Mathematics for Engineers and Scientists

Vinh Phu Nguyen

Mathematics for Engineers and Scientists

Concepts, Applications, and History

Volume 1

Springer

Vinh Phu Nguyen
Faculty of Engineering
Department of Civil Engineering
Monash University
Melbourne, VIC, Australia

ISBN 978-981-97-4630-9 ISBN 978-981-97-4631-6 (eBook)
https://doi.org/10.1007/978-981-97-4631-6

This Springer imprint is published by the registered company Springer Nature Singapore Pte Ltd.
The registered company address is: 152 Beach Road, #21-01/04 Gateway East, Singapore 189721, Singapore

Preface

There are several issues with the traditional mathematics education. First, it focuses too much on technical details. For example, students are asked to routinely apply the formula $-b\pm\sqrt{b^2-4ac}/2a$ to solve many quadratic equations (e.g. $x^2-2x+1=0$, $x^2+5x-10=0$ etc. and the list goes on). Second, the history of mathematics is completely ignored; textbook exposition usually presents a complete reversal of the usual order of developments of mathematics. The main purpose of the textbooks is to present mathematics with its characteristic logical structure and its incomparable deductive certainty. That's why in a calculus class students are taught what is a function, then what is a limit, then what is a derivative and finally applications. The truth is the reverse: Fermat implicitly used derivative in solving maxima problems; Newton and Leibniz discovered it; Taylor, Bernoulli brothers, Euler developed it; Lagrange characterized it; and only at the end of this long period of development, that spans about 200 years, did Cauchy and Weierstrass define it. Third, there is little opportunity for students to discover (rediscover to be exact) the mathematics for themselves. Definitions, theorems are presented at the outset, the students study the proofs and do applications.

Born and grew up in Vietnam in the early 80s, I received such a mathematical education. Lack of books and guidance, I spent most of the time solving countless of mathematical exercises. Even though I remembered enjoying some of them, admittedly the goal was always to get high marks in exams and particularly pass the university entrance examination. Most of the time, it was some clever tricks that are learned, not the true meaning of the mathematical concepts or their applications. Of course why people came up with those concepts and why these concepts are so defined were not discussed by the teachers (and unfortunately I did not ask these important questions). After my bachelor, I enrolled in a master program. Again, I was on the same education route: solving as many problems as possible. And you could guess, after a master was a Ph.D. study in the Netherlands. Though I had time and freedom and resources to do whatever I felt needed, the focus was still to pass yet another form of examination—graduation. This time it is measured by a number of research papers published in a peer-reviewed journal. To pursue an academic career, I took a postdoctoral job of which the main aim is to have as many papers as possible.

As you can imagine, I became technically fluent in a narrow field but on a weak foundation.

Eventually, I got a job in a university in 2016. For the first time in my life, I did not have to "perform" but I am able to really learn things (staff in universities still need to perform to satisfy certain performance criteria which is vital for probation and promotion). This is when I started reading books not on my research field, and I found that very enjoyable.

The turning point was the book called *A Mathematician's Lament* by Paul Lockhart, a professional mathematician turned college teacher. Paul Lockhart describes how maths is incorrectly taught in schools and he provides better ways to teach maths. He continues in *Measurement* by showing us how we should learn maths by "re-discovering maths" for ourselves. That made me to decide to re-learn mathematics. But this time it must be done in a (much) more fun and efficient way. A bit of researching led me to reading the book *Learning How to Learn* by Barbara Oakley and Terry Sejnowski. The biggest lesson taken from Oakley and Sejnowski's book is that you can learn any subject if you do it properly.

So, I started learning mathematics from scratch during my free time. It started probably in 2017. I have read many books on mathematics and physics and books on the history of mathematics. I wrote some notes on my iPad recording what I have learned. Then, it came the COVID-19 pandemic, also known as the coronavirus pandemic which locked down Melbourne–the city I am living in. That was when I decided to put my iPad notes in a book format to have a coherent story which is not only beneficial to me, but it will be helpful to others, hopefully.

This book is a set of notes covering (elementary) algebra, Euclidean geometry, trigonometry, analytic geometry, calculus of functions of single variables, and probability. This covers the main content of the mathematics curriculum for high school students. These are followed by statistics, calculus of functions of more than one variable, differential equations, variational calculus, linear algebra, and numerical analysis. These topics are for undergraduate college students majoring in science, technology, engineering, and mathematics. Very few such books exist, I believe, as the two targeted audiences are too different. This one is different because it was written for me, firstly and mainly. However, I do believe that high school students can benefit from "advanced" topics by seeing what can be applications of high school mathematics and what could be extensions or better explanations thereof. On the other hand, there are college students not having a solid background in mathematics who can use the elementary parts of this book as a review.

Although it is a thick book, I have left out many topics. I do not discuss topology, graph theory, abstract algebra, differential geometry, simply because I do not know them (and plan to learn them when the time is ready). But the book goes beyond a study of mathematics just to apply it to sciences and engineering.

The style of the book, as you might guess, is informal. Mostly because I am not a mathematician and also I like a conversational tonne. This is not a traditional mathematics textbook, so it does not include many exercises. Instead it focuses on the mathematical concepts, their origin (why we need them), their definition (why they are defined like the way they're), their extension. The process leading to proofs and solutions is discussed as most often it is the first step which is hard, all the remaining is mostly labor work (involving algebra usually). And of course, history of mathematics is included by presenting major men in mathematics and their short biographies.

Of course there is no new mathematics in this book as I am not a mathematician; I do not produce new mathematics. The maths presented is standard, and thus I do not cite the exact sources. But, I do mention all the books and sources where I have learned the maths.

And finally, influenced by the fact that I am an engineer, the book introduces programming from the beginning. Thus, young students can learn mathematics and programming at the same time! For now, programming is just to automate some tedious calculations, or to compute an infinite series numerically before attacking it analytically. Or a little bit harder as to solve Newton's equations to analyze the orbit of some planets. But a soon exposure to programming is vital to their future career. Not least, coding is fun! All the code is put in github.[1] at this https://github.com/vinhphunguyen/maths_juliaaddress.

Clayton, Australia Vinh Phu Nguyen
June 2024

[1] GitHub is a website and cloud-based service that helps developers store and manage their code, as well as track and control changes to their code.

Acknowledgments

I was lucky to get help from some people. I would like to thank "anh Bé" who tutored me, for free, on mathematics when I needed help most. To my secondary school math teacher "Thay Dieu", who refused to receive tutor fee, I want to acknowledge his generosity. To my high school math teacher "Thay Son", whose belief in me made me more confident in myself, I would like to say thank you very much. To my friend Phuong Thao, who taught me not to memorize formulas, I want to express my deepest gratitude as this simple advice has changed completely the way I have studied since. To Dr. Han Duc Tran (University of Utah) who replied "XFEM" when I asked him 20 years ago what was the hottest topic in the field of computational mechanics. His answer led me to my fateful encounter with Stephane Bordas (Prof. at Luxembourg University). And finally to Prof. Hung Nguyen-Dang, whose the EMMC master program has changed the course of my life and many other Vietnamese, "em cam on Thay rat nhieu".

In the learning process, I cannot say thank you enough to some amazing YouTube channels such as 3Blue1Brown, Mathologer, blackpenredpen, Dr. Trefor Bazett. They provide amazing, clear animation based explanation for many mathematics topics from which I have learned a lot.

I have received encouragement along this journey, and I would like to thank you Miguel Cervera at Universitat Politècnica de Catalunya whom I have never met, Laurence Brassar at University of Oxford, Haojie Lian at Taiyuan University of Technology, Stéphane Bordas at The University of Luxembourg–who was my master supervisor back in 2013. A special thank you is given to Daniel Alves Paladim–a former colleague at Cardiff University–but now working for the American self-driving car company Cruise LLC. Daniel has shared with me all the fascinating books on various topics that he has read over the years. To my close friend Chi Nguyen-Thanh (Royal HaskoningDHV Vietnam), thank you very much for your friendship and encouragement to this project.

This book was typeset with LaTeX on a MacBook. A majority of figures in the book were created using open source software `Asymptote` and `Tikz`. Other figures were generated using `geogebra, processing, Desmos, Julia, Python`. I want to say thank you to Le Huy Tien, a lecturer at Vietnam National University, Hanoi for his help and encouragement on `Asymptote` and `Tikz`.

I dedicate this book to all mathematicians who have created beautiful mathematics that helps making our world a better place to live.

Clayton, Australia Vinh Phu Nguyen
June 2024

Contents

About the Author

Vinh Phu Nguyen born in 1980 in Hue (Vietnam). Vinh Phu Nguyen enrolled in Hai Ba Trung High School in 1995 and graduated 3 years later. He then went to Hochiminh University of Technology (known as Bach Khoa Sai gon among Vietnamese) in 1998. He graduated in 2003 with a bachelor's in civil engineering. Seeing himself unfit for the industry of civil engineering, he enrolled in the EMMC (European Master in Mechanics of Construction) program, Liège University (Belgium). Upon graduation in 2005, he got a CIFRE scholarship to go to the Ècole nationale d'ingenieurs de Saint-Etienne (ÈNISE; National Engineering School of Saint-Etienne) in France to work with Prof Jean Michel Bergheau. The Ph.D. topic seemed so difficult for him that after 1 year he left France empty-handed.

He got a second chance working with Prof. Bert Sluys at Faculty of Civil Engineering and Geosciences, Delft University of Technology (The Netherlands) in 2007. He graduated with a Ph.D. in computational mechanics in 2011 and went immediately to Johns Hopkins University (USA) for a postdoc. After a 6-month spell in the US he went back to Vietnam to get married and then in 2012 went to Cardiff University (UK) to work with Prof Stéphane Bordas, his master supervisor back in 2004, till 2014. Since 2016, he has been a lecturer at the Department of Civil Engineering, Monash University (Australia).

Dr. Nguyen has done some research on numerical methods for solving partial differential equations in solid mechanics problems (e.g. isogeometric analysis, extended finite element method, material point method) and on damage/fracture models (e.g. phase-field fracture models) for solving fracture mechanics and large deformation mechanics problems. He has published one book *Material Point Method: Theory, Implementations and Applications* with Springer Nature.

Since 2017, he became interested in the history of mathematics and science and the problems of high school education in mathematics.

Chapter 1
Introduction

Without doubt, this is a very difficult chapter to write. It is my attempt to explain what mathematics really is (Sect. 1.1). All learners should know at least about the big picture of the topic they're going to study. In Sect. 1.2, common terminologies in mathematics such as axiom, definition, theorem and proof are introduced. Next, the differences between mathematical exercises and problems are discussed (Sect. 1.3). The message is to focus on problems rather than on routine exercises. Section 1.4 is then devoted to some problem-solving strategies. The role of computers in teaching and learning mathematics is treated in Sect. 1.5.

"I'm just not a math person". We hear it all the time. And we've had enough. I tried in Sect. 1.6 to uncover this myth. You can be better at maths than you're thinking.

Mathematics is not a dead, complete subject, created thousands of years ago. In fact, it is a living subject with many unsolved problems and everyday new mathematics is being created or discovered. Section 1.7 presents some of these unsolved problems. It is these problems that keep mathematicians working late at night.

But how one should learn math? If you're wondering about this, then Sect. 1.8 is for you. In it, I present a list of actionable advice on how you should learn mathematics.

Finally, the organization of the book is outlined in Sect. 1.9.

You do not have to try to understand everything in this chapter if your maths is not solid yet. Just skim through it and sometimes get back to it to see the big picture.

V. P. Nguyen, *Mathematics for Engineers and Scientists*,
https://doi.org/10.1007/978-981-97-4631-6_1

1.1 What Is Mathematics?

In *The Mathematical Experience*—a National Book Award in Science—Philip Davis[1] and Reuben Hersh[2] wrote

> Mathematics consists of true facts about imaginary objects

Mathematicians study imaginary objects that are called mathematical objects. Some examples are numbers, functions, triangles, matrices, groups, and more complicated things such as vector spaces and infinite series. These objects are said imaginary or *abstract* as they do not exist in our physical world. For instance, in geometry, a line does not have thickness and a line is perfectly straight! And certainly, mathematicians don't care if a line is made of steel or wood. There are no such things in the physical world. Similarly, we cannot hold and taste the number 3. When we write 3 on a beach and touch it, we only touch a *representation* of the number 3.

Why working with abstract objects useful? One example from geometry is provided as a simple answer. Suppose that we can prove that the area of a (mathematical) circle is π times the square of the radius, then this fact would apply to the area of a circular field, the cross-section of a circular tree trunk or the floor area of a circular temple. An abstract idea can be applied to a wide range of problems.

Having now in their hands some mathematical objects, how do mathematicians deduce new knowledge? As senses, experimentation and measurement are not sufficient, they rely on *reasoning*. Yes, logical reasoning. This started with the Greek mathematicians. It is obvious that we can not use our senses to estimate the distance from the Earth to the Sun. It would be tedious to measure the area of a rectangular region than measuring just its sides and use mathematics to get the area. It is very time consuming and error prone to design structures by pure experimentation. If a bridge is designed in this way, it would only be fair that the designer be the first to cross this bridge.

What mathematicians are really trying to get from their objects? Godfrey Hardy answered this best[3]:

> A mathematician, like a painter or poet, is a maker of patterns. If his patterns are more permanent than theirs, it is because they are made with ideas.

Implied by Hardy is that *mathematics is a study of patterns of mathematical objects*. Let's confine to natural numbers as the mathematical object. The following is one example of how mathematicians play with their objects. They start with a question: what is the sum of the first n natural numbers? This sum is mathematically written as

---

[1] Philip J. Davis (1923–2018) was an American academic applied mathematician.

[2] Reuben Hersh (1927–2020) was an American mathematician and academic, best known for his writings on the nature, practice, and social impact of mathematics. His work challenges and complements mainstream philosophy of mathematics.

[3] Godfrey Harold Hardy (February 1877–December 1947) was an English mathematician, known for his achievements in number theory and mathematical analysis. In biology, he is known for the Hardy–Weinberg principle, a basic principle of population genetics.

$$S(n) = 1 + 2 + 3 + \cdots + n$$

For example, if $n = 3$, then the sum is $S(3) = 1 + 2 + 3$, which is 6, and if $n = 4$ then the sum is $S(4) = 1 + 2 + 3 + 4$, and so on. Now, mathematicians are lazy creatures, they do not want to compute the sums, again and again, for different values of n. They want to find a single formula for the sum that works for any n. To achieve that they have to see through the problem or to see the pattern. Thus, they compute the sum for a few special cases: for $n = 1, 2, 3, 4$, the corresponding sums are

$$\begin{aligned}
n &= 1 : S(1) = 1 \\
n &= 2 : S(2) = 1 + 2 = 3 = \frac{2 \times 3}{2} \\
n &= 3 : S(3) = 1 + 2 + 3 = 6 = \frac{3 \times 4}{2}
\end{aligned}$$

A pattern emerges and they guess the following formula

$$S(n) = 1 + 2 + 3 + \cdots + n = \frac{n(n+1)}{2} \tag{1.1.1}$$

What is more interesting is how they prove that their formula is true. They write $S(n)$ in the usual form, and they also write it in a reverse order,[4] and then they add the two

$$\begin{aligned}
S(n) &= 1 + 2 + \cdots + (n-1) + n \\
S(n) &= n + (n-1) + \cdots + 2 + 1 \\
2S(n) &= \underbrace{(n+1) + (n+1) + \cdots + (n+1) + (n+1)}_{n \text{ terms}} = n(n+1)
\end{aligned}$$

That little narrative is a (humbling) example of the mathematician's art: asking simple and elegant questions about their imaginary abstract objects, and crafting satisfying and beautiful explanations. Now how did mathematicians know to write $S(n)$ in a reverse order and add the twos? How does a painter know where to put his brush? *Experience, inspiration, trial and error, dump luck*, as Paul Lockhart puts it [41]. (The number in brackets refers to the number of the book quoted in the Bibliography at the end of this book). That is the art of it. There is no systematic approach to maths problems. And that's why it is interesting; if we do the same thing over and over again, we get bored. In mathematics, you won't get bored.

Why there was a pattern in the sum of the first n natural numbers? Are mathematicians the only people that look for patterns? Philip Ball in his fascinating book *Patterns in Nature* answers these questions well. He wrote:

> The world is a confusing and turbulent place, but we make sense of it by finding order. We notice the regular cycles of day and night, the waxing and waning of the moon and tides, and

---

[4] They can do this because of the commutative property of addition: Changing the order of addends does not change the sum. Or, $a + b = b + a$.

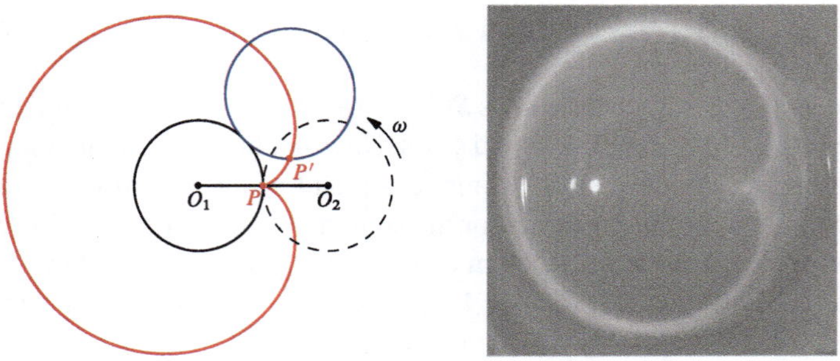

(a) P traces out a cardioid (b)

Fig. 1.1 The cardioid occurs in geometry and in your coffee

the recurrence of the seasons. We look for similarity, predictability, regularity: those have always been the guiding principles behind the emergence of science. We try to break down the complex profusion of nature into simple rules, to find order among what might at first look like chaos. This makes us all pattern seekers.

All high school students know that in mathematics we have different territories: algebra, geometry, analysis, combinatorics, probability, and so on. What they usually do not know is that there is a connection between different branches of mathematics. Quite often a connection that we least expect of. To illustrate the idea, let us play with circles and see what we can get. Here is the game and the question: roll a circle with a marked point around another circle of the same radius, this point traces a curve. What is the shape of this curve? In Fig. 1.1a, we roll the dashed circle around the black circle, and we follow point P and this point traces out a beautiful heart-shaped curve, which is called a cardioid. This beautiful heart-shaped curve shows up in some of the most unexpected places.

Got your coffee? Turn on the flashlight feature of your phone and shine the light into the cup from the side. The light reflects off the sides of the cup and forms a caustic on the surface of the coffee. This caustic is a cardioid (Fig. 1.1b). Super interesting, isn't it[5]?

So far, the cardioid appears in geometry and in real life. Where else? How about times table? We all know that $2 \times 1 = 2$, $2 \times 2 = 4$, $2 \times 3 = 6$, and so on. Let's describe this geometrically and a cardioid will show up! Begin with a circle (of any radius) and mark a certain number (designated symbolically by N) of evenly spaced points around the circle, and number them consecutively starting from zero: $0, 1, 2, ..., N - 1$. Then for each n, draw a line between points n and $2n$ mod N. For example, for $N = 10$, connect 1–2, 2–4, 3–6, 4–8, 5–0 (this is similar to clock: after 12 h the hour hand returns to where it was pointing to), 6–2, 7–4, 8–6, 9–8. Figure 1.2 is the results for $N = 10, 20, 200$, respectively. The envelope of these lines is a cardioid, clearly for large N.

[5] For more detail check https://divisbyzero.com/2018/04/02/i-heart-cardioids/.

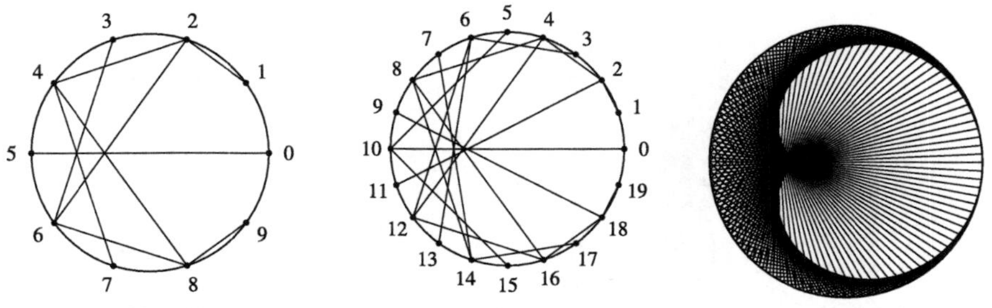

Fig. 1.2 A cardioid emerges from the times table of two when N is big

Let's enjoy another unexpected connection in mathematics. The five most important numbers in mathematics are 0,1 (which are foundations of arithmetic), $\pi = 3.14159\ldots$, which is the most important number in geometry; $e = 2.71828\ldots$, which is the most important number in calculus; and the imaginary number i, with $i^2 = -1$. And yet, they are connected via the following simple relation:

$$e^{i\pi} + 1 = 0$$

which is known as Euler' equation and it is the most *beautiful* equation in mathematics! Why an equation is considered beautiful? Because *the pursuit of beauty in pure mathematics is a tenet*. Neuroscientists in Great Britain discovered that the same part of the brain that is activated by art and music was activated in the brains of mathematicians when they looked at math they regarded as beautiful.

I think these unexpected connections are sufficient for many people to spend time playing with mathematics. People who do mathematics just for fun is called *pure mathematicians*. To get an insight into the mind of a working pure mathematician, there is probably no book better than Hardy's essay *A Mathematician's Apology*. In this essay, Hardy offers a defense of the pursuit of mathematics. Central to Hardy's "apology" is an argument that mathematics has value independent of possible applications. He located this value in the beauty of mathematics.

Below is a mathematical joke that reflects well on how mathematicians think of their field:

> Philosophy is a game with objectives and no rules. Mathematics is a game with rules and *no objectives*.

But, if you are pragmatic, you will only learn something if it is useful. Mathematics is super useful. With it, physicists unveil the secrets of our universe; engineers build incredible machines and structures; biologists study the geometry, topology, and other physical characteristics of DNA, proteins, and cellular structures. The list goes on. People who do mathematics with applications in mind are called *applied mathematicians*.

(a) Un-zoomed image (b) Zoomed in image

Fig. 1.3 Newton fractals for $z^4 - 1 = 0$

And a final note on the usefulness of mathematics. In 1800s, mathematicians worked on wave equations for fun. And in 1864, James Clerk Maxwell–a Scottish physicist—used them to predict the existence of electrical waves. In 1888, Heinrich Rudolf Hertz—a German physicist—confirmed Maxwell's predictions experimentally and in 1896, Guglielmo Giovanni Marconi—an Italian electrical engineer—made the first radio transmission.

Is the above story of radio wave unique? Of course not. We can cite the story of differential geometry (a mathematical discipline that uses the techniques of differential calculus, integral calculus, linear algebra and multilinear algebra to study problems in geometry) by the German mathematician Georg Friedrich Bernhard Riemann in the 19th century, which was used later by the German-born theoretical physicist Albert Einstein in the twentieth century to develop his general relativity theory. And the Greeks studied the ellipse more than a millennium before Kepler used their ideas to predict planetary motions.

The Italian physicist, mathematician, astronomer, and philosopher Galileo Galilei once wrote:

> Philosophy [nature] is written in that *great book* which ever is before our eyes—I mean the universe—but we cannot understand it if we do not first learn the language and grasp the symbols in which it is written. *The book is written in mathematical language*, and the symbols are triangles, circles, and other geometrical figures, without whose help it is impossible to comprehend a single word of it; without which one wanders in vain through a dark labyrinth.

And if you think mathematics is dry, I hope that Fig. 1.3 will change your mind. These images are Newton fractals obtained from considering this equation of one single complex variable $f(z) = z^4 - 1 = 0$. There are four roots corresponding to four colors in the images. A grid of 200×200 points on a complex plane is used as initial guesses in the Newton method of finding the solutions to $f(z) = 0$. The points are colored according to the color of the root they converge to. Refer to Sect. 4.5.4 for detail.

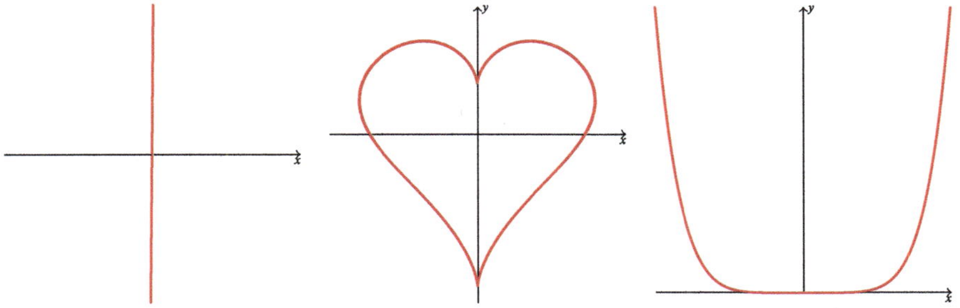

(a) $y = 1000x$ (b) $(x(t), y(t))$ (c) $y = x^6/16$

Fig. 1.4 I love you by mathematicians. The parametric equation for the heart is: $x(t) = 16\sin^3(t)$, $y(t) = 13\cos(t) - 5\cos(2t) - \cos(3t) - \cos(4t)$

Fig. 1.5 Sierpiński triangles

And who said mathematicians are boring, please look at Figs. 1.4, and 1.5, where we start with an equilateral triangle. Subdivide it into four smaller congruent equilateral triangles and remove the central triangle. Repeat step 2 with each of the remaining smaller triangles infinitely. What we obtain are Sierpiński triangles.[6]

Let's now play the "chaos game" and we shall meet Sierpiński triangles again. The process is simple: (1) Draw an equilateral triangle on a piece of paper and draw a *random initial point*, (2) Draw the next point *midway* to one of the vertices of the triangle, *chosen randomly*, (3) Repeat step 2 ad infinitum. What is amazing is when the number of points is large, a pattern emerges, and it is nothing but Sierpiński triangles (Fig. 1.6)! If you are interested in making these stunning images (and those in Fig. 1.7), check Appendix A.11.

To know what is mathematics, there is no better way than to see how mathematicians think and act. And for that I think mathematical jokes are one good way. Mathematicians Andrej and Elena Cherkaev from University of Utah have provided a collection of these jokes at Mathematical humor and I use the following one

> An engineer, a physicist and a mathematician are staying in a hotel. The engineer wakes up and smells smoke. He goes out into the hallway and sees a fire, so he fills a trash can from his room with water and douses the fire. He goes back to bed. Later, the physicist wakes up and smells smoke. He opens his door and sees a fire in the hallway. He walks down the hall

[6] The Polish mathematician Wacław Sierpiński (1882–1969) described the Sierpinski triangle in 1915. But similar patterns already appeared in the thirteenth-century Cosmati mosaics in the cathedral of Anagni, Italy.

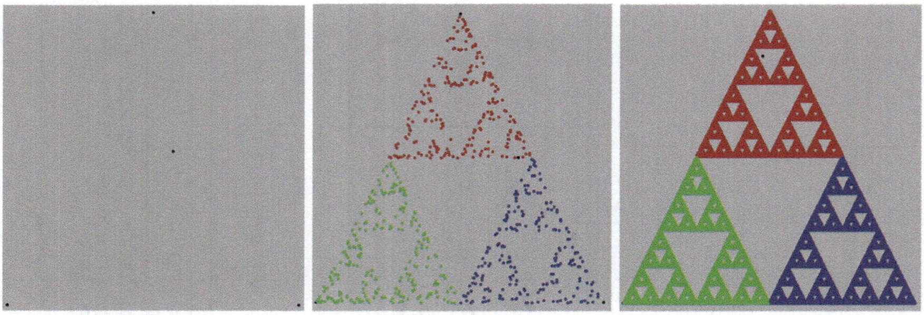

Fig. 1.6 Chaos game and Sierpiński triangles. `Processing` source: check folder `chaos_game_pde` in my github account mentioned in the preface

Fig. 1.7 Chaos game, pentagon, and fractals. `Processing` source: check folder `chaos_game_2`

> to a fire hose and after calculating the flame velocity, distance, water pressure, trajectory, etc. extinguishes the fire with the minimum amount of water and energy needed. Later, the mathematician wakes up and smells smoke. He goes to the hall, sees the fire and then the fire hose. He thinks for a moment and then exclaims, *"Ah, a solution exists!"* and then *goes back to bed*.

to demonstrate that sometimes showing that something exists is just as important as finding itself.

With just pen and paper and reasoning mathematics can help us uncover the hidden secrets of many things from giant objects such as planets to minuscule objects such as bacteria and every other in between. Let's study this fascinating language; the language of our universe.

Hey, but what if someone does not want to become an engineer or scientist, does he/she still have to learn mathematics? I believe he/she should because of the following reasons. According to Greek, mathematics is learning and according to Hebrew it is thinking. So learning mathematics is to learn how to think, how to reason, logically. Réne Descarte once said "I think then I am".

Before delving into the world of mathematics, we first need to get familiar to some common terminologies; terms such as axioms, theorems, definitions, and proofs. The next section is for those topics.

1.2 Axiom, Definition, Theorem, and Proof

Axioms are assumptions that all agree to be true. Or in other words, they are *self-evident truths*. No proof is needed for axioms as they are the *rules of the game*. Actually, we cannot prove axioms and that is why we have to accept them. We cannot prove the axioms because they are the starting points for us to do logic. This may seem disappointing to those who thought that in mathematics everything is proved! After axioms, then come definitions. A definition is to define a word. For example to define an even function, mathematicians write: 'A function $f(x) : \mathbb{R} \to \mathbb{R}$ is an even function if for any $x \in \mathbb{R}$: $f(-x) = f(x)$'. Mathematicians define something when they need it for their work. The following joke is a good example about this:

> One day a farmer called up an engineer, a physicist, and a mathematician and asked them *to fence of the largest possible area with the least amount of fence*. The engineer made the fence in a circle and proclaimed that he had the most efficient design. The physicist made a long, straight line and proclaimed "We can assume the length is infinite..." and pointed out that fencing off half of the Earth was certainly a more efficient way to do it. The Mathematician just laughed at them. He built a tiny fence around himself and said "*I declare myself to be on the outside.*"

Unlike scientists and engineer who study real things in our real world and that's why they are restricted by the laws of nature, mathematicians study objects such as numbers, functions that live in a mathematical world. Thus, mathematicians have more freedom.

Next come theorems. A theorem is a statement about properties of one or more than objects. One can have this theorem regarding even functions: "If $f(x)$ is an even function, then its derivative is an odd function". We need to provide a mathematical proof for a mathematical statement to become a theorem.

The word "proof" comes from the Latin *probare* (to test). The development of mathematical proof is primarily the product of ancient Greek mathematics, and one of its greatest achievements. Thales and Hippocrates of Chios gave some of the first known proofs of theorems in geometry. Mathematical proof was revolutionized by Euclid (300 BCE,[7]) who introduced the *axiomatic method* still in use today. Starting with axioms, the method proves theorems using deductive logic: if A is true, and A implies B, then B is true. Or "All men smoke weed; Sherlock Holmes is a man; therefore, Sherlock Holmes smokes weed".

---

[7] Common Era (CE) and Before the Common Era (BCE) are alternatives to the Anno Domini (AD) and Before Christ (BC) notations used by the Christian monk Dionysius Exiguus in 525. The two notation systems are numerically equivalent: "2022 CE" and "AD 2022" each describe the current year; "400 BCE" and "400 BC" are the same year.

As a demonstration of mathematical proofs, let's consider the following problem. Given $a \geq b \geq c \geq 0$ and $a + b + c \leq 1$, prove that $a^2 + 3b^2 + 5c^2 \leq 1$.

Proof. We first rewrite the term $a^2 + 3b^2 + 5c^2$ as (why? how do we know to do this step?)

$$a^2 + 3b^2 + 5c^2 = a^2 + b^2 + c^2 + 2b^2 + 2c^2 + 2c^2$$

Then using the data that $a \geq b \geq c \geq 0$, we know that $2b^2 = 2bb \leq 2ab$, thus

$$a^2 + 3b^2 + 5c^2 \leq a^2 + b^2 + c^2 + 2ab + 2ca + 2cb$$

Now, we recognize that the RHS[8] is nothing but $(a + b + c)^2$ because of the well known identity $(a + b + c)^2 = a^2 + b^2 + c^2 + 2ab + abc + 2ca$. Thus, we have

$$a^2 + 3b^2 + 5c^2 \leq (a + b + c)^2$$

And if we combine this with the data that $a + b + c \leq 1$, we have proved the problem. ■

To indicate the end of a proof several symbolic conventions exist. While some authors still use the classical abbreviation Q.E.D., which is an initialism of the Latin phrase *quod erat demonstrandum*, meaning "which was to be demonstrated", it is relatively uncommon in modern mathematical texts. Paul Halmos pioneered the use of a solid black square at the end of a proof as a Q.E.D symbol, a practice which has become standard (and followed in this text), although not universal.

The proof is simple because this is a problem for grade 7/8 students. But how about a proof with shapes? See Fig. 1.8 for such a geometry-based proof. Terms like a^2 should be seen as the area of a square of side a. Inside the big square of side $a + b + c$—of which the area is smaller than or equal to 1, we have many smaller squares including one a^2, three b^2, and five c^2, and it is quite obvious that we have the inequality $a^2 + 3b^2 + 5c^2 \leq 1$. Essentially, this geometry-based proof is similar to the previous proof, but everyone would agree it is easier to understand. I recommend the book *Proofs without words* by Roger Nelsen [51] for such elegant proofs.

I present another problem. Let's take the case of a triangle inside a semicircle. If we play with it long enough, we will see one remarkable thing: no matter where on the circle we place the tip of the triangle, it always forms a nice right triangle (Fig. 1.9a). Mathematics is full of surprises like this. But is it true? We need a proof to answer the question *How do we know it?* In the same figure, I present a proof commonly given in high school geometry classes. A complete proof would be more verbose than what I present here. Does it exist a better proof? See Fig. 1.9b: $ACBC'$ is a rectangle and thus ABC is a right triangle! This geometric proof is said to be elegant because it allows us to see why the theorem is true.

---

[8] The expression on the right side of the "=" sign is the right side of the equation and the expression on the left of the "=" is the left side of the equation. For example, in $x + 5 = y + 8$, $x + 5$ is the left-hand side (LHS) and $y + 8$ is the right-hand side (RHS).

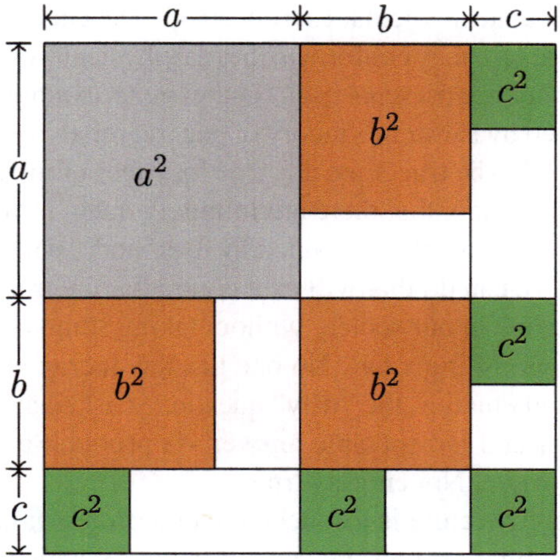

Fig. 1.8 A proof of an algebra problem using geometry

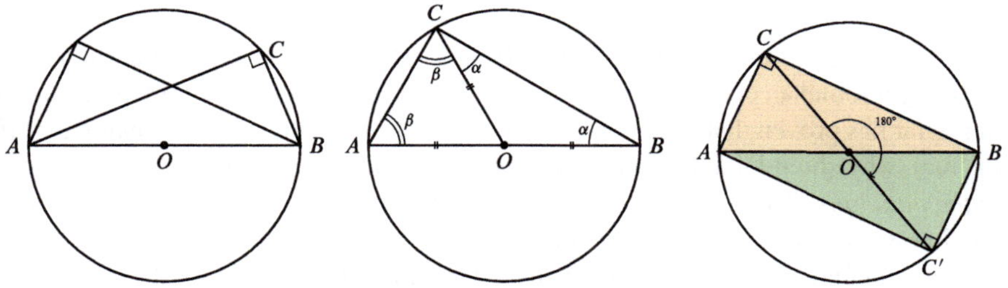

(a) In triangle ABC: $2(\alpha + \beta) = \pi$ (b) $ACBC'$ is a rectangle

Fig. 1.9 The angle inscribed in a semicircle is always a right angle (90°): two proofs. (a) The angles of the same marking are equal, and thus we have $2(\alpha + \beta) = \pi$ or $\alpha + \beta = \pi/2$. The key to the proof is to draw the line OC which does not initially exist in the problem. And (b) Rotate ABC 180° counterclockwise we get a new triangle ABC'. Alternatively, draw OC (as in the first proof) but extend it to meet the circle at C' (to have symmetry). Now the box $ACBC'$ is a parallelogram. But it is a special parallelogram: the two diagonals are equal (both are diameters of the circle). So, $ACBC'$ must be a rectangle. The second proof is considered to be better than the first because we can "see" the result

Not all proofs are as simple as the above ones. For example, in number theory, Fermat's Last Theorem states that no three positive integers a, b, and c satisfy the equation $a^n + b^n = c^n$ for any integer value of n greater than 2. This theorem was first stated as a theorem by Pierre de Fermat around 1637 in the margin of a copy of *Arithmetica*; Fermat added that he had a proof that was too large to fit in the margin. After 358 years of effort by countless number of mathematicians, the first successful proof was released only very recently, in 1994, by Andrew Wiles (1953) an English mathematician. About Wiles' proof, it is 192 pages long.

Proofs are what separate mathematics from all other sciences. In other sciences, we accept certain laws because they conform to the real physical world, but those laws can be modified if new evidence presents itself. One famous example is Newton's theory of gravity was replaced by Einstein's theory of general relativity. But in mathematics, if a statement is proved to be true, then it is true forever. For instance, Euclid proved, over two thousand years ago, that there are infinitely many prime numbers (a prime number is a number that can only be divided by itself and 1 without remainders), and there is nothing that we can do that will ever contradict the truth of that statement.

It is impossible to live in our society without taking some body of knowledge on authority (i.e.,, without asking why). No one has the energy and capacity to check everything. But in mathematics, the "why" question is asked at every stage, with the expectation of a clear and indisputable answer via proofs. So, in mathematics, we have only black and white. Not entirely true...

In mathematics, a conjecture is a conclusion or a proposition which is suspected to be true due to preliminary supporting evidence, but for which no proof or disproof has yet been found. For example, on 7 June 1742, the German mathematician Christian Goldbach wrote a letter to Leonhard Euler in which he proposed the following conjecture: *Every positive even integer can be written as the sum of two primes*. The guess seems to be true: $8 = 5 + 3$, $24 = 19 + 5$, $64 = 23 + 41$ etc. Furthermore, no one has yet found an even number for which this statement does not work out. But no one so far has proven that exceptions cannot exist. Is the pattern real or imaginary? If it is real, why should numbers have this property? *Such mysteries are what motivate mathematicians*. This became Goldbach's conjecture and is one of the oldest and best-known unsolved problems in number theory and all of mathematics.

If we can find one *counterexample*, just one, to a conjecture then we disprove it. To illustrate this point, I present the following example taken from the interesting book *Proofs: A Long-Form Mathematics Textbook* by Jay Cummings [16]. The problem is: suppose we are investigating how many regions are formed if one places n dots randomly on a circle and then connects them with line. We start with $n = 2, 3, 4, 5$ and count the regions, The results given in Fig. 1.10 suggest that the number of regions can be 2^{n-1}, which forms our conjecture. It turns out that this conjecture is, however, wrong. When we put 6 points on a circle and count the number of regions,

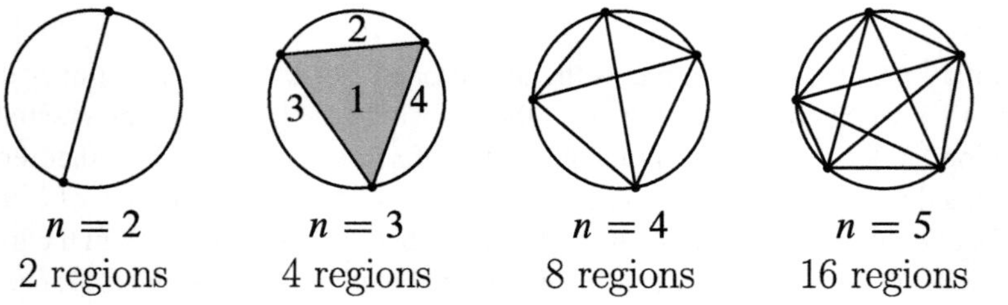

Fig. 1.10 A circle with n random points on it; two points are connected by a line. These lines make regions—space bounded by closed curves

we find that there are only 31 regions, not 32 as suggested by 2^{n-1}; thus disproving the conjecture.

This was an easy situation because we could find a counterexample for such a small value of n. Things become much harder when the conjecture is correct for all values of n that we can verify but we just don't know how to prove it. That is the situation with Goldbach's conjecture. This example also demonstrates why mathematicians are proof-obsessed. We cannot simply believe in something just because it looks correct or because it holds for certain case. We need a proof.

In addition to the above-mentioned terminologies, we also have proposition, lemma, and corollary. A proposition is a less important but nonetheless interesting true statement. A lemma is a true statement used in proving other true statements (that is, a less important theorem that is helpful in the proof of other more important theorems). And finally, a corollary denotes a true statement that is a simple deduction from a theorem or proposition. For example, the sum of the interior angles of any triangle is always 180 degrees. One corollary to that theorem: Each interior angle of an equilateral triangle is 60 degrees.[9]

1.3 Exercises Versus Problems

It is vital to differentiate exercises and problems as we should not spend too much time on the former. It is the latter that is fun and usually leads to interesting things. Roughly speaking, exercises are problems or questions that you know the procedure to solve them (even though you might not be able to finish it). For example, a typical exercise is using the formula $\frac{-b\pm\sqrt{b^2-4ac}}{2a}$ to solve many quadratic equations, e.g., $x^2 - 2x + 1 = 0$, $x^2 + 5x - 10 = 0$. I think that you should solve only three quadratic equations: (i) $(x-1)(x-2) = 0$ (two real solutions), (ii) $(x-1)(x-1) = 0$ (two repeated solutions) and $x^2 + 1 = 0$ (no real solutions) and plot the graphs of them. That's it. Solving fifty more quadratic equations would not bring you any joy and useful things.

On the other hand, problems are questions that we do not know before hand the procedure to solve. There are two types of mathematics questions: ones with known solutions and ones with/without solutions. The latter problems arise in mathematical research. Herein, we focus on the former—problems with known solutions but we do not know them yet. For example, consider this problem of finding the roots of the following fourth-order equation:

$$x^4 - 3x^3 + 4x^2 - 3x + 1 = 0 \tag{1.3.1}$$

How can we solve this equation? There is no formula for x. After many attempts, we have found that dividing this equation by x^2 is a correct direction (actually this was used by Lagrange some 300 years ago):

[9] If we denote by α the angle of an equilateral triangle then the theorem tells us $\alpha + \alpha + \alpha = 180°$. From that we get $\alpha = 60°$.

$$x^2 + \frac{1}{x^2} - 3\left(x + \frac{1}{x}\right) + 4 = 0 \tag{1.3.2}$$

Due to symmetry, we do a change of variable with $u = x + {}^1/_x$, thus we obtain

$$u^2 - 3u + 2 = 0 \Longrightarrow u = 1, \quad u = 2 \tag{1.3.3}$$

If we allow only real solutions, then with $u = 2$, we have $x + 1/x = 2$ which gives $x = 1$.

1.4 Problem-Solving Strategies

Whenever we have solved (successfully) a math problem, should we celebrate our success a bit and move on to another problem as quick as possible? Celebration yes, but moving on to another problem, no. Instead we should perform the fourth step recommended by the Hungarian mathematician George Pólya (1887–1985) in his celebrated book *"How to solve it"* [53]. That is *looking back*, by answering the following equations:

- Can we check the result? Substituting $x = 1$ into the LHS of Eq. 1.3.1 indeed yields zero;
- Can we guess the result? Can we solve it differently? We can, by trial and error, see that $x = 1$ is a solution and factor the LHS as $(x - 1)(x^3 - 2x^2 + 2x - 1)$. And proceed from there.
- Can we use the method for some other problem? Yes, we can use the same technique for equations of this form $ax^4 + bx^3 + cx^2 + bx + a = 0$.

This step of looking back is actually similar to reflection in our lives. We all know that once in a while we should stop doing what we supposed to do to think about what we have done.

Another useful strategy is to get familiar with the problem before solving it. For example, consider these two simultaneous equations:

$$\begin{aligned} 127x + 341y &= 274 \\ 218x + 73y &= 111 \end{aligned} \tag{1.4.1}$$

There is a routine method for solving such equations, which I do not bother you with here. What I want to say here is that if we're asked to solve the following equations by hands, should we just apply that routine method?

$$\begin{aligned} 6751x + 3249y &= 26751 \\ 3249x + 6751y &= 23249 \end{aligned} \tag{1.4.2}$$

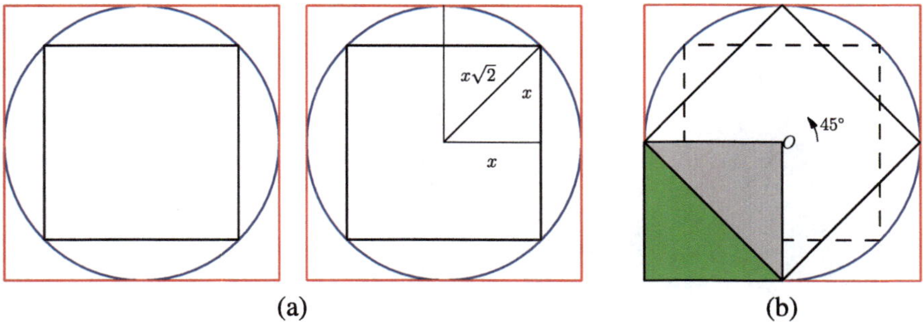

Fig. 1.11 Using symmetry to solve problems

No, we leave that for computers. We're better. Let's spend time with the problem first, and we see something special now:

$$\begin{aligned} 6751x + 3249y &= 26751 \\ 3249x + 6751y &= 23249 \end{aligned} \tag{1.4.3}$$

We see a symmetry in the coefficients of the equations. This guides us to perform operations that *maintain this symmetry*: if we sum the two equations we get $x + y = \ldots$ And if we subtract the first from the second we get $x - y = \ldots$ (we can do the inverse to get $y - x = \ldots$). Now, the problem is very easy to solve.

As another example of exploiting the symmetry of a problem, consider this geometry problem: a square is inscribed in a circle that is inscribed in a square. Find the ratio of the area of the smaller square over that of the large square. We can introduce symbols to the problem and use the Pythagorean theorem to solve this problem (Fig. 1.11a): that ratio is $1/2$. But we can also use symmetry: if we rotate (counterclockwise) the smaller square 45°C with respect to the center of the circle, we get a new problem shown in Fig. 1.11b. And it is obvious that the ratio that we're looking for is $1/2$.

For problem-solving skills, I recommend to read Pólya's book and the book by Paul Zeit, [70]. The latter contains more examples at a higher level than Pólya's book. Another book is âŁ˜'Solving mathematical problems: a personal perspective" by the Australian-American mathematician Tarence Tao (1975). He is widely regarded as one of the greatest living mathematicians. If you want to learn "advanced" mathematics, his blog is worth of checking.

1.5 Computing in Mathematics

Herein I discuss the role of computers in learning mathematics and solving mathematical problems. First, we can use computers to plot complex functions (Fig. 1.12). Second, a computer can help to understand difficult mathematical concepts such as

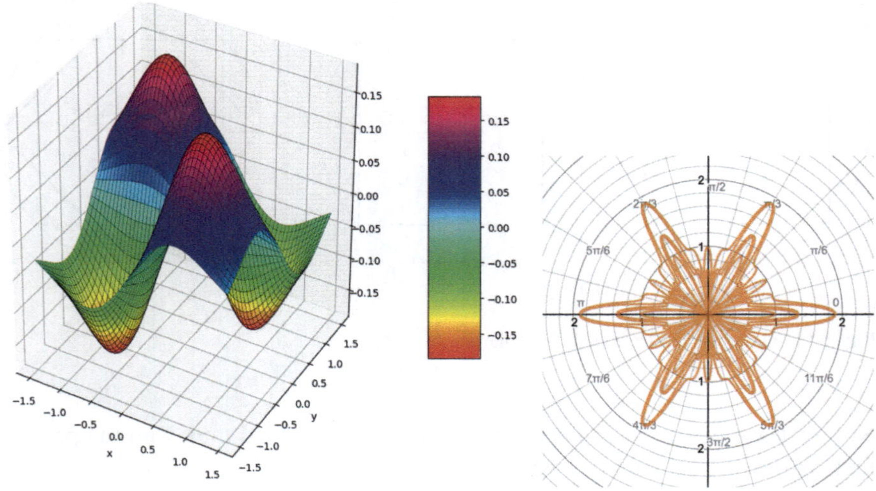

(a) $z = x^2 + 10xy + y^2$

(b) $r = \sin^2(1.2\theta) + \cos(6\theta)^3$

Fig. 1.12 Computers are used to plot complex functions

n	S_n
1	0.5
2	0.75
3	0.875
4	0.9375
5	0.96875
6	0.984375
7	0.992188
⋮	⋮
14	0.999939
15	0.999969

(a)

```
using PrettyTables %you have to install this package first
function geometric_series(n)
  S = 0.
  for k=1:n
    S += 1/2^k
  end
  return S
end

data = zeros(20,2)
for i=1:20
  S = geometric_series(i)
  data[i,1] = i
  data[i,2] = S
end
pretty_table(data, ["n", "S"]) # print the table to terminal
```

(b)

Fig. 1.13 A small `Julia` program to compute the geometric series

limit. To demonstrate this point, let's consider the geometric series $S = 1/2 + 1/4 + 1/8 + \cdots$. Does this series converge (i.e.,, when enough terms are used one gets the same result) and what is the sum? We can program a small code shown in Fig. 1.13b to produce the table shown in Fig. 1.13a. The data shown in this table clearly indicate that the geometric series do converge and its sum is 1.

Third, when it comes to applied mathematics, computers are an invaluable tool. In applied mathematics, problems are not solved exactly by hands, but approximately using some algorithms which are tedious for hand calculations but suitable for computers. To illustrate what applied mathematics is about, let's solve this equation $f(x) = \cos x - x = 0$; i.e.,, finding all values of x such that $f(x) = 0$. Hey, there is no formula similar to $\frac{-b\pm\sqrt{b^2-4ac}}{2a}$ for this equation. That's why Newton developed a method to get approximate solutions. Starting from an initial guess x_0, his method iteratively generates better approximations:

$$x_{n+1} = x_n + \frac{\cos x_n - x_n}{1 + \sin x_n}$$

With only four such calculations, we get $x = 0.73908513$ which is indeed the solution to $\cos x - x = 0$.

And finally, computers are used to build amazing animations to explain mathematics, see for example this YouTube video. Among various open source tools to create such animations, `processing`[10] is an easy to use tool, based on `Java`—a common programming language. Figure 1.6 was made using `processing`.

I have introduced two tools for programming, namely, `Julia` and `processing`. This is because the latter is better suited for making animations while the former is for scientific computing.

For the role of computers in doing mathematics, I refer to the great book *Mathematics by Experiment: Plausible Reasoning in the 21st Century* by Jonathan Borwein and David Bailey [9].

But if you think that computers can replace mathematicians, you are wrong. Even for arithmetic problems, computers are not always better than human. One example is the computation of a sum like this (containing 10^{12} terms)

$$S = \frac{1}{1} + \frac{1}{4} + \frac{1}{9} + \cdots + \frac{1}{10^{24}}$$

Even though a powerful computer can compute this sum by adding term by term, it takes a long time (On my macbook pro, `Julia` crashed when computing this sum!). The result is $S = 1.6449340668482264$.[11] Mathematicians developed smarter ways to compute this sum; for example, this is how Euler computed this sum in the eighteenth century:

$$\begin{aligned} S = &\frac{1}{1} + \frac{1}{4} + \frac{1}{9} + \frac{1}{16} + \frac{1}{25} + \frac{1}{36} + \frac{1}{49} + \frac{1}{64} + \frac{1}{81} \\ &+ \frac{1}{10} + \frac{1}{200} + \frac{1}{6000} - \frac{1}{3 \times 10^6} \end{aligned}$$

a sum of only 13 terms and got 1.644934064499874—a result which is correct up to eight decimals! The story is while solving the Basel problem (i.e.,, what is $S =$

[10] Available for free at https://processing.org.

[11] And this number is exactly $\pi^2/6$. Why π is here? It's super interesting, isn't it? Check this youtube video for an explanation.

$1 + 1/4 + 1/9 + 1/16 + \cdots + 1/k^2 + \cdots$; Sect. 2.21.4), Euler discovered/developed the so-called Euler-Maclaurin summation formula (Sect. 4.18).

Computers can be valuable assistants, but only when a lot of human thought has gone into setting up the computations.

1.6 Mathematical Anxiety or Math Phobia

Mathematical anxiety, also known as math phobia, is anxiety about one's ability to do mathematics. Math phobia, to some researchers, is gained—not from personal experience but from parents and teachers and the textbooks they used. So, to the kids who think you are a math phobia, don't panic. That's not your fault.

To illustrate the problem of textbooks and teachers, I quote the prologue of the classic 1910 text named *Calculus Made Easy* by the eccentric British electrical engineer Silvanus Phillips Thompson (1851–1916):

> Considering how many fools can calculate, it is surprising that it should be thought either a difficult or a tedious task for any other fool to learn how to master the same tricks. Some calculus-tricks are quite easy. Some are enormously difficult. The fools who write the text-books of advanced mathematics—and they are mostly clever fools—*seldom take the trouble to show you how easy the easy calculations are*. On the contrary, *they seem to desire to impress you with their tremendous cleverness by going about it in the most difficult way*. Being myself a remarkably stupid fellow, I have had to unteach myself the difficulties, and now beg to present to my fellow fools the parts that are not hard. Master these thoroughly, and the rest will follow. *What one fool can do, another can.*

And Thompson's view on textbooks was shared by Cornelius Lanczos (1893–1974), a Hungarian-American mathematician and physicist, who wrote in the preface of his celebrated book *The Variational Principles of Mechanics* these words:

> Many of the scientific treaties of today are formulated in a half-mystycal language, as though *to impress the reader with the uncomfortable feeling that he is in the permanent presence of a superman*. The present book is conceived in a humble spirit and is written for humble people.

Paul Lockhart in *A Mathematician's Lament* wrote the following words

> THERE IS SURELY NO MORE RELIABLE WAY TO KILL enthusiasm and interest in a subject than to make it a mandatory part of the school curriculum. Include it as a major component of standardized testing and you virtually guarantee that the education establishment will suck the life out of it. *School boards do not understand what math is; neither do educators, textbook authors, publishing companies, and, sadly, neither do most of our math teachers.*

Talking about teachers, Nobel winning physicist Richard Feynman has once said "If you find science boring, you are learning it from wrong teacher". He implied that if you have a good teacher you can learn any topic.

Let me get back to those kids who thought they fell behind the math curriculum. What should you do? I have some tips for you. First, read *A Mathematician's Lament* of Paul Lockhart. After you have finished that book, you would be confident that if you study maths properly you can enjoy mathematics. Second, spend lots of time (I spent one summer when I fell behind in the 9th grade) to learn maths from scratch.[12] Lockhart's other books (see Sect. 1.8.1) will surely help. And this book (Chaps. 1/2/3 and Appendices A/B) could be useful.

Is math ability genetic? Yes, to some degree. Essentially none of us could ever be as good at math as Terence Tao, no matter how hard we tried or how well we were taught. But here's the thing: We don't have to! For high-school and college math, inborn talent is much less important than hard work, preparation, and self-confidence.

Ok. **What one fool can do, another can**. What a simple sentence but it has a tremendous impact on people crossing it. It has motivated many people to start learning calculus, including Feymann. And we can start learning maths with it.

1.7 Millennium Prize Problems

Is mathematics complete as the way it is presented in textbooks? On the contrary, far from that. There are many mathematical problems of which solutions are still elusive to even the brightest mathematicians on Earth. The most famous unsolved problems are *the Millennium Problems*. They are a set of seven problems for which the Clay Mathematics Institute offered a US$ 7 million prize fund ($ 1 million per problem) to celebrate the new millennium in May 2000. The problems all have significant impacts on their field of mathematics and beyond, and were all unsolved at the time of the offering of the prize.

The Riemann hypothesis—posed by the Germain Bernhard Riemann in 1859 in his paper "Ueber die Anzahl der Primzahlen unter einer gegebenen Grösse" (On the number of primes less than a given magnitude)—is perhaps the most famous unsolved problem in mathematics. It concerns the nontrivial zeroes of the Riemann zeta function, which is defined for $\Re s > 1$ by the infinite sum

$$\zeta(s) = \sum_{s=1}^{n} \frac{1}{n^s}$$

This function has trivial zeroes (that are all s such that $\zeta(s) = 0$) on the negative real line, at $s = -2, -4, -6, \ldots$ The location of its other zeroes is more mysterious; the conjecture is that

> The nontrivial zeroes of the zeta function lie on the line $\Re s = 0.5$

[12] If you're in the middle of a semester, then spend less time on other topics. You cannot have everything!.

Yes, the problem statement is that simple, but its proof is elusive to all mathematicians to date.

In 1900 at the International Congress of Mathematicians in Paris, the Germain mathematician David Hilbert gave a speech which is perhaps the most influential speech ever given to mathematicians, given by a mathematician, or given about mathematics. In it, Hilbert outlined 23 major mathematical problems to be studied in the coming century. And the Riemann hypothesis was one of them. Hilbert once remarked:

> If I were to awaken after having slept for a thousand years, my first question would be: Has the Riemann hypothesis been proven?

Judging by the current rate of progress (on solving the hypothesis), Hilbert may well have to sleep a little while longer.

It is usually while solving unsolved mathematical problems that mathematicians discover new mathematics. The new maths also help to understand the old maths and provide better solution to old problems. Some new maths are also discovered by scientists especially physicists while they are trying to unravel the mysteries of our universe. Then, after about 100 or 200 years some of the new maths come into the mathematics curriculum to train the general public. And the educators, whoever they are, hope that our kids can understand these maths—the mathematics that were once only within the grasp of a few greatest mathematicians!

1.8 How to Learn Mathematics

High school students should probably start this book with this section. Herein I provide some reading materials (Sect. 1.8.1) and how to learn maths tips (Sect. 1.8.2).

1.8.1 Reading Materials

When you're solving problems, working through textbooks, getting into the nitty-gritty details of each topic, it's so easy to *lose the forest for the trees* and *forget why you even became inspired to study the topic that you're learning in the first place*. If you read only the textbooks, you will find the subject dull. Textbooks on mathematics are written for people who already possess a strong desire to study mathematics: they are not written to crease such a desire. Do not begin by reading the subject. Instead, begin by reading around the subject. This is where really, really good (and non-speculative) books on that topic come in handy: they inspire, they encourage, and they help you understand the big picture. For mathematics and physics, the following are among the best (at least to me):

- A Mathematician's Lament by Paul Lockhart;
- Measurement by Paul Lockhart;
- The Joy of x by Steven Strogatz;
- The Feynman Lectures on Physics;
- An Imaginary Tale by Paul Nahin;
- Character of Physical Law by Richard Feynman;
- Evolution of Physics by Einstein and Infeld;
- Letters to a young mathematician by Ian Stewart.

In *A Mathematician's Lament* Paul Lockhart describes how maths is incorrectly taught in schools and he provides better ways to teach maths. He continues in *Measurement* by showing us how we should learn maths by "re-discovering maths" for ourselves. Of course, what Paul suggested works only for self-study people. What if you are a high school student? There are two possibilities. First, if you fortunately have a great teacher, then just stick with her/him. Second, if you do not have such luck, you can ignore her/him and self-study maths with your own pace. Do not forget that mark is not important for deep understanding. Having said that, marks are vital for getting scholarships, sadly.

The Joy of x by Steven Strogatz belongs to a family of maths books that aim to popularize mathematics. In this family, you can also find equally interesting books such as *Journey through Genius* by William Dunham, or *17 equations that changed the world* by Ian Stewart *etc.* It is beneficial at a young age to read these books to realize that mathematics is not a dry, boring topic. On the contrary, it is interesting. Similarly, *An Imaginary Tale: The story of square root of -1* by Paul Nahin is a popular maths book which tells the fascinating story of $\sqrt{-1}$. In the book, I have referred to many other popular math books (see the Reference list).

The Feynman Lectures on Physics[13] by the Nobel winning physicist Richard Feymann is probably the best to learn college level mathematics by studying physics. Bill Gates once said "Feymann is the best teacher I never had". In these lectures, Feymann beautifully introduced various physics topics and the mathematics required to describe them. He also describes how physicists think about problems. Another reason to read these lectures is that *it is good to read books at a level higher than your knowledge*. Feymann lectures were written for Caltech (California Institute of Technology) undergraduates.

Evolution of Physics by the greatest physicist Einstein teaches us how to imagine. Through imaginary thought experiments, the book explains the basic concepts of physics. It is definitely a must read for all students who want to learn physics.

And if you want to become a professional mathematician, read *Letters to a young mathematician* by Ian Stewart [59]. Ian Stewart (born 1945) is a British mathematician who is best known for engaging the public with mathematics and science through his many bestselling books, newspaper and magazine articles, and radio and television appearances.

And don't forget to read the history of mathematics. Henri Poincaré advised,

---

[13] The lectures are freely available at https://www.feynmanlectures.caltech.edu.

> The task of the educator is to make the child's spirit pass again where its forefathers have gone, moving rapidly through certain stages but suppressing none of them. In this regard, the history of science must be our guide.

Here are some books on this topic:

- A history of Mathematics: an introduction by Victor Katz [33];
- A short account of the history of mathematics by W. W. Rouse Ball [6];
- Mathematics and Its History by John Stillwell [64];
- Men of Mathematics: The Lives and Achievements of the Great Mathematicians from Zeno to Poincaré by E. T. Bell [7].

If you prefer watching the history of maths unfold, the BBC Four The story of Maths is excellent. You can find it on YouTube.

How should we read a mathematics textbook?. Of course the first thing to notice is that we cannot read a math book like reading a novel. The second thing is that we should not read it page-by-page, word-by-word from the beginning to the end in one go. The third thing is that maths textbooks are usually many times longer than necessary because they have to include a lot of exercises (at the end of each section or chapter). Why so? Mostly to please the publishers who aim for financial targets not educational ones! As discussed in Sect. 1.3, it is better to spend time-solving problems rather than exercises. It is certain that we first still have to do a few exercises to understand a concept/method. But that's it.

Here is one suggestion on how we should read a math book (based on many recommendations that I have collected from various sources). It is clear that something that works for one person might not work for others, but it can be a start:

- 1st read: skim through a section/chapter first. The idea is to see the forest, not the trees. Knowing all the trees in the first go would be too much;
- 2nd read: read slowly (with paper/pencil) to get know the trees; focus on the motivation, the definition, the theorem;
- 3rd read: read around; read the history of the concept;
- 4th read: pay attention to the proofs; study them carefully and reproduce a proof for yourself.

1.8.2 Learning Tips

> One way to learn a lot of mathematics is by reading the first chapters of many books
> (Paul Halmos)

It is not a surprise that many of us have studied many topics naturally, i.e.,, without understanding how the brain works. We can compensate for that lack of knowledge by reading *Learning How to Learn* by Barbara Oakley and Terry Sejnowski. I do not repeat their advice here, because they're the experts and I am not. Instead, I provide my owns that I have learned and developed over the years (I do not claim they are the best practices, I just feel that I should share what I think useful; these are the tips that I wish I had known them when in school):

- If you have a bad teacher, simply ignore his/her class. There are excellent math teachers online. Learn from them instead. You can listen to the story of Steven Strogatz at https://www.youtube.com/watch?v=SUMLKweFAYk to see how a teacher can change your love to mathematics and then your life;
- If you have questions (any) on maths, you can post them to https://math.stackexchange.com and get answers;
- The best way to learn is to teach. If you do not have such opportunity, you can write about what you know. Similar to this book (which was initially a note for me). Or you can write a blog on maths. Writing is one of the best way to consolidate your understanding of what you have learn (not only maths).[14] You might wonder 'but writing is time consuming'. That is not true if you write just *one page per day* and you're doing that consistently for everyday;
- LaTeX is the best tool (as for now) for writing mathematics. So it is not a bad idea to learn it and use it (for Mathematics Stack Exchange you have to use LaTeX anyway). This book was typeset using LaTeX; If you do not know where to start with LaTeX, check this youtube video out;
- While learning maths, it is a good habit to keep in mind that mathematics is about ideas not formula or numbers. So, first you should be able to *express the idea in your own speaking language*. Then, *translate that to the language of maths*. For example, the idea of convergence of a sequence is expressed in both English and mathematics as follows:

$\forall \epsilon > 0$	$\exists N \in \mathbb{N}$	such that	$\forall n > N$	$\lvert a_n - a \rvert < \epsilon$
however small epsilon is	there is a point in the sequence	such that beyond that point		all the terms are within epsilon of a

- Just like learning any speaking languages, to speak the language of maths you have to study its vocabulary. You should get familiar with Greek symbols like ϵ, δ, $\forall$ etc.;
- And as Euclid told Ptolemy 1st Soter, the first king of Egypt after the death of Alexander the Great 'there is no royal road to geometry', you have to do mathematics. Just as to enjoy swimming you have to jump into the water, by just watching others swimming you will never understand the excitement;
- Knowing the name of something doesn't mean you understand it.[15] There is a way to test whether you understand something or only know the name/definition. It's called the Feynman Technique, and it works like this: "*Without using the new word*

[14] As Dick Guindon once said *Writing is nature's way of letting you know how sloppy your thinking is.*

[15] Feymann's father once told him "See that bird? It's a brown-throated thrush, but in Germany it's called a halzenfugel, and in Chinese they call it a chung ling and even if you know all those names for it, you still know nothing about the bird.".

which you have just learned, try to rephrase what you have just learned in your own language".

- As there is no single book that can covers everything about any topics, it is better to have *a couple of good books for any topics*;
- Read mathematics books very slowly; do not lose the forest for the trees. Study the definitions carefully, why we need them. Then, play with the definitions to see what properties they might possess. Until then, study the theorems. And finally the proofs. If you just want to be a scientist or engineer, then focus less on the proofs;
- Study the history of mathematics. Not only it tells you interesting stories but also it reveals that great mathematicians are also human; they struggled, and they failed many times before succeeded in developing a sound mathematical idea;
- If you fall behind in maths, physics, chemistry (I used to in 8th grade), just focus on improving your maths. Being better at math, you will do fine with physics and chemistry. Remember that math is the language God talks;
- It is impossible to understand algebra if you have not mastered arithmetic. It is impossible to understand calculus if you have not mastered algebra. So, do not rush, it is important to go back to an early stage;
- Maths is a huge subject and it is impossible to be good at everything in maths. If you're not good at geometry it does not mean that you're not good at maths. You might not be good at pure maths, but you can excel in applied maths;
- Be aware of focused vs diffuse mode of thinking. Check the book *Learning How to Learn* for details. In short, diffuse mode is when your mind is relaxed and free. You're thinking about nothing in particular. You're in diffuse mode when you're walking in a park (without a phone of course), having a bath, etc. And usually, it is *when you're in a diffuse mode that you find solutions to problems that you have been struggling to solve.*[16] And of one the best way to get into a diffuse mode is walking. It's not a coincidence that many of the finest thinkers in history were enthusiastic walkers. An old example is Aristotle, the famous Greek philosopher, empiricist, who conducted his lectures while walking the grounds of his school in Athens;
- How long should you fight before giving up to look at the solutions? I admit it is very tempting to look at the solutions when we're stuck. But don't! The best is to play with the exercises for a while (2 h?,[17]) if still no luck, then forget it, do something else. Come back to it later, do the same thing. After 1 or 2 days, still

[16] Archimedes has gone down in history as the guy who ran naked through the streets of Syracuse shouting "Eureka!"—or "I have it!" in Greek. The story behind that event was that Archimedes was charged with proving that a new crown made for Hieron, the king of Syracuse, was not pure gold as the goldsmith had claimed. Archimedes thought long and hard but could not find a method for proving that the crown was not solid gold until he took a bath.

[17] Of course how long before giving up is a personal decision. But I want to use Polya's words about the pleasure of finding something out for yourself: "A great discovery solves a great problem but there is *a grain of discovery in the solution of any problem*. Your problem may be modest; but if it challenges your curiosity and brings into play your inventive faculties, and if you solve it by your own means, you may *experience the tension* and *enjoy the triumph of discovery*".

stuck, then look at the solutions, but only the first step, solve the problem and do self-reflection. If you have to look at the entire solution, then make sure you can repeat all the steps by yourself later. Only then the material is yours. Don't fool yourself by just looking at the solutions and think that you understand the math. No! That is *illusion of competence*—a mental situation where you think you've mastered a set of material but you really haven't. We all watch Messi scoring a goal from a free kick: he just puts the ball into the high left corner so that the goal keeper cannot reach it. But can we repeat that?;

- Do more self-reflection. What is the place that you learn most effectively? When is the time you're most productive? After solving every math question, ask questions like: why the method works, why that answer, is the answer reasonable, does the method work if we modify the question? Why I could not see the solution? Are there other ways to solve the same problem? Only after having answered all these questions, then move to a new math problem;
- Facing a math problem, you should do something: loosen up yourself, draw something, write down something ... And in your head say that "I can solve it, I can solve it". This is called a growth mindset a term presented by Psychologist Dr. Carol Dweck of Stanford University;
- To have a sharp mind and body we do exercises. Similarly your maths will be rusty if you do not use it. I heard that Zdeněk Bažant—a Professor of Civil Engineering and Materials Science at Northwestern University—keeps solving a partial differential equation every week! Note that he is not a mathematician; but he needs maths for his work;
- If you plan to become an engineer or scientist and you were not born with drawing abilities, then practice drawing. Many figures in this book were drawn manually and this was intentional as it is a good way for me to practice drawing;
- Mathematics can be difficult to fully comprehend and that understanding often comes through repeated exposure and practice rather than a sudden moment of insight. Thus, the mathematician John von Neumann once said:

  > In mathematics you don't understand things. You just get used to them.

- Finally, I have collected some learning tips into a document which can be found here.

Feynman's Epilogue. At the end of his famous physics course at Caltech, Feynman said the following words, I quote

> Well, I've been talking to you for two years and now I'm going to quit. In some ways I would like to apologize, and other ways not. I hope—in fact, I know—that two or three dozen of you have been able to follow everything with great excitement, and have had a good time with it. But I also know that "the powers of instruction are of very little efficacy except in those happy circumstances in which they are practically superfluous." So, for the two or three dozen who have understood everything, may I say I have done nothing but shown you the things. For the others, if I have made you hate the subject, I'm sorry. I never taught elementary physics before, and I apologize. I just hope that I haven't caused a serious trouble to you, and that you do not leave this exciting business. I hope that someone else can teach it to you in a way that doesn't give you indigestion, and that you will find someday that, after all, it isn't as horrible as it looks.

> Finally, may I add that *the main purpose of my teaching has not been to prepare you for some examination—it was not even to prepare you to serve industry or the military.* I wanted most to give you some *appreciation of the wonderful world and the physicist's way of looking at it*, which, I believe, is a major part of the true culture of modern times. (There are probably professors of other subjects who would object, but I believe that they are completely wrong.)

This is probably the ideal learning environment that cannot be repeated by other teachers. What is then the solution? Selft studying! With a computer connected to the world wide web, some good books (those books that I've used to write this note are good in my opinion), and amazing free teachers (e.g., 3Blue1Brown, Mathologer, blackpenredpen, Dr. Trefor Bazett), you can learn mathematics (or any topic) in a fun and productive way.

1.9 Organization of the Book

Some wise person has said "writing is learning". While this is not true for many authors, it cannot be more true for me. I have written this note first for myself then for you—our readers (whoever you are). The ultimate goal is to learn mathematics not for grades and examinations but for having a better understanding of the subject and then of the physical world. This goal cannot be achieved if we proceed with pace and shortcuts. So, we will go slowly and starting from basic concepts of numbers and arithmetic operations all the way up to college level mathematics.

In Chap. 2, I discuss all kinds of numbers: natural numbers, integer numbers, rational numbers, irrational numbers, and complex numbers. The presentation is such that these concepts emerge naturally. Next, arithmetic operations on these numbers are defined. Then, linear/quadratic/cubic equations are treated with emphasis on cubic equations, which led to the discovery of imaginary number $i = \sqrt{-1}$. Inverse operations are properly introduced: addition/subtraction, multiplication/division, exponential/logarithm.

Chapter 3 presents a concise summary of Euclidean geometry and trigonometry. Both trigonometry functions such as $\sin x$, $\cos x$, $\tan x$, and $\cot x$ and inverse trigonometry functions, e.g., $\arcsin x$ are introduced. A comprehensive table of trigonometry identities is provided with derivation of all identities. Few applications of this fascinating branch of mathematics in measuring large distances indirectly are also discussed. It is a short chapter as other applications of trigonometry are discussed in other chapters.

Chapter 4 is all about calculus of functions of single variable. Calculus is "the language God talks" according to Richard Feynman—the 1964 Nobel-winning theoretical physicist. Feynman was probably referring to the fact that physical laws are written in the language of calculus. Steven Strogatz in his interesting book *Infinite Powers* [67] wrote "Without calculus, we wouldn't have cell phones, computers, or microwave ovens. We wouldn't have radio. Or television. Or ultrasound for expectant mothers, or GPS for lost travelers. We wouldn't have split the atom, unraveled the human genome, or put astronauts on the moon". Thus it is not surprising that calculus occupies an important part in the mathematics curriculum in both high schools and universities. Sadly, as being taught in schools, calculus is packed with many theorems, formulae and tricks. This chapter attempts to present calculus in an intuitive way by following as much as possible the historical development of the subject. It's a long chapter of more than one hundred pages. This is unavoidable as calculus deals with complex problems. But it mainly concerns the two big concepts: derivative ($f'(x)$ and those dy, dx) and integral ($\int_a^b f(x)dx$).

Chapter 5 presents a short introduction to the mathematical theory of probability. Probability theory started when mathematicians turned their attention to games of chance (e.g., dice rolling). Nowadays, it is used widely in areas of study such as statistics, mathematics, science, finance, gambling, artificial intelligence, machine learning, computer science, game theory, and philosophy to, for example, draw inferences about the expected frequency of events. Probability theory is also used to describe the underlying mechanics and regularities of complex systems.

After calculus of functions of single variable is calculus of functions of multiple variables (Chap. 6). There are two types of such functions: scalar-valued multivariate functions and vector-valued multivariate functions. An example of the former functions is $T = g(x, y, z)$, which represents the temperature of a point in the earth. An example of the latter is the velocity of a fluid particle. We first introduce vectors and vector algebra (rules to do arithmetic with vectors). Certainly, dot product and vector product are the two most important concepts in vector algebra. Then I present the calculus of these two families of functions. For the former, we will have partial derivatives and double/triple integrals. The calculus of vector-valued functions is called vector calculus, which was firstly developed for the study of electromagnetism. Vector calculus then finds applications in many problems: fluid mechanics, solid mechanics etc. In vector calculus, we will meet divergence, curl, line integral, and Gauss's theorem.

In Chap. 7, I discuss what probably is the most important application of calculus: differential equations. These equations are those that describe many physical laws. The attention is on how to derive these equations more than on how to solve them. Derivation of the heat equation $\frac{\partial \theta}{\partial t} = \kappa^2 \frac{\partial^2 \theta}{\partial x^2}$, the wave equation $\frac{\partial^2 u}{\partial t^2} = c^2 \frac{\partial^2 u}{\partial x^2}$, etc. are presented. Also discussed is the problem of mechanical vibrations.

I then discuss in Chap. 8 the calculus of variations, which is a branch of mathematics that allows us to find a function $y = f(x)$ that minimizes a functional $I = \int_a^b G(y, y', y'', x)dx$. For example, it provides answers to questions like "what is the plane curve with maximum area with a given perimeter". You might have correctly guessed the answer: in the absence of any restriction on the shape, the curve is a circle. But calculus of variation provides a proof and more. One notable result of variational calculus is variational methods such as Ritz-Galerkin method, which led to the finite element method. The finite element method is a popular method for numerically solving differential equations arising in engineering and mathematical modeling. Typical problem areas of applications include structural analysis, heat transfer, fluid flow, mass transport, and electromagnetic potential.

Chapter 9 is about linear algebra. Linear algebra is central to almost all areas of mathematics. Linear algebra is also used in most sciences and fields of engineering. Thus, it occupies a vital part in the university curriculum. Linear algebra is all about matrices, vector spaces, systems of linear equations, eigenvectors, you name it. It is common that a student of linear algebra can do the computations (e.g., compute the determinant of a matrix, or the eigenvector), but he/she usually does not know the why and the what. This chapter hopefully provides some answers to these questions.

Chapter 10 is all about numerical methods: how to compute a definite integral numerically, how to interpolate a given data, how to solve numerically and approximately an ordinary differential equation. The basic idea is to use the power of computers to find *approximate solutions* to mathematical problems. This is how Katherine Johnson—the main character in the movie *Hidden Figures*—helped put a man on the moon. She used Euler's method (a numerical method discussed in this chapter) to do the calculation of the necessary trajectory from the earth to the moon for the US Apollo space program. Just that she did by hands.

The book also contains one appendix. In Appendix A, I provide some `Julia` codes that are used in the main text. The idea is to introduce young students to programming as much early as possible.

When we listen to a song or look at a painting we really enjoy the song or the painting much more if we know just a bit about the author and the story about her/his work. In the same manner, mathematical theorems are poems written by mathematicians who are human beings. Behind the mathematics are the stories. To enjoy their poems we should know their stories. The correspondence between Ramanujan—a 23-year-old Indian clerk on a salary of only *£20 per annum* and Hardy—a world renown British mathematician at Cambridge is a touching story. Or the story about the life of Galois who said these final words *Ne pleure pas, Alfred ! J'ai besoin de tout mon courage pour mourir à vingt ans* (Don't cry, Alfred! I need all my courage to die at twenty) to his brother Alfred after being fatally wounded in a duel. His mathematical legacy—Galois theory and group theory, two major branches of abstract algebra—remains with us forever. Because of this, in this book, biographies and some stories of leading mathematicians are provided. But I am not a historian. Thus, I recommend readers to consult MacTutor History of Mathematics Archive. MacTutor is a free online resource containing biographies of nearly 3000 mathematicians and over 2000 pages of essays and supporting materials.

How this book should be read? For those who do not know where to start, this is how you could read this book. Let's start with this chapter. Then proceed with Chapters 2, 3, and 4 that cover more than the high school curriculum. If you're interested in using the maths to do some science projects, check out Chap. 10 where you will find techniques (easy to understand and program) to solve simple harmonic problems (spring-mass or pendulum) and N-body problems (e.g., Sun-Earth problem, Sun-Earth-Moon problem). If you get up to there (and I do not see why you cannot), then feel free to explore the remaining of the books.

Conventions. Equations, figures, tables, theorems are numbered consecutively within each section. For instance, when we're working in Sect. 2.2, the fourth equation is numbered (2.2.4). And this equation is referred to as Eq. (2.2.4) in the text. Same conventions are used for figures and tables. I include many code snippets in the appendix, and the numbering convention is as follows. For instance, Listing B.5 refers to the fifth code snippet in Appendix B. Asterisks (*), daggers (†), and similar symbols indicate footnotes.

Without further ado, let's get started and learn maths in the spirit of Richard Feynman:

> I wonder why. I wonder why
> I wonder why I wonder
> I wonder why I wonder why
> I wonder why I wonder!

Because a curious mind can lead us far. After all, you see, millions saw the apple fall, but only Newton asked why.

And don't forget that ability (intelligence) is malleable via efforts. If a guy in his early 40, married, with two kids, and a full-time job can learn math, you all can too. And you will do it better.

Chapter 2
Algebra

Algebra is generous: she often gives more than is asked for. Jean d'Alembert

Algebra is one of the broad parts of mathematics, together with number theory, geometry, and analysis. In its most general form, algebra is the study of mathematical symbols and the rules for manipulating these symbols; it is a unifying thread of almost all of mathematics. It includes everything from elementary equation solving to the study of abstractions such as groups, rings, and fields. Elementary algebra is generally considered to be essential for any study of mathematics, science, or engineering, as well as such applications as medicine and economics.

This chapter discusses some topics of elementary algebra. By elementary, we meant the algebra in which the commutative of multiplication rule $a \times b = b \times a$ holds. There exists other algebra that violates this rule. There is also matrix algebra that deals with groups of numbers (called matrices) instead of single numbers.

Our starting point is not the beginning of the history of mathematics; instead, we start with the concept of positive integers (or natural numbers) along with the two basic arithmetic operations of addition and multiplication. Furthermore, we begin immediately with the decimal, also called Hindu-Arabic, or Arabic, number system that employs 10 as the base and requiring 10 different numerals, the digits 0, 1, 2, 3, 4, 5, 6, 7, 8, 9. And finally, we take it for granted the liberal use of symbols such as x, y and write $x(10 - x)$ rather than as

> If a person puts such a question to you as: 'I have divided ten into two parts, and multiplying one of these by the other the result was twenty-one;'? then you know that one of the parts is thing and the other is ten minus thing.

from al-Khwarizimi's 'Algebra' (ca. 820 AD).

Our approach is reasonable given the fact that we have limited lifespan and thus it is impossible to trace the entire history of mathematics.

V. P. Nguyen, *Mathematics for Engineers and Scientists*,
https://doi.org/10.1007/978-981-97-4631-6_2

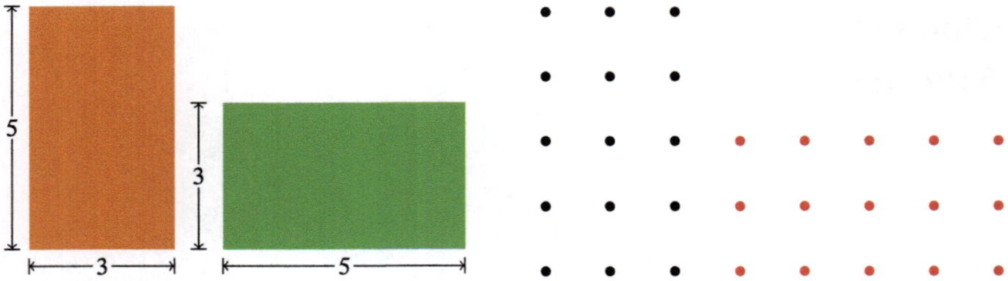

Fig. 2.1 Visual demonstration of the commutativity of multiplication $3 \times 5 = 5 \times 3$

2.1 Natural Numbers

The practice of counting objects led to the development of natural numbers such as 1, 2, 3, ... With these numbers, we can do two basic operations of arithmetic, namely addition (+) and multiplication ($\times$ or $\cdot$). Addition such as $3 + 5 = 8$ (8 is called the sum) and multiplication such as $4 \times 5 = 4 \cdot 5 = 20$ (where 20 is called the product), are easy to understand.

It can be seen that the addition and multiplication operations have the following properties

$$3 + 5 = 5 + 3, \quad 3 \times 5 = 5 \times 3, \quad 3 \times (2 + 4) = 3 \times 2 + 3 \times 4 \ (= 18) \tag{2.1.1}$$

One way to understand why $3 \times 5 = 5 \times 3$ is to use visual representations. Figure 2.1 provides two such representations; for the one on the left, the area of that rectangle does not care whether we place it in on the long or short side, thus 3×5 must be equal to 5×3.[1] For the one on the right, we use a block of rocks (plotted as dots): we can either have a block of 5 rows, each row having 3 rocks, thus in total there are 5×3 rocks, or have a block of 3 rows, each row having 5 rocks, thus in total there are 3×5 rocks. For $3 \times (2 + 4) = 3 \times 2 + 3 \times 4$, see Fig. 2.2.

As there is nothing special about numbers 3, 5, 2, 4 in Eq. (2.1.1); other natural numbers obey these rules too, one can define the following arithmetic rules for natural numbers a, b and c[2]:

---

[1] If you did not know what is area and the formula for the area of simple geometries, check out Sect. 3.1.6.

[2] Note that we do not attempt to prove these rules. We feel that they are reasonable to accept (using devices such as Fig. 2.1). You can consider them as the rules of the game if we want to play with natural numbers. Of course, some mathematicians are not happy with that, and they came up with other axioms (rules) and from those axioms they can indeed prove the rules we are discussing now. If interested you can Google for Peano axioms.

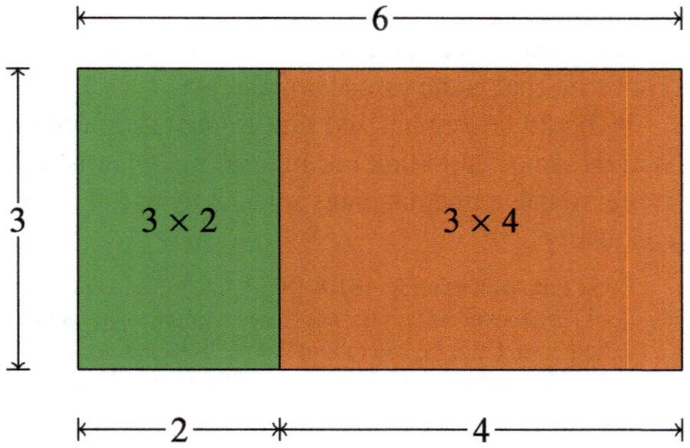

Fig. 2.2 Visual demonstration of the associativity of addition/multiplication $(2 + 4) \times 3 = 2 \times 3 + 4 \times 3$. Try to draw a similar diagram to see that $(4 - 2) \times 3 = 4 \times 3 - 2 \times 3$

$$\begin{aligned}
&\text{(a1) commutative } a + b = b + a \\
&\text{(a2) commutative } ab = ba \\
&\text{(b1) associative} \quad (a + b) + c = a + (b + c) \\
&\text{(b2) associative} \quad (ab)c = a(bc) \\
&\text{(c1) distributive} \quad a(b \pm c) = ab \pm ac
\end{aligned} \tag{2.1.2}$$

where ab means $a \times b$, but 23 means twenty three not $2 \times 3 = 6$: the rule is the multiplication sign is omitted between letters and between a number and a letter (for mathematicians had to write these a lot and they do not want to spend time writing these boring signs, they always focus on the ideas and patterns). We should pause and appreciate the power of these rules. For example, the b1 rule allows us to put the parentheses anywhere we like (or even omit them).

Once we have recognized how numbers behave, we can take advantage of that. For example, to compute $571 \cdot 36 + 571 \cdot 64$ the naive way we need two multiplications and one addition. Using the distributive property we can do: $571 \cdot (36 + 64) = 571 \cdot 100$—one addition and one easy multiplication. That's a humbling example of the power of recognizing the pattern in mathematics. Another example, to compute the sum $1 + 278 + 99$, we can use the b1 rule to proceed as $(1 + 99) + 278 = 100 + 278 = 378$. Note also that the distributive rule c1 can be written as $(b + c)a = ba + ca$, and this is the rule we implicitly use when we write $5a + 7a = 12a$.

We must note here that the introduction of a symbol (say a) to label any natural number is a significant achievement in mathematics, done in about the 16th century. Before that mathematicians only worked with specific concrete numbers (e.g., 2 or 10). With symbols comes the power of generalization; Eq. (2.1.2) covers the behavior of all natural numbers in one go! Note that we have infinity of such numbers and just one short equation can state a property of all of them. But if we think deeply we see that we do it all the times in our daily life. We use "man" and "woman" to represent any man and woman, whereas "John" and "Mary" describe two particular man and woman!

Simply put, we can say that in algebra we are doing arithmetic with just one new feature: we use letters to represent numbers. Thus, instead of focusing on $2 + 3$, we pay attention to $x + 3$, or $x + y$. As these letters (e.g., x, y, z, a, b, c) representing numbers and we can do arithmetic with numbers, so we can do arithmetic with letters. That's it. That is the essence of algebra.

It should be emphasized that *arithmetic is not mathematics*. The fact that $3 + 5$ is eight is not important nor interesting; what is more interesting is $3 + 5 = 5 + 3$. Professional mathematicians are usually bad at arithmetic as the following true story can testify[3]:

> Ernst Eduard Kummer (1810–1893), a German algebraist, was sometimes slow at calculations. Whenever he had occasion to do simple arithmetic in class, he would get his students to help him. Once he had to find 7×9. "Seven times nine," he began, "Seven times nine is er – ah — ah – seven times nine is. . . ." "Sixty-one," a student suggested. Kummer wrote 61 on the board. "Sir," said another student, "it should be sixty-nine." "Come, come, gentlemen, it can't be both," Kummer exclaimed. "*It must be one or the other.*"

However, as future engineers and scientists you have to be very exact with your calculations. The main purpose of me telling this story to emphasize that arithmetic (which is boring) is not mathematics (which is super interesting).

Abstraction and representation. As kids we were introduced to natural numbers too early that most of the time we take them for granted. When we're getting old enough, we should question them by asking questions *where these numbers come from?, why we write 5 for five? etc.* From concrete things in life such as **five** trees, **five** fishes, **five** cows, etc. human beings have developed number five to represent the *five-ness*. This number five is an *abstract* entity in the sense that we never see, hear, feel, or taste it. And yet, it has a definite existence for the rest of our lives. Do not confuse number five and its representation (5 in our decimal number system) as there are many representations of a number (e.g., V in the Roman number system).

We observed a pattern (five-ness) and we created an abstract entity from it. This is called *abstraction*. And this abstract entity is very powerful. While it is easy to explain a collection of five or six objects (using your fingers), imagine how awkward would it be to explain a set of thirty-five objects without using the number 35.

Now that we have the concept of natural numbers, how we are going to represent them? Ancient people used dots to represent numbers, and tallies were also used. But it was soon realized that all these methods are bad at representing large numbers. Just think of how you would write the number "one hundred" using dots and you understand what I meant. Only after a long period that we developed the decimal number system with only 10 digits (0, 1, 2, 3, 4, 5, 6, 7, 8, 9) that can represent any number you can imagine of!

Is the decimal number system the only one? Of course not, the computers only use two digits 0 and 1. Is it true that we're comfortable with the decimal number system because we have ten fingers? I do not know. I posed this question just to demonstrate that even for something as simple as counting numbers, that we have merely taken for granted, there are many interesting aspects to explore. A curious mind can lead us far.

---

[3] Source: https://www.math.utah.edu/~cherk/mathjokes.html.

History of the equal sign. The inventor of the equal sign '=' was the Welsh physician and mathematician Robert Recorde (c. 1512 – 1558). In 1557, in The Whetstone of Witte, Recorde used two parallel lines (he used an obsolete word gemowe, meaning 'twin') to *avoid tedious repetition of the words 'is equal to'*. He chose that symbol because 'no two things can be more equal'. Recorde chose well. His symbol has remained in use for 464 years.

2.2 Doing Some Algebra

With these rules, we can start doing some algebra. What I meant by that is to manipulate algebraic expressions. An algebraic expression is an expression involving numbers, parentheses, operation signs ($+$, $-$, $\times$, $\div$, etc.) and variables (e.g., a, b, x, y). Examples of algebraic expressions are: $3x + 1$ and $5(x^2 + 3x)$. Note that the multiplication sign is omitted between letters and between a number and a letter: so we write $2x$ instead of $2 \times x$. Having understood what an algebraic expression is, we should be able to tell what not. For example, this is not an algebraic expression: $3x+$, as it does not make sense: $3x$ plus what?

As the first example of manipulating algebraic expressions, consider this problem: what is the square of $a + b$, which is $(a + b)(a + b)$? Think of a square of side $a + b$, then its area is $(a + b)(a + b)$. Mathematicians are lazy, so they use the notation $(a + b)^2$ for $(a + b)(a + b)$. For a given natural number a, its square is $a^2 = a \times a$, its cube is $a^3 = a \times a \times a$; they are examples of powers of a. In Sect. 2.19 we will talk more about powers.

Getting back to $(a + b)^2$, we proceed as[4]

$$\begin{aligned}
(a+b)^2 &= (a+b)(a+b) && \text{(definition)} \\
&= a(a+b) + b(a+b) && \text{(distributive c1)} \\
&= a^2 + ab + ba + b^2 && \text{(distributive c1)} \\
&= a^2 + ab + ab + b^2 && \text{(commutative for multiplication a2)} \\
&= a^2 + 2ab + b^2
\end{aligned} \tag{2.2.1}$$

What is the significance of this? It tells us that the two seemingly different expressions $(a + b)^2$ and $a^2 + 2ab + b^2$ are the same! A geometry proof of this so-called *identity* is shown in Fig. 2.3a. This was how ancient Greek mathematicians thought of $(a + b)^2$. They thought in terms of geometry: any quadratic term of the form ab is associated with the area of a certain shape. And this way of geometric thinking is very useful as we will see in this book. I am against memorizing any formula (including this identity); this is because it is understanding that is important.

What is a mathematical identity? An example would answer this question: $(a + b)^2 = a^2 + 2ab + b^2$ is an identity, as it involves equality of two expressions (that

---

[4] One more exercise to practice: $(3a)^2 = (3a)(3a)$—this is just definition. Now $(3a)(3a) = (3)(3)(a)(a)$ because of the associative property b2. Finally, $(3a)^2 = 9a^2$.

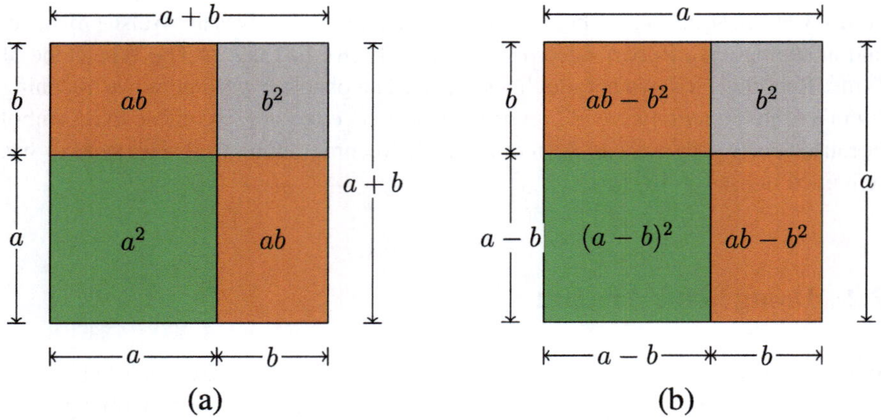

Fig. 2.3 Geometric visualization of $(a+b)^2$ and $(a-b)^2$

contains some variables) for all values of these variables. Plugging $a=2, b=3$ into this identity we surely get $5^2 = 4 + 12 + 9$, which is correct.

What we have done in Eq. (2.2.1) was to expand and simplify an expression. That is, we multiplied out the brackets and then simplified the resulting expression by collecting the like terms (i.e.,, ab and ba).

Let's pause for a moment and think more about $(a+b)^2 = a^2 + 2ab + b^2$. What else does this tell us? Surprisingly a lot! We can think of $(a+b)^2$ as a hamster (in our mathematical world). If we do not touch it, talk to it, it does not talk back. And that's why we just see it as $(a+b)^2$. However, when we talk to it by massaging it, it talks back by revealing its secret: it has another name and it is $a^2 + 2ab + b^2$. So, we can think of mathematicians as magicians (but without a tick), while magicians can get a rabbit out of an empty hat with a tick, mathematicians can too: they poke their numbers and pop out many interesting facts.

But hey, why knowing another name of that hamster is useful? First, mathematicians—as human beings—are curious by nature: they want to know everything about mathematical objects. Second, probably not a very good example, but if the more you know about your enemy, the better, don't you think?

The next thing to do is $(a+b)^3$, which can be computed as

$$\begin{aligned}
(a+b)^3 &= (a+b)(a+b)(a+b) && \text{(definition)}\\
&= (a+b)^2(a+b) && \text{(definition of } (a+b)^2\text{)}\\
&= (a^2+2ab+b^2)(a+b) && \text{(Eq. (2.2.1))}\\
&= a^3+a^2b+2a^2b+2ab^2+b^2a+b^3 && \text{(distributive c1)}\\
&= a^3+3a^2b+3ab^2+b^3 && \text{(collect like terms)}
\end{aligned} \tag{2.2.2}$$

Of course, a geometric interpretation of this expression is available (volumes instead of areas); but I do not think such a diagram would help here. By playing with expressions such as $(a+b)^2$, $(a+b)^3$ etc. that our ancestors came up with the so-called Pascal triangle and the binomial theorem (see Sect. 2.27).

How about $(a+b+c)^2$? Of course, we can do the same way by writing this term as $[(a+b)+c)]^2$, then seeing $a+b=d$ as just a number, then using Eq. (2.2.1) for $(d+c)^2$. However, there is a better way: guessing the result! Our guess is as follows

$$(a+b+c)^2 = a^2+2ab+b^2+c^2+2bc+2ca \tag{2.2.3}$$

The red terms are present when $c=0$, the blue terms are due to the fact that a, b, c are equal: if there is a^2, there must be c^2. By doing this way we're gradually developing a feeling of mathematics.

The FOIL rule of algebra.

In many textbooks on algebra we have seen this identity:

$$(a+b)(c+d) = ac+ad+bc+bd$$

And to help students memorize it someone invented the **FOIL** rule (First-Outer-Inner-Last). I'm against this way of teaching mathematics. This identity is very natural as it comes from the arithmetic rules given in Eq. (2.1.2). Let's denote $c+d=e$ (the sum of two natural numbers is a natural number), so we can write

$$\begin{aligned}
(a+b)(c+d) = (a+b)e &= ae+be && \text{(distributive rule)}\\
&= a(c+d)+b(c+d) && \text{(replace } e \text{ with } c+d\text{)}\\
&= ac+ad+bc+bd && \text{(again, distributive rule)}
\end{aligned}$$

We have, so far, expanded brackets, it's time to do the opposite. The problem is: write a^2-b^2 in another form. We can proceed using geometry as shown in Fig. 2.4. From that, we have the following identity

$$a^2-b^2 = (a+b)(a-b)$$

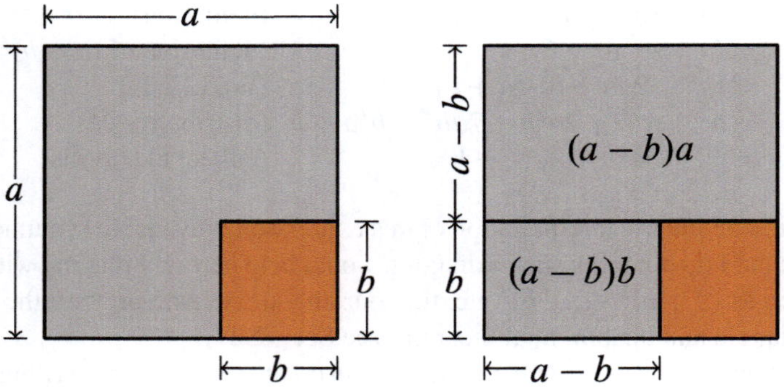

Fig. 2.4 Geometric interpretation of $a^2 - b^2$. We have two squares, one of side a and the other of side b. By our arrangement of these two squares, the term $a^2 - b^2$ is the area of the gray shaded region shown in the left diagram. This figure can be divided into two rectangles, one of area $(a - b)b$ and one of area $(a - b)a$. Adding these two areas and we obtain $(a + b)(a - b)$

which can be verified by going from $(a + b)(a - b)$: it is $a^2 - ab + ba - b^2 = a^2 - b^2$. This identity can help us for example in computing $100\,002^2 - 99\,998^2$ without a calculator nearby. Squaring and subtracting would take quite a while, but the identity is of tremendous help: $100\,002^2 - 99\,998^2 = (100\,002 + 99\,998)(100\,002 - 99\,998) = 4(200\,000)$.

Writing $a^2 - b^2$ as $(a - b)(a + b)$ is called factorizing the term. In mathematics, factorization or factoring consists of writing a number or another mathematical object as a product of several factors, usually *smaller or simpler objects* of the same kind. For what? For a better understand of the original object. For example, 3×5 is a factorization of the integer 15, and $(x - 2)(x + 2)$ is a factorization of the polynomial $x^2 - 4$. We have more to say about factorization in Sect. 2.15. And when we meet other mathematical objects (e.g., matrices) later in the book, we shall see that mathematicians do indeed spend a significant of time just to factor matrices.

2.3 Integer Numbers

This section discusses integer numbers that are whole numbers which can be positive or negative (Sect. 2.3.1). A short introduction of negative numbers is given in Sect. 2.3.2. Then, in Sect. 2.3.3, the arithmetic of negative numbers is given; the answer to questions such as why $-1 \times -1 = 1$ can be found herein.

2.3.1 Negative Numbers

So far so good: addition and multiplication of natural numbers are easy. But what is more important is this observation: adding (or multiplying) two natural numbers gives us another natural number. Mathematicians say that natural numbers are *closed under addition and multiplication*. Why they care about this? Because it ensures security: we never step outside of the familiar world of natural numbers, until... when it comes to subtraction. What is $3 - 5$? Well, we can take 3 from 3 and we have nothing (zero). How can we take away two from nothing? It seems impossible. Shall we only allow subtraction of the form $a - b$ when $a \geq b$ (this is how mathematicians say a is larger than or equal to b)? That sounds restricted, doesn't it?

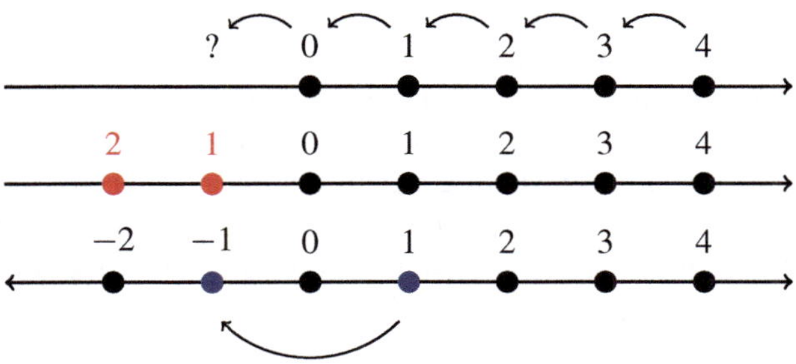

$$4 - 1 = 3$$
$$4 - 2 = 2$$
$$4 - 3 = 1$$
$$4 - 4 = 0$$
$$4 - 5 = ?$$

(2.3.1)

If we imagine a line on which we put zero at a certain place, and on the right of zero we place 1, 2, 3, and so on (see the figure in Eq. (2.3.1)). The English mathematician, John Wallis (1616–1703) is credited with giving some meaning to negative numbers by inventing the number line, which is what I am presenting here. Now, when we do a subtraction, let say $4 - 1$, we start from 4 on this line and walk towards zero one step: we end up at three. Similarly, when we do $4 - 2$ we walk towards zero two steps. Eventually, we reach zero when have walked four steps: $4 - 4 = 0$. What happens then if we walk past zero one step? It is exactly what $4 - 5$ means. We should now be at the position marked by number one but in red (to indicate that this position is on the left side of zero). So, we have solved the problem: $4 - 5 = 1$. Nowadays people write -1 (read 'negative one') instead of using a different color. Thus, $4 - 5 = -1$.

Now we have two kinds of numbers: the ones on the right-hand side of zero (e.g., $1, 2, \ldots$) and the ones on the left-hand side (e.g., $-1, -2, \ldots$). The former are called *positive integers* and the latter *negative integers*; together with zero they form the so-called integers: $\{\ldots, -3, -2, -1, 0, 1, 2, 3, \ldots\}$.[5]

The number line is kind of a two-way street: starting from zero, if we go to the right we go in the positive direction (for we see positive integers), and if we go to the left, we follow the negative direction. For every positive integers a, we have a negative counterpart $-a$. We can think of—as an operation that flips a to the other

[5] Sometimes, we also write positive integers as $+2$.

side of zero. Why we have to start with a positive integer (if all numbers should be treated equal)? If we start with a negative number, let say, $-b$ ($b > 0$), then to flip it to the other side of zero is: $-(-b)$ which is b. So we have $-(-b) = b$ for any integer—positive and negative. If $b > 0$ you can think of this as taken away a debt is an asset.

2.3.2 A Brief History on Negative Numbers

Negative numbers appeared for the first time in history in the *Nine Chapters on the Mathematical Art*, which in its present form dates from the period of the Han Dynasty (202 BC–AD 220). The mathematician Liu Hui (c. third century) established rules for the addition and subtraction of negative numbers. The nine chapters used *red counting rods* to denote positive numbers and *black rods* for negative numbers. During the seventh century AD, negative numbers were used in India to represent debts. The Indian mathematician Brahmagupta, in Brahma-Sphuta-Siddhanta (written c. AD 630), gave rules regarding operations involving negative numbers and zero, such as "A debt cut off from nothingness becomes a credit; a credit cut off from nothingness becomes a debt". He called positive numbers "fortunes", zero "a cipher", and negative numbers "debts".

While we have no problems accepting positive numbers, it is mentally hard to grasp negative numbers. What is negative four cookies? This is because negative numbers are more *abstract* than positive ones. For a long time, negative solutions to problems were considered "false". In Hellenistic Egypt, the Greek mathematician Diophantus, in his book *Arithmetica*, while referring to the equation $4x + 20 = 4$ (which has a negative solution of -4) saying that the equation was *absurd*. This is because Greek mathematics was found on geometrical ideas: a number is a certain length or area or volume of something; thus number is always positive.

2.3.3 Arithmetic of Negative Integers

How arithmetical operations work for negative numbers? The first thing to notice is that the symbol of subtraction is used to indicate negative integers. Why? One way to explain is that: $0 - 2 = -2$ (taken 2 away from zero results in a debt of 2, or going from zero two steps to its left side results in -2).

Using the number line (the one with two directions), or the concept of debts, it is not hard to do the following arithmetical problems:

$$
\begin{array}{ll}
\text{(a) adding two negative integers:} & (-2) + (-3) = -5, \quad (-3) + (-1) = -4 \\
\text{(b) adding a negative and a positive:} & (-2) + 1 = -1, \quad (-2) + 3 = 1 \\
\text{(c) subtracting two negative integers:} & (-3) - (-1) = -2, \quad (-3) - (-4) = 1
\end{array}
\tag{2.3.2}
$$

For (c), using the rule $-(-b) = b$ for any integer—positive and negative. When $b > 0$, this rule can be understood as "taken away a debt is an asset".

Let's study now the arithmetic rules for multiplication of negative numbers. We should always start simple: multiplication of a negative integer and a positive integer. It is obvious to see that

$$(-1) + (-1) + (-1) = -3$$

as, after all, if I borrow you one dollar a week for three weeks, then I own you three dollars.[6] This immediately results in the following

$$(-1) \times 3 := (-1) + (-1) + (-1) = -3 \quad (= 3 \times (-1))$$

So, when a positive number a is multiplied with -1 it is flipped to the other side of zero on the number line at $-a$. That is, $-a = (-1) \times a = (-1)a$ for integer a. And with that we know how to handle -2×10, which is -20, and so on. The general rule is $(-a)(b) = -(ab)$, for positive integers a and b. In English, it reads "the multiplication of a positive and a negative number yields a negative number whose numerical value is the product of the two given numerical values".

But what maths has to do with debts? Can we deduce the rules without resorting to debts, which are very negative. Ok, let's compute $5 \times (3 + (-3))$ in two ways. First, as $3 + (-3) = 0$, we have $5 \times (3 + (-3)) = 0$. But from Eq. (2.1.2), we also have (distributive rule)

$$5 \times (3 + (-3)) = 5 \times 3 + 5 \times (-3) = 0 \Longrightarrow 5 \times (-3) = -15$$

Thus, if we insist that the usual arithmetic rules apply also for negative numbers, we have deduced a rule that is consistent with daily experience. From a mathematical viewpoint, mathematicians always try to have a set of rules that works for as many objects as possible. They have the rules in Eq. (2.1.2) for positive integers, now they gave birth to negative integers. To make positive and negative integers live happily together the negative integers must follow the same rules. (They can have their own rules, that is fine, but they must obey the old rules).

But how about $(-1) \times (-1)$? One way to figure out the result is to look at the following

$$\left.\begin{aligned} (-1) \times 3 &= -3 \\ (-1) \times 2 &= -2 \\ (-1) \times 1 &= -1 \\ (-1) \times 0 &= 0 \end{aligned}\right\} \Longrightarrow (-1) \times (-1) = 1$$

and observe that from top going down, the RHS numbers get increased by one. Thus $(-1) \times 0 = 0$ should lead to $(-1) \times (-1) = 0 + 1 = 1$. This is certainly not a proof

[6] If you prefer thinking of geometry, the the number line is very useful: $(-1) + (-1) + (-1)$ is walking three steps to the negative direction from zero, we must end up at -3.

for we're not sure that the pattern will repeat. This was just one short explanation. If you were not happy with that, then $(-1) \times (-1) = 1$ was a consequence by our choice to maintain the arithmetic rules, the distributive rule, in Eq. (2.1.2):

$$
\begin{aligned}
&1 + (-1) = 0 \Rightarrow [1 + (-1)] \times (-1) = 0: \\
&1 \times (-1) + (-1) \times (-1) = 0 \Longrightarrow (-1) \times (-1) = 1
\end{aligned}
$$

Coincidentally, it is similar to the ancient proverb *the enemy of my enemy is my friend*. If you are struggling with this, it is OK as the great Swiss mathematician Euler (who we will meet again and again in this book) also struggled with the fact that $(-1)(-1) = 1$ too.

Knowing the arithmetic rules of negative integers, we can now study the expression $(a-b)^2$. We proceed as follows

$$
\begin{aligned}
(a-b)^2 &= (a-b)(a-b) && \text{(definition)} \\
&= (a-b)a - (a-b)b && \text{(distributive c1)} \\
&= a^2 - ba - (ab - b^2) && \text{(distributive c1)} \\
&= a^2 - ba - ab + b^2 && \text{(open bracket)} \\
&= a^2 - 2ab + b^2
\end{aligned}
\tag{2.3.3}
$$

A geometry-based proof of it is given in Fig. 2.3b. The open bracket rule is: $-(a - b) = a + b$. That is, if a bracket is preceded by a minus sign, change positive signs within it to negative and vice-versa when removing the bracket. Need a proof? Here it is:

$$
\begin{aligned}
-(a-b) &= (-1) \times (a-b) \\
&= (-1) \times a + (-1) \times (-b) \\
&= -a + (-1) \times [(-1) \times b] \\
&= -a + [(-1) \times (-1)] \times b = -a + b
\end{aligned}
\tag{2.3.4}
$$

Question 1 *How many are there integer numbers?*

2.4 Playing with Natural Numbers

Once human have created numbers, they started playing with them and discovered many interesting properties. For example, some numbers are even and some are odd (Fig. 2.5). Even natural numbers are $2, 4, 6, 8, \ldots$, which can be written as $2, 2 \times 2, 2 \times 3, \ldots$ Then, a generalization can be made, an even number is the one of the form $2k$—it is divisible by 2. And an odd number is $2k + 1$—it is not divisible by 2. To say about the oddness or evenness of a number, mathematicians use the term *parity*.

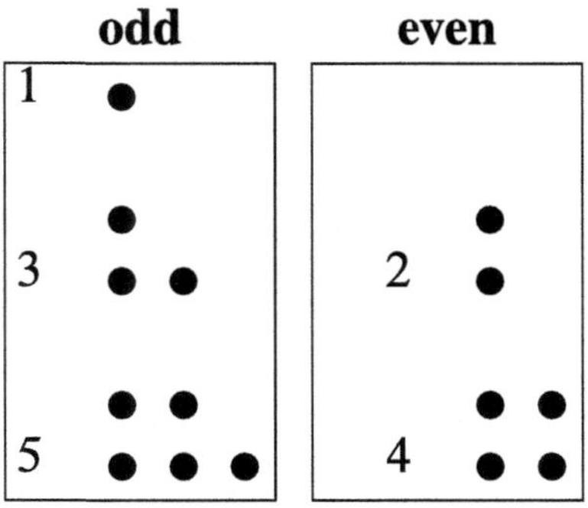

Fig. 2.5 Even and odd numbers. Using this visualization, an odd number is obviously written as $2k + 1$—it is not divisible by 2

To demonstrate the importance of definition in mathematics, below are three definitions of even numbers. What is the one mathematicians use?

- Definition 1: An even number is any integer that can be divided exactly by 2;
- Definition 2: A number that is divisible by 2 and generates a remainder of 0 is called an even number.
- Definition 3: An even number is an integer of the form $n = 2k$, where k is an integer.

They are equivalent, but the definition 3 is the one that is the one that mathematicians adopt. Why? Because they can use this definition to deduce the behavior of even numbers. Precisely, they can manipulate the expression $n = 2k$ in many ways. I present some examples now so that you can see the point. Other definitions are useless as they are all verbal.

Now we have two groups of even and odd numbers, questions about their relation arise. For instance, is there any relation between even/odd numbers? Yes, for example: even times even is even (e.g., $2 \times 6 = 12, 4 \times 6 = 24$) or odd times odd is odd (e.g., $3 \times 5 = 15, 3 \times 7 = 21$). But why so? Below are proofs. Even though these proofs are elementary, I still present them to emphasize that there are proofs everywhere, not just in Euclidean geometry class!

Proof that the product of two even numbers is always an even number According to the definition number of even number, we write $2n$ for the first even number and $2k$ for the second even number, where n and k are integers. Their product is then $(2n)(2k)$. But, according to the associativity rule of multiplication, Eq. (2.1.2), we can write this product as $2(2nk)$, which is nothing but $2m$, where $m = 2nk$ is an integer (why?). We use the definition (of even numbers) again: the product is of the form $2m$, thus it is even. ■

This proof is a bit verbal, so I will try to have a shorter proof:

Proof that the product of two even numbers is always an even number Assume that a and b are two even numbers. According to the definition 3 (of even numbers), we then have $a = 2n$ and $b = 2k$ for integers n, k. Their product is $ab = (2n)(2k)$, which can be rewritten as $2m$, where $m = 2nk$ is an integer. Therefore, by definition 3, ab is even. ■

Here is another property of numbers that we find: if we multiply two consecutive odd numbers (or consecutive even numbers) we get a number that is one less than a perfect square (e.g., $5 \times 7 + 1 = 36 = 6^2$, $10 \times 12 + 1 = 121 = 11^2$). Is this always the case, or can we find two consecutive odd or even numbers for which this phenomenon does not occur? We can check this forever, or we can prove it once and for all. *Mathematicians are lazy*, so they prefer the latter. Here is the proof. The plan of the proof is: (i) translate the English phrase "two consecutive even numbers" into mathematical symbols: if the first even number is $2k$, then the next even number is $2k + 2$,[7] (ii) translate "multiply two consecutive even numbers and add 1" to $(2k)(2k + 2) + 1$. The remaining is simply algebra:

Proof Let's call the first even number by $2k$, then the next even number is $2k + 2$ where $k = 1, 2, 3, \ldots$ The sum of their product and one is:

$$(2k)(2k + 2) + 1 = (2k)^2 + 4k + 1 = (2k)^2 + 2(2k) + 1 = (2k + 1)^2$$

which is obviously a perfect square. You should completely understand the above sequence of steps: the first one is the distributive property c1 in Eq. (2.1.2) (i.e.,, $a(b + c) = ab + ac$ with $a = 2k, b = 2k$ and $c = 2$); the second step is to make appear $(a + b)^2 = a^2 + 2ab + b^2$ with $a = 2k$ and $b = 1$. And why $4k = 2(2k)$? It is so because $4k = (2 \times 2)k$, which is also $2(2k)$ due to property b2 $(ab)c = a(bc)$ in Eq. (2.1.2). ■

Why mathematicians care about perfect squares? One reason: it is super easy to compute the square root of a perfect square.

Here is yet another property of integers: multiplying three successive integers and adding the middle integer to this product always yields a perfect cube! For example, $2 \times 3 \times 4 + 3 = 27 = 3^3$. Why?

Proof Let's denote three successive integers by $k, k + 1, k + 2$, then we can write

$$k(k + 1)(k + 2) + (k + 1) = (k + 1) \underbrace{[k(k + 2) + 1]}_{(k+1)^2} = (k + 1)^3$$

which is a perfect cube. ■

2.4.1 Divisibility

$$\frac{\text{divident}}{\text{divisor}} = \text{quotient}$$

Divisibility is the ability of a number to be *evenly divided* by another number. For example, four divided by two is equal to two, an integer, and therefore we say four

[7] Think of concrete examples such as 2, 4 or 6, 8, and you will see this.

is *divisible* by two. Mathematicians also say that two divides four, which means that "2 divides 4 into 2 parts". I introduce some terminologies now: the number which is getting divided here is called the *dividend*. The number which divides a given number is the *divisor*. And the number which we get as a result is known as *the quotient*. So, in $6 : 3 = 2$, 6 is the dividend, 3 is the divisor and 2 is the quotient. Mathematicians write $3|6$ to say that 3 divides 6 or equivalently 6 is divisible by 3.[8] Thus, $a|b$ means that there exists a whole number k such that $ak = b$.

Here is another property of counting numbers regarding divisibility: a number is divisible by 9 *if and only if* the sum of its digits is divisible by 9. For example, 351 is divisible by 9 and $3 + 5 + 1 = 9$ is clearly divisible by 9. To see why, we write 351 as follows

$$\begin{aligned}
\overset{210}{351} &= 3 \times 100 + 5 \times 10 + 1 \times 1 \\
&= 3 \times 10^2 + 5 \times 10^1 + 1 \times 10^0 \\
&= 3 \times (99 + 1) + 5 \times (9 + 1) + 1 \\
&= 3 \times 99 + 5 \times 9 + \underbrace{3 + 5 + 1}_{\text{sum of the digits}} \\
&= 9(3 \times 11 + 5 + 1)
\end{aligned}$$

Therefore, it is divisible by 9. But this does not make a proof! It, however, give us some hints for a proof. Now, we write a counting number in this form

$$a_n a_{n-1} \cdots a_1 a_0$$

which is a number having $n + 1$ digits whatever n is.[9] Then, we do similarly to what we have done for the number 351:

$$\begin{aligned}
a_n a_{n-1} \cdots a_1 a_0 &= a_n \times 10^n + a_{n-1} \times 10^{n-1} + \cdots + a_1 \times 10 + a_0 \\
&= a_n(10^n - 1 + 1) + a_{n-1}(10^{n-1} - 1 + 1) + \cdots + a_0 \\
&= (a_n + a_{n-1} + \cdots + a_1 + a_0) \\
&+ 9(\underbrace{a_n a_n \cdots a_n}_{n terms} + \underbrace{a_{n-1} a_{n-1} \cdots a_{n-1}}_{n-1 terms} + \cdots + a_1)
\end{aligned}$$

And that concludes the proof of the divisibility of 9: when $a_n + a_{n-1} + \cdots + a_1 + a_0 = 9k$. Note that, in the above equation, we have used

[8] $a|b$ does not mean the same thing as a/b. The latter is a number, the former is a statement about two numbers.

[9] How mathematicians came up with this notation? Let's start simple with two digit number (e.g., 24, 99): they can be written as ab. Then a three digit number is abc. But there are only about 25 alphabets! The solution is to use subscripts a_1,a_2. There are infinite counting numbers, thus we do not have to worry about using them to label anything.

$$10^n - 1 = \underbrace{99\cdots 9}_{nterms} \Longrightarrow a_n(10^n - 1) = a_n \times \underbrace{99\cdots 9}_{nterms}$$
$$= a_n \times 9 \times \underbrace{11\cdots 1}_{nterms} = 9 \times \underbrace{a_n a_n \cdots a_n}_{nterms}$$

So, you have seen that after the property has been discovered, the proof might not be so difficult, especially when we have worked out a proof for a special case.

Remark If you think about my "proof" of the statement "a number is divisible by 9 if and only if the sum of its digits is divisible by 9", you might realize that it is not complete. We come back to this issue in Sect. 2.5.

A good question is how we have discovered the property in the first place? It is simple: by playing with numbers very carefully. For example, we all know the times table for 9. If we not just look at the multiplication, but also the inverse, i.e.,, the division, we see this:

$$\begin{aligned} 9 \times 1 &= 9 & 9 : 9 &= 1 \\ 9 \times 2 &= 18 & 18 : 9 &= 2 \\ 9 \times 3 &= 27 & 27 : 9 &= 3 \\ 9 \times 4 &= 36 & 36 : 9 &= 4 \end{aligned}$$

Then, by looking at the red numbers, the divisibility of a number for 9 was discovered. The lesson is always to look at a problem from different angles. For example, if you see the word "Rivers", it can be a name of a person not just the rivers.

Without double that there are rules regarding the divisibility of a number by 2, 3, 4, 5, and so on. Try to find them for yourself. We shall get back to this problem when know more about mathematics (Sect. 2.30.4).

Here are only a few interesting facts about natural numbers. There are tons of other interesting results. If you have found that they are interesting, study them! The study of natural numbers has gained its reputation as the "queen of mathematics" according to Gauss—the famous German mathematician, and many of the greatest mathematicians have devoted study to numbers. You could become a number theorist (a mathematician who studies natural numbers) or you could work for a bank on the field of information protection—known as "cryptography". Or you could become an amateur mathematician like Pierre de Fermat who was a lawyer but studied mathematics in free time for leisure purposes.

If you do not enjoy natural numbers, that is of course also totally fine. For sciences and engineering, where real numbers are dominant, a good knowledge of number theory is not needed. Indeed, before writing this book, I knew just a little about natural numbers and relations between them.

One of the amazing things about pure mathematics—mathematics done for its own sake, rather than out of an attempt to understand the "real world"—is that sometimes, purely theoretical discoveries can turn out to have practical applications. This happened, for example, when non-Euclidean geometries described by the mathematicians Karl Gauss and Bernard Riemann turned out to provide a model for the relativity between space and time, as shown by Albert Einstein.

Taxicab number 1729. The name is derived from a conversation in about 1919 involving British mathematician G. H. Hardy and Indian mathematician Srinivasa Ramanujan. As told by Hardy:

> I remember once going to see him [Ramanujan] when he was lying ill at Putney. I had ridden in taxi-cab No. 1729, and remarked that the number seemed to be rather a dull one, and that I hoped it was not an unfavorable omen. "No," he replied, "it is a very interesting number"; it is the smallest number expressible as the sum of two [positive] cubes in two different ways.

Ramanujan meant that $1729 = 1^3 + 12^3 = 9^3 + 10^3$!

Let's see some math magics, which, unlike other kinds of magics, can be explained.

Magic numbers.

This magic trick is taken from the interesting book *Alex's Adventures in Numberland* by Alex Bellos [8]. The trick is: "I ask you to name a three-digit number for which the first and last digits differs by at least two. I then ask you to reverse that number to give you a second number. After that, I ask you to subtract the smaller number from the larger number. I then ask you to add this intermediary result to its reverse. The result is 1089, regardless whatever number you have chosen". For instance, if you choose 214, the reverse is 412. Then, 412 – 214 = 198. I then asked you to add this intermediary result to its reverse, which is 198 + 891, and that equals 1089.

The question is why? See some hints at this footnote.[10]

2.4.2 Math Contest Problem

Let $a_1, a_2, \ldots, a_n$ represent an arbitrary arrangement of the numbers $1, 2, \ldots, n$[11]. Prove that if n is odd, the product $(a_1 - 1)(a_2 - 2)(a_3 - 3) \cdots (a_n - n)$ is an even

[10] The proof is to start with a three-digit number abc where $a, b, c \in \mathbb{N}$ and $a \neq 0$. Its reverse is cba, then the difference of cba and abc is $99(c - a)$. Note that $c - a = \{2, 3, 4, 5, 6, 7, 8\}$. Thus, the intermediary number can be only one of $\{198, 297, 396, 495, 594, 693, 792\}$. And we write this number as xyz, with $x + z = 9$ and $y = 9$ and thus its inverse is zyx. Now, adding xyz to zyx will result in $100(xz) + 20y + (x + z) = 100 \times 9 + 20 \times 9 + 9 = 1089$.

[11] If you're not sure what this sentence means, take $n = 3$ for example, then we have three integers $1, 2, 3$. Arrangements of them are $(1, 2, 3)$, $(1, 3, 2)$, $(2, 1, 3)$, and so on.

number. First, note that we do not know what is a_1, a_2, and so on. Second, the question is about the parity of the product of a bunch of integers.

Starting from the fact that odd times odd is odd,[12] we can generalize it to get a new fact that the product of a bunch of odd numbers is odd. And the product of a bunch of integers is even if there is at least one even in those integers. Using this fact, the problem is now amount to proving that among the integers $a_1 - 1, a_2 - 2, a_3 - 3, \ldots, a_n - n$ there is *at least* one even number. What we have achieved? A lot! We have transformed the given problem into an easier problem: instead of dealing with a product of numbers that we do not know, now we just need to find one even number.

Let's make the problem concrete so that it is easier to deal with. We consider the case $n = 5$. Thus, we have the numbers $1, 2, 3, 4, 5$ and $(a_1, a_2, a_3, a_4, a_5)$ can be $(1, 3, 4, 2, 5)$. We have to prove that among the numbers

$$a_1 - 1, a_2 - 2, a_3 - 3, a_4 - 4, a_5 - 5$$

there exists at least one even number. Proving this is hard (because it is not clear which one is even), so we transform the problem to prove that it is impossible that all those numbers are odd. If we can prove that, then at least one of them is even. This technique is called proof by contradiction.

If we assume that all numbers $a_1 - 1, a_2 - 2, a_3 - 3, a_4 - 4, a_5 - 5$ are odd, we get a_1 is even, a_2 is odd, a_3 is even, a_4 is odd and a_5 is even. Thus, there are three even numbers and two odds. But in $1, 2, 3, 4, 5$, there are two evens and three odds! We arrive at a contradiction, thus our assumption is wrong. We have proved the problem, at least for $n = 5$.

Nothing is special about 5, the same argument works for $7, 9, \ldots$ Actually $1, 2, 3, \ldots, n$ starts with 1, an odd number, and thus there are more odd numbers than even ones. But $a_1 - 1, a_2 - 2, \ldots, a_n - n$ starts with an even number and hence has more evens than odd numbers.

It was a good proof, but what do you think of the following proof? Even though the problem concerns a product, let's consider the sum of $a_1 - 1, a_2 - 2, \ldots, a_n - n$:

$$\begin{aligned} S &= (a_1 - 1) + (a_2 - 2) + (a_3 - 3) + \cdots + (a_n - n) \\ &= (a_1 + a_2 + \cdots + a_n) - (1 + 2 + \cdots + n) \end{aligned}$$

Why bother with this sum? Because it is zero whatever the values of $a_1, a_2, \ldots$.[13] Now the sum of an odd number of integers is zero (which is even) leads to the conclusion that one of the number must be even. (Otherwise, the sum would be odd; think of $3 + 5 + 7$ which is odd).

Why mathematicians knew to look at the sum S instead of the product? I do not know the exact answer. One thing is sum, products are familiar things to think of. But if that did not convince you, then the following joke tells it best:

---

[12] To be mathematicians alike, we say that the set of odd integers is closed under multiplication.

[13] This sum is called an invariant of the problem. Thus, this problem-solving technique is to find for invariants in the problem. Check the book by Paul Zeit, [70] for more.

A man walking at night finds another on his hands and knees, searching for something under a streetlight. "What are you looking for?", the first man asks; "I lost a quarter," the other replies. The first man gets down on his hands and knees to help, and after a long while asks "Are you sure you lost it here?". "No," replies the second man, "*I lost it down the street. But this is where the light is.*"

2.5 If and Only If: Conditional Statements

When reading about mathematics, one phrase that regularly shows up is "if and only if". This phrase particularly appears within statements of mathematical theorems. For example, "an integer number is divisible by 9 *if and only if* the sum of its digits is divisible by 9" is one such theorem. But what, precisely, does this statement mean?

To understand "if and only if", we must first know what is meant by a conditional statement. A conditional statement is one that is formed from two other statements, which we will denote by A and B. To form a conditional statement, we say "if A then B". Here is one example from our daily experiences: "If it is raining outside, then I stay inside". By the way, in "if A then B", A is called the hypothesis and B is called the conclusion.

Let's see the following two mathematical statements:

- (S1): A number is divisible by 5 *if* it ends in 5.
- (S2): A number is divisible by 5 *only if* it ends in 5.

These two statements are very similar but different in one keyword "only". Interestingly, one statement is true and the other is false. Here's why. Some of the multiples of 5 are: $5, 10, 15, 20, \ldots$ which convinces us that the statement (S2) "a number is divisible by 5 only if it ends in 5" is incorrect: the number 10 ends in 0, and yet it is divisible by 5.

We now consider the statement "an integer number is divisible by 9 if the sum of its digits is divisible by 9". This statement is true (as proved in Sect. 2.4.1). As remarkable as it is, this theorem does not answer the question: what if the sum of the digits of a number is NOT divisible by 9? So, to have a complete answer about the divisibility of a number by 9, we have to consider the converse: if a number is divisible by 9 then the sum of its digits is divisible by 9.

Given a conditional statement "if A then B", we're also interested in the converse: "if B then A". It is easy to see that the converse is not always true. When the converse is true, we have a biconditional statement:

$$\text{"Something is an } A \text{ if and only if (iff) it is a } B\text{"} \quad (A \Longleftrightarrow B)$$

"If and only if" is sometimes abbreviated to "iff" and symbolized by $\Longleftrightarrow$. For example, the statement "A triangle is equilateral iff its angles all measure 60°" means both "If a triangle is equilateral then its angles all measure 60°" and "If all the angles of a triangle measure 60° then the triangle is equilateral".

To prove a theorem stated in the form $A \iff B$, we have to prove two things. First, $A \implies B$. Second, $B \impliedby A$. Using the divisibility by 9 as an example, we now need to prove 'if a number is divisible by 9 then its digits sum is divisible by 9', as the other part was proved previously in Sect. 2.4.1.

Proof We know that $z = a_n a_{n-1} \cdots a_1 a_0$ can be written as

$$\begin{aligned} a_n a_{n-1} \ldots a_1 a_0 = (a_n + a_{n-1} + \cdots + a_0) + \\ 9(a_n a_n \cdots a_n + a_{n-1} a_{n-1} \cdots a_{n-1} + \cdots + a_1) \end{aligned}$$

If $9|z$, then $z = 9k$, thus we have

$$(a_n + a_{n-1} + \cdots + a_0) + 9(a_n a_n \cdots a_n + a_{n-1} a_{n-1} \cdots a_{n-1} + \cdots + a_1) = 9k$$

which results in $a_n + a_{n-1} + \cdots + a_0 = 9l$, or the sum of the digits is divisible by 9. ■

Remark For the statement $A \iff B$ we say that A is the necessary and sufficient condition for B, and vice versa. Note that the hypothesis that a number ends in 5 is sufficient for it to be divisible by 5, but it is not necessary. Because numbers end in 0 are also divisible by 5.

2.6 Sums of Whole Numbers

In this section, we discuss sums involving the first n whole numbers, e.g., what is $1 + 2 + 3 + \cdots + 1000$? In the process, we introduce proof by induction, which is a common mathematical proof method when there is n involved. Also introduced is the sigma notation for sum, i.e.,, $\sum_{i=1}^{n} i$.

2.6.1 Sum of the First n Whole Numbers

The sum of the first n integers is written as

$$S(n) = 1 + 2 + 3 + \cdots + n \tag{2.6.1}$$

The notation $S(n)$ indicates this is a sum and its value depends on n. The ellipsis ... also known informally as dot-dot-dot, is a series of (usually three) dots that indicates an intentional omission of a word, sentence, or whole section from a text without altering its original meaning. The word (plural ellipses) originates from the Ancient Greek élleipsis meaning "leave out". In the above equation, an ellipsis $\cdots$ (raised to the center of the line) used between two operation symbols (+ here) indicates the omission of values in a repeated operation.

There are different ways to compute this sum. I present three ways to demonstrate that there are usually more than one way to solve a mathematical problem. And the more solutions you can have the better. Among these different ways to a solve a problem, if it can be applied to many different problems, it is a powerful technique that should be studied.

The first strategy is simple: *get your hands dirty* by calculating manually this sum for some cases of $n = 1, 2, 3, 4, \ldots$ and try to *find a pattern*. Then, we propose a formula and if we can prove it, we have discovered a mathematical truth (if it is significant then it will be called theorem, and your name is attached to it forever). For $n = 1, 2, 3, 4$, the corresponding sums are

$$\begin{aligned}
n = 1 &: S(1) = 1 \\
n = 2 &: S(2) = 1 + 2 = 3 && = \frac{2 \times 3}{2} \\
n = 3 &: S(3) = 1 + 2 + 3 = 6 && = \frac{3 \times 4}{2} \\
n = 4 &: S(4) = 1 + 2 + 3 + 4 = 10 && = \frac{4 \times 5}{2}
\end{aligned}$$

From that (the red numbers), we can guess the following formula

$$S(n) = 1 + 2 + 3 + \cdots + n = \frac{n(n+1)}{2} \tag{2.6.2}$$

You should now double-check this formula for other n, and only when you're convinced that it might be correct, then prove it. Why bother? Because if you do not prove this formula for any n, it remains only as a conjecture: it can be correct for all ns that you have manually checked, but who knows whether it holds for others. How are we going to prove this? Mathematicians do not want to prove Eq. (2.6.2) n times; they are very lazy which is actually good as it forces them to come up with clever ways. A technique suitable for this kind of proof is **proof by induction**. The steps are: (1) check $S(1)$ is correct—this is called the basis step, (2) assume $S(k)$ is correct, this is known as the induction hypothesis and (3) prove that $S(k + 1)$ is correct: the induction step. So, the fact that $S(1)$ is valid leads to $S(2)$ is correct, which in turn leads to $S(3)$ and so on. This is similar to the familiar domino effect.

Proof by induction of Eq. (2.6.2) It is easy to see that $S(1)$ is true (Eq. (2.6.2) is simply $1 = 1$). Now, assume that it holds for k—a natural number, thus we have

$$S(k) = 1 + 2 + 3 + \cdots + k = \frac{k(k+1)}{2}$$

Now, we consider $S(k+1)$, which is $1+2+\cdots+k+k+1$, which is $S(k)+(k+1)$. If we can show that $S(k+1)=0.5(k+1)(k+1+1)$, then we're done. Indeed, we have

$$S(k+1) = S(k) + (k+1) = \frac{k(k+1)}{2} + (k+1) = \frac{(k+1)(k+1+1)}{2}$$

■

I now present another way done by the 10-year-old Gauss (who would later become the prince of mathematics and one of the three greatest mathematicians of all time, along with Archimedes and Newton). The story goes like this. As a young student, Gauss and his classmates were asked one day to add up all the integers from 1 to 100. We can imagine the groans. But only a few seconds later Gauss gave the correct answer. How could he possibly have performed such a feat? He had not seen the problem before, nor was he a phenomenal calculator. He was, however, an extraordinary *pattern recognizer*. What he recognized is this: if we add 1 to 100, or 2 to 99, or 3 to 98 etc. we all get 101, and there are 50 such pairs, the sum must be then $101 \times 50 = 5050$:

$$\begin{aligned} S &= 1+2+3+\cdots+100 \\ S &= 100+99+98+\cdots+1 \\ 2S &= 101+101+\cdots+101 = 101 \times 100 \\ S &= \frac{100 \times 101}{2} \end{aligned} \tag{2.6.3}$$

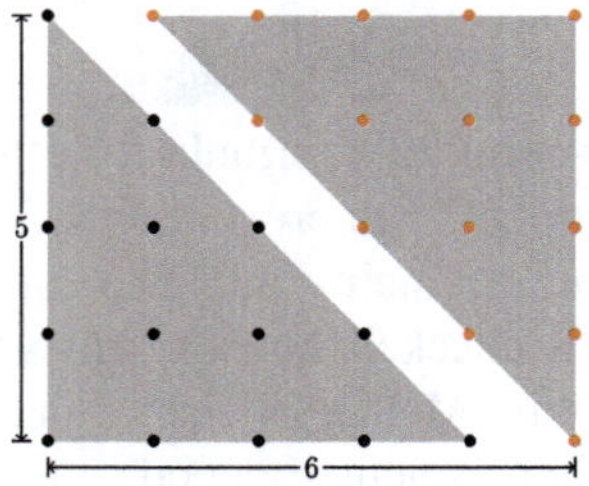

What a great idea! A geometric illustration of Gauss' clever idea is given in the figure: our sum is a triangle, and by adding to this triangle another equal triangle we get a rectangle which is easier to count the dots. Why $1+2+3+\cdots$ makes a triangle? See Fig. 2.6 for the reason. The lesson here is to try to have different views (or representations) of the same problem. In this problem, we move away from the abstract (numbers $1, 2, 3, \ldots$) back to the concrete (rocks or dots) and by playing with the dots, we can see the way to solve the problem.

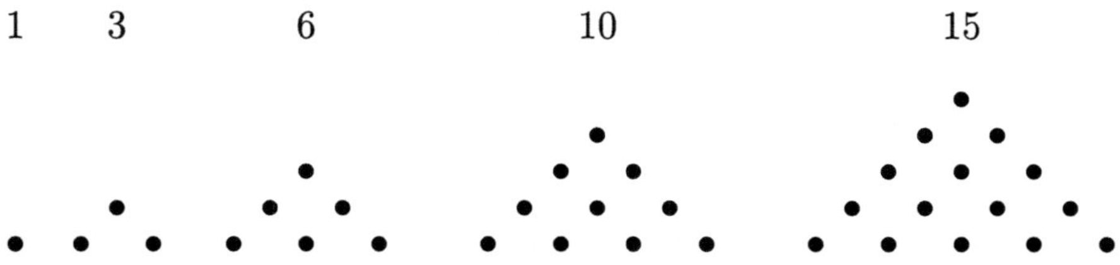

Fig. 2.6 Triangular numbers are those of the form $1 + 2 + 3 + \cdots + n$

Triangular numbers and factorial. The first four triangular numbers are

$$\begin{aligned} 1 &= 1 \\ 3 &= 1 + 2 \\ 6 &= 1 + 2 + 3 \\ 10 &= 1 + 2 + 3 + 4 \end{aligned}$$

If we replace the plus sign by the multiplication, we get factorials[14]: $1 \times 2 = 2!$, $1 \times 2 \times 3 = 3!$ and so on. It is super interesting.

The power of a formula. What is significant about Eq. (2.6.2)? First, it simplifies computation by reducing a large number of additions to three fixed operations: one of addition, one of multiplication and one of division. Second, as we have at our disposal a formula that produces a number if we plug in a number, we can, in theory, to compute $S(5/2)$, it is $35/8$. Of course, it does not make sense to ask the sum of the first $5/2$ integers. Still, formula extends the scope of the original problem to values of the variable other than those for which it was originally defined.[15]

The third way is to write the sum as follows

$$S(n) = \textcircled{1} + 2 + 3 + \cdots + \textcircled{n} = \sum_{k=1}^{n} k \tag{2.6.4}$$

The notation $\sum_{k=1}^{n} k$ reads sigma of k for k ranges from $1, 2, 3,$ to n; k is called the index of summation. It is a dummy variable in the sense that it does not appear in the actual sum. Indeed, we can use any letter we like; we can write $\sum_{i=1}^{n} i$; 1 is the starting point of the summation or the lower limit of the summation; n is the stopping point or upper limit of the summation. And $\sum$ is the capital Greek letter *sigma* corresponding to S for sum. This summation notation was introduced by Fourier in 1820. You will see that mathematicians introduce weird symbols all the times. Usually, they use Greek letters for this purpose. Note that there is no reason to

[14] Refer to Sect. 2.26.2 for detail on factorial.

[15] Believe me, it is what mathematicians do and it led to many interesting and beautiful results; one of them is the factorial of 0.5 or $(1/2)! = \sqrt{\pi}/2$, why π here?, see Sect. 4.20.2.

be scared of them, just like any human languages we need time to get used to these symbols.

Now comes the art. Out of the blue,[16] mathematicians consider this identity $(k-1)^2 = k^2 - 2k + 1$ to get

$$(k-1)^2 = k^2 - 2k + 1 \Longrightarrow \boxed{k^2 - (k-1)^2 = 2k - 1} \tag{2.6.5}$$

The boxed equation is an identity, i.e.,, it holds for $k = 1, 2, 3, \ldots$ Now, we substitute $k = 1, 2, \ldots, n$ in the boxed identity, we get n equations, and if we add these n equations we're led to the following which involves $S(n)$[17]

$$\sum_{k=1}^{n}[k^2 - (k-1)^2] = \sum_{k=1}^{n}(2k-1) = 2\sum_{k=1}^{n} k - n = 2S(n) - n \tag{2.6.6}$$

Now if the sum on the left-hand side can be found, we're done. As it turns out it is super easy to compute this sum, to see that we just need to write out $\sum_{k=1}^{n}[k^2 - (k-1)^2]$ explicitly:

$$\begin{aligned}\sum_{k=1}^{n}[k^2 - (k-1)^2] &= (1^2 - 0^2) + (2^2 - 1^2) + (3^2 - 2^2) + \cdots + (n^2 - (n-1)^2) \\ &= \cancel{1^2} + \cancel{2^2} - \cancel{1^2} + \cancel{3^2} - \cancel{2^2} + \cdots + (n^2 - (n-1)^2) = n^2\end{aligned}$$

This sum is known as *a sum of differences*, and it has a *telescoping property* that its sum depends only on the first and the last term for many terms cancel each other (e.g., the red and blue terms). We will discuss more about sum of differences, when we see that it is a powerful technique (as the sum is so easy to compute).

Introducing the above result into Eq. (2.6.6) we can compute $S(n)$ and the result is identical to the one that we have obtained using Gauss' idea and induction.

2.6.2 Sum of the Squares of the First n Whole Numbers

The sum of the squares of the first n whole numbers is expressed as (note that we use the same symbol $S(n)$ but this time a different sum is concerned. This should not cause any confuse hopefully)

---

[16] If you are really wondering the origin of this magical step, Sect. 2.21.6 provides one answer.

[17] To see why $\sum_{k=1}^{n}(2k-1) = 2\sum_{k=1}^{n} k - n$, go slowly: $\sum_{k=1}^{n}(2k-1) = \sum_{k=1}^{n} 2k - \sum_{k=1}^{n} 1$. Now, $\underbrace{1 + 1 + \cdots + 1}_{n \text{ terms}} = n$, but $1 + 1 + \cdots + 1 = \sum_{k=1}^{n} 1$. For the term $\sum_{k=1}^{n} 2k$, it is $2 \cdot 1 + 2 \cdot 2 + \cdots + 2 \cdot n = 2(1 + 2 + \cdots + n) = 2\sum_{k=1}^{n} k$.

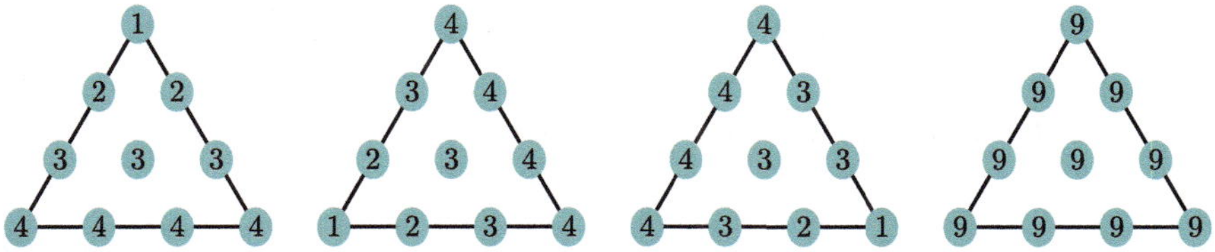

a) *b)*

Fig. 2.7 $S(n) = 1^2 + 2^2 + 3^2 + \cdots n^2$: pictorial explanation of the result $3S(n) = n(n+1)/2(2n+1)$

$$S(n) = 1^2 + 2^2 + 3^2 + \cdots n^2 = \sum_{k=1}^{n} k^2 \tag{2.6.7}$$

Among the previous three ways, which one can be used now? Obviously, the clever Gauss's trick is out of luck here. The tedious way of computing the sum for a few cases, find the pattern, guess a formula, and prove it might work. But it is hard in the step of finding the formula.

So, we adopt the telescope sum technique starting with this identity $(k-1)^3 = k^3 - 3k^2 + 3k - 1$

$$(k-1)^3 = k^3 - 3k^2 + 3k - 1 \Longrightarrow k^3 - (k-1)^3 = 3k^2 - 3k + 1$$

It follows then

$$\sum_{k=1}^{n} [k^3 - (k-1)^3] = 3\sum_{k=1}^{n} k^2 - 3\sum_{k=1}^{n} k + n$$

But, the telescope sum on the right-hand side is n^3, i.e.,, $\sum_{k=1}^{n} [k^3 - (k-1)^3] = n^3$. Thus, we can write

$$3S(n) = n^3 + 3\frac{n(n+1)}{2} - n = \frac{n(n+1)}{2}(2n+1) \tag{2.6.8}$$

where we have used the result from Eq. (2.6.2) for $\sum_{k=1}^{n} k$. Can we understand why the result is as it is? Consider the case $n = 4$, i.e.,, $S(4) = 1 + 4 + 9 + 16$. We can express this sum as a triangle shown in the first picture in Fig. 2.7. As the sum does not change if we rotate this triangle, we consider two rotations (the first rotation is an anti-clockwise 120 degrees about the center of the triangle) shown in the two remaining figures. If we sum these three triangles, i.e.,, $3S(4)$, we get a new triangle shown in the last picture of Fig. 2.7. What is the sum of this triangle? It is $9(1 + 2 + 3 + 4)$, and $9 = 2(4) + 1$, so this triangle gives $(2 \times 4 + 1)(4)(5)/2$, which is the RHS of Eq. (2.6.8).

Why we knew that a rotation would solve this problem? This is because any triangle in Fig. 2.7 is rotationally symmetric.

2.6.3 Sum of the Cubes of the First n Whole Numbers

The sum of the squares of the first n whole numbers is expressed as

$$S(n) = 1^3 + 2^3 + 3^3 + \cdots n^3 = \sum_{k=1}^{n} k^3 \tag{2.6.9}$$

As this point, you certainly know how to tackle this sum. We start with $(k-1)^4$:

$$(k-1)^4 = k^4 - 4k^3 + 6k^2 - 4k + 1 \Longrightarrow k^4 - (k-1)^4 = 4k^3 - 6k^2 + 4k - 1 \tag{2.6.10}$$

So,

$$\sum_{k=1}^{n}[k^4 - (k-1)^4] = 4\sum_{k=1}^{n} k^3 - 6\sum_{k=1}^{n} k^2 + 4\sum_{k=1}^{n} k - n \tag{2.6.11}$$

We know the LHS ($\sum_{k=1}^{n}[k^4 - (k-1)^4] = n^4$), and the second and third sums (from previous problems) in the RHS except the one we are looking for (the red term), so we can compute it as:

$$\begin{aligned} 4S(n) &= n^4 + n(n+1)(2n+1) - 2n(n+1) + n \\ \Longrightarrow S(n) &= \frac{n[n^3 + (n+1)(2n+1) - 2(n+1) + 1]}{4} \\ \Longrightarrow S(n) &= \frac{n^2(n+1)^2}{4} = \left(\frac{n(n+1)}{2}\right)^2 \end{aligned} \tag{2.6.12}$$

Using Eq. (2.6.2) we can see that, the sum of the first n cubes is the square of the sum of the first n natural numbers. Actually, we can see this relation geometrically, as shown in the below figure for the case of $n = 3$: $S(3) = 1 + 8 + 27 = (1 + 2 + 3)^2$.

Let's summarize all the results we have to see if a pattern exists[18]:

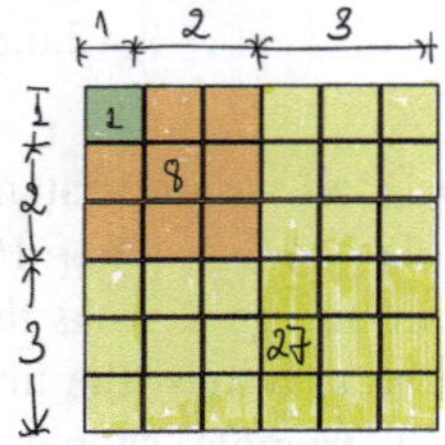

$$\begin{aligned} \sum_{k=1}^{n} k^1 &= \frac{n(n+1)}{2} &&= \frac{n^2}{2} + \frac{n}{2} \\ \sum_{k=1}^{n} k^2 &= \frac{n(n+1)(2n+1)}{6} &&= \frac{n^3}{3} + \frac{3n^2 + n}{6} \\ \sum_{k=1}^{n} k^3 &= \frac{n^2(n+1)^2}{4} &&= \frac{n^4}{4} + \frac{2n^3 + n^2}{4} \end{aligned} \tag{2.6.13}$$

Clearly, we can see a pattern that allows us to write for any whole number p (we believe[19] in the pattern that it will hold when $p = 4, 5, \ldots$)

[18] If you know calculus, this is the younger brother of this $\int x^n dx = x^{n+1}/n+1 + C$.

[19] Yes, many times following your gut is the best way to go. And in maths, patterns are everywhere.

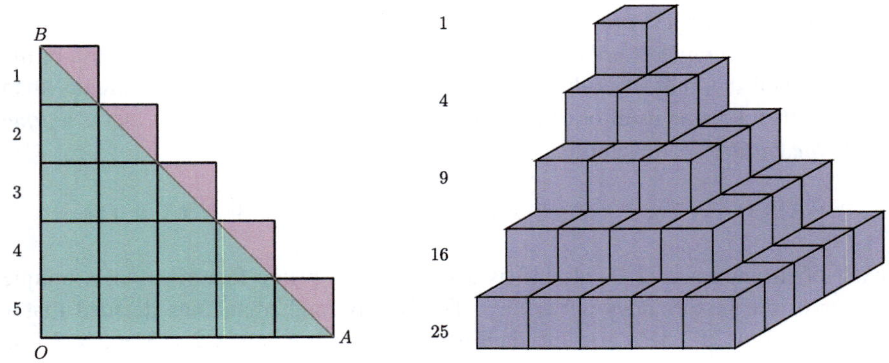

(a) $1 + 2 + \cdots + n$ for $n = 5$ (b) $1^2 + 2^2 + \cdots + n^2$ for $n = 5$

Fig. 2.8 Geometric interpretation of some of Eq. 2.6.13

$$\sum_{k=1}^{n} k^p = \frac{n^{p+1}}{p+1} + R(n) \tag{2.6.13}$$

where the ratio of $R(n)$ over n^{p+1} approaches zero when n is infinitely large; see Sect. 2.22 for a discussion on sequence and limit. This result would become useful in the development of calculus (precisely, in the problem of determining the area under the curve $y = x^p$).

All the sums in Eq. 2.6.13 contain two terms, and we can see why by looking at Fig. 2.8. For $\sum_{k=1}^{n} k^1$, the term $n^2/2$ is the area of the cyan triangle OAB. And the term $n/2$ is the area of the pink staircases (Fig. 2.8a). Similarly, for $\sum_{k=1}^{n} k^2$, the term $n^3/3$ is the volume of the pyramid (Fig. 2.8b). If you're good at geometry you should be able to compute this sum geometrically following this pyramid interpretation. However, for $\sum_{k=1}^{n} k^p$ $p \geq 3$, it is impossible to use geometry while algebra always gives you the result, albeit more involved.

Question 2 *We have found the sums of integral powers up to power of three. One question arises naturally: is there a general formula that works for any power?*

2.7 Prime Numbers

Again by playing with natural numbers long enough and pay attention, we see this:

$$4 = 2 \times 2,\ 6 = 2 \times 3,\ 8 = 2 \times 4,\ 9 = 3 \times 3$$
$$2 = 1 \times 2,\ 3 = 1 \times 3,\ 5 = 1 \times 5,\ 7 = 1 \times 7$$

So, we have two groups of natural number as far as factorizing (expressing a number as a product of other numbers) them is concerned. In one group $(2, 3, 5, 7, \ldots)$, the numbers can only be written as a product of one and itself. Such numbers are called *prime numbers*, or just primes. The other group (4, 6, 8, 9, ...) contains non-prime numbers or *composite numbers*. Primes are central in number theory because of the fundamental theorem of arithmetic stating that every natural number greater than one is either a prime itself or can be factorized as a product of primes that is *unique up to their order*. For example,

$$328\,152 = 2 \times 2 \times 2 \times 3 \times 11 \times 11 \times 113 = 2^3 \times 3^1 \times 11^2 \times 113$$

each of the numbers 2, 3, 11, 113 is a prime. This prime factorization is unique (order of the factors does not count). That's why mathematicians decided that 1 is not a prime. If 1 was a prime then we could write $6 = 1 \times 2 \times 3 = 2 \times 3$: the factorization is not unique! As with matters are made of atoms, numbers are made of prime numbers!

Theorem 2.7.1 (Fundamental Theorem of Arithmetic)

Every integer greater than 1 can be written in the form

$$p_1^{n_1} p_2^{n_2} \cdots p_k^{n_k}$$

where $n_1, n_2, \ldots$ *are positive integers, and* $p_1, p_2, \ldots$ *are primes. The factorization is unique, except possibly for the order of the factors.*

This theorem tells us two things: (1) any integer greater than 1 can be written in this form $p_1^{n_1} p_2^{n_2} \ldots p_k^{n_k}$—this is about the existence of such a factorization and (2) this form is unique—this is about uniqueness of the factorization; no other factorization is possible.

In this apparently introductory exposure to primes, we start with the question of how many primes are there and the beautiful proof of Eulicd (Sect. 2.7.1). Then, we discuss the prime number theorem which is about the distribution of primes (Sect. 2.7.2). Finally, we talk about prime twins and the fascinating of Yitang Zhang (Sect. 2.7.3).

2.7.1 How Many Primes Are There?

Now we have discovered new type of number—the primes—we ask questions about them and usually we discover interesting things. And we create new maths. One question is: how many prime numbers are there? How to answer this question? The easiest way is to count them. We count the number of primes from 1 to N for $N = 100, 1\,000, 10\,000, \ldots$, designated by $\pi(N)$,[20] then we divide it by N to have the density of prime numbers. The result of this analysis is given in Table 2.1.

[20] The Greek letter π makes a "p" sound, and stands for "prime".

Table 2.1 The density of prime numbers

N	$\pi(N)$	$\pi(N)/N$
100	25	0.25
1 000	168	0.168
10 000	1 229	0.123
100 000	9 592	0.096
1 000 000	78 498	0.079

Of course, a computer program was used to get this result.[21] The result shows that $\pi(100) = 25$; that is there are 25 primes in the first 100 integers. Among the first 1 000 integers, there are 168 primes, so $\pi(1000) = 168$, and so on. Note that as we considered the first 100, 1000 and 10 000 integers, the percentage of primes went from 25 to 12.3%. These examples suggest, and the prime number theorem confirms, that the density of prime numbers at or below a given number decreases as the number gets larger.

But if we keep counting for bigger N we see that the list of prime numbers goes on. Indeed, there are infinite prime numbers as proved by Euclid more than 2000 years ago. His proof is one of the most famous, most often quoted, and most beautiful proofs in all of mathematics. Why? It is because the largest prime that Euclid knew was probably a small number, but with reasoning only he could prove that there are infinite primes.

His proof is now known as *proof by contradiction* (also known as the method of *reductio ad absurdum*, Latin for "reduction to absurdity"). To use this technique, we assume the negate of the statement we are trying to prove and use that to arrive at something impossibly correct. So, we assume that there are finite prime numbers namely $p_1, p_2, \ldots, p_n$. And from this assumption we do something to arrive at something absurd, thus invalidating our starting point.

Euclid considered this number p:

$$p = p_1 \times p_2 \times \cdots \times p_n + 1$$

Because we have assumed there are only n primes, p cannot be a prime. Thus, according to the fundamental theorem of arithmetic, p must be divisible by any of p_i $(1 \leq i \leq n)$, but the above equation says that p divides by any p_i always with remainder of 1. A contradiction! So the assumption that there are finite primes is wrong, and thus there are infinite prime numbers.[22]

---

[21] I used the package `Primes.jl` that provides the function `isprime(n)` to check if a given n is prime or not.

[22] A misconception is that p is always a prime. One example $2 \times 3 \times 5 \times 7 \times 11 \times 13 + 1 = 30031 = 59 \times 509$, not a prime.

2.7.2 The Prime Number Theorem

The prime number theorem states that among the first n integers, there are about $n/\log(n)$ primes. This theorem was proved at the end of the 19th century by Hadamard and de la Valle Poussin, and $\log(n)$ is the natural logarithm of n, see Sect. 2.24.

How did mathematicians come up with this amazing theorem? Because they're lazy. Instead of counting the primes one by one, which is tediously slow, this theorem gives us in one go approximately how many primes from 1 to n. But exactly how mathematicians discovered this theorem? A good thing to know is that it took them lots of time to do so. And it was the work of many great men.

The first step is to have a table of all the primes from 1 to n when n is as big as possible (keep in mind that back then there was no computer). An early figure in tabulating primes is John Pell, an English mathematician who dedicated himself to creating tables of useful numbers. Thanks to his efforts, the primes up to 100 000 were widely circulated by the early 1700s. By 1800, independent projects had tabulated the primes up to 1 million.

Having the tables (or data) is one thing and getting something out of it is another. And we need a genius to do that. And that genius was Gauss. In a letter to his colleague Johann Encke about prime numbers, Gauss claimed merely to have looked at the data and seen the pattern; his complete statement reads "I soon recognized that behind all of its fluctuations, this frequency is on the average inversely proportional to the logarithm".

We are not Gauss so we need to visualize the data. We can say $\pi(N)$ is a function and call it the prime counting function. It is a function because when we feed to it a number it returns another number. In Fig. 2.9, the plot[23] of $\pi(N)$ is given for $N = 10^2, 10^3, 10^4$. What can we get from these plots? It is clear that as N get larger and larger $\pi(N)$ can be considered as a smooth function. Among all functions that we know of it is $N/\log N$ that best approximates $\pi(N)$.

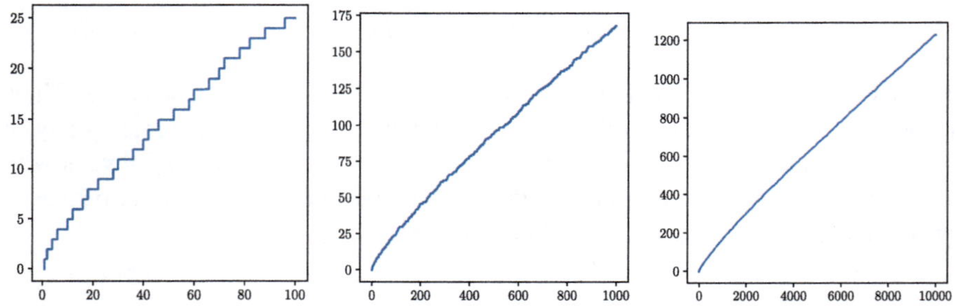

Fig. 2.9 Plot of the prime counting function $\pi(N)$ for $N = 10^2, 10^3, 10^4$

[23] Created using the function `step` of `matplotlib`.

Table 2.2 The density of prime numbers. The fourth col is the difference in third col

N	$\pi(N)$	$N/\pi(N)$	Δ
1000000	78498	12.7392	2.30794
10000000	664579	15.0471	2.30961
10000000	5761455	17.3567	2.30991
100000000	50847534	17.3567	–

But why log? See Table 2.2 and the red numbers. The red number is exactly log 10. In this table, the third column is $N/\pi(N)$ and the first entry in the fourth column is the difference of the second entry and the first entry in the third column. Let $f(N)$ be the mysterious function for $N/\pi(N)$, then we have $f(10N) = f(N) + 2.3$. A function that turns a product into a sum! That can be a logarithm. Indeed, $\log(10N) = \log N + \log 10$, and $\log 10 = 2.3$. This table was probably the one that Gauss merely looked at and guessed correctly the function. Without doubt, he was a genius.

However, Gauss did not prove his conjecture.[24] The theorem was proved independently by Jacques Hadamard and Charles Jean de la Vallée Poussin in 1896 using ideas introduced by Bernhard Riemann (guess what, Riemann was Gauss's student), in particular, the Riemann zeta function.

2.7.3 *Twin Primes and the Story of Yitang Zhang*

It would be incomplete if we did not mention twin primes. The first 25 prime numbers (all the prime numbers less than 100) are

$$\{2, 3, 5, 7, 11, 13, 17, 19, 23, 29, 31, 37, 41, 43, 47, 53, 59, 61, 67, 71, 73, 79, 83, 89, 97\}$$

Mathematicians call the prime pairs $(3, 5)$, $(5, 7)$, $(11, 13)$, etc., the twin primes. Thus, we have the following definition:

Definition 2.7.1 A couple of primes (p, q) are said to be twins if $q = p + 2$.

Note that except $(2, 3)$, 2 is the *smallest* possible distance (or gap) between two primes. Mathematicians then ask the same old question: how many are there twin primes? It is unknown whether there are infinitely many twin primes (the so-called twin prime conjecture) or if there is a largest pair. The breakthrough work of Yitang Zhang in 2013, as well as work by James Maynard, Terence Tao, and others, has made substantial progress towards proving that there are infinitely many twin primes,

[24] I am not sure why. Maybe the maths of his time was not sufficient.

but at present this remains unsolved. For a list of unsolved maths problems check here.

It is usually while solving unsolved mathematical problems that mathematicians discover new mathematics. The new maths also help to understand the old maths and provide better solution to old problems. Then, after about 100 or 200 years, some of the new maths come into the mathematics curriculum to train the general public.

Yitang Zhang (born 5 February 1955). On 17 April 2013, a paper arrived in the inbox of Annals of Mathematics, one of the discipline's top journals. Written by a mathematician virtually unknown to the experts in the field—*a 58 year old*[25] lecturer at the University of New Hampshire named Yitang Zhang—the paper claimed to have taken a huge step forward in solving the twin primes conjecture, one of mathematics' oldest problems. Just 3 weeks later Zhang's paper was accepted. Rumors swept through the mathematics community that a great advance had been made by an unknown mathematician—someone whose talents had been so overlooked after he earned his doctorate in 1991 that he had found it difficult to get an academic job, working for several years as an accountant and *even in a Subway sandwich shop*.[26]

"Basically, no one knows him" said Andrew Granville, a number theorist at the Université de Montréal. "Now, suddenly, he has proved one of the great results in the history of number theory". For Zhang's story, you can watch this documentary movie.

There are many more interesting stories about primes but we stop here, see Fig. 4.73 for a prime spiral.

2.8 Rational Numbers

This section expands our system of number to include the so-called rational numbers—those numbers such as $3/3$, $7/3$, and so on.

Section 2.8.1 presents a definition of rational numbers, including a geometric construction using straightedge and compass. The arithmetic of them is discussed in Sect. 2.8.2, where the rules of addition/subtraction, multiplication/division of rational numbers are given. Fractions of which denominators are 10, 100, 1000, ... can be written using decimals; e.g., $1/10 = 0.1$; and a discussion of decimals is given in Sect. 2.8.3.

---

[25] "No mathematician should ever allow himself to forget that mathematics, more than any other art or science, is a young man's game", Hardy wrote. He also wrote, "I do not know of an instance of a major mathematical advance initiated by a man past fifty".

[26] The pursuit of tenure requires an academic to *publish frequently*, which often means refining one's work within a field, a task that Zhang has no inclination for. He does not appear to be competitive with other mathematicians, or resentful about having been simply a teacher for years while everyone else was a professor. As he did not have to publish many papers he had all the time to focus on big problems. I think his situation is somehow similar to Einstein being a clerk in the Swiss patent office.

2.8.1 *What Is 5/2?*

Sharing is not easy whether you are a kid or an adult. And it is also the case with mathematics. While it is straightforward to recognize that $6/3 = 2$ (six candies equally shared by 3 kids) or $8/2 = 4$, what is $5/2$? From an algebraic point of view, we can say that while the equation $3x = 6$ has one solution: $x = 2$, the equation $2x = 5$ has no integer solution. In other words, integers are not closed under division. Again, it was needed to expand our system of number one more time. This is a modern view of how rational numbers (numbers such as $5/2$) are defined. Historically, it was developed very practically.

While counting objects resulted in the development of natural numbers, it was the practical problem of measurement (measuring length and area) that led to the birth of rational numbers. Similar to counting discrete objects (one bird, two carrots, etc.), one needs to define a unit before measurement can be done. For example, how long is a rod? We can define *a unit of length* to which we assign a value of 1 and then the rod length is expressed in terms of this unit. If the unit is meter, the rod is 5 meters. If the unit is yard, the rod is 5.46807 yards.

One problem arises immediately. Not all quantities can be expressed as integral multiples of a unit. A rod can be one meter and something long. To handle this, we define *a sub-unit*. For example, we can divide 1 meter into 100 equal parts and each part (which we call a centimeter by the way) is now a new unit. The rod length is now 120 centimeters, or $120(1/100)$ meters. We can generalize this by dividing 1 into m equal parts to obtain $1/m$ the measure of our new sub-unit. Any length can then be expressed as an integral multiple of $1/m$ or n/m, a ratio. And that's how mathematicians defined rational[27] numbers.

Definition 2.8.1 A rational number is a number that can be written in the form p/q where p and q are integers and q is not equal to zero; p is named the numerator and q is the denominator.

The requirement that q is not equal to zero comes from the fact that division by zero is meaningless. Because, if we allowed it, we would get absurd results. For instance, as $0 \times 1 = 0 \times 2$, divide both sides by 0, we get $1 = 2$, which is non-sense. It is quite obvious that the fraction p/p is nothing but one.

2.8.2 *Arithmetic with Rational Numbers*

We now need to define addition, multiplication, and division for rational numbers. We first present these rules here (explanations follow immediately):

[27] Rational here does not mean logical or reasonable, it is a ratio of two integers.

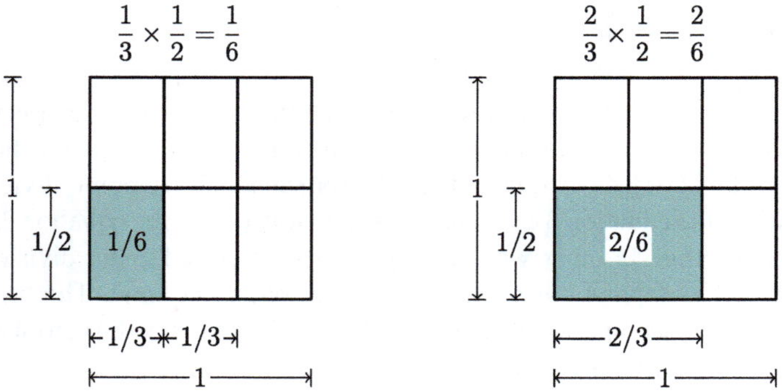

Fig. 2.10 Multiplication of two rational numbers

$$\frac{a}{b} \times \frac{c}{d} = \frac{ac}{bd}, \quad \frac{a}{b} + \frac{c}{d} = \frac{ad+bc}{bd}, \quad \frac{a/b}{c/d} = \frac{a}{b} \times \frac{d}{c} \tag{2.8.1}$$

Surprisingly, the rule for multiplication is easier to grasp than that for addition. We refer to Fig. 2.10 for an illustration. Imagine of a square wooden plate of unit sides. Thus, the area of this plate is 1 (whatever the unit). Now we divide the longer side into 3 equal parts, so each part is 1/3. Similarly, we chop the shorter edge into two halves, so each part is 1/2. Now, the area of one peace is $^1/_3 \times {}^1/_2$ and it must be equal to $^1/_6$ as there are six equal rectangular pieces, and in total they make a unit square of which the area is one. If we take two pieces then the area is $^2/_6$ and they make a rectangle of sides $^2/_3$ and $^1/_2$; so $^2/_3 \times {}^1/_2 = {}^2/_6$. That's why we have the rules of multiplying two rationals: multiplying the numerators and multiplying the denominators, then take the ratio.

We now move to the problem of adding two rationals. First, it is not hard to add two rational numbers when they have the same denominator:

$$\frac{1}{2} + \frac{3}{2} = \frac{1+3}{2} = \frac{4}{2}$$

This is because one-half plus three halves is certainly four halves, which is $4/2$. This is similar to one carrot plus two carrots is three carrots. The unit is just a half instead of 1 carrot. For rational numbers having different denominators, the rule is then to convert them to have the same denominator:

$$\frac{1}{2} + \frac{4}{3} = \frac{1 \times 3}{2 \times 3} + \frac{4 \times 2}{3 \times 2} = \frac{3}{6} + \frac{8}{6} = \frac{11}{6}$$

The conversion is based on the equality of two rational numbers explained in Fig. 2.11.

Finally, we need to define division. What is $(a/b) \div (c/d)$? We use a trick here:

$$\frac{1}{2} = \frac{2}{4} = \frac{3}{6}$$

$$\frac{a}{b} = \frac{c}{d} \Longrightarrow ad = bc$$

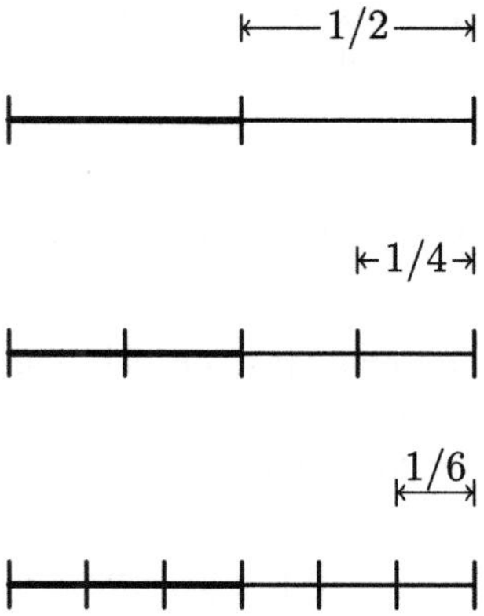

Fig. 2.11 Equality of two rational numbers. The rational 1/2 is said to be in its lowest term as it is impossible to simplify it. On the other hand, 2/4 is not in lowest term. How to know when a/b is in a lowest term? It is when the greatest common divisor of a and b is 1. See Sect. 2.34.1 for a discussion on how to find the greatest common divisor of two integers

$$\frac{\frac{a}{b}}{\frac{c}{d}} = \frac{\frac{a}{b} \times \frac{d}{c}}{\frac{c}{d} \times \frac{d}{c}} = \frac{\frac{a}{b} \times \frac{d}{c}}{1} = \frac{a}{b} \times \frac{d}{c}$$

That is, we multiplied both the numerator and denominator with d/c, to make the denominator equal one. The rule is: divide a fraction is multiply its inverse. For example, we have

$$\frac{1}{1/2} = 1 \times \frac{2}{1} = 2, \quad \frac{2/3}{5/7} = \frac{2}{3} \times \frac{7}{5} = \frac{14}{15}$$

How we use this division rule is correct? We can check. If $6 : 3 = 2$ then $6 = 2 \times 3$. So, we do the same check:

$$\frac{\frac{a}{b}}{\frac{c}{d}} = \frac{ad}{bc} \Longrightarrow \frac{a}{b} = \frac{a\not{d}}{b\not{c}} \times \frac{\not{c}}{\not{d}} = \frac{a}{b}$$

Percentage. In mathematics, a percentage (from Latin per centum meaning "by a hundred") is a ratio expressed as a fraction of 100. It is often denoted using the percent sign ("%"), although the abbreviations "pct.", "pct" and sometimes "pc" are also used. For example, 50% is $50/100 = 1/2$. As a ratio, a percentage is a dimensionless number (pure number); it has no unit of measurement.

Arithmetic is important but this is more important. We have to check whether the rules of integers, stated in Eq. (2.1.2), still hold for the new number—the rationals? It turns out that the rules hold. For example, the addition is still commutative:

$$\frac{a}{b} + \frac{c}{d} = \frac{ad + bc}{bd} = \frac{bc + ad}{bd} = \frac{bc}{bd} + \frac{ad}{bd} = \frac{c}{d} + \frac{a}{b}$$

Note that in the proof we have used $ad + bc = bc + ad$, as these numbers are integers. Why this is important? Because mathematicians want to see $2 = 2/1$—that is an integer is a rational number. Thus, the arithmetic for the rationals must obey the same rules for the integers.

2.8.3 Decimal Notation

Decimals are a convenient and useful way of writing fractions of which denominators are powers of ten, e.g., 10, 100, 1000, ... For example, $3/10$ is written as 0.3. We can understand this notation as follows. Consider for example the number 351

$$\overset{210}{351} = 3 \times 100 + 5 \times 10 + 1 \times 1 = 3 \times 10^2 + 5 \times 10^1 + 1 \times 10^0$$

which means that the units are in the 0 position, the tens in the 1 position and the hundreds in the 2 position, and the position decides the power of tens. Now, $3/10 = 3 \times 10^{-1}$ is zero unit and three tenths, thus the digit 3 must be placed on the -1 position, which is before the units: 03, but we need something to separate the two digits otherwise it is mistaken with 3. So, $3/10$ is written as 0.3. The decimal point separates the units column from the tenths column. Similarly, $3/100$, which is three hundredths, is written as 0.03—the number 3 is at position -2. The number 351.3 is understood as

$$\overset{210-1}{351.3} = 3 \times 10^2 + 5 \times 10^1 + 1 \times 10^0 + 3 \times 10^{-1}$$

The Flemish mathematician Simon Stevin (1548–1620), sometimes called Stevinus, first used a decimal point to represent a fraction with a denominator of ten in 1585. While decimals had been used by both the Arabs and Chinese long before this time, Stevin is credited with popularizing their use in Europe. An English translation of Stevin's work was published in 1608 and titled Disme, The Arts of Tenths or Decimal Arithmetike, and it inspired the third president of the United States Thomas Jefferson to propose a decimal-based currency for the United States (for example, one tenth of a dollar is called a dime).

If we do long division for rationals we see the following decimals

$$\frac{1}{4} = 0.25, \quad \frac{1}{3} = 0.3333\ldots, \quad \frac{1}{7} = 0.142857142857\ldots \tag{2.8.2}$$

First, I introduce some terminologies. In decimals, the number of places filled by the digits after (to the right of) the decimal point is called the *decimal places*. Thus, 0.25 has 2 decimal places and 0.2 has 1 decimal place. That's boring (but we need to know the term to understand other people). Second, what's more interesting lies in Eq. (2.8.2): we can see that there are two types of decimals for rational numbers. The decimal 0.25 is a *terminating decimal*. The (long) division process terminates. On the other hand, $1/3 = 0.3333\ldots$ with infinitely many digits 3 as the division does not terminate. The decimal 0.3333... is called a *recurring decimal*. How about $1/7$? Is it a recurring decimal? Of course it is, you might say. But think about this: how can you sure that the red digits repeat forever? It could be like this: $1/7 = 0.142857142857\ldots142857531\ldots$ But things are not that complicated for rational numbers. Any recurring decimal has the pattern forever. And the reason is not hard to see. Let's look at the following division of integers by 7:

$$\begin{array}{lll}
0 = 0\cdot 7 + 0, & 6 \;= 0\cdot 7 + 6, & 12 = 0\cdot 7 + 5\\
1 = 0\cdot 7 + 1, & 7 \;= 0\cdot 7 + 0, & 13 = 0\cdot 7 + 6\\
2 = 0\cdot 7 + 2, & 8 \;= 0\cdot 7 + 1, & 14 = 0\cdot 7 + 0\\
3 = 0\cdot 7 + 3, & 9 \;= 0\cdot 7 + 2, & 15 = 0\cdot 7 + 1\\
4 = 0\cdot 7 + 4, & 10 = 0\cdot 7 + 3, & 16 = 0\cdot 7 + 2\\
5 = 0\cdot 7 + 5, & 11 = 0\cdot 7 + 4, & 17 = 0\cdot 7 + 3
\end{array}$$

Look at the remainders: there are only six (except 0) of them: $\{0, 1, 2, 3, 4, 5, 6\}$. That's why $1/7 = 0.142857142857\ldots$, which has a cycle of six—the length of the repeating digits.

Sometimes you're asked to find the fraction corresponding to a recurring decimal. For example, what is the fraction of $0.2272727 = 0.2\overline{27}$ where the bar on 27 is to indicate the repeated digits. To this end, we write $0.2\overline{27} = 0.2 + 0.0\overline{27}$. Now, we plan to find the fraction for $0.0\overline{27}$. We start with $y = 0.\overline{27}$, then taking advantage of the repeating pattern, we will find a linear equation in terms of y to solve for it:

$$\begin{aligned}
100y &= 27.\overline{27}\\
99y &= 27 \Longrightarrow y = \frac{27}{99} \Longrightarrow 0.0\overline{27} = \frac{27}{990} \Longrightarrow 0.2\overline{27} = \frac{2}{10} + \frac{27}{990} = \frac{5}{22}
\end{aligned}$$

Is 0.9999... equal to 1? We all know that $1/3 = 0.\overline{3}$, multiplying both sides with 3, we obtain $1 = 0.\overline{9} = 0.9999\ldots$ And there are many other proofs for this. For example, the following proof is common and easy to get:

$$\begin{aligned}
x &= 0.999\ldots\\
100x &= 99.999\ldots\\
99x &= 99 \Longrightarrow x(= 0.999\ldots) = 1
\end{aligned}$$

But what is going on here? The problem is at the equal sign, and the never ending 9999. To fully understand this, we need to go to infinity and this will be postponed until Sect. 2.22.

2.9 Irrational Numbers

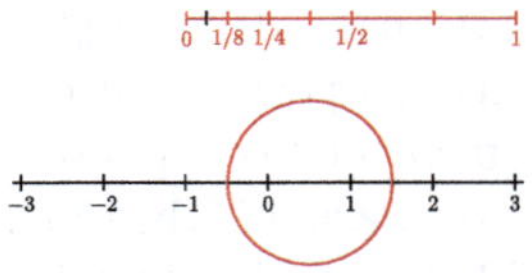

So we have integers and rational numbers. It is easy to see that the average of two rational numbers is also a rational number. Thus, for any two rational numbers, even if they are really close to each other, we can always find another rational number in between them. We might be tempted to conclude that all numbers are rational. And the ancient Greeks believed it for a while. But, surprisingly that's not true. There exists another type of numbers and this section is devoted to the discussion of those numbers that are not the rationals. They are the ... irrationals.

Our story starts with a square and its diagonal (Sect. 2.9.1): it shows that the diagonal of a square of side a has a length of $a\sqrt{2}$ which is an irrational number. We move then to the arithmetic of the irrationals (Sect. 2.9.2). From square root $\sqrt{a}$ we generalize to cube roots $\sqrt[3]{a}$ and to $\sqrt[n]{a}$ (Sect. 2.9.3). Section 2.9.4 is about writing $1/\sqrt{2}$ as $\sqrt{2}/2$ and why $\sqrt{4+2\sqrt{3}} = 1+\sqrt{3}$: this is about rationalizing the denominators and simplifying radicals. Next in our discussion is a special irrational—the golden ratio: $\phi = 1+\sqrt{5}/2$ (Sect. 2.9.5). Section 2.9.6 presents the axioms for the real numbers.

2.9.1 *Diagonal of a Unit Square*

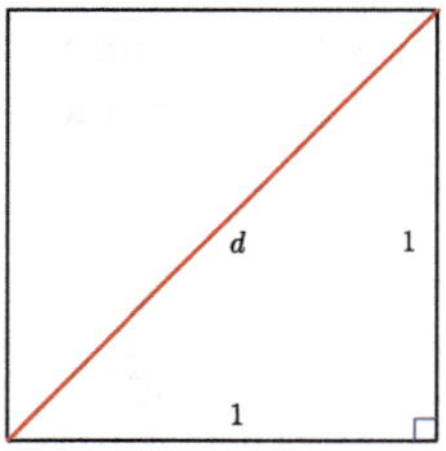

Our journey with the evolution of the number systems continues with a simple but perplexing problem. Consider a unit square as shown in the beside figure, what is the length of the diagonal d? Let's assume that we know the Pythagorean theorem, then it is obvious that $d^2 = 1^2 + 1^2 = 2$ (see Fig. 2.12 for a geometry explanation without resorting to the Pythagorean theorem). But what d exactly is? It turns out that d cannot be expressed as a/b where a and b are integers. In other words, there are no integers a and b such that $a^2/b^2 = 2$. Nowadays, we call such a number an *irrational* number.

It would be inconvenient to refer to d as a number such that d^2 is two. We need a name for it and a symbol as well. Nowadays, we say d is the *square root* of 2, and write it as $d = \sqrt{2}$. We will talk more about roots later in Sect. 2.9.3.

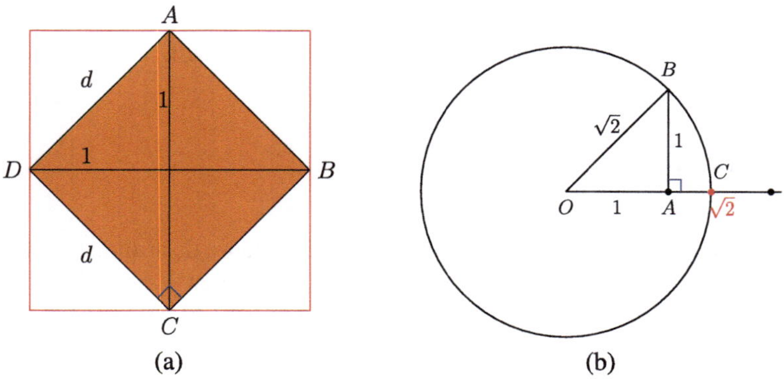

Fig. 2.12 Proof that the diagonal of a unit square has a length of $\sqrt{2}$: Starting with one unit square, we add three more unit squares to the problem, and we suddenly get a symmetrical geometry object. The area of the square $ABCD$ is d^2 and this square is twice as large as the unit square. Thus, $d^2 = 2$ (**a**). A geometric construction of a line segment with length being $\sqrt{2}$ (**b**). We start with the right triangle OAB with $AO = AB = 1$. The Pythagorean theorem then tells us that $OB = \sqrt{2}$. Now using a compass, draw a circle centered at O and with OB as radius we get point C with $OC = \sqrt{2}$. And that point C is where the irrational number $\sqrt{2}$ lives

How are we going to prove that $\sqrt{2}$ is irrational? The only information we have is the definition of an irrational number—the number which is not a/b. So, the goal is to prove that $\sqrt{2} \neq a/b$. Where do we begin? It seems easier if we start with $\sqrt{2} = a/b$, and play with this to see if something comes up. That is we're trying to use proof by contradiction. Let's do it.

Assume that $\sqrt{2}$ is a rational number, i.e.,, $\sqrt{2} = a/b$ or $a^2/b^2 = 2$ where a, b are not both even (if they are, one can always cancel out the factor 2). So, $a^2 = 2b^2$ which is an even number (since it is 2 multiplied by some number). Thus, a is an even number (even though this is rather obvious, as always, prove it). Since a is even, we can express it as $a = 2c$ where $c = 1, 2, 3, \ldots$

$$a = 2c \Longrightarrow a^2 = 4c^2 \Longrightarrow 4c^2 = 2b^2 \Longrightarrow b^2 \text{ is even, or } b \text{ is even}$$

So, we are led to the fact that both a, b are even, which is in contradiction with a, b being not both even. So, the square root of two must be irrational. We used proof by contradiction. To use this technique, we assume the negate of the statement we are trying to prove and use that to arrive at something impossibly correct.

Examples of irrational numbers include square roots of integers that are not complete squares, e.g., $\sqrt{10}$, cube roots of integers that are not cubes, like $\sqrt[3]{7}$, and so on. Multiplying an irrational number by a rational coefficient or adding a rational number to it produces again an irrational number.[28] The most famous irrational number

[28] For example, assume that $\sqrt{2} + r_1 = r_2$ where r_1, r_2 are two rationals, then we get $\sqrt{2} = r_2 - r_1$. But rationals are closed under subtraction, i.e.,, $r_2 - r_1$ is a rational. Thus we arrive at the absurd conclusion that $\sqrt{2}$ is rational. Therefore, $\sqrt{2} + r_1$ must be irrational.

is π—the ratio of a circle circumference to its diameter– $\pi = 3.14159265\ldots$ The decimal portion of π is infinitely long and never repeats itself.

2.9.2 Arithmetic of the Irrationals

Now that we have discovered the irrationals, how should we do arithmetic with them? That's the question we try to answer now. Should we say that $\sqrt{2} + \sqrt{7} = \sqrt{9}$—that is we're saying that the sum of square roots is the square root of the sum. It sounds a nice rule. To know whether this rule is reasonable, we go back (always) to our old friend: the whole numbers. It is easy to see that $\sqrt{4} + \sqrt{9} \neq \sqrt{13}$.[29] We have found one *counterexample* thus we have disproved the aforementioned rule. But then what is $\sqrt{2} + \sqrt{7}$? The answer is $\sqrt{2} + \sqrt{7}$!

Moving now to multiplication, we observe that $\sqrt{4} \times \sqrt{9} = 2 \times 3 = 6 = \sqrt{36}$. And we can use a calculator to see that $\sqrt{a} \times \sqrt{b} = \sqrt{ab}$. For a mathematical proof, check Sect. 2.19. As division is related to multiplication, we should have the same rule. One example: $\sqrt{4}/\sqrt{9} = 2/3 = \sqrt{4/9}$.

In practical applications (in science and engineering for instance), we often approximate irrational numbers by fractions or decimals (e.g., we replace $\sqrt{2}$ by 1.414) because actual physical objects cannot be constructed exactly anyway. Thus, we are happy with $\sqrt{2} \approx 1.414$, and if we need more accuracy, we can use a better approximation $\sqrt{2} \approx 1.414213$.

But then why in mathematics courses, students are asked to compute, let say, $1/1+\sqrt{2}$ without replacing $\sqrt{2}$ with 1.414? There are many reasons. One is that mathematicians love patterns not the answer. For example, the Basel problem asked mathematicians to compute the sum of infinite terms:

$$S = 1 + \frac{1}{4} + \frac{1}{9} + \frac{1}{16} + \cdots$$

Anyone knows that the answer is 1.6449, approximately. But Euler and many other mathematicians were not happy with that: they wanted to find out a formula/expression for S. That problem defied all mathematicians except Euler. He eventually found out that the *exact* answer is $\pi^2/6$. Not only this is a beautiful result in itself, Euler had discovered other mathematical results while working on this problem.

2.9.3 Roots $\sqrt[n]{x}$

A square root of a number x is a number y such that $y^2 = x$; in other words, a number y whose square (the result of multiplying the number by itself, or $y \times y$) is

[29] Because the LHS is 5, and square of 5 is 25 not 13.

x. For example, 4 and -4 are square roots of 16, because $4^2 = (-4)^2 = 16$. Every *nonnegative* real number x has a unique nonnegative square root, called the principal square root, which is denoted by $\sqrt{x}$ where the symbol $\sqrt{\ }$ is called the *radical sign*. The term (or number) whose square root is being considered is known as the *radicand*. In other words, the radicand is the number or expression underneath the radical sign. The radical symbol was first used in print in 1525, in Christoph Rudolff's Coss.[30] It is believed that this was because it resembled a lowercase "r" (for "radix"). The fact that the symbol of square root is $\sqrt{\ }$ is not as important as the concept of square root itself. However, for the communication of mathematics, we have to get to know and use this symbol when it has become standard.

The definition of a square root of x as a number y such that $y^2 = x$ has been generalized in the following way. A cube root of x is a number y such that $y^3 = x$; it is denoted by $\sqrt[3]{x}$. We need a cube root when we know the volume of a box and need to determine its side. Extending to other roots is straightforward. If n is an integer greater than two, an nth root of x is a number y such that $y^n = x$; it is denoted by $\sqrt[n]{x}$.

What is $\sqrt{-4}$? It is a number y such that $y^2 = -4$ which is absurd. So, we only compute square roots of positive numbers, at least for now.

Calculation of square roots. What is the value of $\sqrt{5}$? And you have to find that value without using a calculator. Why bothering with this? Because you could develop an algorithm for calculating a square root of any positive number by yourself. Itself is a big achievement (even though someone had done it before you). Furthermore, this activity is important if you later on follow a career in applied mathematics, sciences and engineering. In these areas people often use approximate methods to solve problems; for example they solve the equation $x = \sin x$ approximately using algorithms similar (in nature) to the one we are discussing in this section. If you are lazy and just use a calculator, you would learn nothing!

Perhaps the first algorithm used for approximating $\sqrt{x}$ is known as the Babylonian method. The method is also known as Heron's method, named after the first-century Greek mathematician Hero of Alexandria who gave the first explicit description of the method in his AD 60 work *Metrica*. So, what is exactly the algorithm? It starts with an initial guess of the square root x_0 and this observation: if x_0 is smaller than the true square root of S, then S/x_0 is larger than the root of S. So, an average of these two numbers might be a better approximation:

$$x_1 = \frac{1}{2}\left(x_0 + \frac{S}{x_0}\right) \tag{2.9.1}$$

And we use x_1 to compute $x_2 = 0.5(x_1 + S/x_1)$. The process is repeated until we get the value that we aim for. How good is it algorithm? Using Julia (see Listing A.1) I

[30] Christoph Rudolff (1499–1545) was the author of the first German textbook on algebra "Coss". Check this.

Table 2.3 Calculation of $\sqrt{5}$ with starting value $x_0 = 2$

n	x_n	Error $e = x_n - \sqrt{5}$
1	2.25	1.00e–2
2	2.2361111	4.31e–5
3	2.2360680	2.25e–8

wrote a small function implementing this algorithm. Using it I computed $\sqrt{5}$ (which is about 2.236067977) with $x_0 = 2$ and the results are given in Table 2.3.

The performance of the algorithm is so good, with three iterations and simple calculations we get a square root of 5 with 6 decimals. However, there are many questions to be asked. For example, where did Eq. (2.9.1) come from?

One derivation of Eq. (2.9.1) is as follows. Assume that x_0 is *close to* $\sqrt{S}$, and e is the error in that approximation, then we have $(x_0 + e)^2 = S$. We can solve for e from this equation:

$$(x_0 + e)^2 = S \Longrightarrow x_0^2 + 2x_0 e + e^2 = S \Longrightarrow e = \frac{S}{2x_0} - \frac{x_0}{2} \tag{2.9.2}$$

where e^2 was omitted as it is negligible. Having obtained e, adding e to x_0 we will get Eq. (2.9.1). Actually, the Babylonian method is an example of a more general method—the Newton method for solving $f(x) = 0$—see Sect. 4.5.4.

How about the calculation of $\sqrt[n]{x}$? Does the Newton method still work? If so, what should be the initial guess? Is the Newton method fast? Using a small program you can investigate all these questions, and discover for yourselves some mathematics.

2.9.4 *Rationalizing Denominators and Simplifying Radicals*

Do you remember that when you wrote $1/\sqrt{2}$ and your strict teacher corrected it to $\sqrt{2}/2$? They are the same, so why bother? I think that the reason is historical. Before calculators, it is easier to compute $\sqrt{2}/2$ (as approximately $1.4142135/2$) than to compute $1/1.4142135$. And thus it has become common to not write radicals in the denominators. Now, we know the why, let's move to the how.

How to rationalize the denominator of this term $1/(1 + \sqrt{2})$? The secret lies in the identity $(a+b)(a-b) = a^2 - b^2$, and thus $(1+\sqrt{2})(1-\sqrt{2}) = -1$, *the radical is gone*. So, we multiply the nominator and denominator by $1 - \sqrt{2}$, which is the *conjugate radical*[31] of $1 + \sqrt{2}$:

[31] The word conjugate comes from Latin and means (literally) "to yoke together", and the idea behind the word is that the things that are conjugate are somehow bound to each other.

$$\frac{1}{1+\sqrt{2}} = \frac{1}{1+\sqrt{2}} \times 1 = \frac{1}{1+\sqrt{2}} \times \frac{1-\sqrt{2}}{1-\sqrt{2}} = \frac{1-\sqrt{2}}{-1} = \sqrt{2}-1$$

And it is exactly the same idea when we have to divide two complex numbers $(a+bi)/(c+di)$. We multiply the nominator and denominator by $c-di$, which is the complex conjugate of $c+di$. This time, doing so eliminates i in the denominator as $i^2=-1$.

In general, the radical conjugate of $a+b\sqrt{c}$ is $a-b\sqrt{c}$. When multiplied together it gives us a^2-b^2c. The principle of rationalizing denominators is as simple as that. But, let's try this problem: simplify the following expression

$$S = \frac{1}{3+2\sqrt{2}} + \frac{1}{2\sqrt{2}+\sqrt{7}} + \frac{1}{\sqrt{7}+\sqrt{6}} + \frac{1}{\sqrt{6}+\sqrt{5}} + \frac{1}{\sqrt{5}+2} + \frac{1}{2+\sqrt{3}}$$

A rush application of the technique would work, but in a tedious way. Let's spend time with the expression and we see something special, a pattern (in the red terms we have $7, 6$ then $6, 5$):

$$S = \frac{1}{3+2\sqrt{2}} + \frac{1}{2\sqrt{2}+\sqrt{7}} + \frac{1}{\sqrt{7}+\sqrt{6}} + \frac{1}{\sqrt{6}+\sqrt{5}} + \frac{1}{\sqrt{5}+2} + \frac{1}{2+\sqrt{3}}$$

So, we rewrite the expression as

$$S = \frac{1}{\sqrt{9}+\sqrt{8}} + \frac{1}{\sqrt{8}+\sqrt{7}} + \frac{1}{\sqrt{7}+\sqrt{6}} + \frac{1}{\sqrt{6}+\sqrt{5}} + \frac{1}{\sqrt{5}+\sqrt{4}} + \frac{1}{\sqrt{4}+\sqrt{3}}$$

Now, we apply the trick to, say, $1/(\sqrt{9}+\sqrt{8})$ and get a nice result of $\sqrt{9}-\sqrt{8}$. Doing the same for other terms, and add them altogether gives us:

$$S = \sqrt{9}-\sqrt{8}+\sqrt{8}-\sqrt{7}+\sqrt{7}-\sqrt{6}+\sqrt{6}-\sqrt{5}+\sqrt{5}-\sqrt{4}+\sqrt{4}-\sqrt{3} = 3-\sqrt{3}$$

where all terms, except the first and last, are canceled leaving us a neat final result of $3-\sqrt{3}$. This is called a *telescoping sum* and we see this kind of sum again and again in mathematics, for instance Sect. Section 2.21.4. The name comes from the old collapsible telescopes you see in pirate movies, the kind of spyglass that can be stretched out or contracted at will. The analogy is the original sum appears in its stretched form, and it can be telescoped down to a much more compact expression.

Another common type of exercise about square/cube roots is to simplify radicals. For example, what is $\sqrt{4+2\sqrt{3}}$ without using a computer. Again, the purpose is not about the result. As we know that the radicand should be a perfect square. This is because the whole exercise would be silly if the radicand was not a perfect square!, we thus assume that $4+2\sqrt{3}=(a+\sqrt{3})^2$, and we're going to find a such that:

$$4+2\sqrt{3} = (a^2+3) + 2a\sqrt{3}$$

From that we have two equations by equating the red and blue terms: $4 = a^2 + 3$ and $2 = 2a$, which gives us $a = 1$. So $\sqrt{4 + 2\sqrt{3}} = 1 + \sqrt{3}$. This technique is called *the method of undetermined coefficients*.

Let's challenge ourselves a bit further. Can we simplify the following radical?

$$\sqrt{104\sqrt{6} + 468\sqrt{10} + 144\sqrt{15} + 2006}$$

The solution is based on the belief that the radicand must be a perfect square, i.e.,, it is of the form $(\cdots)^2$. And this radicand has four terms, we think of the identity $(x + y + z)^2 = x^2 + \cdots$, and this leads to the beautiful compact answer of $13\sqrt{2} + 4\sqrt{3} + 18\sqrt{5}$. Details: we wish $104\sqrt{6} + 468\sqrt{10} + 144\sqrt{15} + 2006$ to be of the form $(x\sqrt{a} + y\sqrt{b} + z\sqrt{c})^2$. The question is: what are a, b, c? Look at 6, 10, 15 and ask why not 6, 10, 14, then you'll see that $a = 2, b = 3, c = 5$. For x, y, z we have $xz = 234$, $yz = 72$, $xy = 52$. A teacher proceeds the reverse with $(x\sqrt{a} + y\sqrt{b} + z\sqrt{c})^2$, and thus she can generate infinitely many problems of this type. But, as a student you just need to do just one.

Some exercises on square roots.

1. Simplify the following

$$\sqrt[6]{26 + 15\sqrt{3}} - \sqrt[6]{26 - 15\sqrt{3}}$$

2. China Mathematical Olympiad, 1998, evaluate the following square root:

$$A = \sqrt{\frac{1998 \times 1999 \times 2000 \times 2001 + 1}{4}}$$

3. Simplify the following

$$\frac{\sqrt{10 + \sqrt{1}} + \sqrt{10 + \sqrt{2}} + \cdots + \sqrt{10 + \sqrt{99}}}{\sqrt{10 - \sqrt{1}} + \sqrt{10 - \sqrt{2}} + \cdots + \sqrt{10 - \sqrt{99}}}$$

The answer is ... 1.41421356... for the first question. For the second question, use the strategy of solving a simpler problem, e.g., $\sqrt{1 \times 2 \times 3 \times 4 + 1}$, which is nothing but $\sqrt{5^2}$, to see the pattern. For the third question, the answer is $1 + \sqrt{2}$, which can be guessed using a short `Julia` script.

Common errors in algebraic expression manipulations. Understanding the rules of rational numbers, we can avoid the following mistakes:

$$\frac{\cancel{3x^2} + 6x^4}{\cancel{3x^2}} = 6x^4, \quad \frac{\cancel{3}x^2 + \cancel{3}x^4 + \sqrt{\cancel{3}x}}{\cancel{3}x^2} = \frac{x^2 + x^4 + \sqrt{x}}{x^2}$$

The correct answer for the first expression is $1 + 2x^2$. It is clear that $(6 + 3)/6$ is definitely not 3! If you're not sure, one example can clarify the confuse.

About the second expression, due to the square root in $\sqrt{3x}$, it is incorrect to cancel 3 inside the square root. This is clear if you think of the last term as $\sqrt{3}/3$, forget the x, and this is definitely not 1!

2.9.5 Golden Ratio

Suppose you want to divide a segment of a line into two parts, so that the ratio of the larger part (a units long) to the smaller part (b units long) is the same as the ratio of the whole (c units long) to the larger part. This ratio is known as the Golden Ratio (also known as the Golden Section, Golden Mean, Divine Proportion). Let's find its value first, which is quite simple:

$$\frac{a}{b} = \frac{c}{a} = \frac{a+b}{a} \Longrightarrow \boxed{\phi = 1 + \frac{1}{\phi}} \quad \text{or} \quad \phi^2 - \phi - 1 = 0 \tag{2.9.3}$$

where $\phi = a/b$. Solving the above quadratic equation for ϕ (Sect. 2.13.2 discusses quadratic equations), we get

$$\left(\phi - \frac{1}{2}\right)^2 = 1 + \frac{1}{4} \Longrightarrow \phi = \frac{1+\sqrt{5}}{2} = 1.618033988 \tag{2.9.4}$$

The number ϕ is irrational.[32] It exhibits many amazing properties. Euclid (325–265 B.C.) in his classic book *Elements* gave the first recorded definition of ϕ. His own words are "A straight line is said to have been cut in extreme and mean ratio when, as the whole line is to the greater segment, so is the greater to the lesser". The German astronomer and mathematician Johannes Kepler once said "Geometry has two great treasures: one is the theorem of Pythagoras, the other the division of a line into extreme and mean ratio. The first we may compare to a mass of gold, the second we may call a precious jewel".

Golden rectangles and mathematical spiral. Let's start with a square of any side, say x (the pink square in Fig. 2.13a), then construct a rectangle by stretching the square horizontally by a factor of ϕ (what else?). What obtained is a *golden rectangle* (b). If we put the (original pink) square over the green rectangle so that the left edges

---

[32] Because $\sqrt{5}$ is irrational.

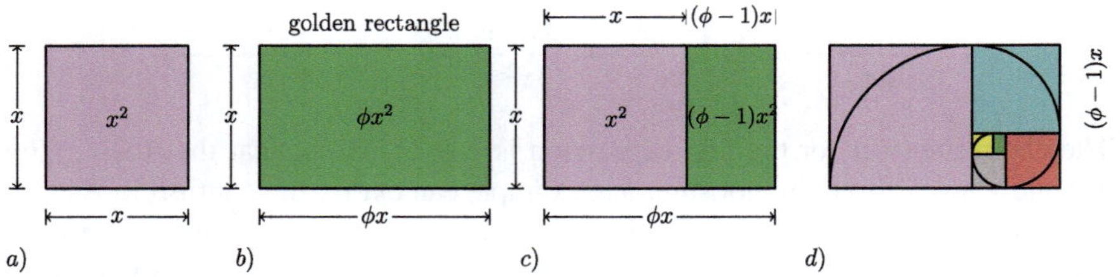

Fig. 2.13 From golden ratio ϕ to golden rectangles (green colored) and mathematical spirals

Fig. 2.14 Spirals occur in various forms in nature

are aligned, you get two areas following the golden ratio (c). In (c), the right (green) rectangle is also a golden rectangle as the ratio of the sides is $^1/_{\phi-1} = \phi$. Now, we split it into a square and a rectangle, then we get another rectangle, and we repeat this infinitely (d). Starting from the left most square, let's draw a circular arc, then another arc for the next square, etc. What you obtain is a spiral which appears in nature again and again (Fig. 2.14).

The golden ratio appears in a pentagon as shown in Fig. 2.15. In this figure, $ABCDE$ is a pentagon of which the sides are one (i.e.,, $|AB| = 1$, the notation $|AB|$ is to denote the length of edge AB), and the diagonals are d (noting that all diagonals of the pentagon are of equal length, see Fig. 2.15a). Looking now at Fig. 2.15b, as the two triangles AHC and EHD (shaded) are similar (using the AA test for they have the same marked angles[33]), we have the ratio of corresponding sides is $^1/_d$. But, we

[33] Check Sect. 3.1 for a presentation of Euclidean geometry.

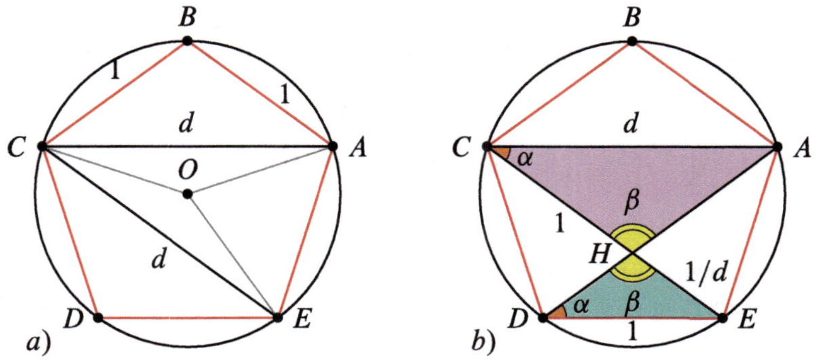

Fig. 2.15 The ratio of a diagonal over a side of a pentagon is the golden ratio

have $|CH| = 1$ (as the triangle CDH is isosceles), thus we must have $|HE| = 1/d$. Now, $|CE| = d = |CH| + |HE| = 1/d + 1$: the short portion of the diagonal CE plus the longer portion equals the diagonal itself. So, $d = \phi$. The flake in Fig. 1.7 is also related to the golden ratio. It's super cool, isn't it?.[34]

2.9.6 *Axioms for the Real Numbers*

Now that we have rational numbers and irrational numbers, it is time to introduce a new term—real numbers. The real numbers include all the rational numbers, such as the integer -5 and the fraction $2/3$, and all the irrational numbers, such as $\sqrt{2}$, π, and so on. The adjective real in this context was introduced in the seventeenth century by René Descartes, who distinguished between real and imaginary roots of polynomials. The set of all real numbers is denoted by $\mathbb{R}$. To do arithmetic with real numbers, we use the following axioms (accepted with faith) for a, b, c being real numbers:

$$
\begin{array}{ll}
\text{Axiom 1: } a + b = b + a & \text{(Commutative law for addition)} \\
\text{Axiom 2: } ab = ba & \text{(Commutative law for multiplication)} \\
\text{Axiom 3: } (a + b) + c = a + (b + c) & \text{(Associative law for addition)} \\
\text{Axiom 4: } (ab)c = a(bc) & \text{(Associative law for multiplication)} \\
\text{Axiom 5: } a + 0 = 0 + a = a & \text{(Existence of additive identity)} \\
\text{Axiom 6: } a + (-a) = 0 & \text{(Existence of additive inverse)} \\
\text{Axiom 7: } a \cdot 1 = 1 \cdot a = a & \text{(Existence of multiplicative identity)} \\
\text{Axiom 8: } a \cdot \frac{1}{a} = 1, \quad a \neq 0 & \text{(Existence of multiplicative inverse)} \\
\text{Axiom 9: } a(b + c) = ab + ac & \text{(Distributivelaw law)}
\end{array}
\tag{2.9.5}
$$

[34] Check this wikipedia for detail.

We use these axioms all the time without realizing that we are actually using them. As an example, below are three results which are derived from the above axioms:

$$\begin{aligned} -a &= (-1)a \\ -(-a) &= +a = a \\ -(a-b) &= -a+b \end{aligned} \tag{2.9.6}$$

The third is known as a rule saying that *if a bracket is preceded by a minus sign, change positive signs within it to negative and vice-versa when removing the bracket.*[35]

Proof First we prove $-a = (-1)a$ using the axioms in Eq. (2.9.5):

$$\begin{array}{ll} -a = -a + 0 & \text{(Axiom 5)} \\ -a = -a + 0 \times a & (a \times 0 = 0) \\ -a = -a + (1 + (-1)) \times a & \text{(Axiom 6)} \\ -a = -a + a + (-1) \times a & \text{(Axiom 9)} \\ -a = (-1) \times a & \text{(Axiom 6)} \end{array}$$

With that result, it is not hard to get $-(-a) = (-1)(-a) = (-1)(-1)(a) = 1a = a$ (axiom 4). For $-(a-b) = -a+b$, we do:

$$\begin{aligned} -(a-b) &= (-1)(a-b) && (-x = (-1)x) \\ &= (-1)a + (-1)(-b) && \text{(Axiom 9)} \\ &= -a + (-1)[(-1)b] && (-x = (-1)x) \\ &= -a + (-1)(-1)b = -a + b \end{aligned}$$

■

You might be thinking: are mathematicians crazy? About these proofs of obvious things George Pólya once said

> Mathematics consists of proving the most obvious thing in the least obvious way
>
> George Pólya

But why they had to do that? The answer is simple: to make sure the axioms selected are minimum and yet sufficient to provide a foundation for the theory they're trying to build.

2.10 Fibonacci Numbers

There is a special relationship between the Golden Ratio and the Fibonacci Sequence to be discussed in this section. The original problem that Fibonacci investigated (in the year 1202) was about how fast rabbits could breed in ideal circumstances. Suppose a newly-born pair of rabbits (one male, one female) are put in a field. Rabbits are

---

[35] Always use one example to check: $-(5-2)$, which is -3, is equal to $-5+3$, which is -2. So, the rule is OK.

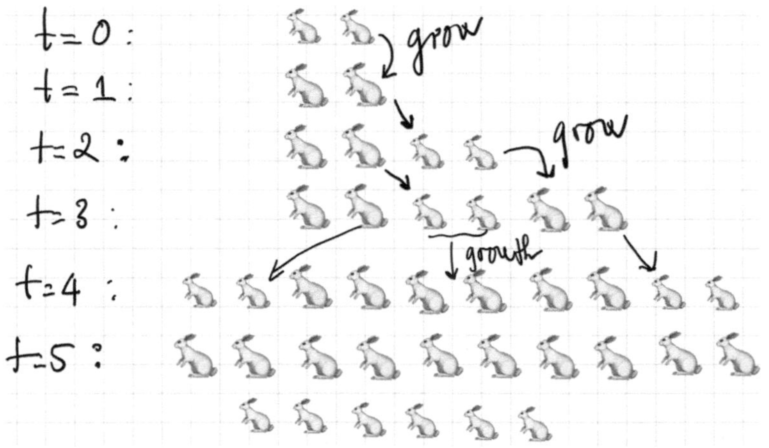

Fig. 2.16 Fibonacci's rabbit problem

mature at the age of one month. And after one month, a mature female can produce another pair of rabbits (male and female). Furthermore, it is assumed that our rabbits never die. The puzzle that Fibonacci posed was: How many pairs will there be in one year?

At the end of the first month, they mate, but there is still one only one pair. At the end of the second month the female produces a new pair, so now there are two pairs of rabbits in the field. At the end of the third month, the original female produces a second pair, making three pairs in all in the field. At the end of the fourth month, the original female has produced yet another new pair, the female born two months ago produces her first pair also, making five pairs (Fig. 2.16).

This led to the Fibonacci sequence $1, 1, 2, 3, 5, 8, 13, 21, 34, \ldots$ which can be defined as follows.

Definition 2.10.1 The Fibonacci sequence starts with 1,1 and the next number is found by adding up the two numbers before it:

$$F_n = F_{n-1} + F_{n-2}, \quad n \geq 2 \tag{2.10.1}$$

For example, $F_2 = F_1 + F_0 = 1 + 1 = 2$, $F_3 = F_1 + F_2 = 1 + 2 = 3$, and so on. Now comes the first surprise: there is a connection between the Fibonacci numbers and the golden ratio. To see the relation between Fibonacci sequence and the golden ratio, we computed some Fibonacci numbers and computed the ratio between two consecutive Fibonacci numbers. The data shown in Table 2.4 indicate that far along

Table 2.4 Ratios of two consecutive Fibonacci numbers approach the golden ratio ϕ

n	F_n	F_{n+1}/F_n
2	2	–
3	3	1.50000000
4	5	1.66666667
$\vdots$	$\vdots$	$\vdots$
19	6765	–
20	10946	1.61803400
21	28657	1.61803399

in the Fibonacci sequence the ratios approach ϕ.[36] But why? Let's denote by x the ratio of consecutive Fibonacci numbers:

$$x = \frac{F_{n+1}}{F_n} = \frac{F_{n+2}}{F_{n+1}} = \frac{F_{n+3}}{F_{n+2}} = \cdots$$

Hence, we can write F_{n+2} in terms of x and F_{n+1}, and F_{n+1} in terms of x and F_n to finally get F_{n+2} in terms of x and F_n:

$$F_{n+2} = xF_{n+1} = x^2F_n$$

Now, in the above equation, we replace $F_{n+2} = F_{n+1} + F_n$, and again replace F_{n+1} by xF_n, we get

$$F_n + F_{n+1} = x^2F_n, \quad F_n + xF_n = x^2F_n$$

Now, divide the last equation by F_n and we get $x^2 = x + 1$: the same quadratic equation that the golden ratio satisfies. That is why the ratio of consecutive Fibonacci numbers is the golden ratio.

There exists another interesting relation between the golden ratio and Fibonacci numbers; it is possible to express the powers of the golden ratios in terms of $a + b\phi$ where a, b are certain Fibonacci numbers. The procedure is as follows:

$$\begin{aligned}
\phi^2 &= 1 + \phi \\
\phi^3 &= \phi\phi^2 = \phi(1+\phi) = \phi + \phi^2 = 1 + 2\phi \\
\phi^4 &= \phi\phi^3 = \phi(1+2\phi) = \phi + 2(1+\phi) = 2 + 3\phi \\
\phi^5 &= 3 + 5\phi
\end{aligned} \tag{2.10.2}$$

[36] Of course, this table was generated by a small `Julia` program. Equation 2.10.1 is a recursive definition, so in this program, we also used that technique. In a program, we define a function and within its definition we use it.

That is, starting with $\phi^2 = 1 + \phi$, which is just the definition of ϕ, we raise the exponent by one to get ϕ^3, and replace ϕ^2 by $1 + \phi$. Then, we use ϕ^3 to get the the fourth-power, and so on. The expression for ϕ^5 was not obtained by detailed calculations, but by guessing, again we believe the pattern we are seeing: the coefficients of the power of the golden ratio are the Fibonacci numbers. In general, we can write:

$$\phi^n = F_{n-2} + F_{n-1}\phi \tag{2.10.3}$$

Note that the equation $\phi = 1 + 1/\phi$ has two solutions, one is ϕ and the other is $\psi = {}^1\!/\!_2(1 - \sqrt{5})$ and these two solutions are linked together by $\phi\psi = -1$. That is the negative solution is $-1/\phi$. If we have Eq. (2.10.3) for ϕ, should we also have something similar for $-1/\phi$—the other golden ratio? Following the same procedure done in Eq. (2.10.2). As $-1/\phi$ is a solution to $\phi = 1 + 1/\phi$, we have

$$-\frac{1}{\phi} = 1 - \phi$$

Squaring the both sides of this, using $\phi^2 = 1 + \phi$ and $\phi = 1 + 1/\phi$:

$$\left(-\frac{1}{\phi}\right)^2 = 1 - 2\phi + \phi^2 = 2 - \phi = 1 - \frac{1}{\phi}$$

And from that we get $(-{}^1\!/\!_\phi)^3$ and so on:

$$\begin{aligned}
\left(-\frac{1}{\phi}\right)^3 &= (1-\phi)(2-\phi) = 3 - 2\phi = 1 - 2\frac{1}{\phi} \\
\left(-\frac{1}{\phi}\right)^4 &= (1-\phi)(2-\phi) = 3 - 2\phi = 2 - 3\frac{1}{\phi} \\
\left(-\frac{1}{\phi}\right)^5 &= (1-\phi)(3-2\phi) = 3 - 5\frac{1}{\phi}
\end{aligned}$$

In all the final equalities, we have used $\phi = 1 + 1/\phi$ so that final expressions are written in terms of $1/\phi$. Now, we're ready to have the following

$$\left(-\frac{1}{\phi}\right)^n = F_{n-2} - F_{n-1}\frac{1}{\phi} \tag{2.10.4}$$

Now comes a nice formula for the Fibonacci sequence, a direct formula not recursive one. If we combine Eqs. (2.10.3) and (2.10.4) we have

$$\left.\begin{aligned}
\phi^n &= F_{n-2} + F_{n-1}\phi \\
\left(-\frac{1}{\phi}\right)^n &= F_{n-2} - F_{n-1}\frac{1}{\phi}
\end{aligned}\right\} \Longrightarrow \phi^n - \left(-\frac{1}{\phi}\right)^n = \left(\phi + \frac{1}{\phi}\right)F_{n-1}$$

And thus, (because $\phi + 1/\phi = \sqrt{5}$)

$$F_{n-1} = \frac{1}{\sqrt{5}}\left[\phi^n - \left(-\frac{1}{\phi}\right)^n\right], \quad F_n = \frac{1}{\sqrt{5}}\left[\phi^{n+1} - \left(-\frac{1}{\phi}\right)^{n+1}\right] \tag{2.10.5}$$

And this equation is now referred to as Binet's Formula in the honor of the French mathematician, physicist, and astronomer Jacques Philippe Marie Binet (1786 – 1856), although the same result was known to Abraham de Moivre a century earlier.

We have one question for you: in Eq. (2.10.5), $\phi = 0.5(1+\sqrt{5})$ is an irrational number, and F_n is always a whole number. Is it possible?

The purpose of this section was to present something unexpected in mathematics. Why on earth the golden ratio (which seems to be related to geometry) is related to a bunch of numbers coming from the sky like the Fibonacci numbers[37]? But there are more. Equation 2.10.1 is now referred to as a difference equation or recurrence equation. And similar equations appear again and again in mathematics (and in science); for example in probability as discussed in Sect. 5.8.7.

History note 2.1: Fibonacci (1170–1240–50)

Fibonacci was an Italian mathematician from the Republic of Pisa, considered to be "the most talented Western mathematician of the Middle Ages". Fibonacci popularized the Hindu–Arabic numeral system in the Western World primarily through his composition in 1202 of Liber Abaci (Book of Calculation). He also introduced Europe to the sequence of Fibonacci numbers, which he used as an example in Liber Abaci.
Although Fibonacci's Liber Abaci contains the earliest known description of the sequence outside of India, the sequence had been described by Indian mathematicians as early as the sixth century.

[37] That is not entirely true as we can see Fibonacci numbers in the petals of flowers. That is, in nearly all flowers, the number of petals is one of the numbers of 3, 5, 8, 13, 21, 34, 55, 89. For example, lilies have three petals, buttercups have five, many delphiniums have eight, marigolds have thirteen, asters have twenty-one, and most daisies have thirty-four, fifty-five, or eighty-nine.

2.11 Continued Fractions

First, continued fractions are fractions of fractions.[38] To see what it means, let's consider a rational number $45/16$ and we first express it as a whole number—called the integer part–plus another fraction a/b, which is smaller than 1. Next, we invert this fraction to the form $1/(b/a)$, and we keep doing this process for the fraction b/a. For example, $45/16 = 2.8125$, thus 2 is its integer part, so we can write $45/16 = 2 + 13/16$. Now $13/16 < 1$, we invert it to get $1/(16/13)$, and the whole process is repeated for $16/13$:

$$\frac{45}{16} = 2 + \frac{13}{16} = 2 + \frac{1}{16/13} = 2 + \frac{1}{1 + \frac{3}{13}} = 2 + \frac{1}{1 + \frac{1}{4 + \frac{1}{3}}}$$

Ok, so continued fraction is just another way to write a number. What's special? Let's explore more. How about an irrational number, like $\sqrt{2}$? How to express the square root of 2 as a continued fraction? Of course, we start by writing $\sqrt{2} = 1 + \cdots$:

$$\sqrt{2} = 1 + \sqrt{2} - 1 = 1 + \frac{1}{1 + \sqrt{2}} \quad \text{(because } (\sqrt{2} + 1)(\sqrt{2} - 1) = 1\text{)}$$

Now, we replace $\sqrt{2}$ in the fraction, i.e.,, $1/(1 + \sqrt{2})$ by the above equation, and doing so gives us:

$$\sqrt{2} = 1 + \frac{1}{1 + \sqrt{2}} = 1 + \frac{1}{2 + \frac{1}{1 + \sqrt{2}}} = 1 + \frac{1}{2 + \frac{1}{2 + \frac{1}{2 + \cdots}}} \tag{2.11.1}$$

We got an *infinite* continued fraction. Note that for $45/16$, a rational number, we got a finite continued fraction. Equation 2.11.1 can be used to compute square roots (in the time with no calculator). The term "continued fractions" was first used by John Wallis in 1653 in his book Arithmetica infinitorum.

Using the same idea, we can write the golden ratio ϕ as an infinite continued fraction

$$\phi = 1 + \frac{1}{\phi} \Longrightarrow \phi = 1 + \frac{1}{1 + \frac{1}{\phi}} = 1 + \frac{1}{1 + \frac{1}{1 + \frac{1}{\phi}}} \tag{2.11.2}$$

And as $\phi = 0.5(1 + \sqrt{5})$, we get this beautiful equation:

---

[38] http://www.maths.surrey.ac.uk/hosted-sites/R.Knott/Fibonacci/cfINTRO.html.

$$\frac{1+\sqrt{5}}{2} = 1 + \cfrac{1}{1 + \cfrac{1}{1 + \cdots}} \tag{2.11.3}$$

And that is not the end. You have probably seen this: $0.5(1+\sqrt{5}) = \sqrt{1+\sqrt{1+\sqrt{1+\cdots}}}$. Here is why:

$$\phi = 1 + \frac{1}{\phi} \Longrightarrow \phi^2 = 1 + \phi \Longrightarrow \phi = \sqrt{1+\phi} \Longrightarrow \phi = \sqrt{1+\sqrt{1+\sqrt{1+\cdots}}}$$

Fixed point iterations. Now, we're going to compute ϕ using its definition: $\phi = 1 + 1/\phi$. We're using a method called fixed point iterations. What is a fixed point of a function? The definition given below answers that question:

Definition 2.11.1 A fixed point x^* of a function $f(x)$ is such a point that $x^* = f(x^*)$.

I am not sure about the origin of this method, but in Sect. 3.7, we shall see how the Persian astronomer al-Kashi (c. 1380 – 1429) in his book *The Treatise on the Chord and Sine*, computed sin 1° to any accuracy. For that problem, he needed to solve a cubic equation of which solution was not available at his time. And he presented a fixed point iteration method to compute sin 1°.

A geometric illustration of a fixed point is shown in Fig. 2.17. Among other things, this concept can be used to solve equations $g(x) = 0$. First, we rewrite the equation in this form $x = f(x)$, then starting with x_0, we compute a sequence $(x_n) = (x_1, x_2, \ldots, x_n)$ with (for now, a sequence is nothing but a list of numbers. In Sect. 2.22, we talk more about sequences)

$$x_{n+1} = f(x_n) \tag{2.11.4}$$

As shown in Fig. 2.18, the sequence (x_n) converges to the solution x^*, if x_0 was chosen properly. Starting from x_0, draw a vertical line that touches the curve $y = f(x)$, then go horizontally until we get to the diagonal $y = x$. The x-coordinate of this point is

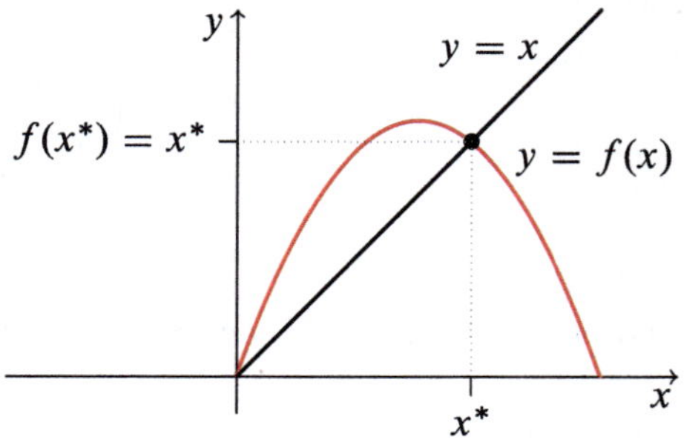

Fig. 2.17 A fixed point of a function $f(x)$ is the intersection of the two curves: $y = f(x)$ and $y = x$

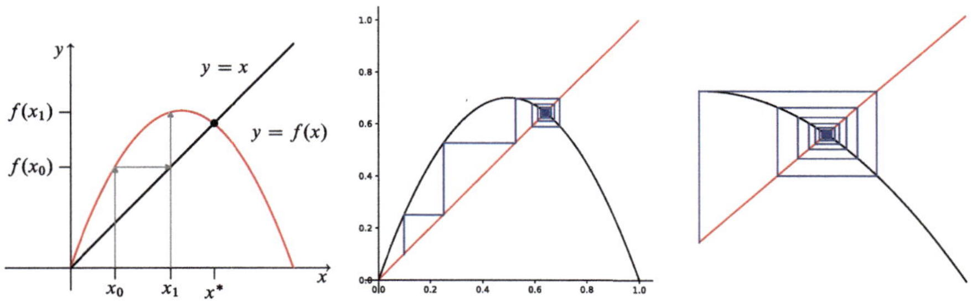

(a) (b) (c)

Fig. 2.18 Fixed point iterations for the function $x = 2.8x(1-x)$. In (b,c), the red line is $y = x$. The code used to generate Figs b and c is in `fixed_point_iter.jl`

Table 2.5 Fixed point iterations for $\phi = 1 + 1/\phi$: $\phi_{n+1} = 1 + 1/\phi_n$

n	ϕ_{n+1}	n	ϕ_{n+1}
1	2.0	1	–1.5
2	1.5	2	0.3333333
3	1.666666	3	3.9999999
4	1.6	4	1.25
5	1.625	5	1.8
⋮	⋮	⋮	⋮
19	1.618034	19	1.618034
20	1.618034	20	1.618034

x_1, and we repeat the process. Fig. 2.18b, c are the results of fixed point iterations for the function $y = 2.8x(1-x)$. What we are seeing is called a cobweb.

I demonstrate how this fixed point iteration scheme works for the golden ratio ϕ. In Table 2.5, I present the data obtained with $\phi_{n+1} = 1 + 1/\phi_n$ with two starting points $\phi_0 = 1.0$ and another $\phi_0 = -0.4$. Surprisingly, both converge to the same solution of 1.618. Thus, the second negative solution of $\phi = 1 + 1/\phi$—which is $\phi = 0.5(1-\sqrt{5})$–escaped. In Fig. 2.19, we can see this clearly.

There are many questions remain to be asked regarding this fixed point method. For example, for what functions the method works, and can we prove that (to be 100% certain) that the sequence (x_n) converges to the solution? To answer these questions, we need calculus and thus I postpone the discussion to Sect. 10.5. In that section you will find the answer to why we could not get the negative solution of $\phi = 1 + 1/\phi$: the first derivative of the function $1 + 1/\phi$ evaluated at $\phi = 0.5(1-\sqrt{5})$ is -2.61 which is larger than one. This also demonstrates the importance of the derivative of a function. It can tell us so many things about a function!

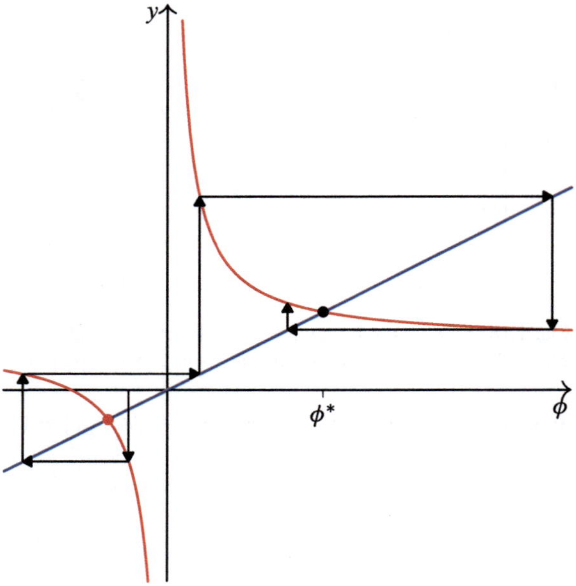

Fig. 2.19 Fixed point iterations for $\phi = 1 + 1/\phi$: $\phi_{n+1} = 1 + 1/\phi_n$ with the starting $\phi_0 = -0.4$. The positive solution (which is the golden ratio) is marked by a black dot whereas the negative solution by a red dot

2.12 Pythagoras' Theorem

Ask any school student to name a famous mathematician, and more often than not they will opt for Pythagoras, if they can think of one. Pythagoras's present-day fame rests on the theorem that bears his name: Pythagoras's theorem or Pythagorean theorem. We have no idea whether Pythagoras actually proved the theorem. In fact, we don't know whether it was his theorem at all. But he got the credit and his name stuck.

Pythagoras' theorem is a fundamental relation in Euclidean geometry among the three sides of a right triangle. It states that the area of the square whose side is the hypotenuse (the side opposite the right angle) is equal to the sum of the areas of the squares on the other two sides. This theorem can be written as an equation relating the lengths of the sides a, b and c, often called the "Pythagorean equation":

$$a^2 + b^2 = c^2 \tag{2.12.1}$$

where c represents the length of the hypotenuse and a and b the lengths of the triangle's other two sides (Fig. 2.20). For example, in a right triangle if two legs are 3 and 4, the hypotenuse is 5 (because $5^2 = 3^2 + 4^2$). This makes the famous $3 - 4 - 5$ right triangle.

The theorem has been given numerous proofs—possibly the most for any mathematical theorem.[39] I present in Fig. 2.21b one proof. And young students are recommended to prove this theorem as many ways as possible.

---

[39] An early twentieth century mathematician named Elisha Scott Loomis collected and published 367 proofs in a book called *The Pythagorean Proposition.*

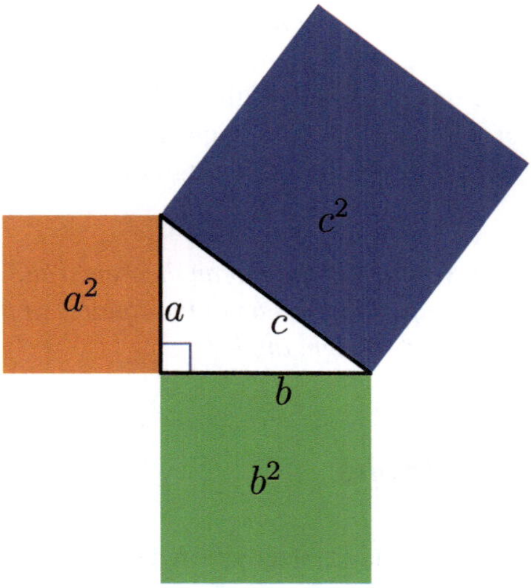

Fig. 2.20 Pythagorean theorem. The sum of the areas of the two squares on the legs (a and b) equals the area of the square on the hypotenuse (c)

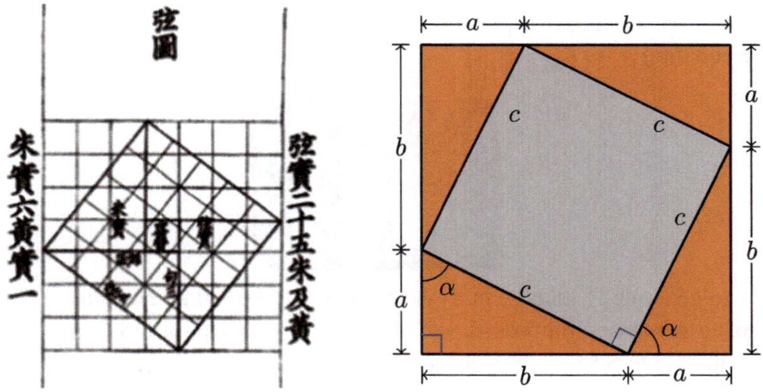

(a) The hsuan-thu diagram

(b) The modern proof

Fig. 2.21 The Chinese proof of the Pythagorean theorem

Proof The proof given in Fig. 2.21b is using the same strategy adopted in Fig. 2.12: we start with one right triangle of sides a, b, c and we add three more, arranged in a special way that together they make a big square of side $a + b$. Then, we compute the area of this big square in two ways. First way is $(a + b)^2$, which is $a^2 + 2ab + b^2$. And the second way is: its area is equal to the sum of areas of the four triangles and the square of side c (why this gray-shaded shape is a square? Look at the angles α). This area is $4(1/2)ab + c^2$. So, we have

$$a^2 + 2ab + b^2 = 2ab + b^2 \implies a^2 + b^2 = c^2$$

William Dunham called this proof the "Chinese Proof" as it was embodied in the *hsuan-thu* diagram of a square tilted in another square, dated from somewhere between 1000BC–1 AD [19]. Note that this diagram was for the specific $3-4-5$ right triangle only. ■

Question 3 *Is the converse of the Pythagorean theorem true? That is, in a triangle, if the square of one side is equal to the sum of the squares of the other two sides, is the angle opposite the first side a right angle?*

2.12.1 Pythagorean Triples

Plimpton 322 is a Babylonian clay tablet, believed to have been written about 1800 BC, has a table of four columns and 15 rows of numbers in the cuneiform script of the period. This table lists two of the three numbers in what are now called Pythagorean triples. One example is $(3, 4, 5)$.

Definition 2.12.1 Integer triples (a, b, c) are called Pythagorean triples if they satisfy the equation $a^2 + b^2 = c^2$.

How to generate Pythagorean triples? There are more than one way and surprisingly using complex numbers is one of them. If you need a brief recall on complex numbers, see Sect. 2.25. Let's start with a complex number $z = u + vi$ where u, v are positive integers and i is the number such that $i^2 = -1$. Its modulus is $|z| = \sqrt{u^2 + v^2}$. The key point is that the modulus of the square of z is $u^2 + v^2$, which is an integer. Indeed, let's compute z^2 and its modulus:

$$z^2 = (u + vi)^2 = u^2 - v^2 + 2uvi \implies |z^2| = \sqrt{(u^2 - v^2)^2 + (2uv)^2} = u^2 + v^2$$

This result indicates that $(u^2 - v^2)^2 + (2uv)^2 = (u^2 + v^2)^2$. Thus, the triple $(u^2 - v^2, 2uv, u^2 + v^2)$ is a Pythagorean triple! We are going to compute some Pythagorean triples using this and Table 2.6 presents the result.

Note that the triples $(3, 4, 5)$ and $(12, 16, 20)$ are related; the latter can be obtained by multiplying the former by 4. The corresponding right triangles are similar. Generally, if we take a Pythagorean triple (a, b, c) and multiply it by some other number d, then we obtain a new Pythagorean triple (da, db, dc). This leads to the so-called primitive Pythagorean triples in which a, b, c have no common factors. A common factor of a, b and c is a number d so that each of a, b and c is a multiple of d. For

Table 2.6 Pythagorean triples $(u^2 - v^2, 2uv, u^2 + v^2)$

(u, v)	$(u^2 - v^2, 2uv, u^2 + v^2)$
(2, 1)	(3, 4, 5)
(4, 2)	(12,16,20)
(3, 2)	(5, 12, 13)
(4, 3)	(7, 24, 25)
(5, 4)	(9, 40, 41)

example, 3 is a common factor of 30, 42, and 105, since $30 = 3 \times 10, 42 = 3 \times 14$, and $105 = 3 \times 35$, and indeed it is their largest common factor. On the other hand, the numbers 10, 12, and 15 have no common factor (other than 1).

If (a, b, c) is a primitive Pythagorean triple, it can be shown that

- a, b cannot be both even;
- a, b cannot be both odd;
- a is odd, b is even ($b = 2k$) and c is odd. See Table 2.6 again.

From the definition of (a, b, c), we write

$$a^2 + b^2 = c^2 \Longrightarrow b^2 = (c - a)(c + a) \tag{2.12.2}$$

As both a and c are odd, so its sum and difference are even. Thus, we can write $c - a = 2m, c + a = 2n$. Equation 2.12.2 becomes

$$(2k)^2 = (2m)(2n) \Longrightarrow k^2 = mn \Longrightarrow m = p^2, \quad n = q^2 \tag{2.12.3}$$

Now, we can solve for a, b, c in terms of p and q, and obtain the same result as before:

$$c = p^2 + q^2, \quad a = q^2 - p^2, \quad b = 2k = 2pq$$

2.12.2 *Fermat's Last Theorem*

In number theory, Fermat's Last Theorem states that no three positive integers a, b, and c satisfy the equation $a^n + b^n = c^n$ for any integer value of n greater than 2. The cases $n = 1$ and $n = 2$ have been known since antiquity to have infinitely many solutions. Recall that $n = 2$, the solutions are Pythagoreas triplets discussed in Sect. 2.12.

We found Fermat's Last Theorem in the margin of his copy of Diophantus' Arithmetica: "It is impossible to separate a cube into two cubes, or a fourth power into two fourth powers, or in general, any power higher than the second, into two like powers". He also wrote that "I have discovered a truly marvelous proof of this proposition which *this margin is too narrow to contain*". This habit of not revealing his calculations or the proofs of his theorems frustrated his adversaries: Descartes came to call him a "braggart", and the Englishman John Wallis referred to him as "that damn Frenchman". About this story, there is a story of another mathematician that goes like this

> A famous mathematician was to give a keynote speech at a conference. Asked for an advance summary, he said he would present a proof of Fermat's Last Theorem—but they should keep it under their hats. When he arrived, though, he spoke on a much more prosaic topic. Afterwards the conference organizers asked why he said he'd talk about the theorem and then didn't. He replied this was his standard practice, just in case *he was killed on the way to the conference.*

The son of a wealthy leather merchant, Pierre de Fermat (1601–1665) studied civil law at the University of Orleans (France) and progressed to a comfortable position in the Parliament of Toulouse, which allowed him to spend his spare time on his great love: mathematics. In the afternoons, Fermat put aside the law and dedicated to mathematics. He studied the treatises of the scholars of classical Greece and combined those old ideas with the new methods of algebra by François Viète.

After 358 years of effort by mathematicians, the first successful proof was released in 1994 by Andrew Wiles (1953) an English mathematician. In 1963, a 10-year-old boy named Andrew Wiles read that story of Fermat's last theorem, was fascinated, and set out to dedicate his life to proving Fermat's Last Theorem. Two decades later, Wiles became a renowned mathematician before deciding to return to his childhood dream. He began to secretly investigate finding the solution to the problem, a task which would take *seven years of his life*.

There are many books written on this famous theorem, for example *Fermat's Last Theorem: The Book* by Simon Singh. I strongly recommend it to young students. About Wiles' proof, it is 192 pages long and I do not understand it at all. Note that I am an engineer not a pure mathematician.

2.12.3 Solving Integer Equations

Find positive integers that satisfy the following equation[40]

$$\sqrt{a} + \sqrt{b} = \sqrt{2009} \tag{2.12.4}$$

[40] The expression $\sqrt{a} + \sqrt{b} = \sqrt{2009}$ is not an identity since it does not hold for all a and b. That's why it's called an equation. We have more to say about equations in Sect. 2.13.

Table 2.7 Solutions to $\sqrt{a}+\sqrt{b}=\sqrt{2009}$

(n, m)	$(0, 7)$	$(1, 6)$	$(2, 5)$	$(3, 4)$
(a, b)	$(0, 7)$	$(41, 1476)$	$(164, 1025)$	$(369, 656)$

How can we solve this? Some hints: (1) a and b are symmetrical so if (a, b) is a solution, so is (b, a); (2) usually squaring is used to get rid of square roots. But we have to first isolate a, b before squaring:

$$\begin{aligned}\sqrt{a} &= \sqrt{2009}-\sqrt{b}\\ a &= 2009+b-2\sqrt{2009b} \quad \text{(squaring the above)}\end{aligned}$$

Thus, we have $\sqrt{2009b}=c$, where c is a positive integer. Now we rewrite $\sqrt{2009b}$ as $7\sqrt{41b}$. Now we know that only the square root of a perfect square is a natural number, thus $\sqrt{41b}$ is a natural number when $b=41m^2$, where $m\in\mathbb{N}$ (this is similar to writing m is a natural number, but shorter, we will discuss about this notation later). Since a and b are playing the same role, we also have $a=41n^2$, $n\in\mathbb{N}$. With these findings, Eq. (2.12.4) becomes:

$$n\sqrt{41}+m\sqrt{41}=7\sqrt{41}\implies n+m=7$$

It is interesting that the scary looking Eq. (2.12.4) is equivalent to this easy equation $n+m=7$, which can be solved by kids of 7 years ago and above by a rude method: trial and error guessing (Table 2.7).

If we're skillful enough and lucky—if we transform the equations in just the right way—we can get them to reveal their secrets. And things become simple. Creativity is required because it often isn't clear which manipulations to perform. And that's why mathematics is exciting. If all problems can be solved by routine procedures, we all get bored!

History note 2.2: Pythagoras (c. 570 BC–495 BC)

Pythagoras was an an ancient Ionian Greek philosopher and the eponymous founder of Pythagoreanism. In antiquity, Pythagoras was credited with many mathematical and scientific discoveries, including the Pythagorean theorem, Pythagorean tuning, the five regular solids, the Theory of Proportions, the sphericity of the Earth, and the identity of the morning and evening stars as the planet Venus. Pythagoras was the first to proclaim his being a philosopher, meaning a 'lover of ideas.' Pythagoras had followers. A whole group of mathematicians signed up to be his pupils, to learn everything he knew, they were called the Pythagoreans. Numbers, Pythagoras believed, were the elements behind the entire universe. The Pythagoreans had sacred numbers. Seven was the number of wisdom, 8 was the number of justice, and 10 was the most sacred number of all. Every part of math was holy. When they solved a new mathematical theorem, they would give thanks to the gods by sacrificing an ox.

2.12.4 From Pythagorean Theorem to Trigonometry and More

Many of the triangles encountered in real life are not right-angled, so the direct applications of the Pythagorean theorem may seem restricted. However, noting that any triangle can be cut into two right-angled ones, and any polygons can be cut into triangles. So right-angled triangles are the key: they show that there is a useful relation between the shape of a triangle and the lengths of its sides. The subject that developed from this insight is trigonometry meaning "triangle measurement". We have an entire Chap. 3 for this topic.

Later in Sect. 3.24.1 we shall see that the Pythagorean theorem provides us a way to compute distance between points defined on a coordinate system.

2.13 Imaginary Number

Our next destination in the universe of numbers is i with the property $i^2 = -1$. To understand the context in which i was developed or discovered, we need to talk about solving equations of one single unknown. We will discuss three types of equation as shown in Eq. (2.13.1):

$$2x = 5 \quad \text{linear equation} \tag{2.13.1a}$$
$$x^2 + 10x = 39 \quad \text{quadratic equation} \tag{2.13.1b}$$
$$x^3 + x^2 + 10x = 39 \quad \text{cubic equation} \tag{2.13.1c}$$

From these three examples, we can see that an equation is a formula that expresses the equality of two expressions (e.g., $2x$ and 5 in Eq. (2.13.1a)), by connecting them with the equals sign $=$. In these three equations, x is called the *unknown* of the equation.

Of course, instead of x, we can equally use y or z, but no mathematicians would use a or b. We shall discuss why later. Our job is to solve the equation: finding x such that it satisfies the equation (make the equation a true statement). For example, $x = 5/2$ is the solution to the first equation for $2(5/2) = 5$.

The presentation on solving these linear/quadratic/cubic equations serves only to introduce the need for mathematicians to consider square root of negative numbers. We shall get back to the topic of solving equations in depth in Sect. 3.25.

History note 2.3: Importance of mathematical notations

There are many major mathematical discoveries but only those which can be understood by others lead to progress. However, the easy use and understanding of mathematical concepts depends on their notation.
The convention we use (letters near the end of the alphabet representing unknowns e.g. x, y, z) was introduced by Descartes in 1637. Other conventions have fallen out of favor, such as that due to Viète who used vowels for the unknowns and consonants for the knowns.

We start off with linear equations (Sect. 2.13.1), then we move to quadratic equations (Sect. 2.13.2). Then, we discuss cubic equations—the context in which imaginary number i first appeared in Sect. 2.13.3. Viète's solution of cubic equation using trigonometry functions is presented in Sect. 2.13.4. And finally, the interesting history of the cubic equation is presented in Eq. (2.13.5).

2.13.1 Linear Equation

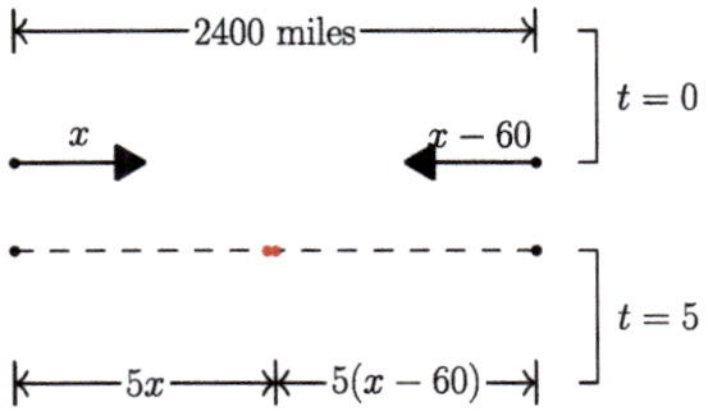

Let us consider the following *word problem*: two planes, which are 2 400 miles apart, fly toward each other. Their speeds differ by 60 miles per hour. They pass each other after 5 hours. Find their speeds. As we shall see, this problem involves a linear equation of the form Eq. (2.13.1a). Let's denote by x (miles/hour) the speed of the left plane. Thus, the speed of the right plane is $x - 60$ (or $x + 60$). The distance traveled by the left plane after 5 hours is $x_1 = 5x$ and that by the right plane is $x_2 = 5(x - 60)$. Then, we get the following equation (see figure)

$$5x + 5(x - 60) = 2400$$

which is the mathematical expression of the fact that the two planes have traveled a total distance of 2400 miles. To solve this *linear* equation, we simply massage it in the way that x is isolated in one side of the equality symbol: $x = \cdots$ So, we do (the algebra is based on the arithmetic rules stated in Eq. (2.1.2); there is nothing to memorize here![41])

$$\begin{aligned}
5x + 5(x - 60) = 2400 &\Leftrightarrow 10x - 300 = 2400 \\
&\Leftrightarrow 10x = 2400 + 300 \Leftrightarrow x = 2700/10 = 270
\end{aligned}$$

Thus, the speed of one plane is 270 miles per hour and the speed of the other plane is 210 miles per hour. In the above equation, the symbol $\Leftrightarrow$ means "equivalent" that is the two sides of this symbol are equivalent: one can go from one side to the other and vice versa; sometimes $\Longleftrightarrow$ is used for the same purpose.

Let's see another problem: adding 6 to a number results in a number that is three times itself; Find the number. Of course, we use x to label the number, then we have: $6 + x = 3x$. Solving this we get $x = 3$. Now, comes one key idea: instead of solving all concrete instances of linear equations (e.g., $2x = 5$ or $5x - 7 = 0$), mathematicians solved the following equation

$$\boxed{ax + b = 0 \Longleftrightarrow x = -\frac{b}{a}} \tag{2.13.2}$$

The equation $ax + b = 0$ is called a linear equation because x is a linear term. Geometrically, if we plot the function $y = ax + b$ on the Cartesian plane we get a line. Section 3.24 discusses Cartesian plane and analytic geometry.

Have you ever wondered why mathematicians did not write Eq. (2.13.2) as $ac + b = 0$ (i.e.,, using c for x)? Noting that in Eq. (2.13.2), a, b, x are all real numbers, but a, b play one role and x plays another role, in the sense that x cannot be arbitrary, it depends on a and b in a specific way. By using the letters near the end of the alphabet, e.g., x, y, z to represent the unknown, we can easily spot out what are unknowns and what are knowns. The expression

$$x = -\frac{b}{a}$$

is called a *formula*. It relates x to given a and b. It allows us to compute/evaluate x. For example, if $a = 2$ and $b = 6$, then $x = -6/2 = -3$. Our world is governed by many formula; probably the most famous one is Einstein's $E = mc^2$.

Usually solving a linear equation in x is straightforward, but the following equation looks hard:

---

[41] Actually there is another obvious rule: if $a = b$, then $a + c = b + c$ for any c. I used it in the third step to remove -300 in $10x - 300$. Many teachers tell students: move a number from one side to the other side of the $=$ symbol, change the sign. Yet another rule to memorize!

$$x + \frac{x}{1+2} + \frac{x}{1+2+3} + \cdots + \frac{x}{1+2+3+\cdots+4021} = 4021$$

The solution is 2021. If you cannot solve it, look at the red term (all the denominator has same form) and ask yourself what it is.

2.13.2 *Quadratic Equation*

We move now to the quadratic equation. It emerged from a very practical problem: if you have a rectangular piece of land, of which one side is longer than the other side 10 units and the area of the land is 39 unit squared, how long are the sides? Already familiar with symbols (we take them for granted), we denote by x the length of the shorter side of the rectangle; then the length of the longer side is $x + 10$. The area of this land is thus $x(x + 10)$ and it is equal to 39. Therefore, we obtain this quadratic equation $x(x + 10) = 39$ or $x^2 + 10x - 39 = 0$. To see the power of symbols, this is how the same equation was referred to in the ninth century: "a square and ten of its roots are equal to thirty-nine"; roots refer to x term, squares to x^2 terms and numbers to constants.

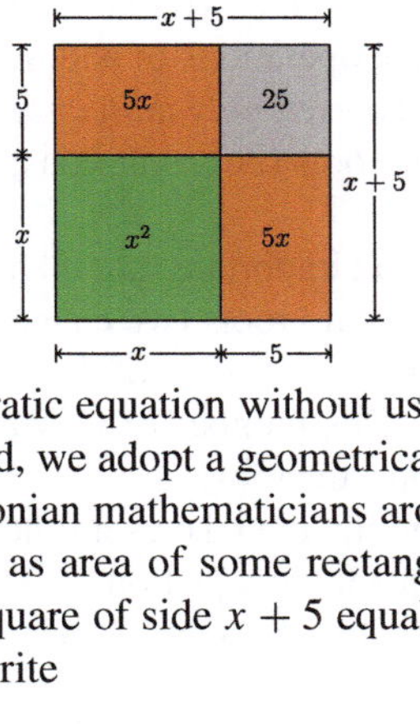

Let's now solve this quadratic equation without using the well-known quadratic formula $-b\pm\sqrt{b^2-4ac}/2a$. Instead, we adopt a geometrical approach typically common in the ancient time by Babylonian mathematicians around 500 BC. By considering all the terms in the equation as area of some rectangles (see next figure), we see that the area of the biggest square of side $x + 5$ equals the sum of areas of smaller rectangles. This leads us to write

$$(x + 5)^2 = x^2 + 10x + 25 = 39 + 25 = 64 = 8^2$$

where, in the second equality, we have replaced $x^2 + 10x$ by 39. The last equality means $x + 5 = 8$, and hence $x = 3$. The ancient mathematicians stopped here, i.e.,, they did not find out the other solution—the negative $x = -13$, because for them numbers simply represent geometric entities (length, area, ...). Note that $8^2 = 64$ and $(-8)^2$ is also 64.

Algebraically, the above is equivalent to using the identity $(a + b)^2 = a^2 + 2ab + b^2$, we have added $25 = 5^2$ to *complete the square* $(x + 5)^2$, which is the area of the square of side $x + 5$. This is also the key to solving cubic equations by a similar *completing a cube* procedure.

Instead of solving many concrete quadratic equations (e.g., $2x^2 - 5x + 3 = 0$, $x^2 + 2x - 6 = 0$, and so on), it was a big mind shift in mathematics when mathematicians considered $ax^2 + bx + c = 0$ with $a \neq 0$. This equation with undefined a, b, c, called *coefficients*, cover all quadratic equations at once. If we can solve it, we solve all quadratic equations! To solve it, first we rewrite it as $x^2 + {}^{b}/_{a}x + {}^{c}/_{a} = 0$,[42] then we complete the square:

$$x^2 + 2\frac{b}{2a}x + \frac{b^2}{4a^2} + \frac{c}{a} - \frac{b^2}{4a^2} = 0 \Longleftrightarrow \left(x + \frac{b}{2a}\right)^2$$
$$= \frac{b^2}{4a^2} - \frac{c}{a} \Longleftrightarrow x + \frac{b}{2a} = \pm\sqrt{\frac{b^2}{4a^2} - \frac{c}{a}}$$

And from that we obtain the famous quadratic formula:

$$\boxed{x = -\frac{b}{2a} \pm \frac{\sqrt{b^2 - 4ac}}{2a}} \tag{2.13.3}$$

This expression is called a *solution by radicals* or an *algebraic solution* for the quadratic. That is, it involves only the original coefficients in the equation—that is, a, b and c—and the algebraic operations of addition, subtraction, multiplication, division, and extraction of roots, *used only finitely often*.[43] That formula is the first formula ever that expresses the roots (i.e.,, solutions) of an equation in terms of its coefficients.

The number $b^2 - 4ac$ plays an important role in the roots: if it is positive, we get two roots, if it is negative we get no roots (we cannot have square root of negative numbers) and if it is zero we have two roots of the same value $-{}^{b}/_{2a}$; this is called a root of multiplicity of two. This number $b^2 - 4ac$ is called the discriminant,[44] designated by Δ: it discriminates the roots.

As a preparation for the cubic equation, I present another way to solve the quadratic equation. Let's consider this quadratic equation $x^2 + bx + c = 0$ where the coefficient of the x^2 term is one to simplify the maths. This method is based on the observation that it is easy to solve this quadratic equation $x^2 - d = 0$ ($d \geq 0$): the solutions are $x = \pm\sqrt{d}$. The question is: can we convert $x^2 + bx + c = 0$ to $x^2 - d = 0$?

The answer is yes: using a change of variable $x = u - b/2$ enables us to get rid of the term bx to obtain this reduced quadratic equation $u^2 = d$, with $d = b^2/4 - c$. How? Just plugging $x = u - b/2$ into the equation $x^2 + bx + c = 0$, do some algebraic manipulations and we get $u^2 = d$. But we can solve easily this reduced equation: $u = \pm\sqrt{d}$ (assuming $d > 0$). From u, we then get x: $x = \pm\sqrt{d} - b/2$. But, how did we know of this change of variable? By trial and errors I guess: starting with

[42] Make sure you understand why we can do this.

[43] Why we need to mention this? Recall that we have seen in Eq. (2.11.1) an expression that involves an infinite number of terms.

[44] The term "discriminant" was coined in 1851 by the British mathematician James Joseph Sylvester.

$x = u + z$, where we do not know what z is. Introducing this x into $x^2 + bx + c = 0$ we obtain:

$$(u + z)^2 + b(u + z) + c = 0 \Longleftrightarrow u^2 + (2z + b)u + bz + z^2 + c = 0$$

Since we do not like the term $(2z + b)u$, we choose z such that $2z + b = 0$. If you still wonder why we know that $x = u + z$, then I do not have a convincing answer. But there is a geometric meaning behind this change of variable. To answer that we need to know Cartesian coordinates and plotting functions. I leave this to Sect. 3.25. With geometry, this change of variable is no more mysterious.

Remark What we have just done? For the equation $x^2 + bx + c = 0$, we used a change of variable to get $u^2 - d = 0$. That is, we removed the term bx, which is the term of order one less than the highest order term x^2. Why is this important? Because it is how mathematicians solved the cubic equation $x^3 + ax^2 + bx + c = 0$. Which term they will remove then?

Remark Irrational numbers arise naturally when solving quadratic equations. The roots of $x^2 - 6x + 4 = 0$ are $3 \pm \sqrt{5}$—two irrational numbers. Al-Khwarizmi referred to irrational numbers as "inaudible" numbers, which was later translated to the Latin *surdus*, which means "deaf" or "mute". From this, we obtained the term *surd*, which is still used, typically for the square roots of non-square integers [56]. That is, we call $\sqrt{5}$ a surd.

Quadratic equations in disguise. Many equations are actually quadratic equations in disguise. For example, $x^4 - 2x^2 + 1 = 0$ is a quadratic equation $t^2 - 2t + 1 = 0$ with $t = x^2$.

To demonstrate unexpected things in maths, let's consider this equation:

$$\sqrt{5 - x} = 5 - x^2$$

To remove the square root, we follow the old rule: squaring both sides of the equation:

$$5 - x = 25 - 10x^2 + x^4$$

Ops! We've got a quartic equation! Now comes the magic of maths, when I first saw this it was like magic. Instead of seeing the equation as a quartic equation in terms of x, how about seeing it as a quadratic equation in terms of 5? With that in mind, we re-write the equation as

$$5 - x = 5^2 - 5(2x^2) + x^4 \Longleftrightarrow 5^2 - (2x^2 + 1)5 + x^4 + x = 0$$

And we solve for 5 using the quadratic formula ${-b \pm \sqrt{b^2 - 4ac}}/{2a}$ (or we can complete a square):

$$5 = \frac{(2x^2 + 1) \pm \sqrt{(2x^2 + 1)^2 - 4(x^4 + x)}}{2} = \frac{(2x^2 + 1) \pm |(2x - 1)|}{2}$$

Simplifying the above, we get two equations:

$$\begin{cases} x^2 + x = 5 \\ x^2 - x + 1 = 5 \end{cases} \Longrightarrow x = \frac{\sqrt{21} - 1}{2}, \quad x = \frac{1 - \sqrt{17}}{2}$$

(We had to ignore two roots that would not make sense for $\sqrt{5 - x} = 5 - x^2$; as the LHS is a non-negative number, so is the RHS. Thus, $-\sqrt{5} \le x \le \sqrt{5}$).

Should we memorize the quadratic formula? The formula ${-b \pm \sqrt{b^2-4ac}}/{2a}$ is significant as it is the first formula ever that expresses the roots of an equation in terms of its coefficients. But young students should not memorize and use it to solve quadratic equations. If we just use the formula all the time we would forget how it was derived (i.e.,, the completing the square method).

Remark Did we completely solve the quadratic equation? If your answer is yes, think again. We have found two solutions given in Eq. (2.13.3). But who ensures that these are the only solutions that the quadratic equation can have? Anyway, we anticipate that a quadratic equation can have at most two solutions. And this thinking eventually led to the fundamental theorem of algebra.

2.13.3 Cubic Equation

The most general form of a cubic equation is written as

$$ax^3 + bx^2 + cx + d = 0, \quad a \neq 0 \tag{2.13.4}$$

The condition $a \neq 0$ is needed, otherwise Eq. (2.13.4) becomes a quadratic equation (suppose that $b \neq 0$). As we can always divide Eq. (2.13.4) by a, it suffices to consider the following cubic equation[45]

$$x^3 + bx^2 + cx + d = 0 \tag{2.13.5}$$

It turned out that solving a full cubic equation Eq. (2.13.5) was not easy. So, in 1545, the Italian mathematician Gerolamo Cardano (1501–1576) presented a solution to the following depressed cubic equation (it is always possible to convert a full cubic equation to the depressed cubic by using this change of variable $x = u - b/3$ to get rid of the quadratic term—the term bx^2[46])

[45] Note that the b, c, d in Eq. (2.13.5) are different from Eq. (2.13.4).

[46] Again, calculus helps to understand why this change of variable: $x = -b/3$ is the x-coordinate of the inflection point of the cubic curve $y = x^3 + bx^2 + cx + d$. Note, however, that at the time of Cardano, calculus has not yet been invented. But with the success of reducing a quadratic equation to the form $u^2 - d = 0$, mathematicians were confident that they should be able to do the same for the cubic equation.

$$x^3 + px = q \tag{2.13.6}$$

of which his solution is

$$\boxed{x = \sqrt[3]{\frac{q}{2} + \sqrt{\frac{q^2}{4} + \frac{p^3}{27}}} - \sqrt[3]{-\frac{q}{2} + \sqrt{\frac{q^2}{4} + \frac{p^3}{27}}}} \tag{2.13.7}$$

It was actually Scipione del Ferro who first discovered this solution. We have more to say about this fascinating story in Sect. 2.13.5. In many books, Cardano's formula is written slightly different: $x = \sqrt[3]{q/2 + \sqrt{q^2/4 + p^3/27}} + \sqrt[3]{q/2 - \sqrt{q^2/4 + \frac{p^3}{27}}}$. Let's first see one example.

Example 2.1 To check his formula, Cardano solved a concrete example: $x^3 + 6x = 20$. We have $p = 6$ and $q = 20$, plugging them into the formula, Eq. (2.13.7) (note that $p^3/27 = (p/3)^3$ to ease the calculation): we obtain

$$x = \sqrt[3]{10 + \sqrt{108}} - \sqrt[3]{-10 + \sqrt{108}}$$

Is this messy expression really the solution? For us, it is easy: using a calculator to check. But Cardano did not have that luxury. What he did then? He noted that $x = 2$ is one solution to $x^3 + 6x = 20$: $2^3 + 6(2) = 20$. If he can show that $\sqrt[3]{10 + \sqrt{108}} - \sqrt[3]{-10 + \sqrt{108}}$ is indeed nothing but 2, then his formula is correct. Noting that $108 = 36 \times 3$, hence $\sqrt{108} = 6\sqrt{3}$. With that, we obtain (with some trial/errors)

$$10 + \sqrt{108} = (1 + \sqrt{3})^3, \quad -10 + \sqrt{108} = (-1 + \sqrt{3})^3$$

Because of this Cardano could get rid of the ugly cube root and got 2!

For the moment, how del Ferro/Cardano came up with Eq. (2.13.7) is not as important as how Eq. (2.13.7) led to the discovery of the imaginary number, now designated by i: $i^2 = -1$. To see that, just consider the following equation

$$x^3 - 15x = 4 \tag{2.13.8}$$

Using Eq. (2.13.7) with $p = -15$ and $q = 4$, we get

$$x = \sqrt[3]{2 + \sqrt{-121}} - \sqrt[3]{-2 + \sqrt{-121}} \tag{2.13.9}$$

As Eq. (2.13.7) was successfully used to solve many depressed cubic equations, it was perplexing that for Eq. (2.13.8) it involves the square root of a negative number, i.e.,, $\sqrt{-121}$.

So, Cardano stopped there and it took almost 30 years for someone to make progress. It was Rafael Bombelli (1526–1572)—another Italian—in 1572 who examined Eq. (2.13.9). He knew that $x = 4$ is a solution to Eq. (2.13.8). Thus, he was about to check the validity of the following identity

$$\sqrt[3]{2+\sqrt{-121}} - \sqrt[3]{-2+\sqrt{-121}} \stackrel{?}{=} 4 \tag{2.13.10}$$

where the LHS is the solution if the cubic formula is correct and 4 is the true solution. In the process, he accepted the square root of negative numbers and treated it as an ordinary number. In his own words, it was a *wild thought* as he had no idea about $\sqrt{-121}$. He computed this term $(2+\sqrt{-1})^3$ as

$$\begin{aligned}(2+\sqrt{-1})^3 &= 8 + 3(2)^2\sqrt{-1} + 3(2)(\sqrt{-1})^2 + (\sqrt{-1})^3 \\ &= 8 + 12\sqrt{-1} - 6 - \sqrt{-1} = 2 + 11\sqrt{-1} = 2 + \sqrt{-121}\end{aligned} \tag{2.13.11}$$

Thus, he knew $\sqrt[3]{2+\sqrt{-121}} = 2+\sqrt{-1}$. Similarly, he also had $\sqrt[3]{-2+\sqrt{-121}} = -2+\sqrt{-1}$. Plugging these into Eq. (2.13.9) indeed gave him four (his intuition was correct):

$$x = \sqrt[3]{2+\sqrt{-121}} - \sqrt[3]{-2+\sqrt{-121}} = 4 \tag{2.13.12}$$

About this, Leibniz—the co-inventor of calculus—expressed his confusion, "I did not understand how ... a quantity could be real, when imaginary or impossible numbers were used to express it". [46]. And he found this so astonishing that among some unpublished papers of him, we found several such expressions, as if he had calculated them endlessly.

Remark Knowing one solution $x = 4$, it is straightforward to find the other solutions using a factorization as

$$x^3 - 15x - 4 = 0 \iff (x-4)(x^2+4x+1) = 0$$

If you're not sure of this factorization, please refer to Sect. 2.29.2. The other solutions can be found by solving the quadratic equation $x^2 + 4x + 1 = 0$. This quadratic equation gives us two solutions. Thus, in total, the considered cubic equation has three solutions. That makes sense; the quadratic equation has two solutions, it is reasonable then to suppose that a cubic equation has three solutions. This reasoning eventually leads to the fundamental theorem of algebra. To conclude, if we can find one solution to the cubic equation, we can find the remaining ones.

del Ferro's method to solve the depressed cubic equation. For unknown reason, he considered the solution $x = u + v$. Putting this into the depressed cubic equation, we get:

$$(u^3+v^3) + (3uv+p)(u+v) = q$$

He needed another equation (as there are two unknowns), so he considered $3uv + p = 0$, or $v = -p/3u$. With this, the above equation becomes $u^3 + v^3 = q$, or[47]

[47] Later on Lagrange called this equation the "resolvent" of the original cubic equation.

$$u^3 - \frac{p^3}{27u^3} = q$$

which is a disguised quadratic equation with $t = u^3$:

$$t^2 - qt - \frac{p^3}{27} = 0 \Longrightarrow t_1 = \frac{q}{2} + \sqrt{\frac{q^2}{4} + \frac{p^3}{27}}, \quad t_2 = \frac{q}{2} - \sqrt{\frac{q^2}{4} + \frac{p^3}{27}}$$

Select only the solution $t_1 = {}^{q}/_{2} + \sqrt{{}^{q^2}/_{4} + \frac{p^3}{27}}$, we get $u = \sqrt[3]{t_1}$. By symmetry, we can do the same thing but with v: we get $v = \sqrt[3]{t_2}$. Finally, the solution to the original depressed cubic equation is $x = u + v$, which is Cardano's formula given in Eq. (2.13.7). Looking back, we can see that del Ferro used this change of variable $x = u - {}^{p}/_{3u}$ to convert a cubic equation to a sixth order equation (but actually a simple quadratic equation).

2.13.4 How Viète Solved the Depressed Cubic Equation

Viète exploited trigonometry to solve the depressed cubic equation. Note the similarity of $\cos^3\theta = {}^3/_4 \cos\theta + {}^1/_4 \cos(3\theta)$ (a trigonometric identity) and the cubic equation $x^3 = px + q$. We put them side by side to see this similarity:

$$\begin{aligned} \cos^3\theta &= \frac{3}{4}\cos\theta + \frac{1}{4}\cos(3\theta) \\ x^3 &= px + q \end{aligned} \tag{2.13.13}$$

It follows that $x = a\cos\theta$ where a and θ are functions of p, q. Substituting this form of x into the cubic equation we obtain

$$\cos^3\theta = \frac{p}{a^2}\cos\theta + \frac{q}{a^3} \tag{2.13.14}$$

As any θ satisfies the above trigonometric identity, we get the following system of equations to solve for a and θ in terms of p and q:

$$\begin{aligned} \frac{p}{a^2} &= \frac{3}{4} \\ \frac{q}{a^3} &= \frac{1}{4}\cos(3\theta) \end{aligned} \Longrightarrow a = \frac{2\sqrt{3p}}{3}, \quad \theta = \frac{1}{3}\cos^{-1}\left(\frac{3\sqrt{3}q}{2p\sqrt{p}}\right) \tag{2.13.15}$$

Thus, the final solution is

$$x = \frac{2\sqrt{3}}{3}\sqrt{p}\cos\left[\frac{1}{3}\cos^{-1}\left(\frac{3\sqrt{3}q}{2p\sqrt{p}}\right)\right] \tag{2.13.16}$$

Does Viète's solution work for the case $p = 15$ and $q = 4$ (the one that caused trouble with Cardano's solution)? Using Eq. (2.13.16) with $p = 15$ and $q = 4$, we get

$$x = 2\sqrt{5}\cos\left[\frac{1}{3}\cos^{-1}\left(\frac{2\sqrt{5}}{25}\right)\right] \tag{2.13.17}$$

which can be evaluated using a computer (or calculator) to give 4 (with angle of 1.3909428270). Note that this equation also gives the other two roots −3.73205 (angle is $1.3909428270 + 2\pi$) and −0.267949 (angle is $1.3909428270 + 4\pi$). And there is no $\sqrt{-1}$ involved! What does this tell us? The same thing (i.e.,, the square root of a negative number) can be represented by i and by cosine/sine functions. Thus, there must be a connection between i and sine/cosine. We shall see this connection later.

Seeing how Viète solved the cubic equation, we can unlock de Ferro's solution. de Ferro used this identity $(u + v)^3 = u^3 + v^3 + 3u^2v + 3uv^2$. We put this identity and the depressed cubic equation altogether

$$\begin{aligned}(u + v)^3 &= 3uv(u + v) + u^3 + v^3\\ x^3 &= -px + q\end{aligned}$$

So, with $x = u + v$ we obtain from the depressed cubic equation $(u + v)^3 = -p(u + v) + q$. Compare this with the identity $(u + v)^3 = \cdots$, we then get two equations to solve for p and q:

$$p = -3uv, \quad q = u^3 + v^3$$

And voilà! We now understand the solution of de Ferro. **Obviously, the algebra of** his solution is easy, what is hard is to think of that identity $(u + v)^3 = u^3 + v^3 + 3u^2v + 3uv^2$ in the first place.

History note 2.4: Caspar Wessel (1745–1818)

François Viète, Seigneur de la Bigotière (1540–1523 February 1603) was a French mathematician whose work on new algebra was an important step towards modern algebra, due to its innovative use of letters as parameters in equations. He was a lawyer by trade, and served as a privy councilor to both Henry III and Henry IV of France.

Viète's most significant contributions were in algebra. While letters had been used to describe an unknown quantity by earlier writers, Viète's was the first to also use letters for the parameters or constant coefficients in an equation. Thus, while Cardano solved particular cubic equations such as

$$x^3 - 5x = 6$$

Viète could treat the general cubic equation

$$A^3 + px = q$$

where p and q are constants. Note that Viète's version of algebra was still cumbersome and wordy as he wrote 'D in R–D in E aequabitur A quad' for $DR - AE = A^2$ in our notation.

2.13.5 History About Cardano's Formula

This section gives a brief historical account on the cubic formula following the interesting book *Journey through Genius* by William Dunham.

Scipione del Ferro (1465–1526), a professor at the University of Bologna, was first to make significant advancement in solving cubic equations. del Ferro could solve the depressed cubic equations—those of the form $x^3 + cx = d$, but he did not publish his discovery. Instead he kept it until his death. The reason is that he had to have secret weapons to be deployed in "mathematical duels"—challenges from other Italian mathematicians. On his dead bed, he passed over his formula to his student Antonio Fior. Although Fior was not so good a mathematician as his mentor in 1535 he leveled a challenge at the noted Brescian scholar Niccolo Fontana (1499–1557), better known as Tartaglia (meaning "the stammerer", after a teenage facial injury from a French soldier's sword).

The self-taught Tartaglia (Fig. 2.22a) discovered how to solve a different form of the cubic—one missing the linear term; that is $x^3 + mx^2 = n$. This set the stage for a mathematical duel between Fior and Tartaglia. In 1535, they exchanged 30 problems with a deadline of a month and a half. Tartaglia sent Fior a variety of problems, whereas the mathematically weaker Fior sent Tartaglia 30 depressed cubics, employing the 'all eggs in one basket' strategy. Just days before the deadline, Tartaglia discovered how to solve the depressed cubic, and quickly finished all 30 problems. Meanwhile, Fior solved none of Tartaglia's problems. News spread throughout Italy of Tartaglia's achievement, and Fior, humiliated, faded from view.

But then entered Gerolamo Cardano (Fig. 2.22b). Cardano tried, and failed, to replicate Tartaglia's success with the cubic, so he began a pressure campaign to convince Tartaglia to share his method, even promising a vow of secrecy:

> I swear to you by the Sacred Gospel, and on my faith as a gentleman, not only never to publish your discoveries, if you tell them to me, but I also promise and pledge my faith as

(a) Tartaglia (b) Cardano

Fig. 2.22 Tartaglia and Cardano: two major figures in the solution of cubic equations

> a true Christian to put them down in cipher so that after my death no one shall be able to understand them.

Eventually, in 1539, Tartaglia relented and shared his knowledge for depressed cubic equations with Cardano, albeit written as a poem. For the clever Cardano, however, just knowing the result, was enough to discover the underlying mathematics. He even discovered how to solve the full cubic equation $ax^3 + bx^2 + cx + d = 0$. Despite his vow to Tartaglia, Cardano shared these results to his talented assistant Ludovico Ferrari. Cardano recognized the importance of these accomplishments and desperately wanted to publish the results. But he could not do so because of his oath.

Then, on a trip to Bologna in 1543, Cardano saw in del Ferro's notebooks the solution of the depressed cubic before Tartaglia. In Cardano's mind, this discovery freed him of his obligation to Tartaglia. Two years later, Cardano published Ars Magna, which contained his and Ferrari's work on cubic and quartic equations.

To describe his formula, re-given below

$$x = \sqrt[3]{\frac{q}{2} + \sqrt{\frac{q^2}{4} + \frac{p^3}{27}}} - \sqrt[3]{-\frac{q}{2} + \sqrt{\frac{q^2}{4} + \frac{p^3}{27}}}$$

Cardano wrote

> Cube one-third the coefficient of x; add to it the square of one-half the constant of the equation; and take the square root of the whole. You will duplicate [repeat] this, and to one of the two you add one-half the number you have already squared and from the other you subtract one-half the same. Then, subtracting the cube root of the first from the cube root of the second, the remainder which is left is the value of x.

2.14 Mathematical Notation

Up to this point, we have used what is called mathematical notations to express mathematical ideas. It is time to discuss these notations. Mathematical notation consists of using symbols for representing operations, unspecified numbers, relations, and any other mathematical objects, and assembling them into expressions and formulas. Beside mathematics, mathematical notation is widely used in science and engineering for representing complex concepts and properties in a concise, unambiguous and accurate way. For instance, Einstein's equation $E = mc^2$ is the representation in mathematical notation of the mass–energy equivalence. Mathematical notation was first introduced by François Viète, and largely expanded during the seventeenth and eighteenth century by René Descartes, Isaac Newton, Gottfried Leibniz, and overall Leonhard Euler.

2.14.1 Symbols

The use of many symbols is the basis of mathematical notation. They play a similar role as words in natural languages. They may play different roles in mathematical notation similarly as verbs, adjective, and nouns play different roles in a sentence.

Lectures as symbols. Letters are typically used for naming mathematical objects. Typically the Latin and Greek alphabets are used, but some letters of Hebrew alphabet are sometimes used. We have seen a, b, α, β, and so on. Obviously, these alphabets are not sufficient: to have more symbols, and for allowing related mathematical objects to be represented by related symbols, diacritics (e.g., f'), subscripts (e.g., x_2) and superscripts (e.g., z^3) are often used. For quadratic equations, we can use x and y to denote its two roots. But it is sometimes better to use x_1 and x_2 (both are x and we can see what is the first and what is the second root). What is more, when we want to talk about the n roots of an n-order polynomial, we have to use $x_1, x_2, \ldots, x_n$. Why because we do not even know what is n.

Other symbols. Symbols are not only used for naming mathematical objects. They can be used for operations ($+, -, \times, \div, \sqrt{}, \ldots$), for relations ($=, >, <, \ldots$), for logical connectives ($\Longrightarrow$, $\Longleftrightarrow, \vee, \ldots$), for quantifiers ($\forall, \exists$) and for other purposes.

What we need to know is that a notation is a personal choice of the particular mathematician who used it for the first time. If interested, you can read *A history of mathematical notations* by the Swiss-American historian of mathematics Florian Cajori (1859–1930) [13].

2.15 Factorization

I have discussed a bit about factorization when presenting the identity $a^2 - b^2 = (a - b)(a + b)$. Herein, we delve into this topic with more depth. Recall that factorization or factoring consists of writing a number or another mathematical object as a product of several factors, usually smaller or simpler objects of the same kind. Factorization was first considered by ancient Greek mathematicians in the case of integers. They proved the fundamental theorem of arithmetic, which asserts that every positive integer may be factored into a product of prime numbers, which cannot be further factored into integers greater than one. For example,

$$48 = 16 \times 3 = 2 \times 2 \times 2 \times 2 \times 3$$

Then comes the systematic use of algebraic manipulations for simplifying expressions (more specifically equations) dated to ninth century, with Al-Khwarizmi's book *The Compendious Book on Calculation by Completion and Balancing*.

The following identities are useful for factorization, for two real numbers a and b:

$$\begin{aligned}
&\text{(a) difference of squares: } && a^2 - b^2 = (a-b)(a+b) \\
&\text{(b) difference of cubes: } && a^3 - b^3 = (a-b)(a^2+ab+b^2) \\
&\text{(c) sum of cubes: } && a^3 + b^3 = (a+b)(a^2-ab+b^2)
\end{aligned} \tag{2.15.1}$$

In using these identities, we need to see 1 as 1^2 or 1^3, then the identity appears. For example, $a^3 - 1$ is $a^3 - 1^3 = (a-1)(a^2+a+1)$. This is similar to in trigonometry we see 1 as $\sin^2 x + \cos^2 x$.

The first method for factorization is finding *a common factor* and using the distributive law $a(b+c) = ab + ac$. For example,

$$6x^3y^2 + 8x^4y^3 - 10x^5y^3 = 2x^3y^2(3 + 4xy - 5x^2y)$$

Another technique is grouping:

$$4x^2 + 20x + 3xy + 15y$$

Then, factorizing each group and a common factor for the entire expression will show up:

$$4x(x+5) + 3y(x+5) = (x+5)(4x+3y)$$

In many cases, we have to look at the expressions carefully so that the identities in Eq. (2.15.1) will appear. For example, let's simplify the following fraction

$$\frac{x^6 + a^2x^3y}{x^6 - a^4y^2}$$

We can process the numerator as $x^3(x^3 + a^2y)$. About the denominator we should see it as $(x^3)^2 - (a^2y)^2$, then things become easy as the denominator becomes $(x^3 + a^2y)(x^3 - a^2y)$. And the fraction is simplified to $x^3/x^3 - a^2y$.[48]

[48] Nothing is mysterious here. We all know that $2 \cdot 3/5 \cdot 3 = 2/5$: number 3 in both the numerator and denominator gone.

Example 2.2 The next exercise about factorization is the following expression:

$$A = \frac{a^3 + b^3 + c^3 - 3abc}{(a-b)^2 + (b-c)^2 + (c-a)^2}$$

Now we make some observations. First, the nominator is of degree three (a^3) and the denominator is of second degree. Second the three variables a, b, c are symmetrical. Thus, if that expression can be factorized into a polynomial, it must be of this form of degree 1 (similar to $a^3/a^2 = a = a^1$)

$$A = pa + qb + rc \Longrightarrow A = p(a + b + c)$$

The fact that $p = q = r$ stems from the symmetry of a, b, c. To find p, just use $b = c = 0$ in the original expression, we find that $p = 0.5$. Thus, one answer might be:

$$A = \frac{a + b + c}{2}$$

And now we just need to check if

$$a^3 + b^3 + c^3 - 3abc = \left(\frac{a + b + c}{2}\right)[(a-b)^2 + (b-c)^2 + (c-a)^2]$$

And it is indeed the case. Thus, the answer is $0.5(a + b + c)$.

The above method is not the usual one often presented in textbooks. Here is the textbook method:

$$\begin{aligned}
(a^3 + b^3) + c^3 - 3abc &= (a+b)^3 - 3ab(a+b) + c^3 - 3abc \\
&= [(a+b)^3 + c^3] - 3ab(a+b+c) \\
&= [(a+b) + c][(a+b)^2 - (a+b)c + c^2] - 3ab(a+b+c) \\
&= (a+b+c)(\ldots)
\end{aligned}$$

where in the third equality we have used the identity $x^3 + y^3 = (x + y)(x^2 - xy + y^2)$. Now you see why in the expression of A, we must have the term $3abc$, not $4abc$ or anything else. It must be $3abc$, otherwise there is nothing to simplify!

Another powerful method to do factorization is to use the identity difference of squares, i.e.,, $(X)^2 - (Y)^2 = (X - Y)(X + Y)$. The thing is we have to make appear the form $(X)^2 - (Y)^2$ called difference of squares. One way is to complete the square *by adding zero to an expression*. For example, suppose that we need to factorize the following expression:

$$A = x^4 + 4$$

We add zero to it so that a square appears:

$$\begin{aligned} A &= \left[(x^2)^2 + 2^2 + 4x^2\right] - 4x^2 \\ &= (x^2+2)^2 - (2x)^2 \\ &= (x^2+2+2x)(x^2+2-2x) \end{aligned}$$

Let's solve one challenging problem in which we will meet a female mathematician and an identity attached to her name. The problem is: compute the following without calculator:

$$A = \frac{(10^4+324)(22^4+324)\cdots(58^4+324)}{(4^4+324)(16^4+324)\cdots(52^4+324)}$$

Observe first that $324 = 4 \cdot 81 = 4 \cdot 3^4$. Then all terms in A have this form: $a^4 + 4b^4$ with $b = 3$. So, let's factorize $a^4 + 4b^4$:

$$\begin{aligned} (a^2)^2 + (2b^2)^2 &= (a^2)^2 + 4a^2b^2 + (2b^2)^2 - 4a^2b^2 \\ &= (a^2+2b^2)^2 - 4a^2b^2 \\ &= (a^2+2b^2+2ab)(a^2+2b^2-2ab) \end{aligned} \tag{2.15.2}$$

This identity is known as the Sophie Germain identity, named after the French mathematician, physicist, and philosopher Marie-Sophie Germain (1776–1831). Despite initial opposition from her parents and difficulties presented by society, she gained education from books in her father's library and from correspondence with famous mathematicians such as Lagrange, Legendre, and Gauss (under the pseudonym of "Monsieur LeBlanc"). Because of prejudice against her sex, she was unable to make a career out of mathematics, but she worked independently throughout her life. Before her death, Gauss had recommended that she be awarded an honorary degree, but that never occurred!

Sophie Germain was born in an era of revolution. In the year of her birth, the American Revolution began. Thirteen years later the French Revolution began in her own country. In many ways, Sophie embodied the spirit of revolution into which she was born. Sophie's interest in mathematics began during the French Revolution when she was 13 years old and confined to her home due to the danger caused by revolts in Paris. She spent a great deal of time in her father's library, and one day *she ran across*

a book in which the legend of Archimedes' death was recounted. Legend has it that "during the invasion of his city by the Romans Archimedes was so engrossed in the study of a geometric figure in the sand that he failed to respond to the questioning of a Roman soldier. As a result he was speared to death". *This sparked Sophie's interest. If someone could be so engrossed in a problem as to ignore a soldier and then die for it, the subject must be interesting!* Thus, she began her study of mathematics.

Using Eq. (2.15.2) with $b = 3$, we have:

$$(a^2)^2 + 324 = (a^2 + 18 + 6a)(a^2 + 18 - 6a) = [a(a+6) + 18][a(a-6) + 18] \tag{2.15.3}$$

Now A is making sense: in the above identity we have $a - 6$ and $a + 6$, and note that the numbers in the nominator and denominator in A differ by 6: 10 and 4, 22, 16 etc. This means that there are many terms that can be canceled. Indeed, with Eq. (2.15.3), we have:

$$\frac{10^4+324}{4^4+324} = \frac{(10\cdot 16+18)(\cancel{10\cdot 4+18})}{(\cancel{4\cdot 10+18})(4\cdot(-2)+18)}$$
$$\frac{58^4+324}{52^4+324} = \frac{(58\cdot 64+18)(\cancel{58\cdot 52+18})}{(\cancel{52\cdot 58+18})(52\cdot 46+18)}$$

Almost all terms cancel each other and we get $A = 373$.

To master factorization we need practice and patience. We need to have a feeling of common algebraic expressions. And one way to achieve that is to play with algebra so that it becomes your friend.

We leave this fraction for you to simplify it[49]

$$\frac{\sqrt{2}+\sqrt{3}+\sqrt{4}}{\sqrt{2}+\sqrt{3}+\sqrt{6}+\sqrt{8}+\sqrt{16}}$$

Why factorization? Because factored expressions are usually more useful than the corresponding un-factored expressions. For example, we use factorization to simplify fractions. We use factorization to solve equations. It is hard to know what is the solution of $x^3 - 6x^2 + 11x - 6 = 0$, but it is easy with $(x-1)(x-2)(x-3) = 0$. Factors can be helpful for checking expressions. For instance, consider a triangle of sides a, b, c, its area is denoted by A, then we have two equivalent expressions for $16A^2$:

$$\begin{aligned}16A^2 &= 2b^2c^2 + 2c^2a^2 + 2a^2b^2 - a^4 - b^4 - c^4\\ &= (a+b+c)(a+b-c)(b+c-a)(c+a-b)\end{aligned}$$

As we know that the triangle area will be zero if $a + b = c$, and thus the factored expression for $16A^2$ reveals this clearly while the un-factored expression does not. By the way, the factored expression above is known as Heron's formula, see Eq. 3.1.1.

---

[49] The answer is $\sqrt{2} - 1$.

As another example to demonstrate the usefulness of factoring, consider the following equation: a rectangle with positive integer side lengths (in cm) has area A (in cm^2) and perimeter P (in cm). The question is can $A + P$ be 102? To solve this problem, let a and b be the two side lengths of the rectangle, where a, b are positive integers. Thus, we need to see if the equation

$$A + P = ab + 2(a + b) = 102 \tag{2.15.4}$$

has positive integer solutions. How to proceed then? Now, we have to factor $A + P$ as follows (the key point again is to add zero):

$$\begin{aligned} A + P &= ab + 2a + 2b = a(b + 2) + 2b + 4 - 4 \\ &= a(b + 2) + 2(b + 2) - 4 = (a + 2)(b + 2) - 4 \end{aligned}$$

Now, Eq. (2.15.4) becomes

$$(a + 2)(b + 2) = 106 = 53 \times 2$$

Now, we can answer the question: since 53 is a prime, and b cannot be zero, there does not exist such a and b. If we replace the number 102 by 104 then the answer is yes: $(a + 2)(b + 2) = 108 = 54 \times 2 = 27 \times 4$. We have $a = 25$ and $b = 2$.

Manipulation of algebraic expressions is a useful skill that can be learned. Herein we discuss some manipulation techniques. Consider this problem: given that the sum of a number and its reciprocal (i.e.,, its inverse) is one, find the sum of the cube of that number and the cube of its reciprocal.

We can proceed as follows. Let's denote by x the number, we then have $x + 1/x = 1$. Solving this quadratic equation we get $x = (1 \pm i\sqrt{3})/2$. Now, to get $x^3 + 1/x^3$ we need to compute x^3, which is $(1 \pm i\sqrt{3})^3/8$, but that would be difficult.[50] There should be a better way. This is what we need

$$S = x^3 + \frac{1}{x^3}$$

and we have $x + 1/x = 1$. Let's cube both sides of $x + 1/x = 1$ and S will show up:

$$\begin{aligned} \left(x + \frac{1}{x}\right)^3 &= \left(x^3 + \frac{1}{x^3}\right) + 3x^2\frac{1}{x} + 3x\frac{1}{x^2} \\ 1^3 &= S + 3(x + \frac{1}{x}) \\ 1 &= S + 3 \times 1 \Longrightarrow S = -2 \end{aligned}$$

[50] Not really, but for maths—as an art form—we aim for beautiful solutions not ugly ones.

We found S without even solving for x. With that success, how about this problem: finding $S = x^{2021} + 1/x^{2021}$ given that $x + 1/x = \sqrt{2}$? Obviously, the solution just presented would not work as no one would dare to expand $(x + 1/x)^{2021}$. To handle such crazy exponent (of 2021), we need complex numbers and de Moivre's formula discussed later in Sect. 2.25.2. The solution is $S = -\sqrt{2}$. Through this exercise, we see again unexpected connections in maths: we use complex numbers (depending on i which we call imaginary) to compute something real.

Let's consider another problem: given two real numbers $x \neq y$ that satisfy

$$\begin{cases} x^2 = 17x + y \\ y^2 = 17y + x \end{cases}$$

What is the value of $S = \sqrt{x^2 + y^2 + 1}$?

The problem is obviously symmetrical, so we will perform symmetrical operations: we sum the two given equations, and we subtract the second from the first one:

$$\begin{cases} x^2 + y^2 = 18(x + y) \\ x^2 - y^2 = 16(x - y) \end{cases} \tag{2.15.5}$$

Then, we multiply the resulting equations, and we can compute $x^2 + y^2$: $x^2 + y^2 = (16)(18)$:

$$(x^2 + y^2)(x^2 - y^2) = (16)(18)(x^2 - y^2)$$

Thus, $S = \sqrt{(16)(18) + 1}$. Another way (a bit slower) is to solve for $x + y$ from the second equation of Eq. (2.15.5), and then put it into the first to solve for $x^2 + y^2$.

2.16 Word Problems and System of Linear Equations

Consider this word problem âĽ˘âĽ˘To complete a job, it takes: Alice and Bob 2 hours, Alice and Charlie 3 hours and Bob and Charlie 4 hours. How long will the job take if all three work together"?

For future scientists and engineers, working with these word problems is more useful than solving a given system of equations (for instance Eq. (2.15.5)). This is because for engineers and scientists setting up the equations is an important task. If they cannot solve the equations, they can ask a computer or a friend from the maths department to do it. We see it quite often; the English theoretical physicist Stephen Hawking (1942–2018) had collaborated with the British mathematician Roger Penrose. Another example is the collaboration between Albert Einstein and Marcel Grossmann. Grossmann (April 9, 1878–September 7, 1936) was a Swiss mathematician and a friend and classmate of Albert Einstein. Einstein told Grossmann: *"You must help me, or else I'll go crazy"*.

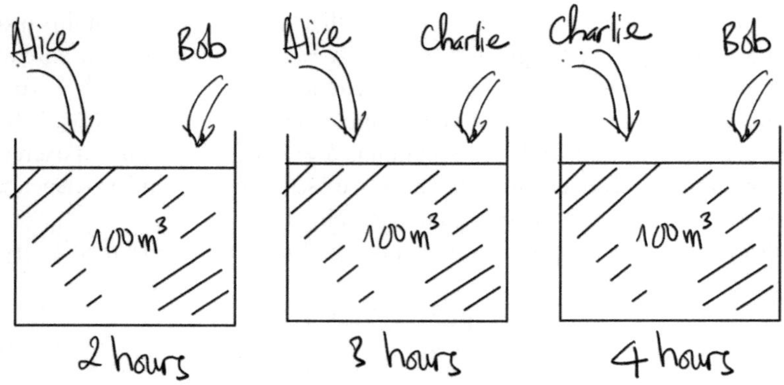

Fig. 2.23 Alice, Bob, and Charlie pouring concrete into a container. Why 100? The idea is not to use small numbers such as 1, 2, ... to avoid working with fractions. If you like, choosing 60 is working fine. But I think 100 is a nice number

Now, let's get back to Alice, Bob, and Charlie. There are *three sentences* that can be translated into *three equations* and solving them gives the three unknowns. This is how word problems work. The question is how to get a correct translation. Many US college students gave an answer of 4.5 hours for this problem. Do you see why that must be wrong? If not, you should *develop a habit of guessing a plausible solution without solving it.* Paul Dirac (1902–1984), an English theoretical physicist who is regarded as one of the most significant physicists of the twentieth century, once said "I consider that I understand an equation when I can predict the properties of its solutions, without actually solving it".

There are many ways to translate the words into equations. But it is probably easy if we think of a specific job of, let say, filling concrete into a container of 100 m^3 (see Fig. 2.23). Let's denote by A, B and C the number of concrete volume (in m^3) that Alice, Bob, and Charlie can pour into the container within 1 h.[51] With this, it is straightforward to translate the sentence "to complete a job, it takes Alice and Bob 2 h" to $2A + 2B = 100$. So, we have this system of equations.[52]

$$\begin{aligned} 2A + 2B &= 100 \\ 3A + 3C &= 100 \\ 4B + 4C &= 100 \end{aligned} \tag{2.16.1}$$

We have a system of three linear equations that is why we call it a system of linear equations. The solution of this system is the three numbers A, B, C that when substituted into the system we get true statements. How are we going to solve it? We know how to solve $ax + b = 0$, so the plan is *to remove/eliminate two unknowns and we're*

[51] You can name them x, y, z or whatever.

[52] We have made an assumption that the efficiency of Alice, Bob and Charlie is constant, even though that it is possible that when working with Bob Charlie might work harder than with Alice. Thus, solving word (or real) problems demands us to ignore unimportant things to make the problem manageable.

left with one unknown. To remove two unknowns, we first remove one unknown. To do that we can use any equation, e.g., $B + C = 25$, write the to-be-removed unknown in terms of the other: for instance $C = 25 - B$. Now C is gone.

We can start removing any unknown, I start with C: from the third equation, we can get $C = 25 - B$, put it into the second equation of Eq. (2.16.1) we get $3A - 3B = 25$. This and the first equation is the new system (with only two unknowns A, B) that we need to solve. We do the same thing again: from $2A + 2B = 100$ we get $B = 50 - A$ (i.e.,, we're removing B), put that into $3A - 3B = 25$: $A = 175/6$. Now we go backward to solve for B and for C. Altogether, the solution is $A = 175/6$, $B = 125/6$ and $C = 25/24$.[53] Then, if the time required for all three people work together to fill the container is t, then the amount of concrete will be $(A + B + C)t$. Thus, we have

$$(A + B + C)t = 100 \Longrightarrow t = \frac{100}{A + B + C} = \frac{24}{13} \text{ hours} \tag{2.16.2}$$

This solution is plausible because it is smaller than the two hours that take Alice and Bob; Charlie should be useful even though he is a bit slower than the other two kids.

Let's consider another word problem taken from *The joy of x* by the American mathematician Steven Strogatz (born 1959). The problem goes like this. If the cold faucet can fill a bathtub in half an hour and the hot faucet fills it in 1 hour, then how long does it take if both faucets are filling together the bathtub? At the age of 10 or 11 Strogatz's answer was 45 minutes when given this problem by his uncle. What's your solution?

Here is his uncle's solution. In one minute, the cold faucet fills $1/30$ of the bathtub and the hot faucet fills $1/60$ of the bathtub. So, together they can fill $1/30 + 1/60 = 1/20$ of the bathtub in one minute. Thus, it takes them 20 minutes. That's the answer. What if we do not know fractions?

Is it possible to get the same answer without using fractions? Yes, using hours instead of minutes! So, in 1 hour, the cold faucet can fill two bathtubs, and the hot faucet fills one bathtub. Together, in 1 hour, they can fill three bathtubs. So, it takes them $1/3$ hour to fill in one bathtub. This is the solution of the older Strogatz. It does not involve fractions but it involves three bathtubs. We could not think of this solution if our mind is fixed with the image of a real bathtub: one bathtub with two faucets. Don't forget maths is a world existing independently with our world. You can do anything you like in this mathematical world!

Let's stretch further, can we solve this problem without doing any maths? Still remember Paul Dirac's above-mentioned quote? This is the way to have deep understanding. Setting up the equations and solving them without doing this step is like a robot.

Let's try. Ok, we know that the cold faucet fills the tub in 30 minutes, so regardless the rate of the hot faucet, together they have to fill in the tub in less than 30 mins.

[53] Did we solve the system? Even though we spent sometime and found A, B, C satisfying the solution, to be honest with you, we have just found one solution. Of course if we can prove that this system has only one solution, then our A, B, C are *the solution*. Can you explain why this system has a unique solution and when such a system does not have solution? And can it have more than one solution?

On the other hand, if the hot faucet rate was the same as the cold one, then together they would do the job in 15 mins. So, without doing any maths, we know the answer t is $15 < t < 30$. What we have just done is, according to Polya in *How to solve it*, considering special cases of the problem that we're trying to solve. We might not be able to solve the original problem, but we can solve at least some simpler problems.

Systems of linear equations in chemistry. Back then in high school I did not know how to balance chemical equations like the following one $C_3H_8 + 5\,O_2 \longrightarrow 3\,CO_2 + 4\,H_2O$. The problem is to find whole numbers x_1, x_2, x_3, x_4 such that

$$x_1\, C_3H_8 + x_2 O_2 \longrightarrow x_3\, CO_2 + x_4\, H_2O$$

That is, to balance the total numbers of carbon (C), hydrogen (H) and oxygen (O) atoms on the left and on the right of the chemical reaction.[54] Now, C, H, and O play similar role of Alice, Bob, and Charlie. There are three atoms, and conservation of each atom gives one equation:

$$\begin{aligned} 3x_1 &= x_3 && \text{(balancing the total numbers of carbon)} \\ 8x_1 &= 2x_4 && \text{(balancing the total numbers of hydrogen)} \\ 2x_2 &= 2x_3 + x_4 && \text{(balancing the total numbers of oxygen)} \end{aligned} \tag{2.16.3}$$

Again, we see a system of linear equations! Solving this is easy: elimination technique. There is one catch: we have four unknowns but only three equations. Let $x_4 = n$, then we can solve for x_1, x_2, x_3 in terms of n: $x_1 = n/4$, $x_3 = 3n/4$, $x_2 = 5n/4$. Take $n = 4$, we get $x_1 = 1, x_2 = 5, x_3 = 3$. If you take $n = 8$ you get another four solutions. Thus, we have infinite number of solutions (which makes sense for $2 = 2$ or $4 = 4$ and so on).

Systems of linear equations. Equation 2.16.1 is one example of a system of linear equations. In these systems, there are n equations for n unknowns $x_1, x_2, \ldots, x_n$ where all equations are linear in terms of x_i $(i = 1, 2, \ldots)$, i.e.,, we will not see nonlinear terms like $x_i x_j$. In what follows, we give examples for $n = 2, 3, 4$:

$$\begin{aligned} 2x_1 + x_2 &= 1 \\ 3x_1 + 2x_2 &= 2 \end{aligned}, \quad \begin{aligned} 2x_1 + 2x_2 &= 10 \\ 3x_1 + 3x_2 &= 2 \\ 4x_2 + 4x_3 &= 4 \end{aligned}, \quad \begin{aligned} 1x_1 + 2x_2 + 4x_3 + 1x_4 &= 1 \\ 2x_1 + 1x_2 + 1x_3 + 7x_4 &= 3 \\ 5x_1 + 1x_2 + 3x_3 + 4x_4 &= 5 \\ 6x_1 + 7x_2 + 2x_3 + 3x_4 &= 2 \end{aligned}$$

If we focus on how to solve these equations, we would come up with the so-called Gaussian elimination method (when we're pressed to solve a system with many unknowns, say $n \geq 6$). On the other hand, if we are interested in the question when such a system has a solution, when it does not have a solution and so on, we could come up with matrices and determinant. For example, we realize that putting all the coefficients in a system of linear equations in an array like

[54] Because atoms are neither destroyed nor created in the reaction.

$$\mathbf{A} = \begin{bmatrix} 1 & 2 & 4 & 1 \\ 2 & 1 & 1 & 7 \\ 5 & 1 & 3 & 4 \\ 6 & 7 & 2 & 3 \end{bmatrix} \tag{2.16.4}$$

and we can play with this array similarly to the way we do with numbers. We can add them, multiply them, subtract them. And we give it a name: **A** is a *matrix*. Matrices, determinants and how to solve efficiently large systems of linear equations (n in the range of thousands and millions) belong to a field of mathematics named *linear algebra*, see Chap. 9.

We're not sure about the original source of systems of linear equations, but systems of linear equations arose in Europe with the introduction in 1637 by René Descartes of coordinates in geometry. In fact, in this new geometry, now called analytical geometry, lines and planes are represented by linear equations, and computing their intersections amounts to solving systems of linear equations.

But if systems of linear equations only come from analytical geometry we would only have systems of three equations (a plane in 3D is of the form $ax + bc + cz = 0$), and life would be boring. Systems of many (ten thousands are not uncommon) linear equations appear again and again in many fields (e.g., physics, biology, economics, and in mathematics itself). For example, in Sect. 10.7.2, when solving the diffusion equation, we shall meet this system of linear equations.

Some word problems.

1. Two dogs, each traveling 10 ft/s, run towards each other from 500 feet apart. As they run, a flea flies from the nose of one dog to the nose of the other at 25 ft/s. The flea flies between the dogs in this manner until it is crushed when the dogs collide. How far did the flea fly?
2. Alok has three daughters. His friend Shyam wants to know the ages of his daughters. Alok gives him first hint: The product of their ages is 72. Shyam says this is not enough information Alok gives him a second hint: the sum of their ages is equal to the number of my house. Shyam goes out and look at the house number and tells 'I still do not have enough information to determine the ages'. Alok admits that Shyam cannot guess and gives him the third hint: my oldest daughter likes strawberry ice-cream.' With this information, Shyam was able to determine all three of their ages. How old is each daughter?

Regarding the daughter-age problem, we have three unknowns and three hints, so it seems to be a good problem. But did you try to set up the equations? There is only one equation, that is $xyz = 72$ if x, y, z are the ages of the daughters. What if the product of their ages is a smaller number, let say, 12? Ah, we can list out the ages as there are only a few cases. If that method works for 12, of course it will work for 72; just a bit extra work. If you still cannot find the solution, check this this website out. What if the product of their ages was a big number?

This is a good exercise to show that we should be flexible. Setting up equations is a good method to solve word problems; but it does not solve all problems. There seems to be a problem that defy all existing mathematics. And it is a good thing as it is these problems that keep mathematicians working late at nights.

Algebra is a language of symbols. Now, if we think again about the word problems, we see that algebra is actually a language—a language of symbols (such as a, or A). What is the advantage of this language? It is comprehensible: it can translate a lengthy verbose problem into a compact form that *the eyes can see quickly and the mind can retain what is going on*. Compare this

> To complete a job, it takes: Alice and Bob 2 hours, Alice and Charlie 3 hours, and Bob and Charlie 4 hours. How long will the job take if all three work together?

and

$$\begin{aligned} 2A + 2B &= 100 \\ 3A + 3B &= 100 \\ 4B + 4C &= 100 \end{aligned}$$

Anyone can undoubtedly recognize the powerful of algebra.

2.17 System of Nonlinear Equations

Contrary to systems of linear equations where we have a systematic method (e.g., Gauss elimination method) to find the solutions, systems of nonlinear equations are harder to solve. But they are less important than systems of linear equations. That's why we have an entire course on linear algebra just to handle systems of linear equations, whereas there is no course on systems of nonlinear equations. I bet you're correctly guessing that one good way to tackle a system of nonlinear equations is to somehow transform it to a system of linear equations.

Let's consider the following two equations:

$$\begin{aligned} x^3 + 9x^2y &= 10 \\ y^3 + xy^2 &= 2 \end{aligned} \tag{2.17.1}$$

Can we eliminate one variable? It might be possible, but we do not dare to follow that path. Try it and you'll see why. There must be a better way. Why? because this is a math exercise! High school students should be aware of this fact: nearly all questions in a test/exam have solutions and it is usually not hard and time consuming (as the test duration is finite!). Furthermore, if there is a hard question, its mark is often low. Thus, *you do not need to spend all of your time to study to get A grades. Use that time to explore the world.*

We present the first solution by considering $(x+3y)^3$. Why this term? Because upon expansion, we will have terms appearing in the two equations:

$$\begin{aligned}(x+3y)^3 &= x^3+9x^2y+27xy^2+27y^3\\ &= x^3+9x^2y+27(xy^2+y^3)\\ &= 10+27\times 2=64\end{aligned}$$

Now, we have $x+3y=4$ or $x=4-3y$. Of course, we substitute x in Eq. (2.17.1) to get an equation in terms of y:

$$y^3+(4-3y)y^2=2 \Longrightarrow y^3-2y^2+1=0 \tag{2.17.2}$$

Recognizing $y=1$ is one solution of the above equation, we can factor its LHS and write[55]

$$(y-1)(y^2-y-1)=0 \Longrightarrow y=1\ \ (x=1),\ y=\frac{1\pm\sqrt{5}}{2}\ \ (x=\frac{5\mp 3\sqrt{5}}{2})$$

Is this solution a good one? Yes, but it is not general as it cannot be used when the second equation is slightly different, e.g., $y^3+5xy^2=2$. We need another solution which works for any coefficients.

What is special about Eq. (2.17.1)? We see x^3, x^2y^1, y^3 and x^1y^2; these terms are all of cubic order! If we do this substitute $y=kx$ (or $x=ky$), all these terms become x^3, kx^3, k^3x^3 and k^2x^3, and thus we can factor out x^3 and thus cancel this x^3 and we have an equation for k. That's the trick:

$$\begin{aligned}x^3+9x^2y=10 &\Longrightarrow x^3(1+9k)=10\\ y^3+xy^2=02 &\Longrightarrow x^3(k^3+k^2)=2\end{aligned}$$

By dividing the first equation by the second one, we get the following cubic equation for k:

$$1+9k=5(k^3+k^2) \Longrightarrow 5k^3+5k^2-9k-1=0 \Longrightarrow (k-1)(5k^2+10k+1)=0$$

with solutions $k=1$ (which results in $x=y=1$) and $k=-1\pm {}^{2\sqrt{5}}/_{5}$. Having k, we can solve for x as

$$x^3=\frac{10}{1+9k}=\frac{25}{-20\pm 9\sqrt{5}} \Longrightarrow x^3=5(9\sqrt{5}+20),\quad x^3=-5(9\sqrt{5}-20) \tag{2.17.3}$$

But we do not know how to (if we did not know the solutions $x=\frac{5\mp 3\sqrt{5}}{2}$) compute x from the above expressions for x^3. It turns out that using $x=ky$ makes our life

[55] This exercise was not about solving cubic equations, so this cubic equation must be easy. That's why guessing one solution is the best technique here.

easier: we can get y from y^3. Try it. I am not sure why this is better, probably because y is simpler than x (Eq. (2.17.2)).

Let's solve another system of radical equations:

$$\begin{aligned} \sqrt{x}+\sqrt{y}&=3\\ \sqrt{x+5}+\sqrt{y+3}&=5 \end{aligned} \tag{2.17.4}$$

We can isolate terms involving y and square to get two equations for x:

$$\begin{cases} \sqrt{x}=3-\sqrt{y}\\ \sqrt{x+5}=5-\sqrt{y+3} \end{cases} \Longrightarrow \begin{cases} x=9+y-6\sqrt{y}\\ x+5=25+y+3-10\sqrt{y+3} \end{cases}$$

which leads to the following equation for y

$$7-5\sqrt{y+3}=-3\sqrt{y}$$

which can be solved for y by squaring both sides (two times). Not bad, but not elegant. What if we can have a change of variable so that

$$x=()^2, \quad x+5=()^2$$

then, we can get rid of the square roots for $\sqrt{x}$ and $\sqrt{x+5}$ easily. Such a change of variable does exist! And it is related to the familiar identity $(p\pm q)^2=p^2\pm 2pq+q^2$:

$$\begin{cases} (p+q)^2=p^2+2pq+q^2\\ (p-q)^2=p^2-2pq+q^2 \end{cases} \Longrightarrow (p+q)^2-(p-q)^2=4pq$$

Dividing both sides by 4, and introducing a new variable r such that $pq=r$, we get:

$$\frac{1}{4}(p+q)^2=\frac{1}{4}(p-q)^2+pq \Longrightarrow \boxed{\frac{1}{4}\left(p+\frac{r}{p}\right)^2=\frac{1}{4}\left(p-\frac{r}{p}\right)^2+r} \tag{2.17.5}$$

And that's what we need: the red term is $x=()^2$, then $x+5=()^2$. So, using the boxed equation in Eq. (2.17.5), we introduce these changes of variables:

$$\begin{cases} x=\dfrac{1}{4}\left(a-\dfrac{5}{a}\right)^2\\ y=\dfrac{1}{4}\left(b-\dfrac{3}{b}\right)^2 \end{cases} \Longrightarrow \begin{cases} x+5=\dfrac{1}{4}\left(a+\dfrac{5}{a}\right)^2\\ y+3=\dfrac{1}{4}\left(b+\dfrac{3}{b}\right)^2 \end{cases}$$

The original system of Eq. (2.17.4) becomes simply as:

$$\begin{cases} \frac{1}{2}\left(a-\frac{5}{a}\right)+\frac{1}{2}\left(b-\frac{3}{b}\right)=3 \\ \frac{1}{2}\left(a+\frac{5}{a}\right)+\frac{1}{2}\left(b+\frac{3}{b}\right)=5 \end{cases} \Longrightarrow \begin{cases} a+b=8 \\ \frac{5}{a}+\frac{3}{b}=2 \end{cases}$$

which can be solved easily. A correct change of variable goes a long way!

Sometimes we can solve a hard equation by converting it to a system of equations that is easier to deal with. As one typical example, let's solve the following equation:

$$\sqrt[3]{14+\sqrt{x}}+\sqrt[3]{14-\sqrt{x}}=4$$

If we look at the terms under the cube roots, we see something special: their sum is constant, i.e.,, without x. So, if we do $u=\sqrt[3]{14+\sqrt{x}}$ and $v=\sqrt[3]{14-\sqrt{x}}$, we have $u^3+v^3=28$. And of course, we also have $u+v=4$ from the original equation. Thus, we have

$$\begin{cases} u+v=4 \\ u^3+v^3=28 \end{cases}$$

which can be solved to have $u=1, v=3$, and from that we get $x=169$. If the equation is slightly changed to $\sqrt[3]{14+\sqrt{x}}+\sqrt[3]{14-a\sqrt{x}}=4$, a is any number, then our trick would not work. Don't worry you will not see that in standardized tests. In real life, probably. But then we can just use a numerical method (e.g., Newton's method, discussed in Sect. 4.5.4, or a graphic method) to find an approximate solution.

2.18 Algebraic and Transcendental Equations

The linear, quadratic, and cubic equations that we have just seen belong to a general class of algebraic equations. Other equations that contain polynomials, trigonometric functions, logarithmic functions, exponential functions etc., are called transcendental equations. For example $x=\cos x$ is a transcendental equation.

Definition 2.18.1 A polynomial equation of the form $f(x)=a_nx^n+a_{n-1}x^{n-1}+a_{n-2}x^{n-2}+\cdots+a_1x+a_0=0$ is called an algebraic equation. An equation that contains polynomials, trigonometric functions, logarithmic functions, exponential functions, etc., is called a transcendental equation.

In Sect. 2.13, we have solved linear/quadratic/cubic equations directly. That is, the solutions of these equations can be expressed as roots of the coefficients in the equations, e.g., $x={}^{-b\pm\sqrt{b^2-4ac}}/_{2a}$ in case of quadratic equations. It is also possible to do the same thing for fourth-order algebraic equations (the formula is too lengthy to be presented here). But, as the French mathematician and political activist Évariste Galois (1811–1832) showed us, polynomials of fifth order and beyond have no closed form solutions using radicals. Why fifth order equations so hard? To answer this question, we need to delve into the so-called abstract algebra—a field about symmetries and groups. I do not know much about this branch of mathematics, so I do not discuss it here. I strongly recommend Ian Stewart's book *Why Beauty Is Truth: The History of Symmetry* [60].

For transcendental equations, we need to use *numerical methods*, i.e.,, those methods that give approximate solutions not exact ones expressed as roots of the coefficients in the equations. For example, a numerical method would give the solution $x = 0.73908513$ to the equation $\cos x - x = 0$. We refer to Sect. 4.5.4 for a discussion on this topic.

Associated with algebraic equations and transcendental equations we have algebraic and transcendental numbers, respectively. An algebraic number is any complex number (including real numbers) that is a root of a non-zero polynomial in one variable with rational coefficients (or equivalently, by clearing denominators, with integer coefficients). All integers and rational numbers are algebraic, as are all roots of integers. Real and complex numbers that are not algebraic, such as π and e, are called transcendental numbers. If you're fascinated by numbers, check out [52].

2.19 Rules of Powers (Exponentiation)

If a square has a side of length x, its area is then x^2, and the volume of a cube of side x is x^3. In general, if n is a positive integer and x is any real number, then x^n corresponds to the repeated multiplication $x \cdot x \cdot x \cdot \dots \cdot x$, where x appears n times in the product. We call this *x raised to the power of n* or *x to the power of n*. Here, x is the base and n is the exponent or the power.

This section is about the arithmetic of such powers x^n. That is, how we can do addition/subtraction, multiplication/division of powers. For example, what is $(2^3) \times (2^5)$. We start humbly with the case $x = 2$ in Sect. 2.19.1. In Sect. 2.19.2, we discuss what is $2^{\sqrt{2}}$.

2.19.1 Powers of 2

We know that $2^2 = 2 \times 2$ and 2^3 is nothing but $2 \times 2 \times 2$. The product of a number (here 2) by itself a number of times is called a power. Herein, we discuss this concept. The two to power four is two multiplied by itself four times, which is expressed as

$$2^4 := \underbrace{2 \times 2 \times 2 \times 2}_{\text{4 times}} \tag{2.19.1}$$

Thus, 2^4 is nothing but a shorthand for $2 \times 2 \times 2 \times 2$. So, for positive integer as exponents, a power is just a repeated multiplication.[56]

We can deduce rules for common operations with powers. I first summarize these rules below, for a reals a, b and m, n being positive integers:

$$\begin{array}{llll}
\text{(a) Product rule 1} & a^m \times a^n = a^{m+n} & \text{(b) Quotient rule 1} & a^m/a^n = a^{m-n} \\
\text{(c) Power of power rule} & (a^m)^n = a^{mn} & \text{(d) Zero exponent rule} & a^0 = 1 \\
\text{(d) Negative exponent rule} & a^{-n} = \frac{1}{a^n} & \text{(e) Rational exponent rule} & a^{m/n} = \sqrt[n]{a^m} \\
\text{(f) Product rule 2} & (ab)^n = a^n b^n & \text{(g) Quotient rule 2} & (a/b)^n = \frac{a^n}{b^n}
\end{array} \tag{2.19.2}$$

And our task is to prove them in what follows. We use a specific number 2 but shall generalize the result to any real number.

For example, multiplication of two powers of two is given by (m, n are positive integers)

$$2^m \times 2^n := \underbrace{(2 \times 2 \times \cdots \times 2)}_{m \text{ times}} \times \underbrace{(2 \times 2 \times \cdots \times 2)}_{n \text{ times}} = 2^{m+n} \tag{2.19.3}$$

which basically says that to multiply two exponents with the same base (2 here), you keep the base and add the powers. And this is the product rule $a^m \times a^n = a^{m+n}$ for $m, n \in \mathbb{N}$.[57]

The next thing is certainly division of two powers. Division of two powers of two is written as

$$\frac{2^m}{2^n} = 2^{m-n} \tag{2.19.4}$$

If that was not clear, we can always check a concrete case. For example,

$$\frac{2^5}{2^3} = \frac{2 \times 2 \times 2 \times 2 \times 2}{2 \times 2 \times 2} = \frac{2 \times 2 \times \cancel{2} \times \cancel{2} \times \cancel{2}}{\cancel{2} \times \cancel{2} \times \cancel{2}} = 2 \times 2 = 2^2 = 2^{5-3}$$

How about raising a power, i.e.,, a power of a power such as $(2^3)^2$? It's $8^2 = 64 = 2^6$. And we generalize this to $(2^m)^n$:

$$(2^m)^n := \underbrace{\underbrace{(2 \times 2 \times \cdots \times 2)}_{m \text{ times}} \times \underbrace{(2 \times 2 \times \cdots \times 2)}_{m \text{ times}} \times \cdots \times \underbrace{(2 \times 2 \times \cdots \times 2)}_{m \text{ times}}}_{n \text{ times}} = 2^{mn} \tag{2.19.5}$$

And we also have this result $(2^m)^n = (2^n)^m$ as both are equal to 2^{mn}.

56 We did the same game before: multiplication (of 2 integers) is a repeated addition. Now, we define a new math object based on repeated multiplication. Why? Because it saves time.

57 Refer to Sect. 2.25.8 for what $\mathbb{N}$ is. Briefly, it is the set (collection) of all natural numbers. Instead of writing the lengthy "n is a natural number", mathematicians write $a \in \mathbb{N}$.

Table 2.8 Powers of 2 with positive, zero and negative index

n	2^n	Value
3	$2 \times 2 \times 2$	8
2	2×2	4
1	2	2
0	2^0	1
-1	2^{-1}	1/2

So far so good, we have rules for powers with positive integer index. How about zero and negative index, e.g., 2^0 and 2^{-1}? To answer these questions, again we follow the rule applied to $-1 \times -1 = 1$: the new rule should be consistent with the old rule. Thus, we compute 2^3, 2^2, 2^1 to see the pattern to deduce what should be 2^0 and 2^{-1}. From the data in Table 2.8: $2^0 = 1$ and $2^{-1} = 1/2$: in this table, while going down from the top row, the value of any row in the third column is obtained by dividing the value of the previous row by two. Thus, $2^{-2} = {}^1\!/_4 = {}^1\!/_{2^2}$. So, we have $2^{-n} = {}^1\!/_{2^n}$.

The next natural question is how to find powers to a rational index, e.g., $2^{1/2}$. It turns out that it is nothing but $\sqrt{2}$. Why? Suppose that we write $\sqrt{2} = 2^a$ and we want to find a. We first write

$$2 = \sqrt{2} \times \sqrt{2} = 2^a \times 2^a$$

Now, we want the product rule to be still valid even a is not integer (we do not know yet what it is). So, we have

$$2 = 2^{a+a} = 2^{2a} \Longrightarrow 2a = 1 \Longrightarrow a = \frac{1}{2}$$

So,

$$2^{1/2} = \sqrt{2} \tag{2.19.6}$$

which reads "2 to the power of 1/2 is the square root of 2", nothing new comes up here. This also indicates that the power of power rule is assumed to hold for rational exponents. We have $(2^{1/2})^2 = 2^{(1/2)2} = 2^1 = 2$, which is true because $2^{1/2}$ is the square root of 2.

In the same manner, $2^{1/3}$ is computed as

$$(2^{1/3})^3 = 2^{(1/3)3} = 2 \Longrightarrow 2^{1/3} = \sqrt[3]{2}$$

And $2^{5/3}$ is computed as

$$(2^{5/3})^3 = 2^{(5/3)3} = 2^5 \Longrightarrow 2^{5/3} = \sqrt[3]{2^5}$$

We can now generalize these results to base a, which is a real number. This was obtained by replacing 2 by a—a real number, as in previous development there is nothing special about 2; what we have done for 2 works exactly for any real number. That's why we have Eq. (2.19.2).

Now that we have defined powers with a rational index $a^{m/n}$. Do all the rules (e.g., the product rule) still apply for such powers? That is do we still have $a^{m/n}a^{p/q} = a^{m/n+p/q}$? To gain insight, we can try few examples. For instance, $3^{1/2} \times 3^{1/2}$ equals 3 (from square), but is also equal 3 from $3^{1/2+1/2} = 3^1$. Now we need a proof, once and for all!

Proof We write $a^{m/n}$ as $a^{qm/qn}$ and $a^{p/q}$ as $a^{pn/qn}$, then it follows

$$a^{\frac{qm}{qn}} a^{\frac{pn}{qn}} = \sqrt[qn]{a^{qm}} \sqrt[qn]{a^{pn}} = \sqrt[qn]{a^{qm}a^{pn}} = \sqrt[qn]{a^{qm+pn}} = a^{\frac{qm+pn}{qn}} = a^{m/n+p/q}$$

■

A bit historical account

A bit of history about notation of exponents is in order. The notation we use today to denote an exponent was first used by Scottish mathematician, James Hume in 1636. However, he used Roman numerals for the exponents. Using Roman numerals as exponents became problematic since many of the exponents became very large so Hume's notation didn't last long. A year later in 1637, Rene Descartes became the first mathematician to use the Hindu-Arabic numerals of today as exponents. It was Newton who first used powers with negative and rational index. Before him, Wallis wrote $1/a^2$ instead of a^{-2}.

2.19.2 Power with an Irrational Index

For a number raised to a fractional exponent, i.e.,, $a^{p/q}$, the result is the denominator-th root of the number raised to the numerator, i.e.,, $\sqrt[q]{a^p}$. Again, we should ask ourselves this question: so what happens when you raise a number to an irrational number? Obviously, it is not so simple to break it down like what we have done in, e.g., Eq. (2.19.6).

What is $2^{\sqrt{2}}$? It cannot be 2 multiplied by itself $\sqrt{2}$ times! So, the definition in Eq. (2.19.1) no longer works. In other words, the starting point that a power is just a repeated multiplication is no longer valid. This situation is similar to multiplication is a repeated addition ($2 \times 3 = 2 + 2 + 2$) does not apply to 2×3.4.

To see what might be $2^{\sqrt{2}}$, we can proceed as follows, without a calculator of course. Otherwise we would not learn anything interesting but a meaningless number.

Table 2.9 Calculation of $2^{\sqrt{2}}$

$2^{1.4}$	$2^{14/10} = \sqrt[10]{2^{14}}$	2.6390158
$2^{1.41}$	$2^{141/100} = \sqrt[100]{2^{141}}$	2.6573716
$2^{1.414}$	$2^{1414/1000} = \sqrt[1000]{2^{1414}}$	2.6647496
$2^{1.4142}$		2.6651190
$2^{1.41421}$		2.6651190
$2^{1.41421356}$		2.6651441383063186
$2^{1.414213562}$		2.665144142000993
$2^{1.4142135623}$		2.665144142555194
$2^{1.41421356237}$		2.6651441426845075

We approximate $\sqrt{2}$ successively by 1.4, 1.41, 1.414, etc. and we compute the corresponding powers (e.g., $2^{1.4} = 2^{14/10} = 2^{7/5} = \sqrt[5]{2^7}$). The results given in Table 2.9 show that as a more accurate approximation of the square root of 2 is used, the powers converge to a value. Note that I have used a calculator to compute each approximation of $2^{\sqrt{2}}$ e.g. $2^{14/10} = \sqrt[10]{2^{14}}$. This is not cheating as the main point here is to get the value of these approximations.

But, how can we be sure that $2^{\sqrt{2}}$ is a number? This can be guaranteed by looking at the function 2^x as shown in Fig. 2.24. There is no hole in this curve or the function is continuous, so there must exist $2^{\sqrt{2}}$.

Are the rules of powers still apply for irrational index? Do we still have $a^x a^y = a^{x+y}$ with x, y being irrational numbers? If so, we say that the power rules work for real numbers, and we're nearly done (if we did not have complex numbers). How to prove this? One easy but not strict way is to say that we can always replace a^x by a^r with r is a rational number close to x, and a^y by a^t. Thus $a^x a^y \approx a^r a^t = a^{r+t}$. To conclude, Eq. (2.19.2) is valid for real numbers a, b and for real exponents (e.g., m can be $\sqrt{2}$).

We have calculated $2^{\sqrt{2}}$ by approximating the square root of 2 with a rational number, e.g. $2^{1414/1000} = \sqrt[1000]{2^{1414}}$. However, calculating the 1000th root is not an easy task. There must be a better way to compute 2^x for any real number x directly and efficiently. For this, we need calculus (Chap. 4). That is, algebra can only help us so far, to go further we need new mathematics.

Adding up powers of two. Let's consider the summation of powers of two starting from 2^1 to 2^{n-1}:

$$S(n) = 1 + 2 + 4 + 8 + \cdots + 2^{n-1} = \sum_{i=0}^{n-1} 2^i = ? \tag{2.19.7}$$

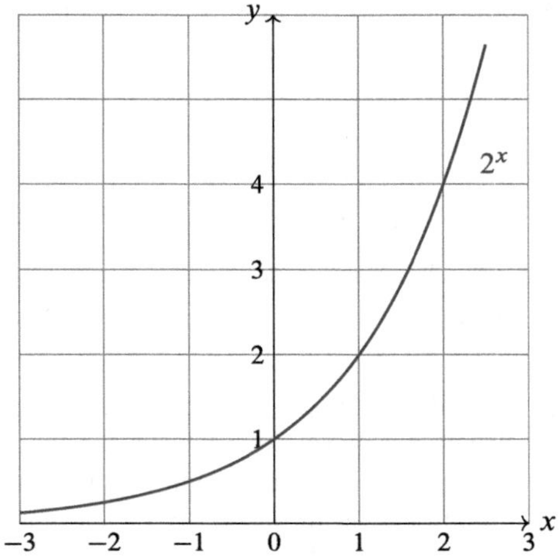

Fig. 2.24 Plot of function 2^x

Table 2.10 $S(n) = 1 + 2 + 4 + 8 + \cdots + 2^{n-1}$

n	$S(n)$	$S(n)$	$S(n)$
1	1	$2 - 1$	$2^1 - 1$
2	3	$4 - 1$	$2^2 - 1$
3	7	$8 - 1$	$2^3 - 1$
4	15	$16 - 1$	$2^4 - 1$

I have added the shorthand notation using the sigma $\sum$ just for people not familiar with this to practice using it. It is useless for our purpose here though. To find the expression for $S(n)$, we need to get **our hands dirty** by computing $S(n)$ for a number of values of n. The results for $n = 1, 2, 3, 4$ are tabulated in Table 2.10. From this data we can find a pattern (see columns 3 and 4 of this table). And this brings us to the following conjecture:

$$S(n) = 1 + 2 + 4 + 8 + \cdots + 2^{n-1} = 2^n - 1 \tag{2.19.8}$$

And if we can prove that this conjecture is correct then we have discovered a theorem.

Proof It is easy to see that $S(1)$ is correct ($1 = 2^1 - 1$). Now, assume that $S(k)$ is correct, or

$$1 + 2 + 4 + 8 + \cdots + 2^{k-1} = 2^k - 1$$

Multiplying this equation by two results in the following

$$2 + 4 + 8 + \cdots + 2^{k-1} + 2^k = 2 \times 2^k - 2$$
$$1 + 2 + 4 + 8 + \cdots + 2^k = 2^{k+1} - 1$$

So, $S(k+1)$ is correct. ■

Why powers?

I think that the concept of power emerged from practical geometry problems. If you have a square of length 2, what is the area? It is 2×2 or two squared. If you have a cube of length 2, the volume is $2 \times 2 \times 2$ or two cubed. The notation 2^3 is just a convenient shortcut for $2 \times 2 \times 2$. Then, mathematicians generalize to a^n for any n.

Becoming mathematician like.

What is $\sqrt{2}^{\sqrt{2}^{\sqrt{2}}}$? It is 2, an integer! You can check this using a calculator and then prove it using the rules of powers that you're now familiar with. Let's go crazy: how about $\pi^{\pi^{\pi}}$?

The second power of x or x squared?

We know that x, x^2, x^3 are called the first, second, and third powers of x. But we also know that x^2 is written/read x squared and x^3 as x cubed. Why? This is because ancient Greek mathematicians see x^2 as the area of a square of side x.

Scientific notation.

When working with very large numbers such as 3 trillion we do not write it as 3 000 000 000 000 as there are too many zeros. Instead, we write it as 3×10^{12} (there are 12 zeros explicitly written). Any number can be written as the product of a number between 1 and 10 and a number that is a power of ten. For example, we can write 257 as 2.57×10^2 and 0.00257 as 2.57×10^{-3}. This system is called the scientific notation.
Doing arithmetic with this notation is easier due to properties of exponents. For example, when we multiply numbers, we multiply coefficients and add exponents:

$$(3 \times 10^6) \times (4 \times 10^8) = (3 \times 4) \times 10^{14} = 12 \times 10^{14} = 1.2 \times 10^{15}$$

The scientific notation immediately reveals how big a number is. We use the order of magnitude to measure a number. Generally, the order of magnitude of a number is the smallest power of 10 used to represent that number. For example, $257 = 2.57 \times 10^2$, so it has an order of magnitude of 2.

2.20 Inequalities

In mathematics, an inequality is a relation that makes a non-equal comparison between two numbers or mathematical expressions. It is used most often to compare two numbers on the number line by their size. There are several different notations used to represent different kinds of inequalities:

- The notation $a < b$ means that a is less than b.
- The notation $a > b$ means that a is greater than b.
- The notation $a \leq b$ means that a is less than or equal b.

Inequalities are governed by the following properties:

$$\begin{array}{lll}
\text{(a)} & \text{transitivity} & \text{if } a \leq b \text{ and } b \leq c \text{ then } a \leq c \\
\text{(b)} & \text{addition} & \text{if } x \geq y \text{ and } a \geq b \text{ then } x + a \geq y + b \\
\text{(c1)} & \text{multiplication} & \text{if } x \geq y \text{ and } a \geq 0 \text{ then } ax \geq ay \\
\text{(c2)} & \text{multiplication} & \text{if } x \geq y \text{ and } a \leq 0 \text{ then } ax \leq ay \\
\text{(d)} & \text{reciprocals} & \text{if } x \geq y \text{ and } xy > 0 \text{ then } 1/x \leq 1/y
\end{array} \tag{2.20.1}$$

I skip the proof of these simple properties herein. But if you find one which is not obvious you should convince yourself by proving it.

Section 2.20.1 presents some simple inequality problems. Section 2.20.2 is about inequalities involving the arithmetic and geometric means. The Cauchy-Schwarz inequality is introduced in Sect. 2.20.3. Next, inequalities concerning absolute values are treated in Sect. 2.20.4. Solving inequalities, e.g., finding x such that $|x - 5| \geq 3$ is presented in Sect. 2.20.5. And finally, how inequality can be used to solve equations is given in Sect. 2.20.6.

2.20.1 Simple Proofs

Let's solve the following inequality problems. Of course, we're forbidden to use a computer/calculator. I repeat that we are not interested in which term is larger; instead we're interested in the mathematical techniques used in solving these inequality problems. Our task is to compare the left-hand side and right-hand side terms (alternatively replacing the question mark by either > or < symbol):

1. $\sqrt{19} + \sqrt{99} \; ? \; \sqrt{20} + \sqrt{98}$
2. $\dfrac{1998}{1999} \; ? \; \dfrac{1999}{2000}$
3. $\dfrac{10^{1999}+1}{10^{2000}+1} \; ? \; \dfrac{10^{1998}+1}{10^{1999}+1}$
4. $40^{39} \; ? \; 39^{40}$

One simple technique is to transform the given inequalities to easier ones. We are going to use this method for the first problem. To this end, we square two sides to get rid of the roots:

$$\begin{aligned} 19 + 99 + 2\sqrt{19 \cdot 99} \quad &? \quad 20 + 98 + 2\sqrt{20 \cdot 98} \\ \sqrt{19 \cdot 99} \quad &? \quad \sqrt{20 \cdot 98} \end{aligned}$$

We square two sides again, of course, and a bit of algebra gives us:

$$\begin{aligned} 19 \cdot 99 \quad &? \quad 20 \cdot 98 = (19 + 1) \cdot 98 \\ 19 \cdot 99 \quad &? \quad 19 \cdot 98 + 98 \\ 19 \quad &? \quad 98 \end{aligned}$$

Now we know ? should be <, thus $\sqrt{19} + \sqrt{99} < \sqrt{20} + \sqrt{98}$.

For the second problem, let's first replace fractions:

$$\frac{1998}{1999} \quad ? \quad \frac{1999}{2000} \Longleftrightarrow 1998 \cdot 2000 \quad ? \quad 1999^2$$

Now come the trick; we replace 1999 by $0.5(1998 + 2000)$, and the solution follows immediately:

$$\begin{aligned} 1998 \cdot 2000 &\quad ? \quad \left(\frac{1998+2000}{2}\right)^2 \\ 4 \cdot 1998 \cdot 2000 &\quad < \quad (1998+2000)^2 \end{aligned}$$

Because $(x+y)^2 \geq 4xy$ for all x, y due to $(x-y)^2 \geq 0$.

Let's now solve this inequality with another way using calculus. Looking at $1998/1999$ and $1999/2000$, they are both of the form $x/1+x$. So, if we consider the function $f(x) = x/(1+x)$, then our problem becomes comparing $f(1998)$ and $f(1999)$. If $f(x)$ is either a monotonically increasing or decreasing function, then we know the answer to the question. To reveal the nature of $f(x)$ we need to massage it a bit: $f(x) = 1/(1+1/x)$. And with this form, $f(x)$ is a monotonically increasing function. Thus, $f(1998) < f(1999)$.

For the third problem, denoting 10^{1998} by x and writing other numbers in terms of x, we will have a simple inequality. I leave it as an exercise for you. We now tackle the final problem. One way to compare two positive numbers a and b is to compare the ratio a/b with one: if this ratio is larger than one then $a > b$ and vice versa. So, we make the ratio $A = 40^{39}/39^{40}$, and massage it a bit:

$$A = \frac{40^{39}}{39^{40}} = \frac{40^{39}}{39^{39} \times 39^1} = \frac{1}{39}\left(\frac{40}{39}\right)^{39}$$

Next, we replace 40 by $1 + 39$, and we then have

$$A = \frac{1}{39}\left(1 + \frac{1}{39}\right)^{39} \approx \frac{2.718}{39} < 1$$

Hey, but why $(1 + 1/39)^{39} \approx 2.718$? The answer is given in Sect. 2.28.

2.20.2 Inequality of Arithmetic and Geometric Means

Starting with $(x-y)^2 \geq 0$, one can get $(x+y)^2 \geq 4xy$. Now consider only non-negative x, y, then by taking the square root for both sides of $(x+y)^2 \geq 4xy$, we have

$$\boxed{\frac{x+y}{2} \geq \sqrt{xy}} \tag{2.20.2}$$

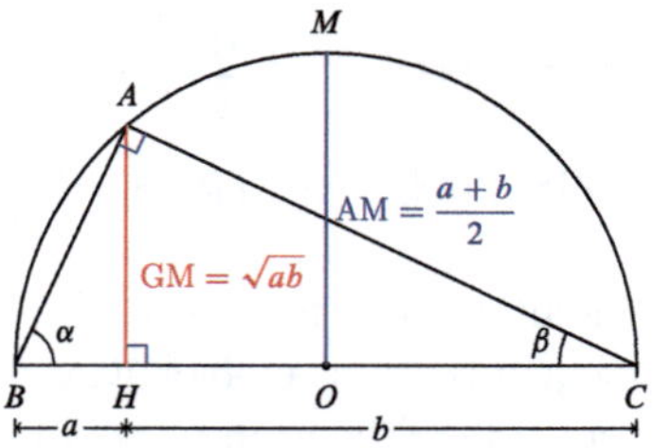

As the left-hand side is the arithmetic mean (AM) of x, y and the RHS is the geometric mean (GM), this inequality is known as the AM-GM inequality. A geometry explanation of this inequality is given in the next figure. Consider a circle with diameter BC and center O. Select a point A on the circle, and the triangle ABC is a right triangle (if this is not clear, check Fig. 3.17). Draw AH perpendicular to BC: $BH = a$ and $HC = b$. Then $(a+b)/2 = OM$, the radius of the circle. It can be shown that $AH = \sqrt{ab}$.[58] It is obvious that when A is traveling on the circle we always have $AH \le OM$.

Now we should ask this question: does this AM-GM inequality holds for more than 2 numbers? For example, do we also have, for example

$$\frac{a+b+c}{3} \ge \sqrt[3]{abc}, \quad \frac{a+b+c+d}{4} \ge \sqrt[4]{abcd}$$

for $a, b, c, d \ge 0$?

Let's first check for the case of 4 numbers as it is easier. Indeed, using the AM-GM for the case of two numbers, we can write

$$\left.\begin{aligned}\frac{a+b}{2} &\ge \sqrt{ab}\\ \frac{c+d}{2} &\ge \sqrt{cd}\end{aligned}\right\} \Longrightarrow \frac{a+b+c+d}{2} \ge \sqrt{ab} + \sqrt{cd}$$

Using again the AM-GM for two numbers $\sqrt{ab}$ and $\sqrt{cd}$, we get what we wanted to verify[59]:

$$\frac{a+b+c+d}{2} \ge 2\sqrt{\sqrt{ab}\sqrt{cd}}, \quad \text{or} \quad \frac{a+b+c+d}{4} \ge \sqrt[4]{abcd}$$

Now, we show that using the AM-GM for 4 numbers, we can get the AM-GM for 3 numbers. The idea is of course to remove d so that only three numbers a, b, c are left. Using $d = (a+b+c)/3$,[60] and the AM-GM inequality for 4 numbers, we have

---

[58] There are different ways to see this. Using trigonometry, from two right triangles ABH and ACH we have $AH = a\tan\alpha$ and $AH = b\tan\beta$. Then, $AH^2 = ab\tan\alpha\tan\beta$. But $\tan\alpha\tan\beta = 1$. In fact, this result is the geometric mean theorem in Euclidean geometry, see Sect. 3.1.10.

[59] If it is not clear to you that $S = \sqrt{\sqrt{xy}} = \sqrt[4]{xy}$, here is the details: $S = ((xy)^{1/2})^{1/2} = (xy)^{1/4}$. See Sect. 2.19.1 if still not clear.

[60] This is the term we need to appear.

$$\frac{a+b+c+\dfrac{a+b+c}{3}}{4} \geq \sqrt[4]{abc\left(\frac{a+b+c}{3}\right)} \tag{2.20.3}$$

which is equivalent to

$$\frac{a+b+c}{3} \geq \sqrt[4]{abc\left(\frac{a+b+c}{3}\right)}$$

and a simple raising to 4th power gives the final result:

$$\frac{a+b+c}{3} \geq \sqrt[3]{abc}$$

Good! Should we aim higher? Of course. We have AM-GM for $n = 2, 4$ and certainly we have similar inequalities for $n = 2^k$ for $k \in \mathbb{N}$. And from $n = 4$ we obtained the AM-GM for $n = 3$. We can start from $n = 32$, and get $n = 31$, so on. It seems that we have the general AM-GM inequality given by

$$\frac{a_1 + a_2 + \cdots + a_n}{n} \geq \sqrt[n]{a_1 a_2 \ldots a_n} \tag{2.20.4}$$

I present a proof of this inequality carried out by the French mathematician, civil engineer, and physicist Augustin-Louis Cauchy (1789–1857) presented in his *Cours d'analyse*. This book is frequently noted as being the first place that inequalities, and $\delta - \epsilon$ arguments were introduced into Calculus. Judith Grabiner wrote Cauchy was "the man who taught rigorous analysis to all of Europe". The AM-GM inequality is a special case of the Jensen inequality discussed in Sect. 4.5.2.

Cauchy used a forward-backward-induction. In the forward step, he proved the AM-GM inequality for $n = 2^k$ for any counting number k. This is a generalization of what we did for the $n = 4$ case. In the backward step, assuming that the inequality holds for $n = k$, he proved that it holds for $n = k - 1$ too.

Proof Assume the inequality holds for $n = k$, we prove that it holds for $n = 2k$. As the inequality is true for k numbers, we can write

$$\frac{a_1 + a_2 + \cdots + a_k}{k} \geq \sqrt[k]{a_1 a_2 \ldots a_k}$$
$$\frac{a_{k+1} + a_{k+2} + \cdots + a_{2k}}{k} \geq \sqrt[k]{a_{k+1} a_{k+2} \ldots a_{2k}}$$

Adding the above inequalities, we get

$$\frac{a_1 + a_2 + \cdots + a_{2k}}{k} \geq \sqrt[k]{a_1 a_2 \ldots a_k} + \sqrt[k]{a_{k+1} a_{k+2} \ldots a_{2k}}$$

And apply the AM-GM for the two numbers in the RHS of the above equation, we obtain

$$\frac{a_1 + a_2 + \cdots + a_{2k}}{k} \geq 2\sqrt{\sqrt[k]{a_1 a_2 \ldots a_k}\sqrt[k]{a_{k+1} a_{k+2} \ldots a_{2k}}} = 2\sqrt[2k]{a_1 a_2 \ldots a_{2k}}$$

■

Proof Cauchy's forward-backward-induction of the AM-GM inequality. Backward step. Assume the inequality holds for $n = k$, we prove that it holds for $n = k - 1$. As the inequality is true for k numbers, we can write

$$\frac{a_1 + a_2 + \cdots + a_k}{k} \geq \sqrt[k]{a_1 a_2 \ldots a_k}$$

To get rid of a_k, we replace it by $a_1+a_2+\cdots+a_{k-1}/k-1$, and the above inequality becomes

$$\frac{a_1 + a_2 + \cdots + a_{k-1} + \dfrac{a_1 + a_2 + \cdots + a_{k-1}}{k-1}}{k} \geq \sqrt[k]{a_1 a_2 \ldots a_{k-1} \frac{a_1 + a_2 + \cdots + a_{k-1}}{k-1}}$$

A bit rearrangement of the LHS gives us

$$\frac{a_1 + a_2 + \cdots + a_{k-1}}{k-1} \geq \sqrt[k]{a_1 a_2 \ldots a_{k-1} \frac{a_1 + a_2 + \cdots + a_{k-1}}{k-1}}$$

Raising two sides of the above inequality to kth power:

$$\left(\frac{a_1 + a_2 + \cdots + a_{k-1}}{k-1}\right)^k \geq a_1 a_2 \ldots a_{k-1} \frac{a_1 + a_2 + \cdots + a_{k-1}}{k-1}$$

And we get what we needed:

$$\frac{a_1 + a_2 + \cdots + a_{k-1}}{k-1} \geq \sqrt[k-1]{a_1 a_2 \ldots a_{k-1}}$$

■

Isoperimetric problems. If $x + y = P$ where P is a given positive number, then Eq. (2.20.2) gives us $xy \leq P^2/4$. And the maximum of xy is attained when $x = y$. In other words, among all rectangles (of sides x and y) with a given perimeter (P), a square has the maximum area. Actually we can also discover this fact using only arithmetic, see Table 2.11. This is a special case of the so-called isoperimetric problems. An isoperimetric problem is to determine a plane figure of the largest possible area whose boundary has a specified length.

The Roman poet Publius Vergilius Maro (70–19 B.C.) tells in his epic Aeneid the story of queen Dido, the daughter of the Phoenician king of the ninth century B.C. After the assassination of her husband by her brother she fled to a haven near Tunis.

Table 2.11 Given two whole numbers such that $n + m = 10$ what is the maximum of nm

n	m	nm
1	9	9
2	8	16
3	7	21
4	6	24
5	5	25

Fig. 2.25 Queen Dido's isoperimetric problem

There she asked the local leader, Yarb, for as much land as could be enclosed by the hide of a bull. Since the deal seemed very modest, he agreed. Dido cut the hide into narrow strips, tied them together and encircled a large tract of land which became the city of Carthage (Fig. 2.25). Dido knew the isoperimetric problem!

Another isoperimetric problem is "Among all planar shapes with the same perimeter the circle has the largest area". How can we prove this? We present a simple "proof":

1. Among triangles of the same perimeter, an equilateral triangle has the maximum area;
2. Among quadrilaterals of the same perimeter, a square has the maximum area;
3. Among pentagon of the same perimeter, a regular pentagon has the maximum area;
4. Given the same perimeter, a square has a larger area than an equilateral triangle;
5. Given the same perimeter, a regular pentagon has a larger area than a square.

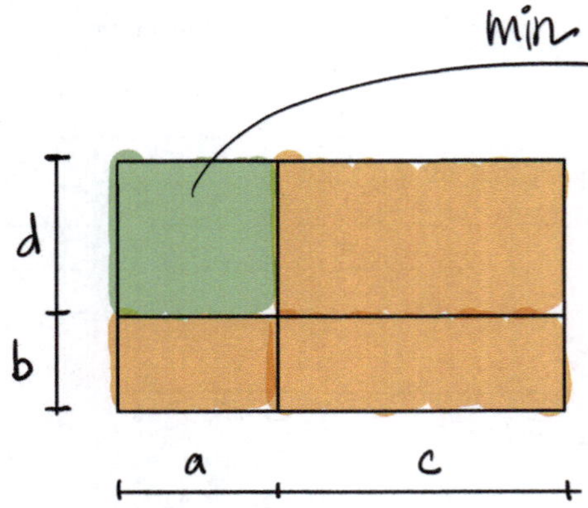

Fig. 2.26 .

We can verify these results. And we can see where this reasoning leads us to: given a perimeter, a regular polygon with infinite sides has the largest area, and that special polygon is nothing but our circle!

Now, let's solve the following problem: assume that a, b, c, d are positive integers with $a + b + c + d = 63$, find the maximum of $ab + bc + cd$. This is clearly an isoperimetric problem. This term $A = ab + bc + cd$ is not nice to a and d in the sense that a and d appear only once. So, let's bring justice to them (or make the term symmetrical): $A = ab + bc + cd + da - da$. A bit of algebra leads to $A = (a + c)(b + d) - da$.

Now we visualize A as in Fig. 2.26. Now the problem becomes maximizing the area of the big rectangle and minimizing the small area ad. The small area is 1 when $a = d = 1$. Now the problem becomes easy.

2.20.3 *Cauchy–Schwarz Inequality*

For a, b, c, x, y, z being real numbers, the Cauchy-Schwarz inequalities read:

$$
\begin{aligned}
(ax + by)^2 &\leq (a^2 + b^2)(x^2 + y^2) \\
(ax + by + cz)^2 &\leq (a^2 + b^2 + c^2)(x^2 + y^2 + z^2)
\end{aligned}
\tag{2.20.5}
$$

The proof of these inequalities is straightforward. Just expand all the terms, and we will end up with: $(ay - bx)^2 \geq 0$ for the first inequality and $(ay - bx)^2 + (az - cx)^2 + (bz - cy)^2 \geq 0$ for the second inequality, which are certainly true. Can we have a geometric interpretation of $(ax + by)^2 \leq (a^2 + b^2)(x^2 + y^2)$? Yes, see Fig.

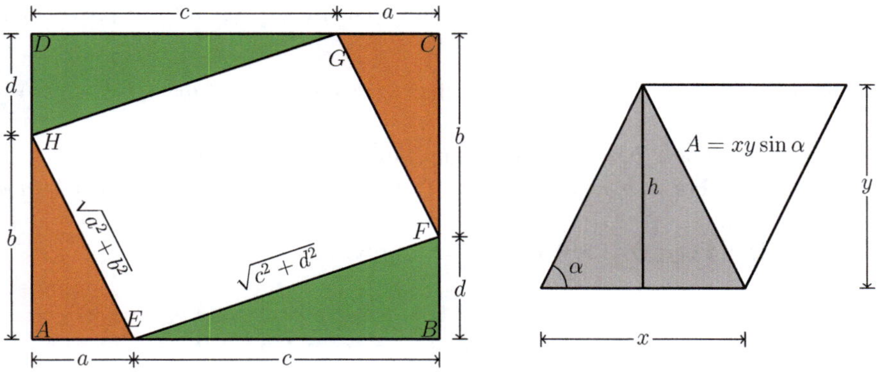

Fig. 2.27 Geometric meaning of $(ax + by)^2 \leq (a^2 + b^2)(x^2 + y^2)$

2.27; the area of the parallelogram $EFGH$ is the area of the big rectangle $ABCD$ minus the areas of all triangles:

$$A = (a + c)(b + d) - (ab + dc) = ad + bc$$

But this area is at most equal to $\sqrt{a^2 + b^2}\sqrt{c^2 + d^2}$ stemming from the fact that the area of a parallelogram is maximum when it is a rectangle (proof in the right figure of Fig. 2.27: $xy \sin\alpha$ attains the maximum value of xy when $\sin\alpha = 1$ or $\alpha = \pi/2$). Thus, we have

$$ad + bc \leq \sqrt{a^2 + b^2}\sqrt{c^2 + d^2} \quad \text{or} \quad (ad + bc)^2 \leq (a^2 + b^2)(c^2 + d^2)$$

Now you might have guessed correctly what we are going to do. We generalize Eq. (2.20.5) to

$$\boxed{\begin{gathered}(a_1b_1 + a_2b_2 + \cdots + a_nb_n)^2 \leq (a_1^2 + a_2^2 + \cdots + a_n^2)(b_1^2 + b_2^2 + \cdots + b_n^2) \\ \left(\sum_{i=1}^{n} a_ib_i\right)^2 \leq \left(\sum_{i=1}^{n} a_i^2\right)\left(\sum_{i=1}^{n} b_i^2\right)\end{gathered}} \tag{2.20.6}$$

And this is the Cauchy-Schwarz inequality, also known as the Cauchy-Bunyakovsky-Schwarz inequality. The inequality for sums, Eq. (2.20.6), was published by Augustin-Louis Cauchy (1821), while the corresponding inequality for integrals was first proved by Viktor Bunyakovsky (1859). The modern proof of the integral version was given by Hermann Schwarz (1888). The Cauchy-Schwarz inequality is a useful inequality in many mathematical fields, such as vector algebra, linear algebra, analysis, probability theory, etc. *It is considered to be one of the most important inequalities in all of mathematics.*

We need to prove Eq. (2.20.6), but let's first use Eq. (2.20.5) to prove some interesting inequalities given below.

1. Example 1. For $a, b, c > 0$, prove that $(a^2b + b^c + c^2a)(ab^2 + bc^2 + ca^2) \geq 9a^2b^2c^2$
2. Example 2. For $a, b, c \geq 0$, prove that $\sqrt{3(a+b+c)} \geq \sqrt{a} + \sqrt{b} + \sqrt{c}$
3. Example 3. For $a, b, c, d > 0$, prove that $1/a + 1/b + 4/c + 16/d \geq 64/(a + b + c + d)$.
4. Example 4. Let $a, b, c > 0$ and $abc = 1$, prove that:

$$\frac{1}{a^3(b+c)} + \frac{1}{b^3(c+a)} + \frac{1}{c^3(a+b)} \geq \frac{3}{2}$$

This is one question from IMO 1995. The International Mathematical Olympiad (IMO) is a mathematical Olympiad for pre-university students and is the oldest of the International Science Olympiads.

Example 1: using Eq. (2.20.5) for $(a^2b + b^c + c^2a)(ab^2 + bc^2 + ca^2)$ to have $(a^2b + b^c + c^2a)(ab^2 + bc^2 + ca^2) \geq \ldots$, and use the 3 variable AM-GM inequality for the ... Example 2: direct application of Eq. (2.20.5) after writing $3(a + b + c)$ as $(1^2 + 1^2 + 1^2)((\sqrt{a})^2 + (\sqrt{b})^2 + (\sqrt{c})^2)$.

About Example 4, even though we know we have to use the AM-GM inequality and the Cauchy-Schwarz inequality, it's very hard to find out the way to apply these inequalities. Then, I thought why I don't reverse engineer this problem, i.e.,, generate it from a fundamental fact. Let's do it and see what happens.

Let $x, y, z > 0$ and $xyz = 1$, using the AM-GM inequality we then immediately have

$$x + y + z \geq 3\sqrt[3]{xyz} = 3$$

I call this the fundamental inequality (for $x, y, z > 0$ and $xyz = 1$). And what we want to do is to do some algebraic manipulations to this fundamental inequality and hope that the IMO inequality will show up. That's the plan. Looking at the IMO problem, it is of the form $S \geq 3$ where S is something that we seek out to find out. To get that, we can start from

$$S(x + y + z) \geq (x + y + z)^2 \quad (\Longrightarrow S \geq x + y + z \geq 3)$$

Re-writing the above as, we see the Cauchy-Schwarz inequality appears:

$$(1x + 1y + 1z)^2 \leq S(x + y + z) \tag{2.20.7}$$

This is because the LHS is in the form $(ax + by + cz)^2$. Of course, we rewrite the 1s by something else, e.g., $1 = \sqrt{z+y}/\sqrt{z+y}$, then the LHS of the above becomes (I label it by A)

$$A := \left(\sqrt{y+z}\frac{x}{\sqrt{y+z}} + \sqrt{z+x}\frac{y}{\sqrt{z+x}} + \sqrt{x+y}\frac{z}{\sqrt{x+y}}\right)^2$$

Applying the Cauchy-Schwarz inequality to A, we get:

$$A \le 2(x+y+z)\left(\frac{x^2}{y+z} + \frac{y^2}{x+z} + \frac{z^2}{x+y}\right) \tag{2.20.8}$$

Now, comparing Eqs. (2.20.7) and (2.20.8) we have found out our S:

$$S = 2\left(\frac{x^2}{y+z} + \frac{y^2}{x+z} + \frac{z^2}{x+y}\right)$$

And since $S \ge 3$, we have a new inequality:

$$\text{Let } x, y, z > 0 \text{ and } xyz = 1, \text{ then } \quad \frac{x^2}{y+z} + \frac{y^2}{x+z} + \frac{z^2}{x+y} \ge \frac{3}{2}$$

And this inequality can be a good exercise for a test. But not for the IMO as it is too obvious with the square terms (which cause us to think of the Cauchy-Schwarz inequality). Now a bit of transformation will give us another inequality. Note that $xyz = 1$, so $x^2 = x^2 \times 1 = x^2(xyz)$, and the same for y^2 and z^2, and voilá cubic terms appear:

$$\begin{aligned} \frac{x^2(xyz)}{y+z} + \frac{y^2(xyz)}{x+z} + \frac{z^2(xyz)}{x+y} &\ge \frac{3}{2} \\ \frac{x^3}{1/y+1/z} + \frac{y^3}{1/x+1/z} + \frac{z^3}{1/x+1/y} &\ge \frac{3}{2} \end{aligned}$$

With $a = 1/x, b = 1/y, c = 1/z$, we get the IMO inequality:

$$\frac{1}{a^3(b+c)} + \frac{1}{b^3(c+a)} + \frac{1}{c^3(a+b)} \ge \frac{3}{2}$$

The cubic terms made this problem hard to prove. And of course, the solution is often presented in a reverse order by starting with $a = 1/x, b = 1/y, c = 1/z$, but, sadly without explaining where that idea came from.

Some special cases. For $b_1 = b_2 = \ldots = b_n = 1$, we have

$$(a_1 + a_2 + \cdots + a_n)^2 \le n(a_1^2 + a_2^2 + \cdots + a_n^2)$$

or re-writing so that the AM appears, we obtain the so-called root-mean square-arithmetic mean inequality:

$$\boxed{\frac{a_1 + a_2 + \cdots + a_n}{n} \le \sqrt{\frac{a_1^2 + a_2^2 + \cdots + a_n^2}{n}}} \tag{2.20.9}$$

This is because the RHS is the root mean square (RMS) that is the square root of the mean square (the arithmetic mean of the squares):

$$\boxed{RMS = Q = \sqrt{\frac{a_1^2 + a_2^2 + \cdots + a_n^2}{n}}} \tag{2.20.10}$$

Proof Now is the time to prove Eq. (2.20.6). Let's start with the simplest case:

$$(a_1 b_1 + a_2 b_2)^2 \le (a_1^2 + a_2^2)(b_1^2 + b_2^2)$$

We consider the following function,[61] which is always non-negative:

$$f(x) = (a_1 x + b_1)^2 + (a_2 x + b_2)^2 \ge 0 \ \text{ for } \forall x$$

We expand this function to write it as a quadratic equation:

$$f(x) = (a_1^2 + a_2^2)x^2 + 2(a_1 b_1 + a_2 b_2)x + (b_1^2 + b_2^2)$$

Now we compute the discriminant Δ of this quadratic equation:

$$\Delta = 4\left[(a_1 b_1 + a_2 b_2)^2 - (a_1^2 + a_2^2)(b_1^2 + b_2^2)\right]$$

As $f(x) = 0$ does not have roots or at most has one root, we have $\Delta \le 0$. And that concludes the proof. For the general case Eq. (2.20.6), just consider this function $f(x) = (a_1 x + b_1)^2 + (a_2 x + b_2)^2 + \cdots + (a_n x + b_n)^2$. ■

What happened to IMO winners? One important point is that the IMO, like almost all other mathematical olympiad contests, is a *timed exam* concerning carefully-designed *problems with solutions*. Real mathematical research is almost never dependent on whether you can find the right idea within the next three hours. In real maths research it might not even be known which questions are the right ones to ask, let alone how to answer them. Producing original mathematics requires *creativity*, *imagination* and *perseverance*, not the mere regurgitation of knowledge and techniques learned by rote memorization.

We should be aware of the phenomenon of 'burn-out', which causes a lot of promising young mathematicians—those who might be privately tutored and entered for the IMO by pushy, ambitious parents—to become disenchanted in mathematics and drop it as an interest before they even reach university. It is best to let the kids follow their interests.

[61] How mathematicians know to consider this particular function? No one knows.

If you enjoy participating in IMO contests, it is absolutely fine. Just be aware of the above comments and you can listen to Terence Tao—who won the IMO when he was only 13 years old—to see how he grew up happily.

2.20.4 Inequalities Involving the Absolute Values

In many maths problems, we need to measure the distance between two points, to know how close or far they are from each other. Note that numbers can be seen as points living on the number line. On this number line, there lives a special number: zero. And we want to quantify the distance from any point x to zero. Thus, mathematicians defined $|x|$—the absolute value of x—as the distance of x from zero. For instance, both –2 and +2 are two units from zero, thus $|2| = |-2| = 2$.

For any real number x, the absolute value or modulus of x is defined as

$$|x| = \begin{cases} x, & \text{if } x \geq 0 \\ -x, & \text{if } x < 0. \end{cases} \tag{2.20.11}$$

For example, the absolute value of 3 is 3, and the absolute value of -3 is also 3. The notation $|x|$, with a vertical bar on each side, was introduced by the German mathematician Karl Weierstrass (1815–1897) in 1841. He was often cited as the "father of modern analysis" and we will have more to say about him in Chap. 4.

With that, let's solve the first absolute value inequality:

$$|x| < 3$$

which means finding all values of x so that the distance of x from zero is less than 3. With a simple picture (Fig. 2.28a), we can see that the solutions are:

$$-3 < x < 3 \quad \text{or} \quad x \in (-3, 3)$$

I have also presented the solutions using set notation $x \in (-3, 3)$. The notation (a, b) indicates all number x such that $a < x < b$. It is called an open bracket as the two ends (i.e.,, -3 and 3) are not included. Then the symbol $\in$ means belong to. We will have more to say about sets in Sect. 2.31.

The next problem is:

$$|2x + 3| \leq 6$$

And by seeing $2x + 3$ as X, the above becomes $|X| \leq 6$, of which solutions are $-6 \leq X \leq 6$. Now, replacing X by the small x, then we have (using the rules in Eq. (2.20.1))

$$-6 \leq 2x + 3 \leq 6 \Longleftrightarrow -6 - 3 \leq 2x \leq 6 - 3 \Longleftrightarrow -9/2 \leq x \leq 3/2$$

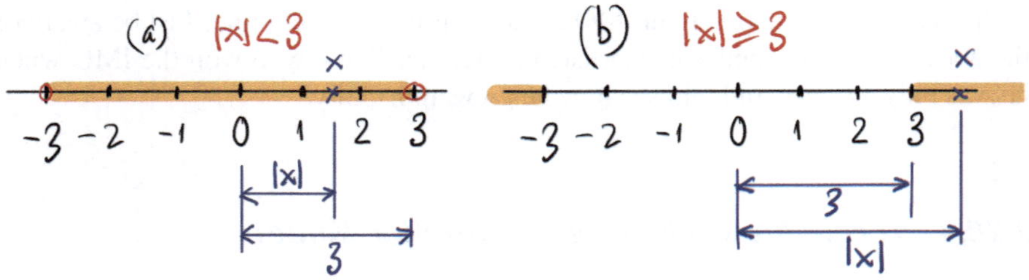

Fig. 2.28 Geometry of $|x|$ as the distance from x to zero

Or using the set notation, we can also express the solution as

$$x \in [-9/2, 3/2]$$

Here, $[a, b]$ indicates a closed bracket.

Let's move to the problem of finding all values of x so that the distance of x from zero is bigger than something:

$$|x| \geq 3$$

Again with a simple picture (Fig. 2.28b), we can see that the solutions are:

$$x \geq 3 \quad \text{or} \quad x \leq -3$$

Or using the set notation, we can also express the solution as

$$x \in (-\infty, -3] \cup [3, \infty]$$

where the symbol $\cup$ in $A \cup B$ means a union of both sets A and B. Noting that it is not necessary to write solutions of inequality problems using set notations. It is clear that we do not gain anything by writing $x \in [-9/2, 3/2]$ instead of $-9/2 \leq x \leq 3/2$. Since set theory is the foundation of [modern] mathematics, thus some people thought that an early exposure to it might be useful. That's why/how set theory entered in high school curriculum.

Triangle inequality. Now comes probably the most important inequality involving absolute values:

$$|a + b| \leq |a| + |b| \tag{2.20.12}$$

This inequality is used extensively in proving results regarding limits, see Sect. 4.10. (We actually used already in Sect. 2.22) Why triangles involved here? It comes from the fact that for a triangle the length of one side is smaller than the sum of the lengths of the other sides. Using the language of vectors, see Sect. 9.1, this is expressed as

$$||\boldsymbol{a} + \boldsymbol{b}|| \leq ||\boldsymbol{a}|| + ||\boldsymbol{b}||$$

Note the similarity of Eq. (2.20.12) compared with the above inequality. That explains its name.

2.20.5 Solving Inequalities

Solving an inequality is to find (real) x such that the inequality holds. For example, find x such that

$$\frac{4x^2}{(1-\sqrt{1+2x})^2} < 2x+9 \tag{2.20.13}$$

First, we determine x such that the inequality makes sense[62]:

$$x \neq 0, \quad 1+2x \geq 0, \quad 2x+9 \geq 0 \Longleftrightarrow x \geq -\frac{1}{2}, \quad x \neq 0$$

Then, we simplify the LHS of Eq. (2.20.13), because we see it is of the form $()^2$ and we can remove the square root in the denominator:

$$\frac{4x^2}{(1-\sqrt{1+2x})^2} = \left(\frac{2x}{1-\sqrt{1+2x}}\right)^2 = \left(\frac{2x(1+\sqrt{1+2x})}{-2x}\right)^2 = 2+2x+2\sqrt{1+2x}$$

Thus, Eq. (2.20.13) is simplified to

$$2+2x+2\sqrt{1+2x} < 2x+9 \Longleftrightarrow x < \frac{45}{8}$$

Combined with the condition of x so that the inequality makes sense, we have the final solution: $-0.5 \leq x < 45/8$ and $x \neq 0$. Alternatively, using the set notation, we can write the solution as (draw a fig like Fig. 2.28 would help):

$$x \in \left[-\frac{1}{2}, 0\right) \cup \left(0, \frac{45}{8}\right)$$

2.20.6 Using Inequalities to Solve Equations

Let's first solve the following equation:

$$x^2+2x+3 = \sqrt{4-x^2}$$

[62] That is, terms under square roots must be non-negative; in this problem, the LHS is non-negative and smaller than the RHS, thus the RHS must be non-negative. We're not interested in the case $x = 0$, because the problem would become $0 < 9$: nothing there to do!

The first approach is to square both sides to get rid of the square root. Doing so results in a fourth order polynomial equation, which is something we should avoid. Let's see if there is an easier way. Note that the RHS is always smaller or equal 2. How about the LHS? It is equal to $(x+1)^2+2$ which is always bigger or equal than 2. So, we have an equation in which the LHS ≥ 2 and RHS ≤ 2. The only case where they are equal is when both of them are equal to two:

$$(x+1)^2+2=2, \quad \sqrt{4-x^2}=2 \Longleftrightarrow x=-1, \quad x=0$$

There is no real solutions! If you prefer a visual solution: the LHS is a parabola facing up with a vertex at $(-1, 2)$ while the RHS is a semi-circle centered at $(0, 0)$ with radius of 2 above the x-axis. These two curves do not intersect! Of course this 'faster' method would not work if number 3 in the LHS is replaced by another number so that the two curves intersect.[63]

2.21 Infinity

This section presents a few things about infinity, the concept of something that is unlimited, endless, without bound. The common symbol for infinity, ∞, was invented by the English mathematician John Wallis in 1655. Mathematical infinities occur, for instance, as the number of points on a continuous line or as the size of the endless sequence of counting numbers: 1, 2, 3, etc.

The symbol ∞ essentially means arbitrarily large or bigger than any positive number. Likewise, the symbol $-\infty$ means less than any negative number.

This section mostly concerns infinite sums, e.g., what is the sum of all positive integers. Such sums are called series. In Sect. 2.21.1, I present arithmetic series (e.g., $2+4+6+\cdots$), in Sect. 2.21.2, I present geometric series (e.g., $1+2+4+\cdots$), and in Sect. 2.21.3 the harmonic series $1+1/2+1/3+\cdots$. In Sect. 2.21.4, the famous Basel problem is presented. Section 2.21.5 is about the first infinite product known in mathematics, and the first example of an explicit formula for the exact value of π.

Why we have to bother with infinite sums? One reason is that many functions can be expressed as infinite sums. For example,

$$f(x) = a_0 + \sum_{n=1}^{\infty}(a_n \cos nx + b_n \sin nx)$$
$$(1-x^2)^{1/2} = 1 - \frac{1}{2}x^2 - \frac{1}{8}x^4 - \frac{1}{16}x^6 + \cdots$$

[63] What we would do then? Either solving a fourth order equation or using Newton's method.

2.21.1 Arithmetic Series

A child's mother gives him 10 cents one day. Everyday thereafter his mom gives him 3 more cents than the previous day. After 10 days, how much does he have?

This simple problem exhibits what is called an *arithmetic series*. After day 1, he has 10 cents. On the second day he gets 13 cents, on the third day 16 cents, and so on. The list of amounts he gets each day

$$10, 13, 16, 19, 22, \ldots,$$

is called a *sequence*. When we add up the terms in this sequence to get the total amount he has at some point

$$10 + 13 + 16 + 19 + 22 + 25 + 28 + 31 + 34 + 37$$

the result is a *series* or precisely a finite series because the number of terms is finite. Shortly, we shall discuss infinite series in which the number of terms is infinite. In this particular case, where each term is separated by a fixed amount from the previous one, both series and sequence are called arithmetic.

The amount that the boy has is simply obtained as a sum of ten terms, it is 235. But we need a smarter way to solve this problem, just in case we face this problem: what is the amount after a year? Doing the sum for 365 terms is certainly a boring task.

What we want here is a formula that gives us directly the arithmetic series. And mathematicians solve this specific problem by considering a general problem (as it turns out it is easier to handle the general problem with symbols). Let's first define a general arithmetic sequence with a being the first term and d being the difference between successive terms. The arithmetic sequence is then

$$a, a + d, a + 2d, \ldots, a + (n - 1)d, \ldots \tag{2.21.1}$$

where the nth term is $a + (n - 1)d$. Now, the sum of the first n terms of this sequence is $a + a + d + a + 2d + \cdots + a + (n - 1)d$. To compute this sum, we follow Gauss, by writing the sum S in the usual order and in a reverse order (for 4 terms only, which is enough to see the point):

$$\begin{array}{llll} S \;= a & + a + d & + a + 2d & + a + 3d \\ S \;= a + 3d & + a + 2d & + a + d & + a \\ \hline 2S = 2a + 3d & + 2a + 3d & + 2a + 3d & + 2a + 3d \end{array}$$

We can see that $2S = 4 \times (2a + 3d)$, or $S = (4/2)(2a + 3d) = (4/2)[(a) + (a + 3d)]$. Now we see the pattern, and thus the general arithmetic series is given by

$$\boxed{a + a + d + \cdots + a + (n-1)d = \frac{\text{num. of terms}}{2}(\text{1st term} + \text{final term})} \tag{2.21.2}$$

Thus, with observation, we have developed a formula that just requires us to do one addition and one multiplication, regardless of the number of terms involved! That's the power of mathematics.

2.21.2 Geometric Series

Suppose that the door is two meters away. To get to it, you must travel half of the distance (one meter), then half of what is left (half a meter), then half of what is left (a quarter of a meter), and so on. In total, you must travel a distance of $1 + S$ with S is the following *infinite sum*:

$$S = \frac{1}{2} + \frac{1}{4} + \frac{1}{8} \cdots = \sum_{i=1}^{\infty} \frac{1}{2^i} \tag{2.21.3}$$

where the ellipsis "..." means "and so on forever". This sum is called a *geometric series*, that is a series with a constant ratio (1/2 for this particular case) between successive terms. Geometric series are among the simplest examples of infinite series with finite sums, although not all of them have this property. Why 'geometric'? I shall explain it shortly.

Hey! What kind of human that walking to a door like that? The story is like this, you might guess correctly that it came from a philosopher. In the fifth century BC, the Greek philosopher Zeno of Elea posed four problems, and the above is one of them passed on to us by Aristotle. Zeno was wondering about the continuity of space and time.

To have an idea what S might be, you can compute it for some concrete values of n to see what the sum might be. I did that for n up to 20 (of course using a small Julia code, Listing A.2) and the result given in Table 2.12 indicates that $S = 1$. Even though the sum involves infinite terms it converges to a finite value of one! And a geometry representation of this sum shown in Fig. 2.29 confirms this. Noting that, in the past, Zeno argued that you would never be able to get to the door! This is because the Greeks had no notion that *an infinite number of terms could have a finite sum.*

Although we have numerical and geometric evidence that the sum is one, we still need a mathematical proof. We need to do some algebra tricks here. The idea is: we do not go to infinity (where is it?), thus we *consider only n terms in the sum,* then we see *what happens to this sum when we let n go to infinity* (the danger is for n not for us, and this works). That's why mathematicians introduce the *partial sum* $S_n = \sum_{i=1}^{n} 1/2^i$. With this symbol, they start doing some algebraic manipulations to it and it reveals its secret to them. First they multiply S_n by $1/2$ and put S_n and $(1/2)S_n$ together to see the connection:

Table 2.12 $S = \sum_{i=1}^{n} \frac{1}{2^i}$

Terms	S
1	0.5
2	0.75
3	0.875
$\vdots$	$\vdots$
10	0.9990234375
20	0.9999990463

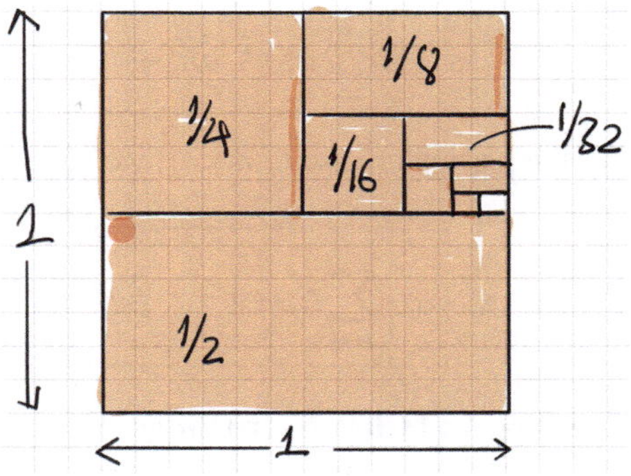

Fig. 2.29 Geometry visualization of $S = \sum_{i=1}^{\infty} \frac{1}{2^i}$

$$\begin{aligned}S_n &= \frac{1}{2} + \frac{1}{4} + \frac{1}{8} + \cdots + \frac{1}{2^{n-1}} + \frac{1}{2^n} \\ \frac{1}{2}S_n &= \qquad\;\; \frac{1}{4} + \frac{1}{8} + \cdots + \frac{1}{2^n} + \frac{1}{2^{n+1}}\end{aligned}$$

What next then? Many terms are identical in S_n and half of it, so it is natural to subtract them from each other to cancel out the common terms:

$$S_n - \frac{1}{2}S_n = \frac{1}{2} - \frac{1}{2^{n+1}} \Longrightarrow \boxed{S_n = 1 - \frac{1}{2^n}}$$

Because the series involves infinite terms, we should now consider the case when n is very large, i.e.,, $n \to \infty$. For such n, the term $1/2^n$—which is the inverse of a giant number—is very very small, and thus S_n is approaching one, which means that S approaches one too:

$$S = 1 \quad \text{when } n \to \infty$$

There is nothing special about $1/2, 1/4, \ldots$ in the series. Thus, we now generalize the above discussion to come up with the following geometric series, with the first term a and the ratio r:

$$S = a + ar + ar^2 + ar^3 + \cdots \tag{2.21.4}$$

Then, we introduce the partial sum S_n (n is the number of terms) and multiply it with r, rS_n, as follows

$$\begin{aligned} S_n &= a + ar + ar^2 + ar^3 + \cdots + ar^{n-1} \\ rS_n &= 0 + ar + ar^2 + ar^3 + \cdots + ar^n \end{aligned}$$

It follows then,

$$(1-r)S_n = a - ar^n \Longrightarrow S_n = \frac{a}{1-r}(1-r^n)$$

Or,

$$a + ar + ar^2 + ar^3 + \cdots + ar^{n-1} = \frac{a}{1-r}(1-r^n) \tag{2.21.5}$$

For the particular case of $a = 1$, we have this result

$$\boxed{\sum_{i=0}^{n-1} r^i = 1 + r + r^2 + r^3 + \cdots + r^{n-1} = \frac{1-r^n}{1-r}} \tag{2.21.6}$$

The Rice And Chessboard Story. There's a famous legend about the origin of chess that goes like this. When the inventor of the game showed it to the emperor of India, the emperor was so impressed by the new game, that he said to the man "Name your reward!". The man responded, "Oh emperor, my wishes are simple. I only wish for this. Give me one grain of rice for the first square of the chessboard, two grains for the next square, four for the next, eight for the next and so on for all 64 squares, with each square having double the number of grains as the square before".

Let's see how many grains would be needed. It can be seen that the total number of grains is a geometric series with $a = 1$ and $r = 2$. Using Eq. (2.21.6), we can compute it:

$$S = 1 + 2 + 4 + \ldots = \frac{1}{1-2}(1-2^{64}) = 18,446,744,073,709,551,615 \tag{2.21.7}$$

The total number of grains equals 18, 446, 744, 073, 709, 551, 615 (eighteen quintillion four hundred forty-six quadrillion, seven hundred forty-four trillion, seventy-three billion, seven hundred nine million, five hundred fifty-one thousand, six hundred and fifteen)! Not only it is a very large number, it is also a prime; the number of grains is the 64th Mersenne number. A Mersenne number is a prime number that is one less than a power of two ($2^n - 1$). This number is named after Marin Mersenne, a French Minim friar, who studied them in the early seventeenth century.

So we have seen two geometric series, one in Eq. (2.21.3) with $r = 1/2 < 1$ and one in the chessboard legend with $r = 2 > 1$. While the first series *converges*, or is convergent (i.e.,, as the number of terms get bigger and bigger the sum does not explode, it settles to a finite value), the second series *diverges* (or is divergent); the more terms result in a bigger sum. The question now is to study when the geometric series converges. Before delving into that question, noting that r can be negative; actually mathematicians want it to be. Because they always aim for a general result.

To see why geometric series with $r < 1$ converge, let's look at Eq. (2.21.5). We have the term $1 - r^n$ which depends on n. But we also know that if $-1 < r < 1$ (or compactly $|r| < 1$ using the absolute value notation), then r^n approaches zero when n is getting bigger and bigger. You can try these numbers 0.5^{10}, 0.5^{11}, 0.5^{12} and you will see that they become smaller and smaller and approaching zero (On a hand calculator, start with 0.5 and press the x^2 button successively, you will get zero). Not a mathematical proof, but for now it is more than enough. For a proof, we need the concept of limit. (Actually we have seen the idea of limit right in Table 2.12).

So, we have for $|r| < 1$, $1 - r^n$ goes to one when n goes to infinity. From Eq. (2.21.5) the geometric series thus becomes

$$a + ar + ar^2 + ar^3 + \cdots = \frac{a}{1-r}, \quad \text{for } |r| < 1 \tag{2.21.8}$$

Note that this formula holds only for $|r| < 1$. If we use it for $|r| > 1$, we would get absurd results. For example, with $r = 2$, this formula gives us

$$1 + 2 + 4 + 8 + \cdots = -1$$

which is absurd. Weird things can happen if we use ordinary algebra to a divergent series! Now we can understand why Niels Henrik Abel[64] said "Divergent series are the devil, and it is a shame to base on them any demonstration whatsoever".

Using the geometric series formula to express repeating decimals. We can use geometric series to prove that a repeating decimal is a rational number. For example,

$$\begin{aligned}
0.22222222\ldots &= 0.2 + 0.02 + 0.002 + \cdots \\
&= \frac{2}{10} + \frac{2}{100} + \frac{2}{1000} + \cdots \quad (a = 2/10, r = 1/10) \\
&= \frac{2}{10} / \frac{9}{10} = \frac{2}{9} \quad \text{using Eq. (2.21.8)}
\end{aligned}$$

And in the same manner, we have this

---

[64] Niels Henrik Abel (1802–1829) was a Norwegian mathematician. His most famous single result is the first complete proof demonstrating the impossibility of solving the general quintic equation in radicals. He was also an innovator in the field of elliptic functions, discoverer of Abelian functions. He made his discoveries while living in poverty and died at the age of 26 from tuberculosis. Most of his work was done in 6 or 7 years of his working life. Regarding Abel, the French mathematician Charles Hermite said: "*Abel has left mathematicians enough to keep them busy for five hundred years*". Another French mathematician, Adrien-Marie Legendre, said: "what a head the young Norwegian has!"). The Abel Prize in mathematics, originally proposed in 1899 to complement the Nobel Prizes, is named in his honor.

$$0.99999\ldots = 0.9 + 0.09 + 0.009 + \cdots = \frac{9}{10}/\frac{9}{10} = 1 \quad (2.21.9)$$

Can you name a number that is larger than 0.999... and smaller than 1? If not, these two numbers are the same!

2.21.3 Harmonic Series

The harmonic series is the divergent infinite series:

$$S = 1 + \frac{1}{2} + \frac{1}{3} + \frac{1}{4} + \cdots = \sum_{k=1}^{\infty} \frac{1}{k} \quad (2.21.10)$$

Why is this series called the harmonic series? We can find the following answer everywhere. It is such called because each term of the series, except the first, is the harmonic mean of its two nearest neighbors. And the explanation stops there!. This response certainly raises more questions than it answers: What is the harmonic mean? To have a complete understanding, we have to trace to the origin.

The harmonic mean. We know the arithmetic mean of two numbers a and b is $A = 0.5(a + b)$. The geometric mean is $G = \sqrt{ab}$. The harmonic mean is $H = 2ab/(a + b)$. Or equivalently, $1/H = 0.5(1/a + 1/b)$; H is the reciprocal[65] of the average of the reciprocals of a and b. So, $1/n$ is the harmonic mean of $1/(n-1)$ and $1/(n+1)$ for $n > 1$. Now, we are going to unfold the meaning of these means.

It is a simple matter to find the average of two numbers. For example, the average of 6 and 10 is 8. When we do this, we are really finding a number x such that 6, x, 10 forms an arithmetic sequence: 6,8,10. In general, if the numbers a, x, b form an arithmetic sequence, then

$$x - a = b - x \Longrightarrow x = \frac{a + b}{2} \quad (2.21.11)$$

Similarly, we can define the geometric mean (GM) of two positive numbers a and b to be the positive number x such that a, x, b forms a geometric sequence. One example is 2, 4, 8 and this helps us to find the formula for GM:

$$\frac{x}{a} = \frac{b}{x} \Longrightarrow x = \sqrt{ab}$$

Now getting back to the harmonic series. What is the value of S? I do not know, so I programmed a small function and let the computer compute this sum. And for

[65] Reciprocal is like inverse. Mathematicians love doing this. For example, instead of saying "perpendicular", they say "orthogonal".

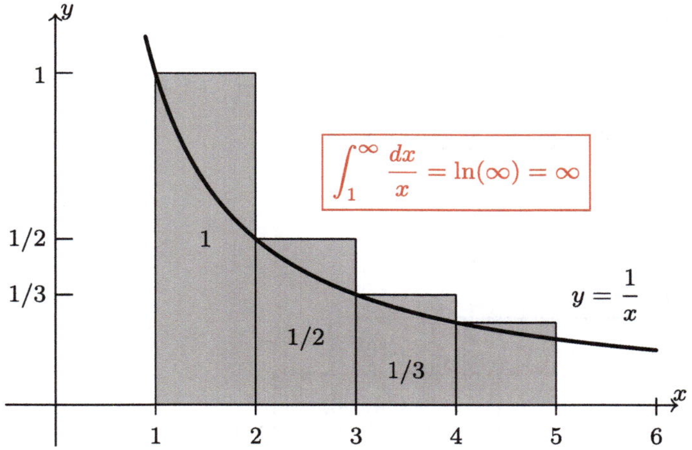

Fig. 2.30 Calculus-based proof of the divergence of the harmonic series. The harmonic series and the area under the curve $y = 1/x$ leads to a famous constant in mathematics. Can you find it?

$n = 10^{10}$ (more than a billion), we got 25.91. Now, we know this sum is infinity, thus called a divergent series. How can we prove that? The divergence of the harmonic series was first proven in the 14th century by the French philosopher of the later Middle Ages Nicole Oresme (1320–1382). Here is what he did:

$$\begin{aligned}
S &= 1 + \frac{1}{2} + \frac{1}{3} + \frac{1}{4} + \cdots \\
S &> 1 + \frac{1}{2} + \left(\frac{1}{4} + \frac{1}{4}\right) + \frac{1}{5} + \frac{1}{6} + \frac{1}{7} + \frac{1}{8} + \cdots \text{ (replace 1/3 by 1/4)} \\
S &> 1 + \frac{1}{2} + \frac{1}{2} + \left(\frac{1}{5} + \frac{1}{6} + \frac{1}{7} + \frac{1}{8}\right) + \cdots \qquad (1/4 + 1/4 = 1/2) \\
S &> 1 + \frac{1}{2} + \frac{1}{2} + \underbrace{\left(\frac{1}{8} + \frac{1}{8} + \frac{1}{8} + \frac{1}{8}\right)}_{1/2} + \cdots \qquad \text{(replace } 1/5, 1/6, 1/7 \text{ by } 1/8\text{)}
\end{aligned} \tag{2.21.12}$$

So Oresme compared the harmonic series with another one which is divergent and smaller than the harmonic series. Thus, the harmonic series must diverge. This proof, which used a *comparison test*, is considered by many in the mathematical community to be a high point of medieval mathematics. It is still a standard proof taught in mathematics classes today. Are there other proofs? How about considering the function $y = 1/x$ and the area under the curve $y = 1/x$? See Fig. 2.30. The area under this curve is infinite and yet it is smaller than the area of those rectangles in this figure. This area of the rectangles is exactly our sum S.

It is interesting to show that one can get the harmonic series from a static mechanics problem of hanging blocks (Fig. 2.31). Let's say that we have two identical blocks

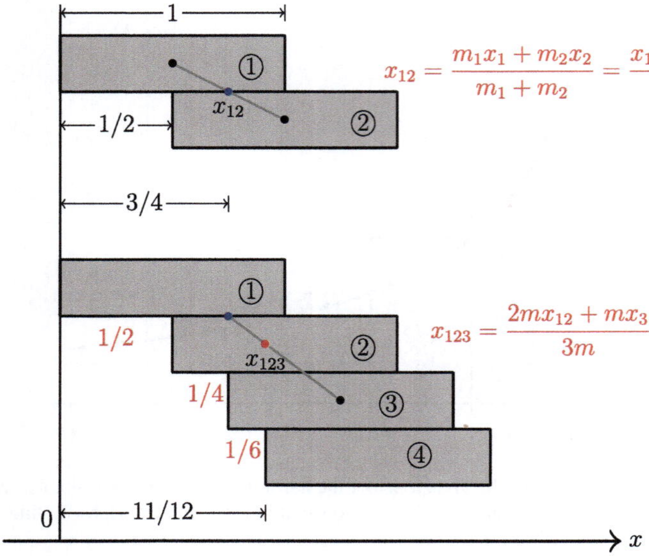

Fig. 2.31 Stacking identical blocks, each has a mass of m, with maximum overhang and its relation to the harmonic series. Without loss of generality, the length of each block is one unit. Noting that x_{123} is the $x-$coordinate of the center of mass of three blocks ①, ②, and ③

and want to position them one on top of the other so that the top one has the largest overhang, but doesn't topple over. The way to do that, using statics, is to place the top block (labeled by ① in the referred figure) precisely halfway across the one underneath (i.e.,, block ②). In this way, the center of mass of the top block falls on the left edge of the bottom block. So, with two blocks, we can have a maximum overhang of $^1/_2$.

With three blocks, we first have to find the center of mass of the two blocks ① and ②. As shown in Fig. 2.31b, this center's $x-$coordinate is $x_{12} = 3/4$ (check Sect. 6.8.7 for a refresh on how to determine the center of mass of an object). Now we place block ③ such that its left edge is exactly beneath that center. From that we can deduce that the overhang for the case of three blocks is $1/2 + 1/4$. Continuing this way, it can be shown that the overhang is given by

$$\frac{1}{2} + \frac{1}{4} + \frac{1}{6} + \cdots = \frac{1}{2}\left(1 + \frac{1}{2} + \frac{1}{3} + \cdots\right)$$

which is half of the harmonic series. Because the harmonic series diverges, it is possible to have an infinite overhang!

To understand why similar series possess different properties, we put the geometric and the harmonic series together below

$$S_{geo} = 0 + \frac{1}{2} + \frac{1}{4} + \frac{1}{8} + \frac{1}{16} + \frac{1}{32} + \frac{1}{64} \cdots$$
$$S_{har} = 1 + \frac{1}{2} + \frac{1}{3} + \frac{1}{4} + \frac{1}{5} + \frac{1}{6} + \frac{1}{7} + \frac{1}{8} \cdots$$

Now we can observe that the terms in the geometric series shrink much faster than the terms in the harmonic series, e.g., the sixth term in the former is 0.015625, while the corresponding term is just $1/7 = 0.142857143$.

2.21.4 Basel Problem

The Basel problem was first posed by the Italian mathematician and clergyman Pietro Mengoli (1626–1686) in 1650 and solved by Leonhard Euler in 1734. As the problem had withstood the attacks of the leading mathematicians of the day (Leibnitz, Bernoulii brothers, Wallis),[66] Euler's solution brought him immediate fame when he was 28. The problem is named after Basel, the hometown of Euler as well as of the Bernoulli family who unsuccessfully attacked the problem.

The Basel problem asks for the precise summation of the reciprocals of the squares of the natural numbers, i.e.,, the precise sum of the infinite series:

$$S = 1 + \frac{1}{4} + \frac{1}{9} + \frac{1}{16} + \cdots + \frac{1}{k^2} + \cdots = ? \tag{2.21.13}$$

Before computing this series, let's see whether it converges. The series converges that can be verified by writing a small code. For a proof, we follow Oresme's idea of using a comparison test. The idea is to compare this series with a larger series that converges. We compare the following series

$$\begin{aligned} S_1 &= 1 + \frac{1}{2\times 2} + \frac{1}{3\times 3} + \frac{1}{4\times 4} + \cdots \quad (S_1 \text{ is nothing but } S) \\ S_2 &= 1 + \frac{1}{1\times 2} + \frac{1}{2\times 3} + \frac{1}{3\times 4} + \cdots \end{aligned} \tag{2.21.14}$$

And if S_2 converges to a finite value, then S_1 should be convergent to some value smaller as $S_1 < S_2$. Indeed, we can re-write the partial sum of the second series as a telescoping sum (without 1)[67]

---

[66] Jakob Bernouilli expressed his eventual frustration at its elusive nature in the comment "If someone should succeed in finding what till now withstood our efforts and communicate it to us, we shall be much obliged to him".

[67] See Sect. 2.21.6 to see why mathematicians think of this way to compute S_2.

$$
\begin{aligned}
S_2(n) - 1 &= \frac{1}{1\times 2} + \frac{1}{2\times 3} + \frac{1}{3\times 4} + \cdots + \frac{1}{n(n+1)} \\
&= \left(1 - \frac{1}{2}\right) + \left(\frac{1}{2} - \frac{1}{3}\right) + \left(\frac{1}{3} - \frac{1}{4}\right) + \cdots + \left(\frac{1}{n} - \frac{1}{n+1}\right) \\
&= \left(1 - \frac{1}{2}\right) + \left(\frac{1}{2} - \frac{1}{3}\right) + \left(\frac{1}{3} - \frac{1}{4}\right) + \cdots + \left(\frac{1}{n} - \frac{1}{n+1}\right) \\
&= 1 - \frac{1}{n+1} \Longrightarrow S_2(n) = 2 - \frac{1}{n+1}
\end{aligned}
\tag{2.21.15}
$$

When n is approaching infinity the denominator in $1/n+1$ is approaching infinity and thus this fraction approaches zero. So, S_2 converges to two. Therefore, S_1 should converge to something smaller than two. Indeed, Euler computed this sum, first by considering the first, say, 100 terms, and found the sum was about 1.6349.[68] Then, using ingenious reasoning, he found that,[69]

$$
S = 1 + \frac{1}{4} + \frac{1}{9} + \frac{1}{16} + \cdots + \frac{1}{k^2} + \cdots = \frac{\pi^2}{6}
$$

How Euler come up with this result? He used the Taylor series expansion of $\sin x$, and the infinite product expansion. See Sect. 4.15.7 for Euler's proof and Sect. 3.15 for Cauchy's proof.

In what follows, I present another proof. This proof is based on the following two lemmas[70]:

- $S = \dfrac{4}{3}\displaystyle\sum_{n=1}^{\infty} \frac{1}{(2n-1)^2}$;
- $\displaystyle\int_1^0 x^m \ln x dx = \frac{1}{(m+1)^2}$

which can be proved straightforwardly. Then, the sum in the Basel problem can be written as

[68] How Euler did this calculation without calculator is another story. Note that the series converge very slow, i.e.,, we need about one billion terms to get an answer with 8 correct decimals. Euler could not do that. But he is a genius; he had a better way. Check Sect. 4.18 for detail.

[69] We have to keep in mind that at that time Euler knew, see Sect. 4.15.4 that another related series has a sum related to π:

$$
\frac{\pi}{4} = 1 - \frac{1}{3} + \frac{1}{5} - \frac{1}{7} \cdots
$$

[70] In mathematics, a lemma (plural lemmas or lemmata) is a generally minor, proven proposition that is used as a stepping stone to a larger result. For that reason, it is also known as a "helping theorem" or an "auxiliary theorem".

$$
\begin{aligned}
S &= \frac{4}{3}\sum_{n=1}^{\infty}\frac{1}{(2n-1)^2} = \frac{4}{3}\sum_{n=0}^{\infty}\frac{1}{(2n+1)^2} = \frac{4}{3}\sum_{n=0}^{\infty}\int_1^0 x^{2n}\ln x dx \\
&= \frac{4}{3}\int_1^0 \ln x\left(\sum_{n=0}^{\infty} x^{2n}\right)dx \quad \text{(the sum is a geometric series)} \\
&= \frac{4}{3}\int_1^0 \frac{\ln x}{1-x^2}dx
\end{aligned}
$$

where in the first equality, we simply changed the dummy variable $2n - 1$ to $2n + 1$ as both represent odd numbers. In the second equality, we used the second lemma with $2n$ plays the role of m. In the third equality, we change the order of sum and integration and finally we computed the sum which is a geometric series $1 + x^2 + x^4 + \cdots$.[71] Why the geometric series appear here in the Basel problem? I do not know, but that is mathematics: when we have discovered some maths, it appears again and again not in maths but also in physics!

Remark And why calculus (i.e.,, integral $\int f(x)dx$) in a class of algebra? Why not? We divide mathematics into different territories (e.g., algebra, number theory, calculus, geometry, etc.). But it is our invention, maths does not care! Most of the times all mathematical objects are somehow related to each other. You can see algebra in geometry and vice versa. That's why I presented this proof here in the chapter about algebra.

2.21.5 Viète's Infinite Product

Viète's formula is the following infinite product of *nested radicals* (a nested radical is a radical expression—one containing a square root sign, cube root sign—that contains or nests another radical expression) representing the mathematical constant π:

$$
\frac{2}{\pi} = \frac{\sqrt{2}}{2}\frac{\sqrt{2+\sqrt{2}}}{2}\frac{\sqrt{2+\sqrt{2+\sqrt{2}}}}{2}\cdots \tag{2.21.16}
$$

Viète formulated the first instance of an infinite product known in mathematics, and the first example of an explicit formula for the exact value of π. Note that this formula does not have any practical application (except that it allows mathematicians to compute π to any accuracy they want[72]; they're obsessed with this task). But in this formula, we see the connection of geometry (π), trigonometry, and algebra.

[71] If not clear this is a geometric series with $a = 1$ and $r = x^2$.

[72] This problem itself does not have any practical application!

Viète's formula may be obtained as a special case of a formula given more than a century later by Leonhard Euler, who discovered that (a proof is given shortly), for n large we have

$$\frac{\sin x}{x} = \cos\frac{x}{2^1}\cos\frac{x}{2^2}\cdots\cos\frac{x}{2^n} \tag{2.21.17}$$

Evaluating this at $x = \pi/2$:

$$\frac{2}{\pi} = \cos\frac{\pi}{4}\cos\frac{\pi}{8}\cos\frac{\pi}{16}\cdots \tag{2.21.18}$$

Starting with $\cos\pi/4 = \sqrt{2}/2$ and using the half-angle formula $\cos\alpha/2 = \sqrt{1+\cos(\alpha)/2}$ (see Sect. 2.25.5 for a proof), the above expression can be computed as

$$\frac{2}{\pi} = \frac{\sqrt{2}}{2}\frac{\sqrt{2+\sqrt{2}}}{2}\frac{\sqrt{2+\sqrt{2+\sqrt{2}}}}{2}\cdots$$

Proof Here is the proof of Eq. (2.21.17). The starting point is the double-angle formula $\sin x = 2\sin\frac{x}{2}\cos\frac{x}{2}$, and repeatedly apply it to $\sin x/2$, then to $\sin x/4$ and so on:

$$\begin{aligned}\sin x &= 2\sin\frac{x}{2}\cos\frac{x}{2}\\ &= 2\left(2\sin\frac{x}{4}\cos\frac{x}{4}\right)\cos\frac{x}{2}\\ &= 2(2)\left(2\sin\frac{x}{8}\cos\frac{x}{8}\right)\cos\frac{x}{4}\cos\frac{x}{2} = 2^3\sin\frac{x}{2^3}\cos\frac{x}{2^1}\cos\frac{x}{2^2}\cos\frac{x}{2^3}\end{aligned} \tag{2.21.19}$$

Thus, after n applications of the double-angle formula for $\sin x$, we get

$$\sin x = 2^n\sin\frac{x}{2^n}\cos\frac{x}{2^1}\cos\frac{x}{2^2}\cdots\cos\frac{x}{2^n} = 2^n\sin\frac{x}{2^n}\prod_{i=1}^{n}\cos\frac{x}{2^i} \tag{2.21.20}$$

where in the last equality, I used the short-hand Pi symbol $\prod$ (it is useless for the proof here, I just wanted to introduce this notation). It is used in mathematics to represent the product of a bunch of terms (think of the starting sound of the word 'product').[73]

Dividing both sides by x gives us

$$\frac{\sin x}{x} = \frac{\sin\frac{x}{2^n}}{x/2^n}\cos\frac{x}{2^1}\cos\frac{x}{2^2}\cdots\cos\frac{x}{2^n}$$

[73] To practice, mathematicians write $\prod_{i=1}^{n} i$ to mean the product of the first n integers (e.g., $(1)(2)\cdots(n)$).

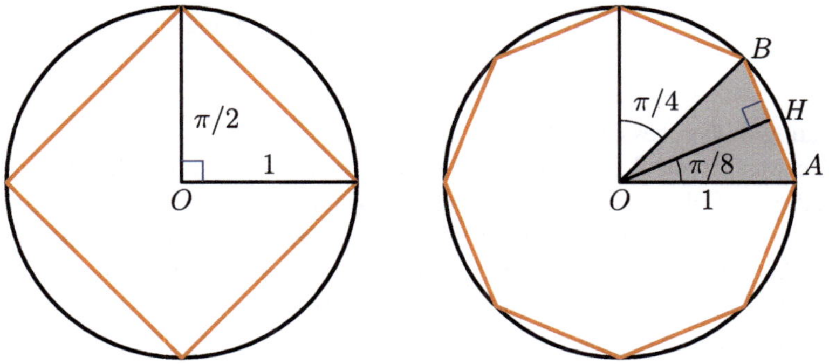

Fig. 2.32 Geometry proof of Viète's formula

As the red term approaches 1 when n is very large (this is the well-known trigonometry limit $\lim_{h\to 0} \sin h/h = 1$ or simply $\sin h \approx h$ when h is small), Euler's formula follows. ■

Viète had a geometry proof, which is now presented. When π is present, there is a circle hidden somewhere. As this formula should be applicable to any circle, let's consider a circle of unit radius. The idea is to compare the area of this circle (which is π) with that of regular polygons inscribed in the circle. Starting with a square, then an octagon, then hexadecagon and so on, see Fig. 2.32. If you do not know about trigonometry, then read Chap. 3 and get back.

For an octagon, its area is eight times the area of the triangle OAB with $OH = \cos \pi/8$ and $AB = 2 \sin \pi/8$:

$$A_8 = (8)\left(\frac{1}{2}\right)\left(2 \sin \frac{\pi}{8}\right) \cos \frac{\pi}{8} = 4 \sin \frac{\pi}{4} = 4\frac{\sqrt{2}}{2}$$

where $\sin 2x = 2 \sin x \cos x$ was used for $x = \pi/8$. And thus, equating this area to the circle area, we get the following equation

$$\pi = 4\frac{\sqrt{2}}{2} \Longrightarrow \frac{2}{\pi} = \frac{\sqrt{2}}{2}$$

which is only a rough approximation for π. We need to use a polygon of more sides. For an hexadecagon, similarly we have

$$A_{16} = (16)\left(\frac{1}{2}\right)\left(2 \sin \frac{\pi}{16}\right) \cos \frac{\pi}{16} = 8 \sin \frac{\pi}{8} = 8\sqrt{\frac{1 - \sqrt{2}/2}{2}} = \frac{4\sqrt{2}}{\sqrt{2 + \sqrt{2}}}$$

And again, equating this area to the circle area gives us

$$\pi = \frac{4\sqrt{2}}{\sqrt{2+\sqrt{2}}} \Longrightarrow \frac{2}{\pi} = \frac{\sqrt{2}}{2}\frac{\sqrt{2+\sqrt{2}}}{2}$$

Continuing this process, we shall get Eq. (2.21.16).

Viète was a typical child of the Renaissance in the sense that he freely mixed the methods of classical Greek geometry with the new algebra and trigonometry. However, Viète did not know the concept of convergence and thus did not worry whether his infinite sequence of operations would blow up or not. That is, one gets different value for π with different numbers of terms adopted (we discuss this issue in Sect. 2.22). As an engineer or scientist of which sloppy engineering mathematics is enough, we just need to write a code to check. But as far as mathematicians are concerned, they need a proof for the convergence/divergence of Viète's formula. And the German and Swiss mathematician Ferdinand Rudio (1856–1929) proved the convergence in 1891.

2.21.6 Sum of Differences

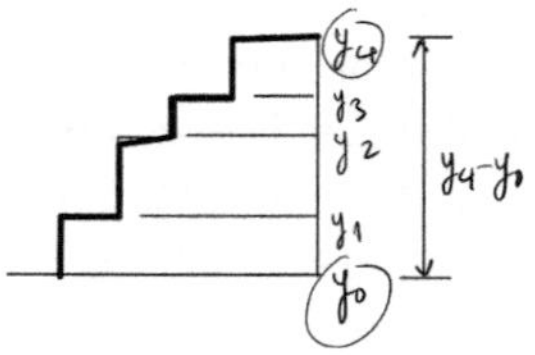

We have seen that a sum of differences (e.g., $\sum_{k=1}^{n}[k^2 - (k-1)^2]$) depends only on the first and the last term. And this is related to actually something we see everyday: a staircase. Assume that someone is climbing a very long and irregular staircase (see figure). And he wants to calculate the total vertical rise from the bottom of the staircase to the top. Of course this is equal to the sum of the heights of all the stairs. And the height of each stair is the difference between the altitude of its top and that of its bottom. So, we have a sum of differences. And this sum should be the same as the difference between the altitude of the top (y_4 in our illustration) and that of the bottom (y_0).

Let's use sum of differences to find the sum of some things. We start with the numbers $0, 1, 2, 3, 4, \ldots$, then we consider the square of these numbers, i.e.,, $0, 1, 4, 9, 16, \ldots$ Now we consider the differences of these squares: $1, 3, 5, 7, 9, \ldots$ (see Table 2.13). Now we can immediately have this result: the sum of the first n odd numbers is n^2:

$$1 + 3 + 5 + 7 + \cdots + (2n+1) = n^2$$

Now, we have discovered another fact about natural numbers: the sum of the first odd numbers is a perfect square. Using dots can you visualize this result, and obtain this fact geometrically? Yes, facts about mathematical objects are hidden in their world waiting to be discovered. And as it turns out many such discoveries have applications in our real world.

Table 2.13 Sum of the first n odd numbers

i	0	1	2	3	4	5
i^2	0	1	4	9	16	25
$i^2-(i-1)^2$		1	3	5	7	9

In Sect. 2.6.1, we considered the sum $S = 1 + 2 + 3 + \cdots + n$ and computed it in three different ways. Now is the fourth way using the fact that the sum of the first n odd numbers is n^2. This sum S consists of the evens and the odds, but we only know the sum of the odds. So, we can transform the evens to the odds: $2 = 1 + 1$, $3 = 2 + 1$ and so on. Without loss of generality, assume $n = 8$, then we can write S as follows

$$\begin{aligned} S &= 1 + ② + 3 + ④ + 5 + ⑥ + \cdots + ⑧ \\ &= 1 + (1+1) + 3 + (3+1) + 5 + (5+1) + 7 + (7+1) \\ &= 2(1+3+5+7) + (1+1+1+1) = 2\left(\frac{8}{2}\right)^2 + \frac{8}{2} = \frac{8 \cdot 9}{2} \end{aligned}$$

I just wanted to show that the motivation for the trick of considering $k^2 - (k-1)^2 = 2k - 1$ presented in Sect. 2.6.1 comes from Table 2.13.

Let's now consider the sum of an infinite series that Huygens asked Leibniz to solve in 1670[74]: the sum of the reciprocals of the triangular numbers[75]:

$$S = \frac{1}{1} + \frac{1}{3} + \frac{1}{6} + \frac{1}{10} + \frac{1}{15} + \cdots$$

Leibniz solved it by first constructing a table similar to Table 2.14. The first row is just the reciprocals of the natural numbers. The second row is, of course, the difference of the first row. Thus, the sum of the second row is

$$\frac{1}{2} + \frac{1}{6} + \frac{1}{12} + \cdots = \frac{1}{2}\left(\frac{1}{1} + \frac{1}{3} + \frac{1}{6} + \cdots\right) = \frac{S}{2}$$

Since this sum is the sum of differences, it is equal to the difference between the first number of the first row, which is one, and the last number (which is zero). But S is twice the sum of the second row, thus $S = 2$.

And who was this Gottfried Wilhelm Leibniz? He would later become the co-inventor of calculus (the other was Sir Isaac Newton). And in calculus that he devel-

74 When Leibniz went to Paris, he met Dutch physicist and mathematician Christiaan Huygens. Once he realized that his own knowledge of mathematics and physics was patchy, he began a program of self-study, with Huygens as his mentor, that soon pushed him to making major contributions to both subjects, including discovering his version of the differential and integral calculus.

75 Refer to Fig. 2.6 for an explanation of triangular numbers.

Table 2.14 Leibniz's table of the reciprocals of natural numbers

i	$\frac{1}{1}$	$\frac{1}{2}$	$\frac{1}{3}$	$\frac{1}{4}$	$\frac{1}{5}$	$\frac{1}{6}$	$\dots$	$\frac{1}{\infty}$
$(i-1)-(i)$		$\frac{1}{2}$	$\frac{1}{6}$	$\frac{1}{12}$	$\frac{1}{20}$	$\frac{1}{30}$		

oped, we have this fact: $\int_a^b f(x)dx = F(b) - F(a)$. What is this? On the LHS we have a sum ($\int$ is the twin brother of Σ) and on the RHS we have a difference! And this fact was discovered by Leibniz—the guy who played with sum of differences. What a nice coincidence!

2.22 Sequences, Convergence, and Limit

Let's start off this section with a discussion of just what a sequence is. A sequence is nothing more than a list of numbers written in a specific order. The list may or may not have an infinite number of terms in them although we will be dealing exclusively with infinite sequences (which are fun to play with). One way to write a sequence is

$$(a_1, a_2, \dots, a_n, a_{n+1}, \dots) \tag{2.22.1}$$

where a_1 is the first term, a_2 is the second term, or generally a_n is the nth term. So, we use a_n to denote the nth term of the sequence and (a_n) to denote the whole sequence (instead of writing the longer expression $a_1, a_2, a_3, \dots$). You will also see this notation $\{a_1, a_2, \dots, a_n, a_{n+1}, \dots\}$ in other books.

Let's study the sequence of the partial sums $S_n = \sum_{i=1}^{n} 1/2^i$ of the geometric series in Eq. (2.21.3). That is the infinite list of numbers: $S_1, S_2, S_3, \dots$ I compute S_n for $n = 1, 2, \dots, 15$ and present the data in Table 2.15 and I plot S_n versus n in Fig. 2.33. What we observe is that as n is getting larger and larger (in this particular example this is when $n > 14$), S_n is getting closer and closer to one. We say that the sequence (S_n) *converges to one and its limit is one*. In symbol, it is written as

$$\lim_{n\to\infty} S_n = 1 \tag{2.22.2}$$

As the limit is finite (in other words the limit exists) the sequence is called convergent. A sequence that does not converge is said to be divergent.

In the previous discussion, our language was not precise as we wrote when n is larger and larger (what measures?) and S_n gets closer to one (how close?). Mathematicians love rigor,[76] so they reword what we have written as to say the limit of the

[76] To see to what certain rigor means to them, this joke says it best: *A mathematician, a physicist, and an engineer were traveling through Scotland when they saw a black sheep through the window of the train. "Aha," says the engineer, "I see that Scottish sheep are black." "Hmm," says the*

Table 2.15 The sequence $(\sum_{i=1}^{n} 1/2^i)$

n	S_n
1	0.5
2	0.75
3	0.875
4	0.9375
⋮	⋮
14	0.999939
15	0.999969

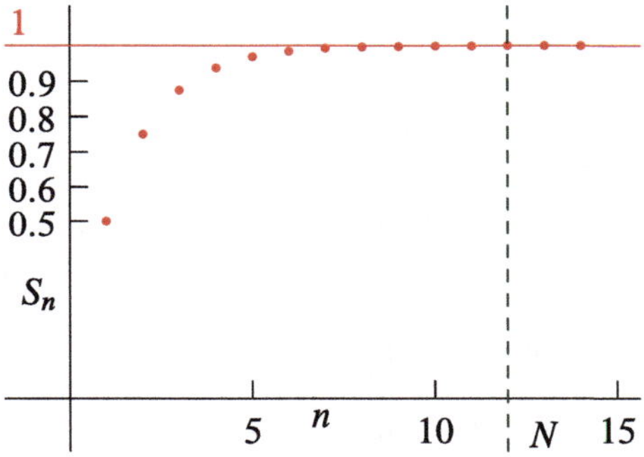

Fig. 2.33 Plot of S_n versus n

sequence (a_n) is a:

$\forall \epsilon > 0$	$\exists N \in \mathbb{N}$	such that	$\forall n > N$	$\|a_n - a\| < \epsilon$
however small epsilon is	there is a point in the sequence	such that beyond that point		all the terms are within epsilon of a

(2.22.3)

So, the small positive number ϵ was introduced to precisely quantify how a_n is close to the limit a. The number N was used to precisely state when n is large enough. The symbol ∀ means "for all" or "for any". The symbol ∃ means 'there exists'.

Now, we can understand why $1 = 0.9999\ldots$ Let

$$S_1 = 0.9, \quad S_2 = 0.99, \quad S_3 = 0.999$$

and so on. S_n will stand for the decimal that has the digit 9 occurring n times after the decimal point. Now, in the sequence $S_1, S_2, S_3, \ldots$, each number is nearer to 1 than the previous one, and by going far enough along, we can make the difference as small as we like. To see this, consider

physicist, "You mean that some Scottish sheep are black." "No," says the mathematician, "All we know is that there is at least one sheep in Scotland, and that at least one side of that one sheep is black!"

$$\begin{aligned}10S_1 &= 9 &&= 10 - 1\\10S_2 &= 9.9 &&= 10 - 0.1\\10S_3 &= 9.99 &&= 10 - 0.01\end{aligned}$$

And thus,

$$\lim_{n\to\infty} 10S_n = \lim_{n\to\infty}\left(10 - \frac{1}{10^{n-1}}\right) = 10 \Longrightarrow \lim_{n\to\infty} S_n = 1 \text{ or } 0.9999\ldots = 1$$

After this brief introduction, we begin with some examples about limit in Sect. 2.22.1. We shall soon realize that there is not much we can do with just the definition (of limit), thus we shall derive some rules that govern limit (Sect. 2.22.2).

2.22.1 Some Examples

Now that we know what is a limit of a sequence, the next thing is to play with some limits to get used to them. Here are a few limit exercises:

1. Prove that $\lim_{n\to\infty} \frac{1}{n} = 0$.
2. Prove that $\lim_{n\to\infty} \frac{1}{n^2} = 0$.
3. Prove that $\lim_{n\to\infty} \frac{1}{n(n-1)} = 0$.
4. Prove that $\lim_{n\to\infty} \frac{n^2}{n^2+1} = 1$.

First, we must ensure that we feel comfortable with the facts that all these limits (except the last one) are zeros. We do not have to make tables and graphs as we did before (as we can do that in our heads now). The first sequence is $1, 1/2, 1/3, \ldots, 1/1\,000\,000, \ldots$ and obviously the sequence converges to 0. For engineers and scientists that is enough, but for mathematicians, they need a proof, which is given here to introduce the style of limit proofs.

The proof is based on the definition of a limit, Eq. (2.22.3), of course. So, what is ϵ, and N? We can pick any value for the former, say $\epsilon = 0.0001$. To choose N we use $|a_n| < 0.0001$ or $1/n < 0.0001$. This occurs for $n > 10\,000$. So we have

$$\text{with } \epsilon = 0.0001, \text{ for all } n > N = 10\,000, \; |a_n| < \epsilon$$

Done! Not really, as $\epsilon = 0.0001$ is just one particular case, mathematicians are not satisfied with this proof; they want a proof that covers all the cases. If we choose $\epsilon = 0.00012$, then $1/\epsilon = 8\,333.333$, not an integer. In our case, we just need $N = 8\,334$. That is when the ceiling function comes in handy: $\lceil x \rceil$ is the least integer greater than or equal to x. If there is a ceiling, then there should be a floor; the floor function is $\lfloor x \rfloor$ which gives the greatest integer smaller than or equal to x.

Here is the complete proof. Let ϵ be *any*[77] small positive number and select N as the least integer greater than or equal to $1/\epsilon$ or $N = \lceil 1/\epsilon \rceil$ using the new ceiling function. Then, for $\forall n > N$, we have $1/n < 1/N < \epsilon$.

And we can prove the second limit in the same way. But, we will find it hard to do the same for the third and fourth limits. In this case, we need to find the rules or the behavior of general limits (using the definition) first, then we apply them to particular cases. Often it works this way. And it makes sense: if we know more about something we can have better ways to understand it. In calculus, we do the same thing: we do not find the derivative of $y = \tan x$ directly but via the derivative of $\sin x$ and $\cos x$ and the quotient rule.

2.22.2 *Rules of Limits*

We have sequences in our own hands, what we're going to do with them? We combine them: we add them, we multiply them, we divide them.[78] And when we do that we discover some laws, similarly with the way people discovered that $a + b = b + a$ if a, b are real numbers.

What is $\lim_{n\to\infty}(1 + 1/n)$? It is one, and it equals 1 plus $\lim_{n\to\infty} 1/n$ (which is zero). And if we see 1 as a sequence $(1, 1, 1, \ldots)$ then we have just discovered the rule stating that *the limit of the sum of two convergent sequences equals the sum of the two limits*. Let's write this formally. Considering two sequences (a_n) and (b_n) and they are convergent with limits a and b, respectively, then we guess that

$$\lim_{n\to\infty}(a_n + b_n) = \lim_{n\to\infty} a_n + \lim_{n\to\infty} b_n = a + b$$

Proof Let ϵ be any small positive number. As (a_n) converges to a, there exists N_1 such that $\forall n > N_1$, $|a_n - a| < \epsilon/2$ (why 0.5ϵ?). Similarly, as (b_n) converges to b, there exists N_2 such that $\forall n > N_2$, $|b_n - b| < \epsilon/2$. Now, let's choose $N = \max(N_1, N_2)$ (so that after N terms, both sequences converge to their corresponding limits), then $\forall n > N$, we have

$$\begin{cases} |a_n - a| < \dfrac{\epsilon}{2} \\ |b_n - b| < \dfrac{\epsilon}{2} \end{cases} \Longrightarrow \underbrace{|(a_n - a) + (b_n - b)| \le |a_n - a| + |b_n - b|}_{\text{triangle inequality}} < \frac{\epsilon}{2} + \frac{\epsilon}{2} = \epsilon$$

So we have $\forall n > N, |(a_n + b_n) - (a + b)| < \epsilon$. Thus, $\lim_{n\to\infty}(a_n + b_n) = \lim_{n\to\infty} a_n + \lim_{n\to\infty} b_n$. If we now reverse the proof particularly the red term, we understand why we selected 0.5ϵ in $|a_n - a| < \epsilon/2$. See also Fig. 2.34 for a better

[77] This is what mathematicians want.

[78] This is exactly what Diego Maradona—the great Argentina footballer—did. He kicks soccer balls. What he did when he saw a tennis ball? He kicks it! Watch this youtube video.

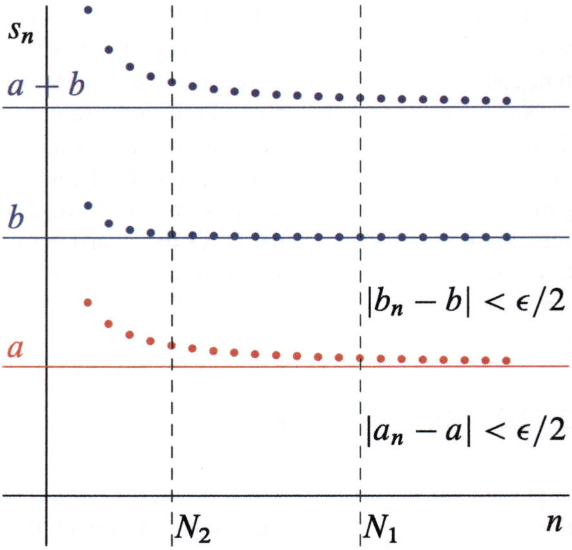

Fig. 2.34 Visual proof of the summation rule of limits

understanding of this proof. As can be seen, in limit proofs, we use extensively the triangle inequality. ■

We are now confident to state the other rules of limits below. The proofs are similar in nature as the proof for the summation rule, but some tricks are required. I refer to textbooks for those proofs.

$$\begin{aligned}
&\text{(a)}\ \lim_{n\to\infty}(a_n \pm b_n) = \lim_{n\to\infty} a_n \pm \lim_{n\to\infty} b_n \\
&\text{(b)}\ \lim_{n\to\infty}(c \cdot a_n) \quad = c \cdot \lim_{n\to\infty} a_n \\
&\text{(c)}\ \lim_{n\to\infty}(a_n b_n) \quad = (\lim_{n\to\infty} a_n)(\lim_{n\to\infty} b_n) \\
&\text{(d)}\ \lim_{n\to\infty}(a_n / b_n) \quad = (\lim_{n\to\infty} a_n)/(\lim_{n\to\infty} b_n)
\end{aligned} \tag{2.22.4}$$

Now equipped with more tools, we can solve other complex limit problems. For example,

$$\begin{aligned}
\lim_{n\to\infty}\frac{n^2}{n^2+1} &= \lim_{n\to\infty}\frac{1}{1+1/n^2} && \text{(algebra)} \\
&= \frac{\lim_{n\to\infty} 1}{\lim_{n\to\infty}(1+1/n^2)} && \text{(quotient rule)} \\
&= \frac{\lim_{n\to\infty} 1}{\lim_{n\to\infty} 1 + \lim_{n\to\infty} 1/n^2} && \text{(summation rule for the denominator)} \\
&= \frac{1}{1+0} = 1
\end{aligned}$$

We just needed to compute one limit: $\lim_{n\to\infty} 1/n^2$. The key step is the first algebraic manipulation step. The limit of n^2/n^2+1 is one, and what does it tell us? It tells us that when n is large, n^2 and $n^2 + 1$ are the same. In fact, if we replace 1 by $a < \infty$, the limit is still one.

2.23 Inverse Operations

Invert, Always Invert.
Carl Gustav Jacob Jacobi

Carl Gustav Jacob Jacobi, nineteenth century mathematician, using the phrase to describe how he thought many problems in math could be solved by looking at the inverse.

It is always a good habit for once in a while to stop doing what we're doing to ponder on the big picture. In this section, we look at carefully the operations we have discussed so far: addition, subtraction, multiplication, division, power, and root.

We start with three numbers a, b, c in which a, c are known. We want to find b such that $a + b = c$; this leads to $b = c - a$. Similarly, finding b such that $ab = c$ gives $b = c/a$. And for $b^a = c$, we get $b = \sqrt[a]{c}$. An example would help for the root operation: which number which is powered to 3 gives 8 (that is $x^3 = 8$). Of course $x = \sqrt[3]{8}$.

We summarize these operations below:

$$
\begin{array}{ll}
\text{(a) addition} & \text{(a}'\text{) subtraction} \\
\quad a + b = c & \quad b = c - a \\
\text{(b) multiplication} & \text{(b}'\text{) division} \\
\quad ab = c & \quad b = c/a \\
\text{(c) polynomial} & \text{(c}'\text{) root} \\
\quad b^a = c & \quad b = \sqrt[a]{c}
\end{array}
\tag{2.23.1}
$$

where the right column is the *inverse* of the operations in the left column. An inverse operation undoes the operation.[79] Starting with the number 2, pressing the x^2 button on a calculator gives you 4, and pressing $\sqrt{x}$ button (on 4) gives you back 2.

This is a powerful way to see subtraction, division and taking roots. For example, we do not have to worry about subtraction as a totally new operation; in fact subtraction is merely the inverse of addition. Later on, when you learn linear spaces, you will see that only addition is defined for linear spaces. This is because $5 - 3$ is simply $5 + (-3)$. Actually, we do inverse operations daily; for example when we put shoes on and take them off.

[79] If musicians can unbreak one's heart, mathematicians can too.

Table 2.16 Logarithm of a geometric progression is an arithmetic progression

x	1	2	4	8	16	32	64	128
$\log_2 x$	0	1	2	3	4	5	6	7

2.24 Logarithm

The question which number which is powered to 2 gives 4 (i.e.,, $x^2 = 4$) gave us the square root. And a similar question, to which index 2 is raised to get 4? (that is find x such that $2^x = 4$), gave us logarithm. We summarize these two questions and the associated operations now

$$\begin{aligned} x^2 &= 4 \Longrightarrow x = \sqrt[2]{4} \\ 2^x &= 4 \Longrightarrow x = \log_2 4 \end{aligned} \tag{2.24.1}$$

Looking at this, we can see that logarithm is not a big deal; it is just the inverse of 2^x in the same manner as square root is the inverse of x^2. For the notation $\log_2 4$ we read *logarithm base 2 of 4*.

You can understand these two equations by using a calculator. Starting with the number 2, pressing the x^2 button gives you 4, and pressing the $\sqrt{x}$ button (on 4) *gives you back 2*—that's why it is an inverse. Similarly, starting with 2, pressing the button 2^x yields 4 and pressing the button $\log_2 x$ returns 2. Historically, logarithm was discovered in an attempt to replace multiplication by summation as the latter is much easier than the former, see Sect. 2.24.3. It was invented by the Scottish mathematician, physicist, and astronomer John Napier (1550–1617) in early seventeenth century.[80]

2.24.1 Rules of Logarithm

After this new $\log_a b$ was discovered, we need to find the rules for them. If you play with them for a while, you will discover the rules for logarithms. One way is to start with the second equation in Eq. (2.24.1): $2^x = 4 \Longrightarrow x = \log_2 4$. For $x = \{0, 1, 2, 3, 4, 5, 6, 7\}$, 2^x yields a geometric progression (GP): 1, 2, 4, 8, 16, 32, 64, 128 (with $r = 2$); the corresponding logarithms (base 2) are an arithmetic progression (AP): 1, 2, 3, 4, 5, 6, 7 (see Table 2.16).

[80] The story is very interesting, see [30] for details. In 1590, James VI of Scotland sailed to Denmark to meet Anne of Denmark—his prospective wife and was accompanied by his physician, Dr John Craig. Bad weather had forced the party to land on Hven, near Tycho Brahe's observatory. Quite naturally, Brahe demonstrated to the party the process of using trigonometry identities to replace multiplication by summation. And Dr Craig happened to have a particular friend whose name is John Napier. With that Napier set out the task of his life: developing a method to ease multiplication. Twenty years later he had succeeded. And we have logarithm.

From this table, we see that $\log_2 4 = 2$, $\log_2 8 = 3$ and $\log_2 32 = 5$. There is a relation between these numbers: $\log_2 32 = \log_2 4 + \log_2 8$. And $32 = 4 \times 8$, thus $\log_2(4 \times 8) = \log_2 4 + \log_2 8$.

In the same manner, we observe that $\log_2 64/2 = \log_2 64 - \log_2 2$. By playing with them long enough, people (and you can too if you're given a chance) discovered the following rules for logarithm (p, m, n are positive integers):

$$
\begin{array}{llll}
\text{(a) Definition} & \log_a a^b = b & \text{(b) Inverse rule} & \log_a \frac{1}{b} = -\log_a b \\
\text{(c) Product rule} & \log_a bc = \log_a b + \log_a c & \text{(d) Quotient rule} & \log_a \frac{b}{c} = \log_a b - \log_a c \\
\text{(e) Power rule 1} & \log_a b^p = p \log_a b & \text{(f) Power rule 2} & \log_a b^{m/n} = \frac{m}{n} \log_a b
\end{array}
\tag{2.24.2}
$$

We are going to prove these rules. The first one $\log_a a^b = b$ is coming from the definition of logarithm $a^x = b \Longrightarrow x = \log_a b$. To prove the product rule, we first show a proof for a particular case $\log_2(4 \times 8)$, to get confident that the rule is correct and use this particular proof for a general proof.

It is obvious that $\log_2(4 \times 8) = 5$ because $2^5 = 32$. We can also proceed as follow[81]

$$\log_2(4 \times 8) = \log_2(2^2 \times 2^3) = \log_2(2^5) = 5 = 2 + 3 = \log_2 2^2 + \log_2 2^3$$

And thus we have proved the product rule for a concrete case of $a = 2$, $b = 4$ and $c = 8$. The key step in this proof was to rewrite $4 = 2^2$ and $8 = 2^3$, i.e.,, expressing 4 and 8 in terms of powers of 2. That is used in the following proof of the product rule $\log_a bc = \log_a b + \log_a c$:

Proof of the product rule Denote $b = a^x$ (thus $x = \log_a b$) and $c = a^y$ (hence $y = \log_a c$), then we can write

$$\log_a bc = \log_a a^x a^y = \log_a a^{x+y} = x + y = \log_a b + \log_a c$$

■

The product rule is for a product of two numbers, but we can generalize it to a product of any number of terms. For example, $\log_a xyz = \log_a(xy)z = \log_a(xy) + \log_a z = \log_a x + \log_a y + \log_a z$. From this, it is easy to see the power rule, when $x = y = z$.

Proof of the power rule 1 The proof of the power rule 1 uses the product rule (first consider the case p is a positive integer):

$$\log_a b^p = \log_a(\underbrace{b \times b \times \cdots \times b}_{p \text{ times}}) = \underbrace{\log_a b + \log_a b + \cdots + \log_a b}_{p \text{ times}} = p \log_a b$$

■

[81] Logarithm is the inverse of powers, so make sure you understand powers.

Table 2.17 Logarithm of a geometric progression is an arithmetic progression

x	$\frac{1}{4}$	$\frac{1}{2}$	1	2	4	8	16
$\log_2 x$	-2	-1	0	1	2	3	4

Table 2.18 Logarithms bases 2 and 3 of 3, 5, 6, 7 and their ratios (last row)

$\log_2 3$	$\log_2 5$	$\log_2 6$	$\log_2 7$
$\log_3 3$	$\log_3 5$	$\log_3 6$	$\log_3 7$
1.58496	1.58496	1.58496	1.58496

Interestingly, this rule also works when p is a negative integer, i.e.,, $p = -q$ and q is a counting number. To see that we need to observe that $\log_a 1/b = -\log_a b$. Why? See Table 2.17. This table is obtained from Table 2.16 by extrapolating what is true to the cases that we're not sure (and want to investigate). (We did this because we believe (again) in patterns.) In this table, if we go from right to left, $\log_2 x$ is reduced by one starting from 4, therefore after 0 we should get -1 and -2. So, we guess that $\log_2 1/4 = -\log_2 4$. Is it true? Yes, because $\log_2 1/4 = -2$ (as $2^{-2} = 1/4$). To prove $\log_a 1/b = -\log_a b$, observe that $0 = \log_a 1 = \log_a(b)(1/b) = \log_a b + \log_a(1/b)$.

The proof of the quotient rule uses the product rule and the power rule 1: $\log_a \frac{b}{c} = \log_a bc^{-1}$.

Proof Proof of the power rule 2 (with rational index). Setting $u = b^{m/n}$, then $u^n = b^m$. Thus,

$$\begin{aligned}\log_a u^n &= n \log_a u\\ \log_a b^m &= n \log_a b^{m/n} \quad (\text{use } u^n = b^m,\, u = b^{m/n})\\ m \log_a b &= n \log_a b^{m/n}\end{aligned}$$

■

It is often the case that we need to change the base of logarithm. Let's find the formula for that. The idea, as always, is to play with the numbers and find a pattern. So, we compute the logarithm with two bases (2 and 3) for some positive integers and put the results in Table 2.18. But hey we do not know how to compute, let say, $\log_3 5$! I was cheating here, I've used a calculator. We shall come back to this question shortly.

From this table, we can see that $\log_2 x/\log_3 x = \alpha$, where α is a constant. We aim to look for this constant. Let's denote $\log_2 x = y$, thus $x = 2^y$, then we can compute $\log_3 x$ in terms of y as

$$\log_3 x = \log_3 2^y = y \log_3 2 \Longrightarrow \log_3 x = \log_3 2 \log_2 x$$

We are a bit cheating here as we have used the power rule for logarithm $\log_a b^p = p \log_a b$ even when p is not a whole number (y is real here). Lucky for us, this rule is valid for the case p is real; but to show that we need calculus (see Chap. 4, Sect. 4.4.14). There is nothing special about a and b here, so we can generalize the above result to arbitrary bases a and b

$$\log_a x = \log_a b \times \log_b x, \quad \text{or} \quad \log_a b = \frac{\log_a x}{\log_b x}, \quad \text{or} \boxed{\log_a b \log_b x = \log_a x} \tag{2.24.3}$$

2.24.2 Some Exercises on Logarithms

Herein I present some exercises on logarithms to become fluent in this new operation. Find x such that $\log_8 2 = x$. We just need to use the definition of logarithm and rules of powers:

$$\log_8 2 = x \Longrightarrow 2 = 8^x = (2^3)^x = 2^{3x}$$

Now, we have $2 = 2^{3x}$, leading to $3x = 1$, the solution is $x = 1/3$. Let's solve another logarithm equation:

$$\log_{\sqrt{3}} \sqrt[3]{9} = x$$

Using the definition of logarithm we then have $(\sqrt{3})^x = \sqrt[3]{9}$. Using rules of powers, we rewrite this in the form $3^{(\dots)} = 3^{(\dots)}$, then we're done:

$$3^{\frac{x}{2}} = 3^{\frac{2}{3}} \Longrightarrow x = \frac{4}{3}$$

The next exercise is to compute $\log_3 1/243$. We need the inverse rule and note that $243 = 3^5$:

$$\log_3 \frac{1}{243} = -\log_3 243 = -\log_3 3^5 = -5$$

Finally, let's compute the following product:

$$P = (\log_2 3)(\log_3 4)(\log_4 5)(\log_5 6)\dots(\log_{31} 32)$$

Looks scary at first, but if you recall the boxed equation in Eq. (2.24.3), where the bases are cancelled, then P is nothing but $\log_2 32$, which is simply 5.

2.24.3 Why Logarithm Useful

In the early seventeenth century, due to colonization by the Europeans, world trade was really taking off. There was intense interest in astronomy, since this increased the chances of a ship coming back with its bounty. Clockmakers were also in great demand. All of this required more and more sophisticated calculation.

Activities like banking and trade resulted in huge volumes of calculation, and accuracy was essential. For example, compound interest and the distances of moons, planets, and stars involved a large number of multiplications and divisions. But this was very tedious and time-consuming, as well as being prone to error.

Surely there had to be a better way?

Logarithms were developed in the early seventeenth century by the Scotsman John Napier and the Englishman Henry Briggs (who later suggested base 10 rather than Napier's strange choice). Their ideas were refined later by Newton, Euler, John Wallis, and Johann Bernoulli towards the end of the seventeenth century.

When the idea of logarithm hit the scene in the early seventeenth century, its impact was substantial and immediate. Modern historians of mathematics, John Fauvel and Jan van Maanen, illustrate this vividly:

> When the English mathematician Henry Briggs learned in 1616 of the invention of logarithms by John Napier, he determined to travel the four hundred miles north to Edinburgh to meet the discoverer and talk to him in person.

A common argument for the use of technology is that it frees students from doing boring, tedious calculations, and they can focus attention on more interesting and stimulating conceptual matters. This is wrong. Mastering 'tedious' calculations frequently goes hand-in-hand with a deep connection with important mathematical ideas. And that is what mathematics is all about, is it not? We use calculators in a physics class or chemistry class, but not in a math class.

To show the usefulness of logarithm assume we have to compute this product $18\,793.26 \times 54\,778.18$ (without a calculator of course). Using logarithm turns this multiplication problem into a summation one (product rule):

$$\log_{10}(18\,793.26 \times 54\,778.18) = \log_{10} 18\,793.26 + \log_{10} 54\,778.18$$

Assume that we know the logs of 18 793.26 and 54 778.18 (we will come to how to compute them in a minute, Briggs provided tables for such values; which became obsolete with the born of modern calculators), then sum them to get A. Finally, the product we are looking for is then simply 10^A (there were/are tables for this and thus we obtain the product just by summing two numbers).

2.24.4 How Henry Briggs Calculated Logarithms in 1617

In 1617, Briggs published a table of logarithms (base 10) followed in 1624 by the more complete Arithmetica Logarithmica. Briggs is viewed by Goldstine[82] as one of the great figures of numerical analysis.

[82] Herman Heine Goldstine (1913–2004) was a mathematician and computer scientist, who worked as the director of the IAS machine at Princeton University's Institute for Advanced Study, and helped

Table 2.19 Successive roots of 10: 10^s or $\sqrt[2^s]{10}$

n	$s = 1/2^n$	$10^s = 1 + \epsilon$	ϵ	s/ϵ	$\frac{10^s-1}{s}$
0		10			
1		3.16227766			
2		1.77827941			
3		1.33352143			
4		1.15478198			
5		1.15478198			
6		1.07460783			2.38745051
7		1.03663293			2.34450742
8		1.01815172			2.32342038
9		1.00903504			2.31297148
10	0.00097656	1.00225115	0.00225115	0.43380638	2.30777050
11	0.00048828	1.00112494	0.00112494	0.43405039	2.30517585
12	0.00024414	1.00056231	0.00056231	0.43417242	2.30387999
13	0.00012207	1.00028112	0.00028112	0.43423345	2.30323242
⋮					
20	9.53674316e–7	1.00000219	0.00000219	0.434294005	2.30258762

Here is what Briggs did, without a calculator.[83] He calculated the successive square roots of 10, i.e.,, $\sqrt{10} = 10^{1/2}$, then $\sqrt{\sqrt{10}} = 10^{1/4}$ etc. We denote this by 10^s with $s = 1/2^n$ ($n = 0, 1, 2, \ldots$) and put them in the third column of Table 2.19. Briggs might have used the algorithm described in Eq. (2.9.1) for this task. We of course used a calculator (as we're not interested in the square root itself herein).

From the third column in Table 2.19, we can see that successive square roots of 10 will be of the form $1 + \epsilon$ where ϵ is a very small positive number. We put ϵ in the fourth column of Table 2.19. Another observation is that the logarithm (base 10) of the number in the third column (which is the second column) is proportional to ϵ; looking at the second column, rows 10 and 11, s is decreased by half, and the corresponding values in the fourth column are decreasing half as well. We can find this proportion by calculating s/ϵ (fifth column).

These two observations allow us to write for a positive number x (not just 10)

$$x^{1/2^n} = 1 + \epsilon, \quad \log(1 + \epsilon) \approx \alpha\epsilon \tag{2.24.4}$$

where $\alpha = 0.43429448190325$ (later on we know that $\alpha = \log_{10} e$, $e = 2.718281$ is the famous number discussed later in Sect. 2.28). And thus, by taking the logarithm

to develop ENIAC, the first of the modern electronic digital computers. He subsequently worked for many years at IBM as an IBM Fellow, the company's most prestigious technical position.

[83] I learned of this in Feynman's Lectures on Physics, Vol I.

of both sides of the first equation in Eq. (2.24.4) and using the second equation, we get

$$\frac{1}{2^n}\log x = \log(1+\epsilon) \approx \alpha\epsilon \Longrightarrow \boxed{\log x \approx 2^n\alpha(x^{1/2^n}-1)} \tag{2.24.5}$$

In summary, Brigg's algorithm for logarithm of x is to calculate successive square roots of x, minus 1, multiplied by α and 2^n. Quite simple, but one question arises: what should be the value for n? It should be large enough so that the approximation $\log(1+\epsilon) \approx \alpha\epsilon$ holds, but it must be small to reduce the numerical error in the calculation of $(x^{1/2^n}-1)$.

With this algorithm, Briggs computed the logarithms of all *prime numbers* smaller than 100; there are only 25 such primes. From this, the logarithms of composite numbers are simply the sum of the logarithms of their prime factors. For example, $\log 21 = \log(3 \times 7) = \log 3 + \log 7$.

Another observation from Brigg's calculations is that, for a small x, we have

$$10^x \approx 1 + kx \tag{2.24.6}$$

which can be seen from the sixth column of Table 2.19. And we have $k = 1/\alpha$. With calculus, we will know that $k = \ln 10$ ($\ln x$ is the logarithm of base e).

History note 2.5: Henry Briggs (1561–1630)

Henry Briggs was an English mathematician notable for changing the original logarithms invented by John Napier into common (base 10) logarithms, which are sometimes known as Briggsian logarithms in his honor. In 1624 his Arithmetica Logarithmica was published, in folio, a work containing the logarithms of thirty thousand natural numbers to fourteen decimal places (1-20,000 and 90,001 to 100,000). Briggs's early research focused primarily on astronomy and its applications to navigation, and he was among the first to disseminate the ideas of the astronomer Johannes Kepler (1571–1630) in England.

2.24.5 *Solving Exponential Equations*

Let's consider the following 'basic' exponential equations:

1. Solving the following equation

$$4^x - 3 \cdot 2^x + 2 = 0$$

2. Solving the following equation

$$4^x + 6^x = 9^x$$

The first one is simply a quadratic equation in disguise ($t^2 - 3t + 2 = 0$ with $t = 2^x$). Here, we use the rule of raising a power saying that $(a^m)^n = a^{mn} = (a^n)^m$. For the second one, we divide the equation by 4^x:

$$1 + \left(\frac{3}{2}\right)^x = \left(\frac{3}{2}\right)^{2x}$$

Then, it is a disguised quadratic equation. Of course, our solution method will not work if, for example, number 4 was replaced by 5 in this equation! Can you find the rules that high school teachers used (probably) to generate this kind of equation? It is $(ab)^x + (b^2)^x = (a^2)^x$, for a, b being positive integers.

Ok, let's solve non-standard exponential equations; they are more fun. One such equation is the following:

$$16^{\frac{x-1}{x}} \cdot 5^x = 100$$

We know that logarithm of a product is the sum of logarithms, and sum is easier to deal with. So, we take logarithm of both sides of the equation: (we do not know which base is the best, so we use a for that)

$$\begin{aligned}
&\frac{x-1}{x} \log_a 16 + x \log_a 5 = \log_a 100 \\
\Longleftrightarrow\ & 4\left(\frac{x-1}{x}\right) \log_a 2 + x \log_a 5 = 2 \log_a 10
\end{aligned}$$

Looking at the red numbers, you see that they are related: 5=10/2. If we pick $a = 10$, we can get nice numbers:

$$\begin{aligned}
&4\left(\frac{x-1}{x}\right) \log_{10} 2 + x \log_{10} \frac{10}{2} = 2 \\
\Longleftrightarrow\ & 4\left(\frac{x-1}{x}\right) \log_{10} 2 + x(1 - \log_{10} 2) = 2 \\
\Longleftrightarrow\ & (1 - \log_{10} 2)x^2 + (4 \log_{10} 2 - 2)x - 4 \log_{10} 2 = 0
\end{aligned}$$

Finally, we get a quadratic equation in terms of x, even though the coefficients are a bit scary. Don't worry, this is an exercise, the answers are usually of a compact form. So, using the quadratic formula, we have:

$$x = \frac{2 - 4\log_{10} 2 \pm 2}{2(1 - \log_{10} 2)} = \begin{cases} 2 \\ \dfrac{4\log_{10} 2}{2\log_{10} 2 - 2} = -\dfrac{\log_{10} 4}{\log_{10} 5} \end{cases}$$

That's it! We used the fundamental property of logarithm to get a quadratic equation. If the numbers 16,5,100 are replaced by others, then still we have a quadratic equation.

Can we have another solution, easier? Yes, if we divide the original equation by 100, factor $100 = 4 \cdot 5^2$. That's the key point as 4 and 5 appear in $16^{x-1/x} \cdot 5^x$. Don't forget that $16 = 4^2$. After that we take logarithm base 10:

$$\begin{aligned} \frac{16^{\frac{x-1}{x}} \cdot 5^x}{100} = 1 &\iff \frac{16^{\frac{x-1}{x}} \cdot 5^x}{4 \cdot 5^2} = 1 \\ &\iff 4^{\frac{x-2}{x}} \cdot 5^{x-2} = 1 \\ &\iff \left(\frac{x-2}{x}\right)\log_{10} 4 + (x-2)\log_{10} 5 = 0 \\ &\iff (x-2)(\log_{10} 4 + x\log_{10} 5) = 0 \end{aligned}$$

No need to use the quadratic formula.

So far so good, but how about the following equation?

$$2^x + 2^{1/x} = 4$$

Well, this is non-standard: we cannot transform it into an algebraic equation. In fact, we have to use something rather special: using the AM-GM inequality, Eq. (2.20.2), is the key as the LHS is always greater than or equal to 4. Make sure you understand how to use the AM-GM inequality for the LHS. The solution is when $2^x = 2^{1/x}$ or $x = 1/x$ or $x = 1$. If the RHS is 5 instead of 4, then we have to use the graphic method (plot the function of the LHS and see where it intersects with the horizontal line $y = 5$) or Newton's method.

Some exercises on non-standard exponential equations.

1. Solving the following equation

$$3^{\frac{x+2}{3x-4}} - 7 = 2 \cdot 3^{\frac{5x-10}{3x-4}}$$

2. Solving the following equation

$$2^x = 3^{\frac{x}{2}} + 1$$

3. Solving the following equation

$$4^{2x} + 2^{-x} + 1 = (129 + 8\sqrt{2})(4^x + 2^{-x} - 2^x)$$

4. Solving the following equation

$$4^x + 9^x + 25^x = 6^x + 10^x + 15^x$$

Solution of the first two equations is $x = 2$. In the first equation, pay attention to the exponents, they're related! In the second one, it is easy to see $x = 2$ is one solution. You need to prove it's the only solution. For the fourth equation, pay attention to the red numbers: on the LHS we have $4 = 2^2$, $9 = 3^2$ and $25 = 5^2$. And on the RHS we have $6 = 2 \cdot 3$, $10 = 2 \cdot 5$ and $15 = 3 \cdot 5$. Thus, all we have are numbers 2, 3, 5: squares of them and products of them. This leads to $a^2 + b^2 + c^2 = ab + bc + ca$. The answer is $x = 0$.

2.25 Complex Numbers

Bombelli's insight into the nature of the Cardano formula broke the mental logjam concerning $\sqrt{-1}$. His work made clear that manipulating $\sqrt{-1}$ using ordinary arithmetic results in perfectly correct results (i.e.,, real numbers come out of expressions involving $\sqrt{-1}$). Despite the success of Bombelli, there still lacked a physical interpretation of $\sqrt{-1}$. Mathematicians of the sixteenth century were tied to the Greek tradition of geometry, and they felt uncomfortable with concepts to which they could not give a geometric meaning (so that they can see it).

To read this section you need some knowledge about trigonometry. Chapter 3 provides a discussion of this topic. You might need to check it before continuing.

We start with a definition of complex numbers and how arithmetic of them is done in Sect. 2.25.1. We then discuss the powers of complex numbers with the well-known de Moivre formula (Sect. 2.25.2). After that, root of complex numbers is treated, because root is simply power with a fractional exponent (Sect. 2.25.3). It is the square root of -1 that led us to i and to complex numbers, we now want to check what the square root of i brings to the table (Sect. 2.25.4). Complex numbers provide an easy way to derive many trigonometric identities, and Sect. 2.25.5 is about this. What is the power of a real number with a complex exponent? This question gives us the most beautiful equation in mathematics: Euler's identity $e^{i\alpha} = \cos\alpha + i \sin\alpha$ (Sect. 2.25.6). What is i^i? Is it a real number or a complex number; these questions are studied in Sect. 2.25.7. Finally, a summary of our system of numbers from integers, rationals, reals to complex numbers is given in Section 2.25.8.

2.25.1 Definition and Arithmetic of Complex Numbers

The idea of a complex number as a point in the complex plane was first described by Caspar Wessel in 1799, although it had been anticipated as early as 1685 in Wallis's A Treatise of Algebra. Wessel's memoir appeared in the Proceedings of the Copenhagen Academy but went largely unnoticed. In 1806, Jean-Robert Argand[84] independently issued a pamphlet on complex numbers and provided a rigorous proof of the fundamental theorem of algebra. Carl Friedrich Gauss had earlier published an essentially topological proof of the theorem in 1797 but expressed his doubts at the time about "the true metaphysics of the square root of -1". It was not until 1831 that he overcame these doubts and published his treatise on complex numbers as points in the plane, largely establishing modern notation and terminology.[85]

In the complex plane, the horizontal axis represents the real (Re) part and the vertical axis represents the imaginary (Im) part. Thus, the number $1 + 2i$ is a point obtained by walking on the real axis one step and then walking along the imaginary two steps upward.

Definition 2.25.1 A complex number z is the one given by $z = a + bi$ where a and b are real numbers and $i = \sqrt{-1}$—the imaginary unit; a is called the real part, and b is called the imaginary part. Geometrically, a complex number is a point in a complex plane, shown in Fig. 2.35.

The adjective *complex* in complex numbers indicates that complex numbers have more than one part, rather than complicated.

As a new number, we need to define arithmetic rules for complex numbers. We first list the rules for addition/subtraction and multiplication as follows

$$\begin{aligned}(a_1 + b_1 i) + (a_2 + b_2 i) &= (a_1 + a_2) + (b_1 + b_2)i \\ (a_1 + b_1 i) - (a_2 + b_2 i) &= (a_1 - a_2) + (b_1 - b_2)i \\ (a_1 + b_1 i)(a_2 + b_2 i) &= a_1 a_2 - b_1 b_2 + (a_1 b_2 + a_2 b_1)i\end{aligned} \tag{2.25.1}$$

How these rules were defined? It depends. In the first way, we can assume that the rule of arithmetic for ordinary numbers also apply for complex numbers, then there is no mystery behind Eq. (2.25.1): we treat i as an ordinary number and whenever we see i^2 we replace that by -1 (hence $i^3 = i^2 \times i = -i$). In the second way, one first defines the addition and multiplication of two vectors. The rule for addition follows the rule of vector addition (known since antiquity from physics), see Fig. 2.36a. It was Wessel's genius to discover/define the multiplication of two vectors: the resulting vector has a length being the product of the lengths of the two vectors

[84] Jean-Robert Argand (1768–1822) was an amateur mathematician. In 1806, while managing a bookstore in Paris, he published the idea of geometrical interpretation of complex numbers known as the Argand diagram and is known for the first rigorous proof of the Fundamental Theorem of Algebra.

[85] Gauss was a star thus people would be more willing to accept his theory than that proposed by Wessel and Argand who were relatively unknown.

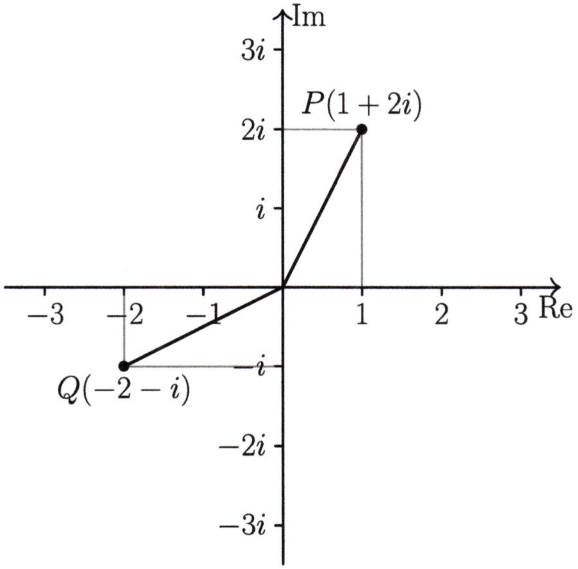

Fig. 2.35 Complex plane

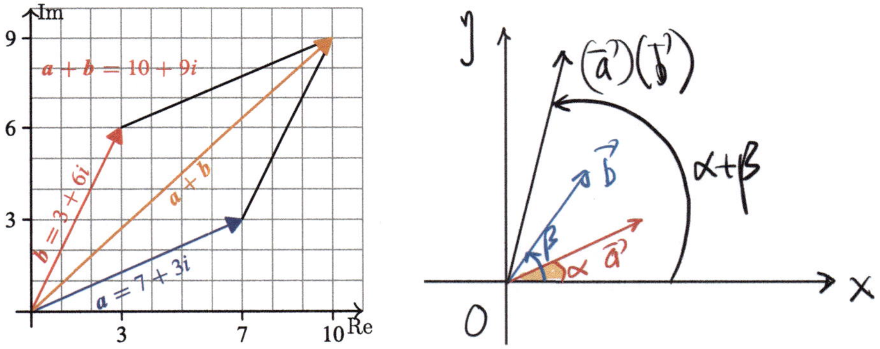

(a) addition (b) multiplication

Fig. 2.36 Addition and multiplication of complex numbers

and a direction being the sum of the direction of the two vectors (with respect to a horizontal line), see Fig. 2.36b. How he got this multiplication rule? As I am not good at geometry I do not want to study his solution. But do not worry, with a new way to represent points on a plane, his rule reveals its mystery to us!

For a point on a plane, there are many ways to define its location. We have used the Cartesian coordinates so far, but we can also use polar coordinates (r, θ); refer to Sect. 4.13.1 if you do not know what polar coordinates are. Using polar coordinates leads to the so-called *polar form* of complex numbers. This polar form is easy to obtain by just relating the Cartesian coordinates (a, b) to (r, θ) using the trigonometric functions: $a = r\cos\theta$ and $b = r\sin\theta$. The polar form of a complex number $z = a + bi$ is given by

Fig. 2.37 .

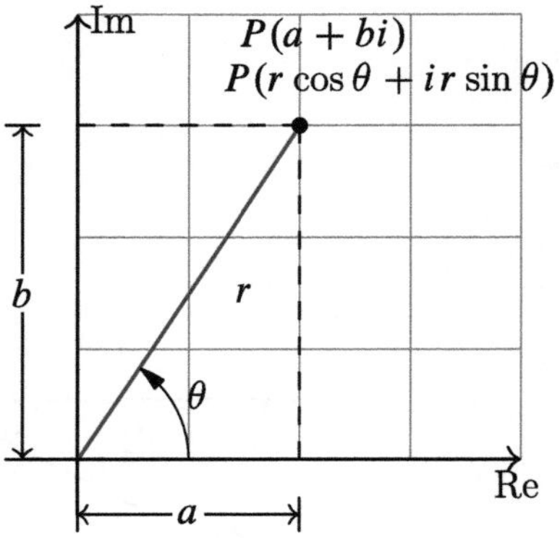

$$z = r(\cos\theta + i\sin\theta) \tag{2.25.2}$$

where $r = \sqrt{a^2 + b^2}$ is called the modulus of z and $\tan\theta = a/b$, θ is the argument of the complex number, see Fig. 2.37. More compactly, people also write $z = r\angle\theta$.

Using the polar form, the multiplication of two complex numbers $z_1 = r_1(\cos\alpha + i\sin\alpha)$ and $z_2 = r_2(\cos\theta + i\sin\theta)$ is written as

$$\begin{aligned} z_1z_2 &= r_1(\cos\alpha + i\sin\alpha) \times r_2(\cos\theta + i\sin\theta) \\ &= r_1r_2[(\cos\alpha\cos\theta - \sin\alpha\sin\theta) + i(\sin\theta\cos\alpha + \sin\alpha\cos\theta)] \\ &= r_1r_2[\cos(\alpha+\theta) + i\sin(\alpha+\theta)] \end{aligned} \tag{2.25.3}$$

From which the geometry meaning of multiplication of two complex numbers is obtained, effortlessly and without any geometric genius insight! With Euler's identity $e^{i\theta} = \cos\theta + i\sin\theta$ (see Sect. 2.25.6),[86] it is even easier to see the geometric meaning of complex number multiplication:

$$\left.\begin{aligned} z_1 &= r_1(\cos\alpha + i\sin\alpha) = r_1e^{i\alpha} \\ z_2 &= r_2(\cos\beta + i\sin\beta) = r_2e^{i\beta} \end{aligned}\right\} \Longrightarrow z_1z_2 = r_1r_2e^{i(\alpha+\beta)}$$

Now we can understand why complex numbers live in the complex plane given in Fig. 2.35. The question is always where $\sqrt{-1}$ lives? Let's represent $\sqrt{-1}$ as $r\angle\theta$ with unknown length and unknown angle. What Wessel know? He defined the multiplication of two vectors, so he used it:

$$(\sqrt{-1})(\sqrt{-1}) = r^2\angle 2\theta \Longleftrightarrow -1 = r^2\angle 2\theta$$

[86] We have anticipated that there must be a link between i and sine/cosine, but we could not expect that e is involved. To reveal this secrete we need the genius of Euler. Refer to Sect. 2.28 for what e is.

But we know where -1 stays; left to the origin at a distance of one. In other words, $-1 = 1\angle 180°$, thus:

$$1\angle 180° = r^2\angle 2\theta \implies \begin{cases} r = & 1 \\ \theta = & 90° \end{cases}$$

And thus $\sqrt{-1}$ is on an axis perpendicular to the horizontal axis and at a unit distance from the origin, here stays $\sqrt{-1}$ which is now designated by the iconic symbol i (standing for imaginary):

$$\boxed{i := \sqrt{-1} = 1\angle 90°}$$

But that is just one i, if we go one around (or a any number of rounds) starting from i we get back to it. So,

$$i = i \sin\left(\frac{\pi}{2} + k2\pi\right), \quad k \in \mathbb{N} \tag{2.25.4}$$

A nice problem about complex numbers. For $j = e^{i\frac{2\pi}{3}}$, let's compute the following product:

$$P = (1 + j)(1 + j^2)(1 + j^3)\cdots(1 + j^{2023})$$

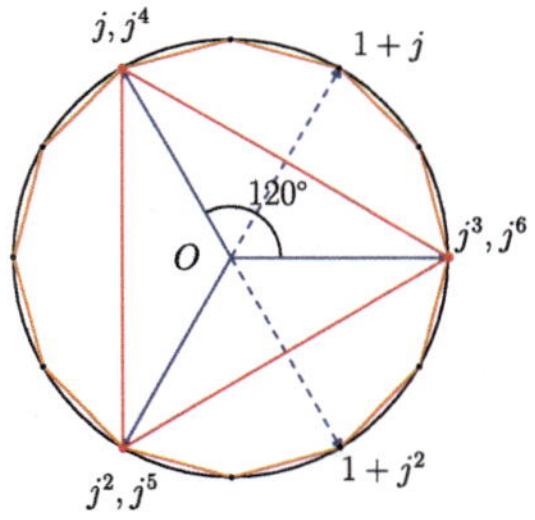

The first thing to do is to notice that j is a point on the complex plane; it is on the unit circle, with an argument of 120°. Then, $(j, j^4, \ldots)$, $(j^2, j^5, \ldots)$ and $(j^3, j^6, \ldots)$ are three vertices of an isolescent triangle (see figure). Thus, we have

$$1 + j = 1 + j^4 = \cdots = e^{i\frac{\pi}{3}}, \quad 1 + j^2 = 1 + j^5 = \cdots = e^{-i\frac{\pi}{3}}$$

And,

$$1 + j^3 = 1 + j^6 = \cdots = 2 \iff 1 + j^{3k} = 2, \ k = 1, 2, \ldots$$

So, $1 + j$ and $1 + j^2$ are complex conjugated, thus its product is a real number; actually it is much nicer: $(1 + j)(1 + j^2) = 1$, and there are 674 terms $1 + j^3, 1 + j^6, 1 + j^9, \ldots$ all being 2, thus $P = 2^{674}$. Here are the detail:

$$P = \underbrace{(1 + j)(1 + j^2)}_{1}\underbrace{(1 + j^3)}_{2}\underbrace{(1 + j^4)(1 + j^5)}_{1}\underbrace{(1 + j^6)}_{2}\cdots(1 + j^{2023})$$

Question 4 *If i rotates a vector in the complex plane, then what will rotate a vector in a 3D space? This was the question that led the Irish mathematician William Hamilton (1805–1865) to the development of quartenion, to be discussed in Sect. 9.1.6.*

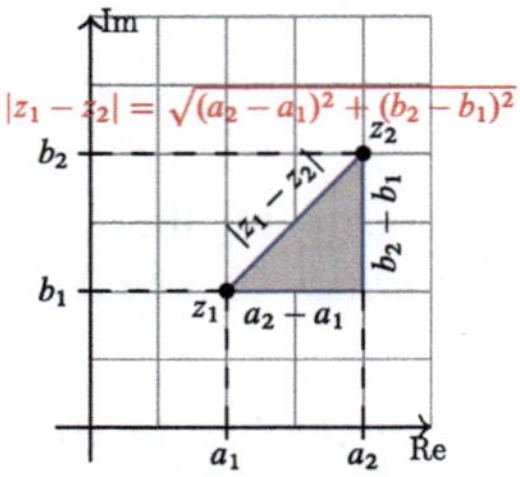

The absolute value of a complex number. In Sect. 2.20.4, we have met the absolute value of a real number x, denoted by $|x|$, which is the distance from x to zero. We extend this concept to complex numbers now. The absolute value of a complex number $z = a + bi$, denoted by $|z|$, is the distance from z to the origin (i.e.,, the point of coordinates $(0, 0)$). Obviously $|z| = r = \sqrt{a^2 + b^2}$ (Fig. 2.37). Usually we're interested in the distance between two complex numbers $z_1 = a_1 + b_1 i$ and $z_2 = a_2 + b_2 i$. The Pythagorean theorem (for the shaded right triangle) gives us that distance: it is simply $\sqrt{(a_2 - a_1)^2 + (b_2 - b_1)^2}$. And it is also $|z_1 - z_2| = |z_2 - z_1|$; the second expression is to show that as the distance between two points is the same if we measure from one point or the other, the distance formula must be *symmetric* with respect to z_1, z_2.

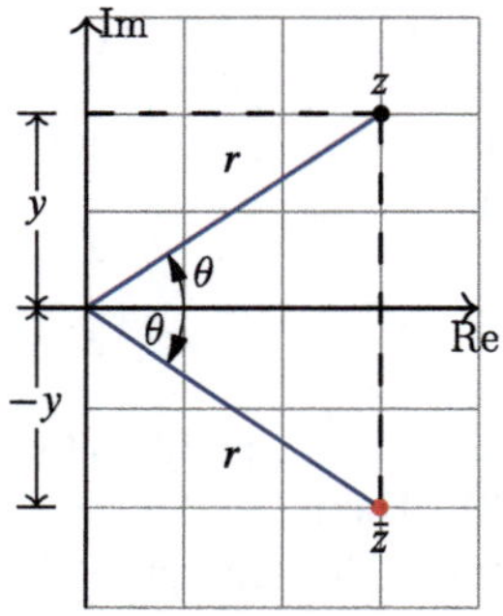

Complex conjugate. If we multiply two complex numbers we get a complex number. But that is not entirely true: $(x + yi)(x - yi) = x^2 + y^2$, which is a real number. For any complex number $z = x + yi$, there is a related complex number of the form $\bar{z} = x - yi$ with the special property that $z\bar{z}$ is a real number. We now define the concept of complex conjugate. The *complex conjugate* of a complex number is the number with an equal real part and an imaginary part equal in magnitude but opposite in sign (see figure). That is, (if x and y are real) then the complex conjugate of $x+yi$ is equal to $x-yi$. The complex conjugate of z is often denoted as $\bar{z}$ (read as z bar). In polar form, the conjugate of $re^{i\phi}$ is $re^{-i\phi}$, which can be shown using Euler's formula. The product of a complex number and its conjugate is a real number $(x + yi)(x - yi) = x^2 + y^2$. In other words, $|z\bar{z}| = |z|^2$.

Below is a summary of some of the properties of the conjugates. Proofs just follow the definition of conjugate.

(a) The complex conjugate of a complex conjugate of z is z: $\overline{\overline{z}} = z$
(b) The complex conjugate of a sum is the sum of the conjugates: $\overline{z+w} = \overline{z} + \overline{w}$
(c) The complex conjugate of the product is the product of the conjugates: $\overline{zw} = \overline{z}\overline{w}$
(d) z is a real number if and only if $\overline{z} = z$.

Since $\overline{\overline{z}} = z$, $|z|$ is called *an involution*. An involution, involutory function, or self-inverse function *is a function f that is its own inverse*. We will see that rules (b) and (c) apply to many mathematical objects.

2.25.2 de Moivre's Formula

Knowing how to multiply complex numbers, we can now get powers of a complex number z. Why bothering with this? Just curiosity. We have been playing with the powers of natural numbers and real numbers. And now we have a new toy (complex number), it is logical to try the old rules with this new kid. More often, interesting things come out (in this case, many useful trigonometric identities can be derived). For example, the two and three powers of $z = r(\cos\alpha + i\sin\alpha)$ are

$$\begin{aligned} z^2 &= zz = r^2[\cos(2\alpha) + i\sin(2\alpha)] \\ z^3 &= z^2 z = r^3[\cos(3\alpha) + i\sin(3\alpha)] \end{aligned} \tag{2.25.5}$$

which can be generalized to $z^n = r^n[\cos(n\alpha) + i\sin(n\alpha)]$ where n is any positive integer. When $r = 1$ this formula is simplified to:

$$\boxed{(\cos\alpha + i\sin\alpha)^n = \cos(n\alpha) + i\sin(n\alpha)} \tag{2.25.6}$$

which is a highly useful formula, known as de Moivre's formula (also known as de Moivre's theorem and de Moivre's identity), named after the French mathematician Abraham de Moivre (1667–1754). Refer to Sect. 2.25.6 to see how it leads to the famous Euler's identity: $e^{i\pi} + 1 = 0$.

It is obvious that the next thing to do is to consider negative powers, e.g., z^{-2}. To do so, let's start simple with z^{-1} which can be computed straightforwardly. We have $z = a + bi = r(\cos\theta + i\sin\theta)$. We can compute z^{-1} using algebra as[87]:

$$z^{-1} = \frac{1}{z} = \frac{1}{a+bi} = \frac{a-bi}{a^2+b^2} = \frac{1}{r}(\cos\theta - i\sin\theta)$$

Thus, we get

[87] If not clear about the last step, see Fig. 2.37, and noting that $a^2 + b^2 = r^2$, and so on.

$$[r(\cos\theta + i\sin\theta)]^{-1} = \frac{1}{r}(\cos\theta - i\sin\theta)$$

which shows that de Moivre's formula still works for $n = -1$.

Alright, we're ready to compute any negative power of a complex number. For example, z^{-2} is given by

$$z^{-2} = (z^{-1})^2 = \frac{1}{r^2}(\cos\theta - i\sin\theta)^2 = \frac{1}{r^2}(\cos 2\theta - i\sin 2\theta) \tag{2.25.7}$$

Now, we're confident that de Moivre's formula holds for any integer. If you want to prove it you can use proof by induction.

2.25.3 Roots of Complex Numbers

Having computed powers of a complex number, it is natural to consider its roots, i.e.,, powers with fractional exponents. The idea is again to use Eq. (2.25.6). But we consider a complex number written as $z = \cos{}^{\alpha}/_{m} + i\sin{}^{\alpha}/_{m}$, for m being a positive integer, then Eq. (2.25.6) gives

$$\left(\cos\frac{\alpha}{m} + i\sin\frac{\alpha}{m}\right)^m = \cos(\alpha) + i\sin(\alpha) \tag{2.25.8}$$

which immediately gives us the formula to compute the m-th root of any complex number

$$\sqrt[m]{r(\cos(\alpha) + i\sin(\alpha))} = \sqrt[m]{r}\left(\cos\frac{\alpha}{m} + i\sin\frac{\alpha}{m}\right) \tag{2.25.9}$$

This is sometimes also referred to as de Moivre's formula.

As the first application of this new formula, we use Eq. (2.25.9) to prove that $\sqrt[3]{2+\sqrt{-121}} = 2 + i$. Why this particular number? It is famous! Still remember Cardano, Bombelli and the cubic equation? This number appeared in Eq. (2.13.9), that led mathematicians into the investigation of imaginary i and ultimately resulted in complex numbers.

Proof First, we write the number under the cube root in polar form of a complex number, then we use Eq. (2.25.9) to get the answer[88]:

$$z = 2 + \sqrt{-121} = 2 + 11i = 11.18034(\cos 1.39094283 + i\sin 1.39094283)$$

$$z^{1/3} = \sqrt[3]{11.18034}(\cos 0.46364761 + i\sin(0.46364761) = 2 + i$$

■

[88] Note that $-121 = 11^2 i^2$, thus $\sqrt{-121} = 11i$.

Doing the same thing, we can also see that $\sqrt[3]{2-\sqrt{-121}} = 2 - i$. But $2 - i$ is the conjugate of $2 + i = \sqrt[3]{2+\sqrt{-121}}$. That is, the conjugate of the kth root is the kth root of the conjugate.

Fifth roots of unity. As another application of Eq. (2.25.9), we are going to compute the fifth roots of unity (one). There are five of them thanks to the fundamental theorem of algebra. We also do that using algebra, and demonstrate that the two approaches yield identical results. First, we write $1 = \cos 2\pi k, k = 0, 1, 2, \ldots$ Then, Eq. (2.25.9) provides the fifth roots:

$$\sqrt[5]{1} = \sqrt[5]{\cos 2\pi k} = \cos\frac{2\pi k}{5} + i\sin\frac{2\pi k}{5}$$

Thus, the four fifth roots of 1 are (note that $k = 0$ gives the obvious answer of 1)

$$\begin{aligned}
k = 1 &: \cos\frac{2\pi}{5} + i\sin\frac{2\pi}{5} = 0.309017 + 0.9510565i \\
k = 2 &: \cos\frac{4\pi}{5} + i\sin\frac{4\pi}{5} = -0.809017 + 0.5877853i \\
k = 3 &: \cos\frac{6\pi}{5} + i\sin\frac{6\pi}{5} = -0.809017 - 0.5877853i \\
k = 4 &: \cos\frac{8\pi}{5} + i\sin\frac{8\pi}{5} = 0.309017 - 0.9510565i
\end{aligned} \tag{2.25.10}$$

As can be seen, these four roots are complex numbers, and together with the real root (i.e.,, 1) they are vertices of a pentagon inscribed in the unit circle,[89] see Fig. 2.38a. And this is true for any nth roots of unity (see Fig. 2.38b for $\sqrt[4]{1}$). What else can we say about them? Among these four complex roots, two are in the upper half of the circle, and the other twos are in the bottom half: they are the conjugates of the ones in the upper half. In Sect. 2.29.2 a proof is provided.

Question 5 *Eq. (2.25.10) are the fifth roots of unity expressed in a transcendental form (because of trigonometric functions). Can you express the fifth roots of unity algebraically? That is writing them using only the four fundamental arithmetic operations* $+, -, \times, \div$ *and extraction of roots. For example, writing* $\cos 2\pi/5 = {}^{-1+\sqrt{5}}/_{4}$ *is writing it algebraically. Why bother? Note that mathematicians are not engineers or scientists, they do not care if* $\cos 2\pi/5 = 0.30901$. *For them,* $\cos 2\pi/5$ *is* ${}^{-1+\sqrt{5}}/_{4}$.

We can also find these roots using algebra. To do so, we solve the equation $z^5 - 1 = 0$, which is equivalent to solving $(z-1)(z^4 + z^3 + z^2 + z + 1) = 0$ thanks to geometric series Eq. (2.21.6).[90] So, we have

---

[89] Why? The unit circle is $x^2 + y^2 = 1$, and the point $\cos\frac{2\pi}{5} + i\sin\frac{2\pi}{5}$ has $x = \cos 2\pi/5$ and $y = \sin 2\pi/5$ satisfying $x^2 + y^2 = 1$.

[90] If not clear, Eq. (2.21.6) gives us: $1 + z + z^2 + z^3 + z^4 = {}^{z^5-1}/_{z-1}$.

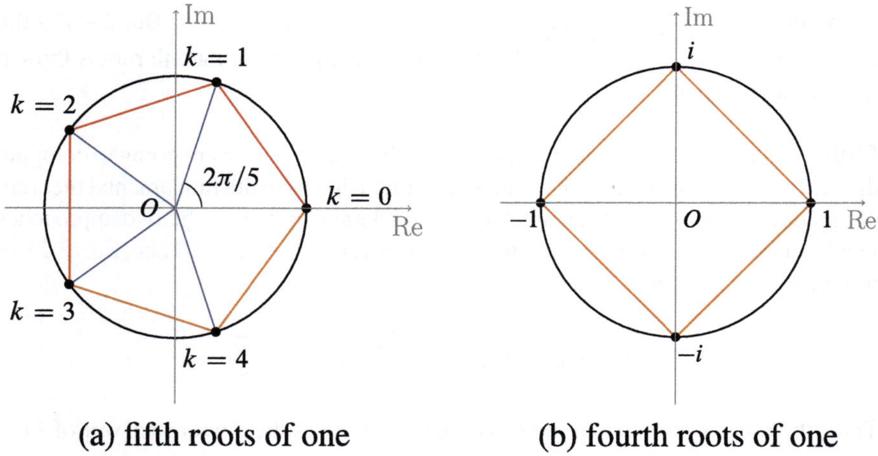

Fig. 2.38 The fifth roots of one—solutions of $z^5 = 1$—are vertices of a pentagon inscribed in the unit circle. The four complex roots, come in pairs: z_1 and z_4 are complex conjugated and similarly for z_2, z_3

$$z^5 - 1 = 0 \Longleftrightarrow (z-1)(z^4 + z^3 + z^2 + z + 1) = 0 \Longrightarrow (z^4 + z^3 + z^2 + z + 1) = 0$$

The last step is possible because we assume $z \neq 1$, i.e.,, we're looking for other roots rather than the boring 1. For the above quartic equation, we use Lagrange's clever trick by dividing this equation by z^2 to get[91]

$$z^2 + z + 1 + \frac{1}{z} + \frac{1}{z^2} = 0 \Longleftrightarrow \left(z^2 + \frac{1}{z^2}\right) + \left(z + \frac{1}{z}\right) + 1 = 0$$

Due to symmetry, we do a change of variable with $u = z + 1/z$, then $z^2 + 1/z^2 = u^2 - 2$ thus the above equation becomes

$$u^2 + u - 1 = 0 \tag{2.25.11}$$

of which solutions are

$$u_1 = \frac{-1 + \sqrt{5}}{2}, \quad u_2 = \frac{-1 - \sqrt{5}}{2}$$

Having obtained u, we can solve for z from $z^2 - uz + 1 = 0$ (which is a quadratic equation again): $z = u \pm \sqrt{u^2 - 4}/2$. Finally, the four solutions are

---

[91] Actually, $z^4 + z^3 + z^2 + z + 1 = 0$ is called a reciprocal polynomial equation for if we put $1/z$ into the equation we get an equation of the same form in which $1/z$ replaces z. A polynomial equation $a_n x^n + a_{n-1} x^{n-1} + \cdots + a_2 x_2 + a_1 x + a_0 = 0$ is a reciprocal equation iff $a_n = a_0$, $a_{n-1} = a_1$, $a_{n-2} = a_2, \ldots$.

$$\begin{aligned} z_1 &= \frac{\sqrt{5}-1}{4} + i\frac{1}{2}\sqrt{\frac{5+\sqrt{5}}{2}}, \quad z_2 = \frac{\sqrt{5}-1}{4} - i\frac{1}{2}\sqrt{\frac{5+\sqrt{5}}{2}} \\ z_3 &= \frac{-\sqrt{5}-1}{4} + i\frac{1}{2}\sqrt{\frac{5-\sqrt{5}}{2}}, \quad z_4 = \frac{-\sqrt{5}-1}{4} - i\frac{1}{2}\sqrt{\frac{5-\sqrt{5}}{2}} \end{aligned} \tag{2.25.12}$$

which are identical to the solutions given in Eq. (2.25.10), if you do the calculation to get the decimals.

Remark Why $u = z + 1/z$ gave us a quadratic equation in terms of u? Why not a cubic or a quartic equation? This is so because the expression $z + 1/z$ can only have two values. Thus, u can have two values, therefore u must be the solutions to a quadratic equation. But why $z + 1/z$ can only have two values? Using Eq. (2.25.12), compute $z_1 + 1/z_1, z_2 + 1/z_2, \ldots$ you'll see that even though there are four zs there are only two $z + 1/z$. But why?

Remark The takeaway point from solving $z^4 + z^3 + z^2 + z + 1 = 0$ is that to find the roots of the 4th-degree equation, we were able to reduce the problem to finding the roots of three quadratic equations—one with integer coefficients, Eq. (2.25.11), and the two whose coefficients included the roots of the first (i.e.,, $z^2 - uz + 1 = 0$). And in the end, the solutions consisted of nested square roots of rational numbers.

We shall have more, a lot more, to say about roots of unity in Sect. 3.25.4. And we shall meet them in discrete Fourier transform and it plays a key role in the fast Fourier transform, which Gilbert Strang once referred to as 'the most important numerical algorithm in our lifetime.'

Some exercises on $z^n - 1 = 0$.

---

1. Solving the following equation algebraically

$$z^6 - 1 = 0$$

2. Solving the following equation algebraically

$$z^7 - 1 = 0$$

If you like it try to solve algebraically $z^n - 1 = 0$ for n up to 11. Great mathematicians in the past including Euler, Lagrange, Vandermonde solved them and paved the way for Gauss—at the age of 19—to crack $z^{17} - 1 = 0$, which is intimately related to the geometry problem of constructing a polygon of 17 sides using only straightedge and compass. It was this achievement that made Gauss decided to choose mathematics (instead of philosophy) as a career.

2.25.4 Square Root of i

It was $\sqrt{-1}$ that led to the development of complex numbers. Then, a natural question is whether a square root of i exists. If so, then we do not need to invent new kinds of number, as with complex numbers we can do all arithmetic operations: addition, multiplication, division, subtraction, power, and root of any number.

We can use de Moivre's formula to compute this root as follows. First, we adopt Eq. (2.25.4) to express i in the polar form $i = i \sin {}^{\pi(1+4k)}/_{2} = \cos {}^{\pi(1+4k)}/_{2} + i \sin {}^{\pi(1+4k)}/_{2}$, noting that $\cos {}^{\pi(1+4k)}/_{2} = 0$, then we use Eq. (2.25.9):

$$\sqrt{i} = \sqrt{\cos\frac{\pi(1+4k)}{2} + i\sin\frac{\pi(1+4k)}{2}} = \cos\frac{\pi(1+4k)}{4} + i\sin\frac{\pi(1+4k)}{4}$$

So, there exist two square roots of i:

$$\begin{aligned} k = 0 &: \cos\frac{\pi}{4} + i\sin\frac{\pi}{4} = \frac{\sqrt{2}}{2} + \frac{\sqrt{2}}{2}i \\ k = 1 &: \cos\frac{5\pi}{4} + i\sin\frac{5\pi}{4} = -\frac{\sqrt{2}}{2} - \frac{\sqrt{2}}{2}i \end{aligned} \tag{2.25.13}$$

We can also use ordinary algebra to get this result. Let's assume that the square root of i is a complex number:

$$\sqrt{i} = a + bi \Longrightarrow i = 0 + i1 = a^2 - b^2 + i2ab$$

Thus, we have a and b satisfying the following system of equations by comparing the real parts and imaginary parts of the two complex numbers[92]

$$a^2 - b^2 = 0, \quad 2ab = 1$$

of which solutions are $a = b = \pm{}^{\sqrt{2}}/_{2}$. And we get the same result. We have used *the method of undetermined coefficients*.

2.25.5 Trigonometry Identities

de Moivre's formula Eq. (2.25.6) can be used to derive various trigonometry identities. For example, with $n = 2$, we can write

[92] We imply that two complex numbers are equal if they have the same real and imaginary parts, which is reasonable.

$$
\begin{aligned}
(\cos\alpha + i\sin\alpha)^2 &= \cos(2\alpha) + i\sin(2\alpha) \\
\cos^2\alpha - \sin^2\alpha + i2\cos\alpha\sin\alpha &= \cos(2\alpha) + i\sin(2\alpha) \\
\Longrightarrow \cos(2\alpha) &= \cos^2\alpha - \sin^2\alpha \\
\Longrightarrow \sin(2\alpha) &= 2\sin\alpha\cos\alpha
\end{aligned}
\tag{2.25.14}
$$

where we have used the fact that if two complex numbers are equal, the corresponding real and imaginary parts must be equal.

And with $n = 3$, we have

$$
\begin{aligned}
(\cos\alpha + i\sin\alpha)^3 &= \cos(3\alpha) + i\sin(3\alpha) \\
\cos^3\alpha - 3\cos\alpha\sin^2\alpha + 3i\cos^2\alpha\sin\alpha - i\sin^3\alpha &= \cos(3\alpha) + i\sin(3\alpha) \\
\Longrightarrow \cos(3\alpha) &= \cos^3\alpha - 3\cos\alpha\sin^2\alpha \\
\Longrightarrow \sin(3\alpha) &= 3\cos^2\alpha\sin\alpha - \sin^3\alpha
\end{aligned}
\tag{2.25.15}
$$

The last two equations can be modified a little bit to get this

$$
\begin{aligned}
\cos(3\alpha) &= \cos^3\alpha - 3\cos\alpha(1-\cos^2\alpha) = 4\cos^3\alpha - 3\cos\alpha \\
\sin(3\alpha) &= 3(1-\sin^2\alpha)\sin\alpha - \sin^3\alpha = 3\sin\alpha - 4\sin^3\alpha
\end{aligned}
\tag{2.25.16}
$$

Can you do the same thing for $\cos(5\alpha)$ in terms of $\cos(\alpha)$? A knowledge of the binomial theorem (Sect. 2.27) might be useful.

In the same manner, Eq. (2.25.9) allows us to write

$$
\sqrt[m]{\cos(\alpha) + i\sin(\alpha)} = \left(\cos\frac{\alpha}{m} + i\sin\frac{\alpha}{m}\right)
$$

which, for $m = 2$ gives

$$
\sqrt{\cos(\alpha) + i\sin(\alpha)} = \left(\cos\frac{\alpha}{2} + i\sin\frac{\alpha}{2}\right)
$$

or, after squaring both sides

$$
\cos(\alpha) + i\sin(\alpha) = \cos^2\frac{\alpha}{2} - \sin^2\frac{\alpha}{2} + 2i\cos\frac{\alpha}{2}\sin\frac{\alpha}{2}
$$

which results in the familiar trigonometry identities

$$
\cos(\alpha) = \cos^2\frac{\alpha}{2} - \sin^2\frac{\alpha}{2}, \quad \sin(\alpha) = 2\cos\frac{\alpha}{2}\sin\frac{\alpha}{2}
\tag{2.25.17}
$$

which yields (from the first of Eq. (2.25.17)) the equivalent half-angle identities

$$\cos\frac{\alpha}{2} = \sqrt{\frac{1+\cos(\alpha)}{2}}, \quad \sin\frac{\alpha}{2} = \sqrt{\frac{1-\cos(\alpha)}{2}}$$

2.25.6 Power of Real Number with a Complex Exponent

The question is: what 2^{3+2i} is? To answer this question, recall de Moivre's formula that reads

$$(\cos\alpha + i\sin\alpha)^n = \cos(n\alpha) + i\sin(n\alpha)$$

And if, we denote $f(\alpha) = \cos\alpha + i\sin\alpha$, then we observe that (thanks to the above equation)

$$f(\alpha) = \cos\alpha + i\sin\alpha \Longrightarrow [f(\alpha)]^n = f(n\alpha)$$

Which function has this property? An exponential! For example,

$$f(x) = 2^x \Longrightarrow [f(x)]^n = (2^x)^n = 2^{nx} = f(nx)$$

With that, it is reasonable to appreciate the following equation (see below for a popular proof)

$$e^{i\alpha} = \cos\alpha + i\sin\alpha \tag{2.25.18}$$

which, when evaluated at $\alpha = \pi$ yields one of the most celebrated mathematical formula, the Euler's theorem:

$$e^{i\pi} + 1 = 0 \tag{2.25.19}$$

which connects the five mathematical constants: $0, 1, \pi, e, i$. You have met numbers 0,1 and i. We will meet the number e in Sect. 2.28. And of course, π the ratio of a circle's circumference to its diameter. This identity is influential in complex analysis. Complex analysis is the branch of mathematical analysis that investigates *functions of complex numbers*. It is useful in many branches of mathematics, including algebraic geometry, number theory, analytic combinatorics, applied mathematics; as well as in physics, including the branches of hydrodynamics, thermodynamics, and particularly quantum mechanics. Refer to Sect. 6.12 for an introduction to this fascinating field.

So, it is officially voted by mathematicians that $e^{i\pi} + 1 = 0$ is the most beautiful equation[93] in mathematics! As one limerick (a literary form particularly beloved by mathematicians) puts it

e raised to the *pi* times *i*,
And plus 1 leaves you nought but a sigh.
This fact amazed Euler
That genius toiler,
And still gives us pause, bye the bye.

It is possible to express the sine/cosine functions in terms of the complex exponential:

$$e^{i\alpha} = \cos\alpha + i\sin\alpha \implies \cos\alpha = \frac{e^{i\alpha} + e^{-i\alpha}}{2} \tag{2.25.20}$$

$$e^{-i\alpha} = \cos\alpha - i\sin\alpha \implies \sin\alpha = \frac{e^{i\alpha} - e^{-i\alpha}}{2i} \tag{2.25.21}$$

Proof Here is one proof of $e^{i\theta} = \cos\theta + i\sin\theta$ if we know the series of e^x, $\sin x$ and $\cos x$. We refer to Sects. 4.15.5 and 4.15.6 for a discussion on the series of these functions.

Start with the series of e^x where x is a real number:

$$e^x = 1 + \frac{x}{1!} + \frac{x^2}{2!} + \frac{x^3}{3!} + \frac{x^4}{4!} + \frac{x^5}{5!} + \cdots$$

Replacing x by $i\theta$, which is a complex number (why can we do this?, see Sect. 6.12):

$$\begin{aligned} e^{i\theta} &= 1 + \frac{i\theta}{1!} + \frac{i^2\theta^2}{2!} + \frac{(i\theta)^3}{3!} + \frac{(i\theta)^4}{4!} + \frac{(i\theta)^5}{5!} + \cdots \\ &= \underbrace{\left(1 - \frac{\theta^2}{2!} + \frac{\theta^4}{4!} - \cdots\right)}_{\cos\theta} + i\underbrace{\left(\frac{\theta}{1!} - \frac{\theta^3}{3!} + \frac{\theta^5}{5!} - \cdots\right)}_{\sin\theta} \\ &= \cos\theta + i\sin\theta \end{aligned}$$

■

93 $e^{i\pi} + 1 = 0$ is actually not an equation. An equation (in a single variable) is a mathematical expression of the form $f(x) = 0$, for example, $x^2 + x - 5 = 0$, which is true only for certain values of the variable, that is, for the solutions of the equation. There is no x, however, to solve for in $e^{i\pi} + 1 = 0$. So, it isn't an equation. It isn't an identity, either, like Euler's identity $e^{i\alpha} = \cos\alpha + i\sin\alpha$, where α is any angle, not just π radians. That's what an identity (in a single variable) is, of course, a statement that is identically true for any value of the variable. There isn't any variable at all, anywhere, in $e^{i\pi} + 1 = 0$: just five constants.

With Euler's identity, it is possible to derive the trigonometry identity for angle summation *without resorting to geometry*; refer to Sect. 3.8 for such geometry-based derivations. Let's denote two complex numbers on a unit circle as $z_1 = \cos\alpha + i\sin\alpha = e^{i\alpha}$, $z_2 = \cos\beta + i\sin\beta = e^{i\beta}$, we then can write the product z_1z_2 in two ways

$$
\begin{aligned}
z_1z_2 &= e^{i(\alpha+\beta)} = \cos(\alpha+\beta) + i\sin(\alpha+\beta) \\
z_1z_2 &= (\cos\alpha + i\sin\alpha)(\cos\beta + i\sin\beta) \\
&= \cos\alpha\cos\beta - \sin\alpha\sin\beta + i(\sin\alpha\cos\beta + \cos\alpha\sin\beta)
\end{aligned}
$$

Equating the real and imaginary parts of z_1z_2 given by both expressions, we can deduce the summation sine/cosine identities, simultaneously!

History note 2.6: Caspar Wessel (1745–1818)

Wessel was born in Norway and was one of thirteen children in a family. In 1763, having completed secondary school at Oslo Cathedral School, he went to Denmark for further studies. He attended the University of Copenhagen to study law, but due to financial pressures, could do so for only a year. To survive, he became an assistant land surveyor to his brother and they worked on the Royal Danish Academy of Sciences and Letters' topographical survey of Denmark. It was the mathematical aspect of surveying that led him to exploring the geometrical significance of complex numbers. His fundamental paper, *Om directionens analytiske betegning*, was presented in 1797 to the Royal Danish Academy of Sciences and Letters. Since it was in Danish and published in a journal rarely read outside of Denmark, it went unnoticed for nearly a century. The same results were independently rediscovered by Argand in 1806 and Gauss in 1831. In 1815, Wessel was made a knight of the Order of the Dannebrog for his contributions to surveying.

Now we can answer the question asked in the beginning of this section: what is $z = 2^{3+2i}$?

$$
\begin{aligned}
z &= 2^{3+2i} = 2^3 \times 2^{2i} = 8 \times 4^i \\
&= 8 \times \left(e^{\ln 4}\right)^i = 8 \times (\cos(\ln 4) + i\sin(\ln 4))
\end{aligned}
$$

And finally, it is possible to compute a logarithm of a negative number. For example, start with $e^{i\pi} = -1$, take the logarithm of both sides:

$$e^{i\pi} = -1 \Longrightarrow \ln(-1) = i\pi$$

Thus, the logarithm of a negative number is an imaginary number. That's why when we first learned calculus, logarithm of negative numbers was forbidden. This should not be the case since we accept the square root of negative numbers! To know more about complex logarithm, check out Sect. 6.12.

In the story of complex numbers, we have not only Wessel but also Jean-Robert Argand (1768–1822), another amateur mathematician. In 1806, while managing a bookstore in Paris, he published the idea of geometrical interpretation of complex numbers known as the Argand diagram and is known for the first rigorous proof of the Fundamental Theorem of Algebra. We recommend the interesting book *An imaginary tale: The story of square root of -1* by Paul Nahin [46] on more interesting accounts on $i = \sqrt{-1}$.

One exercise on complex numbers.

Assume that $f(z) = {}^{z+1}/_{z-1}$, compute $f^{1991}(2+i)$, where $f^3(z) = f(f(f(z)))$. Don't be scared by 1991! Note that this is an exercise to be solved within a certain amount of time after all. Let's compute $f^1(2+i)$, $f^2(2+i)$, and a pattern would appear for a generalization to whatever year that the test is on:

$$\begin{aligned} f(2+i) &= \frac{3+i}{1+i} = 2-i \\ f^2(2+i) &= f(f(2+i)) = f(2-i) = 2+i \\ f^3(2+i) &= f(f(f(2+i))) = f(2+i) = 2-i \end{aligned}$$

So, you see the pattern. 1991 is an odd number so $f^{1991}(2+i) = f(2+i) = 2-i$.

Some complex numbers problems.

---

1. Find the imaginary part of z^6 with $z = \cos 12° + i \sin 12° + \cos 48° + i \sin 48°$.
2. If θ is a constant st $0 < \theta < \pi$ and $x + 1/x = 2\cos\theta$, then find $x^n + 1/x^n$ in terms of n and θ; n is any positive integer.
3. Evaluate

$$\sum_{n=0}^{\infty} \frac{\cos(n\theta)}{2^n}$$

where $\cos\theta = 1/5$.
The answers are 0, $\cos n\theta$ and 6/7, respectively. If it is not clear about the third problem, see below for a similar problem.

We are now going to solve a problem in which we see the interplay between real numbers and imaginary numbers. That's simply amazing. The problem is: Given the complex number $2 + i$, and let's denote a_n and b_n the real and imaginary parts of $(2 + i)^n$, where n is a non-negative integer. The problem is to compute the following sum

$$S = \sum_{n=0}^{\infty} \frac{a_n b_n}{7^n}$$

Let's find a_n and b_n first that seems a reasonable thing to do. Power of an imaginary number? We can use de Moirve's formula. To this end, we need to convert our number $2 + i$ to the polar form:

$$2 + i = \sqrt{5}(\cos\theta + i\sin\theta), \quad \cos\theta = \frac{2}{\sqrt{5}}, \ \sin\theta = \frac{1}{\sqrt{5}}$$

Then, its power can be determined and from that a_n, b_n will appear to us:

$$(2 + i)^n = (\sqrt{5})^n(\cos n\theta + i\sin n\theta) \Longrightarrow a_n = (\sqrt{5})^n \cos n\theta, \quad b_n = (\sqrt{5})^n \sin n\theta$$

Now the sum S is explicitly given by:

$$S = \sum_{n=0}^{\infty} \frac{(\sqrt{5})^n \cos n\theta(\sqrt{5})^n \sin n\theta}{7^n} = \frac{1}{2}\sum_{n=0}^{\infty} \left(\frac{5}{7}\right)^n \sin 2n\theta$$

We did some massage to S to simplify it. Now comes the good part: we leave the real world and move to the imaginary one, by replacing $\sin 2n\theta$ by the imaginary part of $e^{i2n\theta}$:

$$S = \frac{1}{2}\sum_{n=0}^{\infty} \left(\frac{5}{7}\right)^n \operatorname{Im} e^{i2n\theta} \tag{2.25.22}$$

As the sum of the imaginary parts is equal to the imaginary of the sum,[94] we write S as:

---

[94] If not clear, one example is of great help: $(a_1 + b_1 i) + (a_2 + b_2 i) = (a_1 + a_2) + i(b_1 + b_2)$. Thus sum of imaginary parts $(b_1 + b_2)$ equals the imaginary of the sum.

$$S = \frac{1}{2}\text{Im}\sum_{n=0}^{\infty}\left(\frac{5}{7}\right)^n (e^{i2\theta})^n$$

What is the red term? It is a geometric series!, of the form $1, a, a^2, \ldots$ with $a = (5/7)e^{i2\theta}$, and we know its sum $1/(1-a)$[95]:

$$S = \frac{1}{2}\text{Im}\,\frac{1}{1-\frac{5}{7}e^{i2\theta}}$$

We know $e^{i\theta}$, thus we know its square $e^{i2\theta}$, thus the above expression is simply $7/16$. Details are as follows. First, we find the imaginary part of $\frac{1}{1-\frac{5}{7}e^{i2\theta}}$ by:

$$\begin{aligned}
\frac{1}{1-\frac{5}{7}e^{i2\theta}} &= \frac{7}{7-5(\cos 2\theta + i\sin 2\theta)} && (e^{i\alpha} = \cos\alpha + i\sin\alpha) \\
&= \frac{7[7-5\cos 2\theta + i5\sin 2\theta)]}{(7-5\cos 2\theta)^2 + (5\sin 2\theta)^2} && \text{(remove } i \text{ in the denominator)}
\end{aligned}$$

Then, the imaginary part is given by

$$\text{Im}\,\frac{1}{1-\frac{5}{7}e^{i2\theta}} = \frac{35\sin 2\theta}{74-70\cos 2\theta}$$

Thus, S is simplified to

$$S = \frac{1}{2}\frac{35\sin 2\theta}{74-70\cos 2\theta} = \cdots = \frac{7}{16}$$

We have skipped some simple calculations in ...

Is there a shorter solution? Yes, note that S involves $a_n b_n$ as a product, so we do not really need to know a_n and b_n, separately. From the fact that $(2+i)^n = a_n + ib_n$, what we do to get $a_n b_n$? Yes, we square the equation: $(2+i)^{2n} = a_n^2 - b_n^2 + 2ia_n b_n$. Thus, $a_n b_n$ is half of the imaginary part of $(2+i)^{2n}$. Plugging this into S and we fly off to the result in no time.

2.25.7 Power of an Imaginary Number with a Complex Exponent

This section is devoted to i^i. Is it an imaginary or real number? Why bother? Just out of curiosity! We use Euler's identity to write $i = e^{i\pi/2}$ (i has $r = 1$ and $\theta = \pi/2$),

95 Herein we accept that the results on geometric series also apply to complex numbers. Note that a has a modulus of 5/7 which is smaller than 1.

then we raise it to an exponent of i:

$$z = a + bi = re^{i\theta} \Longrightarrow i = e^{i\frac{\pi}{2}} \Rightarrow i^i = (e^{i\frac{\pi}{2}})^i = e^{-\frac{\pi}{2}}$$

So, i^i is a real number! Actually i^i has many values, we have just found one of them[96]:

$$i = e^{i\left(\frac{\pi}{2}+2n\pi\right)} \Longrightarrow i^i = \left[e^{i\left(\frac{\pi}{2}+2n\pi\right)}\right]^i = e^{-\frac{\pi}{2}-2n\pi}$$

Long before Euler wrote $e^{i\theta} = \cos\theta + i\sin\theta$, the Swiss mathematician Johann Bernoulli (1667–1748)—one of the many prominent mathematicians in the Bernoulli family and Euler's teacher—already computed i^i using a clever technique. It is presented here so that we can enjoy it all (assume you know a bit of calculus here). He considered the area of 1/4 of a unit circle:

$$\frac{\pi}{4} = \int_0^1 \sqrt{1-x^2}dx$$

Now comes the clever idea, he used the following 'imaginary' substitution using i (note that if we proceed with the standard substitution $x = \sin\theta$, we will get $\pi/4 = \pi/4$, which is useless; that's why Bernoulli had to turn to i to have something new coming up):

$$x = -iu \Longrightarrow dx = -idu, \quad 1 - x^2 = 1 + u^2$$

Then, the above integral becomes

$$\frac{\pi}{4} = -i\int_0^i \sqrt{1+u^2}du$$

And the red integral can be computed (check Sect. 4.7 if you're not clear):

$$\frac{\pi}{4} = -\frac{i}{2}\left[\sec\theta\tan\theta + \ln(\sec\theta + \tan\theta)\right]_0^{\theta^*}$$

with $\tan\theta^* = i$. Thus, we have

$$\frac{\pi}{4} = -\frac{i}{2}\ln(i)$$

And from that the result $i^i = e^{-\pi/2}$ follows. As we have seen, once accepted i, mathematicians of the seventeenth century played with them with joy and obtained interesting results. And of course other mathematicians did similar things; for exam-

[96] Check Sect. 6.12 for detail.

ple, the Italian Giulio Carlo dei Toschi Fagano (1682–1766) played with a circle but with its circumference, and got the same result as Bernoulli [46]. It is similar to we—ordinary human—soon introduce many new tricks with a new FIFA play station game.

Now comes a surprise. What is 1^π? We have learned that $1^x = 1$, so you might be guessing $1^\pi = 1$. But then you get only one correct answer. To see why just see 1 as a complex number $1 = 1 + 0i = e^{i(2n\pi)}$ with n being an integer, thus

$$1^\pi = (e^{i(2n\pi)})^\pi = e^{i(2n\pi^2)} = \cos(2n\pi^2) + i\sin(2n\pi^2)$$

where in the last equality we have used Euler's identity $e^{i\theta} = \cos\theta + i\sin\theta$. From this we see that only with $n = 0$ we get $1^\pi = 1$, which is real. Other than that we have complex numbers! This is because $\sin(2n\pi^2)$ is always different from zero for all integers not 0. Why that? Because π is irrational, a result by the Swiss polymath Johann Heinrich Lambert (1728–1777). To see why, let's solve $\sin(2n\pi^2) = 0$, of which solutions are

$$\sin(2n\pi^2) = 0 \Longleftrightarrow 2n\pi^2 = m\pi \Longleftrightarrow 2\pi = \frac{m}{n}$$

which cannot happen as π cannot be expressed by m/n because it is an irrational number.

2.25.8 A Summary of Different Kinds of Numbers

We have come a long way starting with counting numbers. We have added more numbers to the list during the journey. Let's summarize what numbers we have:

$$\begin{aligned}
\text{Natural numbers: } & \mathbb{N} = \{0, 1, 2, 3, 4, 5, 6, \ldots\} \\
\text{Integer numbers: } & \mathbb{Z} = \{\ldots, -3, -2, -1, 0, 1, 2, 3, \ldots\} \\
\text{Rational numbers: } & \mathbb{Q} = \left\{\frac{5}{3}, \frac{22}{7}, 1.5, \ldots\right\} \\
\text{Real numbers: } & \mathbb{R} = \{-1, 0, 1, \sqrt{2}, \pi, e, \ldots\} \\
\text{Complex numbers: } & \mathbb{C} = \{2 + 3i, -1, 0, i, 3 + 4i, \ldots\}
\end{aligned}$$

Note that we have introduced different symbols to represent different collections of numbers. Instead of writing "a is a non-negative integer number", mathematicians write $a \in \mathbb{N}$. When they do so, they mean that a is a member of the set (collection) of non-negative integers; this set is symbolically denoted by $\mathbb{N}$. The notation $\mathbb{Z}$ comes from the German word *Zahlen*, which means numbers. The notation $\mathbb{Q}$ is for quotients.

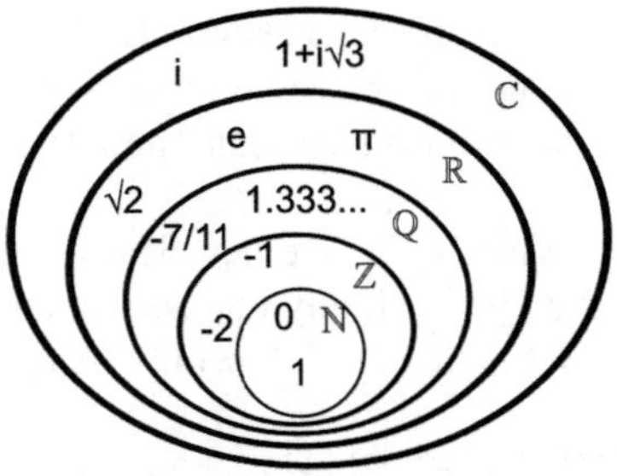

In mathematics, the notion of a number has been extended over the centuries to include 0, negative numbers, rational numbers such as one third ($1/3$), real numbers such as the square root of 5 and π, and complex numbers which extend the real numbers with a square root of -1. Calculations with numbers are done with arithmetical operations, the most familiar being addition, subtraction, multiplication, division, and exponentiation. Besides their practical uses, numbers have cultural significance throughout the world. For example, in Western society, the number 13 is often regarded as unlucky.

The German mathematician Leopold Kronecker (1823–1891) once said, "Die ganzen Zahlen hat der liebe Gott gemacht, alles andere ist Menschenwerk" ("God made the integers, all else is the work of man").

But is that all? Not at all. Complex numbers are cool but after all they are just points on a boring flat plane. Mathematicians wanted to have points in space! And they created other numbers, one of them is quartenions of the form $a + bi + cj + dk$ briefly discussed in Sect. 9.1.6.

2.26 Combinatorics: The Art of Counting

Suppose you're asked to solve the following problems

- At one party each man shook hands with everyone except his spouse, and no handshakes took place between women. If 13 married couples attended, how many handshakes were there among these 26 people?
- How many ordered, nonnegative integer triples (x, y, z) satisfy the equation $x + y + z = 11$?
- A circular table has exactly 60 chairs around it. There are N people seated around this table in such a way that the next person to be seated must sit next to someone. What is smallest possible value of N?

What would you do? While solving them you will see that it involves counting, but it is tedious sometimes to keep track of all the possibilities. There is a need to develop some smart ways of counting. This section presents such counting methods. Later in Sect. 5.2, you will see that to correctly compute probabilities we need to know how to count correctly and efficiently.

We start our discussion with the product rule—which is the basic rule of counting—and tree diagram (Sect. 2.26.1). We then move to factorial (Sect. 2.26.2), permutations (Sect. 2.26.3) and combinations (Sect. 2.26.4). In Sect. 2.26.5, permutations with repetition is treated. Finally, the pigeonhole principle is discussed (Sect. 2.26.6).

2.26.1 Product Rule

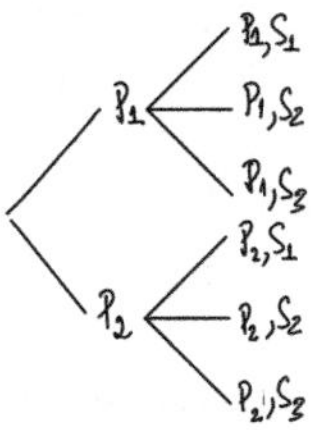

One basic principle of counting is the product rule. Suppose we want to count the number of ways to pick a shirt and pair of pants to wear. If we have 3 shirts and 2 pairs of paints, the total number of ways to choose an outfit is 6. Why? This can be seen by drawing a *tree diagram*. At the first branch we choose a pair of pants, denoted by P_1, and at the second branch we choose the other pair of pants (P_2). Now, in the P_1 branch, we have three sub-branches: one for each shirt. In total, the number of outfits is the number of leaves at the end of the tree. In general, if we have n ways to choose the first and m ways to choose the second, independent of the first choice, there are nm ways—that's why we have the name "product rule"—to choose a pair. And of course no one can stop us to use this rule for more than three things. Imagine that in addition to two pairs of pants and 3 shirts, we have 2 hats to choose. It is quite easy to extend the tree diagram, and see that there are $2 \times 6 = 2 \times 3 \times 2$ ways.

2.26.2 Factorial

Assume that we have to arrange three books on a shelve. The titles of the three books are A, B, and C. The question is there are how many ways to do the arrangement? If we put A on the left most there are two possibilities for B and C: ABC and ACB. If we put B on the left most, then there are also two possibilities: BAC and BCA. Finally, if C is put in the left most, then we have CAB and CBA. In summary, we have six ways of arrangement of three books:

$$ABC \; ACB \; BAC \; BCA \; CAB \; CBA$$

How about arranging four books A, B, C, D? Again, let's put A on the left most position, there are then six ways of arranging the remaining three books (we have just solved that problem!). Similarly, if B is put on the left most position, there are six ways of arranging the other three books. Going along this reasoning, we can see that there are

$$4 \times \text{number of ways to arrange 3 books} = 4 \times 6 = 24$$

ways, and they are

$$
\begin{array}{cccccc}
ABCD & ABDC & ACBD & ACDB & ADBC & ADCB \\
BACD & BADC & BCAD & BCDA & BDAC & BDCA \\
CABD & CADB & CBAD & CBDA & CDAB & CDBA \\
DABC & DACB & DBAC & DBCA & DCAB & DCBA
\end{array}
$$

What if we have to arrange five books? We can see that the number of arrangements is five times the number of arrangements for 4 books. Thus, there are $5 \times 24 = 120$ ways.

There is a pattern here. To see it clearly, let's denote by A_n the number of arrangements for n books ($n \in \mathbb{N}$). We then have $A_5 = 5A_4$,[97] but $A_4 = 4A_3$, we continue this way until A_1—the number of arrangements of only one book, which is one:

$$
\begin{aligned}
A_5 &= 5A_4 \\
&= 5 \times (4A_3) \\
&= 5 \times 4 \times 3A_2 \\
&= 5 \times 4 \times 3 \times 2 \times A_1 = 5 \times 4 \times 3 \times 2 \times 1
\end{aligned}
\tag{2.26.1}
$$

with A_1 being one as there is only one way to arrange one book. We are now able to give the definition of factorial.

Definition 2.26.1 For a positive integer $n \geq 1$, the factorial of n, denoted by $n!$, is defined as

$$
n! = n \times (n-1) \times (n-2) \times \cdots \times 3 \times 2 \times 1 = \prod_{i=1}^{n} i
$$

which is simply a product of the first n natural numbers.

From this definition, it follows that $n! = n(n-1)!$. Using this for $n = 1$, we get $1! = 1 \times 0!$, so $0! = 1$. This is similar to a negative multiplied a negative is a positive. The notation $n!$ was introduced by the French mathematician Christian Kramp (1760–1826) in 1808. We recall the shorthand notation $\prod i$ (called the pi product notation) that was introduced in Eq. (2.21.20).

To understand the notation $n!$, let's compute some factorials: $5! = 120$, $6! = 720$, not so large, but $10! = 3\,628\,800$! How about 50!? It's a number with 65 digits:

$$
50! = 30\,414\,093\,201\,713\,378\,043\,612\,608\,166\,064\,768\,844\,377\,641\,568\,960\,512\,000\,000\,000\,000
$$

No surprise that Kramp used the exclamation mark for the factorial. Note that I have used `Julia` to compute these large factorials. I could not find out the explanation of the name factorial, however.

[97] Just the translation of "the number of arrangements for 5 books is five times the number of arrangements for 4 books".

Question 6 *For a given $n \in \mathbb{N}$, is $n! \in \mathbb{N}$? It is always a good practice, when we have a new function/operation, we should check the result it gives us. Does it give us something not in the number system we're familiar with?*

Factorions. A factorion is a number which is equal to the sum of the factorials of its digits. For example, 145 is a factorion, because

$$145 = 1! + 4! + 5!$$

Can you write a program to find other factorions? The answer is 40 585 and see Listing A.4 for the program.

One problem involving factorial. Let's consider a problem involving factorial: which one of these numbers 50^{99} and 99! is larger? The first attempt is to naturally consider the ratio of these numbers and write out them explicitly (and see if the ratio is smaller than one or not):

$$\frac{50^{99}}{99!} = \frac{50 \times 50 \times \cdots \times 50}{99 \times 98 \times 97 \times \cdots \times 2 \times 1}$$

Now, instead of working directly with 99 terms in the numerator and 99 terms in the denominator, we divide the 99 terms in the numerator into two groups and we're left with one number 50. Similarly, we divide the product in the denominator into two groups and left with 50:

$$\frac{50^{99}}{99!} = \frac{\overbrace{(50 \times 50 \times \cdots \times 50)}^{\text{49 terms}} \times \cancel{50} \times \overbrace{(50 \times 50 \times \cdots \times 50)}^{\text{49 terms}}}{(99 \times 98 \times \cdots \times 51) \times \cancel{50} \times (49 \times 48 \times \cdots \times 2 \times 1)}$$

We can cancel the single 50s, and then combine one term in one group with another term in the other group in the way that 99 is paired with 1, 98 with 2 (why doing that? because $99 + 1 = 100 = 50 \times 2$[98]), and so on:

$$\frac{50^{99}}{99!} = \left(\frac{50^2}{99 \times 1}\right)\left(\frac{50^2}{98 \times 2}\right) \cdots \left(\frac{50^2}{51 \times 49}\right)$$

Now, it is becoming clearer that we just need to compare each term with 1, and it is quite easy to see that all terms are larger than 1 e.g. $50^2/99\times1 > 1$. This is so because we have[99]

$$(a-b)^2 > 0 \Longrightarrow (a+b)^2 > 4ab \Longrightarrow \left(\frac{a+b}{2}\right)^2 > ab$$

[98] Also because pairing numbers is a good technique that we learned from the 10 year old Gauss.

[99] Another way is to write $99 \times 1 = (50+49)(50-49) = 50^2 - 49^2 < 50^2$. In other words, the rectangle 99×1 has an area smaller than that of the square of side 50.

So, 50^{99} is larger than 99!

Another way is to use the AM-GM inequality for $n = 99$ numbers $1, 2, \ldots, 99$. And that proof also gives us a general result that for $n \in \mathbb{N}$ and $n \geq 2$,

$$\left(\frac{n+1}{2}\right)^n > n!$$

Factorial equation. Let's solve one factorial equation: find $n \in \mathbb{N}$ such that

$$n! = n^3 - n$$

Without any clue, we proceed by massage this equation a bit as we see some common thing in the two sides:

$$n(n-1)(n-2)! = n(n-1)(n+1) \Longrightarrow \boxed{(n-2)! = n+1}$$

because n and $n-1$ cannot be zero (as $n = \{0, 1\}$ do not satisfy the equation). At least, now we have another equation, which seems to be less scary (e.g., n^3 gone). What's next then? The next step is to replace $(n-2)!$ by $(n-2)(n-3)!$:

$$(n-2)(n-3)! = n+1 \Longrightarrow (n-3)! = \frac{n+1}{n-2} = \frac{n-2+3}{n-2} = 1 + \frac{3}{n-2}$$

Doing so gives us a direction to go forward: a factorial of a counting number is always a counting number, thus $1 + 3/n-2$ must be a counting number, and that means that $n - 2|3$, or

$$n - 2 = \{1, 3\} \Longrightarrow n = \{3, 5\}$$

It is obvious that $n = 3$ is not a solution, thus the only solution is $n = 5$.

Another solution is to look at the boxed equation $(n-2)! = n+1$ and think about the LHS and the RHS. It is a fact that $n!$ is a very large number, much larger than $n+1$ for n larger than a certain integer. Thus, the two sides are equal only when n is a small integer. Now, we write $(n-2)! = (n-2)(n-3)\ldots(3)(2)(1)$, which is larger than or equal to $2(n-2)$. Now we have,

$$(n-2)! \geq 2(n-2), \quad \text{and} \quad (n-2)! = n+1$$

Therefore, we can deduce that

$$n + 1 \geq 2(n-2) \Longrightarrow n \leq 5 \Longrightarrow n = \{5, 4\}$$

Stirling's approximation is an approximation for factorials. It is named after the Scottish mathematician James Stirling (1692–1770). The factorial of n can be well approximated by:

$$n! \approx \sqrt{2\pi} n^{n+1/2} e^{-n} = \sqrt{2\pi n} \left(\frac{n}{e}\right)^n, \quad \text{for} n = 1, 2, \dots \tag{2.26.2}$$

The need to develop this formula is that it is hard to compute the factorial of a large counting number, especially in Stirling's time. We shall see this shortly.

Proof of Stirling's approximation From Sect. 4.20.2 on the Gamma function, we have the following representation of $n!$:

$$n! = \int_0^\infty x^n e^{-x} dx$$

Using the change of variable $x = ny$, and $y^n = e^{n \ln y}$, the above becomes

$$n! = n^{n+1} \int_0^\infty e^{n(\ln y - y)} dy$$

What is the blue integral? If I tell you it is related to the well-known Gaussian integral $\int_{-\infty}^{\infty} e^{-x^2} dx = \sqrt{\pi}$, do you believe me? If not, plot $e^{n(\ln y - y)}$ for $n = 5$ and $y \in [0, 5]$ you will see that the plot resembles the bell curve. Thus, we need to convert $\ln y - y$ to y^2. And what allows us to do that? Taylor comes to the rescue. Now, we look at the function $\ln y - y$ and plot it, we see that it has a maximum of -1 at $y = 1$ (plot it and you'll see that), thus using Taylor's series we can write $\ln y - y \approx -1 - (y - 1)^2/2$, thus

$$n! = e^{-n} n^{n+1} \int_0^\infty e^{-n(y-1)^2/2} dy$$

Thus, another change of variable $t = \sqrt{n}x/\sqrt{2}$, and the red integral becomes

$$\int_0^\infty e^{-n(y-1)^2/2} dy = \frac{\sqrt{2}}{\sqrt{n}} \int_0^\infty e^{-t^2} dx = \frac{\sqrt{2\pi}}{\sqrt{n}}$$

Why the lower integration bound is zero not $-\infty$ and we still can use $\int_{-\infty}^{\infty} e^{-x^2} dx = \sqrt{\pi}$? This is because the function $e^{n(\ln y - y)}$ quickly decays to zero (plot and you see it), thus we can extend the integration from $[0, \infty]$ to $(-\infty, \infty)$. Actually the method just described to compute the blue integral is called the Laplace method. ■

What is the lesson from Stirling's approximation for $n!$? We have a single object which is $n!$. We have a definition of it: $n! = (1)(2) \cdots (n)$. But this definition is useless when n is large. By having another representation of $n!$ via the Gamma function, we are able to have a way to compute $n!$ for large n's.

Some exercises involve factorials.

1. Compute the following sum

$$S = \frac{3}{1! + 2! + 3!} + \frac{4}{2! + 3! + 4!} + \cdots + \frac{2001}{1999! + 2000! + 2001!}$$

Semifactorial or double factorial. Now we know that $1 \cdot 2 \cdot 3 \cdot 4 \cdot 5 \cdot 6 \cdot 7 \cdot 8 \cdot 9$ is 9! But, in many cases, we encounter this product $1 \cdot 3 \cdot 5 \cdot 7 \cdot 9$, which is 9! without the even factors 2, 4, 6, 8. In 1902, the physicist Arthur Schuster introduced the notation $n!!$ for such products, called it the "double factorial". So,

$$9!! = 9 \cdot 7 \cdot 5 \cdot 3 \cdot 1, \quad 8!! = 8 \cdot 6 \cdot 4 \cdot 2$$

So, the double factorial or semifactorial of a number n, denoted by !!, is the product of all the integers from 1 up to n that have the same parity (odd or even) as n:

$$n!! = \begin{cases} n(n-2)(n-4)\cdots(3)(1) = \prod_{k=1}^{\frac{n+1}{2}} (2k-1), & \text{if} n \text{is odd} \\ n(n-2)(n-4)\cdots(2)(1) = \prod_{k=1}^{\frac{n}{2}} (2k), & \text{if} n \text{is even} \end{cases} \tag{2.26.3}$$

It is obvious that $n!! \neq (n!)!$; e.g., $4!! = (4)(2) = 8$ but $(4!)! = (24)!$.

Double factorials can also be defined recursively. Just as we can define the ordinary factorial by $n! = n(n-1)!$ for $n \geq 1$ with $0! = 1$, we can define the double factorial by

$$n!! = n(n-2)!! \tag{2.26.4}$$

for $n \geq 2$ with initial values $0!! = 1!! = 1$.

2.26.3 *Permutations*

Now we know that there are $n!$ ways to arrange n distinct books. Why just books? Generally, there are $n!$ *permutations* of the elements of a set[100] having n elements. A permutation of a set of n objects is any rearrangement of the n objects. For example, considering this set {1, 2, 3}, we have these arrangements (permutations): {1, 2, 3}, {1, 3, 2}, {2, 1, 3}, {2, 3, 1}, {3, 1, 2} and {3, 2, 1}.

[100] Don't worry about this word "set"; it is what mathematicians use when they mean a collection of things.

We have used the simplest way to count the number of permutations of a set with n elements: we listed all the possibilities. But we can do another way. Imagine that we have n distinct books to be placed into n boxes. For the first box, there are n choices, then for each of these n choices there are $n-1$ choices for the second box, for the third box there are $n-2$ choices and so on. In total, there will be $n(n-1)(n-2)\cdots(3)(2)(1)$ ways. When we multiply all the choices we are actually using the so-called basic rule of counting: the product rule.

There are 5! ways to arrange 5 persons in 5 seats. But, there are how many ways to place five people into two seats? There are only $5 \times 4 = 20$ ways because for the first seat we have 5 choices and for the second seat we have 4 choices. Assuming that the five people are named A, B, C, D, E, then the 20 ways are:

$$\begin{array}{cccccccccc} AB & BC & CD & DE & AC & AD & AE & BD & BE & CE \\ BA & CB & DC & ED & CA & DA & EA & DB & EB & EC \end{array}$$

Now, what we need to do is to find how the result of 20 is related to 5 people and 2 seats. For 5 people and 5 seats, the answer is 5!. So, we expect that 20 should be related to the factorials of 5 and 2—the only information of the problem. Indeed, it can be seen that we can write $20 = 5 \times 4$ in terms of factorials of 5 and 2:

$$5 \times 4 = \frac{5 \times 4 \times 3 \times 2 \times 1}{3 \times 2 \times 1} = \frac{5!}{3!} = \frac{5!}{(5-2)!}$$

We now generalize this. Assume we have a n-set (i.e.,, a set having n distinct elements) and we need to choose r elements from it ($r \leq n$). There are how many ways to do so if order matters? In other words, how many r-permutations? For example considering this 3–set $\{A, B, C\}$ and we choose 2 elements. We have six ways: $\{A, B\}$, $\{B, A\}$, $\{A, C\}$, $\{C, A\}$, $\{B, C\}$, $\{C, B\}$.

The number of r-permutations of an n-element set is denoted by $P(n, r)$ or sometimes by P_n^r, which is defined as:

$$P(n,r) = P_n^r = \frac{n!}{(n-r)!} \tag{2.26.5}$$

And we can write $P(n, r)$ explicitly as:

$$P(n,r) = \frac{n(n-1)(n-2)\cdots(n-r+1)(n-r)!}{(n-r)!} = n(n-1)(n-2)\cdots(n-r+1)$$

This expression is exactly telling us what we have observed. We need to choose r elements; there are n options for the first element, $n-1$ options for the second element, ... and $n-r+1$ options for the last element.

2.26.4 Combinations

In permutations, the order matters: AB is different from BA. Now, we move to combinations in which the order does not matter. Let's use the old example of placing five people into two seats. These are 20 arrangements of five people A, B, C, D, E into two seats (there are 5 options for the 1st seat and 4 options for the second seat):

$$\begin{array}{cccccccccc} AB & BC & CD & DE & AC & AD & AE & BD & BE & CE \\ BA & CB & DC & ED & CA & DA & EA & DB & EB & EC \end{array}$$

And if AB is equal to BA, i.e.,, what matter is who seats next to who not the order, there are only 10 ways. When order does not matter, we are speaking of a *combination*. My fruit salad is a combination of apples, grapes, and bananas. We do not care the order the fruits are in.

We can observe that:

$$10 = \frac{20}{2} = \frac{5!}{(5-2)!2!}$$

which leads to the following r-combinations equation:

$$\binom{n}{r} = C_n^r = \frac{n!}{(n-r)!r!} = \frac{P_n^r}{r!} \tag{2.26.6}$$

The last equality shows the relation between permutation and combination; there are less combinations than permutations due to repetitions. And there are $r!$ repetitions. The notation $\binom{n}{r}$ is read *n choose r*.

$\binom{n}{r}$ is also called the binomial coefficient. This is because the coefficients in the binomial theorem are given by $\binom{n}{r}$ (Sect. 2.27).

Question 7 *The factorial was defined for positive integers. Is it too restrict? If you're feeling this way, that's very good. What is the value of* $(1/2)!$*? The result is surprising; it is not an integer, it is a real number:* $0.5\sqrt{\pi}$.

2.26.5 Generalized Permutations and Combinations

Permutations with repetition. With 3 a's and 2 b's how many 5-letter words can we make? Of course, we do not care about meaningless words. It is clear that we can have these words:

$$aaabb, aabab, abaab, baaab, aabba, ababa, baaba, abbaa, babaa, bbaaa \tag{2.26.7}$$

That is ten words. The question now is how to derive a formula, as listing works only when there are few combinations. First, let's denote by N the number of 5-letter words that can be made from 3 a's and 2 b's. Second (and this is the key to the solution), we *convert this problem to the problem we're familiar with*: permutations without repetition by using a_1, a_2, a_3 for 3 a's and b_1, b_2 for 2 b's. Obviously there are 5! 5-letter words from a_1, a_2, a_3, b_1, b_2. We can get these words by starting with Eq. (2.26.7). For each of them, we add subscripts 1, 2, 3 to the a's (there are 3! ways of doing that), and then we add subscripts 1, 2 to the b's (there are 2! ways). Thus, in total, there are $N3!2!$ 5-letter words. Of course, we have $N3!2! = 5!$, thus

$$N = \frac{5!}{3!2!}$$

Now we generalize the result to the case of n objects which are divided into k groups in which the first group has n_1 identical objects, the second group has n_2 identical objects, ..., the kth group has n_k identical objects. Certainly, we have $n_1 + n_2 + \cdots + n_k = n$. The number of permutations of these n such objects are

$$\frac{n!}{n_1!n_2!\cdots n_k!} \tag{2.26.8}$$

For the special case that $k = 2$, we have one group with r identical elements and one group with $n - r$ elements:

$$\underbrace{aa\cdots a}_{r}\underbrace{bb\cdots b}_{n-r}$$

There are

$$\frac{n!}{r!(n-r)!}$$

permutations of such set. Coincidentally, it is equal to $\binom{n}{r}$:

$$\frac{n!}{r!(n-r)!} = \binom{n}{r} \tag{2.26.9}$$

To remove this confusion between permutations and combinations, we can change how we look at the problem. For example, the problem of making 5-letter words with 3 a's and 2 b's can be seen like this. There are 5 boxes in which we will place 3 a's into 3 boxes. The remaining boxes will be reserved for 2 b's. How many way to select 3 boxes out of 5 boxes? It is $\binom{5}{3}$.

Instead of placing the a's first we can place the b's first. There $\binom{5}{2}$ ways of doing so. Therefore, $\binom{5}{2} = \binom{5}{3}$. Thus, we have the following identity

$$\binom{n}{k} = \binom{n}{n-k} \tag{2.26.10}$$

We can check this identity easily using algebra. But the way we showed it here is interesting in the sense that we do not need any algebra. This is proof by combinatorial interpretation. The basic idea is that we count the same thing twice, each time using a different method and then conclude that the resulting formulas must be equal.

2.26.6 The Pigeonhole Principle

Let's solve this problem. If a Martian has an infinite number of red, blue, yellow, and black socks in a drawer, how many socks must the Martian pull out of the drawer to guarantee that *he has a pair*?

It is obvious that if he pulls out two socks, it is not certain he will get a pair; for example he can get a red and a blue shock. The result is the same if he pulls out three socks or four socks (for this case he might get red/blue/yellow/black socks). Only when he gets out five socks, he certainly will get a pair.

Let's solve another problem. A bag contains 10 red marbles, 10 white marbles, and 10 blue marbles. What is the minimum number of marbles we have to choose randomly from the bag to ensure that we get *four marbles of the same color*?

We can try and see that with 9 marbles we cannot ensure that there are four marbles of the same color. One example is $RRRWWWBBB$ where R stands for red and so on. And that example shows that with 10 marbles it is 100% that we get four marbles of same color. Regardless of the color of the 10th marble, we will get either four red marbles, or four blue or four white ones.

It is clear that these two problems involve the art of counting. And there is a general principle that governs this type of problems. And this principle is related to pigeons.

Suppose that a flock of 10 pigeons flies into a set of 9 pigeonholes to roost. Because there are 10 pigeons but only 9 pigeonholes, at least one of these 9 pigeonholes must have at least two pigeons in it. This illustrates a general principle called the pigeonhole principle,[101] which states that *if there are more pigeons than pigeonholes, then there must be at least one pigeonhole with at least* ***two pigeons*** *in it.* We can use it to easily solve the first problem of picking socks. Here, *the pigeons are the socks* and the *(pigeon-)holes are the sock colors*. Because there are four holes, we need at least

[101] This principle is also known as Dirichlet box principle, named after the German mathematician Johann Peter Dirichlet (1805–1859).

five pigeons so that at least one hole contains two pigeons. As a hole represents a color, when there are two pigeons (socks) in a hole that indicates that there are two socks of the same color!

The second problem of marbles is a generalization of the first problem. In the first problem, we just need at least one hole having two pigeons. In the second problem, we need at least one hole having four marbles. There should be an extended version of the pigeonhole principle.

If we put 11 marbles in 3 holes, then we can either have $\{5, 3, 3\}$ or $\{4, 4, 3\}$; that is there is at least one hole that holds at least 4 marbles. How 4 is related to 11 and 3? It is $\lceil 11/3 \rceil$. Still remember the ceiling function? If not, check Sect. 2.22.1. The extended pigeonhole principle states that *if we put p pigeons in h holes, where $p > h$, then at least one hole must hold at least $\lceil p/h \rceil$ pigeons.* We can try other examples and observe that this extended version holds true. We need a proof, but let's use it to solve the marble problem. The problem can be cast as finding the number of pigeons p to be put in 3 holes (as there are 3 colors) so that one hole has $\lceil p/3 \rceil = 4$. Solving this gives us $p = 10$.

Proof the generalized pigeonhole principle Here is the proof of the extended pigeonhole principle. We use proof by contradiction: first we assume that no hole contains at least $\lceil p/h \rceil$ pigeons and based on this assumption, we're then led to something absurd. If no hole contains at least $\lceil p/h \rceil$, then every hole contains a maximum of $\lceil p/h \rceil - 1$ pigeons. Thus, p holes contain a maximum of

$$(\lceil p/h \rceil - 1)\, h$$

pigeons. We're now showing that this number of pigeons is smaller than p:

$$(\lceil p/h \rceil - 1)\, h < p, \quad (\lceil x \rceil < x + 1)$$

This is impossible because we have started with p pigeons.

■

Some problems to practice the pigeonhole principle.

1. Every point on the plane is colored either red or blue. Prove that no matter how the coloring is done, there must exist two points, exactly a mile apart, that are the same color.
2. Given a unit square, show that if five points are placed anywhere inside or on this square, then two of them must be at most $\sqrt{2}/2$ units apart.

2.26.7 Solutions to Questions

Herein we're going to solve the questions asked at the beginning of the section. The first question is: At one party each man shook hands with everyone except his spouse, and no handshakes took place between women. If 13 married couples attended, how many handshakes were there among these 26 people?.[102]

The number of handshakes is consisted of the number of handshakes between men and the number of handshakes between men and women. The latter is easy to get: it is $13 \times 12 = 156$ for every men can shake hands with 12 women and there are 13 men. For the former—the handshakes between men—there are two ways to get it. One way is to use formula and one way is to start from the scratch. I present the former way first. The first man can shake hands with 12 men, the second with 11, the third with 10, and so on. In total, there are $12 + 11 + \cdots + 2 + 1 = 78$ handshakes between men. But, $1 + 2 + \cdots + 11 + 12 = 0.5(12 \times 13)$, using the formula for the sum of the first n positive integer Eq. (2.6.2). However, $0.5(12 \times 13) = \binom{13}{2}$. And that's exactly the second way to get the number of handshakes between men: we just need to choose two men among 13 men, and there are $\binom{13}{2}$ ways to do just that.

Next, we solve the question "How many ordered, nonnegative integer triples (x, y, z) satisfy the equation $x + y + z = 11$?". A mistake that I made was: there are 12 choices for x, 12 choices for y and 1 choice for z; thus the solution is 144. It is wrong because we cannot use the product rule here as x, y are not independent! Actually, we can solve this equation quite easily by reasoning as follows: fix x $(x = \bar{x})$, then find how many ways we can have for $y + z = 11 - \bar{x}$. Then, we vary $\bar{x}$ from 0 to 11 and count the corresponding ways:

$$\begin{aligned}
\bar{x} = 0 &: \text{ there are 12 choices for } y : y = 0, 1, 2, \ldots, 11 \\
\bar{x} = 1 &: \text{ there are 11 choices for } y : y = 0, 1, 2, \ldots, 10 \\
\ldots & \quad \ldots \\
\bar{x} = 11 &: \text{ there are 1 choices for } y : y = 0
\end{aligned} \tag{2.26.11}$$

Therefore, there are $12 + 11 + \cdots + 1 = 78$ nonnegative triples (x, y, z) such that $x + y + z = 11$. Should we stop here? No! Why 78? How is it related to 11 and x, y, z? If we cannot answer these questions, when facing the problem $x_1 + x_2 + x_3 + x_4 + x_5 = 150$, how can we solve it? The above brute force approach would be tedious.

Lucky for us, we got 78 from $12 + 11 + \cdots + 1$, which is $\binom{13}{2}$. The task now is to understand why $\binom{13}{2}$ is here. To this end, it is much easier to look at the problem from a different angle. Instead of seeing x, y, z as the solutions to the equation $x + y + z = 11$, we think of 11 *objects to be distributed to three bins*. And it is possible to have empty bins. With this view, we can develop a way to *visualize the problem*: we can use stars to represent the objects and bars to denote the bin boundaries. For example, we now see $(0, 0, 11)$ and $(3, 3, 5)$ as (no longer solutions)

[102] This is Problem 16 from the 1990 AHSME (American High School Mathematics Examination) Problems. The AHSME was replaced with the AMC 10 and AMC 12 in 2000.

$$\begin{aligned}(0, 0, 11) &: \big|\big|\star\star\star\star\star\star\star\star\star\star\star \\ (3, 3, 5) &: \star\star\star\big|\star\star\star\big|\star\star\star\star\star\end{aligned}$$

With this representation, the solution appears itself: in the problem there is always 13 places: 11 for 11 stars and 2 for 2 bars, and we just need to select two places of out 13 (to put the bars), which is $\binom{13}{2}$, which is $\binom{11+3-1}{3-1}$. The last formula enables us to generalize to n objects and k bins: $\binom{n+k-1}{k-1}$.

Finally, we're going to solve the problem "A circular table has exactly 60 chairs around it. There are N people seated around this table in such a way that the next person (after N people) to be seated must sit next to someone. What is smallest possible value of N?" Try a smaller problem with 6 chairs, draw a diagram and try to see what arrangement satisfies the condition of the problem. Then, extend it to 60 chairs. The answer is $N = 20$.

2.26.8 *Some Nice Problems*

The first problem is an Amzon interview question that I learned from youtuber Mindyourdecision: Let S be the set of all 5 digit numbers formed using the digits $1, 2, 3, 4, 5$ without repetition. What is the sum of all numbers in S?

If you do not know where to start; then start simple. Change the problem to: S is the set of all 3 digits numbers formed using $1, 2, 3$ (no repetition). Then, we can list all of them:

$$S = \{123, 132, 213, 231, 312, 321\}$$

Having known all the numbers in S, shall we pull out a pocket calculator and do the sum? No, no, no. We do not use that, we do the sum manually and with a delicate observation, we can see the way. Let's denote by s the sum of all numbers in S. The key point is: *There are 2 numbers start with 1, 2 numbers start with 2 and 2 numbers start with 3*. Thus, the sum of S without considering the tenth and unit digits is:

$$s_3 = 1 \times 100 \times 2 + 2 \times 100 \times 2 + 3 \times 100 \times 2 = 2 \times 100 \times (1 + 2 + 3)$$

Similarly, we have the sum of the tenths s_2 and the sum of the units s_1 are

$$\begin{aligned}s_2 &= 1 \times 10 \times 2 + 2 \times 10 \times 2 + 3 \times 10 \times 2 = 2 \times 10 \times (1 + 2 + 3) \\ s_1 &= 1 \times 2 + 2 \times 2 + 3 \times 2 = 2 \times (1 + 2 + 3)\end{aligned}$$

Therefore, the sum of S is:

$$\begin{aligned}s_1 + s_2 + s_3 &= 2 \times (1 + 2 + 3) + 2 \times 10 \times (1 + 2 + 3) \\ &\quad + 2 \times 100 \times (1 + 2 + 3) = 12(1 + 10 + 100)\end{aligned}$$

Now, we can return to the Amazon interview question. The key point is: *There are* 24 *numbers start with 1,* 24 *numbers start with 2,* 24 *numbers start with 3,* 24 *with 4 and* 24 *with 5*. The sum is[103]

$$15 \times 24 \times (1 + 10 + 10^2 + 10^3 + 10^4) = 15 \times 24 \times 11\,111 = \boxed{33\,333 \times 120}$$

The final answer is: 3 999 960. A friend of mine showed me a smarter solution: it is in the boxed equation above. Three is the average number of each digit and there are 120 numbers in S, so the sum is $120 \times 33\,333$.

2.27 Pascal Triangle and the Binomial Theorem

A binomial expansion is one of the form $(a + b)^n$ where a, b are real numbers and n is a positive integer number. With only simple algebra and adequate perseverance one can obtain the following formulae:

$$\begin{aligned}
(a+b)^0 &= 1\\
(a+b)^1 &= a+b\\
(a+b)^2 &= 1a^2 + 2ab + 1b^2\\
(a+b)^3 &= a^3 + 3a^2b + 3ab^2 + b^3\\
(a+b)^4 &= a^4 + 4a^3b + 6a^2b^2 + 4ab^3 + b^4
\end{aligned} \tag{2.27.1}$$

We find the first trace of the Binomial Theorem in Proposition 4 of Book II of Euclid's Elements: "If a straight line be cut at random, the square on the whole is equal to the squares on the segments and twice the rectangle of the segments". This is nothing but $(a + b)^2 = a^2 + b^2 + 2ab$ if the segments are a and b. In $(a + b)^2 = 1a^2 + 2ab + 1b^2$, the numbers 1, 2, 1 are called the coefficient of the binomial expansion. It's a remarkable fact that the coefficients in these binomial expansions make a triangle, which is usually referred to as Pascal's triangle. As shown in Fig. 2.39a, this binomial expansion was known by Chinese mathematician Yang Hui (ca. 1238–1298) long before Blaise Pascal (1623–1662).

The triangle can be constructed as follows (Fig. 2.39b): first placing a 1 along the left and right edges, then the triangle can be filled out from the top by adding together the two numbers just above to the left and right of each position in the triangle. Can you write a small program to build the Pascal triangle? This is a good coding exercise.

Section 2.27.1 is about the binomial theorem, which allows us to expand $(a + b)^n$ for $n \in \mathbb{N}$ without resorting to the Pascal triangle. In the next section, a surprising relation between sum of powers, discussed in Sect. 2.6, and Pascal triangle is presented, together with the so-called Bernoulli numbers (Sect. 2.27.2). Finally, a proof of the binomial theorem is shown in Sect. 2.27.3.

---

[103] Note that $15 = 1 + 2 + 3 + 4 + 5$.

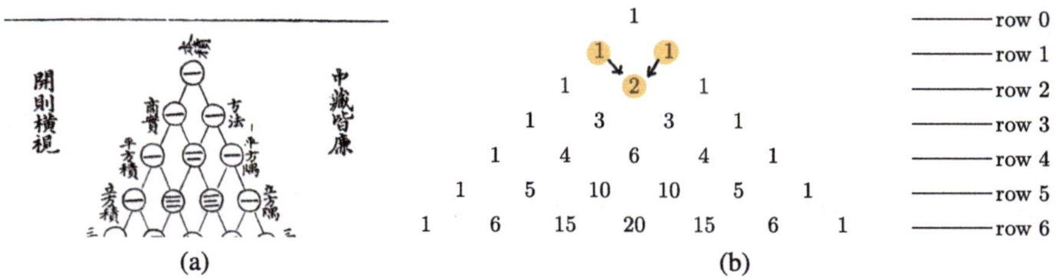

Fig. 2.39 Pascal's triangle for the binomial coefficients

2.27.1 *Binomial Theorem*

Knowing how to draw the Pascal triangle it is then possible to know the coefficients of $(a+b)^{10}$. But it is quite slow. Is there a faster way to know the coefficient of a certain term in $(a+b)^n$ without going through the Pascal triangle? To answer that question, let's consider $(a+b)^3$. We expand it as follows

$$\begin{aligned}(a+b)^3 &= (a+b)(a+b)(a+b)\\ &= (aa+ab+ba+bb)(a+b)\\ &= aaa+aab+aba+abb+baa+bab+bba+bbb\end{aligned}$$

Every term in the last expression has three components containing only a and b (e.g., aba). We also know some of these terms are going to group together; e.g., $aab = aba = baa$, as they are all equal a^2b. Now, there are $\binom{3}{2}$ ways to write a sequence of length three, with only a and b, that has precisely two a's in it. Thus, the coefficient of a^2b is $\binom{3}{2} = 3$. Refer to Sect. 2.26 for a discussion on the notation $\binom{c}{n}$.

Generalization allows us to write the following binomial theorem:

$$(a+b)^n = \sum_{k=0}^{n} \binom{n}{k} a^{n-k} b^k, \quad \binom{n}{k} = \frac{n!}{(n-k)!k!} = \frac{n(n-1)\cdots(n-k+1)}{k!} \tag{2.27.2}$$

As an example, what is the coefficient of a^6b in $(a+b)^7$? We have: $n=7$ and $k=1$, thus the coefficient is $\binom{7}{1} = 7!/6!1! = 7$. You can check the result using the Pascal triangle.

Question 8 *What if the exponent n is not a positive integer? How about $(a+b)^{1/2}$ or $(a+b)^{-3/2}$? To these cases, we have to wait for Newton's discovery of the so-called generalized binomial theorem, see Sect. 4.15.1.*

Question 9 *If we have the binomial theorem for $(a+b)^n$, how about $(a+b+c)^n$? The third power of the trinomial $a+b+c$ is given by $(a+b+c)^3 = a^3+b^3+c^3+3a^2b+3a^2c+3b^2a+3b^2c+3c^2a+3c^2b+6abc$. Is it possible to have a* *formula for the coefficients of the terms in* $(a+b+c)^3$*? And how about* $(x_1+x_2+\cdots+x_m)^n$*?*

2.27.2 Sum of Powers of Integers and Bernoulli Numbers

Now I present a surprising result involving the binomial coefficients. Recall in Sect. 2.6 that we have computed the sums of powers of integers (e.g., $1^3+2^3+\cdots+n^3$). We've considered the sums of powers of one, two and three only. But back in the old days, the German mathematician Johann Faulhaber (1580–1635) did that for powers up to 23. Using that result, Jakob Bernoulli in 1713, and the Japanese mathematician Seki Takakazu (1642–1708), in 1712 independently found a pattern and discovered a general formula for the sum. With $n, m \in \mathbb{N}$ and $m \geq 1$, let

$$S_m(n) := \sum_{k=1}^{n-1} k^m$$

Then, we have[104]

$$S_m(n) = \frac{1}{m+1}\sum_{k=0}^{m}(-1)^k\binom{m+1}{k}B_k n^{m+1-k} \tag{2.27.3}$$

where B_k are now called the Bernoulli numbers. Why not Takakazu numbers, or better Bernoulli-Takakazu numbers? Because history is not what happened, but merely what has been recorded, and most of what has been recorded in English has a distinctly Western bent. This is particularly true in the field of mathematical history. The Bernoulli numbers B_k are

$$B_0 = 1,\ \ B_1 = -\frac{1}{2},\ B_2 = \frac{1}{6},\ \ B_3 = 0,\ \ B_4 = -\frac{1}{30},\ B_5 = 0, \ldots$$

What are the significance of these mysterious numbers? It turns out that, as is often the case in mathematics, the Bernoulli-Takakazu numbers appear in various fields in mathematics, see Sect. 4.17 for more detail.

---

[104] This is how Jacob Bernoulli derived this. He wrote something similar to Eq. 2.6.13 for $m = 1, 2, 3, \ldots, 10$. Then, he looked at the coefficients of $n^{m+1}, n^m, \ldots$ carefully. A pattern emerged, in connection to Pascal's triangle, and he guessed correctly Eq. (2.27.3) believing in the pattern. Thus, he did not prove this formula. It was later proved by Euler.

2.27.3 Binomial Theorem: A Proof

Knowing the binomial theorem, the Pascal triangle is now written using the $\binom{n}{k}$ notation[105]:

$$\binom{0}{0}$$

$$\binom{1}{0} \quad \binom{1}{1}$$

$$\binom{2}{0} \quad \binom{2}{1} \quad \binom{2}{2}$$

$$\binom{3}{0} \quad \binom{3}{1} \quad \binom{3}{2} \quad \binom{3}{3}$$

$$\binom{4}{0} \quad \binom{4}{1} \quad \binom{4}{2} \quad \binom{4}{3} \quad \binom{4}{4}$$

From it we can observe the following identities

$$\binom{2}{1} = \binom{1}{0} + \binom{1}{1}, \binom{3}{1} = \binom{2}{0} + \binom{2}{1}, \ldots$$

which are generally written as

$$\binom{n+1}{k+1} = \binom{n}{k+1} + \binom{n}{k} \quad \text{for } 0 \le k < n \tag{2.27.4}$$

This identity—known as Pascal's rule or Pascal's identity—can be proved algebraically. But that is just an exercise about manipulating factorials. We need a combinatorial proof so that we better understand the meaning of the identity.

The left-hand side (the red term) in Pascal's identity is the number of $(k+1)$-element subsets taken from a set of $n+1$ elements. Now what we want to prove is that the left-hand side is also the number of such subsets. Figure 2.40 shows the proof for the case of $n = 3$ and $k = 1$. I provided only a proof for a special case whereas all textbooks present a general proof. This results in an impression that mathematicians only do hard things. Not at all. In their unpublished notes, they usually had proofs for simple cases!

With this identity, Eq. (2.27.4), we can finally prove the binomial theorem; that is the theorem is correct for any $n \in \mathbb{N}$. The technique we use (actually Pascal did it first) is proof by induction. Observe that the theorem is correct for $n = 1$. Now, we assume that it is correct for $n = k$, that is

---

105 This is typeset using the package tikz.

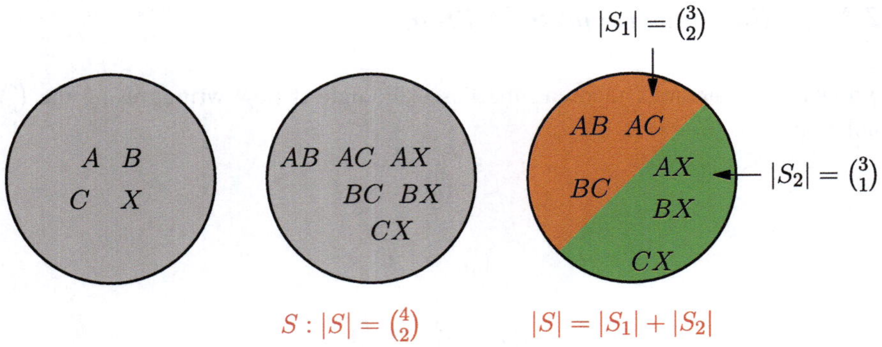

Fig. 2.40 Proof of Pascal's identity for the case of $n = 3$ and $k = 1$. The red term in Eq. (2.27.4) is $\binom{4}{2}$, which is the cardinality of S—a set that contains all subsets of two elements taken from the set $\{A, B, C, X\}$. We can divide S into two subsets: S_1 is the one without X and S_2 is the one with X. Obviously $|S_1| = \binom{3}{2}$

$$(a+b)^k = \sum_{j=0}^{k} \binom{k}{j} a^{k-j} b^j = a^k + \binom{k}{1} a^{k-1} b + \cdots + \binom{k}{k-1} ab^{k-1} + b^k \tag{2.27.5}$$

And our aim is to prove that it is also valid for $n = k + 1$, that is:

$$(a+b)^{k+1} = \sum_{j=0}^{k+1} \binom{k+1}{j} a^{k+1-j} b^j = a^{k+1} + \binom{k+1}{1} a^k b + \cdots + \binom{k+1}{k} ab^k + b^{k+1} \tag{2.27.6}$$

To this end, we compute $(a+b)^{k+1}$ as $(a+b)^k(a+b)$ and use Eq. (2.27.5):

$$\begin{aligned}
(a+b)^{k+1} &= (a+b)^k(a+b) \\
&= \left[a^k + \binom{k}{1} a^{k-1} b + \cdots + \binom{k}{k-1} ab^{k-1} + b^k\right](a+b) \\
&= a^{k+1} + a^k b + \binom{k}{1} a^k b + \binom{k}{1} a^{k-1} b^2 + \cdots + \binom{k}{k-1} ab^k + ab^k + b^{k+1}
\end{aligned}$$

And the rest is some manipulations,

$$\begin{aligned}
(a+b)^{k+1} &= a^{k+1} + \left[\binom{k}{0} a^k b + \binom{k}{1} a^k b\right] + \cdots + \left[\binom{k}{k-1} ab^k + \binom{k}{k} ab^k\right] + b^{k+1} \\
&= a^{k+1} + \left[\binom{k}{0} + \binom{k}{1}\right] a^k b + \cdots + \left[\binom{k}{k-1} + \binom{k}{k}\right] ab^k + b^{k+1} \\
&= a^{k+1} + \binom{k+1}{1} a^k b + \cdots + \binom{k+1}{k} ab^k + b^{k+1} \quad \text{(using (2.27.4))}
\end{aligned}$$

This is exactly Eq. (2.27.6), which is what we wanted to prove.

2.28 Compounding Interest

By the eighteenth century, the only known irrational numbers were those expressed using nth roots (e.g., $\sqrt{5}$ or $\sqrt[3]{7}$). The number e—the subject of this section—was the next number proved to be irrational. And it appears in a surprising place of compound interest.

No other aspect of life has a more mundane character than the quest for financial security. And central to any consideration of money is the concept of interest. The practice of charging a fee for borrowing money goes back to antiquity. For example, a clay tablet from Mesopotamia dated to about 100 B.C. poses the following problem: how long will it take for a sum of money to double if invested at 20% rate compounded annually?

Imagine that you've deposited \$1 000 in a savings account at a bank that pays an incredibly generous interest rate of 100/

But is it the best that we can get? What if we get interest every month? The interest rate is now of course 1/12. Therefore, our money in the bank after the first and second month is:

$$\text{1st month:}\quad 1000 + \frac{1}{12}\times 1000 = \left(1+\frac{1}{12}\right)\times 1000$$

$$\text{2nd month:}\quad \left(1+\frac{1}{12}\right)\times\left(1+\frac{1}{12}\right)\times 1000$$

And the amount of money after 12 months is:

$$\underbrace{\left[\left(1+\frac{1}{12}\right)\times\left(1+\frac{1}{12}\right)\times\cdots\times\left(1+\frac{1}{12}\right)\right]}_{\text{12 times}}\times 1000$$

$$=\left(1+\frac{1}{12}\right)^{12}\times 1000 = 2613.03529$$

which is \$2 613 and better than the annual compounding. Let's be more greedy and try with daily, hourly and minutely compounding. It is a good habit to ask questions "what if" and work hard investigating these questions. It led to new maths in the past! The corresponding calculations are given in Table 2.20.

From this table, we can see that the amount of money increases from \$2 000 and settles at \$2 718,279 242 6. What I have presented was done by Jacob Bernoulli in 1683. But he did not introduce a notation for this number and did not recognize the connection of the number 2.718279 with logarithm. It was Euler in 1731 who introduced the symbol e[106] to represent the rate of continuous compounding:

[106] Was Euler selfish in selecting e for this number? Probably not. Note that it was Euler who adopted π in 1737 and $i^2 = -1$.

Table 2.20 Amounts of money received with yearly, monthly, daily, hourly and minutely compounding

	Formula	Result
Yearly	$(1+1)\times 1000$	2000
Monthly	$(1+1/12)^{12}\times 1000$	2613.035290224676
Daily	$(1+1/365)^{365}\times 1000$	2714.567482021973
Hourly	$(1+1/365/24)^{365\times 24}\times 1000$	2718.1266916179075
Minutely	$(1+1/365/24/60)^{365\times 24\times 60}\times 1000$	2718.2792426663555

$$\boxed{e := \lim_{n\to\infty}\left(1+\frac{1}{n}\right)^n} \tag{2.28.1}$$

The fascinating thing about e is that the more often the interest is compounded, the less your money grows during each period (compare $1+1$ versus $(1+1/12)$ for example). Yet it still amounts to something significant after a year, for it is multiplied over so many periods.

In mathematics, there are three most famous irrational numbers and e is one of them. They are π, ϕ and e. We have met two of them. We will introduce π in Chapter 3 when we talk about geometry.

We have now a definition of e as a limit, but it is not really helpful. What we need is a formula and Sect. 2.28.1 presents exactly such formula. Next, Sect. 2.28.2 proves that e is irrational. And finally, a surprising connection between the Pascal triangle and e is given in Sect. 2.28.3.

2.28.1 How to Compute e?

How we compute e? Looking at its definition, we can think of using the binomial theorem in Eq. (2.27.2) with $a=1$ and $b=1/n$. We compute e as follows

$$\begin{aligned}
\left(1+\frac{1}{n}\right)^n &= \sum_{k=0}^{n}\frac{n!}{k!(n-k)!}1\times\left(\frac{1}{n}\right)^k \\
&= 1+\frac{n!}{(n-1)!n}+\frac{n!}{2!(n-2)!n^2}+\frac{n!}{3!(n-3)!n^3}+\cdots \\
&= 1+1+\frac{1}{2!}\left(1-\frac{1}{n}\right)+\frac{1}{3!}\left(1-\frac{3}{n}+\frac{2}{n^2}\right)+\cdots
\end{aligned} \tag{2.28.2}$$

Thus, taking the limit (i.e.,, when n is very large), we get

$$e := \lim_{n\to\infty}\left(1+\frac{1}{n}\right)^n = 1+\frac{1}{1!}+\frac{1}{2!}+\frac{1}{3!}+\cdots = 2.718281828459045 \quad (2.28.3)$$

because when $n \to \infty$ all the red terms approach one for the terms involving n approach zero. See also for a calculus-based discussion on the fascinating number e in Sect. 4.15.5.

How many terms needed to get *e*? Yes, we have a formula, Eq. (2.28.3), to compute e. One question remains: how many terms should we use? Let's assume that we use only four terms, so

$$e \approx 1+\frac{1}{1!}+\frac{1}{2!}+\frac{1}{3!} = 2.6666666666666665$$

The error of this approximation is of course the terms we have ignored:

$$error = \frac{1}{4!}+\frac{1}{5!}+\frac{1}{6!}+\cdots$$

The task now is to understand this error, because if the error is small then our approximation is good. Surprisingly a bit of massage to it is useful:

$$\begin{aligned} error &= \frac{1}{3!}\left(\frac{1}{4}+\frac{1}{5\times 4}+\frac{1}{6\times 5\times 4}+\cdots\right)\\ &< \frac{1}{3!}\left(\frac{1}{2}+\frac{1}{2\times 2}+\frac{1}{2\times 2\times 2}+\cdots\right)\\ &< \frac{1}{3!}\left(\frac{1}{2}+\frac{1}{4}+\frac{1}{8}+\cdots\right) = \frac{1}{3!} \end{aligned}$$

(Noting that the sum in the bracket is a geometric series being one.) Thus, if we use $n+1$ terms to compute e, the error is smaller than $1/n!$ As $n \to \infty$, the error is approaching zero, and we get a very good approximation for e. With this, you can figure out how many terms needed to get one million digits of e.

2.28.2 *Irrationality of e*

Similar to Euclid's proof of the irrationality of $\sqrt{2}$, we use a proof of contraction here. We assume that e is a rational number and this will lead us to a nonsense conclusion. The plan seems easy, but carrying it out is different. We start with Eq. (2.28.3):

$$1+\frac{1}{1!}+\frac{1}{2!}+\frac{1}{3!}+\cdots = \frac{a}{b}$$

where $a, b \in \mathbb{N}$.

The trick is to make b appear in the LHS of this equation:

$$\left(1+\frac{1}{1!}+\frac{1}{2!}+\cdots+\frac{1}{b!}\right)+\left(\frac{1}{(b+1)!}+\frac{1}{(b+2)!}+\cdots\right)=\frac{a}{b} \tag{2.28.4}$$

We can simplify the two red and blue terms. For the red term, using the fact that $b! = b(b-1)(b-2)\cdots 2!$, we can show that the red term is of this form $c/b!$ where $c \in \mathbb{N}$.

For the second term, we need to massage it a bit:

$$\begin{aligned}\frac{1}{(b+1)!}+\frac{1}{(b+2)!}+\cdots &= \frac{1}{(b+1)!}+\frac{1}{(b+2)(b+1)!}+\cdots\\ &= \frac{1}{(b+1)b!}+\frac{1}{(b+2)(b+1)b!}+\cdots\\ &= \frac{1}{b!}\left(\frac{1}{(b+1)}+\frac{1}{(b+2)(b+1)}+\cdots\right)\end{aligned}$$

Denote by x the blue term, we are going to show that $0 < x < 1/b$. In other words, x is a real number. Indeed,

$$x < \frac{1}{b+1}+\frac{1}{(b+1)^2}+\frac{1}{(b+1)^3}\cdots = \frac{1}{b+1}\left(1+\frac{1}{b+1}+\frac{1}{(b+1)^2}+\cdots\right)=\frac{1}{b}$$

where we used the formula for the geometric series in the bracket.

Now Eq. (2.28.4) becomes as simple as:

$$\frac{a}{b}=\frac{c}{b!}+\frac{1}{b!}x$$

Multiplying this equation with $b!$ to get rid of it, we have:

$$a(b-1)! = c + x$$

And this is equivalent to saying an integer is equal to the sum of another integer and a real number, which is nonsense!

Question 10 *If*

$$e = \lim_{n\to\infty}\left(1+\frac{1}{n}\right)^n$$

Then, what is

$$\lim_{n\to\infty}\left(1-\frac{1}{n}\right)^n = ?$$

Try to guess the result, and check it using a computer.

1	1
1 1	1
1 2 1	2
1 3 3 1	9
1 4 6 4 1	96
1 5 10 10 5 1	2500
1 6 15 20 15 6 1	162000
1 7 21 35 35 21 7 1	26471025

Fig. 2.41 Pascal triangle and the product of all the terms in a row $s_n = \prod_{k=0}^{n} \binom{n}{k}$ for $n = 0, 1, \ldots, 7$

2.28.3 Pascal Triangle and e Number

Let's recall the binomial theorem:

$$(a+b)^n = \sum_{k=0}^{n} \binom{n}{k} a^{n-k} b^k, \quad \binom{n}{k} = \frac{n!}{(n-k)!k!}$$

For a given n if we compute the product of all the binomial coefficients in that row, denoted by s_n, something interesting will emerge. We define s_n as[107]:

$$s_n = \prod_{k=0}^{n} \binom{n}{k} \tag{2.28.5}$$

The first few s_n are shown in Fig. 2.41. The sequence (s_n) grows bigger and bigger. How about the ratio s_n/s_{n-1}?

In Table 2.21, we compute s_n manually for $n = 1, 2, 3, 4, 5, 6$ and automatically (using a `Julia` script) for $n = 89, 90, 91, 899, 900, 901$. The ratios $r_n = s_n/s_{n-1}$ also grows un-boundedly, but the ratio of r_n converges to a value of 2.71677, which is close to e.

So, we suspect that the following is true

$$\lim_{n\to\infty} \frac{r_n}{r_{n-1}} = e, \quad \lim_{n\to\infty} \frac{s_n/s_{n-1}}{s_{n-1}/s_{n-2}} = e, \quad \lim_{n\to\infty} \frac{s_{n+1}/s_n}{s_n/s_{n-1}} = e \tag{2.28.6}$$

Note that when n is very big, n and $n-1$ are pretty the same. That is why in the above equation, we have different expressions they are equivalent.

---

[107] If, instead of a product we consider the sum of all the coefficients in the nth row we shall get 2^n. Check Fig. 2.41, row 3: $1+3+3+1 = 8 = 2^3$.

Table 2.21 $s_n = \prod_{k=0}^{n} \binom{n}{k}$, see Listing A.5 for the code

n	s_n	$r_n = s_n/s_{n-1}$	r_n/r_{n-1}
1	1	1	1
2	2	2	2
3	9	4.5	2.25
4	96	10.67	2.37
5	2500	26.042	2.44
6	162000	64.8	2.49
$\vdots$	$\vdots$	$\vdots$	$\vdots$
89	2.46e+1711		
90	1.77e+1673	5.13e+37	
91	2.46e+1711	1.39e+38	2.70
$\vdots$	$\vdots$	$\vdots$	$\vdots$
899	2.22e+174201		
900	2.17e+174590	9.74e+388	
901	5.74e+174979	2.65e+389	2.71677

Proof Herein we prove that Eq. (2.28.6) is true. First, we compute s_n:

$$s_n = \prod_{k=0}^{n} \binom{n}{k} = \prod_{k=0}^{n} \frac{n!}{(n-k)!k!} = (n!)^{n+1} \prod_{k=0}^{n} \frac{1}{(k!)^2} \tag{2.28.7}$$

To see the last equality, one can work out directly for a particular case. For $n = 3$, we have

$$s_3 = \prod_{k=0}^{3} \frac{n!}{(n-k)!k!} = \frac{3!}{3!0!} \times \frac{3!}{2!1!} \times \frac{3!}{1!2!} \times \frac{3!}{0!3!} = (3!)^4 \prod_{k=0}^{3} \frac{1}{(k!)^2}$$

We can write s_{n+1} and the ratio s_{n+1}/s_n as

$$s_{n+1} = ((n+1)!)^{n+2} \prod_{k=0}^{n+1} \frac{1}{(k!)^2} \Rightarrow \frac{s_{n+1}}{s_n} = \frac{((n+1)!)^n}{(n!)^{n+1}} = \frac{(n+1)^n}{n!} \tag{2.28.8}$$

Therefore, the other ratio s_n/s_{n-1} can be given by

$$\frac{s_n}{s_{n-1}} = \frac{(n)^{n-1}}{(n-1)!} \tag{2.28.9}$$

Finally,

$$\frac{s_{n+1}/s_n}{s_n/s_{n-1}} = \frac{(n+1)^n}{n!}\frac{(n-1)!}{n^{n-1}} = \left(1+\frac{1}{n}\right)^n \tag{2.28.10}$$

Given that $\lim_{n\to\infty}\left(1+\frac{1}{n}\right)^n = e$, the result follows.

■

2.29 Polynomials

A polynomial is an expression consisting of variables (also called indeterminates) and coefficients that involve only the operations of addition, subtraction, multiplication, and non-negative integer exponentiation of variables. An example of a polynomial of a single variable x is $x^2 - x + 2$. An example in three variables is $x^2 + 2xy^3z^2 - yz + 4$. The expression $1/x + x^2 + 3$ is not a polynomial due to the term $1/x = x^{-1}$ (exponent is -1, contrary to the definition).

Polynomials appear in many areas of mathematics and science. For example, they are used to form polynomial equations, they are used in calculus and numerical analysis to approximate other functions. For example, we have Taylor series and Lagrange polynomials to be discussed in Chaps. 4 and 10.

A polynomial in a single indeterminate x can always be written in the form

$$P_n(x) := a_nx^n + a_{n-1}x^{n-1} + \cdots + a_2x^2 + a_1x + a_0 = \sum_{k=0}^{n} a_kx^k, \quad a_k \in \mathbb{R} \tag{2.29.1}$$

The summation notation enables a compact notation (noting $x^0 = 1$). Assume that $a_n \neq 0$, then n is called the *degree of the polynomial* (which is the largest degree of any term with nonzero coefficient). Polynomials of small degree have been given specific names. A polynomial of degree zero is a constant polynomial (or simply a constant); that is, $P_0(x) = a_0$. Polynomials of degree one, two, or three are linear polynomials, quadratic polynomials and cubic polynomials, respectively. For higher degrees, the specific names are not commonly used, although quartic polynomial (for degree four) and quintic polynomial (for degree five) are sometimes used.

This section is about polynomials in a single indeterminate x. We start with the arithmetic of polynomials in Sect. 2.29.1. If we can add numbers, we can add $x^2 + 2x$ and $x^3 + 4x - 3$; why not? When we divide two polynomials, we care about the remainder, and mathematicians dicovered the polynomial reminder theorem while studying these remainders (Sect. 2.29.2). This theorem allows us to factor a polynomial if we know one root of it; for example $x^2 - 5x + 6 = (x-2)(x-3)$. But, how can we guess the root of a polynomial? Descartes gave us a theorem—the rational root theorem (Section 2.29.3). In Sect. 2.25.3, we have observed that the complex roots of $z^n - 1 = 0$ come in conjugate pairs. Now, we can prove that (Section 2.29.4). Section 2.29.5 presents the fundamental theorem of algebra stating that a nth degree polynomial has exactly n roots.

We then turn to the problem of evaluating a given polynomial at a specific point in a most efficient way. For example, given $P_2(x) = x^2 + 5x - 6$, what is $P_2(100)$? Sect. 2.29.6 gives a method discovered by Horner for this task. The section closes with Vieta formula relating the sum of the roots and their product with the coefficients of the polnomial (Sect. 2.29.7) and Girard-Newton indentities (Eq. 2.29.8).

2.29.1 Arithmetic of Polynomials

We can add (subtract), multiply (divide) two polynomials, in the same manner with numbers. Let's start with addition/subtraction of two polynomials:

$$\begin{aligned}(x^3 + 2x^2 - 5x - 1) + (x^2 - 3x - 10) &= (x^3 + 2x^2 - 5x - 1) + (0x^3 + x^2 - 3x - 10)\\ &= x^3 + (2+1)x^2 - (5+3)x - (1+10)\end{aligned}$$

Thus, the sum of two polynomials is obtained by adding together the coefficients of corresponding powers of x. Subtraction of polynomials is the same. And from two polynomials to m polynomials is a breeze thanks to Eq. (2.29.1). To see the power of compact notation, let $\sum_{k=0}^{n} a_k x^k$ be the first polynomial and $\sum_{k=0}^{n} b_k x^k$ be the second, then the sum of them is obviously $\sum_{k=0}^{n}(a_k + b_k)x^k$. It's nice, isn't it?

The next thing is the product of two polynomials:

$$\begin{aligned}(x^2 - 3x - 10)(2x^1 + 3) &= 2x^3 + 3x^2 - 6x^2 - 9x - 20x - 30\\ &= 2x^{cyan3} - 3x^2 - 29x - 30\end{aligned}$$

which comes from the usual arithmetic rules. What is interesting is that for two polynomials p and q, the degree of the product pq is the sum of the degree of p and q:

$$\deg(pq) = \deg(p) + \deg(q)$$

which is nothing but a consequence of the product rule of powers, discussed in Sect. 2.19.1.

The division of one polynomial by another is not typically a polynomial. (This is expected from experience with division of integers.) Instead, such ratios are a more general family of objects, called *rational fractions, rational expressions*, or *rational functions*, depending on context. This is analogous to the fact that the ratio of two integers is a rational number. For example, the fraction $2/(1 + x^3)$ is not a polynomial; it cannot be written as a finite sum of powers of the variable x; or in other words in the form of Eq. (2.29.1).

Let's divide $x^2 - 3x - 10$ by $x + 2$ and $2x^2 - 5x - 1$ by $x - 3$ using long division:

$$\begin{array}{r|l} & \quad x \quad -5, \\ \hline x+2) & x^2-3x-10 \\ & -x^2-2x \\ \hline & \quad -5x-10 \\ & \quad\ \ 5x+10 \\ \hline & \qquad\qquad 0 \end{array} \qquad \begin{array}{r|l} & \qquad 2x+1 \\ \hline x-3) & 2x^2-5x-1 \\ & -2x^2+6x \\ \hline & \qquad\ x-1 \\ & \quad\ -x+3 \\ \hline & \qquad\qquad 2 \end{array}$$

Thus, $x+2$ evenly divides $x^2-3x-10$ (similarly to 2 divides 6), but $x-3$ does not evenly divide $2x^2-5x-1$ for the remainder is non-zero. So, we can write

$$\frac{2x^2-5x-1}{cyanx-3} = 2x+1+\frac{2}{x-3} \iff 2x^2-5x-1 = (x-3)(2x+1)+2$$

The blue term is called the dividend, the cyan term is called the divisor, and the purple term is called the quotient. The red term is called the remainder term. And we want to understand it.

2.29.2 The Polynomial Remainder Theorem

If we "play" with polynomial division we could discover one or two theorems. Let's do it!

1. Divide x^2 by $x-1$. Record the remainder.
2. Divide x^2 by $x-2$. Record the remainder.
3. Divide x^2 by $x-3$. Record the remainder.

If we do all these little exercises (the answers are $1, 4$ and 9), we find out that the remainder of dividing x^2 by $x-a$ is nothing but a^2. Something special is hidden. Let's do a few more exercises to unravel the mystery:

1. Divide x^2+x+1 by $x-a$. Record the remainder.
2. Divide x^2+2x+3 by $x-a$. Record the remainder.

The answers are a^2+a+1 and a^2+2a+3, respectively. What are they? They are exactly the values obtained by evaluating the function at $x=a$. In other words, the remainder of dividing $f(x)=x^2+x+1$ by $x-a$ is simply $f(a)$. Now, no one can stop us from stating the following 'theorem': *if $P(x)$ is a polynomial, then the remainder of dividing $P(x)$ by $x-a$ is $P(a)$*. Of course mathematicians demand a proof, but I leave it as a small exercise. After a proof has been provided, this statement became *the remainder theorem*.

Now we can understand why if the equation $x^3-6x^2+11x-6=0$ has $x=1$ as one solution, we can always factor the equation as $(x-1)(\cdots)=0$. This is due to the remainder theorem that $f(x)=x^3-6x^2+11x-6=(x-1)(\cdots)+f(1)$. But $f(1)=0$ as 1 is one solution of $f(x)=0$, so $f(x)=(x-1)(\cdots)$. To find other solutions of this equation, we need to find out $(\cdots)$, which we denote by $g(x)$. This can be, of course, done via long division. But I really do not like that process. We can do another way using *the method of undetermined coefficients*. First, $g(x)$ must be of the form $x^2 + ax + b$ with a, b being real numbers to be determined. You should understand why $g(x)$ is a quadratic polynomial. Now, we factor $f(x)$:

$$x^3 - 6x^2 + 11x - 6 = (x-1)(x^2 + ax + b) = x^3 + (a-1)x^2 + (b-a)x - b$$

It is quite straightforward to determine a and b: well, we have two expressions for the same $f(x)$, thus the coefficients in the twos must be the same:

$$\begin{aligned} x^3 - 6x^2 + 11x - 6 = x^3 + (a-1)x^2 + (b-a)x - b \Longleftrightarrow a - 1 \\ = -6, \quad b - a = 11, \quad b = 6 \end{aligned}$$

Solving these three equations, we get $a = -5$ and $b = 6$. So, we can finally factor $f(x)$:

$$x^3 - 6x^2 + 11x - 6 = 0 \Longleftrightarrow (x-1)(x^2 - 5x + 6) = 0 \Longleftrightarrow \begin{cases} x - 1 = 0 \\ x^2 - 5x + 6 = 0 \end{cases}$$

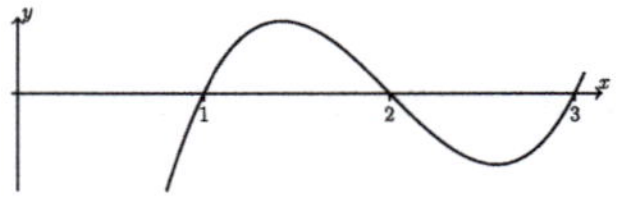

What we have actually achieved with the help of the remainder theorem? We have transformed *a hard cubic equation into two easier equations: one linear and one quadratic equation*! Of course this method works only if we can find out one solution. So, try to find one solution if you have to solve a hard (polynomial) equation. One way is to plot the graph of the polynomial (see figure). But there is an important theme here: it is a powerful idea to convert a hard problem into many smaller easier problems. That is the lesson we need to learn here. If the purpose is to find the solutions to $x^3 - 6x^2 + 11x - 6 = 0$, we can just use a tool to do that (see figure).

Remark While $x^3 - 6x^2 + 11x - 6 = 0$ can be factored as $(x-1)(x^2 - 5x + 6)$, it is impossible to do so for $x^2 + 1$ if we confine ourselves to real numbers. What would you do with this observation? Did you think of the concept of irreducibility of polynomials?

2.29.3 Guessing Roots of a Polynomial the Smart Way

Back in the time of Descartes, there was no calculator to make graphs. Descartes needed a way to guess *easy* roots of polynomial equations. By easy I meant rational roots of the form $x = p/q$ with $p, q \in \mathbb{Z}$. And he discovered what we know call the rational root theorem (or test). Suppose that we have to solve a quadratic equation with integer coefficients (e.g., $x^2 - 5x + 7 = 0$):

$$ax^2 + bx + c = 0, \quad a, b, c \in \mathbb{Z}$$

And we search for rational roots of the form $x = p/q$, where of course $q \neq 0$ and p, q are coprimes (i.e.,, the greatest common divisor of them is one or $gcd(p, q) = 1$). Plugging x into the equation we get $a(p/q)^2 + b(p/q) + c = 0$. Descartes then analyzed this equation to deduce something special about a, b, c and p, q. First, we rewrite this equation as

$$ap^2 + bpq = -cq^2 \Longleftrightarrow p(ap + bq) = -cq^2$$

This equation means that $p| - cq^2$: p divides $-cq^2$. From elementary number theory, we know that this leads to $p|c$. Next, we are doing something similar but for q, we rewrite the above equation as

$$ap^2 = -cq^2 - bpq \Longleftrightarrow ap^2 = -q(cq + bp)$$

This equation means that $q|ap^2$: q divides ap^2. From elementary number theory, we know that this leads to $q|a$. It is straightforward to generalize this to $a_n x^n + a_{n-1}x^{n-1} + \cdots + a_0 = 0$: the rational root $x = p/q$ of this equation, if any, must have $p|a_0$ and $q|a_n$.

The following example demonstrates how to use this test.

Example 2.3 Assume we want to guess the roots of the following quadratic equation with integer coefficients:

$$3x^2 - 4x + 1 = 0$$

If the root is of the form $x = p/q$, then $p|1$ and $q|3$. We now list all the possibilities:

$$p = \{-1, 1\}, \quad q = \{-1, 1, -3, 3\}$$

From that the potential roots should be: $x = \{\pm 1, \pm 1/3\}$. Checking these potential roots, we see that $1/3$ is the root. Then, using the factor theorem, we can find the second root.

2.29.4 *Complex Roots of* $z^n - 1 = 0$ *Come in Conjugate Pairs*

In Sect. 2.25.3, we have observed that the complex roots of $z^n - 1 = 0$ come in conjugate pairs. Now, we can prove that. Suppose that

$$p(x) = a_0 + a_1x + \cdots + a_nx^n, \quad a_i \in \mathbb{R}$$

is a polynomial of real coefficients. Let α be a complex root (or zero) of p, i.e.,, $p(\alpha) = 0$. We need to prove that the complex conjugate of α, i.e.,, $\overline{\alpha}$ is also a root. That is, $p(\overline{\alpha}) = 0$. The starting point is, of course, $p(\alpha) = 0$. So we write $p(\alpha)$

$$p(\alpha) = a_0 + a_1\alpha + \cdots + a_n\alpha^n$$

From that, we compute $p(\overline{\alpha})$:

$$\begin{aligned}
p(\overline{\alpha}) &= a_0 + a_1\overline{\alpha} + \cdots + a_n\overline{\alpha}^n \\
&= \overline{a}_0 + \overline{a}_1\overline{\alpha} + \cdots + \overline{a}_n\overline{\alpha}^n \quad (\overline{a} = a \text{ if } a \text{ is real}) \\
&= \overline{a_0 + a_1\alpha + \cdots + a_n\alpha^n} \quad (\overline{ab} = \bar{a}\bar{b}), \overline{a+b} = \overline{a} + \overline{b} \\
&= \overline{p(\alpha)} = \overline{0} = 0
\end{aligned}$$

2.29.5 The Fundamental Theorem of Algebra

The fundamental theorem of algebra is about the roots of an nth degree polynomial equation. It is easy to state, plausible, and hard to prove. We can guess this theorem reasoning like this. The linear equation $ax + b = 0$ has one root (i.e.,, $x = -b/a$), the quadratic equation $ax^2 + bx + c = 0$ has two roots (can be complex), the cubic equation has three roots. Thus, it is reasonable for Descartes to have stated, in his La Geometrie (1637), that "Every equation can have as many distinct roots as the number of dimensions of the unknown quantity in the equation". Even earlier, his fellow countryman Albert Girard (1590–1632) wrote, in his L'Invention Nouvelle en L'Algebra (1629) that "Every equation of algebra has as many solutions as the exponent of the highest term indicates".

What the theorem says is that every nth degree polynomial equation of the form

$$a_nz^n + a_{n-1}z^{n-1} + \cdots + a_1z + a_0 = 0, \quad a_1, a_2, \ldots \in \mathbb{C}, \quad n \in \mathbb{N}$$

always has exactly n roots in the complex number system. An important detail is that if there are identical roots of multiplicity m then they count as m roots, not just one.

Many famous mathematicians tried to prove this theorem, including d'Alembert, Euler, Lagrange. All failed. The first rigorous proof was published by Argand in 1806. In this book, we take this theorem on faith.

This theorem is useful in many ways. One example is: we know that a polynomial of real coefficients, if it has complex roots, they come in pairs. So, the three roots to a cubic polynomial either must all be real, or there must be one real root and one conjugate pair. There cannot be three complex roots because then there would have to be one complex root without a conjugate mate.

Question 11 *Given a real number a, how many cube roots of a are there? Write down the expression of these roots.*

2.29.6 Polynomial Evaluation and Horner's Method

Given a n-order polynomial $P_n(x) = a_n x^n + a_{n-1}x^{n-1} + \cdots + a_2 x^2 + a_1 x + a_0$, the polynomial evaluation is to compute $P_n(x_0)$ for any given x_0. You might be thinking, why we bother with this. We have the formula for $P_n(x)$, just plug x_0 into it and we're done. Yes, you're almost right, unless that this naive method is slow. We are moving to the land of applied mathematics where we must be pragmatic and speed is very important. Note that usually we need to compute a polynomial many many times. For example, to plot a polynomial we need to evaluate it at many points (because the more points we use the smoother the graph of the function is).

Let's consider a specific cubic polynomial $f(x) = 2x^3 - 6x^2 + 2x + 1$. The first "naive" solution for $f(x_0)$ is

$$f(x_0) = \underbrace{2 \times x_0 \times x_0 \times x_0}_{\text{3 multiplications}} - \underbrace{6 \times x_0 \times x_0}_{\text{2 multiplications}} + \underbrace{2 \times x_0}_{\text{1 multiplications}} + 1$$

which involves 6 multiplications and 3 additions. How about the general $P_n(x_0)$? To count the multiplication/addition, we need to write down the algorithm, Algorithm 1 is such one. Roughly, an algorithm is a set of steps what we follow to complete a task. There are more to say about algorithm in Sect. 2.34. Having this algorithm in hand, it is easy to covert it to a computer program.

Algorithm 1 Polynomial evaluation algorithm.

1: Compute $P = a_0$
2: **for** $k = 1 : n$ **do**
3: $\quad P = P + a_k x_0^k$
4: **end for**

From that algorithm we can count:

$$\begin{aligned} &\text{addition:} && n \\ &\text{multiplication:} && 1 + 2 + \cdots + n = n(n+1)/2 \end{aligned}$$

Can we do better? The British mathematician William George Horner (1786–1837) developed a better method today known as Horner's method. But he attributed to Joseph-Louis Lagrange, and the method can be traced back many hundreds of years to Chinese and Persian mathematicians.

In Horner's method, we massage $f(x_0)$ a bit as:

$$\begin{aligned} f(x_0) &= 2x_0^3 - 6x_0^2 + 2x_0 + 1 \\ &= x_0[2x_0^2 - 6x_0 + 2] + 1 \\ &= x_0[x_0(2x_0 - 6) + 2] + 1 \end{aligned}$$

which requires only 3 multiplications! For $P_n(x_0)$ Horner's method needs just n multiplications. To implement Horner's method, a new sequence of constants is defined recursively as follows:

$$\begin{aligned} b_3 &= \qquad a_3 & b_3 &= 2 \\ b_2 &= x_0 b_3 + a_2 & b_2 &= 2x_0 - 6 \\ b_1 &= x_0 b_2 + a_1 & b_1 &= x_0(2x_0 - 6) + 2 \\ b_0 &= x_0 b_1 + a_0 & b_0 &= x_0(x_0(2x_0 - 6) + 2) + 1 \end{aligned}$$

where the left column is for a general cubic polynomial whereas the right column is for the specific $f(x) = 2x^3 - 6x^2 + 2x + 1$. Then, $f(x_0) = b_0$. As to finding the consecutive b-values, we start with determining b_n, which is simply equal to a_n. We then work our way down to the other b's, using the recursive formula: $b_{n-1} = a_{n-1} + b_n x_0$ (this can be obtained by looking carefully at the left column in the above equation), until we arrive at b_0.

A by-product of Horner's method is that we can also find the division of $f(x)$ by $x - x_0$:

$$f(x) = (x - x_0)Q(x) + b_0, \quad Q(x) = b_3 x^2 + b_2 x + b_1 \tag{2.29.2}$$

And that allows us to evaluate the derivative of f at x_0, denoted by $f'(x_0)$, as

$$f'(x) = Q(x) + (x - x_0)Q'(x), \quad f'(x_0) = Q(x_0)$$

One application is to find all solutions of $P_n(x) = 0$. We use Horner's method together with Newton's method (Sect. 4.5.4). A good exercise to practice coding is to code a small program to solve $P_n(x) = 0$. The input is $P_n(x)$ and press a button we shall get all the solutions, nearly instantly!

Remark Yes, Horner's method is faster than the naive method. But mathematicians are still not satisfied with this. Why? Because there might be another (hidden) method that can be better than Horner's. Imagine that if they can prove that Horner's method is the best then they can stop searching for better ones. And they proved that it is the case. Details are beyond my capacity.

2.29.7 *Vieta's Formula*

We all know the solutions to the quadratic equation $ax^2 + bx + c = 0$; they are:

$$x_{1,2} = \frac{-b \pm \sqrt{b^2 - 4ac}}{2a}$$

What we get if we multiply these roots and sum them?

$$x_1 + x_2 = -\frac{b}{a}, \quad x_1x_2 = \left(\frac{-b + \sqrt{b^2 - 4ac}}{2a}\right)\left(\frac{-b - \sqrt{b^2 - 4ac}}{2a}\right) = \frac{c}{a} \tag{2.29.3}$$

That is remarkable given that the expression for the roots is quite messy: their sum and product are, however, very simple functions of the coefficients of the quadratic equation. And this is known as Vieta's formula discovered by Viète. Not many of high school students (including the author) after knowing the well known quadratic formula asked this question to discover for themselves this formula.

How useful Eq. (2.29.3) is? It can be used to determine the sign of the roots $x_{1,2}$ without solving the equation. If a and c are of different signs, then $ac < 0$, which results in $x_1x_2 < 0$: therefore, we have one negative root and one positive root.

Remark Did you notice something special about Eq. (2.29.3)? Note that $x_1 + x_2$ and x_1x_2 will not change if we switch the roots; i.e.,, $x_2 + x_1$ is exactly $x_1 + x_2$. Is this a coincidence? Of course not. The quadratic equation does not care how we label its roots.

After this, another question should be asked: Do we have the same formula for cubic equations, or for any polynomial equations? Before answering that question, we need to find a better way to come up with Vieta's formula. Because the formula of the roots of a cubic equation is very messy. And we really do not want to even add them not alone multiply them. As x_1 and x_2 are the roots of the quadratic equation, we can write that equation in this form (thanks to the discussion in Sect. 2.29.2)

$$(x - x_1)(x - x_2) = 0 \Longleftrightarrow x^2 - (x_1 + x_2)x + x_1x_2 = 0$$

And this must be equivalent to $x^2 + (b/a)x + c/a = 0$, thus we have $x_1 + x_2 = -b/a$ and $x_1x_2 = c/a$—the same result as in Eq. (2.29.3). This method is nice because we do not need to know the expressions of the roots. With this success, we can attack the cubic equation $x^3 + (b/a)x^2 + (c/a)x + d/a = 0$. Let's denote by x_1, x_2, x_3 its roots, then we write that cubic equation in the following form

$$\begin{aligned}(x - x_1)(x - x_2)(x - x_3) = 0 \Longleftrightarrow x^3 &- (x_1 + x_2 + x_3)x^2 \\ &+ (x_1x_2 + x_2x_3 + x_3x_1)x - x_1x_2x_3 = 0\end{aligned}$$

And thus comesVieta's formula for the cubic equation:

$$x_1 + x_2 + x_3 = -\frac{b}{a}, \quad x_1x_2 + x_2x_3 + x_3x_1 = \frac{c}{a}, \quad x_1x_2x_3 = -\frac{d}{a}$$

Summarizing these results for quadratic and cubic equations, we write (to see the pattern)

$$\begin{aligned}
a_2x^2 + a_1x + a_0 = 0 : \quad & x_1 + x_2 &= -\frac{a_1}{a_2}, \; x_1x_2 &= +\frac{a_0}{a_2} \\
a_3x^3 + a_2x^2 + a_1x + a_0 = 0 : \; & x_1 + x_2 + x_3 &= -\frac{a_2}{a_3}, \; x_1x_2x_3 &= -\frac{a_0}{a_3}
\end{aligned}$$

In the above equation, we see something new: $(x_1 + x_2, x_1 + x_2 + x_3)$, $(x_1x_2, x_1x_2 + x_2x_3 + x_3x_1)$ and $(x_1x_2, x_1x_2x_3)$. If we consider a fourth order polynomial we would see $x_1 + x_2 + x_3 + x_4$, $x_1x_2 + x_2x_3 + x_3x_4 + x_2x_3 + x_2x_4 + x_3x_4$, $x_1x_2x_3 + x_1x_2x_4 + x_1x_3x_4 + x_2x_3x_4$ and $x_1x_2x_3x_4$. As can be seen, these terms are all sums. Moreover, they are symmetric sums (e.g., the sum $x_1 + x_2 + x_3$ is equal to $x_2 + x_1 + x_3$). Now, mathematicians want to define these sums—which they call elementary symmetric sums—precisely. And this is their definition of elementary symmetric sums of a set of n numbers.

Definition 2.29.1 The kth elementary symmetric sum (or polynomial) of a set of n numbers is the sum of all products of k of those numbers $(1 \le k \le n)$. For example, if $n = 4$, and our set of numbers is $\{a, b, c, d\}$, then:

$$\begin{aligned}
\text{1st symmetric sum} &= S_1 = a + b + c + d \\
\text{2nd symmetric sum} &= S_2 = ab + ac + ad + bc + bd + cd \\
\text{3rd symmetric sum} &= S_3 = abc + abd + acd + bcd \\
\text{4th symmetric sum} &= S_4 = abcd
\end{aligned} \tag{2.29.4}$$

With this new definition, we can write the general Vieta's formula. For a nth order polynomial equation

$$a_nx^n + a_{n-1}x^{n-1} + \cdots + a_2x^2 + a_1x + a_0 = 0$$

we have

$$S_j = (-1)^j \frac{a_{n-j}}{a_n}, \quad 1 \le j \le n$$

where S_j is the jth elementary symmetric sum of a set of n roots. With a proper tool, we can have a compact Vieta's formula that encapsulates all symmetric sums of the roots of any polynomial equation!

If we do not know Vieta's formula, then finding the complex roots of the following system of equations:

$$\begin{aligned}
x + y + z &= 2 \\
xy + yz + zx &= 4 \\
xyz &= 8
\end{aligned}$$

would be hard. But it is nothing than this problem: 'solving this cubic equation $t^3 - 2t^2 + 4t - 8 = 0$'!

Some problems using Vieta's formula.

1. Let x_1, x_2 be roots of the equation $x^2 + 3x + 1 = 0$, compute

$$S = \left(\frac{x_1}{x_2 + 1}\right)^2 + \left(\frac{x_2}{x_1 + 1}\right)^2$$

2. If the quartic equation $x^4 + 3x^3 + 11x^2 + 9x + A$ has roots k, l, m and n such that $kl = mn$, find A.

For the first problem, the idea is to use Vieta's formula that reads $x_1 + x_2 = -3$ and $x_1x_2 = 1$. To use $x_1 + x_2$ and x_1x_2 we have to massage S so that these terms show up. For example, for the term x_1/x_2+1, we do (noting that $x_2^2 + 3x_2 + 1 = 0$, thus $x_2^2 + x_2 = -1 - 2x_2$)

$$\frac{x_1}{x_2 + 1} = \frac{x_1x_2}{x_2^2 + x_2} = \frac{x_1x_2}{-1 - 2x_2} = \frac{1}{-1 - 2x_2}$$

Do we need to do the same for the second term? No, we have it immediately once we had the above:

$$\frac{x_2}{x_1 + 1} = \frac{1}{-1 - 2x_1}$$

Now, the problem is easier as S can be written in terms of $x_1 + x_2$ and x_1x_2:

$$S = \frac{1}{(1 + 2x_1)^2} + \frac{1}{(1 + 2x_2)^2} = \frac{(1 + 2x_1)^2 + (1 + 2x_2)^2}{[(1 + 2x_1)(1 + 2x_2)]^2} = \cdots = 18$$

2.29.8 Girard-Newton's Identities

For a quadratic equation $ax^2 + bx + c = 0$ Viète gave us $x_1 + x_2 = -b/a$ and $x_1x_2 = c/a$. From these two, we can compute other things concerning the roots x_1, x_2. For example, we can compute $x_1^2 + x_2^2$ as follows[108]

$$(x_1 + x_2)^2 = x_1^2 + x_2^2 + 2x_1x_2 \Longrightarrow x_1^2 + x_2^2 = (x_1 + x_2)^2 - 2x_1x_2 \quad (2.29.5)$$

[108] I learned this from brilliant.org.

What is special about $x_1^2 + x_2^2$, why not $x_1^2 + x_2^3$? The term $x_1^2 + x_2^2$ is a special case of the so-called *symmetric polynomials*—those polynomials that are invariant when we swap x_1, x_2. For example, $x_1 + x_2 + 2x_1x_2$ is a symmetric polynomial. But $x_1^2 + x_2^3$ is not a symmetric polynomial. Equation 2.29.5 shows that it is possible to express a symmetric polynomial in terms of $x_1 + x_2$ and x_1x_2—the elementary symmetric polynomials. Now, you understand the adjective 'elementary': *the elementary symmetric polynomials are the 'building blocks' for all symmetric polynomials.*[109]

It's interesting, but can we do the same for $x_1^3 + x_2^3$ (what else?)? Let's do that now. We start with $(x_1 + x_2)^3 = x_1^3 + x_2^3 + 3x_1x_2(x_1 + x_2)$, then $x_1^3 + x_2^3$ is

$$\begin{aligned} x_1^3 + x_2^3 &= (x_1 + x_2)^3 - 3x_1x_2(x_1 + x_2) \\ &= (x_1 + x_2)[(x_1 + x_2)^2 - 3x_1x_2] \\ &= (x_1 + x_2)[x_1^2 + x_2^2 - x_1x_2] \end{aligned} \tag{2.29.6}$$

We could have stopped at the second equality where we succeeded in expressing the symmetric polynomial $x_1^3 + x_2^3$ in terms of the elementary symmetric sums. However, we went one step further: we wanted to write $x_1^3 + x_2^3$ in terms of $x_1^2 + x_2^2$.

Now, we put $x_1^2 + x_2^2$ and $x_1^3 + x_2^3$ together to see if there exists a pattern or not:

$$\begin{aligned} x_1^2 + x_2^2 &= (x_1 + x_2)(x_1^1 + x_2^1) - 2x_1x_2 = -\frac{b}{a}(x_1^1 + x_2^1) - \frac{c}{a}\boxed{2} \\ x_1^3 + x_2^3 &= (x_1 + x_2)[x_1^2 + x_2^2 - x_1x_2] = -\frac{b}{a}(x_1^2 + x_2^2) - \frac{c}{a}(x_1^1 + x_2^1) \end{aligned}$$

We're seeing a pattern here (pay attention to the exponents highlighted in red), except the ugly number 2. But, $2 = 1 + 1 = x_1^0 + x_2^0$. Voilà, we have now a nice formula, if we denote $P_i = x_1^i + x_2^i$—the sum of the ith powers of the roots (for example, if $i = 2$, we have $P_2 = x_1^2 + x_2^2$):

$$\boxed{P_i = -\frac{b}{a}P_{i-1} - \frac{c}{a}P_{i-2}, \quad i = 0, 1, 2, \ldots, \quad P_0 = 2, \quad P_1 = -\frac{b}{a}} \tag{2.29.7}$$

If this relation is correct, then we would have

$$P_4 = x_1^4 + x_2^4 = -\frac{b}{a}P_3 - \frac{c}{a}P_2 = -\frac{b}{a}(x_1^3 + x_2^3) - \frac{c}{a}(x_1^2 + x_2^2)$$

And the above is indeed correct! So, we think that Eq. (2.29.7) is correct and we will have to prove it. If you use proof by induction, you will see that it is indeed valid. Why is this formula powerful? Because it allows us to compute the sum of any powers of the roots (without knowing the roots themselves!). How we do this? We start with $P_0 = 2$ and $P_1 = -b/a$, Eq. (2.29.7) then gives us P_2, use it again and we

---

[109] This statement stems from the Fundamental Theorem on Symmetric Polynomials.

get P_3, again to get P_4. You see the point. Can you relate this to something we've met earlier in this chapter?[110]

Some problems using Girard-Newton's identities.

1. Given $x^2 - 2x + 6 = 0$ with two (complex) roots a, b. Compute $a^{10} + b^{10}$.
2. Given $x^3 - 3x^2 + 6x - 9 = 0$ with roots a, b, c. Compute $a^5 + b^5 + c^5$.
3. If the roots of $x^3 + 3x^2 + 4x - 8 = 0$ are a, b, c, compute $a^2(1 + a^2) + b^2(1 + b^2) + c^2(1 + c^2)$. Note[a].

[a] Ans: 7552, 108 and −126. If you are wondering what is the formula for the cubic equation, the answer is simple: derive a counterpart of Eq. (2.29.7) for the cubic equation. If possible, generalize it to polynomial equation of any degree.

2.29.9 *Convolution*

Let's return to the problem of multiplying two polynomials. For example, if we have $f(x) = 1 + 2x + 3x^2$ and $g(x) = 4 + 5x + 6x^2$, then we $h(x) = f(x)g(x)$ is given by:

$$\begin{aligned} h(x) &= 4 + 5x + 6x^2 + 8x + 10x^2 + 12x^3 + 12x^2 + 15x^3 + 18x^4 \\ &= 4 + 13x + 28x^2 + 27x^3 + 18x^4 \end{aligned}$$

There is nothing special here! Be patient. We can represent a polynomial by its coefficient: $f(x)$ by $a = (a_0, a_1, a_2) = (1, 2, 3)$ and $g(x)$ by $b = (b_0, b_1, b_2) = (4, 5, 6)$, and $h(x)$ by $c = (4, 13, 28, 27, 18)$. This is called the coefficient representation of polynomials.[111] Therefore, what we have achieved is to discover that given two lists of numbers (or two sequences) a and b we can make a new list $c = a * b$:

$$\begin{aligned} c_0 &= a_0 b_0 \\ c_1 &= a_0 b_1 + a_1 b_0 \\ c_2 &= a_0 b_2 + a_1 b_1 + a_2 b_0 \\ c_3 &= a_0 b_3 + a_1 b_2 + a_2 b_2, \quad c_4 = a_2 b_2 \end{aligned}$$

Such defined c is called a *convolution* of a and b. The formula for c is:

[110] The answer is Fibonacci sequence.

[111] Do you think of other representations for polynomials?

$$c_k = \sum_{i+j=k} a_i b_j = \sum_{i=0}^{n} a_i b_{k-i} \tag{2.29.8}$$

We shall meet convolution all over place in mathematics: in probability, in Fourier analysis (Fourier transform), in differential equations. If you want to become a electrical engineering, you will meet convolution all the time. It is a very important concept in signal processing.

2.30 Modular Arithmetic

I was writing this section on Friday and suppose I want to know what day of the week it will be 100 days from today. This is how we can solve this problem. We know that every 7 days, we will be back to the same day of the week. So if you start on a Friday, after 7 days it will also be Friday, and after 14 days it will still be Friday, and so on. The smallest multiple of 7 less than 100 is 98, so 98 days from today, it will be a Friday. Therefore, 100 days from today it will be a Sunday.

What we are really doing here is computing the remainder when 100 is divided by 7. That remainder is 2 (because $100 = 7 \times 14 + 2$). So the day of the week it will be 100 days from today is the same as the day of the week it will be 2 days from today. This is the basic concept of modular arithmetic, the art of computing remainders.

The best way to introduce modular arithmetic is to think of the face of a clock. The numbers go from 1 to 12 and then the clock "wraps around": when we get to "16 o'clock", it actually becomes 4 o'clock again. So 16 becomes 4, 14 becomes 2, and so on (Fig. 2.42). This can keep going, so when we get to "28 o'clock", we are actually back to where 4 o'clock is on the clock face (and also where 16 o'clock was too).

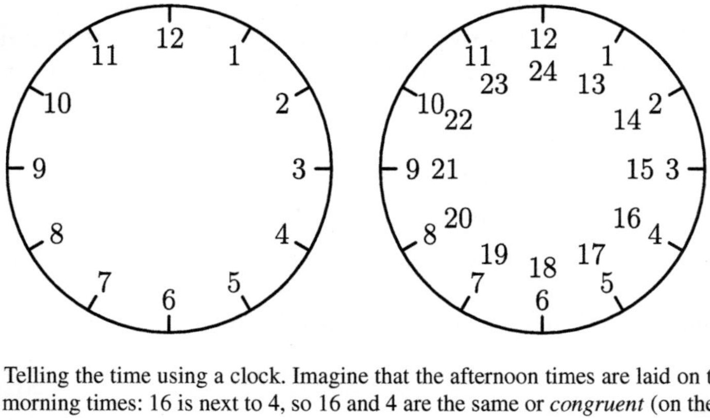

Fig. 2.42 Telling the time using a clock. Imagine that the afternoon times are laid on top of their respective morning times: 16 is next to 4, so 16 and 4 are the same or *congruent* (on the clock)

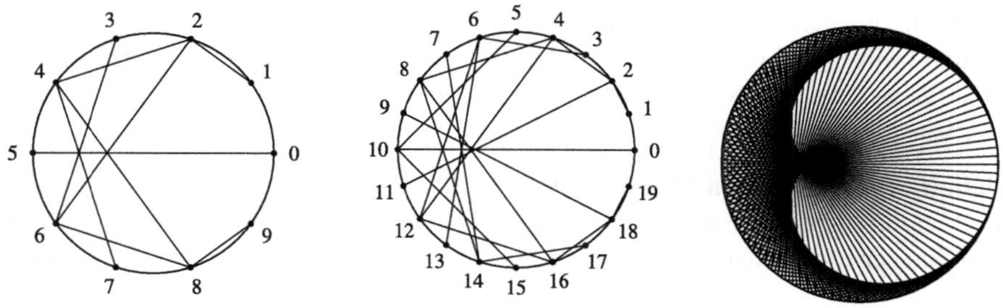

(a) $N = 10$ (b) $N = 20$ (c) $N = 200$

Fig. 2.43 A cardioid emerges from the times table of 2

(a) $N = 200$, table of 8 (b) $N = 200$, table of 51 (c) $N = 200$, 2.5

Fig. 2.44 Interesting things emerge from the times table of 8, 51 and 2.5

So in this clock world, we only care where we are in relation to the numbers 1 to 12. In this world, $1, 13, 25, 37, \ldots$ are all thought of as the same thing, as are $2, 14, 26, 38, \ldots$ and so on.

What we are saying is "$13 = 1 +$ some multiple of 12", and "$26 = 2 +$ some multiple of 12", or, alternatively, "the remainder when we divide 13 by 12 is 1" and "the remainder when we divide 26 by 12 is 2". The way mathematicians express this is:

$$13 \equiv 1 \pmod{12}, \quad 26 \equiv 2 \pmod{12}$$

This is read as "13 is congruent to 1 *mod* (or modulo) 12" and "26 is congruent to 2 mod 12".

But we don't have to work only in mod 12. For example, we can work with mod 7, or mod 10 instead. Now we can better understand the cardioid introduced in Chap. 1, re-given below in Fig. 2.43. Herein, we draw a line from number n to $2n \pmod{N}$ because on the circle we only have N points. For example, $7 \times 2 = 14$ which is congruent to 4 modulo 10. That's why we drew a line from 7 to 4.

Should we stop with the times table of 2? No, of course. We play with times table of three, four, and so on. Figure 2.44a shows the result for the case of eight and 51. How about times table for a non-integer number like 2.5? Why not? See Fig. 2.44c.

So, modular arithmetic is a system of arithmetic for integers, where numbers "wrap around" when reaching a certain value, called the modulus. The modern approach to modular arithmetic was developed by Gauss in his book Disquisitiones Arithmeticae, published in 1801.

Definition 2.30.1 Given an integer $n > 1$, called a modulus, two integers a and b are said to be congruent modulo n, if n is a divisor of their difference (that is, if there is an integer k such that $a - b = kn$ or $a = b + kn$). We then write $a \equiv b \pmod{n}$.

It is certain that there exist rules that govern modular arithmetic, and we discuss them in Sect. 2.30.1. Next, modular arithmetic is used to solve some number theory problems (Sect. 2.30.2). In Sect. 2.30.3, my very first attempt to solve a hard problem from a math contest is presented. Finally, we come back to the problem of divisibility of integers, but via modular arithmetic (Sect. 2.30.4).

2.30.1 Rules of Modular Arithmetic

Now that we have a new kind of arithmetic, the next thing is to find the rules it obey. Actually, some of the rules are simple: if $a_1 \equiv b_1 \pmod{m}$ and $a_2 \equiv b_2 \pmod{m}$, or if $a \equiv b \pmod{m}$ then

(a) addition

$$a_1 \pm a_2 \equiv b_1 \pm b_2 (\text{mod } m)$$

(b) multiplication

$$a_1 a_2 \equiv b_1 b_2 (\text{mod } m)$$

(c) exponentiation

$$a^p \equiv b^p (\text{mod } m), \quad p \in \mathbb{N}$$

(d) scaling

$$ak \equiv bk (\text{mod } m), \quad k \in \mathbb{N}$$

(d) transitivity

$$\text{if } a \equiv b (\text{mod } m) \text{ and } b \equiv c (\text{mod } m) \text{ then } a \equiv c (\text{mod } m) \tag{2.30.1}$$

It is possible to look at the clock on Fig. 2.42 to find examples demonstrating these rules. The proof of these rules is skipped here (only use the definition of $a \equiv b \pmod{m}$); noting that the exponentiation rule simply follows the multiplication rule. Of course, as often the case, we can extend the rules to the case of more numbers. For example, assuming that $a_1 \equiv b_1 \pmod{m}$, $a_2 \equiv b_2 \pmod{m}$ and $a_3 \equiv b_3 \pmod{m}$, then we have

$$a_1 \pm a_2 \pm a_3 \equiv b_1 \pm b_2 \pm b_3 \pmod{m}, \quad a_1 a_2 a_3 \equiv b_1 b_2 b_3 \pmod{m}$$

Note that this list is not meant to be exhaustive.

2.30.2 *Solving Problems Using Modular Arithmetic*

Let's solve some problems using this new mathematics. The first problem is: what is the last digit (also called the units digit) of the sum

$$2403 + 791 + 688 + 4339$$

Of course, we can solve this by computing the sum, which is 8 221, and from that the answer is 1. But, modular arithmetic provides a more elegant solution in which we do not have to add all these numbers.

First, we reduce these big numbers into smaller ones, by noting that[112]

$$2403 \equiv 3 \pmod{10}, \quad 791 \equiv 1 \pmod{10}, \quad 688 \equiv 8 \pmod{10}, \quad 4339 \equiv 9 \pmod{10}$$

Then, the addition rule in Eq. (2.30.1) leads to

$$2403 + 791 + 688 + 4339 \equiv 3 + 1 + 8 + 9 \pmod{10} \equiv 1$$

And the units digit of the sum is one. In this method, we had to only add 3, 1, 8 and 9.

The second problem is: Andy has 44 boxes of soda in his truck. The cans of soda in each box are packed oddly so that there are 113 cans of soda in each box. Andy plans to pack the sodas into cases of 12 cans to sell. After making as many complete cases as possible, how many sodas will he have leftover?

This word problem is mathematically translated as: finding the remainder of the product 44×113—which is the number of soda cans Andy has—when divided by 12. We have

$$44 \equiv 8 \pmod{12}, \quad 113 \equiv 5 \pmod{12}$$

Thus,

$$44 \times 113 \equiv 8 \times 5 \pmod{12} \equiv 40 \pmod{12} \equiv 4$$

So, the number of sodas left over is four.

In the third problem we shall move from addition to exponentiation. The problem is what are the tens and units digits of 7^{1942}? Of course, we find the answers without actually computing 7^{1942}.

Let's consider a much easier problem: what are the two last digits of 1235 using modular arithmetic. We know that $1235 = 12 \times 100 + 35$, thus $1235 \equiv 35 \pmod{100}$. So, we can work with modulo 100 to find the answer. Now, coming back to the original problem with 7^{1942}, of which the strategy is to do simple things first: computing the powers of 7[113] and looking for the pattern:

---

[112] Why using modulus 10?

[113] Do not forget the original problem is about 7^{1942}.

$$
\begin{array}{lll}
7^1 = 7 & : 7^1 \equiv 07 & \pmod{100} \\
7^2 = 49 & : 7^2 \equiv 49 & \pmod{100} \\
7^3 = 343 & : 7^3 \equiv 43 & \pmod{100} \\
7^4 = 2401 & : 7^4 \equiv 01 & \pmod{100} \\
\hline
7^5 = 16807 & : 7^5 \equiv 07 & \pmod{100} \\
7^6 = 117649 & : 7^6 \equiv 49 & \pmod{100} \\
7^7 = \ldots 43 & : 7^7 \equiv 43 & \pmod{100} \\
7^8 = \ldots & : 7^8 \equiv 01 & \pmod{100} \\
\hline
7^9 = \ldots & : 7^9 \equiv 07 & \pmod{100}
\end{array}
\tag{2.30.2}
$$

We're definitely seeing a pattern here, the last two digits of a power of 7 can only be either of 07, 49, 43, 01. Now, as 1942 is an even number, we just focus on even powers that can be divided into two groups: 2, 6, 10, ... and 4, 8, 12, ...[114] The first group can be generally expressed by $2 + 4k$ for $k = 0, 1, 2, \ldots$ Now, solving $2 + 4k = 1942$ gives us $k = 485$. Therefore, the last two digits of 7^{1942} are 49. (Note that if you try with the second group, a similar equation does not have solution, i.e.,, 1942 belongs to the first group).

Although the answer is correct, there is something fishy in our solution. Note that we only computed powers of 7 up to 7^9. There is nothing to guarantee that the pattern repeats forever or at least up to exponent of 1942. Of course, we can prove that this pattern is true using the multiplication rule. We can avoid going that way, by computing 7^{1942} directly by noting that $1942 = 5 \times 388 + 2$. Why this decomposition of 1942? Because $7^5 \equiv 7 \pmod{100}$. With this, we can write

$$
7^{1942} = \left((7^5)^{388}\right)(7^2) \equiv (7^{388})(49) \pmod{100} \tag{2.30.3}
$$

And certainly we play the same game for 7^{388}; writing $388 = 5 \times 77 + 3$, then $77 = 5 \times 15 + 2$, then $15 = 5 \times 3$, we have (do not forget the exponential rule for modular arithmetic)

$$
7^{388} \equiv (7^{77})(7^3) \equiv (7^{15})(7^2)(7^3) \equiv (7^3)(7^2)(7^3) \pmod{100} \tag{2.30.4}
$$

Now using Eq. (2.30.2) to have $7^2 \equiv 49 \pmod{100}$ and $7^3 \equiv 43 \pmod{100}$, Eq. (2.30.3) then becomes $7^{1942} \equiv 7^3 7^2 7^3 7^2$:

$$
7^{1942} \equiv 7^{10} \equiv 49 \pmod{100}
$$

As can be seen the idea is simple: trying to replace the large number (1942) by smaller ones![115]

---

[114] Why these groups? If not clear, look at Eq. (2.30.2): the period of 49 is 4.

[115] Now, consider this problem: finding the tens and units digits of 49^{971}? But wait, isn't it the same problem before? Yes, but you will find that working with powers of 49, instead of 7, is easier.

2.30.3 A Problem from a 2006 Hongkong Math Contest

Let's solve another problem, which is harder than previous problems. Consider a function f that takes a counting number a and returns a counting number obeying this rule

$$f(a) = (\text{sum of the digits of } a)^2$$

The question is to compute $f^{(2007)}(2^{2006})$, i.e.,, f is composed of itself 2007 times.[116] Why 2006? Because this problem is one question from a math contest in Hong Kong happening in the year of 2006.

Before we can proceed, we need to know more about the function f first. Concrete examples are perfect for this. If $a = 321$, then

$$f(321) = (3 + 2 + 1)^2 = 36$$

If we go blindly, this is what we would do: compute 2^{2006} (assuming that we're able to get it), know its digits and sum them and square the sum. Then, applying the same steps for this new number. And do this 2007 times. No, we simply cannot do all of this w/o a calculator. There must be another way.

Because I did not know where to start, I wrote a `Julia` program, shown in Listing 1 to solve it. The answer is then 169.

Listing 2.1 `Julia` program to solve the HongKong problem

```
function ff(x)
 digits_x = digits(x)    # get the digits of x and put in array
 return sum(digits_x)^2
end
function fff(x,n)        # repeatedly applying ff n times
    for i = 1:n
        x = ff(x)
    end
    return x
end
x = big(2)^2006          # have to use big integer as 2^2006 is very big
println(fff(x,2007))
```

But without a computer, how can we solve this problem? If we cannot solve this problem, *let's solve a similar problem but easier*, at least we get some points instead of zero! This technique is known as **specialization**, and it is a very powerful strategy. How about computing $f^{(5)}(2^4)$? That can be done as $2^4 = 16$:

[116] If you do not know what is a function composition, here it is: $f^2(x) = f(f(x))$. Evaluate f at x get the result and put it into the function again.

$$\begin{array}{llll}
f^{(1)}(2^4) = f(16) & = (1+6)^2 & = 49 \\
f^{(2)}(2^4) = f(49) & = (4+9)^2 & = 169 \\
f^{(3)}(2^4) = f(169) & = (1+6+9)^2 & = 256 \\
\hline
f^{(4)}(2^4) = f(256) & = (2+5+6)^2 & = 169 \\
f^{(5)}(2^4) = f(169) & = (1+6+9)^2 & = 256
\end{array}$$

The calculation was simple because 2^4 is a small number. What's important is that we see a pattern. With this pattern it is easy to compute $f^{(n)}(2^4)$ for whatever value of n, $n \in \mathbb{N}$.

So far so good. We made progress because we were able to compute 2^4, which is 16, then we can use the definition of the function f to proceed. For 2^{2006}, it is impossible to go this way. Now, we should ask this question: *why the function f is defined this way*, i.e.,, it depends on the sum of the digits of the input? Why not the product of the digits? Let's investigate the sum of the digits of a counting number. For example,

$$123 \Longrightarrow 1+2+3=6, \quad 4231 \Longrightarrow 4+2+3+1=10$$

If we check the relation between 6 and 123 and between 4231 and 10, we find this:

$$6 \equiv 123 \pmod 9, \quad 10 \equiv 4231 \pmod 9$$

That is: *the sum of the digits of a counting number is congruent to the number modulo 9.*[117] And then, according to the exponentiation rule of modular arithmetic, the square of sum of the digits of a counting number is congruent to the number squared modulo 9. For example, $36 \equiv 123^2 \pmod 9$.

With this useful 'discovery', we can easily do the calculations w/o having to know the digits of 2^4 (in other words w/o calculating this number; note that our actual target is 2^{2006}):

$$\begin{array}{l}
f^{(1)}(2^4) \equiv (2^4)^2 \equiv 4 \pmod 9 \\
f^{(2)}(2^4) \equiv (2^4)^4 \equiv 7 \pmod 9 \\
f^{(3)}(2^4) \equiv (2^4)^8 \equiv 4 \pmod 9 \\
f^{(4)}(2^4) \equiv (2^4)^8 \equiv 7 \pmod 9
\end{array} \tag{2.30.5}$$

Now, if we want to compute $f^{(4)}(2^4)$, we can start with the fact that it is congruent with $7 \pmod 9$. But wait, there are infinite numbers that are congruent with 7 modulo 9; they are $\{7, 16, 25, \ldots, 169, 178, \ldots, \}$. We need to do one more thing; if we can find a smallest upper bound of $f^{(4)}(2^4)$, let say $f^{(4)}(2^4) < M$, we then can remove many options and be able to find $f^{(4)}(2^4)$.

117 Of course, we need to prove this, two examples cannot be a proof. In this case, this is true. The proof is not hard. Actually we have stumbled on this in Sect. 2.4 when we're talking about the divisibility of a counting number with 9.

Now, we can try the original problem. Note that $2^{2006} \equiv 4 \pmod 9$,[118] then by similar reasoning as in Eq. (2.30.5), we get

$$f^{(n)}(2^{2006}) \equiv \begin{cases} 4 \pmod 9, & \text{if } n \text{ is even} \\ 7 \pmod 9, & \text{if } n \text{ is odd} \end{cases} \tag{2.30.6}$$

And we combine this with the following result (to be proved shortly):

$$f^{(n)}(2^{2006}) < 529\ (= 23^2), \quad \forall n \geq 8 \tag{2.30.7}$$

Now, we substitute $n = 2005$ in Eq. (2.30.6), we get $f^{(2005)}(2^{2006}) \equiv 7 \pmod 9$. And because the sum of the digits of a counting number is congruent to the number modulo 9, we now also have (transitivity rule)

$$\text{sum of digits of } f^{(2005)}(2^{2006}) \equiv 7 \pmod 9 \tag{2.30.8}$$

Now, using Eq. (2.30.7) for $n = 2006$, we have

$$f^{(2006)}(2^{2006}) = (\text{sum of digits of } f^{(2005)}(2^{2006}))^2 < 23^2$$

which leads to

$$\text{sum of digits of } f^{(2005)}(2^{2006}) < 23 \tag{2.30.9}$$

Combining the two results on the sum of the digits of $f^{(2005)}(2^{2006})$ given in Eqs. (2.30.8) and (2.30.9), we can see that it can only take one of the following two values:

$$\text{sum of digits of } f^{(2005)}(2^{2006}) = \{7, 16\}$$

which results in 169—the same number as the code returned:

$$f^{(2006)}(2^{2006}) = \{49, 256\} \Longrightarrow f^{(2007)}(2^{2006}) = 13^2 = 169$$

Proof Now is the proof of Eq. (2.30.7). We start with the fact that $2^{2006} < 2^{2007} = 8^{669} < 10^{669}$. In words, 2^{2006} is smaller than a number with 670 digits. By the definition of f, we then have

$$f(2^{2006}) < f(\underbrace{99\cdots 9}_{699 \text{ terms}}) = (9 \times 699)^2 < 10^8$$

This is because $99\ldots9$ with 699 digits is the largest number that is smaller than 10^{699} and has a maximum sum of the digits. Next, we do something similar for $f^{(2)}(2^{2006})$ starting now with 10^8:

---

[118] How can we know this? Simple: by computing the powers of 2 similarly to what we have done for 7 shown in Eq. (2.30.2).

$$f^{(2)}(2^{2006}) < f(\underbrace{99\cdots9}_{\text{8 terms}}) = (9 \times 8)^2 < 10^4$$

Then for $f^{(3)}(2^{2006})$:

$$f^{(3)}(2^{2006}) < f(9999) = (9 \times 4)^2 = 1296$$

And for $f^{(4)}(2^{2006})$:

$$f^{(4)}(2^{2006}) < f(1999) = (28)^2 = 784$$

And for $f^{(5)}(2^{2006})$:

$$f^{(5)}(2^{2006}) < f(799) = (25)^2 = 625$$

Continuing this way and we stop at $f^{(8)}(2^{2006})$:

$$f^{(8)}(2^{2006}) < f(599) = (23)^2 = 529$$

■

2.30.4 Divisibility with Modular Arithmetic

Divisibility by 3 and 9. Even though it is possible to study the divisibility of integers without using modular arithmetic, I want to use it to get familiar with it. Without loss of generality, let n be a four digit integer with digits a, b, c and d. So, we can write $n = 10^3a + 10^2b + 10c + d$. Now, we know that $3|n$ when $n = 3k$ which means that $n \equiv 0 \pmod 3$. Working with mod 3, we have $10 \equiv 1 \pmod 3$, hence $10^2 \equiv 1 \pmod 3$, $10^3 \equiv 1 \pmod 3$. Thus,

$$n = 10^3a + 10^2b + 10c + d \equiv 1a + 1b + 1c + d \pmod 3$$

Thus $3|n$ is when $1a + 1b + 1c + d \equiv 0 \pmod 3$, or when $a + b + c + d$ is a multiple of 3. Following the same argument, a number is a multiple of 9 when the sum of its digits is a multiple of 9.

Divisibility by 2 and 5. Working with modulo 2 (or 5), we have $10 \equiv 0 \pmod{2, 5}$, thus we just need to worry about the last digit of the number. An integer number is a multiple of 2 when its last digit is one of $\{0, 2, 4, 6, 8\}$. And an integer number is a multiple of 5 when its last digit is one of $\{0, 5\}$.

Divisibility by 4 and 8. Let's consider divisibility by 8 first. Working with modulo 8, we have $10 \equiv 2 \pmod 8$, which leads to

$$10^2 \equiv 4 \pmod 8,\ \ 10^3 \equiv 8 \pmod 8 \equiv 0 \pmod 8, \Longrightarrow 10^n \equiv 0 \pmod 8,\ n \geq 3$$

Thus, only the last three digits are responsible for the divisibility. And the rule is: an integer is divisible by 8 if the number represented by its last three digits is a multiple of 8.

Divisibility by 6. The divisibility of a number by 6, which is nothing but 2×3 depends on the divisibility of that number by 2 and 3. We have this result: a number is divisible by 6 iff it is divisible by both 2 and 3. For example, 12 is divisible by 6, and $2|12$ and $3|12$.

History note 2.7: Gauss (30 April 1777–1723 February 1855)

Johann Carl Friedrich Gauss was a German mathematician who made significant contributions to many fields in mathematics and science. Sometimes referred to as the Prince of Mathematics and "the greatest mathematician since antiquity", Gauss had an exceptional influence in many fields of mathematics and science, and is ranked among history's most influential mathematicians. Gauss was born in Brunswick to poor, working-class parents. His mother was illiterate and never recorded the date of his birth. Gauss was a child prodigy. In his memorial on Gauss, Wolfgang von Waltershausen wrote that when Gauss was barely three years old he corrected a math error his father made; and that when he was seven, solved an arithmetic series problem faster than anyone else in his class of 100 pupils.

While at the University of Göttingen, Gauss independently rediscovered several important theorems. His breakthrough occurred in 1796 when he showed that a regular polygon can be constructed by compass and straightedge if the number of its sides is the product of distinct Fermat primes and a power of 2. This was a major discovery in an important field of mathematics; and the discovery ultimately led Gauss to choose mathematics instead of philology as a career. Gauss was so pleased with this result that he requested that a regular heptadecagon[a] be inscribed on his tombstone. The stonemason declined, stating that the difficult construction would essentially look like a circle.

He further advanced modular arithmetic in his textbook *The Disquisitiones Arithmeticae* written when Gauss was 21. This book is notable for having an impact on number theory as it not only made the field truly rigorous and systematic but also paved the path for modern number theory. On 23 February 1855, Gauss died of a heart attack in Göttingen.

[a] A heptadecagon or 17-gon is a seventeen-sided polygon.

2.31 Cantor and Infinity

To start a short discussion on infinity, I use the story adapted from Strogatz's The joy of X:

> A boy is very excited about the number 100. He told me it is an even number and 101 is an odd number, and 1 million is an even number. Then the boy asked this question: "Is infinity even or odd'?'

This is a very interesting question as infinity is something unusual as we have seen in Sect. 2.21. Let's assume that infinity is an odd number, then two times infinity, which is also infinity, is even! So, infinity is neither even nor odd!

This section tells the story of the discovery made by a mathematician named Cantor that there are infinities of different sizes. I recommend the book *To Infinity and Beyond: A Cultural History of the Infinite* by Eli Maor [43] for an interesting account on infinity.

We start with Sect. 2.31.1 that presents a brief introduction to set theory. Next, we discuss finite and infinite sets (Sect. 2.31.2); examples of finite sets are $\{1, 2, 3\}$ and $\{3, 4, 5, 6\}$, and an example of an infinite set is $\{0, 1, 2, 3, \ldots\}$. Cantor showed us that these different sets are of the same size, and they are countable. Finally, Sect. 2.31.3 discusses the set of real numbers, which is uncountable. Now, we have countably infinite sets (such as $\mathbb{N}$, $\mathbb{Z}$, $\mathbb{Q}$) and uncountably infinite sets (such as $\mathbb{R}$).

2.31.1 Sets

Each of you is familiar with the word collection. In fact, some of you may have collections—such as a collection of stamps, a collection of PS4 games. *A set is a collection of things*. For example, $\{1, 2, 5\}$ is a set that contains the numbers 1, 2, and 5. These numbers are called the *elements* of the set. Because the order of the elements in a set is irrelevant, $\{2, 1, 5\}$ is the same set as $\{1, 2, 5\}$. Think of your own collection of marbles; you do not care the location of each invidual marble. And also think of {} as a polythene bag which holds its elements inside in such a way that we can see through the bag to see the elements. Furthermore, an element cannot appear more than once in a set; so $\{1, 1, 2, 5\}$ is equivalent to $\{1, 2, 5\}$.

To say that 2 is a member of the set $\{1, 2, 5\}$, mathematicians write $2 \in \{1, 2, 5\}$ and to say that 6 is not a member of this set, they write $6 \notin \{1, 2, 5\}$.

Of course the next thing mathematicians do with sets is to compare them. Considering two sets: $\{1, 2, 3\}$ and $\{3, 4, 5, 6\}$, it is clear that the second set has more elements than the first. We use the notation $|A|$, called the cardinality, to indicate the number of elements of the set A. The cardinality of a set is the size of this set or the number of elements in the set.

2.31.2 Finite and Infinite Sets

It is obvious that we have finite sets whose cardinalities are finite; for instance $\{1, 2, 3\}$ and $\{3, 4, 5, 6\}$, and infinite sets such as

$$\mathbb{N} = \{0, 1, 2, 3, \ldots\}$$

Things become interesting when we compare infinite sets. For example, Galileo wrote in his *Two New Sciences* about what is now known as Galileo's paradox:

1. Some counting numbers are squares such as $1, 4, 9$ and 16, and some are not squares such as $2, 5, 7$, and so on.
2. The totality of all counting numbers must be greater than the total of squares because the totality of all counting numbers includes squares as well as non-squares.
3. Yet, for every counting number, we can have a one-to-one correspondence between numbers and squares, for example (a doubled headed arrow $\leftrightarrow$ is used for this one-to-one correspondence)

$$\begin{array}{ccccccc} 1 & 2 & 3 & 4 & 5 & 6 & \cdots \\ \updownarrow & \updownarrow & \updownarrow & \updownarrow & \updownarrow & \updownarrow & \cdots \\ 1 & 4 & 9 & 16 & 25 & 36 & \cdots \end{array}$$

4. So, there are, in fact, as many squares as there are counting numbers. This is a contradiction, as we have said in point 2 that there are more numbers than squares.

The following example illustrates the one-to-one correspondence between two sets. Let's assume that a particular stadium can hold up to 50 000 people, and on a certain day, you notice that every seat is taken with no one standing. In this case, you can deduce that there are exactly 50 000 people present in the stadium, *without physically counting them*. This is because when two sets, in this case, the set of seats and the set of people, are in one-to-one correspondence, meaning every individual has a definite seat and every seat has one person, then the number of individuals in each set is the same.

The German mathematician Georg Cantor (1845–1918) solved this problem by introducing a new symbol $\aleph_0$ (pronounced *aleph-null*), using the first letter of the Hebrew alphabet with the subscript 0. He said that $\aleph_0$ was the cardinality of the set of natural numbers $\mathbb{N}$. Every set whose members can be put in a one-to-one correspondence with the natural numbers also has the cardinality $\aleph_0$. How Candor did that? He applied this idea of one-to-one correspondence to an infinite set.

With this new technique, we can show that the sets $\mathbb{N}$ and $\mathbb{Z}$ have the same cardinality. Their one-to-one correspondence is:

$$\begin{array}{cccccccc} 1 & 2 & 3 & 4 & 5 & 6 & 7 & \cdots \\ \updownarrow & \updownarrow & \updownarrow & \updownarrow & \updownarrow & \updownarrow & \updownarrow & \cdots \\ 0 & -1 & 1 & -2 & 2 & -3 & 3 & \cdots \end{array}$$

The next question is how about the set of rational numbers $\mathbb{Q}$? Is this larger or equal the set of natural numbers? Between 1 and 2, there are only two natural numbers, but there are infinitely many rational numbers. Thus, it is tempting for us to conclude that $|\mathbb{Q}| > |\mathbb{N}|$. Again, Cantor proved that we were wrong; $\mathbb{Q} = \aleph_0$!

For simplicity, we consider only positive rational numbers. A positive rational number is a number of this form p/q where $p, q \in \mathbb{N}$ and $q \neq 0$. First, Cantor arranged all positive rational numbers into an infinite array:

$$\begin{array}{cccccc} \frac{1}{1} & \frac{2}{1} & \frac{3}{1} & \frac{4}{1} & \frac{5}{1} & \cdots \\ \frac{1}{2} & \frac{2}{2} & \frac{3}{2} & \frac{4}{2} & \frac{5}{2} & \cdots \\ \frac{1}{3} & \frac{2}{3} & \frac{3}{3} & \frac{4}{3} & \frac{5}{3} & \cdots \\ \frac{1}{4} & \frac{2}{4} & \frac{3}{4} & \frac{4}{4} & \frac{5}{4} & \cdots \\ \frac{1}{5} & \frac{2}{5} & \frac{3}{5} & \frac{4}{5} & \frac{5}{5} & \cdots \\ \vdots & \vdots & \vdots & \vdots & \vdots & \ddots \end{array}$$

where the first row contains all rational numbers with denominator of one, the second row with denominator of two and so on. Note that this array has duplicated members; for instance $1/1, 2/2, 3/3, \ldots$ or $1/2, 3/6, 4/8$.

Next, he devised a zigzag way to traverse all the numbers in the above infinite array, once for each number:

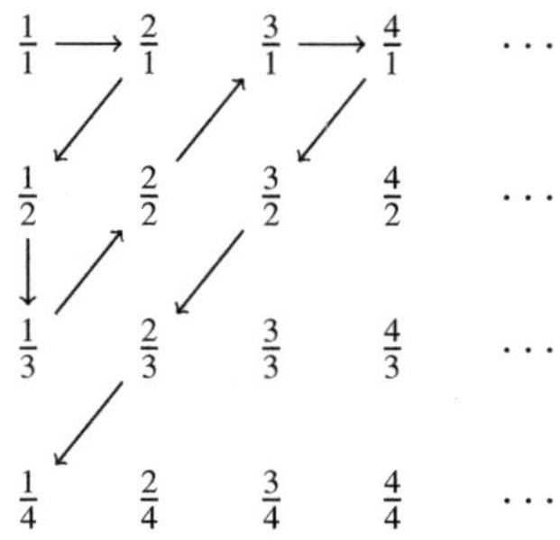

If we follow this zigzag path all along: one step to the right, then diagonally down, then one step down, then diagonally up, then again one step to the right, and so on ad infinitum, we will cover all positive fractions, one by one.[119] In this way we have arranged all positive fractions in a row, one by one. In other words, we can find a one-to-one correspondence for every positive rational with the natural numbers. This discovery that the rational numbers are countable-in defiance of our intuition- left such a deep impression on Cantor that he wrote to Dedekind: "Je le vois, mais je ne le crois pas!" ("I see it, but I don't believe it!").

Thus the natural numbers are countable, the integers are countable and the rationals are countable. It seems as if everything is countable, and therefore all the infinite sets of numbers you can care to mention—even ones our intuition tells contain more objects than there are natural numbers—are the same size.

This is not the case.

2.31.3 Uncountably Infinite Sets

The next thing Cantor showed us is that the set of real numbers $\mathbb{R}$ is uncountable; that is there is no one-to-one correspondence between $\mathbb{N}$ and $\mathbb{R}$. Recall that the real numbers are all the irrationals (those numbers that cannot be written as one integer divided by another: π, $\sqrt{2}$, $e, \ldots$) and the rationals together.

How he proved this? His proof consists of two steps:

- There are exactly the same number of points in any interval $[a, b]$ as in the number line $\mathbb{R}$.
- Using the above result, he proved that for the unit interval $[0, 1]$, there is no one-to-one correspondence between it and the set of natural numbers.

We focus on the second item.[120] You're might be guessing correctly that Cantor used a proof of contradiction. And the proof must go like this. First, he assumed that all

[119] Along our path we will encounter fractions that have already been met before under a different name such as 2/2, 3/3, 4/4, and so on; these fractions we simply cross out and then continue our path as before.

[120] You can prove the first item using ...geometry.

the decimals in [0, 1] is countable. Second he would artificially create a number that is not in those decimals.

The following proof is taken from Bellos' *Alex adventure in numberland*. It is based on Hilbert' hotel—a hypothetical hotel named after the German mathematician David Hilbert that has an infinite number of rooms. One day there are infinite number of guests arriving at the hotel. Each of these guests wears a T-shirt with a never-ending decimal between 0 and 1 (e.g., 0.415783113 . . .). The manager of this hotel is a genius and thus he was able to put all the guests in the rooms:

$$\begin{array}{ll}
\text{room 1: } & 0.4157831134213468\ldots \\
\text{room 2: } & 0.1893952093807820\ldots \\
\text{room 3: } & 0.7581723828801250\ldots \\
\text{room 4: } & 0.7861108557469021\ldots \\
\text{room 5: } & 0.638351688264940\ldots \\
\text{room 6: } & 0.780627518029137\ldots \\
\vdots & \vdots
\end{array}$$

Now what Cantor did was to build one real number that was not in the above list. Cantor used a diagonal method as follows. First, he constructed the number that has the first decimal place of the number in Room 1, the second decimal place of the number in Room 2, the third decimal place of the number in Room 3 and so on. In other words, he was choosing the diagonal digits that are underlined here:

$$\begin{array}{ll}
\text{room 1: } & 0.\underline{\mathbf{4}}157831134213468\ldots \\
\text{room 2: } & 0.1\underline{\mathbf{8}}93952093807820\ldots \\
\text{room 3: } & 0.75\underline{\mathbf{8}}1723828801250\ldots \\
\text{room 4: } & 0.786\underline{\mathbf{1}}108557469021\ldots \\
\text{room 5: } & 0.6383\underline{\mathbf{5}}1688264940\ldots \\
\text{room 6: } & 0.78062\underline{\mathbf{7}}518029137\ldots \\
\vdots & \vdots
\end{array}$$

That number is 0.488157 . . . Second, he altered all the decimals of this number; he added one to all the decimals. The final number is 0.599268 . . .. Now comes the best thing: This number is not in room 1, because its first digit is different from the first digit of the number in room 1. The number is not in room 2 because its second digit is different from the second digit of the number in room 2, and we can continue this to see that the number cannot be in any room n. Although Hilbert Hotel is infinitely large it is not enough for the set of real numbers.

So, now matter how big Hilbert's hotel is it cannot accommodate all the real numbers. The set of real numbers is said to be *uncountable*. Now, we have countably infinite sets (such as $\mathbb{N}$, $\mathbb{Z}$, $\mathbb{Q}$) and uncountably infinite sets (such as $\mathbb{R}$). With the right mathematics, Cantor proved that there are infinities of different sizes.

There are more to say about set theory in Sect. 5.5.

2.32 Number Systems

There is an old joke that goes like this:

> There are only 10 types of people in the world: those who understand binary and those who don't.

If you got this joke you can skip this section and if you don't, this section is for you.

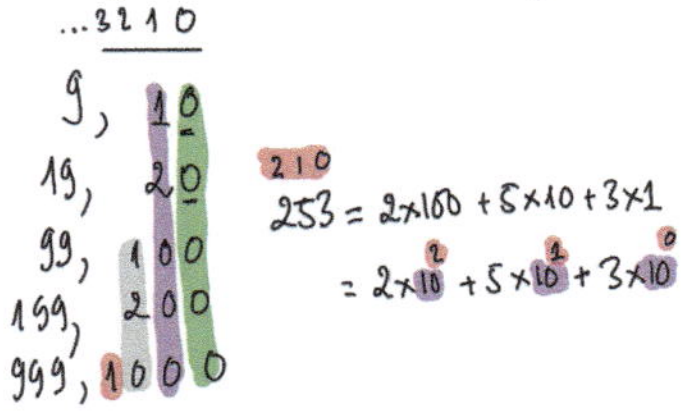

Computers only use two digits: 0 and 1; which are called the **bi**nary digi**t**s from which we have the word "bit". In that binary world, how we write number 2? It is 10. Now, you have understood the above joke. But why $10 = 2$? To answer that question we need to go back to the decimal system. For unknown reason we—human beings—are settled with this system. In this system there are only ten digits: $0, 1, 2, 3, 4, 5, 6, 7, 8, 9$. How we write ten books then? There is no such digit in our system! Note that we're allowed to use only $0, 1, 2, 3, 4, 5, 6, 7, 8, 9$. The solution is written ten as two digits: 10. To understand this more, we continue with eleven (11), twelve (12), until nineteen (19). How about twenty? We do the same thing: 20. Thus, any positive integer is a *combination of powers of 10*. Because of this 10 is called *the base* of the decimal system.

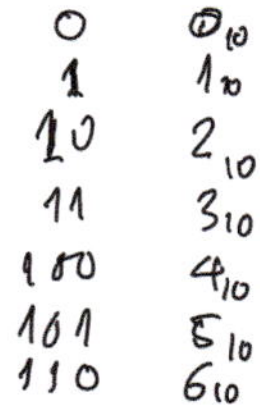

For the binary system, we do the same thing, but with powers of 2 of course. For example, $2_{10} = 10_2$; the subscripts to signify the number system; thus 10_2 is to denote the number ten in the binary system. Refer to the next figure to see the binary numbers for 1 to 6 in the decimal system. With this, it is straightforward to convert from binaries to decimals. For example $111_2 = 1 \times 2^2 + 1 \times 2^1 + 1 \times 2^0 = 7_{10}$. How about the conversion from decimals to binaries? We use the fact that *any binary*

is a combination of powers of two. For example, $75_{10} = 64 + 8 + 2 + 1 = 2^6 + 0 \times 2^5 + 0 \times 2^4 + 2^3 + 0 \times 2^2 + 2^1 + 2^0 = 1001011_2$.

One disadvantage of the binary system is the long binary strings of 1's and 0's needed to represent large numbers. To solve this, the "Hexadecimal" or simply "Hex" number system adopting the base of 16 was developed. Being a base-16 system, the hexadecimal number system therefore uses 16 different digits with a combination of numbers from 0 through to 15. However, there is a potential problem with using this method of digit notation caused by the fact that the decimal numerals of 10, 11, 12, 13, 14, and 15 are normally written using two adjacent symbols. For example, if we write 10 in hexadecimal, do we mean the decimal number ten, or the binary number of two (1 + 0). To get around this tricky problem hexadecimal numbers that identify the values of 10, 11, . . . , 15 are replaced with capital letters of A, B, C, D, E, and F, respectively.

So, let's convert the hex number E7 to the decimal number. The old rule applies: a hex number is a combination of powers of 16. Thus, $E7 = 7 \times 16^0 + 14 \times 16^1 = 231$. Similarly, we can convert a decimal number to a hex number. For example, to convert 1000, we write this number as a combination of powers of 16. That is,

$$1000_{10} = 768 + 232 = 3 \times 16^2 + 14 \times 16^1 + 8 \times 16^0$$

Thus,

$$1000_{10} = 3E8_{16}$$

2.33 Graph Theory

This section is a brief introduction to graph theory. It all started with a mundane entertaining exercise about crossing bridges back in eighteenth century in Europe (Sect. 2.33.1). I then present the problem of map coloring and the famous four color theorem (Sect. 2.33.2). Youngsters who would like to major in computer science should pay attention to graph theory.

2.33.1 The Seven Bridges of Königsberg

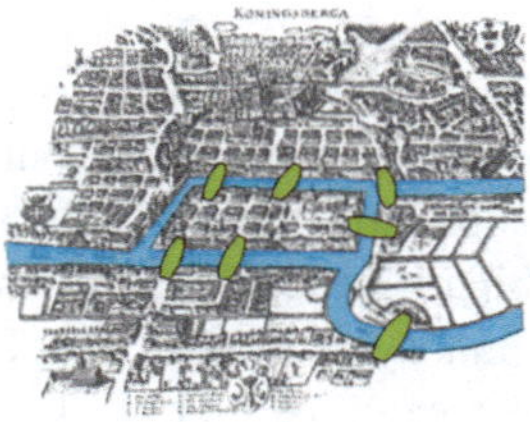

The city of Königsberg in Prussia (now Kaliningrad, Russia) was set on both sides of the Pregel River and included two large islands—Kneiphof and Lomse—which were connected to each other, and to the two mainland portions of the city, by seven bridges. According to lore, the citizens of Königsberg used to spend Sunday afternoons walking around their beautiful city. While walking, the people of the city decided to create a game for themselves: the game is to devise a way in which they could *walk around the city, crossing each of the seven bridges only once*. Furthermore, it is possible to start from one place and end the walk at another place. Even though none of the citizens of Königsberg could invent a route that would allow them to cross each of the bridges only once, still they could not understand why it was impossible. Lucky for them, Königsberg was not too far from St. Petersburg, home of the famous mathematician Leonard Euler.

Carl Leonhard Gottlieb Ehler, mayor of Danzig, asked Euler for a solution to the problem in 1736. And this is what Euler replied (from [31]) seeing no connection between this problem and current mathematics of the time:

> Thus you see, most noble Sir, how this type of solution bears little relationship to mathematics, and I do not understand why you expect a mathematician to produce it, rather than anyone else, for the solution is based on reason alone, and its discovery does not depend on any mathematical principle. Because of this, I do not know why even questions that bear so little relationship to mathematics are solved more quickly by mathematicians than by others.

Even though Euler found the problem trivial, he was still intrigued by it. In a letter written the same year to the Italian mathematician and engineer Giovanni Marinoni, Euler said,

> This question is so banal, but seemed to me worthy of attention in that [neither] geometry, nor algebra, nor even the art of counting was sufficient to solve it.

And as it is often the case, when Euler paid attention to a problem he solved it. Since neither geometry nor algebra (in other words current maths was not sufficient to solve this problem), in the process, he developed a new maths, which we now call graph theory.

The firs thing Euler did was to get rid of things that are irrelevant to the problem. Things such as color of the bridges, of the water, how big the landmasses are are all irrelevant. Thus, he drew a schematic of the problem shown in the left of Fig. 2.45. He labeled the landmasses as A, B, C, D and the bridges a, b, c, d, e, f, g. The problem is just the connection between these entities. Nowadays, we can go further: it is obvious that we do not have to draw the landmasses, we can represent them as dots, and the bridges as lines (or curves). In the right figure of Fig. 2.45, we did that and this is called a graph (denoted by G).

What information can we read from a graph? The first things are: number of vertices and number of edges. Is that all? If so, how can we differentiate one vertex from another? Thus, we have to look at the number of edges that can be drawn from a vertex. To save words, of course mathematicians defined a word for that: it is called the degree of a vertex. For example, vertex C has a degree of five whereas vertices A, B, D both have a degree of three.

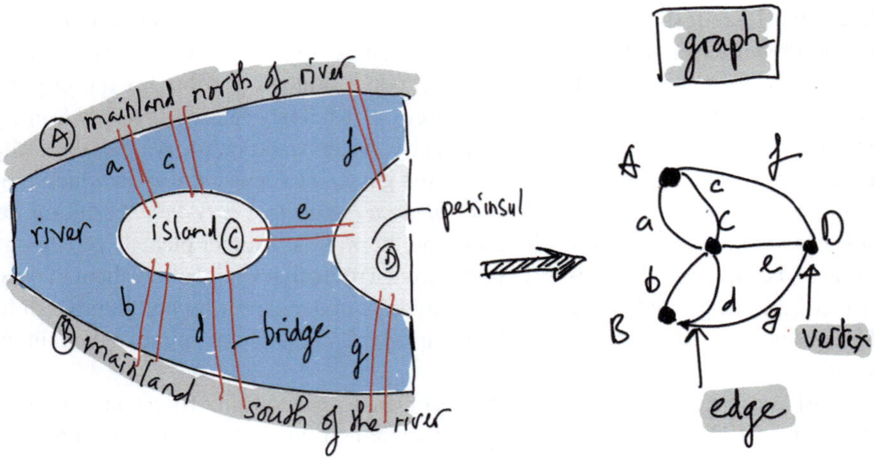

Fig. 2.45 The schematic of the Seven Bridges of Königsberg and its graph

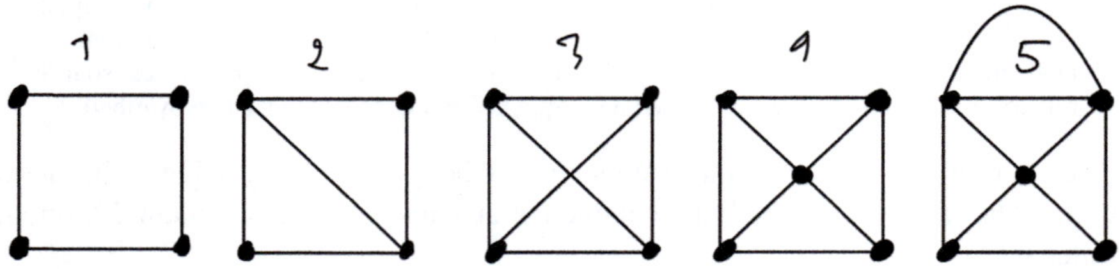

Fig. 2.46 Easy graphs to solve. The number on top each graph is to number them

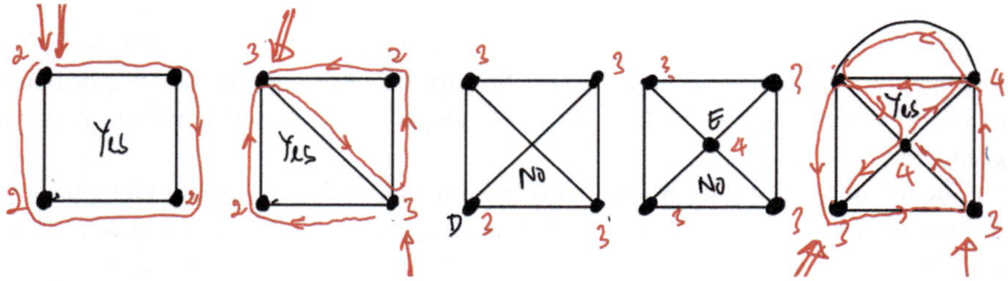

Fig. 2.47 Solution to easy graphs in Fig. 2.46. A single arrow indicates the starting vertex and a double arrow for the finishing vertex

Now, we are going to solve easier graphs and see the pattern. Then we come back to the Seven Bridges of Königsberg. We consider five graphs as shown in Fig. 2.46. Now, try to solve these graphs and fill in a table similar to Table 2.22 and try to see the pattern for yourself before continuing. Based on the solution given in Fig. 2.47, we can fill in the table.

What do we see from Table 2.22? We can only find a solution whenever the number of odd vertices is either 0 or 2. The case of 0 is special: we can start at any

Table 2.22 Results of graphs in Fig. 2.46. An odd vertex is a vertex having an odd degree

Shape	Number of odd vertices	Number of even vertices	Yes/No
1	0	4	Yes
2	2	2	Yes
3	4	0	No
4	4	1	No
5	2	3	Yes

vertex and we end up eventually at exactly the same vertex (Fig. 2.47). For the case of two: we start at an odd vertex, and end up at another odd vertex.

Now we understand why the Seven Bridges problem has no solution: the graph of this problem has 4 vertices with odd degrees, not 2.

What we did with Fig. 2.47 and Table 2.22 is to get familiar with a problem. Thanks to it, we understood the problem better and saw the solution. However, we still haven't explained why the graph must have zero or two vertices with odd degrees in order to start at one vertex and traverse each edge exactly once. To visualize this, let's imagine the vertices as tourist destinations in a sightseeing tour. In any tour, we have a starting point, an ending point, and the remaining points (intermediate points). And of course, the ending point can be the starting point. Once we have classified the points like that, we must ask: what's different about the starting/ending point compared to the intermediate points? *This question is the key that leads us to the solution!* At an intermediate point A, the number of edges must be even: we use 1 edge to get to A, use one edge to leave it; later if we need to return to A, we need two different edges (note that A is an intermediate point, so we must leave it first), and so on. From there, the degree of the intermediate point is even. Consequently, the degree of the starting/ending point can be odd. We have the following cases:

- Starting/ending point with an even degree: graph 1 in Fig. 2.47 belongs to this case. We can start from any vertex, and the tour will end at that vertex.
- Starting/ending point with an odd degree: graphs 2 and 5 in Fig. 2.47 belong to this case. And if there are more than 2 vertices with odd degrees, there is no solution."

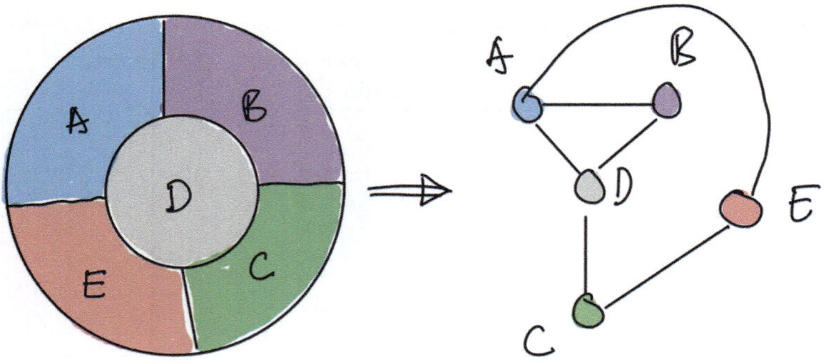

Fig. 2.48 Coloring a map is equivalent to coloring the vertices of its graph

2.33.2 Map Coloring and the Four Color Theorem

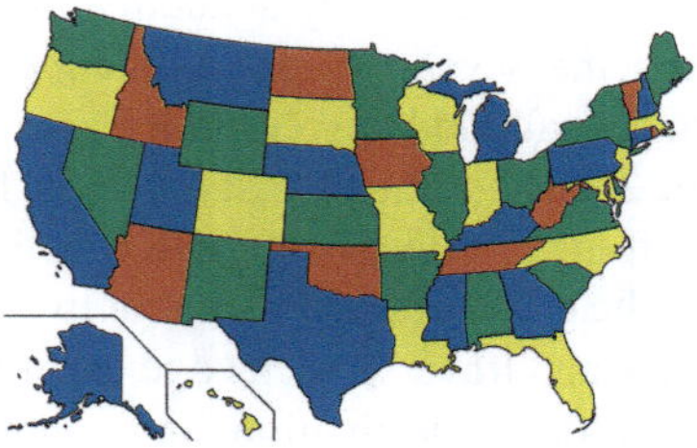

Francis Guthrie,[121] while trying to color the map of counties of England, noticed that only four different colors were needed. It became the four color theorem, or the four color map theorem, which states that no more than four colors are required to color the regions of any map so that no two adjacent regions have the same color.

How is this related to graph theory? Of course it is, this is no different from the seven bridges of Königsberg. Indeed, if we place a vertex in the center of each region (say in the capital of each state) and then connect two vertices if their states share a border, we get a graph. Suddenly, we see that many problems fall under the umbrella of graph theory! *Coloring regions on the map corresponds to coloring the vertices of the graph* (Fig. 2.48). Since neighboring regions cannot be colored the same, our graph cannot have vertices colored the same when those vertices are adjacent.

In general, given any graph G, a coloring of the vertices is called (not surprisingly) a *vertex coloring*. If the vertex coloring has the property that adjacent vertices are colored differently, then the coloring is called proper. Every graph has a proper vertex coloring; for example, you can color every vertex with a different color. But that's boring! Don;t you agree? To make life more interesting, we have to limit the number of colors used to a minimum. And we need a term for that number. The smallest

[121] Francis Guthrie (1831–1899) was a South African mathematician and botanist who first posed the Four Color Problem in 1852.

number of colors needed to get a proper vertex coloring is called the *chromatic number of the graph*, written $\chi(G)$.

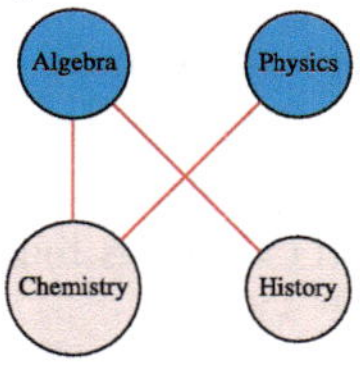

We do not try to prove the four color theorem here. No one could do it without using computers! It was the first major theorem to be proved using a computer (proved in 1976 by Kenneth Appel and Wolfgang Haken). Instead, we present one mundane application of graph coloring: exam scheduling. Suppose algebra, physics, chemistry and history are four courses in a college. And suppose that following pairs have common students: algebra and chemistry, algebra and history, chemistry and physics. If algebra and chemistry exam is held on the same day, students taking both courses then have to miss at least one exam. They cannot take both at the same time. *How do we schedule exams in minimum number of days so that courses having common students are not held on the same day?* You can look at the graphs, where an edge is drawn between classes having common students, and see the solution.[122]

That's all about graph for now. The idea is to inspire young students, especially those who want to major in computer science in the future. If you're browsing the internet, you are using a graph. The story goes like this. In 1998, two Stanford computer science PhD students, Larry Page and Sergey Brin, forever changed the World Wide Web as we know it. They created one of the greatest universal website used daily. Google.com is one of the most successful companies in the world. What was the basis for its success? It was the Google Search Engine that made Larry Page and Sergey Brin millionaires.

The Google Search Engine is based one simple algorithm called PageRank. PageRank is an optimization algorithm based on a simple graph. The PageRank graph is generated by having all of the World Wide Web pages as vertices and any hyperlinks on the pages as edges. To understand how it works we need not only graphs but also linear algebra (Chap. 9), probability (Chap. 5) and optimization theory. Yes, there is no easy road to prosperity and fame.

2.34 Algorithm

To end this chapter, I discuss a bit about algorithms for they are ubiquitous in our world. In Sect. 2.34.1, I present probably the oldest algorithm: the Euclidean algorithm to find greatest common divisor. Then, I show one application of this algorithm for a puzzle from the movie Die hard (Sect. 2.34.2).

[122] We used two colors, so we need 2 days.

2.34.1 Euclidean Algorithm: Greatest Common Divisor

Let's play a game: finding the greatest common divisor/factor (gcd) of two positive integers. The gcd of two integers is the *largest number that divides them both*. The manual solution is: (1) to list all the prime factors of these two numbers and (2) get the product of common factors and (3) that is the gcd, illustrated below for 210 and 84:

$$\begin{aligned} 210 &= ② \times ③ \times 5 \times ⑦ = ㊷ \times 5 \\ 84 &= ② \times ② \times ③ \times ⑦ = ㊷ \times 2 \end{aligned}$$

Thus, the gcd of 210 and 84 is 42: gcd(210, 84) = 42. Obviously if we need to find the gcd of two big integers, this solution is terrible. Is there any better way?

If d is a common divisor to both a and b (assuming that $a > b \geq 0$), then we can write $a = dm$ and $b = dn$ where $m, n \in \mathbb{N}$. Therefore, $a - b = d(m - n)$. What does this mean? It means that $d|(a - b)$ or d is also a divisor of $a - b$.[123] Conversely, if d is a common divisor to both $a - b$ and b, it can be shown that it is a common divisor to both a and b. Therefore, the set of common divisors of a and b is exactly the set of common divisors of $a - b$ and b. Thus, $\gcd(a, b) = \gcd(a - b, b)$. This is a big deal because *we have replaced the original problem with an easier (or smaller) one* for $a - b$ is smaller than a. Coming back to the problem of finding gcd(210, 84): this is how we now proceed: to find gcd(210, 84) we find gcd(126, 84) and to find gcd(126, 84) we find gcd(42, 84), which is equal to gcd(84, 42) which is equal to gcd(42, 42), and this is 42. I summarize what we have just done:

$$\begin{aligned} &\gcd(210, 84) \\ &\gcd(126, 84) \\ &\gcd(42, 84) &= \gcd(84, 42) \\ &\gcd(42, 42) &= 42 \end{aligned}$$

We did not have to do this calculation forever for $\gcd(a, a) = a$ for any integer a. This algorithm is better than the manual solution but it is slow: imagine we have to find the gcd of 1000 and 3, too many subtractions. But if we look at the algorithm we can see many repeated subtractions: for example $210 - 84 = 126$ and $126 - 84 = 210 - 84 - 84 = 42$. We can replace these two subtractions by a single division: $42 = 210 \bmod 84$ or $210 = 2 \times 84 + 42$. So, this is how we proceed:

$$\begin{aligned} &\gcd(210, 84) & &(210 = 2 \times 84 + 42) \\ &\gcd(84, 42) & &(84 = 2 \times 42 + 0) \\ &\gcd(42, 0) &= 42\ &(\gcd(a, 0) = a) \end{aligned}$$

Noting that $\gcd(a, 0) = a$, for $a \neq 0$, since any number is a divisor of 0, and the greatest divisor of a is a.

---

[123] One example: 5|10 and 5|25, and 5|(25 − 10) or 5 is a divisor of 15.

It's time for generalization. The problem is to find gcd(a, b) for $a > b > 0$. The steps are a repeated division: first a divide b to get the remainder r_1, then b divide r_1 to get the remainder r_2 and so on[124]:

$$\begin{array}{lll} \gcd(a,b) & (a = qb + r_1), & 0 \le r_1 < b \\ \gcd(b, r_1) & (b = q_1 r_1 + r_2), & 0 \le r_2 < r_1 \\ \gcd(r_1, r_2) & (r_1 = q_2 r_2 + r_3), & 0 \le r_3 < r_2 \\ \dots & \dots\dots & \end{array}$$

We have obtained a sequence of numbers:

$$b > r_1 > r_2 > r_3 > \cdots > 0$$

Since the remainders decrease with every step but can never be negative, eventually we must meet a zero remainder, at which point the procedure stops. The final nonzero remainder is the greatest common divisor of a and b.

What we have just seen is the Euclidean algorithm, named after the ancient Greek mathematician Euclid, who first described it in his *Elements* (c. 300 BC). It is an example of an algorithm, *a step-by-step procedure* for performing a calculation according to *well-defined rules*, and is one of the oldest algorithms in common use. About it, Donald Knuth wrote in his classic *The Art of Computer Programming*: "The Euclidean algorithm is the granddaddy of all algorithms, because it is the oldest nontrivial algorithm that has survived to the present day".

2.34.2 Puzzle from Die Hard

In the movie Die Hard 3, John McClane had to defuse a bomb by placing exactly 4 gallons of water on a sensor. The problem is, he only has a 5 gallon jug and a 3 gallon jug on hand. This problem may seem impossible without a measuring cup. But McClane solved it (and just in time) and I am sure so do you. What is interesting is how mathematicians solve it.

First, mathematicians consider a general problem not a specific one that McClane solved. This is because they're lazy and do not want to be in McClane's position more than one time. Second, in addition to solving the problem, they also wonder when the problem is solvable. Knowing how to answer this question will save them time when they have to solve this problem "With only a 2 gallon jug and a 4 gallon jug, how to get one gallon of water".

It is interesting to know that solution to this problem lies in the Euclidean algorithm. Take for example the problem of finding gcd(34, 19), using the Euclidean algorithm we do:

---

124 Note that when we do a long division for a/b we get a remainder always smaller than the divisor b.

$$\begin{aligned}
34 &= 19(1) + 15 \ , \ \gcd(19, 15) \\
19 &= 15(1) + 4 \ , \ \gcd(15, 4) \\
15 &= 4(3) + 3 \ , \ \gcd(4, 3) \\
4 &= 3(1) + \textcircled{1} \ , \ \gcd(3, 1) \\
3 &= 3(1) + 0 \ , \ \gcd(1, 0)
\end{aligned} \tag{2.34.1}$$

Thus, gcd(34, 19)=1: they are called coprimes or relatively primes. Now we go backwards, starting from the second last equation with the non-zero remainder of 1 which is the gcd of 34 and 19, we express 1 in terms of 3—the remainder of the previous step, then we do the same thing for 3. The steps are

$$\begin{aligned}
1 &= 4 - (1)3 \\
&= 4 - (1)(15 - 4(3)) = (4)4 - (1)15 && \text{(replaced 3by 3rd eq in (2.34.1))} \\
&= (4)[19 - (15)1] - (1)15 = 4(19) - (5)15 && \text{(replaced 4 by 2nd eq in (2.34.1))} \\
&= 4(19) - (5)[34 - 19(1)] = -5(34) + 5(19) && \text{(replaced 15 by 1st eq in (2.34.1))}
\end{aligned}$$

What did we achieve after all of this boring arithmetic? We have expressed gcd(34, 19), which is 1, as $-5(34) + 5(19)$. This is known as Bézout's identity: $\gcd(a, b) = ax + by$, where $a, b, x, y \in \mathbb{Z}$. In English, the gcd of two integers a, b can be written as an integral linear combination of a and b. (A linear combination of a and b is just a nice name for a sum of multiples of a and multiples of b.)

How does this identity help us to solve McClane's problem? Let $a = 5$ (5 gallon jug) and $b = 3$, then $\gcd(5, 3) = 1$. The Bézout identity tells us that we can always write $1 = 5x + 3y$, which gives us $4 = 5x' + 3y'$ (we need 4 as the problem asked for 4 gallons of water, and if you're wondering what is x', it is $4x$). It is easy to see that the solutions to the equation $4 = 5x' + 3y'$ are $x' = 2$ and $y' = -2$: $4 = 5(2) + (3)(-2)$. This indicates that we need to fill the 5-gallon jug twice and drain out (subtraction!) the 3-gallon jug twice. That's the rule to solving the puzzle.[125]

Now is time for this problem "With only a 2 gallon jug and a 4 gallon jug, how to get one gallon of water". Here $a = 4$ and $b = 2$, we then have $\gcd(4, 2) = 2$. Bézout's identity tells us that $2 = 4x + 2y$ (one solution is $(1, -1)$). But the problem asked for one gallon of water, so we need to find x' and y' so that $1 = 4x' + 2y'$. After having spent quite some time without success to find those guys x' and y', we came to a conjecture that 1 cannot be written as $4x' + 2y'$. And this is true, because the smallest positive integer that can be so written is the $gcd(4, 2)$, which is 2.[126]

---

[125] Details can be seen in the movie or youtube.

[126] Note that $d = \gcd(a, b)$ divides $ax + by$. If $c = ax' + by'$ then $d|c$, or $c = dn \geq d$. Thus d is the smallest positive integer which can be written as $ax + by$.

2.35 Review

We have done lots of things in this chapter. It's time to sit back and think deeply about what we have done. We shall use a technique from Richard Feynman to review a topic. In his famous lectures on physics [23], he wrote (emphasis is mine)

> If, in some cataclysm, all of scientific knowledge were to be destroyed, and only one sentence passed on to the next generations of creatures, what statement would contain the most information in the fewest words? I believe **it is the atomic hypothesis (or the atomic fact, or whatever you wish to call it) that all things are made of atoms**—little particles that move around in perpetual motion, attracting each other when they are a little distance apart, but repelling upon being squeezed into one another. In that one sentence, you will see, there is an enormous amount of information about the world, if just a little imagination and thinking are applied.

I emphasize that using Feynman's review technique is a very efficient way to review any topic for a good understanding of it (and thus useful for exam review). Only few key information are needed to be learned by heart, others should follow naturally as consequences. This avoids rote memorization, which is time consuming and not effective.

I have planned to do a review of algebra starting with just one piece of knowledge, but I soon realized that it is not easy. So I gave up. Instead I provide some observations (or reflection) on what we have done in this chapter (precisely on what mathematicians have done on the topics covered here):

- By observing objects in our physical world and deduce their patterns, mathematicians develop mathematical objects (e.g., numbers, shapes, functions, etc.) which are abstract (we cannot touch them).
- Even though mathematical objects are defined by humans, their properties are beyond us. We cannot impose any property on them, what we can do is just discover them.
- Quite often, mathematical objects live with many forms. For example, let's consider 1, it can be 1^2, 1^3 or $\sin^2 x + \cos^2 x$ etc. Using the correct form usually offers the way to something. And note that we also have many faces too.
- Things usually go in pairs: boys/girls, men/women, right/wrong, etc. They are opposite of each other. In mathematics, we have the same: even/odd numbers, addition/subtraction, multiplication/division, exponential/logarithm, and you will see differentiation/integration in calculus.
- Mathematicians love doing generalization. They first have arithmetic for numbers, then they have arithmetic for functions, for vectors, for matrices. They have two-dimensional and three-dimensional vectors (e.g., a force), and then soon they develop n-dimensional vectors where n can be any positive integer! Physicists only consider a 20-dimensional space. But the boldest generalization we have seen in this chapter was when mathematicians extended the square root of positive numbers to that of negative numbers. So $i = \sqrt{-1}$ was born and soon after complex

numbers $a + bi$. And you know what? Complex numbers play a vital role in quantum mechanics.[127]

- From a practical point of view all real numbers are rational ones. The distinction between rational and irrational numbers are only of value to mathematics itself. Our measurements always yield a terminating decimal, e.g., 3.1456789 which is a rational number.

Is this algebra the only one kind of algebra? No, no, no. Later on we shall meet vectors, and we have vector algebra and its generalization—linear algebra. We also meet matrices, and we have matrix algebra. We have tensors, and we have tensor algebra. Still the list goes on; we have abstract algebra and geometric algebra.

[127] Quantum mechanics is a fundamental theory in physics that provides a description of the physical properties of nature at the scale of atoms and subatomic particles.

Chapter 3
Geometry and Trigonometry

Geometry (means "earth measurement") is one of the oldest branches of mathematics. It is concerned with properties of space that are related to distance, shape, size, and relative position of figures. A mathematician who works in the field of geometry is called a geometer. The first known application of geometry was surveying in Egypt, where annual flooding of the Nile eradicated the demarcations between fields. Geometry provided the survival tools needed to accurately and efficiently redivide the land into individual holdings after the floods.

Trigonometry[1] (from Greek trigonon, "triangle" and metron, "measure") is a branch of mathematics that studies relationships between side lengths and angles of triangles. The field emerged during the third century BC, from applications of geometry to astronomical studies. This is now known as spherical trigonometry as it deals with the study of curved triangles, those triangles drawn on the surface of a sphere. Later, another kind of trigonometry was developed to solve problems in various fields such as surveying, physics, engineering, and architecture. This field is called plane trigonometry or simply trigonometry. And it is this trigonometry that is the main subject of this chapter.

In learning trigonometry in high schools a student often gets confused of the following facts. First, trigonometric functions are defined using a right triangle (e.g. sine is the ratio of the opposite and the hypotenuse). Second, trigonometric functions are later on defined using a unit circle. Third, the measure of angles is suddenly switched from degree to radian without a clear explanation. Fourth, there are two many trigonometry identities. And fifth, why do we have to spend time studying these triangles? In this chapter, I will try to make these issues clear.

My presentation of geometry and trigonometry does not follow its historical development. However, I nevertheless provide some historical perspective to the subject.

[1] Bartholomew Pitiscus (1561–1613) is responsible for the word trigonometry; he chose Trigonometria, a Greek-to-Latin transliteration of "triangle measurement", for the title of his book.

V. P. Nguyen, *Mathematics for Engineers and Scientists*,
https://doi.org/10.1007/978-981-97-4631-6_3

We start with the Euclidean geometry in Sect. 3.1. The topic of the area of curved plane figures is treated in Sect. 3.2. Then, we move to trigonometry. Section 3.3 introduces the trigonometry functions defined using a right triangle (e.g. $\sin x$). Then, trigonometric functions defined on a unit circle are discussed in Sect. 3.4. A presentation of degree versus radian is given in Sect. 3.5.

Section 3.6 lists the very basic properties that trigonometric functions obey (such as $\sin^2\alpha + \cos^2\alpha = 1$). We discuss how to compute the sine for angles between 0 and 360° in Sect. 3.7, without using a calculator of course. Trigonometric identities (e.g. $\sin(x + y) = \sin x \cos y + \sin y \cos x$ for all x, y) are then presented in Sect. 3.8. The identities for triangle angles are given in Sect. 3.9. We shall meet the Pascal triangle again in $\tan n\alpha$ (Sect. 3.10) and we shall enjoy how mathematician used imaginary numbers and calculus to get identities for $\tan n\alpha$ (Sect. 3.11) and for $\cos n\alpha$ and $\sin n\alpha$ (Sect. 3.12) for whatever n (say $n = 100$).

Section 3.13 outlines inverse trigonometric functions e.g. $\arcsin x$. Next, inverse trigonometric identities are treated in Sect. 3.14, where we see Machin's amazing formula to compute π. We devote Sect. 3.15 to trigonometric inequalities, a very interesting topic. Then in Sect. 3.16 we present trigonometric equations and how to solve them. The law of cosines (or generalized Pythagorean theorem) and the law of sines are treated in Sect. 3.17.

Graph of trigonometric functions are discussed in Sect. 3.18. Hyperbolic functions are treated in Sect. 3.19. Some applications of trigonometry are given in Sect. 3.20. A power series for the sine function, as discovered by ancient Indian mathematicians, is presented in Sect. 3.21. With it, it is possible to compute the sine for any angle. An interesting trigonometric identity of the form $\sin\alpha + \sin 2\alpha + \cdots + \sin n\alpha$ is treated in Sect. 3.22.

In Sect. 3.23 I briefly introduce spherical trigonometry as this topic has been removed from the high school curriculum. Analytic geometry in which coordinates and algebra are used to describe geometry is treated in Sect. 3.24 and a brief introduction to non-Euclidean geometries is given in Sect. 3.26. Finally, a brief introduction to CAS (computer algebra system) is given in Sect. 3.27, so that students can get acquaintance early to this powerful tool.

3.1 Euclidean Geometry

Where there is matter, there is geometry. (Johannes Kepler)

I did not enjoy Euclidean geometry in high school. I am not sure why but it might be due to the following reasons. First, it requires a compass and straightedge and everything should be perfect; in contrast, with just a pencil and I can do algebra. Second, a geometry problem is too narrow in the sense that there are too many things (e.g. triangles, circles, points,...) inside a diagram and it's hard to see what is going on.

Fig. 3.1 Geometry in nature: circle, rectangle, and hexagon (from left to right). We also see circles in the ripples on a pond, in the human eye, and on butterflies' wings

But this book would be incomplete without mentioning Euclidean geometry, especially Euclid's *The Elements*. Why? Because Euclid's *Elements* has been referred to as the most successful and influential textbook ever written. It has been estimated to be second only to the Bible in the number of editions published since the first printing in 1482. Moreover, without a proper introduction of Euclid's geometry it would be awkward to talk about trigonometry—a branch of mathematics which is based on geometry.

In *The Elements* Euclid presented what we now call Euclid's geometry, or Euclidean geometry. Written about 300 B.C., it contains the results produced by fine Greek mathematicians such as Thales, Hippias, the Pythagoreans, Hippocrates, and Eudoxus. The Elements begins with plane geometry: lines, circles, triangles, polygons, and so on. These shapes are abstracts of the real geometries we observe in nature (Fig. 3.1). It goes on to the solid geometry of three dimensions. Much of the Elements states results of what are now called algebra and number theory, explained in geometrical language.

Euclidean geometry is an example of *synthetic geometry*, in that it proceeds logically from axioms describing basic properties of geometric objects such as points and lines, to propositions about those objects, all without the use of coordinates to specify those objects. This is in contrast to *analytic geometry*—which was created by René Descartes and Pierre de Fermat in the seventeenth century—which uses coordinates to translate geometric propositions into algebraic formulas. This analytic geometry will be briefly discussed in Sect. 3.24.

This short section cannot present all the details of Euclidean geometry. Instead only a few important topics (the choice is somewhat personal) of it are presented. First, some basic concepts are first introduced (Sect. 3.1.1). Then, the structure of *The Elements* is treated in Sect. 3.1.2. Next, the influence of this masterpiece is discussed (Sect. 3.1.3). The difference between algebraic thinking and geometric thinking is given in Sect. 3.1.4.

I go into detail in the remaining sections. The fifth postulate is studied in Sect. 3.1.5. The area of simple geometrical shapes is treated in Sect. 3.1.6. The topic of congruence and similarity is given in Sect. 3.1.7. Section 3.1.8 presents a discussion related to the question "is square a rectangle"; Sect. 3.1.9 is devoted to interior and exterior angles in convex polygons. Some important theorems about circles are presented in Sect. 3.1.10. This is followed by Sect. 3.1.11 in which some topics on tangent lines to a circle are given.

Section 3.1.12 is about geometric construction—that is the construction of geometric figures using only straight edge and compass. And in that context, the three classical problems of antiquity are discussed: the problem of doubling the cube, the problem of trisecting an angle, and the problem of squaring the circle (Sect. 3.1.13). I mention those problems not only because they are famous problems of more than 2000 years, but also because of their tight connection to algebra.

Tessellation—the mathematics of tiling—is discussed in Sect. 3.1.14. Section 3.1.15 is about regular polyhedra or Platonic solids. And Sect. 3.1.16 discusses Euler's polyhedra formula. This formula is not found in the standard curriculum taught in the schools. Some high school students may know Euler's formula, but most students of mathematics do not encounter this relation until college.

I have read and learned from the following excellent books:

- Paul Lockhart's *Measurement* [42].
- *The Wonder Book of Geometry: A Mathematical Story* by David Acheson [1].
- *Geometry by Its History* by Alexander Ostermann and Gerhard Wanner [3].
- *Tales of Impossibility: The 2000-Year Quest to Solve the Mathematical Problems of Antiquity* by David Richeson [56].
- *The Four Pillars of Geometry* by John Stillwell [63].

If you want to read Euclid's *Elements*, check http://aleph0.clarku.edu/~djoyce/elements/bookI/bookI.html this website out.[2] I still do find solving geometric problems hard. But technology helps; using a colorful iPad really helps and software such as https://www.geogebra.org is very useful.

3.1.1 Basic Concepts

Euclid's geometry operates with basic objects such as points, lines, triangles, circles, and polygons (Fig. 3.2). We are all familiar with the two-dimensional regular polygons. A polygon is regular if every side has the same length and every interior angle has the same measure.[3] The equilateral triangle is the only regular three-sided polygon, the square is the only regular four-sided polygon, and so forth (Fig. 3.3). *There are infinitely many regular polygons*: one n-sided polygon for every integer n greater than two. We paid attention to these regular polygons and circles because beauty is often found in regularity, symmetry, and perfection.

Triangles, rectangles, circles etc.. are two-dimensional (2D) shapes as they live on a plane (of the paper we use to draw them).

---

[2] Link: http://aleph0.clarku.edu/~djoyce/elements/bookI/bookI.html.

[3] Note that this definition is not needed for the equilateral triangles: we can just simply require that either the three sides are congruent or the three angles are congruent. We cannot do that for n-gons for $n \geq 4$. Think of a square and push two opposite vertices a bit, it will become a diamond of which four sides have the same lengths but not the same angles. Because of this triangles are rigid.

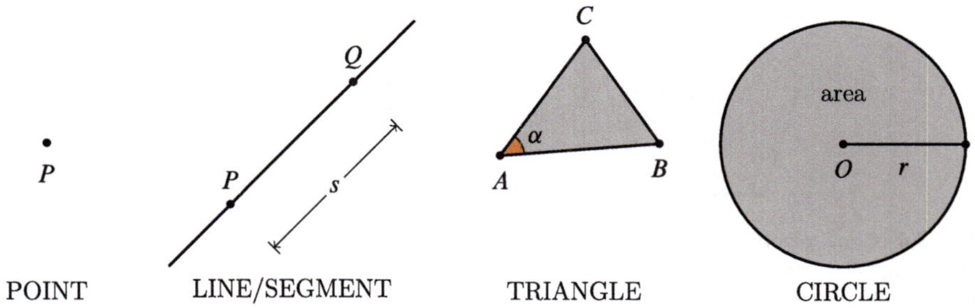

Fig. 3.2 Two-dimensional basic objects in Euclid's geometry. Naming convention: Points are customarily named using capital letters of the alphabet (e.g. $A, B, C, D, \ldots$). Other figures, such as lines, triangles, or circles, are named by listing a sufficient number of points to pick them out unambiguously from the relevant figure, e.g. triangle ABC is a triangle with vertices at points A, B, and C. Angles can be identified by the three points that define them. For example, the angle with vertex A formed by the rays AB and AC (that is, the lines from point A to points B and C) is denoted by $\angle BAC$ (or $\angle CAB$). The symbols α, β, γ etc. are often used to denote the size of angles; e.g. $\alpha = 30°$. To a segment, we assign a positive real number s to quantify its length and to a close figure such as triangles/circles we assign a positive real number called area that quantifies the space inside the figure

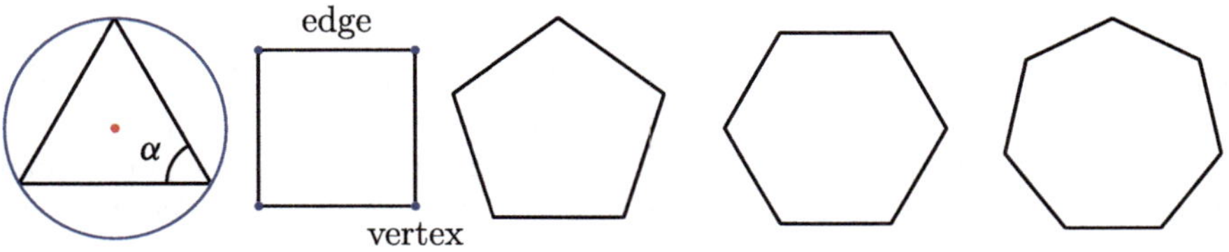

Fig. 3.3 Two-dimensional basic objects in Euclid's geometry: regular polygons with 3, 4, 5, 6 and 7 sides. For an n-gon (i.e., an n-sided polygon) the interior angle α in degrees is: $\alpha = 180°(1 - 2/n)$. For a proof, check Sect. 3.1.9. For a regular polygon, there always exists a circle that passes through all the vertices of the polygon; this circle is called the circumscribed circle or circumcircle (shown for the $n = 3$ case as the blue circle)

All polygons in Fig. 3.3 are special: they are called convex polygons. The line segment between two points of a convex polygon is contained in the union of the interior and the boundary of the polygon (Fig. 3.4). Compared with non-convex polygons, convex polygons are not only more beautiful but also much easier to work with.

Euclidean geometry also considers three-dimensional (3D) solids such as cubes, pyramids, cylinders, cones, and spheres (Fig. 3.5). We call a cube a 3D object as it has three dimensions: width, height, and thickness. It then studies the properties of these 2D/3D objects such as the length of a segment (i.e., a part of a line), the angle between two non-parallel lines, the area of a triangle/circle, the volume of a sphere/cone etc.. This chapter, however, does not discuss the volume problem. I postpone this topic to Sect. 4.3 in Chap. 4 about calculus. This is so because initial

Fig. 3.4 Convex polygons (left) versus non-convex polygons (right). The line segment between two points of a convex polygon is contained in the union of the interior and the boundary of the polygon i.e., the shaded region

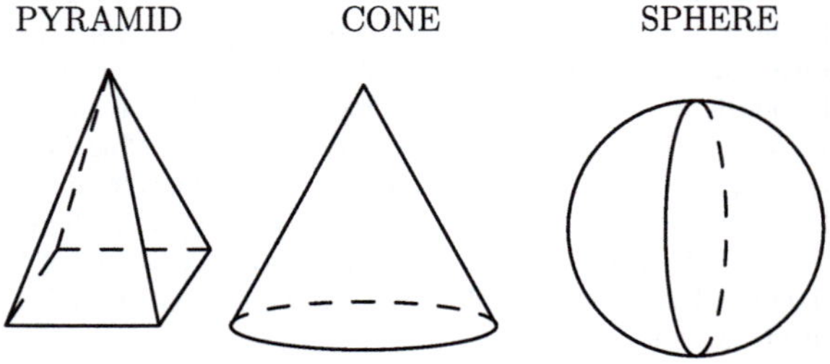

Fig. 3.5 Three-dimensional basic objects in Euclid's geometry: (square based) pyramid, cone, and sphere

ideas of calculus originate from the way ancient mathematicians such as Archimedes calculated the area and volume of curved shapes (i.e., circle, cylinder, sphere).

Similar to numbers—an abstract concept—objects in geometry (e.g. points, lines etc..) are also abstract. For example, a point does not have size. A line does not have thickness and a line in geometry is perfectly straight! And certainly, mathematicians don't care if a line is made of steel or wood. There are no such things in the physical world.

3.1.2 The Structure of Euclid's Elements

> No man can talk well unless he is able first of all to define to himself what he is talking about. Euclid, well studied, would free the world of half its calamities, by banishing half the nonsense which now deludes and curses it. (Abraham Lincoln)

The structure of Euclid's *Elements* is as follows:

1. Twenty-three definitions of the basic concepts: point, line, triangle, circle etc.. Euclid defined a point as: A point is that which has no part (definition 1 in Book I). Definition 23 is about parallel lines: lines that never meet, think of straight railway tracks for example.
2. Five common notions and five postulates on which all subsequent reasoning is based. For example, common notion 1 states that "Things equal to the same thing are equal to each other" (which we now write if $a = b$, $c = b$ then $a = c$).

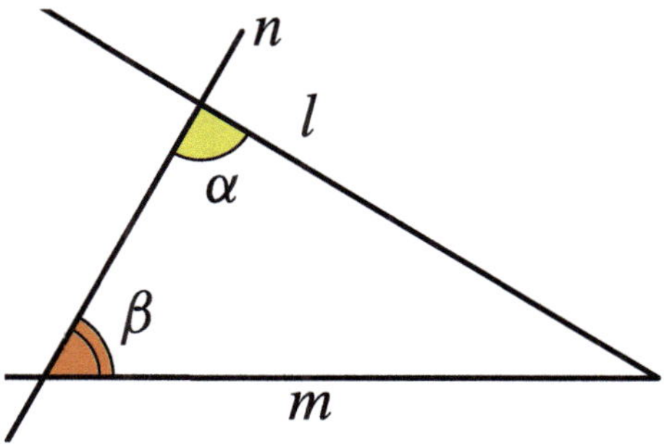

Fig. 3.6 The fifth postulate of Euclidean geometry

3. Using the above definitions, common notions, and postulates, Euclid proceeded to prove many theorems, which he called propositions. Many propositions are proved using other propositions which have been proved earlier.

Thus, *The Elements* is a self-contained logically consistent system for geometry. Now, we discuss the five postulates. Euclid's five postulates are:

- Any two points can be joined by a straight line.
- Any straight line segment can be extended indefinitely in a straight line.
- Given any straight line segment, a circle can be drawn having the segment as radius and one endpoint as center.
- All right angles are congruent (i.e., the same).
- If a straight line n falling on two straight lines (m and n) make the interior angles on the same side less than two right angles (i.e., $\alpha + \beta < 180°$), if produced indefinitely, meet on that side on which are the angles less than the two right angles (Fig. 3.6).

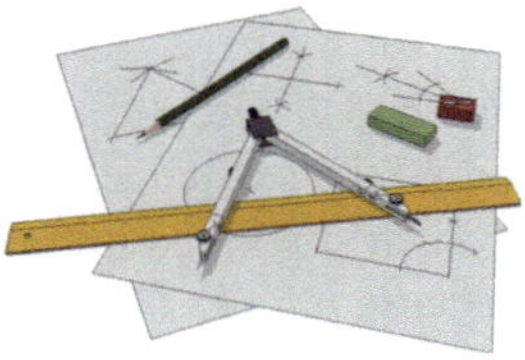

The first postulate tells us that it is possible to construct a straight line given two points. The second postulate tells us that we can always make a line segment longer. That means that we never run out of space; that is, *space is infinite*. The third postulate allows for the *existence* of circles of any size and center—say center A and radius AB. Postulates 1 and 3 set up the "straightedge and compass" framework that was a standard for geometric constructions. A compass, more accurately known as a pair of compasses, is a technical drawing instrument that can be used for drawing circles or arcs. On the other hand, a straightedge is a tool which can draw a perfectly straight line through any two points but which *cannot be used as a ruler to measure lengths*. Why Euclid needed the fourth postulate which basically says that all 90° are equal? This is to indicate that space is homogeneous and isotropic (i.e., everywhere on the plane right angles are the same).

It is clear that the fifth postulate is different from the other four. The first four were simple assertions that few would be inclined to doubt. Far from being instantly self-evident, the fifth postulate was even hard to read and understand (we need to draw a figure). It did not satisfy Euclid and he tried to avoid its use as long as possible—in fact the first 28 propositions of *The Elements* are proved without using it.

Once these postulates are accepted, a lot of facts known as theorems (Euclid called them propositions) about circles, triangles, angles, and so on follows. To illustrate one such proposition, I now present Proposition 1: On a given finite straight line, construct an equilateral triangle. The purpose is to demonstrate how definitions and postulates are used in solving the problem. Furthermore, it is to show that Euclidean geometry is constructive. Postulates 1, 2, 3, and 5 assert the existence and uniqueness of certain geometric figures, and these assertions are of a constructive nature: that is, we are not only told that certain things exist but are also given methods for creating them with no more than a compass and an unmarked straightedge.

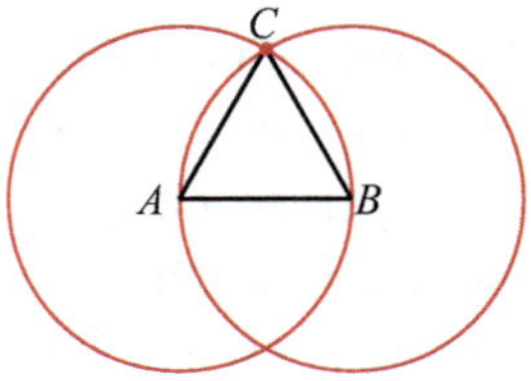

This is how Euclid constructed an equilateral triangle. We start with the line AB, and we need to construct the triangle ABC such that all sides are of equal length (i.e., $AB = BC = CA$[4]). The point is: *where is C*? As $AC = AB$, C must be on the circle centered at A with radius AB. So, we draw this circle (and the reason why we can do this is Postulate 3). In the same manner, $BC = BA$, so C is on the circle centered at B with radius BA. So, we draw this circle. Finally, C is the intersection of these two circles.

It should be, however, mentioned that Euclid states no axiom about the intersection of circles, so he has not justified the existence of the point C used in his very first proposition! There are many such situations in Euclid's *Elements*, in which Euclid assumes something is true because *it looks true in the diagram.*

I now present yet another theorem and the characteristics of a geometry proof. Let's take the case of a triangle inside a semicircle. If we play with it long enough, we will see one remarkable thing: no matter where on the circle we place the tip of the triangle, it always forms a nice right triangle (Fig. 3.7a). But is it true? We need a proof. In the same figure, I present a proof commonly given in high school geometry classes. This proof uses the fact that the sum of angles in a triangle is 180° or π, see Sect. 3.1.5 for proof of this fact. A complete proof would be more verbose than what I present here. Does it exist a better (that means elegant) proof? See Fig. 3.7b: $ACBC'$ is a rectangle and thus ABC is a right triangle! This fact is known as Thales' theorem.

[4] A remark is in order. I use AB to label a segment between two points A and B. But I also used AB to denote its length. Not very rigorous here.

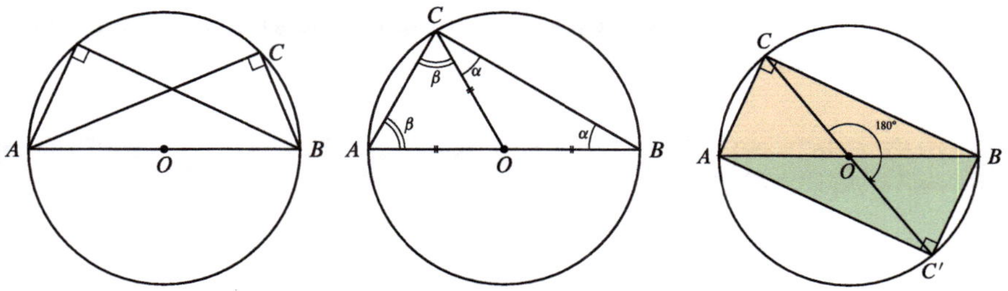

(a) In triangle ABC: $2(\alpha + \beta) = 180°$ (b) $ACBC'$ is a rectangle

Fig. 3.7 The angle inscribed in a semicircle is always a right angle (90°): two proofs. **a** The angles of the same marking are equal, and thus we have $2(\alpha + \beta) = 180°$ or $\alpha + \beta = 90°$. The key to the proof is to draw the line OC which does not initially exist in the problem. And **b** Rotate ABC 180° counterclockwise we get a new triangle ABC'. Alternatively, draw OC (as in the first proof) but extend it to meet the circle at C' (to have symmetry). Now the box $ACBC'$ is a parallelogram. But it is a special parallelogram: the two diagonals are equal (both are diameters of the circle). So, $ACBC'$ must be a rectangle. The second proof is considered to be better than the first because we can "see" the result

In Fig. 3.7 hatch marks are used to denote equal measures of angles and line segments. For example, angles OAC and ACO have the same hatch marks to indicate that they are equal.

3.1.3 Influence of The Elements

The Elements is still considered a masterpiece in the application of logic to mathematics. It has proven enormously influential in many areas of science. Many scientists, the likes of Nicolaus Copernicus, Johannes Kepler, Galileo Galilei, Albert Einstein, and Isaac Newton were all influenced by *The Elements*. When Newton wrote his masterpiece *Philosophiæ Naturalis Principia Mathematica* (Mathematical Principles of Natural Philosophy), he followed Euclid with definition, axioms, and theorems in that order. Albert Einstein recalled a copy of *The Elements* and a magnetic compass as two gifts that had a great influence on him as a boy, referring to *The Elements* as the "holy little geometry book".

The austere beauty of Euclidean geometry has been seen by many in western culture as a glimpse of an otherworldly system of perfection and certainty. Abraham Lincoln kept a copy of Euclid in his saddlebag, and studied it late at night by lamplight; he related that he said to himself, "You never can make a lawyer if you do not understand what demonstrate means; and I left my situation in Springfield, went home to my father's house, and stayed there till I could give any proposition in the six books of Euclid at sight". Thomas Jefferson, a few years after he finished his second term as president, he wrote to his old friend John Adams on 1 January 1812:

"I have given up newspapers in exchange for Tacitus and Thucydides, for Newton and Euclid; and I find myself much happier".

History note 3.1: Euclid (fl. 300 BC)

Euclid (fl. 300 BC) was a Greek mathematician, often referred to as the "founder of geometry" or the "father of geometry". His *Elements* is one of the most influential works in the history of mathematics, serving as the main textbook for teaching mathematics (especially geometry) from the time of its publication until the late 19th or early 20th century. In the Elements, he deduced the theorems of what is now called Euclidean geometry from a small set of axioms. Euclid also wrote works on perspective, conic sections, spherical geometry, number theory, and mathematical rigor.

3.1.4 Algebraic Versus Geometric Thinking

To emphasize the importance of geometry, we turn to the story of Paul Dirac (1902–1984), an English theoretical physicist who is regarded as one of the most significant physicists of the twentieth century. In the thirteen hundred (or so) pages of his published work, Dirac had no use at all for diagrams. He never used them publicly for calculation. His books on general relativity and quantum mechanics contained not a single figure.

Therefore it seems reasonable to assume that Dirac would consider himself an algebraist. On the contrary, he wrote in longhand in his archives something remarkable:

> There are basically two kinds of mathematical thinking, algebraic and geometric. A good mathematician needs to be a master of both. But still he will have a preference for one rather or the other. I prefer the geometric method. Not mentioned in published work because it is not easy to print diagrams. With the algebraic method one deals with equations between algebraic quantities. Even tho I see the consistency and logical connections of the equations, *they do not mean very much to me*. I prefer *the relationships which I can visualize in geometric terms*. Of course with complicated equations one may not be able to visualize the relationships e.g. it may need too many dimensions. But with the simpler relationships one can often get help in understanding them by geometric pictures.

One remarkable thing happened in Dirac's life is that he learned projective geometry early in his life (in secondary school at Bristol). He wrote "This had a strange beauty and power which fascinated me". Projective geometry provided Dirac new insight into Euclidean space and into special relativity.

Of course Dirac could not know that his early exposure to projective geometry would be vital to his future career in physics. We simply can't connect the dots looking forward, as Steven Jobs (February 24, 1955–October 5, 2011)—the Apple co-founder—once said in his famous 2005 commencement speech at Stanford University:

> You can't connect the dots looking forward; you can only connect them looking backwards. So you have to trust that the dots will somehow connect in your future. You have to trust in something—your gut, destiny, life, karma, whatever. This approach has never let me down, and it has made all the difference in my life.

Example 3.1 Pablo, Sofia, and Mia got some candy eggs at a party. Pablo had three times as many eggs as Sofia, and Sofia had twice as many eggs as Mia. Pablo decides to give some of his eggs to Sofia and Mia so that all three will have the same number of eggs. What fraction of his eggs should Pablo give to Sofia? This problem is question five in the 2014-2015 AMC 10.

Algebraic solution. First, I present the algebraic solution. Let m be the number of egss that Mia had; then Sofia had $2m$ egss and Pablo had $6m$ eggs. In total, they have $m + 2m + 6m = 9m$ eggs. After the sharing from Pablo, they have the same number of egss, thus each has $3m$ egss. So, Pablo had to give m eggs to Sofia which constitutes $1/6$ of his egss.

Geometric solution. For a geometric solution we need a diagram, as shown below. The solution comes then with ease. Of course each square is m eggs.

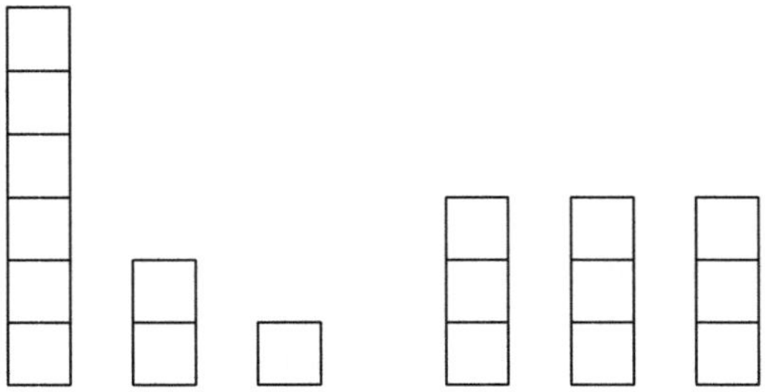

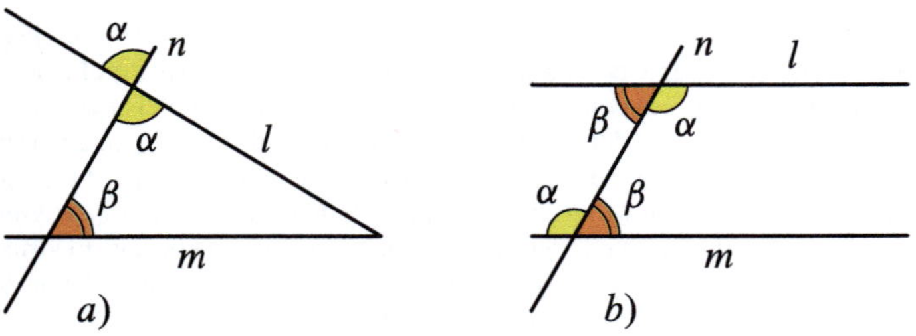

Fig. 3.8 For two parallel lines l and m the alternate angles α (or β) are equal. In the left figure the two angles α are equal and are called vertical angles

3.1.5 The Fifth Postulate and Consequences

Figure 3.8a illustrates the situation described by the fifth postulate, which is what happens when the two lines are *not parallel*. So, the fifth postulate says: if $\alpha + \beta < 180°$ then l intersects m on the right. The negative of this postulate is also true. Therefore, it follows that when $\alpha + \beta = 180°$, then l does not meet m; or l is parallel to m (we express this symbolically as $l \parallel m$). So, if $\alpha + \beta = 180°$ then $l \parallel m$. By introducing alternate interior angles, which are angles that occur on opposite sides of the transversal line n (Fig. 3.8b), we hence also have: if alternate angles are equal then $l \parallel m$. This is Proposition I.27 in *the Elements*.

So, if $\alpha + \beta = 180°$ then $l \parallel m$. How about the converse? Is it true that if $l \parallel m$ then $\alpha + \beta = 180°$? This is true and is Proposition I.29. How to prove it? Proof of contradiction is the way to go. Assume that $\alpha + \beta < 180°$, then by the fifth postulate the lines will meet, which is contradict with our assumption that they are parallel. So, $\alpha + \beta = 180°$. Do we need to consider $\alpha + \beta > 180°$? No.

Playfair's Parallel postulate. Euclid's fifth postulate is wordy. The Scottish mathematician John Playfair rephrased it in 1795 as: Given a line and a point not on that line, there exists *only one* parallel line to the original line going through that point. It now bears the name the Parallel Postulate. We need to make sure that Playfair's Parallel postulate and the fifth postulate are equivalent. To this end, we have to prove two things: (1) from Playfair's Parallel postulate we can get the fifth postulate and vice versa. I omit these proofs for brevity.

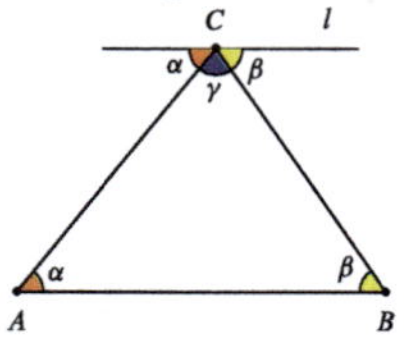

Sum of angles in a triangle is 180°. I know present a beautiful property of triangles that the angle sum of a triangle is 180°. I emphasize that this result is a consequence of the parallel postulate. If α, β and γ are the angles of any triangle then $\alpha + \beta + \gamma = 180°$. To prove this result, draw a line l through C, which is parallel to AB. Then the angle on the left beneath l is alternate to the angle α in the triangle, so it is equal to α. Similarly, the angle on the right beneath l is equal to β. Then, considering the angles at C, we have $\alpha + \beta + \gamma = 180°$.

3.1.6 Area of Simple Geometries

Similar to using a ruler to measure the length of one-dimensional objects (e.g. a stick), to determine the area of a shape, we also need something to compare with. And the conventional choice is the amount of space of a square of unit sides (this square is called a unit square), see Fig. 3.9. So the measurement of area really boils down to the question, how much room is any shape taking up compared to a unit square?

Some areas are relatively easy to measure. And obviously, the easiest area is that of a rectangle as it can fit with the unit square. For example, suppose we have a 5 by 3 rectangle, we can chop it into 15 unit squares. Thus, its area is 15 (Fig. 3.9). So, if the sides of a rectangle are nice whole numbers, *the area is then the product of the length of the two sides*. But what if the sides are not whole numbers? Then use a smaller unit square. The area is still the product of the length of the two sides.

Next comes the triangles. Suppose we have a triangle with base b and height h (see Fig. 3.10a). What is its area? The way to get the triangle's area is to use the area of a rectangle. That is, *using the known to determine the unknown*. We put the triangle inside a rectangle with sides b and h. Then, by dropping a line CH perpendicular to AB, we see that the triangle's area is half of that of the rectangle. Therefore, we get the formula of $1/2bh$.

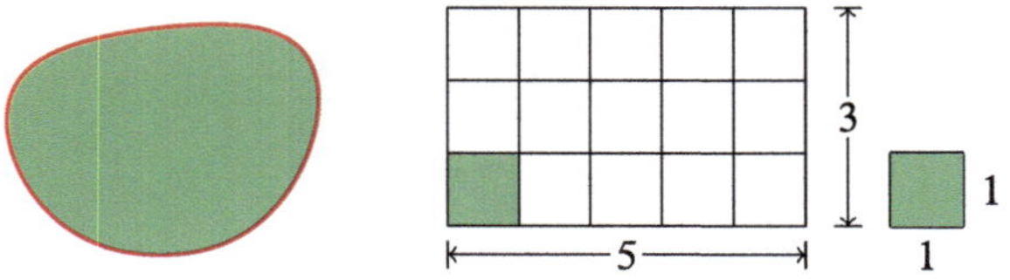

Fig. 3.9 Area of 2D shapes: an area unit is the amount of space taken up by a unit square

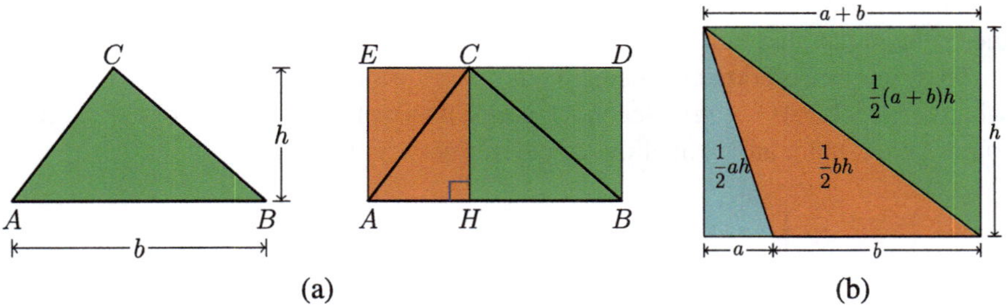

Fig. 3.10 The area of a triangle is related to the area of the bounding rectangle

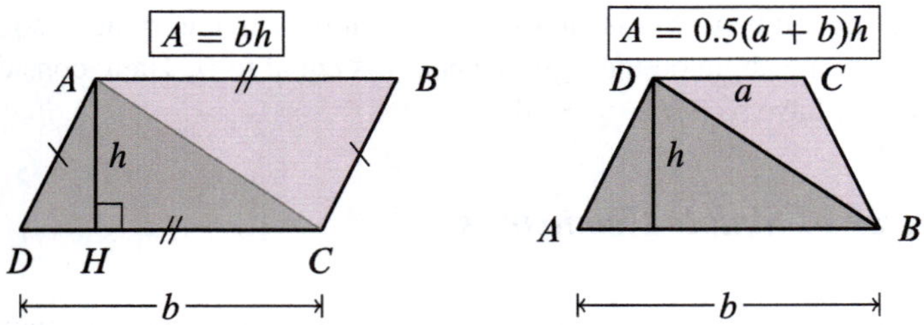

Fig. 3.11 Area of parallelogram and trapezoid. A parallelogram is a quadrilateral with two pairs of parallel sides. The opposite or facing sides of a parallelogram are of equal length. A quadrilateral with at least one pair of parallel sides is called a trapezoid. We find the area of a parallelogram by dividing it into two congruent triangles (shaded in the left diagram); the area of one such triangle is $0.5bh$

Nice. But how about a slanting triangle as the orange triangle in Fig. 3.10b? In this case point C lies outside the bounding rectangle, so the above formula might not work? Again, we use a rectangle of which the area is $(a+b)h$. This area is equal to the sum of areas of the two right triangles and the orange triangle. From that, we can see that the area of the triangle is still $1/2bh$.

Next comes polygons. The area of a polygon is the sum of all the sub-triangles making up the polygon. We can see from this that ancient mathematicians computed the area of new more complex geometries based on the known area of old simpler geometries. See Fig. 3.11 for one example.

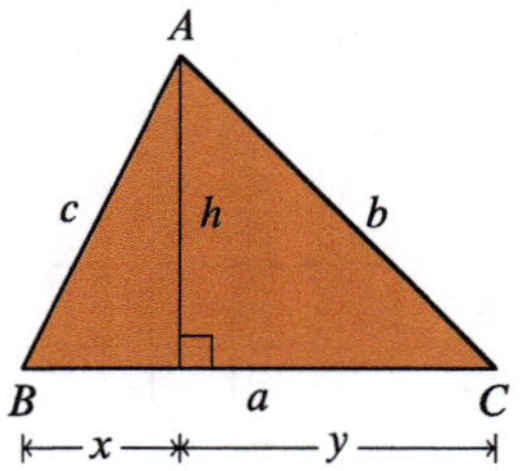

Heron's formula. What is the area of a triangle in terms of its sides a, b, c? The formula is credited to Heron (or Hero) of Alexandria, and a proof can be found in his book, Metrica, written c. CE 60. It has been suggested that Archimedes knew the formula over two centuries earlier. I now present a derivation of this formula using the Pythagorean theorem.

First, the area is computed using the familiar formula "half of the base multiplied with the height": $A = 1/2ah$. Second, the height is expressed in terms of a, b, c. Refer to the figure, there are 3 equations to determine x, y, h:

$$\left.\begin{aligned} x+y&=a \\ x^2+h^2&=c^2 \\ y^2+h^2&=b^2 \end{aligned}\right\} \Longrightarrow x=\frac{a}{2}-\frac{b^2-c^2}{2a}, \quad y=\frac{a}{2}+\frac{b^2-c^2}{2a}, \quad h^2=c^2-x^2$$

As we have h^2, let's compute the square of the area:

$$4A^2 = a^2(c^2 - x^2)$$
$$4A^2 = a^2(c-x)(c+x) = a^2\left(c - \frac{a}{2} + \frac{b^2-c^2}{2a}\right)\left(c + \frac{a}{2} - \frac{b^2-c^2}{2a}\right)$$
$$4A^2 = a^2\left(\frac{2ac - a^2 + b^2 - c^2}{2a}\right)\left(\frac{2ac + a^2 - b^2 + c^2}{2a}\right)$$
$$16A^2 = [b^2 - (a-c)^2][(a+c)^2 - b^2] = (b+a-c)(b-a+c)(a+c+b)(a+c-b)$$

If we introduce $s = 0.5(a+b+c)$—the semi-perimeter of the triangle—the Heron's formula is given by

$$A = \sqrt{s(s-a)(s-b)(s-c)} \tag{3.1.1}$$

The final expression of A is symmetrical with respect to a, b, c and it has a correct dimension (square root of length power 4 is length squared—an area). Thus, it seems correct (if it was $A = \sqrt{s(s-2a)(s-b)(s-c)}$ or $A = \sqrt{s(s-a)^2(s-b)(s-c)}$, then it is definitely wrong). How we know that it's correct? Check it for a triangle of which area we know for sure. Note that using the generalized Pythagorean theorem gives a shorter/easier proof.

What can we do with Heron's formula? We can use it to compute the area of a triangle given the sides a, b, c, of course. The power of symbolic algebra is that we can deduce new information from (Eq. (3.1.1)). We can pose this question: among all triangles of the same perimeter, which triangle has the maximum area? Using the AM-GM inequality (Sect. 2.20), it's straightforward to show that an equilateral triangle (i.e., triangle with three sides equal $a = b = c$) has the maximum area.

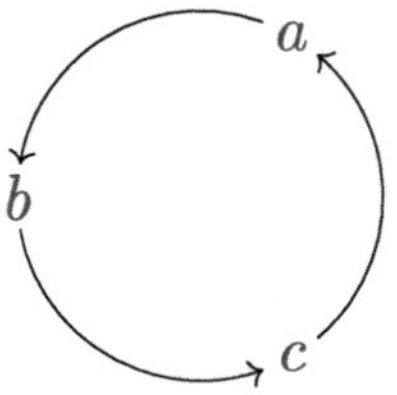

Generalization to quadrilaterals. One of the most beautiful things about Heron's formula is the generalization discovered by the Hindu mathematician Brahmagupta around 620 AD. First, we rewrite Heron's formula as below

$$A = \frac{1}{4}\sqrt{(a+b+c)(b+c-a)(c+a-b)(a+b-c)}$$

Note that starting from the term $a+b+c$ we can get the next term by cycling through a, b, c as shown in the figure. The formula is, however, not symmetrical. In the term $a+b+c$ there is no minus sign! Now comes Brahmagupta: he added $d = 0$ (which is nothing) to the above equation in this form:

$$A = \frac{1}{4}\sqrt{(a+b+c-d)(b+c+d-a)(c+d+a-b)(d+a+b-c)}$$

This equation is fair (or symmetrical) to all quantities involved i.e., a, b, c, d. This beautiful formula then must have a meaning, Brahmagupta argued. Indeed, it is the area of a quadrilateral of sides a, b, c, d inscribed in a circle (such a quadrilateral is called a cyclic quadrilateral).

The following joke describes well the principle of *using the known to determine the unknown*:

> A physicist and a mathematician are sitting in a faculty lounge. Suddenly, the coffee machine catches on fire. The physicist grabs a bucket and leap towards the sink, filled the bucket with water and puts out the fire. Second day, the same two sit in the same lounge. Again, the coffee machine catches on fire. This time, the mathematician stands up, got a bucket, *hands the bucket to the physicist, thus reducing the problem to a previously solved one.*

3.1.7 Congruence and Similarity

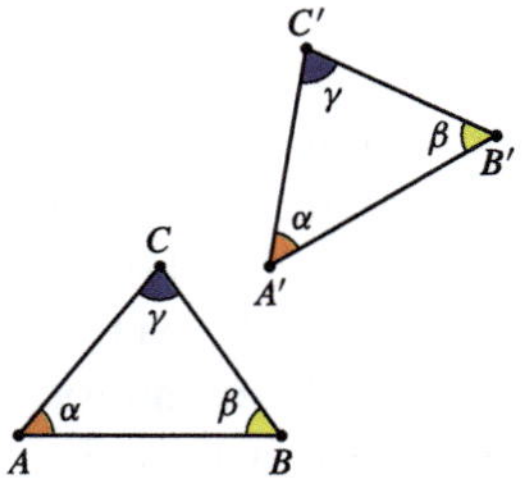

Let's make two identical triangles by paper. Now, separate them apart, and rotate one triangle. What we obtain is similar to the figure next. Obviously, these two triangles are *of equal shape and size*. Mathematicians call them *congruent triangles*. Euclid says that two geometric figures are congruent (he used the word coincide) when one of them can be moved to fit exactly on the other.

The next thing mathematicians have to do is to develop criteria to determine whether two triangles are congruent or not. They do not want to build physical triangles and move them to check this. That's not mathematics!

To say that two triangles ABC and $A'B'C'$ are congruent, we write $\triangle ABC \cong \triangle A'B'C'$. Note that we always write the vertices of the two congruent triangles so that matched vertices and sides can be read off in a natural way. So, the expression $\triangle ABC \cong \triangle A'B'C'$ means that

$$AB = A'B', \quad BC = B'C', \quad CA = C'A', \quad \angle A = \angle A', \quad \angle B = \angle B', \quad \angle C = \angle C'$$

Congruence tests. The three congruence tests for triangles are SAS (side-angle-side), ASA (angle-side-angle), and SSS (side-side-side) as shown in Fig. 3.12. Instead of proving them we consider them as axioms. Even though Euclid proved the SAS as

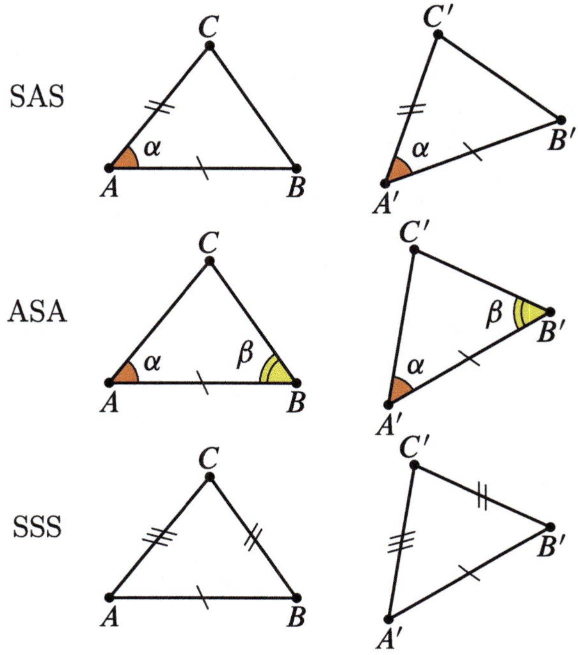

Fig. 3.12 Congruence tests for triangles. To use the SAS, check two sides of the two triangles and the angles included, not other angles

a theorem, his proof relied on motion, which is not easy to be precisely described in elementary mathematics. That's why we accept these tests with faith.

Using congruence we can deduce many interesting facts. For example, in an isosceles triangle, which is a triangle that has two sides of equal length, the angles opposite to these sides are equal. Another result: opposite sides of a parallelogram are equal. The proof always goes like this: find two congruent triangles, for example using the SAS test, then the angles not used in the test are equal. The difficult step is to make the triangles show up!

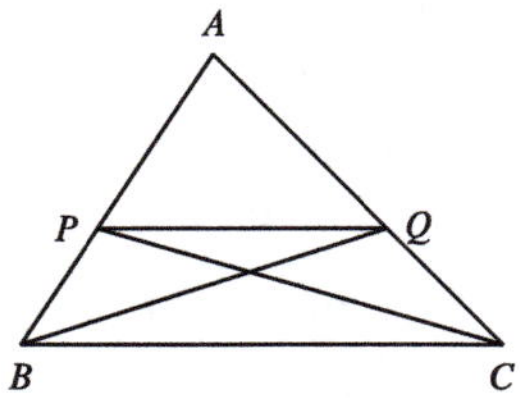

The Thales theorem or the Basic Proportionality Theorem (B.P.T. Theorem). Consider $\triangle ABC$ and a segment PQ that cuts AB and AC in such a way that $PQ \parallel BC$. Thales said that PQ cuts AB and AC proportionally:

$$\text{if } PQ \parallel BC \text{ then } \boxed{|AP|/|PB| = |AQ|/|QC|} \tag{3.1.2}$$

The notation $|AB|$ is to denote the length of AB; in high school math textbooks, it is simply written as AB which is not rigorous. For the proof, we need one elementary result about the area of a triangle: that area is one half the product of the base and the height. To prove the Thales theorem, *draw* (what else?) PC and BQ. Now, we have two new triangles PBQ and QCP and they have the same area (as they have the same base PQ and the same height which is the distance from PQ to BC). Next, we consider triangles APQ and PBQ, their areas are

$$\text{area } APQ = h \times |AP|, \quad \text{area } PBQ = h \times |PB|$$

where h is the distance from Q to AB. From that we get

$$\frac{\text{area } APQ}{\text{area } PBQ} = \frac{|AP|}{|PB|} \tag{3.1.3}$$

In the same manner, if we consider triangles APQ and PQC, we obtain

$$\frac{\text{area } APQ}{\text{area } QCP} = \frac{|AQ|}{|QC|} \tag{3.1.4}$$

From Eqs. (3.1.3) and (3.1.4) and noting that the two triangles PBQ and QCP have the same area, we get what we wanted to prove: $|AP|/|PB| = |AQ|/|QC|$. A bit of algebra then gives us the equivalent result: $|AP|/|AB| = |AQ|/|AC|$.

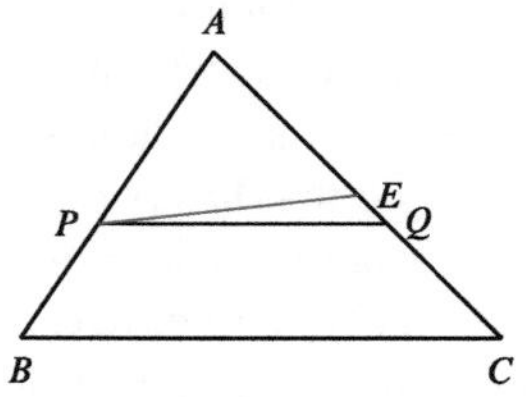

The Thales theorem is: a parallel line implies proportional sides. The converse is: proportional sides imply a parallel. The question is: is this converse true? Yes, it is. The problem is: given $|AP|/|PB| = |AQ|/|QC|$, prove that $PQ \parallel BC$. Introduce a line PE that is parallel to BC. The aim is to show that E is nothing but Q. Since $PE \parallel BC$, we have (due to Thales' theorem):

$$\frac{|AP|}{|PB|} = \frac{|AE|}{|EC|}$$

But $|AP|/|PB| = |AQ|/|QC|$ (given), thus

$$\frac{|AQ|}{|QC|} = \frac{|AE|}{|EC|} \iff \frac{|AQ|}{|QC|} + 1 = \frac{|AE|}{|EC|} + 1 \iff \frac{|AC|}{|QC|} = \frac{|AC|}{|EC|} \iff |QC| = |EC|$$

Similarity of triangles. If we look at the triangles ABC and APQ in the proof of the Thales theorem we see that they have equal angles (Fig. 3.13a). They have

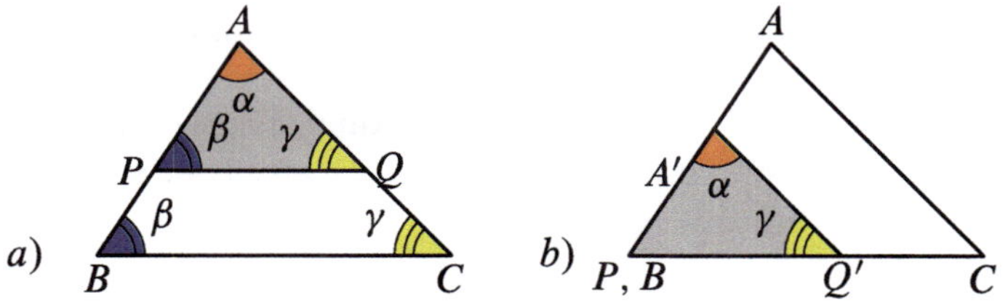

Fig. 3.13 Similar triangles (APQ and ABC) have equal angles, or their sides are proportional

one common angle at A, and the other two angles are the same because $PQ \parallel BC$. These two triangles are *of the same shape* but not of the same size. They are said to be *similar*, denoted by $\triangle ABC \sim \triangle APQ$. From the Thales theorem, we have $|AP|/|AB| = |AQ|/|AC| = k$, where $k > 0$ is any real number. So, two similar triangles have proportional sides (at least for two sides). How about $|PQ|/|BC|$ i.e., the ratio of the remaining sides? That's the question we should ask.

To answer that question we slide $\triangle APQ$ along AB so that P coincides B; we have the situation as shown in Fig. 3.13b. As $A'Q' \parallel AC$, we have $k = |PA'|/|PA| = |PQ'|/|PC|$. But $|PQ'| = |PQ|$ (when we move a triangle its sides do not change) and $|PC| = |BC|$. Thus, $|PQ|/|BC|$ is also k.

To conclude, two triangles are similar if they have equal angles, or their sides are proportional (i.e., if one triangle has sides a, b and c, the other has sides of ka, kb, and kc where $k > 0$ is called the scale factor). When $k = 1$, the two triangles are of course congruent. Thus, congruence is a special case of similarity. What about the areas of the two similar triangles? The ratio of the areas is k^2. One easy proof[5] is to use Heron's formula $A = \sqrt{s(s-a)(s-b)(s-c)}$, with $s = 0.5(a+b+c)$, see Eq. (3.1.1).

Because congruence is a special case of similarity, there are also three similarity tests: SSS, SAS (where the sides are proportional), and AA (as the sum angles is 180°, if two angles are equal the remaining ones must be also equal). For right triangles the AA test becomes A test since we need to check the equality of one acute angle only.

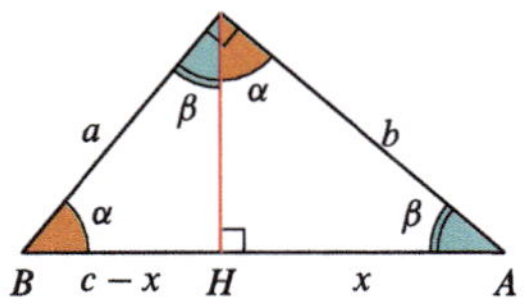

Revisiting the Pythagorean theorem. I now present the second proof of the famous Pythagorean theorem. This proof uses similar triangles. Let's consider the right triangle ABC with sides a, b and hypotenuse c. From C draw CH perpendicular to AB,

[5] Yet another proof is: consider a parallelogram of sides a, b; its area is ab. Now, scale it up by a factor k. The area of the scaled parallelogram is k^2ab. Relating the area of a triangle to that of a parallelogram will give us the result.

and let $AH = x$ (thus $BH = c - x$). This CH divides our triangle into two smaller right triangles CBH and ACH. What are special about these two right triangles? They are similar to $\triangle ABC$ (AA test). This is a skill that high school students must have: to see the similarity of triangles. The rest is algebra, as we shall see. Thus, we have

$$\triangle ABC \sim \triangle CBH \Longrightarrow \frac{|BC|}{|BH|} = \frac{|AB|}{|BC|} \Longrightarrow \frac{a}{c-x} = \frac{c}{a}$$
$$\triangle ABC \sim \triangle ACH \Longrightarrow \frac{|AC|}{|AH|} = \frac{|AB|}{|AC|} \Longrightarrow \frac{b}{x} = \frac{c}{b}$$

From these two, it is easy to get $c^2 = a^2 + b^2$.

3.1.8 Is a Square a Rectangle?

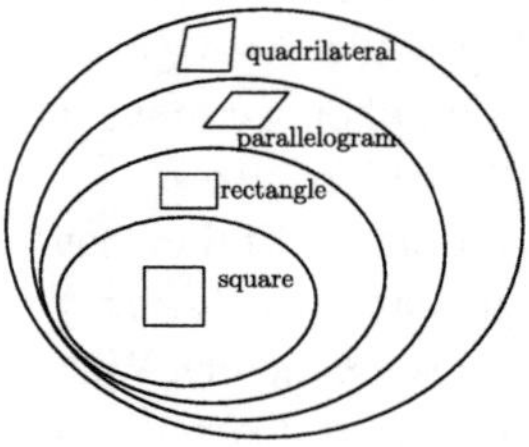

A quadrilateral is a four-sided polygon, having four edges (sides) and four corners (vertices). The word is derived from the Latin words *quadri*, a variant of four, and *latus*, meaning "side". A parallelogram is a simple (non-self-intersecting) quadrilateral with two pairs of parallel sides. A rectangle is a quadrilateral with four right angles. A square is a rectangle with two equal-length adjacent sides. Thus, a square is a regular quadrilateral, which means that it has four equal sides and four equal angles.

3.1.9 Angles in Convex Polygons

An interior angle of a polygon is formed by two sides of the polygon that share an endpoint. It is a well-known fact that the sum of the three interior angles of a triangle is 180°. Referring to Fig. 3.14, this fact is written as $a_1 + a_2 + a_3 = 180°$. Now we move to rectangles, it is obvious that $a_1 + a_2 + a_3 + a_4 = 360°$ because each angle is a right angle (or 90°). From this, we guess that for a n-sided convex polygon, the sum of the angles is $180°(n-2)$.[6] At least it is true for triangles ($n = 3$, angle sum is $(3-2)180° = 180°$) and rectangles ($n = 4$).

[6] This is known as the interior angles theorem.

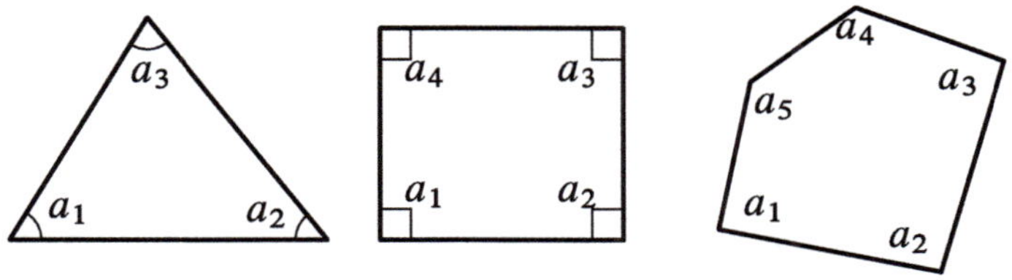

Fig. 3.14 Sum of interior angles in a convex polygon of n sides is $180°(n-2)$?

Fig. 3.15 Exterior angles in a polygon

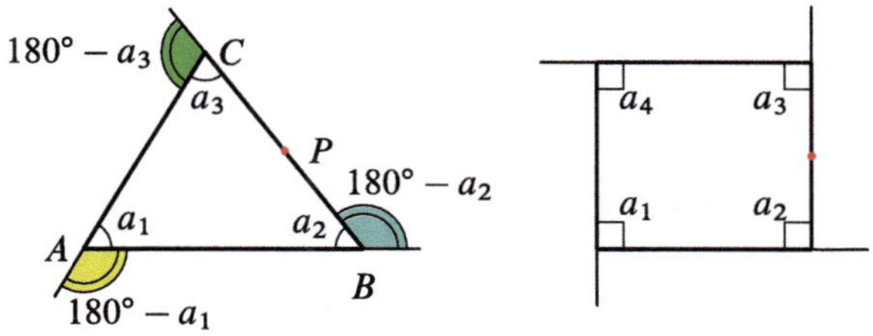

To prove what we have guessed we need to introduce another concept[7]: exterior angles. An exterior angle (also called an external angle or turning angle) is an angle formed by one side of a polygon and a line extended from an adjacent side (Fig. 3.15). If we start at P and travel counterclockwise around the triangle one lap we will get back to P. In doing so, we make one left turn at C of $180° - a_3$, one left turn at A of $180° - a_1$ and one left turn at B of $180° - a_2$. In total we have make a full turn of $(180° - a_1) + (180° - a_2) + (180° - a_3) = 3(180°) - (a_1 + a_2 + a_3)$. One full turn is 360°; if you do not convince, considering a special case of an equilateral triangle with $a_1 = a_2 = a_3 = 60°$, then it is obvious that the total of exterior angles is $120° + 120° + 120° = 360°$.

The same analysis for rectangles yields the same result. So, we guess that

$$(180° - a_1) + (180° - a_2) + \cdots + (180° - a_n) = 360° \tag{3.1.5}$$

If this is correct (and it is as we shall prove), then we obtain the so-called exterior angle theorem that states that the sum of the exterior angles of a polygon is 360°. From this theorem, we get the sum of interior angles:

$$n180° - (a_1 + a_2 + \cdots + a_n) = 2 \times 180° \Longrightarrow \boxed{a_1 + a_2 + \cdots + a_n = (n-2)180°}$$

For a regular n-sided polygon, we have $a_1 = a_2 = \ldots = a_n = \alpha$, the above equation then gives us

---

[7] There exists another proof without using exterior angles. The idea is to decompose a polygon into triangles, and using the fact that the sum of triangle angles is 180°. Assuming that we have an n-sided polygon $A_1A_2...A_n$. From vertex A_1 we draw diagonals $A_1A_3, A_1A_4, \ldots, A_1A_{n-1}$. There are $n-3$ such diagonals which make $n-2$ triangles. Draw a figure to see all of this.

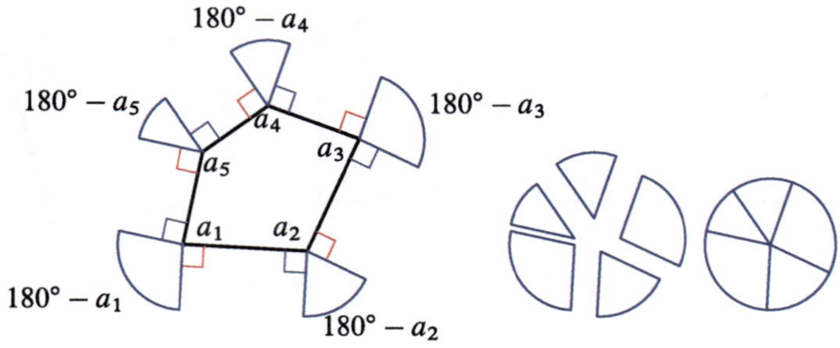

Fig. 3.16 Exterior angles theorem: Polya's proof

$$\boxed{\text{interior angle of a regular } n\text{-sided polygon: } \alpha = 180^\circ(1 - {}^{2}/_{n})} \tag{3.1.6}$$

What can we read from this formula? First, all interior angles in regular polygons are smaller than 180° (because they are 180° multiplied with a factor smaller than one i.e., $1 - 2/n$). Second, the more edges a regular polygon has the bigger its interior angles.[8]

Proof Proof of the exterior angle theorem by George Polya At each corner (of the polygon), draw two line segments pointing outward, one line segment perpendicular to each side (Fig. 3.16). Draw a sector of a unit circle at each vertex with these segments as sides. Now observe that the angle made by these two segments is precisely the exterior angle. Because the sides of each pair of adjacent sectors are parallel, we can reassemble the sectors to form a circle. So, the sum of the exterior angles is 360°. ■

3.1.10 Circle Theorems

iThe Central Angle Theorem states that the central angle from two chosen points A and B on a circle is always twice the inscribed angle from those two points (left diagram in Fig. 3.17). An inscribed angle is the angle whose vertex lies on the circumference. From that we get two immediate results (or corollaries as mathematicians would call) shown in the middle (inscribed angles subtended by the same arc are equal) and right diagram in the referred figure.

Thales's theorem and its converse. The right diagram in Fig. 3.17 shows the Thales theorem: if AB is the diameter of a circle and C is a point (any point) on that circle, then the angle ACB is a right angle. As always in mathematics, we should ask the question: is the converse of the Thales theorem true? That is, given the diameter of a circle AB, and the angle ACB is 90°, is C on the circle or not?

[8] Compare equilateral triangles with angles of 60° and squares with angles of 90°.

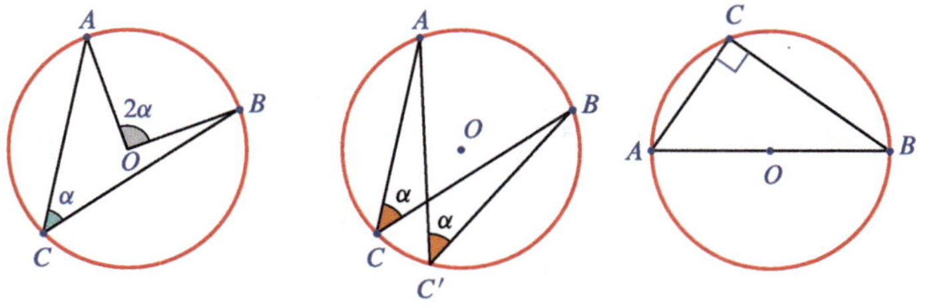

Fig. 3.17 Central angle theorem (left figure): The Central Angle Theorem states that the central angle from two chosen points A and B on the circle is always twice the inscribed angle from those two points. The inscribed angle can be defined by any point along the outer arc AB and the two points A and B. And when AB is the diameter of the circle i.e., $2\alpha = 180°$, we have $C = 90°$. How to prove the central angle theorem? Simply draw the line OC, and we have two isosceles triangles OAC and OCB

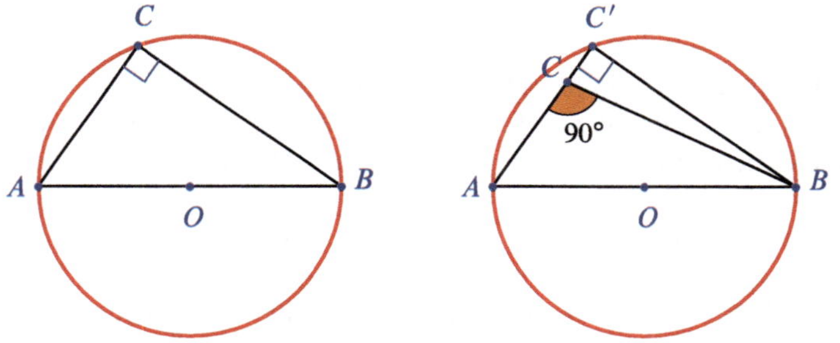

Fig. 3.18 Proof by contradiction for the converse of the Thales theorem: C is inside the circle case

It turns out that the converse of the Thales theorem is also true. And Euclid presented a proof by contradiction (Fig. 3.18). Assume that $\angle ACB$ is 90° *but C is not on the circle*. If we can deduce something absurd from this, then C must be on the circle. That's the gist of proof by contradiction. There are two cases: C is inside the circle and it is outside the circle. Let's consider the former case first. Extend AC so that it cuts the circle at C'. The Thales theorem then tells us that $\angle AC'B = 90°$. Thus we have BC and BC', both perpendicular to AC, are parallel. However, these two parallel lines meet at B! A contradiction: so C cannot be inside the circle. I leave the case C outside the circle for you as an exercise.

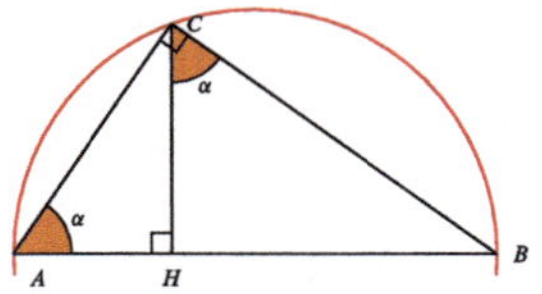

Geometric mean theorem or the right triangle altitude theorem. Given two positive numbers a and b we have met the geometric mean $\sqrt{ab}$ in Sect. 2.21.3. Herein, we see this concept in relation to right triangles. Consider a right triangle ABC with $\angle C = 90°$. (Any right triangle is associated with a circle due to Thales' theorem,

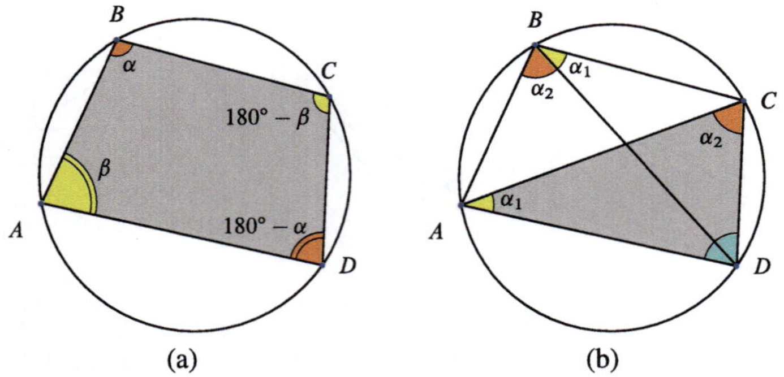

Fig. 3.19 Four points on a circle: opposite angles add up to 180° (**a**) and proof (**b**). The converse is also true: if the opposite angles of a quadrilateral add up to 180°, then the quadrilateral is cyclic. Proof by contradiction, remember the converse of the Thales theorem?

so I intentionally drew a circle, but it is not needed for the theorem). From C, draw $CH \perp AB$: CH is called the altitude on the hypotenuse AB. The triangle HAC is similar to the triangle HCB (AA test), so we have $HC/HB = HA/HC$. Thus,

$$HC^2 = HA \cdot HB \Longrightarrow HC = \sqrt{HA \cdot HB} \tag{3.1.7}$$

It is $HC^2 = HA \cdot HB$ that we can square a rectangle with a ruler and compass, that is to construct a square of equal area to a given rectangle. On the other hand, $HC = \sqrt{HA \cdot HB}$ allows us to construct the square root of $a > 0$ using straight-edge and compass.

Four points on a circle. Consider a circle and four points A, B, C, and D on it. We then have a cyclic quadrilateral $ABCD$. (A cyclic[9] quadrilateral or inscribed quadrilateral is a quadrilateral whose vertices all lie on a circle.) It is interesting that in this quadrilateral opposite angles add up to 180° (Fig. 3.19a).[10] To prove this, we need to make triangles (we know more about triangles than about quadrilaterals). So, we draw the diagonals AC and BD (Fig. 3.19b). In $\triangle ACD$ we have $\alpha_1 + \alpha_2 + \angle D = 180°$. But $\angle ABD = \alpha_2$ (inscribed angles subtended by the same arc are equal) and $\angle DBC = \alpha_1$. Therefore, $\angle B + \angle D = 180°$.

Intersecting chords theorem. A chord of a circle is a line segment whose endpoints both lie on the circle. If two chords AB and CD meet at a point P, then $AP \cdot BP = CP \cdot DP$ (Fig. 3.20a). I could have written this as $|AP| \cdot |BP| = |CP| \cdot |DP|$. But, for the sake of brevity, I adopted the simpler notation. We can rearrange this equation to get ${}^{AP}/{}_{CP} = {}^{DP}/{}_{BP}$. This gives us the idea to prove this theorem, as this new expression is obviously about the proportionality of sides of two triangles APC

[9] The word cyclic is from the Ancient Greek kuklos, which means "circle" or "wheel".

[10] Why do we know this? Imagine a circle and a rectangle with four vertices on the circle, obviously the opposite angles add up to 180°. Generalize this.

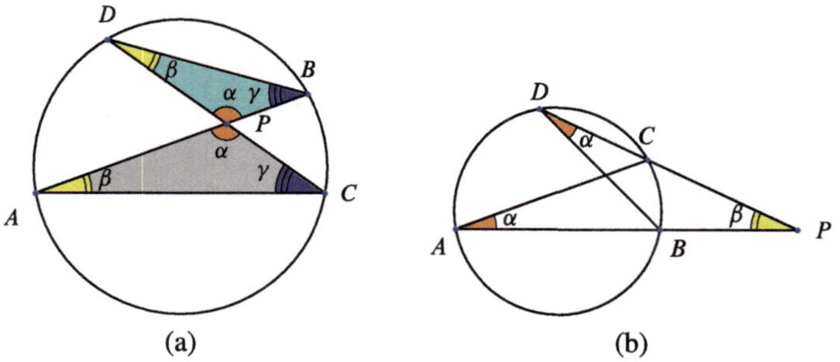

Fig. 3.20 Intersecting chords: $AP \cdot BP = CP \cdot DP$. This result shows that the quantity $AP \cdot BP$ is an invariant (constant) because it does not depend on the position of A and B, with the given circle and point P

and DPB. So, draw BD and AC: we now have triangles in the problem. The two triangles are similar (AAA test), thus their sides are proportional. If the two chords meet outside the circle (Fig. 3.20b), the result still holds.

Ptolemy's theorem. Consider a circle and a cyclic quadrilateral $ABCD$ (Fig. 3.21). Ptolemy discovered the following fact:

$$AC \cdot BD = AB \cdot CD + BC \cdot AD$$

which means that the product of the diagonals is equal to the sum of the products of the opposite sides. How to prove this? It is certain that we have to use similar triangles, but not one pair of two similar triangles but two pairs: because we have $AB \cdot BD$ and $BC \cdot AD$. If we look at Fig. 3.21a, there are two pairs of similar triangles, for instance, triangles labeled as 1 and 1′. But they do not give us what we need (why?).

The key point is to create a new point and thus new triangles. The new point must be on one of the diagonals: we added point E on AC such that $\angle ABE = \angle CBD$ (Fig. 3.21b). Now, $\Delta ABE \sim \Delta DBC$ (AA test), therefore we have

$$\frac{AE}{AB} = \frac{CD}{BD} \Longleftrightarrow AE \cdot BD = AB \cdot CD$$

In the same manner, we have $\Delta EBC \sim \Delta ABD$ (AA test), therefore

$$\frac{EC}{BC} = \frac{AD}{BD} \Longleftrightarrow EC \cdot BD = BC \cdot AD$$

Adding these two results and we're done. The Ptolemy theorem is not a geometry exercise, it provides one of the most important trigonometric identities (to be discussed in Sect. 3.8).

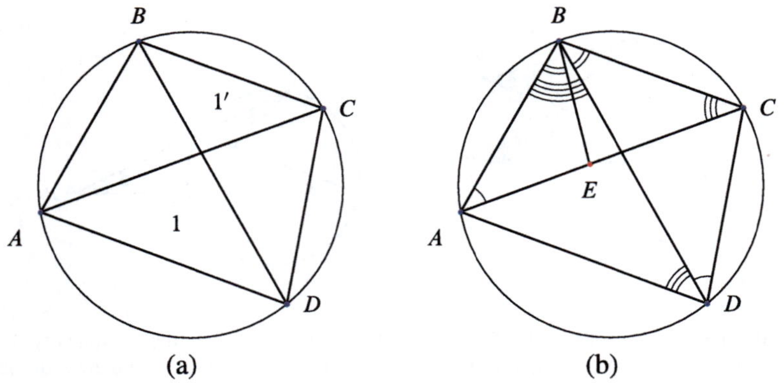

Fig. 3.21 Ptolemy's theorem: $AC \cdot BD = AB \cdot CD + BC \cdot AD$ (**a**) and one proof (**b**). By making $ABCD$ a rectangle, Ptolemy's theorem becomes the Pythagorean theorem ($CD = AB$, $AD = BC$, and $BD = AC$)

3.1.11 Tangents to Circles

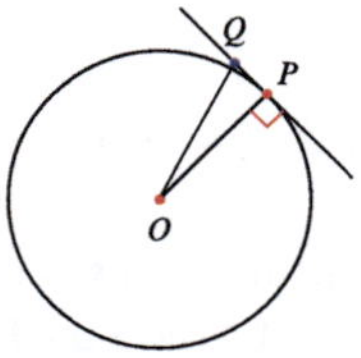

Given a circle and a line there are three possibilities regarding the relative position of them: (1) the line does not intersect the circle—which is boring as there is nothing more to play with, (2) the line intersects the circle at two points and (3) the line touches the circle at exactly one point. We're interested in this last case for it is special. We start with a definition: *A tangent line to a circle is a line that touches the circle at exactly one point*—the point of tangency (P in the figure).
How to draw the tangent line to a circle at P? Draw OP, then draw a line perpendicular to OP, that is the tangent line. But why? Simply consider another point Q on this line, from the Pythagorean theorem we know that the hypotenuse is the largest side in a right triangle[11] (i.e., $OQ > OP$). Thus any point on this line except P is outside the circle.

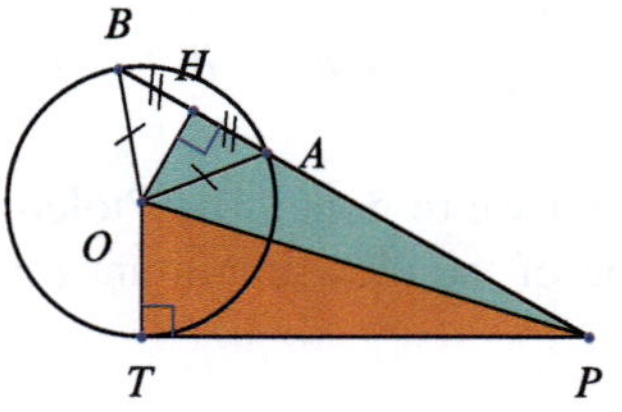

[11] The proof is easy: $a^2 + b^2 = c^2$, thus $c^2 > a^2$ and $c^2 > b^2$.

Tangent-secant theorem. Let P be a point outside the circle, PT be tangent to the circle and PB be a line that cuts the circle at A and B. Then, the tangent-secant theorem says that $PT^2 = PA \cdot PB$. Where does it come from? It comes from Fig. 3.20. Imagine that PA is being rotated counterclockwise around P. While doing so the points A, B come close to each other until they meet at one single point T: PT is then a tangent, and we have $PA = PB = PT$. To prove this theorem, noting that we need PT^2, and that reminds us of the Pythagorean theorem which involves right triangles. So, we draw the triangle OTP which is a right triangle (as PT is tangent to the circle). We need another right triangle: draw OH perpendicular to AB: we then have $HB = HA$ (congruent triangles OHB, OHA). Now, the three right triangles OTP, OHP and OHA give us (r is the radius of the circle):

$$\begin{aligned} r^2 + PT^2 &= OP^2 \\ PH^2 + OH^2 &= OP^2 \\ OH^2 + HA^2 &= r^2 \end{aligned}$$

The rest is algebra: from the first two equations we get $r^2 + PT^2 = PH^2 + OH^2$ (OP is gone), then using the third equation for r^2, we obtain (r is gone too)

$$PT^2 = PH^2 - HA^2 = (PH - HA)(PH + HA) = PA \cdot (PH + HB) = PA \cdot PB$$

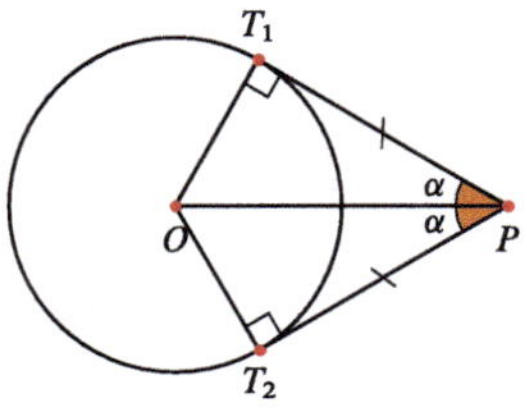

If we have a circle and a point P outside it, we can have a maximum of two tangents from P to the circle: PT_1 and PT_2. From the above result, we know that $PT_1 = PT_2$ and this can be proved using two congruent right triangles OPT_1 and OPT_2 (SSS test, where we have two equal sides, and the third sides are also equal due to the Pythagorean theorem). Thus, OP bisects $\angle T_2PT_1$: it divides this angle into two equal angles. One says that OP is the bisector of $\angle T_12PT_1$. This result will bring us to another interesting property of triangles concerning the so-called inscribed circle or incircle. An incircle is the largest circle which fits inside a triangle just touching the three sides of the triangle.

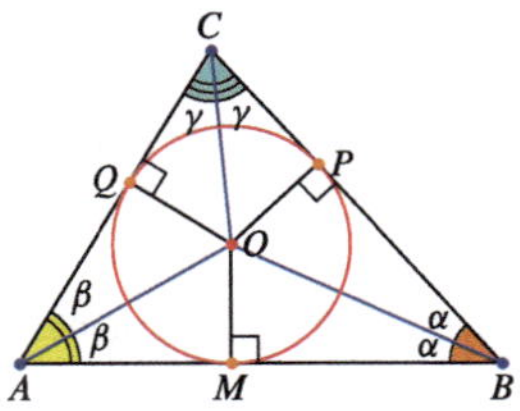

Inscribed circle in a triangle. Consider this problem: given a $\triangle ABC$, draw its inscribed circle. Assume that we have such a circle with a center at O that touches the triangle at P, Q and M, then we have AQ, AM as tangents; BP, BM as tangents. Thus, AO is the bisector of $\angle BAC$; OB is the bisector of $\angle ABC$. Therefore, the center of the incircle of $\triangle ABC$ is the intersection of the bisectors of two of its angles. It follows that the remaining bisector (of $\angle ACB$) also goes through O.[12]

But, how to draw the bisector for a given angle? I am glad you ask. Yes, do not accept anything except the axioms or postulates is the motto.

3.1.12 Geometric Construction

In geometry, geometric construction is the construction of lengths, angles, and other geometric figures using only a straightedge and a pair of compasses. It is important to point out that geometric constructions are not practical problems. The same geometry students could use the protractors and rulers (and even software) to construct many geometric shapes with less difficulties. Furthermore, they are not physical problems neither. They are purely theoretical problems.

All straightedge-and-compass constructions consist of *repeated application of five basic constructions* using the points, lines, and circles that have already been constructed. The five basic constructions are:

- Creating the line through two points;
- Creating the circle that contains one point and has a center at another point;
- Creating the point at the intersection of two (non-parallel) lines;
- Creating the one point or two points in the intersection of a line and a circle (if they intersect);
- Creating the one point or two points in the intersection of two circles (if they intersect).

Collapsing or locking compass. The rules seem to imply that we can't use a locking compass. That is, we are unable to open the compass to a fixed distance, like a pair of dividers, move them to another location on the paper, and draw a circle with that radius. This is troubling because modern compasses are locking compasses and they are time saver when doing geometry problems. Then, are we cheating? No. Euclid, in his second proposition in book I, has shown that collapsing compasses can copy distances. Specially, his second proposition states that given a line segment AB and a point C it is possible to construct a point D so that $CD = AB$.

How to construct D? Let's draw some triangles (if not we just have two boring segments in the diagram): the equilateral triangles ABE and BCF (Fig. 3.22a). (We discussed this problem in Sect. 3.1.2). Now, assume we know where D is, then, for $AE = CD$, $FD = a + b$. This means that D is on the circle centered at F (known)

[12] Proof: two right triangles COQ and COP are congruent (SSS test), hence $\angle QCO = \angle OCP$.

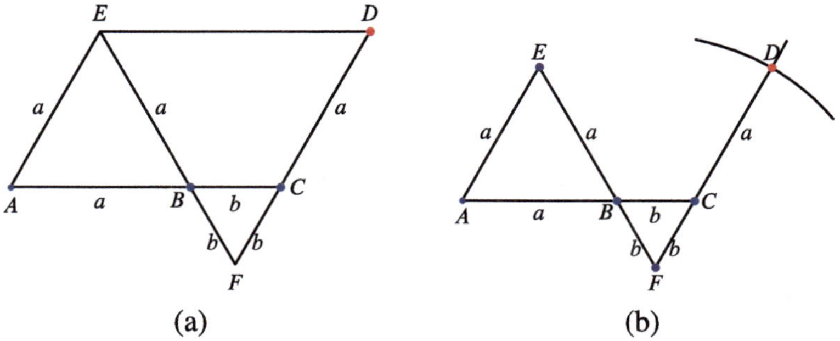

Fig. 3.22 Euclid's second proposition in Book I: using a compass to copy a line segment (AB)

Fig. 3.23 Bisecting a segment. This construction can be applied to construct a line passing through a given point and perpendicular to a given line

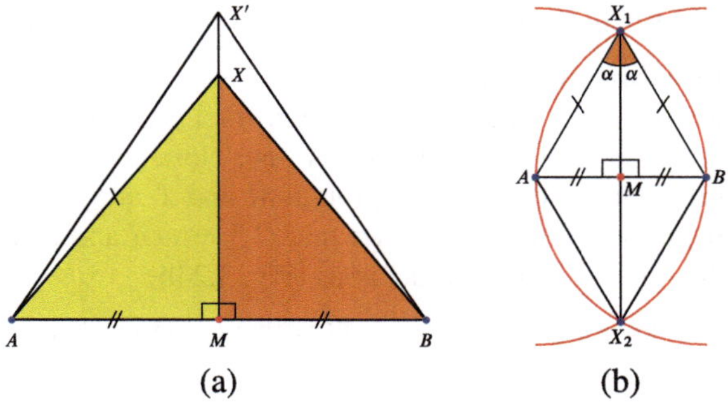

and passes through E (known). And D is also on the line FC. Thus, we know how to construct D (Fig. 3.22b). So, from now on, we can just use the locking compasses.

Bisecting a segment. Given a segment AB, find the point M on it such that $AM = BM$. One way to proceed is: assuming that we can construct such point M, then find its properties and relation with existing figures. As this is geometry, we need to have geometric figures (triangles, circles etc.). Thus, draw $MX \perp AB$, AX and BX (Fig. 3.23a) What is special about X? It is: $AX = BX$. Thus, X is the intersection of two circles: one centered at A with radius AX and the other centered at B with the same radius. But we do not know AX, but we have AB.

So, the construction steps are (Fig. 3.23b):

- Draw a circle centered at A with radius AB;
- Draw a circle centered at B with radius AB;
- These two circles intersect at two points X_1, X_2, we then get X_1 and X_2;
- Draw X_1X_2, it intersects AB at M. That's the point we're after.

More important than the constructions themselves are the proofs that they accomplish what they say they accomplish: we need to prove our construction. Indeed, $\triangle X_2AX_1 \sim \triangle X_2BX_1$ (SSS test), thus we have the same angles at X_1 as indicated, and that leads to $\triangle X_1AM \sim X_1BM$ (SAS), which results in $AM = BM$.

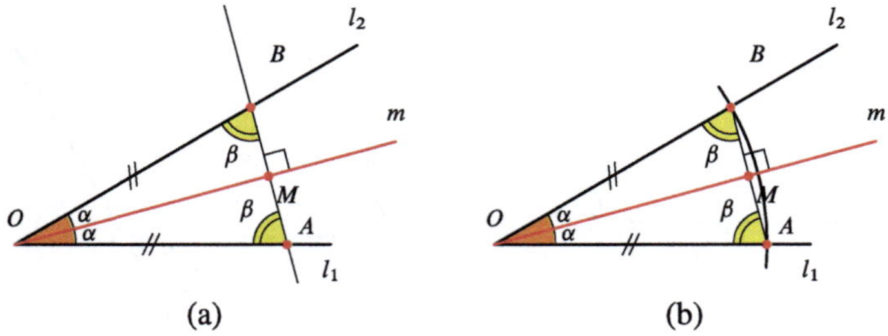

Fig. 3.24 Bisecting an angle

The generalization of this problem is: to divide a segment into n equal parts. To this end, we need to use the Thales theorem on proportion. I have presented this construction when we discuss rational numbers in Definition 2.8.1.

Bisecting an angle. Given an angle, construct a line bisecting this angle. Referring to Fig. 3.24a, what we need to construct is the line m. But a line requires two points, so we need a point M on this line. Now, to make lines l_1 and l_2 involved in the problem, we draw a line through M and $\perp m$.[13] This line cuts l_1 and l_2 at A and B, respectively. It is easy to see that $OA = OB$ and $BM = AM$. And now, we reverse the process and construct m as (Fig. 3.24b)

- Draw a circle centered at O with any radius. This circle intersects l_1 at A and l_2 at B;
- Bisect AB (using the previous construct): then we have M—the mid-point of AB;
- Draw OM. That's the line we're looking for.

Construct regular polygons of 6×2^n edges inscribed in a unit circle. Let's solve this problem: construct a hexagon inscribed in a unit circle. Again, we do the analysis first: assuming that we already have this hexagon (Fig. 3.25a). If we join the center O with the vertices we get six equilateral triangles. For example, $\triangle OAB$ is an equilateral triangle (why?). But we know how to build an equilateral triangle! So, we're able to build a hexagon (Fig. 3.25b).

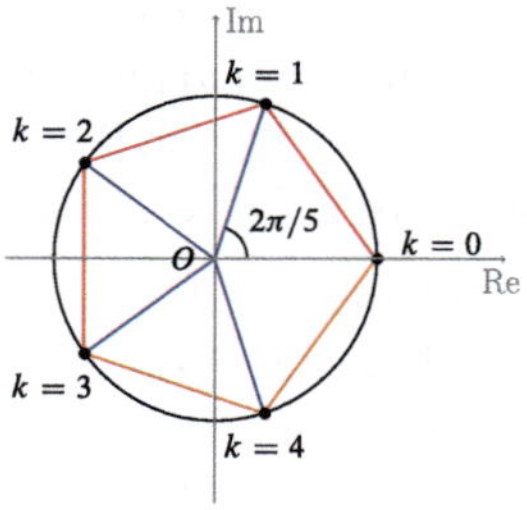

Construct a pentagon inscribed in a unit circle. For me, this problem is hard. I have tried to do the analysis but failed. The reason was that I did not know well of this

[13] Why perpendicular? This is special, because if not, then there are infinite number of lines.

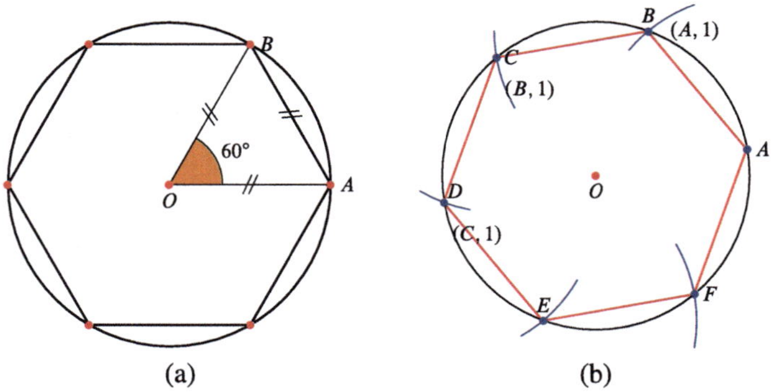

Fig. 3.25 Constructing a hexagon inscribed in a unit circle. Start with a circle with center O, select a point on it, label it A. Draw the circle $(A, 1)$ centered at A with radius 1. This circle cuts the original circle at B. Make a circle $(B, 1)$, it cuts the original circle at C. Repeat this process and we get a hexagon. From this hexagon and the angle bisecting construction, we can build any polygon having 6×2^n edges for $n = \{0, 1, 2, 3. \dots\}$. This successful construction of the hexagon also means that we can trisect a 180°. Trisecting an angle β is to make three equal angles $\beta/3$ from it

pentagon. After a while I realized that if I know the length of one edge of the pentagon, then it can be a good starting point. To find this length, there are many ways, but one way is to use complex numbers, trigonometry, and analytic geometry.[14] In Fig. 2.38, we have met the fifth roots of unity which are the vertices of a pentagon. Now, we know the coordinates of the vertex with $k = 0$ are $(1, 0)$ and the coordinates of $k = 1$ are $(\cos\alpha, \sin\alpha)$ with $\alpha = 2\pi/5$. Therefore, the length of this segment, denoted by a_5, is

$$a_5 = \sqrt{(\cos\alpha - 1)^2 + (\sin\alpha)^2} = \sqrt{2(1 - \cos\alpha)} \tag{3.1.8}$$

But $\cos\alpha = \cos(2\pi/5) = \sqrt{5}-1/4$,[15] thus

$$a_5 = \sqrt{2(1 - \cos\alpha)} = \sqrt{\frac{5}{2} - \frac{\sqrt{5}}{2}} \tag{3.1.9}$$

Now, using this knowledge we will do the analysis as in Fig. 3.26a: we have our circle and our pentagon (assumed it has been constructed). To mimic the coordinate axes, we also draw two perpendicular lines going through the center of the circle. Then, we have $O_1A = 1$. And $\sqrt{5}/2$ can be obtained from a right triangle with sides 1, 1/2 (for $\sqrt{1^2 + (1/2)^2} = \sqrt{5}/2$). Thus, we need a segment of length 1/2. So, we introduce point S—the mid-point of MO, this gives us $SA_1 = \sqrt{5}/2$. We need S' such that $SS' = \sqrt{5}/2$. Why? Then we have $OS' = \sqrt{5}/2 - 1/2$. In the right triangle OA_1S', we have

[14] If you do not know yet trigonometry/analytic geometry, you can return to this after you have mastered these topics.

[15] Computed from $\cos\pi/5$, which is given in Sect. 3.8.

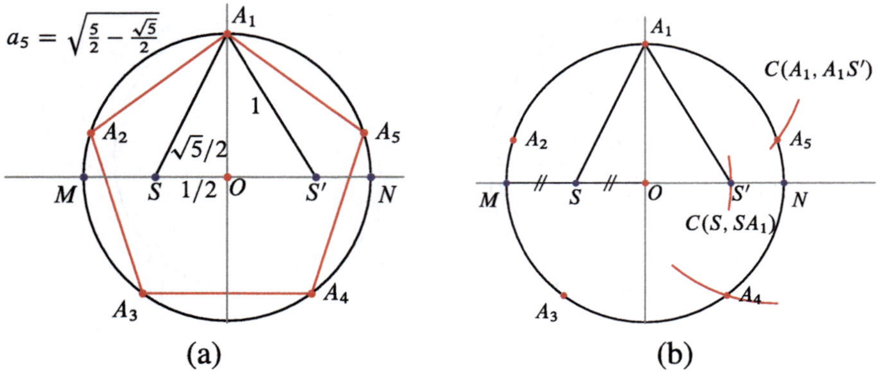

Fig. 3.26 Construction of a pentagon inscribed in a circle

$$A_1S' = \sqrt{1^2 + \left(\sqrt{5}/2 - 1/2\right)^2} = \sqrt{\frac{5}{2} - \frac{\sqrt{5}}{2}}$$

which is nothing but a_5! Therefore, the construction is shown in Fig. 3.26b.[16]

It is time to stop and think about what we have done. We can construct n-gon for $n = \{3, 4, 5, 6\}$. And by angle bisecting construction, we can construct regular polygons of 8, 10, 12, 20, 24, ... edges. How about a 7-gon? How about a 9-gon? This turns out to be an extremely hard problem! For the next 2000 years since Ptolemy, no one had ever succeeded in constructing a regular 7-gon, a regular 9-gon, or a regular 11-gon. Every effort seems to end up in failure. The reason is not because of our incapability but it turns out that *it is impossible to construct those polygons!*

The first mathematician who made a breakthrough in polygon construction was Gauss. On 30 March 1796, when he was 19 years old, Gauss made a major discovery in mathematics: he showed that it is possible to construct a regular 17-gon (called heptadecagon). It is believed that the excitement of this discovery had led Gauss to choose mathematics instead of philosophy as a career. Gauss was so pleased by this result that he requested that a regular 17-gon be inscribed on his tombstone. For some unknown reason, this request was not fulfilled, but on Gauss's memorial stone in Brunswick, we can find a 17-point star.

So, we know that it is possible to construct 3-gon, 5-gon, and 17-gon. What is special about 3, 5 and 17? They are odd, they are primes. What else? Let's touch them so that they reveal their secret: taking one from them we obtain these numbers: 2, 4 and 16, which can be written as 2^0, 2^2, 2^4, which can also be written as 2^{2^0}, 2^{2^1} and 2^{2^2}. These are Fermat prime numbers! Within 5 years Gauss had developed a theory that described exactly which polygons were, and which were not, constructible. In 1801, Gauss presented this result at the end of his now classic *Disquistiones Arithmeticae*,

[16] This construction was given by Ptolemy (c. 100–c. 170), the mathematician, astronomer, and geographer of Greek descent, in the second century in the Almagest, one of the most famous scientific texts of all time.

in which he proves the constructibility of the n-gon for any n that is a Fermat prime (i.e., of the form $2^{2^k}+1$ for $k \in \mathbb{N}$).

How about Gauss's proof? I will discuss it later in Sect. 3.24.3.

3.1.13 The Three Classical Problems of Antiquity

The three classical problems of antiquity are the following:

1. doubling the cube: given a cube, construct another cube which is twice as big,
2. squaring the circle: given a circle, construct a square which has the same area as that of the circle,
3. trisecting the angle: given an angle, divide it into three equal angles.

using only straightedge and compass. Although Euclid solved more than 100 construction problems in the *Elements*, we do not see any of these problems in the *Elements*. The reason is that these problems were not solved until the nineteenth century!

If the cube has a side of a, its volume is then a^3, and the volume of the cube we want to construct is $2a^3$. Therefore, doubling the cube boils down to constructing, using straightedge and compass, a cube with a side of $\sqrt[3]{2}a$. Of course, we can start with $a = 1$, and the problem is to construct a $\sqrt[3]{2}$ segment. This is intimately related to the cubic equation $x^3 - 2 = 0$.

If the circle has a unit radius, its area is then π. A square that has an area of π has a side of which the length is $\sqrt{\pi}$. Therefore, squaring the circle boils down to constructing, using straightedge and compass, a segment of length $\sqrt{\pi}$.

Given an angle 3θ, it is said that we can trisect it if we can construct a segment of length $\cos\theta$. Using the trigonometry identity $\cos 3\theta = 4\cos^3\theta - 3\cos\theta$, the problem seems to become again the business of cubic equations!

For millennia people tried again and again to double a cube and trisect an angle, but they always failed. Such was the status of these problems until 1837 when Pierre Wantzel (1814–1848) published proof that they couldn't be solved (to be discussed in Sect. 3.24.4). The great Gauss had stated that both of these methods were impossible but had never provided proof.

3.1.14 Tessellation: The Mathematics of Tiling

You have probably noticed that floors are usually tiled in squares or sometimes in rectangles. What is so special about these shapes? What are the disadvantages of using other shapes? In mathematics, the term used for tiling a plane (floor in our

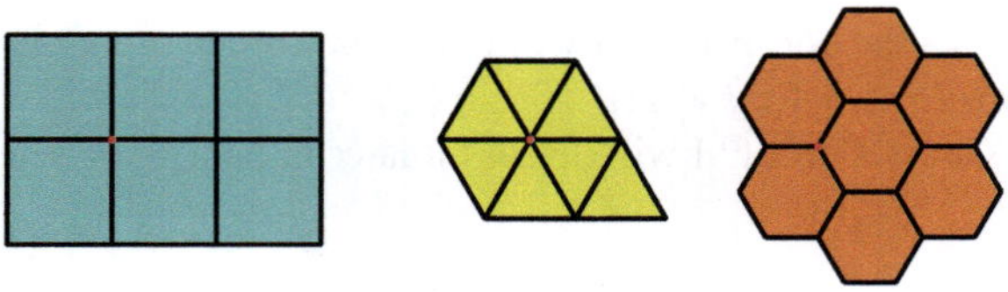

Fig. 3.27 Examples of regular polygons that can tile the floor without gaps and overlaps: square, equilateral triangle, and hexagon. This is known as *regular tessellation*, in which only one type of regular polygons is allowed

context) with no gaps and no overlaps is a tessellation. Tessellations have many real-world examples and are a physical link between mathematics and art.[17]

In order to have no gaps the angle around a point (point far from the boundary of the plane) must be 360°, no more no less. Based on this, we can understand which regular polygons can tessellate a plane. The only regular polygons that can do this are equilateral triangle, square, and hexagon (Fig. 3.27). Equilateral triangles form a regular tessellation because their interior angles are 60° (consequence of Eq. (3.1.6)), and 60 is a divisor of 360: $360 = 6 \times 60$. Squares form a regular tessellation because their interior angles are 90°, and 90 is a divisor of 360 too. Hexagons also form a regular tessellation because their interior angles are 120°, and 120 is a divisor of 360 too.

In order to not have to examine other possibilities, we resort to algebra for a proof. Let's consider a tessellation with p-gons and q such p-gons at a vertex. At this vertex the angle is 360° and this gives us the following equation (using Eq. (3.1.6))

$$180^\circ\left(1-\frac{2}{p}\right)\times q = 360^\circ$$

Dividing both sides of this equation by 180°, we get another equation $(1-2/p)q = 2$. And our task is now to solve this equation—that is finding the tuples (p, q)—noting that $p, q \in \mathbb{N}$. How to do this? Amazingly, a bit of algebraic massage to the equation is a big help:

$$(p-2)q = 2p \iff (p-2)q = 2p-4+4 \iff (p-2)(q-2) = 4 \quad (3.1.10)$$

So, $(p-2, q-2) = \{(1,4), (4,1), (2,2)\}$. Thus, $(p, q) = \{(3,6), (6,3), (4,4)\}$. That's it. The key step was to add $-4+4$, which is nothing, to the RHS of the equation.

That's all we can do with regular tessellation. It is kind of boring so let's relax the requirement a little bit by allowing more than one type of regular polygons in the tessellation. What we then obtain is called *semi-regular tessellation*. See Fig. 3.28 for

[17] I got to know to this topic by reading the interesting book *Mathematics Rebooted: A Fresh Approach to Understanding* by Lara Alcock [2]. Alcock is a Reader in the Department of Mathematics Education at Loughborough University.

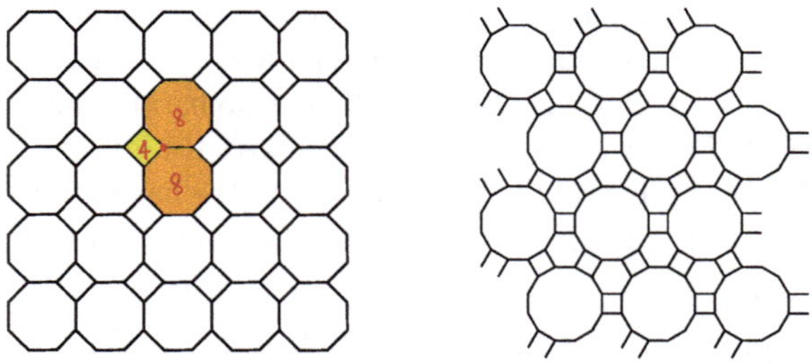

(a) pattern I $(4, 8, 8)$ (b) pattern II $(4, 6, 12)$

Fig. 3.28 Some examples of semi-regular tessellation. The notation (4, 8, 8) indicates the polygons (via its number of edges) surrounding a vertex

some examples. Pattern I is made up of squares, and regular hexagons. Going around each vertex we meet a square (4 edges) and two pentagons (8 edges), we associate the tuple (4, 8, 8) with it. Pattern II is made up of squares, regular hexagons, and regular dodecagons, so we associate the tuple (4, 6, 12) with it. Keep in mind that at every vertex, we must have the same polygons in the same order.

The question that mathematicians are interested in now is: how many semi-regular tessellations are there? It turns out that there are only eight of them. And we're going to prove this. To this end, let the tuple be $(n_1, n_2, \ldots, n_k)$ where each n_i is a positive integer representing the number of edges. Since each polygon has at least three sides, we have $n_i \geq 3$ for every i. We start by showing that $k \leq 6$. That is, there can be no more than six polygons meeting at each vertex. This constraint is due to the fact that at any vertex of tessellation the angle must be 360°. It is obvious that we are going to use Eq. (3.1.6) to obtain:

$$\left(180^\circ - \frac{360^\circ}{n_1}\right) + \left(180^\circ - \frac{360^\circ}{n_2}\right) + \cdots + \left(180^\circ - \frac{360^\circ}{n_k}\right) = 360^\circ$$

A bit of algebraic manipulation (dividing the above equation by 180°) gives us

$$\frac{1}{n_1} + \frac{1}{n_2} + \cdots + \frac{1}{n_k} = \frac{k}{2} - 1 \tag{3.1.11}$$

But we also have $n_i \geq 3$ or $1/n_i \leq 1/3$. So,

$$\frac{1}{n_1} + \frac{1}{n_2} + \cdots + \frac{1}{n_k} \leq \frac{k}{3} \tag{3.1.12}$$

From Eqs. (3.1.11) and (3.1.12) we have $k/2 - 1 \leq k/3$ which leads to $k \leq 6$. Furthermore, note that we also have $k \geq 3$ because at any vertex we need at least three polygons to have 360°.[18] Therefore, there are only four possibilities for k: $k \in \{3, 4, 5, 6\}$.

[18] Why? Because all interior angles in regular polygons are smaller than 180° according to Eq. (3.1.6).

Now we have to consider each case. I show here the case $k = 3$ (as it is the easiest). First, we change the notation as there are only three unknowns. Let's rename n_1, n_2, n_3 as a, b, c. Second, there is no harm in labeling them such that $a \le b \le c$. Now, Eq. (3.1.11) becomes $1/a + 1/b + 1/c = 1/2$. In conjunction with $3 \le a \le b \le c$, we have $1/2 \le 3/a$, or $a \le 6$. We have, therefore, found a: $a = \{3, 4, 5, 6\}$. The next step is to determine b and c, for each a from $1/b + 1/c = 1/2 - 1/a$:

- $a = 3$: finding two positive integers b and c such that $1/b + 1/c = 1/6$. We proceed as (using the same trick as in Eq. (3.1.10))

$$1/b + 1/c = 1/6 \Longleftrightarrow bc - 6(b + c) = 0 \Longleftrightarrow (b - 6)(c - 6) = 36$$

  Thus, $(b - 6, c - 6) = \{(1, 36), (2, 18), (3, 12), (4, 9), (6, 6)\}$. Hence, (b, c) is one of the following: $(7, 42), (8, 24), (9, 18), (10, 15), (12, 12)$.
- $a = 4$: in the same manner, (b, c) is one of the following: $(5, 20), (6, 12), (8, 8)$.
- $a = 5$: (b, c) is: $(5, 10)$.
- $a = 6$: (b, c) is: $(6, 6)$.

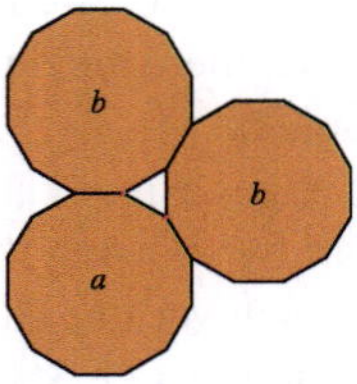

To conclude (a, b, c) is one of the following triples: $(3, 7, 42), (3, 8, 24), (3, 9, 18), (3, 10, 15), (3, 12, 12), (4, 5, 20), (4, 6, 12), (4, 8, 8), (5, 5, 10), (6, 6, 6)$. We recognize that $(6, 6, 6)$ is the regular tessellation with hexagons. For other possibilities, we need to examine them. For example, let's consider $(3, 12, 12)$ shown in the figure but imagine that instead of 12-sided polygons we have a-sided and b-sided polygons. It is then clear that we must have $a = b$. Thus, in the triple (a, b, c) if there is an odd number then the other two must equal.[19] Therefore, $(3, 7, 42), (3, 8, 24), (3, 9, 18), (3, 10, 15), (5, 5, 10)$ are invalid. The only semi-regular tessellations for $k = 3$ are $(3, 12, 12), (4, 8, 8), (4, 6, 12)$.

3.1.15 *Platonic Solids*

The generalization of a polygon in three-dimensional space is called a *polyhedron* (plural is polyhedra). This is a solid with flat faces (or sides) each of which is a polygon. Among the more familiar polyhedra is the square-base pyramid (see Fig. 3.5). And if we generalize the idea of regular polygons to 3D what we get are called *regular polyhedra* or Platonic solids. Figure 3.29 shows the five platonic solids: tetrahedron, cube, octahedron, dodecahedron and icosahedron.

[19] Keep in mind that at every vertex, we must have the same polygons in the same order.

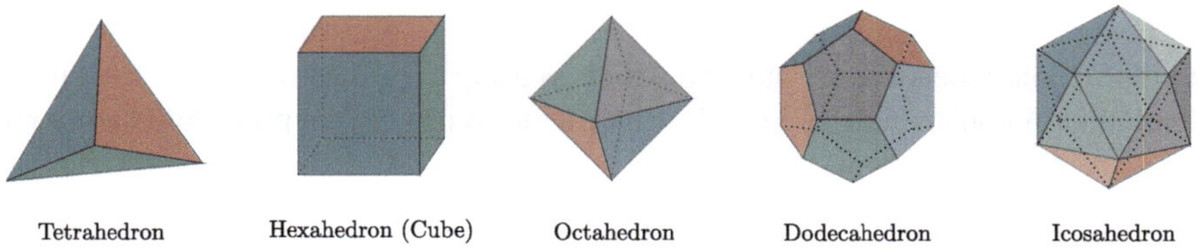

Fig. 3.29 Platonic solids. The tetrahedron has 4 triangular faces, 6 edges, and 4 vertices. The cube or hexahedron has 6 square faces, 12 edges, and 8 vertices. The octahedron has 8 triangular faces, 12 edges, and 6 vertices

What is a platonic solid? A platonic solid is a convex, regular polyhedron in three-dimensional Euclidean space. That is:

- all faces are regular polygons: e.g. a tetrahedron has four faces each of which is an equilateral triangle;
- all faces are congruent i.e., same size and shape;
- the same number of faces meet at each vertex (and this number is larger than or equal to three), and
- the solid is convex (e.g. no indentation).

The name Platonic solids of the five regular polyhedra comes from Plato (423–348 BC), a Greek philosopher born in Athens. Today Plato is best known as a philosopher, but one of his most important contributions was the creation of the Academy. Although Plato did not contribute directly to the body of mathematical knowledge, he was an important figure in the advancement of the subject. He was a lover of mathematics, and mathematics formed the backbone of the curriculum at the Academy. This is clearly illustrated by the inscription found above its entrance: *"Let no one ignorant of geometry enter here"*. Because many mathematicians were trained and nurtured in the Academy, it is said that Plato was not a maker of mathematics, but "a maker of mathematicians" [10].

In his test *Timaeus*—a fictional account of a discussion between Socrates, Hermocrates, Critias, and Timaeus, Plato associated the tetrahedron with fire, the octahedron with air, the icosahedron with water, and the cube with earth; fire, air, water, and earth were the only four elements considered at that time. The fifth solid, the dodecahedron, was considered the shape that encompasses the whole universe and was used for arranging the constellations in heaven. This Greek atomic model was so influential that it was universally accepted until the Irish scientist Robert Boyle (1627–1691) published his book *The Sceptical Chymist* in 1661.

One question arises: Is it just that these five Platonic solids are the only ones we've found? Could it be there are others out there waiting to be discovered[20]? Don't try to find other platonic solids as Euclid's The Elements provided a proof that *there exist no more than five platonic solids.*

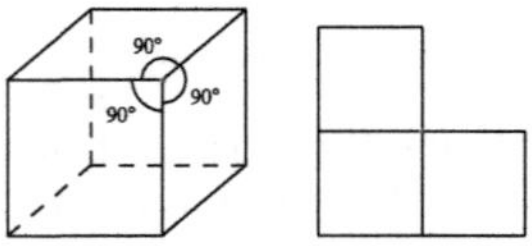

The proof is based on this observation: the sum of the angles of the faces meeting at a vertex on a platonic solid must be less than 2π. As illustrated in the next figure, at a vertex on a cube, there are three faces of which the angles are $90°$: the sum of these angles are $270°$, less than 2π. For a tetrahedron, three faces meet at a vertex of which each face has an angle of $60°$, thus the total angle sum is $180°$, again less than 2π. Now, we're going to use this result to prove that there are only five platonic solids shown in Fig. 3.29.

Proof Let's assume that the platonic solid has as faces n-gons. We know that every n-gon has an interior angle of $\alpha = 180°(1 - 2/n)$. Furthermore, assume that there are m faces meeting at a vertex, noting that $m \geq 3$. Then, the total angle at a vertex is $m\alpha$ or $180°m(1 - 2/n)$. Using the above result we then have the inequality $180°m(1 - 2/n) < 360°$ or $m(1 - 2/n) < 2$. Now, it is simple to proceed. We just need to consider $n = 3, 4, 5, \geq 6$ and see that there are only five platonic solids. For example, if $n = 3$, then we have $m(1 - 2/3) < 2$ or $m < 6$. Thus, $3 \leq m < 6$ or $m = \{3, 4, 5\}$: we have then the tetrahedron, the octahedron, and the icosahedron. For $n \geq 6$ it is impossible to satisfy the inequality $m(1 - 2/n) < 2$ with $m \geq 3$. ∎

3.1.16 Euler's Polyhedra Formula

In a polygon the number of vertices is equal to the number of edges. Thus, there exists a relationship between the number of vertices and the number of edges. The same must hold true for polyhedra, except that the situation is more complicated as we are dealing with faces, vertices, and edges. It is surprising that no mathematician paid attention to this issue until 1639 when Descartes turned away his attention from philosophy to mathematics. He noticed a curious pattern. He probably constructed a table similar to Table 3.1. The way Descartes looked at polyhedra was revolutionary: he was dealing with geometrical objects but he did not calculate the metric of them such as area, volume, or angles. Instead he counted the number of vertices, faces, which seems not a special thing to do. Counting things is, however, a mathematical activity, it is the field of combinatorics.

[20] I used the word discovery because these platonic solids are commonly found in nature. For instance, the carbon atoms in a diamond are arranged in a tetrahedral shape. Common salt and fool's gold (iron sulfide) form cubic crystals, and calcium fluoride forms octahedral crystals.

Table 3.1 Platonic solids: number of faces/vertices/edges. The last two columns serve a later purpose, p denotes the number of edges of a face and q denotes the number of faces meeting at a vertex

Polyhedron	Vertices (V)	Edges (E)	Faces (F)	$F - E + V$	p	q
Tetrahedron	4	6	4	2	3	3
Cube	8	12	6	2	4	3
Octahedron	6	12	8	2	3	4
Dodecahedron	20	30	12	2	5	3
Icosahedron	12	30	20	2	3	5

From this table, Descartes noticed that for any platonic solids we have $F - E + V = 2$[21] if F denotes the number of faces, V the number of vertices and E the number of edges. For unknown reasons, he did not publish this finding. It was Euler who in 1750 re-discovered this result and in 1751 proved it. On 14 November 1750, in a letter to his friend, the number theorist Christian Goldbach (1690–1764), Euler wrote, "It astonishes me that these general properties of stereometry [solid geometry] have not, as far as I know, been noticed by anyone else" [55].

Now I discuss how to build Table 3.1. If you're thinking counting is the way to do it, what's there to discuss, then you're wrong: mathematicians are lazy and they prefer smart ways of counting. Let's assume that each face of a platonic solid is a p-gon, and there are q faces (or q edges) meeting at a vertex. And we assume that we know the number of faces F (this is easy to count). The problem now is to find E and V as functions of F, p, q.

Each face has p edges and there are F faces, so there are pF edges. But we have counted each edge twice as two faces share the same edge. So, in total we have $E = {}^{pF}/_{2}$. Each edge has two vertices, and we have E edges, thus there are $2E$ vertices. But we have counted each vertex q times, so $V = {}^{2E}/_{q}$. We need to check our formula. Take the dodecahedron as an example: $F = 12 p = 5$ and $q = 3$. Thus, $E = {}^{(5)(12)}/_{2} = 30$ and $V = {}^{(2)(30)}/_{3} = 20$.

Getting back to Table 3.1 Euler quickly realized that $F - E + V = 2$ and he proved that this relation holds for any convex polyhedron (not just the Platonic solids). And this equation now bears the name Euler's polyhedra formula, and it ranks as the second most beautiful equation in mathematics. Recall that the most beautiful equation of all maths is Euler's identity $e^{i\pi} + 1 = 0$. Euler is the master of us all, as Laplace puts it.

Using Euler' formula to prove there exist only five Platonic solids. Consider a regular polyhedron with p-gons as faces and there are q faces (or q edges) meeting at a vertex. We have derived the relations:

[21] Historically, Descartes came up with a slightly different relation: he did not use edges but instead used the plane angles.

$$F = \frac{2E}{p}, \quad V = \frac{2E}{q}$$

Now we substitute them into $F - E + V = 2$, we get

$$\frac{2E}{p} - E + \frac{2E}{q} = 2$$

From that we get

$$\frac{2E}{p} - E + \frac{2E}{q} > 0 \Longleftrightarrow \boxed{\frac{1}{p} + \frac{1}{q} > \frac{1}{2}}$$

It can be shown that only the solids in Table 3.1 (see last two columns) satisfy this inequality. How does this proof compare with Euclid's proof? Euclid's proof is geometrical. This proof is called topological. In the proof we just massage some algebraic expressions, nothing about the regularity of the Platonic solids was used.

3.2 Area of Curved Figures

In Sect. 3.2.1, we discuss the area of first curved shape—the lune of Hippocrates. Next, we treat the area under a parabola segment—a problem solved by Archimedes (Sect. 3.2.2). The problem of the circumference and area of circles is given in Sect. 3.2.3. They involve π, and how ancient mathematicians computed π is shown in Sect. 3.2.4.

3.2.1 Area of the First Curved Plane: The Lune of Hippocrates

In geometry, the lune of Hippocrates, Hippocrates of Chios (circa 400 BCE), is a lune bounded by arcs of two circles, the smaller of which has as its diameter a chord spanning a right angle on the larger circle (Fig. 3.30). Equivalently, it is a plane region bounded by one 180-degree circular arc and one 90-degree circular arc. It was the first curved figure to have its exact area calculated mathematically.

Hippocrates wanted to solve the classic problem of squaring the circle, i.e., constructing a square by means of straightedge and compass, having the same area as a given circle. He proved that the lune bounded by the arcs labeled E and F in the figure has the same area as triangle ABO. Again, ancient mathematicians compared

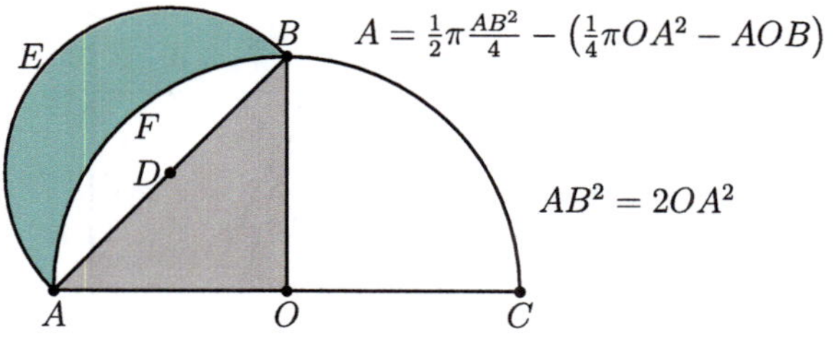

Fig. 3.30 Lune of Hippocrates. The shaded area $AEBF$ is a moon-like crescent shape, and it is called a lune, deriving from the Latin word *luna* for moon. Geometrically a lune is the area between two circular arcs. Let's denote by A the area of the lune. Then, it is shown that it is equal to the area of the triangle AOB, assuming that we know the formula of the area of a circle

new areas to old ones! This afforded some hope of solving the circle-squaring problem, since the lune is bounded only by arcs of circles. Heath[22] concludes that, in proving his result, Hippocrates was also the first to prove that the area of a circle is proportional to the square of its diameter.

3.2.2 Area of a Parabola Segment

Let's see how Archimedes computed the area of a parabola segment without calculus. To simplify the analysis, let's consider the parabola $y = x^2$ which is cut by the horizontal line $y = 1$, see Fig. 3.31. The shaded area is computed as a sum of Δ_1 (the area of the largest triangle), and the left over area. This left over area is approximated by two triangles (Δ_2 and Δ_3) and what is left unaccounted. And the process goes on.

First, the areas of triangles OQR and QBR are identical and equal $1/16$. Thus $\Delta_2 = 1/8$, and therefore $\Delta_2 + \Delta_3 = 1/4$. And note that $\Delta_1 = 1$, so $\Delta_2 + \Delta_3 = 1/4\Delta_1$. So, we can write that

$$A \approx \Delta_1 + \frac{\Delta_1}{4}$$

If we continue the process with the unaccounted region, we get $A \approx \Delta_1 + \Delta_1/4 + \Delta_1/16$. Seeing now the pattern, we write

$$A = \Delta_1 + \frac{\Delta_1}{4} + \frac{\Delta_1}{16} + \cdots = \Delta_1\left(1 + \frac{1}{4} + \frac{1}{16} + \cdots\right) = \frac{4}{3}\Delta_1$$

where use was made of the geometric series (Sect. 2.21.2). It is remarkable that the area of that curved shape is a factor of the area of the largest triangle. From this result, it is simple to deduce that the area below the parabola is $2 - 4/3 = 2/3$.

[22] Sir Thomas Little Heath (1861–1940) was a British civil servant, mathematician, classical scholar, and historian of ancient Greek mathematics. Heath translated works of Euclid of Alexandria, Apollonius of Perga, Aristarchus of Samos, and Archimedes of Syracuse into English.

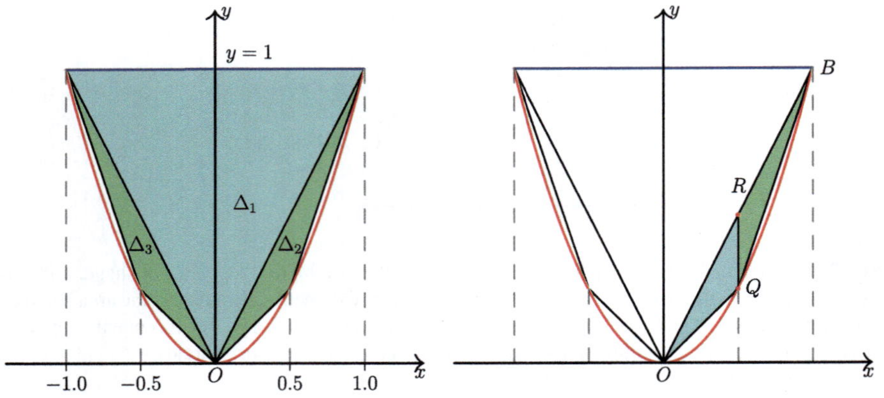

Fig. 3.31 Area of a parabola segment. The coordinates of Q are (0.5, 1/4) and of R are (0.5, 0.5) noting that OB is the line $y = x$. Thus $QR = 1/4$. We have $\Delta_1 = 1$, and Δ_2 is equal to the area of the two triangles OQR and QBR combined

A student in a calculus course would just use integration and immediately obtain the result, as

$$A = 2\int_0^1 (1 - x^2)dx = 2\left(x - \frac{x^3}{3}\right)\Bigg]_0^1 = \frac{4}{3} \tag{3.2.1}$$

This technique has a name because it was widely used by Greek mathematicians: it's called *the method of exhaustion*; as we add more and more triangles they exhaust the area of the parabola segment. There are a lots to learn about Archimedes' solution to this problem. First, he also used the area of simpler geometry (a triangle). Second, and the most important idea, is that he used infinitely many triangles! Only when the number of triangles is approaching infinity, the sum of all the triangle areas approach the area of the parabola segment. This sounds similar to the integral calculus we know of today! But wait, while Eq. (3.2.1) is straightforward, Archimedes' solution requires his genius. For example, how would we know to use the triangles that he adopted?

Even though Archimedes' solution is less powerful than the integral calculus developed much later in the seventeenth century, he and Greek mathematicians were right in going to infinity. The main idea of computing something *finite*, e.g. the area of a certain (curved) shape, is to chop it into many smaller pieces, handle these pieces, and when the number of pieces goes to infinity adding them up will give the answer. This is what Strogatz called the *Principle of Infinity* in his book *The Power of Infinite*. It is remarkable that we see Archimedes' legacy in the modern world, see for instance Fig. 3.32. In computer graphics and in many engineering and science fields, any shape is approximated by a collection of triangles (sometimes quadrilaterals are also used) and tetrahedra for 3D shapes. What is difference is that we do not go to infinity with this process, as we're seeking an approximation. Note that Archimedes was trying to get an exact answer.

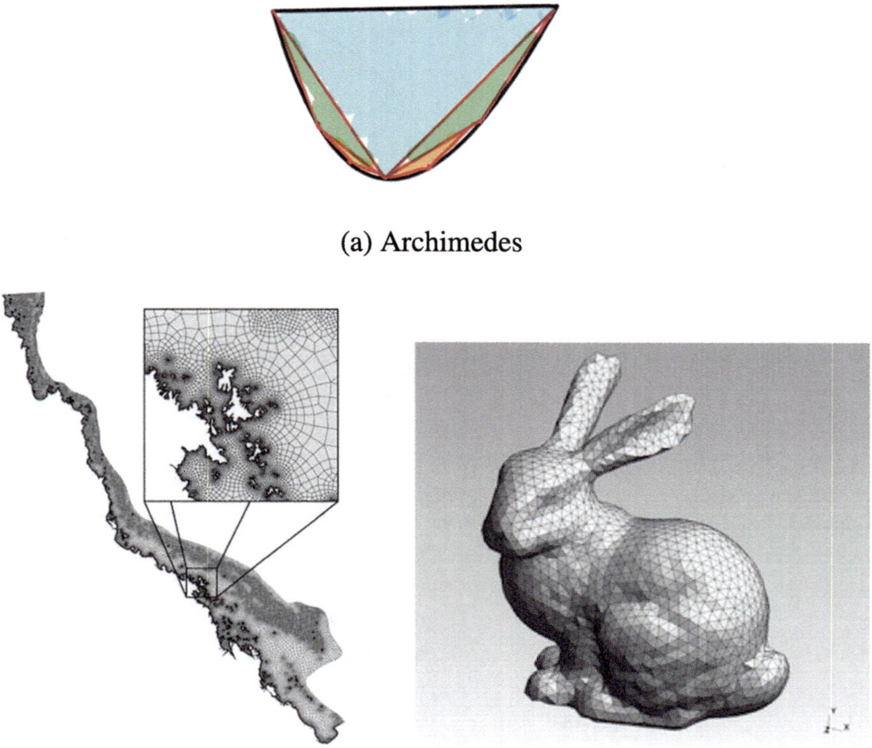

(a) Archimedes

(b) Today

(c) Today

Fig. 3.32 Archimedes' legacy in the modern world: use of triangles and tetrahedra to approximate any 2D and 3D objects (images obtained using `Gmsh` [26])

3.2.3 Circumference and Area of Circles

Below are the facts about circles that ancient mathematicians know:

- the circumference of a circle is proportional to its diameter, so $C = 2\pi_1 r$ assuming that the proportionality is π_1 (which is some number) and r is the circle radius;
- the area of a circle is proportional to the square of its radius, so $A = \pi_2 r^2$ assuming that the proportionality is π_2;
- the circumference C and the area A (of a circle) is related by $A = {}^1\!/_2 Cr$.

The third fact induces that $\pi_1 = \pi_2 = \pi$. A proof of $A = {}^1\!/_2 Cr$ is shown in Fig. 3.33: the area of the circle equals the area of an inscribed regular polygon having infinite sides (in the figure, I illustrate using a pentagon i.e., $n = 8$). The area of this polygon is the sum of the area of all the isosceles triangles OAB; an isosceles triangle is a triangle that has two sides of equal length. These triangles have a height

Fig. 3.33 .

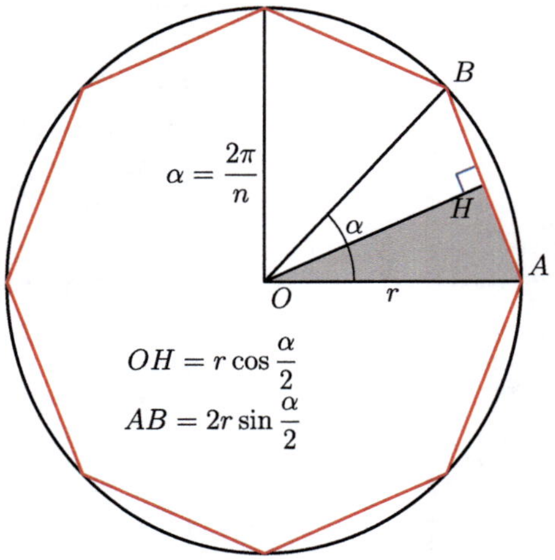

Table 3.2 Proof of $A = 0.5Cr$ with $r = 1$: using regular polygons of 4 to 512 sides

n	A_n	C_n	A_n/C_n
4	2.00000000	5.65685425	0.35355339
8	2.82842712	6.12293492	0.46193977
16	3.06146746	6.24289030	0.49039264
32	3.12144515	6.27309698	0.49759236
64	3.13654849	6.28066231	0.49939773
128	3.14033116	6.28255450	0.49984941
256	3.14127725	6.28302760	0.49996235
512	3.14151380	6.28314588	0.49999059

(OH) equal to the radius of the circle and the sum of their bases equal to the circle's circumference.

If the above reasoning was not convincing enough, here is a better one. Let's consider a regular polygon of n sides inscribed in a circle. Its area is denoted by A_n and its circumference by C_n, from Fig. 3.33, we can get

$$A_n = nr^2 \sin\frac{\pi}{n}\cos\frac{\pi}{n}, \quad C_n = n2r\sin\frac{\pi}{n}$$

Then, we consider the ratio A_n/C_n when n is very large:

$$\frac{A_n}{C_n} = \frac{1}{2}r\cos\frac{\pi}{n} \Longrightarrow \lim_{n\to\infty}\frac{A_n}{C_n} = \frac{1}{2}r$$

See Table 3.2 for supporting data.

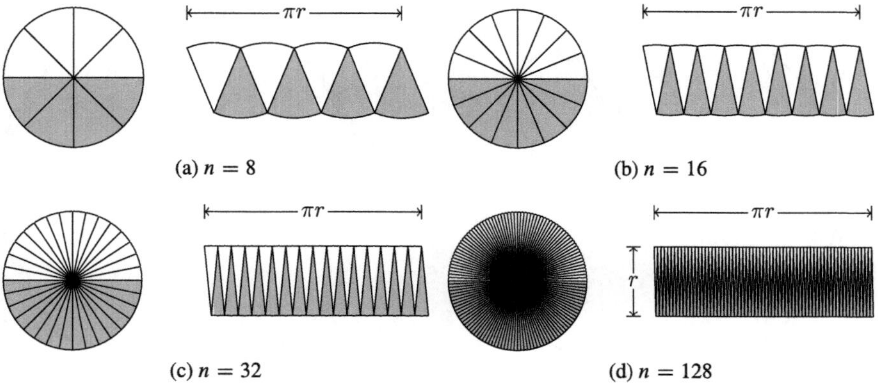

(a) $n = 8$ (b) $n = 16$ (c) $n = 32$ (d) $n = 128$

Fig. 3.34 Quadrating a circle of radius r: dividing the circle into a number of wedges (n). When n is very large what we get is a rectangle of sides r and πr, of which the area is πr^2. And that is also the area of the circle

How ancient mathematicians came up with the formula $A = \pi r^2$? The idea of calculating the area of the circle is the same: *breaking the circle into simpler shapes of which the area is known*. This is what ancient mathematicians did, see Fig. 3.34: they chopped a circle into eight wedge-shaped pieces (like a pizza) and rearranged the wedges. The obtained shape does not look similar to any shape known of. So, they chopped the circle into two wedges: this time 16 pieces. This time, something familiar appears! The wedges together look like a rectangle. Being more confident now, they decided to go extreme: divide the circle into an infinite number of wedges. What they got is a rectangle of sides πr (half of the circle perimeter) and r. Thus, the area of a circle is πr^2. What an amazing idea it was.

3.2.4 Calculation of π

Archimedes was the first to give a method of calculating π to any desired accuracy around 250 BC. It is based on the fact that the perimeter of a regular polygon of n sides inscribed in a circle, denoted by ni, is smaller than the circumference of the circle, whereas the perimeter of an n-polygon circumscribed about the circle, denoted by nc, is greater than its circumference (Fig. 3.35). In other words, the circle circumference $2\pi r$ is squeezed in between ni and nc, and we can always determine ni and nc. By making n sufficiently large, we can determine π accurately, to any degree of approximation we want. That's the essence of Archimedes' method.

Now, we carry out his method. We just need to consider a unit circle (i.e., a circle of unit radius); as π is a constant: it does not depend on the size of the circle. What we need to do is to determine the circumference of n-gon inscribed and circumscribed in this unit circle. To this end, we need to know just one side of these polygons. It

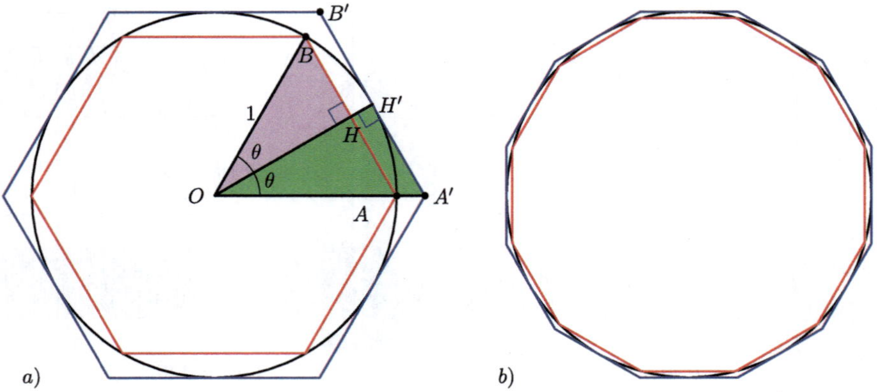

Fig. 3.35 Archimedes's method to determine π. The red hexagon is inscribed in the unit circle ($OB = 1$) whereas the blue hexagon is circumscribed about it. When we increase n-the number of sides-these two polygons go towards the circle (in b, I used $n = 12$)

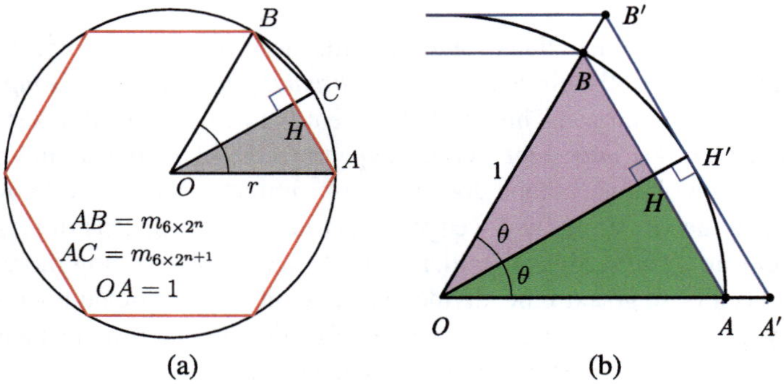

Fig. 3.36 The unit circle with 6×2^n-gon and $6 \times 2^{n+1}$-gon inscribed in the circle (illustrated for $n = 0$). In **b** is depicted the circumscribed 6×2^n-gon (again illustrated for $n = 0$). The diagrams are still valid for other polygons, by simply replacing θ by $\theta/2$

turns out that there is a relation between the polygons in the sense that, if we start with a hexagon inscribed in the unit circle (of which the side is 1), we can determine the side of a 12-gon inscribed in the circle (Fig. 3.36a), and from a 12-gon we can get a 24-gon and so on. Furthermore, we can also compute the side of a circumscribed hexagon (Fig. 3.36b).

We need to have a notation that allows us to differentiate the different n-gons. As we consider only $6, 12, 24, 48, \ldots$, which can be written as $6 \times 2^0, 6 \times 2^1, 6 \times 2^2, 6 \times 2^3, \ldots$, we can use the notation 6×2^n-gon to label a polygon in our collection.

The length of a side of the 6×2^n-gon is denoted by $m_{6\times 2^n}$. The first task is to relate $m_{6\times 2^n}$ to $m_{6\times 2^{n+1}}$. Using the Pythagorean theorem for $\triangle BHC$ and for $\triangle OAH$, we get the following equation (see Fig. 3.36a, illustrated with $N = 6$, AB is the side of the 6×2^n-gon and BC is the side of the $6 \times 2^{n+1}$-gon), noting that $HC = 1 - OH$ and $OH^2 + (M/2)^2 = 1^2$:

$$m_{6\times 2^{n+1}} = \sqrt{\frac{(m_{6\times 2^n})^2}{4} + \left(1 - \sqrt{1 - \frac{(m_{6\times 2^n})^2}{4}}\right)^2} = \sqrt{2 - \sqrt{4 - (m_{6\times 2^n})^2}} \tag{3.2.2}$$

With $n = 0$ we have $m_{6\times 2^n} = m_6$, which is the length of the side of an inscribed hexagon, which is one. This equation gives us $m_{12} = \sqrt{2 - \sqrt{4-1}} = \sqrt{2 - \sqrt{3}}$. From it, we can compute m_{24} and so on.

The second task is to relate the side of an inscribed 6×2^n with that of a circumscribed 6×2^n (denoted by $M_{6\times 2^n}$). Referring to Fig. 3.36b, illustrated again with $N = 6$, AB is the side of the inscribed 6×2^n-gon and $A'B'$ is the side of the circumscribed 6×2^n =gon. From similar triangles OBH and $OB'H'$ (or Thales's theorem due to $BH \parallel B'H'$), we have $H'B'/HB = OH'/OH$. But, $HB = 0.5AB = 0.5m_{6\times 2^n}$, $OH' = 1$ and $OH = \sqrt{1 - (m_{6\times 2^n})^2/4}$. Thus,

$$M_{6\times 2^n} = \frac{2AB}{\sqrt{4 - AB^2}} = \frac{2m_{6\times 2^n}}{\sqrt{4 - (m_{6\times 2^n})^2}} \tag{3.2.3}$$

We start with a hexagon (following Archimedes): $m_6 = 1$ and $M_6 = 2/\sqrt{3}$ after Eq. (3.2.3). Thus, π is

$$6 \times 1 < 2\pi < 6 \times \frac{2}{\sqrt{3}} \Longrightarrow 3 < \pi < 2\sqrt{3}$$

Next, we consider a 12-gon with $m_{12} = \sqrt{2 - \sqrt{3}}$, and Eq. (3.2.3) provides $M_{12} = 2\sqrt{2-\sqrt{3}}/\sqrt{2+\sqrt{3}}$. Therefore,

$$12\left(\frac{\sqrt{2-\sqrt{3}}}{2}\right) < \pi < 12\left(\frac{\sqrt{2-\sqrt{3}}}{\sqrt{2+\sqrt{3}}}\right)$$

Continuing this way until 96-gons, Archimedes stopped:

$$96\left(\frac{\sqrt{2-\sqrt{2+\sqrt{2+\sqrt{2+\sqrt{3}}}}}}{2}\right) < \pi < 96\left(\frac{\sqrt{2-\sqrt{2+\sqrt{2+\sqrt{2+\sqrt{3}}}}}}{\sqrt{2+\sqrt{2+\sqrt{2+\sqrt{2+\sqrt{3}}}}}}\right)$$

Now, Archimedes faced a tedious task: computing those square roots without a calculator, without decimal numbers (without number 0 even). He probably used Eq. (2.9.1) to compute the roots. I used my computer to do these roots to get

$$3.14103195089053 < \pi < 3.1427145996453882 \tag{3.2.4}$$

My computer gave me $\pi = 3.1415926535897$. Therefore, Archimedes got only two correct decimals. But it was a magnificent achievement! This polygonal algorithm dominated for over 1 000 years until infinite series were discovered. I presented one such infinite series for π in Eq. (2.21.16). And there is Machin's formula in Eq. (3.14.3). And we shall see more in this book.

But Archimedes could not write Eq. (3.2.4) as he did not have decimals. Instead, he used fractions and wrote: $3\frac{10}{71} < \pi < 3\frac{1}{7}$.

Pi is a special number and various books are written about it. There is even a day called Pi day (March 14), which is coincidentally also the birthday of Albert Einstein (14 March 1879). People keep calculating more and more digits of this number. Note that no one cares about the decimal digits of $\sqrt{2}$. I recommend the book *A History of Pi* by Petr Beckmann.

History note 3.2: Archimedes (c. 287 - c. 212 BC)

Archimedes of Syracuse was a Greek mathematician, physicist, engineer, inventor, and astronomer. Considered to be the greatest mathematician of ancient history, and one of the greatest of all time, Archimedes anticipated modern calculus and analysis by applying concepts of infinitesimals and the method of exhaustion to derive and rigorously prove a range of geometrical theorems, including: the area of a circle; the surface area and volume of a sphere; area of an ellipse; the area under a parabola; the volume of a segment of a paraboloid of revolution; the volume of a segment of a hyperboloid of revolution; and the area of a spiral. He derived an accurate approximation of π.

Archimedes died during the Siege of Syracuse, where he was killed by a Roman soldier despite orders that he should not be harmed. Cicero describes visiting the tomb of Archimedes, which was surmounted by a sphere and a cylinder, which Archimedes had requested be placed on his tomb to represent his mathematical discoveries.

Liu Hui's algorithm. Liu Hui (third century CE) was a Chinese mathematician and writer who lived in the Three Kingdoms period (220–280) of China. In 263, he edited and published a book with solutions to mathematical problems presented

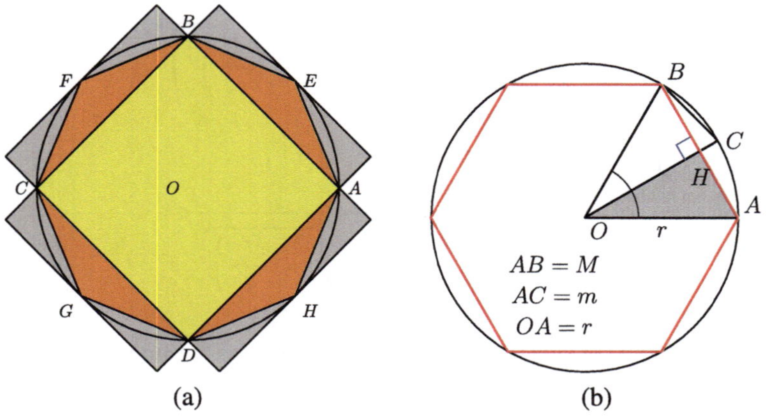

Fig. 3.37 Liu Hui's method for determining π based on the area of inscribed polygons with N and $2N$ sides

in the famous Chinese book of mathematics known as The Nine Chapters on the Mathematical Art, in which he was possibly the first mathematician to discover, understand, and use negative numbers. Along with Zu Chongzhi (429–500), Liu Hui was known as one of the greatest mathematicians of ancient China. In this section I present his method to determine π.

Liu Hui first derived an inequality for π based on the area of inscribed polygons with N and $2N$ sides. In the diagram of Fig. 3.37a, $ABCD$ is an N polygon whereas $AEBFCGDH$ is a $2N$ polygon, both inscribed in the circle. Regarding the areas of these polygons where A_N is the area of the N-polygon, A_{2N} the area of the $2N$-polygon and the circle (of area A_c), we have the following relations:

$$A_N = \text{yellow area} \tag{3.2.5a}$$
$$A_{2N} = \text{yellow area + orange area} \tag{3.2.5b}$$
$$A_{2N} < A_c < A_{2N} + \text{gray area} \tag{3.2.5c}$$
$$\text{gray area} = \text{orange area} \tag{3.2.5d}$$

Therefore, we can deduce that

$$A_{2N} < A_c < 2A_{2N} - A_N \Longrightarrow A_{2N} < \pi < 2A_{2N} - A_N \tag{3.2.6}$$

where the last inequality holds when considering a circle of unit radius.

Liu Hui then computed the area of inscribed polygons with N and $2N$ sides. To that end, he needed a formula relating the side of a $2N$-gon, denoted by m with that of a N-gon, denoted by M. Using the Pythagorean theorem for triangle BHC and for triangle OAH, he derived the following equation (see figure, illustrated with $N = 6$, AB is the side of the N-gon and BC is the side of the $2N$-gon both inscribed in the circle of radius r), noting that $HC = r - OH$ and $OH^2 + (M/2)^2 = r^2$:

$$m = \sqrt{\left(\frac{M}{2}\right)^2 + \left(r - \sqrt{r^2 - \left(\frac{M}{2}\right)^2}\right)^2}$$

Putting $r = 1$ (unit), we have

$$m = \sqrt{\frac{M^2}{4} + \left(1 - \sqrt{1 - \frac{M^2}{4}}\right)^2} \tag{3.2.7}$$

Now, he calculated the area of a N-gon approximately as the sum of the areas of all triangles making the polygon:

$$A_N \approx N \times \left(\frac{1}{2}Mr\right) \approx N \times \left(\frac{1}{2}M\right) \tag{3.2.8}$$

Now comes the complete algorithm: we start with $N = 6$ (hexagon), thus $M = 1$ (as $r = 1$). Then, we do:

- compute m using Eq. (3.2.7)
- compute $A_N = 0.5NM$, compute $A_{2N} = 0.5(2N)m$
- then Eq. (3.2.6) yields $A_{2N} < \pi < 2A_{2N} - A_N$
- next iteration: $M = m$, $N = 2N$.

If we repeat this algorithm four times i.e., using 46-gon and 98-gon, we get this approximation for π: $3.14103195 < \pi < 3.14271370$. And the Chinese astronomer and mathematician Zu Chongzhi (429–500 AD) got $3.141592619365 < \pi < 3.141592722039$ with 12288-gon, a record which would not be surpassed for 1200 years. Even by 1600 in Europe, the Dutch mathematician Adriaan Anthonisz and his son obtained π value of 3.1415929, accurate only to 7 digits.

Ramanujan's pi formula. Srinivasa Ramanujan (22 December 1887–26 April 1920) was an Indian mathematician who lived during the British Rule in India.

Albeit without any formal training in pure mathematics, he has made substantial contributions to mathematical analysis, number theory, infinite series, and continued fractions, including solutions to mathematical problems then considered unsolvable.

Ramanujan initially developed his own mathematical research in isolation: according to Hans Eysenck, a German-born British psychologist: "*He tried to interest the leading professional mathematicians in his work, but failed for the most part. What he had to show them was too novel, too unfamiliar, and additionally presented in unusual ways; they could not be bothered*". Seeking mathematicians who could better understand his work, in 1913 he began a postal correspondence with the English mathematician Godfrey Hardy at the University of Cambridge. Recognizing Ramanujan's work as extraordinary, Hardy arranged for him to travel to Cambridge.

Ramanujan gave us, among many other amazing formulas, the following formula for $1/\pi$

$$\frac{1}{\pi} = \frac{2\sqrt{2}}{9801} \sum_{k=0}^{\infty} \frac{(4k)!}{(k!)^4} \frac{(1103 + 26390k)}{396^{4k}} \tag{3.2.9}$$

With only one term, we get $\pi = 3.1415926535897936$! I do not know the derivation of it. But it is certain that it did not come from the method of ancient mathematicians which relied on geometry. Ramanujan had in his hands the power of twentieth-century mathematics. To know more about Ramanujan, I recommend the 2015 British biographical drama film 'The Man Who Knew Infinity'. The movie is based on the 1991 book of the same name by Robert Kanigel.

History note 3.3: The first letter of Ramanujan to Hardy.

Dear Sir,

I beg to introduce myself to you *as a clerk* in the Accounts Department of the Port Trust Office at Madras on a salary of only *£20 per annum*. I am now about 23 years of age. I have had *no University education* but I have undergone the ordinary school course. After leaving school I have been employing the spare time at my disposal to work at Mathematics. I have not trodden through the conventional regular course which is followed in a University course, but I am striking out a new path for myself. I have made a special investigation of divergent series in general and the results I get are termed by the local mathematicians as 'startling'.

Just as in elementary mathematics you give a meaning to a^n when n is negative and fractional to conform to the law which holds when n is a positive integer, similarly the whole of my investigations proceed on giving a meaning to Eulerian Second Integral for all values of n. My friends who have gone through the regular course of University education tell me that $\int_0^\infty x^{n-1}e^{-x}dx = \Gamma(n)$ is true only when n is positive. They say that this integral relation is not true when is negative. Supposing this is true only for positive values of n and also supposing the definition $n\Gamma(n) = \Gamma(n+1)$ to be universally true, I have given meanings to these integrals and under the conditions I state the integral is true for all values of n negative and fractional.

My whole investigations are based upon this and I have been developing this to a remarkable extent so much so that the local mathematicians are not able to understand me in my higher flights.
Very recently I came across a tract published by you styled Orders of Infinity in page 36 of which I find a statement that no definite expression has been as yet found for the number of prime numbers less than any given number. I have found an expression which very nearly approximates to the real result, the error being negligible. I would request you to go through the enclosed papers. *Being poor, if you are convinced that there is anything of value I would like to have my theorems published.* I have not given the actual investigations nor the expressions that I get but I have indicated the lines on which I proceed. Being inexperienced I would very highly value any advice you give me. Requesting to be excused for the trouble I give you.
I remain, Dear Sir, Yours truly, S. Ramanujan

3.3 Trigonometric Functions: Right Triangles

Considering two similar right-angled triangles (or right triangles) as shown in Fig. 3.38. Now is the time to apply Euclid's geometry: as the two triangles ABC and $A'B'C'$ are similar we have:

$$\frac{AC}{A'C'} = \frac{AB}{A'B'}$$

From that it is simple to deduce

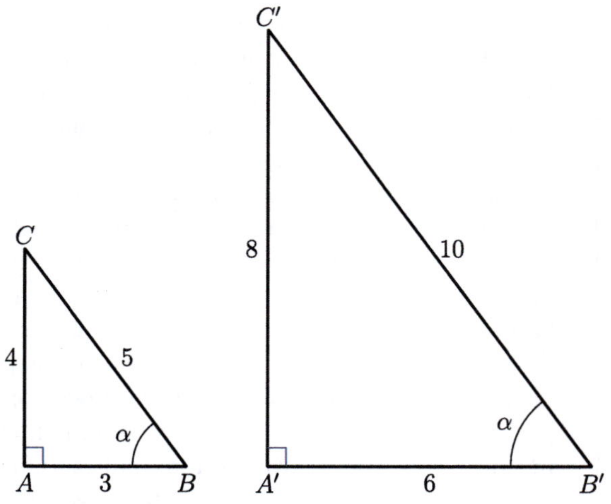

Fig. 3.38 Two similar right-angled triangles ABC and $A'B'C'$: $AC/AB = A'C'/A'B'$

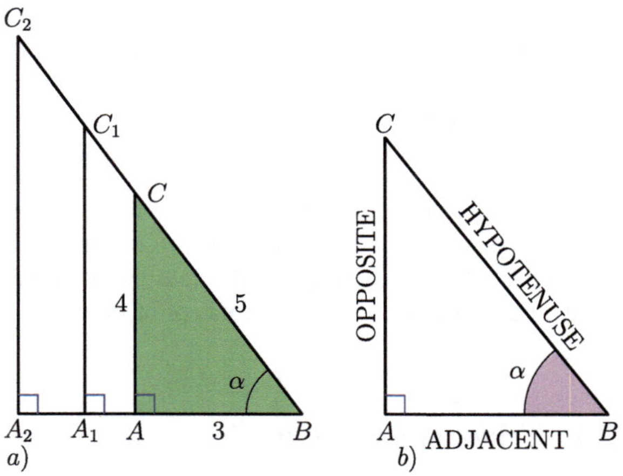

Fig. 3.39 Knowing $\tan\alpha$ allows us to determine $A_1C_1 = A_1B\tan\alpha$

$$\frac{AC}{AB} = \frac{A'C'}{A'B'} \quad \left(= \frac{4}{3}\right)$$

What does this mean? This shows that for all right triangles with an angle α, the side ratios AC/AB are constant.

Now comes the key point. If we can manage to compute the ratio AC/AB for a given angle α, then we can use it to solve any right triangle with the angle at B equal α. Thus, we have our very first trigonometric function—the tangent:

$$\tan\alpha := \frac{AC}{AB}$$

Thus a trigonometric function relates an angle of a right-angled triangle to ratios of two side lengths. And if we have a table of the tangent i.e., for each angle α, we can look up its $\tan\alpha$, we then can solve every right triangle problems; in Fig. 3.39a we can determine $A_1C_1 = A_1B\tan\alpha$. The first trigonometric table was apparently compiled by Hipparchus of Nicaea (180–125 BCE), who is now consequently known as "the father of trigonometry".

Why just the ratio AC/AB? All three sides of a triangle should be treated equally and their ratios are constants for all right triangles with the same angle α. If so, from three sides, we can have six ratios! And voilà, we have six trigonometric functions. Quite often, they are also referred to as six trigonometric ratios. They include: sine, cosine, and tangent and their reciprocals, and are defined as (Fig. 3.39b):

$$\begin{aligned}
\cos\alpha &= \frac{\text{adjacent}}{\text{hypotenuse}} = \frac{AB}{BC}, \quad \sec\alpha = \frac{BC}{AB} = \frac{1}{\cos\alpha}\\
\sin\alpha &= \frac{\text{opposite}}{\text{hypotenuse}} = \frac{AC}{BC}, \quad \csc\alpha = \frac{BC}{AC} = \frac{1}{\sin\alpha}\\
\tan\alpha &= \frac{\text{opposite}}{\text{adjacent}} = \frac{AC}{AB}, \quad \cot\alpha = \frac{AB}{AC} = \frac{1}{\tan\alpha}
\end{aligned}$$

The secant of α is 1 divided by the cosine of α, the cosecant of α is defined to be 1 divided by the sine of α, and the cotangent (cot) of α is 1 divided by the tangent of α. These three functions (secant, cosecant, and cotangent) are the reciprocals of the cosine, sine, and tangent.

Where these names come from is to be explained in the next section.

Remark Loosely speaking a function is a device that receives a number (mostly a real number)—called the input—and returns another number, the output. For example, $\sqrt{x}$ is a function: input x and it returns the square root of x. I refer to Sect. 4.2 for a detailed discussion on the concept of function.

3.4 Trigonometric Functions: Unit Circle

Earlier forms of trigonometry can be traced back to the ancient Greeks, notably to the two mathematicians Hipparchus and Ptolemy. This version of trigonometry was based on chords in a circle. Precisely, Hipparchus' trigonometry was based on the chord subtending a given arc in a circle of fixed radius R, Ptolemy used $R = 60$, see Fig. 3.40. Note that $\mathrm{crd}(\alpha)$ is simply a length. We can see the connection with our modern sine: $\mathrm{crd}(2\alpha) = 2R\sin\alpha$, or $\mathrm{crd}(\alpha) = 2R\sin\alpha/2$. Indian mathematicians inherited this trigonometry and modified it.[23] Instead of using the chord they used half-chord. Therefore, the Indian half-chord is closely related to our sine.

The Sanskrit word for chord-half was jya-ardha, which was sometimes shortened to jiva. This was brought into Arabic as jiba, and written in Arabic simply with two

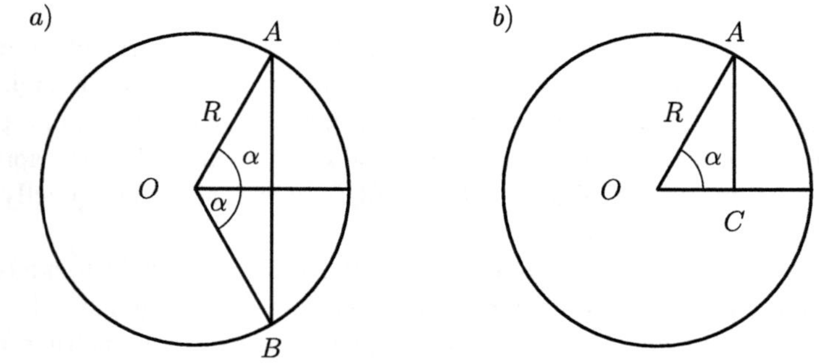

$AB = \mathrm{crd}(2\alpha)$ $\qquad$ $AC = \frac{1}{2}AB$

Fig. 3.40 Greek's chord and Indian's half-chord. When $R = 1$, Indian's half-chord, AC is our sine

[23] Why? Because Ptolemy's definition becomes cumbersome in practice. For instance, for a right triangle with one angle α known and the hypotenuse known, to compute the other side, one has to: (1) doubling the angle 2α (because the definition uses the central angle), (2) table look up to get $\mathrm{crd}(2\alpha)$, and (3) multiplying the result with a factor $c/120$. Too much work.

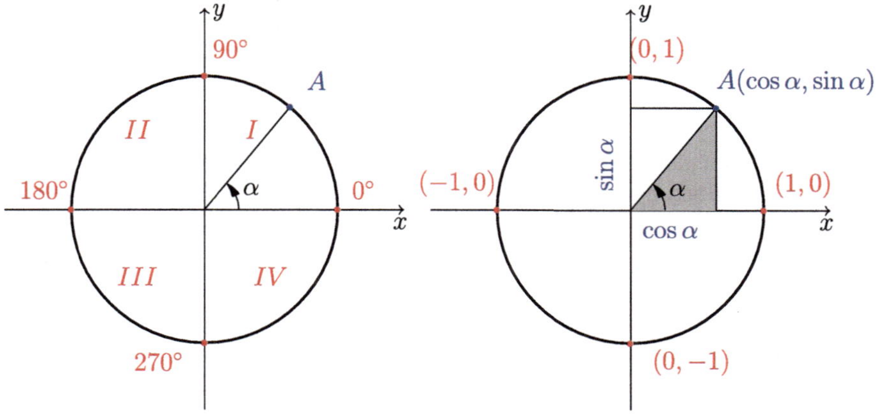

Fig. 3.41 The unit circle: angles are measured counterclockwise

consonants jb, vowels not being written. Later, Latin translators selected the word sinus to translate jb thinking that the word was an Arabic word jaib, which meant breast, and sinus had breast and bay as two of its meanings. In English, sinus was imported as "sine". This word history for sine is interesting because it follows the development path of trigonometry from India, through the Arabic language from Baghdad through Spain, into western Europe in the Latin language, and then to modern languages such as English and the rest of the world.

Right triangles have a serious limitation. They are excellent for angles up to 90°. How about angles larger than that? And how about negative angles? A circle can solve all these limitations.

We consider a unit circle—a circle with a unit radius—centered at the origin of the Cartesian coordinate system (refer to Sect. 3.24.1 for details). It was Euler who made the decision of using the unit circle. Angles are measured from the positive x axis counterclockwise; thus 90° is straight up, 180° is to the left (Fig. 3.41). The circle is divided into four quadrants: the first quadrant is for angles $\alpha \in [0°, 90°]$; those angles are called *acute* angles, the second quadrant is for angles $\alpha \in [90°, 180°]$; those angles are *obtuse* angles. Angles in the third and fourth quadrants are $\alpha \in [180°, 360°]$ and called *reflex* angles.

An angle α is corresponding to a point A on the circle. And the x-coordinate of this point is $\cos\alpha$ whereas the y-coordinate is $\sin\theta$. With this special radius of unity, the definition of trigonometric functions as ratios (Sect. 3.3) and the definition of Indian mathematicians as chord lengths coincide. There is even no triangle with this view: trigonometry becomes analytic. But if we look carefully, in Fig. 3.41, the shaded figure is a right triangle!

Mnemonics in trigonometry. The sine, cosine, and tangent ratios in a right triangle can be remembered by representing them as strings of letters, for instance SOH-CAH-TOA in English:

$$\text{Sine} = \text{Opposite/Hypotenuse}$$
$$\text{Cosine} = \text{Adjacent/Hypotenuse}$$
$$\text{Tangent} = \text{Opposite/Adjacent}$$

One way to remember the letters is to sound them out phonetically.

However, the acronym SOH-CAH-TOA does not help us understand why the ratio is defined this way. Armed only with "SOH-CAH-TOA", it becomes confusing to find the values of trigonometric ratios for angles whose measures are 90° or greater, or negative. The acronym is not a bad tool, but it is insufficient to help understand trigonometric ratios. We believe it is better not to teach students this SOH-CAH-TOA (and the likes). Instead, as the origin of the sine is a chord, which measures the height of a right triangle, the sine must be the ratio of opposite over hypotenuse. Cosine then uses the other side of the triangle, namely the adjacent. And tangent is just a derived quantity: it is the ratio of sine over cosine. There is no need to memorize anything about it.

3.5 Degree Versus Radian

Angles are measured either in degrees or in radians. Even though it is common to have different units for a quantity (we have meters, miles, and yards as units for length), we should understand the origin and benefits of the different units for angles. Angles were first expressed in terms of degree and a full circle is 360°. Why 360? We never know why but it came from Babylonians, probably because we have 365 days in a year. In any case, it fitted well with the Babylonian sexagesimal (base 60) number system. Although this system is now obsolete, the division of a circle into 360 parts has survived, and not only that, we also have the division of an hour into 60 minutes and a minute into 60 s.

The use of degree as the measure for angles is quite arbitrary. Mathematicians found a better way for this purpose: they used the length of an arc. Referring to Fig. 3.42, one radian (1 rad) is the angle corresponding to the arc (AB in the figure) of which the length is the circle radius (denoted by r). If the angle is two radians, the length of the arc is $2r$. Thus, if the angle is α (radians), the length of an arc is αr. As a full circle has a perimeter of $2\pi r$, a full circle is 2π or 360°; therefore $\pi = 180°$. So, an angle of α (in degrees) is $\alpha\pi/180$ in radians: 90° is $\pi/2$, 180° is π etc.

With radian, the mathematics is simpler. For example, if α is in radian, the length of an arc is simply αr, where r is the radius of the circle. On the other hand, the arc length would be $(\pi\alpha/180)r$ if the angle is in degrees. In calculus, derivatives of trigonometric functions are also simpler when angles are expressed in radians. For example, the derivative of $\sin x$ would be $(\pi/180)\cos x$ instead of $\cos x$, see Sect. 4.4.8 for details.

The concept of radian measure, as opposed to the degree of an angle, is normally credited to the English mathematician Roger Cotes (1682–1716) in 1714. He

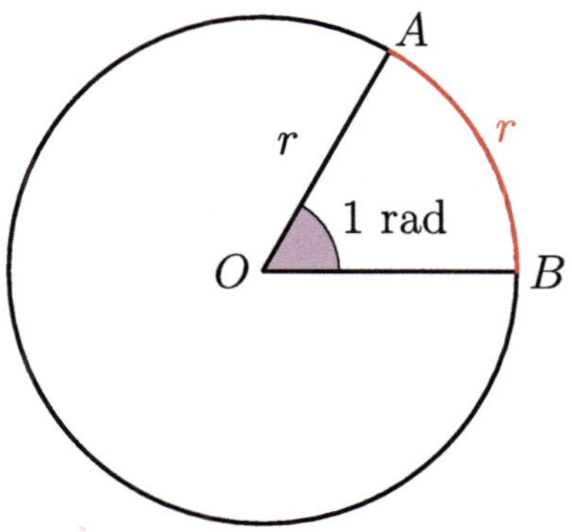

Fig. 3.42 Definition of radian as unit for angle measurement

described the radian in everything but name and recognized its naturalness as a unit of angular measure. Prior to the term radian becoming widespread, the unit was commonly called a circular measure of an angle. The idea of measuring angles by the length of the arc was already in use by other mathematicians. For example, the Persian astronomer and mathematician al-Kashi (c. 1380–1429) used so-called diameter parts as units, where one diameter part was $1/60$ radian.

The term radian first appeared in print on 5 June 1873, in examination questions set by the British engineer and physicist James Thomson (1822–1892) at Queen's College, Belfast.[24] He had used the term as early as 1871, while in 1869, the Scottish mathematician Thomas Muir (1844–1934) was vacillated between the terms rad, radial, and radian. In 1874, after a consultation with James Thomson, Muir adopted radian. The name Radian was not universally adopted for some time after this. Longmans' School Trigonometry still called the radian circular measure when published in 1890.

3.6 Some First Properties

It is obvious that these six trigonometric functions are not independent of each other. One can derive the following relations between them quite straightforwardly. We start with $\sin^2\alpha + \cos^2\alpha = 1$, which is a consequence of the Pythagorean theorem, see the right diagram in Fig. 3.41:

$$\begin{aligned} \sin^2\alpha + \cos^2\alpha &= 1 \\ 1 + \tan^2\alpha &= \frac{1}{\cos^2\alpha} = \sec^2\alpha \\ 1 + \cot^2\alpha &= \frac{1}{\sin^2\alpha} = \csc^2\alpha \end{aligned} \tag{3.6.1}$$

[24] James Thomson's reputation is substantial though it is overshadowed by that of his younger brother William Thomson or Lord Kelvin whose name is used for absolute temperatures.

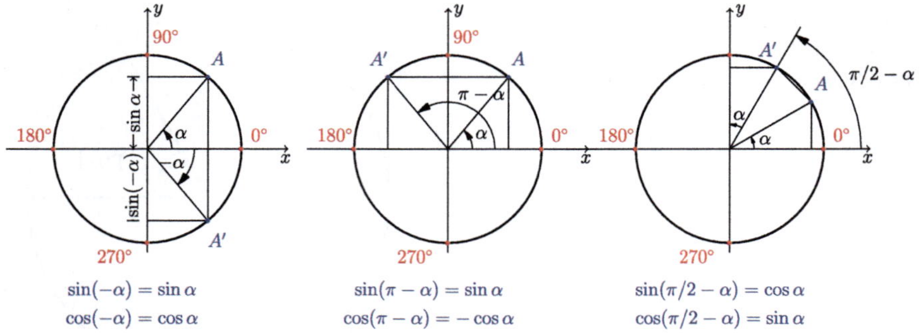

Fig. 3.43 Some simple facts about $\sin x$ and $\cos x$

where the second and third identities are obtained from the first by dividing both side by $\cos^2\alpha$ and $\sin^2\alpha$, respectively.

Without actually doing any calculations, just starting from the definition of trigonometry functions, and some observations (Fig. 3.43), we can see that $-1 \leq \sin\alpha \leq 1$, $-1 \leq \cos\alpha \leq 1$, and

$$\begin{cases} \sin(-\alpha) = -\sin\alpha \\ \cos(-\alpha) = \cos\alpha \end{cases}, \quad \begin{cases} \sin(\pi-\alpha) = \sin\alpha \\ \cos(\pi-\alpha) = -\cos\alpha \end{cases}, \quad \begin{cases} \sin(\pi/2-\alpha) = \cos\alpha \\ \cos(\pi/2-\alpha) = \sin\alpha \end{cases}, \tag{3.6.2}$$

The first means that the function $y = \sin x$ is odd, and the second tells us that $y = \cos x$ is even (see Sect. 4.2.1 for more details). Why bother? Not only because we do like classification but also odd and even functions possess special properties (e.g. $\int_{-\pi}^{\pi} \sin x dx = 0$, a nice result, isn't it?). The second in Eq. (3.6.2) allows us to just compute the sine for angles $0 \leq \theta \leq \pi/2$ (these angles are called first quadrant angles), the sine of $\pi - \theta$ is then simply $\sin\theta$. From the third equation we see that the value of the cosine $\cos(x)$ is equal to the values of $\sin(\pi/2 - x)$ for its complementary angle. And that explains the name "cosine": complementary of sine.

To see why tangent is such called, see Fig. 3.47. And then, it is easy to understand the name cotangent as $\tan(\pi/2 - \alpha) = \cot\alpha$. Note that we did not list identities for tangent and cotangent here for brevity; as $\tan\theta$ and $\cot\theta$ are functions of the sine and cosine.

If we start at a certain point on a circle (this point has an angle of θ) and we go a full round then we're just back to where we started. That means $\sin(\theta + 2\pi)$ is simply $\sin\theta$. But if we go another round we also get back to the starting point. Thus, for n being any whole number, we have

$$\begin{aligned} \sin(\alpha + 2n\pi) &= \sin\alpha \\ \cos(\alpha + 2n\pi) &= \cos\alpha \end{aligned} \tag{3.6.3}$$

That's why when we solve trigonometric equations like $\cos x = \sqrt{2}/2$, the solution is $x = \pm^{\pi}/_{4} + 2n\pi$ with $n \in \mathbb{N}$ not simply $x = \pm^{\pi}/_{4}$.

3.7 Sine Table

Now, we are going to calculate the sines and cosines of angles from 0° to 360°. You might be asking why bother if we all have calculators which are capable of doing this for us? Note that we're not interested herein in the results (i.e., the sine of a given angle), but interested in how trigonometric functions are actually computed. Along the process, new mathematics were invented (or discovered) to finish the job. Astronomers, for example, always need to know the sine/cosine of angles from 0° to 360°. That was why the Greek astronomer, geometer, and geographer Ptolemy in Egypt during the second century AD, produced such a table in Book I of his masterpiece *Almagest.*

Table of sin (angle)

Angle	sin (a)	Angle	sin (a)	Angle	sin (a)	Angle	sin (a)
0.0	0.0	25.0	.4226	46.0	.7193	71.0	.9455
1.0	.0174	26.0	.4384	47.0	.7314	72.0	.9511
2.0	.0349	27.0	.4540	48.0	.7431	73.0	.9563
3.0	.0523	28.0	.4695	49.0	.7547	74.0	.9613
4.0	.0698	29.0	.4848	50.0	.7660	75.0	.9659
5.0	.0872	30.0	.5000	51.0	.7772	76.0	.9703
6.0	.1045	31.0	.5150	52.0	.7880	77.0	.9744
7.0	.1219	32.0	.5299	53.0	.7986	78.0	.9781
8.0	.1392	33.0	.5446	54.0	.8090	79.0	.9816
9.0	.1564	34.0	.5592	55.0	.8191	80.0	.9848
10.0	.1736	35.0	.5736	56.0	.8290	81.0	.9877
11.0	.1908	36.0	.5878	57.0	.8387	82.0	.9903
12.0	.2079	37.0	.6018	58.0	.8480	83.0	.9925
13.0	.2249	38.0	.6157	59.0	.8571	84.0	.9945
14.0	.2419	39.0	.6293	60.0	.8660	85.0	.9962
15.0	.2588	40.0	.6428	61.0	.8746	86.0	.9976
16.0	.2756	41.0	.6561	62.0	.8829	87.0	.9986
17.0	.2924	42.0	.6691	63.0	.8910	88.0	.9994
18.0	.3090	43.0	.6820	64.0	.8988	89.0	.9998
19.0	.3256	44.0	.6947	65.0	.9063	90.0	1.00
20.0	.3420	45.0	.7071	66.0	.9135		
21.0	.3584			67.0	.9205		
22.0	.3746			68.0	.9272		
23.0	.3907			69.0	.9336		
24.0	.4067			70.0	.9397		

As sine and cosine are related via $\sin^2 \alpha + \cos^2 \alpha = 1$ and other trigonometry functions are derived from sine/cosine, knowing the sines is enough. And we set for ourselves a task of *building a sine table for angles from* 0° *to* 90°. We start simple with only whole numbered angles i.e., $1°, 2°, \ldots$ Now, one small observation will save us big time: as $\sin^2 \alpha + \sin^2(\pi/2 - \alpha) = 1$, if we know $\sin \alpha$, we know $\sin(\pi/2 - \alpha)$. Thus the work is cut half.[25]

As such a detailed sine table is of no value today, we present the sines/cosines of few "common" angles in Table 3.3 for reference. Among these values, we just need to calculate the sines for $\theta = \{45°, 60°\}$. The calculation for 45° and 60° is demonstrated in Fig. 3.44; we simply used the definition of sine/cosine, and of course the famous Pythagorean theorem. Because $\theta = \{0°, 90°, 180°, 270°\}$ coincide with the vertices of the unit circle, their sines are easy to obtain.

Ptolemy computed sine for 36° or $\pi/5$ using geometric reasoning based on Proposition 10 of Book XIII of Euclid's *Elements* (we present a different way later). Knowing the sine for 36° we then know the sine of 54°. Using the trigonometry identity $\sin(\alpha \pm \beta) = \sin \alpha \cos \beta \pm \sin \beta \cos \alpha$, to be discussed in Sect. 3.8, we get the sine for 72° with $\alpha = \beta = 36°$, the sine for 18° with $\alpha = 72°, \beta = 54°$ and the sine for

[25] For example, if we know $\sin 44°$, then $\sin 46° = \sqrt{1 - \sin^2 44°}$.

Table 3.3 Sines and cosines of some angles from 0° to 360°

Angle θ [degree]	Angle [rad]	$\sin\theta$	$\cos\theta$
0	0	0	1
30	$\pi/6$	$1/2$	$\sqrt{3}/2$
45	$\pi/4$	$\sqrt{2}/2$	$\sqrt{2}/2$
60	$\pi/3$	$\sqrt{3}/2$	$1/2$
90	$\pi/2$	1	0
180	π	0	–1
270	$3\pi/2$	–1	0
360	2π	0	1

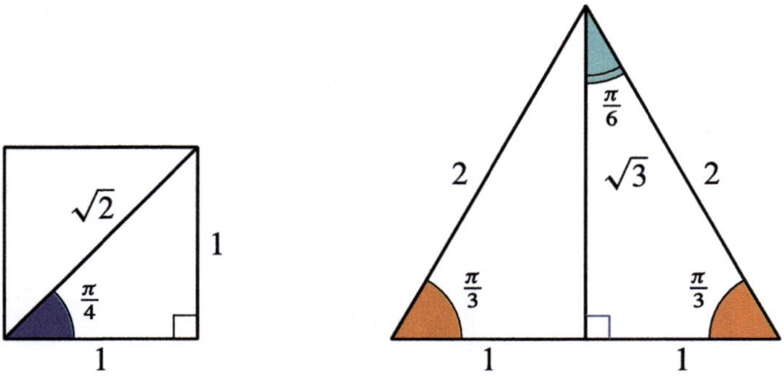

Fig. 3.44 Calculation of sine and cosine for $\theta = \pi/4, \theta = \pi/6$ and $\theta = \pi/3$. For $\pi/4$, we start with the unit square and draw a diagonal of which the length is $\sqrt{2}$, due to the Pythagorean theorem. We then have two identical right triangles. Consider one right triangle, the acute angles are $\pi/4$. We know all the sides of this right triangle, thus we know how to compute the sine/cosine of $\pi/4$. Similarly, for $\pi/3$, we start with an equilateral triangle of side 2 (why?)

75° with $\alpha = 30°, \beta = 45°$. With $\alpha = 75°, \beta = 72°$ we get $\sin 3°$. And from that we can get the sines for all multiples of 3° i.e., 6°, 9°, 12° etc. (for example using the identity $\sin 2x = 2\sin x\cos x$).

If we know $\sin 1°$, then we will know $\sin 2°$, $\sin 6°$, $\sin 5°$. (Note that we already have $\sin 3°$.) etc. and we're done. But Ptolemy could not find $\sin 1°$ directly, he found an approximate method for it (see Sect. 3.15). The Persian astronomer al-Kashi (c. 1380–1429) in his book *The Treatise on the Chord and Sine*, computed $\sin 1°$ to any accuracy. In the process, he discovered the triple angle identity often (wrongly) attributed to François Viète in the sixteenth century.

Using the triple identity $\sin(3\alpha) = 3\sin\alpha - 4\sin^3\alpha$ (to be discussed in the next section[26]), he related $\sin 1°$ with $\sin 3°$ (which he knew) via the following cubic equation:

$$\sin 3° = 3\sin 1° - 4\sin^3 1°$$

[26] Actually we derived this identity in Sect. 2.25.5 using complex numbers.

But the cubic would not be solved for another 125 years by Cardano. Clearly, al-Kashi could not wait that long. What did he do? With $x = \sin 1°$, he wrote

$$\sin 3° = 3x - 4x^3 \implies x = \frac{\sin 3° + 4x^3}{3}$$

What is this? This is the fixed point iterations method discussed in Sect. 2.11. With only four iterations we get sin 1° with accuracy of 12 decimal places:

$$\sin 1° = \sin 0.017453292520 = 0.017452406437$$

al-Kashi gave us sin 1°. Is there anything else? Look at the red numbers, what did you see? It seems that we have $\sin x \approx x$ at least for $x = 1°$. This is even more important than what sin 1° is. Why? Because if it is the case, we can replace $\sin x$—which is a complex function—with a very simple x. We shall see that replacing $\sin x$ by x is making the life of engineers and scientists much easier in many problems (e.g. in solving harmonic oscillation problems). But, $\sin \approx x$ is now just an observation, we need a proof (Sect. 3.15).

3.8 Trigonometric Identities

We have seen some mathematical identities, for example, $(a + b)^2 = a^2 + 2ab + b^2$. This section presents common trigonometric identities—those identities that involve trigonometry functions; for example, $\sin^2 x + \cos^2 x = 1$ is a trigonometric identity as it is correct for all x.

There are several trigonometric identities (actually there are too many unfortunately) and I once learnt them by heart. It was a terrible mistake.[27] Below, we will list the most popular identities and provide proofs for them. So, the point is: do not learn them by heart, instead understand how to construct them from scratch. As will be shown, all identities are derived from one basic identity (I did not count $\sin^2 \alpha + \cos^2 \alpha = 1$)! This identity can be $\sin(\alpha + \beta)$ or $\cos(\alpha - \beta)$.

Even though this section covers many common trigonometry identities, the list given below is not meant to be exhaustive. Later sections will present some interesting but less common identities.

First, I give a list of the so-called angle addition and subtraction identities:

[27] The French-American mathematician Serge Lang (1927–2005) in his interesting book *Math: Encounters with high school students* [39] advised high school students to simultaneously *memorize formula and understand the proof*. He wrote that he himself did that. I think Lang's advice is helpful for tests and exams where time matters. Lang was a prolific writer of mathematical texts, *often completing one on his summer vacation*. Lang's Algebra, a graduate-level introduction to abstract algebra, was a highly influential text.

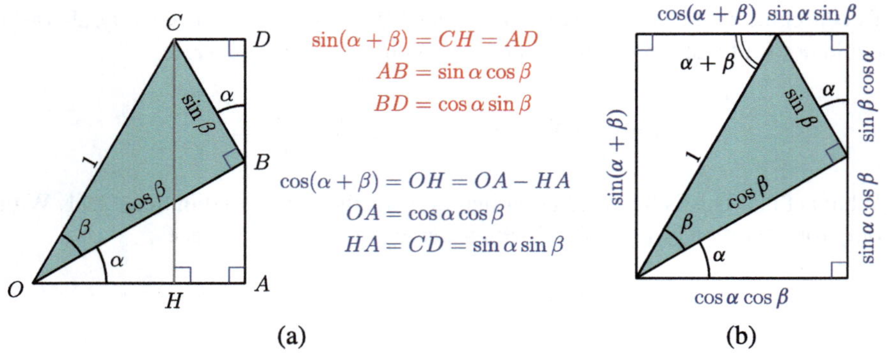

Fig. 3.45 Proof of $\sin(\alpha + \beta)$ and $\cos(\alpha + \beta)$. Noting that $OC = 1$

(angle addition and subtraction)

$$\boxed{\sin(\alpha \pm \beta) = \sin\alpha\cos\beta \pm \sin\beta\cos\alpha}$$
$$\cos(\alpha \pm \beta) = \cos\alpha\cos\beta \mp \sin\alpha\sin\beta$$
$$\tan(\alpha \pm \beta) = \frac{\tan\alpha \pm \tan\beta}{1 \mp \tan\alpha\tan\beta} \tag{3.8.1}$$
$$\cot(\alpha \pm \beta) = \frac{\pm\cot\alpha\cot\beta - 1}{\cot\alpha \pm \cot\beta}$$

The proof of the addition angle formulae for sine and cosine is shown in Fig. 3.45a. The idea is to use the definition of sine and cosine, and thus constructing right triangles containing α and β and their sum. The choice of $OC = 1$ simplifies the calculations. The formula for $\sin(\alpha - \beta)$ can be obtained from the addition angle formula by replacing β with $-\beta$ and noting that $\sin(-\beta) = -\sin\beta$. Or we can prove the formula for $\cos(\alpha - \beta)$ using the unit circle as given in Fig. 3.46.

The identity for the addition angle for the tangent is obtained directly from its definition and the available formulae for sine and cosine:

$$\tan(\alpha + \beta) = \frac{\sin(\alpha + \beta)}{\cos(\alpha + \beta)} = \frac{\sin\alpha\cos\beta + \sin\beta\cos\alpha}{\cos\alpha\cos\beta - \sin\alpha\sin\beta} = \frac{\tan\alpha + \tan\beta}{1 - \tan\alpha\tan\beta} \tag{3.8.2}$$

where in the last step, we divide both the denominator and numerator by $\cos\alpha\cos\beta$ so that tangents will appear.

Remark We just need either $\sin(\alpha \pm \beta)$ or $\cos(\alpha \pm \beta)$ as we can get the latter given the former and vice versa. For example, we can start with $\cos(\alpha + \beta)$, and obtain $\sin(\alpha - \beta)$ by writing $\sin(\alpha - \beta) = \cos(\pi/2 - (\alpha - \beta)) = \cos[(\pi/2 - \alpha) + \beta]$. Then using the addition angle formula for cosine.

Remark I present in Fig. 3.45b a *proof without words* for $\sin(\alpha + \beta)$ and $\cos(\alpha + \beta)$. This type of proof is concise as readers are expected to fill in routine steps for

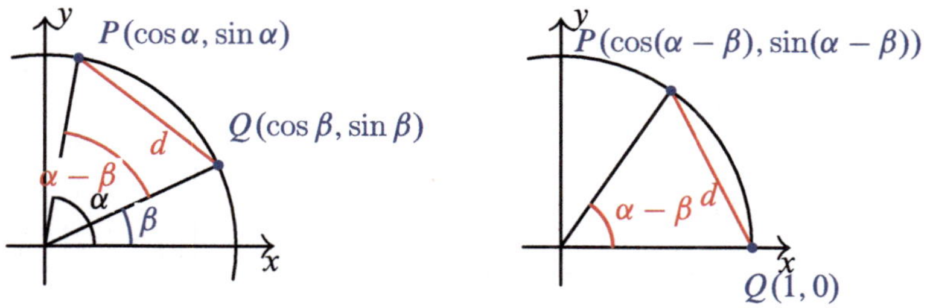

Fig. 3.46 Proof of $\cos(\alpha - \beta)$: expressing the distance d (between points P and Q) two ways (one from the left figure and one from the right figure). Recall $d^2 = (x_1 - x_2)^2 + (y_1 - y_2)^2$ is the squared distance between two points (x_1, y_1) and (x_2, y_2). That exercise gives us $\cos(\alpha - \beta)$

themselves—just as movie characters may suddenly appear in new surroundings without it being necessary to show how they got there. I recommend the book "Proof without words" [51].

Question 1 *But how we knew that* $\sin(\alpha + \beta) = \sin\alpha\cos\beta + \sin\beta + \cos\alpha$ *in the first place? Or, in other words, in which context the identity* $\sin(\alpha + \beta)$ *was discovered? It was Ptolemy, while building a table of sines (chords), that discovered this identity from the Ptolemy theorem. I present his discovery in Sect. 3.17.*

From the addition angle formula for sine, it follows that $\sin(2\alpha) = \sin(\alpha + \alpha) = 2\sin\alpha\cos\alpha$. Similarly, one can get the double-angle for cosine. If you do not like this geometric based derivation, don't forget we have another proof using complex numbers (Sect. 2.25). Thus, we have the following double-angle identities

$$\begin{aligned}
&\text{(double-angle)}\\
&\sin(2\alpha) = 2\sin\alpha\cos\alpha\\
&\cos(2\alpha) = \cos^2\alpha - \sin^2\alpha = 2\cos^2\alpha - 1 = 1 - 2\sin^2\alpha\\
&\tan(2\alpha) = \frac{\sin 2\alpha}{\cos 2\alpha} = \frac{2\sin\alpha\cos\alpha}{1 - 2\sin^2\alpha} = \frac{2\tan\alpha}{1 - \tan^2\alpha}
\end{aligned} \tag{3.8.3}$$

The triple-angle formula for sine can be obtained from the addition angle formula as follows

$$\begin{aligned}
\sin(3\alpha) &= \sin(2\alpha + \alpha)\\
&= \sin(2\alpha)\cos\alpha + \sin\alpha\cos(2\alpha)\\
&= 2\sin\alpha\cos^2\alpha + \sin\alpha(\cos^2\alpha - \sin^2\alpha)\\
&= 2\sin\alpha(1 - \sin^2\alpha) + \sin\alpha(1 - \sin^2\alpha - \sin^2\alpha)
\end{aligned}$$

And the derivation of the triple-angle for tangent is straightforward from the definition of tangent:

$$\tan(3\alpha) = \frac{\sin(3\alpha)}{\cos(3\alpha)} = \frac{3\sin\alpha - 4\sin^3\alpha}{4\cos^3\alpha - 3\cos\alpha} = \frac{3\tan\alpha - \tan^3\alpha}{1 - 3\tan^2\alpha}$$

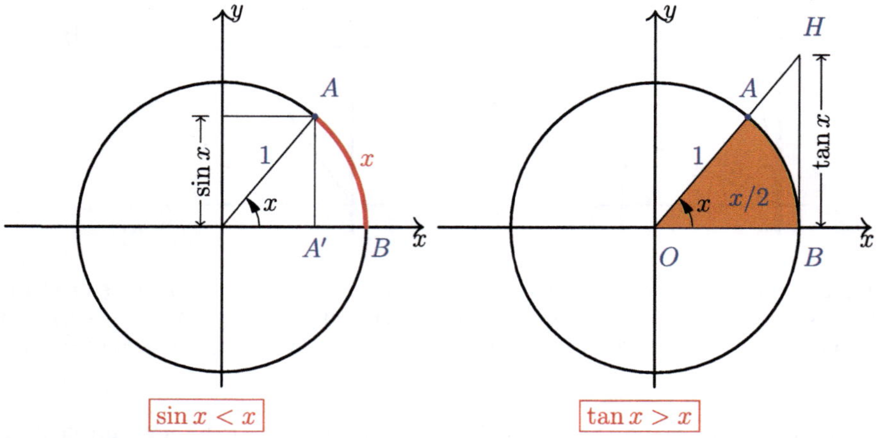

Fig. 3.47 Basic inequalities in trigonometry. Noting that x is measured in radians

where in the last equality, we divided both numerator and denominator by $\cos\alpha$ so that $\tan\alpha$ will appear. Admittedly, we did that because we already know the result. But if you did not know the result, you would also do the same thing. Why? This is so because we believe in pattern: $\sin(3\alpha)$ is expressed in terms of powers of $\sin(\alpha)$ and the same pattern occurs for cosine. Why not tangent? Usually, this belief pays off. Remember the story of Newton discovering the binomial theorem?

The triple angle identities is thus given by

$$\begin{aligned}
&\text{(triple-angle)}\\
&\sin(3\alpha) = 3\sin\alpha - 4\sin^3\alpha\\
&\cos(3\alpha) = 4\cos^3\alpha - 3\cos\alpha\\
&\tan(3\alpha) = \frac{3\tan\alpha - \tan^3\alpha}{1 - 3\tan^2\alpha}
\end{aligned} \tag{3.8.4}$$

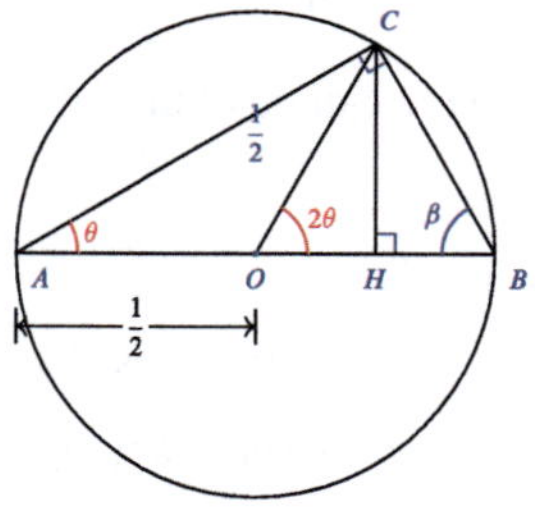

From the double-angle for cosine: $\cos(2\alpha) = \cos^2\alpha - \sin^2\alpha = 2\cos^2\alpha - 1$ we can derive the identity for half-angle. If you prefer a geometry proof for this, it is shown in Fig. 3.47. Consider a circle of radius of $1/2$ and a right triangle ACB with AB being the diameter of the circle. Note that $\sin\beta = \cos\theta$ as $\theta + \beta = \pi/2$. The proof is simple but it requires some knowledge of Euclidean geometry. In particular, we need the central angle theorem. It is this theorem that gives us the angle at O is 2θ in Fig. 3.47. What we need is $AH = \cos^2\theta$, and $AH = AO + OH$:

$$OH = \frac{1}{2}\cos 2\theta, \quad AC = s \in \beta$$
$$AH = AC\cos\theta = \sin\beta\cos\theta = \cos^2\theta$$
$$\cos^2\theta = OA + OH = \frac{1}{2} + \frac{1}{2}\cos 2\theta$$

The half-angle identities for the sine and cosine are

$$\begin{aligned} &\text{(half-angle)} \\ &\cos\alpha = \sqrt{\frac{1+\cos(2\alpha)}{2}} \\ &\sin\alpha = \sqrt{\frac{1-\cos(2\alpha)}{2}} \end{aligned} \tag{3.8.5}$$

Next come the so-called product identities:

$$\begin{aligned} &\text{(Product identities)} \\ &\sin\alpha\cos\beta = \frac{\sin(\alpha+\beta)+\sin(\alpha-\beta)}{2} \\ &\cos\alpha\cos\beta = \frac{\cos(\alpha+\beta)+\cos(\alpha-\beta)}{2} \\ &\sin\alpha\sin\beta = \frac{\cos(\alpha-\beta)-\cos(\alpha+\beta)}{2} \end{aligned} \tag{3.8.6}$$

The product identities $\sin\alpha\sin\beta$ are obtained from the addition/subtraction identities:

$$\left.\begin{aligned} \sin(\alpha+\beta) &= \sin\alpha\cos\beta + \sin\beta\cos\alpha \\ \sin(\alpha-\beta) &= \sin\alpha\cos\beta - \sin\beta\cos\alpha \end{aligned}\right\} \Longrightarrow \sin(\alpha+\beta)+\sin(\alpha-\beta) = 2\sin\alpha\cos\beta$$

Another form of the product identities is the sum-product identities given by

$$\begin{aligned} &\text{(Sum-product identities)} \\ &\sin\alpha+\sin\beta = 2\sin\frac{\alpha+\beta}{2}\cos\frac{\alpha-\beta}{2} \\ &\cos\alpha+\cos\beta = 2\cos\frac{\alpha+\beta}{2}\cos\frac{\alpha-\beta}{2} \\ &\cos\alpha-\cos\beta = -2\sin\frac{\alpha+\beta}{2}\sin\frac{\alpha-\beta}{2} \end{aligned} \tag{3.8.7}$$

which are nothing but Eq. (3.8.6) which are of the product-sum form. Later on in Sect. 7.10 we will use these identities when discussing superposition of waves and the beat phenomenon. When you hear the "wah wah wah" sound of the two pitches, close to each other, you are hearing the sum-to-product identities at work!

And finally are two identities relating sine/cosine with tangent of half-angle[28]:

$$\text{(Sine/cosine in terms of } \tan x/2) \qquad \sin x = \frac{2u}{1+u^2}, \quad \cos x = \frac{1-u^2}{1+u^2} \quad (u = \tan\tfrac{x}{2}) \tag{3.8.8}$$

So, what we have just done? We proved the angle addition (or subtraction) identity, and all the rest were derived (using simple algebra) from it.

Historically, the product identities, Eq. (3.8.6), were used before logarithms were invented to perform multiplication. Here's how you could use the second one. If you want to multiply $x \times y$, use a table to look up the angle α whose cosine is x and the angle β whose cosine is y. Look up the cosines of the sum $\alpha + \beta$ and the difference $\alpha - \beta$. Average those two cosines. You get the product xy! Three table look-ups, and computing a sum, a difference, and an average rather than one multiplication. Tycho Brahe (1546–1601), among others, used this algorithm known as prosthaphaeresis.

Derivation of $\cos(\alpha-\beta) = \cos\alpha\cos\beta + \sin\alpha\sin\beta$ using vector algebra.

If we know vector algebra (Sect. 9.1), we can derive this identity easily. Consider two unit vectors $\boldsymbol{a}$ and $\boldsymbol{b}$. The first vector makes with the horizontal axis an angle α and the second vector an angle β. So, we can express these two vectors as

$$\boldsymbol{a} = \cos\alpha\boldsymbol{i} + \sin\alpha\boldsymbol{j}, \quad \boldsymbol{b} = \cos\beta\boldsymbol{i} + \sin\beta\boldsymbol{j}$$

Then, the dot product of these two vectors can be computed by two ways:

$$\boldsymbol{a}\cdot\boldsymbol{b} = \cos(\alpha-\beta), \quad \boldsymbol{a}\cdot\boldsymbol{b} = \cos\alpha\cos\beta + \sin\alpha\sin\beta$$

So, we have $\cos(\alpha-\beta) = \cos\alpha\cos\beta + \sin\alpha\sin\beta$. How about the similar identity for sine? It seems using the cross product (which involves the sine) of two vectors is the way. You could try it yourself.

[28] The proof goes like this: $\sin x = 2\sin{}^{x}/_{2}\cos{}^{x}/_{2} = 2\tan{}^{x}/_{2}\cos^2{}^{x}/_{2}$. Furthermore, $1 + \tan^2 x/2 = 1/cos^2 x/2$.

Compute sine/cosine of $\pi/5$.

Let's denote $\theta = \pi/5$, or $5\theta = \pi$, so

$$\begin{aligned}
2\theta &= \pi - 3\theta \\
\sin(2\theta) &= \sin(\pi - 3\theta) = \sin(3\theta) \\
2\sin\theta\cos\theta &= 3\sin\theta - 4\sin^3\theta \\
4\cos^2\theta - 2\cos\theta - 1 = 0 &\Longrightarrow \cos\theta = \frac{1+\sqrt{5}}{4}
\end{aligned}$$

Did you notice something familiar? Cosine of 36° is related to the golden ratio ϕ. To see why check Fig. 2.15. I pass the baton to the interested reader to investigate that.

3.9 Trigonometric Identities for Angles of Plane Triangles

For the three angles of a triangle, there are many beautiful trigonometry identities e.g. $\cot\frac{x}{2} + \cot\frac{y}{2} + \cot\frac{z}{2} = \cot\frac{x}{2}\cot\frac{y}{2}\cot\frac{z}{2}$ for $x + y + z = \pi$. This section presents some of them.

Let's consider a plane triangle with three angles denoted by x, y, and z (in many books we will see the notations A, B, and C). We thus have the constraint $x + y + z = \pi$. We have then many identities. For example,

$$\cot\frac{x}{2} + \cot\frac{y}{2} + \cot\frac{z}{2} = \cot\frac{x}{2}\cot\frac{y}{2}\cot\frac{z}{2}$$

Proof The above formula is equivalent to the following

$$\tan\frac{x}{2}\tan\frac{y}{2} + \tan\frac{y}{2}\tan\frac{z}{2} + \tan\frac{z}{2}\tan\frac{x}{2} = 1$$

From $x + y + z = \pi$, we can relate the tangent of $(x + y)/2$ to the tangent of $z/2$, and use the addition angle formula for the tangent, we will arrive at the formula:

$$\begin{aligned}
\tan\left(\frac{x}{2} + \frac{y}{2}\right) &= \cot\frac{z}{2} = \frac{1}{\tan\frac{z}{2}} \\
\frac{\tan\frac{x}{2} + \tan\frac{y}{2}}{1 - \tan\frac{x}{2}\tan\frac{y}{2}} &= \frac{1}{\tan\frac{z}{2}}
\end{aligned}$$

■

We list in what follows some trigonometry identities for triangle angles:

$$\begin{aligned}
&\text{(a) Sine-related identities}\\
&\sin x + \sin y + \sin z = 4\cos\frac{x}{2}\cos\frac{y}{2}\cos\frac{z}{2}\\
&\sin 2x + \sin 2y + \sin 2z = 4\sin x \sin y \sin z\\
&\text{(b) Cosine-related identities}\\
&\cos x + \cos y + \cos z = 4\sin\frac{x}{2}\sin\frac{y}{2}\sin\frac{z}{2} + 1\\
&\cos 2x + \cos 2y + \cos 2z = -4\cos x \cos y \cos z - 1\\
&\text{(c) Tangent-related identities}\\
&\tan x + \tan y + \tan z = \tan x \tan y \tan z\\
&\text{(d) Co-tangent-related identities}\\
&\tan x \tan y + \tan y \tan z + \tan z \tan x = 1
\end{aligned} \tag{3.9.1}$$

Proof follows the same reasoning: using $x + y + z = \pi$ to replace z and use corresponding identities, Sect. 3.8. These identities are remarkable for their symmetry (i.e., the role/form of x, y, z is identical); one might even call them beautiful.

Proof This is a proof for $\cos x + \cos y + \cos z = 4\sin\frac{x}{2}\sin\frac{y}{2}\sin\frac{z}{2} + 1$: From $x + y + z = \pi$, we can relate the cosine of $z/2$ to the cosine of $x + y$, and use the summation formula for cosine to the term $\cos x + \cos y$, we will make appear half angles. Also using the double-angle formula $\cos 2u = 2\cos^2 u - 1$:

$$\begin{aligned}
\cos x + \cos y + \cos z &= 2\cos\left(\frac{x+y}{2}\right)\cos\left(\frac{x-y}{2}\right) - \cos(x+y)\\
&= 2\cos\left(\frac{x+y}{2}\right)\cos\left(\frac{x-y}{2}\right) - 2\cos^2\left(\frac{x+y}{2}\right) + 1\\
&= 2\sin\left(\frac{z}{2}\right)\left[\cos\left(\frac{x-y}{2}\right) - \cos\left(\frac{x+y}{2}\right)\right] + 1
\end{aligned}$$

Using the identity $\cos() - \cos()$ will conclude the proof. ■

3.10 Pascal Triangle Again

If we compute $\tan n\alpha$ in terms of $\tan\alpha$ for $n \in \mathbb{N}$, we get the following (only up to $n = 4$):

$$
\begin{aligned}
\tan\alpha &= t \\
\tan 2\alpha &= \frac{2t}{1-t^2} \\
\tan 3\alpha &= \frac{3t-t^3}{1-3t^2} \\
\tan 4\alpha &= \frac{4t-4t^3}{1-6t^2+1t^4}
\end{aligned} \tag{3.10.1}
$$

And see what? Binomial coefficients multiplying the powers $\tan^m \alpha$ show up. The binomial coefficients, corresponding to the numbers of the row of Pascal's triangle, occur in the expression in a zigzag pattern (i.e., coefficients at positions $1, 3, 5, \ldots$ are in the denominator, coefficients at the positions $2, 4, 6, \ldots$ are in the numerator, or vice versa), following the binomials in row of Pascal's triangle in the same order.

3.11 Bernoulli's Imaginary Trick for tan $n\alpha$

The way we obtained $\tan n\theta$ in terms of $\tan\theta$ works nicely for small n. Is it possible to have a method that works for any n? Yes, Bernoulli presented such a method, but it adopted imaginary number $i^2 = -1$ and the new infinitesimal calculus that Leibniz just invented.[29] Here is what he did:

$$
\left.\begin{aligned} x &= \tan\theta, \\ y &= \tan n\theta \end{aligned}\right\} \Longrightarrow \tan^{-1} y = n\tan^{-1} x
$$

We refer to Sect. 3.14 for a discussion on inverse trigonometric functions (e.g. $\tan^{-1} x$). Briefly, given an angle θ, press the tangent button gives us $\tan\theta$, and pressing the $\tan^{-1}$ button gives us back the angle. Now, he differentiated $\tan^{-1} y = n\tan^{-1} x$ to get

$$
\frac{dy}{1+y^2} = n\frac{dx}{1+x^2}
$$

Then he indefinitely integrated the above equation to get

$$
\int \frac{dy}{1+y^2} = \int n\frac{dx}{1+x^2} \tag{3.11.1}
$$

Now comes the trick of using i:

$$
\boxed{\frac{1}{1+x^2} = \frac{1}{x^2-i^2} = \frac{1}{(x-i)(x+i)} = \frac{1}{2i}\left(\frac{1}{x-i} - \frac{1}{x+i}\right)}
$$

[29] If you do not know calculus yet, skip this. Calculus is discussed in Chap. 4.

So what he did is called factoring into imaginary components, and in the final step, a partial fraction expansion. With that, it's easy to compute the integral $\int dx/1+x^2$:

$$\int \frac{dx}{1+x^2} = \frac{1}{2i}\left(\int \frac{dx}{x-i} - \int \frac{dx}{x+i}\right) = \frac{1}{2i}(\ln|x-i| - \ln|x+i|) = \frac{1}{2i}\ln\left|\frac{x-i}{x+i}\right|$$

With this result, he could proceed with Eq. (3.11.1):

$$\frac{1}{2i}\ln\left|\frac{y-i}{y+i}\right| = n\frac{1}{2i}\ln\left|\frac{x-i}{x+i}\right| + C \Longleftrightarrow \ln\left|\frac{y-i}{y+i}\right| = n\ln\left|\frac{x-i}{x+i}\right| + C' \tag{3.11.2}$$

He found C' with the condition $x = y = 0$ when $\theta = 0$: $C' = \ln[(-1)^{n-1}]$. With this C', Eq. (3.11.2) becomes:

$$\ln\left(\frac{y-i}{y+i}\right) = \ln\left(\frac{x-i}{x+i}\right)^n + \ln[(-1)^{n-1}] = \ln\left[(-1)^{n-1}\left(\frac{x-i}{x+i}\right)^n\right] \tag{3.11.3}$$

Thus, he obtained this

$$\frac{y-i}{y+i} = (-1)^{n-1}\left(\frac{x-i}{x+i}\right)^n \tag{3.11.4}$$

which gave him

$$\begin{aligned} \frac{y-i}{y+i} &= +\left(\frac{x-i}{x+i}\right)^n \quad n = 1, 3, 5, \ldots \\ \frac{y-i}{y+i} &= -\left(\frac{x-i}{x+i}\right)^n \quad n = 2, 4, 6, \ldots \end{aligned}$$

And solving for y (the above equations are just linear equations for y), Bernoulli have obtained a nice formula for y or $\tan n\theta$ with $x = \tan\theta$

$$\begin{aligned} \tan n\theta &= i\frac{(x+i)^n + (x-i)^n}{(x+i)^n - (x-i)^n} \quad n = 1, 3, 5, \ldots \\ \tan n\theta &= i\frac{(x+i)^n - (x-i)^n}{(x+i)^n + (x-i)^n} \quad n = 2, 4, 6, \ldots \end{aligned} \tag{3.11.5}$$

Now, we check this result, by applying it to $n = 2$, and the above equation (the second one of course as $n = 2$) indeed leads to the correct formula of $\tan 2\theta = 2\tan\theta/(1-\tan^2\theta)$.

3.12 $\sin n\alpha$ and $\cos n\alpha$

In Sect. 2.25.5 we have used de Moivre's formula to derive the formula for $\sin 2\alpha$, $\sin 3\alpha$ in terms of powers of $\sin\alpha$. In principle, we can follow that way to derive the formula of $\sin n\alpha$ for any n, but the process is tedious (try with $\sin 5\alpha$ and you'll understand what I meant). There should be an easier way. In what follows, a method that uses i, the binomial theorem, and Euler's identity is presented.

The trick is in Eq. (2.25.21), which we rewrite here:

$$\sin\alpha = \frac{e^{i\alpha} - e^{-i\alpha}}{2i} \tag{3.12.1}$$

Using it for $n\alpha$ we have:

$$\begin{aligned}
\sin n\alpha &= \frac{e^{in\alpha} - e^{-in\alpha}}{2i} = \frac{(e^{i\alpha})^n - (e^{-i\alpha})^n}{2i} \\
&= \frac{(\cos\alpha + i\sin\alpha)^n - (\cos\alpha - i\sin\alpha)^n}{2i} \quad (\text{use } e^{i\alpha} = \cos\alpha + i\sin\alpha) \\
&= \frac{\sum_{k=0}^{n}\binom{n}{k}\cos^{n-k}\alpha(i\sin\alpha)^k - \sum_{k=0}^{n}\binom{n}{k}\cos^{n-k}\alpha(-i\sin\alpha)^k}{2i} \\
&= \sum_{k=0}^{n}\binom{n}{k}\cos^{n-k}\alpha\sin^k\alpha\frac{i^k - (-i)^k}{2i} \\
&= \frac{n}{1!}\cos^{n-1}\alpha\sin\alpha - \frac{n(n-1)(n-2)}{3!}\cos^{n-3}\alpha\sin^3\alpha + \cdots
\end{aligned} \tag{3.12.2}$$

where in the third equality, we have used the binomial theorem to expand $(\cdots)^n$, and the red term is equal to zero for $k = 0, 2, 4, \ldots$ and equal to one for $k = 1, 3, 5, \ldots$

$\cos n\alpha$ **for any** n. If we have something for sine, cosine is jealous. So, we do the same analysis for cosine, and get

$$\begin{aligned}
\cos n\alpha &= \frac{e^{in\alpha} + e^{-in\alpha}}{2} = \sum_{k=0}^{n}\binom{n}{k}\cos^{n-k}\alpha\sin^k\alpha\frac{i^k + (-i)^k}{2} \\
&= \cos^n\alpha - \frac{n(n-1)}{2!}\cos^{n-2}\alpha\sin^2\alpha \\
&\quad + \frac{n(n-1)(n-2)(n-3)}{4!}\cos^{n-4}\alpha\sin^4\alpha + \cdots
\end{aligned} \tag{3.12.3}$$

where in the third equality, we have used the binomial theorem to expand $(\cdots)^n$, and the red term is equal to zero for $k = 1, 3, 5, \ldots$ and equal to one for $k = 0, 4, 8, \ldots$, and equal to minus one for $k = 2, 6, 10, \ldots$

With Eq. (3.12.3), we can write the formula for $\cos(n\alpha)$ for the first few values of n:

$$\begin{aligned}\cos(0\alpha) &= 1\\ \cos(1\alpha) &= \cos\alpha\\ \cos(2\alpha) &= 2\cos^2\alpha - 1\\ \cos(3\alpha) &= 4\cos^3\alpha - 3\cos\alpha\\ \cos(4\alpha) &= 8\cos^4\alpha - 8\cos^2\alpha + 1\\ \cos(5\alpha) &= 16\cos^5\alpha - 20\cos^3\alpha + 5\cos\alpha\end{aligned} \tag{3.12.4}$$

What is the purpose of doing this? The next step is to try to *find a pattern in these formula*. One question is, is it possible to compute $\cos(6\alpha)$ w/o resorting to Eq. (3.12.3)? Let' see how we can get $\cos(2\alpha) = 2\cos^2\alpha - 1$ from $\cos 1\alpha = \cos\alpha$: we can multiply $\cos 1\alpha$ with $2\cos\alpha$ and minus 1, and 1 is $\cos 0\alpha$:

$$\cos(2\alpha) = 2\cos(\alpha) \times \cos(1\alpha) - \cos(0\alpha)$$

Thus, we can compute $\cos(k\alpha)$ from $\cos(k-1)\alpha$ and $\cos(k-2)\alpha$! The formula is[30]

$$\boxed{\cos(k\alpha) = 2\cos\alpha\cos(k-1)\alpha - \cos(k-2)\alpha}$$

Now, we have a recursive formula for $\cos n\alpha$

$$\cos(n\alpha) = \begin{cases} 1, & \text{if } n = 0\\ \cos\alpha, & \text{if } n = 1\\ 2\cos\alpha\cos(n-1)\alpha - \cos(n-2)\alpha, & \text{if } n \geq 2\end{cases} \tag{3.12.5}$$

One application of this formula is to derive the Chebyshev polynomials of the first kind[31] described in Sect. 10.3.2. Why this has to do with polynomials? Note that from the above equation, $\cos(n\alpha)$ is a polynomial in terms of $\cos\alpha$, e.g. $\cos 3\alpha = 4(\cos\alpha)^3 - 3\cos(\alpha)$. That's why. If you forget what is a polynomial, check Sect. 2.29.

3.13 Inverse Trigonometric Functions

For each of the trigonometric functions, there is a corresponding inverse trigonometric function. Let's take the sine function as an example. Start with an angle,

[30] Note that this was how to discover this relation between $\cos(k\alpha)$, $\cos(k-1)\alpha$ and $\cos(k-2)\alpha$. When we know such formula exists, we can prove it in an easier way. I leave it as a trigonometry exercise.

[31] If there are polynomials of first kind, then where are those of second kind? They're polynomials related to $\sin n\alpha$. Those related to the cosine are called the first kind probably because $\cos\alpha$ is the real part of $e^{i\alpha}$.

say $x = \pi/4$, we get $y = \sin(\pi/4) = \sqrt{2}/2$. Then, the inverse sine gives back x: $\pi/4 = \sin^{-1}(\sqrt{2}/2)$. So, the inverse sine function answers the question *what is the angle of which the sine is* y; it is $\sin^{-1} y$. Similarly we have $\cos^{-1}(x)$, $\tan^{-1}(x)$, etc. This notation was introduced by John Herschel (1792–1871), an English polymath, mathematician, astronomer, chemist, inventor, and experimental photographer who invented the blueprint and did botanical work, in 1813.

Another notation exists for the inverse trigonometric functions. The most common convention is to name inverse trigonometric functions using an arc-prefix: $\arcsin(x)$, $\arccos(x)$, $\arctan(x)$, etc. For example, $\pi/4 = \arcsin(\sqrt{2}/2)$. This notation arises from the following geometric relationships. When measuring in radians, an angle of θ radians will correspond to an arc whose length is $r\theta$, where r is the radius of the circle. Thus in the unit circle, the arc whose cosine is x is the same as "the angle whose cosine is x", because the length of the arc of the circle is the same as the measurement of the angle.

3.14 Inverse Trigonometric Identities

To each trigonometry identity we have a corresponding inverse trigonometry identity. Herein I present one such identity and its use to compute π. The trigonometry identity is tangent of the difference of two angles:

$$\tan(\alpha - \beta) = \frac{\tan\alpha - \tan\beta}{1 + \tan\alpha\tan\beta}$$

Starting from this identity and we introduce two new symbols x, y which are tangents of α and β:

$$\begin{cases} x = \tan\alpha \\ y = \tan\beta \end{cases} \Longrightarrow \frac{x - y}{1 + xy} = \tan(\alpha - \beta) \quad \text{[from the above identity]}$$

Thus, we get the following

$$\alpha - \beta = \arctan\frac{x - y}{1 + xy}$$

And by substituting $\alpha = \arctan x$ and $\beta = \arctan y$ into the above equation, we get the following inverse trigonometry identity:

$$\arctan x - \arctan y = \arctan\frac{x - y}{1 + xy}$$

Introducing $x = a_1/b_1$, $y = a_2/b_2$, we get the same identity in a slightly different form (I have included two versions: one for angle addition and one for angle difference):

$$\boxed{\arctan\frac{a_1}{b_1} \pm \arctan\frac{a_2}{b_2} = \arctan\frac{a_1b_2 \pm a_2b_1}{b_1b_2 \mp a_1a_2}} \tag{3.14.1}$$

Machin's formula. John Machin (1686–1751) was a professor of astronomy at Gresham College, London. He is best known for developing a quickly converging series for π in 1706 and using it to compute π to 100 decimal places. He derived the following formula, now known as Machin's formula, using Eq. (3.14.1) (details given later)

$$\frac{\pi}{4} = 4\arctan\frac{1}{5} - \arctan\frac{1}{239} \tag{3.14.2}$$

Then he combined his formula with the Taylor series expansion for the inverse tangent (see Eq. (4.15.15) in Chap. 4) to compute π to 100 decimal places (w/o a calculator of course; he did not have that luxury).[32] Machin's formula remained the primary tool of Pi-hunters for centuries (well into the computer era). For completeness, details are given as follows.

The power series for $\arctan x$ is: $\arctan x = x - \frac{x^3}{3} + \frac{x^5}{5} - \frac{x^7}{7} + \cdots$. Using it in Eq. (3.14.2) gives us a series to compute π:

$$\frac{\pi}{4} = 4\left[\frac{1}{5} - \frac{1}{3\cdot 5^3} + \frac{1}{5\cdot 5^5} - \frac{1}{7\cdot 5^7} + \cdots\right] - \left[\frac{1}{239} - \frac{1}{3\cdot 239^3} + \frac{1}{5\cdot 239^5} - \cdots\right] \tag{3.14.3}$$

With only five terms in the arctangent series, this formula gave us eight correct decimals for π.

Proof Derivation of Machin's formula Eq. (3.14.2) using Eq. (3.14.1). We start with $\arctan\frac{1}{5} + \arctan\frac{1}{5}$ to get $2\arctan\frac{1}{5}$:

$$\begin{aligned} 2\arctan\frac{1}{5} &= \arctan\frac{1}{5} + \arctan\frac{1}{5} \\ &= \arctan\frac{1\cdot 5 + 1\cdot 5}{5\cdot 5 - 1\cdot 1} = \arctan\frac{5}{12} \end{aligned} \tag{3.14.4}$$

Now, with $2\arctan\frac{1}{5} + 2\arctan\frac{1}{5}$ we get $4\arctan\frac{1}{5}$:

$$\begin{aligned} 4\arctan\frac{1}{5} &= 2\arctan\frac{1}{5} + 2\arctan\frac{1}{5} \\ &= \arctan\frac{5}{12} + \arctan\frac{5}{12} && \text{(Eq. (3.14.4))} \\ &= \arctan\frac{5\cdot 12 + 5\cdot 12}{12\cdot 12 - 5\cdot 5} = \arctan\frac{120}{119} && \text{(Eq. (3.14.1))} \end{aligned}$$

Finally, we consider $4\arctan\frac{1}{5} - \frac{\pi}{4}$, writing $\pi/4$ as $\arctan 1/1$:

[32] In passing we note that Brook Taylor was Machin's contemporary at Cambridge University.

$$4\arctan\frac{1}{5}-\frac{\pi}{4}=4\arctan\frac{1}{5}-\arctan\frac{1}{1}=\arctan\frac{120}{119}-\arctan\frac{1}{1}=\arctan\frac{1}{239}$$

■

Compute π from thin air. Machin's formula for π is great, but there is an unbelievable way to get π, from thin air. To be precise from $i^2=-1$. Recall that we have (Sect. 2.25.7):

$$\frac{\pi}{4}=-\frac{i}{2}\ln(i)$$

Now we do a bit of algebra to convert i to a fraction form in the RHS of the above:

$$\frac{\pi}{4}=-\frac{i}{2}\ln(i)=-\frac{i}{2}\ln\left(\frac{1+i}{1-i}\right)=-\frac{i}{2}\left(\ln(1+i)-\ln(1-i)\right) \tag{3.14.5}$$

Now, we use the power series of logarithm, written for a complex number z:

$$\ln(1+z)=z-\frac{z^2}{2}+\frac{z^3}{3}-\frac{z^4}{4}+\cdots$$

Thus, we have

$$\begin{aligned}\ln(1+i)&=i-\frac{i^2}{2}+\frac{i^3}{3}-\frac{i^4}{4}+\cdots\\ \ln(1-i)&=-i-\frac{(-i)^2}{2}+\frac{(-i)^3}{3}-\frac{(-i)^4}{4}+\cdots\end{aligned} \tag{3.14.6}$$

Finally, substituting Eq. (3.14.6) into Eq. (3.14.5), we get π:

$$\frac{\pi}{4}=1-\frac{1}{3}+\frac{1}{5}-\frac{1}{7}+\cdots=\sum_{n=1}^{\infty}(-1)^{n+1}\frac{1}{2n-1}$$

What a great formula: π and all the odd numbers. We got π from $\sqrt{-1}$; a real number from an imaginary one! It seems impossible, so we should check this result. That's why I have provided the last (red) expression, which can be coded on a computer. The outcome of that exercise is that the more terms we use the more close to $\pi/4=0.7853981633\ldots$ we get. However, this series is too slow in the sense that we need too many terms to get an accurate value of π. That's why Machin and other mathematicians developed other formulas.

But still this is not Eq. (3.14.3). Don't worry. The German mathematician Karl Heinrich Schellbach (1809–1890) did this in 1832. He used:

$$\pi=\frac{2}{i}\ln\left[\frac{(5+i)^4(-239+i)}{(5-i)^4(-239-i)}\right]$$

It is certain that Schellbach was aware of Machin's formula, and that was how he could think of the crazy expression in the bracket for i.

Derivation of Eq. (3.14.1) **using complex numbers**. If we consider two complex numbers $b_1 + a_1 i$ with the angle $\theta_1 = \arctan a_1/b_1$ and $b_2 + a_2 i$ with the angle $\theta_2 = \arctan a_2/b_2$, then its product is $b_1 b_2 - a_1 a_2 + (a_1 b_2 - a_2 b_1)i$ with the angle $\theta = \arctan(a_1 b_2 - a_2 b_1)/(b_1 b_2 - a_1 a_2)$. Then Eq. (3.14.1) is nothing but $\theta = \theta_1 + \theta_2$, a property of complex number multiplication. And this is expected as we started from the trigonometry identity for angle difference/addition.

3.15 Trigonometric Inequalities

When Ptolemy was making his sine table, he could not find $\sin 1^\circ$ directly,[33] and he found an approximate method. His idea was to compare $\sin 1^\circ$ with $\sin 3^\circ$ using this inequality $\sin 3^\circ < 3 \sin 1^\circ$. If you remember the trigonometry identity $\sin 3x = 3 \sin x - 4 \sin^3 x$, you'll understand this inequality.

As we now pay attention to inequalities of trigonometry functions, we turn to the unit circle with sine/cosine/tangent, Fig. 3.47, and we discover these inequalities:

$$\sin x < x < \tan x \tag{3.15.1}$$

where the first inequality (i.e., $\sin x < x$) was obtained by comparing the length of the line AA'—which is $\sin x$—versus the length of the arc AB (which is x) in the left figure. The second inequality was obtained by comparing the area of the right triangle OHB, which is $1/2 \tan x$, and the area of the shaded region in the right figure (which is $x/2$).

There is nothing special about $\sin 3^\circ < 3 \sin 1^\circ$, which can be written in this form $\sin 3^\circ/\sin 1^\circ < 3/1$, if we have this, we should have this:

$$\frac{\sin \alpha}{\sin \beta} < \frac{\alpha}{\beta}, \quad \text{for all } \alpha > \beta \in [0, \pi/2] \tag{3.15.2}$$

And of course we need a proof as for now it is just our guess. Before presenting a proof, let's see how Eq. (3.15.2) was used by Ptolemy to compute $\sin 1^\circ$:

$$\begin{aligned}
\alpha = (3/2)^\circ, \beta = 1^\circ &: \sin 1^\circ > \frac{2}{3} \sin(3/2)^\circ \\
\alpha = 1^\circ, \beta = (3/4)^\circ &: \sin 1^\circ < \frac{4}{3} \sin(3/4)^\circ
\end{aligned}$$

[33] The fact that Ptolemy could not find $\sin 1^\circ$ given that $\sin 3^\circ$ is known is due to the fact that it is impossible to trisect an angle using only straightedge and compass.

From $\sin 3^\circ$, we can compute $\sin(3/2)^\circ$ and $\sin(3/4)^\circ$ (of course using half-angle identities). Thus, we get $0.0174512988719154 33 < \sin 1^\circ < 0.01745279409512592$. So, we obtain $\sin 1^\circ = 0.01745$. The accuracy is only 5 decimal places. Can you improve this technique?

Proof We're going to prove Eq. (3.15.2) using algebra and Eq. (3.15.1). There exists a geometric proof of Aristarchus of Samos—an ancient Greek astronomer and mathematician who presented the first known heliocentric model that placed the Sun at the center of the known universe with the Earth revolving around it. Thus, this inequality is known as Aristarchus's inequality. I refer to Wikipedia for the geometry proof.

First, using algebra to transform the inequality to a 'better' form:

$$\begin{aligned}
\frac{\sin\alpha}{\sin\beta} &< \frac{\alpha}{\beta} \\
\beta\sin\alpha &< \alpha\sin\beta \qquad \text{(all quantities are positive)} \\
\beta(\sin\alpha - \sin\beta) &< (\alpha-\beta)\sin\beta \ \text{(add } -\beta\sin\beta \text{ to both sides)} \\
\frac{\sin\alpha-\sin\beta}{\alpha-\beta} &< \frac{\sin\beta}{\beta} \qquad (\alpha-\beta>0)
\end{aligned}$$

The key step is of course the highlighted one where we brought the term $-\beta\sin\beta$ to the game. Why that particular term? Because it led us to this term $\sin\alpha-\sin\beta/\alpha-\beta$. As all steps are equivalent, we just need to prove the final inequality. We use the identity for $\sin\alpha - \sin\beta$ to rewrite the LHS of the last inequality (the blue term) and use $\sin x < x$:

$$\frac{\sin\alpha-\sin\beta}{\alpha-\beta} = \frac{2\sin\left(\frac{\alpha-\beta}{2}\right)\cos\left(\frac{\alpha+\beta}{2}\right)}{\alpha-\beta} < \left(\frac{2}{\alpha-\beta}\right)\frac{\alpha-\beta}{2}\cos\left(\frac{\alpha+\beta}{2}\right) = \cos\left(\frac{\alpha+\beta}{2}\right)$$

The next move is to get rid of α in $\cos(\alpha+\beta/2)$. For that, we need this fact $\cos x < \cos y$ if $x > y$ for $x, y \in [0, \pi/2]$. Because $\alpha > \beta$, then $0.5(\alpha+\beta) > 0.5(\beta+\beta) = \beta$, thus

$$\frac{\sin\alpha-\sin\beta}{\alpha-\beta} < \cos\left(\frac{\alpha+\beta}{2}\right) < \cos\beta \ \left(= \frac{\sin\beta}{\tan\beta}\right)$$

The last step was to convert from $\cos\beta$ to $\sin\beta$ (we need this term) noting that we have a tool not yet used, that is $\tan\beta > \beta$, or $1/\tan\beta < 1/\beta$. ■

Actually, if we know calculus, the proof is super easy (Fig. 3.48); it does not require us being genius. The function $f(x) = \sin x/x$ is a decreasing function for $x \in [0, \pi]$ (compute its first derivative and using $\tan x > x$ to see that the derivative is indeed negative). Now considering two numbers α, β in $[0, \pi]$ such that $\alpha > \beta$, we then have immediately $f(\alpha) < f(\beta)$. Done. Alternatively, if we consider the function $y = \sin x$, we also have the inequality. Comparing with Aristarchus's proof, which was based on circles and triangles, the calculus-based proof is straightforward. Why? Because in old trigonometry, sine was attached to angles of triangles, whereas in calculus it is free of angles/triangles. It is simply a function.

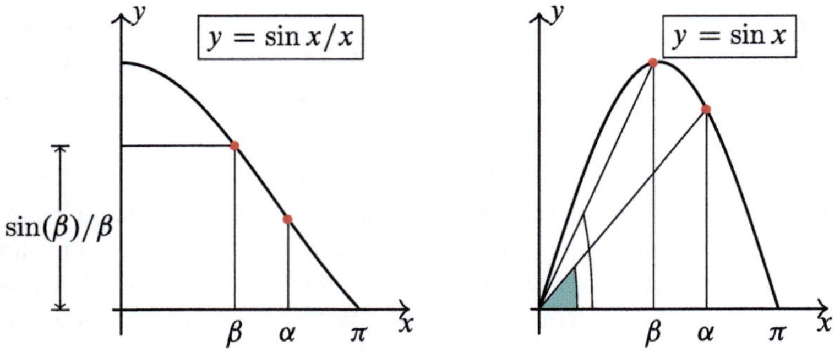

Fig. 3.48 Calculus based proofs of Aristarchus's inequality $\sin\alpha/\sin\beta < \alpha/\beta$. The proof using the $\sin x$ function is based on the fact that the graph is concave. The term $\sin\alpha/\alpha$ is the tangent of the shaded angle. The angle corresponding to α is smaller than that of β.kohktikgto

An approximation of the sine: $\sin x \approx x$. When we were building our sine table, we have discovered that $\sin x \approx x$, at least when $x = 1° = \pi/180$. It turns out that for small x, this is always true. And it stems from Eq. (3.15.1), which we rewrite as

$$\sin x < x < \tan x \Longleftrightarrow \cos x < \frac{\sin x}{x} < 1$$

Now, let x approach zero, then $\cos x$ approaches 1, and thus $1 < \sin x/x < 1$: that is $\sin x/x$ is sandwiched between 1 and 1. This leads to:

$$\lim_{x\to 0} \frac{\sin x}{x} = 1 \tag{3.15.3}$$

Some inequalities for angles of a triangle. Herein I present some well-known inequalities involving angles of a triangle. We label the three angles by A, B, C this time. For all inequalities, equality occurs when $A = B = C$ or when the triangle is equilateral. They are

$$\begin{aligned}
&\text{(a) } \sin A + \sin B + \sin C \le \frac{3\sqrt{3}}{2}\\
&\text{(b) } \cos A + \cos B + \cos C \le \frac{3}{2}\\
&\text{(c) } \cot A \cot B \cot C \le \frac{\sqrt{3}}{9}\\
&\text{(d) } \cot A + \cot B + \cot C \ge \sqrt{3}\\
&\text{(e) } \sin^2 A + \sin^2 B + \sin^2 C \le \frac{9}{4}\\
&\text{(f) } \cot^2 A + \cot^2 B + \cot^2 C \ge 1
\end{aligned} \tag{3.15.4}$$

Proof We prove (a) using the Jensen inequality (check Sect. 4.5.2 if it's new to you) which states that for a convex function $f(x)$, $f((x+y+z)/3) \le (1/3)(f(x)+f(y)+f(z))$. As the function $y = \sin x$ for $0 \le x \le \pi$ is a concave function, we have:

$$\sin\left(\frac{A+B+C}{3}\right) \ge \frac{\sin A + \sin B + \sin C}{3}$$

Thus,

$$\sin A + \sin B + \sin C \le 3\sin\left(\frac{\pi}{3}\right) = 3\frac{\sqrt{3}}{2} \quad \blacksquare$$

Proof You might be thinking the proof of (b) is similar to (a). Unfortunately, the cosine function is harder: its graph consists of two parts, see Fig. 3.51. Only for acute-angled triangles, we can use the Jensen inequality as in (a). Hmm. We need another proof for all triangles. First, we convert the term $\cos A + \cos B + \cos C$ to:

$$\begin{aligned}\cos A + \cos B + \cos C &= 2\cos\frac{A+B}{2}\cos\frac{A-B}{2} + 1 - 2\sin^2\frac{C}{2}\\ &= 2\sin\frac{C}{2}\cos\frac{A-B}{2} + 1 - 2\sin^2\frac{C}{2}\end{aligned}$$

Then, the inequality becomes:

$$E = -2\left(\sin\frac{C}{2}\right)^2 + 2\cos\frac{A-B}{2}\sin\frac{C}{2} - \frac{1}{2}$$

This is a quadratic function in terms of $\sin C/2$ with a negative highest coefficient (i.e., -2). To show that $E \le 0$ for all A, B, C, we just need to check its discriminant Δ. Indeed, the discriminant is always negative, thus E is always below the x-axis, and thus it is always smaller or equal to 0. A consequence of this result is another inequality that reads

$$\sin\frac{A}{2}\sin\frac{B}{2}\sin\frac{C}{2} \le \frac{1}{8}$$

which is obtained from (c) and the identity $\cos A + \cos B + \cos C = 4\sin\frac{A}{2}\sin\frac{B}{2}\sin\frac{C}{2} + 1$. ■

Proof We prove (c) using the Jensen inequality for the convex function $\tan x$. Note that we only need to consider acute angles (because if one angle is not acute, its cotangent is negative whereas the other cotangents are positive, and thus the inequality holds):

$$\tan\left(\frac{A+B+C}{3}\right) \le \frac{\tan A + \tan B + \tan C}{3}$$

Thus,

$$\tan A + \tan B + \tan C \geq 3 \tan\left(\frac{\pi}{3}\right) = 3\sqrt{3}$$

But, we have $\tan A + \tan B + \tan C = \tan A \tan B \tan C$, thus

$$\tan A \tan B \tan C \geq 3\sqrt{3} \iff \frac{1}{\tan A \tan B \tan C} \leq \frac{\sqrt{3}}{9}$$

■

Proof We prove (d) as follows. The key point is the identity $\cot A \cot B + \cot B \cot C + \cot C \cot A = 1$. We start with:

$$(\cot A + \cot B + \cot C)^2 = \cot^2 A + \cot^2 B + \cot^2 C \\ + 2(\cot A \cot B + \cot B \cot C + \cot C \cot A)$$

And we can relate $\cot^2 A + \cot^2 B + \cot^2 C$ with $\cot A \cot B + \cot B \cot C + \cot C \cot A$:

$$\begin{cases} \cot^2 A + \cot^2 B \geq 2 \cot A \cot B \\ \cot^2 B + \cot^2 C \geq 2 \cot B \cot C \\ \cot^2 C + \cot^2 A \geq 2 \cot C \cot A \end{cases}$$

Therefore,

$$\cot^2 A + \cot^2 B + \cot^2 C \geq \cot A \cot B + \cot B \cot C + \cot C \cot A$$

And then,

$$(\cot A + \cot B + \cot C)^2 \geq 3(\cot A \cot B + \cot B \cot C + \cot C \cot A)$$

■

Proof We prove (e) using some algebra and the inequality (b). First, we transform $\sin^2 A + \sin^2 B + \sin^2 C$ to $\cos 2A, \ldots$:

$$\sin^2 A + \sin^2 B + \sin^2 C = \frac{3}{2} - \frac{1}{2}(\cos 2A + \cos 2B + \cos 2C)$$

Then, using Eq. (3.9.1), we get

$$\sin^2 A + \sin^2 B + \sin^2 C = 2 + 2 \cos A \cos B \cos C$$

If one angle (assuming that angle is A without loss of generality) is not acute, then $\cos A < 0$ and $\cos B, \cos C > 0$, thus $\cos A \cos B \cos C < 0$. Therefore, $\sin^2 A +$

$\sin^2 B + \sin^2 C < 2$. If, all angles are acute, $\cos A, \cos B, \cos C > 0$, we can use the AM-GM inequality:

$$\sqrt[3]{\cos A \cos B \cos C} \le \frac{1}{3}(\cos A + \cos B + \cos C)$$

And using the inequality (b), we get

$$\cos A \cos B \cos C \le \frac{1}{27}(\cos A + \cos B + \cos C)^3 \le \frac{1}{27}\frac{27}{4} = \frac{1}{4}$$

And the result follows immediately:

$$\sin^2 A + \sin^2 B + \sin^2 C \le 2 + \frac{1}{2} = \frac{9}{4}$$

■

Proof We can prove (f) using the Cauchy-Swatch inequality and the inequality (d). ■

Cauchy's proof of Basel problem. In Sect. 2.21.4 I have introduced the Basel problem and one calculus-based proof. Herein, I present Cauchy's proof using only elementary mathematics. The plan of his proof goes as:

- The starting point is the trigonometry inequality we all know:

$$\sin\theta < \theta < \tan\theta, \quad 0 < \theta < \pi/2$$

- The above inequality gives us an equivalent one:

$$\cot^2\theta < \frac{1}{\theta^2} < 1 + \cot^2\theta \tag{3.15.5}$$

- Now, he introduced two new positive integer variables n and N such that

$$\theta = \frac{n\pi}{2N+1}, \quad 1 \le n \le N \tag{3.15.6}$$

  This definition of θ comes from the requirement that $\theta < \pi/2$. Now, Eq. (3.15.5) becomes

$$\cot^2\frac{n\pi}{2N+1} < \left(\frac{2N+1}{n\pi}\right)^2 < 1 + \cot^2\frac{n\pi}{2N+1} \tag{3.15.7}$$

  Now that the Basel problem is about the summation of the reciprocals of the squares of the natural numbers i.e., $\sum_n 1/n^2$, he made $1/n^2$ appear:

$$\frac{\pi^2}{(2N+1)^2}\cot^2\frac{n\pi}{2N+1} < \frac{1}{n^2} < \frac{\pi^2}{(2N+1)^2} + \frac{\pi^2}{(2N+1)^2}\cot^2\frac{n\pi}{2N+1} \tag{3.15.8}$$

- The next step is, of course, to introduce $\sum_n {}^1\!/\!_{n^2}$:

$$\sum_{n=1}^{N}\frac{\pi^2}{(2N+1)^2}\cot^2\frac{n\pi}{2N+1} < \sum_{n=1}^{N}\frac{1}{n^2} < \sum_{n=1}^{N}\frac{\pi^2}{(2N+1)^2} + \sum_{n=1}^{N}\frac{\pi^2}{(2N+1)^2}\cot^2\frac{n\pi}{2N+1} \tag{3.15.9}$$

- Now, the Basel sum is an infinite series, we have to consider $N \to \infty$ (then the blue term vanishes); I also introduced S for $\sum_{n=1}^{N} {}^1\!/\!_{n^2}$:

$$\lim_{N\to\infty}\frac{\pi^2}{(2N+1)^2}\sum_{n=1}^{N}\cot^2\frac{n\pi}{2N+1} < S < \lim_{N\to\infty}\frac{\pi^2}{(2N+1)^2}\sum_{n=1}^{N}\cot^2\frac{n\pi}{2N+1} \tag{3.15.10}$$

What Cauchy needed now is to be able to evaluate the red sum.

- The next move is to adopt de Moivre's formula (Eq. (2.25.6)):

$$\cos nx + i\sin nx = (\cos x + i\sin x)^n \tag{3.15.11}$$

And a bit of massage, we get $\cot x$ in the game:

$$\frac{\cos nx + i\sin nx}{\sin^n x} = (\cot x + i)^n \tag{3.15.12}$$

Now, we use the binomial theorem, Eq. (2.27.2), to expand $(\cot x + i)^n$:

$$(\cot x + i)^n = \binom{n}{0}\cot^n x + \binom{n}{1}\cot^{n-1} x i + \cdots + \binom{n}{n-1}\cot x i^{n-1} + \binom{n}{n} i^n$$

And from that we can extract the imaginary part of $(\cot x + i)^n$:

$$\text{Im}(\cot x + i)^n = \binom{n}{1}\cot^{n-1} x - \binom{n}{3}\cot^{n-3} x + \binom{n}{5}\cot^{n-5} x + \cdots$$

Using Eq. (3.15.12) and equating the imaginary parts of the sides, we get

$$\frac{\sin nx}{\sin^n x} = \binom{n}{1}\cot^{n-1} x - \binom{n}{3}\cot^{n-3} x + \binom{n}{5}\cot^{n-5} x + \cdots \tag{3.15.13}$$

This is by itself a trigonometry identity that holds for any $n \in \mathbb{N}$ and $x \in \mathbb{R}$. Now we take this identity, fix a positive integer N and set $n = 2N + 1$ and $x_k = {}^{k\pi}\!/\!_{2N+1}$, for $k = 1, 2, \ldots, N$. Why that? Because the LHS of the identity is zero with this choice: $\sin nx_k = \sin(2N+1)^{k\pi}\!/\!_{2N+1} = \sin k\pi = 0$. Therefore Eq. (3.15.13) becomes

$$0 = \binom{2N+1}{1}\cot^{2N} x_k - \binom{2N+1}{3}\cot^{2N-2} x_k + \cdots + \binom{2N+1}{2N+1}(-1)^N \tag{3.15.14}$$

for $k = 1, 2, \ldots, N$. The numbers x_k are distinct numbers in the interval $0 < x_k < \pi/2$. The numbers $t_k = \cot^2 x_k$ are also distinct numbers in this interval. What Eq. (3.15.14) means is that, the numbers t_k are the roots of the following Nth degree polynomial:

$$p(t) = \binom{2N+1}{1}t^N - \binom{2N+1}{3}t^{N-1} + \cdots + \binom{2N+1}{2N+1}(-1)^N \tag{3.15.15}$$

Now, Vieta's formula (Sect. 2.29.7) links everything together: the sum of all the roots is the negative of the ratio of the second coefficient and the first one:

$$\sum_{k=1}^{N} t_k = \frac{\binom{2N+1}{3}}{\binom{2N+1}{1}} = \frac{(2N)(2N-1)}{6} \tag{3.15.16}$$

Replacing t_k by its definition and noting that $x_k = {}^{k\pi}/_{2N+1}$, we get what is needed in Eq. (3.15.10):

$$\sum_{k=1}^{N} \cot^2 \frac{k\pi}{2N+1} = \frac{(2N)(2N-1)}{6} \tag{3.15.17}$$

With that sum, Eq. (3.15.10) is simplified to:

$$\lim_{N\to\infty} \frac{\pi^2}{(2N+1)^2}\frac{(2N)(2N-1)}{6} < S < \lim_{N\to\infty} \frac{\pi^2}{(2N+1)^2}\frac{(2N)(2N-1)}{6} \tag{3.15.18}$$

Or

$$\frac{\pi^2}{6} < S < \frac{\pi^2}{6}$$

Thus, S is sandwiched between $\pi^2/6$ and $\pi^2/6$, it must be $\pi^2/6$. And we come to the end of the amazing proof due to the great Cauchy.

3.16 Trigonometric Equations

This chapter would be incomplete if we left out trigonometry equations; those similar to finding x such that $\sin x + \cos x = 1$ for example. Basically solving trigonometry functions involves using some trigonometry identities, some algebra tricks, and the fact that $-1 \leq \sin x, \cos x \leq 1$.

Let's start with solving this equation $\tan^2 x + \cos 4x = 0$. There are certainly more than one way to solve this, I present one solution only. The idea is to *look at the equation carefully* and ask why $\tan^2 x$ and not $\tan^3 x$ and why $\cos 4x$ not $\cos 5x$. With that observation, we already half solve this problem: $\tan^2 x$ can be converted to $\cos 2x$, and certainly $\cos 4x$ can also be converted to $\cos 2x$. Eventually we only have an equation of $\cos 2x$:

$$\begin{aligned}
\frac{\sin^2 x}{\cos^2 x} + 2\cos^2 2x - 1 &= 0 \\
\frac{1-\cos 2x}{1+\cos 2x} + 2\cos^2 2x - 1 &= 0 \\
\frac{1-u}{1+u} + 2u^2 - 1 &= 0 \quad (u = \cos 2x) \\
u(u^2 + u - 1) &= 0
\end{aligned}$$

And the remaining is piece of cake, isn't it[34]?

We present another trigonometry equation and that's it, no more. You can enjoy solving them, but honestly they do not teach you more on mathematics (except you would become more fluent in manipulating algebraic expressions, which is an important skill after all). The equation is $\sin^{100} x + \cos^{100} x = 1$. Obviously, as the equation involves too high power (100th power), it requires a special technique.

Of course you do not want to mess up with the left-hand side of the equation (it is too messy doing so). So, we have to play with the RHS which is the number 1. But if you know trigonometry, you know that $1 = \sin^2 x + \cos^2 x$. Thus, the equation becomes, and now you can massage the equation, and hope something useful would appear:

$$\begin{aligned}
\sin^{100} x + \cos^{100} x &= \sin^2 x + \cos^2 x \\
\underbrace{\sin^2 x(1 - \sin^{98} x)}_{\geq 0} + \underbrace{\cos^2 x(1 - \cos^{98} x)}_{\geq 0} &= 0
\end{aligned}$$

When the sum of two non-negative terms is zero, it is only possible when the two terms are both zeros:

$$\sin^2 x(1 - \sin^{98} x) = 0, \quad \cos^2 x(1 - \cos^{98} x) = 0$$

which requires that $\sin x = 0$, $\cos x = \pm 1$ or $\cos x = 0$, $\sin x = \pm 1$. And now you can solve the scary-looking equation $\sin^{2020} x + \cos^{2020} x = 1$.

We think we should pay less attention on solving trigonometry equations because up to this point we still do not know how to compute $\sin x$ for any given x. All we know is just Table 3.3. When we use a calculator and press $\sin 0.1$ to get 0.09983341664, how does the calculator compute that? See Sect. 3.21 for solution, sort of.

[34] Solving that equation yields $u = \{0, (-1 \pm \sqrt{5})/2\}$. As $u = \cos 2x$ is always larger than -1, we do not accept $u = -(1 + \sqrt{5})/2$.

Some trigonometry problems.

1. Compute the sum $\sin^2 10° + \sin^2 20° + \sin^2 30° + \sin^2 40° + \cdots + \sin^2 90°$.
2. Solve $8 \sin x \cos^5 x - 8 \sin^5 x \cos x = 1$.
3. Compute $\cos 36° - \cos 72°$.
4. Solve $\cos^2 x + \cos^2 2x + \cos^2 3x = 1$ (IMO 1962).
5. Prove $\cos \pi/7 - \cos 2\pi/7 + \cos 3\pi/7 = 1/2$ (IMO 1963).

Answers are 5, 7.5° and 0.5, respectively. Hints: for the first problem, follow Gauss (see Sect. 2.6.1 in case you have missed it) by grouping two terms together so that something special appears. For the third problem, do not first find $\cos 36°$ and then $\cos 36° - \cos 72°$. With little massage, you can compute $\cos 36° - \cos 72°$ directly. For the final problem, remember how we computed $\sin \pi/5$?

3.17 Laws of Cosines and Sines

For right (right-angled) triangles, we have the famous Pythagoras theorem $c^2 = a^2 + b^2$. For oblique triangles, the generalized Pythagoras theorem extends the Pythagoras theorem. For a triangle with sides a, b and c, see Fig. 3.49a, the generalized Pythagoras theorem states that

$$\begin{aligned} a^2 &= b^2 + c^2 - 2bc \cos A \\ b^2 &= c^2 + a^2 - 2ca \cos B \\ c^2 &= a^2 + b^2 - 2ab \cos C \end{aligned} \tag{3.17.1}$$

In Fig. 3.49a, the proof for $b^2 = c^2 + a^2 - 2ca \cos B$ is obtained by applying the Pythagoras theorem for the right triangle ADC.[35] Now, we do some checking for the newly derived formula. First, when B is a right angle, its cosine is zero, and we get the familiar $b^2 = a^2 + c^2$ again. Second, the term $2ca \cos B$ has a dimension of length squared, which is correct (if that term was $2a^2 b \cos B$, the formula would be wrong because we cannot add a square of length with a cubic of length. We cannot add area with volume). There is no need to prove the other second formula. As a, b, c are symmetrical, from $b^2 = c^2 + a^2 + 2ca \cos B$ we can get the other two by permuting the variables: $a \to b, b \to c, c \to a$.

The generalized Pythagoras theorem provides us an easy way to prove the converse of the Pythagorean theorem: if $c^2 = a^2 + b^2$, then $\angle C = 90°$. In any triangle,

[35] For this case we get $b^2 = c^2 + a^2 + 2ca \cos(\pi - B)$, but note that $\cos(\pi - B) = -\cos B$.

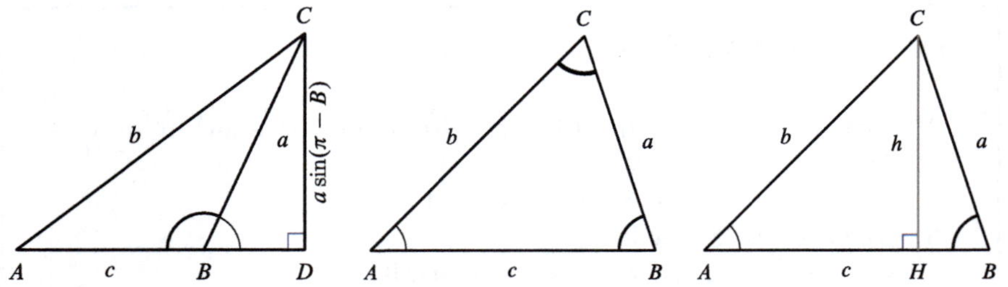

(a) (b) (c)

Fig. 3.49 Proof of the generalized Pythagoras theorem (**a**), the sine law (**b**) and proof (**c**)

we have $c^2 = a^2 + b^2 - 2ab\cos C$, now given that $c^2 = a^2 + b^2$, must then have $2ab\cos C = 0$, and since $a, b > 0$, this leads to $\cos C = 0$ or $\angle C = 90°$.

The generalized Pythagoras theorem is also known as the law of cosines for it relates the lengths of the sides of a triangle to the cosine of one of its angles. If there are law of cosines, then it should exist the law of sines. This law is written as (Fig. 3.49b)

$$\frac{a}{\sin A} = \frac{b}{\sin B} = \frac{c}{\sin C} \tag{3.17.2}$$

The proof is simple: see Fig. 3.49c. In the two right triangles ACH and BCH, computing $h = b\sin A = a\sin B$. Hence, $a/\sin A = b/\sin B$. The law of sines can be used to compute the remaining sides of a triangle when two angles and a side are known—a technique known as triangulation.

In the law of sines we have ${}^{a}/_{\sin A} = {}^{b}/_{\sin B} = {}^{c}/_{\sin C}$, but what is this number ${}^{a}/_{\sin A}$? It must be the length of some segment (why?) and it turns out to be the diameter of the triangle's circumcircle. This result dates back to Ptolemy (see the left diagram in Fig. 3.50 for a proof: considering the right triangle OHB, we have immediately ${}^{c}/_{\sin C} = 2r$). Now, if we assume $2r = 1$, then we get what is shown in the right diagram. Then, the law of sines simply states that each side of a triangle inscribed in a unit circle is equal to the sine of the opposite angle. Now, we can understand why Ptolemy defined sine as the length of a chord.

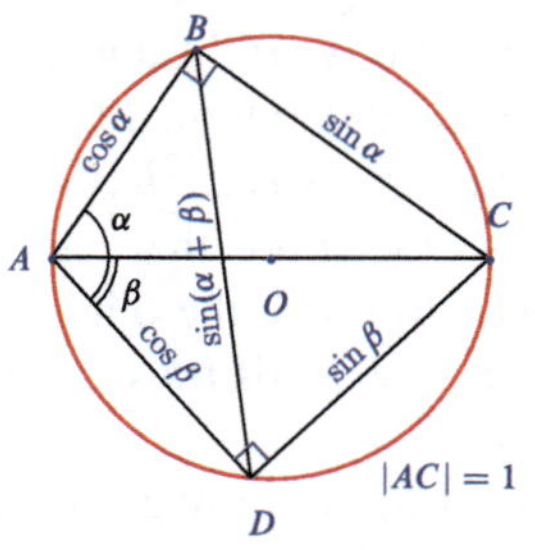

Ptolemy's discovery of the $\sin(a + b)$ **identity**. With the law of sines, and the Ptolemy theorem, we can now explain the discovery of the trigonometric identity $\sin(\alpha + \beta)$. To use the Ptolemy theorem, we consider a unit circle, and a cyclic

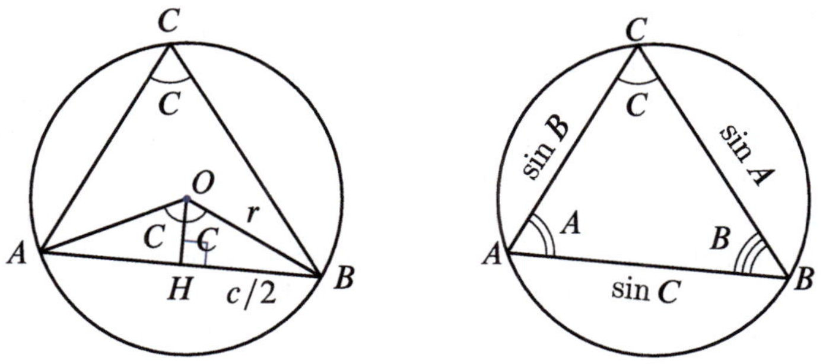

Fig. 3.50 Proof of the law of sines: $c/\sin C = 2r$ (left) and when $2r = 1$, the chords become the sines (right). We have $\angle AOB = 2C$ thanks to the central angle theorem

quadrilateral $ABCD$, but with $AC = 1$ being the circle's diameter (to have right triangles). Let $\angle DAC$ and $\angle CAB$ be α and β, respectively. As $AC = 1$, all the segments in the diagram are associated with either sine or cosine. And the Ptolemy theorem tells us that $AC \cdot BD = AB \cdot CD + BC \cdot AD$, which becomes $\sin(\alpha + \beta) = \sin\alpha\cos\beta + \sin\beta\cos\alpha$. Noting that $BD = \sin(\alpha + \beta)$ is due to the law of sines (Fig. 3.50).

3.18 Graph of Trigonometric Functions

Even though I postpone the discussion on the concept of mathematical functions to Sect. 4.2, I present here the graph of some trigonometry functions, mostly for completeness of this chapter. Loosely speaking a function is a device that receives a number (mostly a real number)—called the input—and returns another number, the output. If we denote by x the input of the sine function, we write $y = \sin x$. By varying x from negative infinity to positive infinity (of course, practically just a finite interval was considered, here is $[-4\pi, 4\pi]$), we compute the corresponding ys, and plot all the points (x, y) to get the graphs shown in Fig. 3.51.

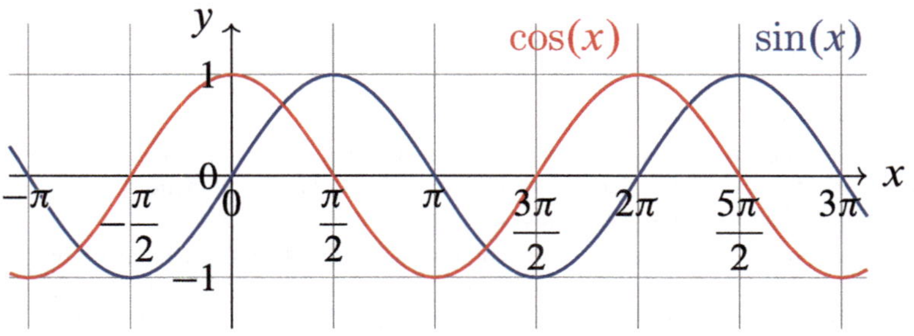

Fig. 3.51 Graphs of sine and cosine functions. Made with `tikz`

OK. We have used technology to do the plotting for us (as it does a better job than human beings), but we should be able to "read" information from the graph. No computer is able to do that. First, the two graphs are confined in the interval $[-1, 1]$ (because both sine and cosine are smaller/equal to 1 and larger/equal to -1). Second, where the sine is maximum or minimum the cosine is zero and vice versa (why? Which identity is responsible for this?). Third, by focusing on the interval $[0, 4\pi]$, one can see that the sine starts with zero, increases to 1 (at $\pi/2$), then decreases to zero (at π), continues decreasing until it gets to -1, then increases back to zero (at 2π). After that, the graph repeats. Thus, sine is a *periodic function*. And its period T is 2π. The cosine function has the same period. The period of a periodic function $f(x)$ is the smallest number T such that[36]

$$f(x+T) = f(x), \quad \forall x \tag{3.18.1}$$

Why smallest? This is because if T is the period, then $2T, 3T, \ldots$ are also periods. So, we just need to use the smallest.

The graph of the tangent function is given in Fig. 3.52. It can be seen that the tangent function is periodic with a period of π i.e., $\tan(x+\pi) = \tan x$, which can be proved using trigonometry identity $\tan(a+b) = {}^{\tan a + \tan b}/_{1 - \tan a \tan b}$. As $\tan x = {}^{\sin x}/_{\cos x}$, the function is not defined for angles $\bar{x}$ such that $\cos\bar{x} = 0$. Solving this equation yields $\bar{x} = \pi/2 + k\pi$, $k = 0, \pm 1, \pm 2, \ldots$ The vertical lines at $\bar{x}$ are the vertical asymptotes of the tangent curve.

We now shift our discussion to the graph of a nice function. It is given by

$$y(x) = \frac{\sin x}{x} = \frac{1}{x}\sin x$$

which is obtained by taking the sine function and the $1/x$ function and multiply them. But hey, why this function? Because it shows up a lot in mathematics.[37] For example, in calculus we need to find the derivative of the sine function. Here is what we do, by considering the function $y = \sin t$:

[36] As is always the case in mathematics, whenever we have a new object (herein the period), we have theorems (facts) on them. Here is one: if $f_1(x)$ and $f_2(x)$ are two functions of the same period T, then the function $\alpha f_1(x) + \beta f_2(x)$ also has the period of T. Prove it.

[37] Actually this function is named the sinc function by the British mathematician Philip Woodward (1919–2018). In his 1952 article "Information theory and inverse probability in telecommunication", in which he said that the function "occurs so often in Fourier analysis and its applications that it does seem to merit some notation of its own". In Sect. 7.12.4, I present a Fourier analysis, routinely done by electrical engineers, in which this function shows up.

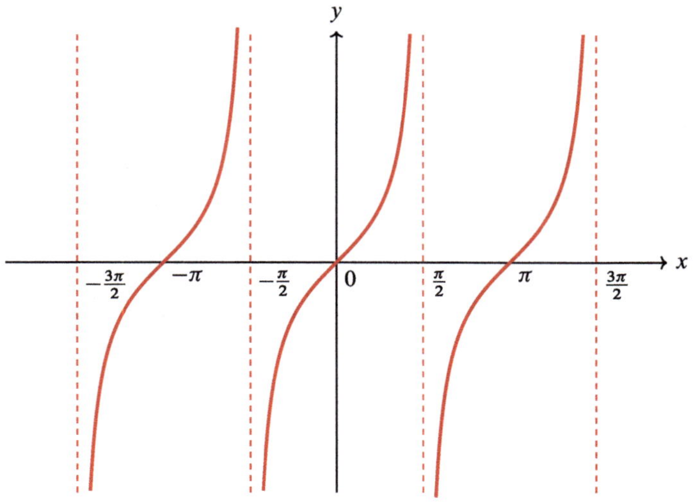

Fig. 3.52 Graphs of the tangent function. The vertical dashed lines at $\bar{x} = \pi/2 + k\pi$ are the vertical asymptotes of the tangent curve. An asymptote is a line that the graph of a function approaches as either x or y goes to positive or negative infinity. There are three types of asymptotes: vertical, horizontal, and oblique

$$
\begin{aligned}
(\sin t)' &= \lim_{x \to 0} \frac{\sin(t+x) - \sin t}{x} \\
&= \sin t \lim_{x \to 0} \left(\frac{\cos x - 1}{x} \right) + \cos t \lim_{x \to 0} \frac{\sin x}{x}
\end{aligned}
$$

I refer to Sect. 4.4.8 if something was not clear. If this is not enough to get your attention, note that the function $\sin x/x$ is very popular in signal processing. So if you are to enroll in an electrical engineering course, you will definitely see it.

In Fig. 3.53 I plot $\sin x$, $1/x$ and $\sin x/x$. What can we observe from the graph of $f(x) = \sin x/x$? First, it is symmetrical with respect to the y-axis (this is because $f(-x) = f(x)$, or as mathematicians call it, it is an even function). Second, similar to $\sin x$, $\sin x/x$ is also oscillatory. However not between -1 and 1. The amplitude of this oscillation is decreasing when $|x|$ gets larger. Can we find how this amplitude depends on x precisely?

Yes, we can:

$$
-1 \leq \sin x \leq 1 \Longrightarrow -\frac{1}{x} \leq \left(\frac{1}{x} \right) (\sin x) \leq \frac{1}{x}
$$

This comes from the fact that if $a \leq b$, and $c > 0$ then $ac \leq bc$. So, the above inequality for $\sin x/x$ works only for $x > 0$. But due to the symmetry of this function, the inequality holds for $x < 0$ as well. Now we see that $\sin x/x$ can never exceed

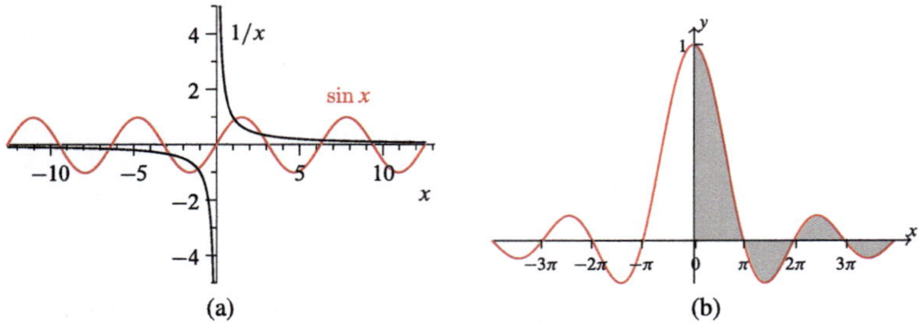

Fig. 3.53 Graph of $\sin x$, $1/x$ (**a**) and graph of $\sin x/x$ (**b**). What is interesting is that the area of the shaded region when $x \to \infty$ is nothing else but exactly half of π. See Dirichlet integral in Eq. (4.7.45) to see why

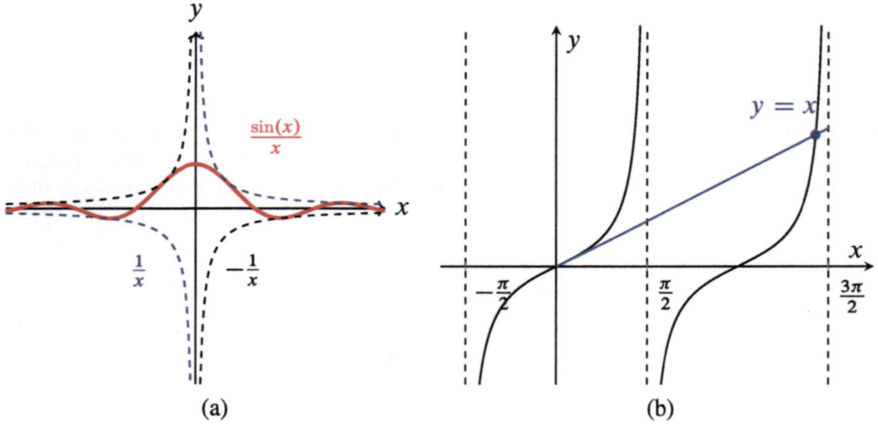

Fig. 3.54 Envelops of $\sin x/x$ are $1/x$ and $-1/x$ (**a**) and solving $\tan x = x$ graphically (**b**)

$1/x$ and $-1/x$; these two functions are therefore called the envelops of $\sin x/x$, see Fig. 3.54a.

Is it everything about $\sin x/x$? No, no. There is at least one more thing: where are the stationary points of this function? To that, we need to use calculus as algebra or geometry is not powerful enough for this task. From calculus we know that at a stationary point the derivative of the function vanishes:

$$f'(x) = \frac{x \cos x - \sin x}{x^2} \implies f'(x) = 0 : \tan x = x$$

How we're going to solve this equation of $\tan x = x$ or $g(x) := \tan x - x = 0$? Well, we do not know. So we fall back to a simple solution: the solutions of $\tan x = x$ are the intersection points of the curve $y = \tan x$ and the line $y = x$. From Fig. 3.54b we see that there is one solution $x = 0$, and infinitely more solutions close to $3\pi/2, 5\pi/2, \ldots$

But the graphical method cannot give accurate solutions. To get them we have to use approximate methods and one popular method is the Newton-Raphson method

Table 3.4 The first four solutions of $\tan x = x$ obtained with the Newton-Raphson method

n	x	x
1	0.00000000	0
2	4.49340946	$2.86\pi/2$
3	7.72525184	$4.92\pi/2$
4	10.9041216	$6.94\pi/2$

described in Sect. 4.5.4, see Eq. (4.5.8). In this method one begins with a starting point x_0, and gets better approximations via:

$$x_{n+1} = x_n - \frac{g(x_n)}{g'(x_n)} = x_n - \frac{\tan x_n - x_n}{1/\cos^2 x_n - 1}, \quad n = 0, 1, 2, \ldots \tag{3.18.2}$$

If you program this and use it you will see that the method sometimes blows up i.e., the solution is a very big number. This is due to the tangent function which is very large for x such that $\cos x = 0$. So, we better off use this equivalent but numerically better $g(x) := x \cos x - \sin x$:

$$x_{n+1} = x_n - \frac{x_n \cos x_n - \sin x_n}{-x_n \sin x_n} \tag{3.18.3}$$

With this and starting points close to 0, $3\pi/2$, $5\pi/2$, and $7\pi/2$ we get the first four solutions given in Table 3.4. The third column gives the solutions in terms of multiples of $\pi/2$ to demonstrate the fact that the solutions get closer to the asymptotes of the graph of the tangent function.

Here are two lessons learned from studying the graph of the nice function $\sin x/x$:

- Not all equations can be solved exactly. However, one can always use numerical methods to solve any equation approximately. Mathematicians do that and particularly scientists and engineers do that all the time;
- Mathematically, $\tan x - x = 0$ is equivalent to $x \cos x - \sin x = 0$. However, it is easier to work with the latter, because all involved functions are defined for all x; $\tan x$ is not a nice function to work with: it is a discontinuous function, at the vertical asymptotes! So some mathematical objects are easier to work with than others, exactly similar to human beings.

Period of $\sin 2x + \cos 3x$. The problem that we're now interested in is what is the period of a sum of trigonometric functions? Specifically, $\sin 2x + \cos 3x$. There is one easy way: plotting the function. Figure 3.55 reveals that the period of this function is 2π.

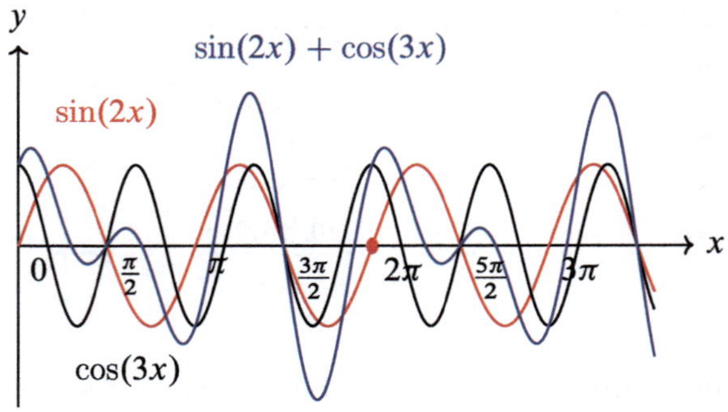

Fig. 3.55 Plot of $\sin 2x$ (red), $\cos 3x$ (black) and $\sin 2x + \cos 3x$ (blue)

Of course, there is another way without plotting the function. We know that the period of $\sin x$ is 2π, and thus the period of $\sin 2x$ is $2\pi/2 = \pi$.[38] Similarly, the period of $\cos 3x$ is $2\pi/3$. Therefore, we have

$$\sin 2x \text{ repeats at:} \pi, 2\pi, 3\pi, \ldots$$
$$\cos 3x \text{ repeats at: } \frac{2\pi}{3}, 2\frac{2\pi}{3}, 3\frac{2\pi}{3}, \ldots$$

Thus, $\sin 2x + \cos 3x$ will repeat the first time (considering positive x only) when $x = 2\pi$, that is the period of this function. What is the point of considering the sum of sine and cosine? In Sect. 4.19 about Fourier series, we do just that. And Fourier series are everywhere in both pure and applied maths.

3.19 Hyperbolic Functions

In this section I present hyperbolic functions $\sinh x$, $\cosh x$ as they are similar to the trigonometric functions discussed previously. There are several ways to introduce these functions, but for now I decided to use the decomposition of a function into an even part and an odd part.

Any function, $f(x)$, can be decomposed into two parts as

$$f(x) = \frac{1}{2}[f(x) + f(-x)] + \frac{1}{2}[f(x) - f(-x)] \tag{3.19.1}$$

in which the first term is an even function, i.e., $g(-x) = g(x)$ and the second is an odd function i.e., $g(-x) = -g(x)$ (see Sect. 4.2.1).

Applying this decomposition to the exponential function $y = e^x$, we have:

[38] If this is not clear, here is one way to explain. The function $y = \sin 2x$ is obtained by horizontally shrinking $y = \sin x$ by a factor of 2 (Fig. 4.8). And thus it has a period as half as that of $\sin x$.

$$e^x = \frac{1}{2}\left[e^x + e^{-x}\right] + \frac{1}{2}\left[e^x - e^{-x}\right] \tag{3.19.2}$$

from that we define the following two functions:

$$\boxed{\begin{aligned}\sinh x &= \frac{1}{2}\left(e^x - e^{-x}\right)\\ \cosh x &= \frac{1}{2}\left(e^x + e^{-x}\right)\end{aligned}} \tag{3.19.3}$$

They are called the hyperbolic sine and cosine functions, which explain their symbols. I explain the origin of these names shortly. First, the graphs of these two functions together with $y = 0.5e^x$ and $y = 0.5e^{-x}$ are shown in Fig. 3.56a. The first thing we observe is that for large x, the hyperbolic cosine function is similar to $y = 0.5e^x$, this is because $0.5e^{-x} \to 0$ when x is large. Second, the hyperbolic cosine curve is always above that of $y = 0.5e^x$. Third, $\cosh x \geq 1$. This can be explained using the Taylor series of e^x and e^{-x} (refer to Sect. 4.15.8 if you're not familiar with Taylor series):

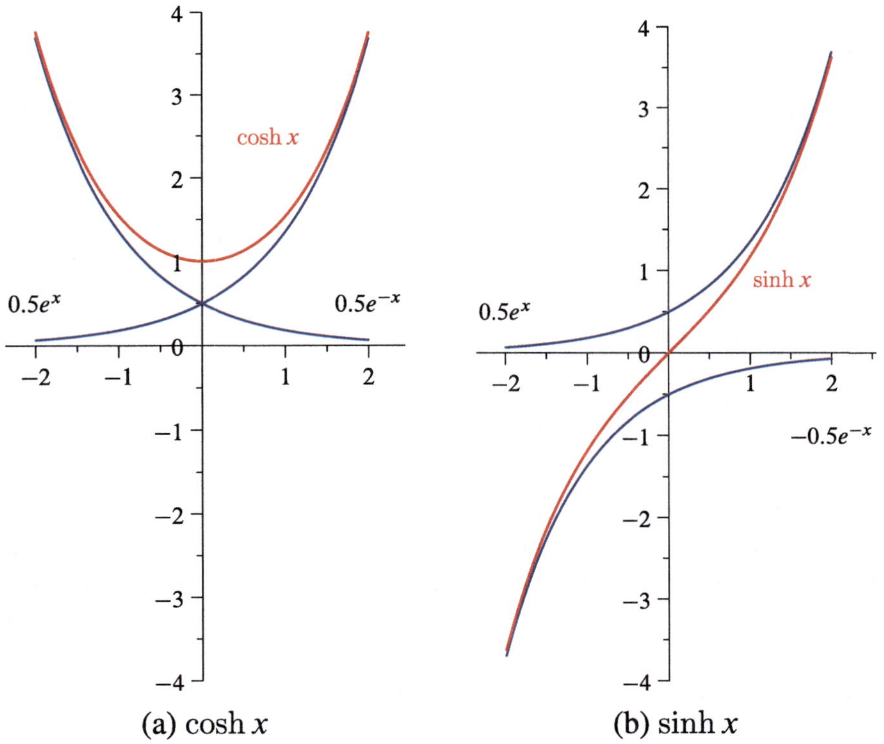

Fig. 3.56 Plot of the hyperbolic sine and cosine functions along with their exponential components

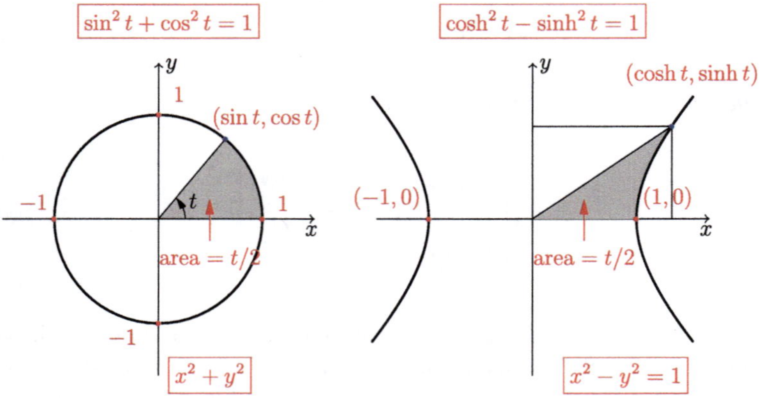

Fig. 3.57 Trigonometric functions $\sin x, \cos x$ are related to a unit circle; they are *circular trigonometry functions*. On the other hand, $\sinh x$ and $\cosh x$ are related to the right hyperbola $x^2 - y^2 = 1$; they are hyperbolic trigonometry functions. The meaning of the parameter t is that it is twice the area of the shaded region. For the circle, that is easy to see. For the hyperbola, we need calculus. Check this youtube channel out if you're interested in the detail

$$\begin{cases} e^x \approx 1 + x + \frac{x^2}{2!} + \frac{x^3}{3!} + \frac{x^4}{4!} + \cdots \\ e^{-x} \approx 1 - x + \frac{x^2}{2!} - \frac{x^3}{3!} + \frac{x^4}{4!} + \cdots \end{cases} \implies \frac{e^x + e^{-x}}{2} \approx 1 + \frac{x^2}{2!} + \frac{x^4}{4!} + \cdots \geq 1$$

From Eq. (3.19.3), it can be seen that $\cosh^2 x - \sinh^2 x = 1$. And we have more identities bearing similarity with trigonometry identities that we're familiar with. For example, we have

$$\begin{aligned} \cosh(a + b) &= \cosh a \cosh b + \sinh a \sinh b \\ \sinh(a + b) &= \sinh a \cosh b + \cosh a \sinh b \end{aligned} \tag{3.19.4}$$

The proof is based on the following results:

$$\begin{cases} e^a = \cosh a + \sinh a \\ e^{-a} = \cosh a - \sinh a \end{cases}, \quad \cosh(a + b) = \frac{e^{a+b} + e^{-a-b}}{2}, \quad \sinh(a + b) = \frac{e^{a+b} - e^{-a-b}}{2}$$

Why called hyperbolic trigonometry? Remember the parametric equation of a unit circle centered at the origin[39]? It is given by $x = \sin t$, $y = \cos t$. Similarly, from the identity $\cosh^2 t - \sinh^2 t = 1$, the hyperbola $x^2 - y^2 = 1$ is parameterized as $x = \cosh t$ and $y = \sinh t$. That explains the name "hyperbolic functions" (Fig. 3.57). Not sure what is a hyperbola? Check out Sect. 4.1.

Another derivation of hyperbolic functions. Start with Euler's identity $e^{i\theta} = \cos\theta + i \sin\theta$ but written for $\theta = x$ and $\theta = -x$:

[39] Refer to Sect. 4.2.6 for a discussion on parametric curves.

$$e^{ix} = \cos x + i \sin x$$
$$e^{-ix} = \cos x - i \sin x$$

We then have (adding the above two equations):

$$\cos x = \frac{e^{ix} + e^{-ix}}{2} \tag{3.19.5}$$

Now we consider a complex variable $z = x + iy$, and use z in the above equation:

$$\begin{aligned}
\cos(x+iy) &= \frac{e^{i(x+iy)} + e^{-i(x+iy)}}{2} \\
&= \frac{e^{ix-y} + e^{-ix+y}}{2} = \frac{e^{ix}e^{-y} + e^{-ix}e^{y}}{2} \\
&= \frac{(\cos x + i \sin x)e^{-y} + (\cos x - i \sin x)e^{y}}{2} \\
&= \cos x \left(\frac{e^{y} + e^{-y}}{2}\right) - i \sin x \left(\frac{e^{y} - e^{-y}}{2}\right)
\end{aligned}$$

And you see the hyperbolic sine/cosine show up! With our definition of them in Eq. (3.19.3), we get this $\cos(x + iy) = \cos x \cosh y - i \sin x \sinh y$. And a similar equation is awaiting for sine:

$$\sin(x + iy) = \sin x \cosh y + i \cos x \sinh y$$

Putting them together,

$$\cos(x + iy) = \cos x \cosh y - i \sin x \sinh y$$
$$\sin(x + iy) = \sin x \cosh y + i \cos x \sinh y$$

And they are quite similar to the real trigonometry identities of $\sin(a + b)$ and $\cos(a + b)$! Now putting $x = 0$ in the above, we get

$$\cos(iy) = \cosh y, \quad \sin(iy) = i \sinh y \tag{3.19.6}$$

which means that the cosine of an imaginary angle is real but the sine of an imaginary angle is imaginary.

Can sine/cosine be larger than one? We all know that for real angles x, $|\sin x| \leq 1$. But for complex angles z, might we have $\cos z > 1$? Let's find z such that $\cos z = 2$. We start with

$$\cos(x + iy) = \cos x \cosh y - i \sin x \sinh y$$

Then, $\cos z = 2$ is equivalent to

$$\cos(x + iy) = 2 \iff \cos x \cosh y - i \sin x \sinh y = 2 + 0i$$

And we obtain the following equations to solve for x, y:

$$\begin{cases} \cos x \cosh y = 2 \\ \sin x \sinh y = 0 \end{cases}$$

From the second equation we get $\sin x = 0$; noting that we're not interested in $\sinh y = 0$ or $y = 0$ as we're looking for complex angles not real ones. With $\sin x = 0$, we then have $\cos x = \pm 1$. But we remove the possibility of $\cos x = -1$, as from the first equation we know that $\cos x > 0$ as $\cosh y > 0$ for all y. So, we have $\cos x = 1$ (or $x = 2n\pi$), and with that we have $\cosh y = 2$:

$$\cosh y = 2 \iff \frac{e^y + e^{-y}}{2} = 2$$

of which solutions are $y = \ln(2 \pm \sqrt{3})$. Finally, the angle we're looking for is

$$z = 2n\pi + i \ln(2 \pm \sqrt{3})$$

These hyperbolic functions are the creation of the human minds, but again they model satisfactorily natural phenomena. For example in Sect. 8.2 we shall demonstrate that the hyperbolic sine is exactly the shape of a hanging chain.[40]

3.20 Applications of Trigonometry

I present some applications of trigonometry in this section. Some materials are borrowed from [37], which is a great book to read. We shall see more applications of trigonometric functions in later chapters. Section 3.20.1 presents how trigonometry is used to measure the radius of the earth. And Sect. 3.20.2 treats the problem of determining the distance between two points on the earth.

40 A hanging chain or cable is a parabola-like shape that a cable assumes under its own weight when supported only at its end.

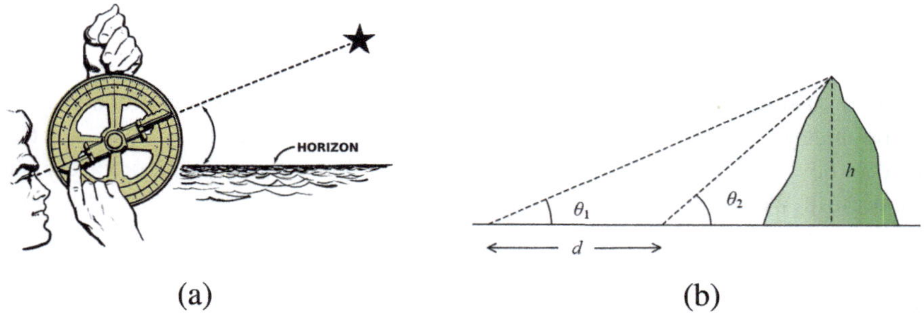

Fig. 3.58 Al-Biruni's measurement of the height of a mountain

3.20.1 Measuring the Earth

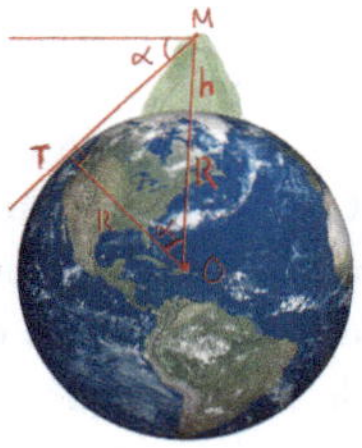

It is obvious that without modern technologies such as GPS, measuring the radius of the earth must be done indirectly. It was Abu Reyhan Al-Biruni, the tenth century Islamic mathematical genius, who combined trigonometry and algebra to achieve this very numerical feat. He first measured the height of a hill near the Fort of Nandana in today's Punjab province of Pakistan. He then climbed the hill to measure the horizon. Using trigonometry and algebra, he got the value equivalent to 3928.77 English miles, which is about 99 percent close to today's radius of the earth. Using the law of sines, Eq. (3.17.2), for the right triangle OTM, we have

$$\frac{h+R}{\sin \pi/2} = \frac{R}{\sin(\pi/2 - \alpha)} \Longrightarrow R = \frac{h \cos \alpha}{1 - \cos \alpha}$$

Al-Biruni tells us that the mountain height is 305.1 meters and the angle α is $0.5667°$. Using these data and a calculator (which he did not have), we find that $R = 6\,238$ km. And the earth circumference is thus 39 194 km. Its actual value is 40 075 km (from Google).

How Al-Biruni measured h and α? First, he measured the angle of elevation of a mountain top at two different points lying on a straight line using an astrolabe, θ_1 and θ_2. Then he measured the distance between these two points d (Fig. 3.58).

Finally, simple application of trigonometry gives us the height h:

$$h = \frac{d \tan \theta_1 \tan \theta_2}{\tan \theta_2 - \tan \theta_1}$$

3.20.2 *Charting the Earth*

To demonstrate how ancient Greek astronomers (also mathematicians) used trigonometry to find distances between two points on the earth surface, we first need to have a coordinate system for the earth so that we can locate precisely any point on it. We know that our earth is a perfect sphere. Two notes here: first how you knew that the earth is a sphere? Because other people said so? It's better to find out why for yourself. Second, the earth is not a perfect sphere; we are making an approximation.

On this sphere there are three special points: the center O, the north pole N and the south pole S. We draw many circles with center at O and pass through S and N (see Fig. 3.59). Each half of such circles is called a *line of longitude* or meridian. Among many such meridians, we define the prime meridian which is the meridian at which longitude is defined to be 0°. The prime meridian divides the sphere into two equal parts: the eastern and western parts.

All points on a meridian have the same longitude, which leads to the introduction of another coordinate. To this end, parallel circles perpendicular to the meridians are drawn on the sphere. One special parallel is the equator which divides the earth sphere into two equal parts: the northern and southern part.

Now we can define precisely what longitude and latitude mean. Referring to Fig. 3.59, we first define a special point A which is the intersection of the equator and the prime meridian. Now, the longitude is the angle AOB in degrees measured from the prime meridian. Thus a longitude is an angle ranging from 0°E to 180°E or 0°W to 180°W. Similarly, the latitude is the angle BOC measured from the equator up (N) or down (S), ranging from 0°N to 90° or 0°S to 90°S.

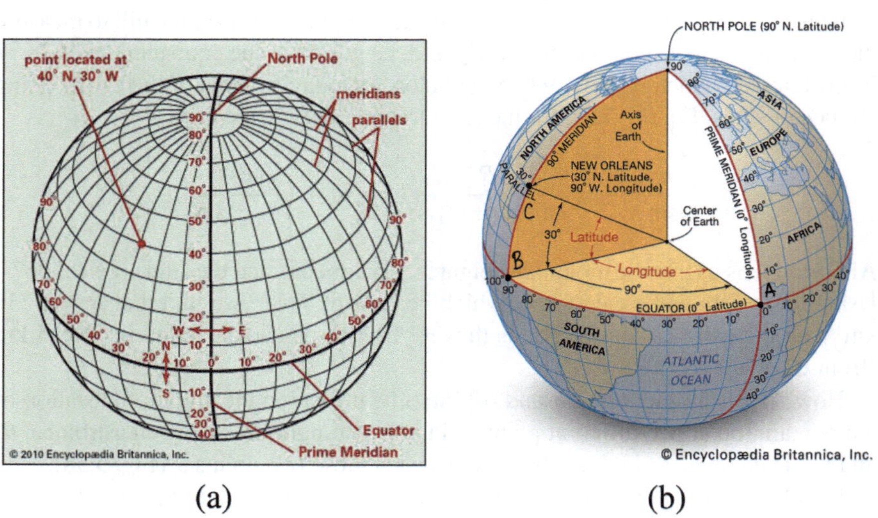

Fig. 3.59 Longitudes and latitudes: *Source* from Britannica

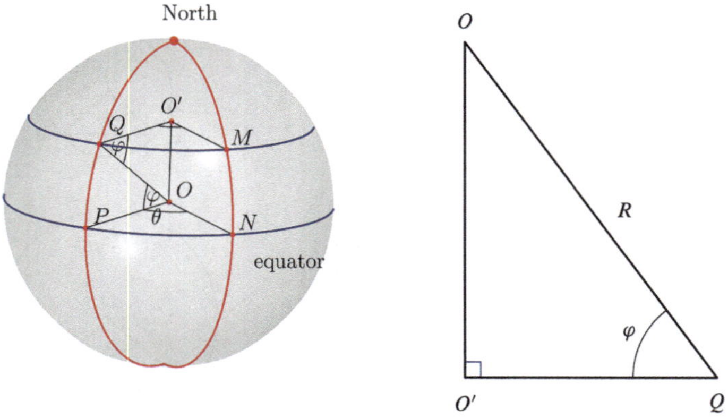

Fig. 3.60 Two cities located at P and Q having the same longitude. The arc PQ, which is part of a great circle. A great circle is the intersection of a sphere with a central plane, a plane through the center of that sphere. Noting that OP is parallel to $O'Q$

Considering two cities located at P and Q having the same longitude, P is on the equator (Fig. 3.60). Now assume that the city located at Q has a latitude of φ. The question we're interested in is: how far is Q from P (how far from the equator)? The answer is the arc PQ, which is part of the great circle of radius R where R being the radius of the earth. Thus:

$$PQ = \pi R \frac{\varphi}{180^\circ}$$

Now considering two cities located at Q and M having the same latitude (Fig. 3.60). What is the distance between them traveling along this latitude? This is the arc QM of the small circle centered at O'. If we can determine the radius of this small circle, then we're done. This radius is $O'Q = R\cos\varphi$ (see the right picture in Fig. 3.60). Then the distance QM is given by

$$QM = \pi O'Q \frac{\theta}{180^\circ} = \pi R \cos\varphi \frac{\theta}{180^\circ}$$

where θ is the difference (assuming that these two points are either on the eastern or western part) of the longitudes of Q and M. But is this distance the shortest path between Q and M? No! The shortest path is the great-circle distance. The great-circle distance or spherical distance is the shortest distance between two points on the surface of a sphere, measured along the surface of the sphere. At this moment, we accept this with faith.

Figure 3.61 illustrates how to find such a great-circle distance. The first step is to find $r = O'Q$ as done before. Then in $\Delta O'QM$ using the cosine law, i.e., Eq. (3.17.1), we can compute the straight line distance between QM, denoted by d:

$$d^2 = r^2 + r^2 - 2r^2\cos\theta = 2r^2(1-\cos\theta) = 2R^2\cos^2\varphi(1-\cos\theta)$$

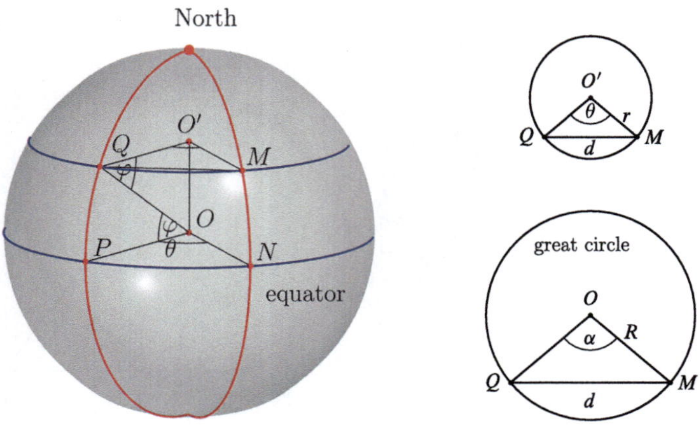

Fig. 3.61 Calculating the great-circle distance between Q and M. The great circle of which QM is an arc lies in the plane going through O and perpendicular to the vector $\overrightarrow{OM} \times \overrightarrow{OQ}$. This great circle is now shown in the sphere on the left picture. Note that $r = O'Q = R\cos\varphi$

Then using the cosine law again but now for the triangle OQM to determine the angle α:

$$d^2 = R^2 + R^2 - 2R^2\cos\alpha \Longrightarrow \alpha = \arccos\left(\frac{2R^2 - d^2}{2R^2}\right) = \arccos\left(1 - (1 - \cos\theta)\cos^2\varphi\right)$$

Knowing the angle of the arc QM in the great circle, it's easy to compute its length:

$$QM = \pi R\frac{\alpha}{180^\circ} = \pi R\frac{\arccos\left(1 - (1 - \cos\theta)\cos^2\varphi\right)}{180^\circ}$$

How about the great-circle distance between any two points on the surface of the earth? We do not know (yet) as it requires spherical trigonometry.

3.21 Infinite Series for Sine

In this section I present how the infinite series for the sine was developed long before calculus. The idea is to consider a sector of a unit circle and calculate its area using two ways. The first is an exact way, and the second is an approximation following Archimedes' idea of exhausting this area by infinitely many triangles (see Sect. 3.2.2 if you're not familiar with the method of exhaustion).

The exact area of the sector $OBAC$ is $\theta/2$ (see Fig. 3.62). This area is approximated as a sum of the area of triangles OBC, ABC, and ABD . We first compute the area of these triangles now. The area of the triangle OBC is easy (recall that the circle has a unit radius):

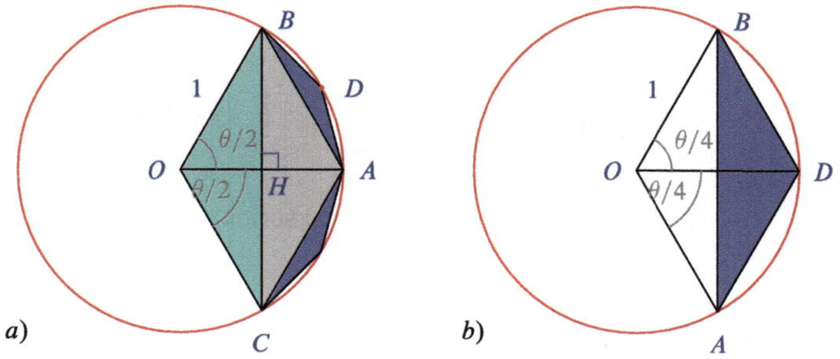

Fig. 3.62 Computing the approximate area of the sector $OBAC$ using the colored triangles. D is the point such that its angle with OA is $\theta/4$

$$OBC = \frac{1}{2}\left(2\sin\frac{\theta}{2}\right)\left(\cos\frac{\theta}{2}\right) = \frac{1}{2}\sin\theta$$

The area of the triangle ABC is also straightforward:

$$\begin{aligned} ABC &= \frac{1}{2}\left(2\sin\frac{\theta}{2}\right)\left(1-\cos\frac{\theta}{2}\right) \\ &= 2\sin\frac{\theta}{2}\sin^2\frac{\theta}{4} \quad \text{(double-angle formula for cosine)} \end{aligned} \tag{3.21.1}$$

We now use the approximation that $\sin x \approx x$ for small x: thus the area of ABC is approximated as $\theta^3/16$.

Next we compute the area of the triangle ABD. If we work with Fig. 3.62a then finding out this area might be hard, but if we rotate OAB $\theta/4$ counterclockwise we get the diagram in Fig. 3.62b, we'll see that $\triangle ABD$ is similar to $\triangle ABC$, but with $\theta/2$, thus its area is

$$ABD \approx \frac{1}{16}\left(\frac{\theta}{2}\right)^3 = \frac{\theta^3}{128}$$

Let's sum the areas of all these triangles (ABD counted twice), and we get

$$A \approx \frac{1}{2}\sin\theta + \frac{\theta^3}{16} + \frac{\theta^3}{64}$$

We can see a pattern here: the denominators are $16, 64 = 16 \times 4$, and thus the final formula for the area of the sector is

$$A \approx \frac{1}{2}\sin\theta + \frac{\theta^3}{16} + \frac{\theta^3}{64} + \frac{\theta^3}{256} + \cdots$$

The added terms account for the areas not considered in our approximation of the sector area. The red term looks familiar: it's a geometric series, so we can compute the red term, and get a more compact formula for A as:

$$\begin{aligned}
A &\approx \frac{1}{2}\sin\theta + \frac{\theta^3}{16} + \frac{\theta^3}{64} + \frac{\theta^3}{256} + \cdots \\
&= \frac{1}{2}\sin\theta + \theta^3\left(\frac{1}{16} + \frac{1}{64} + \frac{1}{256} + \cdots\right) \\
&= \frac{1}{2}\sin\theta + \frac{\theta^3}{12} \quad \text{(geometric series)}
\end{aligned}$$

Now we have two expressions for the same area, so we get the following equation, which leads to an approximation for $\sin\theta$:

$$\frac{\theta}{2} \approx \frac{1}{2}\sin\theta + \frac{\theta^3}{12} \Longrightarrow \boxed{\sin\theta \approx \theta - \frac{\theta^3}{6}}$$

Want to have an even better approximation? Let's apply $\sin x \approx x - x^3/6$ into Eq. (3.21.1) to get $ABC \approx \theta^3/128 - \theta^5/8192$ (the algebra is indeed a bit messy, thus I have used a CAS to help me doing this tedious algebraic manipulation, see Sect. 3.27). And we repeat what we have just done to get

$$\sin\theta \approx \theta - \frac{\theta^3}{6} + \frac{\theta^5}{120}$$

And of course we want to do better. What should be the next term after $\theta^5/120$? It is $-\theta^7/x$ with $x = 5040$:

$$\sin\theta \approx \theta - \frac{\theta^3}{6} + \frac{\theta^5}{120} - \frac{\theta^7}{5040}$$

Are you asking if there is any relation between those numbers in the denominators and those exponents in the nominators? There is! If you just played with factorial (Sect. 2.26.2) enough you would recognize that $6 = 3!$, $120 = 5!$ and of course 5040 must be 7! (pattern again!), thus

$$\boxed{\sin\theta \approx \frac{\theta^1}{1!} - \frac{\theta^3}{3!} + \frac{\theta^5}{5!} - \frac{\theta^7}{7!} + \cdots = \sum_{i=0}^{\infty}(-1)^i\frac{\sin^{2i+1}\theta}{(2i+1)!}} \tag{3.21.2}$$

Can we develop a similar formula for cosine? Of course. But for that we need to wait until the seventeenth century to meet Euler and Taylor who gave us a systematic way to derive infinite series for trigonometry functions. Refer to Sects. 4.15.6 and 4.15.8 if you cannot wait.

Why Eq. (3.21.2) was a significant development in mathematics? Remember that we have built a sine table in Sect. 3.7? It is useful but it is only for integral angles e.g. 30° or 45°. If the angle is not in the table, we have to use interpolation, which is of low accuracy. To have higher accuracy (and thus better solutions to navigation problems in the old days), ancient mathematicians had to find a formula that could give them the value of the sine for any angle. And Eq. (3.21.2) is one such formula; it involves only simple addition/subtraction/multiplication/division.

Remark For many mathematicians Eq. (3.21.2) is the definition of the sine. No triangle, no circle, no side ratio, no chord.

3.22 Unusual Trigonometric Identities

We now all know that the sum of the first n whole numbers is given by

$$1 + 2 + 3 + \cdots + n = \frac{n(n+1)}{2} \tag{3.22.1}$$

And some mathematicians discovered the following identity (the material herein is from the interesting book *Trigonometry delights* of Eli Maor [44]):

$$\sin\alpha + \sin 2\alpha + \sin 3\alpha + \cdots + \sin n\alpha = \frac{(\sin n\alpha/2)(\sin(n+1)\alpha/2)}{\sin\alpha/2} \tag{3.22.2}$$

Even though if you do not believe this identity, one thing clear is a striking similarity between it and the sum in Eq. (3.22.1). We should first test Eq. (3.22.2) for a few α to be confident that it's correct. Since the values we choose for α are random, if Eq. (3.22.2) is correct for them, it should be correct for others and hence for all α such that $\sin\alpha/2 \neq 0$.

Proof Proof of Eq. (3.22.2). Remember how the 10-year-old Gauss computed the sum of the first n whole numbers? We follow him here. Denoting by S the sum on the LHS of Eq. (3.22.2), we write

$$\begin{aligned} S &= \sin\alpha \quad + \sin 2\alpha \qquad\quad + \cdots + \sin(n-1)\alpha + \sin n\alpha \\ S &= \sin n\alpha + \sin(n-1)\alpha + \cdots + \sin 2\alpha \qquad\quad + \sin\alpha \end{aligned}$$

Then, we sum these two equations:

$$\begin{aligned} 2S = &(\sin\alpha + \sin n\alpha) + (\sin 2\alpha + \sin(n-1)\alpha) + \cdots \\ &+ (\sin(n-1)\alpha + \sin 2\alpha) + (\sin n\alpha + \sin\alpha) \end{aligned}$$

And now, of course, we use the sum-to-product trigonometry identity $\sin a + \sin b = 2\sin(a+b)/2\cos(a-b)/2$ for each sum (because it helps for the factorization):

$$\begin{aligned} 2S = &2\sin\frac{(n+1)\alpha}{2}\cos\frac{(1-n)\alpha}{2} + 2\sin\frac{(n+1)\alpha}{2}\cos\frac{(3-n)\alpha}{2} + \cdots + \\ &+ 2\sin\frac{(n+1)\alpha}{2}\cos\frac{(n-3)\alpha}{2} + 2\sin\frac{(n+1)\alpha}{2}\cos\frac{(n-1)\alpha}{2} \end{aligned}$$

A common factor appears, so we factor the above as:

$$2S = 2\sin\frac{(n+1)\alpha}{2}\left[\cos\frac{(1-n)\alpha}{2} + \cos\frac{(3-n)\alpha}{2} + \cdots + \cos\frac{(n-3)\alpha}{2} + \cos\frac{(n-1)\alpha}{2}\right] \tag{3.22.3}$$

So far so good. The next move is the key and we find it thanks to Eq. (3.22.2). So, this is definitely not the way the author of this identity came up with it (because he did not know of this identity before discovering it). In Eq. (3.22.2) we see the term $\sin\alpha/2$, so we multiply Eq. (3.22.3) with it:

$$\begin{aligned} 2S\sin\frac{\alpha}{2} = \sin\frac{(n+1)\alpha}{2}\Bigg[&2\sin\frac{\alpha}{2}\cos\frac{(1-n)\alpha}{2} + 2\sin\frac{\alpha}{2}\cos\frac{(3-n)\alpha}{2} + \cdots \\ &+ 2\sin\frac{\alpha}{2}\cos\frac{(n-3)\alpha}{2} + 2\sin\frac{\alpha}{2}\cos\frac{(n-1)\alpha}{2}\Bigg] \end{aligned}$$

Now we want to simplify the term in the bracket. To this end, we use the product-to-sum trigonometric identity $2\sin\alpha\cos\beta = \sin(\alpha+\beta) + \sin(\alpha-\beta)$:

$$\begin{aligned} 2S\sin\frac{\alpha}{2} = \sin\frac{(n+1)\alpha}{2}\Bigg[&\sin\frac{n\alpha}{2} + \sin\frac{(2-n)\alpha}{2} + \sin\frac{(n-2)\alpha}{2} + \sin\frac{(4-n)\alpha}{2} + \\ &+ \cdots + \sin\frac{(n-2)\alpha}{2} + \sin\frac{(4-n)\alpha}{2} + \sin\frac{(2-n)\alpha}{2} + \sin\frac{n\alpha}{2}\Bigg] \end{aligned}$$

And lucky for us that all terms in the bracket cancel out except the red terms. It's a bit hard to see how other terms are canceled out, one way is to do this for $n = 3$ and $n = 4$ to see that it is indeed the case. Now, the above equation becomes

$$2S\sin\frac{\alpha}{2} = 2\sin\frac{(n+1)\alpha}{2}\sin\frac{n\alpha}{2}$$

And from that we can get our identity. ■

If we have one identity for the sine, we should have one for the cosine and from that one for the tangent:

$$\begin{aligned} \sin\alpha + \sin 2\alpha + \sin 3\alpha + \cdots + \sin n\alpha &= \frac{(\sin n\alpha/2)(\sin(n+1)\alpha/2)}{\sin\alpha/2} \\ \cos\alpha + \cos 2\alpha + \cos 3\alpha + \cdots + \cos n\alpha &= \frac{(\sin n\alpha/2)(\cos(n+1)\alpha/2)}{\sin\alpha/2} \\ \frac{\sin\alpha + \sin 2\alpha + \sin 3\alpha + \cdots + \sin n\alpha}{\cos\alpha + \cos 2\alpha + \cos 3\alpha + \cdots + \cos n\alpha} &= \tan\frac{(n+1)\alpha}{2} \end{aligned} \tag{3.22.4}$$

So, it is quite impressive that we were able to prove Eq. (3.22.2). But how someone could discover this crazy identity? Here might be the way that these identities were discovered. Let's compute the following sum:

$$A = e^{i\alpha} + e^{i2\alpha} + \cdots + e^{in\alpha} \tag{3.22.5}$$

Why this sum is related to Eq. (3.22.4)? This is because Euler's identity tells us that $e^{i\alpha} = \cos\alpha + i\sin\alpha$:

$$\begin{aligned} A &= (\cos\alpha + i\sin\alpha) + (\cos 2\alpha + i\sin 2\alpha) + \cdots + (\cos n\alpha + i\sin n\alpha) \\ &= (\cos\alpha + \cos 2\alpha + \cdots + \cos n\alpha) + i(\sin\alpha + \sin 2\alpha + \cdots + \sin n\alpha) \end{aligned} \tag{3.22.6}$$

The terms in our identities show up both for the sine and cosine! That's the power of complex numbers. Now is the plan: we will compute A in another way, and from that we get the real and imaginary parts of it. Then, we compare that result with Eq. (3.22.6): equating the imaginary parts gives us the sine formula, and equating the real parts gives us the cosine formula.

It can be seen that A is a geometric series, so it's not hard to compute it:

$$A = e^{i\alpha} + e^{i2\alpha} + \cdots + e^{in\alpha} = e^{i\alpha}(1 + e^{i\alpha} + e^{i2\alpha} + \cdots + e^{i(n-1)\alpha}) = \frac{e^{i\alpha}(1 - e^{in\alpha})}{1 - e^{i\alpha}} \tag{3.22.7}$$

Of course, now we bring back sine and cosine (because that's what we need), and A becomes:

$$\begin{aligned} A &= \frac{e^{i\alpha}(1 - e^{in\alpha})}{1 - e^{i\alpha}} = (\cos\alpha + i\sin\alpha)\frac{1 - \cos n\alpha - i\sin n\alpha}{1 - \cos\alpha - i\sin\alpha} \\ &= (\cos\alpha + i\sin\alpha)\frac{(1 - \cos n\alpha - i\sin n\alpha)(1 - \cos\alpha + i\sin\alpha)}{(1 - \cos\alpha - i\sin\alpha)(1 - \cos\alpha + i\sin\alpha)} \\ &= (\cos\alpha + i\sin\alpha)\frac{(1 - \cos n\alpha - i\sin n\alpha)(1 - \cos\alpha + i\sin\alpha)}{2(1 - \cos\alpha)} \end{aligned} \tag{3.22.8}$$

What we have just done in the second equality is the standard way to remove i in the denominator, now we can get the real and imaginary parts of A. Let's focus on the imaginary part:

$$\text{Im}A = \frac{(\sin\alpha + \sin n\alpha) - (\sin\alpha\cos n\alpha + \sin n\alpha\cos\alpha)}{2(1 - \cos\alpha)} = \frac{(\sin n\alpha/2)(\sin(n+1)\alpha/2)}{\sin\alpha/2} \tag{3.22.9}$$

Now comparing Eq. (3.22.6) with Eq. (3.22.9), we can get the sine identity.

Hey, but wait. Euclid would ask where is geometry? We can construct the sum $\sin\alpha + \sin 2\alpha + \ldots$ and $\cos\alpha + \cos 2\alpha + \ldots$ as in Fig. 3.63. To ease the presentation we considered only the case $n = 3$. It can be seen that $\sin\alpha + \sin 2\alpha + \ldots$ equals the y-coordinate of P_3. Now if we can compute d and β, then we're done.

Indeed, O, P_1, P_2, ... are vertices of a regular polygon inscribed in a circle of radius r. This regular polygon has sides of unit length, and thus we can deduce r in terms of α. Indeed, in the triangle OCP_1 (right diagram in Fig. 3.63), we have $1 = 2r\sin\frac{\alpha}{2}$. Or, $2r = 1/\sin 0.5\alpha$. Now, we consider ΔCOP_3, we have

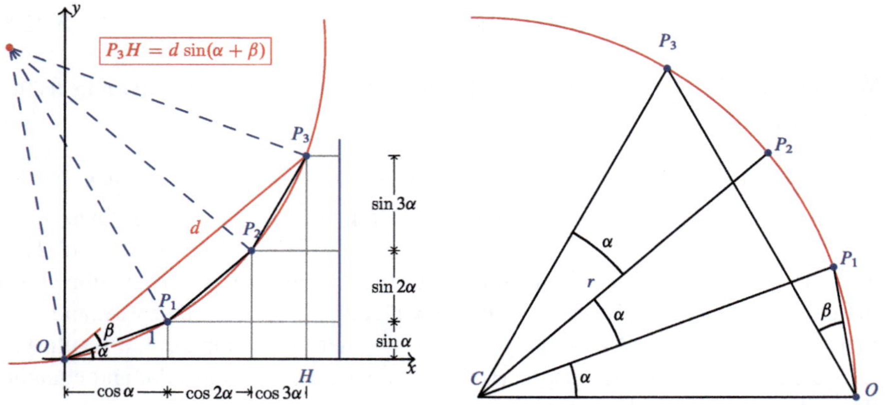

Fig. 3.63 Starting at the origin (which we have labeled as O), we draw a line segment OP_1 of unit length forming an angle α with the positive x-axis. At P_1 we draw a second line segment of unit length forming an angle α with the first segment and thus an angle 2α with the positive x-axis. Continuing in this manner n times, we arrive at the point P_n (which is P_3 in the illustration), whose coordinates we shall denote by X and Y. Obviously, Y is what we're looking for i.e., $Y = \sum_{k=1}^{n} \sin k\alpha$

$$d = 2r \sin\frac{n\alpha}{2} = \frac{\sin n\alpha/2}{\sin\alpha/2}$$

upon substituting $2r$. Now the angle β subtends the chord P_1P_3 and is therefore equal to half the central angle that subtends the same chord (Fig. 3.17):

$$\beta = \frac{1}{2}(n\alpha - \alpha) \Longrightarrow \alpha + \beta = \frac{(n+1)\alpha}{2}$$

Now, we can determine the sum of sines straightforwardly. In the right triangle OP_3H (left diagram), we have $Y = d\sin(\alpha+\beta)$. Substituting $d = {}^{\sin n\alpha/2}/_{\sin\alpha/2}$ and $\alpha + \beta = {}^{(n+1)\alpha}/_{2}$, we obtain Y, which is what we need:

$$\sin\alpha + \sin 2\alpha + \sin 3\alpha + \cdots + \sin n\alpha = d\sin(\alpha+\beta) = \frac{\sin n\alpha/2}{\sin\alpha/2}\sin\frac{(n+1)\alpha}{2}$$

I emphasize that there are no real-life applications of Eq. (3.22.4). If you're asking why we bothered with these formulas, the answer is simple: we had fun playing with them. Is there anything more important than that in life, especially when we're young. Moreover, once again we see the connection between geometry, algebra, and complex numbers. And we saw the telescoping sum again.

Example 3.2 This example is taken from the 2021 Oxford MAT admission test: compute the following sum

$$S = \sin^2(1°) + \sin^2(2°) + \sin^2(3°) + \cdots + \sin^2(89°) + \sin^2(90°) \qquad (3.22.10)$$

One solution is, using $\sin^2 x = 0.5(1 - \cos 2x)$, to write S as

$$S = \frac{1}{2}(90 - S_1)\,,\ S_1 = \cos(2^\circ) + \cos(4^\circ) + \cdots + \cos(180^\circ)$$

then, using the second identity in Eq. (3.22.4) with $n = 90$ and $\alpha = 2^\circ$, we can compute S_1

$$S_1 = \frac{\sin 90^\circ \cos 91^\circ}{\sin 1^\circ} = -1$$

And thus,

$$S = 45.5$$

How to make sure the solution is correct? Write a small program to check!

Admittedly, no one can remember the second identity in Eq. (3.22.4)! There must be another easier way. And indeed, there is. If we write the sum in this way (still remember how the 10-year-old Gauss computed the sum of the first 100 integers?):

$$\begin{aligned} S = &\left[\sin^2(1^\circ) + \sin^2(89^\circ)\right] + \left[\sin^2(2^\circ) + \sin^2(88^\circ)\right] + \cdots \\ &+ \left[\sin^2(44^\circ) + \sin^2(46^\circ)\right] + \sin^2(45^\circ) + \sin^2(90^\circ) \end{aligned} \tag{3.22.11}$$

There are 44 terms of the form $\left[\sin^2(x^\circ) + \sin^2(90^\circ - x^\circ)\right]$, and each term is equal to one (why?), and the red and blue terms are easy, thus

$$S = 44 \times 1 + 1 + \left(\frac{\sqrt{2}}{2}\right)^2 = 45.5$$

3.23 Spherical Trigonometry

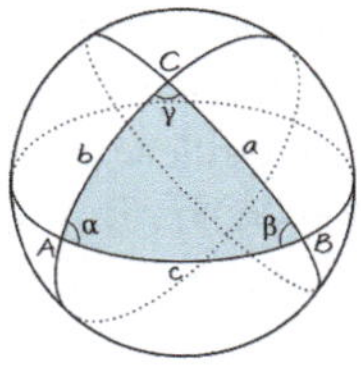

Spherical trigonometry is the study of *curved triangles* called spherical triangles, triangles drawn on the surface of a sphere. The sides of a spherical triangle are arcs of great circles. A great circle is the intersection of a sphere with a central plane, a plane through the center of that sphere. In the figure, you can see three great circles and they make the spherical triangle ABC. The subject is practical, for example, because we live on a sphere. Spherical trigonometry is of great importance for calculations in astronomy, geodesy, and navigation.

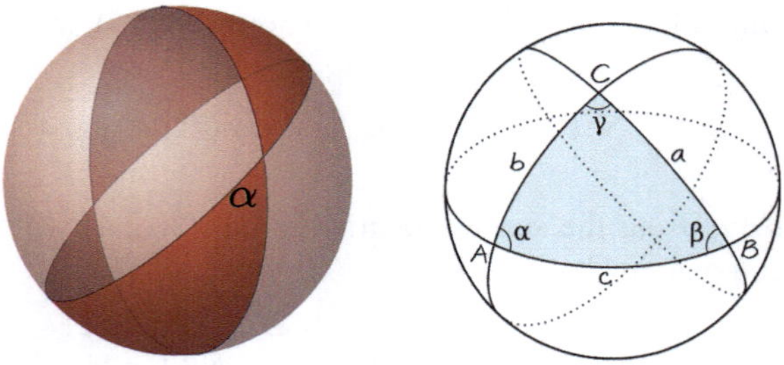

(a) spherical lune (b) spherical triangle

Fig. 3.64 Spherical lune and triangle

This section serves as a brief introduction to the subject. Furthermore, the aim is to understand Legendre's proof of Euler's polyhedral formula. In Sect. 3.23.1, we discuss how to compute the area of a spherical triangle. Then, the area of the spherical polygon is given in Sect. 3.23.2. With that knowledge, we can understand Legendre's proof of Euler's polyhedral formula (Sect. 3.23.3).

To learn more about spherical trigonometry, we refer to the textbook of Glen Van Brummelen [11]. Glen Robert Van Brummelen (born 1965) is a Canadian historian of mathematics specializing in the historical applications of mathematics to astronomy. In his words, he is the "best trigonometry historian, and the worst trigonometry historian" (as he is the only one).

3.23.1 *Area of Spherical Triangles*

The first problem we want to solve in spherical trigonometry is to determine the area of a spherical triangle by knowing its interior angles. An easier problem is to determine the area of a lune. A lune is a two-sided polygon on a sphere where the sides are two halves of great circles (Fig. 3.64a). Let's denote by α the angle of the lune, and by r the radius of the sphere. The area of the lune is then $2\alpha r^2$; when $\alpha = 2\pi$ the lune is simply the entire sphere surface of which the area is $4\pi r^2$.

Now, to compute the area of the spherical triangle ABC shown in Fig. 3.64b, we extend the three sides to get full great circles. Doing so gives us three lunes of which the total surface area is $(2\alpha + 2\beta + 2\gamma)r^2$. But this surface area is also the area of half of the sphere and twice of that of ABC. Thus,

$$(2\alpha + 2\beta + 2\gamma)r^2 = 2\pi r^2 + 2\text{area}(ABC)$$

And from that we can get the formula for the area of ABC:

$$\boxed{\text{area}(ABC) = [(\alpha + \beta + \gamma) - \pi]\, r^2} \tag{3.23.1}$$

In other words, the area is angle sum minus π for unit spheres. Because the sum of the interior angles of every planar triangle is π, we can restate the above formula in yet another way: area = angle sum minus angle sum for a planar triangle.

This formula is known as the Harriot-Girard theorem as it was published by the French mathematician Albert Girard (1595–1632) in 1626 but has also been attributed to the English mathematician Thomas Harriot (1560–1621) who discovered it in 1603 but he had the habit of not publishing anything. Girard's work on trigonometry is the first to use the abbreviations sin, cos, and tan.

3.23.2 Area of Spherical Polygons

Having obtained the formula to compute the area of spherical triangles, it is a small step to consider the area of spherical polygons. We simply divide the polygon into triangles and use the Harriot-Girard theorem for them. Let's start with a four-sided polygon and it is divided into two triangles as shown in Fig. 3.65a. Applying the Harriot-Girard theorem to these triangles and adding up the areas of them we get

$$\text{area} = (a_1 + a_2 + a_3 + a_4 - 2\pi)r^2$$

Doing the same exercise for five-sided polygons (Fig. 3.65b), we get area $= a_1 + a_2 + a_3 + a_4 + a_5 - 3\pi$. Obviously we can see a pattern here, and

$$\boxed{\text{area} = [a_1 + a_2 + \cdots + a_n - (n-2)\pi]r^2} \tag{3.23.2}$$

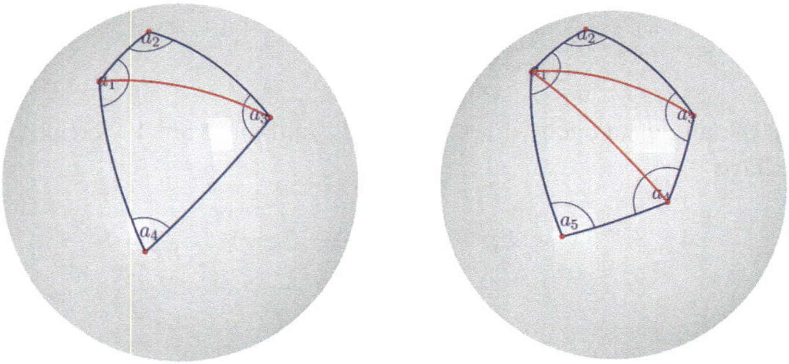

(a) four sided polygon (b) five sided polygon

Fig. 3.65 Surface area of spherical polygons: divide and conquer approach. The polygon is divided into spherical triangles and then the Harriot-Girard theorem is used for all the triangles

But $(n-2)\pi$ is the sum of the angles of a planar n-sided convex polygon. So we can restate the above formula in yet another way for a unit sphere: area = angle sum minus angle sum for a planar convex polygon.

3.23.3 *Legendre's Proof of Euler's Polyhedra Formula*

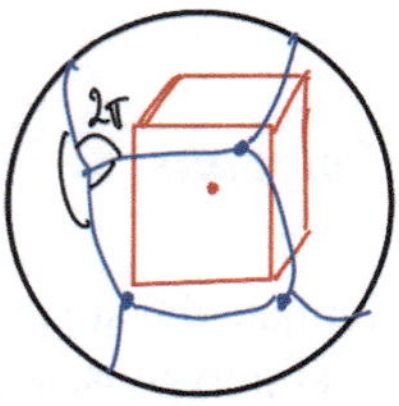

Let's start with a convex polyhedron with V vertices, E edges, and F faces, each of which is a p-gon. Let $*x$ be any point in the interior of the polyhedra. As shown in the figure, construct a sphere centered at $*x$ that surrounds the polyhedron completely. Because the units are irrelevant, we choose them so that the sphere has a radius of one. Project the polyhedron onto the sphere using rays emanating from $*x$. Now imagine that the polyhedron is a wireframed model and that $*x$ is a light bulb. The projection is the shadow of the wire frame on the surface of the encompassing sphere. In this case, the faces of the polyhedron become spherical polygons.

Then, Legendre computed the surface area of the unit sphere in two ways. The first way is of course 4π, and the second way is: this area is the sum of the areas of all the spherical polygons. Now, Eq. (3.23.2) comes into play:

$$4\pi = \sum_{\text{faces}} \left[a_1 + a_2 + \cdots + a_p - (p-2)\pi\right]$$

Noting that the angles at each vertex total to 2π and there are V vertices, and $Fp = 2E$, so we have

$$4\pi = 2\pi V - Fp\pi + 2\pi F \Longrightarrow 2 = V - E + F$$

3.24 Analytic Geometry

> As long as algebra and geometry have been separated, their progress have been slow and their uses limited; but when these two sciences have been united, they have lent each mutual forces, and have marched together towards perfection. (Joseph-Louise Lagrange)

Simply put, analytic geometry or coordinate geometry is geometry with numbers. In analytic geometry, points are represented by an ordered pair of numbers that is called coordinates: e.g. a point is specified by (x, y) where $x, y \in \mathbb{R}$. Curves are represented by equations. For example, the circle of radius one with the center at point $(0, 0)$ is the equation $x^2 + y^2 = 1$. We then manipulate these coordinates and equations to study geometric figures, explore their properties etc..

This section briefly introduces this powerful branch of mathematics. We start with a discussion of the concept of coordinates in Sect. 3.24.1. In Sect. 3.24.2 the equation of lines is derived and we revisit the Thales theorem and prove it using equations. In Sect. 3.24.3 we address the geometric construction regarding what is constructible and what is not. With this preparation, we can prove that it is impossible to double the cube nor trisect the angle (Sect. 3.24.4)—two of the three classical problems of antiquity. In Sect. 3.25, the topic of solving polynomial equations algebraically is treated, which continues what we have left in Chap. 2.

3.24.1 Cartesian Coordinate System

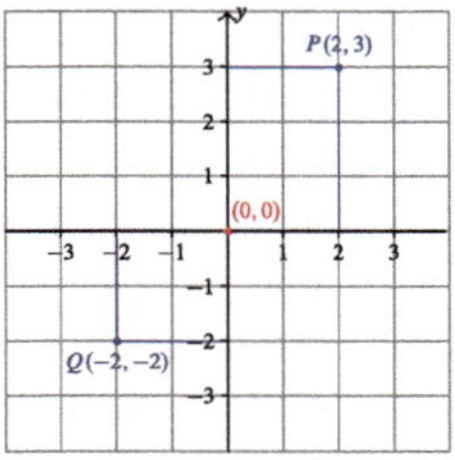

In the seventeenth century, René Descartes (Latinized name: Cartesius) and Pierre de Fermat developed Cartesian coordinates. The name came from Descartes's work La Géométrie published in 1637 as an appendix to Discours de la méthode (Discourse on the Method). This is one of the most influential scientific works of the seventeenth century. For example, it was one of the only two books that the young Isaac Newton owned. Note that Fermat never published his work on this topic. On a plane we draw two fixed perpendicular oriented lines (called axes) meeting at a point called the origin (marked as a dot in the figure). The convention is to call the horizontal axis the x-axis and the vertical axis the y-axis. Mark points along these two axes according to their distance from the origin, like the markings on a ruler: positive numbers to the right and up, negative to the left and down. Then, *every point is uniquely defined by an ordered pair of numerical coordinates*, which are the signed distances to the point from these two axes. For example, $(2, 3)$ is the point with a distance to the y-axis of 2 and a distance to the x-axis of 3. The coordinates of the origin are then $(0, 0)$.

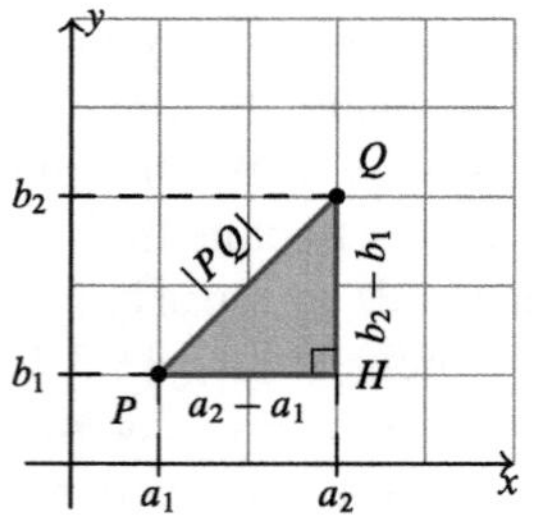

An arbitrary point on the plane is usually denoted by (x, y). The first coordinate (x) is called the abscissa, whereas the second coordinate (y) is called the ordinate.[41] The Pythagorean theorem allows us to determine the distance between two points on this Cartesian plane. For example, the distance between the origin $(0, 0)$ and $(2, 3)$ is the length of the segment joining the two points; that length is $\sqrt{2^2 + 3^2}$. The distance between (a_1, b_1) and (a_2, b_2) is (see the right triangle PQH)

$$\boxed{|PQ| = \sqrt{(a_2 - a_1)^2 + (b_2 - b_1)^2}} \tag{3.24.1}$$

Using the Cartesian coordinate system, geometric shapes (such as curves) can be described by Cartesian equations: algebraic equations involving the coordinates of the points lying on the shape. For example, a circle of radius 2, centered at the origin of the plane, may be described as the set of all points whose coordinates x and y satisfy the equation $\sqrt{x^2 + y^2} = 2$ or $x^2 + y^2 = 4$.

The invention of Cartesian coordinates revolutionized mathematics by providing the first systematic link between Euclidean geometry and algebra. Cartesian coordinates are the foundation of analytic geometry and provide enlightening geometric interpretations for many other branches of mathematics, such as linear algebra, complex analysis, differential geometry, calculus, and more. A familiar example is the concept of the graph of a function. Cartesian coordinates are also essential tools for most applied disciplines that deal with geometry, including astronomy, physics, engineering, and many more. They are the most common coordinate systems used in computer graphics, computer-aided geometric design, and other geometry-related data processing.

[41] This is not so important, but we need to know the terminologies to read books.

History note 3.4: René Descartes (31 March 1596 – 11 February 1650)

René Descartes (Latinized: Renatus Cartesius) was a French philosopher, mathematician, and scientist who spent a large portion of his working life in the Dutch Republic, initially serving the Dutch States Army of Maurice of Nassau, Prince of Orange and the Stadtholder of the United Provinces. One of the most notable intellectual figures of the Dutch Golden Age, Descartes is also widely regarded as *one of the founders of modern philosophy*. His mother died when he was very young, so he and his brothers were sent to live with his grandmother. His father believed that *a good education was important*, so Descartes was sent off to boarding school at a young age.
In 1637, Descartes published his Discours de la methodé in which he explained his rationalist approach to the interpretation of nature. La methodé contained three appendices: La dioptrique, Les météories, and La géométrie. The last of these, The Geometry, was Descartes' only published mathematical work. Approximately 100 pages in length, The Geometry was not a large work, but it presented a new approach in mathematical thinking. Descartes boasted in his introduction that "Any problem in geometry can easily be reduced to such terms that a knowledge of the length of certain straight lines is sufficient for construction." But Descartes' La géométrie was difficult to understand and follow. It was written in French, not the language of scholarly communication at the time, and Descartes' writing style was often obscure in its meaning. In 1649, Frans van Schooten (1615–1660), a Dutch mathematician, published a Latin translation of Descartes' Geometry, adding his own clarifying explanations and commentaries.

3.24.2 Lines and Thales Theorem Revisited

If we look at the x-axis, which is obviously a line, we see that all the points on it have zero y-coordinates. We say that $y = 0$ is the equation of the x-axis. We set a task now to derive the equation for a line passing through two points. There are different ways to do this, but we start humbly with a line going through the origin O and a point $P(x_P, y_P)$ as shown in Fig. 3.66a.

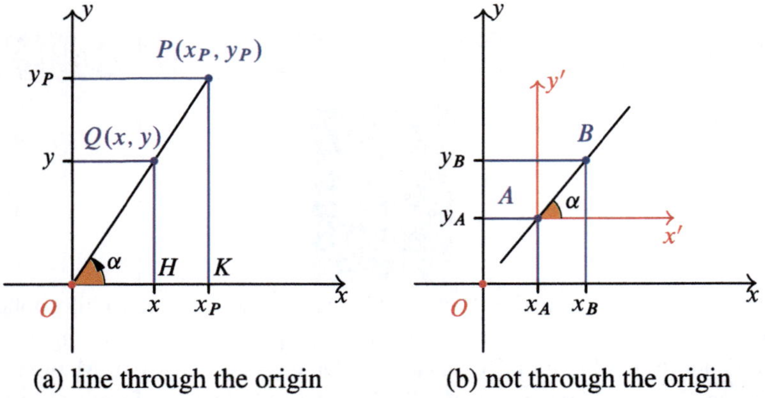

Fig. 3.66 Equation of lines

From the two similar triangles OQH and OPK (AA test), we get $x/x_P = y/y_P$. From that, we get

$$\text{line through the origin and } P(x_P, y_P): \quad y = \frac{y_P}{x_P}x, \quad \text{or} \quad y = ax, \quad a := \frac{y_P}{x_P} \tag{3.24.2}$$

We call a the slope of the line; it is the tangent of the angle (α) between the x-axis and the line measured counterclockwise. Obviously, the slope can be negative. Thus, $y = ax$ describes the line going through the origin and has a slope of a.

Having obtained this, we now move to the general case of a line passing through two arbitrary points $A(x_A, y_A)$ and $B(x_B, y_B)$ as shown in Fig. 3.66b. There is more than one way to derive the equation, but I present a way in which we can reuse Eq. (3.24.2), and in the process, we learn the idea of using whatever coordinate system that makes our life easier.

To reuse Eq. (3.24.2), we introduce a new set of coordinates $x'y'$ with the origin at A. In this coordinate system, the equation of the line AB is $y' = (y'_B/x'_B)x'$ (because it is a line through the origin). We return to the original coordinate system using the relation between the two:

$$x = x' + x_A, \quad y = y' + y_A,$$

Thus,

$$y' = \frac{y'_B}{x'_B}x' \Longrightarrow y - y_A = \frac{y_B - y_A}{x_B - x_A}(x - x_A) \tag{3.24.3}$$

The red term is the slope of the line (Fig. 3.66b). And this line is of the form $y = ax + b$. Having this equation available, it is easy to check if two lines are parallel or perpendicular. If two lines $y = ax + b$ and $y = a'x + b'$ have the same slopes (i.e., $a = a'$) then they are parallel (assuming that $b \neq b'$).

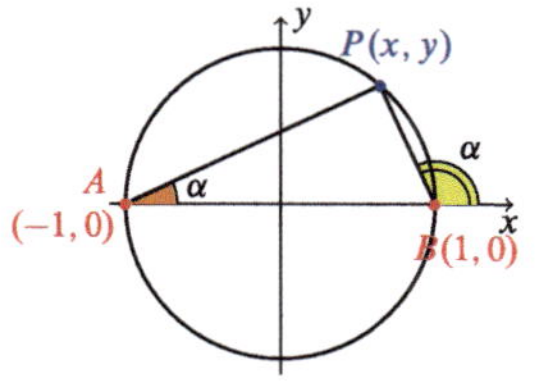

Revisiting the Thales theorem. We have proved the Thales theorem in Fig. 3.17. Now, we prove it using coordinates. To this end, we consider a unit circle centered at the origin of the Cartesian coordinate system. Furthermore, we consider the diameter AB with $A(-1, 0)$ and $B(1, 0)$. Now, let $P(x, y)$ be any point on the circle. Our task is to prove that AP is perpendicular to PB. First, as P is on the circle, its coordinates (x, y) satisfy the equation: $x^2 + y^2 = 1$. We now use the fact that two lines are perpendicular to each other if the product of their slopes is minus one. We thus compute the slopes using Eq. (3.24.3):

$$\text{slope of } APP \text{ is: } \; s_1 = \frac{y-0}{x+1}, \quad \text{slope of } BP \text{ is: } \; s_2 = \frac{y-0}{x-1}$$

Obviously $s_1 > 0$ and $s_2 < 0$. The product of slopes is

$$s_1 s_2 = \frac{y}{x+1}\frac{y}{x-1} = \frac{y^2}{x^2-1} = -1 \quad (x^2 + y^2 = 1)$$

3.24.3 Constructible Numbers

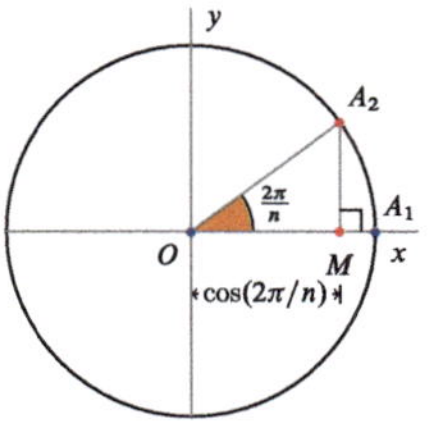

It is time now to see how Gauss proved the constructibility of the 17-gon in 1796 at the age of 19. His work is based on analytic geometry, complex numbers, theory of equations, and number theory. But first, we need a definition of what is a constructible n-gon. Our starting point are: point O with coordinates $(0, 0)$ and point A_1 with coordinates $(1, 0)$. From them we can build the unit circle, and the two coordinate axes Ox and Oy. It is now easy to see that to construct an n-gon we need to create the second vertex A_2. To this end, we just need to construct point M with a distance $OM = \cos(2\pi/n)$. From M, draw a line perpendicular to Ox, this line intersects the unit circle at A_2—the second vertex of our polygon. A circle centered at A_2 with radius being A_1A_2 will intersect the unit circle at A_3—the third vertex and so on.

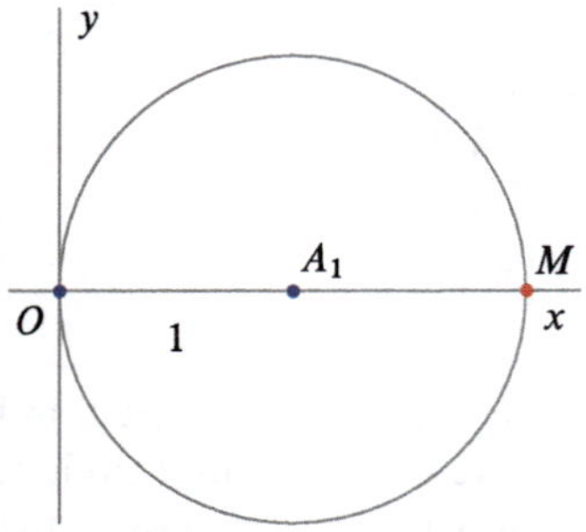

(a) $C(A_1, 1)$ gives $M(2, 0)$

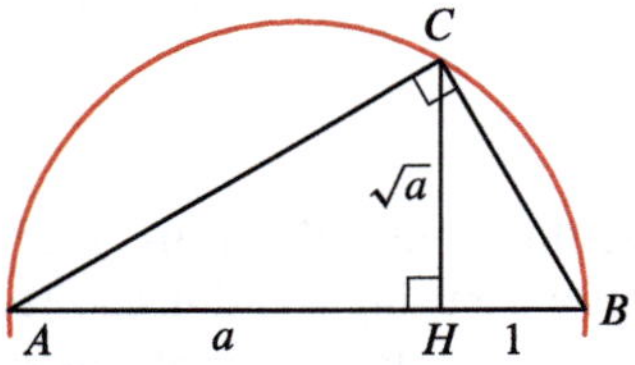

(b) $|AH| = a, |HB| = 1$

Fig. 3.67 Constructing the integers and the square root of $a > 0$. Points A, H, B are there, construct a circle centered at the mid-point of AB with the diameter being AB. Draw then a line through H and perpendicular to AB, this line intersects the circle at C. And CH is exactly $\sqrt{a}$ (thanks to the altitude theorem)

Definition 3.24.1 Constructible n-gon A regular n-gon is said to be constructible if the number $\cos(2\pi/n)$ is constructible.

But, when is point M of a distance $\cos(2\pi/n)$ from O constructible? Let's look at the expression of $\cos 2\pi/n$ for $n = 3, 4, 5, 7$ and see the relation of it with the constructibility of the corresponding n-gon:

$$\begin{aligned}
n = 3 &: \cos\frac{2\pi}{3} = -\frac{1}{2} && \text{(constructible)} \\
n = 4 &: \cos\frac{2\pi}{4} = 1 && \text{(constructible)} \\
n = 5 &: \cos\frac{2\pi}{5} = \frac{\sqrt{5}-1}{4} && \text{(constructible)} \\
n = 7 &: \cos\frac{2\pi}{7} = \sqrt[3]{7/2(1+3\sqrt{-3})} + \cdots && \text{(non-constructible)}
\end{aligned} \tag{3.24.4}$$

I wrote that the 7-gon is not constructible just based on the fact that no one was able to provide such a construction until 1796. Now, this summary of the situation suggests that if $\cos 2\pi/n$ can be written as an expression that involves the four fundamental arithmetic operations (addition/subtraction, multiplication/division) and square roots of positive integers, it then seems that the n-gon is constructible.

Our task now is to see what numbers we can obtain with straightedge/compass construction, from two starting points $(0, 0)$ and $(1, 0)$; points O and A_1 in Fig. 3.67a. First, it is fairly easy to construct points $(2, 0)$, $(3, 0)$ etc. by drawing some circles. For example, a circle centered at A_1 and radius 1 gives us point $M(2, 0)$. Thus, all integers are constructible. Second, how about $\sqrt{5}$? It is constructible thanks to the altitude theorem (Fig. 3.67b). Third, if we apply this square root construction to $b = \sqrt{a}$, we can get $\sqrt{b}$, which is $\sqrt{\sqrt{a}}$.

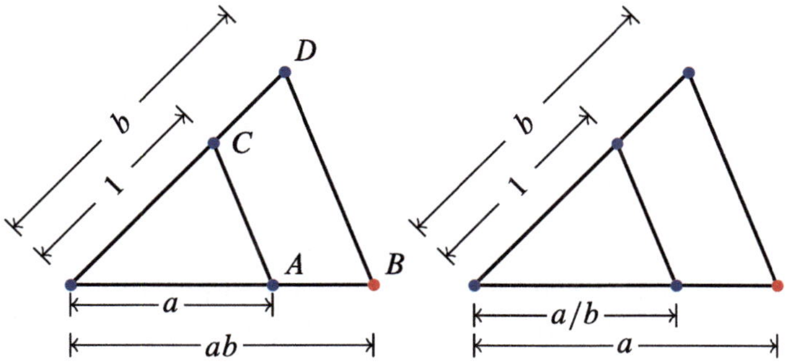

Fig. 3.68 Constructing numbers ab and a/b based on the Thales theorem on similar triangles. Segments a, b and 1 are given. To this end, we just need to construct a line going through a point (DB) and parallel to a given line (AC). How to do that? If we can construct a perpendicular line to a given line, we can construct parallel lines

What about $\sqrt{3}+\sqrt{9}$? It is constructible because we can use straightedge/compass to add two numbers a and b (if they are already constructed). Similarly, we can construct $a-b$. The next question is how about $2\sqrt{3}$? Can we construct this number? And $\sqrt{5}/2$? Yes, we can (Fig. 3.68).

What Gauss did in 1796 was to compute explicitly $\cos {2\pi}/{17}$. He showed that

$$\begin{aligned}\cos\frac{2\pi}{17} &= \frac{1}{16}\left(\sqrt{17}-1+\sqrt{34-2\sqrt{17}}\right)\\ &+\frac{1}{8}\left(\sqrt{17+3\sqrt{17}-\sqrt{34-2\sqrt{17}}-2\sqrt{34+2\sqrt{17}}}\right)\end{aligned} \tag{3.24.5}$$

which involves only addition/subtraction, multiplication/division, and square roots of integers. Thus, the 17-gon is constructible.

Thus we have found a lot of constructible numbers. Every integer, every rational number, and irrational numbers such as $\cos 2\pi/17$ are constructible. But are all real numbers constructible? The answer is no, as you might have guessed correctly, thanks to the following theorem due to Descartes.

Theorem 3.24.1 *Descartes's constructible number theorem A real number is constructible if and only if it can be obtained from the integers using addition, subtraction, multiplication, division, and the extraction of square roots.*

How to prove this theorem? It is based on the equation of line $y=ax+b$ and of circle $(x-a)^2+(y-b)^2=c^2$. An intersection of a line and a circle is then the solution of a quadratic equation, which involves only addition, subtraction, multiplication, division, and the extraction of square roots.

Quadratic irrationals. To be able to do algebra with constructible numbers, we need to find a form for them, similar to an even number is $2n$ for $n \in \mathbb{Z}$. Let a, b, c, d, e, f

be some rationals, then the following constructible numbers are called quadratic irrationals

$$\begin{aligned} &a + b\sqrt{c} && \text{(has an order of one)} \\ &d + \sqrt{a + b\sqrt{c}} && \text{(has an order of two)} \\ &d + e\sqrt{f + \sqrt{a + b\sqrt{c}}} && \text{(has an order of three)} \end{aligned} \tag{3.24.6}$$

where the order of a quadratic irrational is the maximum depth of all of the nested square roots. A rational number is a quadratic irrationality of order zero. Now comes the key idea: any quadratic irrational can be written as $X + Y\sqrt{Z}$ where X, Y and Z are quadratic irrationals and $\sqrt{Z}$ has a higher order than either X or Y. For example, we can write $\sqrt[4]{a+b}$ as

$$\sqrt[4]{a+b} = 0 + \sqrt{0 + \sqrt{a+b}}: \quad X = 0, \quad Y = 1, \quad Z = \sqrt{a+b}$$

Noting that $Z = \sqrt{a+b}$ is a quadratic irrational: $Z = 0 + 1 \times \sqrt{a+b}$.

With this, we're now able to see how Wantzel attacked the doubling the cube and trisecting the angle problems.

3.24.4 Wantzel's Solution on Two Classical Problems of Antiquity

Herein I present Wantzel's proof that it is impossible to double the cube and to trisect the angle using straightedge and compass. Let's first solve the doubling of the cube first. It amounts to whether the cubic equation $x^3 - 2 = 0$ has a constructible solution. Wantzel showed that this equation cannot have such root, thus it is impossible to double the cube, with straightedge and compass.

First, we notice that $x^3 - 2 = 0$ does not have rational roots.[42] Second, assume that this equation has a constructible root, this root must be then of the form $A + B\sqrt{C}$, where A, B and C are quadratic irrationals and $\sqrt{C}$ has a higher order than either A or B. We further assume that $B \neq 0$; we shall see why shortly. Introducing $A + B\sqrt{C}$ into $x^3 - 2 = 0$, we get

$$A^3 + 3A^2B\sqrt{C} + 3AB^2C + B^3C\sqrt{C} - 2 = 0$$

What we do with this equation? On the RHS we have a quadratic irrational, which is not in the right form yet, so we rewrite it

$$(A^3 + 3AB^2C - 2) + (B^3C + 3A^2B)\sqrt{C} = 0$$

This leads to

[42] Remember the rational root theorem of Descartes discussed in Sect. 2.29.3? It is the key here.

$$A^3 + 3AB^2C - 2 = 0, \quad B^3C + 3A^2B^1 = 0$$

If $B = 0$, the above becomes $A^3 - 2 = 0$, which is just the cubic equation that we started with. By looking at the exponents of B in the above two equations, we can see that $-B$ also satisfies them.[43] Thus, if $A + B\sqrt{C}$ is a root, so is $A - B\sqrt{C}$. Using Vieta's theorem about the sum of the roots (Sect. 2.29.7), we then have

$$(A + B\sqrt{C}) + (A - B\sqrt{C}) + x_3 = 0 \Longrightarrow x_3 = -2A$$

But this cubic equation does not have rational roots, so A must be a quadratic irrational. Thus, we can write $x_3 = D + E\sqrt{F}$. Applying the same logic to this root $D + E\sqrt{F}$ we also have $x_4 = -2D$ as a root of the cubic equation. Now, what we have achieved was to show that the cubic equation has four roots

$$x_1 = A + B\sqrt{C}, \quad x_2 = A - B\sqrt{C}, \quad x_3 = -2A, \quad x_4 = -2D$$

These four roots are different for they are of different orders. And a cubic cannot have four different equations. A contradiction! Thus, the cubic equation $x^3 - 2 = 0$ cannot have a constructible root. In other words, it is impossible to double the cube.

Now, we turn our attention to the problem of trisecting the angle. When we say that it is possible to trisect an angle, we mean that we can trisect *any angle*. Therefore, to show that it is impossible to do so, we just need one counterexample. And Wantzel chose an angle of 60°. We need to find the equation for cos 20°:

$$\frac{1}{2} = 4\cos^3 20° - 3\cos 20°$$

With $x = 2\cos 20°$, the above becomes $x^3 - 3x - 1 = 0$. Now, instead of $x^3 - 2 = 0$, we have another cubic equation. Using the same argument, we can conclude that it is impossible to trisect the angle.

3.25 Solving Polynomial Equations Algebraically

It has been shown that the vertices of a regular n-gon are the nth roots of unity. Therefore, the geometric problem of constructing an n-gon is related to solving the polynomial equation $z^n - 1 = 0$ for $z \in \mathbb{C}$. An understanding of geometry hence requires an understanding of equations. This section is a brief introduction to the topic of finding the algebraic solutions to polynomial equations.

To make the terms clear, here is one example. The algebraic solutions of the quadratic equation $ax^2 + bx + c = 0$ are

---

[43] You can plug $-B$ into the first equation, you get the same equation because of B^2. If you plug $-B$ into the second one, you get $-(B^3C + 3A^2B)$, which is also zero.

$$x_{1,2} = \frac{-b \pm \sqrt{b^2 - 4ac}}{2a} \tag{3.25.1}$$

And our aim is to study solutions to the cubics, quartics, and so on. All the solutions must be written in a form similar to Eq. (3.25.1). That is, the solutions are written using only the four fundamental arithmetic operations $+, -, \times, \div$ and extraction of roots.

3.25.1 Solving Quadratic Equations

By plotting the function $y(x) = ax^2 + bx + c$ on a Cartesian plane, as done in Fig. 3.69, we can see that the curve either intersects the x-axis twice or does not intersect at all. Thus, the quadratic equation $ax^2 + bx + c = 0$ has (1) two real solutions or (2) two complex solutions (or no real solutions); do not forget that $x^2 = -1$ has two roots $x = \pm i$. So, any quadratic equation has a maximum of two solutions.

The next step is to examine why the roots are of the form in Eq. (3.25.1). First, there is a special point on the parabola $y(x) = ax^2 + bx + c$: point P—*the lowest point on the curve*. We can find this point by algebra. Using the complete-the-square technique, we can massage the expression $ax^2 + bx + c$ as

$$x^2 + \frac{b}{a}x + \frac{c}{a} = \left(x + \frac{b}{2a}\right)^2 + \frac{c}{a} - \frac{b^2}{4a^2} \geq \frac{c}{a} - \frac{b^2}{4a^2} \quad \forall x$$

Thus, the x-coord of P is $-b/2a$. The parabola is symmetrical with respect to the vertical line drawn from P. Because of this symmetry, the two roots must have the

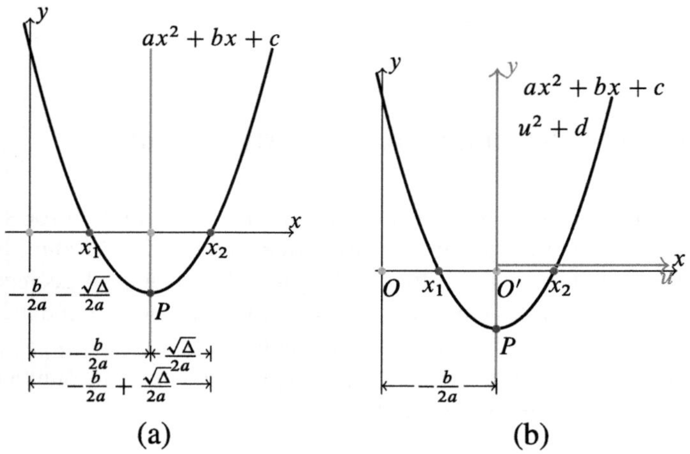

Fig. 3.69 Graph of $y(x) = ax^2 + bx + c$

form $-b/2a$ plus/minus something. We also understand the geometry behind the change of variable $x = u - b/2a$: we use a different coordinate system with the origin at O' (Fig. 3.69b). In this new coordinate system, the parabola is of the simple form $u^2 + d$ (because its lowest point has an x-coordinate of 0).

3.25.2 Solving Cubic Equations

We reconsider the depressed cubic equation given in Eq. (2.13.6)

$$x^3 + px = q \tag{3.25.2}$$

However, our task now is to derive all the three roots of this equation via a formula like Eq. (3.25.1). The approach is of course based on Cardano's solution technique in combination with the fact that an n order polynomial equation has n complex roots.

We start with a change of variable

$$x = u + v, \quad v = -\frac{p}{3u}$$

to get what Lagrange called the resolvent equation of the original cubic equation:

$$u^3 - \frac{p^3}{27u^3} = q \Longleftrightarrow u^6 - qu^3 - \frac{p^3}{27} = 0 \tag{3.25.3}$$

which is a six-degree equation, but disguised as a quadratic equation with $t = u^3$: $t^2 - qt - p^3/27 = 0$. Solving this quadratic equation we obtain

$$t_1 = \frac{q}{2} + \sqrt{\frac{q^2}{4} + \frac{p^3}{27}}, \quad t_2 = \frac{q}{2} - \sqrt{\frac{q^2}{4} + \frac{p^3}{27}} \tag{3.25.4}$$

Do not forget Vietè told us: $t_1 t_2 = -p^3/27$.

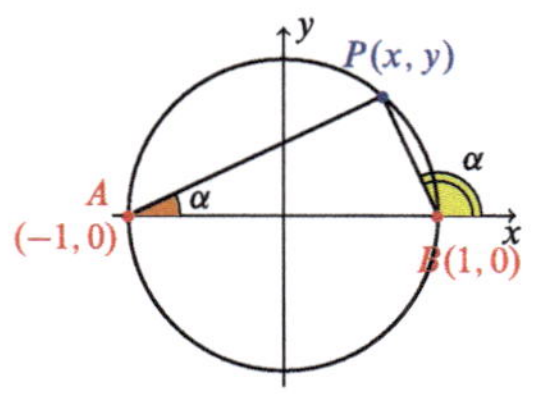

Having now t (two of them), we use $t = u^3$ to get u. Note that we have to be careful not to miss any roots. We must find out six u. From t_1, we get the first real solution $u_1 = \sqrt[3]{t_1}$. How to get the complex roots? Referring to the figure, we see that $u_2 = \sqrt[3]{t_1}\omega$, where $\omega = \cos 2\pi/3 + i \sin 2\pi/3$ (a cubic root of unity). We get another one $u_3 = \sqrt[3]{t_1}\omega^2$. And doing the same thing for t_2, finally, all the six roots of the resolvent equation are

$$\begin{aligned} u_1 &= \sqrt[3]{t_1}, \quad u_2 = u_1\omega, \quad u_3 = u_1\omega^2, \\ u_4 &= \sqrt[3]{t_2}, \quad u_5 = u_4\omega, \quad u_6 = u_4\omega^2 \end{aligned} \tag{3.25.5}$$

We have $u_1u_4 = \sqrt[3]{t_1t_2} = -p/3$. Now, we determine x_1 and x_4, the real roots of the original depressed cubic equation:

$$x_1 = u_1 - \frac{p}{3u_1}, \quad x_4 = u_4 - \frac{p}{3u_4}$$

Using $u_1u_4 = -p/3$, we then have $x_1 = x_4 = u_1 + u_4$. Thus, the two real roots are identical! And this real root is nothing but the sum of u_1 and u_4—the real roots of the resolvent equation. We can get Cardano's formula for this root

$$x_1 = \sqrt[3]{\frac{q}{2} + \sqrt{\frac{q^2}{4} + \frac{p^3}{27}}} + \sqrt[3]{\frac{q}{2} - \sqrt{\frac{q^2}{4} + \frac{p^3}{27}}}$$

Similarly, we have

$$\begin{aligned} x_2 &= u_2 - \frac{p}{3u_2}, \quad u_2u_6 = u_1u_4 = -\frac{p}{3} \Longrightarrow x_2 = u_2 + u_6 = x_6 \\ x_3 &= u_3 - \frac{p}{3u_3}, \quad u_3u_5 = u_1u_4 = -\frac{p}{3} \Longrightarrow x_3 = u_3 + u_5 = x_5 \end{aligned} \tag{3.25.6}$$

What we have just achieved? We have expressed all of the roots of the cubic equation in terms of all of the roots of the resolvent:

$$\begin{aligned} x_1 &= u_1 + u_4 \\ x_2 &= u_2 + u_6 \\ x_3 &= u_3 + u_5 \end{aligned} \tag{3.25.7}$$

Example 3.3 Solving the cubic equation $x^3 - 6x = 4$. This one was solved by Euler in the seventeenth century. We are going to solve this equation in two ways. In the first way, we use Descartes' rational root test to find out easy roots. In the second way, we use the formula in Eq. (3.25.7).

- First way: assuming that we have roots of the form $x = p/q$, then Descartes told us that $p| - 4$ and $q|1$. Thus, we have $p = \{\pm 1, \pm 2, \pm 4\}$ and $q = \{\pm 1\}$. Therefore, $x = \{\pm 1, \pm 2, \pm 4\}$. Plugging those values into the equation we see that $x = -2$ is a root. Now, it is easy to solve the cubic equation using the factor theorem:

$$x^3 - 6x - 4 = 0 \Longleftrightarrow (x + 2)(x^2 - 2x - 2) = 0$$

  Now we just need to solve the quadratic equation $(x^2 - 2x - 2) = 0$. Altogether, we have three real distinct roots:

$$x_1 = -2, \quad x_2 = 1 + \sqrt{3}, \quad x_3 = 1 - \sqrt{3}$$

- Second way: we first compute t_1, t_2 with $p = -6$ and $q = 4$: $t_1 = 2 + 2i$ and $t_2 = 2 - 2i$. (Noting that $q^2/4 + p^3/27 = -4$ is negative) Writing t_1, t_2 in polar form we can then compute u_1 and u_4:

$$u_1 = \sqrt{2}e^{i\pi/12}, \quad u_4 = \sqrt{2}e^{-i\pi/12}$$

Thus, we get the first solution $x_1 = u_1 + u_4 = 2\sqrt{2}\cos\pi/12$. We can compute $\cos\pi/12$ from $\cos\pi/6 = \sqrt{3}/2$.

$$x_1 = 2\sqrt{2}\sqrt{\frac{1}{2} + \frac{\sqrt{3}}{4}} = 1 + \sqrt{3}$$

In the same manner, it is possible to find other roots. For example, $x_2 = u_2 + u_6 = u_1\omega + u_4\omega^2$ with $\omega = \cos 2\pi/3 + i\sin 2\pi/3$.

3.25.3 *Studying Roots of Cubic Using Its Graph*

We have seen from Example 3.3 that when $4p^3 + 27q^2 < 0$ there are three real roots. This section presents an analysis using the tangent line concept (from calculus) to understand the reason why. We consider the depressed cubic equation $x^3 + px = q$. Now, we rewrite this equation as $x^3 = -px + q$, and we study the relation of two curves: one is the cubic $y = x^3$, and one is the line $y = -px + q$. Where these two curves intersect lie the roots of the cubic equation.

First, on a Cartesian plane we plot the simplest cubic $C : y = x^3$ (the red curve in the figure). Now we focus on the tangents to C, because they're special. To this end, consider an arbitrary point P on C with coordinates (x_0, x_0^3); x_0 can be anything from $-\infty$ to ∞. The tangent line to C at P is[44]:

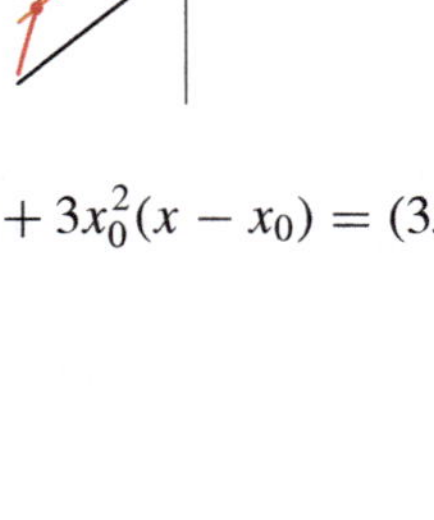

$$T : y = x_0^3 + 3x_0^2(x - x_0) = (3x_0^2)x - 2x_0^3$$

[44] See Sect. 4.4.6 for detail.

Thus the slope of the tangent T is $3x_0^2$ and its intercept is $-2x_0^3$. Given a slope $k > 0$, there are two x_0 such that $3x_0^2 = k$: there are two tangents that are parallel to each other. They are orange lines in the diagram. Why these two tangents are special in our study? Because if a line having the same slope with them and lives inside the domain bounded by them, this line intersects C at three distinct points (the purple line). Now, the line $y = -px + q$ will have the same slope with the tangents if $-p = 3x_0^2$, and to have three distinct roots, we need $|q| < 2x_0^3$. Thus, we have

$$-p = 3x_0^2, \quad |q| < 2x_0^3 \Longrightarrow q^2 < 4x_0^6 = 4\left(-\frac{p}{3}\right)^3 \Longleftrightarrow \boxed{4p^3 + 27q^2 < 0}$$

As the number of roots depends on the sign of $4p^3 + 27q^2$, it is the discriminant of the depressed cubic equation: $\Delta = 4p^3 + 27q^2$. We then have,

- $\Delta < 0$: three real distinct roots;
- $\Delta = 0$:
- $\Delta > 0$: one real root and two complex conjugate roots.

3.25.4 Roots of Unity

We have discussed roots of unity i.e., those solutions to $z^n - 1 = 0$ in Sect. 2.25.3. Herein, we study a bit deeper into this topic. If we let

$$\omega = \cos\left(\frac{2\pi}{n}\right) + i\sin\left(\frac{2\pi}{n}\right) = \exp\left(\frac{2\pi i}{n}\right), \quad n \in \mathbb{N}$$

we will have

$$\omega^k = \cos\left(\frac{2\pi k}{n}\right) + i\sin\left(\frac{2\pi k}{n}\right), \quad k \in \mathbb{N}$$

so that the roots of $z^n - 1 = 0$ will all be expressed by the powers of this quantity ω. That is, these roots are $\omega, \omega^2, \omega^3, \ldots, \omega^n$, of which the last ω^n is always be equal to 1.

3.26 Non-Euclidean Geometries

Out of nothing I have created a strange new universe. (János Bolyai)

The fifth postulate of Euclidean geometry states that "through a given point not on a given line there is exactly one line parallel to the given line and passing through the given point". This postulate is not self-evident and was not accepted by many mathematicians. In fact, to them the fifth postulate is an inelegant axiom. Jordan

Ellenberg (1971), once wrote "An inelegant axiom is like a stain in the corner of the floor; it doesn't get in your way, per se, but it's maddening, and one spends an inordinate amount of time scrubbing and scouring and trying to make the surface nice and clean. In the mathematical context, this amounted to trying to show that the fifth axiom, the so-called parallel postulate, followed from all the others. If that were so, the axiom could be removed from Euclid's list, leaving it shiny and pristine".

After 2000 years of scrubbing, the stain was still there. So, are mathematicians still scrubbing the stain on the floor corner (TD5)? Luckily for them, by the 1820s, Gauss, Bolyai, Lobachevsky, and Riemann had finished scrubbing it. These scholars suggested forgetting about TD5 and instead using a different set of axioms. For example:

> Through a point not on a given line, there is NO line parallel to the given line that passes through the given point.

or

> Through a point not on a given line, there are INFINITELY MANY lines parallel to the given line that pass through the given point.

By doing so, they created a geometry system different from Euclid's. Thus, Gauss named Euclid's system *Euclidean geometry*, and the others became *non-Euclidean geometries* (non-Euclidean geometry). What Gauss, Bolyai, and Lobachevsky discovered is called hyperbolic geometry, and what Riemann found is called elliptic geometry. Why hyperbolic and elliptic? It's a great question, but unfortunately, I cannot answer it here.

It turns out there are many geometrical systems, depending on the rules you play by! The overall picture of these geometries is presented in ??: feel free to create 'premises' as long as they don't contradict each other, and you'll have various geometrical systems.

So, which geometry describes the space we live in? Mathematicians are indifferent to this question. Mathematicians are the tailors (fussy) of scientists. They tailor suits that anyone can wear comfortably. It is the Physicists who are interested. And Einstein let us know that the space we live in follows non-Euclidean geometry, not the familiar Euclidean geometry.

3.27 Computer Algebra Systems

A computer algebra system (CAS) or symbolic algebra system (SAS) is any mathematical software with the ability to manipulate mathematical expressions in a way similar to the traditional manual computations of mathematicians and scientists. The main general-purpose computer algebra systems are `Maple`, `Mathematica` (proprietary software) and `Axiom`, `Maxima`, `Magma`, `SageMath` (free).

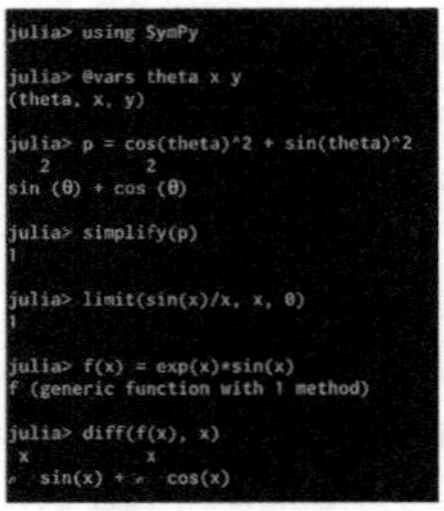

The primary goal of a Computer Algebra system is to *automate tedious and sometimes difficult algebraic manipulation tasks*. Computer algebra systems have not only changed how mathematics is taught at many schools and universities but have provided a flexible tool for mathematicians worldwide. Computer algebra systems can be used to simplify rational functions, factor polynomials, find the solutions to a system of equations, and various other manipulations. In calculus, they can be used to find the limit, derivative, and integrals of functions, all done symbolically. Computer algebra systems began to appear in the 1960s and evolved out of two quite different sources—the requirements of theoretical physicists and research into artificial intelligence.

In the next figure, some use of a CAS using the Python `SymPy` package is illustrated. Appendix A.10 presents a short introduction to this package.

3.28 Review

This chapter has presented Euclidean geometry and trigonometry—the two major topics taught in high schools. The take-home messages for the geometry part are

- Euclid's *Elements* presents a *self-contained logically consistent* system for geometry. It starts with definitions of the objects that it wants to talk about. Then, it states the axioms—rules of the game—that the object obeys. Then, it deduces propositions—facts—about these objects.
- Important theorems in geometry are the Pythagorean theorem, the theorem on similar triangles that similar triangles are proportional, and the central angle theorem. Other theorems are consequences of these basic theorems.
- It is possible to study geometry in more than one way. Euclid presented one way without coordinates and Descartes showed us the second way with coordinates.
- There is a close connection between geometry and algebra. We have seen this for example in the three classical problems of antiquity. The lesson is that mathematics is not simply a set of disconnected pieces of information; rather, it can be thought of as a vast terrain with its own geography, which both its practitioners and originators use to navigate through what would otherwise be an insurmountable wilderness.
- If we change the rules of the game, we can make new games. By removing the fifth postulate of Euclid, mathematicians created other geometries such as hyperbolic geometry and so on.

For trigonometry, below are some key summaries:

- Grew out of astronomy works of ancient mathematicians, there are two types of trigonometry: plane trigonometry and spherical trigonometry.
- All trigonometry identities are derived from a minimum number of identities: $\sin^2 x + \cos^2 x = 1$, $\sin(-x) = -\sin x$, $\cos(-x) = \cos x$, $\cos(\pi/2 - x) = \sin x$ and $\cos(x + y) = \cos x \cos y - \sin x \sin y$. The first four basic identities can be obtained from the unit circle.
- We have learnt a lot by building a sine table with our bare hands (inequalities, fixed iteration method, and so on).
- There are many definitions of the sine: (1) as side ratios in a right triangle, (2) as the y-coordinate of a point on the unit circle, (3) as half-chord length and (4) as $\approx {}^{\theta^1}/_{1!} - {}^{\theta^3}/_{3!} + {}^{\theta^5}/_{5!} - \cdots$ They point to the same object being looked at from different angles.
- What if you do not care about astronomy and right triangles, are sines and cosines still useful? Yes, they play a key role in the so-called Fourier series which is a way to look at periodic function $f(x)$

  $$f(x) = a_0 + (a_1 \cos x + a_2 \cos 2x + \cdots) + (b_1 \sin x + b_2 \sin 2x + \cdots)$$

  The fact that we're paying an affordable phone bill is due to this Fourier series. So, those sines and cosines have a wide range of applications in science and engineering (electronic engineering, earthquake engineering etc.).

Now that we know a bit of algebra and a bit of geometry and trigonometry, it is time to meet calculus. About calculus, the Hungarian-American mathematician, physicist, John von Neumann said

> The calculus was the first achievement of modern mathematics and it is difficult to overestimate its importance. I think it defines more unequivocally than anything else the inception of modern mathematics; and the system of mathematical analysis, which is its logical development, still constitutes the greatest technical advance in exact thinking.

Chapter 4
Calculus

Ancient geometers (i.e., mathematicians working on geometrical problems) was obsessed with two problems: (1) finding the area of planar shapes (e.g. the area of a circle) and (2) finding the tangent to a curve i.e., the line that touches the curve at only one point. Although some results were obtained, mostly by Archimedes with the method of exhaustion, a universal method that can be applied to any curves was not available until the early of the seventeenth century.

The mathematicians of the seventeenth century were equipped with more powerful mathematics; they had the symbolic algebra of Viète and the analytic geometry of Descartes and Fermat. Furthermore, the work of Kepler on the motion of heavenly objects and Galileo on the motion of earthly objects has put the study of motion into the scene. The seventeenth mathematicians no longer saw static objects (such as curves) as motionless. They saw curves as the trajectory of the motion of a particle. With these new tools and dynamics view, they again solved the two above geometrical problems, old results were confirmed and new results were obtained.

The pinnacle of the mathematical developments of this century were the introduction, by Newton and Leibniz, of the two concepts—derivative and integral. The former provides the final answer to the tangent problem and the latter is the solution to the area problem. What is more is that a connection between the derivative and the integral was discovered, what is now we call the fundamental theorem of calculus.

With this new mathematics, called the calculus, problems that once required the genius of Archimedes can be solved by any high school students. A powerful thing was developed. And as in many other cases in mathematics, the calculus turns out to be a very effective tool to solve many other problems; those involve changes. That's why Richard Feynman—the 1964 Nobel-winning theoretical physicist once said "Calculus is the language God talks". Feynman was probably referring to the fact that physical laws are written in the language of calculus. Steven Strogatz in his interesting book *Infinite Powers* [67] wrote "Without calculus, we wouldn't have cell

V. P. Nguyen, *Mathematics for Engineers and Scientists*,
https://doi.org/10.1007/978-981-97-4631-6_4

phones, computers, or microwave ovens. We wouldn't have radio. Or television. Or ultrasound for expectant mothers, or GPS for lost travelers. We wouldn't have split the atom, unraveled the human genome, or put astronauts on the moon".

Thus it is not surprising that calculus occupies an important part in the mathematics curriculum in both high schools and universities. Sadly, as being taught in schools, calculus is packed with many theorems, formulas, and tricks. The truth is, the essence of calculus is quite simple: calculus is often seen as *the mathematics of changes*. The ball rolls down an inclined plane: change of position in time or motion. A curve is something that changes direction. That were the two types of changes that motivated the development of the calculus.

But calculus does not work with all kinds of change. It only works with changes of continuous quantities. Mathematicians (and physicists as well) assume that space and time are continuous. For example, given a length, we can cut it in two halves, cut one half into two halves, and so on to infinity. What do we get from this infinite division? A very very small quantity which is not zero but smaller than any positive real numbers. We call such quantity an infinitesimal.

How does calculus work? It works based on one single principle: the principle of infinity—a term coined by Strogatz. Take as example the problem of finding a tangent to a curve. This curve is divided into infinitely many line segments, each has an infinitesimal length. With that, a tangent to a curve at any point is simply the slope of the line segment connecting that point to the next point (infinitesimally nearby). The slope of a line? We know it.

This chapter is devoted to the calculus of functions of a single variable. I use primarily the following books for the material presented herein:

- *Infinite Powers* by Steven Strogatz[1] [67]. I recommend anyone to read this book before taking any calculus class;
- *Calculus* by Gilbert Strang[2] [65];
- *What is mathematics?: an elementary approach to ideas and methods* by Richard Courant, Herbert Robbins,[3] Ian Stewart [15];

---

[1] Steven Henry Strogatz (born 1959) is an American mathematician and the Jacob Gould Schurman Professor of Applied Mathematics at Cornell University. He is known for his work on nonlinear systems, including contributions to the study of synchronization in dynamical systems, for his research in a variety of areas of applied mathematics, including mathematical biology and complex network theory. Strogatz is probably famous for his writings for the general public, one can cite Sync, The joy of x, Infinite Powers.

[2] William Gilbert Strang (born 1934) is an American mathematician, with contributions to finite element theory, the calculus of variations, wavelet analysis, and linear algebra. He has made many contributions to mathematics education, including publishing seven mathematics textbooks and one monograph.

[3] Herbert Ellis Robbins (1915–2001) was an American mathematician and statistician. He did research in topology, measure theory, statistics, and a variety of other fields. The Robbins lemma, used in empirical Bayes methods, is named after him. Robbins algebras are named after him because of a conjecture that he posed concerning Boolean algebras.

- *The historical development of the calculus* by Charles Edwards [21]
- *Calculus: An Intuitive and Physical Approach* by Moris Kline[4] [36]

Our plan in this chapter is as follows. First, in Sect. 4.1, we briefly discuss the analytic geometry with the introduction of the Cartesian coordinate system, the association of any curve with an equation. Second, the concept of function is introduced (Sect. 4.2). Then, integral calculus of which the most important concept is an integral is treated in Sect. 4.3. That is followed by a presentation of the differential calculus of which the most vital concept is a derivative (Sect. 4.4). We then present some applications of the derivative in Sect. 4.5. The connection between integral and derivative is treated in Sect. 4.6, followed by methods to compute integrals in Sect. 4.7.

Section 4.9 gives some applications of integration. A proper definition of the limit of a function is then stated in Sect. 4.10. Some theorems in calculus are presented in Sect. 4.11. Polar coordinates are discussed in Sect. 4.13. Bézier curves—a topic not provided in high school and even college program—is shown in Sect. 4.14. Infinite series and in particular Taylor series are the topics of Sect. 4.15. Applications of the Taylor series are given in Sect. 4.16. Fourier series are given in Sect. 4.19. Section 4.20

4.1 Conic Sections

Conic sections received their name because they can be represented by a cross section of a plane cutting through a cone (Fig. 4.1a). A conic section is a curve on a plane that is defined by a second-degree polynomial equation in two variables. Conic sections are classified into four groups: parabolas, circles, ellipses, and hyperbolas. The conic sections were named and studied as long ago as 200 B.C.E., when Apollonius of Perga undertook a systematic study of their properties.

Two well-known conics are the circle and the ellipse. They arise when the intersection of the cone and plane is a closed curve (Fig. 4.1a). The circle is a special case of the ellipse in which the plane is perpendicular to the axis of the cone. If the plane is parallel to a generator line of the cone, the conic is called a parabola. Finally, if the intersection is an open curve and the plane is not parallel to the generator lines of the cone, the figure is a hyperbola.

Conic sections are observed in the paths taken by celestial bodies (e.g. planets). When two massive objects interact according to Newton's law of universal gravitation, their orbits are conic sections if their common center of mass is considered to be at rest. If they are bound together, they will both trace out ellipses; if they are moving apart, they will both follow parabolas or hyperbolas (Fig. 4.1b).

Straight lines use $1, x, y$. The next curves use x^2, xy, y^2, which are conics. It is important to see both the **curves** and their **equations**. This section presents the **analytic geometry** of René Descartes and Pierre de Fermat in which the geometry of the curve is connected to the analysis of the associated equation. Numbers are

[4] Morris Kline (1908–1992) was a professor of Mathematics, a writer on the history, philosophy, and teaching of mathematics, and also a popularizer of mathematical subjects.

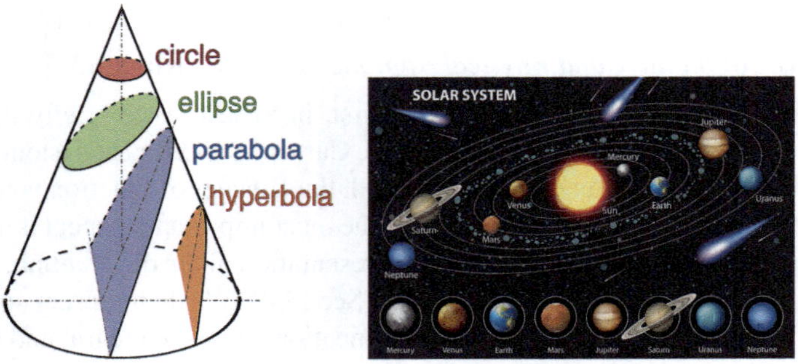

(a) (b)

Fig. 4.1 Conic sections: parabolas, circles, ellipses, and hyperbolas. *Source* Wikipedia Commons (**a**) and `freepik.com` (**b**)

assigned to points, and we speak about the point $(1, 2)$. Euclid and Archimedes might not have understood as Strang put it.

4.1.1 Circles

Definition 4.1 A circle is a set of points whose distance to a special point—the center—is constant.

From this definition, we can derive the equation of a circle. Let's denote the center by (x_c, y_c) and the radius by r, then we consider a point (x, y) on the circle. The fact that the distance from this point to the center is r is written as (using Eq. (3.24.1))

$$\sqrt{(x - x_c)^2 + (y - y_c)^2} = r \Longrightarrow (x - x_c)^2 + (y - y_c)^2 = r^2$$

Upon expansion, we get the following form

$$x^2 + y^2 - 2x_c x - 2y_c y + x_c^2 + y_c^2 - r^2 = 0 \tag{4.1.1}$$

When $x_c = y_c = 0$ i.e., the center of the circle is at the origin, the equation of the circle is much simplified:

$$x^2 + y^2 = r^2 \tag{4.1.2}$$

4.1.2 Ellipses

An ellipse is a plane curve surrounding two focal points, such that for all points on the curve, the sum of the two distances to the focal points is a constant. It generalizes a circle, which is a special type of ellipse in which the two focal points are the same.

Definition 4.2 The ellipse is the set of all points (x, y) such that the sum of the distances from (x, y) to the foci is constant.

We are going to use the definition of an ellipse to derive its equation. Assume that the ellipse is centered at the origin, and its foci are located at $F_1(-c, 0)$ and $F_2(c, 0)$; see Fig. 4.2. The two vertices on the horizontal axis are $A_1(a, 0)$ and $A_2(-a, 0)$.

It is clear that the distances from A_1 (or A_2) to the two foci are $2a$, and that is the constant mentioned in the definition. So, pick any point $P(x, y)$, and compute its distances to the foci $d_1 + d_2$, set it to $2a$ and do some algebraic manipulations, we have

$$\begin{aligned}
d_1 + d_2 &= 2a \quad \text{(definition of ellipse)} \\
\sqrt{(x+c)^2 + y^2} + \sqrt{(x-c)^2 + y^2} &= 2a \quad \text{(definition of distance)} \\
\sqrt{(x+c)^2 + y^2} &= 2a - \sqrt{(x-c)^2 + y^2} \\
(x+c)^2 + y^2 &= 4a^2 + (x-c)^2 + y^2 - 4a\sqrt{(x-c)^2 + y^2} \\
a\sqrt{(x-c)^2 + y^2} &= a^2 - xc \\
(a^2 - c^2)x^2 + a^2y^2 &= a^2(a^2 - c^2) \\
\frac{x^2}{a^2} + \frac{y^2}{a^2 - c^2} &= 1
\end{aligned}$$

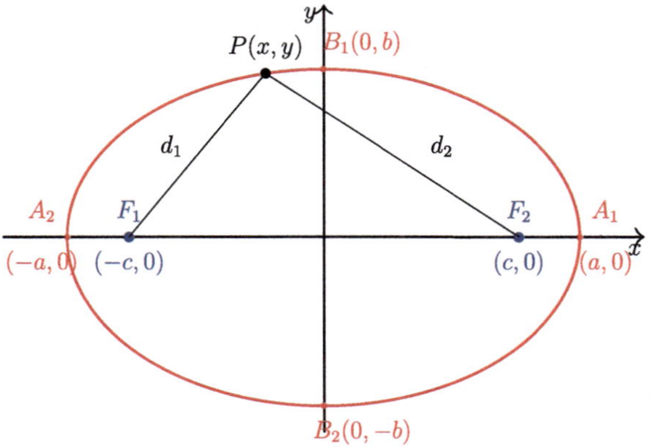

Fig. 4.2 An ellipse centered at the origin. The major axis of an ellipse is its longest diameter: a line segment that runs through the center and both foci, with ends at the widest points of the perimeter. The semi-major axis is one half of the major axis. The semi-minor axis is a line segment that is perpendicular with the semi-major axis and has one end at the center

All steps from the third equality are just algebraic, to remove the square root. Now, the final step is to bring b into play by considering that distances from B_1 to the foci are also $2a$ (from the very definition of an ellipse). This gives us $b^2 + c^2 = a^2$. So, we have

$$\frac{x^2}{a^2} + \frac{y^2}{b^2} = 1, \quad b^2 + c^2 = a^2 \tag{4.1.3}$$

From which an ellipse is reduced to a circle when $a = b$.

Ellipses are common in physics, astronomy and engineering. For example, the orbit of each planet in the solar system is approximately an ellipse with the Sun at one focus point. The same is true for moons orbiting planets and all other systems of two astronomical bodies. The shapes of planets and stars are often well described by ellipsoids.

Area of ellipse. If we know the area of a circle is πr^2, then what is the area of an ellipse? We can get the formula without actually computing it. This area must be in the form $\pi f(a, b)$, and $f(a, b) = f(b, a)$ and $f(a, a) = a^2$. The only form is $f(a, b) = ab$. So, the area of an ellipse is πab.

Reflecting property of ellipses. The ellipse reflection property says that rays of light emanating from one focus, and then reflected off the ellipse, will pass through the other focus. Now, apart from being mathematically interesting, what makes this property so fascinating? Well, there are several reasons. Most notable of which is its significance to physics, primarily optics and acoustics. Both light and sound are affected in this way. In fact there are many famous buildings designed to exploit this property. Such buildings are referred to as whisper galleries or whisper chambers. St. Paul's Cathedral in London, England was designed by architect and mathematician Sir Christopher Wren (1632–1723) and contains one such whisper gallery. The effect that such a room creates is that if one person is standing at one of the foci, a person standing at the other focus can hear even the slightest whisper spoken by the other. We refer to Sect. 4.4.2 for a proof.

4.1.3 Parabolas

When you kick a soccer ball (or shoot an arrow, fire a missile or throw a stone) it arcs up into the air and comes down again ... following the path of a parabola. A parabola is a curve where any point is at an equal distance from a fixed point (called the focus), and from a fixed straight line (called the directrix).

In Fig. 4.3a, we label the focus as F with coordinates (a, b), and a horizontal directrix l with the equation $y = k$ (of course we can have parabolas with a vertical directrix). Now, we pick a point $P(x, y)$ and the definition of a parabola tells us $|PF| = |Pl|$ (i.e., the distance from P to F is equal to the distance from P to the directrix l). Computing $|PF|$ and $|Pl|$ and equate them, we have

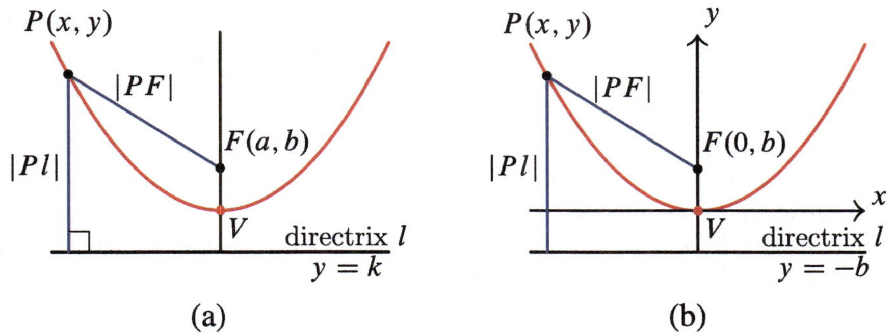

Fig. 4.3 A parabola is a curve where any point is at an equal distance from a fixed point (the focus F), and a fixed straight line (the directrix l); or $|PF| = |Pl|$. To compute the distance from P to l we compute the distance between P and its projection on l which is (x, k). The vertex V is the lowest point on the parabola

$$
\begin{aligned}
\sqrt{(y-k)^2} &= \sqrt{(x-a)^2 + (y-b)^2} \\
y^2 - 2yk + k^2 &= x^2 - 2ax + a^2 + y^2 - 2yb + b^2 \\
y &= \frac{(x-a)^2}{2(b-k)} + \frac{b+k}{2}
\end{aligned}
\tag{4.1.4}
$$

One can see that $b+k/2$ is the ordinate of the vertex of the parabola. To simplify the equation, we can put the origin at V, as done in Fig. 4.3b, then we have $a = 0$ and $k = -b$, thus

$$
y = \frac{x^2}{4b} \quad \text{or} \quad x^2 = 4by
$$

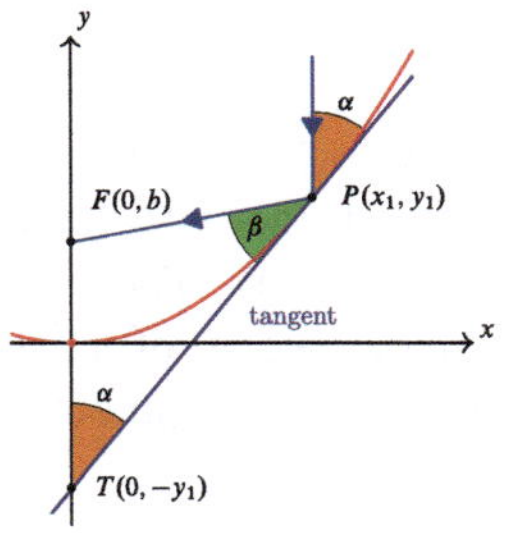

Reflecting property of parabola. The parabola reflection property says that rays of light emanating from one focus, and then reflected off the parabola in a path parallel to the $y-$axis (or vice versa). To prove this property, see the next figure. We consider a parabola with the vertex at the origin. We then consider a point $P(x_1, y_1)$ on the parabola. Through P we draw a tangent line that intersects the $y-$axis at T. We can write the equation for this tangent line (see Sect. 4.4.6), and thus determine the ordinate of T; it is $-y_1$. From optic (Sect. 4.4.2) we know that the light follows the path such that $\alpha = \beta$. So all we need to prove is that PF is making an angle (with the tangent) exactly equal to α (i.e., consistent with the physics of light). This is indeed the case as the triangle TFP is an isosceles triangle (proved by checking that $TF = FP$, all coordinates known).

What are some applications of this nice property of the parabola? A solar collector and a TV dish are parabolic; they concentrate sun rays and TV signals onto a point—a heat cell or a receiver collects them at the focus. Car headlights turn the idea around: the light starts from the focus and emits outward. Is this reflection property related to that of an ellipse? Yes, for the parabola one focus is at infinity.

4.1.4 Hyperbolas

Definition 4.3 A hyperbola is the set of all points (x, y) in a plane such that the difference of the distances between (x, y) and the two foci is a positive constant.

Notice that the definition of a hyperbola is very similar to that of an ellipse. The distinction is that the hyperbola is defined in terms of the difference of two distances, whereas the ellipse is defined in terms of the sum of two distances. So, the equation of a hyperbola is very similar to the equation of an ellipse (instead of a plus sign we have a minus sign):

$$\frac{x^2}{a^2} - \frac{y^2}{b^2} = 1 \tag{4.1.5}$$

What is the graph of a hyperbola looks like? First, we need to rewrite the equation in the usual form $y = f(x)$:

$$y = \pm \frac{b}{a}\sqrt{x^2 - a^2}, \quad |x| \geq a$$

Thus, there are two branches, one for $x \geq a$ and one for $x \leq -a$. When $x \to \infty$, $y \to \infty$. But we can do better than that: we have $x^2 - a^2 \approx x^2$ when $x \to \infty$. Thus, for $x \to \infty$, y approaches $\pm(b/a)x$. These two lines are therefore called the asymptotes of the hyperbola. We can see all of this in Fig. 4.4a for a particular case with $a = 5$ and $b = 3$. When $a = b$, the asymptotes are perpendicular, and we get a rectangular or right hyperbola (Fig. 4.4b).

4.1.5 General Form of Conic Sections

Any conic section namely ellipse, circle, parabola or hyperbola can be generally described by the following equation:

$$Ax^2 + Cy^2 + Dx + Ey + F = 0 \tag{4.1.6}$$

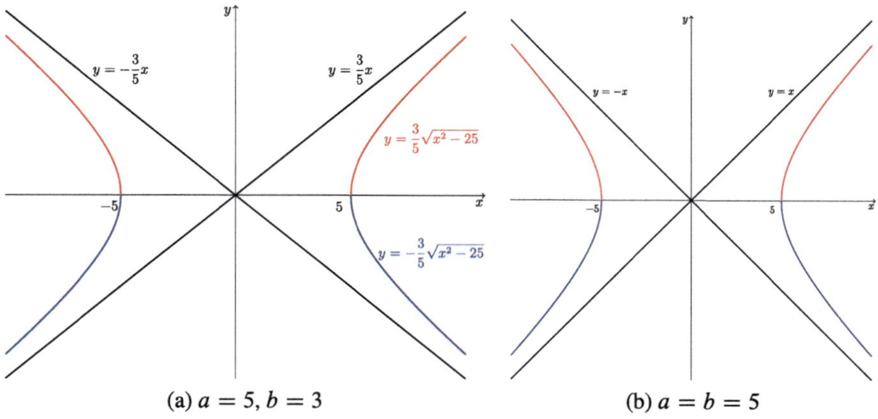

Fig. 4.4 Graph of hyperbolas: when $a = b$, we get a rectangular hyperbola

But this equation is not complete in the sense that it lacks the term xy. Actually, the most general form of a conic section is the following with a xy term (mathematics is fair, isn't it?):

$$Ax^2 + Bxy + Cy^2 + Dx + Ey + F = 0 \tag{4.1.7}$$

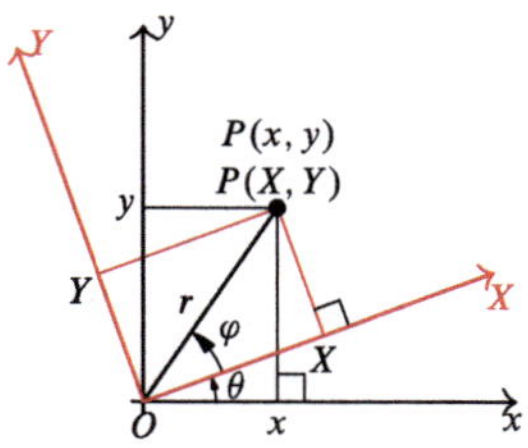

The proof is based on the fact that we can transform Eq. (4.1.7) to Eq. (4.1.6) by a specific rotation of axes to be described in what follows. First we consider axes Ox and Oy. We then rotate these axes an angle θ counterclockwise to have OX and OY. Considering a point P which has coordinates (x, y) in the xy system and (X, Y) in the rotated system. The aim is now to relate these two sets of coordinates. From the figure, we have these results:

$$\begin{cases} X &= r\cos\varphi \\ Y &= r\sin\varphi \end{cases}, \quad \begin{cases} x &= r\cos(\varphi + \theta) \\ y &= r\sin(\varphi + \theta) \end{cases}$$

Using the trigonometry identities for $\sin(a + b)$ and $\cos(a + b)$, we can write x, y in terms of X, Y as

$$\begin{cases} x &= X\cos\theta - Y\sin\theta \\ y &= X\sin\theta + Y\cos\theta \end{cases} \tag{4.1.8}$$

and by solving X, Y, we have

$$\begin{cases} X &= +x\cos\theta + y\sin\theta \\ Y &= -x\sin\theta + y\cos\theta \end{cases}$$

Substituting Eq. (4.1.8) into Eq. (4.1.7) we get an equation in terms of X, Y

$$A(X\cos\theta - Y\sin\theta)^2 + B(X\cos\theta - Y\sin\theta)(X\sin\theta + Y\cos\theta)+$$
$$C(X\sin\theta + Y\cos\theta)^2 + D(X\cos\theta - Y\sin\theta) + E(X\sin\theta + Y\cos\theta) + F = 0 \tag{4.1.9}$$

which has this form

$$A'X^2 + B'XY + C'Y^2 + D'X + E'Y + F = 0$$

and we're interested in the XY term with the coefficient given by

$$B' = B(\cos^2\theta - \sin^2\theta) + 2(C - A)\sin\theta\cos\theta = B\cos 2\theta + (C - A)\sin 2\theta$$

The condition $B' = 0$ (so that no cross term XY is present) gives us

$$B\cos 2\theta + (C - A)\sin 2\theta = 0 \Longrightarrow \boxed{\cot 2\theta = \frac{A - C}{B}}$$

Example 4.1 Now we show that the equation $xy = 1$ is a hyperbola. This is of the form in Eq. (4.1.7) with $A = C = 0$ and $B = 1$. Thus, $\cot 2\theta = 0$, hence $\theta = \pi/4$. With this rotation angle, using Eq. (4.1.8) we can write x, y in terms of X, Y as

$$x = X\frac{\sqrt{2}}{2} - Y\frac{\sqrt{2}}{2}, \quad y = X\frac{\sqrt{2}}{2} + Y\frac{\sqrt{2}}{2}$$

And therefore $xy = 1$ becomes

$$\frac{X^2}{2} - \frac{Y^2}{2} = 1$$

which is obviously a hyperbola.

We also compute A' and C' for another use later:

$$A' = A\cos^2\theta + B\sin\theta\cos\theta + C\sin^2\theta, \quad C' = A\sin^2\theta - B\sin\theta\cos\theta + C\cos^2\theta$$

Isn't is remarkable that even though A', B', C' are different from A, B, C, certain quantities do not. For example, the sum $A' + C'$ is invariant:

$$A' + C' = A + C$$

We also have another invariant—the so-called discriminant of the equation given by $B'^2 - 4A'C'$:

$$\begin{aligned}
B'^2 - 4A'C' &= (B(\cos^2\theta - \sin^2\theta) + 2(C - A)\sin\theta\cos\theta)^2 - \\
&4(A\cos^2\theta + B\sin\theta\cos\theta + C\sin^2\theta)(A\sin^2\theta - B\sin\theta\cos\theta + C\cos^2\theta) \\
&= B^2(\cos^2\theta - \sin^2\theta)^2 + (4BC - 4BA)(\cos^3\theta\sin\theta - \sin^3\theta\cos\theta) + \\
&4(C - A)^2\sin^2\theta\cos^2\theta - \\
&4A^2\cos^2\theta\sin^2\theta + 4AB\cos^3\theta\sin\theta - 4AC\cos^4\theta - 4AB\sin^3\theta\cos\theta + \\
&4B^2\sin^2\theta\cos^2\theta - 4BC\sin\theta\cos^3\theta - 4CA\sin^4\theta + 4CB\sin^3\theta\cos\theta - \\
&4C^2\cos^2\theta\sin^2\theta \\
&= B^2(\cos^4\theta + \sin^4\theta) + 2B^2\cos^2\theta\sin^2\theta - 8AC\sin^2\theta\cos^2\theta \\
&- 4AC(\cos^4\theta + \sin^4\theta) \\
&= (B^2 - 4AC)(\cos^4\theta + \sin^4\theta) + 2(B^2 - 4AC)\sin^2\theta\cos^2\theta \\
&= (B^2 - 4AC)(\cos^4\theta + \sin^4\theta + 2\sin^2\theta\cos^2\theta) \\
&= (B^2 - 4AC)(\cos^2\theta + \sin^2\theta)^2 = B^2 - 4AC
\end{aligned}$$

We have shown the proof to demonstrate the fact that sometimes mathematics can be boring with tedious manipulations of algebraic expressions. But to be able to do something significant, we have to be patient and resilience. Intelligence alone is not enough. Albert Einstein once said "*It is not that I'm so smart. But I stay with the questions* ***much longer***". And Voltaire—one of the greatest of all French Enlightenment thinkers and writers—also said "*no problem can withstand the assault of sustained thinking*".

And yet, there are tools called computer algebra systems such as `Maple`, `Mathematica` and `SageMath` that can do symbolic calculations, so these tools can help us with some tedious symbolic calculations such as the one we have just done. See Sect. 3.27 for details.

Now, you might ask why bother with this term $B'^2 - 4A'C'$? To answer that question, we ask you to tell us which conic section is associated with the equation

$$5x^2 + y^2 + y - 8 = 0$$

Well, you can massage (completing the square technique) the equation to arrive at

$$x^2/5 + (y + 1/2)^2 = 33/4$$

which is an ellipse. Very good! How about $5x^2 - 23xy + y^2 + y - 8 = 0$? As completing the square trick does not work due to the cross term xy, you think of using a program to plot this equation and the type of curve comes out immediately. But, this question is not about the final result but about finding a way (using only paper/pencil) to classify conic sections. Another way (not so good) is to rotate the axes, so that $B' = 0$, then using the complete the square technique. It works, but not so elegantly.

Now you have seen that we can rotate a conic section $Ax^2 + Bxy + Cy^2 + Dx + Ey + F = 0$ to get the same conic but written in this simpler form $A'X^2 + C'Y^2 + D'X + E'Y + F = 0$. And we have shown that $B^2 - 4AC = -4A'C'$. It can be shown that using this $A'X^2 + C'Y^2 + D'X + E'Y + F = 0$, one can deduce the type of the conic based on the sign of $-4A'C'$, thus for the general form of conic $Ax^2 + Bxy + Cy^2 + Dx + Ey + F = 0$, we have this theorem:

$$\begin{cases} B^2 - 4AC > 0: & \text{hyperbola} \\ B^2 - 4AC < 0: & \text{ellipse} \\ B^2 - 4AC = 0: & \text{parabola} \end{cases} \tag{4.1.10}$$

4.2 Functions

Consider now Galileo's experiments on balls rolling down a ramp. He measured how far a ball went in a certain amount of time. If we denote time by t and distance by s, then we have a relation between s and t. As s and t are varying quantities, they are called *variables*. The relation between these two variables is a function. Loosely stated for the moment, a function is a relation between variables.

The most effective mathematical representation of a function is what we call a formula. For example, the distance the ball traveled is written as $s = t^2$. The formula immediately gives us the distance at any time; for example, by plugging $t = 2$ into the formula the distance traveled is 4.[5] As s depends on t, t is an *independent variable* and s a *dependent variable*. And we speak of $s = t^2$ as *s is a function of t*.

As we see more and more functions it is convenient to have a notation specifically invented for functions. Euler used the notation $s = f(t)$, reads f of t, to describe all functions of a single variable t. When the independent variable is not time, mathematicians use $y = f(x)$. You should think of $y = f(x)$ as *f acts on x to yield y*. Note that this short notation $y = f(x)$ represents all functions that take one number x and return another number y! It can be $y = x$, $y = \sin x$ etc.

In the function $y = f(x)$ for each value of x we have a corresponding value for y $(= f(x))$. But what are the possible values of x? That varies from functions to functions. For $y = x$, x can be any real number (mathematicians like to write $x \in \mathbb{R}$ for this). For $y = \sqrt{x}$, x must be any real number that is equal or larger than zero (we do not discuss complex numbers in calculus in this chapter). That's why when

[5] Units are not important here and thus skipped.

Table 4.1 Tabulated values of three functions: $y = x$, $y = x^2$ and $y = 2^x$

x	$y = x$	$y = x^2$	$y = 2^x$
0	0	0	1
1	1	1 (1)	2 (1)
2	2	4 (3)	4 (2)
3	3	9 (5)	8 (4)
4	4	16 (7)	16 (8)
5	5	25 (9)	32 (16)
6	6	36 (11)	64 (32)

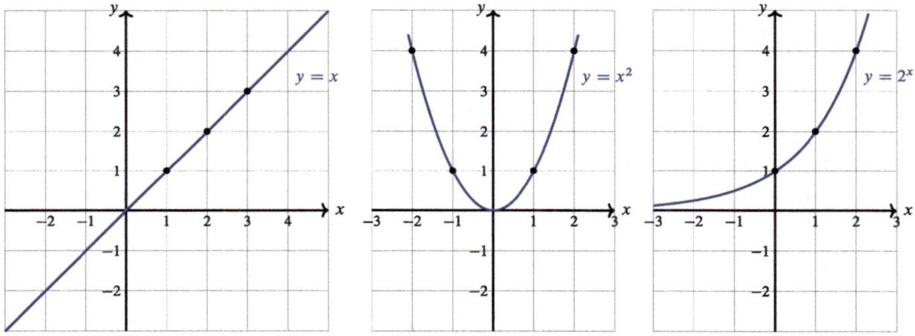

(a) $y = x$ (b) $y = x^2$ (c) $y = 2^x$

Fig. 4.5 Graph of functions $y = x$, $y = x^2$ and $y = 2^x$. The dots are those points x, $f(x)$ which are then connected by lines to get a graph. Of course, many points are used to get smooth curves

we talk about a function we need to be clear about the range of the input (called the domain of a function) and also the range of the output. The notation for that is $f : \mathbb{R} \to \mathbb{R}$ for any function that takes a real number and returns a real number.

Now we consider three common functions: a linear function $y = f(x) = x$, a power function $y = x^2$ and an exponential function $y = 2^x$. For various values of the input x, Table 4.1 presents the corresponding outputs. It is obvious that *it is hard to get something out of this table*, algebra is not sufficient. We need to bring in geometry to get insights. A picture is worth 1000 words. That's why we plot the points $(x, f(x))$ in a Cartesian plane and connect the points by lines and we get the so-called *graphs of functions*. See Fig. 4.5 for the graphs of the three functions under consideration.

With a graph you can actually see how the function is changing, where its zeroes and inflection points are, how it behaves at each point, what are its minima etc. Compare looking at a graph to looking at a picture of someone and looking at an equation to reading a description of that person.

4.2.1 Even and Odd Functions

Just as with numbers where we have even numbers and odd numbers, we also have even and odd functions. If we plot an even function $y = f(x)$ we observe that it is symmetrical with respect to (sometimes, the abbreviated w.r.t is used) the y-axis; the part on one side of the vertical axis is a *reflection* of the part on the other side, see Fig. 4.6. This means that $f(-x) = f(x)$. On the other hand, the graph of an odd function has rotational symmetry with respect to the origin, meaning that its graph remains unchanged after *rotation* of 180 °C about the origin. So, even functions and odd functions are functions which satisfy particular symmetry relations. Mathematicians define even and odd functions as:

Definition 4.4 (a) A function $f(x) : \mathbb{R} \to \mathbb{R}$ is an even function if for any $x \in \mathbb{R}$: $f(-x) = f(x)$.
(b) A function $f(x) : \mathbb{R} \to \mathbb{R}$ is an odd function if for any $x \in \mathbb{R}$: $f(-x) = -f(x)$.

Decomposition of a function. Any function $f(x)$ can be decomposed into a sum of an even function and an odd function, as

$$f(x) = f^e(x) + f^o(x) \tag{4.2.1}$$

where the even/odd functions can be found as

$$\left.\begin{aligned} f(x) &= f^e(x) + f^o(x) \\ f(-x) &= f^e(x) - f^o(x) \end{aligned}\right\} \Longrightarrow \begin{aligned} f^e(x) &= 0.5[f(x) + f(-x)] \\ f^o(x) &= 0.5[f(x) - f(-x)] \end{aligned} \tag{4.2.2}$$

Why such a decomposition is worthy of studying? One example: As integral is defined as area, from Fig. 4.6, we can deduce the following results:

$$\int_{-a}^{a} f^e(x)dx = 2\int_{0}^{a} f^e(x)dx, \quad \int_{-a}^{a} f^o(x)dx = 0 \tag{4.2.3}$$

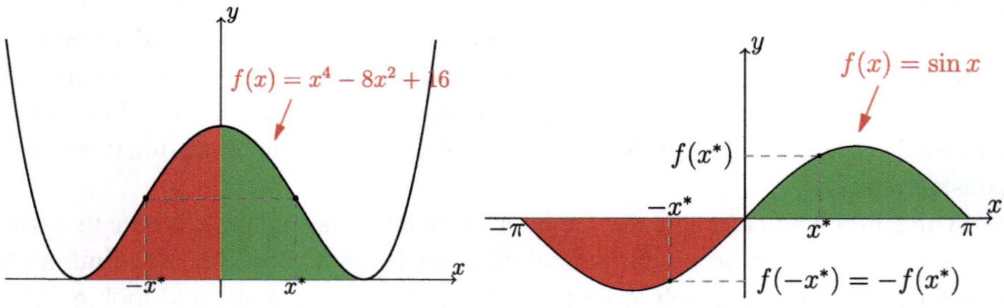

(a) even function (b) odd function

Fig. 4.6 Graphs of some even and odd functions. Typical even functions are $y = x^{2n}$, $y = \cos x$ and typical odd functions are $y = x^{2n+1}$, $y = \sin x$

4.2.2 Transformation of Functions

For ordinary objects, we can do certain kinds of transformation to them such as vertical translation (e.g. bring a book upstairs), horizontal translation. Mathematicians do the same thing, but of course to their mathematical objects which are functions in this discussion. Figure 4.7 shows vertical/horizontal translation of $y = x^2$. And this seemingly useless stuff will prove to be useful when we study waves in Sect. 7.10. To mathematicians, a traveling wave is just a function moving in space, just like what we're doing now to $y = x^2$.

And as we stretch (or squeeze/shrink) a solid object mathematicians stretch and squeeze functions. They can do a horizontal stretching by the transformation $f(cx)$ ($c < 1$) and a vertical stretching with $cf(x)$ ($c > 1$). Figure 4.8 illustrates these scaling transformations for $y = \sin x$.

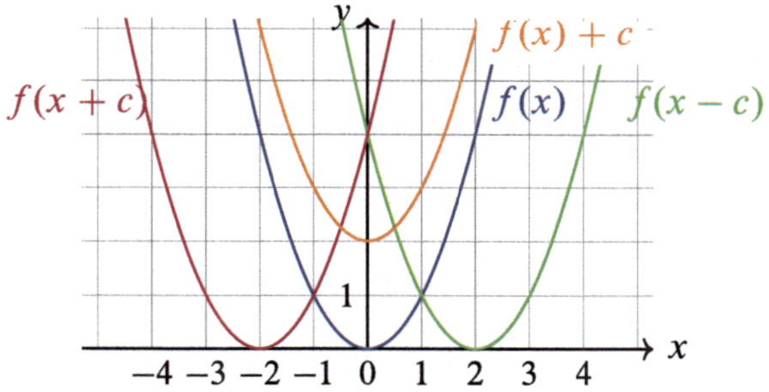

Fig. 4.7 Translation of a function $y = f(x)$: vertical translation $f(x) + c$ displaces the function a distance c upward ($c > 0$), and downward if $c < 0$. Horizontal translation to the right with $f(x - c)$ and to the left with $f(x + c)$ for $c > 0$. Note: the original function, plotted as the blue curve, is $y = x^2$

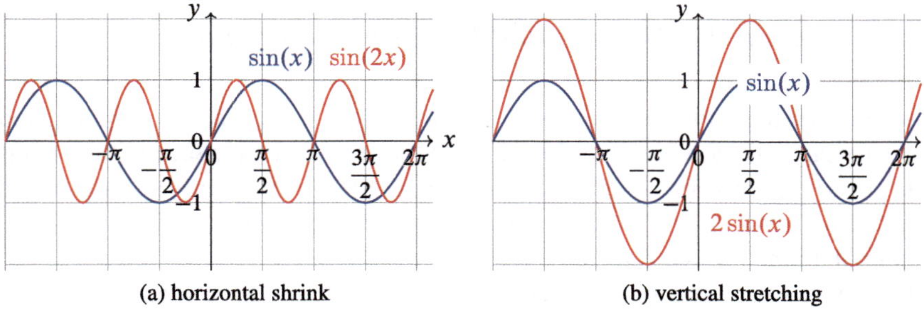

Fig. 4.8 Stretching and shrinking mathematical functions

4.2.3 Function of Function

Function composition (or function of function) is an operation that takes two functions f and g and produces a function h such that $h(x) = g(f(x))$. Intuitively, composing functions is a chaining process in which the output of function f feeds the input of function g. This is one way for mathematicians to create new functions from old ones.

The notation $(g \circ f)(x)$ is used to represent a composition of two functions:

$$(g \circ f)(x) := g(f(x)) \tag{4.2.4}$$

For example, consider two functions: $g(x) = \sin x$ and $f(x) = x^2$, we obtain the composite function $\sin(x^2)$. If we do the inverse i.e., $(f \circ g)(x)$ we get $\sin^2 x$. So, $(g \circ f)(x) \neq (f \circ g)(x)$. Is it interesting to know that, later on in linear algebra course, we will see that this fact is why matrix-matrix multiplication is not commutative (Sect. 9.6).

How about chaining three functions $h(x)$, $g(x)$ and $f(x)$? It is built on top of composing two functions:

$$[h \circ (g \circ f)](x) = h([g \circ f](x)) = h(g[f(x)]) \tag{4.2.5}$$

where we have used Eq. (4.2.4) in the first equality. It can be seen that (verify it for yourself)

$$[h \circ (g \circ f)](x) = [(h \circ g) \circ f](x)$$

That is function composition is not commutative but is associative (similar to $(ab)c = a(bc)$ for reals a, b, c).

4.2.4 Domain, Co-Domain and Range of a Function

If we consider these two functions $y = x^2$ and $y = \sqrt{x}$ we see that the first function is an easy guy who accepts any value of x. On the other hand, the second function is picky; it only accepts non-negative numbers i.e., $x \geq 0$.[6] So, to specify all the possible values of the input, mathematicians invented the concept *domain of a function*. The domain of a function is the set of all inputs.

If we have something for the input, you know that we also have something for the output: that is called the co-domain. The co-domain is the set of outputs. Thus, the co-domain of $y = x^2$ and $y = \sqrt{x}$ is $\mathbb{R}$ i.e., all real numbers. However, we know that $y = \sqrt{x}$ can only output non-negative reals. Ok. Mathematicians invented another

[6] We confine our discussion in this chapter mostly to functions of real numbers. Functions of complex numbers are left to Chap. 7.

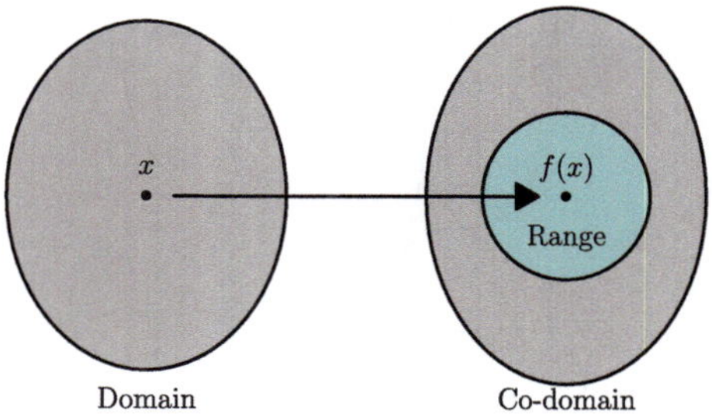

Fig. 4.9 Venn diagram for domain, co-domain, and range of a function

concept: *the range of a function*, which is a sub-set of its co-domain which contains the actual outputs. For example, if $f(x) = x^2$, then its co-domain is all real numbers but its range is only non-negative reals. Using Venn diagrams[7] we can visualize these concepts (Fig. 4.9).

Example 4.2 One example is sufficient to demonstrate how to find the domain of a function:

$$f(x) = \frac{2x - 1}{1 - \sqrt{x - 5}}$$

As we forbid division by zero and only real numbers are considered, the function only makes sense when:

$$\begin{cases} 1 - \sqrt{x - 5} \neq 0 \\ \qquad x - 5 \geq 0 \end{cases} \Longrightarrow x \neq 6 \ \text{ and } x \geq 5$$

To say x is a number that is larger or equal 5 and different from 5, we can write $x \neq 6$ and $x \geq 5$. Mathematicians seems to write it this way: $x \in [5, 6) \cup (6, \infty)$. This is because they're thinking this way: considering the number line starting from 5, and you make a cut at 6 (we do not want it). Thus the line is broken into two peaces $[5, 6)$ and $(6, \infty)$. And the symbol $\cup$ in $A \cup B$ means a union of both sets A and B. The brackets mean that the interval is closed—it includes the endpoints. An open interval (a, b), on the other hand, would not include endpoints a and b, and would be defined as $a < x < b$.

4.2.5 Inverse Functions

Given a function $y = f(x)$ which produces a number y from an input x, the inverse function $x = g(y)$ undoes the function f by giving back the number x that we started with. That is if $y = f(x)$, then $x = g(f(x))$. The inverse function of f is commonly denoted by f^{-1} (not $1/f$). We now illustrate some examples. Corresponding to a

[7] Check Sect. 5.5 if you're not sure of Venn diagrams.

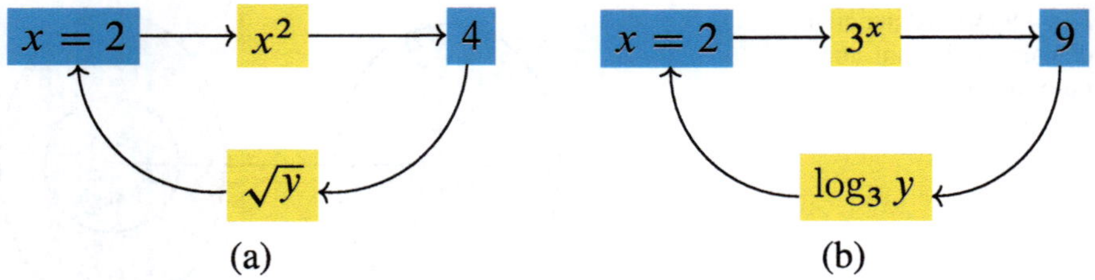

Fig. 4.10 Illustration of some inverse functions: $x^2/\sqrt{y}$ and $3^x/\log_3 y$

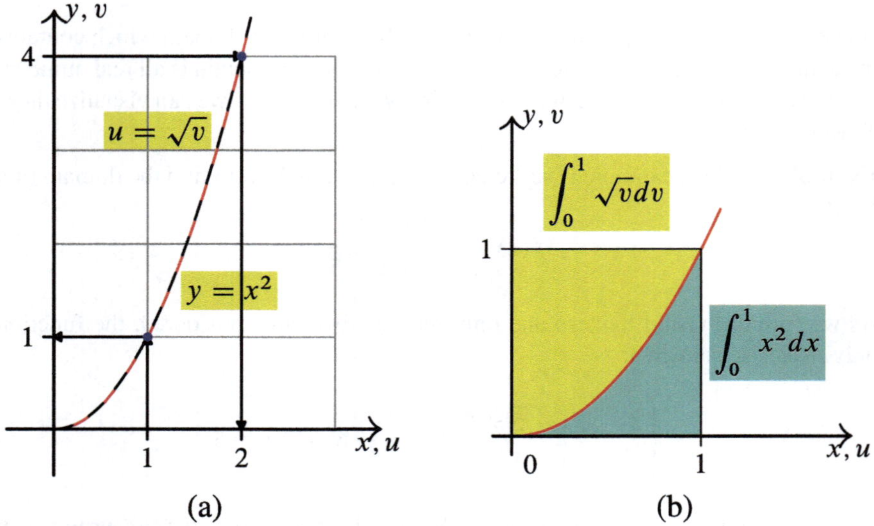

Fig. 4.11 As the sum of $\int_0^1 x^2dx$ and $\int_0^1 \sqrt{v}dv$ is one, knowing the former will give us the latter

power function and an exponential function, we have their inverses: the root function and the logarithm function functions, see Fig. 4.10.

Let's assume that we know the integral of $y = x^2$ between 0 and 1, it is written as $\int_0^1 x^2dx$ (Sect. 4.3.2 will discuss this weird symbol). What is then $\int_0^1 \sqrt{v}dv$? As the two functions x^2 and $\sqrt{u}$ are inverses of each other, it follows that the sum of these integrals equal one (Fig. 4.11). So, knowing one integral yields the other.

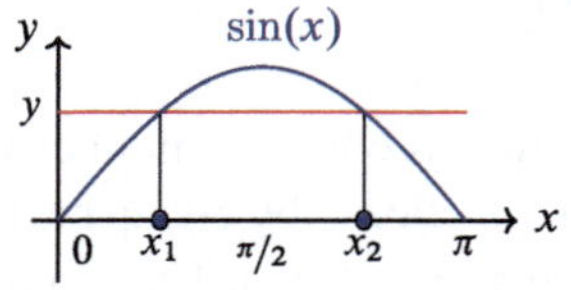

Invertible functions. If we plot the function $y = \sin x$ on the interval $0 \leq x \leq \pi$, then there are two $x-$coordinates, x_1 and x_2, having the same $y-$coordinate of $\sin(y)$ for $0 \leq y \leq 1$. Thus there does not exist the inverse of $y = \sin x$ on the interval $0 \leq x \leq \pi$. This leads to the following definition. A function $f : A \to B$ is **one-to-one** (or injective) if distinct inputs give distinct outputs. That is, f is one-to-one if $x_1 \neq x_2$, then $f(x_1) \neq f(x_2)$.

4.2.6 *Parametric Curves*

When we express the equation of a circle of radius r centered at the origin in the form $y = f(x)$, it reveals one big limitation: we need two equations $y = \pm\sqrt{r^2 - x^2}$ to fully describe the circle. By introducing an additional variable, usually denoted by t, it is possible to express the full circle by just one equation

$$\begin{aligned} x(t) &= r \cos t \\ y(t) &= r \sin t, \quad 0 \leq t \leq 2\pi \end{aligned} \tag{4.2.6}$$

Curves, represented by an equation of the form $(x(t), y(t))$ are called *parametric curves*. And the variable t is called a parameter. Why the symbol t is used for the parameter? Think of a particle moving on a plane, as it moves it traces out a curve; its position is specified by its coordinates which are a function of time.

How to get the graph of a parametric curve? That is simple: for each value of t, we compute $x(t)$ and $y(t)$, which constitute a point in the xy plane. The locus of all such points is that graph. Figure 4.12 shows some parametric curves. We have more to say about parametric curves in Sect. 4.12.

4.2.7 *History of Functions*

M. Kline credits Galileo with the first statements of dependency of one quantity on another, e.g., "The times of descent along inclined planes of the same height, but of different slopes, are to each other as the lengths of these slopes". In 1714, Leibniz already used the word "function" to mean quantities that depend on a variable. The notation we use today, $f(x)$, was introduced by Euler in 1734 [35].

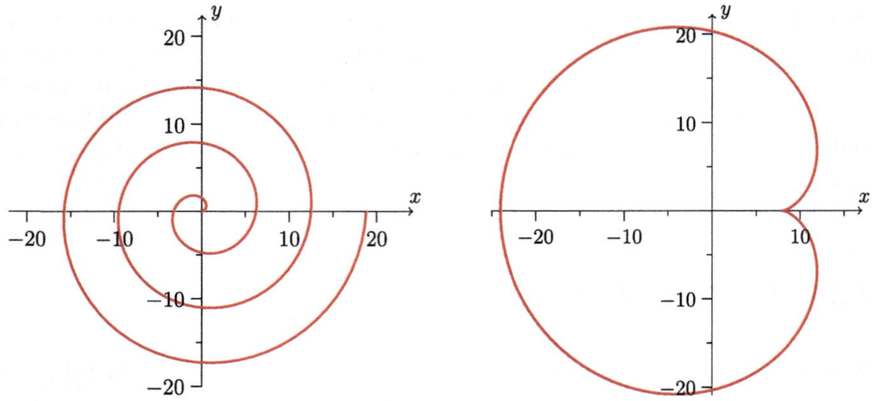

(a) $(t \cos t, t \sin t)$ (b) $a(2\cos t - \cos(2t)), a(2\sin t - \sin(2t))$

Fig. 4.12 Spiral and the cardioid, which we have met in Fig. 1.1. In Sect. 4.13.3 I will discuss how to derive the equation of the cardioid

4.2.8 Some Exercises About Functions

Let's $f(x) : [0, 1] \to \mathbb{R}$, given by $f(x) = {}^{4^x}/_{4^x+2}$. Compute the following sum

$$S = f\left(\frac{1}{40}\right) + f\left(\frac{2}{40}\right) + \cdots + f\left(\frac{39}{40}\right) - f\left(\frac{1}{2}\right)$$

This is from JEE-Advanced 2021 exam. Joint Entrance Examination—Advanced (JEE-Advanced) is an academic examination held annually in India.

Ok. Assume that we're not sitting any exam, and the ultimate goal is to get the sum, then it is super easy. Write a few lines of code and you'll see that $S = 19$. But what if we actually have to do this without a calculator not alone a PC. What we're going to do? We pay attention to the expression of S and we observe a regularity:

$$S = f\left(\frac{1}{40}\right) + f\left(\frac{2}{40}\right) + \cdots + f\left(\frac{38}{40}\right) + f\left(\frac{39}{40}\right) - f\left(\frac{1}{2}\right)$$

That is the pairs of the same color sum to one (e.g. $1/40 + 39/40 = 1$, $2/40 + 38/40 = 1$, $3/40 + 37/40 = 1$ etc.). Let's compute then $f(x) + f(1-x)$, and hope that something good is there. To test this idea, we compute $f(0) + f(1)$ (as these sums are easy), and it gives us 1: very promising. Moving on to $f(x) + f(1-x)$[8]:

[8] Just use the rule $a^{y-z} = {}^{a^y}/_{a^z}$, then $4^{1-x} = {}^{4}/_{4^x}$.

$$f(x) + f(1-x) = \frac{4^x}{4^x+2} + \frac{4^{1-x}}{4^{1-x}+2} = \cdots = 1$$

The sum is also 1. Then, S is consisted of 19 sums of the form $f(x) + f(1-x)$, which is nothing but one, plus $f(20/40)$ minus $f(1/2)$. The final result is thus simply 19.[9]

A second problem we discuss is the following:

$$f_0(x) = \frac{x}{x+1}, \quad f_{n+1}(x) = (f_0 \circ f_n)(x), \quad n = 0, 1, 2, \ldots$$

Find $f_n(x)$. How we're going to solve it? We have a rule to find $f_{n+1}(x)$ for whole numbers. Let's try to compute few of them e.g. $f_1(x), f_2(x), \ldots$ to see what we get

$$\begin{aligned}
n = 0: &\quad f_1(x) = (f_0 \circ f_0)(x) = \frac{\frac{x}{x+1}}{\frac{x}{x+1}+1} = \frac{x}{2x+1}\\
n = 1: &\quad f_2(x) = (f_0 \circ f_1)(x) = \frac{\frac{x}{2x+1}}{\frac{x}{2x+1}+1} = \frac{x}{3x+1}\\
n = 2: &\quad f_3(x) = (f_0 \circ f_3)(x) = \frac{\frac{x}{3x+1}}{\frac{x}{3x+1}+1} = \frac{x}{4x+1}
\end{aligned}$$

What we just did is starting from $f_0(x) = x/x+1$, using $f_{n+1}(x) = (f_0 \circ f_n)(x)$ we compute $f_1(x)$ (recall that $(f_0 \circ f_n)(x)$ is a composite function), then using $f_1(x)$ to compute $f_2(x)$ and so on. Lucky for us that we see a pattern.[10] Observe the red numbers on each equation and we can write

$$f_n(x) = \frac{x}{(n+1)x+1}$$

Now we prove this formula using ... proof by induction (what else?). The formula works for $n = 0$. Now we assume it works for $n = k$:

$$f_k(x) = \frac{x}{(k+1)x+1}$$

And we're going to prove that it's also correct for $n = k+1$ i.e.,

$$f_{k+1}(x) = \frac{x}{(k+2)x+1}$$

You can check this.

---

[9] With this, now you can try this Canadian math problem in 1995. Let $f(x) = 9^x/(9^x+3)$. Compute $S = \sum_{n=1}^{1995} f(n/1996)$.

[10] It has to have a pattern. Why? Because this is a test! It must be answered within a short amount of time.

4.3 Integral Calculus

Integral calculus is originally concerned with *the quadrature of closed curves*. That is the problem of determining the area of a closed curve (e.g. what is the area of a circle of radius r). This problem is certainly very old dating back to the times of Greek mathematicians. But why do mathematicians bother with this problem and what is really the area of a shape? The problem is as simple as this: there are two pieces of lands, one of a rectangular shape and the other of a triangular shape. The question is which piece is larger, and by that we mean which piece takes more space. This measurement is called area.

After area of two-dimensional objects, we move to the volume of three dimensional objects: cube, cone, pyramid etc. (Sect. 4.3.1).

Although ancient mathematicians came up with genius ways to determine the area of various shapes, but the method is not universal: each shape requires a different solution! Mathematicians in the seventeenth century came up with a smarter way to solve this area problem. In Sect. 4.3.2 a definition of a definite integral is given, and in Sect. 4.3.3 we compute the first definite integral using that definition.

4.3.1 Volumes of Simple Solids

After area is volume of simple 3D solids (Fig. 4.13). The volume of a cube of side a is a^3, and that of a cuboid of sides a, b, c is abc. This is simple. But what is the volume of a cylinder? A cylinder is an interesting object. It's kind of round and it's kind of straight.

The method of working out the volume of a cylinder of height h and radius r is the same: chopping it by vertical thin slices (Fig. 4.14). Each slice is a thin cuboid of height h and base area A_i. The volume of the cylinder is the total volume of all the cuboids. As the number of slices gets larger, the area of the bases of these rectangles (e.g. $\sum_i A_i$) approaches the area of the base of the cylinder, which is πr^2. Thus, the volume of the cylinder is $\pi r^2 h$: base area multiplied with the height.

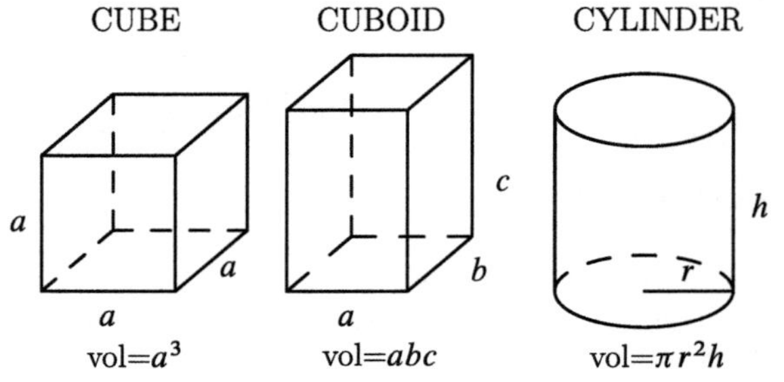

Fig. 4.13 Volume of a cube, a cuboid and a cylinder. The volumes of these solids are always: base area times height

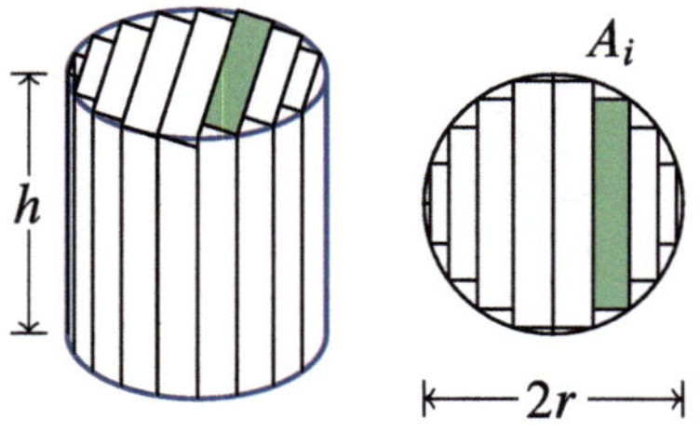

$$\begin{aligned}\text{vol} &= \sum \text{vol of cuboids}\\ &= \sum_{i=1}^{n} h \times A_i\\ &= h\left(\sum_{i=1}^{n} A_i\right)\\ &= h \times \pi r^2\end{aligned}$$

Fig. 4.14 Method of exhaustion to determine the volume of a cylinder with radius r and height h. The volume of the cylinder is $\pi r^2 h$: base area multiplied with the height

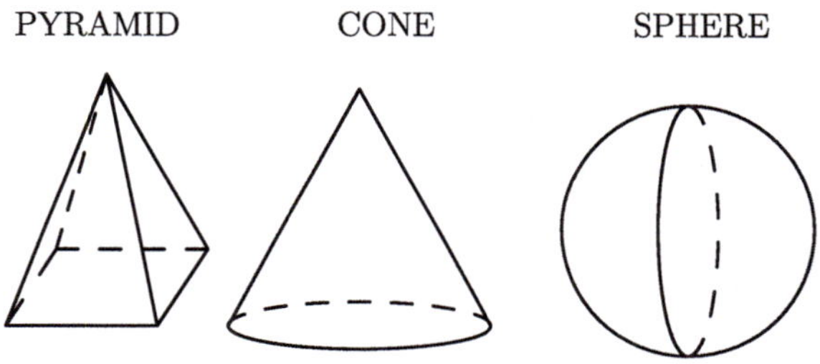

Fig. 4.15 Pyramid, cone and sphere

So far so good. The next problem of ancient mathematics was to determine the volume of pyramid, cone, and sphere (Fig. 4.15). Their arguments to find these volumes are fascinating.

Volume of a pyramid. The pyramid can be seen as a triangle, so to get its volume we can go back to how the area of a triangle was computed. We put a triangle in a rectangular case (Fig. 3.10), and asked how much space of the case the triangle takes up? It turns out that it takes up one half the area of the rectangle. Now, we do the same thing: we put a pyramid inside a case which is a cuboid and we guess the pyramid's volume is a constant times the volume of the cuboid. And we know the volume of the cuboid. What should be the constant? Is it $1/2$? No, it is 3D, the constant is $1/3$. Thus, the volume of a pyramid of base A and height h is $\frac{1}{3}Ah$. But it is not a proof!

Let's get back to 2D and we start with a square of sides a. Its area is a^2. Drawing two diagonals we get four equal triangles, each has an area of $a^2/4$ (Fig. 4.16a). Thus, if the triangle has a base a and a height $a/2$, its area is $\frac{1}{2}(a/2)(a)$. What if the height is not $a/2$? Does the formula still work? Yes, we did that in Fig. 3.10. Thus, if the height is a then the area is $0.5a^2$, which is twice the old area. What we're seeing is that: if we dilate the triangle in the vertical direction by a factor α, the new area would be α times the old one. Finally, we want a general triangle (as shown in

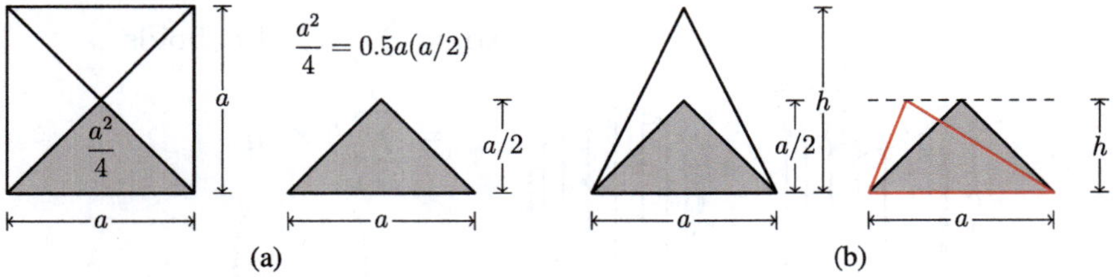

Fig. 4.16 Area of a triangle: effects of dilation and shearing

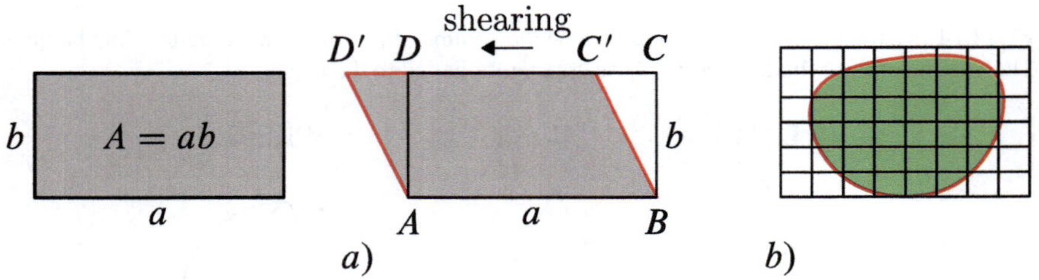

Fig. 4.17 Shearing a shape does not change its area: easy to see for rectangles/squares; as $CC' = DD'$ the area of $ABCD$ and $ABC'D'$ are equal (**a**). For a general shape (**b**), think of it as made of many many tiny squares. When we shear it each square does not change area, thus the area of the shape, which is the total area of all the squares, does not change. What a nice argument! This can be also proved using linear algebra, particularly linear transformation and determinant. See Chap. 9

the right most figure in Fig. 4.16b): the area is still $0.5bh$. What does it tell us? It indicates that shearing a shape does not change its area (Fig. 4.17).

For a pyramid, the argument is exactly the same: see Fig. 4.18.

Volume of a cone. A cone looks similar to a pyramid, so is it volumes also $1/3Ah$? It is. How to prove it? Use the volume of a pyramid. A cone is made of many many pyramids whose the bases making up the base of the cone. All these pyramids have the same height, h. Thus each pyramid has a volume of $1/3A_ih$. And the total volume of these pyramids is $1/3h \sum A_i$, but $\sum A_i = A$, the base of the cone. This is exactly the strategy used to get the volume of a cylinder (see Fig. 4.14).

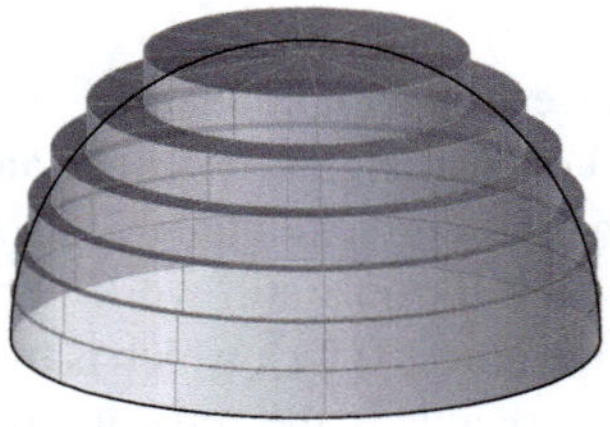

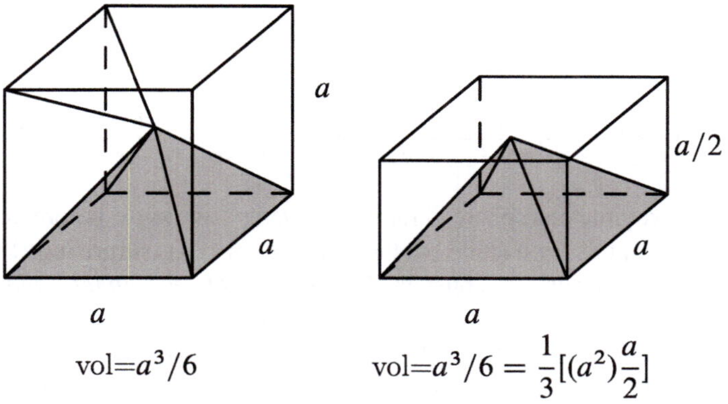

Fig. 4.18 The volume of a pyramid of base A and height h is $1/3Ah$. A cube of side a contains six equal pyramids of which the volume is $a^3/6$. This is $1/3A \times h$ where A is the area of the base and $h = a/2$ is the height. Then a vertical dilation to any h and a shearing leads to the formula $1/3A \times h$ for any pyramid

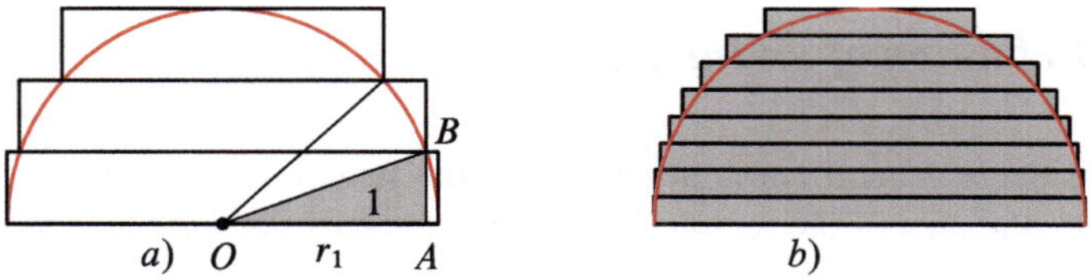

Fig. 4.19 Calculation of the volume of half of a sphere of unit radius. The hemisphere is divided into n cylinders of height $AB = 1/n$. In **a** $n = 3$ and in **b** $n = 8$. The cylinders are numbered from bottom to top starting from 0; cylinder i has a radius $r_i = \sqrt{1-(i/n)^2}$ from the Pythagorean theorem (for the right triangle OAB). Cylinder i has a volume of $\pi r_i^2(1/n)$

Volume of a ball. Finally, we address the volume of a ball of radius one. Our strategy is to divide the top half of this ball into many many cylinders of the same height. We know the volume of cylinder, and we just add up all these volumes (see beside figure). Note that this is not the only way to divide the sphere: we can equally use cylinders that fit inside the sphere. But this does not matter when we consider $n \to \infty$.

Referring to Fig. 4.19a, cylinder i has a volume of $\pi r_i^2(1/n)$. Thus, the volume V of the ball, which is the sum of the volume of n cylinders when n is very large, is

$$V = 2 \lim_{n\to\infty} \pi \left[1 + \left(1 - \left(\frac{1}{n}\right)^2\right) + \left(1 - \left(\frac{2}{n}\right)^2\right) + \cdots + \left(1 - \left(\frac{n-1}{n}\right)^2\right)\right]\frac{1}{n}$$

Now, we have a problem with algebra. Let's first massage the above expression a bit, and we shall see something familiar:

$$V = 2\pi \lim_{n\to\infty} \frac{1}{n}\left[n - \frac{1}{n^2}\left(1^2 + 2^2 + \cdots + (n-1)^2\right)\right]$$

A sum of the first $n-1$ squares! We know how to get this sum, see Eq. (2.6.8). So, we have

$$V = 2\pi \lim_{n\to\infty}\left(1 - \frac{(n-1)(2n-1)}{6n^2}\right) = 2\pi\left(1 - \lim_{n\to\infty}\frac{(n-1)(2n-1)}{6n^2}\right) = \frac{4\pi}{3} \tag{4.3.1}$$

Therefore, the volume of a sphere of radius r is $4/3\pi r^3$. We know beforehand that the volume must be of this form $c\pi r^3$, but the fact that c is $4/3$ is surprising. Archimedes was the first to find out this constant. In his work *On the Sphere and Cylinder*, Archimedes proved that the ratio of the volume of a sphere to the volume of the cylinder that contains it is 2 : 3. In that same work, he also proved that the ratio of the surface area of a sphere to the surface area of the cylinder that contains it, together with its circular ends, is also 2 : 3. Because expressions for the volume and surface area of a cylinder were known before his time, Archimedes' results established the first exact expressions for the volume and surface area of a sphere.

4.3.2 Definition of an Integral

Let's consider a general curve described by the function $y = f(x)$, and we want to calculate the area of the region bounded by this curve and $y = 0$ and $x = a$ and $x = b$. The idea is, following Archimedes, to chop this space into a large number of thin rectangles or slices (Fig. 4.20) and compute the areas of all these slices and add them up. It's then obvious that using more slices results in a better estimation of the area. And when the number of slices goes to infinity (or approaches) the total area of all the slices is exactly the area under the curve.

To make the above statement more precise, let's call n the number of slices, and A_n the total area of n slices. We start with 1 slice, then 2 slices, 3 slices, etc. up to infinity. Thus, we get a sequence $(A_n) = \{A_1, A_2, \ldots, A_n\}$. If A is the area under the curve, then this sequence approaches A when n approaches infinity. So, we define A to be the limit of the area sequence (A_n):

$$A := \lim_{n\to\infty} A_n \tag{4.3.2}$$

What we need to do now is to compute A_n. Luckily that' simple and it should be because it is our choice to make this chop! For simplicity, assume that these rectangles have the same base $\Delta x = (b-a)/n$ (Fig. 4.21). That is we place $n+1$ equally spaced points $x_0, x_1, \ldots$ over the interval $[a, b]$, we have then n sub-intervals $[x_i, x_{i+1}]$. Actually, we have two ways to build the slices: one way is to use the left point x_i[11] of $[x_i, x_{i+1}]$ (similar to an inscribed polygon in a circle); the second way is to use the right point x_{i+1} (similar to circumscribed polygon). The area A is now written as:

[11] Note that i here is not the imaginary unit $i^2 = -1$.

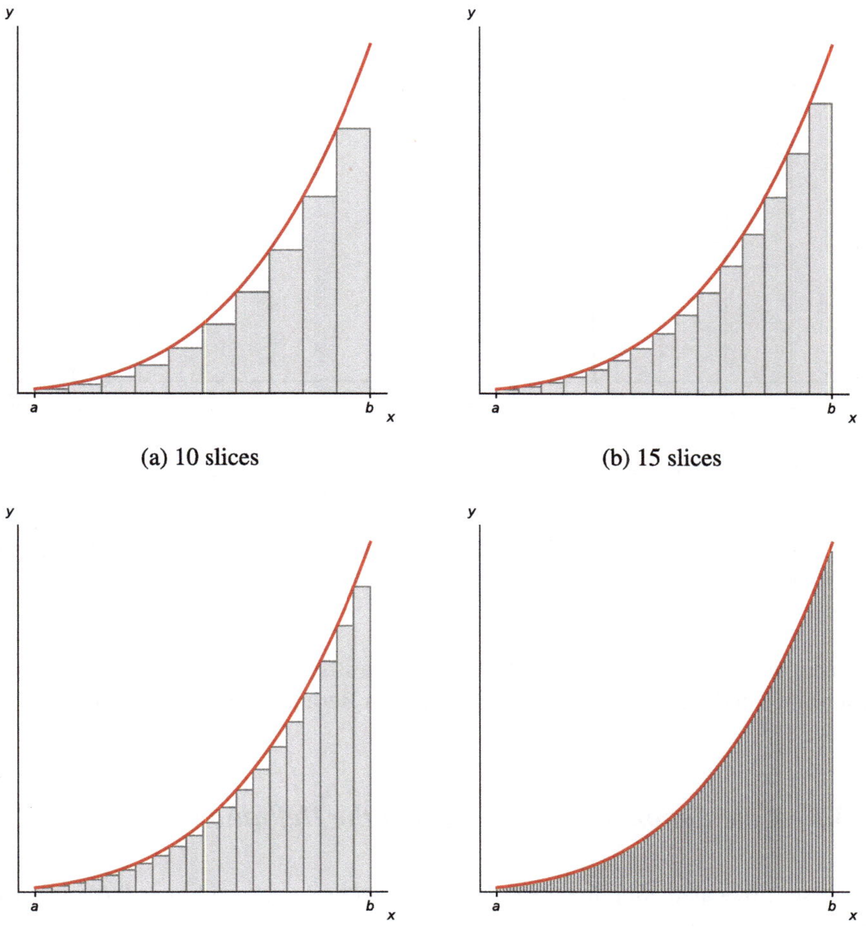

Fig. 4.20 Approximating the area under the curve $y = f(x)$ by many thin rectangles

$$A := \lim_{n\to\infty} \sum_{i=0}^{n-1} \Delta x f(x_i) := \lim_{n\to\infty} \sum_{i=1}^{n} \Delta x f(x_i) = \int_a^b f(x)dx \left(= \int_a^b dA\right) \tag{4.3.3}$$

The notation $\int$ was introduced by Gottfried Wilhelm Leibniz to represent the long S (for sum)[12] The function $f(x)$ under the integral sign is called the integrand. The points a and b are called the limits of integration and $[a, b]$ is called the interval of integration. The modern notation for the definite integral, with limits above and below the integral sign (a and b), was first used by Joseph Fourier in Mémoires of the

[12] He first used the symbol omn, short for *omnia* which is Latin for sum. On the other hand, Newton did not care about notation, thus he did not have a systematic notation for the integral.

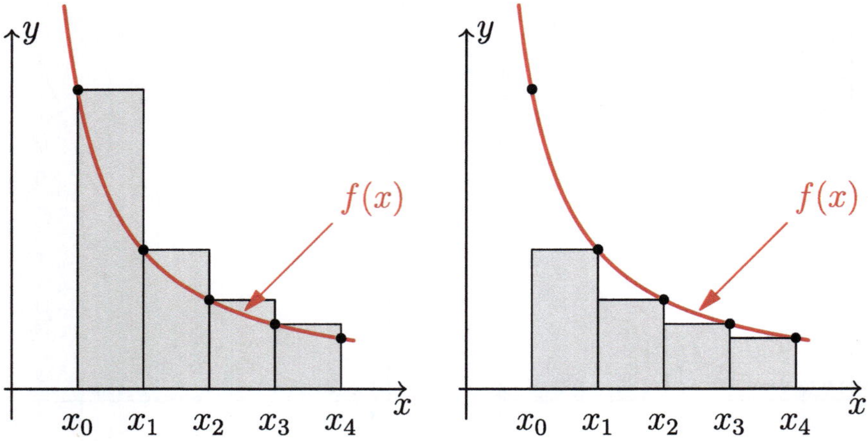

Fig. 4.21 The area of the region bounded by the curve $y = f(x)$ and $y = 0$ and $x = a$ and $x = b$: computed by chopping that region into an infinite number of thin slices. The interval $[a, b]$ is divided into n sub-intervals $[x_i, x_{i+1}]$ where $x_i = a + i(b - a)/n$, $i = 0, 1, \ldots, n$. We can either use the left point (left figure) or the right point of the sub-intervals to define the height of one slice. And of course, we can also pick any point inside a sub-interval to define the slice height

French Academy around 1819–20. The red sum is called the Riemann sum named after nineteenth-century German mathematician Bernhard Riemann.

4.3.3 Calculation of Integrals Using the Definition

Let's start by calculating the area under a parabola $y = x^2$ with $a = 0$ (the easiest area problem) and see if we get the same result that Archimedes got a long time ago. The steps are ($a = 0$, $\Delta x = b/n$, $x_i = ib/n$):

$$
\begin{aligned}
\int_0^b x^2 dx &= \lim_{n\to\infty} \sum_{i=1}^{n} \left(\frac{ib}{n}\right)^2 \frac{b}{n} && \text{(definition)}\\
&= b^3 \lim_{n\to\infty} \frac{1}{n^3} \sum_{i=1}^{n} i^2 && \text{(algebra)}\\
&= b^3 \lim_{n\to\infty} \frac{n(n+1)(2n+1)}{6n^3} && (\textstyle\sum_{i=1}^{n} i^2 \text{in Eq. 2.6.13})\\
&= b^3 \lim_{n\to\infty} \left(\frac{1}{3} + \frac{1}{2n} + \frac{1}{6n^2}\right) = \frac{b^3}{3}
\end{aligned}
$$

The red terms vanish when n approaches infinity; they are infinitely small. The result before going to the limit is quite messy (many terms), but in the limit, a simple result of $b^3/3$ was obtained. This is similar to how ancient mathematicians found the area of the circle (Fig. 3.34). By the way, the red terms account for those small triangles above the curve (the right figure in Fig. 4.21). If $b = 1$, the area is $1/3$ which agrees with Archimedes' finding.

Let's do another integral for $y = x^3$, and hope that we can see a pattern for $y = x^p$ with p being a positive integer (because we do not want to repeat this "boring" procedure for $y = x^4$, $y = x^5$ etc.; mathematics would be less interesting then):

$$\int_0^b x^3 dx = \lim_{n\to\infty} \sum_{i=1}^{n} \left(\frac{ib}{n}\right)^3 \frac{b}{n} = \lim_{n\to\infty} \frac{b^4}{n^4} \sum_{i=1}^{n} i^3 = \lim_{n\to\infty} \frac{b^4}{4} \frac{n^4 + 2n^3 + n^2}{n^4} = \frac{b^4}{4} \tag{4.3.4}$$

and we have used Eq. (2.6.12) to compute $\sum_{i=1}^{n} i^3$. We are seeing a pattern here, and thus for any positive integer p, we have the following results

$$\int_0^b x^p dx = \frac{b^{1+p}}{1+p} \Longrightarrow \int_0^a x^p dx = \frac{a^{1+p}}{1+p} \Longrightarrow \int_a^b x^p dx = \frac{b^{1+p}}{1+p} - \frac{a^{1+p}}{1+p} \tag{4.3.5}$$

which is based on Eq. (2.6.13).

The next step is to do the integration for $y = x^{1/m}$. As we know the integral of $y = x^m$—the inverse of $y = x^{1/m}$, and the sum of these two areas is known, see Fig. 4.11, it is possible to compute $\int_0^b x^{1/m} dx$. That sum is $b \times b^m = b^{1+m}$, and one area is $\int_0^b x^m dx = b^{1+m}/1+m$, and the other area is $\int_0^{b^m} v^{1/m} dv$. So, we have

$$b^{1+m} = \int_0^b x^m dx + \int_0^{b^m} v^{1/m} dv \Longrightarrow \int_0^{b^m} v^{1/m} dv = \frac{m}{1+m} b^{1+m}$$

So, we're able to compute $\int_0^b x^{1/m} dx$:

$$\int_0^b x^{1/m} dx = \frac{m}{1+m} b^{1/m+1}, \quad (m \neq -1) \tag{4.3.6}$$

One way to be sure is to use $m = 2$, $b = 1$ and Fig. 4.11 to check the new result. Obviously the integral of the hyperbola $y = 1/x$ $(m = -1)$ cannot be computed using Eq. (4.3.6) as it involves the nonsense $1/0$.

4.3.4 Rules of Integration

The following rules of integration follow from its definition. Or they can be verified geometrically as shown in Fig. 4.22. They are

$$\begin{aligned}\int_a^b f(x)dx &= \int_a^c f(x)dx + \int_c^b f(x)dx \\ \int_a^a f(x)dx = 0 &\Longrightarrow \int_a^c f(x)dx = -\int_c^a f(x)dx \\ \int_a^b [\alpha f(x) + \beta g(x)]dx &= \alpha \int_a^b f(x)dx + \beta \int_a^b g(x)dx \\ \int_a^b f(x)dx > 0 \quad &\text{if } f(x) > 0 \;\; \forall x \in [a,b]\end{aligned} \tag{4.3.7}$$

The first rule means that we can split the integration interval into sub-intervals and do the integration over the sub-intervals and sum them up. The second rule indicates that if we reverse the integration limits, the sign of the integral change. The third rule is actually a combination of two rules: $\int_a^b \alpha f(x)dx = \alpha \int_a^b f(x)dx$ and $\int_a^b [f(x) + g(x)]dx = \int_a^b f(x)dx + \int_a^b g(x)dx$. The fourth means that if the integrand is positive within an interval, then over this interval the integral is positive.

Another rule (or property) of integrals is the following

$$\text{if } h(x) \le f(x) \le g(x) \;\; (a \le x \le b) \Longrightarrow \int_a^b h(x)dx \le \int_a^b f(x)dx \le \int_a^b g(x)dx \tag{4.3.8}$$

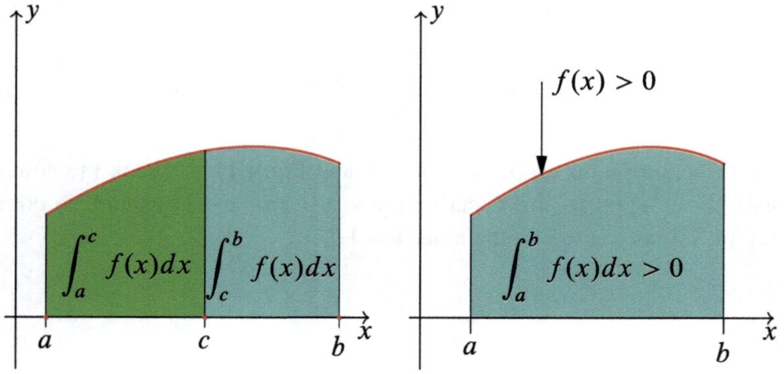

Fig. 4.22 Some properties of the integral $\int_a^b f(x)dx$

One application of Eq. (4.3.8) is to prove $\sin x \leq x$:

$$\cos t \leq 1 \Longrightarrow \int_0^x \cos dt \leq \int_0^x dt \Longrightarrow \sin x \leq x \tag{4.3.9}$$

4.3.5 *Indefinite Integrals*

When the limits of an integral are not fixed, we have an indefinite integral, which is a function. Usually, we assume the lower limit is fixed and the upper limit is a variable:

$$F(x) = \int_a^x f(u)du = \int_a^x f(t)dt \tag{4.3.10}$$

I have used two notations $f(u)du$ and $f(t)dt$ to illustrate that u or t can be thought of dummy variables; any variable (not x) can be used.

That's all we can do with integral calculus, for now. We are even not able to compute the area of a circle using the integral! We need the other part of calculus—differential calculus, which is the topic of the next section.

4.4 Differential Calculus

While integration was an idea from antiquity, the idea of derivative was relatively new. This section presents the basic ideas of differential calculus. I first present in Sect. 4.4.1 how Fermat solved a maxima problem using the idea behind the concept of derivative that we know today. As Fermat (and all the mathematicians of his time) did not know the concept of limit—which is the foundation of calculus, his maths was not rigorous, but it worked in the sense that it provided correct results. The motivation for the inclusion of Fermat's work is to show that mathematics were not developed as it is now presented in textbooks: everything works nicely. Far from that, there are set backs, doubts, criticisms, and so on. Then in Sect. 4.4.3 we talk about uniform and non-uniform speeds as a motivation for the concept of derivative introduced in Sect. 4.4.4. As we have already met the limit concept in Sect. 2.22, I immediately use limit to define the derivative of a function. But I will postpone a detailed discussion of what is a limit until Sect. 4.10 to show that without limits seventeenth century mathematicians with their intuition can proceed without rigor. This way of presentation style will, hopefully, comfort many students. It took hundreds of years for the best mathematicians to develop the calculus that we know today. It is OK for us to be confused, to make mistakes, and to have low grades.

4.4.1 Maxima of Fermat

As an elementary example of a maxima problem that Pierre de Fermat solved back in the seventeenth century we consider this problem: prove that, of all rectangles with a given perimeter, it is the square that has the largest area. This problem belongs to the so-called optimization problems. Fermat's solution, discussed below, presented the first steps in the development of the concept of a derivative.

Let's denote the perimeter by a, and one side of the rectangle by x, the other side is hence $y = a/2 - x$. Therefore, the area—which is xy—is

$$M(x) = \frac{ax}{2} - x^2 \tag{4.4.1}$$

Before presenting Fermat's solution, let's solve it the easy way: $M(x)$ is a concave parabola with an inverted bowl shape thus it has the highest point. We can rewrite M in the following form

$$M = \left(\frac{a}{4}\right)^2 - \left(x - \frac{a}{4}\right)^2 \Longrightarrow M \leq \left(\frac{a}{4}\right)^2 \tag{4.4.2}$$

thus M is the maximum when the red term vanishes or when $x = a/4$. Thus $y = a/4$, and a square has the largest area among all rectangles with a given perimeter. One thing to notice herein is that this algebraic way is working only for this particular problem. We need something more powerful which can be, hopefully, applicable to all optimization problems, not just Eq. (4.4.1).

Fermat's reasoning was that: if x is the one that renders M maximum, then *adding a small number* ϵ *to* x would not change M.[13] This gives us the equation $M(x + \epsilon) = M(x)$, and with Eq. (4.4.1), we get

$$\frac{a(x+\epsilon)}{2} - (x+\epsilon)^2 = \frac{ax}{2} - x^2 \tag{4.4.3}$$

which leads to another equation, by dividing the above equation by ϵ (this can be done because $\epsilon \neq 0$):

$$\frac{a}{2} - 2x + \epsilon = 0 \tag{4.4.4}$$

Then, he used $\epsilon = 0$, to get x

$$\frac{a}{2} - 2x = 0 \Longrightarrow x = \frac{a}{4} \tag{4.4.5}$$

[13] Why? Imagine you're climbing up a hill. When you're not at the top each move increases your altitude. But when you're already at the top, then a move will not change your altitude. Actually, it changes, but only insignificantly (assuming that your step is not giant).

To someone who knows calculus, it is easy to recognize that Eq. (4.4.5) is exactly $M'(x) = 0$ in our modern notation, where $M'(x)$ is the first derivative of $M(x)$. Thus Fermat was very close to the discovery of the derivative concept.

It is important to clearly understand what Fermat did in the above process. First, he introduced a quantity ϵ which is initially non-zero. Second, he manipulated this ϵ as if it is an ordinary number. Finally, he set it to zero. So, this ϵ *is something and nothing simultaneously!* Newton and Leibniz's derivative, also based on similar procedures, thus lacks a rigorous foundation for 150 years until Cauchy and Weierstrass introduced the concept of limit (Sect. 4.10). But Fermat's solution is correct!

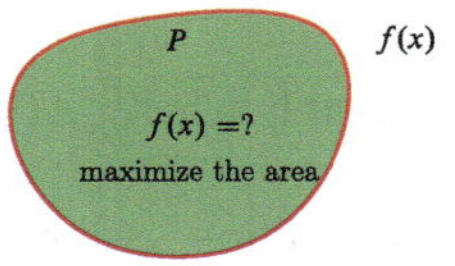

Isoperimetric problems. We have seen that among rectangles with a given perimeter, the square is the curve which has the largest area. A general version of this so-called isoperimetric problems is: if you have a piece of rope with a fixed length, what shape should you make with it in order to enclose the largest possible area? Here we are trying to choose a function $f(x)$ to maximize an integral giving the area enclosed by $f(x)$, given the constraint that the length of the curve (P) is fixed. See beside figure. We used the derivative of a function to find the solution to the rectangle problem. But it is not useful for this general isoperimetric problem as we do not know the function $f(x)$. Solving this requires a new kind of mathematics known as variational calculus developed in the seventeenth century by the likes of Euler, Bernoulli brothers, and Lagrange. See Chap. 8 for details on variational calculus.

As we're at optimization problems, let me introduce another optimization problem in the next section. This is to demonstrate that optimization problems are everywhere. We shall see that not only we try to optimize things (maximize income, minimize cost and so on) but so does nature.

4.4.2 Heron's Shortest Distance

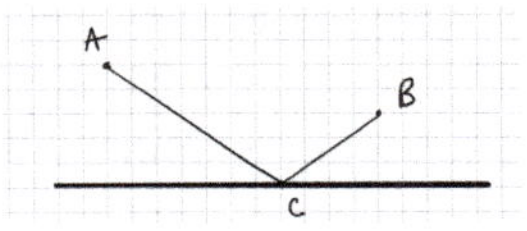

One of the first non-trivial optimization problems was solved by Heron of Alexandria, who lived about 10–75 C.E. Heron's 'Shortest Distance' problem is as follows. Given two points A and B on one side of a straight line, find the point C on the line such

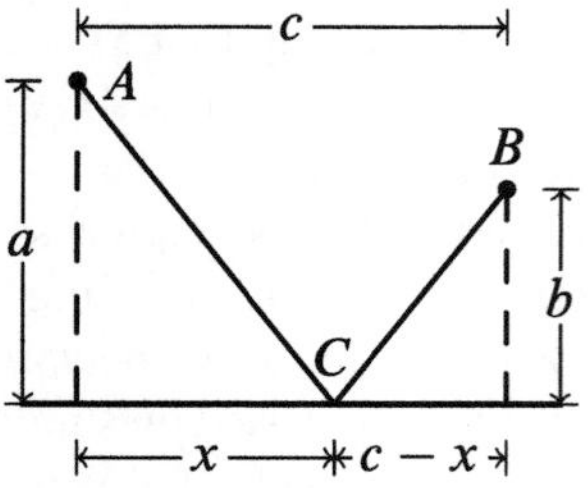

$$AC = \sqrt{a^2 + x^2}$$
$$BC = \sqrt{b^2 + (c - x)^2}$$
$$f(x) = \sqrt{a^2 + x^2} + \sqrt{b^2 + (c - x)^2}$$
$$f'(x_0) = 0 \Rightarrow x_0 \Rightarrow C$$

Fig. 4.23 Heron's shortest distance problem: we need to find the best location for point C. Note that a, b, c are constants, only x—measuring the horizontal distance from A to C—changes

that $|AC| + |CB|$ is as small as possible where $|AC|$ is the distance between A and C.

Before presenting Heron's smart solution, assume that we know calculus, then this problem is simply an exercise of differential calculus. We express the distance $|AC| + |CB|$ as a function of x—the position of point C that we're after, then calculate $f'(x)$ and set it to zero. That's it. The derivative of the distance function is (Fig. 4.23)

$$f'(x) = \frac{x}{\sqrt{a^2 + x^2}} - \frac{c - x}{\sqrt{b^2 + (c - x)^2}}$$

Thus, setting the derivative to zero gives

$$f'(x) = 0 : \implies xb = (c - x)a \implies \frac{a}{x} = \frac{b}{c - x} \tag{4.4.6}$$

What is this result saying? It is exactly the law of light reflection if we see the problem as light is moving from A, reflected on a surface at C and goes to B: angle of incidence (i.e., the angle of which tangent is a/x) equals angle of reflection, which was discovered by Euclid some 300 years earlier.

So what is exactly what Heron achieved? He basically demonstrated that reflected light takes the *shortest* path—or the *shortest* time, assuming light has a finite speed. Why is this a significant achievement? Because this was the first evidence showing that our universe is lazy. When it does something it always selects its way so that a certain quantity (e.g. time, distance, energy, action) is minimum. Think about it for a while then you would be fascinated by this idea. As a human being, we do something in many ways, and from these trials we select the best (optimal) way. Does nature do the same thing? It seems not. Then, why it knows to select the best way? To know more about this topic, I recommend the book *The lazy universe* by Coopersmith [14].[14]

[14] You can also watch this youtube video.

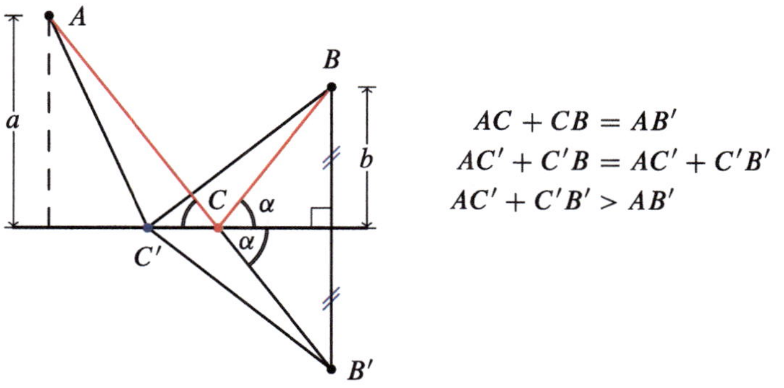

Fig. 4.24 Heron's genius solution. It uses one property of a triangle: the sum of two edges is larger than the remaining one. The crux of the proof is to create point B'

Heron's proof of the shortest distance problem. Referring to Fig. 4.24, Heron created a new point B' which is the reflection of point B through the horizontal line. Then, the solution—point C—is the intersection of line AB' and the horizontal line. An elegant solution, no question. But it lacks generality. On the other hand, the calculus-based solution is universally applicable to almost any optimization problem and it does not require the user to be a genius. With calculus, things become routine.

But wait, how did Heron know to create point B'? Inspiration, experience, trial and error, dumb luck. That's the art of mathematics, creating these beautiful little poems of thought, the sonnets of pure reason.

Algebra vs geometry. This problem illustrates the differences between algebra and geometry. Geometry is intuitive and visual. It appeals to the right side of the brain. With geometry, beginning an argument requires strokes of genius (like drawing the point B'). On the other hand, algebra is mechanical and systematic. Algebra is left-brained.

Proof of reflection property of ellipse. The reflective property of an ellipse is simply this: A ray of light starting at one focus will bounce off the ellipse and go through the other focus. Referring to Fig. 4.25, we need to prove that a light starts from F_1 coming to P, bounces off the ellipse E and gets reflected to F_2. Using the result of Heron's shortest distance property, what we have to prove is: F_1P and F_2P make equal angles with the tangent line to E at P i.e., $\theta_2 = \theta_3$.

Extend F_1P past P to F_2' such that $F_2P = PF_2'$. Let L is the line that bisects the angle $F_2'PF_2$ so that $\theta_1 = \theta_2$. Thus, L is the perpendicular bisector of F_2F_2'. What we need to prove now is: proving L is the tangent to E at P. Why? Because then we have $\theta_1 = \theta_3$ (vertical angles), but $\theta_1 = \theta_2$ (as L is the bisector), so $\theta_2 = \theta_3$.

Proof (Proof of the fact that L is tangent to E at P) We use proof by contradiction. Suppose that L is not tangent to E at P. Then L meets E at another point Q. As

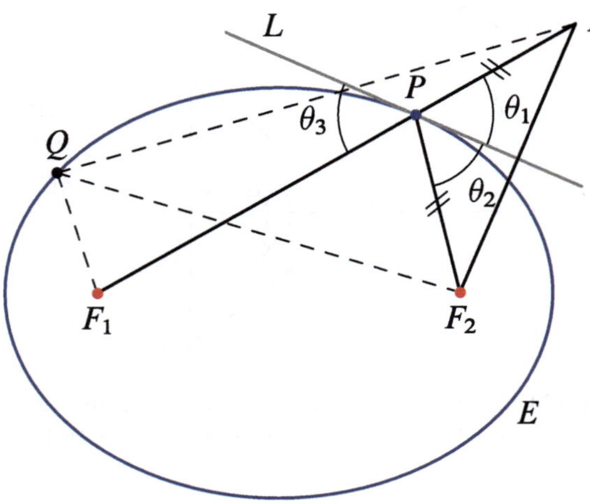

Fig. 4.25 Proof of reflection property of ellipse ($PF_1 + PF_2 = 2a$): $\theta_2 = \theta_3$

Q is on L we have $QF_2 = QF_2'$. By definition of ellipse, we have $F_1Q + F_2Q = F_1P + F_2P$. Consider now the triangle F_1QF_2', we have

$$F_1Q + F_2'Q = F_1Q + QF_2 = F_1P + F_2P = F_1P + PF_2' = F_1F_2'$$

which violates the triangle inequality in the triangle F_1QF_2'. So, L must be the tangent to E at P.

4.4.3 Uniform Versus Non-uniform Speed

To understand the concept of derivative, one can either consider the problem of finding a tangent to a curve at a certain point on the curve or the problem of determining the velocity of an object at a certain moment in time if it is moving with a non-uniform velocity. I take the second approach as there is change inherently in this problem. This was also how Newton developed his fluxions.[15] Note that Newton is not only a mathematician, but also a physicist.

Let's start simple with a car moving with a constant speed. If it has gone 30 km in 1 h, we say that its speed is 30 km per h. To measure this speed, we divide the distance the car has traveled by the elapsed time. If s measures the distance and t measure time, then

[15] The modern definition of a function had not yet been created when Newton developed his calculus. The context for Newton's calculus was a particle "flowing" or tracing out a curve in the $x - y$ plane. The x and y coordinates of the moving particle are *fluents* or flowing quantities. The horizontal and vertical velocities are the *fluxions* (which we call derivatives) of x and y, respectively, associated with the flux of time.

$$\text{uniform speed} = \frac{\text{distance}}{\text{time interval}} = \frac{\Delta s}{\Delta t} \tag{4.4.7}$$

The ratio $\Delta s/\Delta t$ is called *a time rate of change of position* i.e., change of position per unit time. Sometimes it is simply referred to as the rate of change of position. Note that Δ does not stand for any number. Δ stands for "the change in"—that and nothing else. Thus, Δs (read Delta s) is used to indicate a change in s and Δt (read Delta t) is used to indicate a change in t.

But life would be boring if everything is moving at a constant speed. Then, one would need no differential calculus. Luckily, non-uniform motions are ubiquitous. Kepler discovered that the planets moved non-uniformly around their ellipses with the Sun as focus, sometimes hesitating far from the Sun, sometimes accelerating near the Sun. Likewise, Galileo's projectiles moved at ever-changing speeds on their parabolic arcs. They slowed down as they climbed, paused at the top, and then sped up as they fell back to earth. The same was true for pendulums. And a car which travels 30 mi in an hour does not travel at a speed of 30 mi an hour. If its owner lives in a big town, the car travels slowly while it is getting out of the town, and makes up for it by doing 50 on the arterial road in the country.

How could one quantify motions in which speed changed from moment to moment? It was the task that Newton set out for himself. And to answer that question he invented calculus. We are trying here to reproduce his work. We use Galileo's experiment of a ball rolling down an inclined plane (Table 4.2 from $s = t^2$) and seek out to find the ball speed at any time instant, the notation for that is $v(t)$, where v is for velocity.

Let us first try to find out how fast the ball is going after one second. First of all, it is easy to see that the ball *continually goes faster and faster*. In the first second, it goes only 1 foot; in the next second 3 ft; in the third second 5 ft, and so on. As the average speed during the first second is 1 foot per second, the speed of the ball at 1 s must be larger than that. Similarly, the average speed during the second is 3 ft per s, thus the speed of the ball at 1 s must be smaller than that. So, we know $1 < v(1) < 3$.

Can we do better? Yes, if we have a table similar to Table 4.2 but with many more data points not at whole seconds. For example, if we consider 0.9, 1, and 1.1 s (Table 4.3), we can get $1.9 < v(1) < 2.1$. And if we consider 0.99 s, 1 s and 1.01 s, we get $1.99 < v(1) < 2.01$. And if we take a thousandth of a second, we find the speed lies between 1.999 and 2.001. And if we keep refining the time interval, we find that the only speed satisfying this is 2 ft per second. Doing the same thing, we find the speed at whole seconds in Table 4.4. If $s = t^2$, then $v = 2t$.

So the speed at any moment will not differ very much from the average speed during the previous tenth of a second. It will differ even less from the average speed

Table 4.2 Galileo experiment of ball rolling down an inclined plane

Time [second]	0	1	2	3	4	5	6
Distance [feet]	0	1	4	9	16	25	36

Table 4.3 Galileo experiment of ball rolling down an inclined plane with time increments of 0.1 s

Time [second]	0.9	1	1.1
Distance [feet]	0.81	1	1.21

Table 4.4 Galileo experiment of ball rolling down an inclined plane: instantaneous speed

Time [second]	0	1	2	3	4	5	6
Speed [feet/s]	0	2	4	6	8	10	12

Table 4.5 Limit of average speeds when the time interval h is shrunk to zero

h	$(t_0+h)^2-t_0^2/h$	$(t_0-h)^2-t_0^2/-h$
10^{-1}	4.100000000000	3.900000000000
10^{-2}	4.010000000000	3.990000000000
10^{-3}	4.001000000000	3.998999999999
10^{-4}	4.000100000008	3.999900000000
10^{-5}	4.000010000027	3.999990000025
10^{-6}	4.000001000648	3.999998999582

for the previous thousandth of a second. In other words, if we take the average speed for smaller and smaller lengths of time, we shall get nearer and nearer—as near as we like—to the true speed. Therefore, the instantaneous speed i.e., the speed at a time instant is *defined as the value that the sequence of average speeds approaches when the time interval approaches zero*. We show this sequence of average speeds in Table 4.5 at the time instant $t_0 = 2$s. Note that this table presents not only the average speeds from the time instances $t_0 + h$ and t_0, but also from $t_0 - h$ and t_0. And both sequences converge to the same speed of 4, which is physically reasonable. Later on, we know that these correspond to the right and left limits.

But saying "the value that the sequence of average speeds approaches when the time interval approaches zero" is verbose, we have a symbol for that, discussed in Sect. 2.22. Yes, that value (i.e., the instantaneous speed) is the limit of the average speeds when the time interval approaches zero. Thus, the instantaneous speed is defined succinctly as

$$\text{instantaneous speed} s'(t) \text{ or } \dot{s} \equiv \lim_{\Delta t \to 0} \frac{\Delta s}{\Delta t} \tag{4.4.8}$$

where, we recall, the notation Δs is used to indicate a change in s; herein it indicates the distance traveled during Δt. And we use the symbol $s'(t)$ to denote this instantaneous speed and call it the derivative of $s(t)$. Newton's notation for this derivative is $\dot{s}$, and it is still being used especially in physics. This instantaneous speed is the number that the speedometer of your car measures.

4.4.4 The Derivative of a Function

Leaving behind distances and speeds, if we have a function $f(x)$, then its derivative at point x_0, denoted by $f'(x_0)$, is defined as

$$f'(x_0) = \lim_{\Delta x \to 0} \frac{f(x_0 + \Delta x) - f(x_0)}{\Delta x} = \lim_{\Delta x \to 0} \frac{\Delta f}{\Delta x} \tag{4.4.9}$$

In words, the derivative $f'(x_0)$ is the limit of the ratio of change of f (denoted by Δf) and change of x (denoted by Δx) when Δx approaches zero. The term $\Delta f/\Delta x$ is called *a difference quotient.*

Instead of focusing on a specific value x_0, we can determine the derivative of $f(x)$ at an arbitrary point x, which is denoted by $f'(x)$. For an x we have a corresponding number $f'(x)$, thus $f'(x)$ is a function in itself. Often we use h in place of Δx because it is shorter. Thus, the derivative is also written as

$$f'(x) = \lim_{h \to 0} \frac{f(x+h) - f(x)}{h}$$

Notations for the derivative. There are many notations for the derivative: (1) Newton's notation $\dot{f}$, (2) Leibniz's notation for the derivative dy/dx, and (3) Lagrange's notation $f'(x)$. Let's discuss Lagrange's notation first as it is easy. Note that given a function $y = f(x)$, its derivative is also a function, which Lagrange called a *derived function* of $f(x)$. That's the origin of the name "derivative" we use today. Lagrange's notation is short, and thus very convenient.

How about Leibniz's notation? I emphasize that when Leibniz developed the concept of derivative, the concept of limit was not available.[16] Leibniz was clear that the derivative was obtained when Δf and Δx were very small, thus he used df and dx, which he called the infinitesimals (infinitely small quantities) or differentials. An infinitesimal is a hazy thing. It is supposed to be *the tiniest number we can possibly imagine that isn't actually zero*. In other words, an infinitesimal is smaller than everything but greater than nothing (0). On the other hand, the notation dy/dx has these advantages: (i) it reminds us that the derivative is the rate of change $\Delta y/\Delta x$ when $\Delta x \to 0$ (the ds remind us of the limit process), (ii) it reveals the unit of the derivative immediately as it is written as a ratio while $f'(x)$ is not. But, the major

[16] This only came to life about 200 years after Newton and Leibniz!.

advantage is that we can use the differentials dy and dx separately and perform algebraic operations on them just like ordinary numbers.

4.4.5 Infinitesimals and Differentials

To understand Leibniz's infinitesimals, surprisingly a simple question is a big help. We all know that $2^3 = 8$, but what is 2.001^3? We're not interested in the final result but in the structure of the result. Using a calculator, we get

$$2.001^3 = 8.012006001$$

So, it is 8 plus a bit. That makes sense, a tiny change from 2 to 2.001 results in a tiny change from 8 to 8.012006001 (a change of 0.012006001). What is interesting is that we can decompose this change into a sum of three parts as follows

$$0.012006001 = 0.012 + 0.000006 + 0.000000001$$

which is a small plus a super-small plus a super-super small.

We can use algebra to understand this structure of the result. Let's consider x_0 (instead of 2 as we did previously) and a change Δx, then we ask what is $(x_0 + \Delta x)^3$. It is given by (we can multiply out or use Pascal's triangle if we're lazy):

$$(x_0 + \Delta x)^3 = x_0^3 + 3x_0^2\Delta x + 3x_0(\Delta x)^2 + (\Delta x)^3$$

Thus the change $(x_0 + \Delta x)^3 - x_0^3$ is

$$(x_0 + \Delta x)^3 - x_0^3 = 3x_0^2\Delta x + 3x_0(\Delta x)^2 + (\Delta x)^3$$

And putting $x_0 = 2$ into the above equation, we have

$$(2 + \Delta x)^3 - 2^3 = 12\Delta x + 6(\Delta x)^2 + (\Delta x)^3 \tag{4.4.10}$$

Now we can see why the change consists of three parts of different sizes. The small but dominant part is $12\Delta x = 12(.001) = .012$. The remaining parts $6(\Delta x)^2$ and $(\Delta x)^3$ account for the super-small .000006 and the super-super-small .000000001. The more factors of Δx there are in a part, the smaller it is. That's why the parts are graded in size. Every additional multiplication by the tiny factor Δx makes a small part even smaller.

Now comes the power of Leibniz's notation dx and dy. In Eq. (4.4.10), if we replace Δx by dx and call dy the change due to dx, and of course we neglect the super and super-super small parts (i.e., $(dx)^2$ and $(dx)^3$), then we have a nice formula:

$$dy := (2 + dx)^3 - 2^3 = 12dx \tag{4.4.11}$$

which allows us to write

$$\frac{dy}{dx} = 12 \text{ which is nothing but the derivative of} x^3 \text{ at } x = 2$$

For Leibniz dx and dy exist (sometimes he doubted their existence), or in other words, to him, they are fundamental mathematical objects and the derivative just happens to be their ratio. Many mathematicians object to this way of doing calculus. But I do not care. Euler, Bernoullis, and many other mathematicians used these dx and dy successfully. If you want rigor, that is fine, then use limit.

Differential operator. Yet another notation for the derivative of $y = f(x)$ at x_0 is

$$\left. \frac{d}{dx} f(x) \right|_{x=x_0} = \lim_{h \to 0} \frac{f(x_0 + h) - f(x_0)}{h}$$

This notation adopts the so-called *differential operator* $\frac{d}{dx} f(x)$. What is an operator? Think of the square root of a number. Feed in a number x, the operator square root $\sqrt{\Box}$ gives another number $\sqrt{x}$. Similarly, feed in a function $f(x)$, the operator $d/dx \Box$ gives another function—the derivative $f'(x)$. For the time being, just think of this operator as another notation that works better aesthetically (not objective) for functions of which the expression is lengthy. Compare the following two notations and decide for yourself:

$$\left(\frac{x^2 + 3x + 5}{\sqrt{x^3 - 3x + 1}} \right)', \quad \frac{d}{dx} \left(\frac{x^2 + 3x + 5}{\sqrt{x^3 - 3x + 1}} \right)$$

Later on, we shall see that mathematicians consider this operator as a legitimate mathematical object and study its behavior. That is, they remove the functions out of the picture and think of the differentiation process (differentiation is the process of finding the derivative).

Nonstandard analysis. The history of calculus is fraught with philosophical debates about the meaning and logical validity of fluxions and infinitesimals. The standard way to resolve these debates is to define the operations of calculus using the limit concept rather than infinitesimals. And that resulted in the so-called *real analysis*. On the other hand, in 1960, Abraham Robinson[17] developed *nonstandard analysis* that reformulates the calculus using a logically rigorous notion of infinitesimal numbers. This is beyond the scope of the book and my capacity as I cannot afford to learn another kind of number—known as the hyperreals (too many already!).

[17] Abraham Robinson (1918–1974) was a mathematician who is most widely known for development of nonstandard analysis, a mathematically rigorous system whereby infinitesimal and infinite numbers were reincorporated into modern mathematics.

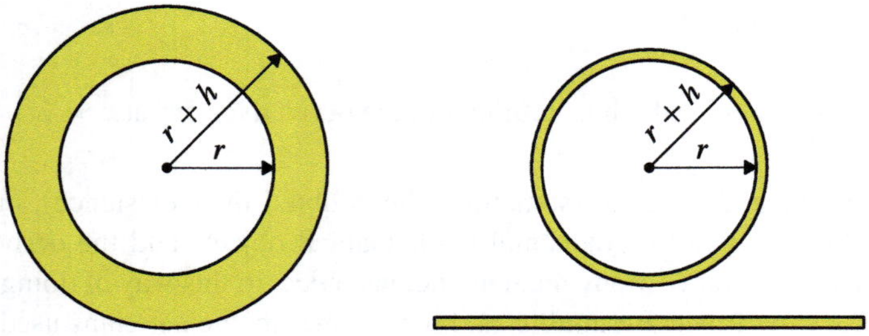

Fig. 4.26 Area of an annulus of radii r and $r + h$: interesting appears when considering small hs

Surface of a ball. Equipped with infinitesimals, we're now able to determine the surface area of a sphere (of radius r). Thanks to Archimedes, we know that it is $4\pi r^2$. But why? To derive this result, let's get back to the circumference of a circle. We want to derive the formula of the circumference given the area formula. The idea is to consider an annulus of radii r and $r + h$ as shown in Fig. 4.26. It is obvious that the area of the annulus is the area of the bigger circle minus that of the smaller circle:

$$A_{\text{annulus}} = \pi(r+h)^2 - \pi r^2 = \pi[(r+h)^2 - r^2] = \pi(2rh + h^2)$$

Now, we consider tiny h, in which case h^2 is negligible, so $A_{\text{annulus}} = (2\pi r)h$. But this thin annulus is nothing but a rectangle of length C—the circumference of our circle and width h. Therefore, $C = 2\pi r$.

We now play the same game for spheres: we consider two spheres of which one has a radius r and the bigger has a radius $r + h$, then we compute the volume of the region between the two spheres:

$$V = \frac{4}{3}\pi(r+h)^3 - \frac{4}{3}\pi r^3 = \frac{4}{3}\pi(3r^2h + 3rh^2 + h^3)$$

When h is super tiny, we then have

$$V = \frac{4}{3}\pi(3r^2h) = (4\pi r^2)h$$

Therefore, the surface area of a sphere is $4\pi r^2$. It is not $4.01\pi r^2$, it is exactly $4\pi r^2$; in other words, the surface area of a sphere is exactly four times the area of its great circle.

4.4.6 The Geometric Meaning of the Derivative

We have used algebra to define the derivative of a function, let's consider the geometric meaning of this important concept. To this end we use Descartes' analytic geometry by plotting the graph of the function $y = f(x)$ on the Cartesian xy plane. We then consider a point P on the curve with coordinates $(x_0, f(x_0))$, cf. Figure 4.27a. To have change, we consider another point Q on the curve with coordinates $(x_0 + h, f(x_0 + h))$. Then we have the average rate of change of the function at P, that is $\Delta f/h$ where $\Delta f = f(x_0 + h) - f(x_0)$. This rate of change is nothing but the slope of the secant PQ (see the shaded triangle). Now, the process of considering smaller and smaller h, to get the derivative, is amount to considering points Q', Q'' which are closer and closer to P (Fig. 4.27b). The secants PQ, PQ', PQ'', ... approach the line PP' which touches the curve $y = f(x)$ at P. *PP' is the tangent to the curve at P*. The average rate of change $\Delta f/h$ approaches df/dx—the derivative of $f(x)$ at x_0.

When h approaches 0, the secants approach the tangent and their slopes approach the derivative. Thus, the derivative of the function at P is the slope of the tangent to $f(x)$ at the same point. That's the geometric meaning of the derivative.

Now we derive the equation for this tangent. It is the line going through the point $P(x_0, y_0)$ and has a slope equal $f'(x_0)$, thus the equation for the tangent is

$$\text{tangent to } y = f(x) \textit{ at } P(x_0, y_0): \quad f(x_0) + f'(x_0)(x - x_0) \tag{4.4.12}$$

And this leads to the so-called linear approximation to a function, discussed later in Sect. 4.5.3. The idea is to replace a curve—which is hard to work with—by its tangent (which is a line and easier to work with).

We now understand the concept of the derivative of a function, algebraically and geometrically. Now, it is the time to actually compute the derivative of functions that we know: polynomials, trigonometric, exponential etc.

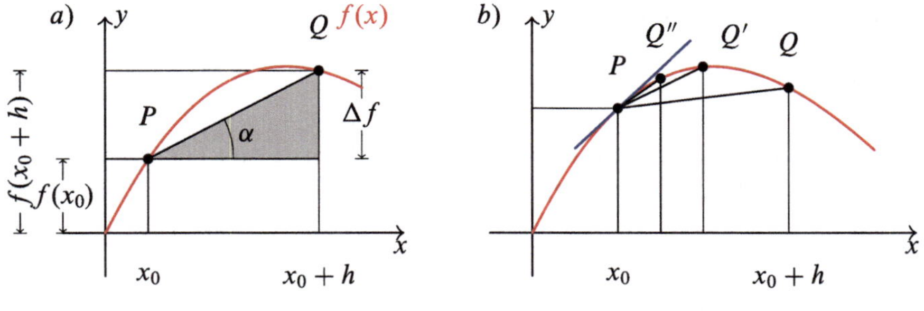

Fig. 4.27 The derivative of a function $y = f(x)$ evaluated at x_0 is the slope of the tangent to the curve at $(x_0, f(x_0))$. The ratio $\Delta f/h$ is the slope of the secant PQ. When $h \to 0$, this ratio becomes the slope of the tangent (blue line) of the curve at P

4.4.7 Derivative of $f(x) = x^n$

We start simple first. Let's compute the derivative of $y = x^2$ at x_0, and we use the definition in Eq. (4.4.9):

$$\begin{aligned}
f'(x_0) = \lim_{h\to 0}\frac{f(x_0+h)-f(x_0)}{h} &= \lim_{h\to 0}\frac{(x_0+h)^2 - x_0^2}{h} \quad \text{(def.)}\\
&= \lim_{h\to 0}\frac{2x_0h + h^2}{h} \quad \text{(purely algebra)}\\
&= \lim_{h\to 0}(2x_0+h) \quad \text{(algebra, } h \text{ is not zero!)}\\
&= 2x_0 \text{ (when } h \text{ is approaching 0)}
\end{aligned} \tag{4.4.13}$$

The algebra was simple but there are some points worthy of further discussion. First, if we used $h = 0$ in the difference quotient $2x_0h+h^2/h$ we would get this form $0/0$—which is mathematically meaningless. This is so because to get the derivative which is a rate of change at least we should allow h to be different from zero (so that some change is happening). That's why the derivative was not defined as the difference quotient when $h = 0$. Instead, it is defined as the limit of this quotient when h approaches zero. Think of the instantaneous speed (Table 4.5), and the thing is clear.

As always, it is good to try to have a geometric interpretation. What we are looking for is what is the change of x^2 if there is a tiny change in x. We think of x^2 immediately as the area of a square of side x (Fig. 4.28). Then, a tiny change dx leads to a change in the area of $2xdx$, because the change $(dx)^2$ is so small that it can be neglected.

So, it's up to you to like the limit approach or the infinitesimal one. If you prefer rigor then using a limit is the way to go. But if you just do not care what is the meaning of infinitesimals (whether they exist for example), then use dx and dy freely like Leibniz, Euler, and many seventeenth century mathematicians did. And the results are the same!

Similar steps give us the derivatives of x^3, x^4:$(x^3)' = 3x^2$, and $(x^4)' = 4x^3$. It is hard to resist to write this general formula for all positive integers n[18]:

$$(x^n)' = nx^{n-1} \tag{4.4.14}$$

How about the derivative when n is negative? Let's start with $f(x) = x^{-1} = 1/x$. Using the definition, we can compute its derivative as

$$\left(\frac{1}{x}\right)' = \lim_{h\to 0}\frac{\dfrac{1}{x+h}-\dfrac{1}{x}}{h} = \lim_{h\to 0}\frac{-1}{x(x+h)} = -\frac{1}{x^2}$$

---

[18] If you need a proof so that you can have a good sleep at night, then follow Eq. (4.4.13) and use the binomial theorem.

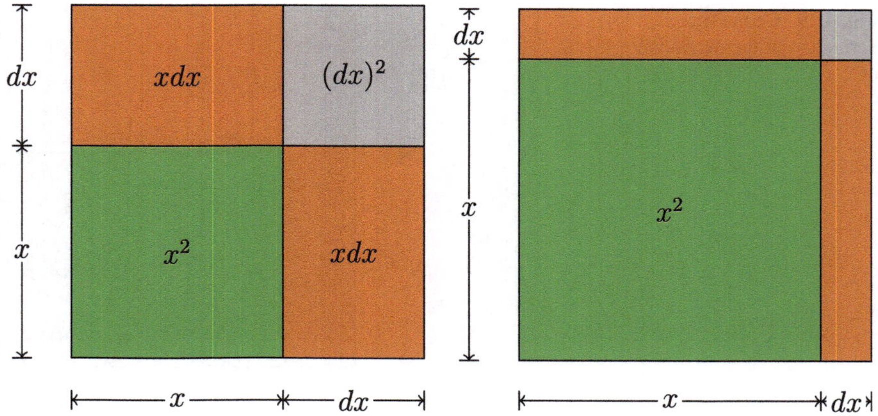

Fig. 4.28 Geometric derivation of the derivative of x^2. The change $(dx)^2$ is super small compared with $2xdx$ and thus it will approach zero when dx is approaching zero

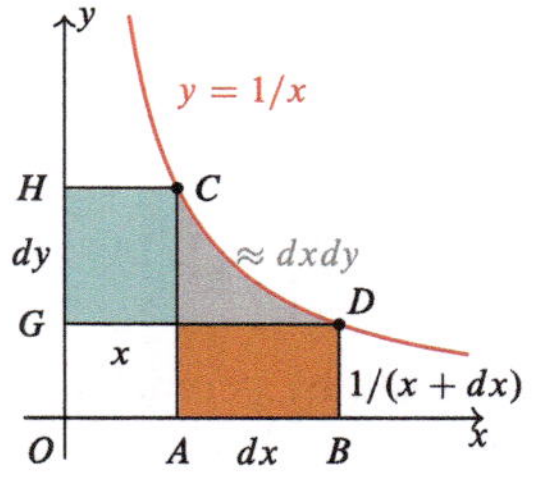

Let's see whether we can have a geometry-based derivation of this result. We plot the function $1/x$ and pick two points on the curve close to each other: one point is $C = (x, 1/x)$ and the other point is $D = (x + dx, 1/(x + dx))$. As the areas of the two rectangles $OACH$ and $OBDG$ are equal (equal 1), the areas of the two rectangle strips (those are shaded) must be equal. We can compute these two areas and equate them, then we will obtain an equation relating dx and dy; just note that $HG = -dy$ because dy is negative. So we can write

$$(-dy)(x) = (dx)\left(\frac{1}{x + dx}\right) \Longrightarrow -x^2dy = dx \Longrightarrow \frac{dy}{dx} = -\frac{1}{x^2}$$

In the algebra, we removed the term $x(dy)(dx)$ as super-super-small quantity in the same manner discussed in Sect. 4.4.5.[19] Geometrically, $dydx$ is proportional to the area of the gray-shaded region in the figure; obviously this area shrinks to zero when dx is tiny. What we just got means that the formula in Eq. (4.4.14) (i.e., $(x^n)' = nx^{n-1}$) still holds for negative powers at least for $n = -1$.

[19] This is to demonstrate that we can use dx and dy as ordinary numbers. But keep in mind that all of this works because of the properties of limit.

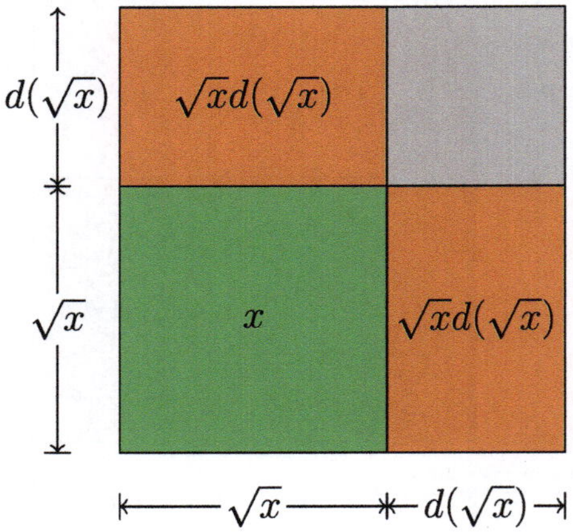

Fig. 4.29 Geometric derivation of the derivative of $\sqrt{x}$

Now, we compute the derivative of the square root function i.e., $\sqrt{x}$. We assume that (once again believe in mathematical patterns) Eq. (4.4.14) still applies for fractional exponents, so we write

$$(\sqrt{x})' = \left(x^{1/2}\right)' = \frac{1}{2}x^{-1/2} = \frac{1}{2\sqrt{x}} \tag{4.4.15}$$

Let's try if we can get the same result by geometry. As $\sqrt{x}$ is the inverse of x^2, we use area concept. We consider a square of side $\sqrt{x}$, its area is thus x. We consider a change in the side $d(\sqrt{x})$, and see how the square area changes, see Fig. 4.29.

Change in the area is dx, and it is written as

$$dx = 2\sqrt{x}d(\sqrt{x}) \Longrightarrow \frac{d(\sqrt{x})}{dx} = \frac{1}{2\sqrt{x}} \tag{4.4.16}$$

4.4.8 Derivative of Trigonometric Functions

I present the derivatives of sine/cosine in this section. Once we have known these derivatives, the derivative of tangent and other trigonometric functions is straightforward to compute as these functions are functions of sine/cosine. Let's start with a direct application of the definition of a derivative for $\sin x$:

$$
\begin{aligned}
(\sin x)' &= \lim_{h\to 0} \frac{\sin(x+h)-\sin x}{h} \\
&= \lim_{h\to 0} \frac{\sin x \cos h + \sin h \cos x - \sin x}{h} \\
&= \sin x \lim_{h\to 0} \left(\frac{\cos h - 1}{h}\right) + \cos x \lim_{h\to 0} \frac{\sin h}{h}
\end{aligned}
$$

We need the following limits (proof of the first will be given shortly, for the second limit, check Eq. (3.15.3))

$$
\lim_{h\to 0} \left(\frac{\cos h - 1}{h}\right) = 0, \quad \lim_{h\to 0} \frac{\sin h}{h} = 1 \tag{4.4.17}
$$

which leads to

$$
(\sin x)' = \cos x \tag{4.4.18}
$$

We can do the same thing to get the derivative of cosine. But we can also use trigonometric identities and the chain rule (to be discussed next) to obtain the cosine derivative:

$$
(\cos x)' = \frac{d}{dx}\left[\sin\left(\frac{\pi}{2} - x\right)\right] = -\cos\left(\frac{\pi}{2} - x\right) = -\sin x \tag{4.4.19}
$$

A geometric derivation of the derivative of $\sin x$, shown in Fig. 4.30, is easier and without requiring the two limits in Eq. (4.4.17).

Using the quotient rule, we can compute the derivative of $\tan x$:

$$
(\tan x)' = \left(\frac{\sin x}{\cos x}\right)' = \frac{\cos^2 x + \sin^2 x}{\cos^2 x} = \frac{1}{\cos^2 x} \tag{4.4.20}
$$

Proof Herein, we prove that the limit of $\cos h - 1/h$ equals zero. The proof is based on the limit of $\sin h/h$ and a bit of algebra:

$$
\lim_{h\to 0} \left(\frac{\cos h - 1}{h}\right) = \lim_{h\to 0} \left(\frac{-\sin^2 h}{h(1+\cos h)}\right) = \lim_{h\to 0} \left(\frac{\sin h}{h}\right) \lim_{h\to 0} \left(\frac{-\sin h}{(1+\cos h)}\right) = 1 \times \frac{0}{2}
$$

4.4.9 Rules of Derivative

Let's summarize what we know about derivative. We know the derivatives of x^n for positive integers n, of $\sqrt{x}$, of x^{-1}, of trigonometric functions, exponential functions a^x or e^x and logarithm functions (to be discussed). What about the derivative of $x^2 \sin x$, $x^2-3/\cos x$? For them, we need to use the rules of differentiation. With these rules, only derivatives of elementary functions are needed, derivative of other func-

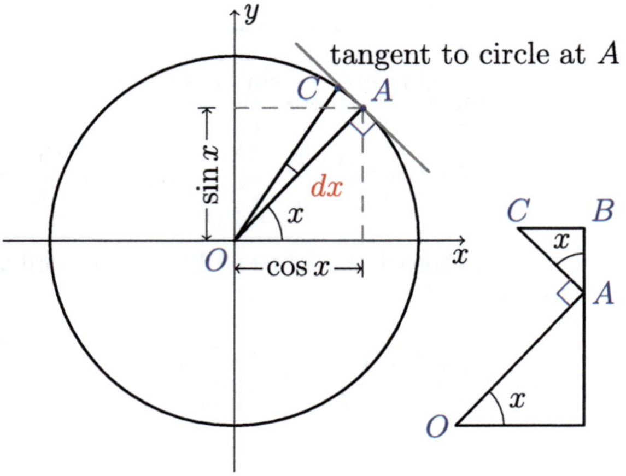

Fig. 4.30 Geometric derivation of the derivative of the sine/cosine functions by considering a unit circle and point A with coordinates $(\cos x, \sin x)$. For a small variation in angle dx, we have $AC = dx$ and $AC \perp OA$ because AC is tangential to the circle. Then, $d(\sin x) = AB = dx \cos x$ from the right triangle ABC. Note that angles are in radians. If it is not the case, $AC = (\pi dx/180)$, and the derivative of $\sin x$ would be $(\pi/180) \cos x$

tions (inverse functions, composite functions) are calculated using these rules. They are first summarized in what follows for easy reference $(a, b \in \mathbb{R})$

$$
\begin{aligned}
[a]' &= 0 && \text{(constant function rule)}\\
[af(x) + bg(x)]' &= af'(x) + bg'(x) && \text{(sum rule)}\\
[f(x)g(x)]' &= f'(x)g(x) + g'(x)f(x) && \text{(product rule)}\\
\left(\frac{1}{f(x)}\right)' &= -\frac{f'}{f^2} && \text{(reciprocal rule)}\\
\left(\frac{f(x)}{g(x)}\right)' &= \frac{f'g - fg'}{g^2} && \text{(quotient rule)}\\
(f(g(x)))' &= f'(g(x))g'(x) && \text{(chain rule)}
\end{aligned}
$$

Among these rules the chain rule is the hardest (and left to the next section), other rules are quite easy. The function $y = a$ is called a constant function for $y = a$ for all x. Obviously we cannot have change with this boring function, thus its derivative is zero.

If we follow Eq. (4.4.13) we can see that the derivative of $3x^2$ is $3(2x)$. A bit of thinking will convince us that the derivative of $af(x)$ is $af'(x)$, which can be verified using the definition of derivative, Eq. (4.4.9). Again, following the steps in Eq. (4.4.13), the derivative of $x^3 + x^4$ is $3x^2 + 4x^3$, and this leads to the derivative of $f(x) + g(x)$ is $f'(x) + g'(x)$: *the derivative of the sum of two functions is the sum of the derivatives*. This can be verified using the definition of derivative, Eq. (4.4.9).

Now, $af(x)$ is a function and $bg(x)$ is a function, thus the derivative of $af(x) + bg(x)$ is $(af(x))' + (bg(x))'$, which is $af'(x) + bg'(x)$. And this is our first rule.[20]

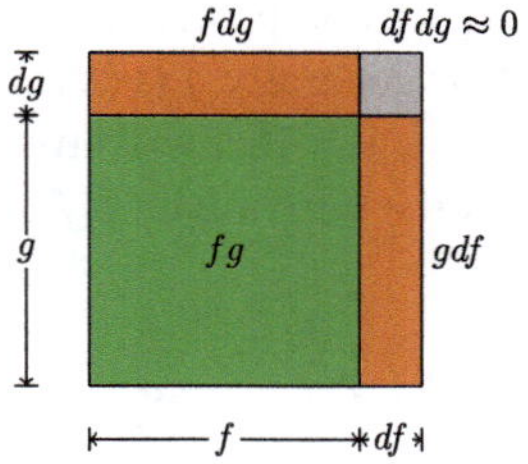

The sum rule says that the derivative of the sum of two functions is the sum of the derivatives. Thus Leibniz believed that the derivative of the product of two functions is the product of the derivatives. It took him no time (with an easy example, let say $x^3(2x+3)$) to figure out that his guess was wrong, and eventually he came up with the correct rule. The proof of the product rule is given in the beside figure. The idea is to consider a rectangle of sides f and g with an area of fg. (Thus implicitly this proof applies to positive functions only). Now assume that we have an infinitesimal change dx, which results in a change in f, denoted by $df = f'(x)dx$ and a change in g, denoted by $dg = g'(x)dx$. We need to compute the change in the area of this rectangle. It is $gdf + fdg + dfdg$, which is $gdf + fdg$ as $(df)(dg)$ is minuscule. Thus the change in the area which is the change in fg is $[gf'(x) + fg'(x)]dx$. That concludes our geometric proof.

The proof of the reciprocal rule starts with this function $f(x) \times {}^{1}/_{f(x)} = 1$. Applying the product rule for this constant function, we get

$$0 = f'(x)\frac{1}{f(x)} + f(x)\frac{d}{dx}\left(\frac{1}{f(x)}\right) \Longrightarrow \frac{d}{dx}\left(\frac{1}{f(x)}\right) = -\frac{f'(x)}{f^2(x)}$$

The quotient rule is obtained from the product rule and the reciprocal rule as shown in Eq. (4.4.21)

$$\begin{aligned}\frac{d}{dx}\left(\frac{f}{g}\right) &= \frac{d}{dx}\left(f\frac{1}{g}\right)\\ &= \frac{df}{dx}\frac{1}{g} + f\frac{d}{dx}\left(\frac{1}{g}\right)\\ &= \frac{df}{dx}\frac{1}{g} - f\left(\frac{dg/dx}{g^2}\right) = \frac{f'g - fg'}{g^2}\end{aligned} \tag{4.4.21}$$

[20] Note that this rule covers many special cases. For example, taking $a = 1$, $b = -1$, we have $[f(x) - g(x)]' = f'(x) - g'(x)$. Again, subtraction is secondary for we can deal with it via addition. Furthermore, even though our rule is stated for two functions only, it can be extended to any number of functions. For instance, $[f(x) + g(x) + h(x)]' = f'(x) + g'(x) + h'(x)$, this is so because we can see $f(x) + g(x)$ as a new function $w(x)$, and we can use the sum rule for the two functions $w(x)$ and $h(x)$.

4.4.10 The Chain Rule: Derivative of Composite Functions

The chain rule is for derivative of composite functions. For example, what is the derivative of $f(x) = \sin(x^2)$? Or generally $f(g(x))$. In the case of $f = \sin(x^2)$, we have $f = \sin(y)$ and $y = x^2$. We know the derivative of f w.r.t y and the derivative of y w.r.t x. But the question is the derivative of f w.r.t x. Using Leibniz's dy and dx, it is easy to derive the rule:

$$\frac{\Delta f}{\Delta x} = \frac{\Delta f}{\Delta y}\frac{\Delta y}{\Delta x} \Longrightarrow \frac{df}{dx} = \frac{df}{dy}\frac{dy}{dx} \tag{4.4.22}$$

which means that derivative of f w.r.t x is equal to the derivative of f w.r.t y multiplied by the derivative of y w.r.t x.

Thus, for $f = \sin(x^2)$, its derivative is: $\cos(x^2)2x$. We should at least check Eq. (4.4.22). For example, considering $y = x^6$, then $y' = 6x^5$. Now, we write $y = x^6 = (x^2)^3$, then the chain rule yields $y' = (3x^4)(2x) = 6x^5$. Not a proof, but for me it is enough.

4.4.11 Derivative of Inverse Functions

We have discussed inverse functions in Sect. 4.2.5. Calculus is always about derivative and integration. In this section, we discuss how to find the derivative of an inverse function. Given a function $y = f(x)$, the inverse function is $x = f^{-1}(y)$. And our aim is to find dx/dy.

We write $x = f^{-1}(y) = f^{-1}(f(x))$ and differentiate (w.r.t x) two sides of this equation. On the RHS, we use the chain rule:

$$x = f^{-1}(f(x)) \Longrightarrow 1 = \frac{df^{-1}}{dy}\frac{df}{dx}$$

So, we have the rule for the derivative of an inverse function:

$$\boxed{\frac{dx}{dy} = \frac{1}{dy/dx}} \tag{4.4.23}$$

Let's check this rule with $y = x^2$ and $x = \sqrt{y}$. We compute dx/dy using Eq. (4.4.23): $dx/dy = 1/(dy/dx) = 1/(2x) = 1/(2\sqrt{y})$. And this result is identical to the derivative of $x = \sqrt{y}$.

4.4.12 Derivatives of Inverses of Trigonometry Functions

Using Eq. (4.4.23) we can compute the derivative of inverse trigonometric functions. We summarize the results in Table 4.6. Proofs follow shortly.

We present the proof of the derivative of $\arcsin x$. Write $y = \sin x$, then we have $dy/dx = \cos x$. The inverse function is $x = \arcsin y$. Using the rule of the derivative of an inverse function:

$$\frac{dx}{dy} = \frac{1}{dy/dx} = \frac{1}{\cos x} = \frac{1}{\sqrt{1-y^2}} \tag{4.4.24}$$

where in the final step we have converted from x to y as dx/dy is a function of y. Now considering the function $y = \arcsin x$, we have $dy/dx = 1/\sqrt{1-x^2}$. Proofs of other trigonometric inverses follow.

4.4.13 Derivatives of a^x and Number e

In this section, we seek the derivative of the exponential functions. To start with, consider the function 2^t. As will be seen, using the definition to analytically find out the derivative of 2^t is quite hard. So, we use computations to see the pattern: we compute the derivative of 2^t at $t = 1, 2, 3, 4$ using $dt = 1$; the results given in Table 4.7 indicate that the derivative at point t is the function itself evaluated at the same point.

Table 4.6 Derivative of inverses of trigonometric functions

$f(x)$	$f^{-1}(x)$	$\frac{df^{-1}}{dx}$
$\sin x$	$\arcsin x$	$\frac{1}{\sqrt{1-x^2}}$
$\cos x$	$\arccos x$	$-\frac{1}{\sqrt{1-x^2}}$
$\tan x$	$\arctan x$	$\frac{1}{1+x^2}$
$\cot x$	$\text{arccot} x$	$\frac{1}{1+x^2}$

Table 4.7 Derivative of 2^t using a finite increment $dt = 1$

t	2^t	$\frac{2^{t+1}-2^t}{dt}$
1	2	2
2	4	4
3	8	8
4	16	16

Table 4.8 Limit of $2^{dt}-1/dt$

dt	$\frac{(2^{dt}-1)}{dt}$
0.1	0.7177346253629313
0.01	0.6955550056718884
0.001	0.6933874625807412
0.00001	0.6931495828199629
0.000001	0.6931474207938493

Table 4.9 $(a^t)' = ka^t$, to find k: $\frac{a^{dt}-1}{dt}$ with $dt = 10^{-7}$

$\frac{(2^{dt}-1)}{dt}$	$\frac{(3^{dt}-1)}{dt}$	$\frac{(4^{dt}-1)}{dt}$	$\frac{(6^{dt}-1)}{dt}$	$\frac{(8^{dt}-1)}{dt}$
0.693147	1.098612	1.386294	1.791760	2.079442
		0.693147 × 2	1.791760	0.693147 × 3

We know that this cannot be true as dt is too big. But it gave us some hint that the derivative should be related to 2^t. So, we use the definition of derivative and do some algebra this time so that 2^t shows up:

$$\frac{d(2^t)}{dt} = \lim_{dt\to 0} \frac{2^{t+dt} - 2^t}{dt} = \lim_{dt\to 0} \frac{2^t(2^{dt}-1)}{dt} = 2^t \lim_{dt\to 0} \frac{2^{dt}-1}{dt}$$

And once again, we face the eternal question of calculus: does the limit in the above equation exist and if so, what is the value? Herein, we pass the first question (we know from textbooks that it does exist), and focus on finding the value. Using different values for dt, we can see that the limit is 0.6931474 (Table 4.8).

So, the derivative of 2^t is 2^t multiplied by a constant. We can generalize this result as there is nothing special about number 2. For the exponent function $y = a^t$, its derivative is given by

$$\frac{d(a^t)}{dt} = ka^t \tag{4.4.25}$$

where k is a constant. To find its value, we compute k for a few cases of $a = 2, 3, 4, 6, 8$ to find a pattern. The results are given in Table 4.9. If this k can be expressed as a function of a, $f(a)$, then we have $f(4) = f(2^2) = 2f(2)$, and $f(8) = f(2^3) = 3f(2)$. What function has this property? A logarithm! But logarithm of which base, we do not know yet.

Instead of finding k, we can turn the problem around and ask if there exists an exponential function such that *its derivative is itself*. In other words, $k = 1$. From Table 4.9, we guess that there exists a number c within $[2, 3]$ that the derivative of c^x is c^x. It turns out that this function is $f(t) = e^t$, where e is the Euler number (its value is approximately 2.78) that we have found in the context of continuously compounding interest (Sect. 2.28). Indeed,

$$\frac{d(e^t)}{dt} = e^t \lim_{dt \to 0} \frac{e^{dt} - 1}{dt} = e^t \tag{4.4.26}$$

Because e is defined as a number that satisfies the following limit

$$\lim_{dt \to 0} \frac{e^{dt} - 1}{dt} = 1 \tag{4.4.27}$$

You can see where this definition of e comes from by looking at Eq. (2.24.6) (in the context that Briggs calculated his famous logarithm tables). It can be shown that this definition is equivalent to the definition of e as the rate of continuously compound interest:

$$e = \lim_{dt \to 0} \left(1 + dt\right)^{1/dt} = \lim_{n \to \infty} \left(1 + \frac{1}{n}\right)^n \tag{4.4.28}$$

which is exactly Eq. (2.28.1).

Now, we can find the constant k in Eq. (4.4.25), $k = \ln a$ where $\ln x$ is the natural logarithm function of base e, which is the inverse function of e^x (we will discuss about $\ln x$ in the next section)

$$\frac{d(a^t)}{dt} = \ln a a^t \tag{4.4.29}$$

Proof The proof of the derivative of a^t is simple. Since we know the derivative of e^x, we write a^x in terms of e^x. So, we write $a = e^{\ln a}$, thus

$$a^t = e^{(\ln a)t} \implies \frac{da^t}{dt} = \frac{d(e^{(\ln a)t})}{dt} = \ln a e^{(\ln a)t} = \ln a a^t$$

where we have used the chain rule of differentiation.

Are there other functions of which derivatives are the functions themselves? No, the only function that has this property is $y = ce^x$. The function $y = e^x$ is the only function of which the derivative and integral are itself. To it, there is a joke that goes like this.

> An insane mathematician gets on a bus and starts threatening everybody: "I'll integrate you! I'll differentiate you!!!" Everybody gets scared and runs away. Only one lady stays. The guy comes up to her and says: "Aren't you scared, I'll integrate you, I'll differentiate you!!!" The lady calmly answers: "No, I am not scared, I am e^x."

4.4.14 Logarithm Functions

So we have discovered (re-discovered to be exact) the number e, and thus we can define the exponential function $y = e^x$ with the remarkable property that its derivative is itself. It is then straightforward to define the natural logarithm function $x = \ln y$ (if

Table 4.10 The area of $y = 1/u$ from 1 to x: $f(x) = \int_1^x du/u$ for some values of x

x	2.0	4.0	8.0	16.0
$f(x)$	0.69314718	1.38629436	2.07944154	2.77258870
$\Delta f(x)$		0.69314718	0.69314718	0.69314718

you always think of inverse functions or operations). Historically, it was not the case because $y = e^x$ was not known at the time when Flemish Jesuit and mathematicians Grégoire de Saint-Vincent (1584–1667) and Alphonse Antonio de Sarasa (1618–1667) discovered the natural logarithm function while working on the quadrature of the rectangular hyperbola $xy = 1$ in 1647.

We start the discussion by noting that the area of the function $y = 1/x$ has defied almost all seventeenth century mathematicians (including giants such as Descartes, Fermat). This indicates that the integral of $y = 1/x$ can be a new function. To find out about this function, we are going to define a function $f(x)$ as

$$f(x) := \int_1^x \frac{1}{u} du \tag{4.4.30}$$

And the task now is to find out what $f(x)$ is. The way to do that is to compute some values of this function and see where they lead us. To this end, I use the definition of integral to compute $f(x)$ for some values of x. The results are given in Table 4.10. In the definition of the integral (Eq. (4.3.3)) we have to go to infinity, but herein, I have used the mid-point rule with 20 000 sub-divisions to compute these integrals. That is $n = 20\,000$ in Eq. (4.3.3); this is similar to how Archimedes computed π.[21] This is obviously not the way Saint-Vincent computed the integral in Eq. (4.4.30) (since he did not have computers). He also used the definition of the integral as the sum of areas of many thin slices, but his slices were special, see this YouTube link for detail.

Is there anything special from this table? Ah yes. In the first row we have a geometric progression 2, 4, 8, 16, and in the second row we have an arithmetic progression (indicated by a constant $\Delta f(x)$ in the last row). Which function has this property? A logarithm![22] If not clear, you might need to check Table 2.16 again. You can check from the values in the table that

$$f(8) = f(4 \times 2) = f(4) + f(2), \quad \int_1^2 du/u = \int_2^4 du/u = \int_4^8 du/u$$

[21] Refer to Sect. 10.4.1 if you're not sure what is the mid-point rule.

[22] By 1661 Huygens understood this relation between the rectangular hyperbola and the logarithm.

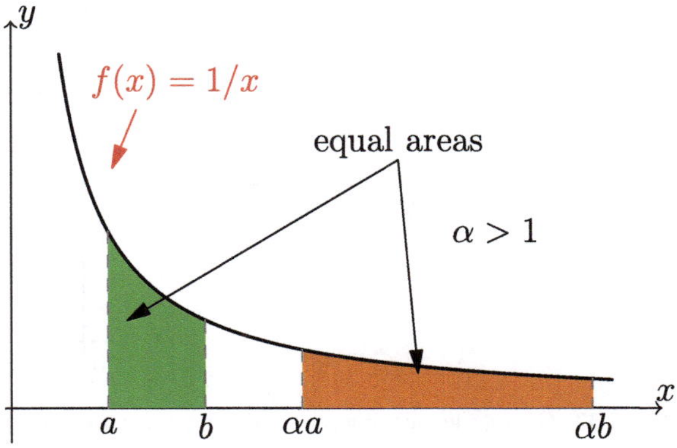

Fig. 4.31 The hyperbola $y = 1/x$

From which we anticipate the following result, see Fig. 4.31:

$$\int_a^b \frac{du}{u} = \int_{\alpha a}^{\alpha b} \frac{du}{u} \tag{4.4.31}$$

How we are going to prove this? Well, it depends on what tools you want to use. If you assume that you know already the chain rule, then a simple substitution would prove that the two integrals in Eq. (4.4.31) are equal. If you assume you were in the sixteenth century, then the proof would be a bit harder. You can use the definition of the integral, Eq. (4.3.3), for $\int_{\alpha a}^{\alpha b} du/u$ and see that α will be canceled out, and thus that integral equals $\int_a^b du/u$.

Ok. So $\int_1^x du/u$ is a logarithm, but what base? Can the base be 2 or 3? Of course not, there is nothing special about these two numbers. To find out the base, let's denote it by y, then we have:

$$\log_y(x) = \int_1^x \frac{1}{u} du$$

The task is now to determine the value of y. Now, we consider small x, and the area under the hyperbola from 1 to $1 + x$, we can then write

$$\log_y(1+x) = \int_1^{1+x} \frac{1}{u} du$$

Why doing this? Because we can compute that integral: $\int_1^{1+x} \frac{1}{u} du \approx x$ (draw a figure and you'll see it). Therefore,

$$\log_y(1+x) \approx x \Rightarrow \lim_{x \to 0} \frac{1}{x} \log_y(1+x) = 1 \Longrightarrow \lim_{x \to 0} \log_y(1+x)^{\frac{1}{x}} = 1 \Longrightarrow \boxed{\lim_{x \to 0} (1+x)^{\frac{1}{x}} = y}$$

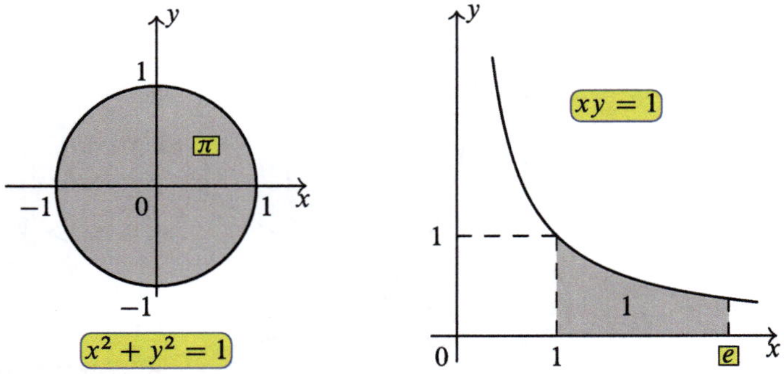

Fig. 4.32 Geometric meaning of π and e: the former is related to the area of a circle and the latter to the area of a hyperbola. Both shapes are conic sections

The boxed formula looks familiar: it is exactly Eq. (2.28.1) that defines e. So, y is nothing but e. Thus, mathematicians define the natural logarithm function $y = \ln x$ as:

$$\boxed{\ln x := \int_1^x du/u} \tag{4.4.32}$$

In 1668 Nicolaus Mercator published Logarithmotechnia where he used the term "natural logarithm" for the first time for logarithms to base e. Before that this logarithm was called hyperbolic logarithm as it comes from the area of a hyperbola. The geometric meaning of e is given in Fig. 4.32, with a comparison with that of π.

And all properties of logarithm (such as $\ln ab = \ln a + \ln b$), that Napier and Briggs discovered (Sect. 2.24) in a completely different way, should follow naturally from this definition. With Fig. 4.33, we can prove $\ln ab = \ln a + \ln b$ as follows:

$$\ln ab = \int_1^{ab} \frac{du}{u} = \int_1^{b} \frac{du}{u} + \int_b^{ab} \frac{du}{u} = \ln b + \ln a \tag{4.4.33}$$

where use was made of Eq. (4.4.31) to convert $\int_b^{ab} du/u$ to $\int_1^a du/u = \ln a$.

I defer the discussion on the derivative of logarithm functions to Sect. 4.4.18. Figure 4.34 presents the graph of the exponential and logarithm functions. Both are monotonically increasing functions. This is so because their derivatives are always positive. A function $f(x)$ is called monotonically increasing if *for all* x and y such that $x \leq y$ one has $f(x) \leq f(y)$. If the order $\leq$ in the definition of monotonicity is replaced by the strict order $<$, one obtains a stronger requirement. A function with this property is called *strictly* increasing (also increasing).

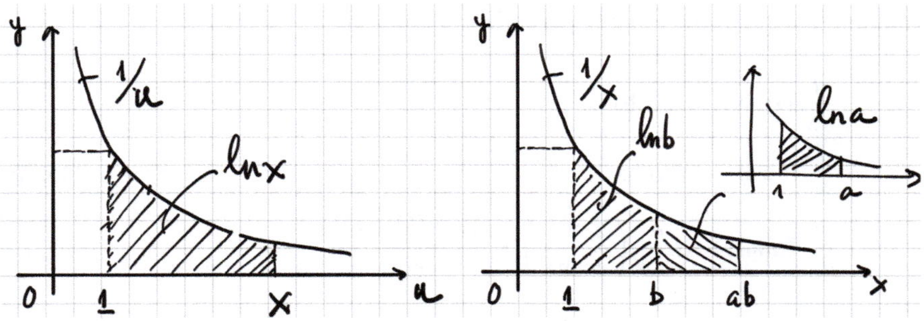

Fig. 4.33 Proof of $\ln ab = \ln a + \ln b$

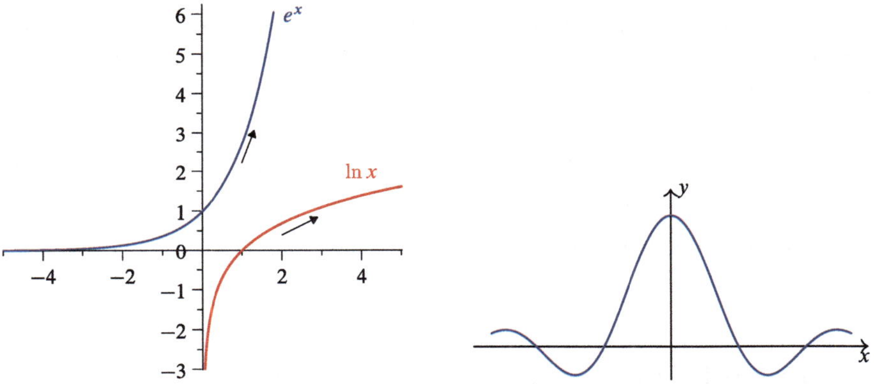

(a) monotonic functions (b) non-monotonic functions

Fig. 4.34 Graph of $y = \ln x$ and $y = e^x$—monotonically increasing functions (a) and graph of $\sin 4x/x$, a non-monotonic function

4.4.15 *Derivative of Hyperbolic and Inverse Hyperbolic Functions*

The hyperbolic sine and cosine functions have been introduced in Sect. 3.19. In what follows, I list all hyperbolic functions in one place for convenience:

$$\begin{aligned}
\sinh x &= \frac{1}{2}\left(e^x - e^{-x}\right) & \operatorname{csch} x &= \frac{1}{\sinh x} = \frac{2e^x}{e^{2x}-1} \\
\cosh x &= \frac{1}{2}\left(e^x + e^{-x}\right) & \operatorname{sech} x &= \frac{1}{\cosh x} = \frac{2e^x}{e^{2x}+1} \\
\tanh x &= \frac{\sinh x}{\cosh x} = \frac{e^{2x}-1}{e^{2x}+1} & \coth x &= \frac{\cosh x}{\sinh x} = \frac{e^{2x}+1}{e^{2x}-1}
\end{aligned} \tag{4.4.34}$$

The first derivatives of some hyperbolic functions are

$$\begin{aligned}\frac{d}{dx}(\sinh x) &= \cosh x\\ \frac{d}{dx}(\cosh x) &= \sinh x\\ \frac{d}{dx}(\tanh x) &= \frac{d}{dx}\left(\frac{\sinh x}{\cosh x}\right) = \frac{\cosh^2 x - \sinh^2 x}{\cosh^2 x} = \frac{1}{\cosh^2 x}\end{aligned} \tag{4.4.35}$$

Note the striking similarity with the circular trigonometric functions. Only sometimes a minus/plus difference.

We're now too familiar with the concept of inverse operators/functions. So it is natural to consider inverse hyperbolic functions. For brevity, we consider only $y = \sinh^{-1} x$ and $y = \cosh^{-1} x$. Let's compute the derivative of $y = \sinh^{-1} x$. We have $x = \sinh y$, and thus $dx/dy = \cosh y = \sqrt{1+x^2}$.[23] So,

$$\frac{d}{dx}\left(\sinh^{-1} x\right) = \frac{1}{dx/dy} = \frac{1}{\sqrt{1+x^2}} \tag{4.4.36}$$

If someone tell you that $\sinh^{-1} x$ is actually a logarithm of x:

$$y = \sinh^{-1} x = \ln(x + \sqrt{1+x^2})$$

Do you believe him? Yes. Because the sine hyperbolic function is defined in terms of the exponential e^x, it is reasonable that its inverse is related to $\ln x$—the inverse of e^x. The proof is simple:

$$x = \sinh y = \frac{e^y - e^{-y}}{2} \Longrightarrow (e^y)^2 - (2x)e^y - 1 = 0 \Longrightarrow e^y = x + \sqrt{1+x^2}$$

4.4.16 High Order Derivatives

Given the function $y = f(x)$, the derivative $f'(x)$ is also a function, so it is natural to compute the derivative of $f'(x)$ which is the derivative of a derivative. What we get is the so-called second derivative, denoted usually by $f''(x)$[24]:

$$f''(x) = \frac{d}{dx}\left(\frac{df}{dx}\right) = \frac{d^2 f}{dx^2} = \lim_{h\to 0} \frac{f'(x+h) - f'(x)}{h} \tag{4.4.37}$$

[23] Using the identity $\cosh^2 x - \sinh^2 x = 1$.

[24] This makes $f'(x)$ the first derivative of $f(x)$.

Consider $y = 2x$, the first derivative is $y' = 2$, and the second derivative is $y'' = 0$. Thus, a line has a zero-second derivative. For a parabola $y = x^2$, we have $y' = 2x$ and $y'' = 2$. A parabola has a constant non-zero second derivative. A line has a constant slope, it does not bend and thus its second derivative is zero. On the other hand, a parabola has a varying slope, it bends and therefore the second derivative is non-zero.

The most popular second derivative is probably the acceleration, which is the second derivative of the position function of a moving object, $x(t)$: $\boldsymbol{a} = \ddot{x}$ following Newton's notation or $\boldsymbol{a} = d^2x/dt^2$ following Leibniz. Equivalently, the acceleration is the derivative of the velocity. Historically, acceleration was the first-second derivative ever. Newton's laws of motions are presented in Sect. 6.10.

Have ever you wondered why mathematicians used the notation d^2f/dx^2 but not df^2/dx^2? In other words, why the number 2 is treated differently in the numerator and denominator? I do not have a rigorous answer. But using the notion of acceleration helps. The acceleration is given by

$$a = \frac{d^2x}{dt^2} \Longrightarrow \text{correct unit: m/s}^2$$

If a was written as $a = dx^2/dt^2$, its unit would be $(\text{m/s})^2$, which is wrong.

Going along this direction, we will have the third derivative e.g. the third derivative of x^3 is 6. And the fourth derivative and so on, usually we denote an n-order derivative of a function $f(x)$ by $f^{(n)}(x)$. But, wait how about derivatives of non-integer order like $d^{1/2}f(x)/dx^{1/2}$? That question led to the development of the so-called *fractional calculus*.

Fractional derivative. Regarding the n-order derivative of a function $f^{(n)}(x)$, in a 1695 letter, l'Hopital asked Leibniz about the possibility that n could be something other than an integer, such as $n = 1/2$. Leibniz responded that "It will lead to a paradox, from which one-day useful consequences will be drawn". Leibniz was correct, but it would not be centuries until it became clear just how correct he was.

There are two ways to think of $f^{(n)}(x)$. The first is the one we all learn in basic calculus: it's the function that we obtain when we repeatedly differentiate f n times. The second is more subtle: we interpret it as an operator whose action on the function $f(x)$ is determined by the parameter n. What l'Hopital is asking is what the behavior is of this operator when n is not an integer. The most natural way to answer this question is to interpret differentiation (and integration) as *transformations* that take f and turn it into a new function.

That's all I know about fractional derivative and fractional calculus. I have presented them here to illustrate the fact that if we break the rules (the order of differentiation is usually a positive integer) we could make new mathematics.

4.4.17 Implicit Functions and Implicit Differentiation

We have discussed the derivative of explicit functions like $y = f(x)$, now it is time to deal with the derivative of implicit functions like $x^2 + y^2 = 25$, what is dy/dx? While it is possible, for this particular case, to write $y = \pm\sqrt{25 - x^2}$ and proceed as usual, it is easy to see that for other implicit functions e.g. $y^5 + xy = 3$ it is impossible to solve y in terms of x. Thus, we need another way known as *implicit differentiation* that requires no explicit expression of $y(x)$.

The best way to introduce implicit differentiation is probably to solve the so-called *related rates* problems. One such problem is given in Fig. 4.35.

We need to relate $z(t)$ to $y(t)$, and then differentiate it with respect to time:

$$[z(t)]^2 = 100^2 + [y(t)]^2 \Longrightarrow 2z\frac{dz}{dt} = 2y\frac{dy}{dt} \Longrightarrow \frac{dz}{dt} = 3\frac{y}{z}$$

When the balloon is above the ground 50 m, $z = 50\sqrt{50}$ m, so at that time, $dz/dt = 3(50)/(50\sqrt{50}) = 3\sqrt{5}/5$ m/s. The problem is easy using the chain rule and it is so because time is present in the problem.

Now we come back to this problem: given $x^2 + y^2 = 25$, what is dy/dx? We can imagine a point with coordinate $(x(t), y(t))$ moving along the circle of radius 5 centered at the origin. Then, we just do the differentiation w.r.t time:

$$[x(t)]^2 + [y(t)]^2 = 25 \Longrightarrow 2x\frac{dx}{dt} + 2y\frac{dy}{dt} = 0 \Longrightarrow 2xdx + 2ydy = 0 \Longrightarrow \boxed{\frac{dy}{dx} = -\frac{x}{y}}$$

Is this result correct? If we write $y = \sqrt{25 - x^2}$ (for the upper part of the circle), then $dy/dx = {-x}/{y}$, the same result obtained using implicit differentiation. You can see that dt disappears. Why ${dy}/{dx} = -{x}/{y}$? This simply means that the tangent to a circle at any point is perpendicular to the line joining the center to that point. You need to draw a figure and using the fact that $({y}/{x})({x}/{y}) = 1$ to see this.

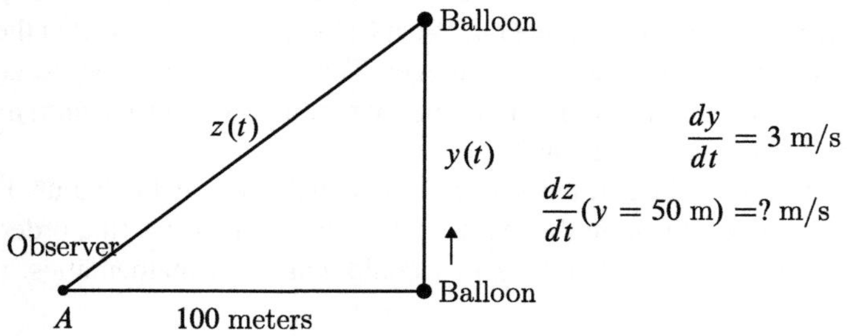

Fig. 4.35 One problem on related rates: a balloon is flying up with a constant speed of 3 m/s. While it is doing so the distance from it to an observer at A, denoted by z, is changing. The question is: find dz/dt when $y = 50$ m

We can apply this to a more complex function[25]:

$$y^5 + xy = 3 \Longrightarrow 5y^4dy + ydx + xdy = 0 \Longrightarrow \frac{dy}{dx} = -\frac{y}{5y^4 + x}$$

As now we're used to this, I have removed dt in the process in the above equation.

4.4.18 Derivative of Logarithms

What is the derivative of $y = \log_a x$? We can use implicit differentiation to find it. Let's first convert to exponential function (as we know how to differentiate this function):

$$y = \log_a x \Longrightarrow x = a^y$$

Differentiating the above, we get

$$dx = \ln a a^y dy \Longrightarrow \frac{dy}{dx} = \frac{1}{\ln a a^y} = \frac{1}{\ln a}\frac{1}{x}$$

With $a = e$, $\ln e = 1$, thus the derivative of $\ln x$ is simply $1/x$.

For sake of convenience, I summarize these two results in the following equation

$$\frac{d}{dx}\left(\log_a x\right) = \frac{1}{\ln a}\frac{1}{x}, \quad \frac{d}{dx}(\ln x) = \frac{1}{x} \tag{4.4.38}$$

Logarithmic differentiation is a useful technique to differentiate complicated functions. For example, what is the derivative of the following function:

$$y = \frac{x^{3/4}\sqrt{x^2+1}}{(3x+2)^5}$$

Taking the natural logarithm of both sides of this equation, we get

$$\ln y = \frac{3}{4}\ln x + \frac{1}{2}\ln(x^2+1) - 5\ln(3x+2)$$

Differentiating this equation we have:

$$\frac{dy}{y} = \frac{3}{4}\frac{dx}{x} + \frac{1}{2}\frac{2xdx}{x^2+1} - \frac{15dx}{3x+2}$$

---

[25] If you do not like dx, dy as independent objects, it is fine to just use dy/dx, and you will get the same result.

And solve this for dy/dx, and replace y by its definition, we get the final result:

$$\frac{dy}{dx} = y\left(\frac{3}{4x} + \frac{x}{x^2+1} - \frac{15}{3x+2}\right) = \frac{x^{3/4}\sqrt{x^2+1}}{(3x+2)^5}\left(\frac{3}{4x} + \frac{x}{x^2+1} - \frac{15}{3x+2}\right)$$

The method of differentiating functions by first taking logarithms and then differentiating is called logarithmic differentiation.

4.5 Applications of Derivative

This section presents some applications of derivative. First, the use of derivative to solve maxima problems (i.e., finding the maximum or minimum value of a function) is given in Sect. 4.5.1. Then, convexity and Jensen's inequality is treated in Sect. 4.5.2. Linear approximation—the idea of replacing a complex function by a linear one is the subject of Sect. 4.5.3. Finally, the Newton method to solve numerically any equation of the form $f(x) = 0$ is presented in Sect. 4.5.4. All these problems are solved thanks to the concept of derivative.

4.5.1 Maxima and Minima

Probably the most important application of derivative is to find minima and maxima of a function. This is an optimization problem—a very important problem in virtually all science and engineering fields. In an optimization problem we have an objective function $y = f(x)$, which we want to minimize. Of course, in the realm of calculus, we consider only continuous functions with at least continuous first derivative.

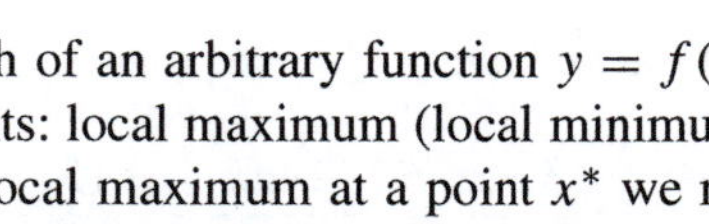

Considering the graph of an arbitrary function $y = f(x)$ in the next figure, we can identify special points: local maximum (local minimum) and global maximum (global minimum). By local maximum at a point x^* we mean the function at that point is largest in a neighborhood of the point: $f(x^*) \geq f(x^* + h)$ for small negative and positive h. You can be the smartest kid in your class (local maximum) but there might be a smarter kid in another class.

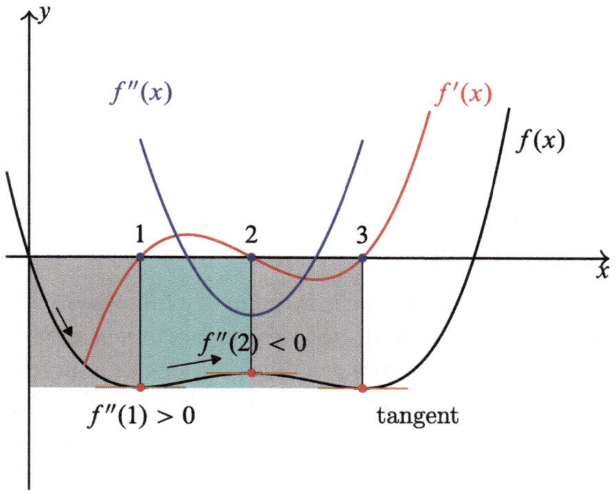

Fig. 4.36 Graph of a fourth-order polynomial $f(x)$ with its first $f'(x)$ and second derivatives $f''(x)$

To discover the rules regarding maxima/minima, let's consider the following fourth-order polynomial:

$$f(x) = \frac{x^4}{4} - 2x^3 + \frac{11x^2}{2} - 6x$$

And we compute the first and second derivative of this function[26]:

$$\begin{aligned} f'(x) &= x^3 - 6x^2 + 11x - 6 = (x-1)(x-2)(x-3) \\ f''(x) &= 3x^2 - 12x + 11 \end{aligned}$$

The graphs of the function $f(x)$, the first derivative $f'(x)$ and the second derivative $f''(x)$ are shown in Fig. 4.36. We can see that:

- The function is decreasing within the interval in which $f'(x) < 0$. This makes sense noting that $f'(x)$ is the rate of change of $f(x)$—when it is negative the function must be decreasing;
- The function is increasing within the interval in which $f'(x) > 0$;
- At point x_0 where $f'(x_0) = 0$, the function is not increasing nor decreasing; it is stationary—*the tangent is horizontal*. There ($x_0 = 1, 2, 3$), the function is either a *local* minimum or a local maximum. It is only a local minimum or maximum for there are locations where the functions can get a larger/smaller value. The

[26] As the first derivative represents the slope of the tangent of the curve, at the minimum or maximum point x_0 the slope is horizontal. In other words, at these points, the first derivative vanish. Thus, it is natural to consider the first derivative of $f(x)$ to study its maxima and minima. The second derivative is needed when it comes to decide at x_0 the function attains a maximum or a minimum. Think of a bowl (similar to $y = x^2$ with second derivative of 2), it has a minimum point (the bottom of the bowl). Now, turn the bowl upside down ($y = -x^2$, with a negative y''), this bowl now has a maximum point. It is as simple as that.

derivative at a point contains local information about a function around the point (which makes sense from the very definition of the derivative);

- A stationary point x_0 is a local minimum when $f''(x_0) > 0$; the tangent is below the function, or the curve is concave up. Around that point the curve has the shape of a cup ∪;
- A stationary point x_0 is a local maximum when $f''(x_0) < 0$; the tangent is above the function, or the curve is concave down. Around that point the curve has the shape of a cap ∩.

Finding a minimum or a maximum of a function is essentially a comparison problem. In principle we need to evaluate the function at all points and pick out the minimum/-maximum. Differential calculus provides us a more efficient way consisting of two steps: (1) finding stationary points where the first derivative of the function is zero, and (2) evaluating the second derivative at these points.

Snell's law of refraction. We now use the derivative to derive the Snell's law of refraction.[27] This law is a formula used to describe the relationship between the angles of incidence and refraction, when referring to light (or other waves) passing through a boundary between two different isotropic media such as water/air. Fermat derived this law in 1673 based on his principle of least time. According to this principle, the light follows a path that minimizes the travel time; or light is lazy.

Referring to Fig. 4.37, we now compute the time required for the light to go from A to B, which is the sum of the time for it to travel from A to O then from O to B:

$$t = t_{AO} + t_{OB} = \frac{\sqrt{a^2+x^2}}{v_1} + \frac{\sqrt{b^2+(d-x)^2}}{v_2}$$

Calculating the first derivative of t and set it to zero gives us

$$\frac{x}{\sqrt{a^2+x^2}v_1} = \frac{d-x}{\sqrt{b^2+(d-x)^2}v_2}, \quad \Longrightarrow \frac{\sin\alpha_1}{v_1} = \frac{\sin\alpha_2}{v_2} \tag{4.5.1}$$

Note that the angles are measured with respect to the normal to the boundary, not to the boundary. Now, introducing the refractive index n, defined as $n = {}^{c}/_{v}$ where c denotes the speed of light in vacuum. Thus, the refractive index describes how fast light travels through a medium. For example, the refractive index of water is 1.333, meaning that light travels 1.333 times slower in water than in a vacuum.

Now, Eq. (4.5.1) becomes:

$$\boxed{\frac{\sin\alpha_1}{v_1} = \frac{\sin\alpha_2}{v_2} \quad \text{or} \quad n_1 \sin\alpha_1 = n_2 \sin\alpha_2} \tag{4.5.2}$$

That's the Snell law of refraction.

[27] Named after the Dutch astronomer and mathematician Willebrord Snellius (1580–1626).

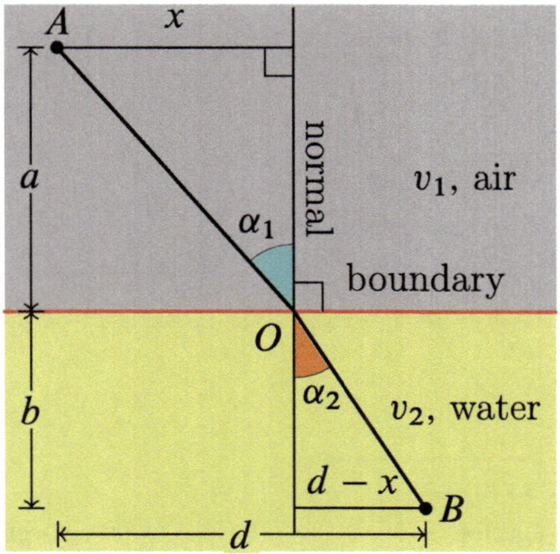

Fig. 4.37 Snell's law of refraction: v_1 and v_2 are the velocities of light in the medium 1 and 2. As the velocity is lower in the second medium, the angle of refraction α_2 is smaller than the angle of incidence α_1. A and B are two points fixed in space (i.e., a, b, d are constants). A light ray travels from A to B can meet the boundary (red line) of the two media at any point. Assuming that the point is O which is x apart from A in the horizontal direction

4.5.2 *Convexity and Jensen's Inequality*

When the second derivative at a stationary point is positive the curve has a ∪ shape. Now we discuss more about this property. We consider a function $y = f(x)$ for $a \le x \le b$ (alternatively, we write that $x \in [a, b]$). The function $f(x)$ is said to be *convex if its graph in the interval* $[a, b]$ *is below the secant line joining the two points* $(a, f(a))$ *and* $(b, f(b))$; they are labeled as A and B in Fig. 4.38a.

To quantify this we consider an arbitrary point x in $[a, b]$; it is given by $x = (1-t)a + tb, t \in [0, 1]$. We now label the point $P(x, f(x))$ on the curve $y = f(x)$. Corresponding to this x we have point Q on the secant AB; the y-coordinate of Q is $(1-t)f(a) + tf(b)$.[28] And the fact that P is below Q is written as:

$$f((1-t)a + tb) \le (1-t)f(a) + tf(b) \tag{4.5.3}$$

A few examples are helpful; with $t = 0.5$ and $t = 2/3$, we have:

$$f\left(\frac{a+b}{2}\right) \le \frac{f(a)+f(b)}{2}, \quad \text{or } f\left(\frac{2}{3}a + \frac{1}{3}b\right) \le \frac{2}{3}f(a) + \frac{1}{3}f(b)$$

We are going to generalize this inequality. We rewrite Eq. (4.5.3) as

$$f(t_1x_1 + t_2x_2) \le t_1 f(x_1) + t_2 f(x_2), \quad t_1 + t_2 = 1 \tag{4.5.4}$$

[28] You need to know the equation of a line/segment given its two points, see Sect. 9.1.3.

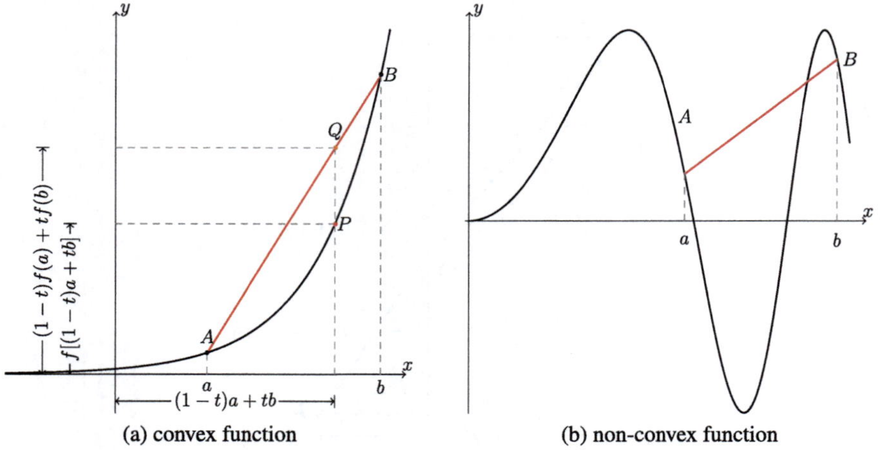

Fig. 4.38 Convex function (left) and non-convex function (right). The function $f(x)$ is said to be convex if its graph in the interval $[a, b]$ is below the second line joining the two end points $(a, f(a))$ and $(b, f(b))$;

And we ask the question: does this nice inequality hold for 3 points? The answer is yes. So, we need to prove this:

$$f(t_1x_1 + t_2x_2 + t_3x_3) \leq t_1 f(x_1) + t_2 f(x_2) + t_3 f(x_3), \quad t_1 + t_2 + t_3 = 1$$

Proof We use Eq. (4.5.4) to prove the above inequality. First, we need to split 3 terms into 2 terms (to apply Eq. (4.5.4)):

$$\begin{aligned} f(t_1x_1 + t_2x_2 + t_3x_3) &= f\left(t_1x_1 + (t_2 + t_3)\frac{t_2x_2 + t_3x_3}{t_2 + t_3}\right) \\ &\leq t_1 f\,(x_1) + (t_2 + t_3) f\left(\frac{t_2x_2 + t_3x_3}{t_2 + t_3}\right) \quad \text{(Eq. 4.5.4)} \end{aligned}$$

After that we apply again Eq. (4.5.4) to $f\ ({t_2x_2+t_3x_3}/{t_2+t_3})$:

$$\begin{aligned} f(t_1x_1 + t_2x_2 + t_3x_3) &\leq t_1 f\,(x_1) + (t_2 + t_3)\left[\frac{t_2}{t_2 + t_3} f(x_2) + \frac{t_3}{t_2 + t_3} f(x_3)\right] \\ &\leq t_1 f\,(t_1x_1) + t_2 f\,(x_2) + t_3 f\,(x_3) \end{aligned}$$

And nothing can stop us to generalize this inequality to the case of n points:

$$\boxed{f\left(\sum_{i=1}^{n} t_i x_i\right) \leq \sum_{i=1}^{n} t_i f\,(x_i)}, \quad \sum_{i=1}^{n} t_i = 1 \tag{4.5.5}$$

And this is known as the Jensen inequality, named after the Danish mathematician Johan Jensen (1859–1925). Jensen was a successful engineer for the Copenhagen Telephone Company and became head of the technical department in 1890. All his mathematics research was carried out in his spare time. Of course if the function is concave, the inequality is reversed.

To avoid explicitly stating $\sum t_i = 1$, another form of the Jensen inequality is

$$\boxed{f\left(\frac{\sum_{i=1}^n a_i x_i}{\sum_{i=1}^n a_i}\right) \leq \frac{\sum_{i=1}^n a_i f(x_i)}{\sum_{i=1}^n a_i}} \tag{4.5.6}$$

where $t_i = a_i / \sum a_i$, and $a_i > 0$ are weights.

Geometric interpretation of Jensen's inequality. For $n = 2$, we have a geometry interpretation of the Jensen inequality (Fig. 4.38). How about the case of more than 2 points? It turns out that there is also a geometry interpretation, but it requires a concept from physics: the center of mass.

Let's consider a convex function $y = f(x)$ on an interval. For a visual demonstration we consider the case $n = 4$. We place four-point masses (with masses m_1, m_2, m_3, m_4, respectively) on the curve of $y = f(x)$, of which the x-coordinates are $x_i, i = 1, \ldots, 4$ (see Fig. 4.39). The coordinates of the center of mass of these four-point masses are (refer to Sect. 6.8.7 if your memory on this was rusty):

$$x_{\text{CM}} = \frac{\sum_{i=1}^4 m_i x_i}{\sum_{i=1}^4 m_i}, \quad y_{\text{CM}} = \frac{\sum m_i f(x_i)}{\sum m_i}$$

Note that in the equation for y_{CM} the limits of summation were skipped for the sake of brevity.

The nice thing is that the center of mass is always inside the polygon with vertices being the point masses—the shaded region in Fig. 4.39 (I refer Sect. 6.8.7 for a proof of this). This leads immediately to $y_{\text{CM}} \geq f(m_i x_i / m)$.

AM-GM inequality. We discussed the AM-GM inequality in Sect. 2.20.2 together with Cauchy's forward-backward-induction-based proof. Now, we show that the AM-GM inequality is simply a special case of the Jensen inequality. The function $y = \log x$ for $x > 0$ is concave (using the fact that $f''(x) < 0$ or looking at its graph), thus applying Eq. (4.5.6) with $a_i = 1$:

$$\log\left(\frac{x_1 + x_2 + \cdots + x_n}{n}\right) \geq \frac{\log(x_1) + \log(x_2) + \cdots + \log(x_n)}{n}$$

Fig. 4.39 Geometric interpretation of the Jensen inequality for the case of more than 2 points

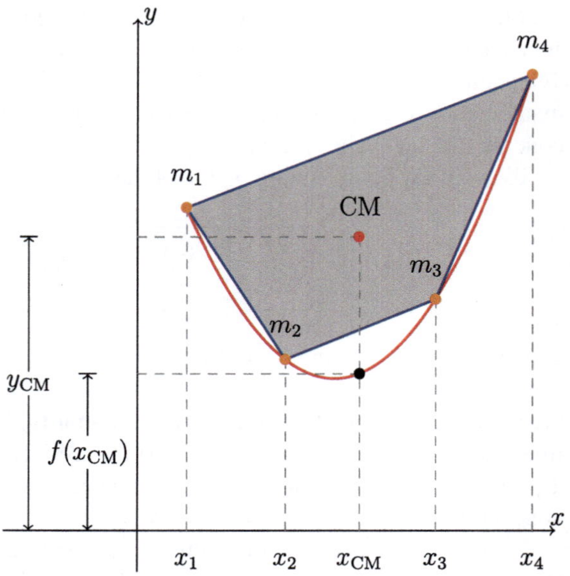

Using the property of logarithm that $\log ab = \log a + \log b$ for the RHS of the above, we are led to the AM-GM inequality:

$$\log\left(\frac{x_1+x_2+\cdots+x_n}{n}\right) \geq \log(x_1x_2\cdots x_n)^{\frac{1}{n}} \implies \frac{x_1+x_2+\cdots+x_n}{n} \geq (x_1x_2\cdots x_n)^{\frac{1}{n}}$$

where use was made of the fact that $\log x$ is an increasing function (Fig. 4.34) i.e., if $\log a \geq \log b$ then $a \geq b$.

Why convex functions important. Convex functions are important because they have nice properties. Given a convex function within an interval, if a local minimum (maximum) is found, it is also the global minimum (maximum). And it leads to convex optimization. Convex optimization is the problem of minimizing a convex function over convex constraints. It is a class of optimization problems for which there are fast and robust optimization algorithms, both in theory and in practice.

Now, you have the tool, let's solve this problem: Given three positive real numbers a, b, c, prove that

$$a^a b^b c^c \geq (abc)^{\frac{a+b+c}{3}}$$

The art of using the Jensen inequality is to use what function. If you know what $f(x)$ to be used, then it becomes easy.

4.5.3 Linear Approximation

Bertrand Russell Although this may seem a paradox, all exact science is dominated by the idea of approximation.

We now discuss a useful application of the derivative: that is to approximate a complex function by a linear function. The problem is that we have a complex function of which the graph is a curve. Focus now on a specific point x_0, and if we zoom in closely at this point we see not a curve but a segment! That segment has a slope equal to $f'(x_0)$. Thus, in the neighborhood of x_0, we replace $f(x)$ (which is usually complex) by a line with the equation of the following form

$$Y(x) = f(x_0) + f'(x_0)(x - x_0) \tag{4.5.7}$$

Of course working with a line is much easier than with a complex curve. Figure 4.40 shows this approximation.

At x_0 we have $Y(x_0) = f(x_0)$, but the approximation gets worse for x far way from x_0 (for we follow the tangent line not the real curve). This is obvious. We need to know the error of this approximation; the error is $e = Y(x) - f(x)$. Let's try with a function and play with the error. We can spot the pattern from this activity. We use $y = \sqrt{x}$ and $x_0 = 100$ (nothing special about this point except its square root is 10). We compute the square root of $100 + h$ for $h = \{1.0, 0.1, 0.01, 0.001\}$ using Eq. (4.5.7), which yields $\sqrt{100 + h} \approx 10 + h/20$, and the error associated with the approximation is $e(h) := Y(100 + h) - \sqrt{100 + h}$.

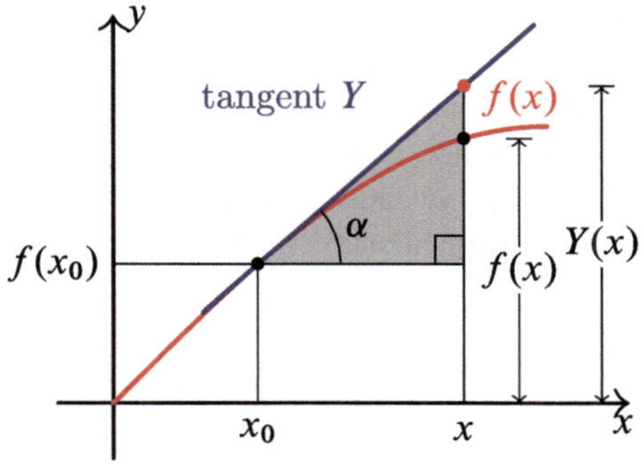

Fig. 4.40 Linear approximation of a function $f(x)$ using the first derivative $f'(x)$. In the shaded right triangle, we have $\tan\alpha = {}^{Y - f(x_0)}/_{x - x_0}$, but $f'(x_0) = \tan\alpha$. And voila we get our linear approximation

Table 4.11 Linear approximations of $\sqrt{x}$ at $x_0 = 100$ for various h

h	$Y = 10 + h/20$	$e(h)$
1.0	10.05	1.243789e-04
0.1	10.005	1.249375e-06
0.01	10.0005	1.249938e-08
0.001	10.00005	1.249987e-10

The results are given in Table 4.11. Looking at this table we can see that $e(h) \sim h^2$. That is when h is decreasing by $1/10$ the error is decreasing by $1/100$. We can also get this error measure by squaring $\sqrt{100+h} \approx 10 + h/20$

$$\sqrt{100+h} \approx 10 + \frac{h}{20} \implies 100 + h \approx 100 + h + \frac{h^2}{400}$$

Some common linear approximations near $x = 0$ are $e^x \approx 1 + x$ and $\sin x \approx x$,[29] and the approximation for the sine function is used in solving the oscillation of a pendulum.

4.5.4 *Newton's Method for Solving* $f(x) = 0$

Newton's method for solving an equation of the form $f(x) = 0$ (e.g. $\sin x + x^2 = 0.5$) uses the first derivative of $f(x)$. The method belongs to a general class of *iterative methods*. In an iterative method, a starting point x_0 is selected, then, a better value of the solution x_1 is computed using the information evaluated at x_0. This iterative process produces a sequence of $x_0, x_1, x_2, \ldots$ which converges to the root x^*.

The idea of the Newton method is illustrated in Fig. 4.41. At point $(x_n, f(x_n))$, we draw a line tangent to the curve and find x_{n+1} as the intersection of this line and the x-axis. Thus, x_{n+1} is determined using x_n, $f(x_n)$ and $f'(x_n)$:

$$\boxed{x_{n+1} = x_n - \frac{f(x_n)}{f'(x_n)}} \tag{4.5.8}$$

Extraction of square roots. We can use Newton's method to extract square roots of any positive number. Let's write $x = \sqrt{a}$ as $f(x) = x^2 - a = 0$. Using Eq. (4.5.8), we have

$$x_{n+1} = x_n - \frac{x_n^2 - a}{2x_n} = \frac{x_n}{2} + \frac{a}{2x_n} = \frac{1}{2}\left(x_n + \frac{a}{x_n}\right) \tag{4.5.9}$$

[29] For $\sin x$, $x_0 = 0$, then $Y = \sin(0) + \cos(0)(x - 0) = x$.

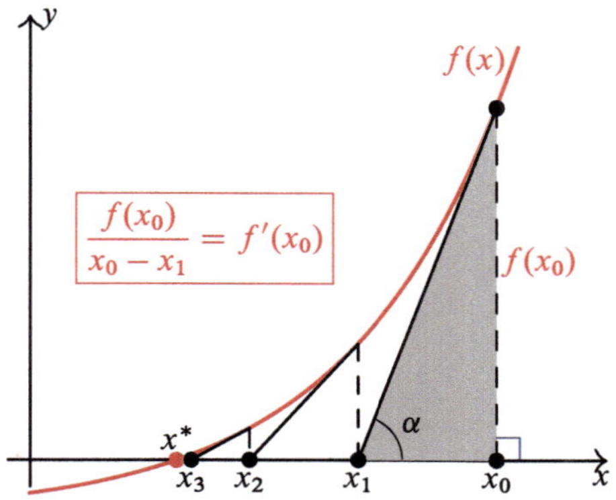

Fig. 4.41 Newton's method of solving $f(x) = 0$ iteratively. We have a curve $y = f(x)$ and it intersects with the x−axis at x^*; that's the solution to $f(x) = 0$. We start with a guess x_0, and we get x_1 using the tangent line to the curve at $(x_0, f(x_0))$. Then we repeat the process until we get to x^*

Table 4.12 Solving $\sqrt{x} = 2$ with $x_0 = 1$

n	x_n	$e(h)$
1	1.0	4.14e-01
2	1.5	−8.58e-02
3	1.416666666	−2.45e-03
4	1.414215686	−2.12e-06
5	1.414213562	−1.59e-12

Note that the final expression was used by Babylonians 1000 years before Newton. The result of the calculation given in Table 4.12 demonstrates that Newton method converges quickly. More precisely it converged quadratically when close to the solution: the last three rows indicate that the error is halved each iteration.

Coding Newton's method. Let's solve this equation $f(x) = \cos x - x = 0$ using a computer. That is we do not compute $f'(x)$ explicitly and use Eq. (4.5.8) to get

$$x_{n+1} = x_n + \frac{\cos x_n - x_n}{1 + \sin x_n}$$

That is too restrictive. We want to write a function that requires a function $f(x)$ and a tolerance. That's it. It will give us the solution for any input function. The idea is to use an approximation for the derivative, see Sect. 10.2.1. The code is given in Listing A.6. In any field (pure or applied maths, science, or engineering), coding has become an essential skill. So, it is better to learn coding when you're young. That's why I have inserted many codes throughout the note.

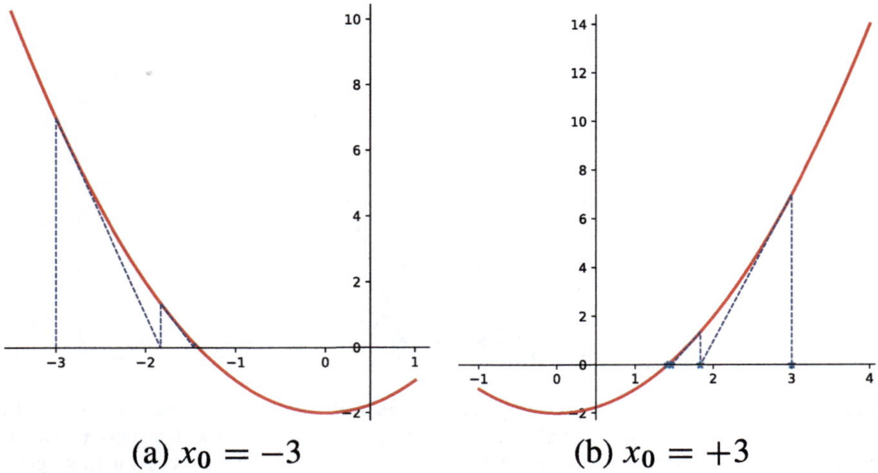

Fig. 4.42 Newton's method is sensitive to the initial guess x_0

Is Newton's method applicable only to one equation with one unknown $f(x) = 0$? No! It is used to solve systems of equations of billion unknowns, see Sect. 6.4. Actually it is used everyday by scientists and engineers. One big application is nonlinear finite element analyses to design machines, buildings, airplanes, you name it.

Exploring Newton's method. With a program implementing the Newton method we can play with it, just to see what happens. For example, in the problem of finding $\sqrt{2}$ by solving $x^2 - 2 = 0$, if we start with $x_0 = -1$, then the method gives us $-\sqrt{2}$. Not that we want! But it is also a root of $x^2 - 2 = 0$. Thus, the method depends on the initial guess (Fig. 4.42). To find a good x_0 for $f(x) = 0$ we can use a graphic method: we plot $y = f(x)$ and locate the points it intersects with the x-axis roughly, and use that for x_0.

Newton's method on the complex plane. We discussed complex numbers in Sect. 2.25, but we seldom use them. Let's see if we can use Newton's method to solve $f(z) = 0$ such as $z^4 - 1 = 0$ where z is a complex number. Just assume that we can treat functions of a complex variable just as functions of a real variable, then

$$\boxed{z_{n+1} = z_n - \frac{f(z_n)}{f'(z_n)}} \tag{4.5.10}$$

Let's solve the simplest complex equation $z^2 + 1 = 0$, this equation has two solutions $z = \pm i$. With the initial guess $z_0 = 1 + 0.5i$ Newton's method converges to $z = i$ (Table 4.13). So, the method works for complex numbers too. Surprise? But happy. If $z_0 = 1 - i$, the method gives us the other solution $z = -i$ (not shown here).

Table 4.13 Solving $z^2 + 1 = 0$ with $z_0 = 1 + 0.5i$. See Listing A.7 for the code

n	z_n
1	$0.1 + 0.45i$
2	$-0.185294 + 1.28382i$
3	$-0.0375831 + 1.02343i$
4	$-0.000874587 + 0.99961i$
5	$3.40826e - 7 + 1.0i$
6	$-1.04591e - 13 + 1.0i$

If we pose this question we can discover something interesting. The question is if $f(z) = 0$ has multiple roots then which initials z_0 converge to which roots? And a computer can help us to visualize this. Assume that we know the exact roots and they are stored in a vector $z_{\text{exact}} = [\bar{z}_1, \bar{z}_2, \ldots]$. Corresponding these exact roots are some colors, one color for each root. Then, the steps are

1. A large number of points on the complex plane is considered.
2. For each of these points with coordinates (x, y), form a complex number $z_0 = x + iy$. Use Eq. (4.5.10) with z_0 to find one root z. Then find the position of z in z_{exact}, thus find the associated color. That point (x, y) is now assigned with that color.
3. Now we have a matrix of which each element is a color, we can plot this matrix as an image.

You can find the code in Listing A.7. Let's apply it to $f(z) = z^3 - 1 = 0$. The roots of $f(z)$ are: 1, $-1/2 + i\sqrt{3}/2$, and $-1/2 - i\sqrt{3}/2$. Three roots and thus three colors. Points in the green color converge to the root $\bar{z}_1 = 1$, those in the purple color to the root $\bar{z}_2 = -1/2 + i\sqrt{3}/2$ and ones in the red color converge to the remaining root. These three domains are separated by a boundary which is known as Newton fractal. We see that complex numbers are very close together, converging to different solutions, arranged in an intricate pattern (Fig. 4.43).

Arthur Cayley (1821–1895) was a prolific British mathematician who worked mostly on algebra. He helped found the modern British school of pure mathematics. In 1879 he published a theorem for the basin of attraction for quadratic complex polynomials. Cayley also considered complex cubics but was unable to find an obvious division for the basins of attraction. It was only later in the early twentieth century that French mathematicians Pierre Joseph Louis Fatou (1878–1929) and Gaston Maurice Julia (1893–1978) began to understand the nature of complex cubic polynomials. With computers, from 1980s mathematicians were able to finally create pictures of the basins of attraction of complex cubic functions.

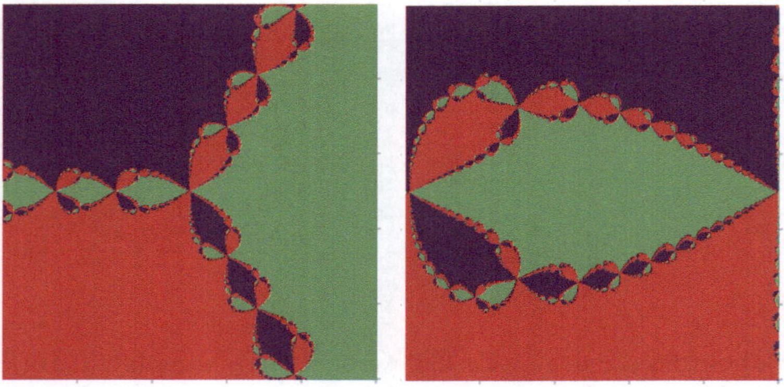

(a) (b)

Fig. 4.43 Newton's fractals for $z^3 - 1 = 0$

4.6 The Fundamental Theorem of Calculus

We have defined the integral of the function $y = f(x)$ over the interval $[a, b]$ as the area under the curve described by $y = f(x)$. By approximating this area by the area of many many thin slices, we have arrived with a definition of the integral as a limit of a sum of many many small parts. But this definition is not powerful, as it only allowed us to compute this simple integral $\int_a^b x^n dx$. How about other functions? How about even the area of a circle?

It turns out that the answer is lying in front of us. We just do not see it. Let's get back to the distance-speed problem. From the distance traveled $s(t)$ we can determine the instantaneous speed by differentiating the distance $s(t)$: $v(t) = ds/dt$. *How about the inverse*? Given the (non-uniform) speed $v(t)$ can we determine the distance? Let's do that.

Assume that the time interval of interest is $[0, T]$ where T is a fixed number (e.g. 5 h). Let's call ds an infinitesimal distance, then the total distance is simply the sum of all these ds, or symbolically $\int_0^T ds$. But $ds = vdt$, so the distance is $\int_0^T vdt$. So, the distance is the area under the speed curve $v(t)$. Actually, this is not unexpected (Fig. 4.44).

If we now consider T a variable, then $s(T)$ is a function of T:

$$s(T) = \int_0^T v(t)dt \tag{4.6.1}$$

But the derivative of $s(T)$ is $v(T)$, thus we have

$$\frac{d}{dT}\int_0^T v(t)dt = v(T) \tag{4.6.2}$$

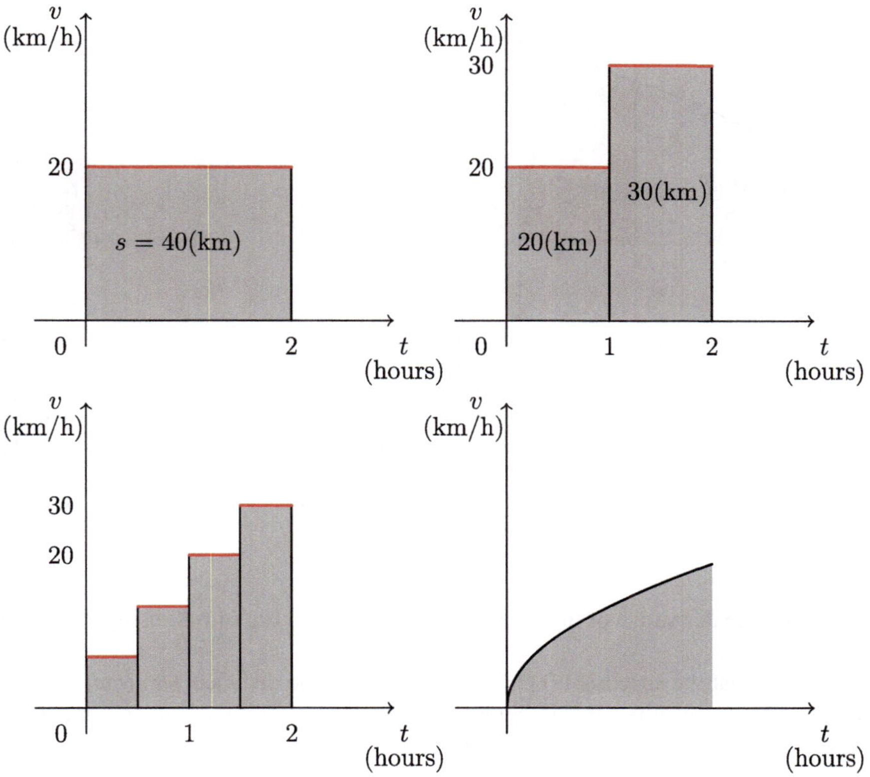

Fig. 4.44 Distance is the area under the speed curve

which says that differentiating an integral of a function gives back the function. In other words, differentiation undoes integration. Actually, we have seen this before[30]:

$$\int_0^x x^2 dx = \frac{x^3}{3}, \quad \frac{d}{dx}\left(\frac{x^3}{3}\right) = x^2$$

Now, we suspect that this relation between differentiation and integration holds for any function. We set the task to examine this. Is the following true?

$$\frac{d}{dx}\int_0^x f(t)dt = f(x) \tag{4.6.3}$$

The integral $\int_0^x f(t)dt$ is the area under the curve $y = f(t)$ from 0 to x. By considering a tiny change dx, we can see that the change in this area is $f(x)dx$ (Fig. 4.45).

[30] For the indefinite integral see Sect. 4.3.3, with b replaced by x.

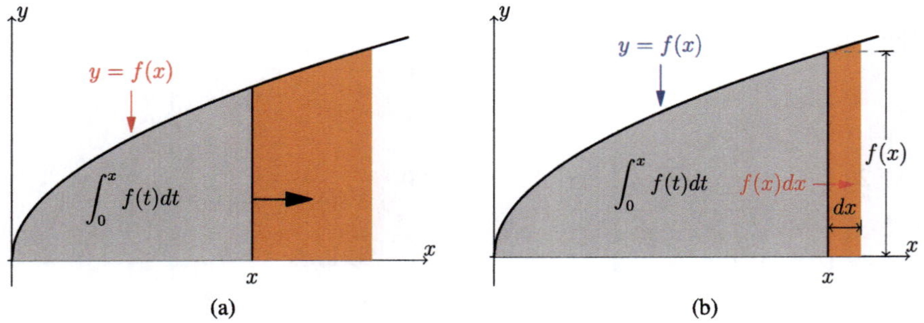

Fig. 4.45 Geometric proof of Eq. (4.6.3). The key point is to think of the area problem dynamically. Imagine sliding x to the right at a constant speed. You could even think of x as time; Newton often did. Then the area of the crossed region changes continuously as x moves. Because that area depends on x, it should be regarded as a function of x. Now considering a tiny change of x, denoted by dx. The area is increased by a tall, thin rectangle of height $f(x)$ and infinitesimal width dx; this tiny rectangle has an infinitesimal area $f(x)dx$. (Actually there is a tiny triangle, but it is nothing compared with the rectangle). Thus, the rate at which the area accumulates is $f(x)$. And this leads to Eq. (4.6.3)

Therefore, the derivative of the area is $f(x)$. We proved Eq. (4.6.3) using the differential dx.

Assume that the speed is $v(t) = 8t - t^2$, what is the distance $\int_0^T v(t)dt$? We do not know how to evaluate this integral (not using the definition of integral of course) but we know that it is a function $s(T)$ such that $ds/dT = v(T) = 8T - T^2$, from Eq. (4.6.2). A function like $s(T)$ is called an *anti-derivative*. We have just met something new here. Before, we are given a function, let say, $y = x^3$, and we're asked (or required) to find its derivative: $(x^3)' = 3x^2$. Now, we're facing the inverse problem: $(?)' = 3x^2$, that is finding the function of which the derivative is $3x^2$. We know that function is x^3. Thus, x^3 is *one anti-derivative* of $3x^2$. I used the word one anti-derivative because we have other anti-derivatives. In fact, there are infinitely many anti-derivatives of $3x^2$, they are $x^3 + C$, where C is called a constant of integration. It is here because the derivative of a constant is zero. Graphically, $x^3 + C$ is just a vertical translation of the curve $y = x^3$, the tangent to $x^3 + C$ at every point has the same slope as those of x^3.

Coming back now to $s(T)$, we can thus write

$$\left.\begin{aligned} \int_0^T (8t - t^2)dt &= s(T) \\ \frac{ds}{dT} &= 8T - T^2 \end{aligned}\right\} \Longrightarrow s(T) = \int_0^T (8t - t^2)dt = 4T^2 - \frac{T^3}{3} + C \tag{4.6.4}$$

To find the integration constant C, we use the fact that $s(0) = 0$, so $C = 0$.

It is straightforward to use Eq. (4.6.4) for determining the distance traveled between t_1 and t_2 (we're really trying to compute the general definite integral $\int_a^b f(x)dx$ here):

$$\begin{aligned}\int_{t_1}^{t_2}(8t-t^2)dt &= s(t_2)-s(t_1)\\ &= \left(4t_2^2-\frac{t_2^3}{3}+C\right)-\left(4t_1^2-\frac{t_1^3}{3}+C\right)\\ &= \left(4t_2^2-\frac{t_2^3}{3}\right)-\left(4t_1^2-\frac{t_1^3}{3}\right)\end{aligned} \tag{4.6.5}$$

There is nothing special about distance and speed, we have, for any function $f(x)$, the following result

$$\int_a^b f(x)dx = F(b)-F(a) \quad \text{with} \quad \frac{dF}{dx}=f(x) \tag{4.6.6}$$

which is known as *the fundamental theorem of calculus*, often abbreviated as FTC. So, to find a definite integral we just need to find one anti-derivative of the integrand, evaluate it at two end points and subtract them. It is this theorem that makes the problem of finding the area of a curve is a trivial exercise for modern high school students. Notice that the same problem once required the genius of the likes of Archimedes.

While it is easy to understand Eq. (4.6.5) as the distance traveled between t_1 and t_2 must be $s(t_2)-s(t_1)$, it is hard to believe that a definite integral which is the sum of all tiny rectangles eventually equals $F(b)-F(a)$; only the end points matter. But this can be seen if we use Leibniz's differential notation:

$$\begin{aligned}\int_a^b f(x)dx = \int_a^b \frac{dF}{dx}dx &= \int_a^b dF\\ &= (\cancel{F_2}-F_1)+(\cancel{F_3}-\cancel{F_2})+(F_4-\cancel{F_3})+\cdots+(F_n-F_{n-1})\\ &= F_n-F_1 = F(b)-F(a)\end{aligned}$$

The term $(F_2-F_1)+(F_3-F_2)+(F_4-F_3)+\cdots+(F_n-F_{n-1})$ is a sum of differences. The same person (Leibniz) who often worked with such sums (see Sect. 2.21.6) was the one who discovered the fundamental theorem of calculus. Newton, on the other hand, discovered the exact same theorem via a different way: the way of motion. And Newton is the father of mechanics—the science of motion!

History note 4.1: Sir Isaac Newton (25 December 1642–20 March 1726/27)

Sir Isaac Newton was an English mathematician, physicist, astronomer, and theologian (described in his own day as a "natural philosopher") who is widely recognized as one of the most influential scientists of all time and as a key figure in the scientific revolution. His book *PhilosophiÃƒÆ’Ã†âŁ™ÃƒâŁ Ã¢â,¬â„¢ÃƒÆ’Ã¢â,¬Å¡ÃƒâŁšÃ‚Â¦ Naturalis Principia Mathematica (Mathematical Principles of Natural Philosophy)*, first published in 1687, established classical mechanics. Newton also made seminal contributions to optics, and shares credit with Gottfried Wilhelm Leibniz for developing the infinitesimal calculus.

Newton was born prematurely in 1642 at his family's home near the town of Grantham, several months after the death of his father, an illiterate farmer. When Newton was three, his mother wed a wealthy clergyman, who didn't want a stepson. Newton's mother went to live with her new husband in another village, leaving behind her young son in the care of his grandparents.

In 1705, Newton was knighted by Queen Anne. By that time, he'd become wealthy after inheriting his mother's property following her death in 1679 and also had published two major works, 1687's "Mathematical Principles of Natural Philosophy"ÃƒÆ’Ã¢â,¬Å¡ÃƒâŁšÃ‚Âž(commonly called the "Principia") and 1704's "Opticks." After the celebrated scientist died at age 84 on March 20, 1727, he was buried in Westminster Abbey, the resting place of English monarchs as well as such notable non-royals as Charles Darwin, Charles Dickens and explorer David Livingstone.

History note 4.2: Gottfried Wilhelm (von) Leibniz (1646-1716)

Gottfried Wilhelm Leibniz as a German philosopher, mathematician, and political adviser, important both as a metaphysician and as a logician and distinguished also for his independent invention of the differential and integral calculus. As a child, he was educated in the Nicolai School but was largely self-taught in the library of his father– Friedrich Leibniz, a professor of moral philosophy at Leipzig–who had died in 1652. At Easter time in 1661, he entered the University of Leipzig as a law student; there he came into contact with the thought of scientists and philosophers who had revolutionized their fields–figures such as Galileo and René Descartes. During the 1670s (slightly later than Newton's early work), Leibniz developed a very similar theory of calculus, apparently completely independently. Within the short period of about two months he had developed a complete theory of differential calculus and integral calculus. Unlike Newton, however, he was more than happy to publish his work, and so Europe first heard about calculus from Leibniz in 1684, and not from Newton (who published nothing on the subject until 1693). When the Royal Society was asked to adjudicate between the rival claims of the two men over the development of the calculus, they gave credit for the first discovery to Newton, and credit for the first publication to Leibniz. However, the Royal Society under the rather biassed presidency of Newton, later also accused Leibniz of plagiarism, a slur from which Leibniz never really recovered. Ironically, it was Leibniz's mathematics that eventually triumphed, and his notation and his way of writing calculus, not Newton's more clumsy notation, is the one still used in mathematics today.

4.7 Integration Techniques

Let's see how many ways we can compute integrals (indefinite or definite) using paper and pencil. The first way is to use the definition of integral as the limit of the sum of all the areas of the small thin rectangles. The fundamental theorem of calculus saves us from going down this difficult track. Therefore, the second way is to find an anti-derivative of the integrand function. Anti-derivatives of many common functions have been determined and tabulated in tables. So, we just do "table look up". Clearly that these tables cannot cover all the functions, so we need a third way (or fourth). This section presents integration techniques for functions of which anti-derivatives are not present in tables.

4.7.1 Integration by Substitution

We will not find the anti-derivative of $\cos(x^2)2x$ in any table. However, $\int \cos(x^2)2xdx$ can be computed quite straightforwardly. Similarly, it is also possible to compute $\int \sqrt{1+x^2}2xdx$. The two integrals are given by

$$\begin{aligned} \int \cos(x^2)2xdx &= \sin(x^2) + C \\ \int \sqrt{1+x^2}2xdx &= \frac{2}{3}(1+x^2)^{3/2} + C \end{aligned} \tag{4.7.1}$$

And you can verify the above equation by differentiating the RHS and you get the integrands in the LHS. If you look at these two integrals carefully, you will recognize that they are of this form:

$$\boxed{\int_a^b f(g(x))g'(x)dx = \int_\alpha^\beta f(u)du, \quad u = g(x)} \tag{4.7.2}$$

So, we do a change of variable $u = g(x)$, which leads to $du = g'(x)dx$, then the LHS of Eq. (4.7.2) becomes the RHS i.e., $\int_a^b f(g(x))g'(x)dx = \int_\alpha^\beta f(u)du$. Of course, $\alpha = g(a)$ and $\beta = g(b)$. Equation (4.7.2) is called *integration by substitution* and it is based on the chain rule of differentiation. Nothing new here, one fact of differentiation leads to another corresponding fact of integration, because they are related.

Now we can understand Eq. (4.7.1). Let's consider the first integral, we do the substitution $u = x^2$, hence $du = 2xdx$, then:

$$\int \cos(x^2)2xdx = \int \cos(u)du = \sin u + C = \sin(x^2) + C$$

Proof Proof of *integration by substitution* given in Eq. (4.7.2). We start with a composite function $F(g(x))$ as we want to use the chain rule. We compute the derivative of this function:

$$\frac{d}{dx}F(g(x)) = F'(g(x))g'(x) \tag{4.7.3}$$

Now we integrate the two sides of the above equation, we get

$$\int_a^b \frac{d}{dx}F(g(x))dx = \int_a^b F'(g(x))g'(x)dx$$

(if we have two identical functions, the areas under the two curves described by these two functions are the same, that's what the above equation means). Now, the FTC tells us that

$$\int_a^b \frac{d}{dx}F(g(x))dx = F(g(b)) - F(g(a)) \tag{4.7.4}$$

Introducing two new numbers $\alpha = g(a)$ and $\beta = g(b)$, then as a result of the FTC, where $u = g(x)$, we have

$$F(\beta) - F(\alpha) = \int_\alpha^\beta F'(u)du \tag{4.7.5}$$

From Eq. (4.7.4) (4.7.5)) we obtain,

$$\int_\alpha^\beta F'(u)du = \int_a^b \frac{d}{dx}F(g(x))dx = \int_a^b F'(g(x))g'(x)dx$$

To make $f(x)$ appear, just introducing $f(x) = F'(x)$, then the above equation becomes

$$\int_a^b f(g(x))g'(x)dx = \int_\alpha^\beta f(u)du$$

So, the substitution rule guides us to replace a hard integral with a simpler one. The main challenge is to find an appropriate substitution. For certain integrals e.g. $\int \sqrt{1-x^2}dx$, the new variable is clear: $x = \sin\theta$ to just get rid of the square root. I present in Sect. 4.7.1 such trigonometry substitutions. For most of the cases, finding a good substitution is a matter in which practice and ingenuity, in contrast to systematic methods, come into their own.

Example 4.3 Let's compute the following integral

$$I = \int_0^\pi \frac{2x^3 - 3\pi x^2}{(1+\sin x)^2}dx$$

which is the 2015 Cambridge STEP 2[31]

[31] Sixth Term Examination Papers in Mathematics, often referred to as STEP, are university admissions tests for undergraduate Mathematics courses developed by the University of Cambridge. STEP papers are typically taken post-interview, as part of a conditional offer of an undergraduate place. There are also a number of candidates who sit STEP papers as a challenge. The papers are designed to test the ability to answer questions similar in style to undergraduate Mathematics.

What change of variable is to be used? After many unsuccessful attempts, we find that $u = \pi - x$ looks promising:

$$u = \pi - x \Longrightarrow du = -dx, \quad 1 + \sin x = 1 + \sin u$$

Now we compute the nominator in terms of u:

$$\begin{cases} x^3 = (\pi - u)^3 = \pi^3 - 3\pi^2 u + 3\pi u^2 - u^3 \\ x^2 = (\pi - u)^2 = \pi^2 - 2\pi u + u^2 \end{cases} \Longrightarrow 2x^3 - 3\pi x^2 = -\pi^3 + 3\pi u^2 - 2u^3$$

And the integration limits do not change, therefore I becomes

$$I = \int_0^\pi \frac{-\pi^3 + 3\pi u^2 - 2u^3}{(1 + \sin u)^2} du = -\pi^3 \int_0^\pi \frac{du}{(1 + \sin u)^2} - \int_0^\pi \frac{2u^3 - 3\pi u^2}{(1 + \sin u)^2} du$$

And what is the red term? It is I, so we have an equation for I and solving it gives us a new form for I:

$$I = -\frac{\pi^3}{2} \int_0^\pi \frac{du}{(1 + \sin u)^2}$$

We stop here, as the new integral seems solvable. What we want to say here is that this integral was designed so that the substitution $u = \pi - x$ works. If we slightly modify the integral as follows

$$I_1 = \int_0^{\pi/2} \frac{2x^3 - 3\pi x^2}{(1 + \sin x)^2} dx, \quad I_2 = \int_0^\pi \frac{2x^3 - 3x^2}{(1 + \sin x)^2} dx, \quad I_3 = \int_0^\pi \frac{3x^3 - 3\pi x^2}{(1 + \sin x)^2} dx$$

Our substitution would not work! That's why it was just a trick; even though a favorite one of examiners. How do we integrate these integrals then? We fall back to the very definition of integral as the sum of many many thin rectangles, but we use the computer to do the boring sum. This is called numerical integration (see Sect. 10.4 if you're interested in, that's how scientists and engineers do integrals).

4.7.2 Integration by Parts

Integration by parts is based on the product rule of differentiation that reads

$$[u(x)v(x)]' = u'(x)v(x) + v'(x)u(x)$$

Integrating both sides of the above equation gives us

$$\boxed{u(x)v(x) = \int u'(x)v(x)dx + \int v'(x)u(x)dx} \tag{4.7.6}$$

So, instead of calculating the integral $\int u'(x)v(x)dx$, we compute $\int v'(x)u(x)dx$ which should be simpler. Basically we transfer the derivative from u to v. The hard thing is to recognize which should be $u(x)$ and $v(x)$. Some examples are provided to see how to use this technique.

Example 4.4 We want to determine $\int \ln x dx$. Start with $x \ln x$ and differentiate that (then $\ln x$ will show up), and we're done:

$$(x \ln x)' = \ln x + 1 \implies \int \ln x dx = x \ln x - x + C$$

Next, considering this integral $\int x \cos x dx$. Start with $x \sin x$ (for $(\sin x)' = \cos x$),

$$(x \sin x)' = \sin x + x \cos x \implies \int x \cos x dx = x \sin x - \int \sin x dx = x \sin x + \cos x + C$$

Example 4.5 Let's consider $\int x^2 e^x dx$. This one is interesting as we will need to do integration by parts two times. First, recognize that the derivative of e^x is itself, so we consider the function $x^2 e^x$, its derivative will make appear $x^2 e^x$ (the integrand), and another term with a lower power of x (which is good). So,

$$(x^2 e^x)' = 2xe^x + x^2 e^x \implies \int x^2 e^x dx = x^2 e^x - 2 \int xe^x dx$$

Now, we have an easier problem to solve: the integral of xe^x. Repeat the same step, we write

$$(xe^x)' = e^x + xe^x \implies \int xe^x dx = xe^x - \int e^x dx = xe^x - e^x$$

And, voilà, the result is

$$\int x^2 e^x dx = x^2 e^x - 2xe^x + 2e^x + C \tag{4.7.7}$$

Should we stop here and move to other integrals? If we stop here and someone come to ask us to compute this integral $\int x^5 e^x dx$ or even $\int x^{20} e^x dx$, we would struggle to solve these integrals. There is a structure behind Eq. (4.7.7), which we will come back to in Sect. 4.7.4.

4.7.3 Trigonometric Integrals: Sine/Cosine

This section presents the following integrals (on the left column are particular examples of the integrals in the right column):

$$\begin{array}{lll}
\int \sin^2 x \cos^3 x dx & \int \sin^p x \cos^q x dx & (p \text{ or } q \text{ is odd}) \\
\int \sin^5 x dx & \int \sin^p x \cos^q x dx & (p \text{ is odd}, q = 0) \\
\int \sin^2 x \cos^2 x dx & \int \sin^p x \cos^q x dx & (p \text{ or } q \text{ is even}) \\
\int_0^{2\pi} \sin 8x \cos 6x dx & \int_0^{2\pi} \sin px \cos qx dx & (p, q = 0, 1, 2, \dots) \\
\int_0^{2\pi} \sin 8x \sin 7x dx & \int_0^{2\pi} \sin px \sin qx dx & (p \neq q) \\
\int_0^{2\pi} \sin^2 8x dx & \int_0^{2\pi} \sin px \sin qx dx & (p = q)
\end{array}$$

We restrict the discussion in this section to non-negative p and q. The next section is devoted to negative exponents, and you can see it is about the integration of tangents and secants. The integrals in the last three rows are very important; they aren't exercises on integrals. They are the basics of the Fourier series (Sect. 4.19).

Before computing these integrals, we would like to calculate the last one without actually calculating it. We know immediately that $\int_0^{2\pi} \sin^2 8x dx = \pi$. Why? This is because

$$\int_0^{2\pi} \sin^2 8x dx + \int_0^{2\pi} \cos^2 8x dx = \int_0^{2\pi} dx = 2\pi \tag{4.7.8}$$

And $\int_0^{2\pi} \sin^2 8x dx = \int_0^{2\pi} \cos^2 8x dx$ because of symmetry.

Example 4.6 Let's compute $\int \sin^2 x \cos^3 x dx$. As $\sin^2 x + \cos^2 x = 1$, we can always replace an even power of cosine ($\cos^2 x$) in terms of $\sin^2 x$. We are left with $\cos x dx$ which is fortunately $d(\sin x)$. So,

$$\begin{aligned}
\sin^2 x \cos^3 x dx &= \sin^2 x \cos^2 x d(\sin x) \\
&= \sin^2 x (1 - \sin^2 x) d(\sin x) = (\sin^2 x - \sin^4 x) d(\sin x)
\end{aligned} \tag{4.7.9}$$

Therefore (the following is actually a substitution with $u = \sin x$)

$$\int \sin^2 x \cos^3 x dx = \int (\sin^2 x - \sin^4 x) d(\sin x) = \frac{1}{3} \sin^3 x - \frac{1}{5} \sin^5 x + C \tag{4.7.10}$$

Example 4.7 How about $\int \sin^5 x dx$? The same idea: $\sin^5 x = \sin^4 x \sin x$, and $\sin x dx = -d(\cos x)$. Details are

$$\begin{aligned}\sin^5 x dx &= -\sin^4 x d(\cos x)\\ &= -(1-\cos^2 x)^2 d(\cos x) = (-1+2\cos^2 x - \cos^4 x) d(\cos x)\end{aligned}$$

Then, substitution now with $u = \cos x$ gives

$$\int \sin^5 x dx = \int (-1+2\cos^2 x - \cos^4 x) d(\cos x) = -\cos x + \frac{2}{3}\cos^3 x - \frac{1}{5}\cos^5 x + C$$

These two examples cover the integral $\int \sin^p x \cos^q x dx$ where $p, q \geq 0$ and either p or q is odd. Next example is a case where the exponent is even.

Example 4.8 Consider this integral $\int \cos^4 x dx$. We can do an integration by parts or use trigonometric identities to lower the exponent. Here is the second approach:

$$\cos^4 x = \left(\frac{1+\cos 2x}{2}\right)^2 = \frac{1+2\cos 2x + \cos^2 2x}{4} = \frac{1}{4} + \frac{\cos 2x}{2} + \frac{1+\cos 4x}{8}$$

Thus, the integral is given by

$$\int \cos^4 x dx = \int \left(\frac{1}{4} + \frac{\cos 2x}{2} + \frac{1+\cos 4x}{8}\right) dx = \frac{3x}{8} + \frac{\sin 2x}{4} + \frac{\sin 4x}{32} + C$$

Example 4.9 Let's compute the integral $\int \sin^2 x \cos^2 x dx$. Again, we use trigonometric identities to lower the powers, this time for both sine and cosine:

$$\begin{aligned}\sin^2 x \cos^2 x &= \frac{1-\cos 2x}{2}\frac{1+\cos 2x}{2} = \frac{1-\cos^2(2x)}{4}\\ &= \frac{1}{8} - \frac{\cos(4x)}{8} \Longrightarrow \int \sin^2 x \cos^2 x dx = \frac{x}{8} - \frac{\sin(4x)}{32} + C\end{aligned} \tag{4.7.11}$$

Example 4.10 Consider the integral $\int_0^{2\pi} \sin 8x \cos 6x dx$. The best way is to use the product-to-sum identity, see Eq. (3.8.6), to replace a product of sines with a sum of two sines:

$$\begin{aligned}\sin 8x \cos 6x &= \frac{1}{2}\Big(\sin 14x + \sin 2x\Big)\\ \Longrightarrow \int_0^{2\pi} \sin 8x \cos 6x dx &= \frac{1}{2}\int_0^{2\pi} (\sin 14x + \sin 2x) dx = 0\end{aligned}$$

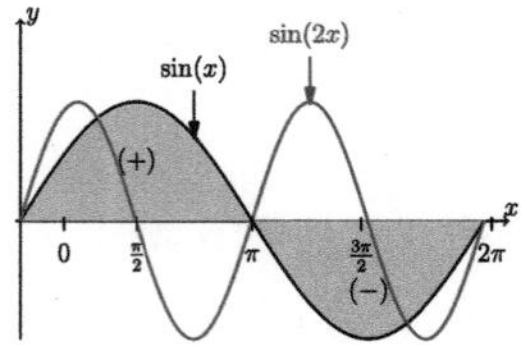

The result is zero because of the nature of the sine function, see the figure. We always have $\int_0^{2\pi} \sin nx dx = 0$ for any positive integer n. This is because the plus area is equal to the negative area. The last integral we consider is $\int \sin 8x \sin 6x dx$. At this moment, this integral is a piece of cake: since $\sin 8x \sin 6x = 1/2(\cos 2x - \cos 14x)$, so we have

$$\int_0^{2\pi} \sin 8x \sin 6x dx = \frac{1}{2}\int_0^{2\pi} (\cos 2x - \cos 14x) dx = 0$$

4.7.4 Repeated Integration by Parts

To illustrate the technique, let's consider $\int \sin^n x dx$, $n \in \mathbb{N}$. By using integration by parts, we obtain a recursive formula for this integral

$$\begin{aligned}
&(\sin^{n-1} x \cos x)' = (n-1)\sin^{n-2} x \cos^2 x - \sin^n x \\
&\Longrightarrow \int \sin^n x dx = (n-1)\int \sin^{n-2} x \cos^2 x dx - \sin^{n-1} x \cos x \\
&\Longrightarrow \int \sin^n x dx = (n-1)\int \sin^{n-2} x(1-\sin^2 x) dx - \sin^{n-1} x \cos x \\
&\Longrightarrow \int \sin^n x dx = \frac{n-1}{n}\int \sin^{n-2} x dx - \frac{1}{n}\sin^{n-1} x \cos x
\end{aligned} \tag{4.7.12}$$

Thus, each integration by part lowers the power of $\sin x$ from n to $n-2$, another integration by parts gets to $\int \sin^{n-4} dx$. We proceed until we get either $\int \sin x dx$ if n is odd or $\int dx$ if n is even.

Now, we show that Eq. (4.7.12) can lead to an infinite product for π. Using the above but with integrations limits 0 and $\pi/2$, we have (the term $[\frac{1}{n}\sin^{n-1} x \cos x]_0^{\pi/2} = 0$)

$$\int_0^{\pi/2} \sin^n x dx = \frac{n-1}{n}\int_0^{\pi/2} \sin^{n-2} x dx \tag{4.7.13}$$

Now, consider two cases: n is even and n is odd. For the former case ($n = 2m$), repeated application of Eq. (4.7.13) gives us[32]

[32] To find out the numbers 3/4 and 1/2 in the last equality, just use $m = 3$. The number $\pi/2$ is nothing but $\int_0^{\pi/2} dx$ when m has been reduced to 1.

$$
\begin{aligned}
\int_0^{\pi/2} \sin^{2m} x dx &= \frac{2m-1}{2m} \int_0^{\pi/2} \sin^{2m-2} x dx \\
&= \left(\frac{2m-1}{2m}\right)\left(\frac{2m-3}{2m-2}\right) \int_0^{\pi/2} \sin^{2m-4} x dx \\
&= \left(\frac{2m-1}{2m}\right)\left(\frac{2m-3}{2m-2}\right)\left(\frac{2m-5}{2m-4}\right) \cdots \left(\frac{3}{4}\right)\left(\frac{1}{2}\right)\frac{\pi}{2}
\end{aligned}
\tag{4.7.14}
$$

And for odd powers $n = 2m + 1$, we have

$$
\int_0^{\pi/2} \sin^{2m+1} x dx = \left(\frac{2m}{2m+1}\right)\left(\frac{2m-2}{2m-1}\right) \cdots \left(\frac{4}{5}\right)\left(\frac{2}{3}\right) \tag{4.7.15}
$$

From Eqs. (4.7.14) and (4.7.15), we obtain by dividing the former equation by the latter equation

$$
\frac{\pi}{2} = \frac{2 \times 2}{1 \times 3} \frac{4 \times 4}{3 \times 5} \cdots \frac{2m \times 2m}{(2m-1) \times (2m+1)} \tag{4.7.16}
$$

where we used the fact that $\int_0^{\pi/2} \sin^{2m} x dx / \int_0^{\pi/2} \sin^{2m+1} x dx = 1$ when m approaches infinity (a proof is due in what follows). And this infinite product for $\pi/2$ is known as Wallis' infinite product, as John Wallis was the first who discovered it (and what I am presenting is basically his derivation). In Sect. 4.15.7, I will present Euler's derivation of this infinite product.

Proof We're proving the following result

$$
\lim_{m \to \infty} \frac{\int_0^{\pi/2} \sin^{2m} x dx}{\int_0^{\pi/2} \sin^{2m+1} x dx} = 1
$$

As $0 \le x \le \pi/2$, we have

$$
0 \le \sin x \le 1 \Longrightarrow \sin^{2m+1} x \le \sin^{2m} x \le \sin^{2m-1} x
$$

Thus, integrating these functions from 0 to $\pi/2$ we get (see Eq. (4.3.8) if not clear)

$$
\int_0^{\pi/2} \sin^{2m+1} x dx \le \int_0^{\pi/2} \sin^{2m} x dx \le \int_0^{\pi/2} \sin^{2m-1} x dx
$$

A bit of arrangement leads to

$$
1 \le \frac{\int_0^{\pi/2} \sin^{2m} x dx}{\int_0^{\pi/2} \sin^{2m+1} x dx} \le \frac{\int_0^{\pi/2} \sin^{2m-1} x dx}{\int_0^{\pi/2} \sin^{2m+1} x dx}
$$

Now, let's denote the ratio on the RHS of the above equation by A and we want to compute it. First, Eq. (4.7.12) is used to get

$$\int_0^{\pi/2} \sin^{2m+1} x dx = \frac{2m}{2m+1} \int_0^{\pi/2} \sin^{2m-1} x dx$$

Thus, A is given by

$$\frac{\int_0^{\pi/2} \sin^{2m-1} x dx}{\int_0^{\pi/2} \sin^{2m+1} x dx} = \frac{2m+1}{2m} \frac{\int_0^{\pi/2} \sin^{2m-1} x dx}{\int_0^{\pi/2} \sin^{2m-1} x dx} = 1 + \frac{1}{2m}$$

From that A approaches 1 when m approaches infinity.

4.7.5 What is $\int_0^\infty x^4 e^{-x} dx$?

In Eq. (4.7.7), using integration by parts twice, we got the following result

$$\int x^2 e^x dx = +x^2 e^x - 2x e^x + 2e^x \tag{4.7.17}$$

We can see the structure in the RHS: $x^2 \to 2x \to 2$; that is the result of the repeated differentiation of x^2. The alternating signs $+/-/+$ are due to the minus sign appearing in each integration by parts.

With this understanding, without actually doing the integration, we know that

$$\int x^4 e^x dx = x^4 e^x - 4x^3 e^x + 12x^2 e^x - 24x e^x + 24e^x$$

We can check this using SymPy.

Now we move to the integral $\int_0^\infty x^4 e^{-x} dx$. First, replacing e^x by e^{-x} we have the following results:

$$\int x^2 e^{-x} dx = -x^2 e^{-x} - 2x e^{-x} - 2e^{-x}$$
$$\int x^4 e^{-x} dx = -x^4 e^{-x} - 4x^3 e^{-x} - 12x^2 e^{-x} - 24x e^{-x} - 24e^{-x}$$

Focus now on the second integral, but now with special integration limits of 0 and infinity, we have:

$$\int_0^\infty x^4 e^{-x} dx = \left[-x^4 e^{-x} - 4x^3 e^{-x} - 12x^2 e^{-x} - 24x e^{-x}\right]_0^\infty - 4! e^{-x}|_0^\infty \tag{4.7.18}$$

All the terms in the brackets are zeroes and $e^{-x}|_0^\infty = -1$, thus we obtain a very interesting result:

$$\boxed{\int_0^\infty x^4 e^{-x} dx = 4!} \tag{4.7.19}$$

This is a stunning result. Can you see why? We will come back to it later in Sect. 4.20.2.

4.7.6 Trigonometric Integrals: Tangents and Secants

This section discusses integrals of $\int \tan^m x dx$ and $\int \sec x dx$. Why do these two functions together? Because they are related via $\sec^2 x - \tan^2 x = 1$.

The plan is to start simple with $m = 1$, $m = 2$ then $m = 3$ and use the pattern observed in these results, hopefully to get the integral for any $m \geq 3$. For $m = 1$, that is $\int \sin x/\cos x dx$, which is of the form $\int \sin^p \cos^q dx$ with $p = 1$, and $q = -1$ (both are odd). So, in the same manner of what discussed in Sect. 4.7.3 we proceed as follows

$$\int \tan x dx = \int \frac{\sin x}{\cos x} dx = -\int \frac{d(\cos x)}{\cos x} = -\ln|\cos x| + C \tag{4.7.20}$$

Now, we move to square of the tangent, and we relate it to the secant (using $1 + \tan^2 x = \sec^2 x$ and $(\tan x)' = \sec^2 x$):

$$\int \tan^2 x dx = \int (\sec^2 x - 1) dx = \tan x - x + C \tag{4.7.21}$$

How about $\int \tan^3 x dx$? We write $\tan^3 x = \tan^2 x \tan x = (\sec^2 x - 1)\tan x$, thus we can relate this integral to $\int \tan x dx$, which we know, and something that is easy:

$$\begin{aligned}
\int \tan^3 x dx &= \int (\sec^2 x - 1)\tan x dx = \int \sec^2 x \tan x dx - \int \tan x dx \\
&= \int \tan x d(\tan x) - \ln|\cos x| \\
&= \frac{\tan^2 x}{2} - \ln|\cos x| \quad (\text{substitution} u = \tan x)
\end{aligned} \tag{4.7.22}$$

Now, we see the way and can do the general integral $\int \tan^m x dx$:

$$\begin{aligned}
\int \tan^m x dx &= \int \tan^2 x \tan^{m-2} x dx = \int (\sec^2 x - 1)\tan^{m-2} x dx \\
&= \int \sec^2 x \tan^{m-2} x dx - \int \tan^{m-2} dx = \frac{\tan^{m-1} x}{m-1} dx - \int \tan^{m-2} dx
\end{aligned} \tag{4.7.23}$$

That is, we have a formula for $\int \tan^m x dx$ that requires $\int \tan^{m-2} x dx$, which in turn involves $\int \tan^{m-4} x dx$ and so on. Depending on m being odd or even, this leads us to either $\int \tan x dx$ or $\int \tan^2 x dx$, which we know how to integrate.

Ok. Let's move to the secant function. How we're going to compute the following integral $\int \sec x dx$? Replacing $\sec x = 1/\cos x$ would not help. Think of its friend $\tan x$, we do this:

$$\int \sec x dx = \int \frac{\sec x}{1} dx = \int \frac{\sec x}{\sec^2 x - \tan^2 x} dx$$

We succeeded in bringing in the two friends. Now the next is just algebra:

$$\int \sec x dx = \int \frac{\sec x}{(\sec x - \tan x)(\sec x + \tan x)} dx = \frac{1}{2} \int \left[\frac{1}{\sec x - \tan x} + \frac{1}{\sec x + \tan x} \right] dx$$

Now, we switch to $\sin x$ and $\cos x$, as we see something familiar when doing so:

$$\begin{aligned}
\int \sec x dx &= \frac{1}{2} \int \left[\frac{\cos x}{1 - \sin x} + \frac{\cos x}{1 + \sin x} \right] dx \\
&= \frac{1}{2} \int \left[-\frac{d(1 - \sin x)}{1 - \sin x} + \frac{d(1 + \sin x)}{1 + \sin x} \right] \\
&= \frac{1}{2} (\ln(1 + \sin x) - \ln(1 - \sin x)) = \frac{1}{2} \ln \frac{1 + \sin x}{1 - \sin x}
\end{aligned}$$

We can stop here. However, we can further simplify the result, noting that

$$\begin{aligned}
\frac{1 + \sin x}{1 - \sin x} &= \frac{\sin^2 x/2 + \cos^2 x/2 + 2 \sin x/2 \cos x/2}{\sin^2 x/2 + \cos^2 x/2 - 2 \sin x/2 \cos x/2} \\
&= \left(\frac{\sin x/2 + \cos x/2}{\sin x/2 - \cos x/2} \right)^2 = \left(\frac{1 + \sin x}{\cos x} \right)^2 = (\sec x + \tan x)^2
\end{aligned}$$

And finally, the integral of $\sec x$ is

$$\int \sec x dx = \frac{1}{2} \ln \frac{1 + \sin x}{1 - \sin x} = \frac{1}{2} \ln (\sec x + \tan x)^2 = \ln |\sec x + \tan x| + C$$

This one was hard, but $\int \sec^2 x dx$ is easy. It is $\int (1 + \tan^2 x) dx$. How about $\int \sec^3 x dx$? We do the same thing we did for $\int \cos^3 x dx$:

$$\begin{aligned}
\int \sec^3 x dx &= \int \sec^2 x \sec x dx \\
&= \int (1 + \tan^2 x) \sec x dx = \int \sec x dx + \int \tan^2 x \sec x dx
\end{aligned}$$

For the integral $\int \tan^2 x \sec x dx$, we use integration by parts with $u = \sec x$ and $v = \tan x$. Finally,

$$\int \sec^3 x dx = 0.5(\sec x \tan x + \ln|\sec x + \tan x|) + C \tag{4.7.24}$$

Why bother with this integral? But this integral is the answer to the problem of calculating the length of a segment of a parabola (Sect. 4.9.1).

4.7.7 Integration by Trigonometric Substitution

Trigonometric substitutions are useful to deal with integrals involving terms such as $\sqrt{a^2 - x^2}$, $\sqrt{x^2 - a^2}$, or $1/x^2+a^2$ or $\sqrt{x^2 + a^2}$. These substitutions remove the square roots and usually lead to simpler integrals.

Example 4.11 For example, consider the following definite integral

$$\int_0^4 \frac{dx}{\sqrt{16 - x^2}}$$

With this substitution $x = 4\sin\theta$, we have

$$x = 4\sin\theta \Longrightarrow \begin{cases} dx & = 4\cos\theta d\theta \\ \sqrt{16 - x^2} & = \sqrt{16(1 - \sin^2\theta)} = 4\cos\theta \\ 0 \le \theta \le \frac{\pi}{2} \end{cases}$$

And thus the integral becomes

$$\int_0^4 \frac{dx}{\sqrt{16 - x^2}} = \int_0^{\pi/2} \frac{4\cos\theta}{4\cos\theta} d\theta = \frac{\pi}{2}$$

Example 4.12 Simple. But, how about the following?

$$\int_4^8 \frac{dx}{\sqrt{x^2 - 16}}$$

The substitution $x = 4\sin\theta$ would not work: $\sqrt{x^2 - 16} = \sqrt{16(\sin^2\theta - 1)}$ which is meaningless, as the radical is negative. So, we use another trigonometric function: the secant function. The details are

$$x = 4\sec\theta \Longrightarrow \begin{cases} dx & = 4\tan\theta\sec\theta d\theta \\ \sqrt{x^2 - 16} & = \sqrt{16(\sec^2\theta - 1)} = 4\tan\theta \\ 0 \le \theta \le \frac{\pi}{3} \end{cases} \tag{4.7.25}$$

And the original integral is simplified to

$$\int_4^8 \frac{dx}{\sqrt{x^2-16}} = \int_9^{\pi/3} \frac{4\tan\theta\sec\theta}{4\tan\theta} d\theta = \int \sec\theta d\theta = \ln(\sec\theta + \tan\theta)]_0^{\pi/3}$$

Example 4.13 Now comes another trigonometric substitution using the tangent function. The following integral

$$\int_0^\infty \frac{dx}{16+x^2} \tag{4.7.26}$$

with

$$x = 4\tan\theta \Longrightarrow \begin{cases} dx & = 4\sec^2\theta d\theta \\ 16+x^2 & = 16(1+\tan^2\theta) = 16\sec^2\theta \\ 0 \le \theta \le \frac{\pi}{2} \end{cases} \tag{4.7.27}$$

is simplified to

$$\int_0^\infty \frac{dx}{16+x^2} = \int_0^{\pi/2} \frac{4\sec^2\theta}{16\sec^2\theta} d\theta = \frac{1}{4}\theta\bigg]_0^{\pi/2} = \frac{\pi}{8}$$

Sometimes we see an integral which is a disguised form of $\int_0^\infty \frac{dx}{16+x^2}$, for example:

$$\int \frac{dx}{5x^2-10x+25}$$

In this case, we just need to complete the square i.e., $5x^2 - 10x + 25 = \Box^2 + c$, c is a constant. Then, the substitution of $x = c\tan\theta$ is used. So, the steps are

$$\begin{aligned} \int \frac{dx}{5x^2-10x+25} &= \frac{1}{5}\int \frac{dx}{x^2-2x+5} = \frac{1}{5}\int \frac{d(x-1)}{(x-1)^2+4} \\ &= \frac{1}{5}\int \frac{du}{u^2+4} = \frac{1}{10}\tan^{-1}\left(\frac{x-1}{2}\right) + C \end{aligned}$$

The second step is completing the square, the third step is to rewrite it in the familiar form of Eq. (4.7.26).

Example 4.14 Now we consider the integrand of the form $\sqrt{x^2+a^2}$. The following integral

$$I = \int \sqrt{x^2+a^2} dx, \quad a > 0 \tag{4.7.28}$$

with

$$x = a\tan\theta \Longrightarrow \begin{cases} dx & = a\sec^2\theta d\theta \\ \sqrt{x^2+a^2} & = a\sec\theta \end{cases}$$

So, I becomes $a^2 \int \sec^3\theta d\theta$, using the result in Eq. (4.7.24), we then have

$$\int \sqrt{x^2+a^2}dx = a^2 \int \sec^3\theta d\theta = 0.5a^2(\sec\theta\tan\theta + \ln|\sec\theta + \tan\theta|)$$

Now, we need to call x back noting that $\tan\theta = {}^x/_a$ and $\sec\theta = {}^{\sqrt{x^2+a^2}}/_a$:

$$\int \sqrt{x^2+a^2}dx = \frac{1}{2}x\sqrt{x^2+a^2} + \frac{a^2}{2}\ln|x+\sqrt{x^2+a^2}| + C \tag{4.7.29}$$

Is $x = a\tan\theta$ the only substitution for this problem? Euler did not think so. This is what he used:

$$\sqrt{x^2+a^2} = x + t \Longrightarrow x = \frac{a^2-t^2}{2t} \Longrightarrow dx = \Box dt \Longrightarrow I = \int\left(\frac{a^2-t^2}{2t}+t\right)\Box dt$$

Now, the original integral in terms of x turns into an integral in terms of t. And this integral can be computed with ease; the result is some $g(t)$. What is interesting is to prove that $g(t)$ is identical to Eq. (4.7.29). That's a nice exercise on algebra. And it also demonstrates the beauty of mathematics: something as messy as $g(t(x))$ is nothing but a nice function in Eq. (4.7.29).

Example 4.15 I present the final trigonometric substitution so that we can evaluate integrals of any rational function of $\sin x$ and $\cos x$. For example,

$$\int \frac{dx}{3-5\sin x}, \quad \int \frac{dx}{1+\sin x - \cos x}$$

The substitution is (discovered by the Germain mathematician Karl Weierstrass (1815–1897))

$$u = \tan\frac{x}{2}, \quad dx = \frac{2du}{1+u^2}$$

This is because, as given in Eq. (3.8.8), we can express $\sin x$ and $\cos x$ in terms of u:

$$\sin x = \frac{2u}{1+u^2}, \quad \cos x = \frac{1-u^2}{1+u^2}$$

Then, $\int \frac{dx}{3-5\sin x}$ becomes:

$$\int \frac{dx}{3-5\sin x} = 2\int \frac{du}{3u^2-10u+3} \tag{4.7.30}$$

This integral is of the form $P(u)/Q(u)$ and we discuss how to integrate it in the next section.

Table 4.14 Summary of trigonometric substitutions

Form	Substitution	dx	New form
$\sqrt{a^2-x^2}$	$x=a\sin\theta$	$a\cos\theta d\theta$	$a\cos\theta$
$\sqrt{x^2-a^2}$	$x=a\sec\theta$	$a\tan\theta\sec\theta d\theta$	$a\tan\theta$
a^2+x^2	$x=a\tan\theta$	$a\sec^2\theta d\theta$	$a^2\sec^2\theta$
$\frac{1+\sin x}{1+\cos x}$	$u=\tan x/2$	$2du/(1+u^2)$	$\frac{P(u)}{Q(u)}$

It is always a good idea to stop doing what we're doing and summarize the achievement. I provide such a summary in Table 4.14.

4.7.8 *Integration of $P(x)/Q(x)$ Using Partial Fractions*

This section is about the integration of rational functions, those of the form $P(x)/Q(x)$ where $P(x)$ and $Q(x)$ are polynomials (Sect. 2.29). The most important thing is that we can always integrate these rationals using elementary functions that we know.

We start off with this observation: it is not hard to evaluate the following indefinite integral

$$\int\left[\frac{1}{x-2}+\frac{3}{x+2}-\frac{4}{x}\right]dx=\ln|x-2|+3\ln|x+2|-4\ln|x|+C$$

However, it is not obvious how to do the following integral $\int\frac{-4x+16}{x^3-4x}dx$. The basic idea is that we can always transform ${-4x+16}/{x^3-4x}$ into a sum of simpler fractions (called partial fractions):

$$\frac{-4x+16}{x^3-4x}=\frac{-4x+16}{x(x-2)(x+2)}=\frac{A}{x}+\frac{B}{x-2}+\frac{C}{x+2}$$

where each partial fraction is of the form $p(x)/q(x)$ where *the degree of the nominator is one less than that of the denominator*. This is called the method of Partial Fraction Decomposition.[33] To find the constants A, B, C, we just convert the RHS into the form of the LHS:

$$\frac{A}{x}+\frac{B}{x-2}+\frac{C}{x+2}=\frac{(A+B+C)x^2+2(B-C)x-4A}{x^3-4x}$$

[33] The concept was discovered independently in 1702 by both Johann Bernoulli and Gottfried Leibniz.

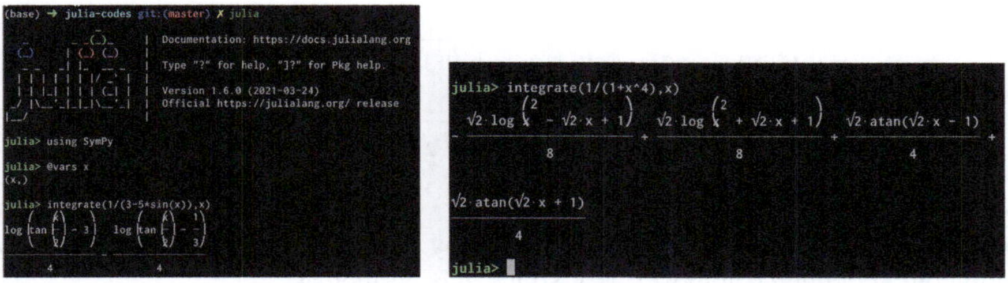

Fig. 4.46 Symbolic evaluation of integrals using the library `SymPy` in `Julia`. `SymPy` is actually a Python library, so we can use it directly not necessarily via `Julia`

As this fraction is equal to ${-4x+16}/{x^3-4x}$, the two nominators must be the same, thus we have $(A+B+C)x^2+2(B-C)x-4A \equiv -4x+16$, which leads to

$$A+B+C=0,\quad 2(B-C)=-4,\quad -4A=16 \Longrightarrow A=-4,\ B=1,\ C=3$$

Now $\int \frac{-4x+16}{x^3-4x}dx$ can be computed with ease:

$$\int \frac{-4x+16}{x^3-4x}dx = \int \left[\frac{1}{x-2}+\frac{3}{x+2}-\frac{4}{x}\right]dx \tag{4.7.31}$$

With this new tool we can finish the integral $\int \frac{dx}{3-5\sin x}$, see Eq. (4.7.30):

$$\begin{aligned}\int \frac{dx}{3-5\sin x} &= 2\int \frac{du}{3u^2-10u+3} = \frac{1}{4}\left[\int \frac{du}{u-3} - \int \frac{du}{u-1/3}\right]\\ &= \frac{1}{4}\left(\ln|u-3| - \ln|u-1/3|\right) = \frac{1}{4}\left(\ln|\tan x/2-3| - \ln|\tan x/2-1/3|\right)\end{aligned}$$

And we can check our result using a CAS (Fig. 4.46).

If you were attentive you would observe that the two integrals that we have just considered are of the form $P(x)/Q(x)$ where the degree of the denominator is larger than that of the nominator. These particular rationals are called *proper rationals*. And we just need to pay attention to them only, as the other case can always be rewritten in this form, for example[34]:

$$\frac{2x^2-5x-1}{x-3} = 2x+1+\frac{2}{x-3}$$

You should have also noticed that in the considered rationals, $Q(x)$ has distinct roots i.e., it can be factored as $Q(x)=(a_1x+b_1)(a_2x+b_2)\cdots(a_nx+b_n)$ where n is the degree of $Q(x)$. In this case, the partial fraction decomposition is

[34] If you do like polynomials, think of $5/3=(3+2)/3=1+2/3$.

$$\begin{aligned}\frac{P(x)}{Q(x)} &= \frac{P(x)}{(a_1x+b_1)(a_2x+b_2)\cdots(a_nx+b_n)} \\ &= \frac{A_1}{a_1x+b_1} + \frac{A_2}{a_2x+b_2} + \cdots + \frac{A_n}{a_nx+b_n}\end{aligned} \tag{4.7.32}$$

And it's always possible to find A_i when $P(x)$ is a polynomial of degree less than n, which is the case for proper rationals. Why is it possible? Note that $P(x)$ is a polynomial of degree at most $n-1$, so it can be written as

$$P(x) = a_{n-1}x^{n-1} + \cdots + a_2x^2 + a_1x + a_0$$

Therefore, there are n unknowns A_i and n known coefficients a_i. So, it is possible to find $A_1, A_2, \ldots$

The case where $Q(x)$ has repeated roots. Now we consider the case where $Q(x)$ has repeated roots, for example the following integral

$$\int \frac{x^2+15}{(x+3)^2(x^2+3)}dx$$

A polynomial is an irreducible polynomial if it cannot be factored. As the blue term x^2+3 does not have real roots, we cannot factor it. It is of an irreducible form. So, we leave it as it is and focus on the red term, which is obviously $(x+3)(x+3)$, and is a reducible form. Note that $Q(x)=0$ has a repeated root of -3.

The decomposition in this case is a little bit special:

$$\frac{x^2+15}{(x+3)^2(x^2+3)} = \frac{Ax+B}{x^2+3} + \frac{C}{(x+3)^1} + \frac{D}{(x+3)^2} \tag{4.7.33}$$

where the red terms follow this rule: *for the term $(ax+b)^k$ (in the denominator) we need a partial fraction for each exponent from 1 up to k*. As I am lazy, I have used `SymPy` to do this decomposition for me:

$$\frac{x^2+15}{(x+3)^2(x^2+3)} = \frac{1-x}{2(x^2+3)} + \frac{1}{2(x+3)} + \frac{2}{(x+3)^2} \tag{4.7.34}$$

Now, it is possible to compute the integral:

$$\begin{aligned}\int \frac{x^2+15}{(x+3)^2(x^2+3)}dx &= \int\left(\frac{1-x}{2(x^2+3)} + \frac{1}{2(x+3)} + \frac{2}{(x+3)^2}\right)dx \\ &= \frac{1}{2\sqrt{3}}\arctan\frac{x}{\sqrt{3}} - \frac{1}{4}\ln(x^2+3) + \frac{1}{2}\ln(x+3) - \frac{2}{x+3} + E\end{aligned}$$

Explanation of the rule for $(ax+b)^k$.

To understand the decomposition when $Q(x)$ has repeated roots, consider the rational $1/(x+3)^2$. Our first attempt to decompose it is

$$\frac{1}{(x+3)^2} = \frac{A}{x+3} + \frac{Bx+C}{(x+3)^2}$$

Obviously, this would not work as we have three unknowns A, B, C but there are only two equations! Surprisingly, if we use a new variable $u = x+3$, it works:

$$\frac{1}{u^2} = \frac{A}{u} + \frac{Bu+C}{u^2} = \left(\frac{A}{u} + \frac{B}{u}\right) + \frac{C}{u^2} = \frac{A_1}{u} + \frac{A_2}{u^2}$$

Now, there are only two unknowns A_1, A_2.

Example 4.16 To wrap up this section, let's compute the following integral

$$I = \int \frac{dx}{1+x^4}$$

We need first to factor $1+x^4$ as $1+x^4+2x^2-2x^2 = (1+x^2)^2 - (\sqrt{2}x)^2$, then we have

$$\frac{1}{1+x^4} = \frac{1}{(1+x^2)^2 - (\sqrt{2}x)^2} = \frac{1}{(x^2+\sqrt{2}x+1)(x^2-\sqrt{2}x+1)}$$

The next step is to do a partial fraction decomposition for this, and we're done. See Fig. 4.46 for the result, done by a CAS.

4.7.9 Tricks

This section presents a few tricks to compute some interesting integrals. If you're fascinated by difficult integrals, you can consult YouTube channels by searching for "MIT integration bee" and the likes.[35] Or you can read the book *Inside Interesting Integrals* of Paul Nahin [49].

The first example is the following integral

[35] One example from the MIT integration bee: $\int \sqrt{x\sqrt{x\sqrt{x\ldots}}}dx$.

$$\int_{-1}^{1} \frac{\cos x}{1+e^{1/x}} dx$$

You should ask why the integration limits are -1 and 1, not -1 and 2? Note that $\int_{-a}^{a} f(x)dx = 0$ if $f(x)$ is an odd function. So, we decompose the integrand function into an even and an odd part (see Eq. (4.2.1) if not clear):

$$\begin{aligned}\frac{\cos x}{1+e^{1/x}} &= \frac{1}{2}\left(\frac{\cos x}{1+e^{1/x}} + \frac{\cos x}{1+e^{-1/x}}\right) + \frac{1}{2}\left(\frac{\cos x}{1+e^{1/x}} - \frac{\cos x}{1+e^{-1/x}}\right) \\ &= \frac{1}{2}\cos x + \frac{1}{2}\left(\frac{\cos x}{1+e^{1/x}} - \frac{\cos x}{1+e^{-1/x}}\right)\end{aligned}$$

And we do not care about the odd part, because its integral is zero, anyway. So,

$$\int_{-1}^{1} \frac{\cos x}{1+e^{1/x}} dx = \int_{0}^{1} \cos x dx = \sin(1)$$

Feymann's trick. This trick is based on the Leibniz rule that basically says:

$$I(t) = \int_a^b f(x,t)dx \Longrightarrow \frac{dI(t)}{dt} = \int_a^b \frac{\partial f(x,t)}{\partial t} dx \tag{4.7.35}$$

We refer to Sect. 6.8.7 for a discussion leading to this rule. The symbol $\frac{\partial f(x,t)}{\partial t}$ is a partial derivative of $f(x,t)$ with respect to t while holding x constant.

As the first application of this rule, we can generate new integrals from old ones. For example, we know the following integral (integrals with one limit that goes to infinity are called improper integrals and they are discussed in Sect. 4.8)

$$I = \int_0^\infty \frac{dx}{x^2+a^2} = \left[\frac{1}{a}\tan^{-1}\left(\frac{x}{a}\right)\right]_0^{\pi/2} = \frac{\pi}{2a} \tag{4.7.36}$$

And by considering a as a variable playing the role of t in Eq. (4.7.35), we can write:

$$I(a) = \int_0^\infty \frac{dx}{x^2+a^2} \Longrightarrow \frac{dI}{da} = \int_0^\infty \frac{-2a}{(x^2+a^2)^2} dx \tag{4.7.37}$$

And from Eq. (4.7.36)—which says $I = \pi/2a$—we can easily get $dI/da = -\pi/2a^2$, and thus we get the following new integral:

$$\int_0^\infty \frac{-2a}{(x^2+a^2)^2} dx = \frac{\pi}{2a} \Longrightarrow \int_0^\infty \frac{dx}{(x^2+a^2)^2} = \frac{\pi}{4a^3}$$

Of course, we can go further by computing d^2I/da^2 and get new integrals. But we stop here to do something else.

Suppose we need to evaluate this integral (of which anti-derivative cannot be found in elementary functions)

$$\int_0^1 \frac{x^2-1}{\ln x}dx \tag{4.7.38}$$

So, we introduce a parameter b, to get

$$I(b) = \int_0^1 \frac{x^b-1}{\ln x}dx \Longrightarrow \frac{dI}{db} = \int_0^1 \frac{d}{db}\left(\frac{x^b-1}{\ln x}\right)dx = \int_0^1 x^b dx = \frac{1}{1+b} \tag{4.7.39}$$

So, we were able to compute dI/db as the integral became simpler! Another integration will give us $I(b)$:

$$\frac{dI}{db} = \frac{1}{1+b} \Longrightarrow I(b) = \ln|1+b| + C \tag{4.7.40}$$

To find C, we just look for a special value of b such that $I(b)$ can be easily evaluated. It can be seen that $I(0) = 0 = \ln 1 + C$, so $C = 0$. And now we come back to the original integral in Eq. (4.7.38)—which is nothing but $I(2)$, but $I(2) = \ln 3$. This trick is very cool. I did not know this in high school, and only became aware of it by reading Nahin's book [49].

Let's consider another integral: $\int_0^\infty e^{-x^2}\cos(5x)dx$. We consider the following integral and do the now familiar procedure

$$I(b) = \int_0^\infty e^{-x^2}\cos(bx)dx \Longrightarrow \begin{cases} \dfrac{dI}{db} = \displaystyle\int_0^\infty -xe^{-x^2}\sin(bx)dx \\ \quad\;\; = -\dfrac{b}{2}\displaystyle\int_0^\infty e^{-x^2}\cos(bx)dx \\ \quad\;\; = -\dfrac{b}{2}I(b) \end{cases} \tag{4.7.41}$$

in which we have used integration by parts to arrive at the final equality. Now, we get an equation to determine $I(b)$, this is in fact an ordinary differential equation

$$\frac{dI}{db} = -\frac{b}{2}I(b) \tag{4.7.42}$$

Following a variable separation (that is, isolate the two variables I and b on two sides of the equation) we can get $I(b)$ by integration:

$$\frac{dI}{I} = -\frac{b}{2}db \Longrightarrow \ln|I| = -\frac{b^2}{4} + D \Longrightarrow I = Ce^{-b^2/4} \quad (C = e^D) \tag{4.7.43}$$

Again, we need to find C and with $b = 0$, we have $I(0) = C = \int_0^\infty e^{-x^2} dx = \sqrt{\pi}/2$.[36] So, we get a nice result for our original integral and many more corresponding with different values of b:

$$\begin{aligned} I(5) &= \int_0^\infty e^{-x^2} \cos(5x) dx = \frac{\sqrt{\pi}}{2} e^{-25/4} \\ I(2) &= \int_0^\infty e^{-x^2} \cos(2x) dx = \frac{\sqrt{\pi}}{2e} \end{aligned} \tag{4.7.44}$$

Dirichlet integral. Another interesting integral is $\int_0^\infty \sin x/x dx$. Let us introduce the parameter b in such a way that differentiating the integrand will give us a simpler integral:

$$I(b) = \int_0^\infty \frac{\sin bx}{x} dx \Longrightarrow \frac{dI}{db} = \int_0^\infty \cos(bx) dx = \left. \frac{\sin bx}{b} \right]_0^\infty$$

Unfortunately, we got an improper integral. So, we need to find another way. We need a function of which the derivative has x. That can be e^{bx}. But due to the limit of infinity, we have to use e^{-bx} with $b \geq 0$. Thus, we consider the following integral

$$I(b) = \int_0^\infty \frac{\sin x}{x} e^{-bx} dx$$

From which $\int_0^\infty \sin x/x dx = I(0)$. Let's differentiate this integral w.r.t b:

$$\frac{dI}{db} = -\int_0^\infty \sin x e^{-bx} dx = -A$$

We can evaluate this integral A using integration by parts:

$$\left.\begin{aligned} A = \int_0^\infty \sin x e^{-bx} dx = \frac{1}{b} \int_0^\infty \cos x e^{-bx} dx \\ \int_0^\infty \cos x e^{-bx} dx = \frac{1 - A}{b} \end{aligned}\right\} \Rightarrow A = \frac{1}{1 + b^2}$$

Now, we have $I'(b)$, and a simple integration gives us $I(b)$

$$I'(b) = -\frac{1}{1 + b^2} \Longrightarrow I(b) = -\int \frac{db}{1 + b^2} = -\tan^{-1} b + C$$

[36] How to compute the integral $\int_0^\infty e^{-x^2} dx$ is another story, see Sect. 5.11.3.

We have to find C: with $b = \infty$, we can compute $I(b)$, $\tan^{-1} b$ and thus we can get C:

$$I(\infty) = \int_0^\infty \frac{\sin x}{x} e^{-\infty x} dx = 0 = -\tan^{-1}\infty + C \Longrightarrow C = \frac{\pi}{2}$$

So,

$$\boxed{\int_0^\infty \frac{\sin x}{x} dx = \frac{\pi}{2}} \tag{4.7.45}$$

Another amazing result is about π. The graph of $\sin x/x$ is given in Fig. 4.47b. And the area of the shaded region is nothing else but exactly half of π.

Mathematicians defined the following function[37]:

$$Si(x) := \int_0^x \frac{\sin t}{t} dt \tag{4.7.46}$$

And our task is to derive an expression for $Si(x)$. We have just showed that we cannot compute the integral directly, the Feynman technique only works for definite integrals in which the limits are numbers not variables. But we have another way, from Newton: we can replace $\sin t$ by its Taylor series, then we can integrate $\sin t/t$ easily:

$$\sin(t) = t - \frac{1}{3!}t^3 + \frac{1}{5!}t^5 - \cdots \Longrightarrow \frac{\sin t}{t} = 1 - \frac{t^2}{6} + \frac{t^4}{5!} - \cdots$$

Thus, we can write

$$\begin{aligned}\int_0^x \frac{\sin t}{t} dt &= \int_0^x \left(1 - \frac{t^2}{3!} + \frac{t^4}{5!} - \cdots\right) dt = \left[t - \frac{t^3}{3 \times 3!} + \frac{t^5}{5 \times 5!} - \cdots\right]_0^x \\ &= \frac{x^1}{1 \times 1!} - \frac{x^3}{3 \times 3!} + \frac{x^5}{5 \times 5!} - \cdots\end{aligned}$$

And voilà, the $Si(x)$ function is written as

$$Si(x) := \int_0^x \frac{\sin t}{t} dt = \sum_{i=0}^{\infty} (-1)^i \frac{x^{2i+1}}{(2i+1)(2i+1)!} \tag{4.7.47}$$

With this result we can visualize the function, see Fig. 4.47 where the graph of $\sin x/x$ is also shown. Given the graph, we now can understand Eq. (4.7.45), when $x \to \infty$, the function approaches 1.57079, which is $\pi/2$.

---

[37] Why they did so? There is no elementary function whose derivative is $\sin x/x$. However, anti-derivatives of this function come up moderately frequently in applications, for example in signal processing. So it has been convenient to give one of its anti-derivatives, $\int_0^x \frac{\sin t}{t} dt$, a name.

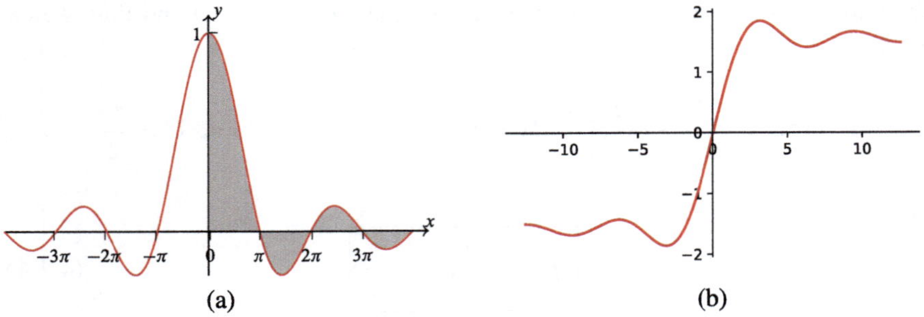

Fig. 4.47 Graph of $\sin x/x$ (**a**) and graph of $Si(x) = \int_0^x \frac{\sin t}{t} dt$ (**b**)

4.8 Improper Integrals

Improper integrals refer to those integrals where one or both of the integration limits are infinity e.g. $\int_1^\infty dx/x^2$ or the integrand goes to infinity $\int_0^1 dx/\sqrt{x}$.

To see how we compute improper integrals, let's consider one simple integral:

$$I = \int_1^\infty \frac{dx}{x^2}$$

We do not know how to evaluate this integral, but we know how to compute $I(b) = \int_1^b dx/x^2$. It is $I(b) = 1 - 1/b$. And by considering different values for b (larger than 1 of course), we have a sequence of integrals, see Fig. 4.48. Let's denote this by $(I_1, I_2, \ldots, I_n)$. It's obvious that this sequence converges to 1 when n approaches infinity. In other words, the area under the curve $y = 1/x^2$ from 1 to infinity is one. Therefore, we define

$$I = \int_1^\infty \frac{dx}{x^2} := \lim_{b\to\infty} \int_1^b \frac{dx}{x^2} = 1$$

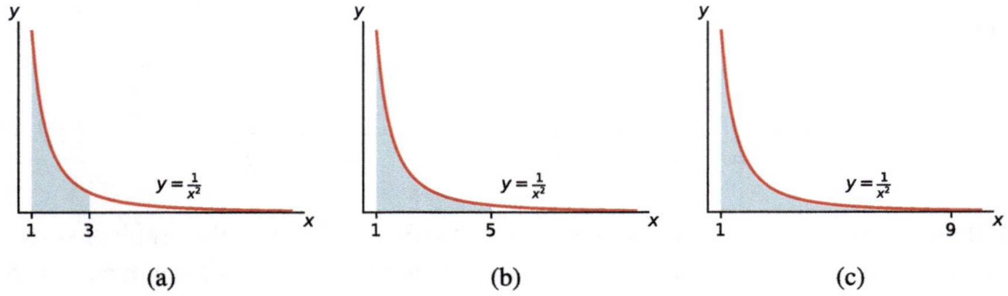

Fig. 4.48 Integral $\int_1^b dx/x^2$ for different values of b. Source code for the figures: `plot-calculus-area.jl`

In the same manner, if the lower integration limit is minus infinity, we have this definition:

$$I = \int_{-\infty}^{b} f(x)dx := \lim_{a\to-\infty} \int_{a}^{b} f(x)dx$$

The next improper integral to be discussed is certainly the one with both integration limits being infinite, like the following

$$I = \int_{-\infty}^{\infty} \frac{dx}{1+x^2}$$

The strategy is to split this into two improper integrals of the form we already know how to compute:

$$I = \int_{-\infty}^{a} \frac{dx}{1+x^2} + \int_{a}^{\infty} \frac{dx}{1+x^2}$$

To ease the computation we will select $a = 0$, just because 0 is an easy number to work with. The above split does not, however, depend on a (as I will show shortly). With the substitution $x = \tan\theta$, see Table 4.14, we can compute the two integrals and thus I as

$$I = [\theta]_{-\pi/2}^{0} + [\theta]_{0}^{\pi/2} = \frac{\pi}{2} + \frac{\pi}{2} = \pi$$

Now to show that any value for a is fine, we just use a, and compute I as

$$I = [\theta]_{-\pi/2}^{\arctan a} + [\theta]_{\arctan a}^{\pi/2} = \left(\arctan a + \frac{\pi}{2}\right) + \left(\frac{\pi}{2} - \arctan a\right) = \pi$$

And what we have done for this particular integral applies for $\int_{-\infty}^{\infty} f(x)dx$.

4.9 Applications of Integration

Some elementary applications of integration are given herein. We start with the computation of length of planar curves in Sect. 4.9.1, of area of 2D geometric figures and volume of 3D figures in Sect. 4.9.2, and the surface area and volume of solids of revolution in Sect. 4.9.3. An application in physics—the gravitation of distributed masses is presented in Sect. 4.9.4. Finally, the use of integrals to compute the limit of sum is the topic of Sect. 4.9.5.

4.9.1 Length of Plane Curves

As the first application of integral, we consider the problem of calculating the length (or arc length) of a plane curve expressed by the equation $y = f(x)$. Determining the length of a curve is also called rectification of a curve. This is because when rectified, the curve gives a straight line segment with the same length as the curve's length. For much of the history of mathematics, even the greatest thinkers (e.g. Descartes) considered it impossible to compute the length of a curve. The advent of infinitesimal calculus led to a general formula that provides closed-form solutions in some cases.

The idea is simple: take a very small segment ds, of which the length is certainly $ds = \sqrt{dx^2 + dy^2}$, then integrating/summing the length of all these segments, we get the total length (Fig. 4.49). In symbols, we write (as $dy = f'(x)dx$)

$$\int_a^b ds = \int_a^b \sqrt{1 + [f'(x)]^2} dx \tag{4.9.1}$$

What does this equation mean? It tells us that the length of a plane curve C is the area of another curve C' with the function $\sqrt{1 + y'}$! And this arc length problem exposes the essence of the calculus in a simplest way: differential calculus allows us to compute the length of a small part of the curve (ds), and then integral calculus gives us a tool to integrate all these small lengths ($\int ds$). And this is used again and again in sciences and engineering.

If you're not convinced with this approach of physicists; for example where is the Riemann sum that is in the definition of a definite integral? For those readers, look at the right picture in Fig. 4.49. You see that, we divide the interval $[a, b]$ into n sub-intervals. For a sub-interval of length Δx_i, $\Delta s_i = \sqrt{1 + f'(x_i)} \Delta x_i$, then

$$L = \lim_{n \to \infty} \sum_{i=1}^{n} \sqrt{1 + f'(x_i)} \Delta x_i$$

And the Riemann sum appears, and that's why we have the integral in Eq. (4.9.1).

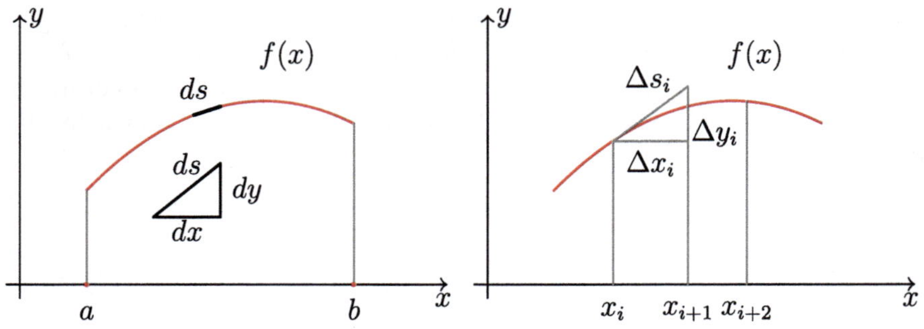

Fig. 4.49 Arc length of a curve $y = f(x)$ for $x \in [a, b]$

Perimeter of a circle. We only have a separate functions of each of the four quarters of a circle, so we compute the length of the first quarter. We write the circle's equation as $y = \sqrt{1-x^2}$, then a direct application of Eq. (4.9.1) gives

$$\int_0^1 \sqrt{1+\frac{x^2}{1-x^2}}dx = \int_0^1 \frac{dx}{\sqrt{1-x^2}} = \int_0^{\pi/2} d\theta = \frac{\pi}{2} \quad (x = \sin\theta)$$

Perimeter of an ellipse. Compute the perimeter of 1/4 of the ellipse given by $y^2 + 2x^2 = 2$. We write the ellipse as $y = \sqrt{2-2x^2}$, then an application of Eq. (4.9.1) results in

$$\int_0^1 \sqrt{\frac{1+x^2}{1-x^2}}dx$$

Unfortunately, we cannot compute this integral unless we use numerical integration (Sect. 10.4). Be careful that the integrand is infinity at $x = 1$ and thus not all numerical integration methods can be used. There is no simple exact closed formula for the perimeter of an ellipse! We will come back to this problem of the determination of the ellipse perimeter shortly.

Arc-length of a parabola. We find the arc length of a parabola $y = x^2$ for $0 \leq x \leq a$:

$$s = \int_0^a \sqrt{1+4x^2}dx$$

You know how to compute this integral (Sect. 4.7.1). Herein we're interested in finding C', which is given by $y = \sqrt{1+4x^2}$ or $y^2 = 1+4x^2$. And this is a hyperbola. So the length of a parabola is nothing new, it's the area of its cousin-a hyperbola.

Elliptic integrals. Consider an ellipse given by $x^2/a^2 + y^2/b^2 = 1$, with $a > b$, its length is given by

$$C = 4\int_0^{\pi/2} \sqrt{a^2\cos^2 t + b^2 \sin^2 t}dt$$

With $k = \sqrt{a^2-b^2}/a$, we can rewrite the above integral as

$$C = 4aE(k), \quad E(k) = \int_0^{\pi/2} \sqrt{1-k^2\sin^2 t}dt$$

The integral $E(k)$ is known as an *elliptic integral*. The name comes from the integration of the arc length of an ellipse. As there are other kinds of elliptic integral, the precise name is the elliptic integral of the second kind. What is then the elliptic integral of the first kind? It is defined as

$$E(k) = \int_0^{\pi/2} \frac{dt}{\sqrt{1 - k^2 \sin^2 t}}$$

It is super interesting to realize that this integral appears again and again in physics. And we will see it in the calculation of the period of a simple pendulum (Sect. 7.8.6).

4.9.2 Areas and Volumes

Herein I present the application of integration in computing the area and volume of some geometrical shapes. First, let's consider a circle of radius r (centered at the origin of the coordinate system). We can write the equation for one-quarter of the circle as $y = \sqrt{r^2 - x^2}$, and thus the area of one-quarter of the circle is

$$\int_0^1 \sqrt{r^2 - x^2} dx = r^2 \int_0^{\pi/2} \cos^2 \theta d\theta = \frac{\pi r^2}{4}$$

And voilà, the circle area is πr^2, a result once required the genius of Archimedes and the likes. This corresponds to the traditional way of slicing a region by thin rectangular strips. For circles which possess rotational symmetry, a better way is to divide the circle into many wedges, see Fig. 4.50:

$$\int_0^{2\pi} \frac{r^2}{2} d\theta = \pi r^2$$

And it gives directly the area of the full circle and no trigonometric substitution is required.

Next, we compute the volume of a cone with radius r and height h. We approximate the cone as a series of thin slices of thickness dy parallel to the base, see Fig. 4.51. The volume of each slide is $\pi R^2 dy$ with $R = r(1 - y/h)$, and thus the volume of the cone is

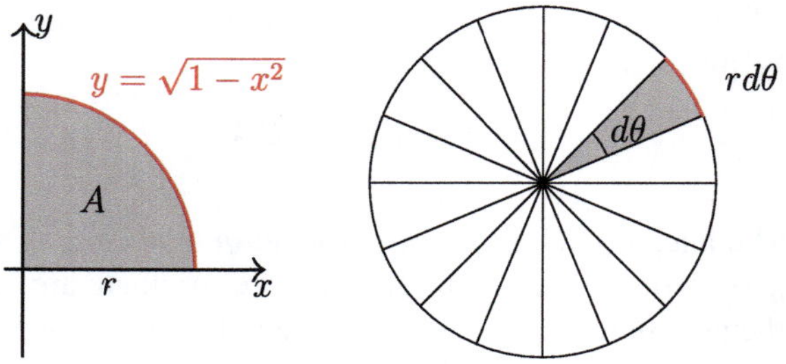

Fig. 4.50 Area of a circle: two ways of integration. In the second way, the area of one wedge is $0.5r(rd\theta)$ as a wedge is a triangle of base $rd\theta$ and height r

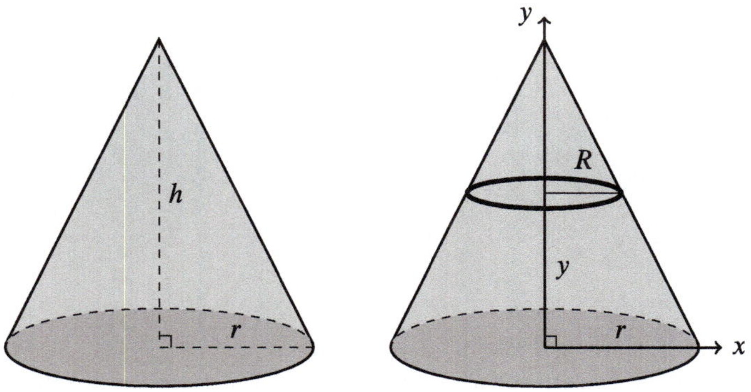

Fig. 4.51 Volume of a cone by integration. The thick slice has a thickness dy and a radius $R = r(1 - y/h)$ where y is the vertical coordinate of its center

$$\int_0^h \pi R^2 dy = \int_0^h \pi r^2 \left(1 - \frac{y}{h}\right)^2 dy = \frac{\pi r^2 h}{3} = \frac{1}{3} Ah, \quad A = \pi r^2$$

Therefore, the volume of a cone is one-third of the volume of a cylinder of the same height and radius, in agreement with the finding of Greek mathematicians.

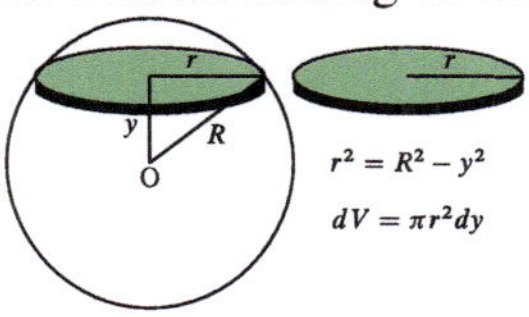

In the same manner, we compute the volume of a sphere of radius R as follows (see figure). We consider a slice of thickness dy—which is nothing but a cylinder—of which the volume is $\pi r^2 dy$, with $r^2 = R^2 - y^2$ where R is the sphere's radius and y is the distance from the origin to the slice. Thus, the total volume, which is the sum of the volumes of all these slices, is

$$2 \int_0^R \pi (R^2 - y^2) dy = \frac{4\pi R^3}{3}$$

We needed a factor of 2 because we considered only slices of the top half of the sphere. Can you see the similarity with Eq. (4.3.1)? In that equation, I presented a pre-calculus method to determine the volume of a sphere.

4.9.3 Area and Volume of a Solid of Revolution

Starting with a curve $y = f(x)$ for $a \leq x \leq b$ and we revolve it around an axis. That produces a solid of revolution (Fig. 4.52). This section presents how to use integra-

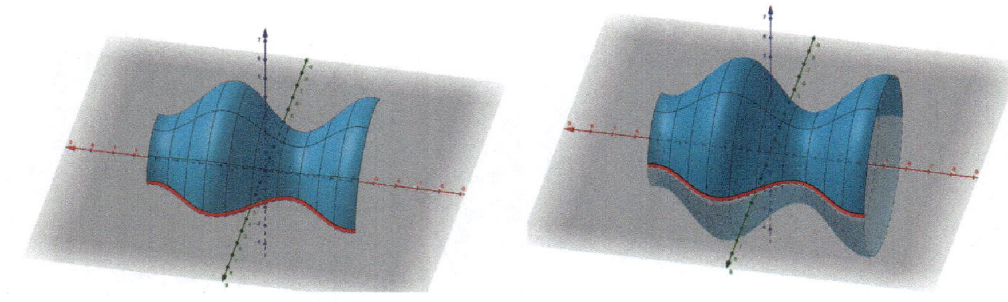

(a) (b)

Fig. 4.52 Solid of revolution: revolving the red curve $y = f(x)$ around an axis (the red axis) one full round (360°) and we get a solid revolution. Generated using the geogeba software

tion to compute the volume of such solids and also the area of the surface of such solids.

Volume of a solid of revolution. Assume we have a solid of revolution that is symmetrical around the $x-$axis. This solid is divided into many slices, each slice is similar to a pizza with area πy^2, $y = f(x)$, and thickness dx, thus the volume of the solid is then given by

$$\text{volume of solid of revolution around } x\text{-axis} = \int_a^b \pi[f(x)]^2 dx \tag{4.9.2}$$

Area of the surface of a solid of revolution. Using the idea of calculus, to find the area of a surface of revolution, we need to divide this surface into many tiny pieces, the area of each piece can be computed. Then, we sum these areas up. When the number of pieces is approaching infinity we get the surface area. We divide the surface into many thin bands shown in Fig. 4.53. As the band is thin, it is actually a truncated cone.

To find the area of a truncated cone, we start from a cone of radius r and slant s. We find its surface area, then we cut the cone by a plane and we get a truncated cone and a smaller cone. We know the surface area of the original cone and that of the smaller cone, thus we're done. That's the plane. To find the surface area of a cone, we flatten the cone out and get a fraction of a circle, see Fig. 4.54a. The area of this flattened cone is $\pi r s$. The area of a truncated cone is therefore $\pi r_1 s_1 - \pi r_2 s_2$ where r_1 is the radius of the original cone and r_2 is the radius of the circle at the cutting plane. It can be seen that this area also equals $2\pi r \Delta s$ where $r = 0.5(r_1 + r_2)$ (Fig. 4.54b) and Δs is the thickness of the band shown in Fig. 4.53.[38]

[38] We have $\pi(r_1 + r_2)\Delta s = \pi(r_1 + r_2)(S_1 - S_2) = \pi r_1 S_1 - \pi r_2 S_2$. This is so because $r_1 S_2 = r_2 S_1$.

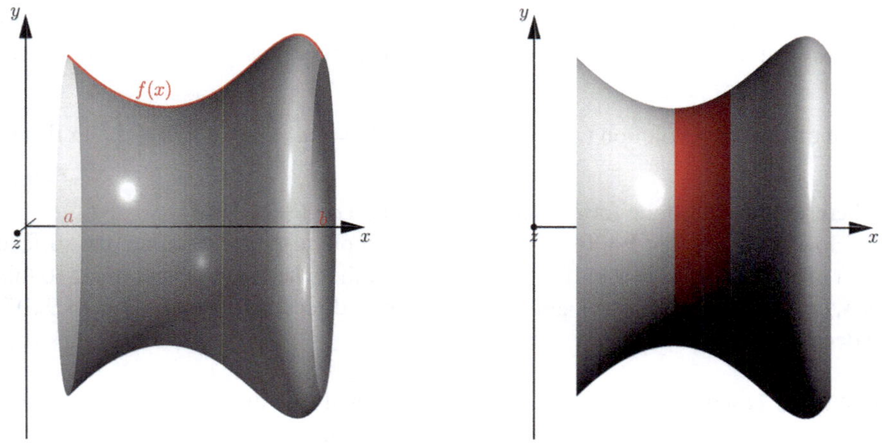

Fig. 4.53 A surface of revolution obtained by revolving a curve $y = f(x)$, $x \in [a, b]$, around the x-axis 360°. To find the surface area, we divide the surface into many tiny bands (one is shown highlighted in red color). The band was intentionally magnified for visualization purposes. We then find the surface area of one band and integrate it to get the total surface

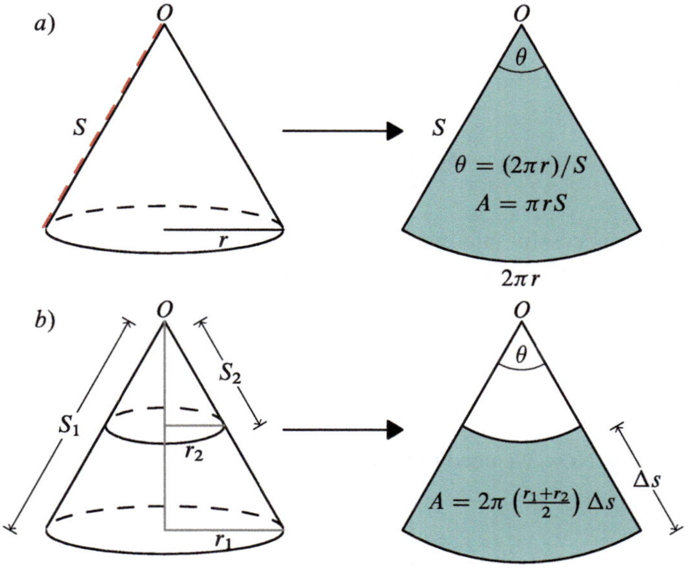

Fig. 4.54 Surface area of a truncated cone is $2\pi r \Delta s$ where r is the average radius and Δs is the width. The dashed line indicates where the cut is. The angle of the flattened cone in **a** is $\theta = 2\pi r/S$. From that we can determine its area as shown in the picture

Of course now we let Δs approach zero, then $\Delta s \approx ds$ and $r = y = f(x)$. The surface area of the solid of revolution is then the integral of those little areas $2\pi f(x)ds$:

$$\text{area of surf. of revolution } (x\text{-axis}) = \int_a^b 2\pi y ds = 2\pi \int_a^b f(x)\sqrt{1+[f'(x)]^2}dx \tag{4.9.3}$$

where we have used the formula for the arc length ds (Sect. 4.9.1).

How we know that Eq. (4.9.3) is correct? Simple: use it to compute the surface area of a sphere of radius r. If the result is $4\pi r^2$, then the formula is correct. To this end, consider $y = \sqrt{r^2 - x^2}$ and $x \in [0, r]$, then a direct application of Eq. (4.9.3)[39] indeed gives us $4\pi r^2$.

Gabriel's Horn is a surface of revolution by revolving the function $y = 1/x$ for $x \geq 1$ around the $x-$axis. This surface has a name because it has a special property: the volume inside Gabriel's Horn is finite but the surface area is infinite. We're first going to prove these results.

The volume is given by using Eq. (4.9.2)

$$V = \int_1^\infty \pi \frac{dx}{x^2} = \pi$$

And the surface is after Eq. (4.9.3):

$$A = 2\pi \int_1^\infty \frac{1}{x}\sqrt{1+\frac{1}{x^4}}dx = ???$$

After many unsuccessful attempts, we realized that it is not easy to compute this integral directly. How about an indirect way? That is, we compare this integral with another easier integral. The easier integral is

$$2\pi \int_1^\infty \frac{dx}{x}$$

which is infinity. Now, we need to find a relation between the two integrals, or these two functions

$$f(x) := \frac{1}{x}\sqrt{1+\frac{1}{x^4}}, \quad g(x) := \frac{1}{x}$$

We can see that $f(x) > g(x)$, thus:

$$\int_1^\infty f(x)dx > \int_1^\infty g(x)dx \Longrightarrow 2\pi \int_1^\infty f(x)dx > 2\pi \int_1^\infty g(x)dx$$

[39] But do not forget to multiply the result with two.

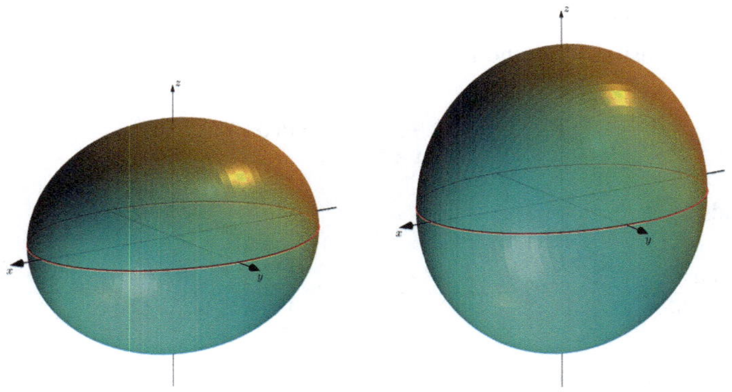

(a) revolution around $x-$axis (b) revolution around $y-$axis

Fig. 4.55 Ellipsoid: revolving an ellipse around an axis

And since $2\pi \int_1^\infty \frac{dx}{x} = \infty$, we also have $2\pi \int_1^\infty f(x)dx = \infty$. In other words, the area of the surface of Gabriel's Horn is infinite. Ok, enough with the maths (which is actually nothing particularly interesting).

Associated with Gabriel's Horn is a painter's paradox. Here it is. One needs an infinite amount of paint to cover the interior (or exterior) of the horn, but only a finite amount of paint is needed to fill up the interior space of the horn. So, either the math is wrong or this paradox is wrong. Of course this paradox is wrong! Can you see why?

Area of ellipsoid. Let's consider an ellipse $x^2/a^2 + y^2/b^2 = 1$; if we revolve it around the $x-$axis or $y-$axis we will get an ellipsoid, see Fig. 4.55. In this section we're interested in the area of such an ellipsoid.

We just need to consider one-quarter of the ellipse in the first quadrant and we revolve it around the $x-$axis $360°$. We parameterize the ellipse by

$$\left.\begin{array}{l} x = a\cos\theta \\ y = b\sin\theta \end{array}\right\} \Longrightarrow \left.\begin{array}{l} dx = -a\sin\theta \\ dy = b\cos\theta \end{array}\right\} \Longrightarrow ds = \sqrt{dx^2 + dy^2} = \sqrt{a^2\sin^2\theta + b^2\cos^2\theta}d\theta$$

Now, we just apply Eq. (4.9.3) to get

$$A = 2\int_0^{\pi/2} 2\pi b\sin\theta\sqrt{a^2\sin^2\theta + b^2\cos^2\theta}d\theta$$

Using this substitution $u = \cos\theta$ the above integral becomes:

$$A = 4\pi b\int_0^1 \sqrt{a^2 - (a^2 - b^2)u^2}du \tag{4.9.4}$$

Now, we assume that $a > b$ (we have to assume this or $a < b$ to use the appropriate trigonometry substitution), and thus we use the following substitution:

$$u = \frac{a}{\sqrt{a^2 - b^2}} \sin \alpha$$

which leads to

$$A = \frac{4\pi b a^2}{\sqrt{a^2 - b^2}} \int_0^{\arcsin \sqrt{a^2-b^2}/a} \frac{1 + \cos 2\alpha}{2} d\alpha = 2\pi \left[b^2 + \frac{a^2 b}{\sqrt{a^2 - b^2}} \arcsin \frac{\sqrt{a^2 - b^2}}{a} \right]$$

Ok, if we now apply this result to concrete cases $a = \ldots$ and $b = \ldots$, then it's fine. But we will miss interesting things. Let's consider the case $a < b$ to see what happens.

Now, consider the case $a < b$, then we write A in a slightly different form (Eq. (4.9.4)):

$$A = 2\int_0^1 2\pi b\sqrt{a^2 + (b^2 - a^2)u^2 du} = 4\pi b\sqrt{b^2 - a^2} \int_0^1 \sqrt{u^2 + c^2 du}, \; c^2 = \frac{a^2}{b^2 - a^2}$$

The red integral is exactly the one we met in Eq. (4.7.29), so we get

$$A = 2\pi \left[b^2 + \ln \left(\frac{b + \sqrt{b^2 - a^2}}{a} \right) \frac{a^2 b}{\sqrt{b^2 - a^2}} \right]$$

Now comes a nice observation. The area of an ellipsoid does not care about the magnitude of a and b. But, then why we have two different expressions for the same thing? This is because we do not allow square root of negative numbers. But hey, we know imaginary numbers. Why don't use them to have a unified expression? Let's do it.

First, define the following:

$$\sin \psi = \frac{\sqrt{a^2 - b^2}}{a}, \quad \cos \psi = \frac{b}{a}$$

Then, we can write:

$$\frac{\sqrt{b^2 - a^2}}{a} = \frac{\sqrt{(a^2 - b^2)(-1)}}{a} = \frac{\sqrt{(a^2 - b^2)i^2}}{a} = \frac{i\sqrt{a^2 - b^2}}{a} = i \sin \psi$$

With this, we have two expressions for A:

$$
\begin{aligned}
A &= 2\pi\left[b^2 + \frac{a^2 b}{\sqrt{a^2-b^2}}\psi\right] \quad (a > b) \\
A &= 2\pi\left[b^2 + \frac{a^2 b}{i\sqrt{a^2-b^2}}\ln\left(\cos\psi + i\sin\psi\right)\right]
\end{aligned}
\tag{4.9.5}
$$

And of course the second terms in the above should be the same:

$$
\ln\left(\cos\psi + i\sin\psi\right) = i\psi
$$

And this is obviously related to Euler's identity $e^{i\psi} = \cos\psi + i\sin\psi$. This logarithmic version of Euler's identity was discovered by the English mathematician Roger Cotes (1682–1716), who was known for working closely with Isaac Newton by proofreading the second edition of the Principia. He was the first Plumian Professor at Cambridge University from 1707 until his early death. About Cotes' death, Newton once said "If he had lived, we might have known something". The above analysis was inspired by [46].

4.9.4 Gravitation of Distributed Masses

In 1687 N published his work on gravity in his classic *Mathematical Principles of Natural Philosophy*. He stated that every object is pulling other object and two objects are pulled by a force that is proportional to the product of their masses and inversely proportional to the square of the distance between them. In mathematical symbols, his law is expressed as:

$$
F = \frac{GMm}{r^2} \tag{4.9.6}
$$

where G is a constant, called the *universal gravitational constant*, that has been experimentally measured by Cavendish about 100 years after Newton's death. It is $G = 6.673 \times 10^{-11}$ Nm^2/kg^2. This section presents how to use integrals to compute F for distributed masses e.g. a rod.

Gravitational pull of a thin rod 1. Consider a thin rod of length L, its mass M is uniformly distributed along the length. Ahead one end of the rod along its axis is placed a small mass m at a distance a (see Fig. 4.56). Calculate the gravitational pull of the rod on m.

Let's consider a small segment dx, its mass is $dm = (M/L)dx$. So, this small mass dm will pull the m with a force dF given by Newton's gravitational theory. The pull of the entire rod is then simply the pull of all these small dF:

$$
dF = \frac{GMm}{L}\frac{dx}{(L+a-x)^2} \Longrightarrow F = \frac{GMm}{L}\int_0^L \frac{dx}{(L+a-x)^2} = \frac{GMm}{a(L+a)} \tag{4.9.7}
$$

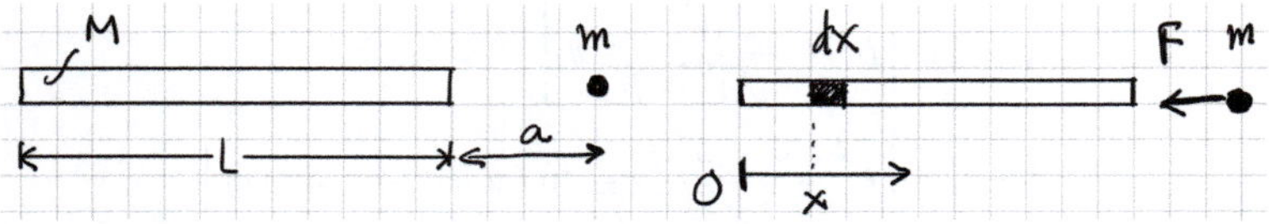

Fig. 4.56 Figure 4.56

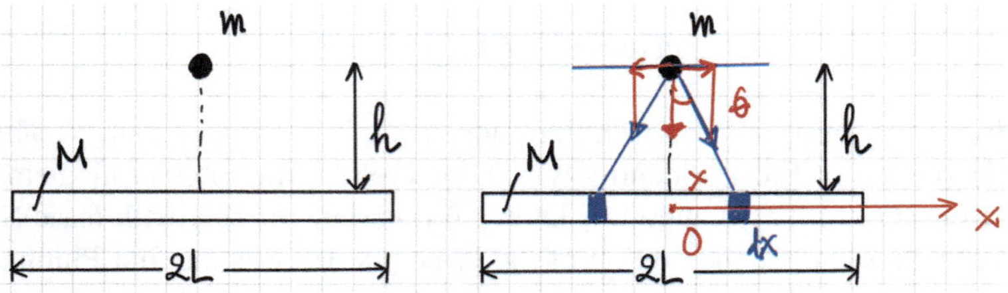

Fig. 4.57 Figure 4.57

Gravitational pull of a thin rod 2. Consider a thin rod of length $2L$, its mass M is uniformly distributed along the length. Above the center of the rod at a distance h is placed a small mass m (see Fig. 4.57). Calculate the gravitational pull of the rod on m.

Due to symmetry the horizontal component of the gravitational pull is zero. Thus, only the vertical component counts. This component for a small segment dx can be computed, and thus the total force is just a sum of all these tiny forces, which is of course an integral:

$$dF = \frac{GMm}{2L}\frac{dx}{h^2+x^2}\cos\theta \Longrightarrow F = \frac{GMm}{L}h\int_0^L \frac{dx}{(h^2+x^2)^{3/2}} \tag{4.9.8}$$

To evaluate the integral $\int_0^L \frac{dx}{(h^2+x^2)^{3/2}}$, we use the trigonometric substitution:

$$x = h\tan\theta \Longrightarrow \begin{cases} dx & = h\sec^2\theta d\theta \\ h^2+x^2 & = h^2\sec^2\theta \\ 0 \le \theta \le \tan^{-1}(L/h) \end{cases} \tag{4.9.9}$$

Thus, the integral becomes

$$\begin{aligned}\int_0^L \frac{dx}{(h^2+x^2)^{3/2}} &= \int_0^{\tan^{-1}(L/h)} \frac{h\sec^2\theta d\theta}{h^3\sec^3\theta} = \frac{1}{h^2}\int_0^{\tan^{-1}(L/h)} \cos\theta d\theta \\ &= \frac{1}{h^2}\left[\sin\theta\right]_0^{\tan^{-1}(L/h)} = \frac{1}{h^2}\sin\left[\tan^{-1}\left(\frac{L}{h}\right)\right] = \frac{L}{h^2\sqrt{h^2+L^2}}\end{aligned}$$

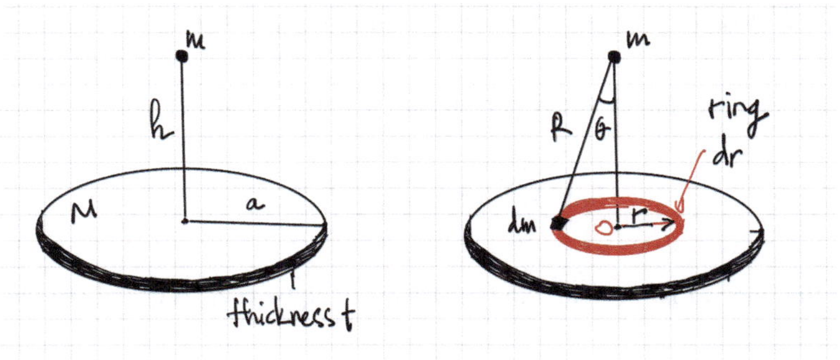

Fig. 4.58 Figure 4.58

And finally, the gravitational force is

$$F = \frac{GMm}{h\sqrt{h^2 + L^2}}$$

Gravitational pull of a thin disk. Consider a thin disk of radius a, its mass M is uniformly distributed. Above the center of the disk at a distance h is placed a small mass m (see Fig. 4.58). Calculate the gravitational pull of this disk on m.

We consider a ring located at distance r from the center, this ring has a thickness dr. We first compute the gravitational pull of this ring on m. Then, we integrate this to get the total pull of the whole disk on m. Again, due to symmetry, only a downward pull exists. Consider a small dm on this ring, we have

$$dF = \frac{dmGm}{R^2}\cos\theta \Longrightarrow F_{\text{ring}} = \int dF = \frac{Gm\cos\theta}{R^2} m_{\text{ring}} \tag{4.9.10}$$

This is because R and $\cos\theta$ are constant along the ring. The mass of the ring is $m_{\text{ring}} = tM\pi[(r+dr)^2 - r^2] = 2\pi rtMdr$. So, the pull of the ring on m is

$$F_{\text{ring}} = \frac{Gm\cos\theta}{R^2} 2\pi rtMdr = GMmt2\pi h \frac{rdr}{\sqrt{h^2+r^2}} \tag{4.9.11}$$

And thus, the total pull of the disk on m is

$$F_{\text{disk}} = \int F_{\text{ring}} = GMmt2\pi h \int_0^a \frac{rdr}{\sqrt{h^2+r^2}} = 2\pi GMmt\left(1 - \frac{h}{\sqrt{h^2+a^2}}\right) \tag{4.9.12}$$

4.9.5 Using Integral to Compute Limits of Sums

We know that we can use the Riemann sum to approximate an integral. We have

$$\lim_{n\to\infty}\sum_{i=1}^{n} f(x_i^*)\Delta x_i = \int_a^b f(x)dx$$

When the interval is [0, 1] and the intervals are equal (i.e., $\Delta x_i = 1/n$), the above becomes

$$\lim_{n\to\infty}\frac{1}{n}\sum_{i=1}^{n} f\left(\frac{i}{n}\right) = \int_0^1 f(x)dx \tag{4.9.13}$$

Now that we know all the techniques to compute definite integrals, we can use integral to compute limits of sum. For example, compute the following limit:

$$\lim_{n\to\infty}\sum_{i=1}^{n}\frac{n}{n^2+i^2}$$

The plan is to rewrite the above sum in the form of a Riemann sum, then Eq. (4.9.13) allows us to equate it to an integral. Compute that integral and we're done. So, we write $n/n^2+i^2 = (1/n)1/1+(i/n)^2$. Thus,

$$\lim_{n\to\infty}\sum_{i=1}^{n}\frac{n}{n^2+i^2} = \lim_{n\to\infty}\frac{1}{n}\sum_{i=1}^{n}\frac{1}{1+(i/n)^2} = \int_0^1 \frac{1}{1+x^2}dx = \cdots = \frac{\pi}{4}$$

Example 4.17 Evaluate the following limit:

$$\lim_{x\to 0^+}\sum_{k=1}^{\infty}\frac{2x}{1+k^2x^2}$$

4.10 Limits

The calculus was invented in the seventeenth century and it is based on limit—a concept developed in the eighteenth century. That's why I have intentionally presented the calculus without precisely defining what a limit is. This is mainly to show how mathematics was actually evolved. But we cannot avoid working with limits, that's why we finally discuss this concept in this section.

Let's consider the quadratic function $y = f(x) = x^2$, and we want to define the derivative of this function at x_0. We consider a change h in x with a corresponding change in the function $\Delta f = (x_0 + h)^2 - x_0^2$. We now know that Newton, Leibniz

Table 4.15 $\lim_{h\to 0} \Delta f/h$

h	$\Delta f/h$
10^{-1}	4.100000000000
10^{-2}	4.010000000000
10^{-3}	4.001000000000
10^{-4}	4.000100000008
10^{-5}	4.000010000027
10^{-6}	4.000001000648

and their fellows defined the derivative as the value that the ratio $\Delta f/h$ tends to when h approaches zero. Here what they did

$$f'(x_0) = \frac{(x_0+h)^2 - x_0^2}{h} = \frac{2x_0h + h^2}{h} = 2x_0 + h = 2x_0$$

The key point is in the third equation where h is not zero and in the final equation where h is zero. Due to its loose foundation h were referred to as "The ghosts of departed quantities" by Bishop George Berkeley of the Church of England (1685–1753) in his attack on the logical foundations of Newton's calculus in a pamphlet entitled *The Analyst* (1734).

Leibniz realized this and solved the problem by saying that h is a differential—a quantity that is non-zero but smaller than any positive number. Because it's non-zero, the third equation in the above is fine, and because it is a super small number, it's nothing compared with $2x_0$, thus we can ignore it.

Was Leibniz correct? Yes, Table 4.15 confirms that. This table is purely numerics, we computed $\Delta f/h$ for many values of h getting smaller and smaller (and we considered $x_0 = 2$ as we have to give x_0 a value).

Now we're ready for the presentation of the limit of a function. The key point here is to see $\Delta f/h$ as a function of h; thus the derivative of $y = f(x)$ at x_0 is the limit of the function $g(h) := \Delta f/h$ when h approaches zero:

$$f'(x_0) = \lim_{h\to 0} \frac{(x_0+h)^2 - x_0^2}{h} = \lim_{h\to 0}(2x_0 + h)$$

And what is this limit? As can be seen from Fig. 4.59, as h tends to zero $2x_0 + h$ is getting closer and closer to $2x_0$. And that's what we call the limit of $2x_0 + h$.

In the preceding discussion we have used the symbol h to denote the change in x when defining the derivative of $y = f(x)$. This led to the limit of another function $g(h)$ with h being the independent variable. It's possible to restate the problem so that the independent variable is always x. We choose a fixed point x_0. And we consider another point x, then we have

$$f'(x_0) = \lim_{x\to x_0} \frac{x^2 - x_0^2}{x - x_0} = \lim_{x\to x_0} x + x_0 = 2x_0$$

Fig. 4.59 $\lim_{h\to 0}(2x_0 + h)$

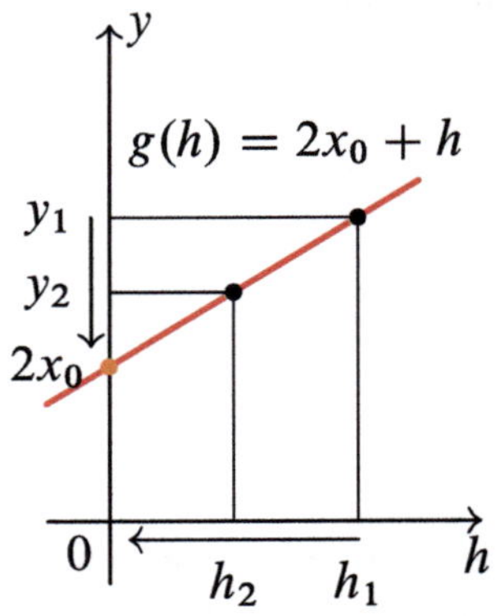

4.10.1 Definition of the Limit of a Function

Now we can forget the derivative and focus on the limit of a function. Let's denote by $y = f(x)$ any function, and we're interested in the limit of this function when x approaches a. Intuitively, we know that $\lim_{x\to a} f(x) = L$ means that we can make x getting closer and closer to a so that $f(x)$ gets closer to L as much as we want. See Table 4.15 again, we have stopped at $h = 10^{-6}$ but we can get $\Delta f/h$ much closer to four by using smaller h.

How to mathematically describe 'x gets closer and closer to a'? Given a positive small number δ, we say that x is close to a when $a - \delta < x < a + \delta$ i.e., $x \in (a - \delta, a + \delta)$ (in plain English, x is in a neighborhood of a). We can write it more compactly as $|x - a| < \delta$. Similarly, $f(x)$ gets closer to L means $|f(x) - L| < \epsilon$, ϵ is yet another small positive number. Cauchy and Bernard Bolzano (1781–1848) were the first who used these ϵ and δ.

There is a little detail here before we present the definition of the limit of a function. We always say that a limit of a function when x approaches a. This implies that *we do not care what happens when $x = a$*. For example, the function $y = {}^{x-1}/_{x^2-1}$ is not defined at $x = 1$, but it is obvious that $\lim\limits_{x\to 1} f(x) = 0.5$.[40] But the classic example is a circle and an n-polygon inscribed in it. When we say a limit of this $n-$polygon when n approaches infinity is the circle, we mean that n is a very large number. But it is meaningless if n is actually infinity. Because in that case we would have a polygon of which each side is of vanished length.

Thus, x close to a and not equal to a is written mathematically as:

$$0 < |x - a| < \delta$$

40 You can try by plotting this function or making a table similar to Table 4.15 to confirm this.

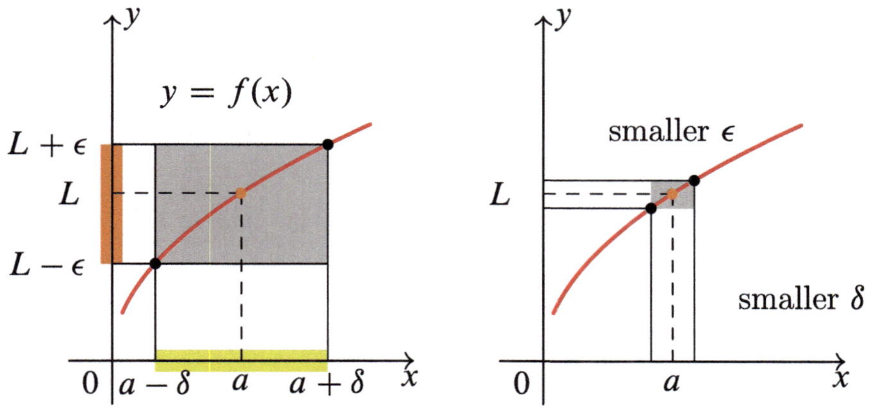

Fig. 4.60 Visualization of the $\epsilon - \delta$ definition of the limit of a function

Definition 4.5 We denote the limit of $f(x)$ when x approaches a by $\lim_{x \to a} f(x)$, and this limit is L i.e.,

$$\lim_{x \to a} f(x) = L$$

when, for any $\epsilon > 0$, there exists a $\delta > 0$ such that

$$\text{if} \quad 0 < |x - a| < \delta \quad \text{then} \, |f(x) - L| < \epsilon$$

This definition was given by the German mathematician Karl Theodor Wilhelm Weierstrass (1815–1897) who was often cited as the "father of modern analysis".

The key point here is that ϵ is the input that indicates the level of accuracy we need for $f(x)$ to approach L and δ is the output (thus depends on ϵ). Figure 4.60 illustrates this; for a smaller ϵ, we have to make x closer to a and thus a smaller δ.

What is analysis by the way? Analysis is the branch of mathematics dealing with limits and related theories, such as differentiation, integration, measure, infinite series, and analytic functions. These theories are usually studied in the context of real and complex numbers and functions. Analysis evolved from calculus, which involves the elementary concepts and techniques of analysis.

One-sided limits. If we want to find the limit of this function $\sqrt{x-1}$ when x approaches 1, we'll see that we need to consider only $x \geq 1$, and this leads to the notion of one-sided limit:

$$\lim_{x \to 1^+} \sqrt{x-1} = \lim_{x \downarrow 1} \sqrt{x-1}$$

which is a right-hand limit when we approach 1 from above, as indicated by the notation $\downarrow 1$, even though this is not popular. And of course, if we have right-hand limit, we have left hand limit e.g. $\lim_{x\to 1^-} \sqrt{1-x}$.

If the limit of $f(x)$ when x approaches a exists, it means that the left-hand and right-hand one-sided limits exit and equal:

$$\text{if } \lim_{x\to a} f(x) = L \text{ then } \quad \lim_{x\to a^+} f(x) = \lim_{x\to a^-} f(x) = L$$

Infinite limits. If we consider the function $y = 1/x^2$ we realize that y is very large for x near 0. Thus, we say that:

$$\lim_{x\to 0} \frac{1}{x^2} = \infty$$

And this is called an infinite limit which is about the limit of a function which is very large near $x = 0$. We can generalize this to have

$$\lim_{x\to a} f(x) = \infty, \quad \lim_{x\to a} f(x) = -\infty, \quad \lim_{x\to a^+} f(x) = \infty$$

Figure 4.61 illustrates some of infinite limits and we can see that the lines $x = a$ are the vertical asymptotes of the graphs. This figure suggests the following definition of infinite limits.

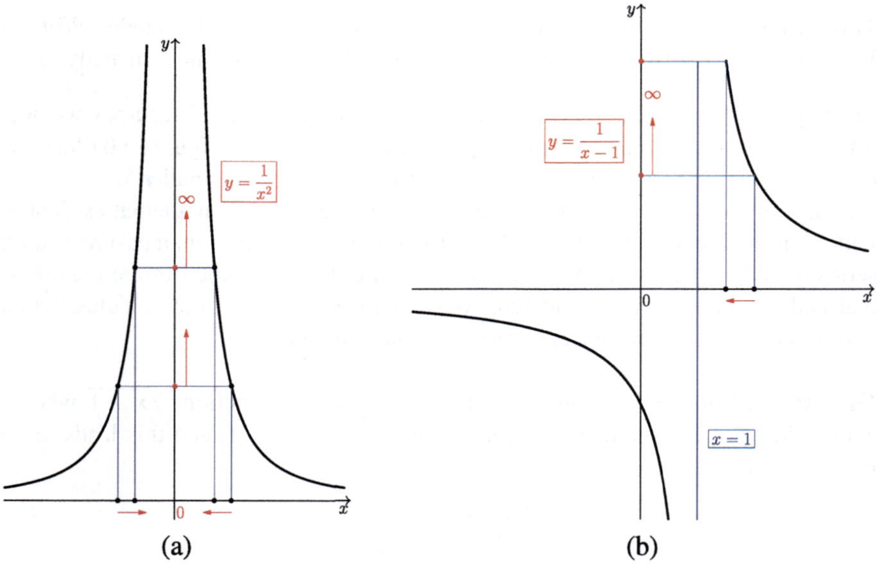

Fig. 4.61 Infinite limits and vertical asymptotes

Definition 4.6 The limit of $y = f(x)$ when x approaches a is infinity, written as,

$$\lim_{x \to a} f(x) = \infty$$

when, for any large number M, there exists a $\delta > 0$ such that

$$\text{if} \quad 0 < |x - a| < \delta \quad \text{then } f(x) > M$$

Limits when x approaches infinity. Again considering the function $y = 1/x^2$ but now focus on what happens when x approaches infinity i.e., x is getting bigger and bigger or when it gets smaller and smaller. It's clear that $1/x^2$ is then getting smaller and smaller. We write $\lim_{x \to +\infty} {}^1\!/_{x^2} = \lim_{x \to -\infty} {}^1\!/_{x^2} = 0$.

Definition 4.7 The limit of $y = f(x)$ when x approaches ∞ is finite, written as,

$$\lim_{x \to \infty} f(x) = a$$

when, for any $\epsilon > 0$, there exists a number $M > 0$ such that

$$\text{if} \quad x > M \quad \text{then } |f(x) - a| < \epsilon$$

We can use this definition to prove that $\lim_{x \to +\infty} {}^1\!/_{x^2} = 0$; select $M = 1/\epsilon$ then $1/x$ will be near to ϵ.

We soon realize that the definition of the limit of a function is not as powerful as it seems to be. For example, with the definition of limit, we're still not able to compute the following limit

$$\lim_{t \to 0} \frac{\sqrt{t^2 + 9} - 3}{t^2}$$

The situation is similar to differentiation. We should now try to find out the rules that limits obey, then using them will enable us to evaluate the limits of complex functions.

4.10.2 *Rules of Limits*

Considering two functions $y = f(x)$ and $y = g(x)$ and assume that $\lim_{x \to a} f(x)$ and $\lim_{x \to a} g(x)$ exit, then we have the following rules:

$$
\begin{aligned}
&\text{(a: constant function rule)} && \lim_{x\to a} c && = c \\
&\text{(b: sum/diff rule)} && \lim_{x\to a}(f(x) \pm g(x)) && = \lim_{x\to a} f(x) \pm \lim_{x\to a} g(x) \\
&\text{(c: linearity rule)} && \lim_{x\to a}(cf(x)) && = c \cdot \lim_{x\to a} f(x) \\
&\text{(d: product rule)} && \lim_{x\to a}(f(x)g(x)) && = (\lim_{x\to a} f(x))(\lim_{x\to a} g(x)) \\
&\text{(e: quotient rule)} && \lim_{x\to a}(f(x)/g(x)) && = (\lim_{x\to a} f(x))/(\lim_{x\to a} g(x)) \\
&\text{(f: power rule)} && \lim_{x\to a}[f(x)]^n && = [\lim_{x\to a} f]^n
\end{aligned}
\tag{4.10.1}
$$

The sum rule basically states that the limit of the sum of two functions is the sum of the limits. And this is plausible: near $x = a$ the first function is close to L_1 and the second function to L_2, thus $f(x) + g(x)$ is close to $L_1 + L_2$. And of course when we have this rule for two functions, we also have it for any number of functions! Need a proof? Here it is

$$\lim_{x\to a}(f + g + h) = \lim_{x\to a}[(f + g) + h] = \lim_{x\to a}(f + g) + \lim_{x\to a} h = \lim_{x\to a} f + \lim_{x\to a} g + \lim_{x\to a} h$$

Similar to multiplication leads to exponents e.g. $2 \times 2 \times 2 = 2^3$, the product rule $\lim_{x\to a} fg = (\lim_{x\to a} f)(\lim_{x\to a} g)$ will lead to $\lim_{x\to a}[f(x)]^n = [\lim_{x\to a} f]^n$ for n being a positive integer. This result also holds for negative integer n by combining it with the quotient rule.

Proof (*Proof of the sum rule of limit*) We assume that

$$\lim_{x\to a} f(x) = L_1, \quad \lim_{x\to a} g(x) = L_2$$

And we need to prove that

$$\lim_{x\to a}[f(x) + g(x)] = L_1 + L_2$$

And this is equivalent to proving the following (using the definition of limit)

$$|f(x) + g(x) - L_1 - L_2| < \epsilon \quad \text{when } |x - a| < \delta$$

Now we use our assumption about the limits of f and g to have ($\delta_1, \delta_2, \epsilon$ are positive real numbers):

$$
\begin{aligned}
|x - a| < \delta_1 \quad &\text{then } |f(x) - L_1| < \frac{\epsilon}{2} \\
|x - a| < \delta_2 \quad &\text{then } |g(x) - L_2| < \frac{\epsilon}{2}
\end{aligned}
\tag{4.10.2}
$$

Then, define $\delta = \min(\delta_1, \delta_2)$, we thus have the two above inequalities:

$$|x - a| < \delta \Longrightarrow |f(x) - L_1| < \frac{\epsilon}{2}, \quad |g(x) - L_2| < \frac{\epsilon}{2}$$

Now using the triangle inequality $|a+b| < |a| + |b|$:

$$|f(x) + g(x) - L_1 - L_2| < |f(x) - L_1| + |g(x) - L_2| < \frac{\epsilon}{2} + \frac{\epsilon}{2} = \epsilon$$

Now you know why we have used $\epsilon/2$ as the accuracy in Eq. (4.10.2). To summarize, the whole proof uses (1) the triangle inequality $|a+b| < |a| + |b|$ and (2) a correct accuracy (e.g. $\epsilon/2$ here). Do we need another proof for the difference rule? No! This is because $a - b$ is simply $a + (-b)$. If you're still not convinced, we can do this:

$$\lim_{x \to a}(f - g) = \lim_{x \to a}[f + (-1)g] = \lim_{x \to a} f + \lim_{x \to a}(-1)g = \lim_{x \to a} f - \lim_{x \to a} g$$

where we used the rule $\lim_{x \to a} cf = c \lim_{x \to a} f$ with $c = -1$.

Proof (*Proof of the product rule of limit*) We assume that

$$\lim_{x \to a} f(x) = L, \quad \lim_{x \to a} g(x) = M$$

It's possible to prove the product rule in the same way as the sum rule, but it's hard. We follow an easier path. First we massage a bit fg[41]:

$$fg = (f - L)(g - M) - LM + Mf + Lg$$

Thus, the limit of fg is

$$\begin{aligned}
\lim_{x \to a} fg &= \lim_{x \to a}(f - L)(g - M) - LM + \lim_{x \to a} Mf + \lim_{x \to a} Lg \\
&= \lim_{x \to a}(f - L)(g - M) - LM + LM + LM \\
&= \lim_{x \to a}(f - L)(g - M) + LM
\end{aligned}$$

Now if we can prove that $\lim_{x \to a}(f - L)(g - M) = 0$ then we're done. Indeed, we have

$$\left.\begin{aligned}
0 < |x - a| < \delta_1 &\Longrightarrow |f - L| < \sqrt{\epsilon} \\
0 < |x - a| < \delta_2 &\Longrightarrow |g - M| < \sqrt{\epsilon}
\end{aligned}\right\}$$

With $\delta = \min(\delta_1, \delta_2)$, we then have

$$0 < |x - a| < \delta \Longrightarrow |f - L||g - M| < \epsilon \text{ or } |(f - L)(g - M) - 0| < \epsilon$$

---

[41] This is the crux of the whole proof. This transform the original problem to this problem: prove $\lim_{x \to a}(f - L)(g - M) = 0$, which is much more easier.

Proof (*Proof of the quotient rule*) First, we prove a simpler version:

$$\lim_{x\to a}\frac{1}{g(x)} = \frac{1}{\lim_{x\to a} g(x)} \tag{4.10.3}$$

Then, it is simple to prove the original rule:

$$\begin{aligned}
\lim_{x\to a}\frac{f(x)}{g(x)} &= \lim_{x\to a}\left[f(x)\frac{1}{g(x)}\right] \\
&= \left[\lim_{x\to a} f(x)\right]\left[\lim_{x\to a}\frac{1}{g(x)}\right] \quad \text{(using product rule)} \\
&= \frac{\lim_{x\to a} f(x)}{\lim_{x\to a} g(x)} \qquad \text{(Eq. (4.10.3))}
\end{aligned}$$

To prove Eq. (4.10.3), let's denote $M = \lim_{x\to a} g(x)$. Then, what we have to prove is that

$$\left|\frac{1}{g(x)} - \frac{1}{M}\right| < \epsilon \quad \text{when } 0 < |x-a| < \delta$$

Or this

$$\frac{1}{|M|}\frac{1}{|g(x)|}|g(x) - M| < \epsilon \quad \text{when } |x-a| < \delta \tag{4.10.4}$$

Now we need to find $\frac{1}{|g(x)|} <?$ and $|g(x) - M| <?$. Because $\lim_{x\to a} g(x) = M$, when $0 < |x-a| < \delta_1$ we have

$$|g - M| < |M|/2$$

We can always select δ_1 so that the above inequality holds. You can draw a picture, similar to Fig. 4.60 to convince yourself about this. Thus,

$$\begin{aligned}
|M| &= |M - g(x) + g(x)| \\
&\le |M - g(x)| + |g(x)| \quad \text{(triangle inequality)} \\
&\le |g(x) - M| + |g(x)| \le \frac{|M|}{2} + |g(x)| \implies \frac{1}{g(x)} < \frac{2}{|M|}
\end{aligned}$$

Now based on Eq. (4.10.4), we need $|g(x) - M| < (M^2/2)\epsilon$. And of course we have it at our disposal because the limit of g is M. This holds true when $0 < |x-a| < \delta_2$. Now, with $\delta = \min(\delta_1, \delta_2)$, we have

$$\frac{1}{g(x)} < \frac{2}{|M|},\quad |g(x) - M| < \frac{M^2}{2}\epsilon \implies \frac{1}{|M|}\frac{1}{|g(x)|}|g(x) - M| < \frac{1}{|M|}\frac{2}{|M|}\frac{M^2}{2}\epsilon = \epsilon$$

Sadly that in many textbooks, the proof is written in a reversal way, which makes students believe that they look stupid. We emphasize again that finding a proof is hard and involves many setbacks. When a proof has been found, the author presents it not in a way the proof was found.

Using the definition of limit, we can see that:

$$\lim_{x \to a} x = a \tag{4.10.5}$$

Combined with the power rule in Eq. (4.10.1), we have

$$\lim_{x \to a} x^n = a^n \tag{4.10.6}$$

If we look at again these two results, we see that the function $y = x^n$ has this nice property: $\lim_{x \to a} f(x) = f(a)$, that is the limit when x approaches a equals the function value at a. We're now turning our discussion to the functions that have this special property: continuous functions.

4.10.3 Continuous Functions

I mentioned in the introduction of this chapter that the essence of calculus is quite simple: calculus is often seen as *the mathematics of changes*. But calculus does not work with all kinds of change. It only works with change of continuous quantities (or continuous functions). Now, we finally can define precisely what is a continuous function.

A function is continuous if we can draw its graph without lifting our pencil off the paper. The functions $\sin x$, $\cos x$, e^x and so on are such functions (Fig. 4.62a). However, if you draw the graph of $1/x$ you have to lift your pencil off the paper as the graph has two branches, one for $x \in (-\infty, 0)$ and one for $x \in (0, \infty)$, see Fig. 4.62b. At $x = 0$ where we have to lift off the pencil we say that there is a discontinuity. This discontinuity is called an *infinite discontinuity*.

To see what else do we have as far as discontinuous functions are involved, we now consider functions that are defined by more than one expression. We break apart the function domain into two or more disjoint pieces, we then use different functions to calculate the output for each x where the function used is based upon the piece into which that particular x falls. Let's start with the function $y = |x|$, which is explicitly written as

$$y = |x| = \begin{cases} x, & \text{if } x \geq 0 \\ -x, & \text{if } x < 0 \end{cases}$$

The graph of this function has two pieces, one defined over $(-\infty, 0)$ with the expression $y = -x$ and one over $(0, \infty)$ with the expression $y = x$ (Fig. 4.62c). But the

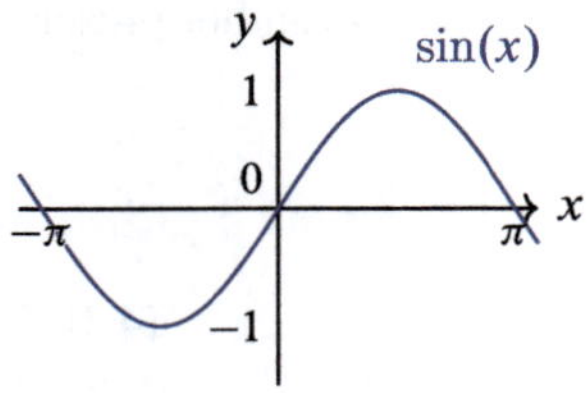

(a) continuous everywhere

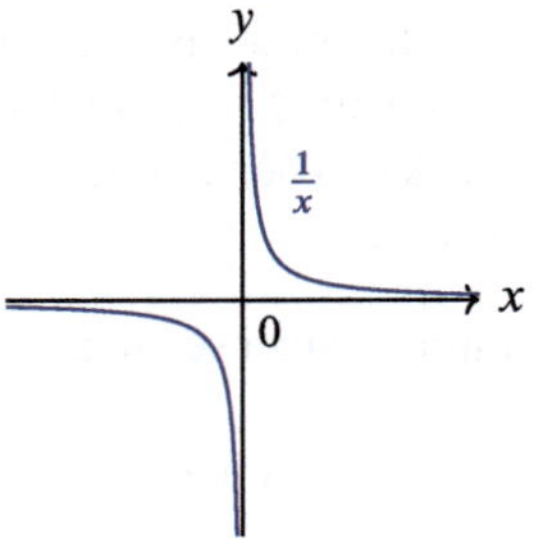

(b) discontinuous at $x = 0$

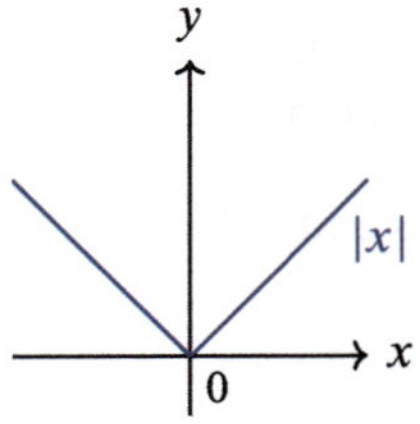

(c) a piecewise function

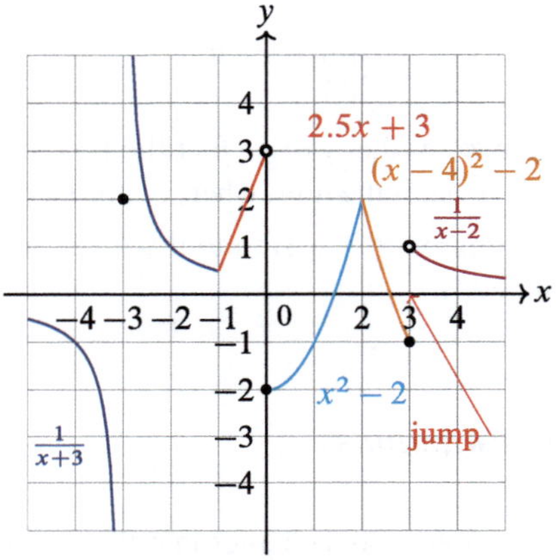

(d) a piecewise functions with discontinuities

Fig. 4.62 Graph of continuous and discontinuous functions

function is definitely continuous everywhere. Understandably, mathematicians call such functions piecewise functions.

The piecewise function in Fig. 4.62d is more interesting. At $x = -3$, the function is defined: $f(-3) = 2$ as indicated by the dot (or closed circle). Still, the function is discontinuous at $x = -3$. At $x = -1$ and $x = 2$, the function is continuous. However, $x = 0$ is special: at this point we can go with either of the two expressions and of course only one of them is chosen. Herein, $x^2 - 2$ is selected and to indicate that a dot is drawn at $(0, f(0))$ or $(0, -2)$ and a hollow (open) circle is drawn at $(0, 3)$:

$$f(x) = \begin{cases} 2.5x + 3, & \text{if } -1 \le x < 0 \\ x^2 - 2, & \text{if } 0 \le x \le 2 \end{cases}$$

As we had to lift off the pencil at $x = 0$, the function is discontinuous there (even though the function is well defined here). As we have a jump from 3 to -2 at $x = 0$, $x = 0$ is called a *jump discontinuity*. Similarly, the function is also discontinuous at $x = 3$.

A function is continuous if we can draw its graph without lifting our pencil off the paper. That's correct, but it is not a definition that mathematicians want. They need a definition that they can use, one with symbols, so that they can manipulate them. From Fig. 4.62 and the surrounding discussion we see that for a function to be continuous at $x = a$ it must be defined at that point; that is, $f(a)$ must exist. But this condition is not enough (jump discontinuity is one example). We also need that $\lim_{x\to a} f(x)$ exists.

Definition 4.8 A function $y = f(x)$ is continuous at point $x = a$ when the limit of $f(x)$ as x approaches a equals the function value at a:

$$\lim_{x\to a} f(x) = f(a) \tag{4.10.7}$$

With that definition of the continuity of a function at a single point, we have another definition. A function is continuous over an interval if it is continuous everywhere in that interval.

It is not hard to discover these rules for continuity of functions:

$$\begin{aligned}
&\text{(a: sum/diff rule) if } f(x) \text{ and } g(x) \text{ are continuous then } f \pm g \text{ is continuous}\\
&\text{(b: linearity rule) if } f(x) \text{ is continuous then cf is continuous}\\
&\text{(c: product rule) if } f(x) \text{ and } g(x) \text{ are continuous then } fg \text{ is continuous}\\
&\text{(d: quotient rule) if } f(x) \text{ and } g(x) \text{ are continuous then } f/g \text{ is continuous}
\end{aligned} \tag{4.10.8}$$

We skip the proof: it's a combination of the definition of continuity and the limit rules in Eq. (4.10.1). Now we're in a position to establish the continuity of many functions we know of.

We start with polynomials, those of the form

$$P(x) = a_n x^n + a_{n-1}x^{n-1} + \cdots + a_2x^2 + a_1x + a_0 = \sum_{i=0}^{n} a_i x^i \tag{4.10.9}$$

They are continuous everywhere. This is because each term $a_i x^i$ is continuous (this in turn is due to $y = x^n$ is continuous and cx^n is also continuous).

Next is rational functions $y = P(x)/Q(x)$; they are continuous due to the quotient rule in Eq. (4.10.8). Of course they're only continuous where $Q(x) \neq 0$. Then, trigonometry functions, logarithm functions, and exponential functions are all continuous.

How about composite functions e.g. $\sin(x^2)$ or e^{-x^2}? Our intuition tells us that they are continuous. We can confirm that by drawing them and see that their graphs are continuous (Fig. 4.63). Therefore, we have

$$\lim_{x\to 1} \sin(x^2) = \sin(1), \quad \lim_{x\to 1} e^{-x^2} = e^{-1}$$

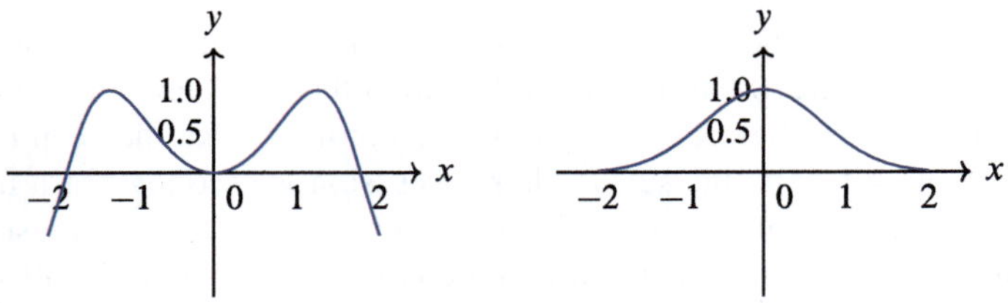

Fig. 4.63 Two composite functions $y = f(g(x))$: $\lim_{x \to 1} f(g(x)) = f(g(1))$

Theorem 4.1 *Limit of a composite functionlimit-compose-func Considering a composite function $y = f(g(x))$ with $\lim_{x \to a} g(x) = b$ and $f(x)$ is continuous at b, then:*

$$\lim_{x \to a} f(g(x)) = f(b) \tag{4.10.10}$$

We're now finally in a position ready to compute some interesting limits. For example,

$$\begin{aligned}
\lim_{t \to 0} \frac{\sqrt{t^2+9}-3}{t^2} &= \lim_{t \to 0} \frac{t^2}{t^2(\sqrt{t^2+9}+3)} = \lim_{t \to 0} \frac{1}{\sqrt{t^2+9}+3} && \text{(algebra)} \\
&= \frac{1}{\lim_{t \to 0}(\sqrt{t^2+9}+3)} && \text{(quotient rule with } f(x)=1) \\
&= \frac{1}{\lim_{t \to 0}(\sqrt{t^2+9})+3} && \text{(sum rule)} \\
&= \frac{1}{\sqrt{\lim_{t \to 0}(t^2+9)}+3} = \frac{1}{\sqrt{9}+3} = 1/6 && \text{(Eq. 4.10.10)}
\end{aligned} \tag{4.10.11}$$

where the first step is to convert the form $0/0$ to something better.

4.10.4 Indeterminate Forms

Let's recall that the derivative of a function $y = f(x)$ at $x = a$ is

$$f'(a) = \lim_{h \to 0} \frac{f(a+h) - f(a)}{h}$$

Even though now we know the quotient rule of limits, we cannot compute $f'(a)$ as

$$f'(a) = \lim_{h \to 0} \frac{f(a+h) - f(a)}{h} = \frac{\lim_{h \to 0} f(a+h) - f(a)}{\lim_{h \to 0} h}$$

Table 4.16 Indeterminate forms

Intermediate form	Conditions
$\frac{0}{0}$	$\lim_{x\to a} f(x) = 0,\ \lim_{x\to a} g(x) = 0$
$\frac{\infty}{\infty}$	$\lim_{x\to a} f(x) = \infty,\ \lim_{x\to a} g(x) = \infty$
$0\cdot\infty$	$\lim_{x\to a} f(x) = 0,\ \lim_{x\to a} g(x) = \infty$
$\infty-\infty$	$\lim_{x\to a} f(x) = \infty,\ \lim_{x\to a} g(x) = \infty$

because it is of the form $0/0$ which is not defined. Limit of the form $0/0$ is called an indeterminate form and we list other indeterminate forms in Table 4.16.

How to compute indeterminate forms then? The first method is to do algebraic manipulations to convert an indeterminate form to a normal form. We actually have used this method in Eq. (4.10.11). We give another example of the form ∞/∞: $\lim_{x\to\infty} {4x^2+x}/{2x^2+x}$. We convert the indeterminate form by dividing both the nominator and denominator by x^2—the highest power:

$$\lim_{x\to\infty}\frac{4x^2+x}{2x^2+x} = \lim_{x\to\infty}\frac{4+\dfrac{1}{x}}{2+\dfrac{1}{x}} = \frac{\lim_{x\to\infty} 4+\dfrac{1}{x}}{\lim_{x\to\infty} 2+\dfrac{1}{x}} = \frac{4}{2} = 2$$

Why we divided both the nominator and denominator by x^2? This is because we know that for a very large x, x is nothing (or negligible) compared with $4x^2$ and $2x^2$, so we can write (not mathematically precise but correct):

$$\lim_{x\to\infty}\frac{4x^2+x}{2x^2+x} = \lim_{x\to\infty}\frac{4x^2}{2x^2} = 2$$

So to say x is nothing is equivalent to convert it to the form $1/x$, and that's why we did the division of x^2. *And there is no value in doing more limits of this form*, as we can guess (note that generalization is a good thing to do) the following result for the ratio of any two polynomials:

$$\lim_{x\to\infty}\frac{P_n(x)}{Q_m(x)} = \begin{cases} 0 & \text{if } n < m \\ \infty & \text{if } n > m \\ \frac{a_n}{b_n} & \text{if } n = m \end{cases}$$

which is nothing but the fact that this limit depends on whether the nominator or denominator overtakes the other. If the denominator overtakes the nominator, the limit is zero.

L'Hopital's rule. The method of using algebra does not apply for this limit: $\lim_{x\to 0} {}^{\sin x}/_{x}$. To deal with this one we had to use geometry, but isn't it against the spirit of calculus? We need to find a mechanical way so that everyone is able to compute this limit and similar limits without resorting to geometry (which is always requiring some genius idea).

What do you think if you see someone doing this?

$$\lim_{x\to 0}\frac{\sin x}{x} = \lim_{x\to 0}\frac{\cos x}{1} = \frac{\lim_{x\to 0}\cos x}{1} = \frac{1}{1} = 1$$

First, the result is correct, and second it is purely algebra. What a magic! Could you guess the formula for this? It is the L'Hopital rule that states: if $f(a) = g(a) = 0$, then

$$\lim_{x\to a}\frac{f(x)}{g(x)} = \frac{f'(a)}{g'(a)} \tag{4.10.12}$$

Actually it is not hard to guess this rule. Recall that for x near a, we have the following approximations for $f(x)$ and $g(x)$:

$$f(x) \approx f'(a)(x-a), \quad g(x) \approx g'(a)(x-a)$$

Thus,

$$\lim_{x\to a}\frac{f(x)}{g(x)} = \lim_{x\to a}\frac{f'(a)(x-a)}{g'(a)(x-a)} = \frac{f'(a)}{g'(a)}$$

What is the limit of $^{x^n}/_{n!}$ when $n \to \infty$? Why bother with this? Because it is involved in the Taylor theorem (Sect. 4.15.10), which is a big thing. Let's start simple and concrete with $x = 2$:

$$\lim_{n\to\infty}\frac{2^n}{n!} = ?$$

A bit of algebraic manipulation goes a long way (of course we assume $n > 2$ as we're interested in the case n goes to infinity):

$$\frac{2^n}{n!} = \frac{2\times 2\times 2\times\cdots\times 2}{1\times 2\times 3\times\cdots\times n} = \frac{2}{1}\times\frac{2}{2}\times\frac{2}{3}\times\frac{2}{4}\times\cdots\times\frac{2}{n}$$

As the red terms are all smaller than one, we're multiplying a constant (the blue term) repeatedly with factors smaller than one, we guess that as n approaches infinity, the limit is zero. But, to be precise, we polish our expression a bit more:

$$\frac{2^n}{n!} = \underbrace{\frac{2}{1}\times\frac{2}{2}\times\frac{2}{3}\times\frac{2}{4}}_{\text{4 terms}}\times\underbrace{\frac{2}{5}\times\frac{2}{6}\times\cdots\times\frac{2}{n}}_{n-4\text{ terms}}$$

What is nice with this new form is that all terms in red are smaller than $1/2$, thus we immediately have

$$\frac{2^n}{n!} < \underbrace{\left(\frac{2}{1} \times \frac{2}{2} \times \frac{2}{3} \times \frac{2}{4}\right)}_{k} \left(\frac{1}{2}\right)^{n-4} = 2^4 k \frac{1}{2^n}$$

Now, it's obvious that the limit is zero:

$$\lim_{n\to\infty} \frac{2^n}{n!} < \lim_{n\to\infty} 2^4 k \frac{1}{2^n} = 2^4 k \lim_{n\to\infty} \frac{1}{2^n} = 2^4 k \frac{1}{\lim_{n\to\infty} 2^n} = 2^4 k \frac{1}{\infty} = 0$$

This proof holds for $x = 3, 4, \ldots$ or even negative integers of which the absolute is larger than one. But how about $x = 3.123$? We just see that

$$\frac{3.123^n}{n!} < \frac{4^n}{n!}$$

And then, we know that for all $x \in \mathbb{R}$, we have $\lim_{n\to\infty} x^n/n! = 0$.

4.10.5 Differentiable Functions

This section is about the differentiability of a function. Usually, we're given a function and asked to compute its derivative. We have done that a lot. And because of that we have the misunderstanding that any continuous function can be differentiated at any point on its domain. That's not true, and thus we have to define the differentiability concept.

Definition 4.9 A function $y = f(x)$ defined on an interval I is differentiable at point $x = a \in I$ if the derivative:

$$f'(a) = \lim_{h\to 0} \frac{f(a+h) - f(a)}{h}$$

exists. If x is an end point of I then the limit in this definition is replaced by an appropriate one-sided limit. The function $f(x)$ is differentiable on I if it is differentiable at each point of I.

If a function is differentiable at a point, it is continuous at that point:

$$\lim_{h\to 0} f(a+h) - f(a) = \left(\lim_{h\to 0} \frac{f(a+h) - f(a)}{h}\right)\left(\lim_{h\to 0} h\right) = f'(a) \times 0 = 0$$

However, if a function is continuous at a point, it might be that it is non-differentiable at that point. The simplest example is the function $y = |x|$, which is continuous

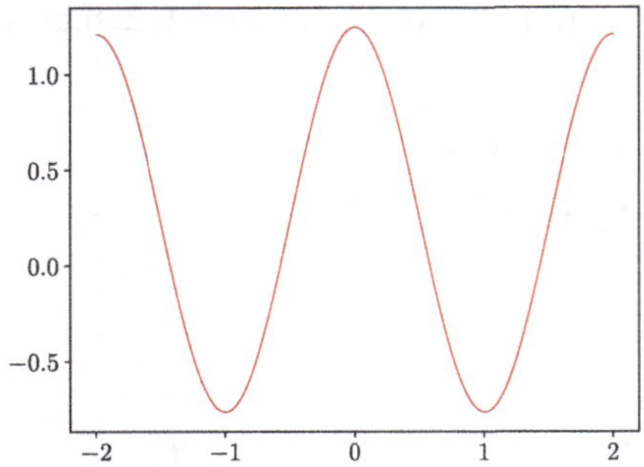

(a) $a = 0.2, b = 0.1, n = 3$

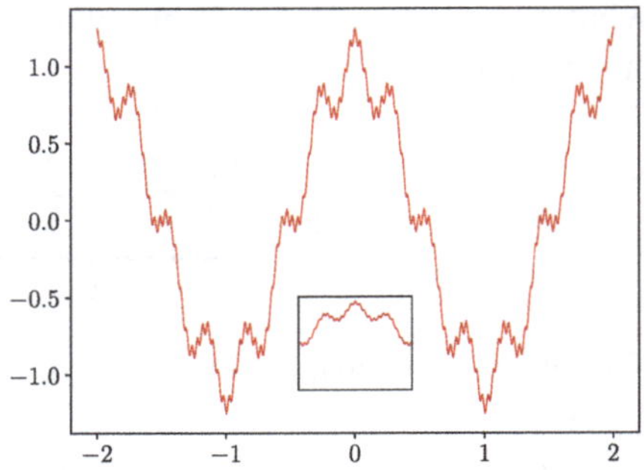

(b) $a = 0.2, b = 7, n = 3$

Fig. 4.64 Weierstrass function:

everywhere, but non-differentiable at $x = 0$. We can try to compute the derivative of $y = |x|$ at $x = 0$ to see this. Or we can think geometrically. At the corner $(0, 0)$, there does not exist a single well-defined tangent to the point.

The most famous example of a continuous function but not differentiable everywhere is the Weierstrass function defined as

$$f(x) = \sum_{n=0}^{\infty} a^n \cos(b^n \pi x), \quad a \in (0, 1) \tag{4.10.13}$$

Figure 4.64 gives the plots of two cases: (i) $a = 0.2, b = 0.1, n = 3$ and (ii) $a = 0.2, b = 7, n = 3$

Definition 4.10 A function $f : (a, b) \to \mathbb{R}$ is continuously differentiable on (a, b), written $f \in C^1(a, b)$, if it is differentiable on (a, b) and $f' : (a, b) \to \mathbb{R}$ is continuous.

For example, the function $y = x^2$ is a C^1 function for all x because it is differentiable everywhere and its first derivative is $2x$, a continuous function.

Definition 4.11 A function $f : (a, b) \to \mathbb{R}$ is said to be k-times continuously differentiable on (a, b), written $f \in C^k(a, b)$, if its derivatives of order j, where $0 \leq j \leq k$, exist and are continuous functions.

4.11 Some Theorems on Differentiable Functions

This section presents some commonly used theorems regarding continuous functions. They are: (1) Extreme value theorem, Fig. 4.65a; (2) Intermediate value theorem, Fig. 4.65b; (3) Rolle's theorem, Fig. 4.66a; (4) Mean value theorem, Fig. 4.66b; and (5) Mean value theorem for integrals.

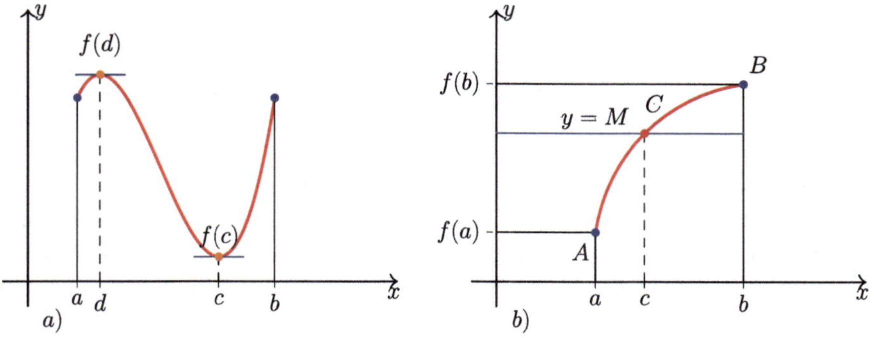

Fig. 4.65 Extreme value theorem and intermediate value theorem

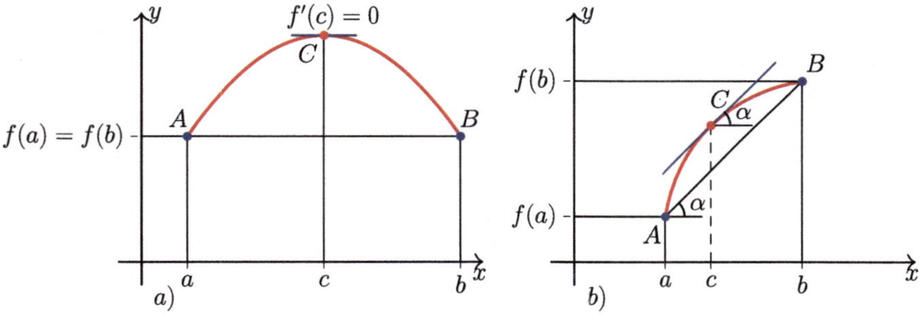

Fig. 4.66 Rolle's theorem and mean value theorem

4.11.1 Extreme Value and Intermediate Value Theorems

As can be seen the first two theorems stem from the property of continuity. The extreme value theorem states that on a closed interval $[a, b]$, a continuous function attains a maximum value at some point c and a minimum value at some point d. Figure 4.65a illustrates only the case where $c, d \in (a, b)$. The intermediate value theorem states that if we draw a horizontal line $y = M$ where $f(a) \leq M \leq f(b)$, then it will always intersect the continuous curve described by the function $y = f(x)$ at c and $a < c < b$. Note that the theorem does not tell us what c is. It just tells that there is such a point only.

Applications. As an application of the intermediate value theorem, let's consider this problem: "prove that the equation $x^3 + x - 1 = 0$ has solutions". Let's denote by $f(x) = x^3 + x - 1$, we then have $f(0) = -1$ and $f(1) = 1$. According to the intermediate value theorem, there exists a point $c \in (0, 1)$ such that $f(c) = 0$ because 0 is an intermediate value between $f(0) = -1$ and $f(1) = 1$.

4.11.2 *Rolle's Theorem and the Mean Value Theorem*

About Rolle's theorem (Fig. 4.66a), we consider a differentiable function $y = f(x)$ defined on a closed interval $[a, b]$ and $f(a) = f(b)$. In this figure, starting from $f(a)$ the function increases to a maximum and then decreases to $f(b) = f(a)$. Of course it attains a maximum at a certain point $c \in (a, b)$, and at c we have $f'(c) = 0$ (from maxima/minima problems). We have another case, where $f(x)$ first decreases to a minimum and then increases back to the starting level. And finally we also have functions that are increasing/decreasing multiple times within $[a, b]$, again the theorem is still true.

We can use Rolle's theorem for this kind of problem "prove that the equation $x^3 + x - 1 = 0$ has exactly one real solution". We use proof by contradiction by assuming that this equation has two roots a and b. That means we have $f(a) = f(b) = 0$. And since $f(x)$ is continuous, according to Rolle's theorem there exists c such that $f'(c) = 0$. But this is impossible because $f'(x) = 3x^2 + 1 > 0$ for all x.[42]

The most important application of Rolle's theorem is to prove the mean value theorem. Rolle's theorem has one restriction that $f(a) = f(b)$. But if we rotate Fig. 4.66a a bit counterclockwise we get Fig. 4.66b, which is the mean value theorem:

$$\exists c \in (a, b) \text{ s.t } f'(c) = \frac{f(b) - f(a)}{b - a} \quad \text{or } f(b) - f(a) = f'(c)(b - a) \tag{4.11.1}$$

This theorem was formulated by the Italian mathematician and astronomer Joseph-Louis Lagrange (1736–1813).

From a geometry point of view, Fig. 4.66b shows that there exists a point $c \in (a, b)$ such that the tangent at c has the same slope as the slope of the secant AB. Let's turn the attention to motion, and consider a continuous motion $s = f(t)$, then the average speed within an interval $[a, b]$ is $f(b)-f(a)/b-a$. The mean value theorem then indicates that there is a time instant t_0 during the interval that the instantaneous speed is equal to the average (or mean) speed.

Proof (*Proof of the mean value theorem*) We need to construct a function $y = g(x)$ such that $g(a) = g(b)$, then Rolle's theorem tells us that there exists a point $c \in (a, b)$ such that $g'(c) = 0$, and that leads to the mean value theorem. So, we must have

$$g'(c) = 0 \Longrightarrow f'(c)(b - a) - (f(b) - f(a)) = 0$$

From that, we know $g'(x)$, and then $g(x)$

$$g'(x) = f'(x)(b - a) - (f(b) - f(a)) \Longrightarrow g(x) = f(x)(b - a) - (f(b) - f(a))x$$

---

[42] Can you think of a function with $f(a) = f(b)$ but there is no $f'(c) = 0$?.

The proof is then as follows. Build the following $g(x)$

$$g(x) = f(x) - \frac{f(b) - f(a)}{b - a}x$$

which is differentiable and $g(a) = g(b)$, thus there exists $c \in (a, b)$ so that $g'(c) = 0$. And that leads to the mean value theorem.

Michel Rolle (1652–1719) was a French mathematician. He is best known for Rolle's theorem. Rolle, the son of a shopkeeper, received only an elementary education. In spite of his minimal education, Rolle studied algebra and Diophantine analysis (a branch of number theory) on his own. Rolle's fortune changed dramatically in 1682 when he published an elegant solution to a difficult, unsolved problem in Diophantine analysis. In 1685 he joined the Académie des Sciences. Rolle was against calculus and ironically the theorem bearing his name is essential for basic proofs in calculus. Among his several achievements, Rolle helped advance the currently accepted size order for negative numbers. Descartes, for example, viewed -2 as smaller than -5. Rolle preceded most of his contemporaries by adopting the current convention in 1691. Rolle's 1691 proof covered only the case of polynomial functions. His proof did not use the methods of differential calculus, which at that point in his life he considered to be fallacious. The theorem was first proved by Cauchy in 1823 as a corollary of a proof of the mean value theorem. The name "Rolle's theorem" was first used by Moritz Wilhelm Drobisch of Germany in 1834 and by Giusto Bellavitis of Italy in 1846.

Analysis of fixed point iterations. In Sect. 2.11 we have seen the fixed point iteration method as a means to solve equations written in the form $x = f(x)$. In the method, we generate a sequence starting from x_0: $(x_n) = \{x_1, x_2, \dots, x_n\}$ using the formula $x_{n+1} = f(x_n)$. We have demonstrated that these numbers converge to x^* which is the solution of the equation. Now, we're going to prove this using the mean value theorem. The whole point of the proof is that if the method works, then the distance from the points $x_1, x_2, \dots$ to x^* must decrease. So, we compute one such distance $x_n - x^*$:

$$x_n - x^* = f(x_{n-1}) - f(x^*) = f'(\xi)(x_{n-1} - x^*), \quad \xi \in [x_n, x^*]$$

Now there are two cases. First, if $|f'(\xi)| \le 1$, then $|x_n - x^*| \le |x_{n-1} - x^*|$, that is, the distance between x_n and x^* is smaller than x_{n-1} and x^*. And that tells us that x_n converges to x^*. Thus, if we start close to x^* i.e., $x_0 \in I = [x^* - \alpha, x^* + \alpha]$, and the absolute of the derivative of the function is smaller than 1 in that interval I, the method works. Table 4.13 discusses more on this topic.

4.11.3 *Average of a Function and the Mean Value Theorem of Integrals*

Let us recall that for n numbers $a_1, a_2, \ldots, a_n$, the ordinary average number is defined as $(a_1+a_1+\cdots+a_n)/n$. But what is the average of all real numbers within [0, 1]? There are infinite numbers living in that interval! Don't worry, integral calculus is capable of handling just that. Finding an answer to that question led to the concept of the average of a function.

The idea is to use integration. Assume we want to find the average of a function $f(x)$ for $a \le x \le b$. We divide the interval $[a, b]$ into n equal sub-intervals of spacing $\Delta x = (b-a)/n$. For each interval we locate a point x_i, so we have

$$\begin{aligned} f_{\text{average}} &= \frac{f(x_1) + f(x_2) + \cdots + f(x_n)}{n} \\ &= \frac{f(x_1)\Delta x + f(x_2)\Delta x + \cdots + f(x_n)\Delta x}{n\Delta x} \quad \text{(multiplying deno/nominator by } \Delta x) \\ &= \frac{\sum f(x_i)\Delta x}{b-a} = \frac{1}{b-a}\int_a^b f(x)dx \end{aligned} \tag{4.11.2}$$

In the final step, we get an integral when n goes to infinity. So, the average of a continuous function is its area divided by $b-a$, which is the *average height of the function.*

Example. Let's compute the average of these functions: $y = x$ in [0, 1], $y = x^2$ in [−1, 1] and $y = \sin^2 x$ in $[0, \pi]$. They are given by

$$f_{\text{average}} = \int_0^1 x dx = \frac{1}{2}, \quad f_{\text{average}} = \frac{1}{2}\int_{-1}^1 x^2 dx = \frac{1}{3}, \quad f_{\text{average}} = \frac{1}{\pi}\int_0^\pi \sin^2 x dx = \frac{1}{2}$$

Looking at Fig. 4.67, it is obvious to see that there always exists a point c in $[a, b]$ such that $f(c)$ is the average height of the function (the horizontal line $y = f_{\text{average}}$ always intersects the curve $y = f(x)$). And this is the mean value theorem of an integral:

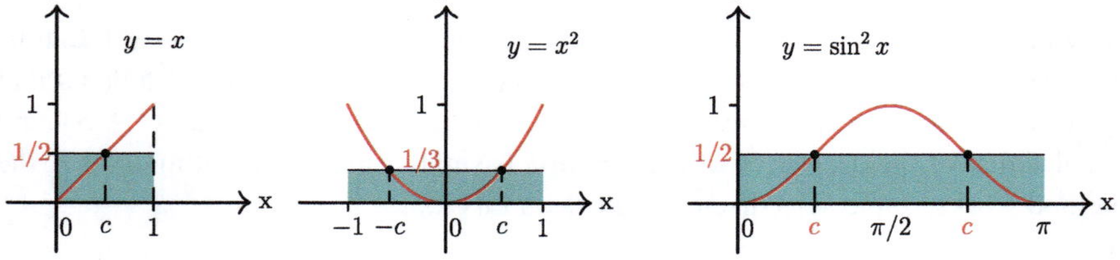

Fig. 4.67 Average of functions: $y = x$ in [0, 1], $y = x^2$ in [−1, 1] and $y = \sin^2 x$ in $[0, \pi]$

$$\exists c \in (a,b) \text{ s.t } f(c) = \frac{1}{b-a}\int_a^b f(x)dx \tag{4.11.3}$$

But some examples do not make a proof. Think about whether we have $f(a) \leq f(c) \leq f(b)$ or not. And what is $f(b)(b-a)$? How it is compared with $\int_a^b f(x)dx$. That's the proof.

For $y = x^2$ we have $c = \pm 1/\sqrt{3}$. They are Gauss points in the Gauss quadrature method to numerically evaluate integrals, see Sect. 10.4 for details.

4.12 Parametric Curves

In Sect. 4.2.6 we have met parametric curves—those which are described by $(x(t), y(t))$ where t is called a parameter. We can think of a curve lying on the plane and a particle is moving on this curve, at time t the position of the particle is given by $(x(t), y(t))$. An equation $y = f(x)$ tells the shape of the path, but not the speed of it.

This section discusses the calculus of parametric curves: tangent to a parametric curve (Sect. 4.12.1), how to calculate its length and area (Sect. 4.12.2). Finally, a presentation of the cycloid—a fascinating curve with many interesting properties—is given (Sect. 4.12.3).

4.12.1 *Tangents*

The slope of the tangent to the curve $y = f(x)$ at point $(x_0, f(x_0))$ is the first derivative of f at x_0 i.e., $f'(x_0)$. We now want to find the slope of the tangent to a parametric curve without having to eliminate the parameter (to convert to the familiar $y = f(x)$). There are many ways to get this slope. First, we can use Leibniz notation to write dy/dx as

$$\frac{dy}{dx} = \frac{(dy/dt)}{(dx/dt)} \tag{4.12.1}$$

4.12.2 *Length and Area of Parametric Curves*

I present two methods to calculate the arc length of a parametric curve. In the first method, we start with the arc length formula of $y = f(x)$ that we know:

$$L = \int_a^b \sqrt{1 + \left(\frac{dy}{dx}\right)^2}\, dx \tag{4.12.2}$$

Now, we replace ${}^{dy}/_{dx}$ and dx by

$$\frac{dy}{dx} = \frac{y'(t)dt}{x'(t)dt} = \frac{y'(t)}{x'(t)}, \quad dx = x'(t)dt$$

Therefore, Eq. (4.12.2) becomes

$$L = \int_{t_1}^{t_2} \sqrt{1 + \left(\frac{y'(t)}{x'(t)}\right)^2} x'(t)dt = \int_{t_1}^{t_2} \sqrt{[x'(t)]^2 + [y'(t)]^2} dt \tag{4.12.3}$$

That's the formula for the arc length of a parametric curve.

In the second way, we start with $\Delta s = \sqrt{\Delta x^2 + \Delta y^2}$, and $\Delta x \approx x'(t)\Delta t$ and $\Delta y \approx y'(t)\Delta t$, hence

$$\Delta s = \sqrt{[x'(t)]^2 + [y'(t)]^2} \Delta t$$

To calculate the area of a parametric curve, we start from the area of $y = f(x)$:

$$A = \int_a^b y(x)dx$$

But $x = x(t)$, hence $dx = x'(t)dt$. So, the area formula is

$$A = \int_{t_1}^{t_2} y(t)x'(t)dt \tag{4.12.4}$$

If $y(t) > 0$ for $t \in [t_1, t_2]$ (i.e., the curve is above the $x-$axis), the area must be positive. But that depends on the sign of $x'(t)$. If $x'(t) > 0$, we traverse the curve from t_1 to t_2, and if $x'(t) < 0$, the curve is traversed from t_2 to t_1. This is the case of an ellipse $x = a \cos t$, $y = b \sin t$. In that case, we need to put a minus sign before the integral in Eq. (4.12.4).

Arc-length of ellipse. We consider again the perimeter of $1/4$ of the ellipse $y^2 + 2x^2 = 2$. Using Eq. (4.12.3), we do

$$\left.\begin{aligned} x &= \cos t \\ y &= \sqrt{2} \sin t \end{aligned}\right\} \Longrightarrow \left\{\begin{aligned} dx/dt &= -\sin t \\ dy/dt &= \sqrt{2} \cos t \end{aligned}\right. \Longrightarrow \int_0^{\pi/2} \sqrt{\sin^2 t + 2\cos^2 t} dt$$

Of course we cannot find an anti-derivative for this integral. Compared to Eq. (4.9.1), this one is better as the integrand does not blow up at the integration limits. Using any numerical quadrature method, we can evaluate this integral easily. This is how an applied mathematician, engineer, or scientist would approach the problem. If they cannot find the answer exactly, they adopt numerical methods. But pure mathematicians do not do that. They will invent new mathematics to deal with integrals that cannot be solved using existing (elementary) functions. Recall that they invented negative integers so that we can solve for $5 + x = 2$, and $i^2 = -1$, and so on.

4.12.3 Cycloid

In geometry, a cycloid is the curve traced by a point on a circle as it rolls along a straight line without slipping. The cycloid has been called "The Helen of Geometers" as it caused frequent quarrels among seventeenth century mathematicians. Galileo, Descartes, Pascal, Fermat, Roberval, Newton, Leibniz, and the Bernoullis, as well as the architect, Christopher Wren, all wrote on various aspects of the cycloid. This section discusses this fascinating curve.

It is possible to mark a point on a bicycle wheel and observe this point closely when someone rides the bike.[43] We cannot do this on paper, so we have to follow a different way. We draw a horizontal line and a circle of radius a. We mark three points: the center O, point A that touches the line and point B such that the arc AB is half the circle (see the leftmost picture in Fig. 4.68). After the circle has rolled for a while point A will be at the highest point on the circle (middle picture) we have the distance $AB = \pi a$ because it is exactly the length of the arc AB (think of straightening the arc AB into the segment AB). Finally, point A will be at the lowest point on the circle again (right picture). So, we know at least three points on the cycloid (red dots) for one period.

Referring to Fig. 4.69, the coordinates of A on the cycloid are (computed from the coordinates of the center)

$$x = a\theta - a\sin\theta, \quad y = a - a\cos\theta \tag{4.12.5}$$

where the derivation was for $\theta \in [0, \pi/2]$ but you can check that it works for any angle. Referring Fig. 4.68 we can see that one arch of the cycloid comes from one full rotation of the circle i.e., $\theta \in [0, 2\pi]$.

Now, we turn to the tangent to the cycloid. The slope of the tangent is Eq. (4.12.1):

$$\frac{dy}{dx} = \frac{\sin\theta}{1 - \cos\theta}$$

From that we can determine points at which the tangent is horizontal: $\sin\theta = 0$ and $1 - \cos\theta \neq 0$; the solutions are $\theta = (2n - 1)\pi$ (that is $\pi, 3\pi, \ldots$). A plot of the cycloid and its tangents is given in Fig. 4.70. How about the tangent at $\theta = 0$? As dy/dx has the form $0/0$ we use the L'Hopital rule:

[43] Blaise Pascal thought that the curve that he saw most in daily life was the cycloid, second only to the circle. Perhaps the large and slowly moving carriage wheels of the seventeenth century were more easily observed than those of our modern automobile.

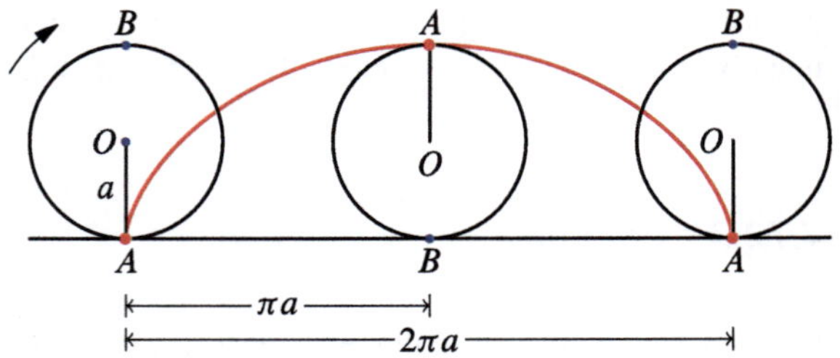

Fig. 4.68 A circle of radius a rolls on a horizontal line. The marked point A traces out a curve called a cycloid. Noting that the circle is traveling on a horizontal line

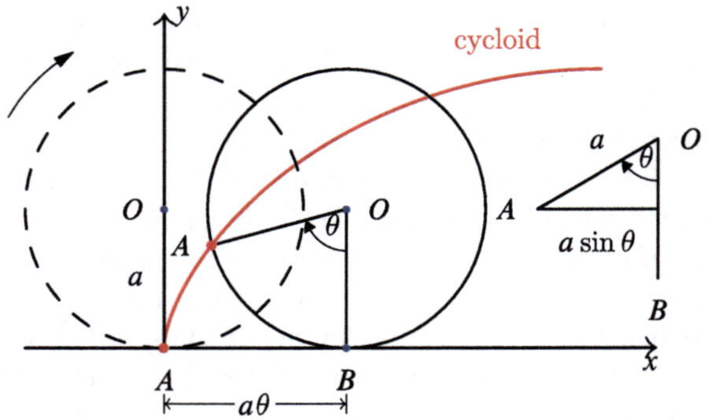

Fig. 4.69 Derivation of the parametric equation of the cycloid. The dashed circle is the starting circle. A Cartesian coordinate system is chosen with the origin at the lowest point A. The coordinates of the center O are $(a\theta, a)$

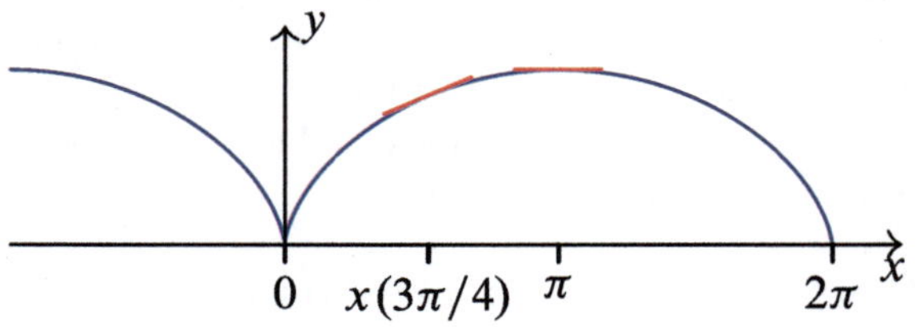

Fig. 4.70 Cycloid with $a = 1$: tangents at $3\pi/4$ and π (slope=0)

$$\lim_{x\to 0^+} \frac{dy}{dx} = \lim_{x\to 0^+} \frac{\sin\theta}{1-\cos\theta} = \lim_{x\to 0^+} \frac{\cos\theta}{\sin\theta} = \frac{1}{0} = \infty$$

In the same manner, $\lim_{x\to 0^-} \frac{dy}{dx} = -\infty$. Thus, there is a vertical tangent at $\theta = 0$ or $\theta = 2n\pi$ where the cycloid touches the $x-$ axis, as we have seen from its graph. These points are called the cusps of the cycloid.

Length and area of the cycloid. Using Eq. (4.12.3) we can compute the length of one arch of the cycloid:

$$x'(\theta) = a(1-\cos\theta), \quad y'(\theta) = a\sin\theta \Longrightarrow ds = a\sqrt{2(1-\cos\theta)}d\theta$$

Therefore, that length is

$$L = \int_0^{2\pi} a\sqrt{2(1-\cos\theta)}d\theta = 2a\int_0^{2\pi} \sin\frac{\theta}{2}d\theta = 8a$$

So, the length of one arch of the cycloid is eight times the radius of the circle which generates the cycloid.

Using Eq. (4.12.4) we can determine the area of one arch of the cycloid: it is $3\pi a^2$; which is three times the area of the circle which generates the cycloid.[44]

The cycloid is the solution to the Brachistochrone problem. Suppose a particle is allowed to slide freely and frictionlessly along a wire under gravity from a point A to point B. Furthermore, assume that the beads start with a zero velocity. Find the curve $y = y(x)$ that minimizes the travel time. Such a curve is called a brachistochrone curve (from Ancient Greek brákhistos khrónos 'shortest time'). And the cycloid is that curve as shown in Sect. 8.4.2. Herein, we compute that minimum time.

I refer to Sect. 8.2 for detail:

$$T = \int dt = \int \frac{ds}{v} = \int \frac{ds}{\sqrt{2gy}} = \int_0^\pi \frac{a\sqrt{2(1-\cos\theta)}d\theta}{\sqrt{2ga(1-\cos\theta)}} = \pi\sqrt{\frac{a}{g}}$$

Thus, the time is equal to π times the square root of the radius (of the circle which generates the cycloid) over the acceleration of gravity.

The cycloid is a tautochrone curve. A tautochrone or isochrone curve (from Greek prefixes tauto- meaning same or iso- equal, and chrono time) is the curve for which the time taken by an object sliding without friction in uniform gravity to its lowest point is independent of its starting point on the curve. Interestingly, the cycloid is such a curve. And we need to prove this.

The time is (similar to the case where the particle is starting at the beginning of the cycle, except now that it begins at a point with ordinate y_0 and the corresponding parameter is θ_0):

$$T = \int \frac{ds}{\sqrt{2g(y-y_0)}} = \int_{\theta_0}^\pi \frac{a\sqrt{2(1-\cos\theta)}d\theta}{\sqrt{2ga(\cos\theta_0-\cos\theta)}} = \sqrt{\frac{a}{g}}\int_{\theta_0}^\pi \frac{\sqrt{1-\cos\theta}d\theta}{\sqrt{\cos\theta_0-\cos\theta}}$$

The task is now to compute the above definite integral, we use the following trigonometric identities:

$$\sqrt{1-\cos\theta} = \sqrt{2}\sin\frac{\theta}{2}, \quad \cos\theta = 2\cos^2\frac{\theta}{2}$$

[44] Galileo approached this area problem empirically by cutting the shape out of a uniform sheet of material and weighing it. He found that the shape weighed the same as three circular plates of the same material cut with the radius of the wheel used to draw the curve. Galileo gave the name "cycloid" to the curve, although it has also been known as a "roulette" and a "trochoid". Historically, an equation for the cycloid was not written down until after its length and area has been discovered by geometry, engineering, and mechanics.

Hence, T becomes

$$T = \sqrt{\frac{a}{g}} \int_{\theta_0}^{\pi} \frac{\sin\frac{\theta}{2}}{\sqrt{\cos^2\frac{\theta_0}{2} - \cos^2\frac{\theta}{2}}} d\theta$$

With $u = \cos(\theta/2)$, we have

$$T = 2\sqrt{\frac{a}{g}} \int_{0}^{\cos\frac{\theta_0}{2}} \frac{du}{\sqrt{\cos^2\frac{\theta_0}{2} - u^2}} = \pi\sqrt{\frac{a}{g}}$$

It is clear T is independent of θ_0 or y_0: the time taken by an object sliding without friction in uniform gravity to its lowest point is independent of its starting point on the cycloid.

4.13 Polar Coordinates

Cartesian coordinates are not the only one way to specify points on a plane. This section presents polar coordinates (Sect. 4.13.1). Also discussed are conic sections written in terms of polar coordinates (Sect. 4.13.2). We have met the cardioid in Fig. 1.1, now is the time for us to derive its equation (Sect. 4.13.3). Finally, the length and area of polar curves are presented (Sect. 4.13.4).

4.13.1 Polar Coordinates and Polar Graphs

The polar coordinate system is a two-dimensional coordinate system in which points are given by an angle and a distance from a central point known as the pole (equivalent to the origin in the more familiar Cartesian coordinate system), cf. Figure 4.71. The polar coordinate system is used in many fields, including mathematics, physics, engineering, navigation and robotics. It is especially useful in situations where the relationship between two points is most easily expressed in terms of angles and distance. For instance, let's consider a unit circle centered at the origin: it is described by two equations $y = \pm\sqrt{1 - x^2}$ in Cartesian coordinates (plus sign for the upper half), but simply $r = \cos\theta$ in polar coordinates.

The full history of polar coordinates is described in *Origin of Polar Coordinates* of the American mathematician and historian Julian Lowell Coolidge (1873–1954). A brief history is given here. The Flemish mathematician Grégoire de Saint-Vincent (1584–1667) and Italian mathematician Bonaventura Cavalieri (1598–1647) independently introduced the concepts at about the same time. In *Acta eruditorum* (1691), Jacob Bernoulli used a system with a point on a line, called the pole and polar axis, respectively. Coordinates were specified by the distance from the pole and the angle

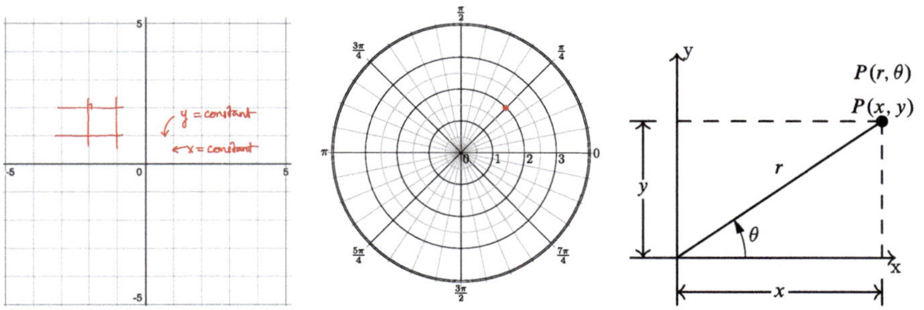

(a) Cartesian coordinates (b) Polar coordinates (c) Relation

Fig. 4.71 Cartesian and polar coordinates: $x = r\cos\theta$ and $y = r\sin\theta$

from the polar axis. The actual term polar coordinates have been attributed to the Italian mathematician Gregorio Fontana (1735–1803). The term appeared in English in George Peacock's 1816 translation[45] of Lacroix's *Differential and Integral Calculus.*[46]

In the Cartesian coordinate system, we lay a grid consisting of horizontal and vertical lines that are at right angles. Two lines are special as their intersection marks the origin from which other points are located. In a polar coordinate system, we also have two axes with an origin. Concentric circles centered at the origin are used to mark constant distances r from the origin. Also, lines starting from the origin are drawn; every points on such a line has a constant angle θ. So, a point is marked by (r, θ) (in Fig. 4.71b, the marked point has a coordinate $(2, \pi/4)$). And of course, there is a relation between Cartesian coordinates (x, y) and polar coordinates (r, θ), see Fig. 4.71c.

Curves are described by equations of the form $y = f(x)$ in the Cartesian coordinate system. Similarly, polar curves are written as $r = f(\theta)$. Let's start with the unit circle. Using Cartesian coordinates, it is written as $x^2 + y^2 = 1$. Using polar coordinates, it is simply as $r = 1$! Fig. 4.72 presents a nice polar curve—a polar rose with as many petals as we want, and a more realistic rose.

What do you think of Fig. 4.73? It is a spiral, from prime numbers! It was created by plotting points $(r, \theta) = (p, p)$, where p is prime numbers beneath 20 000. That is the radius and angle (in radians) are both prime numbers.

---

[45] George Peacock (1791–1858) was an English mathematician and Anglican cleric. He founded what has been called the British algebra of logic.

[46] Sylvestre François Lacroix (1765–1843) was a French mathematician. Lacroix was the writer of important textbooks in mathematics and through these, he made a major contribution to the teaching of mathematics throughout France and also in other countries. He published a two-volume text Traité de calcul differéntiel et du calcul intégral (1797–1798) which is perhaps his most famous work.

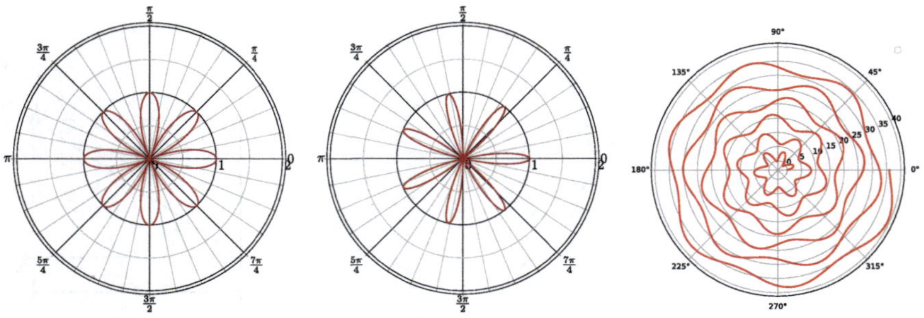

(a) $k = 4$ (b) $k = 7$ (c) real rose

Fig. 4.72 Polar rose $r(\theta) = a \cos k\theta$ with $a = 1$. It is a k-petaled rose if k is odd, or a $2k$-petaled rose if k is even. The variable a represents the length of the petals of the rose. In **c** is a more real rose with $r = \theta + 2\sin(2\pi\theta)$

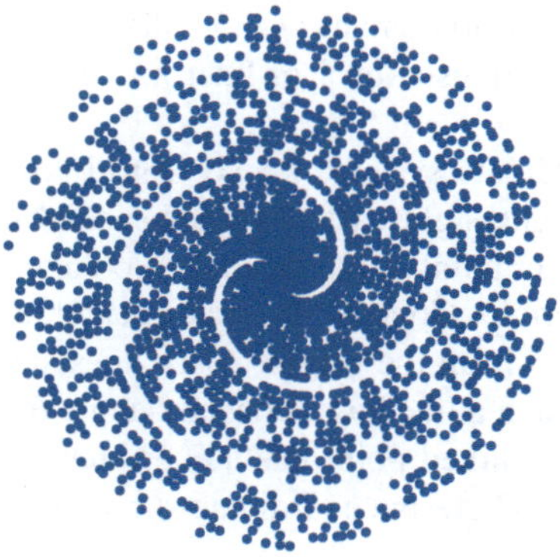

Fig. 4.73 Prime numbers from 1 to 20000 plotted on a polar plane. Generated using `Julia` package `Primes`: the function `primes(n)` returns all primes from 1 to n. Then for every number p in that list, I computed the coordinates $(p \cos p, p \sin p)$. Finally, I plot all these points. Source code: `prime-spiral.jl`

4.13.2 Conic Sections in Polar Coordinates

Herein we derive the equation for conic sections in polar coordinates where the origin is one of the two foci. When using Cartesian coordinates, we define the parabola using the focus and the directrix whereas the ellipse and hyperbola are defined using the distance to the two foci. With polar coordinates, we can define conic sections in a unified way using the focus and the directrix.

Considering Fig. 4.74 where F—the focus—is at the origin, the directrix is the line parallel to the $y-$axis and at a distance d from F. Let's denote by e the eccentricity and by P a point in the conic section with coordinates $(x, y) = (r\cos\theta, r\sin\theta)$, then a conic section is defined as $PF/PE = e$, which leads to the equation:

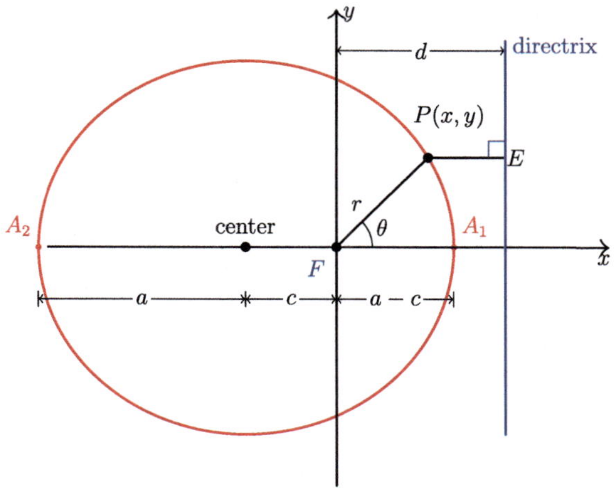

Fig. 4.74 A conic section is defined as $PF/PE = e$. Illustrated with an ellipse in mind

$$r = e(d - r\cos\theta) \Longrightarrow r = \frac{ed}{1 + e\cos\theta} \tag{4.13.1}$$

You might be not convinced that this equation is a conic section. We can check that by either using software to draw this equation and see what we get or we can transform this back to a Cartesian form (which we already know the result). We do the latter now. Why bother doing all of this? This is because, for certain problems, polar coordinates are more convenient to work with than Cartesian coordinates. Later on, we shall use the result in this section to prove Kepler's first law that the orbit of a planet around the Sun is an ellipse (Sect. 6.10.9).

From Eq. (4.13.1), we have $r = e(d - r\cos\theta) = e(d - x)$ for $x = r\cos\theta$, now we square this equation and use $r^2 = x^2 + y^2$, we get

$$r = e(d - x) \Longrightarrow x^2 + y^2 = e^2(d - x)^2 = e^2(d^2 - 2dx + x^2)$$

And a bit of massage to it, we obtain

$$x^2 + \frac{2e^2d}{1 - e^2}x + \frac{y^2}{1 - e^2} = \frac{e^2d^2}{1 - e^2}$$

Knowing already the Cartesian form of an ellipse ($(x/a)^2 + (y/b)^2 = 1$), we now complete the square for x[47]:

$$\begin{aligned}
\left(x + \frac{e^2d}{1 - e^2}\right)^2 + \frac{y^2}{1 - e^2} &= \frac{e^2d^2}{1 - e^2} + \frac{e^4d^2}{(1 - e^2)^2} \quad \text{(complete the square)}\\
\left(x + \frac{e^2d}{1 - e^2}\right)^2 + \frac{y^2}{1 - e^2} &= \frac{e^2d^2}{(1 - e^2)^2} \quad \text{(algebra)}
\end{aligned}$$

[47] If we just learnt by heart the quadratic equation we would forget how to complete a square!

The next step is, of course, to introduce a and b, and h (now we need $e < 1$)

$$a^2 = \frac{e^2d^2}{(1-e^2)^2}, \quad b^2 = \frac{e^2d^2}{1-e^2}, \quad h = \frac{e^2d}{1-e^2} \tag{4.13.2}$$

With these new symbols, our equation becomes, which is the familiar ellipse:

$$\frac{(x+h)^2}{a^2} + \frac{y^2}{b^2} = 1$$

But what is h? You might be guessing correctly that it should be related to c. Indeed, we know that, from Sect. 4.1, the distance from the center of an ellipse to one focus is c and it is defined by $c^2 + b^2 = a^2$, thus

$$c^2 = a^2 - b^2 = \frac{e^2d^2}{(1-e^2)^2} - \frac{e^2d^2}{1-e^2} = \frac{e^4d^2}{(1-e^2)^2} = h^2 \quad \text{[use Eq. (4.13.2)]}$$

The fact that $c = h$ should not be a surprise as we just move the origin from the center of the ellipse to its focus, these two points are separated by a distance c.

Theorem 4.2 *A polar equation of the form*

$$r = \frac{ed}{1 \pm e\cos\theta} \quad or \quad r = \frac{ed}{1 \pm e\sin\theta}$$

represents conic section with eccentricity e. The conic is an ellipse if $e < 1$, a parabola if $e = 1$, and a hyperbola if $e > 1$.

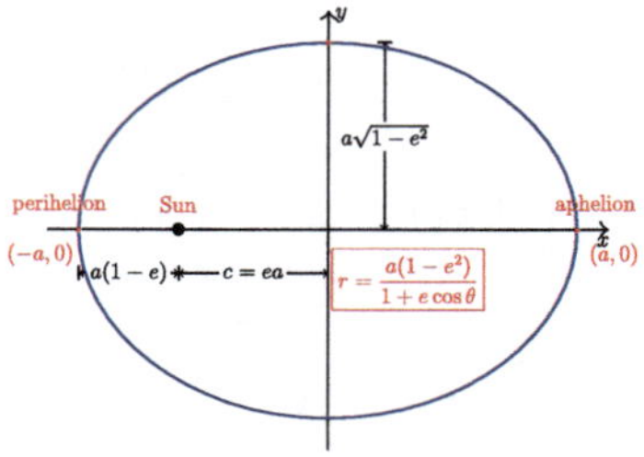

Ellipse in terms of major axis and eccentricity. It is possible to write the ellipse in terms of a—the semi-major axis and e. Using Fig. 4.74, and considering the points A_1 and A_2. Applying the definition of an ellipse to these two points we have

$$a - c = ed - ea + ec, \quad a + c = ed + ea + ec$$

Adding these two equations we get $ed = a - ec$, and substituting this into the first equation, we obtain $c = ea$. Thus, we can specify an ellipse using a and e. We can compute b from $b^2 = a^2 - c^2 = a^2(1 - e^2)$. As planetary orbits around the Sun are ellipses, we pay attention to ellipses and compute the distance from the center of the ellipse to one focus: $c = ea$. Then, we can determine the distance from the Sun to the *perihelion*—the point nearest to the Sun in the planet's orbit, it is $a(1 - e)$. Similarly $a(1 + e)$ is the distance from the Sun to the *aphelion*—the point in the orbit of a planet most distant from the Sun. We shall use these in Sect. 10.6.4 when we simulate the orbit of planets around the Sun.

4.13.3 Cardioid

Let's roll a circle of radius with a marked point around another circle of the same radius, this point traces a curve. What is the shape of this curve? Surprisingly it is of a heart shape, and it is called a cardioid, and a plot of this curve is given in Fig. 1.1. Now, we're going to derive the parametric equation of this fascinating curve.

The plan is to use analytic geometry, put two circles on a Cartesian plane, and consider a point on the rolling circle. Finding the coordinates of this point in terms of the rolling velocity gives us the parametric equation of the cardioid.

Referring to Fig. 4.75, one can determine the location of P' relative to $O_3(2a\cos\alpha, 2a\sin\alpha)$, and then to O_1:

$$\begin{bmatrix} x(\alpha) \\ y(\alpha) \end{bmatrix} = \begin{bmatrix} 2a\cos\alpha \\ 2a\sin\alpha \end{bmatrix} + \begin{bmatrix} \cos\alpha & -\sin\alpha \\ \sin\alpha & \cos\alpha \end{bmatrix} \begin{bmatrix} -a\cos\alpha \\ -a\sin\alpha \end{bmatrix} = \begin{bmatrix} a(2\cos\alpha - \cos 2\alpha) \\ a(2\sin\alpha - \sin 2\alpha) \end{bmatrix} \tag{4.13.3}$$

The matrix in the above is the rotation matrix from the red coordinate system to the blue one (see b). In Sect. 4.1.5 we have discussed rotation of axes (noting that the rotation direction herein is different from that in Sect. 4.1.5). Now, we just replace α by t, and we obtain the parametric equation of a cardioid.

Polar equation. It is more convenient to put the origin at P (see Fig. 4.75a), then just subtract from Eq. (4.13.3) for x an amount of a:

$$\begin{bmatrix} x(\alpha) \\ y(\alpha) \end{bmatrix} = 2a(1 - \cos\alpha) \begin{bmatrix} \cos\alpha \\ \sin\alpha \end{bmatrix} \implies r^2 = x^2 + y^2 = 4a^2(1 - \cos\alpha^2) \implies r = 2a(1 - \cos\alpha) \tag{4.13.4}$$

4.13.4 Length and Area of Polar Curves

Herein we revisit the old problem of calculation of arc length and area of planar curves, but the curves are now written in terms of polar coordinates. Certainly we shall have new formula. But do not worry, the principle is the same. For example, to

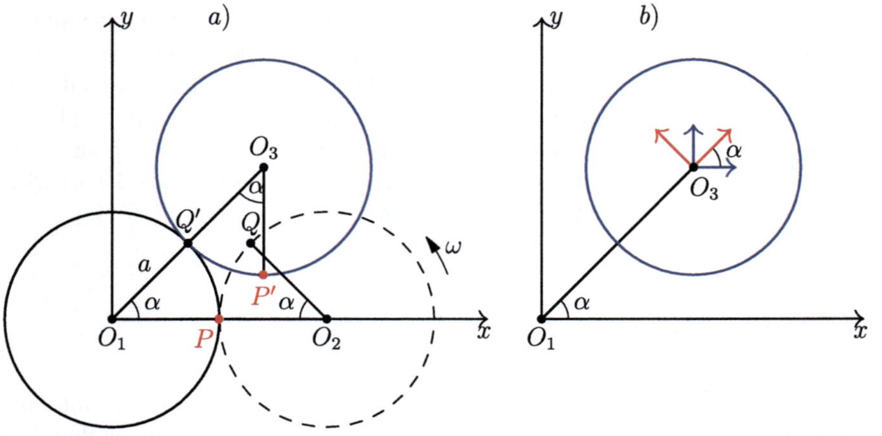

Fig. 4.75 Derivation of the cardioid parametric equation: the center of the fixed circle is $O_1(0, 0)$ and it has a radius of a. At the beginning, the center of the rolling (dashed) circle O_2 is at $(2a, 0)$. The point we pay attention to is P. After Δt, the dashed circle has moved to a new location with O_3 as the center. The segment O_1O_3 makes an angle $\alpha = \omega\Delta t$ with the x–axis, if ω is the rolling angular speed. The point P moves to P'. The key point: $PQ' = P'Q' = PQ$

compute the arc length of a curve, we compute a small portion of that length ds, and integrate that to get the total length.

The arc length problem is: given a polar curve $r = f(\theta)$ with $\theta \in [\theta_1, \theta_2]$, compute its length. We start with the familiar Cartesian coordinates where we write $ds = \sqrt{dx^2 + dy^2}$, and the arc length of a curve $y = f(x)$ is then given by, see Sect. 4.9.1

$$L = \int_1^2 ds, \quad ds = \sqrt{dx^2 + dy^2}$$

As we now work with polar coordinates, we need to convert (x, y) to (r, θ):

$$x = r\cos\theta, \quad y = r\sin\theta, \quad r = f(\theta) \tag{4.13.5}$$

And that allows us to compute dx, dy in terms of dr, $d\theta$ and noting that $dr = f'd\theta$:

$$\begin{aligned} dx &= \cos\theta dr - r\sin\theta d\theta = (\cos\theta f' - f(\theta)\sin\theta)d\theta \\ dy &= \sin\theta dr + r\cos\theta d\theta = (\sin\theta f' + f(\theta)\cos\theta)d\theta \end{aligned} \tag{4.13.6}$$

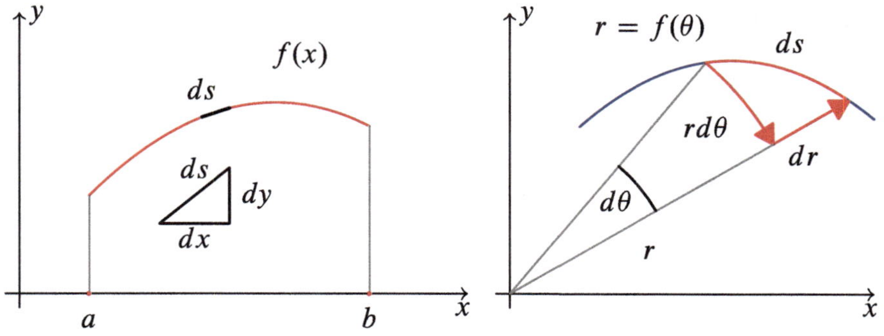

Fig. 4.76 Arc length of polar curves: the idea is exactly the same as for Cartesian curves. We consider two nearby points: one point at (r, θ) and the other at $(r + dr, \theta + d\theta)$. Then, using the Pythagorean theorem noting that as dr and $d\theta$ are infinitesimals, we have a right triangle with sides $dr, rd\theta$ and ds

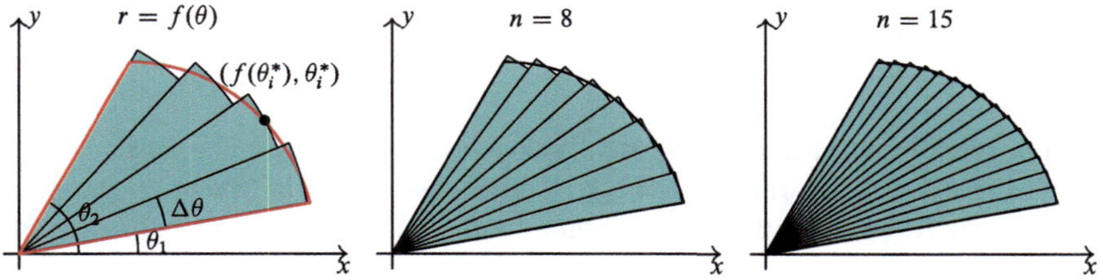

Fig. 4.77 Area of the region bounded by the polar curve $r = f(\theta)$ and by the rays $\theta = \theta_1$ and $\theta = \theta_2$ (the area bounded by the three red curves in the left figure)

And with that, we can now determine ds[48] and the arc length:

$$ds = \sqrt{(f')^2 + f^2}d\theta \Longrightarrow L = \int_{\theta_1}^{\theta_2} \sqrt{[f(\theta)]^2 + [f'(\theta)]^2}d\theta \tag{4.13.7}$$

That derivation is purely algebraic. Many people prefer geometry. With Cartesian coordinates, we consider a point $A(x, y)$ and a nearby point $B(x + dx, y + dy)$, then $ds = AB = \sqrt{dx^2 + dy^2}$ from the Pythagorean theorem. In the same manner, we consider a point $A(r, \theta)$ and a nearby point $B(r + dr, \theta + d\theta)$. Figure 4.76 shows that $ds^2 = (rd\theta)^2 + (dr)^2$, which is exactly what we have obtained using algebra. **The area problem** is: let R be the region bounded by the polar curve $r = f(\theta)$ and by the rays $\theta = \theta_1$ and $\theta = \theta_2$, and $0 \leq \theta_2 - \theta_1 \leq 2\pi$, see the left picture in Fig. 4.77. Compute the area of R. The idea is the same as that was used to determine the area of a circle: we chopped the circle into many many wedges (Fig. 3.34).

[48] Some details are omitted: using Eq. (4.13.6) in $ds^2 = dx^2 + dy^2$ and the identity $\sin^2\theta + \cos^2\theta = 1$.

We divide the angular interval $\theta_2 - \theta_1$ into n sub intervals of equal angle $\Delta\theta = {}^{(\theta_2-\theta_1)}/_n$. For sub-interval i, we consider a point θ_i^* inside it. In the left picture of Fig. 4.77, θ_i^* was chosen such as to bisect the angle $\Delta\theta$. Now, we build a sector of radius $r = f(\theta_i^*)$ and angle $\Delta\theta$, the area of this sector is easy to compute: $1/2[f(\theta_i^*)]^2\Delta\theta$. Therefore, the area that we're seeking is the sum of the area of all the sectors when n approaches infinity (see the middle and right pictures of Fig. 4.77):

$$A = \lim_{n\to\infty}\sum_{i=1}^{n}\frac{1}{2}[f(\theta_i^*)]^2\Delta\theta = \int_{\theta_1}^{\theta_2}\frac{1}{2}[f(\theta)]^2 d\theta$$

4.14 Bézier Curves: Fascinating Parametric Curves

Parametric curves are briefly mentioned in Sect. 4.2.6. Recall that a planar parametric curve is given by

$$C(t) : x(t), y(t), \quad a \leq t \leq b \tag{4.14.1}$$

In high school and in university, usually students are given a parametric curve in which the expressions for $x(t)$ and $y(t)$ are thrown at their faces e.g. the spiral $(t\cos t, t\sin t)$, and asked to do something with that: draw the curve for example. A boring task! The students thus missed a fascinating type of curves called Bézier curves.

Bézier curves are ubiquitous in computer graphics. For instance, one of the most common uses of Bézier curves is in the design of fonts. Cubic Bézier curves are used in Type 1 fonts, and quadratic Bézier curves are used in True Type fonts. Cubic Bézier curves are also used in the TEX fonts designed by Donald Knuth,[49] and one of the clearest explanations is in his book *MetaFont: the Program*. This section presents a brief introduction to these curves.

Starting with two points $\boldsymbol{P}_1$ and $\boldsymbol{P}_2$, the segment $\boldsymbol{P}_1\boldsymbol{P}_2$ is written as[50] (this is a vector equation, the bold symbols are used for the points)

$$\boldsymbol{P}_1\boldsymbol{P}_2 = (1-t)\boldsymbol{P}_1 + t\boldsymbol{P}_2, \quad t \in [0, 1] \tag{4.14.2}$$

This is neither interesting nor new. Do not worry it is just the beginning. If we now have three points $\boldsymbol{P}_1$, $\boldsymbol{P}_2$ and $\boldsymbol{P}_3$, we get a quadratic curve. de Casteljau developed

[49] Donald Ervin Knuth (born 10 January 1938) is an American computer scientist, mathematician, and professor emeritus at Stanford University. He is the 1974 recipient of the ACM Turing Award, informally considered the Nobel Prize of computer science. He has been called the *"father of the analysis of algorithms"*. He contributed to the development of the rigorous analysis of the computational complexity of algorithms. In addition to fundamental contributions in several branches of theoretical computer science, Knuth is the creator of the TEX computer typesetting system by which this book was typeset.

[50] If this is not clear, check Sect. 9.1.3.

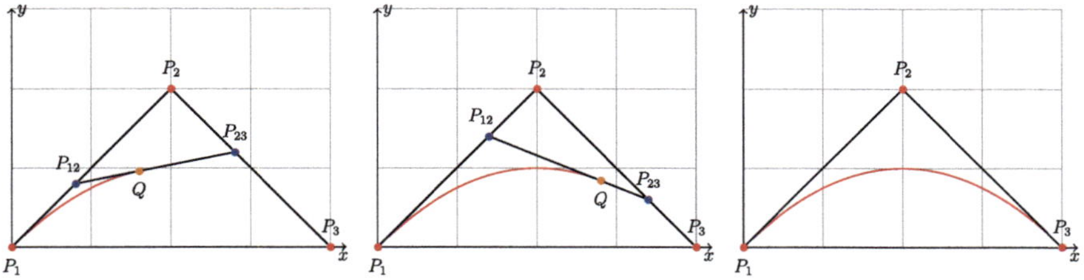

Fig. 4.78 A quadratic Bézier curve (red curve) determined by three control points P_1, P_2, P_3

a recursive algorithm to get that curve.[51] For a given t fixed, using Eq. (4.14.2) to determine two new points $\boldsymbol{P}_{12}$ and $\boldsymbol{P}_{23}$, then using Eq. (4.14.2) again with the two new points to get $\boldsymbol{Q}$ (Fig. 4.78a). When t varies from 0 to 1, this point $\boldsymbol{Q}$ traces a quadratic curve passing $\boldsymbol{P}_1$ and $\boldsymbol{P}_2$ (Fig. 4.78c). The points $\boldsymbol{P}_k$, $k = 1, 2, 3$ are called the control points. They are so-called because the control points control the shape of the curve.

Indeed, the maths gives us:

$$\begin{aligned}\boldsymbol{Q} &= (1-t)\boldsymbol{P}_{12} + t\boldsymbol{P}_{23} \\ &= (1-t)[(1-t)\boldsymbol{P}_1 + t\boldsymbol{P}_2] + t[(1-t)\boldsymbol{P}_2 + t\boldsymbol{P}_3] \quad \text{(Eq. (4.14.2))} \\ &= (1-t)^2\boldsymbol{P}_1 + 2t(1-t)\boldsymbol{P}_2 + t^2\boldsymbol{P}_3\end{aligned} \tag{4.14.3}$$

What we see here is that the last equation is a linear combination of some polynomials (the red terms) and some constant coefficients being the control points.

Moving on to a cubic curve with four control points (Fig. 4.79). The procedure is the same, and the result is

$$\boldsymbol{Q} = (1-t)^3\boldsymbol{P}_1 + 3t(1-t)^2\boldsymbol{P}_2 + 3t^2(1-t)\boldsymbol{P}_3 + t^3\boldsymbol{P}_4 \tag{4.14.4}$$

Animation of the construction of Bézier curves helps the understanding. A coding exercise for people who likes coding is to write a small program to create Fig. 4.79. If you do not like coding, check out geogbra where you can drag and move the control points to see how the curve changes. And this allows *free form geometric modeling*.

To see the pattern (for the generalization to curves of higher orders), let's put the quadratic and cubic Bézier curves together:

$$\begin{aligned}\text{quadratic Bézier curve:} &= 1(1-t)^2\boldsymbol{P}_1 + 2t(1-t)\boldsymbol{P}_2 + 1t^2(1-t)^0\boldsymbol{P}_3 \\ \text{cubic Bézier curve:} &= 1(1-t)^3\boldsymbol{P}_1 + 3t(1-t)^2\boldsymbol{P}_2 + 3t^2(1-t)\boldsymbol{P}_3 + 1t^3(1-t)^0\boldsymbol{P}_4\end{aligned}$$

[51] Paul de Casteljau (born 19 November 1930) is a French physicist and mathematician. In 1959, while working at Citroën, he developed an algorithm for evaluating calculations on a certain family of curves, which would later be formalized and popularized by engineer Pierre Bézier, leading to the curves widely known as Bézier curves.

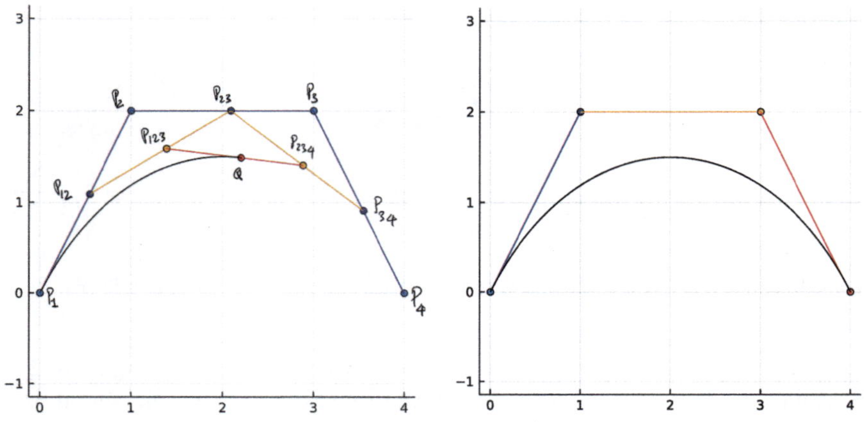

(a) (b)

Fig. 4.79 A cubic Bézier curve determined by four control points. Created using the file `bezier.jl`

And what are we seeing here? Pascal's triangle! And with that we can guess (correctly) that the expression for an n degree Bézier curve determined by $n+1$ control points $\boldsymbol{P}_k$ ($k = 0, 1, 2, \ldots, n$) is

$$B(t) = \sum_{k=0}^{n} \binom{n}{k} (1-t)^{n-k} t^k \boldsymbol{P}_k = \sum_{k=0}^{n} B_{k,n}(t) \boldsymbol{P}_k \tag{4.14.5}$$

where $B_{k,n}$ is the Bernstein basis polynomial,[52] given by

$$B_{k,n}(t) = \binom{n}{k} (1-t)^{n-k} t^k, \quad 0 \le t \le 1 \tag{4.14.6}$$

The Bernstein basis polynomial possess some nice properties: they are non-negative, their sum is one i.e., $\sum_{k=1}^{n} B_{k,n}(t) = 1$,[53] see Fig. 4.80a. Because of these two properties, the point $B(t)$, which is a weighted average of the control points, hence lies inside the convex hull of those points (Fig. 4.80c).

You might be asking: where are calculus stuff? Ok, let's differentiate the cubic curve $B(t)$ to see what we get

$$B'(0) = 3(\boldsymbol{P}_1 - \boldsymbol{P}_0), \quad B'(1) = 3(\boldsymbol{P}_3 - \boldsymbol{P}_2)$$

---

[52] Sergei Natanovich Bernstein (5 March 1880–26 October 1968) was a Soviet and Russian mathematician of Jewish origin known for contributions to partial differential equations, differential geometry, probability theory, and approximation theory.

[53] Why? The binomial theorem is the answer.

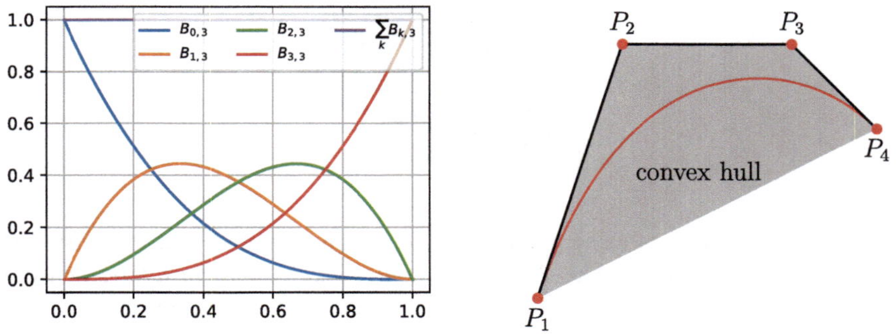

(a) Bernstein cubic polynomials (b) Convex hull property

Fig. 4.80 Bernstein cubic polynomials and convex hull property of Bézier curves

What is this equation telling us? It indicates that the tangent to the curve at $\boldsymbol{P}_1$ (or $t = 0$) is proportional to the line $\boldsymbol{P}_1\boldsymbol{P}_0$. And the tangent to the curve at $\boldsymbol{P}_2$ is proportional to the line $\boldsymbol{P}_3\boldsymbol{P}_2$. This should be not a surprise as we have actually seen this in Fig. 4.79b. Because of this, and the fact that the curve goes through the starting and ending points i.e., $B(0) = \boldsymbol{P}_0$ and $B(1) = \boldsymbol{P}_3$, we say that a cubic Bézier curve is completely determined by four numbers: the values of the curve at the two end points and the slopes of the curve at these points. And this is where Bézier curves look similar to Hermite interpolation.

The vectors extending from $\boldsymbol{P}_0$ to $\boldsymbol{P}_1$ and from $\boldsymbol{P}_3$ to $\boldsymbol{P}_2$ are called *handles* and can be manipulated in graphics programs like Adobe Photoshop and Illustrator to change the shape of the curve. That explains the term free form modeling.

Bézier curves, CAD, and cars. The mathematical origin of Bézier curves comes from a 1912 mathematical discovery: Bernstein discovered (or invented) the now so-called Bernstein basis polynomial, and used it to define the Bernstein polynomial. What was his purpose? Only to prove Weierstrass's approximation theorem (Sect. 10.3.1). We can say that Bernstein polynomials had no practical applications until ... 50 years later. In 1960s, through the work of Bézier[54] and de Castelijau, Bernstein basis polynomials come to life under the form of Bézier curves.

[54] Pierre Étienne Bézier (1 September 1910–25 November 1999) was a French engineer at Renault. Bezier came from a family of engineers. He followed in their footsteps and earned degrees in mechanical engineering from École nationale supérieure d'arts et métiers and electrical engineering from École supérieure d'électricité. At the age of 67 he earned a doctorate in mathematics from Pierre-and-Marie-Curie University.

de Casteljau's idea of using mathematics to design car bodies met with resistance from Citroën. The reaction was: *Was it some kind of joke? It was considered nonsense to represent a car body mathematically. It was enough to please the eye, the word accuracy had no meaning ....* Eventually de Casteljau's insane persistence led to an increased adoption of computer-aided design methods in Citroën from 1963 onward. About his time at Citroën in his autobiography de Casteljau wrote

> My stay at Citroën was not only an adventure for me, but also an adventure for Citroën!

Thanks to people like de Casteljau that now we have a field called computer-aided design (CAD) in which mathematics and computers are used to help the design of all things that you can imagine of cars, buildings, airplanes, phones, and so on.

4.15 Infinite Series

This section presents infinite series, such as

$$\begin{aligned}\sin(x) &= x - \frac{1}{3!}x^3 + \frac{1}{5!}x^5 - \frac{1}{7!}x^7 + \cdots \\ \cos(x) &= 1 - \frac{1}{2!}x^2 + \frac{1}{4!}x^4 - \frac{1}{6!}x^6 + \cdots\end{aligned}$$

This is how computers compute trigonometric functions, exponential functions, logarithms etc. It is amazing that to compute something finite we have to use infinity. Moreover, the expressions have a nice pattern. That's why maths is beautiful. Another theme here is function approximation: a complex function (e.g. $\sin x$) is replaced by a simpler function, e.g. a polynomial $x - {}^1/_{3!}x^3 + {}^1/_{5!}x^5$, which is easier to work with (easier to differentiate and integrate).

Regarding the organization, first, ingenious ways to obtain such infinite series are presented and second, a systematic method, called Taylor's series, is given.

4.15.1 The Generalized Binomial Theorem

We all know that $(1+x)^2 = 1 + 2x + x^2$, but what about $(1+x)^{1/2}$? Newton's discovery of the binomial series gave answer to negative and fractional powers of binomials. Newton were working on the area of curves of which equations are of the form $(1-x^2)^{n/2}$. For $n=1$, this is the problem of calculating the area of a circle segment.

He considered calculating the following integrals

$$f_n(x) = \int_0^x (1-u^2)^{n/2} du \tag{4.15.1}$$

When n is even $f_n(x)$ can be found explicitly since he knows from Wallis[55] that

$$\int_0^x u^p du = \frac{x^{p+1}}{p+1}$$

Hence,

$$\begin{aligned}
f_0(x) &= \int_0^x du = 1\left(\frac{x}{1}\right) \\
f_2(x) &= \int_0^x (1-u^2)du = 1\left(\frac{x}{1}\right) + 1\left(-\frac{x^3}{3}\right) \\
f_4(x) &= \int_0^x (1-u^2)^2 du = 1\left(\frac{x}{1}\right) + 2\left(-\frac{x^3}{3}\right) + 1\left(\frac{x^5}{5}\right) \\
f_6(x) &= \int_0^x (1-u^2)^3 du = 1\left(\frac{x}{1}\right) + 3\left(-\frac{x^3}{3}\right) + 3\left(\frac{x^5}{5}\right) + 1\left(-\frac{x^7}{7}\right)
\end{aligned} \tag{4.15.2}$$

You can see that the red numbers follow Pascal's triangle (Sect. 2.27). These results for even n can be generalized to have the following

$$f_n(x) = \sum_{m=0}^{\infty} a_{mn} \left[(-1)^m \frac{x^{2m+1}}{2m+1}\right] \tag{4.15.3}$$

where a_{mn} denotes the red coefficients in Eq. (4.15.2), they are called *Integral binomial coefficients*[56] and $(-1)^m$ is either +1 or -1 and is used to indicate the alternating plus/minus signs appearing in Eq. (4.15.2). And Newton believed that this formula also works for odd integers $n = 1, 3, 5, \ldots$ So he collected the red coefficients in Eq. (4.15.2) in a table (Table 4.17). And his goal was to find the coefficients for $n = 1, 3, 5, \ldots$ i.e., the boxes in this table. With those coefficients, we know the integrals in Eq. (4.15.1) and by term-wise differentiation we would get the series for $(1-x^2)^n$ for $n = 1/2, 3/2$ etc.

A complete table for integral binomial coefficients is given in Table 4.18. And we determine $a, b, c, d, \ldots$ by equating the m-th row in Table 4.18 with the corresponding row in Table 4.17, but only for columns of even n.

---

[55] John Wallis (1616–1703) was an English clergyman and mathematician who is given partial credit for the development of infinitesimal calculus.

[56] For example, if $n=0$ and $m=0$, then $a_{mn}=1$ by looking at the first in Eq. (4.15.2).

Table 4.17 Integral binomial coefficients. The row of $m = 0$ is all 1, follow Eq. (4.15.2) (coefficient of x term is always 1). The rule of this table is (because a_{mn} follows the Pascal's triangle): $a_{m,n+2} = a_{m,n} + a_{m-1,n}$ for $m \geq 1$ (see the three circled numbers for one example). Note that $a_{1n} = n/2$ for even ns, and Newton believed it is also the case for odd ns. That's why he put 1/2, 3/2 and 5/2 in the row of $m = 1$ for odd ns

	n						
m	0	1	2	3	4	5	6
0	①	1	1	1	1	1	1
1	⓪	1/2	①	3/2	2	5/2	3
2	0	□	0	□	1	3	3
3	0	□	0	□	0	1	1
4	0	□	0	□	0	0	0
5	0	□	0	□	0	0	0

Table 4.18 Integral binomial coefficients

	n					
m	0	1	2	3	4	5
0	a	a	a	a	a	a
1	b	$a+b$	$2a+b$	$3a+b$	$4a+b$	$5a+b$
2	c	$b+c$	$a+2b+c$	$3a+3b+c$	$6a+4b+c$	$10a+5b+c$
3	d	$c+d$	$b+2c+d$	$a+3b+3c+d$	$4a+6b+4c+d$	$10a+10b+5c+d$

For example, considering the third row (the red numbers in Table 4.18), we have the following equations

$$\left.\begin{aligned} c &= 0 \\ a + 2b + c &= 0 \\ 6a + 4b + c &= 1 \end{aligned}\right\} \Longrightarrow c = 0,\ a = \frac{1}{4},\ b = -\frac{1}{8} \Longrightarrow \begin{cases} a_{21} = b + c = -\dfrac{1}{8} \\ a_{23} = 3a + 3b + c = \dfrac{3}{8} \\ a_{25} = 10a + 5b + c = \dfrac{15}{8} \end{cases}$$

Similarly, considering now the fourth row, we have

$$\left.\begin{aligned} d &= 0 \\ b + 2c + d &= 0 \\ 4a + 6b + 4c + d &= 0 \\ 20a + 15b + 6c + d &= 1 \end{aligned}\right\} \Longrightarrow \begin{cases} a = 1/8 \\ b = -1/8 \\ c = 1/16 \\ d = 0 \end{cases} \Longrightarrow \begin{cases} a_{31} = c + d = \dfrac{1}{16} \\ a_{33} = a + 3b + 3c + d = -\dfrac{1}{16} \\ a_{35} = 10a + 10b + 5c + d = \dfrac{5}{16} \end{cases}$$

So, we can write $f_1(x)$ and $f_2(x)$ as

$$f_1(x) = \int_0^x (1-u^2)^{1/2}du = x + \frac{1}{2}\left(-\frac{x^3}{3}\right) - \frac{1}{8}\left(\frac{x^5}{5}\right) + \frac{1}{16}\left(\frac{-x^7}{7}\right) + \cdots$$
$$f_3(x) = \int_0^x (1-u^2)^{3/2}du = x + \frac{3}{2}\left(-\frac{x^3}{3}\right) + \frac{3}{8}\left(\frac{x^5}{5}\right) - \frac{1}{16}\left(\frac{-x^7}{7}\right) \cdots$$

Now, we differentiate the two sides of the above equations; for the LHS the fundamental theorem of calculus is used to obtain directly the result, and for the RHS, a term-wise differentiation is used:

$$\begin{aligned}(1-x^2)^{1/2} &= 1 - \frac{1}{2}x^2 - \frac{1}{8}x^4 - \frac{1}{16}x^6 - \cdots \\ (1-x^2)^{3/2} &= 1 - \frac{3}{2}x^2 + \frac{3}{8}x^4 + \frac{1}{16}x^6 + \cdots\end{aligned} \tag{4.15.4}$$

Verification. To test his result, Newton squared the series for $(1-x^2)^{1/2}$ and observed that it became $1-x^2$ plus some remaining terms which will vanish. Precisely, Newton squared the quantity $1 - 1/2x^2 - 1/8x^4 - 1/16x^6 - 5/128x^8 + R(x)$ and obtained $1 - x^2 + Q(x)$ where $Q(x)$ contains the lowest order of 10 i.e., very small. Today, we can do this verification easily using Sympy.

Now comes the surprising part. We all know the binomial theorem, discussed in Sect. 2.27.1, which says, for $n \in \mathbb{N}$, $(1+x)^n = \sum_{k=0}^{n} \binom{n}{k} x^k$. The LHS of Eq. (4.15.4) are of the same form only with rational exponents. The question is: can Eq. (4.15.4) still be written in the same form of the binomial theorem? That is, for m being a positive rational number, we have

$$(1-x^2)^m = \sum_{k=0}^{\infty} (-1)^k \binom{m}{k} x^{2k}, \quad \binom{m}{k} = \frac{m(m-1)\cdots(m-k+1)}{k!} \tag{4.15.5}$$

The answer is yes. The only difference compared with the integral exponent case is that the binomial expansion is now an infinite series when m is a rational number. With that, I can now state the generalized binomial theorem: for x, y and $r \in \mathbb{R}$, we can expand $(x+y)^r$ as follows

$$(x+y)^r = \sum_{k=0}^{\infty} \binom{r}{k} x^{r-k} y^k, \quad \binom{r}{k} = \frac{r(r-1)\cdots(r-k+1)}{k!} \tag{4.15.6}$$

Root extraction. The generalized binomial theorem Eq. (4.15.6) provides an easy way to extract roots. For example, using it we can write

$$\sqrt{1-x} = (1-x)^{1/2} = 1 - \frac{1}{2}x - \frac{1}{8}x^2 - \frac{1}{16}x^3 - \frac{5}{128}x^4 - \cdots \tag{4.15.7}$$

And this can be used to compute roots. For example, $\sqrt{7} = \sqrt{9-2} = \sqrt{9(1-2/9)} = 3\sqrt{1-2/9}$. Using Eq. (4.15.7) with $x = 2/9$, we then have

$$\sqrt{7} \approx 3\left(1 - \frac{1}{2}\frac{2}{9} - \frac{1}{8}\frac{4}{81} - \frac{1}{16}\frac{8}{729} - \frac{5}{128}\frac{16}{6561}\right) \approx 2.6458047553726565 \tag{4.15.8}$$

With only five terms we got a result that differs from the true value of $\sqrt{7}$ by only 0.00005.

Newton computed π. He considered the first quarter of a unit circle and calculated its area in two ways: the first way is from Greek mathematicians it is $\pi/4$ and the second way is from calculus. And by equating these twos Newton was able to compute π very accurately using simple arithmetics. The function of the first quarter of a unit circle is $y = \sqrt{1-x^2}$, and thus its area is

$$A = \int_0^1 \sqrt{1-x^2}dx$$

Now comes the power of Eq. (4.15.4): Newton replaced $\sqrt{1-x^2}$ by its power series, and with $A = \pi/4$, he obtained:

$$\begin{aligned}
\frac{\pi}{4} &= \int_0^1 \left(1 - \frac{1}{2}x^2 - \frac{1}{8}x^4 - \frac{1}{16}x^6 - \frac{5}{128}x^8 - \cdots\right) dx \\
\frac{\pi}{4} &= \left[x - \frac{1}{2}\frac{x^3}{3} - \frac{1}{8}\frac{x^5}{5} - \frac{1}{16}\frac{x^7}{7} - \frac{5}{128}\frac{x^9}{9} - \cdots\right]_0^1 \\
\pi &= 4\left(1 - \frac{1}{2}\frac{1}{3} - \frac{1}{8}\frac{1}{5} - \frac{1}{16}\frac{1}{7} - \frac{5}{128}\frac{1}{9} - \cdots\right)
\end{aligned}$$

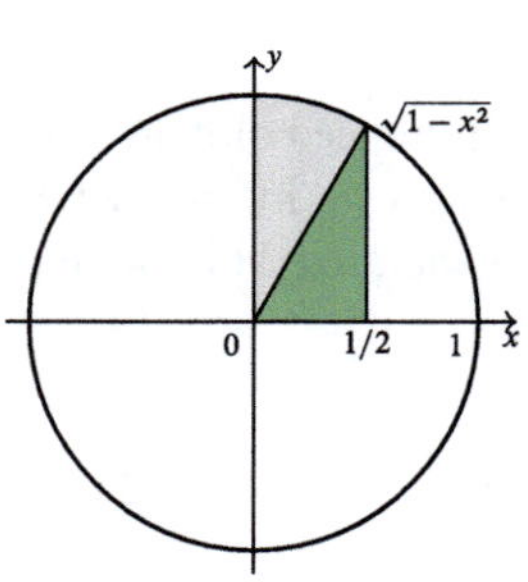

However, he realized that this series converged quite slowly.[57] Why does this series converge slowly? Because in the terms x^n/n, we substituted $x = 1$. If 1 was replaced by a number smaller than 1, then x^n/n would be much smaller, and the series

[57] That is, the series needs lots of terms to get accurate π.

would converge faster. And that is exactly what Newton did: he only integrated to 0.5, and obtained this series (see next figure)

$$\frac{\pi}{12}+\frac{\sqrt{3}}{8}=\frac{1}{2}-\frac{1}{6}\frac{1}{8}-\frac{1}{40}\frac{1}{32}-\frac{1}{112}\frac{1}{128}-\frac{5}{1152}\frac{1}{512}-\cdots$$

with which he managed to compute π with at least 15 digits. At one point, commenting on such approximations, he somewhat sheepishly confessed (at the age of 23) when he wrote, "I am ashamed to tell you to how many figures I carried these computations, *having no other business at the time*". [18].

As you can see, having the right tool, the calculation of π became much easier than the polygonal method of Archimedes.

4.15.2 *Series of* $1/(1+x)$ *or Mercator's Series*

This section presents Newton's work on the function $y=(1+x)^{-1}$. He wanted to compute the area under this curve. The idea is the same: first computing the following integrals

$$f_n(x)=\int_0^x(1+u)^n du \tag{4.15.9}$$

for $n=0,1,2,3,4,\ldots$, then finding a pattern and finally interpolating it to $n=-1$. First thing first, here are $f_n(x)$ for non-negative integers from 0 to 4:

$$\begin{aligned}
f_0(x)&=1(x)\\
f_1(x)&=1(x)+1\left(\frac{x^2}{2}\right)\\
f_2(x)&=1(x)+2\left(\frac{x^2}{2}\right)+1\left(\frac{x^3}{3}\right)\\
f_3(x)&=1(x)+3\left(\frac{x^2}{2}\right)+3\left(\frac{x^3}{3}\right)+1\left(\frac{x^4}{4}\right)\\
f_4(x)&=1(x)+4\left(\frac{x^2}{2}\right)+6\left(\frac{x^3}{3}\right)+4\left(\frac{x^4}{4}\right)+1\left(\frac{x^5}{5}\right)
\end{aligned} \tag{4.15.10}$$

Pascal's triangle shows up again! Now we put all the coefficients from Eq. (4.15.10) in Table 4.19 (left) and want to find the coefficients for column $n=-1$ assuming that the rules work for $n=-1$ as well. The rule of this table is: $a_{m,n+1}=a_{m,n}+a_{m-1,n}$. It follows that the coefficient for $n=-1$ given in the right table ensures this rule.

Table 4.19 Integral binomial coefficients. The row of $m = 0$ is all 1, follow Eq. (4.15.10) (coefficient of x term is always 1). The rule of this table is $a_{m,n+1} = a_{m,n} + a_{m-1,n}$

	n								n						
m	–1	0	1	2	3	4	5	m	–1	0	1	2	3	4	5
0	□	1	1	1	1	1	1	1 0	+1	1	1	1	1	1	1
1	□	0	1	2	3	4	5		−1	0	1	2	3	4	5
2	□	0	0	1	3	6	10	2	+1	0	0	1	3	6	10
3	□	0	0	0	1	4	10	3	−1	0	0	0	1	4	10
4	□	0	0	0	0	1	5	4	+1	0	0	0	0	1	5
5	□	0	0	0	0	0	1	5	−1	0	0	0	0	0	1

Therefore, we can get the integral, and term-wise differentiation gives the series:

$$\int_0^x \frac{du}{1+u} = x - \frac{x^2}{2} + \frac{x^3}{3} - \frac{x^4}{4} + \cdots \Longrightarrow \frac{1}{1+x} = 1 - x + x^2 - x^3 + x^4 - \cdots$$

And we obtain the geometric series! And that confirms Newton was correct.

4.15.3 Geometric Series and Logarithm

From a geometric series and integration, we can obtain interesting series for logarithm. And these series are practical ways to compute the logarithm of any real positive number.

Consider the following geometric series

$$1 + x + x^2 + x^3 + \cdots = \frac{1}{1-x}, \quad |x| < 1 \tag{4.15.11}$$

And by integrating both sides, we get the logarithm series:

$$\int (1 + x + x^2 + x^3 + \cdots)dx = \int \frac{dx}{1-x} \Longrightarrow x + \frac{x^2}{2} + \frac{x^3}{3} + \cdots = -\ln(1-x) \tag{4.15.12}$$

Similarly, this geometric series $1 - x + x^2 - x^3 + \cdots$ gives us $\ln(1+x)$

$$1 - x + x^2 - x^3 + \cdots = \frac{1}{1+x} \Longrightarrow \ln(1+x) = x - \frac{x^2}{2} + \frac{x^3}{3} - \frac{x^4}{4} + \cdots \tag{4.15.13}$$

With this, it is the first time that we are able to compute $\ln 2$ directly using only simple arithmetic operations: $\ln 2 = \ln(1+1) = 1 - 1/2 + 1/3 - 1/4 + \cdots$. Using a calculator we know that $\ln 2 = 0.6931471805599453$. Let's see how the series in

Table 4.20 Convergence rate of two series for ln 2.

n	ln 2 with Eq. (4.15.13)	ln 2 with Eq. (4.15.14)
1	1.0	0.666667
2	0.5	0.666667
$\vdots$	$\vdots$	$\vdots$
11	0.736544	0.693147
$\vdots$	$\vdots$	$\vdots$
1000	0.692647	0.693147

Eq. (4.15.13) performs. The calculation in Table 4.20 (of course done by a Julia code) indicates that this series is practically not useful as it converges too slowly. See column 2 of the table, with 1000 terms, and still the value is not yet close to ln 2.

How can we get a series with a better convergence? The issue might be in the alternating $+/-$ sign in the series. By combining the series for $\ln(1+x)$ and $-\ln(1-x)$, we can get rid of the terms with negative sign:

$$\left\{\begin{aligned} \ln(1+x) &= x - \frac{x^2}{2} + \frac{x^3}{3} - \frac{x^4}{4} + \cdots \\ -\ln(1-x) &= x + \frac{x^2}{2} + \frac{x^3}{3} + \frac{x^4}{4} + \cdots \end{aligned}\right. \Longrightarrow \ln\frac{1+x}{1-x} = 2\left(x + \frac{x^3}{3} + \frac{x^5}{5} + \ldots\right) \tag{4.15.14}$$

Using $x = 1/3$, we have $\ln 2 = 2(1/3 + (1/3)^3/3 + \ldots)$ The data in column 3 in Table 4.20 confirms that these series converge much better: only 11 terms give us 0.693147. What is more, is that while Eq. (4.15.13) cannot be used to compute $\ln e$ (because of the requirement $|x| < 1$), Eq. (4.15.14) can. For any positive number y, $x = {}^{y-1}/_{y+1}$ satisfies $|x| < 1$.

4.15.4 Geometric Series and Inverse Tangent

Let's consider the following geometric series

$$\begin{aligned} 1 + x^2 + x^4 + x^6 + \cdots &= \frac{1}{1-x^2} \\ 1 - x^2 + x^4 - x^6 + \cdots &= \frac{1}{1+x^2} \end{aligned}$$

From the second series we can get the series of the inverse tangent:

$$\int (1 - x^2 + x^4 - x^6 + \cdots)dx = \int \frac{dx}{1 + x^2}$$
$$x - \frac{x^3}{3} + \frac{x^5}{5} - \frac{x^7}{7} \cdots + = \tan^{-1} x \tag{4.15.15}$$

With this, we can derive the magical formula for π discovered by Gregory and Leibniz (actually re-discovered as 200 years before Leibniz some Indian mathematician found it). The angle $\pi/4$ has tangent of 1, thus $\tan^{-1} 1 = \pi/4$, with $x = 1$ we have:

$$\frac{\pi}{4} = 1 - \frac{1}{3} + \frac{1}{5} - \frac{1}{7} \cdots \tag{4.15.16}$$

As π is involved, is there any hidden circle in Eq. (4.15.16)? The answer is yes, and to see that, check https://www.youtube.com/watch?v=NaL_Cb42WyY&feature=youtu.be. This series converges slowly i.e., we need to use lots of terms (and thus lots of calculations) to get a more accurate result for π. However, this series is theoretically interesting as it provided a new way of calculating π (compared with Archimedes' method). To have a better series for π, we should use $x < 1$ in Eq. (4.15.15). For example, note that $\tan \pi/6 = 1/\sqrt{3}$, so with $x = 1/\sqrt{3}$, we can compute a better approximation for π.

4.15.5 Euler's Work on Exponential Functions

From Brigg's work on logarithm, see Sect. 2.24.4, we know the following approximation for a^ϵ where ϵ is a small number:

$$a^\epsilon = 1 + k\epsilon \tag{4.15.17}$$

Now, Euler introduced a^x (this is what we need) with $x = N\epsilon$ where N is a very big number so that ϵ is a very small number. Now, we can write $a^x = a^{N\epsilon} = (a^\epsilon)^N$ and use Eq. (4.15.17):

$$\begin{aligned} a^x &= (1 + k\epsilon)^N \\ &= 1 + N(k\epsilon) + \frac{N(N-1)}{2!}(k\epsilon)^2 + \frac{N(N-1)(N-2)}{3!}(k\epsilon)^3 + \cdots \quad \text{(binomial theorem)} \\ &= 1 + Nk\frac{x}{N} + \frac{N(N-1)}{2!}k^2\frac{x^2}{N^2} + \frac{N(N-1)(N-2)}{3!}k^3\frac{x^3}{N^3} + \cdots \\ &= 1 + \frac{1}{1!}kx + \frac{1}{2!}(kx)^2 + \frac{1}{3!}(kx)^3 + \cdots \end{aligned} \tag{4.15.18}$$

The last equality is due to the fact that $N = N - 1 = N - 2$ as N is very large. Now, we evaluate Eq. (4.15.18) at $x = 1$ to get an equation between a and k:

$$a = 1 + \frac{k}{1!} + \frac{1}{2!}k^2 + \frac{1}{3!}k^3 + \cdots$$

Euler defined e as the number for which $k = 1$:

$$e = 1 + \frac{1}{1!} + \frac{1}{2!} + \frac{1}{3!} + \cdots \tag{4.15.19}$$

The series on the RHS indeed converges because $n!$ gets bigger and bigger and $1/n!$ becomes close to zero. A small code computing this series gives us $e = 2.718281828459045$. With $k = 1$, Eq. (4.15.18) allows us to write e^x as

$$e^x = \left(1 + \frac{x}{N}\right)^N = 1 + \frac{1}{1!}x + \frac{1}{2!}x^2 + \frac{1}{3!}x^3 + \cdots \tag{4.15.20}$$

4.15.6 Euler's Trigonometric Functions

This section presents Euler's derivation of the power series of the sine and cosine functions. He started with the formula $(\cos\alpha \pm i\sin\alpha)^n = \cos(n\alpha) \pm i\sin(n\alpha)$ to get $\cos(n\alpha)$ in terms of $(\cos\alpha + i\sin\alpha)^n$ and $(\cos\alpha - i\sin\alpha)^n$. Then, he used the binomial theorem to expand these two terms. Finally, he replaced n by N a very large positive number, and $\alpha n = \alpha N = x$ so that α is small and $\cos\alpha = 1$.

Let's start with

$$\begin{aligned}(\cos\alpha + i\sin\alpha)^n &= \cos(n\alpha) + i\sin(n\alpha)\\ (\cos\alpha - i\sin\alpha)^n &= \cos(n\alpha) - i\sin(n\alpha)\end{aligned}$$

to get

$$\begin{aligned}\cos(n\alpha) &= \frac{1}{2}\left[(\cos\alpha + i\sin\alpha)^n + (\cos\alpha - i\sin\alpha)^n\right]\\ i\sin(n\alpha) &= \frac{1}{2}\left[(\cos\alpha + i\sin\alpha)^n - (\cos\alpha - i\sin\alpha)^n\right]\end{aligned}$$

Using the binomial theorem: $(a+b)^n = \sum_{k=0}^{n}\binom{n}{k}a^{n-k}b^k$, we can expand the terms $(\cos\alpha + i\sin\alpha)^n$ as

$$\begin{aligned}(\cos\alpha + i\sin\alpha)^n &= \cos^n\alpha + in\cos^{n-1}\alpha\sin\alpha - \frac{n(n-1)}{2!}\cos^{n-2}\alpha\sin^2\alpha\\ &- i\frac{n(n-1)(n-2)}{3!}\cos^{n-3}\alpha\sin^3\alpha + \frac{n(n-1)(n-2)(n-3)}{4!}\\ &\cos^{n-4}\alpha\sin^4\alpha + i\frac{n(n-1)(n-2)(n-3)(n-4)}{5!}\cos^{n-5}\sin^5\alpha + \cdots\end{aligned}$$

and similarly for $(\cos\alpha - i\sin\alpha)^n$ as

$$\begin{aligned}(\cos\alpha - i\sin\alpha)^n &= \cos^n\alpha - in\cos^{n-1}\alpha\sin\alpha - \frac{n(n-1)}{2!}\cos^{n-2}\sin^2\alpha \\ &+ i\frac{n(n-1)(n-2)}{3!}\cos^{n-3}\alpha\sin^3\alpha + \frac{n(n-1)(n-2)(n-3)}{4!} \\ &\cos^{n-4}\alpha\sin^4\alpha - i\frac{n(n-1)(n-2)(n-3)(n-4)}{5!}\cos^{n-5}\alpha\sin^5\alpha + \cdots\end{aligned}$$

Therefore,

$$\begin{aligned}\cos(n\alpha) &= \cos^n\alpha - \frac{n(n-1)}{2!}\cos^{n-2}\alpha\sin^2\alpha + \frac{n(n-1)(n-2)(n-3)}{4!} \\ &\cos^{n-4}\alpha\sin^4\alpha + \cdots \sin(n\alpha) = n\cos^{n-1}\alpha\sin\alpha - \frac{n(n-1)(n-2)}{3!}\cos^{n-3}\alpha\sin^3\alpha \\ &+ \frac{n(n-1)(n-2)(n-3)(n-4)}{5!}\cos^{n-5}\alpha\sin^5\alpha + \cdots\end{aligned} \tag{4.15.21}$$

Now comes the magic of Euler. In Eq. (4.15.21), he used N for n where N is a very large positive integer. He introduced a new variable x such that $\alpha N = x$ (or $\alpha = x/N$). Obviously α is very small, hence we have $\cos\alpha \approx 1$, and $\sin\alpha \approx \alpha$. Hence, $\cos(n\alpha)$ becomes

$$\begin{aligned}\cos(x) &= 1 - \frac{N(N-1)}{N^2}\frac{1}{2!}x^2 + \frac{N(N-1)(N-2)(N-3)}{N^4}\frac{1}{4!}x^4 - \cdots \\ &= 1 - \frac{1}{2!}x^2 + \frac{1}{4!}x^4 - \cdots\end{aligned}$$

where we have used the fact that for a very large integer $N = N-1 = N-2 = N-3 = \ldots$ and get rid of all coefficients involving this arbitrary number N. In the same manner, the series for $\sin x$ can be obtained. Putting them together, we have

$$\begin{aligned}\sin(x) &= x - \frac{1}{3!}x^3 + \frac{1}{5!}x^5 - \frac{1}{7!}x^7 + \cdots = \sum_{i=1}^{\infty}(-1)^{i-1}\frac{1}{(2i-1)!}x^{2i-1} \\ \cos(x) &= 1 - \frac{1}{2!}x^2 + \frac{1}{4!}x^4 - \frac{1}{6!}x^6 + \cdots = \sum_{i=0}^{\infty}(-1)^{i}\frac{1}{(2i)!}x^{2i}\end{aligned} \tag{4.15.22}$$

I have included the formula using the sigma notation. It is not for beauty, that formula is translated directly to our `Julia` code, see Listing A.3. Even though this was done by the great mathematician Euler, we have to verify them for ourselves. Let's compute $\sin\pi/4$ using the series. With only 5 terms, we got 0.707106781 (same as $\sqrt{2}/2$ computed using trigonometry from high school maths)! Why so fast convergence?

With Eq. (4.15.22) we can see that the derivative of sine is cosine: just differentiating the first series and you will obtain the second. Can we also obtain the identity $\sin^2 x + \cos^2 x = 1$ from these series? Of course, otherwise it was not called sine/-cosine series. Some people are skillful enough to use Eq. (4.15.22) to prove this identity. It is quite messy. We can go the other way:

$$g(x) = \sin^2 x + \cos^2 x \Longrightarrow g'(x) = 2\sin x \cos x - 2\cos x \sin x = 0 \Longrightarrow g(x) = \text{constant}$$

But, we know $g(0) = \sin^2 0 + \cos^2 0 = 1$ (using Eq. (4.15.22) of course). So $g = 1$. What a nice proof. We still have to relate the sine/cosine series to the traditional definition of sine/cosine based on a right triangle. And finally, the identity $\sin(x + y) = \sin x \cos y + \sin y \cos x$ and so on (all of this can be done, but that's enough to demonstrate the idea).

You might ask why bothering with all of this. This is because if we can do so, then you can see that trigonometric functions can be defined completely without geometry! Why is that useful? Because it means that trigonometric functions are more powerful than we once thought. Indeed later on we shall see how these functions play an important role in many physical problems that have nothing to do with triangles!

4.15.7 Euler's Solution of the Basel Problem

This section presents Euler's solution to the Basel problem. Recall that the Basel problem involves the sum of the reciprocals of the squares of natural numbers: $1 + \frac{1}{4} + \frac{1}{9} + \frac{1}{16} + \cdots$. This problem has defied all mathematicians including Leibniz who once declared that he could sum any infinite series that converges whose terms follow some rule. And Euler found the sum is $\pi^2/6$.

Euler's proof was based on the power series of $\sin x$ (see the previous section), and the fact that if $f(x) = 0$ has solutions $x_1 = a$, $x_2 = b$, etc. then we can factor it as $f(x) = (a - x)(b - x)(c - x)\cdots = (1 - x/a)(1 - x/b)(1 - x/c)\cdots$ if all of the solutions are different from zero.

Euler considered the function $f(x) = \sin x/x$. From the power series of $\sin(x)$ in Eq. (4.15.22), we can write $f(x)$ as

$$f(x) = \frac{\sin x}{x} = 1 - \frac{x^2}{3!} + \frac{x^4}{5!} + \cdots \tag{4.15.23}$$

As the non-zero solutions of $f(x) = 0$ are $\pm\pi$, $\pm 2\pi$, $\pm 3\pi$, etc., we can also write it as

$$f(x) = \left(1-\frac{x}{\pi}\right)\left(1+\frac{x}{\pi}\right)\left(1-\frac{x}{2\pi}\right)\left(1+\frac{x}{2\pi}\right)\cdots$$
$$= 1 - \left(\frac{1}{\pi^2}+\frac{1}{4\pi^2}+\frac{1}{9\pi^2}+\cdots\right)x^2+\cdots \tag{4.15.24}$$

By equating the coefficient for x^2 in Eq. (4.15.23) and (4.15.24), we obtain

$$\frac{1}{\pi^2}+\frac{1}{4\pi^2}+\frac{1}{9\pi^2}+\cdots = \frac{1}{3!} \Longrightarrow 1+\frac{1}{4}+\frac{1}{9}+\cdots = \frac{\pi^2}{6} \tag{4.15.25}$$

It is easy to verify this result by writing a small code to calculate the sum of $\sum_{i=1}^{n} 1/i^2$, for example with $n = 1000$ and see that the sum is indeed equal to $\pi^2/6$. And with this new toy, Euler continued and calculated the following sums (note that all involve even powers)

$$1+\frac{1}{4}+\frac{1}{9}+\cdots = \frac{\pi^2}{6} \quad \text{(power 2)}$$
$$1+\frac{1}{16}+\frac{1}{81}+\cdots = \frac{\pi^4}{90} \quad \text{(power 4)}$$

But Euler and no mathematicians after him is able to crack down the sum with odd powers. For example, what is $1+\frac{1}{2^3}+\frac{1}{3^3}+\frac{1}{4^3}\cdots$? Can it be π^3/n? No one knows.

Wallis' infinite product Euler's method simultaneously leads us to Wallis' infinite product regarding π. The derivation is as follows

$$\frac{\sin x}{x} = \left(1-\frac{x}{\pi}\right)\left(1+\frac{x}{\pi}\right)\left(1-\frac{x}{2\pi}\right)\left(1+\frac{x}{2\pi}\right)\cdots = \left(1-\frac{x^2}{\pi^2}\right)\left(1-\frac{x^2}{4\pi^2}\right)\left(1-\frac{x^2}{9\pi^2}\right)\cdots$$

Evaluating the above at $x = \pi/2$ results in Wallis' infinite product

$$\frac{2}{\pi} = \left(1-\frac{1}{4}\right)\left(1-\frac{1}{16}\right)\left(1-\frac{1}{36}\right)\cdots = \left(\frac{3}{4}\right)\left(\frac{15}{16}\right)\left(\frac{35}{36}\right)\cdots$$
$$= \left(\frac{1\times 3}{2\times 2}\right)\left(\frac{3\times 5}{4\times 4}\right)\left(\frac{5\times 7}{6\times 6}\right)\cdots \Longrightarrow \boxed{\frac{\pi}{2} = \left(\frac{2\cdot 2}{1\cdot 3}\right)\left(\frac{4\cdot 4}{3\cdot 5}\right)\left(\frac{6\cdot 6}{5\cdot 7}\right)\cdots}$$

Harmonic series and Euler's constant. Up to now, we have met the three famous numbers in mathematics: π, e, and i. Now is the time to meet the fourth number: $\gamma = 0.577215\ldots$ While Euler did not discover π, e and i he gave the names to two of them (π and e). Now that he discovered γ he did not name it.

Recall that $S(n)$—the n-th harmonic number—is the sum of the reciprocals of the first n natural numbers:

$$S(n) := 1 + \frac{1}{2} + \frac{1}{3} + \cdots + \frac{1}{n} = \sum_{i=1}^{n} \frac{1}{i} \tag{4.15.26}$$

Now, define the following quantity

$$A(n) := S(n) - \ln(n) \tag{4.15.27}$$

The sequence $A(1), A(2), \ldots$ converges because it is a decreasing sequence i.e., $A(n+1) < A(n)$ and it is bounded below because $A(n) > 0$. And the limit of this sequence is called the Euler-Mascheroni constant or Euler's constant γ:

$$\gamma := \lim_{n\to\infty} \left(\sum_{i=1}^{n} \frac{1}{i} - \ln n \right) \tag{4.15.28}$$

Using a computer, with $n = 10^7$, I got $\gamma = 0.577215$, correct to six decimals. In 1734, Euler computed γ to five decimals. A few years later he computed γ up to 16 digits.

But hey! How did Euler think of Eq. (4.15.27)? If someone told you to consider this sequence, you could write a code to compute $A(n)$ and see for yourself that it converges to a value of 0.577215. And you would discover γ. Now you see the problems with how mathematics is currently taught and written. For details on the discovery of γ, I recommend the book *Gamma: exploring Euler's constant* by Julian Havil[58] [30] for an interesting story about γ. There are many books about the great incomparable Euler e.g. *Euler: The master of us all* by Dunham William[59] [20] or Paul Nahin's *Dr. Euler's Fabulous Formula: Cures Many Mathematical Ills* [47].

Is Euler's constant irrational? If so, is it transcendental? No one knows. This is one of unsolved problems in mathematics.

[58] Julian Havil (born 1952) is an educator and author working at Winchester College, Winchester, England. The famous English-American theoretical physicist and mathematician Freeman Dyson was one student of Havil.

[59] William Wade Dunham (born 1947) is an American writer who was originally trained in topology but became interested in the history of mathematics and specializes in Leonhard Euler. He has received several awards for writing and teaching on this subject.

History note 4.3: Euler (1707-1783)

Euler was a Swiss mathematician and physicist. He worked in almost all areas of mathematics: geometry, algebra, calculus, trigonometry, number theory and graph theory. He was the first to write $f(x)$ to denote the function of a sinle variable x. He introduced the modern notation for the trigonometric functions, the letter e for the base of the natural logarithm, $\sum$ for summations, i for the imaginary unit i.e., $i^2 = -1$. Euler was one of the most eminent mathematicians of the 18th century and is held to be one of the greatest in history. A statement attributed to Pierre-Simon Laplace expresses Euler's influence on mathematics: "*Read Euler, read Euler, he is the master of us all.*" He is also widely considered to be the most prolific, as his collected works fill 92 volumes, more than anyone else in the field. He spent most of his adult life in Saint Petersburg, Russia, and in Berlin, then the capital of Prussia. Euler's eyesight worsened throughout his mathematical career. He becaome almost totally blind at the age of 59. Euler remarked on his loss of vision, "Now I will have fewer distractions." Indeed, his condition appeared to have little effect on his productivity, as he compensated for it with his mental calculation skills and exceptional memory. Many of those pages were written while he was blind, and for that reason, Euler has been called the Beethoven of mathematics. *Beethoven could not hear his music. Likewise, Euler could not see his calculations.*

4.15.8 Taylor's Series

In previous sections, we have seen the series representation of various functions: exponential function e^x, trigonometric functions $\sin x$ and $\cos x$, logarithm functions and so on. In all cases, a function $f(x)$ is written as a power series in the following form

$$f(x) = a_0 + a_1 x + a_2 x^2 + a_3 x^3 + \cdots = \sum_{n=0}^{\infty} a_n x^n \tag{4.15.29}$$

where a_n are the coefficients that vary from function to function.

Brook Taylor found a systematic way to find these coefficients a_n for any differentiable functions. His idea is to match the function value and all of its derivatives (first derivative, second derivative, etc.) at $x = 0$. Thus, we have the following equations to solve for a_n:

$$\begin{aligned} f(0) &= a_0 \\ f'(0) &= a_1 \\ f''(0) &= 2!a_2 \\ f'''(0) &= 3!a_3 \\ &\vdots \\ f^{(n)}(0) &= n!a_n \end{aligned} \tag{4.15.30}$$

And putting these coefficients into Eq. (4.15.29), we obtain the Taylor's series of any function $f(x)$[60]:

$$f(x) = f(0) + \frac{f'(0)}{1!}x + \frac{f''(0)}{2!}x^2 + \frac{f'''(0)}{3!}x^3 + \cdots = \sum_{n=0}^{\infty} \frac{f^{(n)}(0)}{n!}x^n \tag{4.15.31}$$

where the notation $f^{(n)}(x)$ denotes the n-order derivative of $f(x)$; for $n = 0$ we have $f^{(0)}(x) = f(x)$ (i.e., the 0th derivative is the function itself). See Fig. 4.81 for a demonstration of the Taylor series of $\cos x$. The more terms we include the better approximation of $\cos x$ we get. What is interesting is that we use information of $f(x)$ only at $x = 0$, yet the Taylor series (with enough terms) match the original function for many more points. Taylor series expanded around 0 is sometimes known as the Maclaurin series, named after the Scottish mathematician Colin Maclaurin (1698–1746).

There is nothing special about $x = 0$. And we can expand the function at the point $x = a$:

$$f(x) = \sum_{n=0}^{\infty} \frac{f^{(n)}(a)}{n!}(x-a)^n \tag{4.15.32}$$

[60] Actually not all functions but smooth functions that have derivatives.

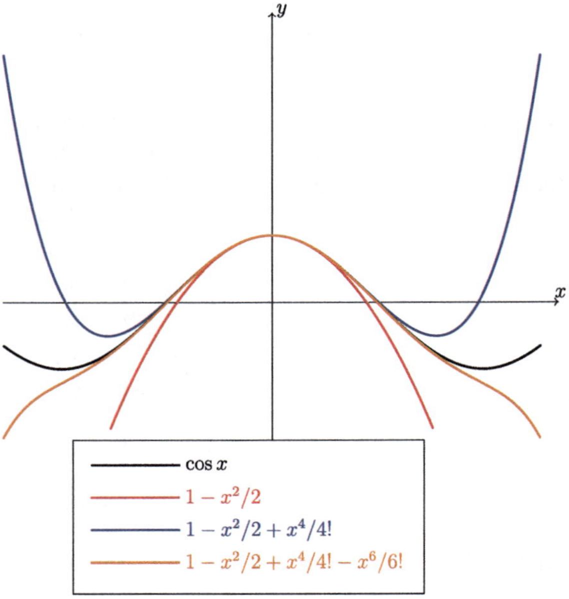

Fig. 4.81 The graph of $\cos x$ and some of its Taylor expansions: $1 - x^2/2$, $1 - x^2/2 + x^4/4!$ and $1 - x^2/2 + x^4/4! - x^6/6!$.

History note 4.4: Taylor (1685-1731)

Brook Taylor was an English mathematician who added to mathematics a new branch now called the 'calculus of finite differences', invented integration by parts, and discovered the celebrated formula known as Taylor's expansion. Brook Taylor grew up not only to be an accomplished musician and painter, but he applied his mathematical skills to both these areas later in his life. As Taylor's family were well off they could afford to have private tutors for their son and in fact this home education was all that Brook enjoyed before entering St John's College Cambridge on 3 April 1703. By this time he had a good grounding in classics and mathematics. The year 1714 marks the year in which Taylor was elected Secretary to the Royal Society. The period during which Taylor was Secretary to the Royal Society marks what must be considered his most mathematically productive time. Two books which appeared in 1715, *Methodus incrementorum directa et inversa* and *Linear Perspective* are extremely important in the history of mathematics. The first of these books contains what is now known as the Taylor series.

4.15.9 Common Taylor Series

Equipped with Eq. (4.15.31) it is now an easy job to develop power series for trigonometric functions, exponential functions, logarithm functions etc. We put commonly used, Taylor series in the following equation:

$$
\begin{aligned}
e^x &= 1 + \frac{1}{1!}x + \frac{1}{2!}x^2 + \frac{1}{3!}x^3 + \cdots &&= \sum_{n=0}^{\infty} \frac{x^n}{n!} && x \in \mathbb{R} \\
\sin x &= x - \frac{1}{3!}x^3 + \frac{1}{5!}x^5 - \frac{1}{7!}x^7 + \cdots &&= \sum_{n=0}^{\infty} (-1)^n \frac{x^{2n+1}}{(2n+1)!} && x \in \mathbb{R} \\
\cos x &= 1 - \frac{1}{2!}x^2 + \frac{1}{4!}x^4 - \frac{1}{6!}x^6 + \cdots &&= \sum_{n=0}^{\infty} (-1)^n \frac{x^{2n}}{(2n)!} && x \in \mathbb{R} \\
\arctan x &= x - \frac{x^3}{3} + \frac{x^5}{5} - \frac{x^7}{7} + \cdots &&= \sum_{n=0}^{\infty} (-1)^n \frac{x^{2n+1}}{(2n+1)} && x \in [-1, 1] \\
\ln(1+x) &= x - \frac{x^2}{2} + \frac{x^3}{3} - \frac{x^4}{4} + \cdots &&= \sum_{n=1}^{\infty} (-1)^{n+1} \frac{x^n}{n} && x \in (-1, 1) \\
\frac{1}{1-x} &= 1 + x + x^2 + x^3 + \cdots &&= \sum_{n=0}^{\infty} x^n && x \in (-1, 1)
\end{aligned}
$$

If we look at the Taylor series of $\cos x$ we do not see odd powers. Why? This is because $\cos(-x) = \cos(x)$ or cosine is an even function. Similarly, in the series of the sine, we do not see even powers. In the above equation, for each series, a condition e.g. $x \in [-1, 1]$ was included. This is to show for which values of x that we can use the Taylor series to represent the origin functions. For example, we cannot use $x - {x^3}/3 + {x^5}/5 - {x^7}/7 + \cdots$ to replace $\arctan x$ for $|x| > 1$.

In Fig. 4.82 we plot e^x and $\ln(1 + x)$ and their Taylor series of different numbers of terms n. We see that the more terms used the more accurate the Taylor series are. But how accurate exactly? You might guess the next thing mathematicians will do is to find the error associated with a truncated Taylor series (we cannot afford to use large n so we can only use a small number of terms, and thus we introduce error and we have to be able to quantify this error). Section 4.15.10 is devoted to this topic.

Taylor's series of other functions. For functions made of elementary functions, using the definition of Taylor's series is difficult. We can find Taylor's series for these functions indirectly. For example, to find Taylor's series of the following function

$$
f(x) = \ln(\cos x), \quad x \in \left(-\frac{\pi}{2}, \frac{\pi}{2}\right)
$$

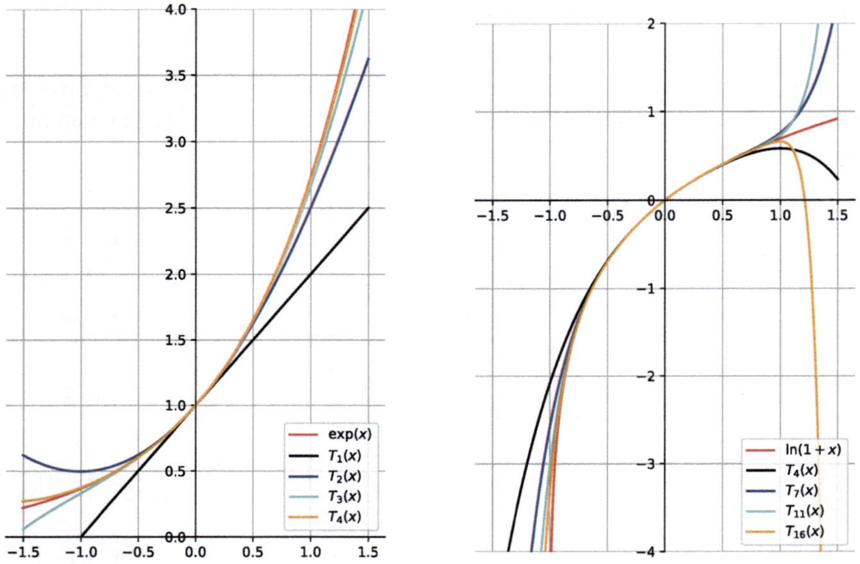

(a) $\exp(x)$ (b) $\ln(1+x)$

Fig. 4.82 Truncated Taylor's series for e^x and $\ln(1+x)$.

we first rewrite $f(x)$ in the form $\ln(1+t)$ so that Taylor's series is available:

$$\begin{aligned} f(x) &= \ln(1 + (\cos x - 1)) \\ &= (\cos x - 1) - \frac{(\cos x - 1)^2}{2} + \frac{(\cos x - 1)^3}{3} - \frac{(\cos x - 1)^4}{4} + \cdots \end{aligned} \tag{4.15.33}$$

Now we use Taylor's series for $\cos x$:

$$\cos x - 1 = -\frac{1}{2!}x^2 + \frac{1}{4!}x^4 - \frac{1}{6!}x^6 + \cdots \tag{4.15.34}$$

Next, we substitute Eq. (4.15.34) into Eq. (4.15.33),

$$\begin{aligned} f(x) = &\left(-\frac{1}{2!}x^2 + \frac{1}{4!}x^4 - \frac{1}{6!}x^6 + \cdots\right) - \frac{1}{2}\left(-\frac{1}{2!}x^2 + \frac{1}{4!}x^4 - \frac{1}{6!}x^6 + \cdots\right)^2 \\ &+ \frac{1}{3}\left(-\frac{1}{2!}x^2 + \frac{1}{4!}x^4 - \frac{1}{6!}x^6 + \cdots\right)^3 + \cdots \end{aligned}$$

Assume that we ignore terms of order 8 and above, we can compute $f(x)$ as

$$\begin{aligned}\ln(\cos x) &= \left(-\frac{1}{2!}x^2 + \frac{1}{4!}x^4 - \frac{1}{6!}x^6 + \cdots\right) - \frac{1}{2}\left(-\frac{1}{2!}x^2 + \frac{1}{4!}x^4\right)^2 + \frac{1}{3}\left(-\frac{1}{2!}x^2\right)^3 \\ &= -\frac{x^2}{2} - \frac{x^4}{12} - \frac{x^6}{45} + \mathcal{O}(x^8)\end{aligned}$$

Big O notation. In the above equation, I have introduced the big O notation ($\mathcal{O}(x^8)$). In that equation, because we neglected terms of order of magnitude equal and greater than eight, the notation $\mathcal{O}(x^8)$ is used. Let's see one example: the sum of the first n positive integers is

$$1 + 2 + 3 + \cdots + n = \frac{n(n+1)}{2} = \frac{n^2}{2} + \frac{n}{2}$$

When n is large, the second term is much smaller relatively than the first term; so the order of magnitude of $1 + 2 + \cdots + n$ is n^2; the factor $1/2$ is not important. So we write

$$1 + 2 + 3 + \cdots + n = \mathcal{O}(n^2)$$

To get familiar with this notation, we write, in below, the full Taylor's series for e^x, and two truncated series

$$\begin{aligned}e^x &= 1 + \frac{1}{1!}x + \frac{1}{2!}x^2 + \frac{1}{3!}x^3 + \cdots = 1 + \frac{1}{1!}x + \mathcal{O}(x^2) \\ e^x &= 1 + \frac{1}{1!}x + \frac{1}{2!}x^2 + \frac{1}{3!}x^3 + \cdots = 1 + \frac{1}{1!}x + \frac{1}{2!}x^2 + \mathcal{O}(x^3)\end{aligned}$$

The notation $\mathcal{O}(x^2)$ allows us to express the fact that the error in $e^x = 1 + x$ is smaller in absolute value than some constant times x^2 if x is close enough to 0.[61] The big O notation is also called Landau's symbol named after the German number theoretician Edmund Landau (1877–1938) who invented the notation. The letter O is for order.

4.15.10 Taylor's Theorem

We recall that it is possible to write any function $f(x)$ as a power series, see Eq. (4.15.32). In Fig. 4.82, we have examined how a power series can approximate a given function. To that end, we varied the number of terms in the series, and we

---

[61] You can play with some values of x close to zero, compute e^x using the exponential function in a calculator, and compute its approximation $1 + x$, the difference between the two values is proportional to x^2.

have seen that the more terms used the more accurately the series approximates the function. To quantify the error of this approximation, mathematicians introduce the concept of the remainder of a Taylor series. That is they divide the series into two sums:

$$f(x)=\sum_{n=0}^{\infty}\frac{f^{(n)}(a)}{n!}(x-a)^n=\underbrace{\sum_{i=0}^{n}\frac{f^{(i)}(a)}{i!}(x-a)^i}_{T_n(x)}+\underbrace{\sum_{i=n+1}^{\infty}\frac{f^{(i)}(a)}{i!}(x-a)^i}_{R_n(x)} \tag{4.15.35}$$

The first sum (has a finite term) is a polynomial of degree n and thus called *a Taylor polynomial*, denoted by $T_n(x)$. The remaining term is called, understandably, the remainder, $R_n(x)$.

It is often that scientists/engineers do this approximation: $f(x) \approx T_n(x)$. This is because it's easy to work with a polynomial (e.g. differentiation/integration, root finding of a polynomial is straightforward). In this case, $R_n(x)$ becomes the error of this approximation. If only two terms in the Taylor series are used, we get

$$T_2(x) = f(a) + f'(a)(x-a)$$

which is the linear approximation we have discussed in Sect. 4.5.3.

How to quantify $R_n(x)$? From Fig. 4.82 we observe that close to $x = a$ the approximation is very good, but far from a the approximation is bad. The following theorem helps to quantify $R_n(x)$.[62]

Theorem 4.3

$$R_n(x) = \frac{f^{(n+1)}(c)}{(n+1)!}(x-a)^{n+1} \tag{4.15.36}$$

Example 4.18 The Taylor series for $y = e^x$ at $a = 0$ with the remainder is given by

$$e^x = 1 + \frac{x}{1!} + \frac{x^2}{2!} + \cdots + \frac{x^n}{n!} + R_n(x), \quad R_n(x) = \frac{e^c}{(n+1)!}x^{n+1}$$

where $0 < c < x$. The nice thing with e^x is that $R_n(x)$ approaches zero as n goes large. Note that we have $|c| < |x|$ and e^x is an increasing function, thus

$$|R_n(x)| \le \frac{e^{|x|}}{(n+1)!}|x^{n+1}| \Longrightarrow \lim_{n\to\infty}|R_n(x)| < e^{|x|}\lim_{n\to\infty}\frac{|x^{n+1}|}{(n+1)!} = 0$$

See Sect. 4.10.4 if you're not clear why the final limit is zero.

[62] I could not find an easy motivating way to come up with this theorem, so I accepted it.

4.16 Applications of Taylor' Series

Herein we present some applications of Taylor series: (1) Evaluate integrals, (2) Evaluate limits, and (3) Evaluate series.

4.16.1 Integral Evaluation

Let's compute the following integral which we could not do before: $\int e^{-x^2} dx$. The idea is to replace the integrand by its Taylor's series:

$$\begin{aligned}
e^x &= 1 + \frac{1}{1!}x + \frac{1}{2!}x^2 + \frac{1}{3!}x^3 + \cdots \\
e^{-x^2} &= 1 + \frac{1}{1!}(-x^2) + \frac{1}{2!}(-x^2)^2 + \frac{1}{3!}(-x^2)^3 + \cdots
\end{aligned}$$

Then, term-wise integration gives us

$$\begin{aligned}
\int e^{-x^2} dx &= \int_0^1 \left[1 + \frac{1}{1!}(-x^2) + \frac{1}{2!}(-x^2)^2 + \frac{1}{3!}(-x^2)^3 + \cdots\right] dx \\
&= x - \frac{1}{1!3}x^3 + \frac{1}{2!5}x^5 - \frac{1}{3!7}x^7 + \cdots = \sum_{n=0}^{\infty}(-1)^n \frac{1}{n!}\frac{x^{2n+1}}{2n+1}
\end{aligned}$$

Now I present another interesting formula for π. This is the formula:

$$\pi = \frac{3\sqrt{3}}{4}\sum_{n=0}^{\infty}\frac{(-1)^n}{8^n}\left(\frac{2}{3n+1} + \frac{1}{3n+2}\right) \tag{4.16.1}$$

First, write a small `Julia` code to verify this formula (using $n = 100$ and compute the RHS to see if it matches $\pi = 3.1415\ldots$). How on earth mathematicians discovered this kind of equation? They started with a definite integral of which the integral involves π:

$$\int_0^{1/2} \frac{dx}{x^2 - x + 1} = \frac{\pi}{3\sqrt{3}}$$

If you cannot evaluate this integral: using a completing a square for $x^2 - x + 1$, then using a trigonometry substitution ($\tan\theta$). That's not interesting. Here is the great stuff:

$$1 + x^3 = (1 + x)(x^2 - x + 1)$$

Thus,

$$I = \int_0^{1/2} \frac{dx}{x^2 - x + 1} = \int_0^{1/2} \frac{x+1}{1+x^3} dx = \int_0^{1/2} \frac{xdx}{1+x^3} + \int_0^{1/2} \frac{dx}{1+x^3}$$

Of course, now we replace the integrands with the corresponding power series. Starting with the geometric series:

$$\frac{1}{1-x} = 1 + x + x^2 + x^3 + \cdots$$

We then have:

$$\begin{aligned} \frac{1}{1+x^3} &= 1 - x^3 + x^6 - x^9 + \cdots \\ \frac{x}{1+x^3} &= x - x^4 + x^7 - x^{10} + \cdots \quad \text{(obtained from the above times } x) \end{aligned}$$

Now, the integral I can be evaluated using these series:

$$\begin{aligned} I &= \int_0^{1/2} (x - x^4 + x^7 - x^{10} + \cdots) dx + \int_0^{1/2} (1 - x^3 + x^6 - x^9 + \cdots +) dx \\ &= \frac{1}{4} \left[\frac{1}{2} \cdot \frac{1}{8^0} - \frac{1}{5} \cdot \frac{1}{8} + \frac{1}{8} \cdot \frac{1}{8^2} - \cdots \right] + \frac{1}{2} \left[1 \cdot \frac{1}{8^0} - \frac{1}{4} \cdot \frac{1}{8} + \frac{1}{7} \cdot \frac{1}{8^2} - \cdots \right] \end{aligned}$$

Now we can understand Eq. (4.16.1).

Mysterious function x^x. What would be the graph of the mysterious function $y = x^x$? Can it be defined for negative x? Is it an increasing/decreasing function? We leave that for you, instead we focus on the integration of this function. That is we consider the following integral:

$$I := \int_0^1 x^x dx$$

Here is how John Bernoulli computed it in 1697. He converted x^x to $e^{\cdots}$:

$$x^x = (e^{\ln x})^x = e^{x \ln x}$$

Then, he used the power series for e^x to replace $e^{x \ln x}$ by a power series:

$$e^{x \ln x} = 1 + \frac{1}{1!} x \ln x + \frac{1}{2!} (x \ln x)^2 + \frac{1}{3!} (x \ln x)^3 + \cdots = \sum_{n=0}^{\infty} \frac{(x \ln x)^n}{n!}$$

Then, the original integral becomes

$$I = \int_0^1 \sum_{n=0}^{\infty} \frac{(x \ln x)^n}{n!} dx = \sum_{n=0}^{\infty} \frac{1}{n!} \int_0^1 (x \ln x)^n dx$$

For the red integral, integration by parts is the way to go:

$$[x^{n+1}(\ln x)^n]' = (n+1)(x \ln x)^n + nx^n(\ln x)^{n-1}$$

Therefore,

$$\begin{aligned}\int_0^1 (x \ln x)^n dx &= \frac{1}{n+1}[x^{n+1}(\ln x)^n]_0^1 - \frac{n}{n+1} \int_0^1 x^n(\ln x)^{n-1} dx \\ &= -\frac{n}{n+1} \int_0^1 x^n(\ln x)^{n-1} dx\end{aligned}$$

This is because $\lim_{x\to 0} x^{n+1}(\ln x)^n = 0$. Now if we repeatedly apply integration by parts to lower the power in $(\ln x)^{n-1}$, we obtain

$$\begin{aligned}\int_0^1 (x \ln x)^n dx &= \left(-\frac{n}{n+1}\right) \int_0^1 x^n(\ln x)^{n-1} dx \\ &= \left(-\frac{n}{n+1}\right)\left(-\frac{n-1}{n+1}\right) \int_0^1 x^n(\ln x)^{n-2} dx \\ &= \left(-\frac{n}{n+1}\right)\left(-\frac{n-1}{n+1}\right)\left(-\frac{n-2}{n+1}\right) \int_0^1 x^n(\ln x)^{n-3} dx\end{aligned}$$

Doing this until $(\ln x)^0$, we're then done:

$$\int_0^1 (x \ln x)^n dx = (-1)^n \frac{n!}{(n+1)^{n+1}}$$

And finally the integral is given by

$$I = \sum_{n=0}^{\infty} \frac{(-1)^n}{(n+1)^{n+1}} = 1 - \frac{1}{2^2} + \frac{1}{3^3} - \frac{1}{4^4} + \cdots$$

4.16.2 Limit Evaluation

Move on to the problems of evaluation of limits, let us consider the following limit

$$\lim_{x\to 0} \frac{x^2 e^x}{\cos x - 1}$$

And again, the idea is to replace e^x and $\cos x$ by its Taylor's series, and we will find that the limit will come easily:

$$A = \frac{x^2(1 + \frac{1}{1!}x + \frac{1}{2!}x^2 + \frac{1}{3!}x^3 + \cdots)}{-\frac{1}{2!}x^2 + \frac{1}{4!}x^4 - \cdots} = \frac{x^2 + \frac{1}{1!}x^3 + \frac{1}{2!}x^4 + \frac{1}{3!}x^5 + \cdots}{-\frac{1}{2!}x^2 + \frac{1}{4!}x^4 - \cdots} \Longrightarrow \lim_{x\to 0} A = -2$$

4.16.3 Series Evaluation

Taylor's series can be used to compute series. For example, what does the following series converge to?

$$1 - \frac{1}{1!} + \frac{1}{2!} - \frac{1}{3!} + \cdots = ?$$

We can recognize that the above series is e^{-x} evaluated at $x = 1$, so the series converges to $1/e$.

4.17 Bernoulli Numbers

In Sect. 2.27 we have discussed the story of Jakob Bernoulli in 1713, and Seki Takakazu in 1712 independently discovered a general formula for the sum of powers of counting numbers (e.g. $1^3 + 2^3 + 3^3 + \cdots$). The formula introduces the now so-called Bernoulli numbers. This section elaborates on these amazing numbers.

Series of $1/(e^x - 1)$ and Bernoulli numbers. Let's start with

$$e^x - 1 = \frac{1}{1!}x + \frac{1}{2!}x^2 + \frac{1}{3!}x^3 + \frac{1}{4!}x^4 + \frac{1}{5!}x^5 + \mathcal{O}(x^6)$$

Then, we can write $1/(e^x - 1)$ as

$$\frac{1}{e^x-1}=\frac{1}{\frac{1}{1!}x+\frac{1}{2!}x^2+\frac{1}{3!}x^3+\frac{1}{4!}x^4+\frac{1}{5!}x^5+\mathcal{O}(x^6)}$$
$$=\frac{1}{x}\left(1+\underbrace{\frac{1}{2}x+\frac{1}{6}x^2+\frac{1}{24}x^3+\frac{1}{120}x^4}_{y}+\mathcal{O}(x^5)\right)^{-1}$$

Now, using the Taylor series for $1/(1+y)=1-y+y^2-y^3+y^4$ (we stop at y^4 as we skip terms of powers higher than 4), and also using SymPy, we get

$$\frac{x}{e^x-1}=1-\frac{x}{2}+\frac{x^2}{12}-\frac{x^4}{720}+\cdots=1\frac{x^0}{0!}-\frac{1}{2}\frac{x^1}{1!}+\frac{1}{6}\frac{x^2}{2!}-\frac{1}{30}\frac{x^4}{4!}+\cdots=\sum_{n=0}^{\infty}B_n\frac{x^n}{n!} \tag{4.17.1}$$

The second equality is to introduce $n!$ into the formula as we want to follow the pattern of the Taylor series. With that, we obtain a nice series for x/e^x-1 in which the Bernoulli numbers show up again! They are

$$B_0=1,\ B_1=-\frac{1}{2},\ B_2=\frac{1}{6},\ B_3=0,\ B_4=-\frac{1}{30},\ B_5=0,\ B_6=\frac{1}{42},\ B_7=0,\ldots$$

Recurrence relation between Bernoulli numbers. Recall that we have met Fibonacci numbers, and they are related to each other. Then, we now ask whether there exists a relation between the Bernoulli numbers. The answer is yes, that's why mathematics is super interesting. The way to derive this relation is also beautiful. From Eq. (4.17.1), we can compute x in terms of e^x-1 and $\sum_{n=0}^{\infty}B_n\frac{x^n}{n!}$:

$$x=(e^x-1)\sum_{n=0}^{\infty}B_n\frac{x^n}{n!}=\left(\frac{x}{1!}+\frac{x^2}{2!}+\cdots\right)\left(\sum_{n=0}^{\infty}B_n\frac{x^n}{n!}\right)$$
$$=\left(\sum_{m=1}^{\infty}\frac{x^m}{m!}\right)\left(\sum_{n=0}^{\infty}B_n\frac{x^n}{n!}\right)=\left(\sum_{m=0}^{\infty}\frac{x^{m+1}}{(m+1)!}\right)\left(\sum_{n=0}^{\infty}B_n\frac{x^n}{n!}\right)$$

The last equality was to convert the lower limit of summation of $\sum_{m=1}^{\infty}x^m/m!$ from 1 to zero, to apply the Cauchy product. Now, we use the Cauchy product for two series[63] to get

$$x=\sum_{n=0}^{\infty}\left(\sum_{k=0}^{n}\frac{x^{n-k+1}}{(n-k+1)!}B_k\frac{x^k}{k!}\right)=\sum_{n=0}^{\infty}\sum_{k=0}^{n}\binom{n+1}{k}B_k\frac{x^{n+1}}{(n+1)!}$$

[63] Refer to Eq. (6.12.2) for derivation.

Replacing $n+1$ by n, with n starts from 1 instead of 0, we have

$$x = \sum_{n=1}^{\infty}\sum_{k=0}^{n-1}\binom{n}{k}B_k\frac{x^n}{n!}$$

Thus, we can conclude that,[64]

$$B_0 = 1, \quad \sum_{k=0}^{n-1}\binom{n}{k}B_k = 0 \text{ for } n > 1 \tag{4.17.2}$$

Explicitly, we have

$$\begin{aligned}
1 &= B_0\\
0 &= B_0 + 2B_1\\
0 &= B_0 + 3B_1 + 3B_2\\
0 &= B_0 + 4B_1 + 6B_2 + 4B_3\\
0 &= B_0 + 5B_1 + 10B_2 + 10B_3 + 5B_4
\end{aligned}$$

You can see the Pascal triangle here!

Cotangent and Bernoulii numbers. If we consider the function $g(x) = {}^{x}/_{e^x-1} - B_1 x$, we get; check Sect. 5.17 for detail,

$$g(x) := \frac{x}{e^x - 1} - B_1 x = \frac{x}{2}\frac{e^{x/2} + e^{-x/2}}{e^{x/2} - e^{-x/2}} = \sum_{n=0}^{\infty}\frac{B_{2n}}{(2n)!}x^{2n}$$

If we know hyperbolic trigonometric functions, see Sect. 3.19, then it is not hard to see that the red term is coth $({}^{x}/_{2})$; and thus we're led to

$$\frac{x}{2}\coth\left(\frac{x}{2}\right) = \sum_{n=0}^{\infty}\frac{B_{2n}}{(2n)!}x^{2n}$$

And to get from coth to cot, just replace x by ix, and we get the series for the cotangent function:

$$\cot x = \sum_{n=0}^{\infty}(-1)^n\frac{2B_{2n}}{(2n)!}(2x)^{2n-1}$$

in terms of the Bernoulli numbers!

---

64 This is similar to $x = b_0 x + (b_1 + b_2)x^2$ for all x, then we must have $b_0 = 1$ and $b_1 + b_2 = 0$.

4.18 Euler-Maclaurin Summation Formula

We have discussed the connection between the sums of powers of integers and the Bernoulli numbers in Sect. 2.27. Recall that we have defined $S_m(n)$ as the sum of the mth power of the first n positive integers:

$$S_m(n) := \sum_{k=1}^{n} k^m$$

Now, to simplify the notation, we simply use S_m for $S_m(n)$. And for later use, we list the first few sums[65]:

$$\begin{aligned}
S_0 &= 1^0 + 2^0 + 3^0 + \cdots + n^0 = B_0 n \\
S_1 &= 1^1 + 2^1 + 3^1 + \cdots + n^1 = \frac{1}{2}\left(B_0 n^2 - 2B_1 n\right) \\
S_2 &= 1^2 + 2^2 + 3^2 + \cdots + n^2 = \frac{1}{3}\left(B_0 n^3 - 3B_1 n^2 + 3B_2 n\right) \\
S_3 &= 1^3 + 2^3 + 3^3 + \cdots + n^3 = \frac{1}{4}\left(B_0 n^4 - 4B_1 n^3 + 6B_2 n^2 + 4B_3 n\right) \\
S_4 &= 1^4 + 2^4 + 3^4 + \cdots + n^4 = \frac{1}{5}\left(B_0 n^5 - 5B_1 n^4 + 10B_2 n^3 + 10B_3 n^2 + 5B_4 n\right)
\end{aligned} \tag{4.18.1}$$

The Euler-Maclaurin summation formula involves the sum of a function $y = f(x)$ evaluated at integer values of x from 1 to n. For example, considering $y = x^2$, and this sum

$$S := f(1) + f(2) + \cdots + f(n) = 1^2 + 2^2 + 3^2 + \cdots + n^2$$

which is nothing but the S_2 we're familiar with. Considering another function $y = x^2 + 3x + 2$, and the sum $S := f(1) + f(2) + \cdots + f(n)$, which is nothing but $S_2 + 3S_1 + 2S_0$. To conclude, for polynomials, S can be written in terms of $S_0, S_1, \ldots$ And we know how to compute $S_0, S_1, \ldots$ using Eq. (4.18.1).

Moving on now to non-polynomial functions such as $\sin x$ or e^x. Thanks to Taylor, we can express these functions as a power series, and we return back to the business of dealing with polynomials. For an arbitrary function $f(x)$—which is assumed to be able to have a Taylor's expansion, we can then write

$$f(x) = c_0 + c_1 x + c_2 x^2 + \cdots$$

Thus, we can compute $S = \sum_{i=1}^{n} f(i)$ in the same manner as we did for polynomials, only this time we have an infinite sum:

---

[65] Two conventions are used in the literature regarding the Bernoulli numbers, in one convention $B_1 = -1/2$ and in the other $B_1 = 1/2$. In this book, I use $B_1 = -1/2$.

$$S := \sum_{i=1}^{n} f(i) = c_0 S_0 + c_1 S_1 + c_2 S_2 + c_3 S_3 + \cdots$$

Substituting $S_0, S_1, \ldots$ in Eq. (4.18.1) into S, we obtain

$$\begin{aligned} S = c_0 B_0 n &+ c_1 \frac{1}{2}\left(B_0 n^2 - 2B_1 n\right) + c_2 \frac{1}{3}\left(B_0 n^3 - 3B_1 n^2 + 3B_2 n\right) + \\ &+ c_3 \frac{1}{4}\left(B_0 n^4 - 4B_1 n^3 + 6B_2 n^2 + 4B_3 n\right) + \\ &+ c_4 \frac{1}{5}\left(B_0 n^5 - 5B_1 n^4 + 10B_2 n^3 + 10B_3 n^2 + 5B_4 n\right) + \cdots \end{aligned}$$

Now, we need to massage S a bit so that it tells us the hidden truth; we group terms with $B_0, B_1, \ldots$:

$$\begin{aligned} S = B_0 &\left(c_0 n + \frac{1}{2} c_1 n^2 + \frac{1}{3} c_2 n^3 + \frac{1}{4} c_3 n^4 + \cdots \right) + \\ &- B_1 (c_1 n + c_2 n^2 + c_3 n^3 + c_4 n^4 + \cdots) + \\ &+ B_2 (c_2 n + \frac{3}{2} c_3 n^2 + 2c_4 n^3 + \cdots) + B_3(\cdots) + \cdots \end{aligned}$$

Now come the magic, the red term is the integral of $f(x)$,[66] the blue term is the first derivative of $f(x)$ at $x = n$ minus $f'(0)$, and the third term is $f''(n) - f''(0)$ and so on, so we have

$$S = \int_0^n f(x)dx - B_1 \left(f(n) - f(0)\right) + \frac{B_2}{2!}\left(f'(n) - f'(0)\right) + \frac{B_3}{3!}\left(f''(n) - f''(0)\right) + \cdots$$

Noting that B_{2n+1} are all zeros except B_1 and $B_1 = -1/2$, we can rewrite the above equation as

$$\sum_{i=1}^{n} f(i) = \int_0^n f(x)dx + \frac{f(n) - f(0)}{2} + \sum_{k=1}^{\infty} \frac{B_{2k}}{(2k)!}\left(f^{(2k-1)}(n) - f^{(2k-1)}(0)\right)$$

Why can this formula be useful when we replace a finite sum by a definite integral (which can be done) and an infinite sum? You will see that this is a powerful formula to compute sums, both infinite sums and finite sums. That was the powerful weapon that Euler used to compute $\sum_{k=1}^{\infty} 1/k^2$ in the Basel problem. But first, we need to polish our formula, because there is an asymmetry in the formula: on the LHS we start from 1, but on the RHS, we start from 0. If we add $f(0)$ to both sides, we get a nicer formula:

---

[66] $\int_0^n (c_0 + c_1 x + c_2 x^2 + \cdots)dx = (c_0 x + c_1 x^2/2 + c_2 x^3/3 + \cdots)|_0^n$.

$$\sum_{i=0}^{n} f(i) = \int_0^n f(x)dx + \frac{f(n)+f(0)}{2} + \sum_{k=1}^{\infty} \frac{B_{2k}}{(2k)!}\left(f^{(2k-1)}(n) - f^{(2k-1)}(0)\right)$$

Now if we ask why start from 0? What if $f(0)$ is undefined (e.g. for $f(x) = 1/x^2$)? We can start from any value smaller than n. Let's consider $m < n$, and we compute two sums:

$$\sum_{i=0}^{n} f(i) = \int_0^n f(x)dx + \frac{f(n)+f(0)}{2} + \sum_{k=1}^{\infty} \frac{B_{2k}}{(2k)!}\left(f^{(2k-1)}(n) - f^{(2k-1)}(0)\right)$$

$$\sum_{i=0}^{m} f(i) = \int_0^m f(x)dx + \frac{f(m)+f(0)}{2} + \sum_{k=1}^{\infty} \frac{B_{2k}}{(2k)!}\left(f^{(2k-1)}(m) - f^{(2k-1)}(0)\right)$$

Now, we subtract the first formula from the second one, we then have a formula which starts from m nearly (note that on the LHS, we start from $m+1$ because $f(m)$ was removed):

$$\sum_{i=m+1}^{n} f(i) = \int_m^n f(x)dx + \frac{f(n)-f(m)}{2} + \sum_{k=1}^{\infty} \frac{B_{2k}}{(2k)!}\left(f^{(2k-1)}(n) - f^{(2k-1)}(m)\right)$$

Using the same trick of adding $f(m)$ to both sides, we finally arrive at

$$\sum_{i=m}^{n} f(i) = \int_m^n f(x)dx + \frac{f(n)+f(m)}{2} + \sum_{k=1}^{\infty} \frac{B_{2k}}{(2k)!}\left(f^{(2k-1)}(n) - f^{(2k-1)}(m)\right) \tag{4.18.2}$$

And this is the Euler-Maclaurin summation formula, usually abbreviated as EMSF, about which D. Pengelley[67] wrote *the formula that dances between continuous and discrete*. This is the form without the remainder term. This is because in the formula we do not know when to truncate the infinite series.[68]

Basel sum. Now we use the EMSF to compute the Basel sum, tracing the footsteps of the great Euler. We write the sum of the second powers of the reciprocals of the positive integers as

$$\sum_{k=1}^{\infty} \frac{1}{k^2} = \sum_{k=1}^{N-1} \frac{1}{k^2} + \sum_{k=N}^{\infty} \frac{1}{k^2} \tag{4.18.3}$$

---

[67] A professor Emeritus in Mathematical Sciences at New Mexico State University.

[68] This derivation was based on Youtuber mathologer.

Now, the first sum with a few terms, we compute it explicitly (i.e., add term by term) and for the second term, we use the EMSF in Eq. (4.18.2). We can compute the red term as, with $f(x) = 1/x^2$

$$\sum_{k=N}^{\infty} \frac{1}{k^2} = \frac{1}{N} + \frac{1}{2N^2} + \frac{1}{6N^3} - \frac{1}{30N^5} + \frac{1}{42N^7} + \cdots$$

For example with $N = 10$, we have (with only four terms in the above series)

$$\sum_{k=1}^{\infty} \frac{1}{k^2} = \sum_{k=1}^{9} \frac{1}{k^2} + \frac{1}{N} + \frac{1}{2N^2} + \frac{1}{6N^3} - \frac{1}{30N^5}$$

An infinite sum was computed using only a sum of 13 terms! How about the accuracy? The exact value is $\pi^2/6 = 1.6449340668482264$, and the one based on the EMSF is 1.644934064499874; an accuracy of eight decimals. If we did not know the EMSF, we would have had to compute 1 billion terms to get an accuracy of 8 decimals! Note that $\sum_{k=1}^{9} 1/k^2$ is only 1.539767731166540.

4.19 Fourier Series

This section is our first encounter with the so-called Fourier series. We will discuss the origin of the Fourier series in Sect. 7.11 and 7.9 where differential equations are discussed. Herein, I present briefly what are Fourier series and how to determine the coefficients of this series. To make a long story short, the story goes like this. While investigating the phenomenon of heat transfer in solids, Fourier discovered that it is possible to write a periodic function[69] $\phi(x)$ as an infinite series of sines and cosines. For example,

$$\phi(x) = \sum_{n=1}^{\infty} B_n \sin(n\pi x) \tag{4.19.1}$$

In Sect. 4.19.1, I present the method to determine B_n for periodic functions of a period of 2π once we have accepted that it is possible to write $\phi(x)$ as in Eq. (4.19.1). Then, in Sect. 4.19.2, we extend the discussion to periodic functions of arbitrary period. Section 4.19.3 gives the complex form of the Fourier series, which is more compact.

[69] We have discussed periodic functions in Sect. 3.18.

4.19.1 Periodic Functions with Period 2π

Before delving into Fourier series, we need some preparing results, for $m, n \in \mathbb{N}$:

$$\int_{-\pi}^{\pi} \cos mx dx = 0, \quad \int_{-\pi}^{\pi} \sin nx \cos mx dx = 0, \quad \int_{-\pi}^{\pi} \cos nx \cos mx dx = \begin{cases} 0 & m \neq n \\ \pi & m = n \end{cases} \tag{4.19.2}$$

of which proof is not discussed here, but check Example 4.10. We once had a thought that, in a boring calculus class, why we spent a significant amount of our Youth to compute these seemingly useless integrals like the above? It is interesting to realize that these integrals play an important role in mathematics and then in our lives. If you need a geometric meaning of Eq. (4.19.2): this equation means that the functions $\cos x, \cos 2x, \sin x, \sin 2x, \ldots$ are perpendicular to each other. To fully understand why two functions can be perpendicular we need linear algebra to be discussed in Chap. 9.

Now, Fourier believed that it is possible to expand any periodic function $f(x)$ with period 2π as a trigonometric infinite series (as mentioned, refer to Sects. 7.11 and 7.9 to see why Fourier came up with this idea; once the idea is there, the remaining steps are usually not hard, as I can understand them):

$$\begin{aligned} f(x) &= a_0 + (a_1 \cos x + a_2 \cos 2x + \cdots) + (b_1 \sin x + b_2 \sin 2x + \cdots) \\ &= a_0 + \sum_{n=1}^{\infty} (a_n \cos nx + b_n \sin nx) \end{aligned} \tag{4.19.3}$$

We do not have b_0 because $\sin 0x = 0$. This trigonometric infinite series is called a Fourier series and the coefficients a_n, b_n are called the *Fourier coefficients*. Our goal now is to determine these coefficients.

To determine a_0, we just integrate two sides of Eq. (4.19.3) from $-\pi$ to π,[70] we get

$$\int_{-\pi}^{\pi} f(x) dx = \int_{-\pi}^{\pi} a_0 dx + \sum_{n=1}^{\infty} \left[a_n \int_{-\pi}^{\pi} \cos nx dx + b_n \int_{-\pi}^{\pi} \sin nx dx \right] \tag{4.19.4}$$

Now the "seemingly useless" integrals in Eq. (4.19.2) come into play: the red integrals are all zeroes, so

[70] The results do not change if we integrate from 0 to 2π. In fact, if a function $y = f(x)$ is T-periodic, then

$$\int_a^{a+T} f(x) dx = \int_b^{b+T} f(x) dx$$

Drawing a picture of this periodic function, and note that integral is an area, and you will see why this equation holds.

$$\int_{-\pi}^{\pi} f(x)dx = 2\pi a_0 \Longrightarrow a_0 = \frac{1}{2\pi}\int_{-\pi}^{\pi} f(x)dx \tag{4.19.5}$$

For a_n with $n \geq 1$, we multiply Eq. (4.19.3) with $\cos mx$ and integrate two sides of the resulting equation. Doing so gives us

$$\int_{-\pi}^{\pi} f(x)\cos mxdx = a_0 \int_{-\pi}^{\pi} \cos mxdx + \\ + \sum_{n=1}^{\infty}\left[a_n \int_{-\pi}^{\pi} \cos mx \cos nxdx + b_n \int_{-\pi}^{\pi} \cos mx \sin nxdx\right]$$

Again, the integrals in Eq. (4.19.2) help us a lots here: the red integrals vanish. We're left with this term

$$A = \sum_{n=1}^{\infty} a_n \int_{-\pi}^{\pi} \cos mx \cos nxdx$$

As the blue integral is zero when $n \neq m$ and it is equal to π when $n = m$, the above term should be equal $a_m\pi$.[71] Thus,

$$\int_{-\pi}^{\pi} f(x)\cos mxdx = a_m\pi \Longrightarrow a_m = \frac{1}{\pi}\int_{-\pi}^{\pi} f(x)\cos mxdx \tag{4.19.6}$$

Similarly, for b_n we multiply Eq. (4.19.3) with $\sin mx$ and integrate two sides of the resulting equation. Doing so gives us

$$b_m = \frac{1}{\pi}\int_{-\pi}^{\pi} f(x)\sin mxdx \tag{4.19.7}$$

Example 4.19 As the first application of Fourier series, let's try the square wave function given by

$$f(x) = \begin{cases} 0 & \text{if } -\pi \leq x < 0 \\ 1 & \text{if} 0 \leq x < \pi \end{cases}, \quad f(x + 2\pi) = f(x) \tag{4.19.8}$$

Square waves are often encountered in electronics and signal processing, particularly digital electronics and digital signal processing. Mathematicians call the function in Eq. (4.19.8) a *piecewise continuous function*. This is because the function is consisted

---

[71] If it is still not clear, one specific case of $m = 3$ helps. We have

$$A = a_1\int_{-\pi}^{\pi} \cos 3x \cos 1xdx + a_2\int_{-\pi}^{\pi} \cos 3x \cos 2xdx + a_3\int_{-\pi}^{\pi} \cos 3x \cos 3xdx + a_4\int_{-\pi}^{\pi} \cos 3x \cos 4xdx + \cdots$$

All red integrals are zeros. Therefore, $A = a_3\pi$.

of many pieces, each piece is defined on a sub-interval. Within a sub-interval the function is continuous, but at some points between two neighboring sub-intervals there is a jump.

The determination of the Fourier coefficients for this function is quite straightforward:

$$\begin{aligned}
a_0 &= \frac{1}{2\pi}\int_{-\pi}^{\pi} f(x)dx = \frac{1}{2\pi}\int_{0}^{\pi} dx = \frac{1}{2} \\
a_n &= \frac{1}{\pi}\int_{-\pi}^{\pi} f(x)\cos nx dx = \frac{1}{\pi}\int_{0}^{\pi} \cos nx dx = 0 \\
b_n &= \frac{1}{\pi}\int_{-\pi}^{\pi} f(x)\sin nx dx = \frac{1}{\pi}\int_{0}^{\pi} \sin nx dx = -\frac{1}{n\pi}(\cos n\pi - 1)
\end{aligned}$$

Noting that b_n is non-zero only for odd n. In that case, $\cos n\pi = -1$. Thus, $b_n = {}^2/_{n\pi}$ for $n = 1, 3, 5, \ldots$ Finally, the Fourier series of this square wave is

$$f(x) = \frac{1}{2} + \frac{2}{\pi}\sin x + \frac{2}{3\pi}\sin 3x + \cdots = \frac{1}{2} + \sum_{n=1}^{\infty} \frac{2}{(2n-1)\pi}\sin(2n-1)x \tag{4.19.9}$$

We have spent some time to compute the Fourier coefficients without knowing for sure that the square wave can indeed be represented by Fourier series. But, we can check this now. Figure 4.83 plots the square wave along with some of its Fourier series with 1, 3, 5, 7 and 15 terms. With more than 7 terms, a good approximation is obtained. Note that Taylor series cannot do this!

It is possible to write a discontinuous function as a sum of sines and cosines which are continuous functions! Lagrange and Laplace just didn't think that was possible.

Let's have some fun with this new toy and we will re-discover an old series. For $0 \le x < \pi$, $f(x) = 1$, so we can write $1 = {}^1/_2 + {}^2/_\pi \sin x + {}^2/_{3\pi} \sin 3x + \cdots$. Then, a bit of algebra, and finally choosing $x = \pi/2$, we see again the well-known series for $\pi/4$:

$$\begin{aligned}
\frac{1}{2} &= \frac{2}{\pi}\sin x + \frac{2}{3\pi}\sin 3x + \cdots \\
\frac{\pi}{4} &= \sin x + \frac{1}{3}\sin 3x + \frac{1}{5}\sin 5x + \cdots \\
\frac{\pi}{4} &= 1 - \frac{1}{3} + \frac{1}{5} - \frac{1}{7} + \cdots \quad \text{(evaluating the above equation at} x = \pi/2)
\end{aligned}$$

Remark 4.1 Similar to 32 457 = 3 × 7 × 11 × 983 (a large composite number can be decomposed into many smaller prime numbers), we can also decompose a periodic function into simpler components (the sines and cosines). These fall into the grand theme: complex things can be built on top of simpler things.

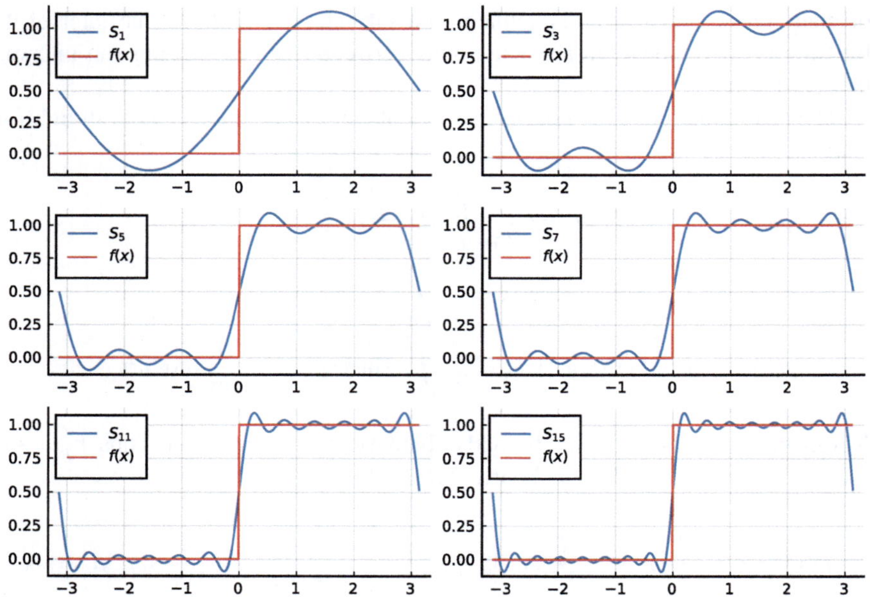

Fig. 4.83 Representing a square wave function by a Fourier series $S_n = 1/2 + 2/\pi \sin x + \cdots + 2/n\pi \sin nx$ for $n = 2k - 1$. Source: `fourier-square-wave.jl`

4.19.2 Functions with Period 2L

Suppose we have a periodic function $f(x)$ with a period $2L$ rather than 2π. Our goal is to derive its Fourier series. We can do the same thing that we did in the previous section. But we do not want to repeat that; instead, we use a change of variable: $t = \pi x/L$

$$f(x) = f\left(\frac{Lt}{\pi}\right) = g(t)$$

As $g(t)$ is a periodic function of 2π period, we have $g(t) = a_0 + \sum_{n=1}^{\infty}(a_n \cos nt + b_n \sin nt)$ with the Fourier coefficients given by

$$a_0 = \frac{1}{2\pi}\int_{-\pi}^{\pi} g(t)dt, \quad a_n = \frac{1}{\pi}\int_{-\pi}^{\pi} g(t)\cos nt dt, \quad b_n = \frac{1}{\pi}\int_{-\pi}^{\pi} g(t)\sin nt dt$$

With $t = \pi x/L$, we have $dt = \pi/L dx$ and thus:

$$f(x) = a_0 + \sum_{n=1}^{\infty}\left(a_n \cos\frac{n\pi x}{L} + b_n \sin\frac{n\pi x}{L}\right) \tag{4.19.10}$$

where

$$a_0 = \frac{1}{2L}\int_{-L}^{L} f(x)dx, \quad a_n = \frac{1}{L}\int_{-L}^{L} f(x)\cos\frac{n\pi x}{L}dx, \quad b_n = \frac{1}{L}\int_{-L}^{L} f(x)\sin\frac{n\pi x}{L}dx$$

Remark 4.2 In many books, you would see the Fourier series written slightly different. For example, instead of $2L-$periodic functions, we can consider $l-$periodic (i.e., $2L = l$), then the Fourier series is written as[72]

$$f(x) = \frac{a_0}{2} + \sum_{n=1}^{\infty}\left(a_n \cos\frac{2n\pi x}{l} + b_n \sin\frac{2n\pi x}{l}\right) \tag{4.19.11}$$

where

$$a_n = \frac{1}{L}\int_{-l/2}^{l/2} f(x)\cos\frac{2n\pi x}{l}dx \ (n = 0, 1, \ldots), \quad b_n = \frac{2}{l}\int_{-l/2}^{l/2} f(x)\sin\frac{2n\pi x}{l}dx$$

Example 4.20 In this example, we consider a triangular wave defined by

$$f(x) = |x|, \quad -1 \le x \le 1, \quad f(x+2) = f(x) \tag{4.19.12}$$

The determination of the Fourier coefficients for this function is also straightforward:

$$\begin{aligned}
a_0 &= \frac{1}{2}\int_{-1}^{1} |x|dx = \frac{1}{2\pi}\int_0^{\pi} dx = \frac{1}{2}\\
a_n &= \int_{-1}^{1} |x|\cos nx dx = 2\int_0^1 x\cos nx dx = \frac{2}{n^2\pi^2}(\cos n\pi - 1)\\
b_n &= \int_{-1}^{1} |x|\sin nx dx = 0 \quad (|x|\sin nx \text{ is an odd function })
\end{aligned}$$

Of course, I've used integration by parts to compute a_n. Noting that a_n is non-zero only for odd n. In that case, $\cos n\pi = -1$. Thus, the Fourier series of this triangular wave is

$$f(x) = \frac{1}{2} - \frac{4}{\pi^2}\cos\pi x - \frac{4}{9\pi^2}\cos 3\pi x + \cdots = \frac{1}{2} - \sum_{n=1}^{\infty}\frac{4}{(2n-1)^2\pi^2}\cos(2n-1)\pi x \tag{4.19.13}$$

A plot of some Fourier series of this function is given in Fig. 4.84. Only four terms and we obtain a very good approximation.

[72] Why a factor 1/2 for a_0? Just not to differentiate a_0 and a_n.

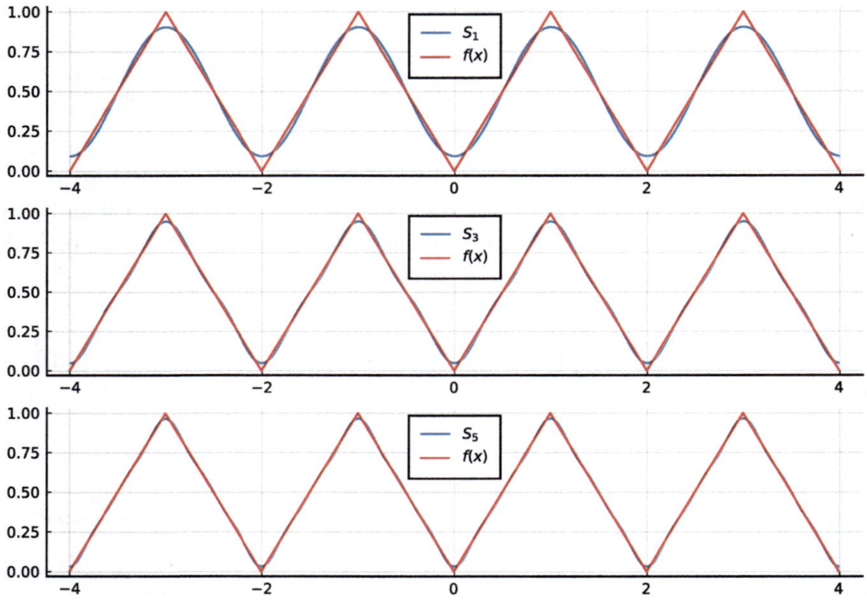

Fig. 4.84 Representing a triangular wave function by a Fourier series $S_n = \frac{1}{2} - \sum_{n=1}^{\infty} {}^{4}/{}_{(2n-1)^2\pi^2} \cos(2n-1)\pi x$

Similarly to Example 4.19, we can also get a nice series related to π^2 by considering $f(x)$ and its Fourier series at $x = 0$:

$$
\begin{aligned}
f(x) &= \frac{1}{2} - \frac{4}{\pi^2}\cos \pi x - \frac{4}{9\pi^2}\cos 3\pi x - \frac{4}{25\pi^2}\cos 5\pi x \\
\frac{1}{2} &= \frac{4}{\pi^2} + \frac{4}{9\pi^2} + \frac{4}{25\pi^2} + \cdots \Longrightarrow \frac{\pi^2}{8} = \frac{1}{1} + \frac{1}{9} + \frac{1}{25} + \cdots
\end{aligned}
$$

Do you see the connection of ${}^{\pi^2}/{}_8 = 1 + 1/9 + 1/125 + \cdots$ to the Basel problem (Sect. 2.21.4)? Actually, using this result, we can compute the Basel sum quite straightforwardly:

$$
\begin{aligned}
S &= \frac{1}{1^2} + \frac{1}{2^2} + \frac{1}{3^2} + \frac{1}{4^2} + \frac{1}{5^2} + \cdots = \left(1 + \frac{1}{3^2} + \frac{1}{5^2} + \cdots\right) + \left(\frac{1}{2^2} + \frac{1}{4^2} + \frac{1}{6^2} + \cdots\right) \\
&= \frac{\pi^2}{8} + \frac{1}{4}\left(1 + \frac{1}{4} + \frac{1}{9} + \cdots\right) = \frac{\pi^2}{8} + \frac{1}{4}S \Longrightarrow S = \frac{\pi^2}{6}
\end{aligned}
$$

Now, what is important to consider is the difference between the Fourier series for the square wave and the triangular wave. I put these two series are side by side now

$$\text{square wave: } f(x) = \frac{1}{2} + \sum_{n=1}^{\infty} \frac{2}{(2n-1)\pi} \sin(2n-1)$$

$$\text{triangular wave: } f(x) = \frac{1}{2} - \sum_{n=1}^{\infty} \frac{4}{(2n-1)^2\pi^2} \cos(2n-1)\pi x$$

Now we can see why we need less terms in the Fourier series to represent the triangular wave rather than the square wave. The difference lies in the red number. The terms in the triangular series approach zero faster than the terms in the square series. And by looking at the shape of these waves, it is obvious that smoother waves (the square wave has discontinuities) are easier for Fourier series to converge.

But, wait. Why only periodic functions? How about non-periodic functions? Can we still decompose non-periodic functions into simpler components? The answer is yes via Fourier transform.

4.19.3 Complex Form of Fourier Series

Herein we derive the complex form of the Fourier series. The idea is to use Eq. (4.19.3) and replace $\cos nx$ and $\sin nx$ with complex exponential using Euler's formula:

$$\cos nx = \frac{e^{inx} + e^{-inx}}{2}, \quad \sin nx = \frac{e^{inx} - e^{-inx}}{2i}$$

Doing so gives us (d is simply a_0):

$$\begin{aligned} f(x) &= d + \sum_{n=1}^{\infty} \left[a_n \left(\frac{e^{inx} + e^{-inx}}{2} \right) + b_n \left(\frac{e^{inx} - e^{-inx}}{2i} \right) \right] \\ &= d + \sum_{n=1}^{\infty} \left[\left(\frac{a_n - ib_n}{2} \right) e^{inx} + \left(\frac{a_n + ib_n}{2} \right) e^{-inx} \right] \end{aligned} \tag{4.19.14}$$

which can be written as

$$f(x) = \sum_{n=-\infty}^{\infty} c_n e^{inx}$$

where the coefficients c_n are given by

$$c_n = \begin{cases} d, & \text{if } n = 0 \\ \frac{a_n - ib_n}{2}, & \text{if } n = 1, 2, 3, \ldots \\ \frac{a_{-n} + ib_{-n}}{2}, & \text{if } n = -1, -2, -3, \ldots \end{cases} \tag{4.19.15}$$

We need the formula for c_n and we're done. Recall that (assuming 2π−periodic functions)

$$d = a_0 = \frac{1}{2\pi}\int_{-\pi}^{\pi} f(x)dx,\ a_n = \frac{1}{\pi}\int_{-\pi}^{\pi} f(x)\cos nx dx,\ b_n = \frac{1}{\pi}\int_{-\pi}^{\pi} f(x)\sin nx dx$$

Thus, for $n > 0$, c_n is written as

$$c_n := \frac{a_n - ib_n}{2} = \frac{1}{2\pi}\int_{-\pi}^{\pi} f(x)(\cos nx - i\sin nx)dx = \frac{1}{2\pi}\int_{-\pi}^{\pi} f(x)e^{-inx}dx$$

And this formula holds for $n \leq 0$ as well.

We now get the complex form of the Fourier series, written as for any periodic functions of $2L$:

$$\boxed{f(x) = \sum_{n=-\infty}^{\infty} c_n e^{in\pi x/L}, \quad c_n = \frac{1}{2L}\int_{-L}^{L} f(x)e^{-in\pi x/L}dx} \tag{4.19.16}$$

Having another way to look at the Fourier series is itself something significant. Still, we can see the benefits of the complex form: instead of having a_0, a_n and b_n and the sines and cosines, now we just have c_n and the complex exponential.

We have more, a lot more, to say about Fourier series e.g. Fourier transforms, discrete Fourier transform, fast Fourier transforms etc. (Sect. 7.12). We still do not know the meanings of the a's and b's (or c_n). We do not know which functions can have a Fourier series. To answer these questions, we need more maths such as linear algebra. I have introduced the Fourier series as early as here for the following reasons. First, we've learned about the Taylor series (which allows us to represent a function with a power series). Now, we have something similar: Fourier series where a function is represented as a trigonometric series. Second, something like the identity $\int_{-\pi}^{\pi} \sin nx \cos mx dx = 0$ looks useless, but it is not.

About Fourier's idea of expressing a function as a trigonometry series, the German mathematician Bernhard Riemann once said:

> Nearly fifty years has passed without any progress on the question of analytic representation of an arbitrary function, when an assertion of Fourier threw new light on the subject. Thus *a new era began* for the development of this part of Mathematics and this was heralded in a stunning way by major developments in mathematical physics.

4.20 Special Functions

4.20.1 Elementary Functions

Before presenting special functions, we need to define the non-special or elementary functions. Those are the familiar functions that we know how to differentiate and integrate:

(a) Powers of x: x, x^2, x^3, etc.
(b) Roots of x: $\sqrt{x}, \sqrt[3]{x}, x^{1/5}$, etc.
(c) Exponentials: e^x
(d) Logarithms: $\log x$.
(e) Trigonometric functions: $\sin x, \cos x, \tan x$, etc.
(f) Inverse trigonometric functions: $\arcsin x, \arccos x, \arctan x$, etc.
(g) Composite functions of the previous six functions: $\log(\sin x)x, \cos^2 x$, etc.
(h) All functions are obtained by adding, subtracting, multiplying, and dividing any of the above seven types a finite number of times. Examples are

$$x^2 + \sin x^3 - \log x$$

4.20.2 Factorial of 1/2 and the Gamma Function

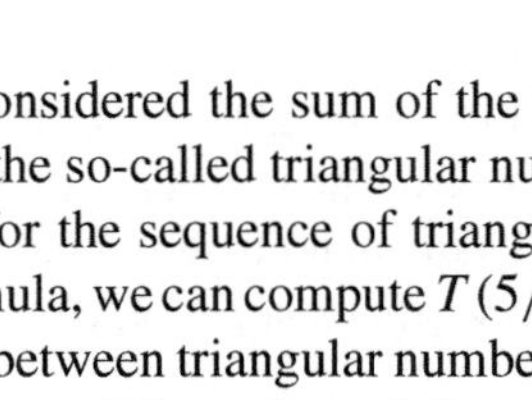

In Sect. 2.6.1, we have considered the sum of the first n counting numbers e.g. $1 + 2 + 3$, and encountered the so-called triangular numbers $1, 3, 6, 10, 15, \ldots$ It is possible to have a formula for the sequence of triangular numbers: $T(n) = n(n+1)/2$. Because we have a formula, we can compute $T(5/2)$ *whatever it means*. In other words, we can interpolate in between triangular numbers. However, for the factorials $1, 2, 6, 24, 120, \ldots$ there is no such formula, and thus we cannot interpolate between factorials of natural numbers. This was the problem proposed by Goldbach in 1720. In the quest for such a formula for $n!$, mathematicians like Euler, Daniel Bernoulli, and Gauss invented the gamma function. As can be seen from the figure, it is possible to draw a curve passing through $(n, n!)$ for $n = 0, 1, 2\ldots$ Thus, theoretically there should exist at least one function $f(x)$ such that $f(n) = n!$ for $n \in \mathbb{N}$.

Recall that in Eq. (4.7.19), we have obtained the following result (as an exercise of integral calculus if we bypass the motivation of the function x^4e^{-x})

$$\int_0^\infty x^4e^{-x}dx = 4!$$

And from that we get (I changed the dummy variable from x to t)

$$n! = \int_0^\infty t^n e^{-t}dt$$

The Gamma function (the notation Γ was due to Legendre) is defined as

$$\boxed{\Gamma(x) := \int_0^\infty t^{x-1}e^{-t}dt} \quad (4.20.1)$$

Therefore,

$$\Gamma(x) = (x-1)! \quad (4.20.2)$$

And with this integral definition of factorial, we are no longer limited to factorials of natural numbers. Indeed, we can compute (0.5)! as[73]

$$\left(\frac{1}{2}\right)! = \Gamma\left(\frac{3}{2}\right) = \int_0^\infty t^{1/2}e^{-t}dt = \frac{\sqrt{\pi}}{2} \quad (4.20.3)$$

$$\left(-\frac{1}{2}\right)! = \Gamma\left(\frac{1}{2}\right) = \int_0^\infty t^{-1/2}e^{-t}dt = \sqrt{\pi} \quad (4.20.4)$$

4.21 Review

It was a long chapter. This is no surprise for we have covered the mathematics developed during a time span of about 200 years. But as is always the case: try to do not lose the forest for the trees. The core of calculus is simple, and I am trying to summarize that core now. Understand that and others will follow quite naturally (except the rigorous foundation—that's super hard).

- The calculus is the mathematics of change: it provides us notions and symbols and methods to talk about changes precisely;
- What is better than motion as an example of change? For motion, we need three notions: (1) position $x(t)$—to quantify the position (that answers the question of where an object is at a particular time), (2) velocity $v(t)$—to quantify the speed

[73] For the final integral, change of variable $u = t^{1/2}$ and we get a new integral $2\int u^2e^{-u^2}du$.

(that answers the question how fast our object is moving), and (3) acceleration $a(t)$—to quantify how fast the object changes its speed.

- Going from (1) to (2) to (3) is called "taking the derivative": the derivative gives us a way to quantify a time rate of change. The velocity is the rate of change of the position per unit time. That's why we have the symbols dx, dt and $^{dx}/_{dt}$;
- Going from (3) to (2) to (1) is called "taking the integral": $x(t) = \int_0^t v dt$. Knowing the speed $v(t)$ and considering a very small time interval dt during which the distance the object has traveled is $v(t)dt$, finally adding up all those tiny distances and we get the total distance $x(t)$;
- So, calculus is the study of derivatives and integrals. But they are not two independent things, they are the inverse of each other like negative/positive numbers, men/women, war/peace, and so on;
- When we studied counting numbers we discovered many rules (e.g. odd + odd = even). The same pattern is observed here: the new toys of mathematicians—the derivative and the integral—have their own rules. For example, the derivative of a sum is the sum of the derivatives. Thanks to this rule, we know how to determine the derivative of $x^{10} + x^5 + 23x^3$, for example, we know how to differentiate each term.
- Calculus does to algebra what algebra does to arithmetic. Arithmetic is about manipulating numbers (addition, multiplication, etc.). Algebra finds patterns between numbers e.g. $a^2 - b^2 = (a - b)(a + b)$. Calculus finds patterns between varying quantities;
- Historically Fermat used derivatives in his calculations without knowing it. Later, Newton and Leibniz discovered it. Other mathematicians such as Brook, Euler, and Lagrange developed and characterized it. Only at the end of this long period of development, which spans about 200 years, did Cauchy and Weierstrass define it.
- Confine to the real numbers, the foundation of the calculus is the concept of limit. This is so because, with limits, mathematicians can prove all the theorems in calculus rigorously. That branch of mathematics is called *analysis*. This branch focuses not on the computational aspects of calculus (e.g. how to evaluate an integral or how to differentiate a function), instead, it focuses on why calculus works.

In the beginning of this chapter, I quoted Richard Feynman saying that "Calculus is the language God talks", and Steven Strogatz writing "Without calculus, we wouldn't have cell phones, computers, or microwave ovens. We wouldn't have radio. Or television. Or ultrasound for expectant mothers, or GPS for lost travelers. We wouldn't have split the atom, unraveled the human genome, or put astronauts on the moon". But for that, we need to learn multivariable calculus and vector calculus (Chap. 6)—the generalizations of the calculus discussed in this chapter and differential equations (Chap. 7). This is obvious: our world is three dimensions and the things we want to understand depend on many other things. Thus, $f(x)$ is not sufficient. But the idea of multivariable calculus and vector calculus is still the mathematics of changes: a small change in one thing leads to a small change in another thing.

Consider a particle of mass m moving under the influence of a force F, then Newton gave us the following equation $m d^2x/dt^2 = F$, which, in conjunction with the data about the position of the particle at $t = 0$, can pinpoint exactly the position of the particle at any time t. This is probably the first differential equation—those equations that involve the derivatives—ever. This is the equation that puts men on the Moon.

Leaving behind the little bits dx, dy and the sum $\int$, our next destination in the mathematical world is a place called probability. Let's go there to see dice, roulette, lotteries—game of chances—to see how mathematicians develop mathematics to describe random events, how they can see through the randomness to reveal its secrets.

Chapter 5
Probability

Games of chance are common in our world—including lotteries, roulette, slot machines, and card games. Thus, it is important to know a bit about the mathematics behind them which is known as probability theory.

Gambling led Cardano—our Italian friend whom we met in the discussion on cubic equations—to the study of probability, and he was the first writer to recognize that *random events are governed by mathematical laws*. Published posthumously in 1663, Cardano's Liber de ludo aleae (Book on Games of Chance) is often considered the major starting point of the study of mathematical probability.

Since then the theory of probability has become a useful tool in many problems. For example, meteorologists use weather patterns to predict the probability of rain. In epidemiology, probability theory is used to understand the relationship between exposures and the risk of health effects. Another application of probability is with car insurance. Companies base your insurance premiums on your probability of having a car accident. To do this, they use information on the frequency of having a car accident by gender, age, type of car, and number of kilometers driven each year to estimate an individual person's probability (or risk) of a motor vehicle accident.

Indeed probability is so useful that the famous French mathematician and astronomer (known as the "Newton of France") Pierre-Simon Marquis de Laplace once wrote:

> We see that the theory of probability is at bottom only common sense reduced to calculation; it makes us appreciate with exactitude what reasonable minds feel by a sort of instinct, often without being able to account for it....It is remarkable that this science, which originated in the consideration of games of chance, should have become the most important object of human knowledge... The most important questions of life are, for the most part, really only problems of probability.

This chapter is an introduction to probability and statistics. It was written based on the following excellent books:

V. P. Nguyen, *Mathematics for Engineers and Scientists*,
https://doi.org/10.1007/978-981-97-4631-6_5

- *The Unfinished game: Pascal, Fermat and the letters* by Keith Devlin[1] [17]
- *Introduction to Probability, Statistics, and Random Processes* by Hossein Pishro-Nik[2]
- *A first course in Probability* by Sheldon Ross[3] [57]
- *The history of statistics: the measurement of uncertainty before 1900* by Stefen Stigler[4] [62].
- *A History of Probability and Statistics and Their Applications before 1750*, by Anders Hald[5] [28].

I did not like gambling and did not pay attention to probability. I performed badly in high school and university when it came to classes on probability; I actually failed the unit. But I do have companies. In 2012, 97 Members of Parliament in London were asked: "If you spin a coin twice, what is the probability of getting two heads?" The majority, 60 out of 97, could not give the correct answer.

I did not plan to re-learn probability but the COVID pandemic came. People are talking about the probability of getting the COVID, etc. And I wanted to understand what they mean. Furthermore, probability is used in machine learning, data science, and many science and engineering fields. Therefore, I decided to study the theory of probability again. I do not have to be scared as this time I will not have to take any exam about probability! I just have fun.

---

[1] Keith J. Devlin (born 16 March 1947) is a British mathematician and popular science writer. His current research is mainly focused on the use of different media to teach mathematics to different audiences.

[2] The book is freely available at https://www.probabilitycourse.com/.

[3] Sheldon Ross (30 April 1943) is the Daniel J. Epstein Chair and Professor at the USC Viterbi School of Engineering. He is the author of several books in the field of probability. In 1978, he formulated what became known as Ross's conjecture in queuing theory, which was solved 3 years later by Tomasz Rolski at Poland's Wroclaw University.

[4] Stephen Mack Stigler (born 10 August 1941) is Ernest DeWitt Burton Distinguished Service Professor at the Department of Statistics of the University of Chicago. He has authored several books on the history of statistics. Stigler is also known for Stigler's law of eponymy which states that no scientific discovery is named after its original discoverer (whose first formulation he credits to sociologist Robert K. Merton).

[5] Anders Hjorth Hald (1913–2007) was a Danish statistician. He was a professor at the University of Copenhagen from 1960 to 1982. While a professor, he did research in industrial quality control and other areas, and also authored textbooks. After retirement, he made important contributions to the history of statistics.

5.1 A Brief History of Probability

This brief historical account is taken from *Calculus*, Volume II by Tom Apostol (2nd edition, John Wiley & Sons 1969).

A gambler's dispute in 1654 led to the creation of a mathematical theory of probability by two famous French mathematicians, Blaise Pascal and Pierre de Fermat. Antoine Gombaud, Chevalier de Méré, a French nobleman with an interest in gaming and gambling questions, called Pascal's attention to an apparent contradiction concerning a popular dice game. The game consisted in throwing a pair of dice 24 times; the problem was to decide whether or not to bet even money on the occurrence of at least one "double six" during the 24 throws. A seemingly well-established gambling rule led de Méré to believe that betting on a double six in 24 throws would be profitable, but his own calculations indicated just the opposite.[6]

This problem and others posed by de Méré led to an exchange of letters between Pascal and Fermat in which the fundamental principles of probability theory were formulated for the first time. Although a few special problems on games of chance had been solved by some Italian mathematicians in the fifteenth and sixteenth centuries, no general theory was developed before this famous correspondence. The correspondence between Pascal and Fermat was told in the interesting book *The Unfinished game: Pascal, Fermat and the letters* by Keith Devlin.

The Dutch scientist Christian Huygens, a teacher of Leibniz who co-invented calculus with Newton, learned of this correspondence and shortly thereafter published the first book on probability in 1657; entitled *De Ratiociniis in Ludo Aleae* or *The Value of all Chances in Games of Fortune*, it was a treatise on problems associated with gambling. Because of the inherent appeal of games of chance, probability theory soon became popular, and the subject developed rapidly during the eighteenth century. The major contributors during this period were Jakob Bernoulli with *Ars Conjectandi* in 1713, and Abraham de Moivre with his classic *The Doctrine of Chances* in 1718.

In 1812, Pierre de Laplace (1749–1827) introduced a host of new ideas and mathematical techniques in his book, *Théorie Analytique des Probabilités* or Analytical Theory of Probability. Before Laplace, probability theory was solely concerned with developing a mathematical analysis of games of chance. Laplace applied probabilistic ideas to many scientific and practical problems. The theory of errors, actuarial

---

6 The probability of getting at least one double six in 24 throws is $1 - (35/36)^{24} = 0.4914$, which is smaller than 0.5.

mathematics, and statistical mechanics are examples of some of the important applications of probability theory developed in the nineteenth century.

Like so many other branches of mathematics, the development of probability theory has been stimulated by the variety of its applications. Conversely, each advance in the theory has enlarged the scope of its influence. Mathematical statistics is one important branch of applied probability; other applications occur in such widely different fields as genetics, psychology, economics, and engineering. Many workers have contributed to the theory since Laplace's time; among the most important are Chebyshev, Markov, von Mises, and Kolmogorov.

One of the difficulties in developing a mathematical theory of probability has been to arrive at a definition of probability that is precise enough for use in mathematics, yet comprehensive enough to be applicable to a wide range of phenomena. The search for a widely acceptable definition took nearly three centuries and was marked by much controversy. The matter was finally resolved in the twentieth century by treating probability theory on an axiomatic basis. In 1933, the Russian mathematician A. Kolmogorov outlined an axiomatic approach that forms the basis for the modern theory.[7] Since then the ideas have been refined somewhat and probability theory is now part of a more general discipline known as measure theory.

5.2 Classical Probability

Probability is a mathematical theory that helps us to quantify random events or experiments. A random event or experiment is the one whose outcomes we can't predict with certainty. For instance, if we flip a coin, we're not sure whether we get a head or a tail.

Although we cannot tell in advance whether a head or a coin will show up, we are, however, able to tell how likely that a head (or a tail) is to occur. Here is how. The possible outcomes of this coin tossing experiment are either a head (H) or tail (T); two possible outcomes. And out of these two outcomes, we have one chance to get H, thus the probability that we get a H is $1/2 = 0.5$. With equal probabilities, we say that getting a H and getting a T are equally likely to occur.

Now suppose that we throw two coins simultaneously. What are the possible outcomes? They are $\{(H, H), (H, T), (T, H), (T, T)\}$. In the first case, we get heads twice, in the last case tails twice, whereas the two intermediate cases lead to the same result since it does not matter to us in which coin heads or tails appear. Thus, we say that the chance of getting heads twice is 1 out of 4 or $1/4$, the chance of getting tails twice is also $1/4$, whereas the chance of one head and one tail is 2 out of 4 or $2/4 = 1/2$.

Similar to other mathematical objects, there exist rules that probability obeys. It is interesting, isn't it? There are regularities behind random events! Here are some.

[7] Kolmogorov's monograph is available in English translation as Foundations of Probability Theory, Chelsea, New York, 1950.

Table 5.1 Tossing a coin three times: all possible outcomes

1st toss	H	H	H	H	T	T	T	T
2nd toss	H	H	T	T	H	H	T	T
3rd toss	H	T	H	T	H	T	H	T
Category	I	II	II	III	II	III	III	IV
P	1/8	3/8		3/8				1/8

Tossing a fair coin and the probability of getting a head is 0.5, for brevity, we write $P(H) = 0.5$, where P stands for probability and the notation $P(H)$ borrows the concept of functions $f(x)$ that we're familiar with from calculus. And we also have $P(T) = 0.5$. Obviously, $P(H) + P(T) = 1$. What does that mean? It indicates that it is 100% sure that either we get a head or a tail. Thus, unity in the theory of probability means a certainty. For the experiment of tossing two coins, again $1/4 + 1/2 + 1/4 = 1$, meaning that we are certain to get one of the 3 possible combinations (two Hs, two Ts, one H/one T).

Let's do a more interesting experiment of tossing a coin three times. The possible outcomes and the probability of some scenarios (called events in the theory of probability) are shown in Table 5.1.

From the three experiments we have discussed, we can see that

$$\begin{aligned}
\text{toss 1 coin 1 time: } P(H) &= 1/2 \\
\text{toss 1 coin 2 times: } P(2H) &= 1/4 \\
\text{toss 1 coin 3 times: } P(3H) &= 1/8
\end{aligned}$$

which indicates that the probability of getting heads twice is equal to the product of the probabilities of getting it separately in the first and in the second tossing; in fact, $1/4 = (1/2)(1/2)$. Similarly, the probability of getting three heads in succession is the product of probabilities of getting it separately in each tossing ($1/8 = (1/2)(1/2)(1/2)$). Thus, if somebody asks you what is the chance of getting head each time in ten tossings, you can easily give the answer: $(1/2)^{10}$. The result will be 0.00098, indicating that the chance is very low: about one chance out of a thousand! Here we have the rule of *multiplication of probabilities*, which states that if you want several different things, you may determine the probability of getting them by multiplying the probabilities of getting the several individual ones. And that chance is usually low as we're asking too much! I am joking, the chance is low because we multiply numbers smaller than one.

If there is a rule of multiplication, there should be another rule: the rule of *addition of probabilities*. This rule states that if we want only *one of several things* (no matter which one), the probability of getting it is the sum of probabilities of getting individual items on our list. For example, when flipping a coin twice, the outcomes are $\{(H, H), (H, T), (T, H), (T, T)\}$. Thus, the chance to get at least one head is

$3/4$, which is equal to $1/4 + 1/4 + 1/4$. Note that to get at least one head, we either need (H, H) or (H, T) or (T, H), each has a probability of 1/4.

What we have discussed is known as classical probability or theoretical probability. It started with the work of Cardano. The classical theory of probability has the advantage that it is conceptually simple for many situations. However, it is limited, since many situations do not have finitely many equally likely outcomes. Tossing a weighted die is an example where we have finitely many outcomes, but they are not equally likely. Studying people's incomes over time would be a situation where we need to consider infinitely many possible outcomes since there is no way to say what a maximum possible income would be, especially if we are interested in the future.

5.3 Empirical Probability

What do we mean actually when we're saying that the chance of getting a head when we toss a coin is $1/2$? What we mean by this is that if we toss a coin 1000 times we would expect to get a head about 500 times. We could do the real experiment of tossing a coin 1000 times. But we ask our computer to do that for us. See Listing A.15 for the `Julia` code for this virtual experiment in which we flip a coin n times and count the number of times that a head shows, labeled by $n(H)$. Then we define the probability of getting a head as its relative frequency of occurrence, mathematically by $n(H)/n$. The result given in Table 5.2 tells us many things. One thing is that when n is large the probability is indeed about 0.5, and this led to the following definition of the probability of an event E:

$$P(E) := \lim_{n\to\infty} \frac{n(E)}{n} \tag{5.3.1}$$

As we have to carry out experiments this theory of probability is called *empirical probability*. Moreover, as the probability is defined based on the relative frequency of the event, this theory of probability is also referred to as *Frequentist probability*.

Limitations of the empirical probability. The limit of this theory of probability lies in Eq. (5.3.1). How do we know that $n(E)/n$ will converge to some constant lim-

Table 5.2 Virtual experiment of tossing a coin n times

n	$n(H)$	P
10	6	0.6
100	48	0.48
1000	492	0.492
2000	984	0.492
10000	5041	0.5041

iting value that will be the same for each possible sequence of repetitions of the experiment? Table 5.2 obviously indicates that the term $n(E)/n$ is actually oscillating.

There is a need of another theory of probability. Axiomatic probability is such a theory, it unifies different probability theories. Similar to Euclidean geometry, the axiomatic probability starts with three axioms called Kolmogorov's Three Axioms, named after the Soviet mathematician Andrey Nikolaevich Kolmogorov (1903–1987).

5.4 Buffon's Needle Problem and Monte Carlo Simulations

5.4.1 Buffon's Needle Problem

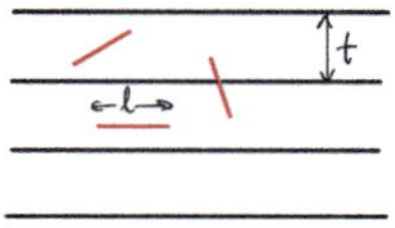

In 1777, Buffon[8] posed (and solved) this problem: Let a needle of length l be thrown at random onto a horizontal plane ruled with parallel straight lines separated by a distance t which is greater than l. What is the probability that the needle will intersect one of these lines?

We use trigonometry and calculus to solve this problem. First, it is sufficient to consider just two lines.[9] To locate the position of the needle, we just need two variables: its center O and its orientation by θ (Fig. 5.1). To specify O, we use d—the distance from O to the nearest line, then $0 \leq d \leq t/2$. For the orientation, $0 \leq \theta \leq \pi/2$. Now, we can specify when the needle cuts the lines:

$$d \leq \frac{l}{2} \sin\theta$$

Now, we plot the function $d = \frac{l}{2}\sin\theta$ on the $\theta - d$ plane. The cut condition is then the area of the shaded region in Fig. 5.1. The probability that the needle will intersect one of these lines is then:

$$P = \frac{\displaystyle\int_0^{\pi/2} l/2 \sin\theta d\theta}{\frac{t}{2}\frac{\pi}{2}} = \frac{2l}{\pi t}$$

8 Georges-Louis Leclerc, Comte de Buffon (1707–1788) was a French naturalist, mathematician, cosmologist, and encyclopédiste.

9 This is the most important step; without it, we cannot proceed further. Why 2? Because 1 line is not enough and there are infinitely many lines in the problem, then 2 is sufficient.

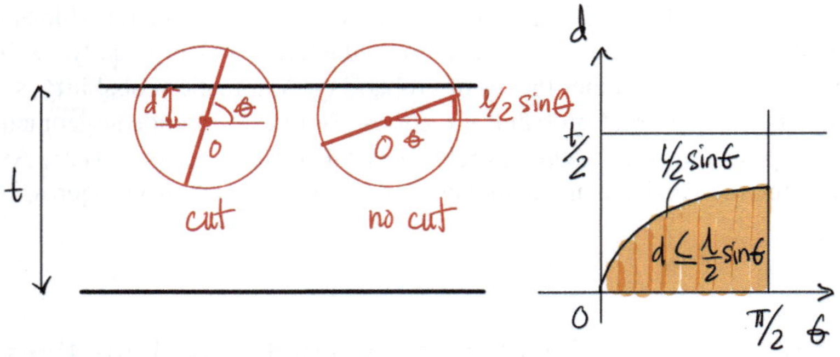

Fig. 5.1 Buffon's needle problem

It is expected that P is proportional to l (the longer the needle the more chance it hits the lines) and inversely proportional to t—the distance between the lines. However, it is unexpected that π shows up in this problem. No circles involved! We discuss this shortly.

In 1886, the French scholar and polymath Marquis Pierre-Simon de Laplace (1749–1827) showed that the number π can be approximated by repeatedly throwing a needle onto a lined sheet of paper N times and counting the number of intersected lines (n):

$$\frac{2l}{\pi t} = \frac{n}{N} \Longrightarrow \boxed{\pi = \frac{2l}{t}\frac{N}{n}}$$

In 1901, the Italian mathematician Mario Lazzarini performed Buffon's needle experiment. Tossing a needle 3408 times with $t = 3$ cm, $l = 2.5$ cm, he got 1808 intersections. Thus, he obtained the well-known approximation $355/113$ for π, accurate to six significant digits. However, Lazzarini's "experiment" is an example of confirmation bias, as it was set up to replicate the already well-known approximation of π, that is $355/113$. Here're the details:

$$\pi = \frac{2l}{t}\frac{N}{n} = \frac{2\cdot 25}{30}\frac{3408}{1808} = \frac{5}{3}\frac{71\cdot 3\cdot 16}{113\cdot 16} = \frac{355}{113} \approx 3.14159292$$

Guessing Buffon's formula. Herein, we're trying to guess the solution without actually solving it. This is a very important skill. (However, I admit that we're doing it only after we have known the result.) As the problem has only two parameters: the needle length l and the distance t between two lines, the result must be of this form $P = c\,(l/t)$ where c is a dimensionless number (refer to Sect. 7.7.1 for detail on dimensions and units). To find out c, we reason that the result should not depend on the shape of the needle. If so, we can consider a needle of the form of a circle of radius r. The length of this circular needle is $2\pi r$ and it must be equal to l, thus its diameter is $d = l/\pi$. The probability is therefore $2l/\pi t$ noting that a circular needle cuts a line twice.

5.4.2 *Monte Carlo Method*

The calculation of π based on the Buffon needle experiment can be considered the first instance of the so-called Monte Carlo method. We can replicate this experiment on a computer, that's the essence of the Monte Carlo method. The underlying concept of the MC method is to use randomness to solve problems that might be deterministic in principle. Throwing randomly n needles is translated to generating n random number $d \in [0, t/2]$ and n random number $\theta \in [0, \pi/2]$. Then we test for a needle cutting a line, if it cuts we record it. The code is given in Listing 5.1.

Listing 5.1 Julia code for Buffon's needle experiment to compute π.

```
1  function buffon_needle(t,l,n)
2    cut = 0
3    for i=1:n
4      d     = (t/2) * rand()
5      theta = 0.5*pi* rand()
6      if ( 0.5*l*sin(theta) >= d ) cut += 1 end
7    end
8    return (2*l/t)*(n/cut) # this is pi
9  end
10 t = 2.0; l = 1.0
11 data = zeros(10,2)
12 data[:,1] = [500 3408 5000 6000 8000 10000 12000 14000 15000 20000]
13 for i=1:size(data,1)
14   data[i,2] = buffon_needle(t,l,data[i,1])
15 end
```

This was probably the simplest introduction to Monte Carlo methods. But if you look at line 5 in the code, you see that we have used $\pi = 3.14159\ldots$ in a program that is about to determine π. Thus, this program is circular.

Another Monte Carlo way for calculating π is presented here. The area of 1/4 of a unit circle is $\pi/4$. We can compute this area by generating N points (x, y) such that $0 \le x \le 1$ and $0 \le y \le 1$.[10] A point is within the area if $x^2 + y^2 \le 1$ (see Fig. 5.2, blue points). We denote the total number of hits by n, then the area is approximately n/N, and thus

$$\pi = 4\frac{n}{N}$$

A Julia code (Listing A.13) was written and the results are given in Table 5.3 for various N. These Monte Carlo methods for approximating π are very slow compared to other methods (e.g., those presented in Sect. 3.2.4), and do not provide any information on the exact number of digits that are obtained. Thus, they are never used to approximate π when speed or accuracy is desired.

[10] Mathematicians like to write this as short as this $\boldsymbol{x} \in [0, 1]^2$, which is to indicate a point inside the square of side 1. But this notation is general enough to cover points inside a unit cube: simply change it to $\boldsymbol{x} \in [0, 1]^3$.

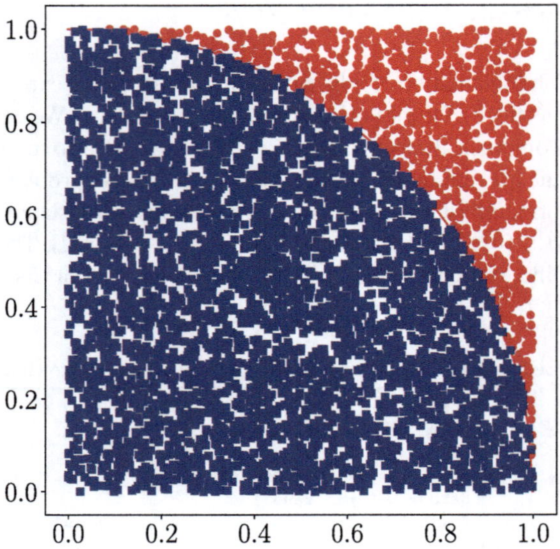

Fig. 5.2 Monte Carlo calculation of π

Table 5.3 Monte Carlo calculation of π

N	$\pi \approx 4\frac{n}{N}$
100	3.40000000
200	3.14000000
400	3.16000000
800	3.14000000
5600	3.16642857

And this MC method can be used to compute numerically any integrals. Section 10.4.5 will present this topic.

5.5 A Review of Set Theory

Probability theory uses the language of sets. Thus, we now briefly review some basic concepts from set theory that are used in this chapter. We discuss set notations, definitions, and operations such as intersections and unions. This section may seem somewhat theoretical and thus less interesting than the rest of the chapter, but it lays the foundation for what is to come.

A set is a collection of things (called elements).[11] We can either explicitly write out the elements of a set as in the set of natural numbers

[11] Why then mathematicians did not use collection? I am not sure, but a collection usually contains a quite large number of things. In maths, a set can be empty or holds one element only.

$$\mathbb{N} = \{1, 2, 3, \ldots\}$$

or, we can also define a set by stating the properties satisfied by its elements. For example, we may write

$$A = \{x \in \mathbb{N} | x \geq 4\}, \quad \text{or} \quad A = \{x \in \mathbb{N} : x \geq 4\}$$

The symbols | and : are read "such that". Thus, the above set contains all counting numbers equal to or greater than four. Because the order of the elements in a set is irrelevant, $\{2, 1, 5\}$ is the same set as $\{1, 2, 5\}$. Furthermore, an element cannot appear more than once in a set; so $\{1, 1, 2, 5\}$ is equivalent to $\{1, 2, 5\}$.

Ordered sets. Let A be a set. An order on A is a relation denoted by $<$ with the following two properties:

- If $x \in A$ and $y \in A$, then one and only one of the following is true:

$$x < y, \quad x = y, \quad x > y$$

- If $x, y, z \in A$, then

$$x < y, \quad y < z \Longrightarrow x < z$$

An ordered set is a set on which an order is defined. For instance, the set of natural numbers is an ordered set.

5.5.1 Subset, Superset, and Empty Set

Set A is a subset of set B if every element of A is also an element of B. We write $A \subset B$ where the symbol $\subset$ indicates "subset". Inversely, B is a superset of A, which we write as $B \supset A$.

A set with no elements is called an *empty set*. How many empty sets there are? To answer that question we need to define when two sets are equal. Two sets are equal if they have the same elements. For example, $\{1, 2, 3\}$ and $\{3, 2, 1\}$ are equal. Now, assume there are two empty sets. If they are not equal, then one set must contain a member that the other does not (otherwise they would be equal). But both sets contain nothing. Thus, they must be equal. *There is only one empty set*: the empty (or null) set is designated by $\emptyset$. This null set is similar to number zero in number theory.

A universal set is the collection of all objects in a particular context. We use the notation S to label the universal set. Its role is similar to the number line in number theory. When we refer to a number we visualize it as a point on the number line. In the same manner, we can visualize a set on the background of the universal set.

The Cartesian product of two sets A and B, denoted by $A \times B$, is defined as the set consisting of all ordered pairs (a, b) for which $a \in A$ and $b \in B$. For example, if $A = x, y$ and $B = \{3, 6, 9\}$, then $A \times B = \{(x, 3), (x, 6), (x, 9), (y, 3), (y, 6), (y, 9)\}$. Note that because the pairs are ordered so $A \times B \neq B \times A$. An important example of sets obtained using a Cartesian product is $\mathbb{R}^n$, where $n \in \mathbb{N}$. For $n = 2$, we have

$$\mathbb{R}^2 = \mathbb{R} \times \mathbb{R} = \{(x, y)|x \in \mathbb{R}, y \in \mathbb{R}\}$$

Thus, $\mathbb{R}^2$ is the set consisting of all points in the two-dimensional plane. Similarly, $\mathbb{R}^3$ is the set of all points in the three-dimensional space that we're living in.

Lower bound and upper bound. Given a set $X \in \mathbb{R}$ (e.g., $X = [0, 5]$), then

- u is an upper bound for X if: $u \geq x$ for $\forall x \in X$;
- l is a lower bound for X if: $l \leq x$ for $\forall x \in X$;
- X is bounded above if there exists an upper bound for X;
- X is bounded below if there exists a lower bound for X;

Sups and Infs. Suppose that X is bounded above, there exist infinite upper bounds.[12] One can define the smallest among the upper bounds. The supremum of X, denoted by $\sup X$, is *the smallest upper bound* for X; that is

- $\sup X \geq x \quad \forall x \in X$ ($\sup X$ is an upper bound);
- $\forall \epsilon > 0, \exists x$ such that $x > \sup X - \epsilon$ ($\sup X$ is the smallest upper bound))

Suppose that X is bounded below, there exist infinite lower bounds. One can define the largest among the lower bounds. The infimum of X, denoted by $\inf X$, is *the largest lower bound* for X; that is

- $\inf X \leq x \quad \forall x \in X$ ($\inf X$ is a lower bound);
- $\forall \epsilon > 0, \exists x$ such that $x < \inf X + \epsilon$ ($\inf X$ is the largest lower bound))

Maximum Versus supremum. Is maximum and supremum of an ordered set the same? Examples can show the answer. Example 1: consider the set $A = \{x \in \mathbb{R}|x < 2\}$. Then, the maximum of A is not 2, as 2 *is not a member of the set*; in fact, the maximum is not well defined. The supremum, though is well defined: 2 is clearly the smallest upper bound for the set. Example 2: $B = \{1, 2, 3, 4\}$. The maximum is 4, as that is the largest element. The supremum is also 4, as 4 is the smallest upper bound.

Venn diagrams. Venn diagrams are useful in visualizing relation between sets. Venn diagrams were popularized by the English mathematician, logician, and philosopher John Venn (1834–1923) in the 1880s. See Fig. 5.3 for one example of Venn diagrams. In a Venn diagram, a big rectangle is used to label the universal set, whereas a circle is used to denote a set.

---

[12] For example, if $X = [0, 5]$, then $6, 7, \ldots$ are all upper bounds of X.

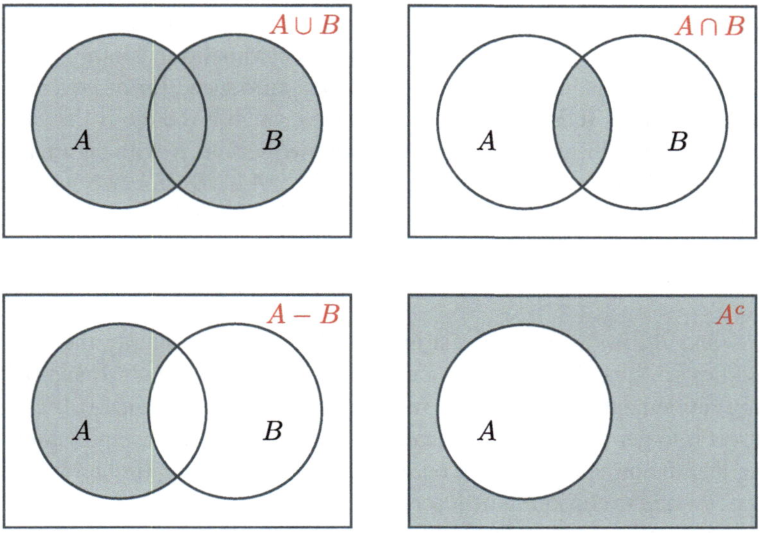

Fig. 5.3 Venn diagrams for set operations

5.5.2 Set Operations

Sets can be combined (if we can combine numbers via arithmetic operations, we can do something similar for sets) via set operations (Fig. 5.3). We can combine two sets in many different ways. First, the union of two sets A and B is a set, labeled as $A \cup B$, containing all elements that are in A or in B. For example, $\{1, 3, 4\} \cup \{3, 4, 5\} = \{1, 3, 4, 5\}$. If we have many sets $A_1, A_2, \ldots, A_n$, the union is written as[13]:$\bigcup_{i=1}^{n} A_i$.

Second, the intersection of two sets A and B, denoted by $A \cap B$, consists of all elements that are both in A and B. For instance, $\{1, 2\} \cap \{2, 3\} = \{2\}$. When the intersection of two sets is empty, i.e., $A \cap B = \emptyset$, the two sets are called *mutually exclusive* or disjoint. We now extend this to more than two sets. If we have n sets $A_1, A_2, \ldots, A_n$, these sets are disjoint if they are pairwise disjoint:

$$A_i \cap A_j = \emptyset \quad \text{for all } i \neq j$$

Third, the difference of two sets A and B is denoted by $A - B$ and is a set consists of elements that are in A but not in B.

Finally, we have another operation on set, but this operation applies to one single set. The complement of a set A, denoted by A^c, is the set of all elements that are in the universal set S but are not in A. The Venn diagrams for the presented set operations are shown in Fig. 5.3.

[13] Note the similarity to the sigma notation $\sum_{i=1}^{n} x_i = x_1 + x_2 + \cdots + x_n$.

Cardinality. The cardinality of a set is basically the size of the set, it is denoted by $|A|$. For finite sets (e.g., the set $\{1, 3, 5\}$), its cardinality is simply the number of elements in A. Again, once a new object was introduced (or discovered) in the mathematical world, there are rules according to which it (herein is the cardinality of a set) obeys. For instance, we can ask given two sets A, B with cardinalities $|A|$ and $|B|$, what is the cardinality of their union, i.e., $|A \cup B|$? For two sets A and B, we have this rule called the *inclusion-exclusion principle* or PIE:

$$|A \cup B| = |A| + |B| - |A \cap B| \tag{5.5.1}$$

When A and B are disjoint, the cardinality of its union is simply the sum of the cardinalities of A and B. When they are not disjoint, when we add $|A|$ and $|B|$, we're counting the elements in $A \cap B$ twice (a Venn diagram would help here), thus we need to subtract it to get the correct cardinality. The name (of the principle) comes from the idea that the principle is based on over-generous inclusion (red term), followed by compensating exclusion (blue term).

Then, mathematicians certainly generalize this result to the union of n sets. For simplicity, we just extend this principle to the case of three sets:

$$|A \cup B \cup C| = |A| + |B| + |C| - |A \cap B| - |B \cap C| - |C \cap A| + |A \cap B \cap C| \tag{5.5.2}$$

Example 5.1 How many integers from 1 to 100 are multiples of 2 or 3? Let A be the set of integers from 1 to 100 that are multiples of 2, then $|A| = 50$ (why?). Let B be the set of integers from 1 to 100 that are multiples of 3, then $|B| = 33$.[14] Our question is amount to computing $|A \cup B|$. Certainly, we use the PIE:

$$|A \cup B| = |A| + |B| - |A \cap B|$$

We need then $A \cap B$ which is the set of integers from 1 to 100 that are multiples of both 2 and 3 or multiples of 6, we have $|A \cap B| = 16$. Thus, $|A \cup B| = 50 + 33 - 16 = 67$.

Generalized principle of inclusion-exclusion. Now we extend the PIE to the case of n sets for whatever n. First, we put the two identities for $n = 2$ and $n = 3$ together to see the pattern:

$$\begin{aligned}
|A \cup B| &= |A| + |B| - |A \cap B| \\
|A \cup B \cup C| &= |A| + |B| + |C| - |A \cap B| - |B \cap C| - |C \cap A| + |A \cap B \cap C|
\end{aligned}$$

To see the pattern, let x belong to all three sets A, B, C. It is then counted in every term in the RHS of the second equation: 4 times added (the red terms) and 3 times subtracted, adding up to 1. As a preparation for the move to n sets, we no longer use

[14] A number that is a multiple of 3 if it can be written as $3m$, then $1 \leq 3m \leq 100$, thus $m = \lfloor 100/3 \rfloor = 33$.

A, B, C, instead we adopt $A_1, A_2, \ldots$[15] Now we write the second equation with the new symbols

$$\left|\bigcup_{i=1}^{3} A_i\right| = |A_1 \cup A_2 \cup A_3| = \sum_{i=1}^{3} |A_i| - \sum_{1\le i<j\le 3} |A_i \cap A_j| + \left|\bigcap_{i=1}^{3} A_i\right|$$

And with that it is not hard to get the general formula:

$$\left|\bigcup_{i=1}^{n} A_i\right| = \sum_{i=1}^{n} |A_i| - \sum_{1\le i<j\le n} |A_i \cap A_j| + \sum_{1\le i<j<k\le n} |A_i \cap A_j \cap A_k| - \cdots + (-1)^{n-1} \left|\bigcap_{i=1}^{n} A_i\right| \tag{5.5.3}$$

Note that the first term is adding up elements in each set, the second term deals with pairs, the third term deals with triplets and so on. The lth term has $\binom{n}{l}$ summands (to see this just return to the case $n = 3$, the second term in the RHS of the formula in this case is

$$-|A \cap B| - |B \cap C| - |C \cap A|$$

which has $\binom{3}{2}$ summands. Did mathematicians stop with Eq. (5.5.3)? No that equation is not in a best form yet. Note that the RHS of that equation involves n terms and each term in turn involves a sum of terms. Mathematicians want to write it as $\sum_i^n(\sum|\cdots|)$. The key to this step is to discard the subscripts i, j, k and replace them by *subscripts with subscripts*: $i_1, i_2, \ldots$

$$\left|\bigcup_{i=1}^{n} A_i\right| = \sum_{k=1}^{n} (-1)^{k+1} \left(\sum_{1\le i_1<\cdots<i_k\le n} |A_{i_1} \cap A_{i_2} \cdots \cap A_{i_k}| \right) \tag{5.5.4}$$

Still they are not happy with this (the blue term, particularly). So they went a further step with defining $A_I = \bigcap_{i\in I} A_i$, $I \subset \{1, 2, ..., n\}$:

$$\boxed{\left|\bigcup_{i=1}^{n} A_i\right| = \sum_{k=1}^{n} (-1)^{k+1} \left(\sum_{\substack{I\subset\{1,2,\ldots,n\}\\ |I|=k}} |A_I| \right), \quad A_I = \bigcap_{i\in I} A_i} \tag{5.5.5}$$

The second sum runs over all subsets I of the indices $1, 2, \ldots, n$[16] which contain exactly k elements (i.e., $|I| = k$). At this moment, mathematicians stop because that form is compact.

If you play with the Venn diagrams you will definitely discover many more identities on sets similar to Eq. (5.5.2). For example, $A = (A \cap B^c) \cup (A \cap B)$. As is

[15] Obviously we will run out of alphabets and moreover subscripts allow for compact notation: we can write $A_1 + A_2 + \cdots = \sum_i A_i$. With A, B, ... we simply cannot.

[16] One example clarifies everything, assume $n = 3$, $k = 2$, then $I = \{1, 2\}$, $I = \{1, 3\}$, $I = \{2, 3\}$.

always in mathematics, this seemingly pointless identity will be useful in other contexts.

Definition 5.5.1 Set A is called countable if one of the following is true

(a) if it is a finite set, i.e., $|A| < \infty$, or
(b) it can be put in a one-to-one correspondence with natural numbers. In this case the set is said to be countably infinite.

A set is called uncountable if it is not countable. One example is the set of real numbers $\mathbb{R}$.

You can check again Sect. 2.31 on Georg Cantor and infinity if anything mentioned in this definition is not clear.

de Morgan's laws state that the complement of the union of two sets is equal to the intersection of their complements and the complement of the intersection of two sets is equal to the union of their complements. The laws are named after Augustus de Morgan (1806–1871)—a British mathematician and logician. He formulated de Morgan's laws and introduced the term mathematical induction, making its idea rigorous. For any two finite sets A and B, the laws are

$$(A \cup B)^c = A^c \cap B^c, \quad (A \cap B)^c = A^c \cup B^c$$

We can draw some Venn diagrams to see that the laws are valid, but that's not enough as we know that the laws might hold true for $n > 2$ sets, in that case, no one can use Venn diagram for a check. The generalized version of de Morgan's first law is

$$(A_1 \cup A_2 \cup \cdots \cup A_n)^c = A_1^c \cap A_2^c \cdots \cap A_n^c, \quad \text{or} \quad \left(\bigcup_{i=1}^{n} A_i\right)^c = \bigcap_{i=1}^{n} A_i^c$$

Proof of de Morgan's first law for two sets The plan is to pick an element x in $(A \cup B)^c$ and prove it is also an element of $A^c \cap B^c$ and vice versa. Let $P = (A \cup B)^c$ and $Q = A^c \cap B^c$. Now, consider $x \in P$, we're going to prove that $x \in Q$, which means that $P \subset Q$. As $x \in (A \cup B)^c$, it is not in $A \cup B$:

$$\begin{aligned}
&\Longrightarrow x \notin (A \cup B) \\
&\Longrightarrow (x \notin A) \text{ and } (x \notin B) \\
&\Longrightarrow (x \in A^c) \text{ and } (x \in B^c) \\
&\Longrightarrow x \in (A^c \cap B^c) : x \in Q \Longrightarrow P \subset Q
\end{aligned}$$

Doing something similar with $y \in Q$ and then showing $y \in P$, we get $Q \subset P$. Now we have $P \subset Q$ and $Q \subset P$. What does it mean? It means $P = Q$. You can use proof by induction to prove the generalized version. ∎

5.6 Random Experiments, Sample Space and Event

Before rolling a die we do not know the result. This is an example of a random experiment. Usually, we carry out a random experiment multiple times; each time (called a trial) we get a result which we call an outcome. *The set of all possible outcomes of a random experiment is called the sample space*. Since this sample space is the biggest space as far as the experiment is concerned, it is our universal set S. An event is a subset of the sample space. Some examples are:

- Random experiment: toss a coin; sample space is $S = \{H, T\}$ (H for head and T for tail), and one event is $E = \{H\}$ or $E = \{T\}$;
- Random experiment: roll a six-sided die; sample space is $S = \{1, 2, 3, 4, 5, 6\}$, and one event can be $E = \{2, 4, 6\}$ if we're interested in the chance of getting an even number;
- Random experiment: toss a coin two times and observe the sequence of heads/tails; the sample space is

$$S = \{(H, H), (H, T), (T, H), (T, T)\}$$

  One event can be $E_1 = \{(H, H), (T, T)\}$.

Note that events are subsets of the sample space S.

5.7 Probability and Its Axioms

Now comes the moment that we can talk about probability in a mathematically precise way. A probability of an event A, denoted by $P(A)$, is a number that is assigned to A. The axiomatic probability theory is based on the following three axioms:

Box 5.1: Three axioms of the theory of probability.

- Axiom 1: The probability of every event is at least zero. For any event A, $P(A) \geq 0$.
- Axiom 2: The probability of the sample space is 100%; $P(S) = 1$.
- Axiom 3: If two events are disjoint, the probability that either of the two events happens is the sum of the probabilities that each happens:

$$P(A \cup B) = P(A) + P(B) \quad \text{if } A \cap B = \emptyset \tag{5.7.1}$$

It is important to take a few moments to understand each axiom thoroughly, because we cannot prove them, we can only accept them and proceed from that. The first axiom states that probability cannot be negative and the smallest probability is zero. When $P(A) = 0$, the event A will never occur. As the axiomatic theory of probability is based on the theory of measure (which I do not know much!), we can think of $P(A)$ as the area of a domain A, and thus the third axiom is something like: the area of two disconnected domains is the sum of the area of each domain. Axiom 2 is just providing a scale. It says that the area of the sample space S—which is the maximum area—is one. All other formulas, results, theorems, whatever, are derived from these three axioms!

Union and intersection of events. As events are sets, we can apply set operations on events. When working with events, intersection means "and", and union means "or". The probability of the intersection of A and B, $P(A \cap B)$ is sometimes written as $P(AB)$ or $P(A, B)$.

- Probability of intersection:

$$P(A \cap B) = P(AB) = P(A \text{ and } B)$$

- Probability of union:

$$P(A \cup B) = P(A \text{ or } B)$$

Example 5.2 We roll a fair six-sided die, what is the probability of getting 1 or 5? So, the event is $E = \{1, 5\}$ and the sample space is $S = \{1, 2, 3, 4, 5, 6\}$. We use the three axioms to compute $P(E)$. First, as the die is fair, the chance of getting any number from 1 to 6 is equal:

$$P(1) = P(2) = P(3) = P(4) = P(5) = P(6)$$

where $P(1)$ is short for $P(\{1\})$. Note that probability is defined only for sets not for numbers. Now, we use axioms 2 and 3 together to write[17]

$$1 \overset{(2)}{=} P(S) \overset{(3)}{=} P(1) + P(2) + \cdots + P(6)$$

which results in the probability of getting any number from 1 to 6 is $1/6$. Then, using the axiom 3 again for E, we have

$$P(\{1, 5\}) = P(1) + P(5) = \frac{1}{6} + \frac{1}{6} = \frac{1}{3}$$

Note that, $1/3 = 2/6$, we can deduce an important formula:

$$P(\{1, 5\}) = \frac{1}{3} = \frac{2}{6} = \frac{|\{1, 5\}|}{|S|}$$

[17] The symbol (2) above the equal sign to indicate that axiom 2 is being used.

Therefore, for a finite sample space S with equally likely outcomes, the probability of an event A is the ratio of the cardinality of A over that of S:

$$P(A) = \frac{|A|}{|S|}$$

Example 5.3 Using the axioms of probability, prove the following:

(a) For any event A, $P(A^c) = 1 - P(A)$. This is the rule of complementary probability.
(b) The probability of the empty set is zero, i.e., $P(\emptyset) = 0$.
(c) For any event A, $P(A) \leq 1$.
(d) $P(A - B) = P(A) - P(A \cap B)$
(e) $P(A \cup B) = P(A) + P(B) - P(A \cap B)$

Proof of $\boldsymbol{P(A^c) = 1 - P(A)}$ Referring back to Fig. 5.3, we know that $A \cup A^c = S$ and A and A^c are disjoint, thus

$$\begin{aligned} P(S) &= P(A \cup A^c) \\ 1 &= P(A) + P(A^c) \Longrightarrow P(A) = 1 - P(A^c) \end{aligned}$$

where use was made of axiom 2 ($P(S) = 1$) and axiom 3 ($P(A \cup A^c) = P(A) + P(A^c)$). We also get $P(A) \leq 1$ from $P(A) = 1 - P(A^c)$ for $P(A^c) \geq 0$ due to axiom 1. ■

Proof of $\boldsymbol{P(A - B) = P(A) - P(A \cap B)}$ See the figure below for the proof. It is based on set properties and axiom 3. ■

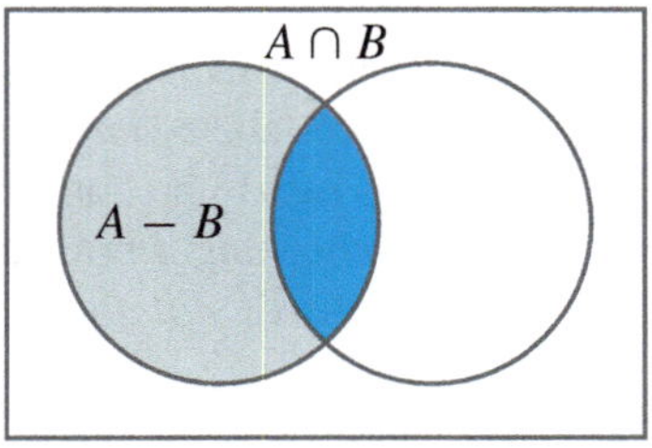

$$(A - B) \cup (A \cap B) = A$$

$$(A - B) \cap (A \cap B) = \emptyset$$

$$P(A - B) + P(A \cap B) = P(A)$$

Proof of $\boldsymbol{P(A \cup B) = P(A) + P(B) - P(A \cap B)}$ Recall the inclusion-exclusion principle that $|A \cup B| = |A| + |B| - |A \cap B|$, $P(A \cup B) = P(A) + P(B) - P(A \cap B)$ is the version of that principle for probability. From the figure below, we have $P(A \cup B) = P(A) + P(B - A)$. Now we replace $P(B - A)$ by $P(B) - P(A \cap B)$—a result from (d). ■

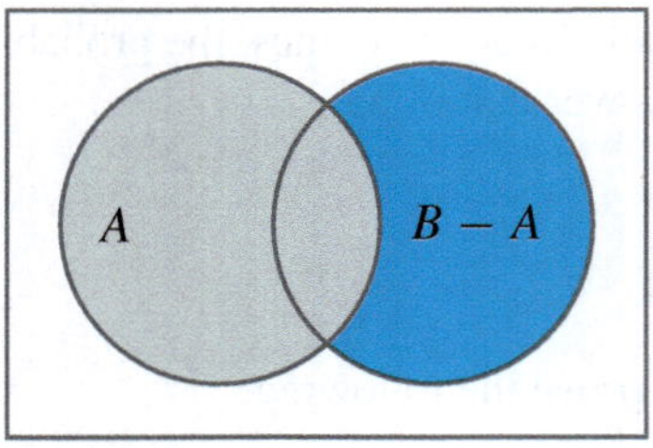

$$A \cap (B - A) = \emptyset$$

$$P(A \cup (B - A)) = P(A) + P(B - A)$$

$$P(A \cup B) = P(A) + P(B - A)$$

The rule (a) can be referred to as the rule of complementary probability. It is very simple and yet powerful for problems in which finding $P(A)$ is hard and finding $P(A^c)$ is much easier. We will use this rule quite often.

Corresponding to the principle of inclusion-exclusion in Eq. (5.5.3), we have the probability version:

$$P\left(\bigcup_{i=1}^{n} A_i\right) = \sum_{i=1}^{n} P(A_i) - \sum_{i<j} P(A_i \cap A_j) + \sum_{i<j<k} P(A_i \cap A_j \cap A_k) - \cdots + (-1)^{n-1} P\left(\bigcap_{i=1}^{n} A_i\right)$$

Or the compact form in Eq. (5.5.5)

$$P\left(\bigcup_{i=1}^{n} A_i\right) = \sum_{k=1}^{n} (-1)^{k+1} \left(\sum_{\substack{I \subset \{1,2,\ldots,n\} \\ |I|=k}} P(A_I) \right), \quad A_I = \bigcap_{i \in I} A_i \tag{5.7.2}$$

Example 5.4 Now we consider a classic example that uses the inclusion-exclusion principle. Assume that a secretary has an equal number of pre-labeled envelopes and business cards (denoted by n). Suppose that she is in such a rush to go home that she puts each business card in an envelope at random without checking if it matches the envelope. What is the probability that *each of the business cards will go to a wrong envelope?*

Always start simple, so we now assume that $n = 3$, and we define the following events:

A_1 : 1st card in correct envelope

A_2 : 2nd card in correct envelope

A_3 : 3rd card in correct envelope

Now let E be the event that each of the business cards will go to a wrong envelope. We want to compute $P(E)$. E occurs only when none of A_1, A_2, A_3 has happened. Thus,

$$E = A_1^c \cap A_2^c \cap A_3^c = (A_1 \cup A_2 \cup A_3)^c \text{ (de Morgan's laws)}$$

Now, we can compute $P(E)$ as

$$P(E) = 1 - P(E^c) = 1 - P(A_1 \cup A_2 \cup A_3)$$

The next step is to use the PIE to get the red term, and thus $P(E)$ is given by

$$P(E) = 1 - \left(\sum_i^3 P(A_i) - \sum_{i<j} P(A_i \cap A_j) + P(A_1 \cap A_2 \cap A_3) \right)$$

Now we compute all the probabilities, and we get something special:

$$P(A_i) = \frac{(3-1)!}{3!} = \frac{1}{3}, \quad P(A_i \cap A_j) = \frac{(3-2)!}{3!} = \frac{1}{6}, \quad P(A_1 \cap A_2 \cap A_3) = \frac{(3-3)!}{3!} = \frac{1}{6}$$

How to get the above probabilities? Note that for $n = 3$ there are a total of 3! outcomes, and to have 2 cards in correct envelops we just need to care about the remaining cards (which is $3 - 2$), and for them, there are of course $(3 - 2)!$ ways.

What is special about this problem is that $P(A_I)$ in Eq. (5.7.2) is equal and it is $p_k = {}^{(n-k)!}/_{n!}$ and each term in Eq. (5.7.2) has $\binom{n}{k}$ summands. Therefore, we're able to go for the general case n with Eq. (5.7.2), and $P(E)$ is given by

$$P(E) = 1 - \sum_{k=1}^n (-1)^{k+1} \binom{n}{k} p_k = 1 - \sum_{k=1}^n \frac{(-1)^{k+1}}{k!} = \sum_{k=0}^n \frac{(-1)^k}{k!}, \quad p_k = \frac{(n-k)!}{n!}$$

To check this theoretical result, we can perform an experiment using the Monte Carlo method. To practice Monte Carlo methods, you're encouraged to implement it for this problem. If need help, check the code `monte-carlo-pi.jl`, function `MC_secretary_prob`, on my GitHub account. You'll see that the theoretical result matches the MC result.

5.8 Conditional Probabilities

A conditional probability is the likelihood of an event occurring given that another event has already happened. Conditional probabilities allow us to evaluate how prior information affects probabilities. When we incorporate existing facts into the calculations, it can change the probability of an outcome. Probability is generally counter-intuitive, but conditional probability is the worst! Conditioning can subtly alter probabilities and produce unexpected results.

This section will introduce the famous Bayes' theorem. But, we first start with a simple example for motivation.

5.8.1 What Is a Conditional Probability?

Example 5.5 Consider a family that has two children. We are interested in the children's genders. Our sample space is $S = \{(G, G), (G, B), (B, G), (B, B)\}$. Also assume that all four possible outcomes are equally likely; that is $P(G, G) = P(G, B) = \cdots = 1/4$.

- What is the probability that both children are girls?
- What is the probability that both children are girls *given that the first child is a girl*?
- What is the probability that both children are girls *given that we know at least one of them is a girl*?

Of course, the probability that both children are girls is $1/4$. The two remaining probabilities are more interesting and new; and most of us would say the answer is $1/2$ for both. Let's denote by A the event that both children are girls and by B the event that the first child is a girl. That is $B = \{(G, G), (G, B)\}$. Now, the chance to have two girls is therefore $1/2$. Let's denote by C the event that one of the children is a girl. That is $C = \{(G, G), (G, B), (B, G)\}$. Now, the chance to have two girls is $1/3$.

The probability that both children are girls (event A) given that the first child is a girl (event B) is called a *conditional probability*. And it is written as $P(A|B)$; the vertical line | is read "given that". This example clearly demonstrates that when we incorporate existing facts into the calculations, it can change the probability of an outcome. The sample space is changed!

The next thing we need to do is to find a formula for $P(A|B)$.

Because B has occurred it becomes the sample space, and the only way that A can happen is when the outcome belongs to the set $A \cap B$, we thus have $P(A|B)$ as

$$P(A|B) = \frac{|A \cap B|}{|B|}$$

Now we can divide the denominator and the numerator by $|S|$, the cardinality of the original sample space, to have

$$P(A|B) = \frac{|A \cap B|/|S|}{|B|/|S|} = \frac{P(A \cap B)}{P(B)} \tag{5.8.1}$$

Of course as B has occurred, $P(B) > 0$, so there is no danger in dividing something by it. Note that Eq. (5.8.1) was derived for sample spaces with equally likely outcomes only. For other cases, take it as a definition for conditional probability.

5.8.2 $P(A|B)$ Is Also a Probability

As $P(A|B)$ is a probability it must satisfy the three axioms of probability (it can have other properties). I list them first, and then we shall prove them:

- Axiom 1: The conditional probability of every event is at least zero: $P(A|F) \geq 0$.
- Axiom 2: The conditional probability of the sample space is 100%: $P(S|F) = 1$.
- Axiom 3: If two events are disjoint, the conditional probability that either of the two events happens is the sum of the conditional probabilities that each happens; $P(A \cup B|F) = P(A|F) + P(B|F)$ if $A \cap B = \emptyset$.

Proof The proof of axiom 2 goes as simple as (based on the fact $S \cap F = F$)

$$P(S|F) = \frac{P(SF)}{P(F)} = \frac{P(F)}{P(F)} = 1 \quad (SF = S \cap F = F)$$

The proof of axiom 3 goes like this, I go from the LHS to the RHS, it's just a personal taste:

$$\begin{aligned} P(A \cup B|F) &= \frac{P((A \cup B)F)}{P(F)} = \frac{P(AF \cup BF)}{P(F)} \\ &= \frac{P(AF) + P(BF)}{P(F)} \quad (AF \cap BF = \emptyset) \\ &= \frac{P(AF)}{P(F)} + \frac{P(BF)}{P(F)} = P(A|F) + P(B|F) \end{aligned}$$

So, the proof used the given information that A and B are disjoint, thus AF and BF are also disjoint (why?). ■

The generalized version of axiom 3 is

$$P\left(\bigcup_{i=1}^{\infty} A_i|F\right) = \sum_{i=1}^{\infty} P(A_i|F)'$$

You should prove it. The proof is exactly the same as the one I presented for two events A_1 and A_2!

If we define $Q(E) = P(E|F)$, then $Q(E)$ may be regarded as a probability function on the events of S because it satisfies the three axioms. Hence, all of the propositions previously proved for probabilities apply to $Q(E)$. For example, all results from Example 5.3 hold for conditional probabilities:

(a) For any event A, $P(A^c|F) = 1 - P(A|F)$.
(b) $P(A \cup B|F) = P(A|F) + P(B|F) - P(AB|F)$

Proof of $\boldsymbol{P(A^c|F) = 1 - P(A|F)}$ The proof is based on the fact that $F = AF \cup A^cF$, thus $P(F) = P(AF) + P(A^cF)$, noting that AF and A^cF are disjoint (a

Venn diagram will show all this). The rest is to use the definition of the conditional probability and some algebraic manipulations:

$$1 - P(A|F) = 1 - \frac{P(AF)}{P(F)} = \frac{P(F) - P(AF)}{P(F)} = \frac{P(AF) + P(A^cF) - P(AF)}{P(F)}$$
$$= \frac{P(A^cF)}{P(F)} = P(A^c|F).$$

■

5.8.3 Multiplication Rule for Conditional Probability

A simple massage to Eq. (5.8.1) results in (note that $P(A \cap B) = P(B \cap A)$ because $A \cap B = B \cap A$)

$$P(A|B) = \frac{P(A \cap B)}{P(B)} \Longrightarrow \boxed{P(A \cap B) = P(B)P(A|B)}, \quad \text{or} \quad P(A \cap B) = P(A)P(B|A) \tag{5.8.2}$$

This formula, known as the multiplication rule of probability, is particularly useful in situations when we know the conditional probability, but we are interested in the probability of the intersection. In words, this formula states that the probability that both A and B occur is equal to the probability that B (or A) occurs multiplied with the conditional probability of A (or B) given that B (or A) occurred. Note that we're not talking about causality (with a direction), only about statistical dependency.

We can generalize the multiplication rule to more than two events. Let's start with three events A, B and C. We can write $A \cap B \cap C = (A \cap B) \cap C$, thus

$$P(A \cap B \cap C) = P((A \cap B) \cap C) = P(A \cap B|C)P(C)$$

To find $P(A \cap B|C)$, we use $P(A \cap B) = P(B)P(A|B)$ and condition on C:

$$P(A \cap B) = P(B)P(A|B) \Longrightarrow P(A \cap B|C) = P(B|C)P(A|B, C) \tag{5.8.3}$$

Thus,

$$P(A \cap B \cap C) = P(C)P(B|C)P(A|B, C)$$

Nothing can stop mathematicians to extend this rule to n events. How should they name the events now? No longer $A, B, C, \ldots$ as there are less than 30 symbols! They now use subscripts for that: $E_1, E_2, \ldots, E_n$. The generalized multiplication rule is

$$P(E_1E_2E_3 \ldots E_n) = P(E_1)P(E_2|E_1)P(E_3|E_1E_2) \cdots P(E_n|E_1E_2 \ldots E_{n-1}) \tag{5.8.4}$$

You wanna a proof? It is simple: application of the definition of conditional probability to all terms, except the first one $P(E_1)$ in the RHS of Eq. (5.8.4):

$$\cancel{P(E_1)} \frac{\cancel{P(E_1E_2)}}{\cancel{P(E_1)}} \frac{\cancel{P(E_1E_2E_3)}}{\cancel{P(E_1E_2)}} \cdots \frac{E_1E_2E_3\ldots E_n}{\cancel{E_1E_2E_3\ldots E_{n-1}}}$$

where all the terms cancel each other except the final numerator, which is the LHS of Eq. (5.8.4).

5.8.4 *Bayes' Formula*

Starting with this fact for two events E and F (you need to draw a Venn diagram to convince yourself):

$$E = EF \cup EF^c$$

Then, we can express $P(E)$ in terms of $P(E)$, $P(F)$ and so on:

$$\begin{aligned} P(E) &= P(EF) + P(EF^c) && (EF \cap EF^c = \emptyset) \\ &= P(E|F)P(F) + P(E|F^c)P(F^c) && \text{(conditional prob.)} \\ &= P(E|F)P(F) + P(E|F^c)[1 - P(F)] && \text{(complementary prob.)} \end{aligned} \tag{5.8.5}$$

which simply states that the probability of event E is the sum of the conditional probabilities of event E given that event F has (red term) or has not occurred. This formula is extremely useful when it is difficult to compute the probability of an event (E) directly, but it is *straightforward to compute it once we know whether or not some second event (F) has occurred*. The following example demonstrates how to use this formula.

Example 5.6 An insurance company believes that people can be divided into two classes: those who are accident-prone and those who are not. The company's statistics show that an accident-prone person will have an accident at some time within a fixed 1-year period with probability 0.4, whereas this probability decreases to 0.2 for a person who is not accident-prone. If we assume that 30% of the population is accident-prone, what is the probability that a new policyholder will have an accident within a year of purchasing a policy?

Solution. Let's denote by E the event a new policyholder will have an accident within a year of purchasing a policy. We need to find $P(E)$. This person is either accident-prone or not. Let's call F the event that a new policyholder is accident-prone, then F^c is the event that this person is not accident-prone. Then, we have $P(F) = 0.3$ and $P(F^c) = 0.7$, $P(E|F) = 0.4$ and $P(E|F^c) = 0.2$, then Eq. (5.8.5) gives:

$$P(E) = (0.4)(0.3) + (0.2)(0.7)$$

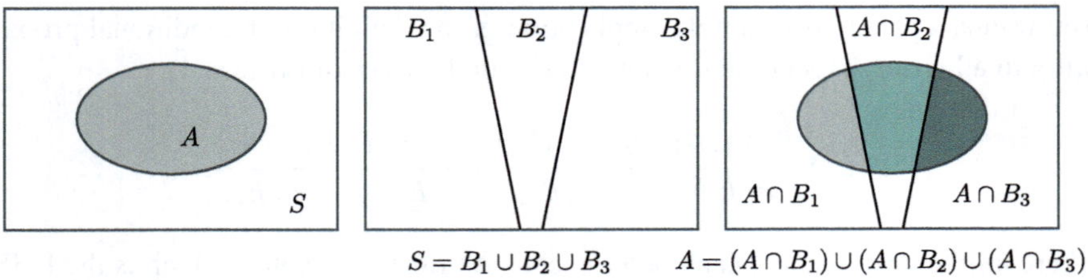

Fig. 5.4 The sample space S is partitioned into three disjoint events B_1, B_2, B_3. Then, we have $A = (A \cap B_1) \cup (A \cap B_2) \cup (A \cap B_3)$

Now we generalize Eq. (5.8.5). How? Note that in that formula, we have two events F and F^c, which are two disjoint events that together fill completely the sample space. Now, what we have to do is to just generalize this to n events. First, for simplicity, assume that we can partition the sample space S into three disjoint sets B_1, B_2, and B_3. Then, we have, see Fig. 5.4

$$A = (A \cap B_1) \cup (A \cap B_2) \cup (A \cap B_3)$$

and $A \cap B_1$, $A \cap B_2$, and $A \cap B_3$ are mutually disjoint. Thus, we can write $P(A)$ as

$$\begin{aligned} P(A) &= P((A \cap B_1) \cup (A \cap B_2) \cup (A \cap B_3)) \\ &= P(A \cap B_1) + P(A \cap B_2) + P(A \cap B_3) \qquad \text{(axiom 3)} \\ &= P(A|B_1)P(B_1) + P(A|B_2)P(B_2) + P(A|B_3)P(B_3) \quad \text{(Eq. (5.8.2))} \end{aligned}$$

With that, we have this general result, with $B_i, i = 1, 2, \ldots, n$ partition S:

$$P(A) = \sum_{i=1}^{n} P(A|B_i)P(B_i) \tag{5.8.6}$$

which is referred to as *the law of total probability*. This formula states that $P(A)$ is equal to a weighted average of $P(A|B_i)$, each term being weighted by the probability of the event on which it is conditioned.

We're now deriving Bayes' formula or Bayes' rule that relates $P(A|B)$ to $P(B|A)$. We start with the conditional probability:

$$P(A|B)P(B) = P(B|A)P(A) \quad (= P(A \cap B) = P(B \cap A))$$

Dividing this equation by $P(A) > 0$, we get the Bayes' formula:

$$P(B|A) = \frac{P(A|B)P(B)}{P(A)} \tag{5.8.7}$$

This formula is referred to as Bayes' theorem or Bayes' Rule or Bayes' Law and is the foundation of the field of Bayesian statistics. Bayes Theorem is also widely used in the field of machine learning. For sure, it is one of the most useful results in conditional probability. The rule is named after eighteenth-century British mathematician Thomas Bayes. The term $P(B|A)$ is referred to as the *posterior probability* and $P(B)$ is referred to as the *prior probability*.

We can use Eq. (5.8.6) to compute $P(A)$, and thus obtained the extended form of Bayes' formula:

$$P(B_j|A) = \frac{P(A|B_j)P(B_j)}{\sum_{i=1}^{n} P(A|B_i)P(B_i)} \tag{5.8.8}$$

Example 5.7 A certain disease affects about *1 out of 10000 people*. There is a test to check whether the person has the disease. The test is quite accurate. In particular, we know that the probability that the test result is *positive* (i.e., the person has the disease), given that the person does not have the disease, is only 2%; the probability that the test result is *negative* (i.e., the person does not have the disease), given that the person has the disease, is only 1%. A random person gets tested for the disease and *the result comes back positive*. What is the probability that the person has the disease?

Solution. A person either gets the disease or not. So the sample space is partitioned into two sets: D for having the disease and D^c for not. We have $P(D) = 0.0001$ and $P(D^c) = 1 - 0.0001$. Let's denote by A the event that the test result is positive. The problem is asking us to compute $P(D|A)$. From the problem description, we have $P(A|D^c) = 0.02$ and $P(A^c|D) = 0.01$ which also yields $P(A|D) = 1 - 0.01$ (complementary probability). Now, it is just an application of Bayes' formula, i.e., Eq. (5.8.8)[18]

$$P(D|A) = \frac{P(A|D)P(D)}{P(A|D)P(D) + P(A|D^c)P(D^c)} = \frac{(1-0.01)(0.0001)}{(1-0.01)(0.0001) + (0.02)(1-0.0001)}$$

which is 0.0049, which indicates that there is less than half a percent chance that the person has the disease. This might seem counter-intuitive because the test is quite accurate. The point is that the disease is very rare. Thus, there are two competing forces here, and since the rareness of the disease (1 out of 10,000) is stronger than the accuracy of the test (98 or 99%), there is still good chance that the person does not have the disease.

Example 5.8 The Monty Hall problem is a probability puzzle, loosely based on the American television game show *Let's Make a Deal* and named after its original host, Monty Hall. The problem was originally posed and solved in a letter by Steve Selvin

[18] Herein, $B_1 = D$ and $B_2 = D^c$.

Table 5.4 The Monty Hall problem: listing all possible outcomes. We chose door 1

Door 1	Door 2	Door 3	Stay at door 1	Switch to offered door
Car	Goat	Goat	WIN	LOSS
Goat	Car	Goat	LOSS	WIN
Goat	Goat	Car	LOSS	WIN

to the American Statistician in 1975. In the problem, you are on a game show, being asked to choose between three doors. A car is behind one door and two goats behind the other doors. You choose a door. The host, Monty Hall, picks one of the other doors, which he knows has a goat behind it, and opens it, showing you the goat. (You know, by the rules of the game, that Monty will always reveal a goat.) Monty then asks whether you would like to switch your choice of door to the other remaining door. Assuming you prefer having a car more than having a goat, *do you choose to switch or not to switch?*

Vos Savant's response was that the contestant should switch to the other door. Many readers of vos Savant's column refused to believe switching is beneficial and rejected her explanation. After the problem appeared in Parade, approximately 10000 readers, including nearly 1000 with PhDs, wrote to the magazine, most of them calling vos Savant wrong. Even when given explanations, simulations, and formal mathematical proofs, many people still did not accept that switching is the best strategy. Paul Erdős[19] remained unconvinced until he was shown a computer simulation demonstrating vos Savant's predicted result.

First, we solve this problem using a computer simulation. The code of a computer simulation of this problem is given in Listing 5.2. The result shows that the probability of not switching is $1/3$, which is making sense, and the probability of switching is $2/3$, that is twice higher. The code assumes that the car is behind door 1 without loss of generality. Note that the host will choose a door that the player did not select and that does not contain a car and reveal this to us.

Another way to see the solution is to explicitly list out all the possible outcomes, and count how often we get the car if we stay versus switch. Without loss of generality, suppose our selection was door 1. Then the possible outcomes can be seen in Table 5.4. In two out of three cases, we win the car by changing our selection after one of the doors is revealed.

[19] Paul Erdős (1913–1996) was a renowned Hungarian mathematician. He was one of the most prolific mathematicians and producers of mathematical conjectures of the twentieth century. He devoted his waking hours to mathematics, even into his later years—indeed, his death came only hours after he solved a geometry problem at a conference in Warsaw. Erdős published around 1500 mathematical papers during his lifetime, a figure that remains unsurpassed. He firmly believed mathematics to be a social activity, living an itinerant lifestyle with the sole purpose of writing mathematical papers with other mathematicians.

Listing 5.2 Monte Carlo simulation of the Monty Hall problem. Source: `monty_hall.jl`

```julia
using  Random
function monty_hall_one_trial(changed)
  # assume that door 1 has the car, changed=1: switching; changed=0: no switching
  # the contestant select one door, can be any of (1,2,3,...)
  number_of_doors = 3
  chosen_num      = rand(1:number_of_doors)
  # if the contestant decided to change
  if changed == 1
    if chosen_num == 1 revealed_num = rand(2,3) end
    if chosen_num == 2 revealed_num = 3 end # revealed_num: by the host
    if chosen_num == 3 revealed_num = 2 end
    #  switch to the remaining door
    avai_doors = setdiff(1:number_of_doors, (chosen_num,revealed_num))
    chosen_num = rand(avai_doors)
  end
  return chosen_num == 1
end
N                  = 10000 # Monte Carlo trials
prob_changed       = sum([monty_hall_one_trial(1) for _ in 1:N])/N # => ~2/3
prob_no_changed = sum([monty_hall_one_trial(0) for _ in 1:N])/N # => ~1/3
```

5.8.5 *The Odds Form of the Bayes' Rule*

The odds of an event A, denoted by $O(A)$ are defined by

$$O(A) := \frac{P(A)}{P(A^c)} = \frac{P(A)}{1 - P(A)} \tag{5.8.9}$$

That is, the odds of an event A tell how much more likely it is that A occurs than it is that it does not occur. For instance, if $P(A) = 2/3$, then $P(A) = 2P(A^c)$, so the odds are 2. If the odds are equal to α, then it is common to say that the odds are α to 1, or $\alpha : 1$ in favor of the hypothesis.

Having defined the odds of an event, we now write the Bayes' formula in the odds form. To this end, consider now a hypothesis H that is true with probability $P(H)$, and suppose that new evidence E is introduced (or equivalently, new data is introduced). Then the conditional probabilities, given the evidence E, that H is true and that H is not true are respectively given by (from Eq. (5.8.7))

$$P(H|E) = \frac{P(E|H)P(H)}{P(E)}, \quad P(H^c|E) = \frac{P(E|H^c)P(H^c)}{P(E)} \tag{5.8.10}$$

Therefore, the new odds after the evidence E has been introduced are, obtained by taking the ratio of $P(H|E)$ and $P(H^c|E)$

$$\frac{P(H|E)}{P(H^c|E)} = \frac{P(H)}{P(H^c)} \frac{P(E|H)}{P(E|H^c)} \tag{5.8.11}$$

That is, the new value of the odds of H is the old value, multiplied by the ratio of the conditional probability of the new evidence given that H is true to the conditional probability given that H is not true.

Example 5.9 Suppose there are two bowls of cookies. Bowl 1 contains 30 vanilla cookies and 10 chocolate cookies. Bowl 2 contains 20 of each. Now suppose you choose one of the bowls at random and, without looking, select a cookie at random. The cookie is vanilla. What is the probability that it came from Bowl 1?

Solution. Let's denote by H the event that the cookie comes from Bowl 1, and E the event that the cookie is a vanilla. We have $P(H) = P(H^c) = 1/2$ (without the information that the chosen cookie was a vanilla, the probability for it to come from either of the two bowls is 50%). We also have $P(E|H)$, the probability that the cookie is a vanilla given that it comes from Bowl 1, which is $30/40 = 3/4$. Similarly, we have $P(E|H^c) = 20/40 = 1/2$. Then, using the odds form of Bayes' rule, we have

$$\frac{P(H|E)}{P(H^c|E)} = \frac{P(H)}{P(H^c)}\frac{P(E|H)}{P(E|H^c)} = \left(\frac{1/2}{1/2}\right)\left(\frac{3/4}{1/2}\right) = \frac{3}{2}$$

Therefore, $P(H|E)$ is $3/5$. Of course, we can also find this probability w/o using the odds form of Bayes' rule: Eq. (5.8.8) gives us

$$P(H|E) = \frac{P(E|H)P(H)}{P(E|H)P(H) + P(E|H^c)P(H^c)} = \cdots = \frac{3}{5}$$

And this is not unexpected as the two formula are equivalent. The odds form is still useful, as demonstrated in the next example, for cases that we do not know how to compute the prior odds.

Now, we introduce a new term—Bayes factor—to express Eq. (5.8.11) in a simpler form, easier to memorize. For a hypothesis H and evidence (or data) E, the Bayes factor is the ratio of the likelihoods:

$$\text{Bayes factor} := \frac{P(E|H)}{P(E|H^c)} \tag{5.8.12}$$

With this definition, Eq. (5.8.11) can be succinctly written as

$$\text{posterior odds} = \text{prior odds} \times \text{Bayes factor} \tag{5.8.13}$$

From this formula, we see that the Bayes factor (BF) tells us whether the evidence/data provides evidence for or against the hypothesis:

- If $BF > 1$ then the posterior odds are greater than the prior odds. So the data provides evidence for the hypothesis.
- If $BF < 1$ then the posterior odds are less than the prior odds. So the data provides evidence against the hypothesis.

- If $BF = 1$ then the prior and posterior odds are equal. So the data provides no evidence either way.

The two forms are summarized in Box 5.2.

Box 5.2: Summary of important formulae of conditional probability.

- conditional probability $P(A|B)$

$$P(A|B) = \frac{P(AB)}{P(B)}$$

- the law of total probability

$$P(A) = \sum_{i=1}^{n} P(A|B_i)P(B_i)$$

- for two events A and B with $P(A) \neq 0$, we have

$$P(B|A) = \frac{P(A|B)P(B)}{P(A)}$$

- if $B_1, B_2, \ldots, B_n$ form a partition of the sample space S, and $P(A) \neq 0$, then we have

$$P(B_j|A) = \frac{P(A|B_j)P(B_j)}{\sum_{i=1}^{n} P(A|B_i)P(B_i)}$$

- the odds form:

$$\frac{P(H|E)}{P(H^c|E)} = \frac{P(H)}{P(H^c)} \frac{P(E|H)}{P(E|H^c)} \tag{5.8.14}$$

Example 5.10 Here is another problem from MacKay's *Information Theory, Inference, and Learning Algorithms*: Two people have left traces of their own blood at the scene of a crime. A suspect, Oliver, is tested and found to have type "O" blood. The blood groups of the two traces are found to be of type "O" (a common type in the local population, having frequency 60%) and of type "AB" (a rare type, with frequency 1%). Do these data [bloods of type "O" and "AB" found at the scene] give evidence in favor of the proposition that Oliver was one of the people who left blood at the scene?

Solution. Let's call H the hypothesis (or proposition) that Oliver was one of the people who left blood at the scene. And let E be the evidence that there are bloods of type "O" and "AB" found at the scene. The only formula we have is the odds form of Bayes' rule:

$$\frac{P(H|E)}{P(H^c|E)} = \frac{P(H)}{P(H^c)} \frac{P(E|H)}{P(E|H^c)}$$

It is obvious that we cannot compute $P(H)/P(H^c)$. In fact, we do not need it, because the question is not about the actual probability that Oliver was one of the people who left blood at the scene! If we can compute the Bayes factor and based on whether it is larger than or smaller than one, we can have a conclusion. What is $P(E|H)$? When H happens, Oliver left his blood of type "O" at the scene, the other people has to have type "AB" blood with a probability of 0.01. Thus, $P(E|H) = 0.01$. For $P(E|H^c)$, we have then two random people at the scene, and we want the probability that they have type "O" and "AB" blood. Thus, $P(E|H^c) = 0.6 \times 0.01 \times 2$.[20]

So, the Bayes factor is

$$\frac{P(E|H)}{P(E|H^c)} = \frac{0.01}{0.6 \times 0.01 \times 2} = 0.83333333$$

Since the Bayes factor is smaller than 1, the evidence does not support the proposition that Oliver was one of the people who left blood at the scene.

Another suspect, Alberto, is found to have type "AB" blood. Do the same data give evidence of the proposition that Alberto was one of the two people at the scene?

$$\frac{P(E|H)}{P(E|H^c)} = \frac{0.6}{0.6 \times 0.01 \times 2} = 50$$

Since the Bayes factor is a lot larger than 1, the data provides strong evidence in favor of Alberto being at the crime scene.

History note 5.1: Thomas Bayes (1701–1761)

Thomas Bayes was an English statistician, philosopher and Presbyterian minister who is known for formulating a specific case of the theorem that bears his name: Bayes' theorem. Bayes never published what would become his most famous accomplishment; his notes were edited and published posthumously by Richard Price as *Essay towards solving a problem in the doctrine of chances* published in the Philosophical Transactions of the Royal Society of London in 1764.

[20] Note that we have assumed that the blood types of two people are independent (so that we can just multiply the probabilities). And why 2?

5.8.6 Independent Events

It is obvious that usually $P(A|B)$ is different from $P(A)$, but when they are equal, i.e., the probability of event A is not changed by the occurrence of event B, we say that event A is *independent* of event B. Using the conditional probability definition, Eq. (5.8.1), we can show that $P(A|B) = P(A)$ leads to $P(AB) = P(A)P(B)$:

$$P(A|B) := \frac{P(AB)}{P(B)} = P(A) \Longrightarrow \boxed{P(AB) = P(A)P(B)}$$

Note that this equation is symmetric with respect to A and B, thus if A is independent of B, B is also independent of A. Thus, we have this definition of two independent events A, B:

$$A \text{ and } B \text{ are independent when} \quad P(AB) = P(A)P(B) \tag{5.8.15}$$

What this formula says is that for two independent events A and B, the chance that both of them happen at the same time is equal to the product of the chance that A happens and the chance that B happens. And this is the multiplication rule of probability that Cardano discovered, check Sect. 5.2.

Example 5.11 Suppose that we toss 2 fair six-sided dice. Let E_1 denote the event that the sum of the dice is 6, E_2 be the event that the sum of the dice equals 7, and F denote the event that the first die equals 4. The questions are: are E_1 and F independent and are E_2 and F independent?

Solution. We just need to check whether the definition of independence of two events, i.e., $P(AB) = P(A)P(B)$ holds. We have

$$P(E_1)P(F) = \frac{5}{36} \times \frac{6}{36} = \frac{5}{216}$$

and

$$P(E_1F) = P(\{(4,2)\}) = \frac{1}{36}$$

Thus, $P(E_1F) \neq P(E_1)P(F)$: the two events E_1 and F are not independent; we call them dependent events.

In the same manner, we compute

$$P(E_2)P(F) = \frac{6}{36} \times \frac{6}{36} = \frac{1}{36}$$

and

$$P(E_2F) = P(\{(4,3)\}) = \frac{1}{36}$$

Thus, $P(E_2F) = P(E_2)P(F)$: the two events E_2 and F are independent. Shall we move on to other problems? No, we had to compute many probabilities to get the answers. Can we just use intuitive guessing? Let's try. To get a sum of six (event E_1), the first die must be one of any of $\{1, 2, 3, 4, 5\}$; the first die cannot be six. Thus, E_1 depends on the outcome of the first die. On the other hand, to get a sum of seven (event E_2), the first die can be anything of $\{1, 2, 3, 4, 5, 6\}$; all the possible outcomes of a die. Therefore, E_2 does not depend on the outcome of the first die.

Independent events Versus disjoint events. Are disjoint events independent or not? If A and B are two disjoint events, then $AB = \emptyset$, thus $P(AB) = 0$, whereas $P(A)P(B) \neq 0$. So, $P(AB) \neq P(A)P(B)$. Two disjoint events are dependent.

Some rules of independent events. Given that A and B are two independent events, what can we say about their complements or unions? Regarding the complementary events, we have this result: If A and B are independent then

- A and B^c are independent;
- A^c and B are independent;
- A^c and B^c are independent;

Thus, if A and B are independent events, then the probability of A's occurrence is unchanged by information as to whether or not B has happened.

Now we are going to generalize the definition of independence of two events to more than two events. Let's start simple with three events, and with one concrete example. It motivates our definition of the independence of three events.

Example 5.12 Two fair 6-sided dice are rolled, one red and one blue. Let A be the event that the red die's result is 3. Let B be the event that the blue die's result is 4. Let C be the event that the sum of the rolls is 7. Are A, B, C mutually independent?

Solution. It's clear that A and B are independent. From Example 5.11, we also know that A, C are independent and B, C are also independent. We're now checking whether $P(ABC) = P(A)P(B)P(C)$; this is our guess based on generalization of the case of two independent events. First,

$$P(A)P(B)P(C) = \frac{1}{6} \times \frac{1}{6} \times \frac{6}{36} = \frac{1}{216}$$

Second, using Eq. (5.8.4) we can compute $P(ABC)$:

$$P(ABC) = P(A)P(B|A)P(C|AB) = \frac{1}{6} \times \frac{1}{6} \times 1 = \frac{1}{36}$$

Three events A, B, and C are independent if all of the following conditions hold

$$\begin{aligned} P(AB) &= P(A)P(B) \\ P(BC) &= P(B)P(C) \\ P(CA) &= P(C)P(A) \\ P(ABC) &= P(A)P(B)P(C) \end{aligned} \tag{5.8.16}$$

And from three to n events is a breeze even n is infinite.

5.8.7 The Gambler's Ruin Problem

Two gamblers, A and B, are betting on the tosses of a fair coin. At the beginning of the game, **player** A has 1 coin and **player** B has 3 coins. So there are 4 coins between them. In each play of the game, **a fair coin is tossed**. If the result of the coin toss is head, player A collects 1 coin from player B. If the result of the coin toss is tail, player A pays player B 1 coin. The game continues until one of the players has all the coins (or one of the players loses all his/her coins). What is the probability that player A ends up with all the coins?

This is a simple version of the classic gambler's ruin problem: two players begin with fixed stakes, transferring points until one or the other is "ruined" by getting to zero points. The earliest known mention of the gambler's ruin problem is a letter from Blaise Pascal to Pierre Fermat in 1656 (2 years after the more famous correspondence on the problem of points).

A solution to the general version of the gambler's ruin problem is presented now. Let's denote by E the event that player A ends up with all the coins when he starts with i coins, $i = 0, 1, 2, \ldots, N$ where N is the total coins of both players. And to show the dependence on i, we use the notation $P_i = P(E)$. Now, we compute $P(E)$ by conditioning *on the event that the first coin lands on head or tail*. Using the law of total probability we can write:

$$P(E) = P(H)P(E|H) + P(H^c)P(E|H^c)$$

where $P(H) = p$ is the probability that the coin lands on head and $P(H^c) = q$ is the probability that the coin land on tail. What is $P(E|H)$ and $P(E|H^c)$? Now, as the first coin lands on head (i.e., H), A has $i + 1$ coins. Since successive flips are assumed to be independent, we just have the same game in which A starts with $i + 1$ coins. Therefore, $P(E|H) = P_{i+1}$. Similarly, if the first coin shows tail, $P(E|H^c) = P_{i-1}$. With that, we can write

$$P_i = pP_{i+1} + qP_{i-1} \tag{5.8.17}$$

Now, using the fact that $p + q = 1$, we have $P_i = (p + q)P_i = pP_i + qP_i$. Replace P_i in the above equation with this, we obtain

$$pP_i + qP_i = pP_{i+1} + qP_{i-1} \Longrightarrow P_{i+1} - P_i = \frac{q}{p}(P_i - P_{i-1}), \quad i = 1, 2, 3, \ldots, N-1$$

Now, we will explicitly write out this equation for $i = 1, 2, 3, \ldots$, with the so-called boundary condition that $P_0 = 0$, i.e., we assume that if player A starts with zero coin, he will lose:

$$\begin{aligned}
i = 1 \quad &: \cancel{P_2} - \boxed{P_1} = \frac{q}{p}(P_1 - P_0) = \frac{q}{p}P_1 \\
i = 2 \quad &: \cancel{P_3} - \cancel{P_2} = \frac{q}{p}(P_2 - P_1) = \left(\frac{q}{p}\right)^2 P_1 \quad \text{(use the result from above row)} \\
i = 3 \quad &: \boxed{P_4} - \cancel{P_3} = \frac{q}{p}(P_3 - P_2) = \left(\frac{q}{p}\right)^3 P_1 \\
\vdots \quad &: \vdots \qquad \vdots \\
i = k - 1 &: P_k - P_{k-1} = \frac{q}{p}(P_{k-1} - P_{k-2}) = \left(\frac{q}{p}\right)^{k-1} P_1
\end{aligned}$$

What we do next? We sum up all the above equations, because we see a telescoping sum $(P_2 - P_1) + (P_3 - P_2) + \cdots$; for the first three rows, only P_4 and P_1 are left without being canceled out, for $k = 1, 2, 3, \ldots, N$[21]:

$$P_k = P_1 \left(1 + \frac{q}{p} + \left(\frac{q}{p}\right)^2 + \left(\frac{q}{p}\right)^3 + \cdots + \left(\frac{q}{p}\right)^{k-1}\right)$$

And what is the infinite sum on the RHS? It is a geometric series, recall from Eq. (2.21.5) that

$$a + ar + ar^2 + ar^3 + \cdots + ar^{n-1} = \frac{a}{1-r}(1 - r^n)$$

Thus, we have a geometric series with $a = 1$, $r = q/p$, thus for $i = 1, 2, 3, \ldots, N$ (I switched back to i instead of k)

$$P_i = \begin{cases} iP_1, & \text{if } q/p = 1 \\ P_1 \dfrac{1 - (q/p)^i}{1 - q/p}, & \text{if } q/p \neq 1 \end{cases}$$

All is good, but still we do not know P_1. Now, we use another boundary condition $P_N = 1$, and then we're able to determine P_1 and then P_i. Plug $i = N$ into the above equation, we can determine P_1[22]:

---

[21] I have moved P_1 to the RHS.

[22] Note that $q/p = 1$ is equivalent to $p = 1/2$.

$$P_1 = \begin{cases} \dfrac{1}{N}, & \text{if } p = 1/2 \\ \dfrac{1 - q/p}{1 - (q/p)^N}, & \text{if } p \neq 1/2 \end{cases}$$

And with that, P_i is given by

$$P_i = \begin{cases} \dfrac{i}{N}, & \text{if } p = 1/2 \\ \dfrac{1 - (q/p)^i}{1 - (q/p)^N}, & \text{if } p \neq 1/2 \end{cases} \tag{5.8.18}$$

What are the outcomes of this gambler's ruin game? First outcome is player A wins, the second is player B wins. Is that all? Is it possible that the game is never ending? To check that we need to compute the probability that player B wins when A starts with i coins, this probability is designated by Q_i. And if $P_i + Q_i = 1$, then the game will definitely end with either A wins or B wins.

By symmetry, we can get the formula for Q_i from P_i by replacing i with $N - i$, which is the amount of coins that player B starts, and p with q:

$$Q_i = \begin{cases} \dfrac{N-i}{N}, & \text{if } p = 1/2 \\ \dfrac{1 - (p/q)^{N-i}}{1 - (p/q)^N}, & \text{if } p \neq 1/2 \end{cases}$$

The sum $P_i + Q_i$ is 1 for $p = 1/2$. For $p \neq 1/2$, we have

$$P_i + Q_i = \frac{1 - (q/p)^i}{1 - (q/p)^N} + \frac{1 - (p/q)^{N-i}}{1 - (p/q)^N} = \cdots = 1$$

Some details were skipped for sake of brevity. Thus, the game will end with either A wins or B wins. Let's pause a bit and see what we have seen: we have seen a telescoping sum, and a geometric series in a game of coin tossing! Isn't mathematics cool?

Solution using difference equations. Equation (5.8.17) is a (linear) difference equation (or a recurrence equation) which involves the differences between successive values of a function of a discrete variable. In that equation we have the difference between P_i, P_{i+1} and P_{i-1}, all are values of a function of i—a discrete variable. (A discrete variable is a variable of which values can only be integers.) Note that a difference equation is the discrete analog of a differential equation discussed in Chap. 7.

To solve Eq. (5.8.17), we rewrite it as follows

$$pP_{i+1} - P_i + qP_{i-1} = 0 \tag{5.8.19}$$

Now, we guess that the solution of this equation is of the form Ar^i[23]:

$$P_i = q^2 P_{i-2} = q^3 P_{i-3} = \cdots = q^i P_0$$

Thus, the solution is of an exponential form.:

$$P_i = Ar^i \Longrightarrow P_{i+1} = Ar^{i+1}, \quad P_{i-1} = Ar^{i-1}$$

Substituting these into Eq. (5.8.19) results in

$$Ar^{i-1}(pr^2 - r + q) = 0 \Longrightarrow \boxed{pr^2 - r + q = 0} \tag{5.8.20}$$

This is a quadratic equation, thus, for the case $p \neq 1/2$ it has two roots (note that $q = 1 - p$):

$$r_1 = 1, \quad r_2 = \frac{q}{p}$$

Thus, $A_1 r_1^i + A_2 r_2^i$ is the general solution to Eq. (5.8.19), so we can write

$$P_i = A_1 + A_2 \left(\frac{q}{p}\right)^i \tag{5.8.21}$$

Now, we determine A_1 and A_2 using the two boundary conditions: $P_0 = 0$ and $P_N = 1$:

$$A_1 + A_2 = 0, \quad A_1 + A_2 \left(\frac{q}{p}\right)^N = 1$$

Solving these two equations, we obtain A_1 and A_2:

$$A_1 = \frac{1}{1 - (q/p)^N}, \quad A_2 = -A_1$$

Substituting $A_{1/2}$ into Eq. (5.8.21) gives the final solution:

$$P_i = A_1 \left(1 - \left(\frac{q}{p}\right)^i\right) = \frac{1 - (q/p)^i}{1 - (q/p)^N}$$

Let's see what are the odds playing in a casino. Assume that $N = 10000$ Units. Using Eq. (5.8.18), the odds are calculated for different initial wealth. The results shown in Table 5.5 are all bad news. As we cannot have more money than the casino, we look at the top half of the table, and the odds are all zero (do not look at the

[23] Why this form? If we start with this simpler equation $P_i = qP_{i-1}$, then, we have.

Table 5.5 Probabilities of player A breaking the bank with total initial wealth $N = 10000$ Units

A's initial wealth	Fair game	Craps	Roulette
i	$p = 0.5$	$p = 0.493$	$p = 0.474$
100	0.0100	0.0000	0.0000
500	0.0500	0.0000	0.0000
1000	0.1000	0.0000	0.0000
5000	0.5000	0.0000	0.0000
6000	0.6000	0.0000	0.0000
9000	0.9000	0.0000	0.0000
9950	0.9950	0.2466	0.0055
9990	0.9990	0.7558	0.3531

column with $p = 0.5$; that's just for reference). One way to improve our odds is to be bold: instead of betting 1 dollar, betting 10 dollars, for example.

If $N = 100$ dollars, and player A starts with 10 dollars, what is his chance if he bets 10 dollars per game? Think of 1 coin is 10 dollars, then we can just use Eq. (5.8.18) with $i = 1$ and $N = 10$: $P_1 = {}^{1-(q/p)^1}/_{1-(q/p)^{10}}$.

5.9 The Secretary Problem or Dating Mathematically

The statement of the secretary problem goes as follows

> You are the HR manager of a company and need to hire the best secretary out of a given number *N of candidates*. You can interview them *one by one, in random order*. However, the decision of appointing or rejecting a particular applicant must be taken *immediately after the interview*. If nobody has been accepted before the end, the last candidate is chosen. What strategy do you use to maximize the chances to *hire the best applicant*?

The first thing we need to do is to translate the problem into mathematics. Let's assign a counting number to each candidate. Thus, four candidates John, Sydney, Peter and Laura would be translated to a list of integers: $(1, 7, 3, 9)$, an integer can be thought of as the score of a candidate. In general we denote by $(a_1, a_2, \ldots, a_N)$ this list. The problem now is to find the maximum of this list, denoted by $a_{\max}$.

If you're thinking, this is easy, I pick Laura as the best applicant for 9 is the maximum of $(1, 7, 3, 9)$. No, you cannot do this for one simple reason: *you cannot look ahead*. Think of your dating, you cannot know in advance who will you date in the future! Thus, at the time the HR manage is interviewing Peter (3) she does not know that there is a better candidate waiting for her. Note that she has to make a decision (rejecting or accepting) immediately after the interview. That is the rule of this problem. It might not be real, but mathematicians do not care.

Ok. I pick the last applicant! But the probability of getting the best is only $1/N$, if N is large then that probability is slim. So, we cannot rely on luck, we need some

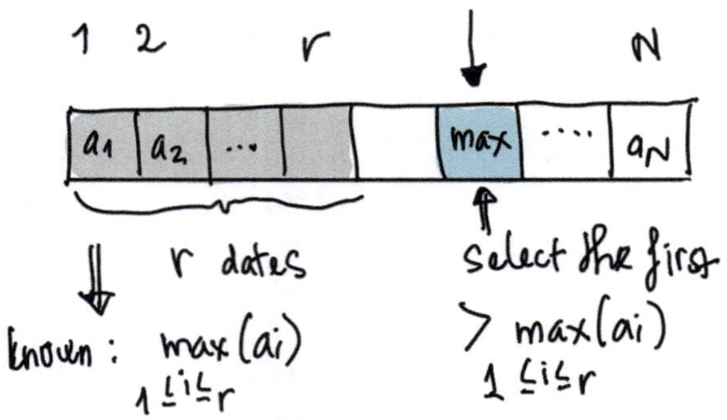

Fig. 5.5 Secretary problem: scan the first r candidates, record the maximum score (i.e., $a^* = \max a_i$, $1 \le i \le r$, and select the first candidate whose score is larger than a^*

strategy here. Again think of dating, what is the strategy there? The strategy most adults adopt—insofar as they consciously adopt a strategy—is *to date around for a while*, gain some experience, figure out one's options, and then *choose the next best thing* that comes around.

We adopt that strategy here. Thus, we scan the first r candidates, record the maximum score, denoted by a^*, (i.e., $a^* = \max a_i$, $1 \le i \le r$), and then select the first candidate whose score is larger than a^* (Fig. 5.5). Now, we're going to compute what is the probability if we do this. Obviously, that probability depends on r; if we have that probability, labeled by $P(r)$, then we can find r that maximizes $P(r)$; such an r is called the optimal r. Assume that optimal r is five, then the optimal strategy (for dating) is: date 5 persons, discard all of them and marry the next person who is better than the best among your five old lovers.

Let n be the nth candidate (after r rejected or scanned candidates) of which the score is maximum. Of course we need to have $n \ge r + 1$ (if not, we would lose the best among the rejected r candidates). The first candidate with a score higher than a^* is the best candidate (i.e., $a_n = a_{\max}$) only happens when the second best is in r candidates. Therefore, $P(r)$ is

$$\begin{aligned}
P(r) &= \sum_{n=r+1}^{N} P(\text{1st} > a^* \text{ and} a_{\max}) \\
&= \sum_{n=r+1}^{N} P(n\text{th is the best and the second best is in } r \text{ candidates}) \\
&= \sum_{n=r+1}^{N} P(n\text{th is the best}) \times P(\text{ the second best is in } r \text{ candidates out of } n-1) \\
&= \sum_{n=r+1}^{N} \frac{1}{N}\frac{r}{n-1} = \frac{r}{N}\sum_{n=r+1}^{N} \frac{1}{n-1} = \frac{r}{N}\sum_{n=r}^{N-1}\frac{1}{n}
\end{aligned} \tag{5.9.1}$$

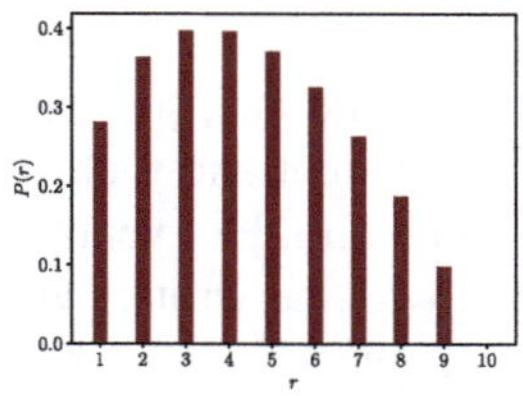

The question now is what should be the value of r so that $P(r)$ is maximum? To answer that question we need to know more about $P(r)$ and there is nothing better than a picture. So, choose $N = 10$ and for $r = 1, 2, 3, 4, 5, 6, 7, 8, 9, 10$ we compute ten $P(r)$ using Eq. (5.9.1), plot them and we obtain a plot shown in the figure. This plot tells us that there is indeed a value of r such that $P(r)$ is maximum, and there is only one such r. Because of that, the next $r+1$ has a lower probability. So, we just need to find r such that $P(r+1) \le P(r)$:

$$P(r+1) \le P(r) \Longleftrightarrow \frac{r+1}{N}\sum_{n=r+1}^{N-1}\frac{1}{n} \le \frac{r}{N}\sum_{n=r}^{N-1}\frac{1}{n} \Longleftrightarrow \sum_{n=r+1}^{N-1}\frac{1}{n} \le 1$$

Recognizing the red sum is related to the n harmonic number,[24] we rewrite the above sum as

$$\sum_{n=1}^{N-1}\frac{1}{n} - \sum_{n=1}^{r}\frac{1}{n} \approx \ln(N-1) - \ln r = \ln\frac{N-1}{r}$$

where the first sum in the left most term is the $(N-1)$th harmonic number H_{N-1}, the second term is the rth harmonic number, and noting that we can approximate $H_n \approx \ln n + \gamma + \mathcal{O}(n)$ where γ is the Euler-Mascheroni constant defined in Eq. (4.15.28). When N is very large, $N - 1 = N$, and thus we need to find r such that

$$\ln\frac{N}{r} \approx 1 \Longleftrightarrow \frac{N}{r} \approx e \Longrightarrow r \approx \frac{N}{e} \approx 0.37N, \quad (e = 2.718281\ldots)$$

What this formula tells us is that we should discard 37% of the total number of candidates, then select the next person that comes along who is better than all of those discarded.

Kepler and the marriage problem. In 1611, after losing his first wife, Barbara, to cholera, the great astronomer and mathematician Johannes Kepler wanted to remarry. His first marriage was an arranged one and not so happy so he decided to find a suitable second wife with care. Now we know how Kepler went about the selection process because he documented it in great detail to Baron Strahlendorf on 23 October 1613. In his process, Kepler had considered 11 different matches over 2 years. The fourth woman was nice to look at—of "tall stature and athletic build", but

[24] If needed, check Sect. 4.15.7 for refresh on harmonic numbers.

Kepler wanted to check out the next one, who, he'd been told, was "modest, thrifty, diligent and [said] to love her stepchildren", so he hesitated. He hesitated so long, that both No. 4 and No. 5 got impatient and took themselves out of the running, leaving him with No. 6, who scared him. He eventually returned to the fifth match, 24-year-old Susanna Reuttinger, who, he wrote, "won me over with love, humble loyalty, economy of household, diligence, and the love she gave the stepchildren. On 30 October 1613, Kepler married Reuttinger who was a wonderful wife and both she and Kepler were very happy.

5.10 Discrete Probability Models

Consider a sample space S, and if S is a countable set, this refers to a discrete probability model. As S is countable, we can list all the elements in S:

$$S = \{s_1, s_2, s_3, \ldots$$

Now if A is an event, we have $A \subset S$, then A is also countable. By the third axiom (of probability), we have

$$P(A) = P\left(\bigcup_{s_j \in A} s_j\right) = \sum_{s_j \in A} P(s_j) \tag{5.10.1}$$

Thus, in a countable sample space, to find the probability of an event, all we need to do is to sum the probability of individual elements in that set. How can we find the probability of individual elements then? We answer this question next.

Finite sample spaces with equally likely outcomes. An important special case of discrete probability models is when we have a finite sample space S, where each outcome is equally likely to occur, i.e.,

$$S = \{s_1, s_2, s_3, \ldots, s_N\}, \quad \text{where } P(s_i) = P(s_j) \text{ for all } i, j \in \{1, 2, \ldots, N\} \tag{5.10.2}$$

Examples are tossing a fair coin or rolling a fair die.

From the second axiom we have $P(S) = 1$, and by denoting $P = P(s_1) = P(s_2) = \cdots = P(s_N)$, we have

$$1 = P(S) = \sum_{i=1}^{N} P(s_i) = NP$$

Therefore,

$$P(s_i) = \frac{1}{N} \quad \text{for all } i = \{1, 2, ..., N\}$$

Next, we're going to calculate $P(A)$ for event A with $|A| = M$, we write

$$P(A) = P\left(\bigcup_{s_j \in A} s_j\right) = \sum_{s_j \in A} P(s_j) = \frac{M}{N} = \frac{|A|}{|S|} \tag{5.10.3}$$

Thus, finding the probability of A reduces to a counting problem in which we need to count how many elements are in A and S. We get the results that Cardano had discovered.[25] And do we know how to count things...efficiently? Yes, we do (Sect. 2.26). If your understanding of factorial, permutations and combinations is not solid (yet), you have to study them again before continuing with probability.

The birthday problem deals with the probability that in a set of n randomly selected people, *at least* two people share the same birthday. This problem is often referred to as the birthday paradox because the probability is counter-intuitively high: with only 23 people, the probability is 50% that at least two people share the same birthday, and with 50 people that chance is about 90%. The first publication of a version of the birthday problem was by Richard von Mises[26] in 1939.

Equipped with probability theory, we're going to solve this problem. But, we need a few assumptions. First, we disregard leap year, which simplifies the math, and it doesn't change the results by much. We also assume that all birthdays have an equal probability of occurring.[27]

Because leap years are not considered, there are only 365 birthdays. And we use this formula $P(A^c) = 1 - P(A)$. That is, instead of working directly, we approach the problem indirectly by asking what is the probability that *none people share the same birthday*. This is because doing so is much easier (note that in the direct problem, handling "at least" two people is not easy as there are two many possibilities).

Let A be the event that none people share the same birthday. We're going to determine $P(A)$, which is not hard. The answer to the problem is simply $1 - P(A)$. The sample space is $\{1, 2, \ldots, 365\}^n$, which has a cardinality of 365^n. For the first person of n people, there are 365 choices, for the second person, there are only 364 choices (to not have the same birthday with the first person), third person 363 choices (to not share the birthday with the first two persons). And for the nth person, there are $365 - n + 1$ choices. Thus, the probability that none people share the same birthday is

---

[25] Note that Cardano could not prove this formula, and we could, starting from Kolmogorov's three axioms.

[26] Richard Edler von Mises (1883–1953) was an Austrian scientist and mathematician who worked on solid mechanics, fluid mechanics, aerodynamics, aeronautics, statistics and probability theory. In solid mechanics, von Mises made an important contribution to the theory of plasticity by formulating what has become known as the von Mises yield criterion. If you want to become a civil/mechanical/aeorspace engineer, you will encounter his name.

[27] The second assumption is not true. But for the first attack to this problem, do not bother too much.

$$P(A) = \frac{(365)(364)\cdots(365-n+1)}{365^n}$$

Therefore, the probability we're looking for is

$$P(n) = 1 - \frac{(365)(364)\cdots(365-n+1)}{365^n} \tag{5.10.4}$$

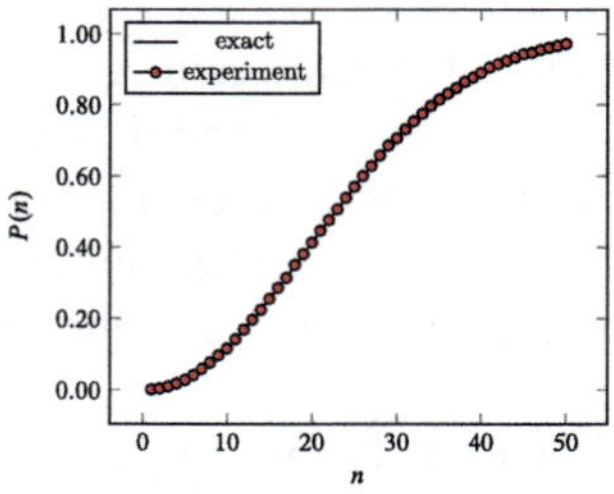

Now, we compute $P(n)$ for different values of n from 1 to 50. And we also carry out virtual experiments (see Appendix A.6). The idea is to check the exact solution—that is Eq. (5.10.4)—with the solution of empirical probability. With 10^5 trials, the empirical solutions match well with the analytical solutions. And we have $P(23) \approx 0.5$, thus with just 23 people in a room, there is 50% chance that at least two of them have the same birthday. With about 50 people that chance is increased to about 90%.

The main reason that this problem is called a paradox is that if you are in a group of 23 and you compare your birthday with the others, *you think you're making only 22 comparisons*. This means that there are only 22 chances of sharing the birthday with someone. However, we don't make only 22 comparisons. That number is much larger and it is the reason that we perceive this problem as a paradox. Indeed, the comparisons of birthdays will be made between every possible pair of individuals. With 23 individuals, there are $\binom{23}{2} = (23 \times 22)/2 = 253$ pairs to consider, which is well over half the number of days in a year (182.5 or 183).

Now, we consider the inverse problem of the birthday problem: how many people (i.e., n =?) so that at least two people will share a birthday with a probability of 0.5? It seems easy, we just need to solve the following equation for n

$$1 - \frac{(365)(364)\cdots(365-n+1)}{365^n} = 0.5$$

Hmm. How to solve this equation? It is interesting to realize that a bit massage to $P(n)$ will be helpful. We rewrite $P(n)$ as follows by $365^n = 365 \times 365 \times \cdots \times 365$, and pair each 365 with one number in the nominator:

$$\begin{aligned} P(n) &= 1 - \left(\frac{365}{365}\right)\left(\frac{364}{365}\right)\cdots\left(\frac{365-n+1}{365}\right) \\ &= 1 - \left(\frac{365}{365}\right)\left(1-\frac{1}{365}\right)\left(1-\frac{2}{365}\right)\cdots\left(1-\frac{n-1}{365}\right) \end{aligned} \tag{5.10.5}$$

Now comes the art of approximation, recall that for small x close to zero, we have[28]

$$e^x \approx 1 + x \Longrightarrow e^{-x} \approx 1 - x$$

(Note that Eq. (5.10.5) has terms of the form $1 - x$). Thus, Eq. (5.10.5) becomes[29]

$$\begin{aligned} P(n) &\approx 1 - \left(e^{-\frac{1}{365}}\right)\left(e^{-\frac{2}{365}}\right)\cdots\left(e^{-\frac{n-1}{365}}\right) \\ &\approx 1 - \exp\left(-\frac{1+2+\cdots+n-1}{365}\right) \\ &\approx 1 - \exp\left(-\frac{n(n-1)}{2\times 365}\right) \approx 1 - \exp\left(-\frac{n^2}{2\times 365}\right) \end{aligned} \tag{5.10.6}$$

where use was made of the sum of the first counting numbers formula (Sect. 2.6.1).

With this approximation, it is easy to find the n such that $P(n) = 0.5$:

$$1 - \exp\left(-\frac{n^2}{2\times 365}\right) = 0.5 \Longrightarrow \frac{n^2}{2\times 365} = \ln 2 \Longrightarrow n = \sqrt{\ln 2 \times 730} = 22.49$$

And from that we get $n = 23$.

5.10.1 *Discrete Random Variables*

Frequently, when an experiment is performed, we are interested mainly in some function of the outcome as opposed to the actual outcome itself. For example, in rolling two dice, we are often interested in the sum of the two dice and are not really concerned about the separate values of each die. That is, we may be interested in knowing that the sum is 4 and may not be concerned over whether the actual outcome was (1, 3), (2, 2) or (3, 1). These quantities of interest are known as *random variables*.

Usually the notation X is used to denote a random variable. And the notation x is used to denote a value of X.

Back to the dice rolling, if we roll two six-sided dice the sample space is shown in Table 5.6. This space is the Cartesian product of $\{1, 2, 3, 4, 5, 6\}$ and $\{1, 2, 3, 4, 5, 6\}$. Now, we can define many random variables. For instance, let's define X as the total number of points on the two dice; a few values of X are 2, 3, 12. The event A that $X = 2$ is the one with two dice showing one on their faces, i.e., $(1, 1)$. The event B that $X = 3$ is the one with $(1, 2)$ or $(2, 1)$. Precisely, such X is called a *discrete random variable* because its possible values are countable.

---

[28] If this is not clear check Taylor series in Sect. 4.15.8. It is hard to live without calculus!

[29] Now we understand why mathematicians need two notations for the exponential function: e^x and $\exp(x)$, the latter is for lengthy inputs.

Table 5.6 Sample space Ω of rolling two six-sided dice: $|\Omega| = 36$

	1	2	3	4	5	6
1	(1, 1)	(1, 2)	(1, 3)	(1, 4)	(1, 5)	(1, 6)
2	(2, 1)	(2, 2)	(2, 3)	(2, 4)	(2, 5)	(2, 6)
3	(3, 1)	(3, 2)	(3, 3)	(3, 4)	(3, 5)	(3, 6)
4	(4, 1)	(4, 2)	(4, 3)	(4, 4)	(4, 5)	(4, 6)
5	(5, 1)	(5, 2)	(5, 3)	(5, 4)	(5, 5)	(5, 6)
6	(6, 1)	(6, 2)	(6, 3)	(6, 4)	(6, 5)	(6, 6)

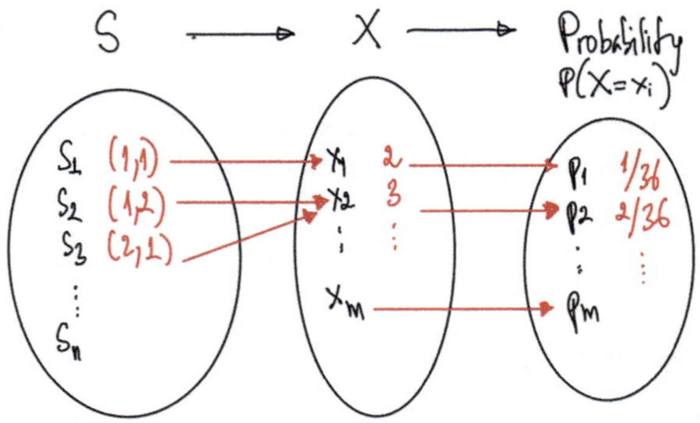

Fig. 5.6 A random variable is a real-valued function from the sample space S to $\mathbb{R}$

Because the value of a random variable X is determined by the outcome of an experiment, we want to assign probabilities to the possible values of the random variable. This is achieved with defining a probability mass function discussed shortly in Sect. 5.10.2.

Mathematically, a random variable X is a real-valued function that maps a set $s \in S$ to a real number. See Fig. 5.6 for a graphical illustration. Because of this, it is a bit confusing when we use the term variable. However, as we will define other functions depending on X, it is called a variable in that sense.

There also exits the so-called continuous random variables. For example, the heights of randomly selected people from a population is a continuous random variable. A continuous variable is so-called because we cannot list it as we do for discrete random variable. Still remember Hilbert's hotel with infinite rooms and Georg Cantor? This section is confined to a discussion of discrete random variables only.

5.10.2 Probability Mass Function

Let's consider a discrete random variable (RV) X of which the range[30] is given by

$$R_X = \{x_1, x_2, x_3, \dots\}$$

where $x_1, x_2, \dots$ are possible values of the random variable X. If we know the probability that X gets a value x_k for all x_k in R_X we will know its probability distribution. The probability of the event $\{X = x_k\}$ is called the probability mass function (PMF) of X. Following Pishro-Nik, the notation for it is $P_X(x_k)$; the subscript X is needed as we shall deal with more than one random variables, each has its own PMF.

Example 5.13 Tossing a coin twice and let X be the number of heads observed. Find the probability mass function P_X. The sample space is $S = \{(H, H), (H, T), (T, H), (T, T)\}$. So, the number of heads X is

$$X = \{0, 1, 2\}$$

Now, we compute $P(X = x_k)$ for $k = 1, 2, 3$:

$$\begin{aligned} P_X(0) &= P(X = 0) = P(T, T) &&= 1/4 \\ P_X(1) &= P(X = 1) = P((H, T), (T, H)) &&= 1/2 \\ P_X(2) &= P(X = 2) = P((H, H)) &&= 1/4 \end{aligned}$$

So, the probability mass function of a random variable X is the function that takes a number $x \in \mathbb{R}$ as input and returns the number $P(X = x)$ as output. (Note that we included continuous random variables in this discussion).

Because a PFM is a probability, it has to satisfy the following two properties:

$$0 \le P_X(x) \le 1, \quad \sum_{x \in R_X} P_X(x) = 1 \tag{5.10.7}$$

To better visualize the PMF, we can plot it. Figure 5.7 shows the PMF of the above random variable X; the plot on the right is known as a bar plot. As we see, the random variable can take three possible values 0, 1 and 2. The figure also clearly indicates that the event $X = 1$ is twice as likely as the other two possible values.

[30] Check Sect. 4.2.4 if you need a refresh on what is a range of a function.

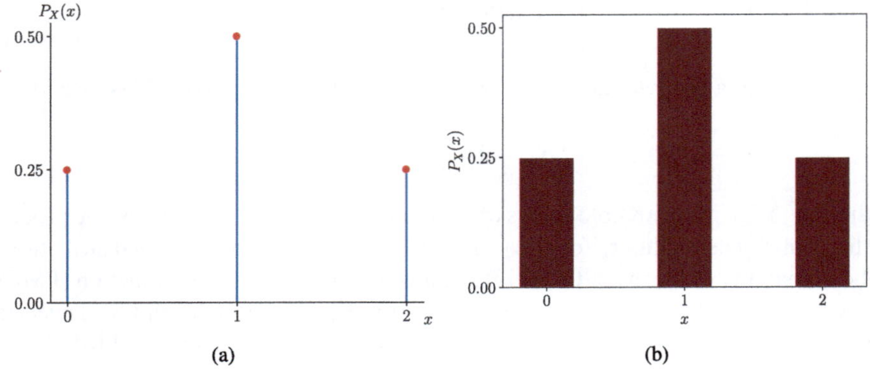

Fig. 5.7 Visualization of the probability distribution of a discrete random variable. Source code: `probability_plots.jl`

5.10.3 Special Distributions

Bernoulli distribution. A Bernoulli random variable is a discrete random variable that can only take two possible values, usually 0 and 1. This random variable models random experiments that have two possible outcomes: "success" and "failure". Here are some examples:

- You take a pass-fail exam. You either pass (resulting in $X = 1$) or fail (resulting in $X = 0$).
- You toss a coin. The outcome is either head or tail.
- A child is born. The gender is either male or female.
- Rolling two dice. You either get a double six (with probability of $1/36$) or not a double six (with a chance of 35/36).

Definition 5.10.1 A random variable X is said to be a Bernoulli random variable with parameter p, denoted by $X \sim Bernoulli(p)$, if its PMF is given by

$$P_X(x) = \begin{cases} p, & \text{if } x = 1 \\ 1 - p, & \text{if } x = 0 \\ 0, & \text{otherwise} \end{cases} \tag{5.10.8}$$

Geometric distribution. Assume that we have an unfair coin for which $P(H) = p$, where $0 < p < 1$ and $p \neq 0.5$. We toss the coin repeatedly until we observe *a head for the first time*. Let X be the total number of coin tosses. Find the distribution of X.

First, we see that $X = \{1, 2, 3, \ldots, k, \ldots\}$. To find the distribution of X is to find $P_X(k) = P(X = k)$ for $k = 1, 2, 3$, and so on. These probabilities are (as all tosses are independent, the probability of, let say, TH is just the product of the probabilities of getting T and H)

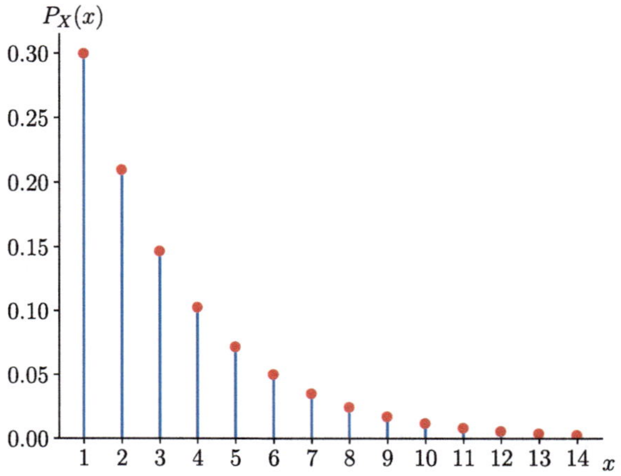

Fig. 5.8 Geometric distribution with $p = 0.3$

$$\begin{aligned}
P_X(1) &: P(H) &&= p \\
P_X(2) &: P(TH) &&= (1-p)p \\
P_X(3) &: P(TTH) &&= (1-p)(1-p)p = (1-p)^2 p \\
&\vdots \quad \vdots && \vdots \\
P_X(k) &: P(TT\dots H) &&= (1-p)^{k-1} p
\end{aligned}$$

If we list these probabilities we obtain this sequence $p, qp, q^2 p, \dots$ with $q = 1 - p$, which is a geometric sequence with $a = p$ and $r = q$ (Sect. 4.15). For that reason, X is called a geometric distribution, denoted by $X \sim Geometric(p)$. See Fig. 5.8 for a plot of this distribution (for $p = 0.3$). Check file `probability_plots.jl` on my GitHub account to see how this plot was generated. Note that, we can think of the experiment behind the geometric distribution as *repeating independent Bernoulli trials until observing the first success.*

Binomial distribution. Suppose that we have a coin for which $P(H) = p$ and thus $P(T) = 1 - p$. We toss it five times. What is the probability that we observe exactly k heads and $5 - k$ tails?[31] To solve this problem, we start with a concrete case: Let A be the event that we observe exactly three heads and two tails. What is $P(A)$?

Because A is the event that we observe exactly three heads and two tails, we can write

$$A = \{HHHTT, TTHHH, THHHT, \dots\}$$

It can be shown that the probability of each member of A is $p^3(1-p)^2$. As there are $|A|$ such members, the probability of A is

$$P(A) = |A| p^3 (1-p)^2$$

[31] Of course $k = 0, 1, 2, 3, 4, 5$.

But from Sect. 2.26.5, we know that $|A| = \binom{5}{3}$, so

$$P(A) = \binom{5}{3} p^3 (1-p)^2$$

With this, we have the following definition of a binomial distribution.

Definition 5.10.2 A random variable X is said to be a binomial random variable with parameters n and p, shown as $X \sim Binomial(n, p)$, if its PMF is given by[32]

$$P_X(k) = \binom{n}{k} p^k (1-p)^{n-k} \quad \text{for } k = 0, 1, 2, \cdots, n \tag{5.10.9}$$

Example 5.14 What is the probability that among five families, each with six children, at least three of the families have four or more girls? Of course, we assume that the probability to have a boy is 0.5.

To solve this problem, first note that the five families are the five trials. And each trial is a success if that family has at least four girls. And if we denote by p_0 the probability of a family to have at least four girls, the probability that at least three of the families have four or more girls is

$$\binom{5}{3} p_0^3 (1-p_0)^2 + \binom{5}{4} p_0^4 (1-p_0) + \binom{5}{5} p_0^5 \tag{5.10.10}$$

To find p_0, we realize that to get six children, each family has to perform six Bernoulli trials with $p = 0.5$ to get a boy or a girl, thus:

$$p_0 = \binom{6}{4}(0.5)^6 + \binom{6}{5}(0.5)^6 + \binom{6}{6}(0.5)^6 = \frac{11}{32}$$

Plugging this p_0 into Eq. (5.10.10) we get the answer to this problem. But that number is not important than the solution process.

We can generalize what we have found in the above example, to have a formula for calculating the probability of $a \le X \le b$:

$$\boxed{P(a \le X \le b) = \sum_{k=a}^{b} \binom{n}{k} p^k (1-p)^{n-k}} \tag{5.10.11}$$

To have a better understanding of the binomial distribution, we plot some of them in Fig. 5.9. The curve has an ascending branch starting from $k = 0$ to k_{max}, and a descending branch with $k \ge k_{max}$. It is possible to determine the value for k_{max}. First,

[32] How to make sure that this is indeed a PMF? Equation (5.10.7) is the answer.

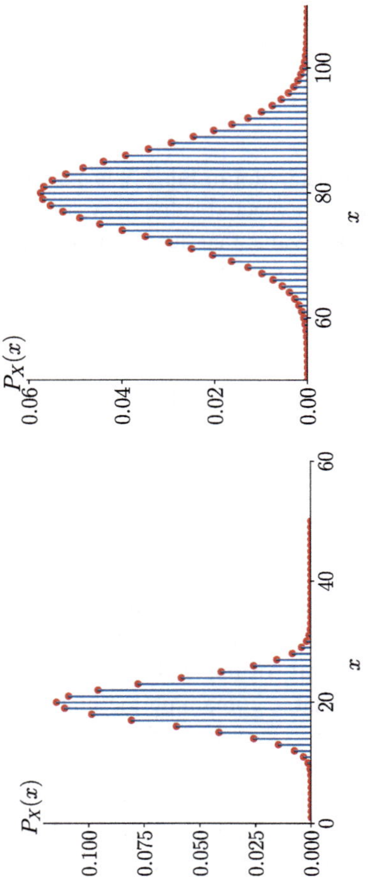

(a) $X \sim Binomial(50, 0.4)$

(b) $X \sim Binomial(100, 0.4)$

Fig. 5.9 Visualization of two binomial distributions. Observe that the curves peak at around np

let's denote $b_n(k) = P_X(k)$, and we need to compute the ratio of two successive terms:

$$\begin{aligned}\frac{b_n(k)}{b_n(k-1)} &= \left(\frac{n!}{(n-k)!k!}p^k(1-p)^{n-k}\right) / \left(\frac{n!}{(n-k+1)!(k-1)!}p^{k-1}(1-p)^{n-k+1}\right) \\ &= \frac{(n-k+1)p}{kq}\end{aligned} \tag{5.10.12}$$

To find the peak of the binomial distribution curve, we find k such that the ratio $b_n(k)/b_n(k-1)$ is larger than or equal to one:

$$\frac{b_n(k)}{b_n(k-1)} \geq 1 \Longleftrightarrow (n+1)p \geq k \Longrightarrow k_{\max} \approx np \tag{5.10.13}$$

Now we can understand why each plot in Fig. 5.9 has a peak near np. And why np is at the peak? Because it is the expected value of X, i.e., it is the average value of X. And it should be the average value that has the highest probability.

Having the ratio between successive terms, it is possible to compute $b_n(k)$ recursively. That is, we compute the first term, i.e., $b_n(0)$, then use it to compute the second term $b_n(1)$ and so on:

$$\begin{aligned}b_n(0) &= (1-p)^n && \text{(from Eq. (5.10.9))} \\ b_n(1) &= \frac{np}{1-p}b_n(0) && \text{(Eq. (5.10.9) with } k=1, q=1-p) \\ b_n(2) &= \frac{(n-1)p}{2(1-p)}b_n(1) \\ \cdots &= \cdots\end{aligned} \tag{5.10.14}$$

John Arbuthnot and Willem Jacob 's Gravesande. In 1710 John Arbuthnot (1667–1735) presented a paper titled *An Argument for Divine Providence* to the London Royal Society, which is a very early example of statistical hypothesis testing in social science. The paper presents a table containing the number of baptized children in London for the previous 82 years. One seemingly spectacular feature of this data was that in each of these 82 years the number of boys was higher than that of the girls. Willem Jacob 's Gravesande (1688–1742)[33] set out a task to find out why.

's Gravesande first found a representative year by taking the average number of births over the 82 years in question, which was 11429. For each year, he then scaled the numbers of births per sex to that average number. In this scaled data, Gravesande found that the number of boys had always been between 5745 and 6128.

33 Willem Jacob 's Gravesande was a Dutch mathematician and natural philosopher, chiefly remembered for developing experimental demonstrations of the laws of classical mechanics and the first experimental measurement of kinetic energy. As professor of mathematics, astronomy, and philosophy at Leiden University, he helped to propagate Isaac Newton's ideas in Continental Europe.

Now, seeing a birth as a Bernoulli trial with $p = 0.5$, he used Eq. (5.10.11) to compute the probability of the number of male births falling within this range in a given year as

$$P = \sum_{k=5745}^{6128} \binom{11429}{k} \left(\frac{1}{2}\right)^{11429} \tag{5.10.15}$$

How did's Gravesande compute this P in 1710[34]? First, he re-wrote it as follows (using the fact that the sum of the coefficients of the nth row in Pascal's triangle is 2^n, check Sect. 2.28.3, he wrote 2^{11429} as $\sum_{k=0}^{11429} \binom{11429}{k}$)

$$P = \frac{\sum_{k=5745}^{6128} \binom{11429}{k}}{2^{11429}} = \frac{\sum_{k=5745}^{6128} \binom{11429}{k}}{\sum_{k=0}^{11429} \binom{11429}{k}} \tag{5.10.16}$$

The problem now boils down to how to handling the coefficients (and sum of them) in a row of Pascal's triangle when n is large. To show how Gravesande did that, just consider the case $n = 5$ (noting that 11429 is an odd number):

$$\binom{5}{0} \binom{5}{1} \binom{5}{2} \boxed{\binom{5}{3}} \binom{5}{4} \binom{5}{5}, \quad \text{or } 1\ 5\ 10\ 10\ 5\ 1 \tag{5.10.17}$$

Since we have the following identity between adjacent binomial coefficients in any row (of the Pascal triangle),

$$\binom{n}{k+1} = \binom{n}{k} \frac{n-k}{k+1} \tag{5.10.18}$$

we now assign the middle term (i.e., $\binom{5}{3}$) to any value; say $\binom{5}{3} = a$, and we then compute the next term $\binom{5}{4}$ in terms of a, then the next term $\binom{5}{5}$ in terms of a. Adding all these three terms and multiplying the result by two,[35] we get the sum of all the coefficients in terms of a. Details are as follows. If we set $a = \binom{5}{3}$, then we have (using Eq. (5.10.18)):

$$\binom{5}{4} = a\left(\frac{n-3}{3+1}\right), \quad \binom{5}{5} = \binom{5}{4}\left(\frac{n-4}{4+1}\right) = a\left(\frac{n-3}{3+1}\right)\left(\frac{n-4}{4+1}\right)$$

[34] Today we would write a few lines of code and get the result of 0.2873. But doing so would not improve our math ability.

[35] Because the row is symmetric.

Thus, P will be:

$$P = \frac{\sum_{k=3}^{5} \binom{5}{k}}{2\sum_{k=3}^{5} \binom{5}{k}} = \frac{1}{2}\left[\frac{a + a\binom{n-3}{3+1} + a\binom{n-3}{3+1}\binom{n-4}{4+1}}{a + a\binom{n-3}{3+1} + a\binom{n-3}{3+1}\binom{n-4}{4+1}}\right] = \frac{1}{2}\left[\frac{1 + \binom{n-3}{3+1} + \binom{n-3}{3+1}\binom{n-4}{4+1}}{1 + \binom{n-3}{3+1} + \binom{n-3}{3+1}\binom{n-4}{4+1}}\right]$$

The *tricky* a has been eliminated!

In this way, Gravesande constructed a table[36] containing half of the coefficients in $(a+b)^{11429}$ starting from the middle term 5715–5973. Note that the coefficients are decreasing from the middle term, and from 5973 on, the coefficients are negligible.

Although 's Gravesande was able to solve this computationally challenging binomial related problem, he stopped there. Thus, 's Gravesande was not a systematic mathematician but rather a good problem solver and a number cruncher.

Abraham de Moivre (1667–1754) was a French-born mathematician who pioneered the development of analytic geometry and the theory of probability. While he was young de Moivre read mathematics texts in his own time. In particular he read Huygens' treatise on games of chance *De ratiociniis in ludo aleae*. He moved to England at a young age due to the religious persecution of Huguenots in France. After arriving in London he became a private tutor of mathematics, visiting the pupils whom he taught and also teaching in the coffee houses of London. As he traveled from one pupil to the next he read Newton's Principia. In 1718 he published *The Doctrine of Chance: A method of calculating the probabilities of events in play*.

First, de Moivre calculates $b_n(k_{\max})$ approximately using Stirling's approximation of $n!$ (which he also derived). Using Eq. (2.26.2) for factorials and noting that $p + q = 1$, he calculates $b_n(k_{\max})$ as follows:

$$b_n(k_{\max}) = \frac{n!}{k_{\max}!(n-k_{\max})!} p^{k_{\max}} q^{n-k_{\max}} \approx \frac{1}{\sqrt{2\pi npq}} \tag{5.10.19}$$

Then, he calculates $b_n(k_{\max}+l)$ $(l = 1, 2, 3, \ldots)$ in terms of $b_n(k_{\max})$. This is similar to 's Gravesande's method, but de Moivre's approach is much better. For example, when $l = 3$, he calculates as follows:

[36] If you like coding write a program to reconstruct Gravesande's table and compute P.

$$\frac{b_n(k_{\max})+3}{b_n(k_{\max})} = \left[\frac{b_n(k_{\max})+1}{b_n(k_{\max})}\right]\left[\frac{b_n(k_{\max})+2}{b_n(k_{\max})+1}\right]\left[\frac{b_n(k_{\max})+3}{b_n(k_{\max})+2}\right]$$

He uses Eq. (5.10.12) to determine the ratio of consecutive numbers (note that $k_{\max} \approx np$):

$$\begin{aligned}\frac{b_n(k_{\max}+i)}{b_n(k_{\max}+i-1)} &= \frac{(n-k_{\max}-i+1)p}{(k_{\max}+i)q} \\ &\approx \frac{(nq-i)p}{(np+i)q} = \frac{1-i/(nq)}{1+i/(np)}\end{aligned} \tag{5.10.20}$$

The logarithm of the final ratio is (using the approximation $\ln(1+x) \approx x$ for small x)[37]:

$$\ln\left(1-\frac{i}{nq}\right) - \ln\left(1+\frac{i}{np}\right) \approx -\frac{i}{nq} - \frac{i}{np} = -\frac{i}{npq} \tag{5.10.21}$$

For $l \geq 1$ and $k_{\max} + l \leq n$, we can calculate $\ln {}^{b_n(k_{\max}+l)}/_{b_n(k_{\max})}$ using Eq. (5.10.21) as follows:

$$\begin{aligned}\ln\frac{b_n(k_{\max}+l)}{b_n(k_{\max})} &= \ln\left(\frac{b_n(k_{\max}+1)}{b_n(k_{\max})} \times \frac{b_n(k_{\max}+2)}{b_n(k_{\max}+1)} \times \cdots \times \frac{b(k_{\max}+l)}{b_n(k_{\max}+l-1)}\right) \\ &= \ln\frac{b_n(k_{\max}+1)}{b_n(k_{\max})} + \ln\frac{b_n(k_{\max}+2)}{b_n(k_{\max}+1)} + \cdots + \frac{b_n(k_{\max}+l)}{b_n(k_{\max}+l-1)} \\ &\approx -\frac{1+2+\cdots+l}{npq} \approx -\frac{1}{2}\frac{l^2}{npq} \quad \text{(Sum of the first } l \text{ integers } = l(l+1)/2)\end{aligned}$$

Thus, $b_n(k_{\max}+l)$ is exponentially related to $b_n(k_{\max})$:

$$b_n(k_{\max}+l) \approx b_n(k_{\max}) \exp\left(-\frac{l^2}{2npq}\right) \tag{5.10.22}$$

At this point, replacing $b_n(k_{\max})$ with Eq. (5.10.19), de Moirve obtains the result:

$$b_n(k_{\max}+l) \approx \frac{1}{\sqrt{2\pi npq}} \exp\left(-\frac{l^2}{2npq}\right) \tag{5.10.23}$$

[37] Hence, this theory works only for small i.

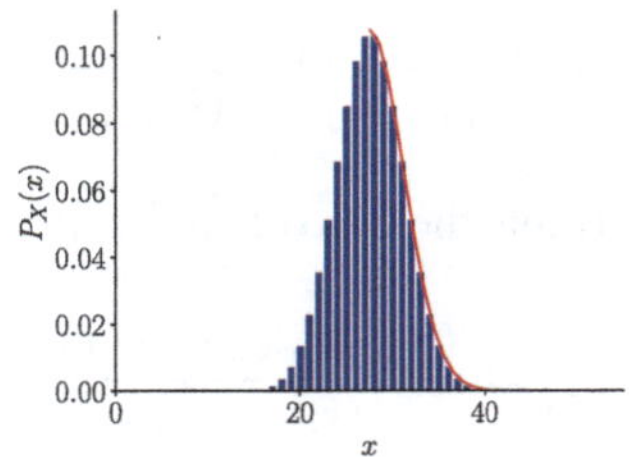

This formula is called de Moivre's approximation of the binomial distribution or the De Moivre-Laplace Central Limit Theorem. Notably, two famous constants in mathematics, $\pi = 3.1415\ldots$ and e, appear in this formula! How good is this formula? The answer is shown in the figure (the red curve calculated from de Moivre's formula). Although de Moivre never drew a red curve, he referred to the "curve" in his work *The Doctrine of Chances*, published in 1738 (when he was 71).[38] He even calculated the two inflection points of the curve. This is perhaps the first appearance of the standard curve.

Why is de Moivre's formula remarkable? Because we no longer have to worry about $n!$; instead, we have the lovely e^{-t^2} that we know how to calculate (using Taylor series). With Eq. (5.10.23), de Moivre can approximate the probability $P(n/2 \le X \le n/2 + d)$ as follows (he assumes $p = q = 0.5$):

$$P(n/2 \le X \le n/2+d) \approx \frac{2}{\sqrt{2\pi n}} \sum_{l=0}^{d} \exp\left(-\frac{2l^2}{n}\right) \approx \frac{2}{\sqrt{2\pi n}} \int_0^d \exp\left(-\frac{2x^2}{n}\right) dx \tag{5.10.24}$$

Thus, to avoid summation, de Moivre uses integration. And to perform this integration, he replaces e^x with the Taylor series: $e^x = \sum_{k=0}^{\infty} x^k/k!$. Thus, he has

$$P(n/2 \le X \le n/2+d) \approx \frac{2}{\sqrt{2\pi n}} \int_0^d \left(1 - \frac{2x^2}{n} + \frac{4x^4}{2!n^2} - \frac{8x^6}{3!n^3} + \frac{16x^8}{4!n^4} - \cdots\right) dx$$

This integral is as easy as eating candy; we have

$$P(n/2 \le X \le n/2+d) \approx \frac{2}{\sqrt{2\pi n}} \left(d - \frac{2d^3}{3n} + \frac{4d^5}{2\cdot 5n^2} - \frac{8d^7}{6\cdot 7n^3} + \frac{16d^9}{24\cdot 9n^4} - \cdots\right)$$

Then what did he do next? What would you do? Here we have d, then n, it looks very complicated, but if you pay attention, the first term is $2/\sqrt{2\pi}$ multiplied by $d/\sqrt{n}$. The second term is proportional to $(d/\sqrt{n})^3$. At this point, de Moivre changes the variable $d = s\sqrt{n}$,

38 He wrote *If the terms of the binomial expression are regarded as standing upright, evenly spaced perpendicular to and above a straight line, then the tops of the terms form a curve.*

$$P(n/2 \leq X \leq n/2 + s\sqrt{n}) \approx \frac{2}{\sqrt{2\pi}}\left(s - \frac{2s^3}{3} + \frac{4s^5}{2\cdot 5} - \frac{8s^7}{6\cdot 7} + \frac{16s^9}{24\cdot 9} - \cdots\right) \tag{5.10.25}$$

With a sword in hand, he used it. He starts with $s = 0.5$,[39] substitutes $s = 0.5$ into the formula above, with 4 terms, and he calculates the number 0.341238. Multiply by 2, and de Moivre has found what we now call the 68 − 95 − 99.7 rule: For a dataset approximately following a normal distribution, values within one standard deviation of the mean account for about 68% of the dataset.

Later, we will know that the standard deviation of the binomial distribution is $\sqrt{npq} = \sqrt{n}/2$. Thus, de Moivre is the first to use the concept of deviation (which he calls *modulus*) as the unit by which distances from the center are measured.

Regarding de Moivre, a friend of mine, Newton, in his later years, told those who came to inquire about mathematics, "Go to Mr. De Moivre; he understands these things better than I do". Despite de Moivre's scientific eminence his main income was as a private tutor of mathematics and he died in poverty. Desperate to get a chair in Cambridge he begged Johann Bernoulli to persuade Leibniz to write a supporting letter for him. Bernoulli did so in 1710 explaining to Leibniz that de Moivre was living a miserable life of poverty. Indeed Leibniz had met de Moivre when he had been in London in 1673 and tried to obtain a professorship for de Moivre in Germany, but with no success. Even his influential English friends like Newton and Halley could not help him obtain a university post.

He was unmarried, and spent his closing years in peaceful study. De Moivre, like Cardano, is famed for predicting the day of his own death. He found that he was sleeping 15 minutes longer each night and summing the arithmetic progression, calculated that he would die on the day that he slept for 24 hours. He was right!

Negative Binomial Distribution. Suppose that we have a coin with $P(H) = p$. We toss the coin until we observe m heads, where $m \in \mathbb{N}$. We define X as the total number of coin tosses in this experiment. Then X is said to have Pascal distribution with parameter m and p. We write $X \sim Pascal(m, p)$. Note that $Pascal(1, p) = Geometric(p)$. Note that by our definition the range of X is given by $R_X = \{m, m+1, m+2, \ldots\}$. This is because we need to toss at least m times to get m heads.

Our goal is to find $P_X(k)$ for $k \in R_X$. It's easier to start with a concrete case, say $m = 3$. What is $P_X(4)$? In other words, what is the probability that we have to toss the coin 4 times to get 3 heads? *The fact that we had to toss the coin 4 times indicating that in the first three tosses we only got 2 heads.* This observation is the key to the solution of this problem. And in the final toss (the fourth one) we got a head. Thus,

$$P_X(4) = P(2 \text{ heads from 3 tosses}) \times P(1 \text{ head for the last toss})$$

[39] Why this number? Note that he needs a number smaller than 1 to only need a few terms in the infinite series.

The problem has become familiar, and we can compute $P_X(4)$:

$$P_X(4) = \binom{3}{2} p^2(1-p)^1 \times p = \binom{3}{2} p^3(1-p)^1$$

And with that, it is just one small step to get the general result:

$$P_X(k) = \binom{k-1}{m-1} p^m(1-p)^{k-m}, \quad k = m, m+1, \ldots \tag{5.10.26}$$

Binomial distribution versus Pascal distribution. A binomial random variable *counts the number of successes in a fixed number of independent trials*. On the other hands, a negative binomial random variable *counts the number of independent trials needed to achieve a fixed number of successes*.

Poisson's distribution. Herein, we're going to present an approximation to the binomial distribution when n is large, p is small and np is finite. Let's introduce a new symbol λ such that $np = \lambda$. We start with $b_n(0)$, and taking advantage of the fact that n is large, we will use some approximations:

$$b_n(0) = (1-p)^n = \left(1 - \frac{\lambda}{n}\right)^n \tag{5.10.27}$$

Now, taking the natural logarithm of both sides of the above equation, and we get

$$\ln b_n(0) = n \ln\left(1 - \frac{\lambda}{n}\right) \tag{5.10.28}$$

Now, we use an approximation for $\ln(1-x)$, check Taylor's series in Sect. 4.15.8 if this was not clear:

$$\ln(1-x) = -x - \frac{x^2}{2} - \frac{x^3}{3} - \frac{x^4}{4} + \cdot$$

With that approximation, we now can write $\ln b_n(0)$ as (with $x = \lambda/n$)

$$\ln b_n(0) = -\lambda - \frac{\lambda^2}{2n} - \frac{\lambda^3}{3n^2} - \cdots \tag{5.10.29}$$

For very large n's, we get a good approximation of $b_n(0)$ by omitting terms with n in the denominator:

$$\ln b_n(0) \approx -\lambda \Longrightarrow b_n(0) \approx e^{-\lambda} \tag{5.10.30}$$

And of course, we use the recursive formula, Eq. (5.10.14), to get the next term $b_n(1)$ and so on. But first, we also need an approximation (when n is large) for the ratio $b_n(k)/b_n(k-1)$; using Eq. (5.10.13) with $p = \lambda/n$, $q = 1 - p$:

$$\frac{b_n(k)}{b_n(k-1)} = \frac{(n-k+1)p}{kq} \approx \frac{\lambda}{k}$$

Now, starting with $b_n(0)$, we obtain $b_n(1)$, $b_n(2)$ and so on:

$$\begin{aligned}
b_n(0) &\approx e^{-\lambda} \\
b_n(1) &\approx \lambda e^{-\lambda} \\
b_n(2) &\approx \frac{\lambda^2}{2} e^{-\lambda} \\
b_n(3) &\approx \frac{\lambda^3}{2 \times 3} e^{-\lambda}
\end{aligned}$$

Thus, we have a formula for any k:

$$b_n(k) \approx \frac{\lambda^k e^{-\lambda}}{k!}$$

And this is now known as Poisson distribution, named after the French mathematician Siméon Denis Poisson (1781–1840). A random variable X is said to be a Poisson random variable with parameter λ, shown as $X \sim Poisson(\lambda)$, if its range is $R_X = \{0, 1, 2, 3, ...\}$, and its PMF is given by

$$P_X(k) = \frac{\lambda^k e^{-\lambda}}{k!} \quad \text{for } k \in R_X \tag{5.10.31}$$

What should we do next after we have discovered the Poisson approximation to the binomial distribution? We should at least do two things[40]:

1. Check the accuracy of the Poisson approximation. We can do this by computing $Binomial(n, p)$ and $Poisson(\lambda)$, with $\lambda = np$, for some values of n and p, and for different x's. I skip this step here for brevity.
2. Justify the need of the Poisson approximation. We're going to do this with one example next.

Suppose you're trying to get something to happen in a video game that is rare; maybe it happens 1% of the time you do something. You'd like to know how likely it is to happen **at least once** if you try, say, 100 times. Here we have $p = 1/100$, $n = 100$. So the binomial distribution gives us an exact answer, namely

$$P = 1 - \left(1 - \frac{1}{100}\right)^{100}$$

[40] Actually we need to check whether $\sum_{k=0}^{\infty} P_X(k) = 1$, or $\sum_{k=0}^{\infty} \frac{\lambda^k e^{-\lambda}}{k!} = 1$.

The result is 0.63396 with a calculator, of course. Using the Poisson approximation with $\lambda = np = 1$, that probability is (easier)

$$P = 1 - e^{-1} = 0.632120.$$

5.10.4 Cumulative Distribution Function

The PMF is one way to describe the probability distribution of a discrete random variable. As we will see later on, the PMF cannot be defined for continuous random variables, because the PMF for a $x \in \mathbb{R}$ would be zero! Why is that? This is because there are infinite values of x (within any interval $[a, b]$ there are infinite real numbers, still remember Hilbert's hotel?), the probability of getting one value of x is zero (for division by infinity is zero). And this is consistent with daily observations. We know that all measurements have a degree of uncertainty regardless of precision and accuracy. This is caused by two factors, the limitation of the measuring instrument (systematic error) and the skill of the experimenter making the measurements (random error). Thus, it is meaningless to say that I measure the length of my son and get exactly 1.43 m.

If we cannot have $P(X = x)$ for real x, then the only option left is $P(a \leq x \leq b)$—the probability that x falls within a range. And the cumulative distribution function (CDF) of a random variable is what we need to describe the distribution of (continuous) random variables. The advantage of the CDF is that it can be defined for any kind of random variable being it a discrete, continuous, and mixed one.

The cumulative distribution function (CDF) of random variable X is defined as

$$\boxed{F_X(x) := P(X \leq x)}, \qquad \text{for all } x \in \mathbb{R}$$

Example. A fair coin is flipped twice. Let X be the number of observed heads. Find the CDF of X. Note that here $X \sim Binomial(2, 1/2)$, the range of X is $RX = \{0, 1, 2\}$ and its PMF is given by

$$P_X(0) = P(X = 0) = \frac{1}{4}, \quad P_X(1) = P(X = 1) = \frac{2}{4}, \quad P_X(2) = P(X = 2) = \frac{1}{4}$$

To find the CDF, we argue as follows. First, note that if $x < 0$, then $F_X(x) = P(X \leq x) = 0$. Second, if $x \geq 2$, then $F_X(x) = P(X \leq x) = 1$. Next, for $0 \leq x < 1$, $F_X(x) = P(X \leq x) = P(X = 0) = 1/4$ and so on. To summarize, the CDF of X is

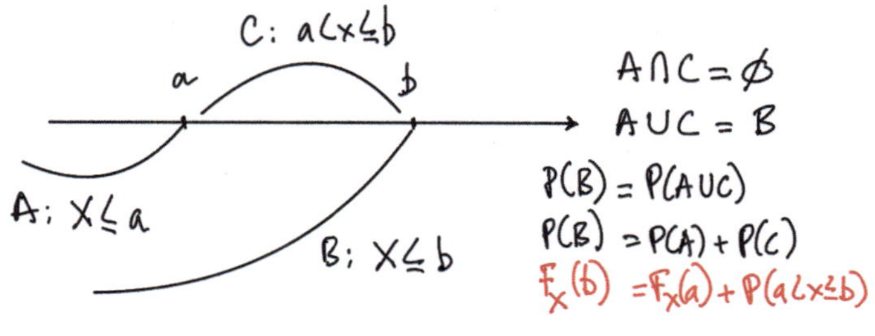

Fig. 5.10 Proof of Eq. (5.10.32)

$$F_X(x) = \begin{cases} 0, & \text{if } x < 0 \\ \frac{1}{4}, & \text{if } 0 \le x < 1 \\ \frac{3}{4}, & \text{if } 1 \le x < 2 \\ 1, & \text{if } x \ge 2 \end{cases}$$

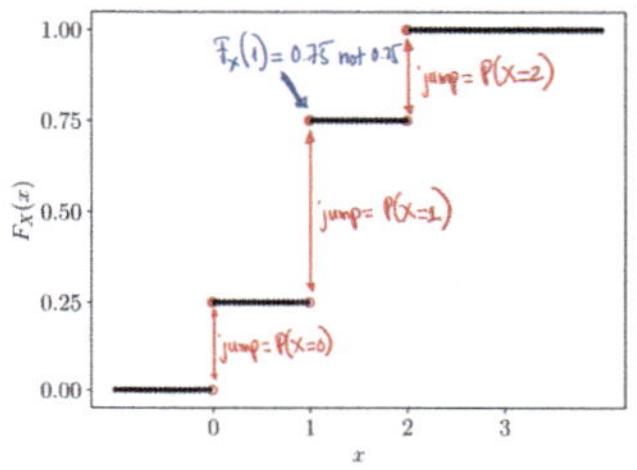

Now that we have seen a CDF, it's time to talk about its properties. By looking at the graph of this CDF, we can tell that

1. The range of $F_X(x)$ is $[0, 1]$;
2. When x approaches $-\infty$ then $F_X(x)$ approaches 0;
3. When x approaches $+\infty$ then $F_X(x)$ approaches 1;
4. The CFD is a non-decreasing function.

The first property is just a consequence of the second and third properties. The second property is just another way of saying that the probability of X smaller than $-\infty$ is zero. Similarly, the third property is the fact that the probability of something in the sample space occurs is one, as any X must be smaller than infinity! About the last property, as we're adding up probabilities, the CDF must be non-decreasing. But we can prove it rigorously using the following result: for $a, b \in \mathbb{R}$ such that $a < b$:

$$P(a < x \le b) = F_X(b) - F_X(a) \tag{5.10.32}$$

of which a proof is given in Fig. 5.10. As probability is always non-negative, the above results in $F_X(b) - F_X(a) \ge 0$ or $F_X(b) \ge F_X(a)$.

5.10.5 Expected Value

A roulette wheel has 36 slots—numbered from 1 to 36, half of them colored red, half black—and two zeros, colored green. Casinos offer even odds for betting on red or black at roulette. Suppose we bet \$100 on red. The question is: on average how much we win or lose per game.

First, we compute the probability of getting a red; it is $18/38 = 9/19$. Thus, the probability of not getting a red is $1 - 9/19 = 10/19$. Now, suppose we play 100 games. In 100 games, we will win in (9/19)(100) games and lose in (10/19)(100) games, thus the amount of money we gain or lose in 100 games is

$$\left(\frac{9}{19}\right)(100)(\$100) + \left(\frac{10}{19}\right)(100)(-\$100) = -(\$5.26)(100)$$

Thus, per game, we will lose \$5.26. What does this number mean? Obviously for each game, we either win \$100 or lose \$100. But in a long run when we have played many games, on average we would have lost \$5.26 per game.

We can see that this average amount can be computed by adding the product of the probability of winning \$100 and \$100 to the product of the probability of losing \$100 and -\$100:

$$\left(\frac{9}{19}\right)(\$100) + \left(\frac{10}{19}\right)(-\$100) = -\$5.26$$

Let's consider another example of rolling a die N times. Assume that among these N times, we observe 1 n_1 times, we observe 2 n_2 times, we observe 3 n_3 times, and so on. Now we compute the average of all the numbers observed:

$$\begin{aligned} x &= \frac{\underbrace{(1+1+\cdots+1)}_{n_1} + \underbrace{(2+2+\cdots+2)}_{n_2} + \cdots + \underbrace{(6+6+\cdots+6)}_{n_6}}{N} \\ &= \frac{(1)(n_1) + (2)(n_2) + \cdots + (6)(n_6)}{N} \end{aligned}$$

Now, assume that N is large, then $n_i/N = 1/6$, which is the probability that we observe i for $i = 1, 2, \ldots$ Thus,

$$\begin{aligned} x &= (1)\left(\frac{n_1}{N}\right) + (2)\left(\frac{n_2}{N}\right) + \cdots + (6)\left(\frac{n_6}{N}\right) = (1+2+3+4+5+6) \times \frac{1}{6} \quad (n_i/N = 1/6) \\ &= \frac{21}{6} = \frac{7}{2} = 3.5 \end{aligned}$$

Thus, the averaged value of rolling a die is $7/2$.

Notice that in both examples the averaged number is the sum of the products of the random variable times its probability. This leads to the following definition for *the expected value*.

Definition 5.10.3 If X is a discrete random variable with values of $\{x_1, x_2, \ldots, x_n\}$ and its PFM is $P_X(x_k)$, then the expected value of X, denoted by $\mathbb{E}[X]$, is defined as

$$\mathbb{E}[X] = x_1 P_X(x_1) + x_2 P_X(x_2) + \cdots = \sum_k x_k P_X(x_k) \tag{5.10.33}$$

History note 5.2: Blaise Pascal (1623–1662)

Blaise Pascal was the third of Étienne Pascal's children. Pascal's mother died when he was only three years old. Pascal's father had unorthodox educational views and decided to teach his son himself. Étienne Pascal decided that Blaise was not to study mathematics before the age of 15 and *all mathematics texts were removed from their house*. Curiosity raised by this, Pascal started to work on geometry himself at the age of 12. He discovered that the sum of the angles of a triangle are two right angles and, when his father found out, *he relented and allowed Blaise a copy of Euclid*. About 1647 Pascal began a series of experiments on atmospheric pressure. By 1647 he had proved to his satisfaction that a vacuum existed. Rene Descartes visited Pascal on 23 September. His visit only lasted two days and the two argued about the vacuum which Descartes did not believe in. Descartes wrote, rather cruelly, in a letter to Huygens after this visit that Pascal *...has too much vacuum in his head*.

Now, we're deriving another formula for the expected value of X, but in terms of the probability of the members of the sample space:

$$\mathbb{E}[X] = \sum_{s \in S} X(s) p(s) \tag{5.10.34}$$

We shall prove the important and useful result that the expected value of a sum of random variables is equal to the sum of their expectations (i.e., $\mathbb{E}[X+Y] = \mathbb{E}[X] + \mathbb{E}[Y]$ for two RVs X and Y) using Eq. (5.10.34).

Proof of Eq. (5.10.34) Let's denote by S_i the event that $X(S_i) = x_i$ for $i = 1, 2, \ldots$ That is,

$$S_i = \{s : X(s) = x_i\}$$

For example, in tossing two dice, and let X be the total number of faces, we have $x_1 = 2$ and $x_2 = 3$, with $S_2 = \{(1, 2), (2, 1)\}$ are the outcomes that led to x_2. Moreover, let $p(s) = P(s)$ be the probability that s is the outcome of the experiment. The proof then starts with the usual definition of $\mathbb{E}[X]$ and replaces $X = x_i$ by S_i, Fig. 5.6 can be helpful to see the connection between s, S and X:

$$\mathbb{E}[X] = \sum_i x_i P_X(x_i) = \sum_i x_i P_X(X = x_i) = \sum_i x_i P(S_i)$$

We continue with replacing $P(S_i)$ by $\sum_{s \in S_i} p(s)$ (that is using the third axiom),

$$\mathbb{E}[X] = \sum_i x_i \sum_{s \in S_i} p(s) = \sum_i \sum_{s \in S_i} x_i p(s) = \sum_i \sum_{s \in S_i} X(s) p(s)$$

And finally, because $S_1, S_2, \ldots$ are disjoint or mutually exclusive, $\sum_i \sum_{S_i}$ is just $\sum_{s \in S}$, thus

$$\mathbb{E}[X] = \sum_{s \in S} X(s) p(s)$$

which concludes the proof. Below, I will elaborate some steps which involves $\sum \sum$. For example, $\sum_i x_i \sum_{s \in S_i} p(s) = \sum_i \sum_{s \in S_i} x_i p(s)$, just use one concrete case:

$$\sum_i x_i \sum_{s \in S_i} p(s) = \sum_i x_i \left(p(s_1) + p(s_2)\right) = \sum_i \left(x_i p(s_1) + x_i p(s_2)\right) = \sum_i \sum_{s \in S_i} x_i p(s).$$

■

5.10.6 *Functions of Random Variables*

Consider this simple statistics problem: we want to find the average height μ of the students at a school. To this end, we randomly select n students and measure their heights; we get $x_1, x_2, \ldots, x_n$, which are random variables. Now, it is reasonable to estimate μ by $\bar{x} = (x_1 + \cdots + x_n)/n$. Such $\bar{x}$ is a function of random variables. Since we cannot be certain that $\bar{x}$ is an exact estimation for μ, we need to find the probability distribution of $\bar{x}$.

Assume now that we are given a discrete RV X along with its probability mass function and that we want to compute the expected value of some function of X, say, $g(X)$. How can we accomplish this? One way is as follows: As $g(X)$ is itself a discrete random variable, it has a probability mass function, which can be determined from the PMF of X, see Fig. 5.11. Once we have determined the PFM of $g(X)$, we can compute $\mathbb{E}[g(X)]$ by using the definition of the expected value.

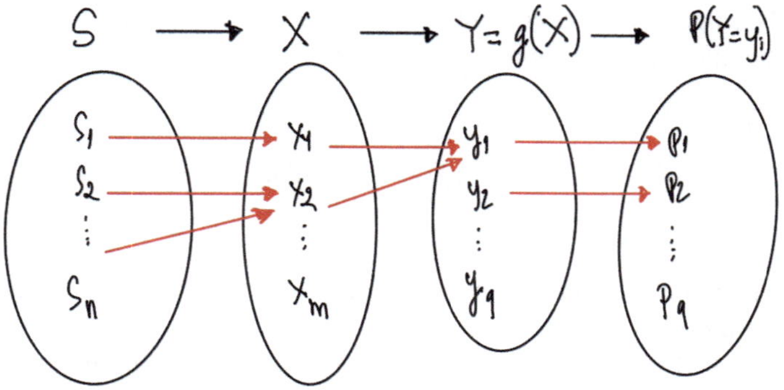

Fig. 5.11 Pictorial presentation of sample space S, RV X, and function of a RV $Y = g(X)$ and its PFM

Example 5.15 Let X be a RV that takes on any values $-1, 0, 1$ with respective probabilities

$$P(X = -1) = 0.2, \quad P(X = 0) = 0.5, \quad P(X = 1) = 0.3$$

Compute $\mathbb{E}[X^2]$; so $g(X) = X^2$ in this example.

First, we compute the PMF of $Y = X^2$ whose range is $\{0, 1\}$:

$$\begin{aligned} P(Y = 0) &= P(X = 0) = 0.5 \\ P(Y = 1) &= P(X = -1) + P(X = 1) = 0.5 \end{aligned}$$

Second, the expected value of Y is computed:

$$\mathbb{E}[X^2] = \mathbb{E}[Y] = (1)(0.5) + (0)(0.5) = 0.5 \tag{5.10.35}$$

But there is a faster way of doing this. The expected value of $g(X)$, $\mathbb{E}[g(X)]$, is simply given by

$$\mathbb{E}[g(X)] = \sum_i g(x_i) P_X(x_i) \tag{5.10.36}$$

And this result is known as *the law of the unconscious statistician*, or LOTUS. This a theorem used to calculate the expected value of a function $g(X)$ of a random variable X when one knows the probability distribution of X but one does not know the distribution of $g(X)$. The name comes from the fact that some statisticians present Eq. (5.10.36) as the definition of the expected value rather than a theorem.

Before proving this result, let's check that it is in accord with the results obtained directly using the definition of $\mathbb{E}[X^2]$ for the above example. Applying Eq. (5.10.36), we get

$$\mathbb{E}[X^2] = (-1)^2(0.2) + (0)^2(0.5) + (1)^2(0.3) = 0.5$$

which is the same as the direct result. To see why the same result was obtained, we can do some massage to the above expression:

$$\begin{aligned}\mathbb{E}[X^2] &= (-1)^2(0.2) + (0)^2(0.5) + (1)^2(0.3)\\ &= (1)(0.2+0.3) + (0)(0.5) \quad \text{(grouping terms with equal } g(x_i))\\ &= (1)(0.5) + (0)(0.5)\end{aligned}$$

The last expression is exactly identical to Eq. (5.10.35). The proof of Eq. (5.10.36) proceeds similarly.

Proof of Eq. (5.10.36) We start with $\sum_i g(x_i)P_X(x_i)$, then group terms with the same $g(x_i)$, and then transform it to $\sum_j y_j P_Y(y_j)$ which is $\mathbb{E}[g(X)]$ with y_j are all the (different) values of Y:

$$\begin{aligned}\sum_i g(x_i)P_X(x_i) &= \sum_j \sum_{i:g(x_i)=y_j} g(x_i)P_X(x_i) \text{ (grouping step)}\\ &= \sum_j \sum_{i:g(x_i)=y_j} y_j P_X(x_i) \quad \text{(replacing } g(x_i) = y_j)\\ &= \sum_j y_j \sum_{i:g(x_i)=y_j} P_X(x_i) = \sum_j y_j P_Y(y_j)\end{aligned}$$

The notation $\sum_{i:g(x_i)=y_j} g(x_i)P_X(x_i)$ means that the sum is over i but only for i such that $g(x_i) = y_j$, and that is achieved by the subscript $i : g(x_i) = y_j$ under the sum notation. ■

5.10.7 Linearity of the Expectation

In this section we shall discuss some properties of the expectation of random variables. For the motivation, let's consider an example first.

Expected value of sum of two random variables. Let's roll two dice and denote by S the sum of faces. If we denote by X the face of the first die and by Y the face of the second die, then $S = X + Y$. Obviously S is a discrete RV, and we can compute its PFM. Thus, we can compute its expected value. First, we list all possible elements of S:

$$\begin{aligned}
S &= 2 \;: (1,1)\\
S &= 3 \;: (1,2), (2,1)\\
S &= 4 \;: (1,3), (3,1), (2,2)\\
S &= 5 \;: (1,4), (4,1), (2,3), (3,2)\\
S &= 6 \;: (1,5), (5,1), (2,4), (4,2), (3,3)\\
S &= 7 \;: (1,6), (6,1), (2,5), (5,2), (3,4), (4,3)\\
S &= 8 \;: (2,6), (6,2), (3,5), (5,3), (4,4)\\
S &= 9 \;: (3,6), (6,3), (4,5), (5,4)\\
S &= 10 : (4,6), (6,4), (5,5)\\
S &= 11 : (5,6), (6,5)\\
S &= 12 : (6,6)
\end{aligned}$$

Now, we can compute $P(S = x_j)$ for $x_j = \{2, 3, \ldots, 12\}$, and then using Eq. (5.10.33) to compute the expected value:

$$\begin{aligned}
\mathbb{E}[S] = 2 \times \frac{1}{36} + 3 \times \frac{2}{36} + 4 \times \frac{3}{36} + 5 \times \frac{4}{36} + 6 \times \frac{5}{36}\\
+ 7 \times \frac{6}{36} + 8 \times \frac{5}{36} + 9 \times \frac{4}{36} + 10 \times \frac{3}{36} + 11 \times \frac{2}{36} + 12 \times \frac{1}{36} = \frac{252}{36} = 7
\end{aligned}$$

You might be asking what is special about this problem? Is it just another application of the concept of expected value? Hold on. Look at the result of 7 again. Rolling one die and the expected value is $7/2$,[41] now rolling two dice and the expected value is 7. We should suspect that

$$\mathbb{E}[X + Y] = \mathbb{E}[X] + \mathbb{E}[Y] \tag{5.10.37}$$

which implies that the expected value of the sum of two random variables is equal to the sum of their individual expected values, regardless of whether they are independent. In calculus, we have the derivative of the sum of two functions is the sum of the derivatives. Here in the theory of probability, we see the same rule.

Proof of Eq. (5.10.37) Let X and Y be two random variables and $Z = X + Y$. We're now using Eq. (5.10.34) for the proof:

$$\begin{aligned}
\mathbb{E}[Z] &= \sum_s Z(s)p(s) = \sum_s [X(s) + Y(s)]p(s)\\
&= \sum_s [X(s)p(s) + Y(s)p(s)] = \sum_s X(s)p(s) + \sum_s Y(s)p(s) = \mathbb{E}[X] + \mathbb{E}[Y]
\end{aligned}$$

■

[41] Check the paragraph before Definition 5.10.3 if this was not clear.

This proof also reveals that the property holds not only for two RVs but for any number of RVs. Thus, for $n \in \mathbb{N}$, we can write

$$\mathbb{E}[nX] = \mathbb{E}[\underbrace{X + X + \cdots + X}_{n} \text{ terms}] = \mathbb{E}[X] + \cdots + \mathbb{E}[X] = nE[X]$$

And from that, it is a short step to guess that[42]

$$\mathbb{E}[aX + b] = a\mathbb{E}[X] + b \quad (a, b \in \mathbb{R}) \tag{5.10.38}$$

Proof of Eq. (5.10.38) Let X be a RV and $Y = g(X) = aX + b$. We're now using Eq. (5.10.36)—the LOTUS—for the proof.

$$\begin{aligned}
\mathbb{E}[g(X)] &= \sum_x g(x)P_X(x) = \sum_x (ax + b)P_X(x) \\
&= \sum_x ax P_X(x) + \sum_x bP_X(x) = a\sum_x xP_X(x) + b\sum_x P_X(x) \\
&= a\mathbb{E}[X] + b \quad \left(\sum_x P_X(x) = 1\right).
\end{aligned}$$

■

5.10.8 Variance and Standard Deviation

It is easy to see that the expected value is not sufficient to describe a probability distribution. For example, consider the three distributions shown in Fig. 5.12. Although they all have an expected value of zero, they are quite different: the values of the third distribution vary a lot. We need to define a measure for this spread. And it is called the variance. To see the motivation behind the definition of the variance, we just need to know that the spread of a probability distribution indicates how far a value is from the expected value. To say how far a number a is from a number b, we can either use $|a - b|$ or $(a - b)^2$.[43]

Now, consider a RV X with $\mathbb{E}[X]$ now being denoted by μ. The variance of a RV X, designated by $\text{Var}(X)$, is defined as the *average value of the squares of the difference from* X *to the mean value*, i.e., $(X - \mu)^2$. Thus, it is given by

$$\text{Var}(X) := \mathbb{E}[(X - \mu)^2] \tag{5.10.39}$$

[42] First, from the fact that $\mathbb{E}[nX] = nE[X]$ we generalize to $\mathbb{E}[aX] = a\mathbb{E}[X]$. We have seen mathematicians did this many time (e.g., check Eq. (2.24.2)).

[43] Of course we prefer working with power functions, and $(a - b)^2$ is the lowest power function.

Fig. 5.12 Three distributions of the same expected value but difference variances

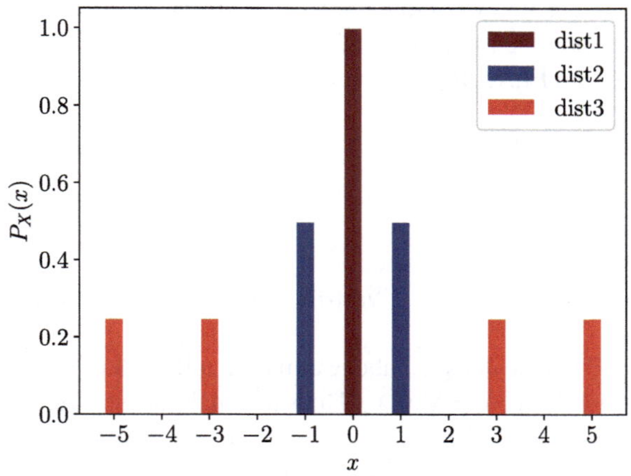

Why square? Squaring always gives a positive value, so the variance will not be zero.[44] A natural question is: the absolute difference also has this property, why we can't define the variance as $\mathbb{E}[|X - \mu|]$? Yes, you can! The thing is that the definition in Eq. (5.10.39) prevails because it is mathematically easier to work with x^2 than to work with $|x|$. Again, just think about differentiating these two functions and you will see what we mean by that statement.

Note that Var(X) has a different unit than X. For example, if X is measured in meters then Var(X) is in meters squared. To solve this issue, another measure, called the *standard deviation*, is defined. The standard deviation, usually denoted by σ_X, is simply the square root of the variance.

Instead of using the definition of the variance directly to compute it, we can use LOTUS to have a nicer formula for it (recall that $\mu = \sum_x x P_X(x)$):

$$\begin{aligned}
\text{Var}(X) &= \mathbb{E}[(X-\mu)^2] = \sum_x (x-\mu)^2 P_X(x) \\
&= \sum_x (x^2 - 2\mu x + \mu^2) P_X(x) \\
&= \sum_x x^2 P_X(x) - 2\mu \sum_x x P_X(x) + \mu^2 \sum_x P_X(x) \\
&= \mathbb{E}[X^2] - \mu^2 = \mathbb{E}[X^2] - (\mathbb{E}[X])^2
\end{aligned} \tag{5.10.40}$$

This formula is useful as we know $\mathbb{E}[X]$ (and thus its squared) and we know how to compute $\mathbb{E}[X^2]$ using the LOTUS. If you want to translate this formula to English, it is: the variance is the mean of the square minus the square of the mean. Eventually, nothing new is needed, it is just a combination of all the things we know of!

---

[44] You're encouraged to think of an example to see this.

Let's now compute $\text{Var}(aX + b)$. Why? To see if the variance is a linear operator or not. Denoting $Y = aX + b$, then $\mu_Y = a\mu + b$, which is the expected value of Y (from Eq. (5.10.38)). Now, we can write

$$\begin{aligned}\text{Var}(Y) &= \mathbb{E}[(Y - \mu_Y)^2] = \mathbb{E}[(aX + b - a\mu - b)^2] \\ &= \mathbb{E}[a^2(X - \mu)^2] = a^2\mathbb{E}[(X - \mu)^2] = a^2\text{Var}(X)\end{aligned} \tag{5.10.41}$$

Thus, we have

$$\text{Var}(aX + b) = a^2\text{Var}(X) \neq a\text{Var}(X) + b \tag{5.10.42}$$

What else does the above equation tell us? Let's consider $a = 1$, that is $Y = X + b$, then $\text{Var}(Y) = \text{Var}(X)$. Does this make sense? Yes, noting that $Y = X + b$ is a translation of X (Sect. 4.2.2), and a translation does not distort the object (or the function), thus the spread of X is preserved.

Sample variance. Herein we shall meet some terminologies in statistics. For example, if we want to find out how much the average Australian earns, we do not want to survey everyone in the population (too many people), so we would choose a small number of people in the population. For example, you might select 10000 people. And that is called a sample. Why 10000, you're asking? It is not easy to answer that question. That's why a whole field called design of experiments was developed, just to have unbiased samples. This is not discussed here.

Ok. Suppose now that we have already a sample with n observations (or measurements) $x_1, x_2, \ldots, x_n$. The question now is: what is the variance for this sample? You might be surprised to see the following[45]

$$S^2 = \frac{1}{n-1}\sum_{i=1}^{n}(x_i - \bar{x})^2, \quad \bar{x} = \frac{1}{n}\sum_{i=1}^{n} x_i$$

Why $n - 1$ but not n? In statistics, this is called Bessel's correction, named after Friedrich Bessel. The idea is that we need S^2 to match the population variance σ^2, to have an unbiased estimator of σ^2. As shown below, with n in the denominator, we cannot achieve this. And what's why $n - 1$ was used.[46]

Proof First, we have the following identity (some intermediate steps were skipped, noting that $\sum_i x_i = n\bar{x}$)

[45] When work with the samples, we do not know the probabilities p_i, and thus we cannot use the definition of mean and expected value directly. Instead we just include each output x as often as it comes. We get the *empirical mean* instead of the expected mean. Similarly, we get the empirical variance.

[46] Another explanation that I found is: one degree of freedom was accounted for in the sample mean. But I do not understand this.

$$\sum_{i=1}^{n}(x_i-\bar{x})^2=\sum_{i=1}^{n}(x_i^2-2x_i\bar{x}+\bar{x}^2)=\sum_{i=1}^{n}x_i^2-n\bar{x}^2 \tag{5.10.43}$$

Applying that identity to $y_i = (x_i - \mu)$ (with $\bar{y} = \bar{x} - \mu$) we have

$$\sum_{i=1}^{n}[(x_i-\mu)-(\bar{x}-\mu)]^2=\sum_{i=1}^{n}(x_i-\mu)^2-n(\bar{x}-\mu)^2$$

Now, we compute the expected value of the LHS of the above equation:

$$\begin{aligned}\mathbb{E}\left[\sum_{i=1}^{n}[(x_i-\mu)-(\bar{x}-\mu)]^2\right]&=\mathbb{E}\left[\sum_{i=1}^{n}(x_i-\mu)^2-n(\bar{x}-\mu)^2\right]\\&=\sum_{i=1}^{n}\mathbb{E}\left[(x_i-\mu)^2\right]-nE[(\bar{x}-\mu)^2]\ \ (\mathbb{E}[X+Y]=\mathbb{E}[X]+\mathbb{E}[Y])\\&=\sum_{i=1}^{n}\text{Var}(x_i)-n\text{Var}(\bar{x})\end{aligned} \tag{5.10.44}$$

Now, we can compute the expected value of S^2:

$$\begin{aligned}\mathbb{E}[S^2]&=\frac{1}{n-1}\mathbb{E}\left[\sum_{i=1}^{n}(x_i-\bar{x})^2\right]=\frac{1}{n-1}\mathbb{E}\left[\sum_{i=1}^{n}((x_i-\mu)-(\bar{x}-\mu))^2\right]\\&=\frac{1}{n-1}\left[\sum_{i}\text{Var}(x_i)-n\text{Var}(\bar{x})\right]\quad\text{(used Eq. (5.10114))}\end{aligned}$$

Note that as $x_1, x_2, \ldots, x_n$ are a random sample from a distribution with variance σ^2, thus (check Eq. (5.12.19) for the second result)

$$\text{Var}(x_i)=\sigma^2,\quad \text{Var}(\bar{x})=\frac{\sigma^2}{n}$$

Substituting these into $\mathbb{E}[S^2]$, we obtain

$$\mathbb{E}[S^2]=\frac{1}{n-1}\left[\sum_{i=1}^{n}\sigma^2-\sigma^2\right]=\frac{1}{n-1}(n\sigma^2-\sigma^2)=\sigma^2$$

Thus the sample variance coincides with the population variance, which justifies the Bessel correction. ■

5.10.9 Expected Value and Variance of Special Distributions

Having established the two vital concepts: expected value and variance, we are now going to compute these parameters for the various discrete distributions considered in this section (e.g., the binomial distribution). To summarize the results, Table 5.7 lists these quantities for the Bernoulli, binomial, geometric, Pascal and Poisson distributions.

How they were computed? Of course using the definition of the expected value and variance, massage the algebraic expression until the simplest form is achieved. I am going to give one example.

Example 5.16 Determine the expected value for the geometric distribution with the PMF given by $q^{k-1}p$ for $k = 1, 2, \ldots$ Using Eq. (5.10.33), we can straightforwardly write $\mathbb{E}[X]$ as

$$\mathbb{E}[X] = \sum_{x_k \in R_X} x_k P_X(x_k) = \sum_{k=1}^{\infty} kq^{k-1}p = p\sum_{k=1}^{\infty} kq^{k-1}$$

Now, the trouble is the red sum. To attack it, we need to use the geometric series,

$$\sum_{k=0}^{\infty} x^k = \frac{1}{1-x} \Longrightarrow \frac{d}{dx}\left(\sum_{k=0}^{\infty} x^k\right) = \sum_{k=0}^{\infty} kx^{k-1} = \frac{1}{(1-x)^2}$$

Thus, we handled the red term, now come back to $\mathbb{E}[X]$:

$$\mathbb{E}[X] = p\sum_{k=1}^{\infty} kq^{k-1} = p\frac{1}{(1-q)^2} = \frac{1}{p}.$$

Table 5.7 Expected value, variance and SD of special distributions

X	Meaning	$\mathbb{E}[X]$	$\mathrm{Var}(X)$	σ
$Bernoulli(p)$		p	p	
$Binomial(n, p)$	n coin toss, X is # of heads observed	np	npq	$\sqrt{npq}$
$Geometric(p)$	X is # of coin toss until a H is observed	$\frac{1}{p}$	p	
$Pascal(m, p)$	X is # of coin toss until m heads observed	$\frac{m}{p}$	npq	
$Poisson(\lambda)$		λ	λ	

5.11 Continuous Probability Models

Whereas a discrete variable is a variable whose value is obtained by counting (e.g., the number of marbles in a jar, the numbers of boys in a class and so on), a continuous variable is a variable whose value is obtained by measuring. For examples, the height of students in class, the weight of students in class, the time it takes to get to school. This section is about probability models for these continuous random variables. The central concept is of a probability density function, that function which gives us the probability of a continuous random variable (Sect. 5.11.1). Then, expected value and variance are discussed in Sect. 5.11.2. Finally, common contiuous distributions (e.g., normal distribution) are given in Sect. 5.11.3.

5.11.1 Probability Density Function

The table below (Table 5.8) gives the heights of fathers and their sons, based on a famous experiment by Karl Pearson[47] around 1903. The number of cases is 1078. Random noise was added to the original data, to produce heights to the nearest 0.1 inch.[48]

One good way to analyze a continuous data sample (such as the one in Table 5.8) is to use a histogram. A histogram is built as follows. First, denote the range of the data, e.g., fathers' heights by $[l, m]$, where l and m represent the minimal and maximal value of the data. Second, we "bin" (or "bucket") the range of values—that is, divide the entire range of values into a series of intervals. Mathematically, the interval $[l, m]$ is partitioned into a finite set of bins $\mathcal{B}_1, \mathcal{B}_2, \ldots, \mathcal{B}_L$. Third, the relative frequency in each bin is recorded. To this end, let's denote by n the number of data observations (in the case of Pearson's data, it is 1078), and for bin j, its frequency f_j is defined (as it should be) the ratio of how many data is in this bin to n. Using symbols, f_j is written as

Table 5.8 Karl Pearson's height data

Row	Father	Son
1	65.00	59.80
2	63.30	63.20
3	65.00	63.30
⋮	⋮	⋮
1077	70.70	69.30
1078	70.00	67.00

[47] Karl Pearson (1857–1936) was a British statistician, leading founder of the modern field of statistics.

[48] You can download the data at https://www.randomservices.org/random/data/Pearson.html.

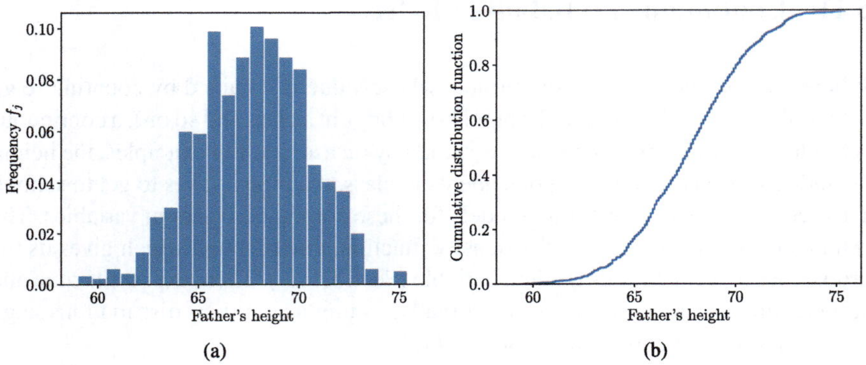

Fig. 5.13 Fathers' height: probability histogram and cumulative distribution function

$$f_j = \frac{1}{n}\sum_{i=1}^{n} \mathbf{1}\{x_i \in \mathcal{B}_j\}, \qquad \text{for } j = 1, 2, \ldots, L \tag{5.11.1}$$

where $\mathbf{1}\{x_i \in \mathcal{B}_j\}$ returns 1 if x_i is in bin $\mathcal{B}_j$ and 0 otherwise.

The final step is to plot the bins and f_j. A bar plot where the X-axis represents the bin ranges and the Y-axis gives information about frequency is used for this. Figure 5.13a presents a histogram for the fathers' heights.[49]

It is useful to assume that the CDF of a continuous random variable is a continuous function, see Fig. 5.13b to see why. Then, recall from Eq. (5.10.32) that

$$P(a < x \le b) = F_X(b) - F_X(a)$$

And from the fundamental theorem of calculus (Chap. 4), we know that

$$F_X(b) - F_X(a) = \int_a^b f_X(x)dx, \quad \text{where } f_X(x) = \frac{dF_X(x)}{dx} \tag{5.11.2}$$

Thus, we can find the probability that x falls within an interval $[a, b]$ in terms of the new function $f_X(x)$:

$$P(a < x \le b) = \int_a^b f_X(x)dx, \quad \text{or} \quad \boxed{P(a \le x \le b) = \int_a^b f_X(x)dx} \tag{5.11.3}$$

The function $f_X(x)$ is called *the probability density function* or PDF. Why that name? This is because $f_X(x) = {}^{dF_X(x)}/_{dx}$, which is probability per unit length. Note

49 See Listing B.1 for the code. I used `Julia` packages to compute and plot the histogram. You're encouraged to code Eq. (5.11.1) if you want to learn programming.

that for a continuous RV writing $P(a < x \le b)$ or $P(a \le x \le b)$ is the same because $P(x = a) = 0$. Actually we have seen something similar (i.e., probability is related to an integral) in Eq. (5.10.24).

The probability density function satisfies the following two properties (which is nothing but the continuous version of Eq. (5.10.7))

1. Probabilities are non-negative:

$$f_X(x) \ge 0 \quad \text{for } \forall x \in \mathbb{R}$$

2. Probabilities sum to one:

$$\int_{-\infty}^{+\infty} f_X(x)dx = 1 \tag{5.11.4}$$

Geometrically, this equation indicates that the area under any PDF curve is one.

5.11.2 Expected Value and Variance

Recall that the expected value and the variance of a discrete RV X are defined as

$$\mathbb{E}[X] := \sum_k x_k P_X(x_k), \quad \text{Var}(X) := \mathbb{E}[(X - \mu)^2]$$

And from that we have the continuous counterparts, where sum is replaced by integral and the PDF replacing the PMF

$$\boxed{\mathbb{E}[X] = \int_{-\infty}^{\infty} x f_X(x)dx, \quad \text{Var}(X) = \int_{-\infty}^{\infty} (x - \mu)^2 f_X(x)dx.} \tag{5.11.5}$$

5.11.3 Special Continuous Distributions

Uniform distribution. This is the simplest type of continuous distributions. What we want is a PDF that is constant (i.e., uniform) in the interval $[a, b]$. Because the PDF is constant, it has a rectangular shape of width $b - a$, and of height $1/b-a$. Why? Because the area under any PDF curve is one (Eq. (5.11.4)). Thus, a continuous random variable X is said to have a Uniform distribution over the interval $[a, b]$, shown as $X \sim Uniform(a, b)$, if its PDF is given by

$$f_X(x) = \begin{cases} \frac{1}{b-a}, & \text{if } a < x < b \\ 0, & \text{otherwise} \end{cases} \tag{5.11.6}$$

Standard normal distribution. de Moivre had derived an approximation to the binomial distribution and it involves the exponential function of the form e^{-x^2}. Thus, there is a need to evaluate the following integral (see Eq. (5.10.24)):

$$I = \int_{-\infty}^{\infty} e^{-x^2} d$$

Unfortunately it is impossible to find an antiderivative of e^{-x^2}. Note that if the integral was $\int 2xe^{-x^2} dx$, then life would be easier. The key point is the factor x in front of e^{-x^2}. If we go to 2D, then, we can make this factor appear. Let's compute I^2 instead[50]:

$$I^2 = \left(\int_{-\infty}^{\infty} e^{-x^2} dx\right)\left(\int_{-\infty}^{\infty} e^{-y^2} dy\right) = \iint_{-\infty}^{\infty} e^{-(x^2+y^2)} dxdy$$

The next step is to switch to polar coordinates in which $dxdy$ will become $rdrd\theta$ (see Sect. 6.8.2), and voilà:

$$I^2 = \int_0^{2\pi} \left[\int_0^{\infty} e^{-r^2} rdr\right] d\theta = \pi \Longrightarrow I = \int_{-\infty}^{\infty} e^{-x^2} dx = \sqrt{\pi}$$

With that, we can define what is called a *standard normal variable* as follows. A continuous random variable Z is said to be a standard normal (or standard Gaussian) random variable, denoted by $Z \sim N(0, 1)$, if its PDF is given by[51]

$$\boxed{Z \sim N(0,1): \quad f_Z(z) = \frac{1}{\sqrt{2\pi}} \exp\left(-\frac{z^2}{2}\right)} \tag{5.11.7}$$

Why this form? Why not this form $(1/\sqrt{\pi})e^{-z^2}$? This one is also a legitimate PDF, actually it is the form that Gauss used. However, the one in Eq. (5.11.7) prevails simply because with it, the variance is one (this is to be shown shortly)—which is a nice number.

The CDF of a standard normal distribution is

$$F_Z(z) = P(Z \le z) = \frac{1}{\sqrt{2\pi}} \int_{-\infty}^{z} \exp\left(-\frac{u^2}{2}\right) du := \Phi(z) \tag{5.11.8}$$

The integral in Eq. (5.11.8) does not have a closed form solution.[52] Nevertheless, because of the importance of the normal distribution, the values of this integral have

[50] Yes, sometimes making a problem harder and we can find the solution to the simpler problem.

[51] The factor $1/\sqrt{2\pi}$ before the exponential function is required because of Eq. (5.11.4).

[52] This means that there is no antiderivative written in elementary functions. The situation is similar to there is no formula for the roots of a polynomial of high degree, e.g., five. This was proved by the French mathematician Joseph Liouville (1809–1882).

Table 5.9 Table for $\Phi(z)$ for $z \geq 0$

z	0.00	0.01	0.02	0.03	0.04	0.05	0.06	0.07	0.08
0.0	0.5000	0.5040	0.5080	0.5120	0.5160	0.5199	0.5239	0.5279	0.5319
0.1	0.5398	0.5438	0.5478	0.5517	0.5557	0.5596	0.5636	0.5675	0.5714
0.2	0.5793	0.5832	0.5871	0.5910	0.5948	0.5987	0.6026	0.6064	0.6103
⋮	⋮	⋮	⋮	⋮	⋮	⋮	⋮	⋮	⋮
2.1	0.9821	0.9826	0.9830	0.9834	0.9838	0.9842	0.9846	0.9850	0.9854

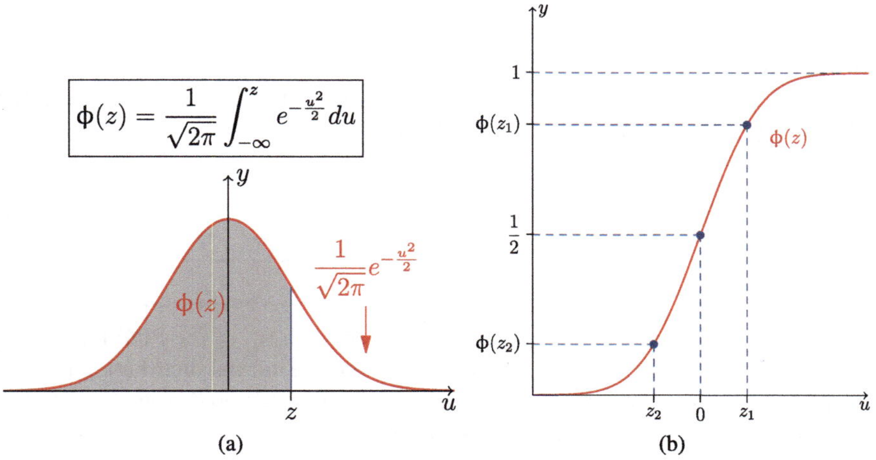

Fig. 5.14 Plot of the standard normal curve (a) and plot of the CDF which is the area underneath the normal curve from $-\infty$ to z. As the total area under the normal curve is one, half of the area is 0.5, thus $\Phi(0) = 1/2$. Another property: $\Phi(-z) = 1 - \Phi(z)$. This property is useful as we only need to make table of $\Phi(z)$ for $z \geq 0$. Why we have this property? Plot the normal curve, mark two points z and $-z$ on the horizontal axis. Then, $1 - \Phi(z)$ is the area under the curve from z to ∞ while $\Phi(-z)$ is the area from $-\infty$ to $-z$. The normal curve is symmetric, thus the two areas must be equal

been tabulated; see Table 5.9 for such a table.[53] Nowadays, it is available in calculators and in many programming languages. Moreover, mathematicians introduced the short notation Φ to replace the lengthy integral expression. Figure 5.14 plots both $f_Z(z)$ and $\Phi(z)$.

To explain the shape of the bell curve, we use calculus. Let's compute the first and second derivatives of $f_Z(z)$:

$$f_Z(z) = \frac{1}{\sqrt{2\pi}} \exp\left(-\frac{z^2}{2}\right) \implies f_Z'(z) = -z f_Z(z), \quad f_Z''(z) = (z^2 - 1) f_Z(z$$

[53] Why we need this table? It is useful for inverse problems where we need to find z^* such that $\Phi(z^*) = a$ where a is a given value. This table was generated automatically (even the LaTeX code to typeset it) using a `Julia` script. For me it was simply a coding exercise for fun.

Thus, $f_Z(z)$ has a maximum at $z = 0$, and $f_Z''(z) < 0$ for $|z| < 1$: the curve is concave here. And $f_Z''(z) > 0$ for $|z| > 1$: the curve is convex here. That's why the curve is of a bell shape.

Now, using Eq. (5.11.5) we're going to find the expected value and the variance of $N(0, 1)$. It can be shown that if $Z \sim N(0, 1)$ then[54]

$$\mathbb{E}[Z] = \int_{-\infty}^{\infty} z^1 f_Z(z)dz = \frac{1}{\sqrt{2\pi}} \int_{-\infty}^{\infty} z^1 e^{-z^2/2} dz = 0$$
$$\text{Var}(Z) = \int_{-\infty}^{\infty} z^2 f_Z(z)dz = \frac{1}{\sqrt{2\pi}} \int_{-\infty}^{\infty} z^2 e^{-z^2/2} dz = 1$$

Normal distribution. A continuous random variable X is said to be a normal (or Gaussian) random variable, denoted by $X \sim N(\mu, \sigma^2)$, where μ is the expected value and σ^2 is the variance of X, if its PDF is given by

$$\boxed{X \sim N(\mu, \sigma^2): \quad f_X(x) = \frac{1}{\sigma\sqrt{2\pi}} \exp\left(-\frac{(x-\mu)^2}{2\sigma^2}\right)} \tag{5.11.9}$$

How did mathematicians come up with the above form of the PDF for the normal distribution? Here is one way. The standard normal distribution has a mean of zero and a variance of one and the graph is centered around $z = 0$. Now, to have a distribution of the same shape (exponential curve) but with mean μ and variance (σ^2) different from one, we need to translate and scale the standard normal curve (Sect. 4.2.2). This is achieved with $X = \sigma Z + \mu$ where $Z \sim N(0, 1)$. We can see that

$$\mathbb{E}[X] = \mathbb{E}[\sigma Z + \mu] = \sigma\mathbb{E}[Z] + \mu = \sigma \times 0 + \mu = \mu$$
$$\text{Var}(X) = \text{Var}(\sigma Z + \mu) = \sigma^2\text{Var}(Z) = \sigma^2$$

So far so good, now to get Eq. (5.11.9), we start with the CDF of X:

$$F_X(x) = P(X \le x) = P(\sigma Z + \mu \le x) = P\left(Z \le \frac{x-\mu}{\sigma}\right) = \Phi\left(\frac{x-\mu}{\sigma}\right)$$

From that we can determine the PDF of X:

$$f_X(x) = \frac{d}{dx}F_X(x) = \frac{d}{dx}\Phi\left(\frac{x-\mu}{\sigma}\right) = \frac{1}{\sigma}\Phi'\left(\frac{x-\mu}{\sigma}\right) = \frac{1}{\sigma}f_Z\left(\frac{x-\mu}{\sigma}\right)$$

---

54 The first integral is zero because the integrand is an even function. For the second integral, using integration by parts.

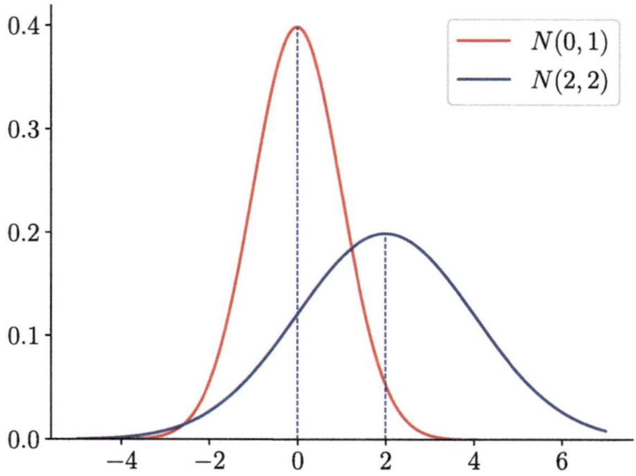

Fig. 5.15 Transformation of the standard normal curve to get a normal curve with $\mu \neq 0$ and $\sigma \neq 1$

Figure 5.15 shows this translating—with $(x - \mu)$—and scaling—with $({}^{x-\mu}/_{\sigma})$. Now we can write the CDF:

$$F_X(x) = P(X \leq x) = \Phi\left(\frac{x - \mu}{\sigma}\right) \tag{5.11.10}$$

And thus we can compute $P(a \leq X \leq b)$ as

$$P(a \leq X \leq b) = \Phi\left(\frac{b - \mu}{\sigma}\right) - \Phi\left(\frac{a - \mu}{\sigma}\right). \tag{5.11.11}$$

5.12 Joint Discrete Distributions

So far we have dealt with only one single variable, now is the time to consider more than one variable. Life is complicated and more often we have to deal with many variables. For example, we might measure the height and weight of horses, or the IQ and birth weight of children, or the frequency of exercise and the rate of heart disease in adults.

In such situations the random variables have a joint distribution that allows us to compute probabilities of events involving both variables and understand the relationship between the variables. The situation is simplest when the variables are independent. When they are not, we use covariance and correlation as measures of the dependence between them.

5.12.1 Two Jointly Discrete Variables

Suppose that we have two discrete random variables X and Y, and that X takes values $\{x_1, x_2, \ldots, x_n\}$ and Y takes values $\{y_1, y_2, \ldots, y_m\}$. The pair (X, Y) take values in the Cartesian product $\{(x_1, y_1), (x_1, y_2), \ldots, (x_n, y_m)\}$; that is the joint range of X and Y: R_{XY}. Now, we're interested in the probability of the event $X = x_i$ and $Y = y_j$, for x_i, y_j in the ranges of X, Y, respectively. We can define the so-called *joint probability mass function*, denoted by $P_{XY}(x_i, y_j)$ as

$$P_{XY}(x_i, y_j) = P(X = x_i, Y = y_j) = P((X = x_i) \text{ and } (Y = y_j)) \tag{5.12.1}$$

which gives us the probability of the joint outcome $X = x_i, Y = y_j$.

Table 5.10 gives an example of a joint PFM. The joint PMF contains all the information regarding the distributions of X and Y. This means that, for example, we can obtain the PMF of X and Y from the joint PMF. For example, the probability that $X = 129$ is computed as, using the law of total probability (Eq. (5.8.6))[55]

$$P(X = 129) = P(X = 129, Y = 15) + P(X = 129, Y = 16) = 0.12 + 0.08 = 0.20$$

Similarly, we computed $P(X = 130)$, and $P(X = 131)$. And we put them in the margins of the original joint PFM table (Table 5.11). Because of this, the probability mass functions for X and Y are often referred to as the *Marginal Distributions* for X and Y.

With that example, we now give the definition of the marginal distribution for X (the one for Y is similar):

$$P_X(x) := \sum_{y_j \in R_Y} P_{XY}(x, y_j) \quad \text{for any } x \in R_X \tag{5.12.2}$$

A joint probability mass function is a probability, thus it has to satisfy the two properties in Eq. (5.10.7):

Table 5.10 Example of a joint PFM

y	x		
	129	130	131
15	0.12	0.42	0.06
16	0.08	0.28	0.04

55 If not clear, see $Y = 15$ as B_1 and $Y = 16$ as B_2.

Table 5.11 Marginal PFM from joint PFM

y	x			
	129	130	131	
15	0.12	0.42	0.06	**0.60**
16	0.08	0.28	0.04	**0.40**
	0.20	**0.70**	**0.10**	

$$0 \le P_{XY}(x_i, y_j) \le 1, \quad \sum_{i=1}^{n}\sum_{j=1}^{m} P_{XY}(x_i, y_j) = 1 \tag{5.12.3}$$

Recall that the cumulative distribution function of random variable X is defined as

$$F_X(x) := P(X \le x), \quad \text{for all } x \in \mathbb{R}$$

From this, we can, in a similar manner, define the joint cumulative distribution function for X and Y:

$$F_{XY}(x, y) := P(X \le x, Y \le y), \quad \text{for all } x, y \in \mathbb{R} \tag{5.12.4}$$

And of course, from the joint CDF $F_{XY}(x, y)$ we can determine the marginal CDFs for X and Y:

$$\begin{aligned} F_X(x) &:= P(X \le x, Y \le \infty) = \lim_{y\to\infty} F_{XY}(x, y) \\ F_Y(y) &:= P(X \le \infty, Y \le y) = \lim_{x\to\infty} F_{XY}(x, y) \end{aligned} . \tag{5.12.5}$$

5.12.2 *Conditional PMF and CDF*

Suppose that we roll a die. Let X be the observed number. Find the probability mass function of X given that we know the observed number was less than 5. Such a PMF is called a conditional PMF. Specifically, the conditional PMF of X given event A is defined as

$$P_{X|A}(x_i) = P(X = x_i|A) = \frac{P(X = x_i \text{ and } A)}{P(A)} \tag{5.12.6}$$

where Eq. (5.8.1) was used in the last equality. Now, instead of event A, we consider the case it is the event that $Y = y_j$, then we have a conditional PMF of X given Y:

$$P_{X|Y}(x_i|y_j) = \frac{P(X = x_i \text{ and } Y = y_j)}{P(Y = y_j)} = \frac{P_{XY}(x_i, y_j)}{P_Y(y_j)}. \tag{5.12.7}$$

5.12.3 Independence

Roll two dice. Let X be the number on the first die and let Y be the number on the second die. Then both X and Y take values 1 to 6 and the joint pmf is $P_{XY}(i, j) = 1/36$ for all i and j between 1 and 6. The joint probability table is shown in Table 5.12. It is obvious that the two events X and Y are independent. Now, look at the mentioned table, we can observe that

$$P_{XY}(x = x_i, y = y_j) = P_X(x_i)P_Y(y_j), \quad i, j = 1, 2, \ldots, 6 \tag{5.12.8}$$

$(1/36 = (1/6)(1/6))$. So, we suspect that for two random variables X and Y to be independent, we should have[56]

$$P_{XY}(x = x_i, y = y_j) = P_X(x_i)P_Y(y_j) \ \ \forall(x_i, y_j$$

To check this definition of independence of two discrete RVs, we are going to use it to prove that the conditional PMF is equal to the marginal PMF. In other words, knowing the value of Y does not provide any information about X. That is we need to prove $P_{X|Y}(x_i|y_j) = P_X(x_i)$. Indeed,

$$P_{X|Y}(x_i|y_j) = \frac{P_{XY}(x_i, y_j)}{P_Y(y_j)} = \frac{P_X(x_i)\cancel{P_Y(y_j)}}{\cancel{P_Y(y_j)}} = P_X(x_i).$$

Table 5.12 Joint PFM of rolling two dice

y	x						
	1	2	3	4	5	6	$P_Y(y_j)$
1	1/36	1/36	1/36	1/36	1/36	1/36	1/6
2	1/36	1/36	1/36	1/36	1/36	1/36	1/6
3	1/36	1/36	1/36	1/36	1/36	1/36	1/6
4	1/36	1/36	1/36	1/36	1/36	1/36	1/6
5	1/36	1/36	1/36	1/36	1/36	1/36	1/6
6	1/36	1/36	1/36	1/36	1/36	1/36	1/6
$P_X(x_i)$	1/6	1/6	1/6	1/6	1/6	1/6	

[56] Actually we know that for two independent events A and B, $P(A \cap B) = P(A)P(B)$.

5.12.4 Conditional Expectation

We have probability and conditional probability. It is thus natural to have conditional expectation. Given a random variable X and an event A that has occured, the conditional expectation of X, denoted by $\mathbb{E}[X|A]$, is defined as

$$\mathbb{E}[X] = \sum_k x_k P_X(x_k) \Longrightarrow \mathbb{E}[X|A] = \sum_k x_k P_{X|A}(x_k) \tag{5.12.9}$$

which is nothing but the ordinary expectation in which the normal PMF is replaced by the conditional PMF. If A is the event $Y = y$, then we have $\mathbb{E}[X|Y = y]$:

$$\mathbb{E}[X|Y = y] = \sum_k x_k P_{X|Y}(x_k|y) \tag{5.12.10}$$

Obviously $\mathbb{E}[X|Y = y]$ is a real number, and it depends on y. Suppose now that $Y = \{y_1, y_2, \ldots, y_m\}$, then for each y_i we have a corresponding $\mathbb{E}[X|Y = y_i]$. So, $\mathbb{E}[X|Y]$ is a function of Y. As Y is a random variable, so is $\mathbb{E}[X|Y]$. It is natural to consider the expected value of $\mathbb{E}[X|Y]$, that is $\mathbb{E}[\mathbb{E}[X|Y]]$ and so on. Let's consider one example first to see what is $\mathbb{E}[\mathbb{E}[X|Y]]$.

Example 5.17 Considering the table below. Let $Z = \mathbb{E}[X|Y]$.

- find the conditional PMF of X given $Y = 0$ and $Y = 1$, i.e., $P_{X|Y}(x|0)$ and $P_{X|Y}(x|1)$
- find the PMF of Z
- find the expected value of Z, i.e., $\mathbb{E}[Z]$

y	x		
	0	1	$P_Y(y_j)$
0	1/5	2/5	3/5
1	2/5	0	2/5
$P_X(x_i)$	3/5	2/5	

- Using Eq. (5.12.7), we compute $P_{X|Y}(0|0)$ as

$$P_{X|Y}(0|0) = \frac{P_{XY}(0,0)}{P_Y(0)} = \frac{1/5}{3/5} = \frac{1}{3}$$

And from that $P_{X|Y}(1|0) = 1 - 1/3 = 2/3$. In the same manner, we have

$$P_{X|Y}(0|1) = \frac{P_{XY}(0,1)}{P_Y(1)} = \frac{2/5}{2/5} = 1, \quad P_{X|Y}(1|1) = 0$$

- Using Eq. (5.12.10) and noting that $Y = \{0, 1\}$, we can obtain

$$Z = \mathbb{E}[X|Y] = \begin{cases} 2/3, & \text{with probability } 3/5 \\ 0, & \text{with probability } 2/5 \end{cases}$$

Therefore, we can write $P_Z(z)$ as

$$P_Z(z) = \begin{cases} 3/5, & \text{if } z = 2/3 \\ 2/5, & \text{if } z = 0 \\ 0, & \text{otherwise} \end{cases}$$

And compute $\mathbb{E}[Z] = (2/3)(3/5) + 0(2/5) = 2/5$. Noting that $\mathbb{E}[X]$ is also $2/5$. Thus, $\mathbb{E}[\mathbb{E}[X|Y]] = \mathbb{E}[X]$, at least for this example.

What we have seen in this example that $\mathbb{E}[X] = \mathbb{E}[\mathbb{E}[X|Y]]$ is known as the law of iterated expectation:

$$\text{The law of iterated expectation}: \quad \mathbb{E}[X] = \mathbb{E}[\mathbb{E}[X|Y]] \tag{5.12.11}$$

Proof We go from $\mathbb{E}[\mathbb{E}[X|Y]]$ to $\mathbb{E}[X]$ using the definition of the conditional expectation and conditional probability:

$$\begin{aligned}
\mathbb{E}[\mathbb{E}[X|Y]] &= \sum_y \mathbb{E}[X|Y = y] P_Y(Y = y) && \text{(def.)} \\
&= \sum_y \sum_x x P_{X|Y}(x|y) P_Y(y) && \text{(def of red term).} \\
&= \sum_y \sum_x x \frac{P_{XY}(x, y)}{P_Y(y)} P_Y(y) && \text{(Eq. (5.12.7))} \\
&= \sum_y \sum_x x P_{XY}(x, y) = \sum_x \sum_y x P_{XY}(x, y) && \text{(switch sum index)} \\
&= \sum_x x \left(\sum_y P_{XY}(x, y) \right) = \sum_x x P_X(x) && \text{(Eq. (5.12.2))}
\end{aligned}$$

If the step in which I switched the sums was not clear, just consider a concrete case in which $X = \{x_1, x_2\}$ and $Y = \{y_1, y_2\}$, then you will see that $\sum_y \sum_x \square = \sum_x \sum_y \square$. Noting that $\sum_x x P_X(x) = \mathbb{E}[X]$. That's the end of our proof. ■

And from Eq. (5.12.11) we get a new way to compute $\mathbb{E}[X]$. Suprisingly it is similar to the law of total probability, see Eq. (5.8.6):

$$\text{The law of total probability}: \quad \mathbb{E}[X] = \sum_y \mathbb{E}[X|Y = y] P_Y(Y = y) \tag{5.12.12}$$

Expectation for independent random variables. The law of unconsscious statistician for two discrete random variables is

$$\mathbb{E}[g(X,Y)] = \sum_{(x_i,y_j)\in R_{XY}} g(x_i,y_j)P_{XY}(x_i,y_j) = \sum_i\sum_j g(x_i,y_j)P_{XY}(x_i,y_j) \tag{5.12.13}$$

Now, we use the LOTUS to derive this result: if X, Y are two independent RVs, then $\mathbb{E}[XY] = \mathbb{E}[X]\mathbb{E}[Y]$.

Proof The proof adopts Eq. (5.12.13) for the function $g = XY$, then uses $P_{XY}(x_i, y_j) = P_X(x_i)P_Y(y_j)$ for the two independent RVs X and Y:

$$\begin{aligned}\mathbb{E}[XY] &= \textstyle\sum_i\sum_j x_i y_j P_{XY}(x_i,y_j)\\ &= \textstyle\sum_i\sum_j x_i y_j P_X(x_i)P_Y(y_j)\\ &= \left(\textstyle\sum_i x_i P_X(x_i)\right)\left(\textstyle\sum_j y_j P_Y(y_j)\right) = \mathbb{E}[X]\mathbb{E}[Y]\end{aligned} \tag{5.12.14}$$

■

Definition 5.12.1 Random variables $X_1, X_2, \ldots, X_n$ are said to be independent and identically distributed (i.i.d) if they are independent and they have the same (marginal) distributions:

$$F_{X_1}(x) = F_{X_2}(x) = \ldots = F_{X_n}(x), \quad \forall x \in \mathbb{R}.$$

5.12.5 Covariance

For two jointly distributed real-valued random variables X and Y, we know that $\mathbb{E}[X+Y] = \mathbb{E}[X] + \mathbb{E}[Y]$. The question is how about the variance of the sum, i.e., $\text{Var}(X+Y)$? Let's see what we get. We start with the definition and use the linearity of the expectation[57]:

$$\begin{aligned}\text{Var}(X+Y) &= \mathbb{E}[((X+Y) - \mathbb{E}[X+Y])^2] \quad \text{(def. Eq. (5.10.39))}\\ &= \mathbb{E}[(X+Y-\mathbb{E}[X]-\mathbb{E}[Y])^2] \quad \text{(linearity of } \mathbb{E}[X+Y])\\ &= \mathbb{E}[(X-\mathbb{E}[X])^2] + \mathbb{E}[(Y-\mathbb{E}[Y])^2] + 2\mathbb{E}[(X-\mathbb{E}[X])(Y-\mathbb{E}[Y])]\\ &= \text{Var}(X) + \text{Var}(Y) + 2\mathbb{E}[(X-\mathbb{E}[X])(Y-\mathbb{E}[Y])]\end{aligned}$$

What can we conclude then? It is this: The variance of a sum is the sum of variances plus something new—the red term. Let's massage it and see what we get (recalling that $\mathbb{E}[aX] = a\mathbb{E}[X]$ from Eq. (5.10.38)):

[57] The step from the second equality to the third is $(X+Y-\mathbb{E}[X]-\mathbb{E}[Y])^2 = [(X-\mathbb{E}[X]) + (Y-\mathbb{E}[Y])]^2 = \cdots$ using $(a+b)^2 = a^2+b^2+2ab$; finally, the linearity of expected value $\mathbb{E}[X+Y] = \mathbb{E}[X] + \mathbb{E}[Y]$ is used again.

$$\mathbb{E}[(X - \mathbb{E}[X])(Y - \mathbb{E}[Y])] \stackrel{(a)}{=} \mathbb{E}[XY - X\mathbb{E}[Y] - Y\mathbb{E}[X] + \mathbb{E}[X]\mathbb{E}[Y]]$$
$$\stackrel{(b)}{=} \mathbb{E}[XY] - \mathbb{E}[Y]\mathbb{E}[X] - \mathbb{E}[X]\mathbb{E}[Y] + \mathbb{E}[X]\mathbb{E}[Y]$$
$$\stackrel{(c)}{=} \mathbb{E}[XY] - \mathbb{E}[X]\mathbb{E}[Y]$$

Steps (a,c) are just algebraic. Step (b) used $\mathbb{E}[aX] = a\mathbb{E}[X]$. For example, $\mathbb{E}[X\mathbb{E}[Y]] = \mathbb{E}[Y]\mathbb{E}[X]$ with $a = \mathbb{E}[Y]$, and $\mathbb{E}[c] = c$ with $c = \mathbb{E}[X]\mathbb{E}[Y]$. Now, if $X = Y$, then the above becomes the variance of Y (or of X, if not clear, check Eq. (5.10.40)). And we know that, see Eq. (5.12.14), if X, Y are independent, then $\mathbb{E}[XY] = \mathbb{E}[X]\mathbb{E}[Y]$, and the red term vanishes. So, what we call the red term? We call it the *covariance of* X and Y, denoted by $\text{Cov}(X, Y)$ or σ_{XY}:

$$\boxed{\text{Cov}(X, Y) = \sigma_{XY} = \mathbb{E}[(X - \mathbb{E}[X])(Y - \mathbb{E}[Y])]} = \mathbb{E}[XY] - \mathbb{E}[X]\mathbb{E}[Y] \tag{5.12.15}$$

Thus, the variance is a measure of the spread of one single variable with respect to its mean. And the covariance is a measure of two variables. The covariance is in units obtained by multiplying the units of the two variables. What are we going to do now? Compute some covariance? That's important but not interesting: the software Microsoft Excel can do that. As usual in maths, we will deduce properties of the covariance before actually computing it!

Properties of the covariance. The covariance can be seen as an operator with two inputs and it looks similar to the dot product of two vectors. If we look at the properties of the dot product in Box 9.2 we guess the following are true (the last one not coming from dot product though):

$$\begin{array}{lll}
\text{(a):} & \text{Cov}(X, X) & = \text{Var}(X) \\
\text{(b): commutative law} & \text{Cov}(X, Y) & = \text{Cov}(Y, X) \\
\text{(d): distributive law} & \text{Cov}(Z, X + Y) & = \text{Cov}(Z, X) + \text{Cov}(Z, Y) \\
\text{(d):} & \text{Cov}(\alpha X, Y) & = \alpha\text{Cov}(X, Y) \\
\text{(e):} & \text{Cov}(X + c, Y) & = \text{Cov}(X, Y)
\end{array} \tag{5.12.16}$$

The proof is skip as it is 100% based on the definition of the covariance, i.e., Eq. (5.12.15). The first property is: if Y always takes on the same values as X, we have the covariance of a variable with itself (i.e., σ_{XX}), which is nothing but the variance.

Example 5.18 Covariance of two discrete random variables We consider the data given in Table 5.10 and use Eq. (5.12.15) to compute σ_{XY}. First, we need the sample means: $\bar{X} = (129 + 130 + 131)/3 = 130$ and $\bar{Y} = (15 + 16)/2 = 15.5$. Then, σ_{XY} can be computed as, using the LOTUS Eq. (5.12.13):

$$
\begin{aligned}
\sigma_{XY} &= \sum_{i=1}^{3}\sum_{j=1}^{2}(X_i - \bar{X})(Y_j - \bar{Y})P_{ij} \\
&= (129-130)(15-15.5)(0.12) + (129-130)(16-15.5)(0.08) + \cdot \\
&+ (130-130)(15-15.5)(0.42) + (130-130)(16-15.5)(0.28) \\
&+ (131-130)(15-15.5)(0.06) + (131-130)(16-15.5)(0.04)
\end{aligned}
$$

5.12.6 Variance of a Sum of Variables

Suppose we have a sum of several random variables, in particular $Y = X_1 + \cdots + X_n$. The question is: what is $\text{Var}(Y)$? If Y is just the sum of two variables, then we know that,[58]

$$\text{Var}(X_1 + X_2) = \text{Var}(X_1) + \text{Var}(X_2) + 2\text{Cov}(X_1, X_2)$$

With that, it is only a small step to go to the general case $Y = \sum_{i=1}^{n} X_i$. It might help if we go slowly with $n = 3$ or $Y = X_1 + X_2 + X_3$, then $\text{Var}(Y) = \text{Cov}(Y, Y)$ can be written as

$$
\begin{aligned}
\text{Var}(Y) &= \text{Cov}\,(X_1 + X_2 + X_3, X_1 + X_2 + X_3) \\
&= \text{Cov}(X_1, X_1 + X_2 + X_3) + \text{Cov}(X_2, X_1 + X_2 + X_3) + \text{Cov}(X_3, X_1 + X_2 + X_3) \\
&= \text{Cov}(X_1, X_1) + \text{Cov}(X_1, X_2) + \text{Cov}(X_1, X_3) + \text{Cov}(X_2, X_1 + X_2 + X_3) \\
&+ \text{Cov}(X_3, X_1 + X_2 + X_3) \\
&= \text{Var}(X_1) + \text{Var}(X_2) + \text{Var}(X_3) + 2\text{Cov}(X_1, X_2) + 2\text{Cov}(X_1, X_3) + 2\text{Cov}(X_2, X_3)
\end{aligned}
\tag{5.12.17}
$$

where in the second equality, we used the distributive property in Eq. (5.12.16). Then, this property is used again in the third equality. Doing the same thing for $\text{Cov}(X_2, X_1 + X_2 + X_3)$ and $\text{Cov}(X_3, X_1 + X_2 + X_3)$ we then obtain the final expression for $\text{Var}(Y)$. Now, we can go to the general case (just Eq. (5.12.17) but with $\sum$ notation):

$$
\begin{aligned}
\text{Var}(Y) = \text{Cov}(Y, Y) = \text{Cov}\left(\sum_{i=1}^{n} X_i, \sum_{j=1}^{n} X_j\right) &= \sum_{i=1}^{n}\sum_{j=1}^{n} \text{Cov}(X_i, X_j) \\
&= \sum_{i=1}^{n} \text{Var}(X_i) + 2\sum_{i<j} \text{Cov}(X_i, X_j), \quad \text{with } Y = X_1 + \cdots + X_n
\end{aligned}
\tag{5.12.18}
$$

---

[58] We can get this formula by this: $\text{Var}(Y) = \text{Cov}(Y, Y)$ and use the distributive law of the covariance operator.

If X_i are uncorrelated all the $\text{Cov}(X_i, X_j)$ terms vanish, and thus we get the nice identity[59]

$$\text{Var}\left(\sum_{i=1}^{n} X_i\right) = \sum_{i=1}^{n} \text{Var}(X_i) \tag{5.12.19}$$

This statement is called the Bienaymé[60] formula and was discovered in 1853. From that we can deduce that $\text{Var}(\bar{X}) = \sigma^2/n$.

For any constants $a_1, \ldots, a_n$ and $b_1, \ldots, b_m$, we have

$$\text{Cov}\left(\sum_{i=1}^{n} a_i X_i, \sum_{j=1}^{m} b_j Y_j\right) = \sum_{i=1}^{n}\sum_{j=1}^{m} a_i b_j \text{Cov}(X_i, Y_j).$$

5.12.7 Correlation Coefficient

It is better to work with dimensionless quantities, we thus introduce the following dimensionless variables U and V:

$$U = \frac{X - \mathbb{E}[X]}{\sigma_X}, \quad V = \frac{Y - \mathbb{E}[Y]}{\sigma_Y}$$

Then, we compute the covariance of U, V, i.e., $\text{Cov}(U, V)$ and give it a name and a symbol:

$$\rho_{XY} := \text{Cov}(U, V) = \text{Cov}\left(\frac{X - \mathbb{E}[X]}{\sigma_X}, \frac{Y - \mathbb{E}[Y]}{\sigma_Y}\right) \overset{\text{(a)}}{=} \text{Cov}\left(\frac{X}{\sigma_X}, \frac{Y}{\sigma_Y}\right) \overset{\text{(b)}}{=} \frac{\text{Cov}(X, Y)}{\sigma_X \sigma_Y}$$

where, for (a) I used property (e) in Eq. (5.12.16). And for (b), I used the definition of the covariance in Eq. (5.12.15) and the property of $\mathbb{E}[\alpha X] = \alpha\mathbb{E}[X]$. The symbol ρ_{XY} denotes the correlation coefficient of X and Y. It is a coefficient as it is dimensionless. So, ρ_{XY} is the covariance between the standardized versions of X and Y.

Now, we are going to show that $-1 \leq \rho_{XY} \leq 1$. The proof uses Eq. (5.12.18) to compute the variance of ${}^{X}/_{\sigma_X} \pm {}^{Y}/_{\sigma_Y}$:

---

[59] In English this rule is familiar: the var of sum is the sum of var. We have similar rules for the derivative, the limit, etc. It is interesting, isn't it?

[60] Irénée-Jules Bienaymé (1796–1878) was a French statistician. He built on the legacy of Laplace generalizing his least squares method. He contributed to the fields of probability and statistics, and to their application to finance, demography, and social sciences.

$$
\begin{aligned}
\text{Var}\left(\frac{X}{\sigma_X} \pm \frac{Y}{\sigma_Y}\right) &= \text{Var}\left(\frac{X}{\sigma_X}\right) + \text{Var}\left(\frac{Y}{\sigma_Y}\right) \pm 2\text{Cov}\left(\frac{X}{\sigma_X}, \frac{Y}{\sigma_Y}\right) \\
&\overset{\text{(a)}}{=} \frac{1}{\sigma_X^2}\text{Var}\,(X) + \frac{1}{\sigma_Y^2}\text{Var}\,(Y) \pm \frac{2}{\sigma_X \sigma_Y}\text{Cov}\,(X, Y) \\
&= 2 \pm 2\rho_{XY} \quad (\text{def. of} \rho_{XY})
\end{aligned}
\tag{5.12.20}
$$

where, for (a) I used Eq. (5.10.42) (i.e., $\text{Var}(aX + b) = a^2\text{Var}(X)$).

But, the variance of $X/\sigma_X \pm Y/\sigma_Y$ is non-negative, thus

$$
0 \le 2 \pm 2\rho_{XY} \Longrightarrow \boxed{-1 \le \rho_{XY} \le 1}.
$$

5.12.8 Covariance Matrix

If we have two variables X, Y we have one single $\text{Cov}(X, Y)$, what if we have more than two variables? Let's investigate the case of three variables X, Y and Z. Of course, we would have $\text{Cov}(X, Y), \text{Cov}(X, Z), \text{Cov}(Y, Z)$, and so on. And if we put all of them in a matrix, we get the so-called *covariance matrix*:

$$
\mathbf{C} = \begin{bmatrix} \text{Cov}(X, X) & \text{Cov}(X, Y) & \text{Cov}(X, Z) \\ \text{Cov}(X, Y) & \text{Cov}(Y, Y) & \text{Cov}(Y, Z) \\ \text{Cov}(X, Z) & \text{Cov}(Y, Z) & \text{Cov}(Z, Z) \end{bmatrix}
$$

Note that the diagonal terms are the variances of the variables. Obviously this matrix $\mathbf{C}$ is a symmetric matrix. Is that all we know about it? It turns out that there is also another property hidden there. Let's investigate it. A 2×2 covariance matrix is sufficient to reveal the secret. Without loss of generality, we consider only discrete random variables X, Y with means $\bar{X}$ and $\bar{Y}$, respectively. Thus, we have

$$
\mathbf{C} = \begin{bmatrix} \text{Cov}(X, X) & \text{Cov}(X, Y) \\ \text{Cov}(X, Y) & \text{Cov}(Y, Y) \end{bmatrix}, \quad \begin{cases} \text{Cov}(X, X) = \sum_i P_i (X_i - \bar{X})^2 \\ \text{Cov}(X, Y) = \sum_i \sum_j P_{ij} (X_i - \bar{X})(Y_j - \bar{Y}) \end{cases}
$$

There is a non-symmetry in the formula of $\text{Cov}(X, X)$ and $\text{Cov}(X, Y)$: there is no P_{ij} in the former! Let's make it appear and something wonderful will show up (this is due to $P_i = \sum_j P_{ij}$, check the marginal probability if this was not clear):

$$
\text{Cov}(X, X) = \sum_i P_i (X_i - \bar{X})^2 = \sum_i \sum_j P_{ij} (X_i - \bar{X})^2
$$

With that, we can have a beautiful formula for **C**, in which **C** is a sum of a bunch of matrices, each matrix is multiplied by a non-negative number (i.e., P_{ij}):

$$\mathbf{C} = \sum_i \sum_j P_{ij} \begin{bmatrix} (X_i - \bar{X})^2 & (X_i - \bar{X})(Y_j - \bar{Y}) \\ (X_i - \bar{X})(Y_j - \bar{Y}) & (Y_i - \bar{Y})^2 \end{bmatrix}$$

What is special about the red matrix? It is equal to $\mathbf{U}\mathbf{U}^\top$, where $\mathbf{U} = (X_i - \bar{X}, Y_i - \bar{Y})$. So what? Every matrix $\mathbf{U}\mathbf{U}^\top$ is positive semidefinite.[61] Thus, **C** combines all these positive semidefinite matrices with weights $P_{ij} \geq 0$: it is also positive semidefinite. This turns out to be a useful property and exploited in principal component analysis—which is an important tool in statistics. Check Sect. 9.15 for a discussion on this topic.

Sample covariance. If we have n samples and each sample has two measurements X and Y, hence we have $X = (x_1, \ldots, x_n)$ and $Y = (y_1, \ldots, y_n)$, then the sample covariance between X and Y is defined as (noting the Bessel's correction $n - 1$ in the denominator)

$$\text{Cov}(X, Y) = \frac{1}{n-1} \sum_{i=1}^{n} (x_i - \bar{x})(y_i - \bar{y}) \tag{5.12.21}$$

What does that actually mean? Assume that X denotes the number of hours studied for a subject and Y is the marks obtained in that object. We can use real data to compute the covariance, and assume that the value is 90.34. What does this value mean? A positive value of covariance indicates that both variables increase or decrease together, e.g., as the number of hours studied increase, the grades also increase. A negative value, on the other hand, means that while one variable increases the other decreases or vice versa. And if the covariance is zero, the two variables are uncorrelated.

Now, we derive the formula for the covariance matrix for the whole data. We start with the sample mean:

$$\begin{array}{l} X : x_1 \;\; x_2 \;\; \cdots \;\; x_n : \bar{x} = 1/n(\sum_i x_i) \\ Y : y_1 \;\; y_2 \;\; \cdots \;\; y_n : \bar{y} = 1/n(\sum_i y_i) \end{array}$$

Then, we subtract the data from the mean, to center the data

$$\mathbf{A} = \begin{bmatrix} x_1 \; x_2 \cdots x_n \\ y_1 \; y_2 \cdots y_n \end{bmatrix} \Longrightarrow \mathbf{A} = \begin{bmatrix} x_1 - \bar{x} \; x_2 - \bar{x} \cdots x_n - \bar{x} \\ y_1 - \bar{y} \; y_2 - \bar{y} \cdots y_n - \bar{y} \end{bmatrix}$$

[61] Check Sect. 9.10.6 for quadratic forms and positive definiteness of matrices. The proof goes: $\boldsymbol{x}^\top(\mathbf{U}\mathbf{U}^\top)\boldsymbol{x} = \|\mathbf{U}^\top\boldsymbol{x}\|^2 \geq 0$.

And the covariance matrix is given by

$$\mathbf{C} = \frac{1}{n-1}\mathbf{A}\mathbf{A}^\top.$$

5.13 Joint Continuous Variables

Similar to the probability of a single random variable, see Eq. (5.11.3), for two continuous RVs X and Y, we have (Fig. 5.16)

$$P(a \le x \le b, c \le y \le d) = \int_a^b \int_c^d f_{XY}(x, y)dxdy, \quad P((X, Y) \in A) = \iint_A f_{XY}(x, y)dxdy \tag{5.13.1}$$

And we call $f_{XY}(x, y)$ the joint density probability function.

5.14 Transforming Density Functions

Suppose that we have a continuous random variable X with known cdf $F_X(x)$ and pdf $f_X(x)$. Assume that we're interested in a derived random variable Y defined as $Y = h(X)$ for some function $y = h(x)$. The question is: what is the pdf of Y, i.e., $f_Y(y)$?

There are two ways to answer this question. In the first way, we first determine the cdf of Y in terms of $F_X(x)$, then we differentiate it to get $f_Y(y)$. Assuming now that $y = h(x)$ is a strictly increasing function. Thus, we have the inverse $x = g(y)$. Then, we can write $F_Y(y)$ as

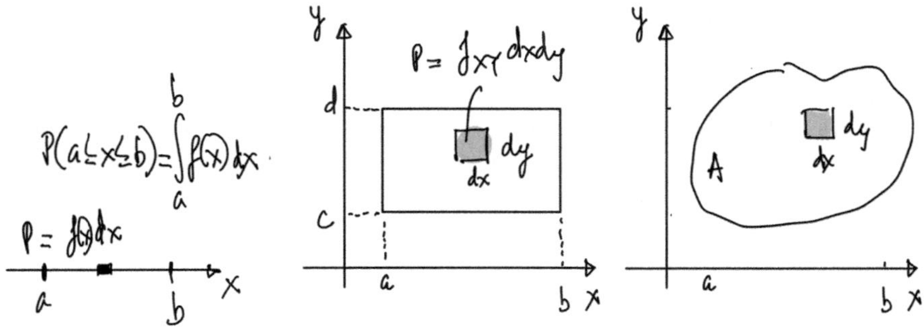

Fig. 5.16 From a probability density function $f_X(x)$ to a joint probability density function $f_{XY}(x, y)$

$$\begin{aligned} F_Y(y) &\stackrel{\text{def}}{=} P(Y \le y) = P(h(X) \le y) \\ &= P(X \le g(y)) \quad (y = h(x) \text{ is an increasing func.}) \\ &\stackrel{\text{def}}{=} F_X(g(y)) \end{aligned} \tag{5.14.1}$$

Next, we're going to differentiate $F_Y(y)$ to get $f_Y(y)$

$$\begin{aligned} f_Y(y) &\stackrel{\text{def}}{=} \frac{d}{dy} F_Y(y) = \frac{d}{dy} F_X(g(y)) \quad \text{(Eq. (5.14.1))} \\ &= F'_X(g(y)) g'(y) \quad \text{(chain rule)} = f_X(g(y)) g'(y) \end{aligned} \tag{5.14.2}$$

If the function $y = h(x)$ is a decreasing function, then by repeating this procedure, we will get $f_Y(y) = -f_X(g(y))g'(y)$ which is non-negative as $g'(y) \le 0$. So,

$$Y : Y = h(X) \Longrightarrow \boxed{f_Y(y) = f_X(g(y))|g'(y)|} \tag{5.14.3}$$

I present now the second way that uses the fact that $P(a \le X \le b) = \int_a^b f_X(x)dx$. Again suppose that $y = h(x)$ is such a function that for $x \in [a, b]$, then $y \in [h(a), h(b)]$. We have

$$P(h(a) \le Y \le h(b)) = P(a \le X \le b) = \int_a^b f_X(x)dx \tag{5.14.4}$$

Now, comes the change of variable $x = g(y)$, and substitute it into the above integral we obtain

$$P(h(a) \le Y \le h(b)) = \int_{h(a)}^{h(b)} f_X(g(y))g'(y)dy. \tag{5.14.5}$$

5.15 Inequalities in the Theory of Probability

There is an adage in probability that says that behind every limit theorem lies a probability inequality (i.e., a bound on the probability of some event happening). Since a large part of probability theory is about proving limit theorems, mathematicians have developed a bewildering number of inequalities. This section presents some common inequalities in probability.

5.15.1 Markov and Chebyshev Inequalities

Let X be any *non-negative* continuous random variable with PDF $f_X(x)$, we can write $\mathbb{E}[X]$ as

$$\begin{aligned}
\mathbb{E}[X] &= \int_{-\infty}^{\infty} x f_X(x)dx = \int_0^{\infty} x f_X(x)dx \text{ (as } X \geq 0) \\
&\geq \int_a^{\infty} x f_X(x)dx && \text{(for any } a > 0) \\
&\geq \int_a^{\infty} a f_X(x)dx && \text{(since } x \geq a) \\
&\geq a \int_a^{\infty} f_X(x)dx \geq aP(X \geq a)
\end{aligned}$$

Thus, we have proved the so-called Markov's inequality:

$$\text{Markov's inequality:} X \text{is any non-negative RV}: \boxed{P(X \geq a) \leq \frac{\mathbb{E}[X]}{a}}$$

Markov's inequality says that for a non-negative random variable X and any positive real number a, the probability that X is greater than or equal to a is less than or equal to the expected value of X divided by a. This is a *tail bound* because it imposes an upper limit on how big the right tail at a can be.

Example 5.19 Suppose that an individual is randomly extracted from a population of individuals having an average yearly income of \$60000. What is the probability that the extracted individual's income is greater than \$200000? In the absence of more information about the distribution of income, we can still use Markov's inequality to calculate an upper bound to this probability:

$$P(X \geq 200000) \leq \frac{60000}{200000} = 0.3$$

Therefore, the probability of extracting an individual having an income greater than \$200000 is less than 30%.

Now, we apply Markov's inequality to get Chebyshev's inequality. Motivation: Markov's inequality involves the expected value. *Where is the variance?* It is involved in Chebyshev's inequality. Can we guess the form of this inequality? The variance is about the spread of X with respect to the mean. So, we would get something similar to $P(|X - \mathbb{E}[X]| \geq b) \leq g(\text{Var}(X), b)$. Note that because of symmetry when talking about spread, we have to have two tails involved: that's where the term $|X - \mathbb{E}[X]| \geq b$ comes into play.

Now, to get the Chebyshev inequality, we consider the non-negative RV $Y = (X - \mathbb{E}[X])^2$. We then have $P(Y \geq b^2) \leq \mathbb{E}[Y]/b^2$ according to Markov. But, $\mathbb{E}[Y] = \text{Var}(X)$, thus we obtain the Chebyshev inequality:

$$\text{Chebyshev's ineq.:} b > 0 : P(|X - \mathbb{E}[X]| \geq b) \leq \frac{\text{Var}(X)}{b^2}$$

Now, it is more convenient to consider $b = z\sigma$ ($z \in \mathbb{N}$), and we obtain the following form of the Chebyshev inequality (with $\mu = \mathbb{E}[X]$ and $\text{Var}(X) = \sigma^2$):

$$P(|X - \mu| \geq z\sigma) \leq \frac{\sigma^2}{z^2\sigma^2} = \frac{1}{z^2}$$

With this form, we can see that, no matter what the distribution of X is, the bulk of the probability is in the interval "expected value plus or minus a few SDs", or symbolically $\mu \pm z\sigma$. Indeed, using $P(A) = 1 - P(A^c)$, we have $P(|X - \mu| < z\sigma) = 1 - P(|X - \mu| \geq z\sigma) \geq 1 - 1/z^2$. I present some concrete cases for $z = 2, 3, 4$:

$$\begin{aligned}
P(\mu - 2\sigma < X < \mu + 2\sigma) &\geq 1 - 1/2^2 = 75\% \\
P(\mu - 3\sigma < X < \mu + 3\sigma) &\geq 1 - 1/3^2 = 88.88\% \\
P(\mu - 4\sigma < X < \mu + 4\sigma) &\geq 1 - 1/4^2 = 93.75\%
\end{aligned}$$

This result is similar to the $68 - 95 - 99.7$ rule, which applies specifically to the normal distribution. Chebyshev's inequality is a more general rule, stating that at least 75% of the values lie within two standard deviations from the expected value, and 88.89% lie within three standard deviations for a wide range of different probability distributions.

5.15.2 Chernoff's Inequality

Now we use moment generating functions (MGF) to derive another inequality. Refer to Sect. 5.17.3 for a discussion on MGF. As the MGF $m_X(t)$ is defined to be $\mathbb{E}[e^{tX}]$, then $X \geq c$ is equivalent to $e^{tX} \geq e^{tc}$ for a fixed $t > 0$ (as the exponential function is an increasing and non-negative function). Now, we can write

$$\begin{aligned}
P(X \geq c) &= P(e^{tX} \geq e^{tc}) \\
&\leq \frac{\mathbb{E}[e^{tX}]}{e^{tc}} && \text{(Markov's inequality for } e^{tX}\text{)} \\
&= \frac{m_X(t)}{e^{tc}} = m_X(t)e^{-tc} && \text{(definition of } m_X(t)\text{)}
\end{aligned}$$

Thus, we obtain the Chernoff's bound on the right tail:

$$P(X \geq c) \leq \min_{t>0} m_X(t)e^{-tc}.$$

5.16 Limit Theorems

In this section, we will discuss two important theorems in probability: (1) the law of large numbers (LLN) and (2) the central limit theorem (CLT). The LLN basically states that the average of a large number of i.i.d. random variables converges to the expected value. The CLT states that, under some conditions, the sum of a large number of random variables has an approximately normal distribution.

5.16.1 The Law of Large Numbers

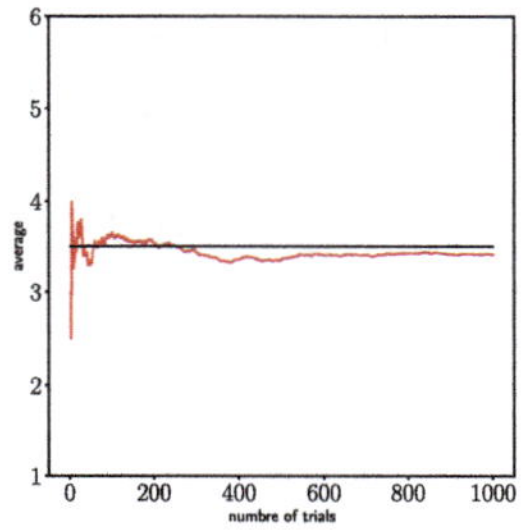

In Sect. 5.10.5 we have discussed the expected value of a random variable. Considering rolling a die, and we say that the expected value is 3.5. What this means is that, if we roll a die for a large number of times then in average we will get 3.5. We can use a MC simulation to verify this. See the figure. As the number of rolls in this run increases, the average of the values of all the results approaches 3.5. The law of large numbers (LLN) is a theorem that describes the result of performing the same experiment a large number of times. According to the law, the average of the results obtained from a large number of trials should be close to the expected value and tends to become closer to the expected value as more trials are performed.

A sequence of independent and identically distributed random variables $X_1, X_2, \ldots, X_n$ is considered. (If you do not like it, then think of the income of a population, and n data of this income.) Then, for each n, we compute the sample mean

$$\bar{X}_n = \frac{1}{n}\sum_{i=1}^{n} X_i$$

Then, we have a sequence of sample means: $\bar{X}_1, \bar{X}_2, \ldots$ If the mean of each random variables is μ, then the law of large number claims that the sequence $\{\bar{X}_n\}_{n=1}^{\infty}$ converges to μ.[62]

There are two different versions of the law of large numbers. They are called the strong law of large numbers and the weak law of large numbers. We cover the weak law only.

Theorem 5.16.1 (Weak law of large numbers) *Let $X_1, X_2, \ldots, X_n$ be iid random variables with expected value $\mu < \infty$. Then, for any $\epsilon > 0$*

$$\lim_{n\to\infty} P(|\bar{X} - \mu| \geq \epsilon) = 0$$

Proof It is obvious that we shalle use Chebyshev's inequality and assume that the variance is finite, i.e., $\text{Var}(X) = \sigma^2 < \infty$. Then, we have

$$P(|\bar{X} - \mu| \geq \epsilon) \leq \frac{\text{Var}(\bar{X})}{\epsilon^2} = \frac{\sigma^2}{n\epsilon^2}$$

Therefore, we have

$$0 \leq \lim_{n\to\infty} P(|\bar{X} - \mu| \geq \epsilon) \leq \lim_{n\to\infty} \frac{\sigma^2}{n\epsilon^2} = 0.$$

■

5.16.2 Central Limit Theorem

We recall that de Moirve, while solving some probability problem concerning the binomial distribution $b(k; n, p)$, had derived a normal approximation for the binomial distribution when n is large. As a binomial variable can be considered as a sum of n independent Bernoulli variables, we suspect that the sum of n independent and identically distributed random variables is approximately normally distributed. Now, we are going to check this hypothesis.

---

[62] A special form of the LLN (for a binary random variable) was first proved by Jacob Bernoulli. It took him over 20 years to develop a sufficiently rigorous mathematical proof which was published in his Ars Conjectandi (The Art of Conjecturing) in 1713. He named this his "Golden Theorem" but it became generally known as "Bernoulli's theorem".

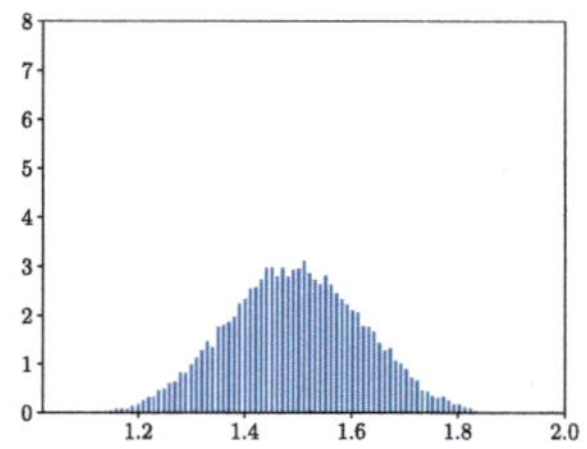

(a) $n = 5$

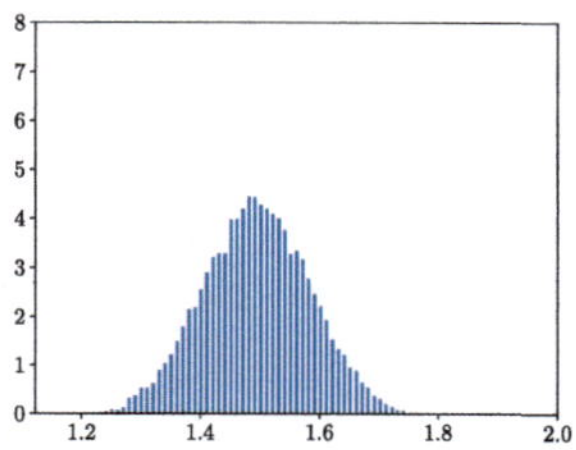

(b) $n = 10$

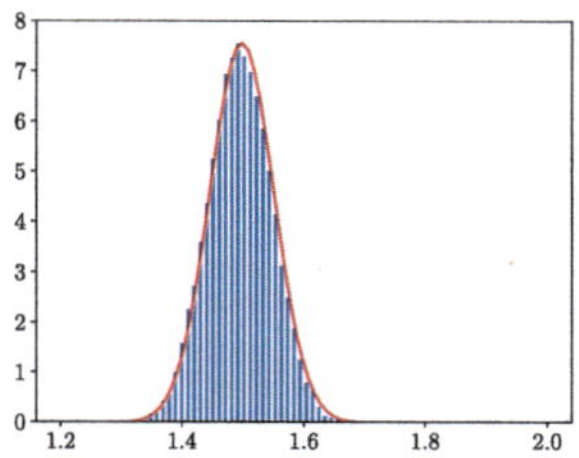

(c) $n = 30$

Fig. 5.17 The mean of n uniformly distributed RVs $X_i \sim Uniform(1, 2)$. Note that each X_i has an expected value of 1.5 and a SD of $1/12$

Mean of n uniformly distributed variables. In this test we consider n uniformly distributed RVs, i.e., $X_i \sim Uniform(1, 2)$ for $i = 1, 2, \ldots, n$. Note that each X_i has an expected value of 1.5 and a SD of $1/12 = 0.08333333$. We now define a new variable Y as the mean of X_i:

$$Y := \frac{X_1 + X_2 + \cdots + X_n}{n} := \bar{X} \tag{5.16.1}$$

What we want to see is whether Y is approximately normally distributed when n is sufficiently large. Three cases are considered: $n = 5$, $n = 10$ and $n = 30$ and the results are given in Fig. 5.17.[63] A few observations can be made from these figures. First, the mean variable Y has an expected value of 1.5—similar to that of X_i—and a SD of $0.08333333/\sqrt{n}$. Second, and what is more is that even though each and every of X_i has a uniform distribution (with a rectangle PDF), but the distribution of their mean has a bell-shaped curve of the normal distribution, see the red curve in Fig. 5.17c which is the normal curve with $\mu = 1.5$ and $\sigma = 0.08333333/\sqrt{30}$.

It is quite simple to verify the observations on the expected value and SD of Y. Indeed, we can compute $\mathbb{E}[Y]$ and $\text{Var}(Y)$ using the linearity of the expected value and the property of the variance. Let's denote by μ and σ^2 the expected value and variance of X_i (all of them have the same). Then,

$$\mathbb{E}[Y] = \mathbb{E}[X_1/n] + \mathbb{E}[X_2/n] + \cdots + \mathbb{E}[X_n/n] = n\left(\frac{1}{n}\mu\right) = \mu \tag{5.16.2}$$

and,

$$\text{Var}(Y) = \text{Var}\left(\frac{X_1 + X_2 + \cdots + X_n}{n}\right) = n\text{Var}\left(\frac{X_i}{n}\right) = \frac{\sigma^2}{n} \tag{5.16.3}$$

[63] See A.19 if you're interested in how this was done.

where in the second equality, the Bienaymé formula, i.e., Eq. (5.12.19) was used to replace the variance of a sum with the sum of variances.

About the bell-shaped curve of Y when n is large, it is guaranteed by the central limit theorem (CLT). According to this theorem (of which proof is given in Sect. 5.17.3), $Y \sim N(\mu, \sigma^2/n)$. Therefore, we have, for large ns (Eq. (5.11.11)):

$$P(a \le Y \le b) = \Phi\left(\frac{b-\mu}{\sigma/\sqrt{n}}\right) - \Phi\left(\frac{a-\mu}{\sigma/\sqrt{n}}\right) \tag{5.16.4}$$

When n is sufficiently large? Another question that comes to mind is how large n should be so that we can use the CLT. The answer generally depends on the distribution of the X_is. Nevertheless, as a rule of thumb it is often stated that if n is larger than or equal to 30, then the normal approximation is very good.

Theorem 5.16.2 (Central limit theorem) *Let $X_1, X_2, \ldots, X_n$ be iid random variables with expected value μ and variance σ^2. Then, the random variable*

$$Z_n = \frac{\bar{X}-\mu}{\sigma/\sqrt{n}} = \frac{X_1 + X_2 + \cdots + X_n - n\mu}{\sqrt{n}\sigma}$$

converges in distribution to the standard normal random variable as n goes to infinity. That is,

$$\lim_{n\to\infty} P(Z_n \le x) = \Phi(x) \quad \text{for all } x \in R$$

Example 5.20 Test scores of all high school students in a state have mean 60 and variance 64. A random sample of 100 ($n = 100$) students from one high school had a mean score of 58. Is there evidence to suggest that this high school is inferior than others?

Let $\bar{X}$ denote the mean of $n = 100$ scores from a population with $\mu = 60$ and $\sigma^2 = 64$. We know from the central limit theorem that $(\bar{X} - \mu)/(\sigma/\sqrt{n})$ is a standard normal distribution. Thus,

$$P(\bar{X} \le 58) = \Phi\left(\frac{58-\mu}{\sigma/\sqrt{n}}\right) = \Phi\left(\frac{58-60}{8/\sqrt{100}}\right) = \Phi(-2.5) = 1 - \Phi(2.5) = 0.006$$

History note 5.3: Pierre-Simon Laplace (1749–1827)

Pierre-Simon, marquis de Laplace was a French scholar and polymath whose work was important to the development of engineering, mathematics, statistics, physics, astronomy, and philosophy. Laplace is remembered as one of the greatest scientists of all time. Sometimes referred to as the French Newton or Newton of France, he has been described as possessing a phenomenal natural mathematical faculty superior to that of any of his contemporaries. He was Napoleon's examiner when Napoleon attended the École Militaire in Paris in 1784. Laplace became a count of the Empire in 1806 and was named a marquis in 1817, after the Bourbon Restoration.
Laplace attended a Benedictine priory school in Beaumont-en-Auge, as a day pupil, between the ages of 7 and 16. At the age of 16 Laplace entered Caen University. As he was still intending to enter the Church, he enrolled to study theology. However, during his two years at the University of Caen, Laplace discovered his mathematical talents and his love of the subject. Credit for this must go largely to two teachers of mathematics at Caen, C Gadbled and P Le Canu of whom little is known except that they realized Laplace's great mathematical potential. Once he knew that mathematics was to be his subject, *Laplace left Caen without taking his degree, and went to Paris*. He took with him a letter of introduction to d'Alembert from Le Canu.

5.17 Generating Functions

In Sect. 4.17 we've got to know the so-called Bernoulli numbers:

$$B_0 = 1,\ B_1 = -\frac{1}{2},\ B_2 = \frac{1}{6},\ B_3 = 0,\ B_4 = -\frac{1}{30},\ B_5 = 0,\ B_6 = \frac{1}{4},\ B_7 = 0, \ldots$$

There are infinity of them, and it seems impossible to understand them. But, with Euler's definition, in 1755, of the Bernoulli numbers in terms of the following function

$$\boxed{\frac{x}{e^x - 1} = \sum_{n=0}^{\infty} B_n \frac{x^n}{n!}} \tag{5.17.1}$$

we have discovered the recurrence relation between B_n, Eq. (4.17.2). The function x/e^x-1 is called a *generating function. It encodes the entire Bernoulli numbers sequence*. Roughly speaking, generating functions transform problems about sequences into problems about functions. And by fooling around with this function we can explore the properties of the sequence it encodes. This is because we've got piles of mathematical machinery for manipulating functions (e.g., differentiation and integration).

Now, we give another example showing the power of generating functions. If, we observe carefully we will see that, except $B_1 = -1/2$, all the odd numbers B_{2n+1} for $n > 1$ are zeros. Why? Let's fool with the function[64]:

$$g(x) := \frac{x}{e^x - 1} - B_1 x = \frac{x}{e^x - 1} + \frac{x}{2} = \frac{x}{2}\frac{e^x + 1}{e^x - 1} = \frac{x}{2}\frac{e^x + 1}{e^x - 1}\frac{e^{-x/2}}{e^{-x/2}} = \frac{x}{2}\frac{e^{x/2} + e^{-x/2}}{e^{x/2} - e^{-x/2}}$$

We added the red term so that we can have a symmetric form ($e^x + 1$ is not symmetric but $e^{x/2} + e^{-x/2}$ is). It's easy to see that $g(-x) = g(x)$, thus it is an even function. Therefore, with Eq. 5.17.1

$$g(x) = 1 + \frac{B_2}{2!}x^2 + \frac{B_3}{3!}x^3 + \cdots \text{ is an even function} \implies B_{2n+1} = 0$$

George Pólya wrote in his book *Mathematics and plausible reasoning* in 1954 about generating functions:

> A generating function is a device somewhat similar to a bag. Instead of carrying many little objects detachedly, which could be embarrassing, we put them all in a bag, and then we have only one object to carry, the bag.

5.17.1 Ordinary Generating Function

The ordinary generating function for the sequence $(a_0, a_1, a_2, \ldots)$ is the power series[65]:

$$\boxed{G(a_n; x) = \sum_{n=0}^{\infty} a_n x^n} \tag{5.17.2}$$

The pattern here is simple: the nth term in the sequence (indexing from 0) is the coefficient of x^n in the generating function. There are a few other kinds of generating

---

64 Why this function?

65 I wrote this part based on the lecture notes of the MIT course *Mathematics for Computer Science*.

functions in common use (e.g., x/e^x-1, which is called an exponential generating function), but ordinary generating functions are enough to illustrate the power of the idea, so we will stick to them and from now on, generating function will mean the ordinary kind.

Remark A generating function is a "formal" power series in the sense that we usually regard x as a placeholder rather than a number. Only in rare cases will we actually evaluate a generating function by letting x take a real number value, so we generally ignore the issue of convergence.

Just looking at the definition in Eq. (5.17.2), there is no reason to believe that we've made any progress in studying anything. We want to understand a sequence $(a_0, a_1, a_2, \ldots)$; how could it possibly help to make an infinite series out of these! The reason is that frequently there's a simple, closed form expression for $G(a_n; x)$. The magic of generating functions is that we can carry out all sorts of manipulations on sequences by performing mathematical operations on their associated generating functions. Let's experiment with various operations and characterize their effects in terms of sequences.

Example 5.21 The generating function for the sequence $1, 1, 1, \ldots$ is $1/1-x$. This is because (if you still remember the geometric series)

$$\frac{1}{1-x} = 1 + x + x^2 + x^3 + \cdots \quad \text{where the coefs. of all } x^n \text{ is } 1$$

We can create different generating functions from this one. For example, if we replace x by $3x$, we have

$$\frac{1}{1-3x} = 1 + 3x + 9x^2 + 27x^3 + \cdots \quad \text{which generates } 1, 3, 9, 27, \ldots$$

Multiplying this with x, we get

$$\frac{x}{1-3x} = 0 + x + 3x^2 + 9x^3 + 27x^4 + \cdots \quad \text{which generates } 0, 1, 3, 9, 27, \ldots$$

which right-shift the original sequence (i.e., $1, 3, 9, 27, \ldots$) by one. We can multiply the GF by x^k to right-shift the sequence k times.

Solving difference equations. Assume that we have this sequence $1, 3, 7, 15, 31, \ldots$ which can be defined as

$$a_0 = 1, \quad a_1 = 3, \quad a_n = 3a_{n-1} - 2a_{n-2} \quad (n \geq 2)$$

The question is: what is the generating function for this sequence? Let's denote by $f(x)$ that function, thus we have (by definition of a generating function)

$$f(x) = 1 + 3x + 7x^2 + 15x^3 + 31x^4 + \cdots \tag{5.17.3}$$

Now, the recurrent relation ($a_n = 3a_{n-1} - 2a_{n-2}$) can be rewritten as $a_n - 3a_{n-1} + 2a_{n-2} = 0$, and we will multiply $f(x)$ by $-3x$ and also multiply $f(x)$ by $2x^2$ and add all up including $f(x)$:

$$\begin{aligned}
f(x) &= 1 + 3x + 7x^2 + 15x^3 + 31x^4 + \cdots + a_n x^n + \cdots \\
-3xf(x) &= 0 - 3x - 9x^2 - 21x^3 - 45x^4 + \cdots - 3a_{n-1}x^n + \cdots \\
2x^2 f(x) &= 0 + 0x + 2x^2 + 06x^3 + 14x^4 + \cdots + 2a_{n-2}x^n + \cdots
\end{aligned}$$

$$f(x)[1 - 3x + 2x^2] = 1 \Longrightarrow f(x) = \frac{1}{1 - 3x + 2x^2}$$

where all the columns add up to zero except the first one, because of the recurrence relation $a_n - 3a_{n-1} + 2a_{n-2} = 0$.

But, why having the generating function is useful? Because it allows us to find a formula for a_n; thus, we no longer need to use the recurrence relation to get a_n starting from $a_0, a_1, \ldots$ all the way up to a_{n-2}.[66] The trick is to rewrite $f(x)$ in terms of simpler functions (using the partial fraction decomposition discussed in Sect. 4.7.8) and then replace these simpler functions by their corresponding power series. Now, we can decompose $f(x)$ easily with the "`apart`" function in `SymPy`[67]

$$f(x) = \frac{1}{1 - 3x + 2x^2} = \frac{-1}{1 - x} + \frac{2}{1 - 2x}$$

Next, we write the series of these two fractional functions[68]:

$$\begin{aligned}
\frac{-1}{1 - x} &= -1 - x - x^2 - x^3 - \cdots \quad \Longrightarrow b_n = -1 \\
\frac{2}{1 - 2x} &= 2 + 4x + 8x^2 + 16x^3 + \cdots \Longrightarrow c_n = 2^{n+1}
\end{aligned}$$

Thus, we can determine a_n:

$$a_n = 2^{n+1} -$$

To conclude, generating functions provide a systematic method to solving recurrence/difference equations. At this step, I recommend you to apply this method to the Fibonacci sequence, discussed in Sect. 2.10, to re-discover the Binet's formula and many other properties of this famous sequence.

---

[66] So, we want a Ferrari instead of a Honda CRV.

[67] Check Sect. 3.27 if you're not sure about `SymPy`.

[68] Note that we also know the series of $1/(1 - 3x + 2x^2)$, but that series is simply the RHS of Eq. (5.17.3).

Evaluating sums. Suppose we need to evaluate this sum $s_n = \sum_{k=0}^{n} a_k$, which is the sum of n elements of a sequence $(a_0, a_1, \ldots)$. Herein, I present a technique to compute such sums using generating functions. To this end, we need to use the Cauchy product formula for two power series. Recall the Cauchy product formula for two power series, see Eq. 5.17.1 and the surrounding for a discussion:

$$\left(\sum_{n=0}^{\infty} a_n x^n\right)\left(\sum_{m=0}^{\infty} b_m x^m\right) = \sum_{n=0}^{\infty}\left(\sum_{k=0}^{n} a_k b_{n-k}\right) x^n$$

Now, we consider the product of two sequences. For two sequences and their associated generating functions,

$$(a_0, a_1, \ldots) \longleftrightarrow A(x), \quad (b_0, b_1, \ldots) \longleftrightarrow B(x)$$

we will obtain a new sequence by multiplying the two given sequences:

$$(c_0, c_1, \ldots) \longleftrightarrow A(x)B(x), \quad c_n = \sum_{k=0}^{n} a_k b_{n-k}$$

using the above mentioned Cauchy product formula.

As we aim for $\sum_{k=0}^{n} a_k$ not $\sum_{k=0}^{n} a_k b_{n-k}$, we need to work with $B(x)$ such that $b_k = 1$ for all k. That special $B(x)$ is $B(x) = 1/1-x$, all b_i's equal one, and thus we have (move from c to s for sum)

$$(s_0, s_1, \ldots) \longleftrightarrow \frac{A(x)}{1-x}, \quad \boxed{s_n = \sum_{k=0}^{n} a_k} \tag{5.17.4}$$

This is called the summation rule as it allows us to compute the sum in the box: that sum is simply the coefficient of the term x^n of the function $A(x)/1-x$. We have turned the problem of sum evaluation to a problem of finding the coefficient of a function! Math is super cool, isn't it? I provide one example below to demonstrate the idea.

Example 5.22 Suppose we want to compute the sum of the first n squares

$$s_n = \sum_{i=0}^{n} i^2$$

All we need to do is (1) to determine $A(x)$, which is the generating functions for the sequence $(0, 1, 4, 9, \ldots)$, and (2) multiply $A(x)$ with $1/1-x$, (3) find the coefficients of that function. First, we have

$$A(x) = \frac{x(1+x)}{(1-x)^3}$$

Therefore,

$$(s_0, s_1, \ldots) \longleftrightarrow \frac{x(1+x)}{(1-x)^4}$$

Which means that $\sum_{i=0}^{n} i^2$ is nothing but the coefficient of x^n in $F(x) = x(1+x)/(1-x)^4$. This is a complicated function, let's break it into simpler ones:

$$F(x) = \frac{x(1+x)}{(1-x)^4} = \frac{x}{(1-x)^4} + \frac{x^2}{(1-x)^4}$$

Now, we observe that the coefficient of x^n in $F(x)$ is the sum of the coefficient x^{n-1} in $1/(1-x)^4$, and the coefficient x^{n-2} in $1/(1-x)^4$. The problem boils down to finding the coefficient of x^k in $G(x) = 1/(1-x)^4$. How to do that? Taylor series is the answer. Recall that

$$f(x) = f(0) + \frac{f'(0)}{1!}x + \frac{f''(0)}{2!}x^2 + \cdots + \frac{f^{(n)}(0)}{n!} + \cdots$$

What does it tell us? It says that the nth coefficient of $f(x)$ is $f^{(n)}(0)/n!$ Now, we need the nth derivative of $G(x)$. Let's compute the first, second derivatives and so on to find the pattern:

$$\left.\begin{aligned} G'(x) &= \frac{4}{(1-x)^5} \\ G''(x) &= \frac{4 \cdot 5}{(1-x)^6} \\ G'''(x) &= \frac{4 \cdot 5 \cdot 6}{(1-x)^7} \end{aligned}\right\} \Longrightarrow G^{(n)}(x) = \frac{(n+3)!}{6(1-x)^{n+4}}$$

Hence, the nth coefficient of $G(x)$ is

$$\frac{G^{(n)}(0)}{n!} = \frac{(n+3)!}{6n!} = \frac{(n+3)(n+2)(n+1)}{6}$$

And from that we can determine the $n-1$th and $n-2$th coefficients, and finally s_n:

$$s_n = \frac{n(n+1)(n+2)}{6} + \frac{(n-1)n(n+1)}{6} = \frac{n(n+1)(2n+1)}{6}$$

Same result as in Eq. (2.6.8).

5.17.2 Counting with Generating Functions

Using the binomial theorem (Eq. (2.27.2)), we can express $(1+x)^k$ as follows:

$$(1+x)^k = \binom{k}{0}x^0 + \binom{k}{1}x^1 + \binom{k}{2}x^2 + \cdots + \binom{k}{k}x^k \tag{5.17.5}$$

Therefore, the coefficient of x^n in $(1+x)^k$ is $\binom{k}{n}$, where $\binom{k}{n}$ represents the number of ways to choose n objects from a set of k elements. With this, we can use the concept of generating functions to solve counting problems.

Example 1. How many integer solutions does the equation $a+b+c=6$ have such that $-1 \le a \le 2$ and $1 \le b, c \le 4$? Note that we can solve this problem using the stars and bars method. However, the solution would not be particularly elegant. Here, we will use generating functions. Since a can only take values in $\{-1, 0, 1, 2\}$, its generating function is $x^{-1}+x^0+x^1+x^2$, i.e., the possible values of a are the exponents. Similarly, the generating function for b is $x^1+x^2+x^3+x^4$, and for c (as c has the same range as b). Thus, the generating function for a, b, c is the product of these functions:

$$G(x) = (x^{-1}+x^0+x^1+x^2)(x^1+x^2+x^3+x^4)(x^1+x^2+x^3+x^4)$$

The solution is the coefficient of x^6 in $G(x)$. If you're feeling lazy, you can use `SymPy` (the `series` command) and immediately see $\cdots + 12x^6 + \cdots$. Therefore, the answer is 12, which matches the solution obtained using the stars and bars method! If you're adventurous enough, we can do it using our brain: first, write $x^{-1}+x^0+x^1+x^2 = (1/x)(1+x+x^2+x^3)$, and do the same for $x^1+x^2+x^3+x^4$[69]:

$$G(x) = \left(\frac{1}{x}\right)(1+x+x^2+x^3)[x(1+x+x^2+x^3)]^2 = x(1+x+x^2+x^3)^3 = x\left(\frac{1-x^4}{1-x}\right)^3$$

Note that $(1-x^4)^3$ can be expanded easily (because it's simple), and for $(1-x)^{-3}$ we can use the generalized binomial theorem:

$$\begin{aligned} G(x) &= x\left(\frac{1-x^4}{1-x}\right)^3 = x(1-3x^4+3x^8-x^{12})(1-x)^{-3} \\ &= (x-3x^5+3x^9-x^{13})(() + \Box_1 x^1 + \cdots + \Box_2 x^5 + ()x^6 + \cdots) \end{aligned}$$

Remember: don't forget our goal is the coefficient of x^6 (only that, no one else is needed) in $(x-3x^5+3x^9-x^{13})(1-x)^{-3}$. Therefore, we write $G(x)$ as

[69] Why do this? Because we want to use $1+x+x^2 = {}^{1-x^3}/_{1-x}$.

$$G(x) = (1x - 3x^5 + 3x^9 - x^{13})(() + \Box_1 x^1 + \cdots + \Box_2 x^5 + ()x^6 + \cdots)$$
$$= \cdots + 1\Box_2 x^6 + (-3)\Box_1 x^6 + \cdots$$

What gives us these coefficients $\Box_1, \Box_2$? Don't forget Newton! The solution is the generalized binomial theorem, spsexrefeq:generalizedspsbinomial, which gives:

$$(1)(-C_5^{-3}) + (-3)C_1^{-3} = 12, \quad C_5^{-3} = \frac{(-3)(-3-1)(-3-2)(-3-3)(-3-4)}{5!}$$

5.17.3 *Moment Generating Functions*

There are several reasons to study moments and moment generating functions. One of them is that moment generating functions make our life easier and the second is that these functions can be used to prove the central limit theorem.

Moments, central moments. To motivate the introduction of moments in probability, let's look at how the expected value and the variance were defined:

$$\mu = \mathbb{E}[X] = \mathbb{E}[X^1], \quad \text{Var}(X) = \mathbb{E}[(X - \mu)^2]$$

It is then logical to define the kth moment of a random variable X as $\mu_k = \mathbb{E}[X^k]$. Why? Because the first moment is then the expected value, the second moment does not give us the variance directly, but indirectly: $\sigma^2 = \mu_2 - \mu_1^2$. And mathematicians also define the kth *central moment* of a random variable X as $\mathbb{E}[X - \mu]^k$. Thus, the variance is simply the second central moment.

Moment generating functions. The moment generating function (MGF) of a random variable (discrete or continuous) X is simply the expected value of e^{tX}:

$$\boxed{m(t) = \mathbb{E}[e^{tX}]} \tag{5.17.6}$$

Now, we will elaborate $m(t)$ to reveal the reason behind its name (and its definition). The idea is to replace e^{tX} by its Taylor series, then applying the linearity of the expected value and we shall see all the moments μ_k:

$$\begin{aligned} m(t) &= \mathbb{E}\left[1 + tX + \frac{(tX)^2}{2!} + \cdots + \frac{(tX)^k}{k!} + \cdots\right] \\ &= 1 + t\mathbb{E}[X] + \frac{t^2\mathbb{E}[X^2]}{2!} + \cdots + \frac{t^k\mathbb{E}[X^k]}{k!} + \cdots \\ &= 1 + \mu_1 t + \frac{\mu_2}{2!}t^2 + \cdots + \frac{\mu_k}{k!}t^k + \cdots \end{aligned}$$

Compared with Eq. (5.17.2), which is the ordinary generating function for the sequence $(a_0, a_1, a_2, \ldots)$, we can obviously see why $m(t)$, as defined in Eq. (5.17.6), is called the moment generating function; it encodes all the moments μ_k of X. How? Simply by differentiating $m(t)$ and evaluate it at $t = 0$, we can retrieve any moment. For example, $m'(0) = \mu_1$, and $m''(0) = \mu_2$.

We can now give a full definition of the moment generating function of either a discrete or continuous random variable:

$$\begin{aligned}\text{Discrete RV} \quad m(t) &= \sum_x e^{tx} P(x) \\ \text{Continous RV} \; m(t) &= \int_{-\infty}^{\infty} e^{tx} f(x)dx\end{aligned} \tag{5.17.7}$$

Next, we are going to consider some examples to see how powerful the MGFs are.

Example 5.23 We consider the geometric series, compute its moment generating function, and see what we can get from it. First, $m(t)$ is given by[70]

$$m(t) = \sum_{k=1}^{\infty} e^{tk} q^{k-1} p = \frac{p}{q} \sum_{k=1}^{\infty} (e^t q)^k = \frac{p}{q} \frac{e^t q}{1 - qe^t} = \frac{pe^t}{1 - qe^t}$$

Now, it is easy to compute the expected value and variance:

$$\mathbb{E}[X] = \mu_1 = m'(0) = \frac{p}{(1-q)^2} = \frac{1}{p}, \quad \left(m'(t) = \frac{pe^t}{(1-qe^t)^2}\right)$$

You can compare this and the procedure in Example 5.16 and conclude for yourself which way is easier. You can use this method to fill in Table 5.7 for other distributions.

Example 5.24 We now determine the MGF of a standard normal variable Z. We use the definition, Eq. (5.17.7), to compute it:

$$\begin{aligned} m_Z(t) &= \int_{-\infty}^{\infty} e^{tz} \frac{1}{\sqrt{2\pi}} e^{-z^2/2} dz = e^{t^2/2} \int_{-\infty}^{\infty} \frac{1}{\sqrt{2\pi}} e^{-1/2(z^2 - 2tz + t^2)} dz \\ &= e^{t^2/2} \int_{-\infty}^{\infty} \frac{1}{\sqrt{2\pi}} e^{-1/2(z-t)^2} dz = e^{t^2/2} \end{aligned}$$

Noting the red integral is simply one: it is the probability density function of $N(t, 1)$.

[70] The red term is a geometric series.

5.17.4 Properties of Moment Generating Functions

Consider $Y = X_1 + X_2$, where X_1, X_2 are independent random variables. Now we compute the moment generating function of Y:

$$m_Y(t) = \mathbb{E}[e^{tY}] = \mathbb{E}[e^{t(X_1+X_2)}] = \mathbb{E}[e^{tX_1}e^{tX_2}] \stackrel{(a)}{=} \mathbb{E}[e^{tX_1}]\mathbb{E}[e^{tX_2}] = m_{X_1}(t)m_{X_2}(t) \tag{5.17.8}$$

which says that the MGF of the sum $\sum_i X_i$ is the product of the MGFs of X_i. This is known as the convolution rule of the MGF. Note that we have used $\mathbb{E}[XY] = \mathbb{E}[X]\mathbb{E}[Y]$ for two independent RVs X and Y in step (a). It is obvious that we're going to generalize this result:

$$Y = \sum_{i=1}^{n} c_i X_i \Longrightarrow m_Y(t) = \prod_{i=1}^{n} m_{X_i}(c_i t) \tag{5.17.9}$$

This result gives us a powerful tool for determining the distribution of the sum of independent random variables.

Now, we derive another property of the MGF. Consider now a transformation $Y = aX + b$, and see what is the MGF of Y, especially how it is related to the MGF of X:

$$m_Y(t) = \mathbb{E}[e^{tY}] = \mathbb{E}[e^{t(aX+b)}] = \mathbb{E}[e^{taX}e^{bt}] = e^{bt}\mathbb{E}[e^{(at)X)}] = e^{bt}m_X(at) \tag{5.17.10}$$

We will use all these results in the next section when we prove the central limit theorem. I am not sure if they were developed for this or not. But note that, for mathematicians considering the sum of X_1, X_2 or $aX + b$ are something very natural to do.

From Example 5.24 and Eq. (5.17.10) we can determine the MGF for $Y \sim N(\mu, \sigma^2)$. The idea is to use

$$\begin{aligned} X \sim N(0, 1) &: m_X(t) = e^{t^2/2} \\ Y = aX + b &: m_Y(t) = e^{bt}m_X(at) = e^{bt}e^{a^2t^2/2} \end{aligned}$$

And we know from Sect. 5.11.3 that we have to use this transformation $Y = \sigma X + \mu$ to have $Y \sim N(\mu, \sigma^2)$. Therefore, we obtain

$$\boxed{Y \sim N(\mu, \sigma^2): \quad m_Y(t) = \exp\left(\mu t + \frac{\sigma^2 t^2}{2}\right)} \tag{5.17.11}$$

With this result and Eq. (5.17.9), we have the following theorem.

Theorem 5.17.1 *Let $X_1, X_2, \ldots, X_n$ be independent normal variables with expected values $\mu_1, \ldots, \mu_n$ and variances $\sigma_1^2, \ldots, \sigma_n^2$. Then, the random variable Y which is a linear combination of X_i, i.e.,*

$$Y = \sum_{i=1}^{n} c_i X_i$$

follows the normal distribution:

$$N\left(\sum_{i=1}^{n} c_i \mu_i, \sum_{i=1}^{n} c_i^2 \sigma_i^2\right)$$

This theorem indicates that the sum of independent normal random variables is itself a normal random variable.

Proof Certainly we have to use Eq. (5.17.9) and (5.17.11). To ease the presentation let's assume $n = 2$, then

$$\begin{aligned} m_Y(t) &= \exp\left(c_1\mu_1 t + \frac{c_1^2\sigma_1^2 t^2}{2}\right) \exp\left(c_2\mu_2 t + \frac{c_2^2\sigma_2^2 t^2}{2}\right) \\ &= \exp\left(c_1\mu_1 t\right) \exp\left(\frac{c_1^2\sigma_1^2 t^2}{2}\right) \exp\left(c_2\mu_2 t\right) \exp\left(\frac{c_2^2\sigma_2^2 t^2}{2}\right) \\ &= \exp((c_1\mu_1 + c_2\mu_2)t) \exp\left((c_1^2\sigma_1^2 + c_2^2\sigma_2^2)\frac{t^2}{2}\right). \end{aligned}$$

■

5.17.5 *Proof of the Central Limit Theorem*

The central limit theorem is probably the most beautiful result in the mathematical theory of probability. Thus, if I cannot somehow see its proof, I feel something is missing. Surprisingly, the proof is based on the moment generating function concept. First, the CLT is recalled now. Let $X_1, X_2, \ldots, X_n$ be iid random variables with expected value μ and variance σ^2. Then, the random variable

$$S_n^* = \frac{S_n - n\mu}{\sqrt{n}\sigma}, \quad S_n = X_1 + X_2 + \cdots + X_n$$

converges in distribution to the standard normal random variable as n goes to infinity.

The plan of the proof: (i) compute the MGF of S_n^*, (ii) show that when n is large this MGF is approximately the MGF of $N(0, 1)$, i.e., $e^{t^2/2}$ (according to Example 5.24) and (iii) if two variables have the same MGFs, then they have the same probability distribution (we need to prove this, but it is reasonable so I accept it). Quite a simple plan, for a big theorem in probability.

The first thing is to write S_n^* as the sum of something:

$$\begin{aligned} S_n^* &= \frac{X_1 + X_2 + \cdots + X_n - n\mu}{\sqrt{n}\sigma} = \frac{(X_1 - \mu) + (X_2 - \mu) + \cdots + (X_n - \mu)}{\sqrt{n}\sigma} \\ &= \sum_{i=1}^{n} \frac{1}{\sqrt{n}} \left(\frac{X_i - \mu}{\sigma} \right) = \sum_{i=1}^{n} \frac{1}{\sqrt{n}} X_i^*, \quad X_i^* = \frac{X_i - \mu}{\sigma} \end{aligned}$$

Why this particular form? Because we have the convolution rule that works for a sum. Using Eq. (5.17.8), the MGF of S_n^* is simply:

$$\begin{aligned} m_{S_n^*}(t) &= \left(m_{X^*/\sqrt{n}}(t)\right)^n \quad \text{(Eq. (5.17.8))} \\ &= \left(m_{X^*}(t/\sqrt{n})\right)^n \quad \text{(Eq. (5.17.10) with } a = 1/\sqrt{n}, b = 0) \end{aligned} \tag{5.17.12}$$

Now, we use Taylor's series to approximate $m_{X^*}(t/\sqrt{n})$ when n is large:

$$m_{X^*}(t/\sqrt{n}) \approx m_{X^*}(0) + m'_{X^*}(0)\frac{t}{\sqrt{n}} + m''_{X^*}(0)\frac{t^2}{2n} = 1 + \frac{t^2}{2n}$$

because $m'_{X^*}(0) = \mathbb{E}[X^*] = 0$ and $m''_{X^*}(0) = \mathbb{E}[(X^*)^2] = 1$. Therefore,[71]

$$m_{S_n^*}(t) \approx \left(1 + \frac{t^2}{2n}\right)^n \approx e^{t^2/2}$$

So, we have proved that when n is large the MGF of S_n^* is approximately the MGF of $N(0, 1)$. Thus, S_n^* has a standard normal distribution. Q.E.D.

[71] This is because $(1 + a/n)^n \to e^a$ when $n \to \infty$. Check Eq. (4.15.20) if this is not clear.

History note 5.4: The Bernoullis

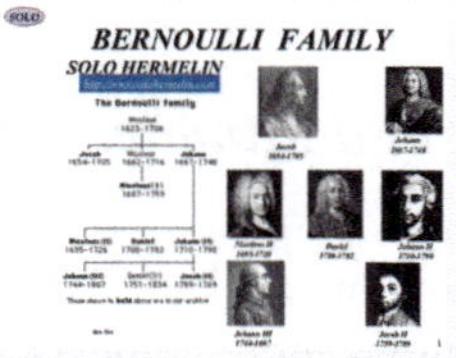

The city of Basel in Switzerland was one of many free cities in Europe and by the 17th century had become an important center of trade and commerce/ The University of Basel became a noted institution largely through the same of an extraordinary family–the Bernouillis. This family had come from Antwerp (Belgium) to Basel and the founder of the mathematical dynasty was Nicholas Bernoulli (1687-1759). He had three sons, two of whom, James or Jacob (1654-1705) and Johann or John (1667–1748), became famous mathematicians. Both were pupils of Leibnitz. James was a professor at Basel until his death in 1705. John, who had been a professor at Groningen, replaced him. It was John who gave Euler special instruction on Saturdays when Euler was young. Johann had also a most famous private pupil Guillaume François Antoine, Marquis de l'Hospital wrote the first ever calculus textbook, which was actually the notes that he had made from his lesson with Bernoulli.
John Bernoulli had three sons. Two of them, Nicolas II and Daniel were mathematicians who befriended Euler. Both went to St. Petersburg in 1725 and Daniel secured a position for Euler at the Russian Academy.
The Bernoulli family had a habit of re-using first names through the generations and this leads to a great deal of confusion amongst people trying to understand the history of 18th century mathematics and physics (me included).

5.18 Multivariate Normal Distribution

The normal distribution is very important in probability theory and it shows up in many different applications. We have discussed a single normal random variable previously in Sect. 5.11.3. In this section we shall now talk about two or more normal random variables. Two important concepts are multivariate normal distribution and the special case for two variables: the bivariate normal distribution. The bivariate distribution helps us understand more the general multivariate case, especially with the use of 3D plots and contour plots.

You need some knowledge of linear algebra to read this section; check Chap. 9 if needed. In what follows $\square^\top$ denotes the transpose of $\square$. And $\boldsymbol{x}^\top \boldsymbol{y}$ is the dot product of two n-vectors $\boldsymbol{x}$ and $\boldsymbol{y}$. That is $\boldsymbol{x}^\top \boldsymbol{y} = \sum_{i=1}^n x_i y_i$. The zero vector is denoted by $\mathbf{0} = (0, 0, .., 0)^\top$ and the identity matrix by $\mathbf{I}$.

5.18.1 Random Vectors and Random Matrices

When dealing with multiple random variables, it is useful to use vector and matrix notations. This makes the formulas more compact and lets us use facts from linear algebra. A random vector $\boldsymbol{X} \in \mathbb{R}^n$ is a vector $(X_1, X_2, \ldots, X_n)^\top$ of scalar random variables X_i.

The expected value vector or the mean vector of the random vector $\boldsymbol{X}$ is simply the vector containing the expected value of each random variable:

$$\mathbb{E}[\boldsymbol{X}] = (\mathbb{E}[X_1], \mathbb{E}[X_2], \ldots, \mathbb{E}[X_n])^\top \tag{5.18.1}$$

All the properties of the expected value work for random vectors. For example, we know that $\mathbb{E}[aX + b] = a\mathbb{E}[X] + b$, which is Eq. (5.10.38). We also have that,

$$\mathbb{E}[\mathbf{A}\boldsymbol{X} + \boldsymbol{b}] = \mathbf{A}\mathbb{E}[\boldsymbol{X}] + \boldsymbol{b}, \quad \mathbf{A} \in \mathbb{R}^{mn}, \ \boldsymbol{b} \in \mathbb{R}^m \tag{5.18.2}$$

If you need a proof, then consider the ith element of the vector $\mathbf{A}\boldsymbol{X} + \boldsymbol{b}$, compute its expected value, using $\mathbb{E}[aX + b] = a\mathbb{E}[X] + b$. Details: $\mathbb{E}(A_{ij}X_j + b_i) = A_{ij}\mathbb{E}[X_j] + b_i$. Note that: a vector is just a bunch of numbers glued together!

If we have random vectors, then we have random matrices. Not a big deal. Actually we have met one random matrix: the covariance matrix. Now, we provide a formal definition of this matrix. Consider again the random vector $\boldsymbol{X} = (X_1, X_2, \ldots, X_n)^\top$, we then have (all are n vectors)

$$\boldsymbol{X} = \begin{bmatrix} X_1 \\ X_2 \\ \vdots \\ X_n \end{bmatrix}, \quad \mathbb{E}[\boldsymbol{X}] = \begin{bmatrix} \mathbb{E}[X_1] \\ \mathbb{E}[X_2] \\ \vdots \\ \mathbb{E}[X_n] \end{bmatrix}, \quad \boldsymbol{X} - \mathbb{E}[\boldsymbol{X}] = \begin{bmatrix} X_1 - \mathbb{E}[X_1] \\ X_2 - \mathbb{E}[X_2] \\ \vdots \\ X_n - \mathbb{E}[X_n] \end{bmatrix}$$

Next, we build the matrix $\mathbf{A} = (\boldsymbol{X} - \mathbb{E}[\boldsymbol{X}])(\boldsymbol{X} - \mathbb{E}[\boldsymbol{X}])^\top$, which is an $n \times n$ matrix:

$$\mathbf{A} = \begin{bmatrix} (X_1 - \mathbb{E}[X_1])^2 & (X_1 - \mathbb{E}[X_1])(X_2 - \mathbb{E}[X_2]) & \cdots & (X_1 - \mathbb{E}[X_1])(X_n - \mathbb{E}[X_n]) \\ (X_2 - \mathbb{E}[X_2])(X_1 - \mathbb{E}[X_1]) & (X_2 - \mathbb{E}[X_2])^2 & \cdots & (X_2 - \mathbb{E}[X_2])(X_n - \mathbb{E}[X_n]) \\ \vdots & \vdots & \ddots & \\ (X_n - \mathbb{E}[X_n])(X_1 - \mathbb{E}[X_1]) & (X_n - \mathbb{E}[X_n])(X_2 - \mathbb{E}[X_2]) & \cdots & (X_n - \mathbb{E}[X_n])^2 \end{bmatrix}$$

Finally, the covariance matrix of $\boldsymbol{X}$, denoted by $\mathbf{C}_X$, is simply defined as $\mathbb{E}[\mathbf{A}]$:

$$\mathbf{C}_X = \mathbb{E}[(\boldsymbol{X} - \mathbb{E}[\boldsymbol{X}])(\boldsymbol{X} - \mathbb{E}[\boldsymbol{X}])^\top] = \begin{bmatrix} \text{Var}(X_1) & \text{Cov}(X_1, X_2) & \cdots & \text{Cov}(X_1, X_n) \\ \text{Cov}(X_2, X_1) & \text{Var}(X_2) & \cdots & \text{Cov}(X_2, X_n) \\ \vdots & \vdots & \ddots & \vdots \\ \text{Cov}(X_n, X_1) & \text{Cov}(X_n, X_2) & \cdots & \text{Var}(X_n) \end{bmatrix}$$

Let $\boldsymbol{X}$ be an n-dimensional random vector and the random vector $\boldsymbol{Y} \in \mathbb{R}^m$ be defined as $\boldsymbol{Y} = \mathbf{A}\boldsymbol{X} + \boldsymbol{b}$, where $\mathbf{A}$ is a fixed $m \times n$ matrix and $\boldsymbol{b}$ is a fixed m-dimensional vector. And we want to compute $\mathbf{C}_Y$—an $m \times m$ matrix—in terms of $\mathbf{C}_X$. The result is

$$\mathbf{C}_Y = \mathbf{A}\mathbf{C}_X\mathbf{A}^\top = \mathbf{A}^\top\mathbf{C}_X\mathbf{A} \tag{5.18.3}$$

Proof First we compute $\mathbb{E}[\boldsymbol{Y}] = \mathbf{A}\mathbb{E}[\boldsymbol{X}] + \boldsymbol{b}$, then using the definition of the covariance matrix to get

$$\begin{aligned}
\mathbf{C}_Y &= \mathbb{E}[(\boldsymbol{Y} - \mathbb{E}[\boldsymbol{Y}])(\boldsymbol{Y} - \mathbb{E}[\boldsymbol{Y}])^\top] \\
&= \mathbb{E}[(\mathbf{A}\boldsymbol{X} + \boldsymbol{b} - \mathbf{A}\mathbb{E}[\boldsymbol{X}] - \boldsymbol{b})(\mathbf{A}\boldsymbol{X} + \boldsymbol{b} - \mathbf{A}\mathbb{E}[\boldsymbol{X}] - \boldsymbol{b})^\top] \\
&= \mathbb{E}[(\mathbf{A}(\boldsymbol{X} - \mathbb{E}[\boldsymbol{X}]))(\mathbf{A}(\boldsymbol{X} - \mathbb{E}[\boldsymbol{X}]))^\top] \\
&= \mathbb{E}[\mathbf{A}(\boldsymbol{X} - \mathbb{E}[\boldsymbol{X}])(\boldsymbol{X} - \mathbb{E}[\boldsymbol{X}])^\top\mathbf{A}^\top] \quad (\text{for } (\mathbf{AB})^\top = \mathbf{B}^\top\mathbf{A}^\top) \\
&= \mathbf{A}\mathbb{E}[(\boldsymbol{X} - \mathbb{E}[\boldsymbol{X}])(\boldsymbol{X} - \mathbb{E}[\boldsymbol{X}])^\top]\mathbf{A}^\top \quad (\text{linearity of } \mathbb{E})
\end{aligned}$$

If the last equality was not clear, the proof is similar to how we have proved Eq. (5.18.1): it is always based on the fact that $\mathbb{E}[aX + b] = a\mathbb{E}[X] + b$. ■

We have proved in Sect. 5.12.8 that the covariance matrix is semi-positive definite for discrete random variables. Herein, we prove this for any random variables.

Theorem 5.18.1 *Let X be a random vector of n elements, then its covariance matrix is always symmetric and semi-positive definite (PSD).*

Proof A matrix $\mathbf{A}$ is PSD if $\boldsymbol{x}^\top\mathbf{A}\boldsymbol{x} \geq 0$ for all $\boldsymbol{x}$. We use this and the fact that $\mathbb{E}[Y^2] \geq 0$ for all Y to prove the theorem. So, let's define $Y = \boldsymbol{b}^\top(\boldsymbol{X} - \mathbb{E}[\boldsymbol{X}])$ with $\boldsymbol{b}$ being any fixed vector, then

$$\begin{aligned}
0 \leq \mathbb{E}[Y^2] = \mathbb{E}[YY^\top] &= \mathbb{E}[\boldsymbol{b}^\top(\boldsymbol{X} - \mathbb{E}[\boldsymbol{X}])(\boldsymbol{X} - \mathbb{E}[\boldsymbol{X}])^\top\boldsymbol{b}] \\
&= \boldsymbol{b}^\top\mathbb{E}[(\boldsymbol{X} - \mathbb{E}[\boldsymbol{X}])(\boldsymbol{X} - \mathbb{E}[\boldsymbol{X}])^\top]\boldsymbol{b} \quad (\text{linearity of } \mathbb{E}) \\
&= \boldsymbol{b}^\top\mathbf{C}_X\boldsymbol{b}
\end{aligned}$$

Thus, the covariance matrix $\mathbf{C}_X$ is always symmetric and semi-positive definite. ■

5.18.2 Functions of Random Vectors

Let $\boldsymbol{X}$ be an n-dimensional random vector with the joint PDF $f_{\boldsymbol{X}}(\boldsymbol{x})$. Let $G : \mathbb{R}^n \to \mathbb{R}^n$ be a continuous and invertible function with continuous partial derivatives, and let $H = G^{-1}$. Suppose that the random variable $\boldsymbol{Y}$ is given by $\boldsymbol{Y} = G(\boldsymbol{X})$, or $\boldsymbol{X} = G^{-1}(\boldsymbol{Y}) = H(\boldsymbol{Y})$. The question is: what is $f_{\boldsymbol{Y}}(\boldsymbol{y})$?

$$\boldsymbol{X} = \begin{bmatrix} X_1 \\ X_2 \\ \vdots \\ X_n \end{bmatrix} = \begin{bmatrix} H_1(Y_1, Y_2, \ldots, Y_n) \\ H_2(Y_1, Y_2, \ldots, Y_n) \\ \vdots \\ H_n(Y_1, Y_2, \ldots, Y_n) \end{bmatrix}$$

To answer that question we use the method described in Sect. 5.14 and the Jacobian in Sect. 6.8.6. The basic idea is the same:

$$P(\boldsymbol{Y} \in A') = P(\boldsymbol{X} \in A) = \int_A f_{\boldsymbol{X}}(\boldsymbol{x}) dA = \int_{A'} f_{\boldsymbol{X}}(H(\boldsymbol{y}))|J| dA \tag{5.18.4}$$

where J is the Jacobian, which is given by

$$J = \det \begin{bmatrix} \frac{\partial H_1}{\partial y_1} & \frac{\partial H_1}{\partial y_2} & \cdots & \frac{\partial H_1}{\partial y_n} \\ \frac{\partial H_2}{\partial y_1} & \frac{\partial H_2}{\partial y_2} & \cdots & \frac{\partial H_2}{\partial y_n} \\ \vdots & \vdots & \ddots & \vdots \\ \frac{\partial H_n}{\partial y_1} & \frac{\partial H_n}{\partial y_2} & \cdots & \frac{\partial H_n}{\partial y_n} \end{bmatrix}$$

The red term is the pdf of $\boldsymbol{Y}$. So, we have

$$f_{\boldsymbol{Y}}(\boldsymbol{y}) = f_{\boldsymbol{X}}(H(\boldsymbol{y}))|J|, \quad J = \det \mathbf{A}, \quad A_{ij} = \frac{\partial H_i}{\partial y_j} \tag{5.18.5}$$

Example 5.25 Let $\boldsymbol{X}$ be an n-dimensional random vector with $f_{\boldsymbol{X}}(\boldsymbol{x})$ and the random vector $\boldsymbol{Y} \in \mathbb{R}^m$ be defined as $\boldsymbol{Y} = \mathbf{A}\boldsymbol{X} + \boldsymbol{b}$, where $\mathbf{A}$ is a fixed $m \times n$ matrix and $\boldsymbol{b}$ is a fixed m-dimensional vector. And we want to compute $f_{\boldsymbol{Y}}(\boldsymbol{y})$.

First, we have $\boldsymbol{X} = \mathbf{A}^{-1}(\boldsymbol{Y} - \boldsymbol{b}) = H(\boldsymbol{Y})$. Equation (5.18.5)

$$f_{\boldsymbol{Y}}(\boldsymbol{y}) = |\det \mathbf{A}^{-1}| f_{\boldsymbol{X}}(\mathbf{A}^{-1}(\boldsymbol{y} - \boldsymbol{b})) = \frac{1}{|\det \mathbf{A}|} f_{\boldsymbol{X}}(\mathbf{A}^{-1}(\boldsymbol{y} - \boldsymbol{b})).$$

5.18.3 Multivariate Normal Distribution

Let $\boldsymbol{X} = (X_1, X_2, \ldots, X_n)^\top$ be a random vector and $\boldsymbol{t} = (t_1, t_2, \ldots, t_n)^\top$ is a vector in $\mathbb{R}^n$. The moment generating function for $\boldsymbol{X}$ is defined by (this is also called the joint moment generating function)

$$M_{\boldsymbol{X}}(\boldsymbol{t}) = \mathbb{E}[e^{\boldsymbol{t}^\top \boldsymbol{X}}] = \mathbb{E}[\exp(t_1 X_1 + \cdots + t_n X_n)] = \mathbb{E}\left[\exp\left(\sum_{i=1}^{n} t_i X_i\right)\right] \tag{5.18.6}$$

The plan of developing the multivariate normal distribution is to repeat what we have done for the univariate normal distribution. Let me recall what we have done:

$$\begin{aligned} X &\sim N(0,1) : m_X(t) = e^{t^2/2} \\ Y &= aX + b : m_Y(t) = e^{bt} m_X(at) = e^{bt} e^{a^2t^2/2} \end{aligned}$$

And with this transformation $Y = \sigma X + \mu$, to have $Y \sim N(\mu, \sigma^2)$, we obtain the MGF of $N(\mu, \sigma^2)$

$$Y \sim N(\mu, \sigma^2) : \quad m_Y(t) = \exp\left(\mu t + \frac{\sigma^2 t^2}{2}\right)$$

Let's carry out that plan. The first step is to consider n independent standard normal variables $Z_1, Z_2, \ldots, Z_n$, i.e., $Z_i \sim N(0, 1)$. Let $\boldsymbol{Z} = (Z_1, Z_2, \ldots, Z_n)^\top$ be a random vector and $\boldsymbol{t} = (t_1, t_2, \ldots, t_n)^\top$. Now, we are going to compute $M_{\boldsymbol{Z}}(\boldsymbol{t})$ using Eq. (5.18.6):

$$\begin{aligned} M_{\boldsymbol{Z}}(\boldsymbol{t}) &= \mathbb{E}[e^{\boldsymbol{t}^\top \boldsymbol{Z}}] = \mathbb{E}[e^{t_1 Z_1 + \cdots + t_n Z_n}] \\ &= \mathbb{E}[e^{t_1 Z_1} \times e^{t_2 Z_2} \times \cdots \times e^{t_n Z_n}] \\ &= \mathbb{E}[e^{t_1 Z_1}]\mathbb{E}[e^{t_2 Z_2}] \cdots \mathbb{E}[e^{t_n Z_n}] \qquad (Z_1, \ldots, Z_n \text{ are independent}) \\ &= e^{t_1^2/2} \cdots e^{t_n^2/2} = \exp\left(\frac{t_1^2}{2} + \cdots + \frac{t_n^2}{2}\right) \text{ (Example. 5.24)} \\ &= \exp\left(\frac{1}{2}\boldsymbol{t}^\top \boldsymbol{t}\right) \end{aligned} \tag{5.18.7}$$

This is nothing but a generalized result of Example 5.24.

The second step, is of course, to consider this variable $\boldsymbol{X} = \mathbf{A}\boldsymbol{Z} + \boldsymbol{\mu}$ with $\boldsymbol{\mu} \in \mathbb{R}^n$ and $\mathbf{A}$ is an $n \times n$ matrix, which is nothing but the generalization of $X = aZ + b$. In what follows, we compute $M_{\boldsymbol{X}}(\boldsymbol{t})$:

$$\begin{aligned} M_X(t) &= e^{t^\top \mu} M_Z(\mathbf{A}^\top t) \qquad \text{(Eq. (5.17.10))} \\ &= e^{t^\top \mu} e^{1/2(\mathbf{A}^\top t)^\top (\mathbf{A}^\top t)} \quad \text{(Eq. (5.18.7))} \\ &= \exp\left(t^\top \mu + \frac{1}{2} t^\top \boldsymbol{\Sigma} t\right), \quad \boldsymbol{\Sigma} = \mathbf{A}\mathbf{A}^\top \end{aligned} \tag{5.18.8}$$

Based on Eq. (5.18.8), it is not a surprise to see the following definition of *a multivariate normal distribution*.

Definition 5.18.1 Let $\boldsymbol{\mu} \in \mathbb{R}^n$ and let $\boldsymbol{\Sigma}$ be an $n \times n$ semi-positive definite matrix. A random vector $\boldsymbol{X} = (X_1, X_2, \ldots, X_n)^\top$ is said to have a multivariate normal distribution with parameters $\boldsymbol{\mu}$ and $\boldsymbol{\Sigma}$ if its multivariate moment generating function is

$$M_X(t) = \exp\left(t^\top \boldsymbol{\mu} + \frac{1}{2} t^\top \boldsymbol{\Sigma} t\right) \tag{5.18.9}$$

The notation is: $\boldsymbol{X} \sim N(\boldsymbol{\mu}, \boldsymbol{\Sigma})$.

This definition is reduced to a single variate normal distribution in Eq. (5.17.11) when $n = 1$. That is, $\boldsymbol{\mu}$ is simply a number μ and the matrix $\boldsymbol{\Sigma}$ is simply the variance σ^2 and $\boldsymbol{t}$ is simple t.

It is clear that, for $\mathbf{Z} = (Z_1, Z_2, \ldots, Z_n)^\top$ with n independent standard normal variables $Z_1, Z_2, \ldots, Z_n$, $\mathbf{Z} \sim N(\mathbf{0}, \mathbf{I})$. Just compare Eq. (5.18.7) with Eq. (5.18.9). And with $\boldsymbol{X} = \mathbf{A}\boldsymbol{Z} + \boldsymbol{\mu}$, we have $\boldsymbol{X} \sim N(\boldsymbol{\mu}, \boldsymbol{\Sigma})$. Just the same old story but now for random vectors.

5.18.4 Mean and Covariance of Multivariate Normal Distribution

First, let's consider $\mathbf{Z} \sim N(\mathbf{0}, \mathbf{I})$, i.e., $\mathbf{Z} = (Z_1, Z_2, \ldots, Z_n)^\top$ where the Z_i are independent $N(0, 1)$ random variables.[72] Then,

$$\mathbb{E}[\mathbf{Z}] = (\mathbb{E}[Z_1], \mathbb{E}[Z_2], \ldots, \mathbb{E}[Z_n])^\top = \mathbf{0} \tag{5.18.10}$$

Thus, the mean vector is the zero vector. Moving on to the covariance matrix, its ijth element is given by

$$\mathbb{E}[(Z_i - \mathbb{E}[Z_i])(Z_j - \mathbb{E}[Z_j])] = \mathbb{E}[Z_i Z_j] \tag{5.18.11}$$

[72] At this point, when reading this you should have in your mind this: $\mathbb{E}[Z_i] = 0$ and $\text{Var}Z_i = 1$.

To proceed, first consider the case $i = j$, then we have $\mathbb{E}[Z_i Z_j] = \mathbb{E}[Z_i^2] = 1$ (why?). But, if $i \neq j$, then we have $\mathbb{E}[Z_i Z_j] = \mathbb{E}[Z_i]\mathbb{E}[Z_j] = 0 \times 0 = 0$. So, we have

$$\mathbb{E}[(Z_i - \mathbb{E}[Z_i])(Z_j - \mathbb{E}[Z_j])] = \begin{cases} 1, & \text{if } i = j \\ 0, & \text{if } i \neq j \end{cases} \tag{5.18.12}$$

So, $\mathbf{C_Z} = \mathbf{I}$—the identity matrix.

Second, we consider $\boldsymbol{X} \sim N(\boldsymbol{\mu}, \boldsymbol{\Sigma})$. Of course, $\boldsymbol{X} = \mathbf{AZ} + \boldsymbol{\mu}$. Hence,

$$\mathbb{E}[\boldsymbol{X}] = \mathbb{E}[\mathbf{AZ} + \boldsymbol{\mu}] = \mathbf{A}\mathbb{E}[\mathbf{Z}] + \boldsymbol{\mu} = \boldsymbol{\mu} \tag{5.18.13}$$

And for the covariance, we use Eq. (5.18.3):

$$\text{Cov}(\boldsymbol{X}) = \text{Cov}(\mathbf{AZ} + \boldsymbol{\mu}) = \mathbf{A}\text{Cov}(\mathbf{Z})\mathbf{A}^\top = \mathbf{AA}^\top = \boldsymbol{\Sigma}. \tag{5.18.14}$$

5.18.5 The Probability Density Function for $N(\boldsymbol{\mu}, \boldsymbol{\Sigma})$

First, we determine the pdf for $\mathbf{Z} \sim N(\mathbf{0}, \mathbf{I})$. Then, we consider the transformation $\boldsymbol{X} = \mathbf{AZ} + \boldsymbol{\mu}$ and then use the result in Example 5.25 to get the pdf of $N(\boldsymbol{\mu}, \boldsymbol{\Sigma})$. That's the plan.

Noting that Z_i are independent $N(0, 1)$ variables, the pdf of $\mathbf{Z}$ is simply the product of the pdf of all Z_i, which we know:

$$f_{\mathbf{Z}}(z) = \prod_{i=1}^{n} f_{Z_i}(z_i) = \prod_{i=1}^{n} \frac{1}{\sqrt{2\pi}} e^{-z_i^2/2} = \frac{1}{(2\pi)^{n/2}} \exp\left(-\frac{1}{2} z^\top z\right) \tag{5.18.15}$$

Now, we consider the transformation $\boldsymbol{X} = \mathbf{AZ} + \boldsymbol{\mu}$, and the result from Example 5.25 gives us

$$\begin{aligned}
f_{\boldsymbol{X}}(\boldsymbol{x}) &= |\det \mathbf{A}^{-1}| f_{\mathbf{Z}}(\mathbf{A}^{-1}(\boldsymbol{x} - \boldsymbol{\mu})) \\
&= |\det \mathbf{A}^{-1}| \frac{1}{(2\pi)^{n/2}} \exp\left(-\frac{1}{2}(\boldsymbol{x} - \boldsymbol{\mu})^\top \left(\mathbf{A}^{-1}\right)^\top \mathbf{A}^{-1}(\boldsymbol{x} - \boldsymbol{\mu})\right) \quad \text{(Eq. (5.18.15))} \\
&= |\det \mathbf{A}^{-1}| \frac{1}{(2\pi)^{n/2}} \exp\left(-\frac{1}{2}(\boldsymbol{x} - \boldsymbol{\mu})^\top \boldsymbol{\Sigma}^{-1}(\boldsymbol{x} - \boldsymbol{\mu})\right) \quad \left(\left(\mathbf{A}^{-1}\right)^\top \mathbf{A}^{-1} = \boldsymbol{\Sigma}^{-1}\right) \\
&= \frac{1}{(2\pi)^{n/2}\sqrt{\det \boldsymbol{\Sigma}}} \exp\left(-\frac{1}{2}(\boldsymbol{x} - \boldsymbol{\mu})^\top \boldsymbol{\Sigma}^{-1}(\boldsymbol{x} - \boldsymbol{\mu})\right)
\end{aligned} \tag{5.18.16}$$

where I have used the following relation between $\det \mathbf{A}^{-1}$ and $\det \boldsymbol{\Sigma}$:

$$|\det \mathbf{A}^{-1}| = \frac{1}{\sqrt{\det \mathbf{AA}^\top}} = \frac{1}{\sqrt{\det \boldsymbol{\Sigma}}}$$

Proof The proof is based on the following facts:

$$\det \mathbf{A}^{-1} = \frac{1}{\det \mathbf{A}}, \quad \det \mathbf{AB} = (\det \mathbf{A})(\det \mathbf{B}), \quad \det \mathbf{A}^\top = \det \mathbf{A}$$

Now, we can write

$$\det(\mathbf{AA}^\top) = (\det \mathbf{A})(\det \mathbf{A}^\top) = (\det \mathbf{A})^2 \Longrightarrow |\det \mathbf{A}| = \sqrt{\det(\mathbf{AA}^\top)}.$$

■

5.18.6 The Bivariate Normal Distribution

The multivariate normal distribution becomes a bivariate normal distribution for $n = 2$. In this case, we have $\boldsymbol{X} = (X_1, X_2)^\top$ and

$$\boldsymbol{\mu} = \begin{bmatrix} \mu_1 \\ \mu_2 \end{bmatrix}, \quad \boldsymbol{\Sigma} = \begin{bmatrix} \mathrm{Var}(X_1) & \mathrm{Cov}(X_1, X_2) \\ \mathrm{Cov}(X_1, X_2) & \mathrm{Var}(X_2) \end{bmatrix} = \begin{bmatrix} \sigma_1^2 & \rho\sigma_1\sigma_2 \\ \rho\sigma_1\sigma_2 & \sigma_2^2 \end{bmatrix}$$

where ρ is the correlation coefficient introduced in Sect. 5.12.7. Thus, the bivariate normal distribution is determined by five scalar parameters: $\mu_1, \mu_2, \sigma_1, \sigma_2$ and ρ.

Now, we compute all quantities involved in the final expression in Eq. (5.18.16):

$$\det \boldsymbol{\Sigma} = (1 - \rho^2)\sigma_1^2\sigma_2^2, \quad \boldsymbol{\Sigma}^{-1} = \frac{1}{\det \boldsymbol{\Sigma}} \begin{bmatrix} \sigma_2^2 & -\rho\sigma_1\sigma_2 \\ -\rho\sigma_1\sigma_2 & \sigma_1^2 \end{bmatrix}$$

and,

$$(\boldsymbol{x} - \boldsymbol{\mu})^\top \boldsymbol{\Sigma}^{-1}(\boldsymbol{x} - \boldsymbol{\mu}) = \frac{1}{1 - \rho^2} \left(\frac{(x_1 - \mu_1)^2}{\sigma_1^2} + \frac{(x_2 - \mu_2)^2}{\sigma_2^2} - \frac{2\rho(x_1 - \mu_1)(x_2 - \mu_2)}{\sigma_1\sigma_2} \right)$$

This term obviously measures the distance (not Euclidean) from $\boldsymbol{x}$ to $\boldsymbol{\mu}$, and it is called the squared Mahalanobis distance.[73]

Finally, the pdf of a bivariate normal distribution is given by

$$f(x_1, x_2) = \frac{1}{2\pi\sigma_1\sigma_2\sqrt{1 - \rho^2}} \times \exp\left[-\frac{1}{2(1 - \rho^2)} \left(\frac{(x_1 - \mu_1)^2}{\sigma_1^2} + \frac{(x_2 - \mu_2)^2}{\sigma_2^2} - \frac{2\rho(x_1 - \mu_1)(x_2 - \mu_2)}{\sigma_1\sigma_2} \right) \right] \tag{5.18.17}$$

[73] Named after Prasanta Chandra Mahalanobis (1893–1972), an Indian scientist and statistician.

For the special case that $\rho = 0$, we then have

$$
\begin{aligned}
f(x_1, x_2) &= \frac{1}{2\pi\sigma_1\sigma_2}\exp\left[-\frac{1}{2}\left(\frac{(x_1-\mu_1)^2}{\sigma_1^2}+\frac{(x_2-\mu_2)^2}{\sigma_2^2}\right)\right] \\
&= \frac{1}{\sqrt{2\pi}\sigma_1}\exp\left[-\frac{1}{2}\left(\frac{(x_1-\mu_1)^2}{\sigma_1^2}\right)\right]\times\frac{1}{\sqrt{2\pi}\sigma_2}\exp\left[-\frac{1}{2}\left(\frac{(x_2-\mu_2)^2}{\sigma_2^2}\right)\right] \\
&= f_{X_1}(x_1)f_{X_2}(x_2)
\end{aligned}
\tag{5.18.18}
$$

Now, for $\mu_1 = \mu_2 = 0$ and $\sigma_1 = \sigma_2 = 1$, from Eq. (5.18.18) we get the following PDF (note that I used (x, y) instead of (x_1, x_2))

$$
f(x, y) = \frac{1}{2\pi}\exp\left[-\frac{1}{2}(x^2 + y^2)\right]
$$

of which the graph is a circular Gaussian surface (see Fig. 5.18a): it is the 3D version of the well-known bell curve. Using Eq. (5.18.17) with $\mu_1 = \mu_2 = 0$, $\sigma_1 = \sigma_2 = 1$ but $\rho = 0.8$, we get Fig. 5.18b.

The distributions plotted in Fig. 5.18 have the following covariance matrices:

$$
\boldsymbol{\Sigma}_1 = \begin{bmatrix} 1 & 0 \\ 0 & 1 \end{bmatrix}, \quad \boldsymbol{\Sigma}_2 = \begin{bmatrix} 1 & 0.8 \\ 0.8 & 1 \end{bmatrix}
\tag{5.18.19}
$$

There are two methods of plotting the bivariate normal distribution. One method is to plot a 3D graph (Fig. 5.18) and the other method is to plot a contour graph (Fig. 5.19). A contour graph is a way of displaying three dimensions on a 2D plot. A 3D plot is sometimes difficult to visualize properly. The contour plot shows only two dimensions, i.e., x_1, x_2. The third dimension is defined by the color. If two points

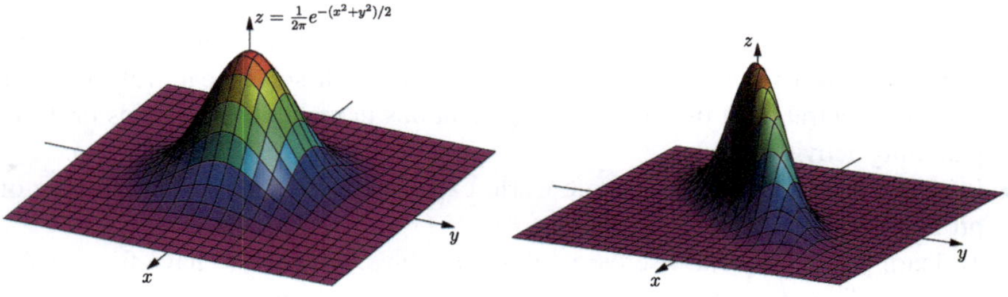

(a) $\rho = 0$ (b) $\rho = 0.8$

Fig. 5.18 Visualization of the PDF of the bivariate normal distributions: generated using `Asymptote`. `Asymptote` is a descriptive vector graphics (programming) language that provides a natural coordinate-based framework for technical drawing. Labels and equations are typeset with LaTeX, the de-facto standard for typesetting mathematics

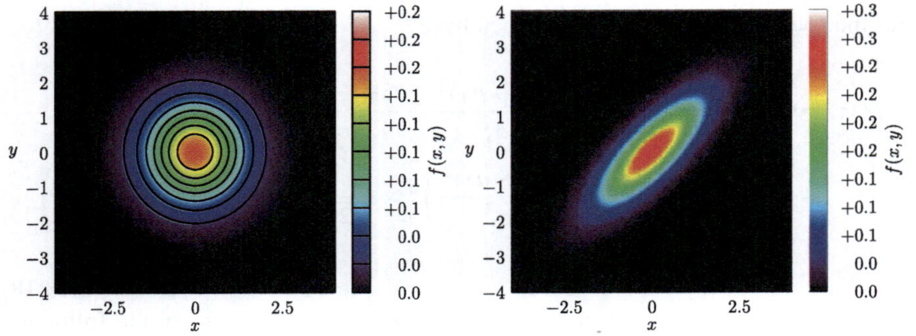

(a) $\rho = 0$ (b) $\rho = 0.8$

Fig. 5.19 Contour plot of the PDF of the bivariate normal distributions: generated using `Asymptote`

have the same color in the contour plot, then they have equal values for their third dimension. A contour plot is usually accompanied by a legend relating the colors to values.

And why the contour of the bivariate normal distribution is an ellipse? The answer is in the square Mahalanobis distance: it is of the form $\boldsymbol{y}^\top \mathbf{A} \boldsymbol{y}$, which is a quadratic form. And in Sect. 9.10.6, we know that $\boldsymbol{y}^\top \mathbf{A} \boldsymbol{y} = c$ is an ellipse.

5.19 Review

I had a bad experience with probability in university. It is quite unbelievable that I have now managed to learn it at the age of 42 to a certain level of understanding. Here are some observations that I made

- Probability had a humbling starting point in games of chances. But mathematicians turned it into a rigorous branch of mathematics with some beautiful theorems (e.g., the central limit theorem) with applications in many diverse fields far from gambling activities;
- It is beneficial to carry out Monte Carlo experiments to support the learning of probability;
- To learn probability for discrete random variables, we need to have first a solid understanding of counting methods (e.g., factorial, permutations, and so on).
- Probability is very counter-intuitive so rote memorization does not help. Facing a probability we should sit back and solve it slowly.
- Probability theory is one of the pillars of machine learning and statistics. That can be a good motivation to learn it.

Chapter 6
Multivariable Calculus

In Chap. 4 we have studied the calculus of functions of one variable ,e.g., functions expressed by $y = f(x)$. Basically, we studied curves in a 2D plane, the tangent to a curve at any point on the curve (first derivative) and the area under the curve (integral). Now is the time to the real world: functions of multiple variables. We will discuss functions of the form $z = f(x, y)$ known as *scalar-valued functions of two variables*. A plot of $z = f(x, y)$ gives a surface in a 3D space. Of course, we are going to differentiate $z = f(x, y)$ and thus partial derivatives $\frac{\partial f}{\partial x}, \frac{\partial f}{\partial y}$ naturally emerge. We also compute integrals of $z = f(x, y)$, the double integrals $\iint f(x, y)dxdy$ which can be visualized as the volume under the surface $f(x, y)$. And triple integrals $\iiint f(x, y, z)dxdydz$ appear when we deal with functions of three variables $f(x, y, z)$. All of these are merely an extension of the calculus we know from Chap. 4. If there are some difficulties, they are just technical not mentally as when we learned about the spontaneous speed of a moving car.

Then comes *vector-valued functions* used to describe vector fields. For example, if we want to study the motion of a moving fluid, we need to know the velocity of all the fluid particles. The velocity of a fluid particle is a vector field and is mathematically expressed as a vector-valued function of the form $\boldsymbol{v}(x, y) = (g(x, y), h(x, y))$ in two dimensions. The particle position is determined by its coordinates (x, y) and its velocity by two functions: $g(x, y)$ for the horizontal component of the velocity and $h(x, y)$ for the vertical component.

And with vector fields, we shall have vector calculus that consists of differential calculus of vector fields and integral calculus of vector fields. In differential calculus of vector fields, we shall meet the gradient vector of a scalar field ∇f, the divergence of a vector field $\nabla \cdot \boldsymbol{C}$ and the curl of a vector field $\nabla \times \boldsymbol{C}$. In the integral calculus, we have the line integral $\int_\Gamma \boldsymbol{F} \cdot d\boldsymbol{s}$, surface integrals $\int_S \boldsymbol{C} \cdot \boldsymbol{n} dA$ and volume integrals. And these integrals are linked together via Green's theorem, Stokes' theorem, and Gauss's theorem. They are generalizations of the fundamental theorem of calculus (Table 6.1).

V. P. Nguyen, *Mathematics for Engineers and Scientists*,
https://doi.org/10.1007/978-981-97-4631-6_6

Table 6.1 Integral calculus of vector fields: a summary

Theorem	Formula	
FTC	$\int_a^b \frac{df}{dx}dx = f(b) - f(a)$	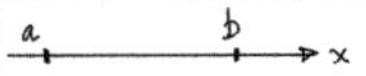
FTC of line integrals	$\int_{1 \text{ along } C}^{2} \nabla\psi \cdot \boldsymbol{ds} = \psi(2) - \psi(1)$	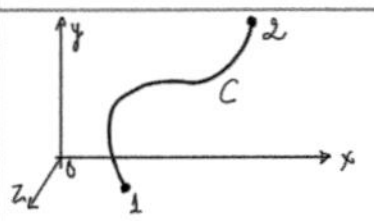
Green's theorem	$\int_S \left(\frac{\partial C_y}{\partial x} - \frac{\partial C_x}{\partial y}\right) dA = \oint_\Gamma (C_x dx + C_y dy)$	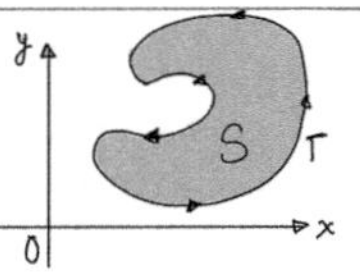
Stokes' theorem	$\int_S (\nabla \times \boldsymbol{C}) \cdot \boldsymbol{n} dA = \oint_\Gamma \boldsymbol{C} \cdot \boldsymbol{ds}$	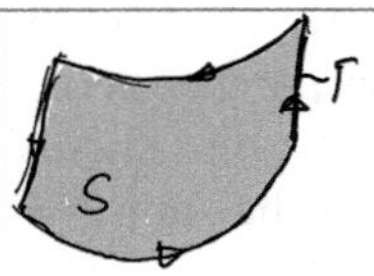
Gauss's theorem	$\int_S \boldsymbol{C} \cdot \boldsymbol{n} dA = \int_V \nabla \cdot \boldsymbol{C} dV$	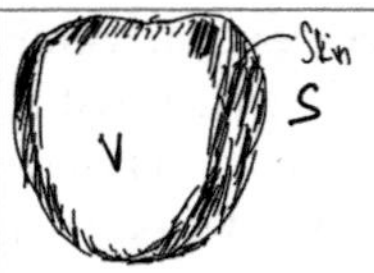

This chapter starts with a presentation of multivariable functions in Sect. 6.1. The derivatives of these functions are discussed in Sect. 6.2. Section 6.3 presents tangent planes and linear approximations. Then, Newton's method for solving a system of nonlinear equations is treated in Sect. 6.4. The gradient of a scalar function and the directional derivative are given in Sect. 6.5. The chain rules are introduced in Sect. 6.6. The problem of finding the extrema of functions of multiple variables is given in Sect. 6.7. Two and three-dimensional integrals are given in Sects. 6.8, and 6.9 discusses parametric surfaces. Newtonian mechanics is briefly discussed in Sect. 6.10. Then comes a big chapter on vector calculus (Sect. 6.11). A short introduction to the wonderful field—complex analysis—is provided in Sect. 6.12.

Some knowledge on vector algebra and matrix algebra are required to read this chapter. Section 9.1 in Chap. 9 provides an introduction to vectors and matrices.

I have used primarily the following books for the material presented herein:

- *Calculus* by Gilbert Strang [65];
- *Calculus: Early Transcendentals* by James Stewart[1] [61];
- *The Feynman Lectures on Physics* by Feynman [23];
- *Vector calculus* by Jerrold Marsden[2] and Anthony Tromba[3] [45].

6.1 Multivariable Functions

The concept of a function of one variable can be easily generalized to the case of functions of two or more variables. If a function returns a scalar we call it *a scalar-valued function*, e.g., $\sin(x + y)$. We discuss such functions in Sect. 6.1.1. On the other hand, if a function returns a vector (multiple outputs), then it is called *a vector-valued function*, see Sect. 6.1.4. One example is a helix curve given by $(\sin t, \cos t, t)$.

6.1.1 Scalar-Valued Multivariable Functions

In the case of a function of two variables, we consider the set of ordered pairs (x, y) where x and y are both real numbers. If there is a law according to which each pair (x, y) is assigned to a single value of z, then we say about a function of two variables. Usually, this function is denoted by $z = f(x, y) : \mathbb{R}^2 \to \mathbb{R}$. Similarly, we can have a function taking three real numbers and produce a real number, mathematically written as $T = g(x, y, z) : \mathbb{R}^3 \to \mathbb{R}$. For example, $T = g(x, y, z)$ is the temperature of a point in the earth.

Visualizing functions of two variables is more difficult by hand as they represent surfaces in a 3D space; these surfaces are formed by the set of all the points (x, y, z) where $z = f(x, y)$, i.e., the set of points $(x, y, f(x, y))$. We need to use software for this task. For example, Fig. 6.1 shows a plot of the function $z = x^2 + y^2$ (which is a paraboloid, a generalization of the 2D parabola) and $x^2 + y^2 + z^2 = 1$ which is a sphere. In Fig. 6.2a, we plot another function $z = f(x, y)$ in which the surface is colored according to the value of z. In this way, it is easy to see where is the highest/lowest points of the surface. Furthermore, it is the only way to visualize $T = f(x, y, z)$ (Fig. 6.2b). This is because the graph of a function $f(x, y, z)$ of three

[1] James Drewry Stewart (1941–2014) was a Canadian mathematician, violinist, and professor emeritus of mathematics at McMaster University. Stewart is best known for his series of calculus textbooks used for high school-, college-, and university-level courses.

[2] Jerrold Eldon Marsden (1942–2010) was a Canadian mathematician. Marsden, together with Alan Weinstein, was one of the world's leading authorities in mathematical and theoretical classical mechanics. He has laid much of the foundation for symplectic topology. The Marsden-Weinstein quotient is named after him.

[3] Anthony Joseph Tromba (born 10 August 1943, Brooklyn, New York City) is an American mathematician, specializing in partial differential equations, differential geometry, and the calculus of variations.

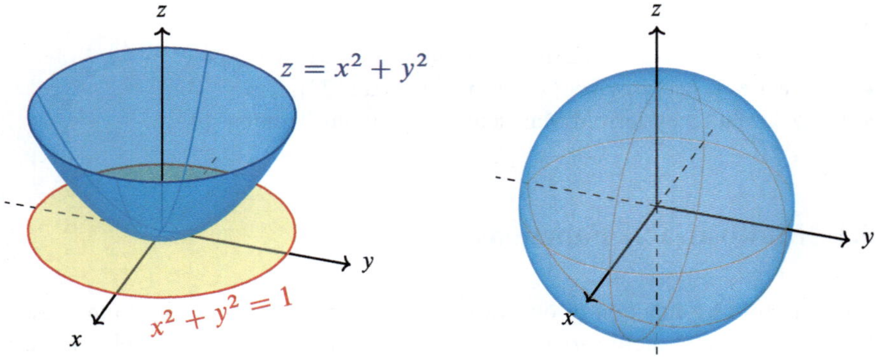

(a) Paraboloid $z = x^2 + y^2$ (b) Sphere

Fig. 6.1 Graph of a paraboloid and a sphere (drawn with `Tikz`). *Source* tikz.net

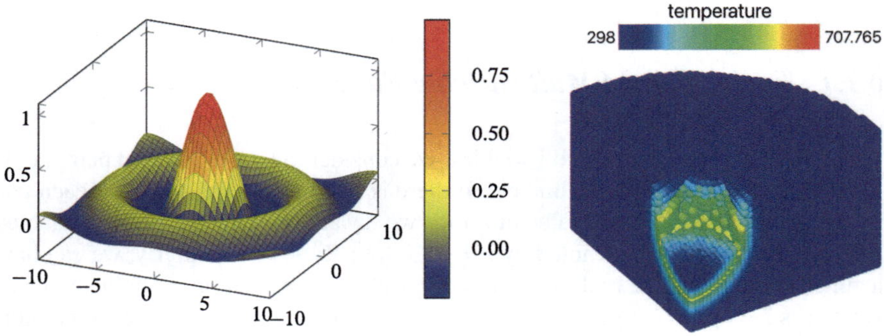

(a) a surface (b) a solid

Fig. 6.2 Graph of the surface $z(x, y) = \sin(\sqrt{x^2 + y^2})/\sqrt{x^2 + y^2}$ in which the surface is colored according to the value of z (**a**) and a 3D solid colored by the temperature inside

variables would be the set of points $(x, y, z, f(x, y, z))$ in four dimensions, and it is difficult to imagine what such a graph would look like.

6.1.2 Level Curves, Level Surfaces, and Level Sets

Three-dimensional plots are more difficult to draw and visualize than two-dimensional plots. Therefore, we now discuss another way of visualizing a multivariable function, borrowed from mapmakers: using level sets, i.e., the set of points in the domain of a function where the function is constant. The nice part of level sets is that they live in the same dimensions as the domain of the function.

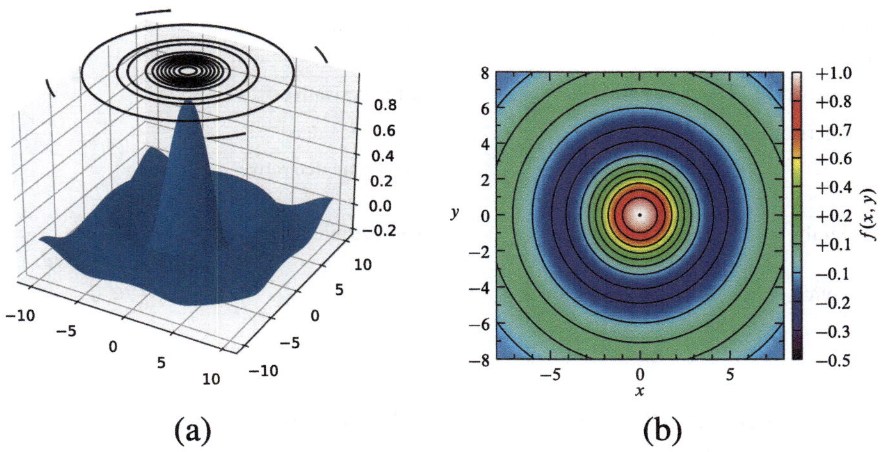

Fig. 6.3 Graph of the surface $z(x, y) = \sin(\sqrt{x^2+y^2})/\sqrt{x^2+y^2}$ and its level curves (**a**). In (**b**) only level curves are plotted and they are just 2D curves. Looking at the level curves, we can say something about the function: the surface is steep where the level curves are close together. And it is flatter when they're father apart

Definition 6.1.1 Level set
A level set of a real-valued function f of n real variables is a set where the function takes on a given constant value c (in the range of f), that is

$$L_c(f) = \{(x_1, x_2, \ldots, x_n) | f(x_1, x_2, \ldots, x_n) = c\} \tag{6.1.1}$$

When the number of independent variables is two, a level set is called a level curve, also known as contour line or isoline; so a level curve is the set of all real-valued solutions of an equation in two variables x, y. When $n = 3$, a level set is called a level surface (or isosurface); so a level surface is the set of all real-valued roots of an equation in three variables x, y, z. For higher values of n, the level set is a level hypersurface.

A level set of a function of two variables $f(x, y)$ is a curve in the two-dimensional xy-plane, called a level curve (Fig. 6.3). A level set of a function of three variables $f(x, y, z)$ is a surface in three-dimensional space, called a level surface. For a constant value c in the range of $f(x, y, z)$, the level surface of f is the implicit surface given by the graph of $c = f(x, y, z)$.

Domain, co-domain, and range of a function. For the function $z = f(x, y) : \mathbb{R}^2 \to \mathbb{R}$, we say that the domain of this function is the entire 2D plane, i.e., $\mathbb{R}^2$. Thus, the domain of a function is the set of all inputs. We also say that the co-domain is $\mathbb{R}$: the co-domain is the set of outputs. And finally, the range of a function is a sub-set of its co-domain which contains the actual outputs. For example, if $f(x, y) = x^2 + y^2$, then its co-domain is all real numbers but its range is only non-negative reals.

Table 6.2 Multivariate calculus is simply an extension of univariate calculus

$f(x)$		$f(x, y)$	
1st derivative	df/dx	Partial derivatives	$\frac{\partial f}{\partial x}, \frac{\partial f}{\partial y}$
2nd derivative	$d^2 f/dx^2$	Second par. der.	$\frac{\partial^2 f}{\partial x^2}, \frac{\partial^2 f}{\partial y^2}, \frac{\partial^2 f}{\partial y \partial x}, \frac{\partial^2 f}{\partial y \partial x}$
		Directional derivative	$D_{\boldsymbol{u}} f(x, y) = u_x \frac{\partial f}{\partial x} + u_y \frac{\partial f}{\partial y}$
Integral	$\int_a^b f(x)dx$	Double/triple integrals	$\iint f(x, y)dxdy$, $\iiint f(x, y)dxdydz$
Extrema	$f_x(x_0) = 0$		$\frac{\partial f}{\partial x} = \frac{\partial f}{\partial y} = 0$

6.1.3 *Multivariate Calculus: An Extension of Univariate Calculus*

If we keep one variable, say y constant, then, from $z = f(x, y)$, we obtain a function of a single variable x, see Fig. 6.1b. We can then apply the calculus we know from Chap. 4 to this function. That leads to partial derivatives. Using these two partial derivatives, we will have directional derivative $D_{\boldsymbol{u}}$ that gives the change in $f(x, y)$ along the direction $\boldsymbol{u}$. Other natural extensions of Chap. 4's calculus are summarized in Table 6.2. We will discuss them, but as you have seen, they are merely extensions of calculus of functions of single variable.

6.1.4 *Vector-Valued Multivariable Functions*

Functions of the form $\boldsymbol{r}(t) = (f(t), g(t), h(t))$ having one variable for input and a vector for output, are called single-variable vector-valued functions. The functions $f(t)$, $g(t)$ and $h(t)$, which are the component functions of $\boldsymbol{r}(t)$, are each a single-variable real-valued function. Single-variable vector-valued functions can be denoted as $\boldsymbol{r} : \mathbb{R} \to \mathbb{R}^n$. The graph of such functions is a 3D curve, see Fig. 6.4a for one example.

A function of the form:

$$f\left(\begin{bmatrix} x \\ y \end{bmatrix}\right) = \begin{bmatrix} f(x, y) \\ g(x, y) \end{bmatrix} \tag{6.1.2}$$

is a multivariable vector-valued function, which maps a point in 2D space to another point in the same space. In Fig. 6.4b, we show such a function: $f(x, y) = (x^2 - y^2, 2xy)$. The black lines are the standard grid lines in a 2D Cartesian plane, and we apply this function to those lines to obtain the red lines. As can be seen, a square was transformed to a curved shape. This function is thus called *a transformation*.

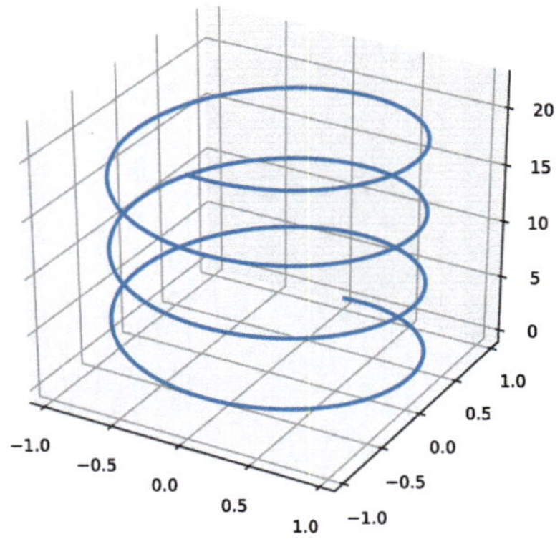

(a) single-variable vector-valued function

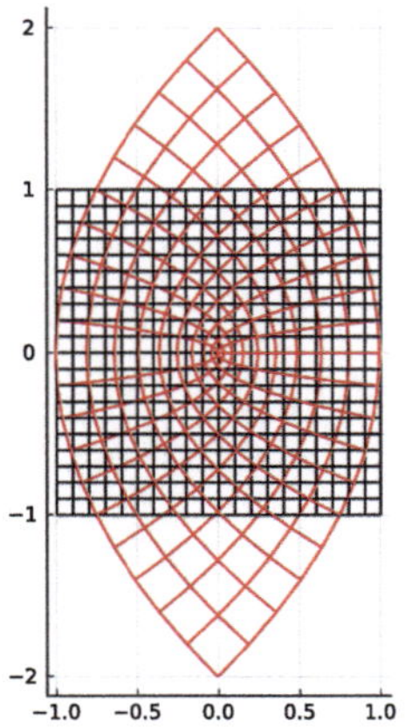

(b) multi-variable vector-valued function

Fig. 6.4 Vector-valued multivariable functions

6.2 Derivatives of Multivariable Functions

For functions of one variable, $y = f(x)$, the derivative was defined as the ratio of the change in the function and the change in x when this change is approaching zero. For functions of two variables $f(x, y)$, it is natural to consider changes in x and in y separately. And this leads to two derivatives: one with respect to x when y is held constant, and the other with respect to y when x is held constant.

For example, consider $f(x, y) = x^2 + y^2$. When x is changed to $x + \Delta x$ (and y is held constant), the corresponding change in f is Δf given by

$$\Delta f = (x + \Delta x)^2 + y^2 - (x^2 + y^2) = 2x\Delta x + (\Delta x)^2$$

And thus $\Delta f/\Delta x = 2x + \Delta x$. The derivative with respect to x, denoted by $\frac{\partial f}{\partial x}$, is therefore $2x$.

So, we are ready to give the formal definition of the partial derivatives of functions of two variables:

$$\begin{aligned} \frac{\partial f}{\partial x} &= \lim_{\Delta x \to 0} \frac{f(x + \Delta x, y) - f(x, y)}{\Delta x} \\ \frac{\partial f}{\partial y} &= \lim_{\Delta y \to 0} \frac{f(x, y + \Delta y) - f(x, y)}{\Delta y} \end{aligned} \tag{6.2.1}$$

In other words, the partial derivative w.r.t. x is the ordinary derivative while holding other variables (y) constant. Sometimes, people write f_x for $\partial f/\partial x$. For $y = f(x)$ the derivative $f'(x_0)$ has the geometrical meaning of being the slope of the tangent to

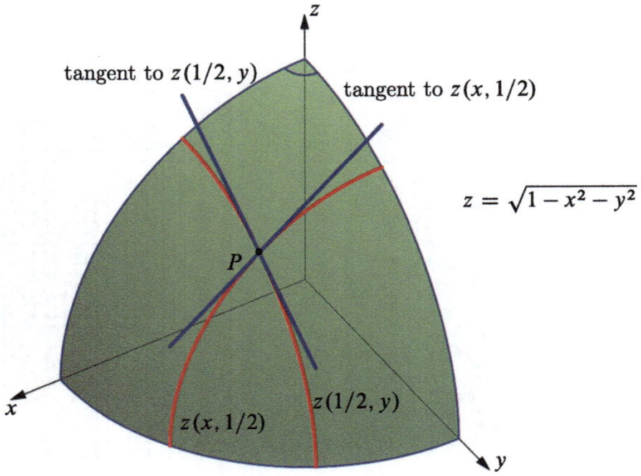

Fig. 6.5 Graph of $z(x, y) = \sqrt{1 - x^2 - y^2}$—which is the first quarter of the unit sphere. Now, let's consider $P(1/2, 1/2, \sqrt{2}/2)$ on this sphere. We consider two curves $C_1 : z(x, 1/2)$ and $C_2 : z(1/2, y)$; the former is the intersection of the surface with the plane $y = 1/2$ and the latter is the intersection of the surface with the plane $x = 1/2$. Then C_1 is given by $g(x) = z(x, 1/2)$, and thus $z_x(1/2, 1/2)$ is the slope of the tangent to C_1 at P. Similarly, $z_y(1/2, 1/2)$ is the slope of the tangent to C_2 at P

$y = f(x)$ at the point $(x_0, f(x_0))$. Do we have something similar for $z = f(x, y)$? Of course, yes. See Fig. 6.5.

And of course, nothing can stop us from moving to second derivatives. From $\partial f/\partial x$, we have $\partial^2 f/\partial x^2$—derivative w.r.t. x of $\partial f/\partial x$ and $\partial^2 f/\partial x \partial y$—derivative w.r.t. y of $\partial f/\partial x$. And from $\partial f/\partial y$, we have $\partial^2 f/\partial y^2$ and $\partial^2 f/\partial y \partial x$. To summarize, we write

$$\begin{aligned}
\frac{\partial f}{\partial x} &\rightarrow \frac{\partial}{\partial x}\left(\frac{\partial f}{\partial x}\right) := \frac{\partial^2 f}{\partial x^2} \ (\text{or}\, f_{xx}), \quad \frac{\partial}{\partial y}\left(\frac{\partial f}{\partial x}\right) := \frac{\partial^2 f}{\partial x \partial y} \ (\text{or}\, f_{xy}) \\
\frac{\partial f}{\partial y} &\rightarrow \frac{\partial}{\partial x}\left(\frac{\partial f}{\partial y}\right) := \frac{\partial^2 f}{\partial y \partial x} \ (\text{or}\, f_{yx}), \quad \frac{\partial}{\partial y}\left(\frac{\partial f}{\partial y}\right) := \frac{\partial^2 f}{\partial y^2} \ (\text{or}\, f_{yy})
\end{aligned} \tag{6.2.2}$$

where f_{xy} and f_{yx} are *cross-derivatives* or *mixed derivatives*. The origin of partial derivatives was partial differential equation such as the wave equation (Sect. 7.5.1). Briefly, eighteenth-century mathematicians and physicists such as Euler, d'Alembert, and Daniel Bernoulli were investigating the vibration of strings (to understand music), and there was a need to consider partial derivatives.

Example 1 Let's consider the function $f(x, y) = x^2y^2 + xy + y$, its first and second (partial) derivatives are

$$\begin{aligned}
f_x &= 2xy^2 + y \quad & f_{xx} &= 2y^2 \quad f_{xy} = 4xy + 1 \\
f_y &= 2x^2y + x + 1 \quad & f_{yy} &= 2x^2 \quad f_{yx} = 4xy + 1
\end{aligned}$$

The calculations were nothing special, but one thing special is $\partial^2 f/\partial y \partial x = \partial^2 f/\partial y \partial x$. Is it luck? Let's see another example.

Example 2 Let's consider this function $f(x, y) = e^{xy^2}$, its first and second derivatives are

$$\begin{aligned} f_x &= y^2 e^{xy^2} & f_{xx} &= y^4 e^{xy^2} & f_{xy} &= 2ye^{xy^2} + 2xy^3 e^{xy^2} \\ f_y &= 2xye^{xy^2} & f_{yy} &= 2xe^{xy^2} + 4x^2y^2e^{xy^2} & f_{yx} &= 2ye^{xy^2} + 2xy^3 e^{xy^2} \end{aligned}$$

Again, we get $\partial^2 f/\partial y \partial x = \partial^2 f/\partial y \partial x$. Actually, there is a theorem called Schwarz's Theorem or Clairaut's Theorem[4] which states that mixed derivatives are equal if they are continuous.

6.3 Tangent Planes, Linear Approximation, and Total Differential

When we considered functions of one variable, we used the first derivative $f'(x_0)$ to get the equation of the tangent line to $f(x)$ at $(x_0, f(x_0))$. The equation of the tangent line is $y = f(x_0) + f'(x_0)(x - x_0)$. And the tangent line led to linear approximation: near x_0 we can use the tangent line instead of the curve. Now, we're doing the same thing for $f(x, y)$. But, instead of tangent lines, we have tangent planes. The equation of a plane passing through the point (x_0, y_0, z_0) is given by (see Sect. 9.1.3 to know why)

$$a(x - x_0) + b(y - y_0) + c(z - z_0) = 0, \quad \text{or} \quad z = z_0 + A(x - x_0) + B(y - y_0) \tag{6.3.1}$$

Our task is now to determine the coefficients A and B in terms of f_x and f_y (we believe in the extension of elementary calculus to multi-dimensions). To determine A, we consider the plane $y = y_0$. The intersection of this plane and the surface $z = f(x, y)$ is a curve in the $x - z$-plane, see Fig. 6.5 for one example. The tangent to this curve at (x_0, y_0) is $z = z_0 + f_x(x_0, y_0)(x - x_0)$ and thus $A = f_x(x_0, y_0)$. Similarly, consider the plane $x = x_0$, and we get $B = f_y(x_0, y_0)$. The tangent plane is now written as

$$z = z_0 + f_x(x_0, y_0)(x - x_0) + f_y(x_0, y_0)(y - y_0) \tag{6.3.2}$$

Linear approximation. Around the point (x_0, y_0), we can approximate the (complicated) function $f(x, y)$ by a simpler function—the equation of the tangent plane:

[4] Alexis Claude Clairaut (13 May 1713–17 May 1765) was a French mathematician, astronomer, and geophysicist. He was a prominent Newtonian whose work helped to establish the validity of the principles and results that Sir Isaac Newton had outlined in the Principia of 1687.

$$f(x, y) \approx f(x_0, y_0) + f_x(x_0, y_0)(x - x_0) + f_y(x_0, y_0)(y - y_0) \tag{6.3.3}$$

which is called a linear approximation of a $f(x, y)$. Compared to a linear approximation of $f(x)$: $f(x_0) + f_x(x_0)(x - x_0)$, we can see the analogy. And we can also guess this is coming from a Taylor's series for $f(x, y)$ where higher order terms are omitted.

Seeing the pattern from functions of single variable to functions of two variables, we can now generalize the linear approximation of functions of n variables ($n \in \mathbb{N}$ and $n \geq 2$):

$$f(\boldsymbol{x}_0) + (\boldsymbol{x} - \boldsymbol{x}_0)^\top \nabla f(\boldsymbol{x}_0) \tag{6.3.4}$$

We will discuss the notation ∇f shortly. Note that vector notation is being used: $\boldsymbol{x} = (x_1, x_2, \ldots, x_n)$ is a point in an n-dimensional space, refer to Sect. 9.1 in Chap. 9 for a discussion on vectors.

Total differential. On the curve $y = f(x)$, a finite change in x is Δx, and if we climb on the curve, we move an amount Δy. But if we move an infinitesimal along x that is dx, and we follow the tangent to the curve, then we move an amount $dy = f'(x)dx$. Now we do the same thing, but we're now climbing on a surface. Using Eq. (6.3.2), we write

$$\boxed{dz = \frac{\partial f}{\partial x}dx + \frac{\partial f}{\partial y}dy} \tag{6.3.5}$$

and we call dz a total differential.

6.4 Newton's Method for Solving Two Equations

We have used the linear approximation to solve $f(x) = 0$. The same idea works for a system of two equations of the following form:

$$\begin{bmatrix} f(x, y) = 0 \\ g(x, y) = 0 \end{bmatrix} \tag{6.4.1}$$

Their linear approximations lead to (where (x_0, y_0) is the starting point):

$$\begin{aligned} f(x_0, y_0) + f_x(x_0, y_0)\Delta x + f_y(x_0, y_0)\Delta y = 0 \\ g(x_0, y_0) + g_x(x_0, y_0)\Delta x + g_y(x_0, y_0)\Delta y = 0 \end{aligned} \tag{6.4.2}$$

which is a system of linear equations for two unknowns Δx and Δy. Formally, we can express the solutions as

$$\begin{bmatrix} \Delta x \\ \Delta y \end{bmatrix} = -\begin{bmatrix} f_x(x_0, y_0) & f_y(x_0, y_0) \\ g_x(x_0, y_0) & g_y(x_0, y_0) \end{bmatrix}^{-1} \begin{bmatrix} f(x_0, y_0) \\ g(x_0, y_0) \end{bmatrix} \tag{6.4.3}$$

With, we update the solution as $x_0 + \Delta x$ and $y_0 + \Delta y$. And the iterative process is repeated until convergence. This Newton method has been applied to solve practical problems that involve millions of unknowns. In the above, $\mathbf{A}^{-1}$ means the inverse of matrix $\mathbf{A}$. We refer to Chap. 9 for details.

6.5 Gradient and Directional Derivative

For functions of one single variable $y = f(x)$, there is only one derivative that measures the rate of change of the function with respect to the change in x. The x-axis is the **only** direction we can go! With functions of two variables $z = f(x, y)$, there are infinite directions to go; on a plane, we can go any direction. We can go along the west-east direction, to have f_x. Or, we can go the north-south direction to have f_y. They are described by partial derivatives. We can go along a direction $\boldsymbol{u} = (u_1, u_2)$ and the change in $f(x, y)$ is described by the so-called *directional derivative*:

$$
\begin{aligned}
\text{change in } x: &\quad \frac{\partial f}{\partial x} = \lim_{\Delta x \to 0} \frac{f(x + \Delta x, y) - f(x, y)}{\Delta x}\\
\text{change in } y: &\quad \frac{\partial f}{\partial y} = \lim_{\Delta y \to 0} \frac{f(x, y + \Delta y) - f(x, y)}{\Delta y}\\
\text{change in } x, y: &\quad D_{\boldsymbol{u}} f = \lim_{h \to 0} \frac{f(x + hu_1, y + hu_2) - f(x, y)}{h}
\end{aligned}
$$

where $\boldsymbol{u}$ is a unit vector (as only its direction is important). Ok, so we have the definition of a new kind of derivative. It generalizes the old partial derivatives: when $\boldsymbol{u} = \boldsymbol{i} = (1, 0)$, we get back to f_x and when $\boldsymbol{u} = \boldsymbol{j} = (0, 1)$, we get back to f_y. But how can we actually compute the directional derivative at a certain point (x_0, y_0) and a given $\boldsymbol{u}$?

One example will reveal the secret. What is simpler than $f(x, y) = x^2 + y^2$? Let's compute $D_{\boldsymbol{u}}$ of this simple function at (x_0, y_0):

$$
\begin{aligned}
D_{\boldsymbol{u}} f(x_0, y_0) &= \lim_{h \to 0} \frac{(x_0 + u_1 h)^2 + (y_0 + u_2 h)^2 - x_0^2 - y_0^2}{h}\\
&= \lim_{h \to 0} \frac{2x_0 u_1 h + 2y_0 u_2 h + (u_1 h)^2 + (u_2 h)^2}{h}\\
&= 2x_0 u_1 + 2y_0 u_2
\end{aligned}
$$

What is this result? It is $f_x(x_0, y_0)u_1 + f_y(x_0, y_0)u_2$. This result makes sense as it is reduced to the old rules of partial derivatives. Indeed, when $\boldsymbol{u}$ are is unit vector, e.g., $\boldsymbol{u} = (1, 0)$, we get the familiar result of $f_x(x_0, y_0)$. And with $\boldsymbol{u} = (0, 1)$, we get the familiar result of $f_y(x_0, y_0)$. So, we guess it is correct for general cases. But, we need a proof.

Proof [Proof of $D_{\boldsymbol{u}}(x_0, y_0)f = f_x(x_0, y_0)u_1 + f_y(x_0, y_0)u_2$.] Of course, we start of with the definition of the directional derivative evaluated at a particular point (x_0, y_0) as

$$D_{\boldsymbol{u}} f(x_0, y_0) = \lim_{h \to 0} \frac{f(x_0 + hu_1, y_0 + hu_2) - f(x_0, y_0)}{h} \tag{6.5.1}$$

Then, we're stuck as there is no concrete expression for $f(x, y)$ for us to manipulate. Here we need a change of view. Note that in the above equation x_0, y_0 and u_1, u_2 are all fixed numbers, only h is a variable. Thus, we can define a new function of a single variable $g(z)$ as

$$g(z) := f(x(z), y(z)), \quad x(z) = x_0 + u_1 z; \quad y(z) = y_0 + u_2 z$$

What we are going to do with this new function? We differentiate it, using the chain rule (Sect. 6.6):

$$g'(z) = f_x \frac{dx}{dz} + f_y \frac{dy}{dz} = f_x u_1 + f_y u_2$$

We're on good track as we have obtained $f_x u_1 + f_y u_2$—the suspect that we're looking for. From this, we have

$$g'(0) = f_x(x_0, y_0)u_1 + f_y(x_0, y_0)u_2$$

Now, we just need to prove that $g'(0)$ is nothing but the RHS of Eq. (6.5.1). That is,

$$g'(0) \stackrel{?}{=} \lim_{h \to 0} \frac{f(x_0 + hu_1, y_0 + hu_2) - f(x_0, y_0)}{h}$$

Indeed, we can compute $g'(0)$ using the definition of derivative and replacing g with f (we need it to appear now):

$$g'(0) = \lim_{h \to 0} \frac{g(h) - g(0)}{h} = \lim_{h \to 0} \frac{f(x_0 + hu_1, y_0 + hu_2) - f(x_0, y_0)}{h}$$

■

The French mathematician, theoretical physicist, engineer, and philosopher of science Henri Poincaré (1854–1912) once said "Mathematics is the art of *giving the same name to different things*". Herein, we see the same expression of $D_{\boldsymbol{u}} f$ but as the normal derivative of $g(z)$. That's the art. This can also be seen in the following joke

> A team of engineers were required to measure the height of a flag pole. They only had a measuring tape, and were getting quite frustrated trying to keep the tape along the pole. It kept falling down, etc. A mathematician comes along, finds out their problem, and proceeds to remove the pole from the ground and measure it easily. When he leaves, one engineer says to the other: "Just like a mathematician! We need to know the height, and he gives us the length!"

We now have a rule to compute the directional derivative for any functions. But there is one more thing in its formula: $\partial f/\partial x u_1 + \partial f/\partial y u_2$ is actually the dot product[5] between the vector $\boldsymbol{u}$ and a vector, which we do not know, with components f_x, f_y.

We now give the rule for a directional derivative for a function $f(x, y, z)$ and define the *gradient vector*, denoted by ∇f (read nabla f or del f):

$$D_{\boldsymbol{u}} f = \nabla f \cdot \boldsymbol{u}, \quad \nabla f = \frac{\partial f}{\partial x}\boldsymbol{i} + \frac{\partial f}{\partial y}\boldsymbol{j} + \frac{\partial f}{\partial z}\boldsymbol{k} \tag{6.5.2}$$

In words, the gradient of a function $f(x, y, z)$ at any point is a 3D vector with components (f_x, f_y, f_z). The gradient vector of a scalar function is significant as it gives us the *direction of steepest ascent*. That is because the directional derivative indicates the change of f in a direction given by $\boldsymbol{u}$. Among many directions, due to the property of the dot product, this change is maximum when $\boldsymbol{u}$ is parallel to ∇f (note that $||\boldsymbol{u}|| = 1$):

$$D_{\boldsymbol{u}} f = \nabla f \cdot \boldsymbol{u} = ||\nabla f|| ||\boldsymbol{u}|| \cos\theta \Longrightarrow \max D_{\boldsymbol{u}} f = ||\nabla f|| \text{ when } \theta = 0$$

where θ is the angle between $\boldsymbol{u}$ and ∇f; the notation $||\nabla f||$ means the Euclidean length of ∇f—which is $||\nabla f|| = \sqrt{f_x^2 + f_y^2 + f_z^2}$.

Let's see how the gradient vector looks like geometrically. For the function $f(x, y) = x^2 + y^2$, we plot its gradient field[6] $2x\boldsymbol{i} + 2y\boldsymbol{j}$ superimposed with the level curves of $f(x, y)$ on Fig. 6.6a. We can see that the *gradient vectors are perpendicular to the level curves*. This is because going along a level curve does not change f: $D_{\boldsymbol{u}} f = \nabla f \cdot \boldsymbol{u} = 0$ when $\boldsymbol{u}$ is perpendicular to ∇f.

So far, we have considered functions of two variables only. How about functions of three variables $w = f(x, y, z)$? We believe that at any point $P = (x_0, y_0, z_0)$ on the level surface $f(x, y, z) = c$ the gradient ∇f is perpendicular to the surface. By this, we mean it is perpendicular to the tangent to any curve that lies on the surface and goes through P (see Fig. 6.6b).

Proof Let's consider a level surface S given by $f(x, y, z) = c$, and a point $P(x_0, y_0, z_0)$ on this surface. Then, we consider a curve C lying on the surface and passing through P (see Fig. 6.6b). We describe C parametrically by $C : (x(t), y(t), z(t))$. Now, for any point on C, we have

$$f(x(t), y(t), z(t)) = c$$

(because points on C are also on S). Using the chain rule, we're going to differentiate the above equation

[5] Refer to Sect. 9.1.2 if you need a refresh on the concept of the dot product of two vectors.

[6] What is a gradient field? At a point (x_0, y_0) the gradient of $f(x, y)$ is a vector. Hence, for every point on the 2D plane, we have a bunch of gradient vectors; together they make a vector field, called the gradient field. We have more to say about fields in Sect. 6.11.

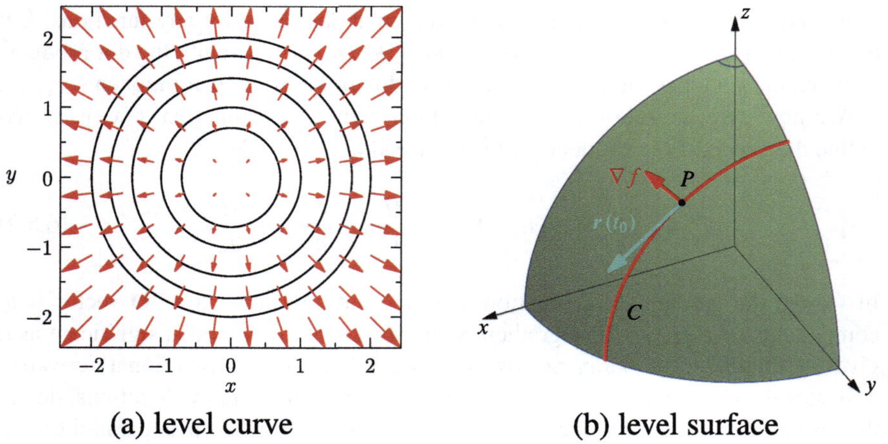

Fig. 6.6 The gradient vector ∇f, those red arrows in (**a**), of a level curve $f(x, y)$ is perpendicular to it (**a**) and the gradient vector is also perpendicular to level surfaces (**b**)

$$\frac{\partial f}{\partial x}\frac{dx}{dt} + \frac{\partial f}{\partial y}\frac{dy}{dt} + \frac{\partial f}{\partial z}\frac{dz}{dt} = 0$$

So, the dot product of ∇f and the vector $(dx/dt, dy/dt, dz/dt)$ is zero, hence ∇f is perpendicular to the vector $\boldsymbol{r} = (dx/dt, dy/dt, dz/dt)$, which is nothing but the tangent vector to the curve C. ■

Why (f_x, f_y, f_z) makes a vector?

It is not true that every three numbers make a vector. For example, we cannot make a vector from this (f_{xx}, f_y, f_z). How to prove that (f_x, f_y, f_z) is indeed a vector? We use the fact that the dot product of two vectors is a scalar. To this end, we consider two nearby points $P_1(x, y, z)$ and $P_2(x + \Delta x, y + \Delta y, z + \Delta z)$. Assume that the temperature at P_1 is T_1 and the temperature at P_2 is T_2. Obviously, T_1 and T_2 are scalars: they are independent of the coordinate system we use. The difference of temperature ΔT is also a scalar, it is given by

$$\Delta T = \frac{\partial T}{\partial x}\Delta x + \frac{\partial T}{\partial y}\Delta y + \frac{\partial T}{\partial z}\Delta z$$

Since ΔT is a scalar, and $(\Delta x, \Delta y, \Delta z)$ is a vector (joining P_1 to P_2), we can deduce that (T_x, T_y, T_z) is a vector.

6.6 Chain Rules

Chain rules are for function composition. With multiple variables, there are many possibilities. Without loss of generality, we consider the following three cases:

1. $f(z)$ with $z = g(x, y)$. We need f_x and f_y;
2. $f(x, y)$ with $x = x(t)$, $y = y(t)$. We need df/dt, as f is just a function of t.
3. $f(x, y)$ with $x = x(u, v)$, $y = y(u, v)$. We need f_u and f_v.

Case 1: $f(z)$ with $z = g(x, y)$. For example, with $f(z) = e^z$ and $z = x^2 + y^2$ we get $f(x, y) = e^{x^2+y^2}$. Thus, $f_x = 2xe^{x^2+y^2}$ and $f_y = 2ye^{x^2+y^2}$. The rule is

$$\frac{\partial f}{\partial x} = \frac{\partial f}{\partial z}\frac{\partial z}{\partial x}, \quad \frac{\partial f}{\partial y} = \frac{\partial f}{\partial z}\frac{\partial z}{\partial y} \tag{6.6.1}$$

Case 2: $f(x, y)$ with $x = x(t)$, $y = y(t)$. This is simple: $df/dt = f_x dx/dt + f_y dy/dt$. When t changes Δt, both x and y change by $\Delta x \approx (dx/dt)\Delta t$ and $\Delta y \approx (dy/dt)\Delta t$, respectively. These changes lead to a change in f:

$$\Delta f \approx f_x \Delta x + f_y \Delta y = \frac{\partial f}{\partial x}\frac{dx}{dt}\Delta t + \frac{\partial f}{\partial y}\frac{dy}{dt}\Delta t$$

Dividing by Δt and letting it go to zero, we get the formula: $df/dt = f_x dx/dt + f_y dy/dt$.

Case 3: $f(x, y)$ with $x = x(u, v)$, $y = y(u, v)$. By holding v constant and using the chain rule in case 2, we can write $\frac{\partial f}{\partial u} = \frac{\partial f}{\partial x}\frac{\partial x}{\partial u} + \frac{\partial f}{\partial y}\frac{\partial y}{\partial u}$. Doing the same thing for $\frac{\partial f}{\partial v}$, and putting these two together, we have

$$\begin{aligned}
\frac{\partial f}{\partial u} &= \frac{\partial f}{\partial x}\frac{\partial x}{\partial u} + \frac{\partial f}{\partial y}\frac{\partial y}{\partial u} \\
\frac{\partial f}{\partial v} &= \frac{\partial f}{\partial x}\frac{\partial x}{\partial v} + \frac{\partial f}{\partial y}\frac{\partial y}{\partial v}
\end{aligned} \tag{6.6.2}$$

This rule can be rewritten in a matrix form as

$$\begin{bmatrix} \dfrac{\partial f}{\partial u} \\ \dfrac{\partial f}{\partial v} \end{bmatrix} = \begin{bmatrix} \dfrac{\partial x}{\partial u} & \dfrac{\partial y}{\partial u} \\ \dfrac{\partial x}{\partial v} & \dfrac{\partial y}{\partial v} \end{bmatrix} \begin{bmatrix} \dfrac{\partial f}{\partial x} \\ \dfrac{\partial f}{\partial y} \end{bmatrix} \tag{6.6.3}$$

We can generalize this to the case of a function of n variables, $f(x_1, x_2, \ldots, x_n)$ and the variables depend on m other variables, $x_i = x_i(u_1, u_2, \ldots, u_m)$ for $i = 1, 2, \ldots, n$, then we have

$$\frac{\partial f}{\partial u_j} = \frac{\partial f}{\partial x_1}\frac{\partial x_1}{\partial u_j} + \frac{\partial f}{\partial x_2}\frac{\partial x_2}{\partial u_j} + \cdots + \frac{\partial f}{\partial x_n}\frac{\partial x_n}{\partial u_j} \quad (1 \le j \le m)$$

$$= \sum_{i=1}^{n} \frac{\partial f}{\partial x_i}\frac{\partial x_i}{\partial u_j} = \frac{\partial f}{\partial x_i}\frac{\partial x_i}{\partial u_j} \quad \text{(Einstein's summation rule on dummy index } i\text{)}$$

6.7 Minima and Maxima of Functions of Two Variables

This section discusses the problem of finding the minima/maxima of the function $z = f(x, y)$. We shall see that this problem is similar to the case of $y = f(x)$ with only one change: partial derivatives are used. Stationary points are discussed in Sect. 6.7.1. The Taylor series of multivariate functions is given in Sect. 6.7.2, and the multi-index notation is presented in Sect. 6.7.3. This notation provides us a convenient way to write partial derivatives and thus we can write the Taylor series of any order for a function of n variables. Quadratic forms are then discussed in Sect. 6.7.4. For a spring of stiffness k its potential energy is $1/2kx^2$, this is a quadratic form even though the simplest one. For unknown reason, quadratic forms appear again and again in physics and mathematics. And finally, constrained minimization problems (e.g., finding the minimum of $x^2 + y^2$ subjecting to $x + y = 5$) using the Lagrange multiplier method is the topic of Sect. 6.7.5.

6.7.1 Stationary Points and Partial Derivatives

Consider a general function of two variable $z = f(x, y)$, of which the graph of such a function is shown in Fig. 6.7. Similar to functions of one variable, at stationary points (which can be a local minimum, local maximum, absolute minimum, etc.) the tangent planes are horizontal. So, at a stationary point (x_0, y_0) the two first partial derivatives are zero (check Eq. (6.3.2) for the equation of a plane if this is not clear):

$$f_x(x_0, y_0) = f_y(x_0, y_0) = 0 \tag{6.7.1}$$

Saddle point. If we consider the function $z = y^2 - x^2$, the stationary point is $(0, 0)$ using Eq. (6.7.1). But this point cannot be a minimum or a maximum point, see Fig. 6.8. We can see that $f(0, 0) = 0$ is a maximum along the x-direction but a minimum along the y-direction. Near the origin the graph has the shape of a saddle and so $(0, 0)$ is called a saddle point of f.

Minimum or maximum or saddle point. For $y = f(x)$, we need to use the second derivative at the stationary point x_0, $f''(x_0)$, to decide if x_0 is a minimum or maximum or inflection point. How did the second derivative help? It decides whether the curve $y = f(x)$ is below the tangent at x_0 (i.e., if $y''(x_0) < 0$ then x_0 is a maximum point

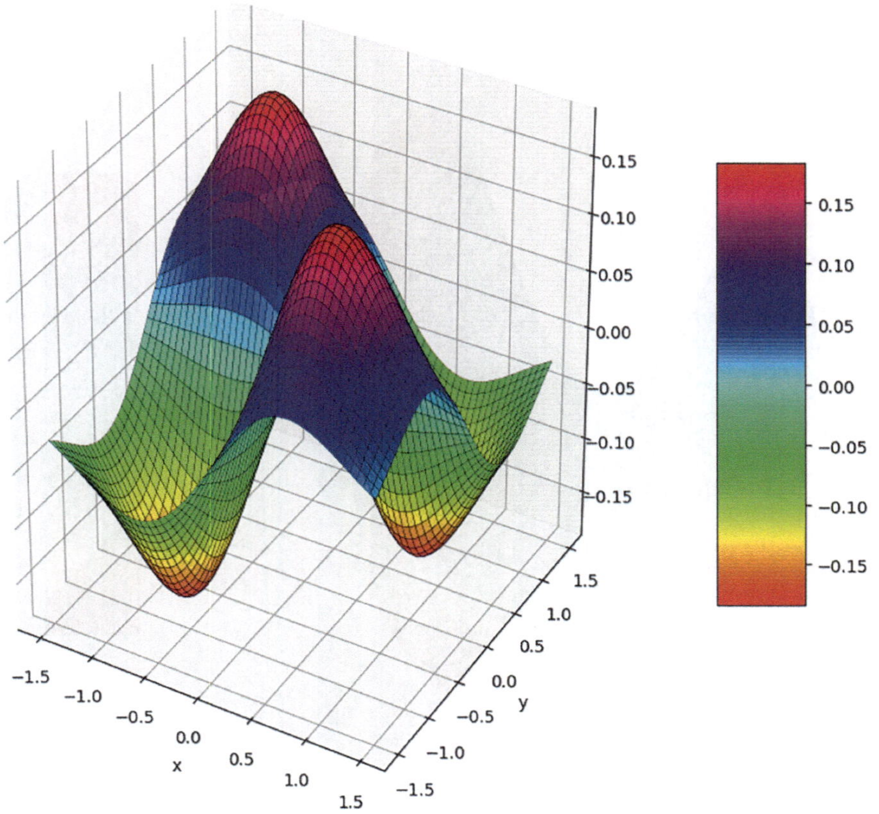

Fig. 6.7 Graph of a function of two variables $z = f(x, y)$ with a colorbar representing the height z. Using a colorbar is common in visualizing functions, especially functions of three variables. We can quickly spot the highest/lowest points based on the color

as we're going downhill) or it is above the tangent (i.e., if $y''(x_0) > 0$ then x_0 is a minimum point). We believe this reasoning also applies for $f(x, y)$. The difficulty is that we now have three second derivatives f_{xx}, f_{yy}, f_{xy} not one!

The idea is to replace the general function $f(x, y)$ by a quadratic function of the form $ax^2 + bxy + cy^2$ to which finding its extreme is straightforward (using only algebra). The means to do this is Taylor's series expansion of $f(x, y)$, see Sect. 6.7.2, around the stationary point (x_0, y_0) up to the second order (as the bending of a surface depends on second order terms only):

$$\begin{aligned}f(x, y) \approx & f(x_0, y_0) + f_x(x_0, y_0)(x - x_0) + f_y(x_0, y_0)(y - y_0) + \\ & + \frac{f_{xx}(x_0, y_0)}{2}(x - x_0)^2 + \frac{f_{yy}(x_0, y_0)}{2}(y - y_0)^2 + f_{xy}(x_0, y_0)(x - x_0)(y - y_0)\end{aligned}$$

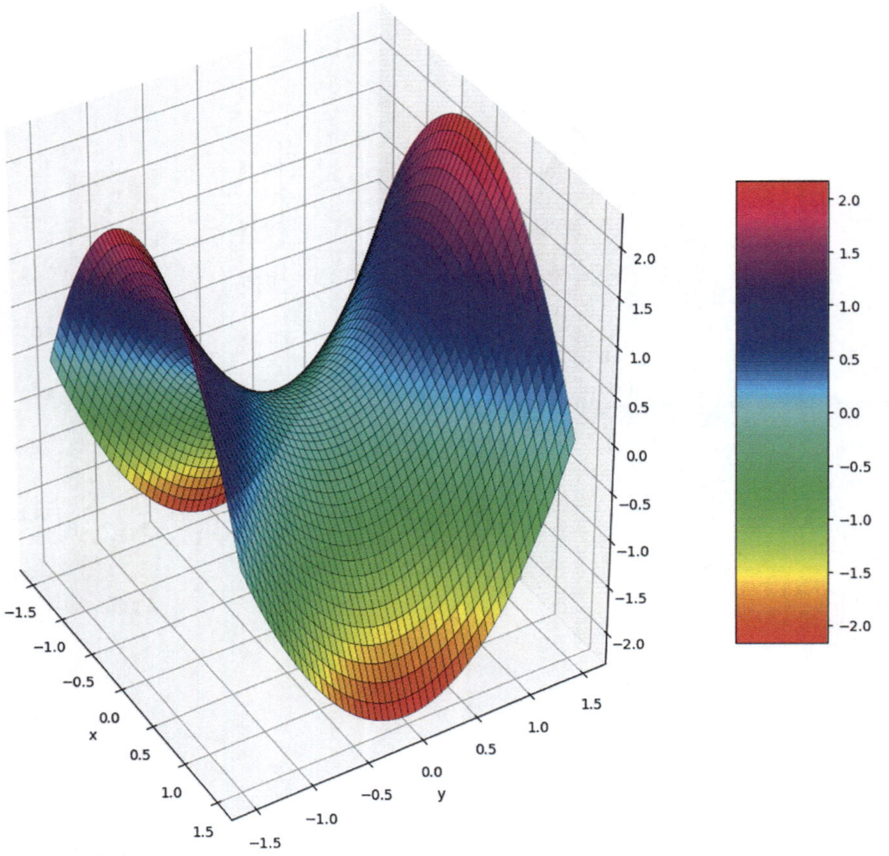

Fig. 6.8 Graph of $z = y^2 - x^2$ which has a saddle point at (0, 0)

At stationary point (x_0, y_0), the first two partial derivatives are zero, so the above is simplified to

$$f(x, y) \approx \frac{f_{xx}(x_0, y_0)}{2}(x - x_0)^2 + \frac{f_{yy}(x_0, y_0)}{2}(y - y_0)^2 + f_{xy}(x_0, y_0)(x - x_0)(y - y_0)$$

where the constant term $f(x_0, y_0)$ was skipped as it does not affect the characteristic of the stationary point (it does change the extremum value of the function, but we're not interested in that here).

Now, we can write $f(x, y)$ in the following quadratic form (we have multiplied the above equation by two, and assume $(x_0, y_0) = (0, 0)$, as if it is not the case we can always do a translation to make it so):

$$f(x, y) = ax^2 + 2bxy + cy^2, \quad a = f_{xx}(0, 0), \quad b = f_{xy}(0, 0), \quad c = f_{yy}(0, 0) \tag{6.7.2}$$

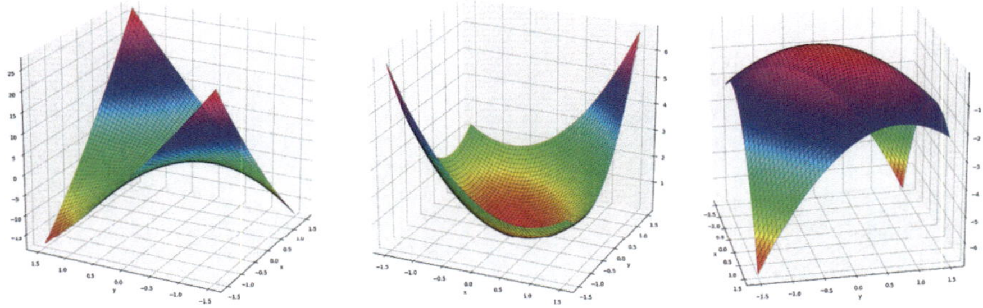

(a) $z = x^2 + 10xy + y^2$ (b) $z = x^2 + xy + y^2$ (c) $z = -x^2 + xy - y^2$

Fig. 6.9 Examples to verify the second derivatives test

which can be rewritten as

$$f(x, y) = a\left[\left(x + \frac{by}{a}\right)^2 + y^2\left(\frac{ac - b^2}{a^2}\right)\right] \tag{6.7.3}$$

from which we can conclude:

$$\begin{array}{lll} \text{if } a > 0 \text{ and } ac > b^2: & f(x, y) > 0\ \forall x, y & \text{minimum at } (0, 0) \\ \text{if } a < 0 \text{ and } ac > b^2: & f(x, y) < 0\ \forall x, y & \text{maximum at } (0, 0) \\ \text{if } ac < b^2: & \text{the parts have opposite signs} & \text{saddle point at } (0, 0) \end{array}$$

This is called a second derivatives test. Figure 6.9 confirms this test. It is helpful to examine the contour plot of the surfaces in Fig. 6.9 to understand geometrically when a function has a min/max/saddle point. Figure 6.10 tells us that around a max/min point the *level curves are oval*, because going any direction will decrease/increase the function. On the other hand, around a saddle point, the level curves are hyperbolas ($xy = c$).

Often, as a means to remember the condition on the sign of $D = ac - b^2$, D is written as the determinant of the following 2×2 matrix containing all the second partial derivatives of f:

$$D = \det\begin{bmatrix} f_{xx} & f_{xy} \\ f_{xy} & f_{yy} \end{bmatrix} = f_{xx}f_{yy} - f_{xy}^2 = ac - b^2 \tag{6.7.4}$$

This matrix is special as it stores all the second derivatives of $f(x, y)$. It must have a special name. It is called a Hessian matrix, named after the German mathematician Ludwig Otto Hesse (22 April 1811–4 August 1874).

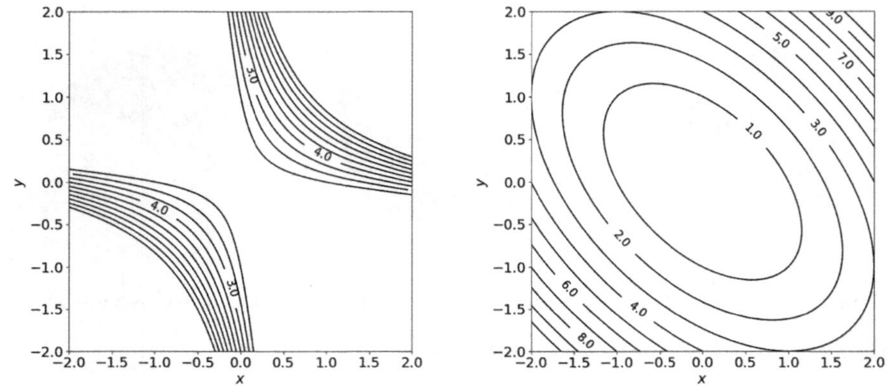

(a) $z = x^2 + 10xy + y^2$ (b) $z = x^2 + xy + y^2$

Fig. 6.10 Contour plots of $z = x^2 + 10xy + y^2$ (saddle point) and $z = x^2 + xy + y^2$ (minimum point)

6.7.2 Taylor's Series of Scalar-Valued Multivariate Functions

We have Taylor's series of functions of single variable $y = f(x)$. Of course, we also have Taylor's series for multivariate functions, $z = f(x, y)$ or $h = f(x, y, z)$, etc. First, we develop a second-order Taylor's polynomial that approximates a function of two variables $z = f(x, y)$.

The second-order Taylor's polynomial has this general form $T(x, y) = a + bx + cy + dxy + ex^2 + fy^2$. We find the coefficients $a, b, c \ldots$ by matching the function at (0, 0) and all derivatives up to second order at (0, 0). The same old idea we have met in univariate calculus:

$$\begin{aligned}
f(0,0) &= T(0,0) &&\Longrightarrow a = f(0,0)\\
f_x(0,0) &= T_x(0,0) &&\Longrightarrow b = f_x(0,0)\\
f_y(0,0) &= T_y(0,0) &&\Longrightarrow c = f_y(0,0)\\
f_{xy}(0,0) &= T_{xy}(0,0) &&\Longrightarrow d = f_{xy}(0,0)\\
f_{xx}(0,0) &= T_{xx}(0,0) &&\Longrightarrow e = \tfrac{1}{2} f_{xx}(0,0)\\
f_{yy}(0,0) &= T_{yy}(0,0) &&\Longrightarrow f = \tfrac{1}{2} f_{yy}(0,0)
\end{aligned}$$

Thus, the second-order Taylor's series for $z = f(x, y)$ is written as

$$f(\boldsymbol{x}) \approx f(\boldsymbol{0}) + f_x(\boldsymbol{0})x + f_y(\boldsymbol{0})y + \frac{1}{2}\left[f_{xx}(\boldsymbol{0})x^2 + 2f_{xy}(\boldsymbol{0})xy + f_{yy}(\boldsymbol{0})y^2\right] \tag{6.7.5}$$

Now, we rewrite this equation using vector-matrix notation, the advantage is that the same equation holds for functions of more than two variables:

$$\begin{aligned} f(\boldsymbol{x}) &\approx f(\mathbf{0}) + \begin{bmatrix} f_x(\mathbf{0}) & f_y(\mathbf{0}) \end{bmatrix} \begin{bmatrix} x \\ y \end{bmatrix} + \frac{1}{2} \begin{bmatrix} x & y \end{bmatrix} \begin{bmatrix} f_{xx}(\mathbf{0}) & f_{xy}(\mathbf{0}) \\ f_{xy}(\mathbf{0}) & f_{yy}(\mathbf{0}) \end{bmatrix} \begin{bmatrix} x \\ y \end{bmatrix} \\ &\approx f(\mathbf{0}) + \nabla f^\top(\mathbf{0})\boldsymbol{x} + \frac{1}{2}\boldsymbol{x}^\top \mathbf{H}(\mathbf{0})\boldsymbol{x} \quad (\boldsymbol{x} = \begin{bmatrix} x \\ y \end{bmatrix}) \\ &\approx f(\mathbf{0}) + \boldsymbol{x}^\top \nabla f(\mathbf{0}) + \frac{1}{2}\boldsymbol{x}^\top \mathbf{H}(\mathbf{0})\boldsymbol{x} \end{aligned} \tag{6.7.6}$$

where the linear part $f_x(\mathbf{0})x + f_y(\mathbf{0})y$ is rewritten as the dot product[7] of the gradient vector and $\boldsymbol{x}$. The quadratic term is rewritten as $\boldsymbol{x}^\top \mathbf{H}\boldsymbol{x}$, as any quadratic form (to be discussed in Sect. 6.7.4) can be written in this form. As the dot product of two vectors is symmetric, we can write the linear term in another way as in the final expression.

6.7.3 Multi-index Notation

In the previous section, I intentionally wrote the Taylor series for $y = f(x, y)$ up to second-order terms. This was simply because it would be tedious to include higher order terms. For functions of a single variable, we are able to write the Taylor series:

$$f(x) = \sum_{n=0}^{\infty} \frac{f^n(x_0)}{n!}(x - x_0)^n$$

The question now is: can we have the same formula as above for $y = f(x_1, x_2, \ldots, x_n)$? The answer is yes and, to that end, mathematicians have developed the so-called multi-index, which generalizes the concept of an integer index to an ordered tuple of indices.

A multi-index $\alpha = (\alpha_1, \alpha_2, \ldots, \alpha_n)$ is an n-tuple of non-negative integers. For example, $\alpha = (1, 2, 3)$, or $\alpha = (2, 3, 6)$. The norm of a multi-index α is defined to be

$$|\alpha| = \alpha_1 + \alpha_2 + \cdots + \alpha_n$$

For a vector $\boldsymbol{x} = (x_1, x_2, \ldots, x_n) \in \mathbb{R}^n$, define

$$\boldsymbol{x}^\alpha = \prod_{i=1}^{n} x_i^{\alpha_i}$$

Also, define the factorial of a multi-index α by

$$\alpha! = \prod_{i=1}^{n} \alpha_i!$$

[7] Refer to Sect. 9.1.2 if you need a refresh on the concept of the dot product of two vectors.

Finally, to write partial derivatives we define the differential operator

$$\partial^\alpha = D^\alpha = \left(\frac{\partial}{\partial x_1}\right)^{\alpha_1}\left(\frac{\partial}{\partial x_2}\right)^{\alpha_2}\cdots\left(\frac{\partial}{\partial x_n}\right)^{\alpha_n}, \quad \left(\frac{\partial}{\partial x_i}\right)^{\alpha_i} := \frac{\partial^{\alpha_i}}{\partial x_i^{\alpha_i}} \tag{6.7.7}$$

With this multi-index notation, the Taylor series for a function $y = f(x_1, \ldots, x_n)$ is given by

$$f(\boldsymbol{x}) = \sum_{|\alpha|=0}^{\infty} \frac{D^\alpha f(\boldsymbol{x}_0)}{\alpha!}(\boldsymbol{x} - \boldsymbol{x}_0)^\alpha$$

To understand this, let's consider $y = f(x_1, x_2)$. The first term in the sum $\sum_{|\alpha|=0}$ is when $|\alpha| = 0$, which is $\alpha = (0, 0)$. Then, $\alpha! = 0!0! = 1$, and $D^\alpha f = f$ according to Eq. (6.7.7). The second term is when $|\alpha| = 1$, which can be $\alpha = (1, 0)$ or $\alpha = (0, 1)$. For the former, we then have $D^\alpha f = \frac{\partial f}{\partial x_1}$, and for the latter $D^\alpha f = \frac{\partial f}{\partial x_2}$. The third term, $|\alpha| = 2$: $\alpha = (2, 0), \alpha = (1, 1)$ and $\alpha = (0, 2)$. We then have, using Eq. (6.7.7)

- If $\alpha = (2, 0)$, then $D^\alpha f = f_{x_1x_1}$;
- If $\alpha = (0, 2)$, then $D^\alpha f = f_{x_2x_2}$;
- If $\alpha = (1, 1)$, then $D^\alpha f = f_{x_1x_2}$;

6.7.4 *Quadratic Forms*

Quadratic forms are homogeneous polynomials of second degree. Let's denote by x_1, x_2, x_3 the variables, then the following are quadratic forms in terms of x_1, x_2, x_3 ($a_1, a_2, \ldots$ are real constants):

$$\begin{aligned}
Q(x_1) &= a_1x_1^2 \\
Q(x_1, x_2) &= a_1x_1^2 + a_2x_1x_2 + a_3x_2^2 \\
Q(x_1, x_2, x_3) &= a_1x_1^2 + a_2x_1x_2 + a_3x_1x_3 + a_4x_2x_3 + a_5x_2^2 + a_6x_3^2
\end{aligned} \tag{6.7.8}$$

Note that we have not used the conventional x, y, z; instead, we have used x_1, x_2, x_3. This is because if we generalize our quadratic forms to the case of, let say 100, variables we will run out of symbols using $x, y, z, \ldots$

Now, we rewrite this quadratic form $Q(x_1, x_2) = a_1x_1^2 + a_2x_1x_2 + a_3x_2^2$ as follows:

$$Q(x_1, x_2) = a_{11}x_1^2 + a_{12}x_1x_2 + a_{21}x_2x_1 + a_{22}x_2^2 = \sum_{i=1}^{2}\sum_{j=1}^{2} a_{ij}x_ix_j$$

so that it can be written using matrix-vector as[8]:

[8] If you're not familiar with matrices, refer to Chap. 9.

$$Q(x_1, x_2) = \begin{bmatrix} x_1 & x_2 \end{bmatrix} \begin{bmatrix} a_{11} & a_{12} \\ a_{21} & a_{22} \end{bmatrix} \begin{bmatrix} x_1 \\ x_2 \end{bmatrix} = \boldsymbol{x}^\top \mathbf{A} \boldsymbol{x} \quad \left(\boldsymbol{x} := \begin{bmatrix} x_1 \\ x_2 \end{bmatrix} \right)$$

So we have just demonstrated that any quadratic form can be expressed in this form $\boldsymbol{x}^\top \boldsymbol{A}\boldsymbol{x}$. Let's do that for this particular quadratic form $Q(x_1, x_2) = x_1^2 + 5x_1x_2 + 3x_2^2$:

$$Q(x_1, x_2) = \begin{bmatrix} x_1 & x_2 \end{bmatrix} \begin{bmatrix} 1 & 1 \\ 4 & 3 \end{bmatrix} \begin{bmatrix} x_1 \\ x_2 \end{bmatrix} = \begin{bmatrix} x_1 & x_2 \end{bmatrix} \begin{bmatrix} 1 & 5/2 \\ 5/2 & 3 \end{bmatrix} \begin{bmatrix} x_1 \\ x_2 \end{bmatrix}$$

It is certain that we prefer the red matrix—which is symmetric, i.e., $a_{12} = a_{21} = 5/2$—than the non-symmetric matrix (the blue one). So, *any quadratic form* can be expressed in this form $\boldsymbol{x}^\top \mathbf{A}\boldsymbol{x}$, where $\mathbf{A}$ is a symmetric matrix. We need a proof because we used the strong word *any* quadratic form, while we just had one example.

Proof Suppose $\boldsymbol{x}^\top \mathbf{B}\boldsymbol{x}$ is a quadratic form where $\mathbf{B}$ is not symmetric. Since it is a scalar, we get the same thing when we transpose it:

$$\boldsymbol{x}^\top \mathbf{B} \boldsymbol{x} = \boldsymbol{x}^\top \mathbf{B}^\top \boldsymbol{x}$$

Thus, we can write

$$\boldsymbol{x}^\top \mathbf{B} \boldsymbol{x} = \frac{1}{2} \left(\boldsymbol{x}^\top \mathbf{B} \boldsymbol{x} + \boldsymbol{x}^\top \mathbf{B}^\top \boldsymbol{x} \right) = \boldsymbol{x}^\top \frac{1}{2} (\mathbf{B} + \mathbf{B}^\top) \boldsymbol{x}$$

The red matrix is our symmetric matrix $\mathbf{A}$. ■

Why quadratic forms? Because, for unknown reasons, they show up again and again in mathematics, physics, engineering, and economics. The simplest example is $1/2kx^2$, which is the energy of a spring of stiffness k. There are more to say about quadratic forms in Sect. 9.10.6, such as positive definiteness of a quadratic form.

6.7.5 Constraints and Lagrange Multipliers

Now we consider a constrained minimization problem: find the minima (or maxima) of a function $z = f(x, y)$ subject to the constraint $g(x, y) = k$. You might be thinking: we can definitely solve it by first solving for x in terms of y using the constraint equation, then substituting that x into the function $f(x, y)$ to get a function of one variable (y). There are, however, many issues with this approach. First, for complex $g(x, y) = k$, it is not possible to solve x in terms of y and vice versa. Second, why eliminating x not y? We are destroying the symmetry of the problem. Third, this technique is hard to apply (or impossible) for problems with *many constraints*. Ok, we need a new way. And it is quite often true that new problems require new mathematics.

Joseph-Louis Lagrange (1736–1813), an Italian mathematician and astronomer, found that new mathematics and herein we reproduce his method. Let's start with an

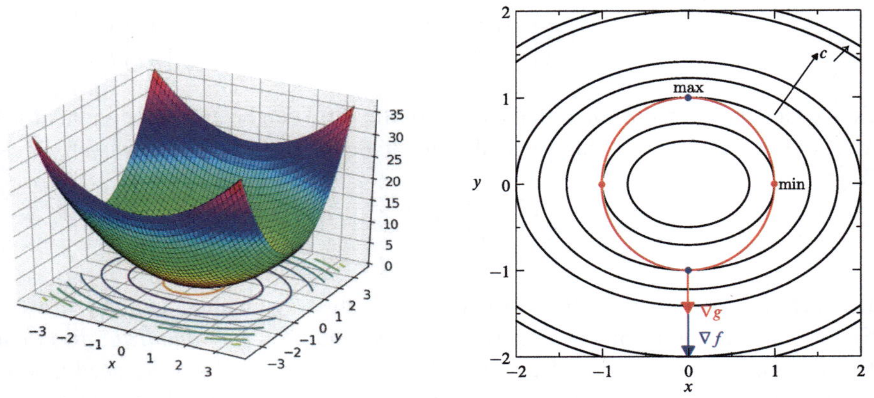

(a) $z = x^2 + 2y^2$ and its level curves

(b) $x^2 + 2y^2 = c$ & $x^2 + y^2 = 1$

Fig. 6.11 Graph of $z = x^2 + 2y^2$ and its level curves (**a**) and the level curves and the constraint (red) curve $x^2 + y^2 = 1$ on a plane (**b**). So we start with the smallest ellipse $x^2 + 2y^2 = c_1$ in (b), c_1 cannot be the minimum value of $z(x, y)$ as the constraint is not satisfied. So, we keep climbing the mountain $z(x, y)$ until we're at the second smallest ellipse. Let's say $x^2 + 2y^2 = c_2$ which touches the constraint curve. Here we have two minima marked by "min" both giving the minimum of c_2 of $z(x, y)$. If we continue going up we will touch the constraint curve the second time: that's where $z(x, y)$ attains the maximum value

example of finding the minima/maxima of $z(x, y) = x^2 + 2y^2$ with the constraint $x^2 + y^2 = 1$. Wherever the extremum points are they must be on the unit circle centered at the origin in the xy-plane. On this plane, we also plot the level curves $x^2 + 2y^2 = c$, which are ellipses, where c ranges from zero to infinity.[9] It is then clear that the extremum points of $z(x, y)$ are where the level curve touches the constraint curve. From Fig. 6.11, we know that they are the points $(\pm 1, 0)$ and $(0, \pm 1)$.

At the touching point of two curves, the tangents are the same. In other words, the normal vectors are parallel:

$$\nabla f(x, y) = \lambda \nabla g(x, y), \quad \text{or} \quad \begin{cases} f_x = \lambda g_x \\ f_y = \lambda g_y \end{cases} \tag{6.7.9}$$

where λ is a real number. These are the two equations to solve for the three unknowns x, y, λ. But do not forget the constraint $x^2 + y^2 = 1$. Three equations for three unknowns. Perfect.

Without constraints, the necessary condition for a function $f(x, y)$ to be stationary at (x_0, y_0) is $\nabla f(x_0, y_0) = \mathbf{0}$. With the constraint $g(x, y) = 0$, we have instead Eq. (6.7.9). With a bit of algebra, we can see the old criterion of zero gradient. Let's introduce a new function $L(x, y, \lambda)$ as (L is called the Lagrange function)

$$L(x, y, \lambda) := f(x, y) - \lambda(g(x, y) - k) \tag{6.7.10}$$

---

[9] Noting that $z = x^2 + 2y^2 \geq 0 \; \forall x, y$.

And we now compute the gradient of L:

$$\nabla L = (f_x - \lambda g_x, f_y - \lambda g_y, -g + k)$$

The condition $\nabla L = \mathbf{0}$ then resembles Eq. (6.7.9) and $g(x, y) = k$. So, by adding one more unknown λ to the problem, and building a new function $L(x, y, \lambda)$, Lagrange *turned a constrained minimization problem into an unconstrained minimization problem!* This method is now known as the Lagrange multiplier method and λ is called a *Lagrange multiplier*. Once Eq. (6.7.10) has been solved, we get possibly a few solutions $(\bar{x}_i, \bar{y}_i)$; the maximum of $f(\bar{x}_i, \bar{y}_i)$ is the maximum we're looking for, and minimum of $f(\bar{x}_i, \bar{y}_i)$ is the minimum we sought for.

As an example, we consider the problem given in Fig. 6.11. Equation (6.7.9) and the constraint gives us the following system of equations to solve for x, y, λ:

$$2x = 2\lambda x, \quad 4y = 2\lambda y, \quad x^2 + y^2 = 1$$

From the first equation, we either get $x = 0$ (which leads to $y = \pm 1$ from the constraint) or $\lambda = 1$. From the second equation, we obtain either $y = 0$ (which leads to $x = \pm 1$ from the constraint) or $\lambda = 2$. So, we have four points $(0, -1, 2)$, $(0, 1, 2)$, $(-1, 0, 1)$, and $(1, 0, 1)$. These points are exactly the ones we have found graphically (shown in Fig. 6.11b). Evaluating f at these four points:

$$f(0, -1) = 2, \quad f(0, 1) = 2, \quad f(-1, 0) = 1, \quad f(1, 0) = 1$$

So the maximum of f is 2 at $(0, \pm 1)$ and the minimum of f is 1 at $(\pm 1, 0)$.

Justifying Lagrange multipliers. We need a proof in addition to the above geometrical derivation. The constraint curve $g(x, y) = k$ can be parametrized as $C(t)$: $(x(t), y(t))$. Now, assuming that $P(x_0, y_0)$ is a point on the surface $z(x, y)$ where the function is a (local) maximum. Of course, P is on the constraint curve C, it is $C(t_0)$. We build the function $h(t) = f(x(t), y(t))$. This function has a maximum at t_0, so we have

$$h'(t_0) = 0 \Longleftrightarrow \frac{\partial f}{\partial x}(x_0, y_0)\frac{dx}{dt}(t_0) + \frac{\partial f}{\partial y}(x_0, y_0)\frac{dy}{dt}(t_0) = 0 \Longleftrightarrow \nabla f(P) \cdot \boldsymbol{r}(P) = 0$$

Hence, at P, ∇f is perpendicular to the constraint curve.[10] But ∇g is also perpendicular to the curve, thus ∇f and ∇g are parallel at P.

The geometric interpretation also reveals that the method still works for the function of three variables $f(x, y, z)$ and the constraint $g(x, y, z) = k$. Instead of level curves, we have level surfaces. And only when the level surface $f(x, y, z) = c$

[10] This result is nothing but the fact that the directional derivative along the constraint curve is zero at P.

touches the level surface $g(x, y, z) = k$, we have a critical point at that touching point.

Meaning of the multiplier.

Two constraints. After one constraint is of course two constraints, and then multiple constraints. For two constraints, we have to move to functions of three variables. Otherwise, two constraints $g(x, y) = c_1$ and $h(x, y) = c_2$ already decide what is the critical point. Nothing left for Lagrange to do!

We start with a concrete example. Consider the function $f(x, y, z) = x^2 + y^2 + z^2$ and two constraints $g(x, y, z) = x + y + z = 9$ and $h(x, y, z) = x + 2y + 3z = 20$. Find the maximum/minimum of f. The two constraints are two planes and they meet at a line C. Now we consider different level surfaces of $f(x, y, z) = x^2 + y^2 + z^2 = c$; they are concentric spheres of radius c (centered at $(0, 0, 0)$). When we increase c from 0, we have expanding spheres, and one of them will *touch* the line C at a point P. At that special point P, we have (for the reason, check Fig. 6.6b):

$$\begin{gathered}\text{the gradient of } f \text{ is perpendicular to } C\\ \text{the gradient of } g \text{ is perpendicular to } C\\ \text{the gradient of } h \text{ is perpendicular to } C\end{gathered}$$

Therefore, all three vectors $\nabla f, \nabla g, \nabla h$ are in the same plane perpendicular to C. Thus, ∇f is a linear combination of ∇g and ∇h. In other words,

$$\nabla f = \lambda_1 \nabla g + \lambda_2 \nabla h \tag{6.7.11}$$

With one constraint $g(x, y) = k$, we have $\nabla f = \lambda_1 \nabla g$. You can see the pattern here. The method still works for any level surfaces $g(x, y, z) = c_1$ and $h(x, y, z) = c_2$, not just planes. Why? Because the gradient vector of a level surface at a point on the surface is perpendicular to it at that point. Now, we build the Lagrange function $L(x, y, z, \lambda_1, \lambda_2) = f(x, y, z) - \lambda_1(g(x, y, z) - c_1) - \lambda_2(h(x, y, z) - c_2)$. Setting the gradient of L to zero: $\nabla L = 0$ will give us Eq. (6.7.11) and the constraints $g(x, y, z) = c_1$ and $h(x, y, z) = c_2$.

Let's get back to the specific case of $f(x, y, z) = x^2 + y^2 + z^2$ and two constraints $g(x, y, z) = x + y + z = 9$ and $h(x, y, z) = x + 2y + 3z = 20$. Equation (6.7.11) gives us

$$2x = \lambda_1 + \lambda_2, \quad 2y = \lambda_1 + 2\lambda_2, \quad 2z = \lambda_1 + 3\lambda_2$$

Substitute these x, y, z into $g(x, y, z) = x + y + z = 9$ and $h(x, y, z) = x + 2y + 3z = 20$, we obtain two equations to solve for λ_1, λ_2: $\lambda_1 = 2$ and $\lambda_2 = 2$. Then, we get $(x, y, z) = (2, 3, 4)$; that's the point P we are looking for.

Proof of the AM-GM inequality. Still remember the AM-GM inequality that states

$$\frac{x_1 + x_2 + \cdots + x_n}{n} \geq \sqrt[n]{x_1 x_2 \ldots x_n}$$

which was proved by Cauchy with his genius backward-forward induction method. Well, with Lagrange and calculus, the proof is super easy. We demonstrate the proof for $n = 3$.

We consider the following function, with the constraint:

$$f(\boldsymbol{x}) = \sqrt[3]{x_1 x_2 x_3} \quad \text{s.t} \;\; x_1 + x_2 + x_3 = c$$

Then we construct the Lagrange function $L(x)$:

$$L(\boldsymbol{x}) := \sqrt[3]{x_1 x_2 x_3} - \lambda(x_1 + x_2 + x_3 - c)$$

And then, we can compute the derivatives of L with respect to x_1, x_2, x_3, then $\nabla L = \boldsymbol{0}$ gives us

$$\begin{aligned}
L_{x_1} &= \frac{1}{3}(x_1 x_2 x_3)^{-2/3} x_2 x_3 - \lambda = 0 \\
L_{x_2} &= \frac{1}{3}(x_1 x_2 x_3)^{-2/3} x_1 x_3 - \lambda = 0 \\
L_{x_3} &= \frac{1}{3}(x_1 x_2 x_3)^{-2/3} x_1 x_2 - \lambda = 0
\end{aligned}$$

Solving this system of equations (easy) gives us $x_1 = x_2 = x_3$, then from the constraint $x_1 + x_2 + x_3 = c$, we get

$$x_1 = x_2 = x_3 = \frac{c}{3}$$

Therefore, the maximum of $f(\boldsymbol{x})$ is $\sqrt[3]{(c/3)^3}$ which is $c/3$ or $1/3(x_1 + x_2 + x_3)$. In other words,

$$\sqrt[3]{x_1 x_2 x_3} \leq \frac{x_1 + x_2 + x_3}{3}$$

6.8 Integration of Multivariable Functions

This section is about integration of $z = f(x, y)$ and $z = f(x, y, z)$. We start with double integral $\iint_D f(x, y)dxdy$ in Sect. 6.8.1. Then, we move to double integrals in polar coordinates (Sect. 6.8.2). Next is triple integral $\iiint_D f(x, y, z)dxdydz$ (Sect. 6.8.3). Triple integrals using spherical and cylindrical coordinates are treated in Sect. 6.8.4. The most important triple integral in physics—Newton's shell theorem—is given in Sect. 6.8.5. It is this theorem that we can treat massive objects such as the

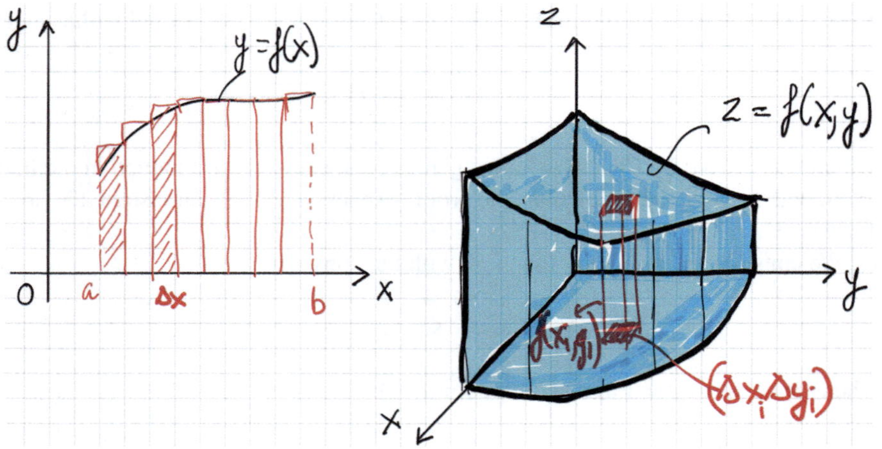

Fig. 6.12 Double integrals as volumes under a surface $z = f(x, y)$

sun as a point mass with all mass concentrated at its center. Then, change of variables and the Jacobian are discussed in Sect. 6.8.6.

One application of double/triple integrals is center of mass and moment of inertia, which is important in physics and engineering. Section 6.8.7 is devoted to such topics. And finally, barycentric coordinates—which are closely related to center of mass—are the topic of Sect. 6.8.8.

6.8.1 Double Integrals

In elementary calculus, we know that $\int_a^b f(x)dx$ can be geometrically seen as the area of a 2D region defined by $x = a$, $x = b$, $y = 0$ and $y = f(x)$. Its natural extension is a double integral that measures a volume of a 3D region. To define this region, on the xy-plane, consider a region R, and for every point x, y in R, we compute $f(x, y)$—the height of the surface at this point.

For 1D integrals, we divide the interval $[a, b]$ into many sub-intervals and compute the area as a sum of the area of all the rectangles (Fig. 6.12). We do the same thing here: the region R is divided into many rectangles $\Delta x_i \Delta y_i$. For a point (x_i, y_i) inside this rectangle, we compute the base $f(x_i, y_i)$ of a box (the 3D counterpart of a rectangle in 2D). Then, the volume is approximated as the sum of all the volumes of these boxes; that is sum of $f(x_i, y_i)\Delta x_i \Delta y_i$. When there are infinitely many such boxes, we get the true volume and define it as a *double*[11] integral:

[11] And that is how mathematicians use the notation with two integral signs.

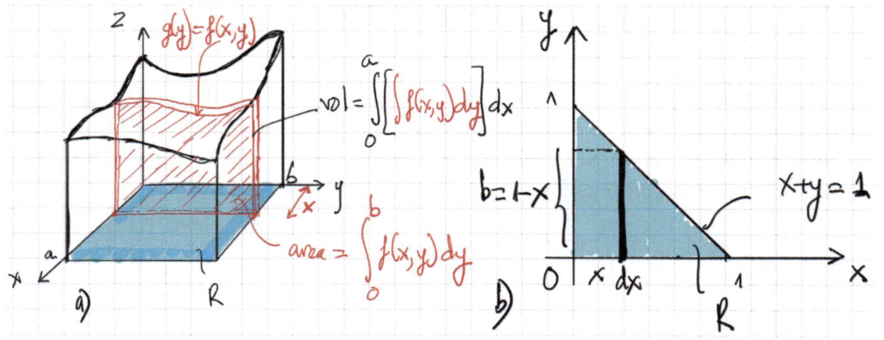

Fig. 6.13 Double integral as an integral (with x) of an integral with y

$$\text{volume} = \lim_{n\to\infty}\sum_{i=1}^{n} f(x_i, y_i)\Delta x_i \Delta y_i = \iint_R f(x, y)dxdy \tag{6.8.1}$$

To compute a double integral, we proceed as shown in Fig. 6.13. First, we consider the plane perpendicular to the x-axis and we fix this plane, this plane intersects with the 3D region of which the volume we're trying to determine. The area of the intersection plane (crossed area in the referred figure) is $A(x) = \int f(x, y)dy$. Multiply this area with the thickness dx we get a volume $A(x)dx$, and integrate this we get the sought-for volume:

$$\iint_R f(x, y)dxdy = \int_0^a \left[\int_0^b f(x, y)dy\right] dx = \int_0^b \left[\int_0^a f(x, y)dx\right] dy \tag{6.8.2}$$

And of course, we can do the other way around. That is why I also wrote the second formula. Noting that the process has been simplified by considering a rectangle for R. In a general case, the integration limits a and b are functions of y and x. The next example is going to show how to handle this situation.

Example 6.1 Compute the volume under $f(x, y) = x - 2y$ and the base triangle (see Fig. 6.13b). Using Eq. (6.8.2), we can write

$$\iint_R (x - 2y)dxdy = \int_0^1 \left[\int_0^{1-x} (x - 2y)dy\right] dx = \int_0^1 \left[xy - y^2\right]_0^{1-x} dx$$

And finally,

$$\iint_R (x - 2y)dxdy = \int_0^1 (-2x^2 + 3x - 1)dx = -1/6$$

6.8.2 Double Integrals in Polar Coordinates

To demonstrate the fact that sometimes it is difficult to compute certain double integrals using x and y, let's compute the mass of a semi-circular plate of unit radius and unit density (refer to Sect. 6.8.7 if you're not familiar with how a mass of a continuous object is computed):

$$m = \iint_R dxdy = 2\int_0^1 \left[\int_0^{\sqrt{1-x^2}} dy\right] dx = 2\int_0^1 \sqrt{1-x^2}dx = 2\int_0^{\pi/2} \cos^2 u du = \frac{\pi}{2}$$

Not hard but still a bit of work. Using polar coordinates (which suitable for circles) is so much more easier. Using polar coordinates, double integrals are given by

$$\boxed{\iint_R f(x,y)dxdy = \iint_S f(r\cos\theta, r\sin\theta)rdrd\theta} \tag{6.8.3}$$

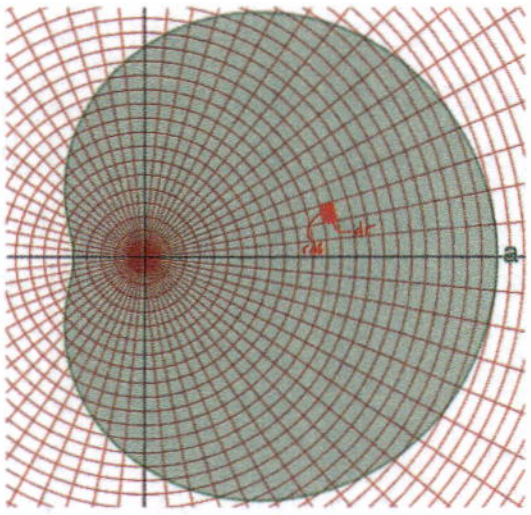

The key point is going from $dxdy$ to $rdrd\theta$ not $drd\theta$. A not-rigorous proof is given here. We chop the integration domain R into infinitely many tiny *polar rectangles*, the integral is then written as the Riemann sum of the area of these polar rectangles multiplied with the integrand evaluated at the centers of the rectangles. So, we just need to compute the area of one polar rectangle, which is $rdrd\theta$ (see figure).

Getting back to the problem of determining the mass of the semi-circular plate, it is much easier with polar coordinates:

$$m = \iint rdrd\theta = \int_0^1 rdr \int_0^{\pi} d\theta = \frac{\pi}{2}$$

As another example of the usefulness of polar coordinates, let's consider the following integral:

$$A = \int_{-\infty}^{\infty} e^{-x^2} dx = \sqrt{\pi}$$

which was computed using polar coordinates, see Sect. 5.11.3 for details.

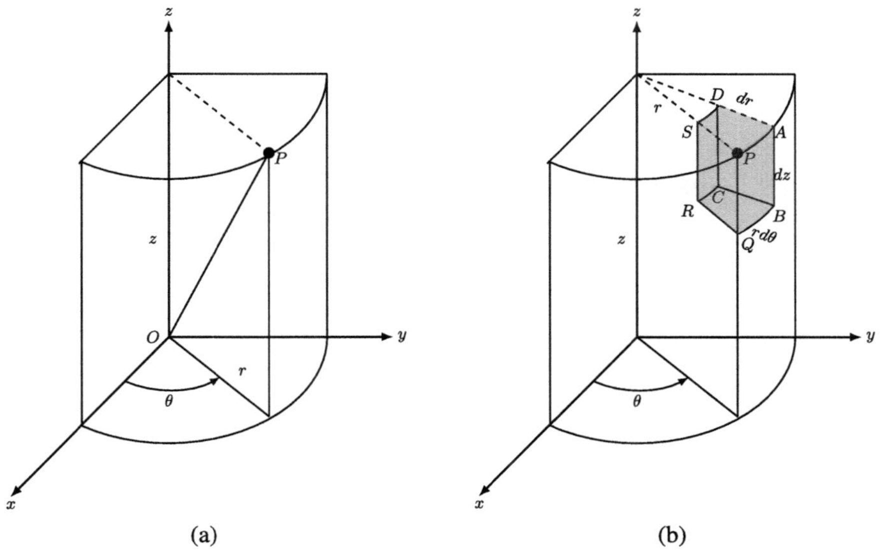

Fig. 6.14 Cylindrical coordinates (r, θ, z): simply a 3D version of the polar coordinates. The differential volume is $rdrd\theta dz$, which is of dimension of length cubed (r, dr, dz)

6.8.3 Triple Integrals

At this point, we have seen $\int f(x)dx$ with differential line dx, $\iint f(x, y)dxdy$ with differential area $dxdy$. We can naturally to write $\iiint f(x, y, z)dxdydz$, with differential volume $dxdydz$. We cannot associate this triple integral as a volume, but if $f(x, y, z) = 1$, we get $\iiint dxdydz$ which is a volume.

6.8.4 Triple Integrals in Cylindrical and Spherical Coordinates

Certain integrals are easier to deal with using not Cartesian coordinates but cylinder and spherical coordinates. This section presents these two coordinates. In a cylindrical coordinate system, we specify a point by (r, θ, z), see Fig. 6.14. The differential volume $dxdydz$ becomes $rdrd\theta dz$. Without the z-component the cylindrical coordinates are polar coordinates. So, I do not treat double integrals using polar coordinates explicitly.

In a spherical coordinate system, we specify a point by (ρ, ϕ, θ), see Fig. 6.15; ρ is the distance from the origin (similar to r in polar coordinates, and ρ is used instead of r to avoid confusion with cylindrical coordinates), θ is the same as the angle in polar coordinates and ϕ is the angle between the z-axis and the line from the origin to the point. If you still remember Sect. 3.20.2, then the angle θ is the longitude, which

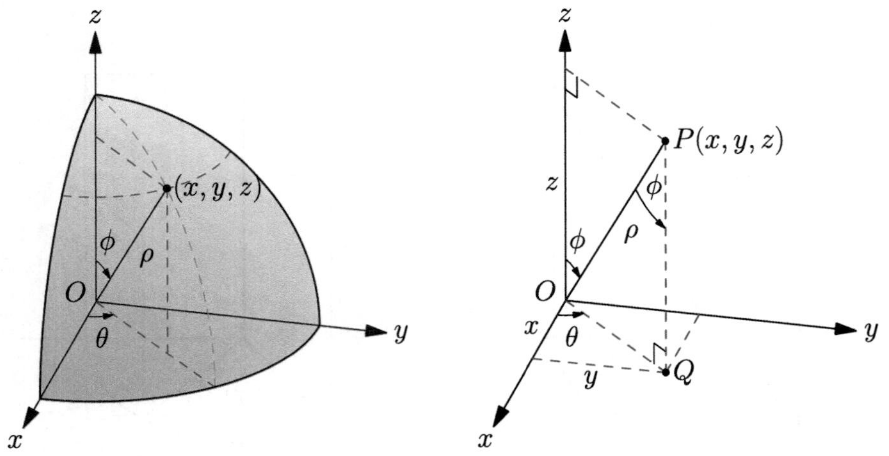

Fig. 6.15 Spherical coordinates (ρ, ϕ, θ)

increases as we travel east around the Equator. The angle ϕ equals 0 at the North Pole and π at the South Pole.

Referring to the middle figure in Fig. 6.15, we're now going to derive the relation between Cartesian coordinates and spherical coordinates. In the right triangle OPQ, we have $PQ = \rho\cos\phi$, and PQ is the z-coordinate of P. We also have $OQ = \rho\sin\phi$. Then, in the Oxy-plane that we're familiar with, we can compute x and y (I also include z):

$$x = OQ\cos\theta = \rho\sin\phi\cos\theta, \quad y = \rho\sin\phi\sin\theta, \quad z = \rho\cos\phi$$

That's how to convert between Cartesian and spherical coordinates. How we know the above is correct? Check $x^2 + y^2 + z^2 = \rho^2$ is the way.

The differential volume $dxdydz$ becomes $\rho^2\sin\phi d\rho d\theta d\phi$, see Fig. 6.16. This differential volume is of dimension of length cubed (ρ^2, $d\rho$).

Volume of a sphere of radius R. It is now quite straightforward to compute the volume of a sphere using spherical coordinates and a triple integral for the function $f(x, y, z) = 1$:

$$V = \iiint \rho^2\sin\phi d\rho d\theta d\phi = \int_0^R \rho^2 d\rho \int_0^\pi \sin\phi d\phi \int_0^{2\pi} d\theta = \frac{R^3}{3}(2)(2\pi) = \frac{4\pi R^3}{3}$$

What's next is one of the most important triple integrals in the history of physics: Newton's gravitational attraction.

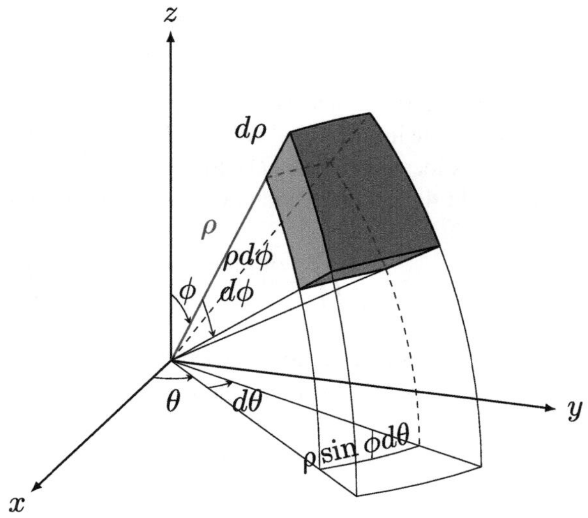

Fig. 6.16 Spherical coordinates. The differential volume $dxdydz$ becomes $\rho^2 \sin\phi d\rho d\theta d\phi$

6.8.5 Newton's Shell Theorem

Newton's gravitation theory states that every object is pulling other object and for two objects of masses M and m they are pulled by a force that is proportional to the product of their masses and inversely proportional to the square of the distance r between them. In mathematical symbols, his law is expressed as (Sect. 6.10.8)

$$F = \frac{GMm}{r^2}$$

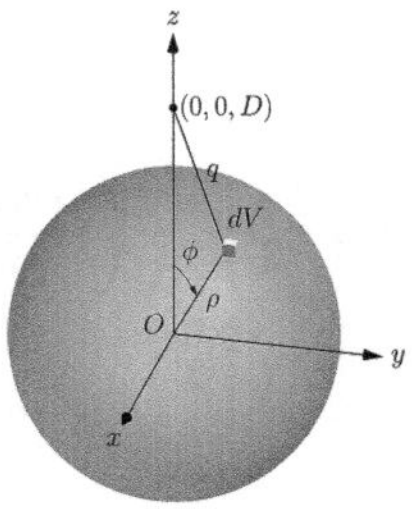

So, when applied to massive objects (such as the sun and planets), it is implied by this theory that regardless of the massive size of the sun (and any planets) it acts as if the mass of the Sun was all concentrated at its center. And that is the Newton shell theorem. Newton proved that the gravitational pull of a sphere (has a total mass M) on a point mass m locating at a distance D from its center is given by GMm/D^2 as if the mass of the sphere was all concentrated at its center. Herein, we use triple integral with spherical coordinates to prove this theorem. For simplicity, we prove the equivalent theorem regarding the potential energy U. That is, we're going to prove that $U = -GMm/D$, see Eq. (6.11.16).

Proof Denoting the density of the sphere by $\bar{\rho}$ (i.e., $\bar{\rho} = M/v$ with $V = 4/3\pi R^3$) and consider an infinitesimal volume dV locating at a distance q from the point mass m, which is located at $(0, 0, D)$, the gravitational potential energy is given by (refer to Sect. 6.11.4 for a discussion on the concept of gravitational potential energy)

$$U_{\text{sphere}} = -\iiint \frac{Gm\bar{\rho}dV}{q} = -Gm\bar{\rho}\iiint \frac{\rho^2 \sin\phi d\rho d\theta d\phi}{q} \tag{6.8.4}$$

where in the second equality, we used spherical coordinates. Using the law of cosines or the generalized Pythagorean theorem, we can compute q in terms of D, ρ, and ϕ:

$$q^2 = D^2 + \rho^2 - 2D\rho\cos\phi = u \tag{6.8.5}$$

Assume for now that ρ is fixed, from Eq. (6.8.5), we have $du = 2\rho D \sin\phi d\phi$ and $q = \sqrt{u}$. With this, Eq. (6.8.4) becomes

$$\begin{aligned} U_{\text{sphere}} &= -\frac{Gm\bar{\rho}}{2D}\iiint \frac{du\rho d\rho d\theta}{\sqrt{u}} \\ &= -\frac{Gm\bar{\rho}}{2D}\int_0^R \left[\int \frac{du}{\sqrt{u}}\int_0^{2\pi} d\theta\right]\rho d\rho \\ &= -\left(\frac{Gm\bar{\rho}}{D}\right)(2\pi)\int_0^R [(D+\rho)-(D-\rho)]\,\rho d\rho \end{aligned} \tag{6.8.6}$$

where 2π comes from $\int_0^{2\pi} d\theta$, the for the integral $\int du/\sqrt{u}$, the limits are $(D-\rho)^2$ with $\phi = 0$ and $(D+\rho)^2$ with $\phi = \pi$; see Eq. (6.8.5). Finally, we do integration along the ρ direction:

$$U_{\text{sphere}} = -\left(\frac{Gm\bar{\rho}}{D}\right)(4\pi)\int_0^R \rho^2 d\rho = -\left(\frac{Gm\bar{\rho}}{D}\right)(4\pi)\frac{R^3}{3} = -\frac{GMm}{D} \tag{6.8.7}$$

■

6.8.6 Change of Variables and the Jacobian

Assume that we need to evaluate this double integral $\iint_R (3x+6y)^2 dA$, where R is the shaded region bounded by the four straight lines shown in Fig. 6.17 (left). Even though it is possible to directly calculate this integral, it is tedious. We can use a change of variables as shown in the figure to simplify the integral. Indeed, the integration limits are now constants.

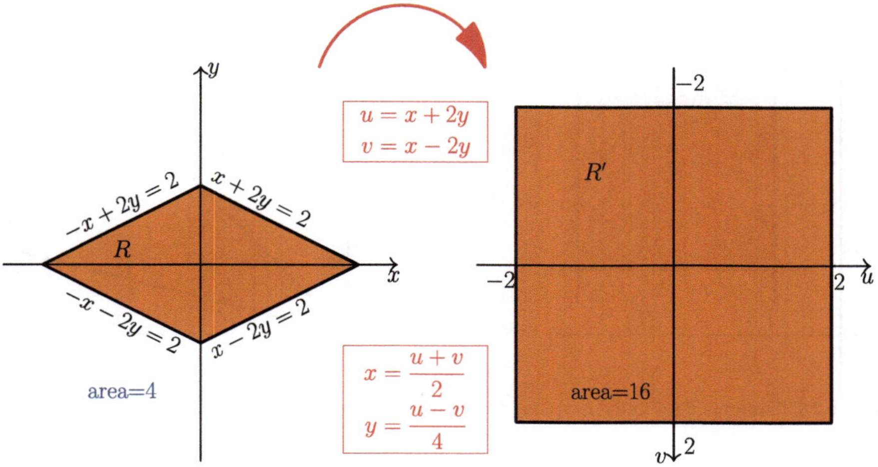

Fig. 6.17 Change of variables $x = f(u, v)$, $y = g(u, v)$

Another example of change of variables is given in Fig. 6.18. This is to demonstrate that straight edges in the uv-plane can be transformed into curves in the xy-plane. What are the transformations? The four curves in the xy-plane are given by

$$xy = 1, \quad xy = 4, \quad \frac{x}{y} = 1, \quad \frac{x}{y} = 4$$

Thus, the region R' is $R' : \{(x, y) : 1 \le x/y \le 4,\ 1 \le xy \le 4\}$. Therefore, if we consider the following transformation:

$$u = xy, \quad v = \frac{x}{y}$$

we can transform R' to R in the uv-plane shown in Fig. 6.18. You can consult this geogebra link to play with 2D transformation.

Actually, we have seen change of variables before: integration by substituion in Sect. 4.7.1 and double integrals using polar coordinates in Sect. 6.8.2. We again believe in patterns and search for a formula for double and triple integrals based on single integrals and the polar coordinates Eq. (6.8.3). So, we put them together in the below equation:

$$\begin{aligned}
\int_{x=g(a)}^{x=g(b)} F(x)dx &= \int_a^b F(g(u))g'(u)du, \quad x = g(u) \\
\iint_R F(x, y)dxdy &= \iint_{R'} F(f(u, v), g(u, v))\square dudv \\
\iiint_R F(x, y, z)dxdydz &= \iiint_{R'} F(f(u, v, w), g(u, v, w), , \quad h(u, v, w))\square dudvdw
\end{aligned}$$

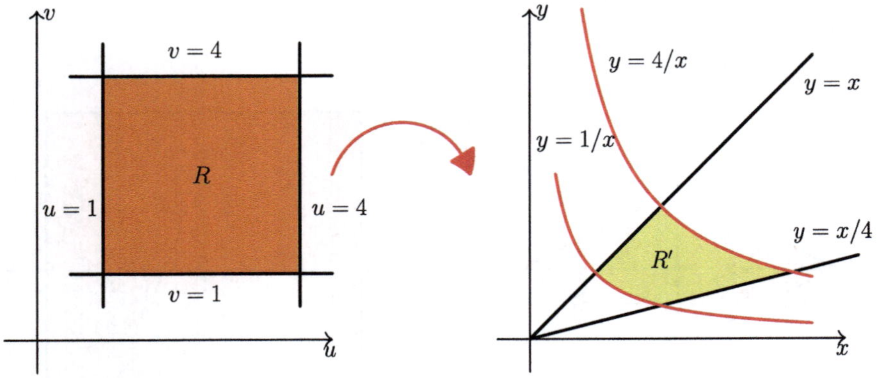

Fig. 6.18 Straight edges in the uv-plane can be transformed to curved edges in the xy-plane

And our task is to find the unknown red box which plays the role of $g'(u)$ when we replace dx by du. For double integrals, this quantity is denoted by J_{uv} and called the Jacobian of the transformation from the uv-plane to the xy-plane. What should J_{uv} be? From the first equation in the above for $\int_a^b f(x)dx$, we guess J_{uv} should be a function of f_u, f_v, g_u, g_v, i.e., all the first derivatives of f and g. If you know linear algebra, precisely linear transformations (Sect. 9.6), you'll see that J_{uv} is the determinant of a matrix containing all these first derivatives. In what follows we explain where this matrix comes from. We note in passing that for completeness we have included triple integrals, but we do not have to consider double and triple integrals separately. What works for double integrals will work for triple integrals.

Local linearity of transformations and the Jacobian matrix. Let's come back to the transformation in Fig. 6.17. That is a linear transformation from a square in the uv-plane to a rhombus in the xy-plane (check Sect. 9.6 if that term, i.e., linear transformation is new to you), and the equation of the transformation is

$$\begin{bmatrix} x \\ y \end{bmatrix} = \begin{bmatrix} \frac{1}{2} & +\frac{1}{2} \\ \frac{1}{4} & -\frac{1}{4} \end{bmatrix} \begin{bmatrix} u \\ v \end{bmatrix} \tag{6.8.8}$$

Thus, from linear algebra, the area of the rhombus is the area of the square (which is 16) scaled with the absolute of the determinant of the red transformation matrix (which is $|-1/4|$): that area is then $16 \times 1/4 = 4$, which is correct (computed by the standard way of plane geometry, see Fig. 6.17-left).

But most of the usual transformations are nonlinear (Fig. 6.18 is one of them: lines are transformed to curves). In that case, how can we use linear transformations to find the area? The answer is: linear approximations turn a curve to a line (tangent), a square to a parallelogram, and then the theory of linear transformations can be used.

Let's consider the following transformation:

$$\begin{bmatrix} x \\ y \end{bmatrix} = \begin{bmatrix} f(u,v) \\ g(u,v) \end{bmatrix} = \begin{bmatrix} u^2 - v^2 \\ 2uv \end{bmatrix}$$

Now we consider small (i.e., infinitesimal) changes in u and v, namely Δu and Δv, and see how x and y change:

$$\begin{bmatrix} (u+\Delta u)^2 - (v+\Delta v)^2 \\ 2(u+\Delta u)(v+\Delta v) \end{bmatrix} - \begin{bmatrix} u^2 - v^2 \\ 2uv \end{bmatrix} \approx \begin{bmatrix} 2u\Delta u - 2v\Delta v \\ 2u\Delta v + 2v\Delta u \end{bmatrix} = \begin{bmatrix} 2u & -2v \\ 2v & 2u \end{bmatrix} \begin{bmatrix} \Delta u \\ \Delta v \end{bmatrix}$$

As can be seen, since for infinitesimal changes $(\Delta u)^2$ and $(\Delta v)^2$ are negligible, we have obtained an approximation to a change in f and g in terms of a matrix containing the four partial derivatives: $f_u = 2u$, $f_v = -2v$, $g_u = 2v$, $g_v = 2u$. This matrix is special and it has a name: the *Jacobian matrix*, named after the German mathematician Carl Gustav Jacob Jacobi (1804–1851). Generally, we then have

$$\begin{bmatrix} dx \\ dy \end{bmatrix} = \begin{bmatrix} f_u & f_v \\ g_u & g_v \end{bmatrix} \begin{bmatrix} du \\ dv \end{bmatrix} \tag{6.8.9}$$

where the matrix is the Jacobian matrix. Globally the transformation is nonlinear but locally (when we zoom in) the transformation is linear.

To find J_{uv}, considering a point (u_0, v_0) and a rectangle of sides du and dv with one vertex at (u_0, v_0), see Fig. 6.19. The vector $(du, 0)$ becomes $(f_u du, g_u du)$ according to Eq. (6.8.9) whereas the vector $(0, dv)$ becomes $(f_v dv, g_v dv)$. The rectangle in the uv-plane has an area of $dudv$, whereas the transformed rectangle, which is a parallelogram, has an area of $(f_u g_v - f_v g_u)dudv$. Thus,

$$\boxed{J_{uv} = \left|\det \begin{bmatrix} f_u & f_v \\ g_u & g_v \end{bmatrix}\right| = \left|\frac{\partial f}{\partial u}\frac{\partial g}{\partial v} - \frac{\partial g}{\partial u}\frac{\partial f}{\partial v}\right|} \tag{6.8.10}$$

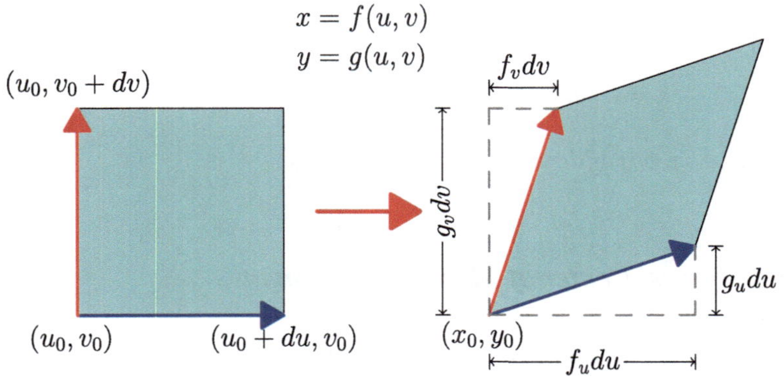

Fig. 6.19 Finding the Jacobian of the transformation J_{uv}

As the determinant can be positive, zero, and negative, we needed to use its absolute value.

Ok. How are we sure that our J_{uv} is correct? The answer is easy: just apply it to a case that we're familiar with polar coordinates. In polar coordinates, we use r, θ which are u, v:

$$\left.\begin{aligned} x &= r\cos\theta \\ y &= r\sin\theta \end{aligned}\right\} \Longrightarrow J_{uv} = \left|\det\begin{bmatrix}\cos\theta & -r\sin\theta \\ \sin\theta & r\cos\theta\end{bmatrix}\right| = r$$

Thus $dxdy = rdrd\theta$.

We come back to the problem in Fig. 6.17. The determinant of the transformation is given by

$$\det\begin{bmatrix}1/2 & 1/2 \\ 1/4 & -1/4\end{bmatrix} = -\frac{1}{4}$$

Therefore, $J_{uv} = 1/4$ and,

$$\iint_R (3x+6y)^2 dxdy = \iint_{R'} 9u^2|J_{uv}|dudv = \frac{9}{4}\int_{-2}^{2}\int_{-2}^{2} u^2 dudv = 48$$

For 2D integrals, J_{uv} is related to the determinant of a 2×2 matrix, and thus for 3D integrals, it is related to the determinant of a 3×3 matrix containing all the nine first partial derivatives:

$$\boxed{J_{uv} = \left|\det\begin{bmatrix}f_u & f_v & f_w \\ g_u & g_v & g_w \\ h_u & h_v & h_w\end{bmatrix}\right|} \tag{6.8.11}$$

which should not be a surprise. And of course we check the correctness of this 3D J_{uv} by applying it to triple integrals using spherical coordinates. We don't provide details, one just needs to know how to compute the determinant of a 3×3 matrix. That determinant in Eq. (6.8.11) for spherical coordinates, i.e.,

$$x = \rho\sin\phi\cos\theta, \quad y = \rho\sin\phi\sin\theta, \quad z = \rho\cos\phi$$

should be $\rho^2\sin\phi d\rho d\theta d\phi$.

6.8.7 Masses, Center of Mass, and Moments

To introduce the concept of center of mass, let's first consider a system of two masses. We can write Newton's second law (check Sect. 6.10 for detail) for these two masses as

$$\begin{aligned}\text{Newton's 2nd law for mass 1: } \boldsymbol{F}_1^{\text{ext}} + \boldsymbol{F}_{12} = \dot{\boldsymbol{p}}_1 \\ \text{Newton's 2nd law for mass 2: } \boldsymbol{F}_2^{\text{ext}} + \boldsymbol{F}_{22} = \dot{\boldsymbol{p}}_2\end{aligned} \tag{6.8.12}$$

where $\boldsymbol{F}_1^{\text{ext}}$ is the external force applied on mass 1; $\boldsymbol{F}_2^{\text{ext}}$ is the external force applied on mass 2; $\boldsymbol{F}_{12}$ is the force applied on mass 1 due to mass 2; $\boldsymbol{p}_1$ is the linear momentum of mass 1. By summing these two equations, we obtain the total momentum of the system (and its time rate of change):

$$\boldsymbol{p} = \boldsymbol{p}_1 + \boldsymbol{p}_2 \Longrightarrow \dot{\boldsymbol{p}} = \dot{\boldsymbol{p}}_1 + \dot{\boldsymbol{p}}_2 \tag{6.8.13}$$

Using Eq. (6.8.12) and Newton's third law which states that $\boldsymbol{F}_{12} = -\boldsymbol{F}_{21}$, these two forces cancel out leaving us only the external forces in $\dot{\boldsymbol{p}}$:

$$\dot{\boldsymbol{p}} = \boldsymbol{F}^{\text{ext}}, \quad (\boldsymbol{F}^{\text{ext}} := \boldsymbol{F}_1^{\text{ext}} + \boldsymbol{F}_2^{\text{ext}}) \tag{6.8.14}$$

We can generalize this to a system of any number of masses to have $\dot{\boldsymbol{p}} = \boldsymbol{F}^{\text{ext}}$.

If we throw a basketball in the air, we will see that it falls in a parabola trajectory. If we throw a bunch of balls connected by strings, we also observe a parabola. But what is moving along this parabola? The concept of center of mass answers this question.

Assuming that the mass of n balls are $m_1, m_2, \ldots, m_n$ and the position vector of them are $\boldsymbol{r}_1, \boldsymbol{r}_2, \ldots, \boldsymbol{r}_n$. If we define $\boldsymbol{R}_{\text{CM}}$—the position vector of the *center of mass* of all basket balls—as

$$\boldsymbol{R}_{\text{CM}} = \frac{m_1\boldsymbol{r}_1 + m_2\boldsymbol{r}_2 + \cdots + m_n\boldsymbol{r}_n}{m_1 + m_2 + \cdots + m_n} = \frac{\sum_{i=1}^n m_i\boldsymbol{r}_i}{\sum_{i=1}^n m_i} \tag{6.8.15}$$

we then can write the system momenta as if *all the mass is concentrated on this center of mass*:

$$\begin{aligned}\boldsymbol{p} &= m_1\dot{\boldsymbol{r}}_1 + m_2\dot{\boldsymbol{r}}_2 + \cdots + m_n\dot{\boldsymbol{r}}_n \\ &= M\dot{\boldsymbol{R}}_{\text{CM}}, \quad M = \sum_i m_i\end{aligned} \tag{6.8.16}$$

By differentiating $\boldsymbol{p} = M\dot{\boldsymbol{R}}_{\text{CM}}$ with respect to time, we get $\boldsymbol{F}^{\text{ext}} = M\ddot{\boldsymbol{R}}_{\text{CM}}$. This equation is significant as it implies that the center of mass moves exactly as if it were a single particle of mass M (the mass of the whole system), subject to the net external force on the system. This is why we can treat extended objects such as planets as if they were point particles.

Even though the math is simple, how we know beforehand the introduction of the center of mass will be useful? We might not know. The idea is to reduce a complicated problem (involving many particles for example) to the simple problem of a single particle that we're familiar with.

In a Cartesian coordinate system, the (position) of the *center of mass* is given by

$$\boldsymbol{R}_{\text{CM}} = \frac{\sum_i m_i \boldsymbol{r}_i}{M} \Longrightarrow \begin{cases} x_{\text{CM}} = \dfrac{m_i x_i}{M} \\ y_{\text{CM}} = \dfrac{m_i y_i}{M} \\ z_{\text{CM}} = \dfrac{m_i z_i}{M} \end{cases} \tag{6.8.17}$$

We can appreciate the usefulness of vector notation; an equation using this notation is really three equations, one for each of the three directions. We note by passing that we have used the Einstein summation in writing $m_i y_i/M$ without the $\sum$ symbol, see Sect. 9.2 for detail. Mathematicians call $\boldsymbol{R}_{\text{CM}}$ a convex combination (less jargon is a weighted average of $\boldsymbol{r}_i$).

Let's play with Eq. (6.8.17) and surely something fun will come to us. We now shall consider only the $x-$direction, because if we can understand that one, we can understand the other two. Now, assume that the object is divided into little pieces (N such pieces), all of which has the same mass m. Then,

$$x_{\text{CM}} = \frac{\sum_i m_i x_i}{M} = \frac{m \sum_i x_i}{mN} = \frac{\sum_i x_i}{N}$$

In other words, x_{CM} is the average of all the x's, if the masses are equal. Now, suppose we have only two masses, and one mass is $2m$ and the other is m. Then we have $x_{\text{CM}} = (2x_1 + 1x_2)/3$. In other words, every mass being counted a number of times proportional to the mass. From that, it can be seen that x_{CM} is somewhere larger than the smallest x and smaller than the largest x. That holds for y_{CM} and z_{CM}. Thus, the CM lies within the envelope of the masses (Fig. 6.20).

Center of mass of solids. What is the center of mass of a continuous object; e.g., a steel disk? Of course, integral calculus is the answer. The sums in Eq. (6.8.17) become integrals

$$\boldsymbol{x}_{\text{CM}} = \frac{1}{M} \iiint \boldsymbol{x} \underbrace{(\rho dx dy dz)}_{dm}, \quad M = \iiint (\rho dx dy dz) \tag{6.8.18}$$

where ρ is the density. Thus for objects with density that does not vary from point to point, the geometric centroid and the center of mass coincide.

Recall that for a particle of mass m, its moment of inertia with respect to an axis is $I = mr^2$, see Sect. 9.1.5. Extending this to a system of N particles, we will have $I = \sum_\alpha m_\alpha r_\alpha^2$ and to a continuum we have $dI = r^2 dm$, and thus:

$$I_z = \int_{\mathcal{B}} \rho(x^2 + y^2) dV = \iiint \rho(x^2 + y^2) dx dy dz \tag{6.8.19}$$

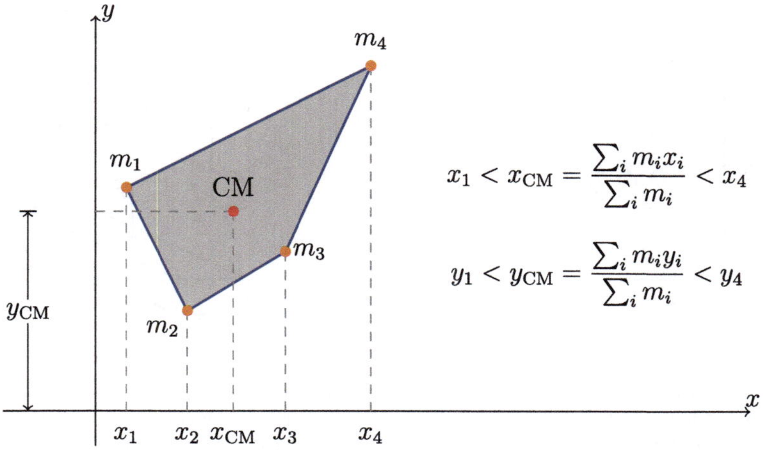

Fig. 6.20 The center of mass of n masses lie within the envelope of the masses

And this is the moment of inertia of a solid $\mathcal{B}$ when it is rotating w.r.t. the z-axis. Similarly, w.r.t. these other two axes, we have

$$\begin{aligned} I_x &= \int_{\mathcal{B}} \rho(y^2 + z^2)dV \\ I_y &= \int_{\mathcal{B}} \rho(x^2 + z^2)dV \end{aligned} \tag{6.8.20}$$

Now, if we consider plane figures, i.e., objects of which the thickness is negligible compared with other dimensions, we can see that $z = 0$ in Eq. (6.8.20), and thus

$$I_z = \int_{\mathcal{B}} \rho(x^2 + y^2)dA = \int_{\mathcal{B}} \rho x^2 dA + \int_{\mathcal{B}} \rho y^2 dA = I_y + I_x \tag{6.8.21}$$

All are two-dimensional integrals as signified by dA. When $\rho = 1$, we have

$$I_x = \int_{\mathcal{B}} y^2 dA, \quad I_y = \int_{\mathcal{B}} x^2 dA \tag{6.8.22}$$

which are known as the second moments of inertia. The second moment of area is a measure of the “efficiency” of a shape to resist bending caused by loading perpendicular to the beam axis (Fig. 6.21). It appeared for the first time in Euler-Bernoulli theory of slender beams.

Example 6.2 Determine the center of gravity and moment of inertia of a semicircular disk of radius a made of a material with a constant density ρ.

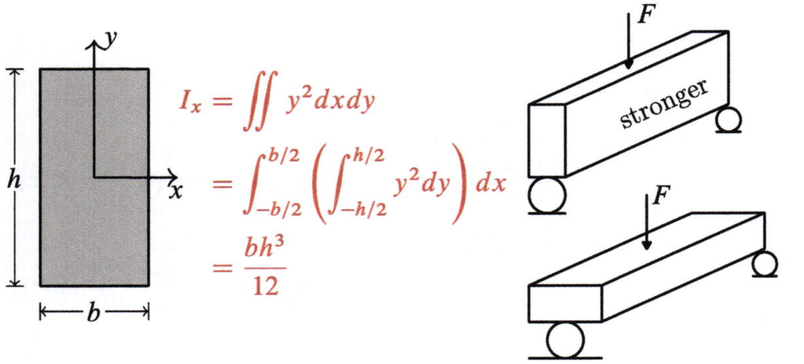

Fig. 6.21 The second moment of area is a measure of the "efficiency" of a shape to resist bending caused by loading perpendicular to the beam axis. For example, a beam of rectangular cross section has $I_x = bh^3/12$, thus the height of the beam h decides the bending resistance. That's why the top beam is stronger than the bottom one

First, we compute the mass. It is given by (Eq. (6.8.18) and use polar coordinates)

$$M = \iint \rho r d\theta dr = \rho \int_0^a r dr \int_0^\pi d\theta = \rho \frac{\pi a^2}{2}$$

Then we determine the center of gravity (due to symmetry, only the y-component is non-zero)

$$y_{\text{CM}} = \frac{1}{M} \iint \rho y r d\theta dr = \frac{1}{M} \iint \rho r^2 \sin\theta d\theta dr = \frac{\rho}{M} \int_0^a r^2 dr \int_0^\pi \sin\theta d\theta = \frac{4a}{3\pi}$$

And the moment of inertia is given by

$$\begin{aligned} I_y &= \rho \iint x^2 r dr d\theta = \rho \iint r^3 \cos^2\theta dr d\theta \\ &= \rho \int_0^a r^3 dr \int_0^\pi \frac{1+\cos 2\theta}{2} d\theta = \rho \frac{\pi a^4}{8} \end{aligned}$$

Figure 6.22 presents a summary of how to determine the center of mass for discontinuous (a) and continuous objects (b). Particularly interesting is the way how the center of mass of a compound object is determined. In Fig. 6.22c, we have an object consisting of two rectangles. As we can treat each rectangle as a point mass with its center of mass already known, Fig. 6.22c, the CM of the compound object can be computed using Eq. (6.8.17). When the thickness of the compound object (t) is constant and it has a uniform density (i.e., the density is the same everywhere), we can convert from mass to area (A), and obtain the following equation:

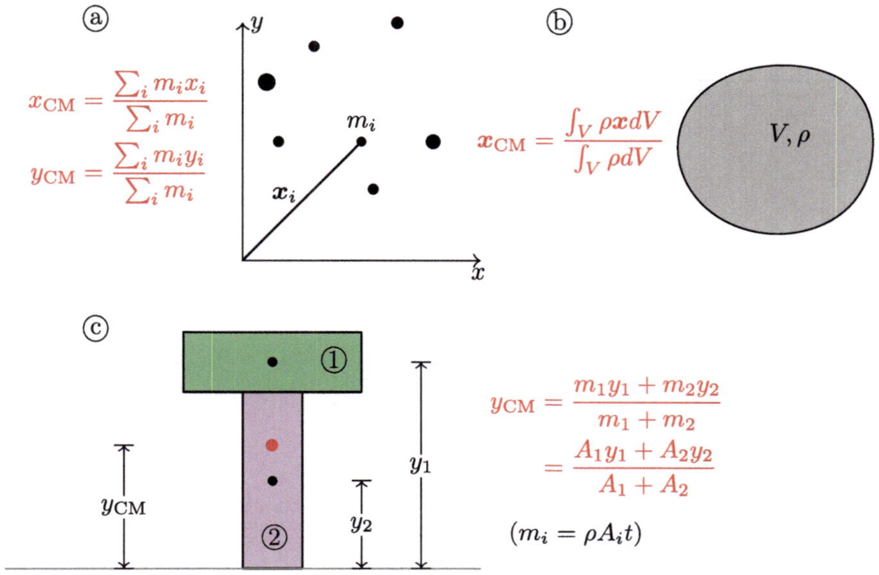

Fig. 6.22 Center of mass: from particles (**a**) to continuous objects (**b**) and compound objects (**c**)

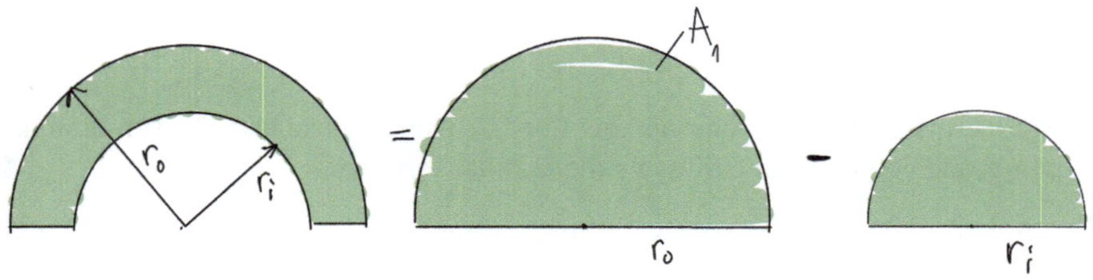

Fig. 6.23 Center of mass of objects obtained by material removal

$$x_{CM} = \frac{\sum_i x_i A_i}{\sum_i A_i} \tag{6.8.23}$$

for the CM of any 2D compound solid. The shape in Fig. 6.22c is the cross section of a T-beam (or tee beam), used in civil engineering. Thus, civil engineers use Eq. (6.8.23) frequently.

In many cases, we remove material from a shape to make a new one, see Fig. 6.23. In that case, the CM of the object is given by

$$x_{CM} = \frac{x_1 A_1 - x_2 A_2}{A_1 - A_2} \tag{6.8.24}$$

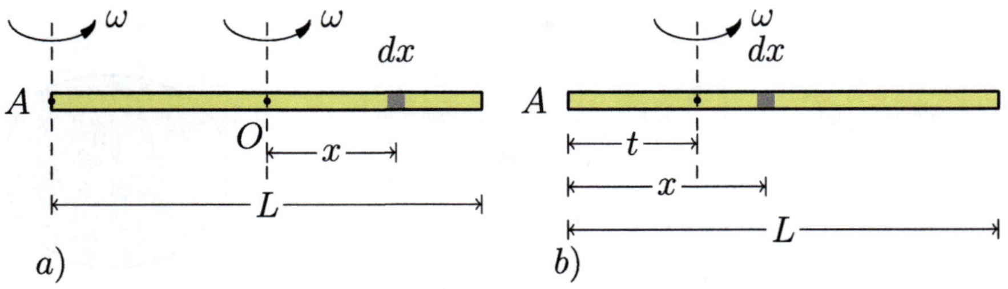

Fig. 6.24 Moment of inertia of a thin rod of length L

Proof of Eq. (6.8.23). Assume that we have a compound object V which is consisted of two sub-domains: V_1 with m_1 and V_2 with m_2. Now, we can write the center of mass of the compound object as

$$\begin{aligned}
\boldsymbol{x}_{\text{CM}} &= \frac{\int_V \rho \boldsymbol{x} dV}{m_1 + m_2} = \frac{\int_{V_1} \rho \boldsymbol{x} dV + \int_{V_2} \rho \boldsymbol{x} dV}{m_1 + m_2} \\
&= \frac{1}{m_1 + m_2} \int_{V_1} \rho \boldsymbol{x} dV + \frac{1}{m_1 + m_2} \int_{V_2} \rho \boldsymbol{x} dV \\
&= \frac{m_1}{m_1 + m_2} \frac{\int_{V_1} \rho \boldsymbol{x} dV}{m_1} + \frac{m_2}{m_1 + m_2} \frac{\int_{V_2} \rho \boldsymbol{x} dV}{m_2}
\end{aligned}$$

The red terms are CMs of sub-domains 1 and 2. Thus, we can treat the sub-domains as points and compute the CM using the familiar formula. ■

Example 2. Determine the moment of inertia of a rod of length L with $\rho = 1$ with respect to various points: the left extreme A and the center O (Fig. 6.24). Could you guess which case has a lower moment of inertia?

As the rod is very thin, we only have 1D integrals. So, the moments of inertia w.r.t. A and O are, see Fig. 6.24a:

$$I_A = \int_0^L x^2 dx = \frac{L^3}{3}, \quad I_O = \int_{-L/2}^{L/2} x^2 dx = \frac{L^3}{12} \tag{6.8.25}$$

And the fact that $I_A > I_0$ indicates it is easier to turn the rod around O—its center of gravity. This is consistent with our daily experiences.

Now, if we ask the following question various interesting things would show up. About which point along the rod, the moment of inertia is minimum? Let's denote $I(t)$ the moment of inertia w.r.t. a point located at a distance t from A. We can compute $I(t)$ as, Fig. 6.24b:

$$I(t) = \int_0^L (x-t)^2 dx = \int_0^L x^2 dx + \int_0^L t^2 dx - 2\int_0^L xt dx = \frac{L^3}{3} + t^2 L - tL^2$$

And differential calculus helps us to find t such that $I(t)$ is minimum:

$$\frac{dI(t)}{dt} = 2tL - L^2 = 0 \Longrightarrow t = \frac{L}{2} \tag{6.8.26}$$

The first thing to notice is that instead of integrating and then differentiating, we can do the reverse. That is we differentiate the function in the integral and then do the integration:

$$\begin{aligned}\frac{dI(t)}{dt} &= \int_0^L \frac{d(x-t)^2}{dt}dx \\ &= -2\int_0^L (x-t)dx = -L^2 + 2tL\end{aligned}$$

And we have got the same result. So, there must be a theorem about this. It is called the Leibnitz rule for differentiating under the integral sign:

$$I(t) = \int_a^b f(x,t)dx \Longrightarrow \frac{dI(t)}{dt} = \int_a^b \frac{df(x,t)}{dt}dx \tag{6.8.27}$$

Parallel axis theorem. In the problem of the calculation of the moment of inertia of a rod of length L, we have $I_A = L^3/3$ and $I_O = L^3/12$. If we ask this question: what is the relation between these two quantities, we will get something interesting. Let's first compute the difference between them:

$$I_A - I_O = \frac{L^3}{3} - \frac{L^3}{12} = \frac{L^3}{4}$$

And this difference must depend on the distance between A and O which is $L/2$, thus we write

$$I_A - I_O = \frac{L^3}{4} = \left(\frac{L}{2}\right)^2 \times L$$

Now, we anticipate the following result: if O' is at a distance d from the CM O, the moment of inertia w.r.t. to O' is given by

$$I_{O'} = I_O + d^2 \times L$$

Next, we extend this result to 3D objects and obtain the so-called parallel axis theorem, which facilitates the calculation of the moment of inertia of a solid about an arbitrary axis: we just need to compute the moment of inertia w.r.t. the CM and use this theorem if we need the moment of inertia w.r.t. any axis.

We consider an object $\mathcal{B}$ with density ρ (Fig. 6.25). A set of coordinate axes is used where O is at the origin. In this coordinate system, the center of mass of the object is located at (x_{CM}, y_{CM}, z_{CM}). Let I_{CM} be the moment of inertia of $\mathcal{B}$ with respect to an axis passing through CM. Now we're determining the moment of inertia I_z w.r.t.

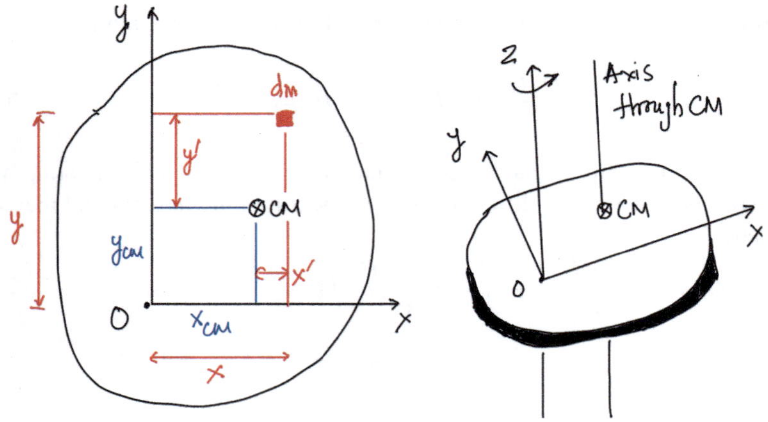

Fig. 6.25 Parallel axis theorem: two parallel axes, one passing through the CM and the other is a distance d away: $d^2 = x^2_{\text{CM}} + y^2_{\text{CM}}$

an axis passing through O by considering an infinitesimal $dm = \rho dV$ locating at (x, y):

$$
\begin{aligned}
I_z &= \int_{\mathcal{B}} \rho(x^2 + y^2)dV \\
&= \int_{\mathcal{B}} \rho\left[(x_{\text{CM}} + x')^2 + (y_{\text{CM}} + y')^2\right] dV \\
&= \int_{\mathcal{B}} \rho(x^2_{\text{CM}} + y^2_{\text{CM}})dV + \int_{\mathcal{B}} \rho(x'^2 + y'^2)dV + \int_{\mathcal{B}} 2\rho x_{\text{CM}} x' dV + \int_{\mathcal{B}} 2\rho y_{\text{CM}} y' dV \\
&= Md^2 + I_{\text{CM}} + 0 + 0
\end{aligned}
\tag{6.8.28}
$$

And that gives us the parallel axis theorem that states:

$$
I_z = I_{\text{CM}} + Md^2 \tag{6.8.29}
$$

You can find I_{CM} for many common solids in textbooks, and from that, the parallel axis theorem allows us to compute the moment of inertia about an arbitrary axis.

But wait why the blue integrals in Eq. (6.8.28) are zero? This is due to one property of the CM:

$$
x_{\text{CM}} = \frac{\displaystyle\int_{\mathcal{B}} \rho x dV}{\displaystyle\int_{\mathcal{B}} \rho dV} \Longrightarrow \int_{\mathcal{B}} \rho(x - x_{\text{CM}})dV = 0 \Longrightarrow \int_{\mathcal{B}} \rho x' dV = 0
$$

Actually, we know this result without realizing it, see Table 6.3.

Table 6.3 $\sum_i (x_i - \bar{x}) = 0$ where $\bar{x}$ is the arithmetic average of x_is

x_i	$\bar{x}$	$x_i - \bar{x}$
1.0	3.0	-2.0
2.0	3.0	-1.0
3.0	3.0	0.0
4.0	3.0	1.0
5.0	3.0	2.0

6.8.8 Barycentric Coordinates

In this section, the barycentric coordinates, discovered by the German mathematician and theoretical astronomer August Ferdinand Möbius (1790–1868), are presented. These coordinates are based on the concept of the center of mass in physics. Let's consider three point masses m_A, m_B, and m_C placed at the three vertices of the triangle ABC with coordinates $\boldsymbol{x}_A, \boldsymbol{x}_B, \boldsymbol{x}_C$. We know that its center of mass is point P:

$$\boldsymbol{x}_P = \frac{m_A}{M}\boldsymbol{x}_A + \frac{m_B}{M}\boldsymbol{x}_B + \frac{m_C}{M}\boldsymbol{x}_C$$

with $M = m_A + m_B + m_C$.

Conversely, given a triangle ABC, what masses/weights must be put at the vertices to balance at some point Q? The solution to this problem defines a new coordinate system relative to the given positions A, B and C: it is possible to locate a point P on a triangle with three numbers (ξ_1, ξ_2, ξ_3). These three numbers are called the barycentric coordinates of P. The barycentric coordinates of a point relative to a triangle are the masses that we would have to place at the vertices of the triangle for its center of mass to be at that point.

We, then have:

$$\begin{aligned} 1 &= \xi_1 + \xi_2 + \xi_3 \\ x &= \xi_1 x_A + \xi_2 x_B + \xi_3 x_C \\ y &= \xi_1 y_A + \xi_2 y_B + \xi_3 y_C \end{aligned} \tag{6.8.30}$$

The second and third equations convert the barycentric coordinates to Cartesian coordinates. They are just Eq. (6.8.17).

Now, we need to determine the barycentric coordinates of the three vertices. It is straightforward to see that the barycentric coords of A is $(1, 0, 0)$: using Eq. (6.8.30) with $(1, 0, 0)$ results in (x_A, y_A). Another way to see this is: the only way so that the center of mass is at A is when m_A is very large compared with m_B and m_C; thus $\xi_1 = {}^{m_A}/_{M} = {}^{m_A}/_{m_A} = 1$. Similarly, the coords of B are $(0, 1, 0)$ and of C are $(0, 0, 1)$.

From that, we can see that every point on the edge BC has $\xi_1 = 0$ (this makes sense as the only case where the center of mass is on BC is that the mass at A is zero).

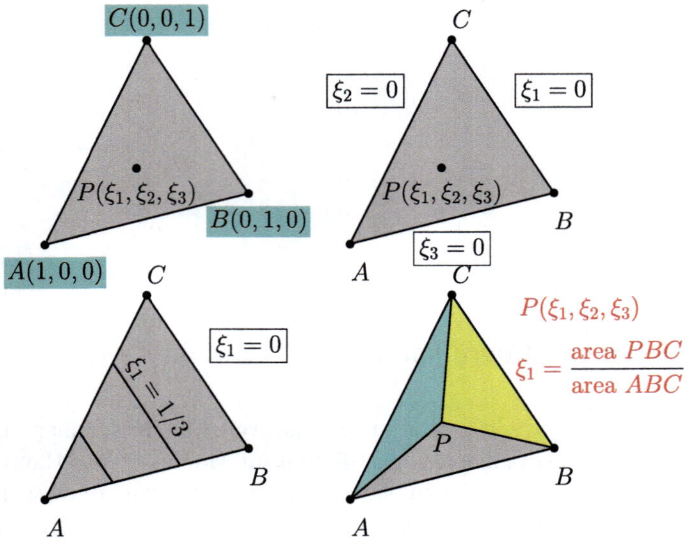

Fig. 6.26 Barycentric coordinates (ξ_1, ξ_2, ξ_3) of points in a triangle

The point is within the triangle if $0 \le \xi_1, \xi_2, \xi_3 \le 1$. If any one of the coordinates is less than zero or greater than one, the point is outside the triangle. If any of them is zero, P is on one of the lines joining the vertices of the triangle. See Fig. 6.26.

Next, we're showing that the line $\xi_1 = a$ (e.g., $\xi_1 = 1/3$) is parallel to the edge BC or the line $\xi = 0$. Using Eq. (6.8.30) with $\xi_1 = 1/3$ (hence $\xi_2 + \xi_3 = 2/3$), we can obtain (x, y) as

$$\begin{aligned} x &= \frac{1}{3}x_A + \xi_2 x_B + \xi_3 x_C = \frac{1}{3}x_A + \frac{2}{3}x_C + \xi_2(x_B - x_C) \\ y &= \frac{1}{3}y_A + \xi_2 y_B + \xi_3 y_C = \frac{1}{3}y_A + \frac{2}{3}y_C + \xi_2(y_B - y_C) \end{aligned} \tag{6.8.31}$$

We have learnt in Sect. 9.1.3 that the above line has the direction vector $\boldsymbol{x}_B - \boldsymbol{x}_C$, which is edge BC. Therefore, the line $\xi_1 = 1/3$ is parallel to BC.

Now, we carry out some algebraic manipulations to $\boldsymbol{x}_P$ to show that there is nothing entirely new about barycentric coordinates. To this end, we replace ξ_1 by $1 - \xi_2 - \xi_3$, and we compute $\boldsymbol{x}_P - \boldsymbol{x}_A$ which is the relative position of P w.r.t. A:

$$\boldsymbol{x}_P - \boldsymbol{x}_A = \left[(1 - \xi_2 - \xi_3)\boldsymbol{x}_A + \xi_2\boldsymbol{x}_B + \xi_3\boldsymbol{x}_C\right] - \boldsymbol{x}_A = \xi_2(\boldsymbol{x}_B - \boldsymbol{x}_A) + \xi_3(\boldsymbol{x}_C - \boldsymbol{x}_A)$$

Or,

$$\overrightarrow{AP} = \xi_2\overrightarrow{AB} + \xi_3\overrightarrow{AC} \tag{6.8.32}$$

So, if we use the vertex A as the origin and two edges AB and AC as the two basic vectors, we have an oblique coordinate system, and in this system, any point P is specified with two coordinates (ξ_2, ξ_3) is simply a linear combination of these two basic vectors with the coefficients being ξ_1 and ξ_2.

One question arises: why don't we just use Eq. (6.8.32)? If we look at this equation carefully, one thing comes to us: it is not symmetric! Why A is the origin? How about B and C? On the other hand, with the barycentric coordinates (ξ_1, ξ_2, ξ_3), everything is symmetric. There is no origin!

Geometrical meaning. The point P divides the triangle ABC into three sub-triangles PBC, PAB and PAC. It can be shown that the barycentric coordinates (ξ_1, ξ_2, ξ_3) are actually the ratio of the areas of these sub-triangles with that of the big triangle:

$$\xi_1 = \frac{\text{area of } PBC}{\text{area of } ABC}, \quad \xi_2 = \frac{\text{area of } PAC}{\text{area of } ABC}, \quad \xi_3 = \frac{\text{area of } PAB}{\text{area of } ABC}$$

One way to prove this is to use Eq. (9.1.21) to compute the areas of PBC and ABC noting that the Cartesian coords of P is $\xi_1 x_A + \xi_2 x_B + \xi_3 x_C$. Because of this property that (ξ_1, ξ_2, ξ_3) are also called the *areal coordinates.*

6.9 Parametric Surfaces

Functions of the form $\boldsymbol{r}(t) = (f(t), g(t), h(t))$ having one variable for input and a vector for output, are called single-variable vector-valued functions. Such single-variable vector-valued functions can be denoted as $\boldsymbol{r} : \mathbb{R} \to \mathbb{R}^3$. And the graph of such functions is a 3D curve. One can see that this function actually transforms a line segment lying on the number line to a curve in a 3D space (Fig. 6.27).

And in the same manner, two-variable vector-valued functions defined as $\boldsymbol{r} : \mathbb{R}^2 \to \mathbb{R}^3$ describes a surface. This function transform a domain in the $u - v$-plane to a surface living in a 3D space (Fig. 6.27). We actually do this kind of transformation in our daily lives when we roll a paper to make a cylinder, see for example (Fig. 6.28). If we use a ruled paper, i.e., one with two sets of perpendicular lines, in the transformed paper, these lines become curves which are called grid curves. The surface functions take two variables, u, and v, because a parametric surface can be seen as a "warped" version of a rectangular grid. The vector function "warps" this grid into a three-dimensional surface.

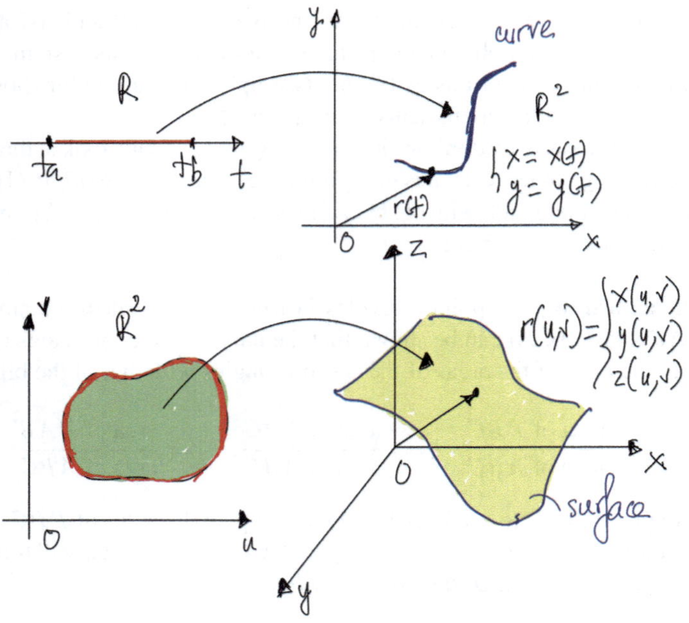

Fig. 6.27 Parametric curves and parametric surfaces

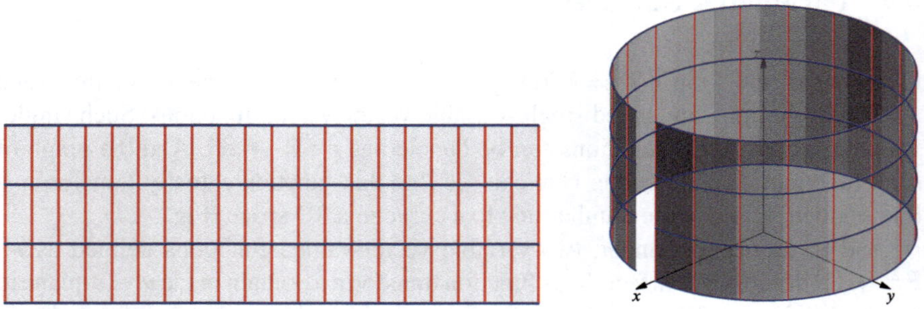

(a) (b)

Fig. 6.28 Paper rolling is a transformation $\boldsymbol{r} : \mathbb{R} \to \mathbb{R}^3$

6.9.1 Parametric Representation of Common Solids

Let's consider a cylinder with a radius a and height h in which the base is in the xy−plane (Fig. 6.28b). The parametric representation of such cylinder is

$$\begin{aligned} x(u, v) &= a \cos u \\ y(u, v) &= a \sin u \\ z(u, v) &= v \end{aligned} \tag{6.9.1}$$

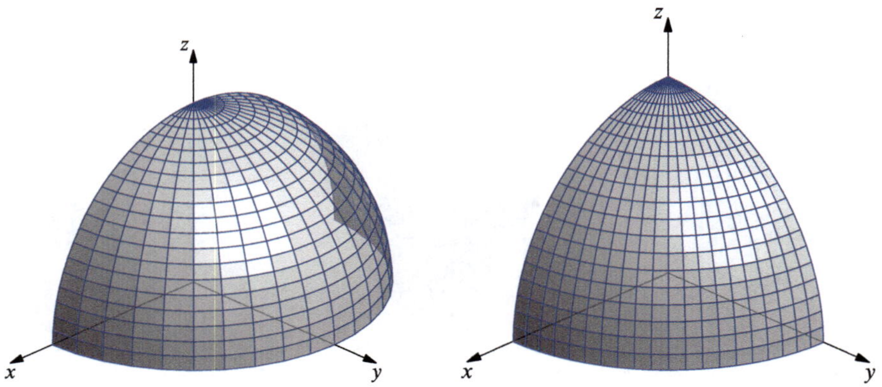

(a) $0 \le \phi \le \pi/2, 0 \le \theta \le \pi$ (b) $0 \le \phi \le \pi/2, 0 \le \theta \le \pi/2$

Fig. 6.29 Plots of sphere using a parametric representation

which is an extension of the parametric representation of circles. The domain of this function is $[0, 2\pi] \times [0, h]$ for a full cylinder. One advantage of the parametric representation is that it is easy to generate the surface: simply limit u to $0, \pi$ then we will have a half cylinder.

From spherical coordinates, we immediately have the parametric representation for a sphere of radius ρ:

$$\begin{aligned} x(\phi, \theta) &= \rho \sin\phi \cos\theta \\ y(\phi, \theta) &= \rho \sin\phi \sin\theta, \quad \phi \in [0, \pi], \quad \theta \in [0, 2\pi] \\ z(\phi, \theta) &= \rho \cos\phi \end{aligned} \tag{6.9.2}$$

Some plots of sphere are given in Fig. 6.29. Sometimes you see $x(\phi, \theta) = \rho \sin\pi u \cos 2\pi v$ with $(u, v) \in [0, 1] \times [0, 1]$. This form is obtained by normalizing the two angle parameters, so that they vary in the unit interval.

Our next destination is the paraboloid $z = x^2 + y^2$. The first parametric representation is obtained when we *use x and y as the parameters*:

$$\begin{aligned} x(u, v) &= u \\ y(u, v) &= v \\ z(u, v) &= u^2 + v^2, \quad u, v \in [-a, a] \end{aligned}$$

And a plot using this equation is given in Fig. 6.30a. It is not nice at all! Furthermore, it is impossible to plot part of the paraboloid. We need a better parametric representation. Noting that this paraboloid is a surface of revolution obtained by revolving the curve $z = y^2$ (or $z = x^2$) around the $z-$axis 2π. The level curves are circles of the form $x^2 + y^2 = r$, thus the parametric representation is

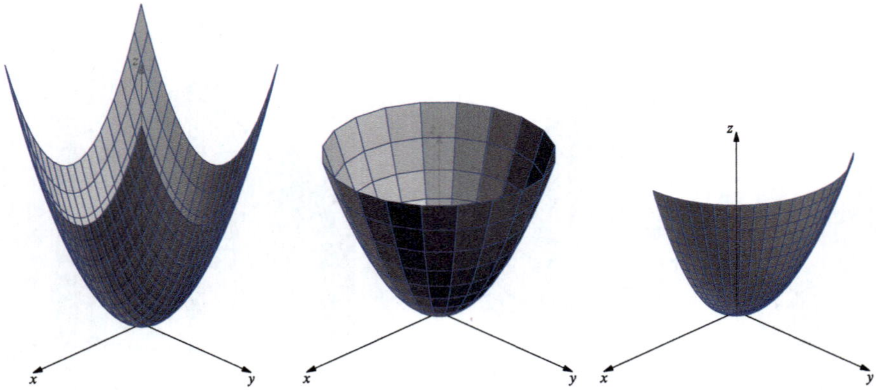

(a) parametric 1 (b) parametric revol (c) parametric revol

Fig. 6.30 Plots of the paraboloid $z = x^2 + y^2$ using an explicit formula and a parametric representation. Obviously, all surfaces can be parameterized in more than one way

$$\begin{aligned}
x(u, v) &= v \cos u \\
y(u, v) &= v \sin u \\
z(u, v) &= v^2, \quad u \in [0, 2\pi], \quad v \in [0, a]
\end{aligned}$$

In Fig. 6.30b and c, I plot two paraboloids using this parametric representation.

We now consider an ellipsoid obtained by revolving the ellipse $x^2/a^2 + y^2/b^2 = 1$ around the $x-$axis. First, we parametrize the ellipse as $x = a \cos u$ and $y = b \sin u$. As the revolution is around the $x-$axis, we consider a point P on the ellipse $x^2/a^2 + y^2/b^2 = 1$; this point is $P(a \cos u, b \sin u)$ for, let say, $u = \pi/3$ as shown in Fig. 6.31. Now, we rotate this point P around the $x-$axis a full round. What we obtain is a circle of radius $b \sin u$. The parametric equation of this circle is thus $(b \sin u) \cos v$, $(b \sin u) \sin v$. Therefore, the parametric equation of this ellipse is

$$\begin{aligned}
x(u, v) &= a \cos u \\
y(u, v) &= b \sin u \sin v \\
z(u, v) &= b \sin u \cos v, \quad u \in [0, \pi], \quad v \in [0, 2\pi]
\end{aligned}$$

See Fig. 6.31 for an illustration of the meaning of u, v. It is obvious that this parametric representation works for any parametric curve written as $(f(u), g(u))$. If we rotate this curve around the $x-$axis, then we have

$$x(u, v) = f(u), \quad y(u, v) = g(u) \sin v, \quad z(u, v) = g(u) \cos v$$

See Fig. 6.31b for a surface of revolution obtained with $y = 2 + \cos x$.

We move now to doughnuts. In geometry, a torus (plural tori) is a surface of revolution generated by revolving a circle in three-dimensional space about an axis

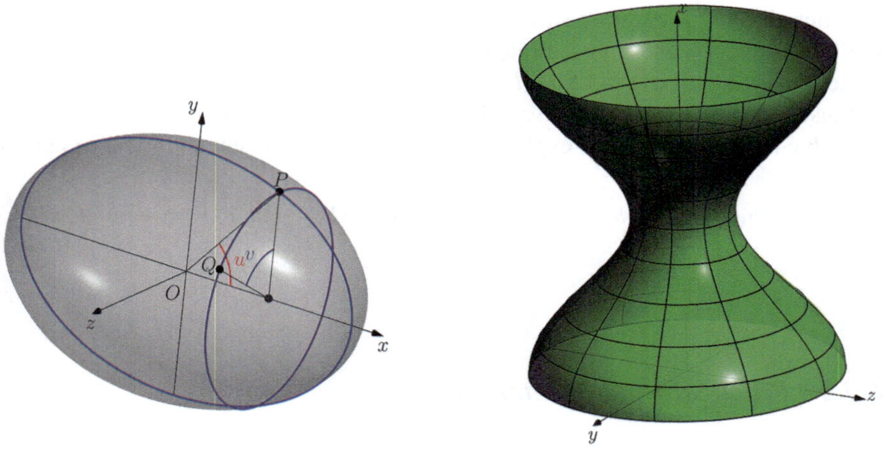

(a) (b)

Fig. 6.31 Parametric representation of surface of revolution: **a** the ellipsoid obtained by revolving the ellipse $x^2/a^2 + y^2/b^2 = 1$ around the $x-$axis. The ellipse $x^2/a^2 + y^2/b^2 = 1$ is the blue curve; **b** $y = 2 + \cos x$

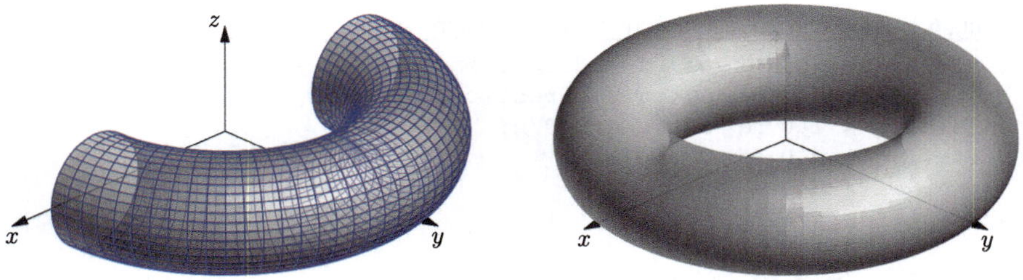

(a) $0 \leq v \leq \pi$ (b) $0 \leq v \leq 2\pi$

Fig. 6.32 Torus. The mesh lines are shown in (**a**) only

that is coplanar with the circle. It is thus a surface of revolution, so we know how to derive a parametric representation of it. Let's consider a circle in the zx-plane with a center at $(a, 0)$ and radius b. The parametric equation of this circle is

$$x = b\cos u + a, \quad z = b\sin u, \quad 0 \leq u \leq 2\pi$$

Now we rotate this circle around the $z-$axis and we get a torus:

$$x = (b\cos u + a)\cos v, \quad y = (b\cos u + a)\sin v, \quad z = b\sin u, \quad u, v \in [0, 2\pi] \tag{6.9.3}$$

See Fig. 6.32 for an illustration with $a = 3$ and $b = 1$.

After all these interesting surfaces we now discuss how to parametrize a plane. Let's consider a plane going through a point $\boldsymbol{P}_0 = (x_0, y_0, z_0)$ and contains two vectors $\boldsymbol{a} = (a_x, a_y, a_z)$ and $\boldsymbol{b} = (b_x, b_y, b_z)$ (these two vectors must be not parallel or must be independent). From Sect. 9.1.3, we know that the plane can be written as $\boldsymbol{P} = \boldsymbol{P}_0 + u\boldsymbol{a} + v\boldsymbol{b}$. Written explicitly, the parametric representation of the plane is

$$x(u, v) = x_0 + ua_x + vb_x, \quad y(u, v) = y_0 + ua_y + vb_y, \quad z(u, v) = z_0 + ua_z + vb_z, \tag{6.9.4}$$

6.9.2 Practical Implementation Detail

This section discusses how computers generate the plot of parametric surfaces. The idea is simple: a number of points on the surface are generated, these points are then connected to make the so-called grid curves. If needed, the surface is shaded. The algorithm is given in Listing 6.1. Note that the parameters are normalized to fall within the unit interval.

Listing 6.1 Algorithm to generate points on a parametric surface.

```
m = number of steps in v direction;
n = number of steps in u direction;
delta_v = 1.0/m; delta_u=1.0/n;
for(int i = 0; i < m; ++i){
  v = i * delta_v;
  for(int j = 0; j < n; ++j){
    u = j * delta_u;
    P = (x(u,v), y(u,v), z(u,v));
  }
}
```

6.9.3 Tangent Plane and Normal Vector

To find the tangent plane at a point $P(u_0, v_0)$ on a parametric surface S, we need to find the two tangent vectors of S at P and using the cross-product to get the normal. First, we fix u, and thus get a curve C_v lying on S, the tangent to this curve at P is (Fig. 6.33)

$$\boldsymbol{r}_v = \frac{\partial x}{\partial v}(u_0, v_0)\boldsymbol{i} + \frac{\partial y}{\partial v}(u_0, v_0)\boldsymbol{j} + \frac{\partial z}{\partial v}(u_0, v_0)\boldsymbol{k} \tag{6.9.5}$$

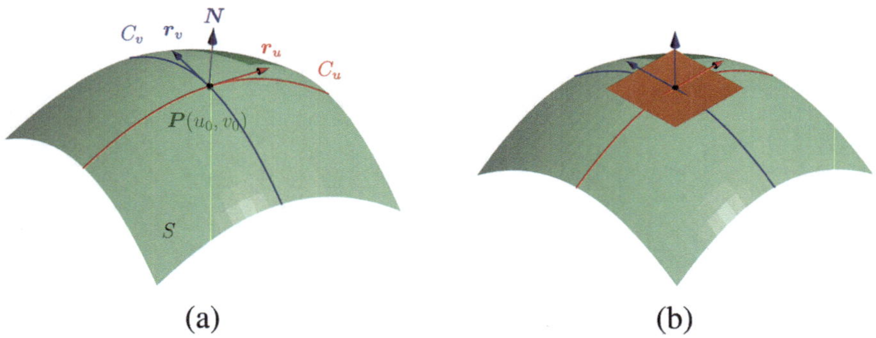

Fig. 6.33 Tangent plane and normal at P on a parametric surface S. The plan's equation is obtained with Eq. (6.9.4) where $\boldsymbol{a} = \boldsymbol{r}_u$ and $\boldsymbol{b} = \boldsymbol{r}_v$

Second, we fix v, and get a curve C_u lying on S, the tangent to this curve at P is

$$\boldsymbol{r}_u = \frac{\partial x}{\partial u}(u_0, v_0)\boldsymbol{i} + \frac{\partial y}{\partial u}(u_0, v_0)\boldsymbol{j} + \frac{\partial z}{\partial u}(u_0, v_0)\boldsymbol{k} \tag{6.9.6}$$

Finally, the normal $\boldsymbol{N}$ to the surface at $P(u_0, v_0)$ is obtained by taking the cross-product of the two tangent vectors:

$$\boldsymbol{N} = \boldsymbol{r}_u \times \boldsymbol{r}_v$$

I refer to Fig. 6.33 for a demonstration, in which the surface is actually a Bézier surface discussed in Sect. 6.9.5.

6.9.4 Surface Area and Surface Integral

The problem we're now interested in is: compute the area of a general parametric surface. Assuming that the parameter domain is a rectangle which is divided into sub-rectangles of sides Δu, Δv (Fig. 6.34a). We call these sub-rectangles the patches. On the surface S the marked patch transfroms to the patch S_{ij}, which is a curved surface (Fig. 6.34b). If we can compute the area of this patch (approximately), then we sum up the areas of all the patches, and when the number of patches tends to infinity, we shall obtain the sought-for surface area. That's the plan.

The area of S_{ij} turns out to be easy for we know the tangents $\boldsymbol{u}$ and $\boldsymbol{r}_v$. We replace S_{ij} by a parallelogram of sides $||\Delta u\boldsymbol{u}||$ and $||\Delta v\boldsymbol{v}||$. This parallelogram lies in the tangent plane to the surface at P_{ij} (see Fig. 6.33b). Thus, the area of S_{ij} is

$$\text{area of one patch} = \|\boldsymbol{r}_u \times \boldsymbol{r}_v\| \Delta u \Delta v$$

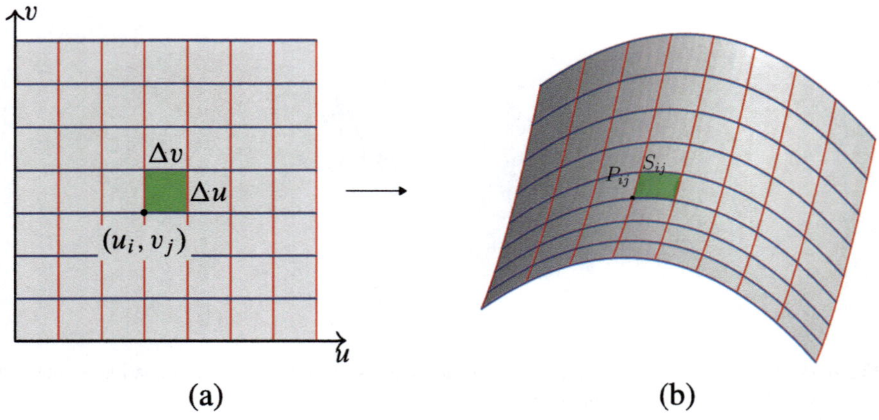

Fig. 6.34 Area of a parametric surface S

We now sum up all the areas:

$$\text{area of surface} = \sum \|\boldsymbol{r}_u \times \boldsymbol{r}_v\| \Delta u \Delta v$$

And we replace the above Riemann sum by a double integral:

$$\text{area of surface} = \iint \|\boldsymbol{r}_u \times \boldsymbol{r}_v\| du dv \tag{6.9.7}$$

6.9.5 Bézier Surfaces

In Sect. 4.14, we have met Bézier curves. It is time to extend them to Bézier surfaces, which are very powerful parametric surfaces as they are extensively used in practice. Recall that a Bézier curve, denoted by $B_n(u)$, with $n+1$ control points $\boldsymbol{P}_k$, $k = 0, 1, ..., n$ is described by the following parametric equation:

$$B_n(u) = \sum_{k=0}^{n} \binom{n}{k} (1-u)^{n-k} u^k \boldsymbol{P}_k = \sum_{k=0}^{n} B_{k,n}(u) \boldsymbol{P}_k \tag{6.9.8}$$

where $B_{k,n}(u)$ is the Bernstein basis polynomial. (Note that I have intentionally changed from t to u for the parameter). The idea now is to introduce another Bézier curve $B_m(v)$ with control points $\boldsymbol{Q}_l$, $l = 0, 1, ..., m$. We then build a Bézier surface $S(u, v)$ by multiplying $B_n(u)$ and $B_m(v)$:

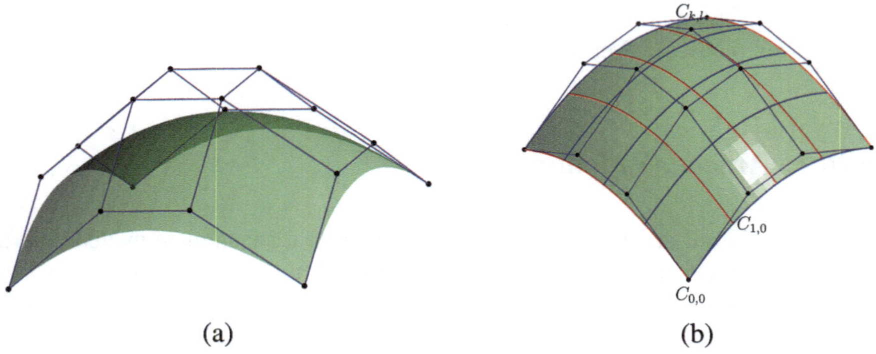

Fig. 6.35 A bi-cubic Bézier surface with 4 × 4 control points

$$S(u, v) = \left(\sum_{k=0}^{n} B_{k,n}(u)\boldsymbol{P}_k\right)\left(\sum_{l=0}^{m} B_{l,m}(v)\boldsymbol{Q}_l\right)$$

We put the control points $\boldsymbol{P}_k$ and $\boldsymbol{Q}_l$ into one single 2D array $\boldsymbol{C}_{k,l}$, which is a matrix $n \times m$ with entries are 3D points:

$$S(u, v) = \sum_{k=0}^{n}\sum_{l=0}^{m} B_{k,n}(u)B_{l,m}(v)\boldsymbol{C}_{k,l}$$

As an example, we consider a bi-cubic Bézier surface with 4×4 control points (Fig. 6.35).

6.10 Newtonian Mechanics

Back in the fifteenth century, it was still debatable about whether the earth orbits the sun or vice versa. The discussion was purely philosophical. It was Tycho Brahe who came up with the idea of making accurate observations of the orbit of the planets. Based on these observations, it should be easy to deduce what is orbiting about what. So, Tycho Brahe spent many years to measure the positions of planets. And he hired Kepler to be his assistant. After Brahe's death, Kepler, based on his former boss's data, has discovered the three Kepler's law of planetary motions. Inspired by these laws, Newton came up with his universal gravitation theory and derived Kepler's laws as consequences of his gravitation theory.

This section is a brief introduction to Newtonian mechanics. The aim is to introduce some applications of differential and integral calculus in the description of motion. And the section also presents how the French astronomer Urbain Le Verrier discovered Neptune with only paper, pencil and of course mathematics. In Francois Arago's apt phrase, Le Verrier had *discovered a planet "with the point of his pen"*.

6.10.1 Aristotle's Motion

Aristotle believed that there are two kinds of motion for inanimate matter, natural and unnatural. Unnatural (or "violent") motion is when something is being pushed, and in this case the speed of motion is proportional to the force of the push. This was probably deduced from watching boats and oxcarts. Natural motion is when something is seeking its natural place in the universe, such as a stone falling, or fire rising.

For the natural motion of heavy objects falling to earth, Aristotle asserted that the speed of fall was proportional to the weight, and inversely proportional to the density of the medium the body was falling through. However, these remarks are very brief and vague, and certainly not quantitative.

Actually, these views of Aristotle did not go unchallenged even in ancient Athens. Thirty years or so after Aristotle's death, Strato pointed out that a stone dropped from a greater height had a greater impact on the ground, suggesting that the stone picked up more speed as it fell from the greater height.

6.10.2 Galileo's Motion

Galileo set out his ideas about falling bodies, and about projectiles in general, in a book called "Two New Sciences". The two were the science of motion, which became the foundation-stone of physics, and the science of materials and construction, an important contribution to engineering.

A biography by Galileo's pupil Vincenzo Viviani stated that Galileo had dropped balls of the same material, but different masses, from the Leaning Tower of Pisa to demonstrate that their time of descent was independent of their mass. It is an amazing feeling to see this ourselves, and you can go to this YouTube webpage, see also the next figure.

History note 6.1: Galileo Galilei (1564–1642)

Galileo di Vincenzo Bonaiuti de' Galilei was an Italian astronomer, physicist, and engineer, sometimes described as a polymath, from Pisa. Galileo has been called the "father of observational astronomy", the "father of modern physics", the "father of the scientific method", and the "father of modern science". Although Galileo considered the priesthood as a young man, at his father's urging he instead enrolled in 1580 at the University of Pisa for a medical degree. In 1581, when he was studying medicine, he noticed a swinging chandelier swinging in larger and smaller arcs. By comparison with his heartbeat, he observed that the chandelier took the same amount of time to swing back and forth, no matter how far it was swinging. At home, he *set up two pendulums of equal length and swung one with a large sweep and the other with a small sweep and found that they kept time together*. Up to this point, Galileo had deliberately been kept away from mathematics, since a physician earned a higher income than a mathematician. However, *after accidentally attending a lecture on geometry, he decided to study mathematics and natural philosophy instead of medicine*.

6.10.3 Kepler's Laws

Based on the data that Brahe had collected and his own genius Kepler has discovered the following laws of planetary motion. Kepler's three laws describe the orbits of planets around the Sun. The laws modified the heliocentric theory of Nicolaus Copernicus, replacing its circular orbits with elliptical trajectories, and explaining how planetary velocities vary. The three laws state that (see Fig. 6.36):

- **Law 1**: Each planet orbits in an ellipse with one focus at the sun (Fig. 6.36a);
- **Law 2**: The vector from the sun to a planet sweeps out an area at a steady state: $dA/dt =$ constant; the shaded areas in Fig. 6.36b are equal.
- **Law 3**: The length of the planet's year (or period) is $T = ka^{3/2}$ where a is the maximum distance from the center, and $k = 2\pi/\sqrt{GM}$ is the same for all planets; G is a constant, called the *universal gravitational constant*, M is the mass of the sun (see Sect. 6.10.8 for detail).

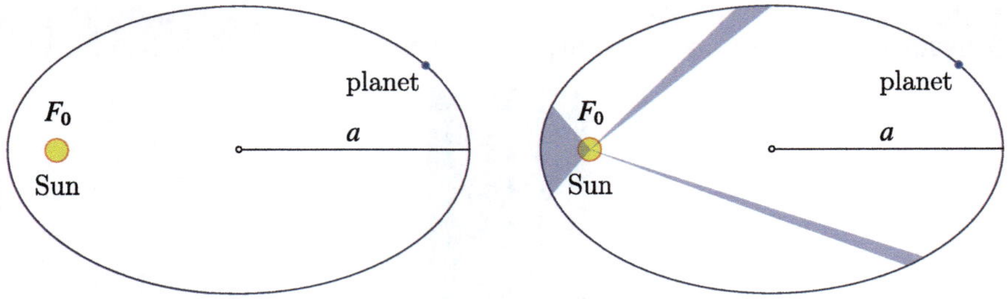

(a) Kepler's 1st law (b) Kepler's 2nd law

Fig. 6.36 Keplers' laws of planetary motion: illustration of the first and second law

The elliptical orbits of planets were indicated by calculations of the orbit of Mars. From this, Kepler inferred that other bodies in the Solar System, including those farther away from the Sun, also have elliptical orbits. The second law helps to establish that when a planet is closer to the Sun, it travels faster. Unlike the first and second laws that describe the motion characteristics of a single planet, the third law makes a comparison between the motion characteristics of different planets. The comparison being made is that the ratio of the squares of the periods (i.e., T^2) to the cubes of their average distances (i.e., a^3) from the sun is the same for every one of the planets. Thus, the third law expresses that the farther a planet is from the Sun, the slower its orbital speed, and vice versa.

History note 6.2: Johannes Kepler (1571–1630)

Johannes Kepler was a German astronomer, mathematician, and astrologer. He is a key figure in the seventeenth-century scientific revolution, best known for his laws of planetary motion, and his books Astronomia nova, Harmonices Mundi, and Epitome Astronomiae Copernicanae. These works also provided one of the foundations for Newton's theory of universal gravitation. He was introduced to astronomy at an early age and developed a strong passion for it that would span his entire life. At age six, he observed the Great Comet of 1577, writing that he "was taken by his mother to a high place to look at it. In 1580, at age nine, he observed another astronomical event, a lunar eclipse, recording that he remembered being "called outdoors" to see it and that the moon "appeared quite red".

6.10.4 Newton's Laws of Motion

Kepler had three laws for planetary motions and Newton also developed his three own laws of motion. We learnt them by heart in high school. And on the surface, they look very simple and indeed the equation is so simple ($F = ma$). The three laws of motion are:

- **Law 1**: states that if a body is at rest or moving at a constant speed in a straight line, it will remain at rest or keep moving in a straight line at constant speed *unless it is acted upon by a force*.
- **Law 2**: is a quantitative description of the changes that a force can produce on the motion of a body. It states that the time rate of change of the momentum of a body is equal in both magnitude and direction to the force imposed on it. The momentum of a body is equal to the product of its mass and its velocity. In symbols, this law is written as $\boldsymbol{F} = m\boldsymbol{a}$.
- **Law 3**: states that when two bodies interact, they apply forces to one another that are equal in magnitude and opposite in direction. The third law is also known as the law of action and reaction.

The first law is known as the law of inertia and was first formulated by Galileo Galilei. This law is very counter-intuitive: if we go shopping with a cart and we stop pushing it it goes for a short distance and stop. The law of inertia is wrong! As explained in the wonderful book *Evolution of Physics* by Einstein and Infeld, only with the imagination that Galilei resolved the problem: there is actually friction acting on the cart. If we can remove it (by having a very smooth road for example) the cart would go indeed further. And with an ideally perfectly smooth road, it goes forever.

We focus now on the second law, which is written fully as

$$\begin{aligned}
F_x &= ma_x = m\frac{d^2x}{dt^2} = m\frac{dv_x}{dt} \\
F_y &= ma_y = m\frac{d^2y}{dt^2} = m\frac{dv_y}{dt} \\
F_z &= ma_z = m\frac{d^2z}{dt^2} = m\frac{dv_z}{dt}
\end{aligned} \tag{6.10.1}$$

How are we going to use it? First, we need to know the force, we then resolve it into three components F_x, F_y and F_z, and finally, we solve Eq. (6.10.1). How to do that is the subject of the next section.

Equation (6.10.1) are what mathematicians refer to as *ordinary differential equations* with the well-known abbreviation ODEs. Precisely they are second-order ODEs as they contain the second time derivative $^{d^2x}/_{dt^2}$. Scientists like to call them dynamical equations because they describe the evolution in time (i.e., dynamics) of the system. Chapter 7 discusses differential equations in detail.

Newton gave us the second law which requires force so he had to give us some forces. And he did. In Sect. 6.10.8, I present his force of gravitation. For other forces, he gave us the third law which in many cases helps us to remove interaction forces (usually unknown) between bodies.

6.10.5 Dynamical Equations: Meaning and Solutions

To illustrate what Eq. (6.10.1) can predict, we consider an object of mass m locating at a height h above the earth. The mass of the earth is M and its radius is R. According to Newton's theory of gravitation presented in Sect. 6.10.8, the earth is pulling the object with a force F pointing to the center of the earth and has a magnitude of

$$F = \frac{GMm}{(R+h)^2}$$

Since h is tiny compared with R, we can approximate $(R+h)^2$ as $R^2 + 2Rh + h^2 \approx R^2$. Thus,

$$F = \frac{GM}{R^2}m = mg, \quad g = \frac{GM}{R^2}$$

where g is called the *acceleration of gravity*. The quantity mg is called the weight of the object, which is how hard gravity is pulling on it. With

$$G = 6.673 \times 10^{-11}\ \text{Nm}^2/\text{kg}^2, \quad M = 5.972 \times 10^{24}\ \text{kg}, \quad R = 6.37 \times 10^6\ \text{m}$$

one can determine that $g = 9.81\ \text{m/s}^2$.

With the gravitational force known, let's solve the first real problem using calculus. The problem is: we are shooting a basket ball or firing a gun; describe its motion. These projectile motions occur in a plane. Let's use the xy-plane with x being horizontal and y vertical. For simplicity, the initial position of the object (with mass m) is at the origin. The initial velocity of the object is $(v_0 \cos\alpha, v_0 \sin\alpha)$ (Fig. 6.37). Our task now is to solve the following dynamical equations:

$$0 = m\frac{d^2x}{dt^2}, \quad -mg = m\frac{d^2y}{dt^2}$$

Solving the first equation for $x(t)$, we get

$$v_x(t) = v_0 \cos\alpha, \quad x(t) = (v_0 \cos\alpha)t \tag{6.10.2}$$

which agrees with the law of inertia: no force on the x direction, the velocity (in the horizontal direction) is then constant. Now, solving the second equation for $y(t)$, we get

Fig. 6.37 Projectile motion. [Generated using Asymptote.]

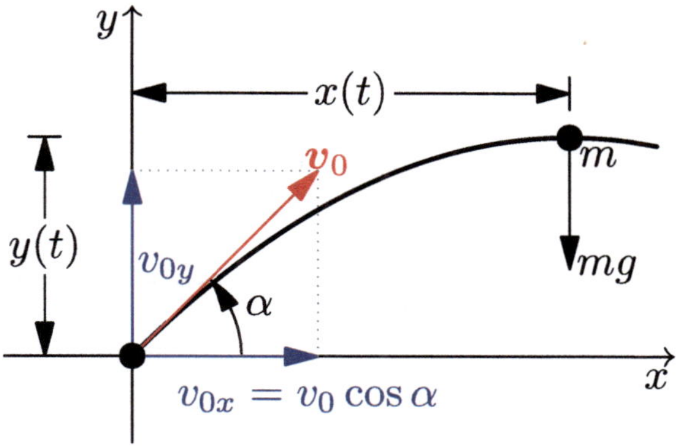

$$\frac{d^2y}{dt^2} = -g \Longrightarrow v_y(t) = -gt + v_0 \sin\alpha \Longrightarrow y(t) = (v_0 \sin\alpha)t - \frac{1}{2}gt^2 \quad (6.10.3)$$

Putting together $x(t)$ and $y(t)$, we get the complete trajectory of the projectile:

$$x(t) = (v_0 \cos\alpha)t, \quad y(t) = (v_0 \sin\alpha)t - \frac{1}{2}gt^2 \quad (6.10.4)$$

What this equation provides us is that: start with the initial position (which is (0, 0) in this particular example) and initial velocity, this equation predicts the position of the projectile at any time instant t. One question here is: what is the shape of the trajectory? Eliminating t will reveal that. From Eq. (6.10.2), we have $t = {}^{x}/_{v_0 \cos\alpha}$, and substitute that into Eq. (6.10.3), we get

$$y = (\tan\alpha)x - \frac{1}{2}\frac{g}{v_0^2 \cos^2\alpha}x^2 \quad (6.10.5)$$

A parabola! We can do a few more things with this: determining when the object hits the ground, and how far. The power of Newton's laws of motions is in the prediction of the motion of planets, see Sect. 6.10.9 for detail.

6.10.6 *Motion Along a Curve (Cartesian)*

With calculus of functions of one variable, we have studied motion along a straight line. An extension along this line is to study motion along a curve path. For example, what is the trajectory of a rocket when time is passing? Such trajectory is defined by a position vector $\boldsymbol{R}(t)$ given by

$$\boldsymbol{R}(t) = x(t)\boldsymbol{i} + y(t)\boldsymbol{j} + z(t)\boldsymbol{k} \quad (6.10.6)$$

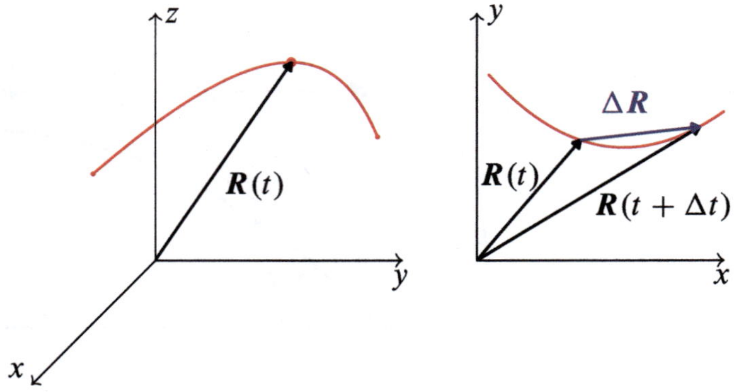

Fig. 6.38 Position vector $\boldsymbol{R}(t)$ and a change in position vector $\Delta\boldsymbol{R}(t)$. Sometimes the notation $\boldsymbol{r}(t)$ is also used

The position vector gives the position of the object in motion at any time instance (Fig. 6.38). If the motion is in a plane, we just omit the third term in the above equation. Such a position vector is mathematically called *a vector-valued function* as we assign to all number t a vector $\boldsymbol{R}$.

Knowing the function, the first step is to do the differentiation; which gives us the velocity vector $\boldsymbol{v}(t)$. To this end, we consider two time instants: at t the position vector is $\boldsymbol{R}(t)$ and at $t+\Delta t$ the position vector is $\boldsymbol{R}(t+\Delta t)$. Then, the velocity is computed as (one note about the notation is in order: vectors are typeset by italic boldface minuscule characters like $\boldsymbol{a}$)[12]

$$\begin{aligned}\boldsymbol{v}(t) &= \lim_{\Delta t\to 0}\frac{\Delta\boldsymbol{R}}{\Delta t}\\ &= \lim_{\Delta t\to 0}\frac{[x(t+\Delta t)-x(t)]\boldsymbol{i}+[y(t+\Delta t)-y(t)]\boldsymbol{j}+[z(t+\Delta t)-z(t)]\boldsymbol{k}}{\Delta t}\\ &= \frac{dx}{dt}\boldsymbol{i}+\frac{dy}{dt}\boldsymbol{j}+\frac{dz}{dt}\boldsymbol{k}\end{aligned} \tag{6.10.7}$$

What does this equation tell us? It tells us that differentiating a vector-valued function is amount to differentiating the three component functions (they are ordinary functions of a single variable). The formula is simple because the unit vectors (i.e., $\boldsymbol{i},\boldsymbol{j},\boldsymbol{k}$) are fixed. As we shall see later, this is not the case with polar coordinates, and the velocity vector has more terms.

The speed (of the object) is then given by $\|\boldsymbol{v}(t)\|$, the length of the velocity vector. The direction of motion is given by the tangent vector $\boldsymbol{T}(t)$, which is given by $\boldsymbol{v}/\|\boldsymbol{v}\|$. The tangent is a unit vector, as we're only interested in the direction.

[12] Implicitly we used the rule of limit: limit of sum is sum of limits.

The acceleration is just the derivative of the velocity:

$$\boldsymbol{a}(t) = \frac{d\boldsymbol{v}}{dt} = \frac{d^2\boldsymbol{R}}{dt^2} = \frac{d^2x}{dt^2}\boldsymbol{i} + \frac{d^2y}{dt^2}\boldsymbol{j} + \frac{d^2z}{dt^2}\boldsymbol{k} \tag{6.10.8}$$

Now, we generalize the rules of differentiation of ordinary functions to vector functions. Let's consider two vector-valued functions $\boldsymbol{u}(t)$ and $\boldsymbol{v}(t)$ and a scalar function $f(t)$, we have the following rules:

$$\begin{aligned}
&\text{(a)}\ \frac{d}{dt}[\boldsymbol{u}+\boldsymbol{v}] = \boldsymbol{u}' + \boldsymbol{v}' \\
&\text{(b)}\ \frac{d}{dt}[f(t)\boldsymbol{u}] = f'(t)\boldsymbol{u} + f(t)\boldsymbol{u}' \\
&\text{(c)}\ \frac{d}{dt}[\boldsymbol{u}\cdot\boldsymbol{v}] = \boldsymbol{u}'\cdot\boldsymbol{v} + \boldsymbol{u}\cdot\boldsymbol{v}' \\
&\text{(d)}\ \frac{d}{dt}[\boldsymbol{u}\times\boldsymbol{v}] = \boldsymbol{u}'\times\boldsymbol{v} + \boldsymbol{u}\times\boldsymbol{v}'
\end{aligned} \tag{6.10.9}$$

These rules can be verified quite straightforwardly. These rules are just some maths exercises, but amazingly we shall use the rule (d) to prove that the orbit of the earth around the sun is a plane curve.

And with all of this, we can study a variety of motions. In what follows, we present uniform circular motion as an example of application of the maths.

Uniform motion along a circle. Uniform circular motion can be described as the motion of an object in a circle at a *constant speed*. This might be a guest on a carousel at an amusement park, a child on a merry-go-round at a playground, a car with a lost driver navigating a round-about or "rotary", a yo-yo on the end of a string, a satellite in a circular orbit around the Earth, or the Earth in a (nearly) circular orbit around our Sun.

At all instances, the object is moving tangentially to the circle. Since the direction of the velocity vector is the same as the direction of the object's motion, the velocity vector is directed tangent to the circle as well. As an object moves in a circle, it is constantly changing its direction. Therefore, it is accelerating (even though the speed is constant).

Let's denote by ω the angular velocity of the object (the SI unit of angular velocity is radians per second). Then, we can write its position vector, and differentiating this vector gives us the velocity vector, which is then differentiated to give us the acceleration vector (assuming that the radius of the circular path is r):

$$\boldsymbol{R}(t) = \begin{bmatrix} r\cos\omega t \\ r\sin\omega t \end{bmatrix} \Longrightarrow \boldsymbol{v}(t) = \begin{bmatrix} -r\omega\sin\omega t \\ +r\omega\cos\omega t \end{bmatrix} \Longrightarrow \boldsymbol{a}(t) = \begin{bmatrix} -r\omega^2\cos\omega t \\ -r\omega^2\sin\omega t \end{bmatrix} \tag{6.10.10}$$

The speed—the length of $\boldsymbol{v}$—is thus $r\omega$. The acceleration vector $\boldsymbol{a}(t)$ can also be written as $-\omega^2\boldsymbol{R}(t)$: that explains the term centripetal acceleration. The word *cen-*

Fig. 6.39 .

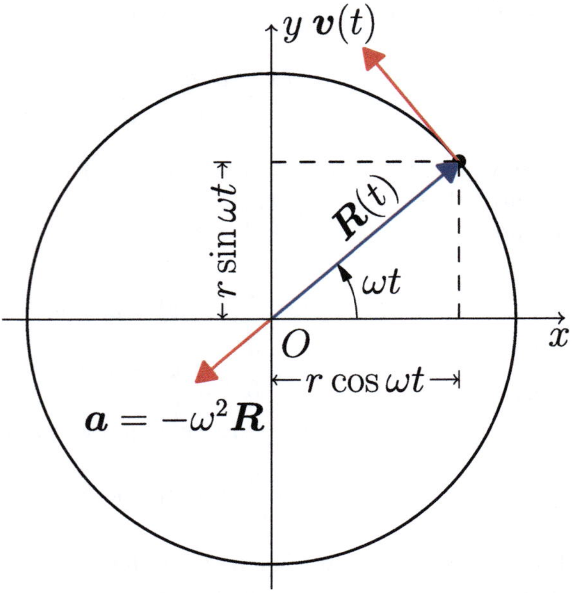

tripetal comes from the Latin words *centrum* (meaning center) and *petere* (meaning to seek). Thus, centripetal takes the meaning "center seeking". Without this acceleration, the object would move in a straight line, according to Newton's laws of motion. About the magnitude, we have $a = v^2/r$ (for $a = ||\boldsymbol{a}|| = r\omega^2$ and $v = r\omega$). We plot the position vector, velocity vector, and acceleration vector in Fig. 6.39.

6.10.7 Motion Along a Curve (Polar Coordinates)

We have described motion along a curve in which space is mathematically represented by a Cartesian coordinate system. Herein, we do the same thing but with polar coordinates. A point in this system is written as (r, θ), and similar to $\boldsymbol{i}$ and $\boldsymbol{j}$—the unit vectors in a Cartesian system, we also have $\hat{\boldsymbol{r}}$, the unit vector in the radial direction and $\hat{\boldsymbol{\theta}}$, the unit vector in the angular direction (Fig. 6.40).

The unit vector in the radial direction $\hat{\boldsymbol{r}}$ is given by

$$\hat{\boldsymbol{r}} := \frac{\boldsymbol{r}}{||\boldsymbol{r}||} = \frac{x}{\sqrt{x^2+y^2}}\boldsymbol{i} + \frac{y}{\sqrt{x^2+y^2}}\boldsymbol{j} = \cos\theta \boldsymbol{i} + \sin\theta \boldsymbol{j} \tag{6.10.11}$$

Knowing $\hat{\boldsymbol{r}}$ allows us to determine the unit vector in the tangential direction $\hat{\boldsymbol{\theta}}$ as the two vectors are perpendicular to each other. Collectively, they are written as

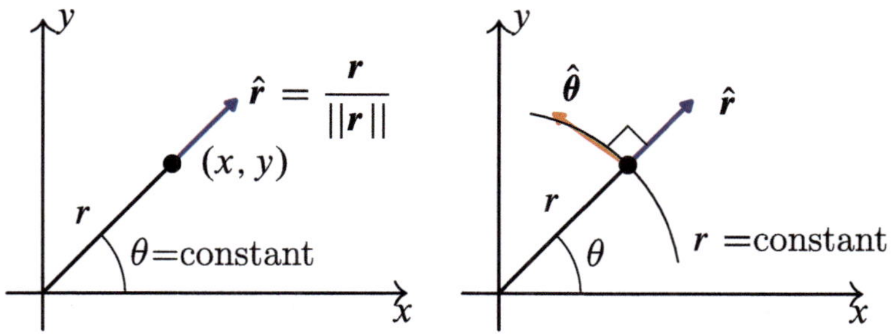

Fig. 6.40 Unit vectors in polar coordinate system. The most important observation is that while $\hat{\boldsymbol{r}}$ and $\hat{\boldsymbol{\theta}}$ are constant in length (because they are both unit vectors), they are not constant in direction. In other words, they are vector-valued functions that change from point to point. Note that $||\boldsymbol{r}|| = \sqrt{x^2 + y^2}$

$$\begin{aligned}\hat{\boldsymbol{r}} &= +\cos\theta\boldsymbol{i} + \sin\theta\boldsymbol{j}\\ \hat{\boldsymbol{\theta}} &= -\sin\theta\boldsymbol{i} + \cos\theta\boldsymbol{j}\end{aligned} \tag{6.10.12}$$

As both of them are functions of θ only, their derivatives with respect to r are zeros. We need their derivatives w.r.t. θ:

$$\begin{aligned}\frac{d\hat{\boldsymbol{r}}}{d\theta} &= -\sin\theta\boldsymbol{i} + \cos\theta\boldsymbol{j} = \hat{\boldsymbol{\theta}}\\ \frac{d\hat{\boldsymbol{\theta}}}{d\theta} &= -\cos\theta\boldsymbol{i} - \sin\theta\boldsymbol{j} = -\hat{\boldsymbol{r}}\end{aligned} \tag{6.10.13}$$

We're now ready to compute the derivative of these unit vectors w.r.t. time (following Newton, use the notation $\dot{f}$ to denote the time derivative of $f(t)$):

$$\begin{aligned}\frac{d\hat{\boldsymbol{r}}}{dt} &= \frac{d\hat{\boldsymbol{r}}}{d\theta}\frac{d\theta}{dt} = \dot{\theta}\hat{\boldsymbol{\theta}}\\ \frac{d\hat{\boldsymbol{\theta}}}{dt} &= \frac{d\hat{\boldsymbol{\theta}}}{d\theta}\frac{d\theta}{dt} = -\dot{\theta}\hat{\boldsymbol{r}}\end{aligned} \tag{6.10.14}$$

Now, we proceed to determine the velocity and acceleration. First, the velocity is

$$\boldsymbol{r} = r\hat{\boldsymbol{r}} \Longrightarrow \frac{d\boldsymbol{r}}{dt} = \dot{r}\hat{\boldsymbol{r}} + r\frac{d\hat{\boldsymbol{r}}}{dt} = \dot{r}\hat{\boldsymbol{r}} + r\dot{\theta}\hat{\boldsymbol{\theta}} \tag{6.10.15}$$

And therefore, the acceleration is

$$
\begin{aligned}
\frac{d^2\boldsymbol{r}}{dt^2} &= \frac{d}{dt}\left(\dot{r}\hat{\boldsymbol{r}} + r\dot{\theta}\hat{\boldsymbol{\theta}}\right) \\
&= \ddot{r}\hat{\boldsymbol{r}} + \dot{r}\frac{d\hat{\boldsymbol{r}}}{dt} + \dot{r}\dot{\theta}\hat{\boldsymbol{\theta}} + r\ddot{\theta}\hat{\boldsymbol{\theta}} + r\dot{\theta}\frac{d\hat{\boldsymbol{\theta}}}{dt} \\
&= (\ddot{r} - r\dot{\theta}^2)\hat{\boldsymbol{r}} + (2\dot{r}\dot{\theta} + r\ddot{\theta})\hat{\boldsymbol{\theta}}
\end{aligned}
\tag{6.10.16}
$$

where use was made of Eq. (6.10.14).

So, Newton's second law in polar coordinates is written as

$$
\begin{aligned}
F_r &= m(\ddot{r} - r\dot{\theta}^2) \\
F_\theta &= m(2\dot{r}\dot{\theta} + r\ddot{\theta})
\end{aligned}
\tag{6.10.17}
$$

Another way to come up with the velocity and acceleration using $\boldsymbol{r} = re^{i\theta}$.

Using complex exponential, we can write

$$
\hat{\boldsymbol{r}} = e^{i\theta}, \quad \hat{\boldsymbol{\theta}} = ie^{i\theta} \tag{6.10.18}
$$

As multiplying with i is a 90° rotation, it is clear that $\hat{\boldsymbol{r}}$ is perpendicular to $\hat{\boldsymbol{\theta}}$. Now, we can differentiate $\boldsymbol{r} = re^{i\theta}$ w.r.t. time:

$$
\boldsymbol{r} = re^{i\theta} \Longrightarrow \frac{d\boldsymbol{r}}{dt} = \dot{r}e^{i\theta} + ire^{i\theta}\dot{\theta} = \dot{r}\hat{\boldsymbol{r}} + r\dot{\theta}\hat{\boldsymbol{\theta}}
$$

which is exactly what we obtained in Eq. (6.10.15). For the acceleration, doing something similar as

$$
\frac{d\boldsymbol{r}}{dt} = \dot{r}e^{i\theta} + ire^{i\theta}\dot{\theta} \Longrightarrow \frac{d^2\boldsymbol{r}}{dt^2} = \ddot{r}e^{i\theta} + \dot{r}ie^{i\theta}\dot{\theta} + i\dot{r}e^{i\theta}\dot{\theta} + ir\dot{\theta}ie^{i\theta}\dot{\theta} + ire^{i\theta}\ddot{\theta}
$$

and we got Eq. (6.10.16).

6.10.8 *Newton's Gravitation*

In 1687, Newton published his work on gravity in his classic *Mathematical Principles of Natural Philosophy*. He stated that every object is pulling other object and for two objects they are pulled by a force that is proportional to the product of their masses and inversely proportional to the square of the distance between them. In mathematical symbols, his law is expressed as

$$F = \frac{GMm}{r^2} \tag{6.10.19}$$

where G is a constant, called the *universal gravitational constant*, that has been experimentally measured by Cavendish[13] about 100 years after Newton's death. The value of this constant is $G = 6.673 \times 10^{-11}$ Nm^2/kg^2. In physics, Eq. (6.10.19) is known as an *inverse square law*. Why such a name? Because electric forces also obey this kind of law. Again, you see that the same mathematics apply for different physical phenomena.

How Newton came up with Eq. (6.10.19)? It can be based on Kepler's third law as shown in the box below.

Assume that a planet of mass m orbits the sun in a circle of radius r with uniform speed v. This is not correct that a planet is orbiting with a uniform the speed. But note that we are just trying to guess what the form of a law looks like. Note also that Newton never knew G in his own equation Eq. (6.10.19)! The period T of the planet, which is the time for it to complete one travel around the sun, is given by $T = {}^{2\pi r}/_{v}$ (nothing but time = distance/speed), and we need T^2:

$$T = \frac{2\pi r}{v} \Longrightarrow T^2 = \frac{4\pi^2 r^2}{v^2}$$

We then determine v in terms of the force F using Newton's second law and $a = v^2/r$ (check Eq. (6.10.10) and the discussion below it):

$$F = ma, \quad a = \frac{v^2}{r} \Longrightarrow v^2 = ar = \frac{Fr}{m}$$

Thus, the squared period T^2 becomes

$$T^2 = \frac{4\pi^2 r^2}{v^2} = \frac{4\pi^2 mr}{F} \tag{6.10.20}$$

And Kepler's third law says $T^2 \propto r^3$, so

$$\frac{4\pi^2 mr}{F} \propto r^3 \Longrightarrow F \propto \frac{m}{r^2} \tag{6.10.21}$$

But the planet also pulls the sun of mass M with the same force (Newton's third law), thus F should be proportional to M too. Thus, $F \propto \frac{Mm}{r^2}$. Eventually, $F = \text{constant} \times \frac{Mm}{r^2}$, and that constant is G—for gravity. Mathematics cannot give you G; for that we need physicists.

[13] Henry Cavendish (1731–1810) was an English natural philosopher, scientist, and an important experimental and theoretical chemist and physicist.

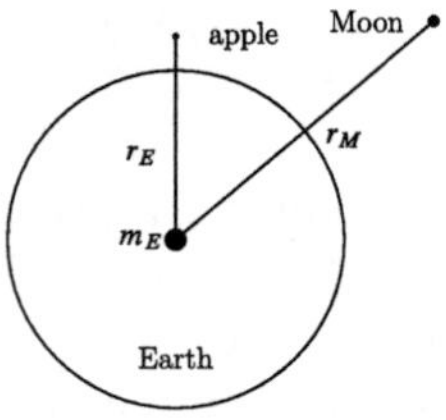

Why Eq. (6.10.19) is called the universal law of gravitation? That is because it is the force described by Eq. (6.10.19) that governs the motion of planets around the sun, the orbit of the moon around the earth and the fall of an apple from a tree to the earth surface. How Newton proved this? He used the following facts: (i) the earth can be approximated as a mass concentrated at its center (obtained via his Shell theorem proved in Sect. 6.8.5) and (ii) the acceleration due to gravity near the earth surface g is 9.81 m/s^2, probably determined by Galileo.

So, using Newton's second law and Eq. (6.10.19), we can compute the acceleration of an apple and the moon as

$$\begin{aligned}\text{Apple:}\quad a_a &= \frac{F}{m_a} = \frac{Gm_E}{r_E^2}\\ \text{Moon:}\quad a_m &= \frac{Gm_E}{r_M^2}\end{aligned}$$

where r_E is the radius of the earth (a known quantity), r_M is the distance from the earth center to the moon (also known), and m_E is the mass of the earth (we do not really need it as it will be canceled in the algebraic manipulations); m_a is the mass of the apple and note that as the apple is near to the earth surface the distance from the center of the earth to the apple is simply r_E. We can thus compute a_m:

$$a_m = \left(\frac{r_E}{r_M}\right)^2 a_a = \left(\frac{6.37 \times 10^6 \text{ m}}{3.84 \times 10^8 \text{ m}}\right)^2 \times 9.81 = \boxed{2.7 \times 10^{-3} \text{ m/s}^2} \tag{6.10.22}$$

The acceleration of the moon can also be computed using another way (Eq. (6.10.20)):

$$a_m = \frac{v^2}{r_M} = \frac{4\pi^2 r_M^2}{T^2 r_M} = \frac{4\pi^2 r_M}{T^2} = \frac{(4\pi^2)3.84 \times 10^8 \text{ m}}{(2.36 \times 10^6 \text{ s})^2} = \boxed{2.72 \times 10^{-3} \text{ m/s}^2} \tag{6.10.23}$$

where $T \approx 27$ days is the period of the moon. The amazing agreement of the two values of the moon acceleration proved the universality of Newton's law of gravity.

All planets orbit the sun on a plane. Using (d) in Eq. (6.10.9), we compute

$$\frac{d}{dt}(\boldsymbol{r} \times \boldsymbol{v}) = \boldsymbol{v} \times \boldsymbol{v} + \boldsymbol{r} \times \boldsymbol{a} = \boldsymbol{0} + \boldsymbol{r} \times \boldsymbol{a} = \boldsymbol{r} \times \boldsymbol{a}$$

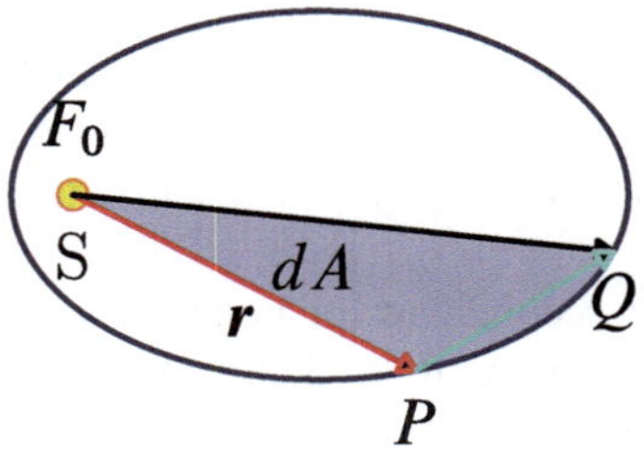

$$\overrightarrow{PQ} = \boldsymbol{v}dt$$

$$dA = \frac{1}{2}||\boldsymbol{r} \times \boldsymbol{v}dt||$$

$$\frac{dA}{dt} = \frac{1}{2}||\boldsymbol{r} \times \boldsymbol{v}|| = \frac{1}{2m}||\boldsymbol{r} \times \boldsymbol{p}||$$

Fig. 6.41 Proof of Kepler's second law. For illustration purposes, the area dA was magnified. The angular momentum is $\boldsymbol{l} = \boldsymbol{r} \times \boldsymbol{p}$

Now comes one observation: $\boldsymbol{r} \times \boldsymbol{a} = \boldsymbol{0}$ because $\boldsymbol{r}$ and $\boldsymbol{a}$ are parallel. Why that? From Newton's second law (i.e., $\boldsymbol{F} = m\boldsymbol{a}$) we deduce that $\boldsymbol{F}$ and $\boldsymbol{a}$ are parallel. But Newton's gravitation tells us that $\boldsymbol{F}$ is parallel to $\boldsymbol{r}$. To conclude, $\frac{d}{dt}(\boldsymbol{r} \times \boldsymbol{v}) = \boldsymbol{0}$, or $\boldsymbol{r} \times \boldsymbol{v}$ is a constant vector $\boldsymbol{H}$. And $\boldsymbol{r}$ stays always in the plane perpendicular to $\boldsymbol{H}$. See Fig. 4.1b for a beautiful picture of this plane.

6.10.9 From Newton's Universal Gravitation to Kepler's Laws

Proof of second law. I provide two proofs of Kepler's second law. Recall that Kepler's second law simply means that dA/dt is constant, see Fig. 6.41. Let's put the origin (of the coordinate system) at the sun which leads to the net torque acting on a planet P is zero. This is because the sun's gravitational pull is a central force and the cross-product of two parallel vectors is a zero vector. If the net torque is zero, then the angular momentum is constant (Sect. 9.1.5). It can be seen that dA/dt is proportional to the length of the angular momentum (which is constant), and thus dA/dt is constant.

As the second proof, we compute the angular momentum. We start with the angular momentum $\boldsymbol{l} = \boldsymbol{r} \times \boldsymbol{p}$, and $\boldsymbol{p} = m\boldsymbol{v}$, $\boldsymbol{v} = \dot{r}\hat{\boldsymbol{r}} + r\dot{\theta}\hat{\boldsymbol{\theta}}$ according to Eq. (6.10.15):

$$\begin{aligned}
\boldsymbol{l} &= \boldsymbol{r} \times \boldsymbol{p} = mr(\hat{\boldsymbol{r}} \times \boldsymbol{v}), \quad (\boldsymbol{r} = r\hat{\boldsymbol{r}}) \\
&= mr\left[\hat{\boldsymbol{r}} \times (\dot{r}\hat{\boldsymbol{r}} + r\dot{\theta}\hat{\boldsymbol{\theta}})\right] \quad \text{(Eq. (6.10.15))} \\
&= mr^2\dot{\theta}(\hat{\boldsymbol{r}} \times \hat{\boldsymbol{\theta}}), \quad (\hat{\boldsymbol{r}} \times \hat{\boldsymbol{r}} = \boldsymbol{0})
\end{aligned} \tag{6.10.24}$$

From this, we can determine the length of the angular momentum as $l = mr^2\omega$, where $\omega = \dot{\theta}$; because $||\hat{\boldsymbol{r}} \times \hat{\boldsymbol{\theta}}|| = 1$ for two perpendicular unit vectors. From Fig. 6.41 and following the steps in Eq. (6.10.24) but without the mass m, we get

$$dA/dt = 0.5\|\boldsymbol{r} \times \boldsymbol{v}\| = \frac{1}{2m}\|\boldsymbol{r} \times \boldsymbol{p}\| = 0.5r^2\dot{\theta}(\hat{\boldsymbol{r}} \times \hat{\boldsymbol{\theta}}) = 0.5r^2\omega = {}^{l}/_{2m}$$

Since the angular momentum l is conserved, we arrive at the conclusion that dA/dt is constant. This proof shows us that as the planet is orbiting the sun, when it is close to the sun (r is small), it speeds up (ω is bigger as $l = mr^2\omega$ is constant).

Proof of second law. We use Newton's second law in polar coordinates, i.e., Eq. (6.10.17) together with Newton's universal gravity to deduce Kepler's first law. The only force is the Sun's gravitational pull written as

$$\boldsymbol{F} = -\frac{GMm}{r^2}\hat{\boldsymbol{r}} \tag{6.10.25}$$

Introducing this force into Eq. (6.10.17), we get the following system of two equations:

$$\begin{aligned} \ddot{r} - r\dot{\theta}^2 &= -\frac{GM}{r^2} \\ 2\dot{r}\dot{\theta} + r\ddot{\theta} &= 0 \end{aligned} \tag{6.10.26}$$

Solution of this system of equations is the orbit of the planet and it should be an equation for an ellipse (but we need to prove this). From the second equation in Eq. (6.10.26), we have

$$d/dt(r^2\dot{\theta}) = 0 \Longleftrightarrow r^2\dot{\theta} = h = \text{constant}$$

The trick is to use a new variable $q = 1/r$.[14] In terms of q, $r^2\dot{\theta} = h$ becomes $hq^2 = \dot{\theta}$. Let's compute dr/dt first:

$$r = \frac{1}{q} \Longrightarrow \dot{r} = -\frac{\dot{q}}{q^2} = -\frac{1}{q^2}\frac{dq}{d\theta}\frac{d\theta}{dt} = -h\frac{dq}{d\theta}$$

Then, using the above expression of $\dot{r}$ we compute $\ddot{r}$, which is what we want, see Eq. (6.10.26):

$$\frac{d^2r}{dt^2} = -h\frac{d}{dt}\left(\frac{dq}{d\theta}\right) = -h\frac{d}{d\theta}\left(\frac{dq}{dt}\right) = -h\frac{d}{d\theta}\left(\frac{dq}{d\theta}\frac{d\theta}{dt}\right) = -h^2q^2\frac{d^2q}{d\theta^2} \tag{6.10.27}$$

where, in the last equality, we used the result that $hq^2 = \dot{\theta}$.

We're now ready to rewrite the first equation of Eq. (6.10.26) in terms of h, q, θ:

$$h^2q^2\frac{d^2q}{d\theta^2} + \frac{1}{q}(hq^2)^2 = GMq^2 \Longrightarrow \boxed{\frac{d^2q}{d\theta^2} + q = C}, \quad (C = {}^{GM}/_{h^2}) \tag{6.10.28}$$

[14] Don't ask me why this new variable. I have no idea.

The boxed equation is a so-called differential equation (DE). We have more to say about differential equations in Chap. 7, but briefly, a DE is an equation that contains derivatives of some function that we're trying to find, e.g., $f(x) + f'(x) = 2$. How are we going to solve the above-boxed equation? Solving DEs is not easy, but in this case, it turns out that the solution is something we know. *What is the boxed equation saying to us?* It tells us to find a function (i.e., q) such that its second derivative equals minus itself (the constant C is not important). *We know that* $\cos\theta$ *is such a function.* So, the solution to this equation is $q = C - D\cos\theta$. Now, forget q—it's just a means to an end—we need r which is

$$r = \frac{1}{C - D\cos\theta}$$

But this is the equation of a conic section (Sect. 4.13.2). We need astronomical data to determine C and D and from that to deduce that this is indeed the equation of an ellipse.

At this moment, you might be thinking "but the orbit of planets around the Sun was known to be an ellipse thanks to Kepler". It is indeed easier to work on a problem of which solution we knew beforehand. But, Newton's universal gravity theory is more powerful than that. It can predict things that we never know of.

6.10.10 Discovery of Neptune

To understand why Newton's universal gravity law is considered one of the best of humankind, let's see how it helped to predict the planet Neptune before Neptune was directly observed.

By 1847, the planet Uranus had completed nearly one full orbit since its discovery by William Herschel in 1781, and astronomers had detected a series of irregularities in its path that could not be entirely explained by Newton's universal gravitation theory. So, either was Newton wrong or there was a mysterious planet that people at that time did not know of. In 1845, the French astronomer Urbain Le Verrier (1811–1877) and the British mathematician and astronomer John Couch Adams (1819–1892)—an undergraduate at Cambridge University, who both believed in Newton, separately began calculations to determine the nature and position of this unknown planet. On 23 September 1846, Johann Galle head of the Berlin Observatory used Le Verrier's calculations to find Neptune only 1° off Le Verrier's predicted position. The planet was then located 12° off Adams' prediction. In Francois Arago's apt phrase, Le Verrier had discovered a planet "with the point of his pen".

Remark Although Newton's theory of gravity was a great achievement Newton himself could not explain the mechanism of his own theory. As the sun and the earth are separated by a large distance and there is nothing in between, how the force of gravity communicates is the big issue. And this is only resolved hundreds of years later when Albert Einstein came up with his theory of general relativity.

6.10.11 Newton and the Great Plague of 1665–1666

Between the summer of 1665 and the spring of 1667, Isaac Newton at the age of 22 made two extended visits to his family farm in Woolsthorpe to escape the plague affecting Cambridge. The bubonic "Great Plague" of 1665–1666 was the worst outbreak of plague in England since the black death of 1348. It spread rapidly throughout the country. London lost roughly 15% of its population, and the villagers of Eyam, Derbyshire, became famous for their heroic quarantine to halt the spread of the disease.

Many town-dwellers, like Newton, retreated to the relative safety of the countryside. What is different is how he set his mind to work in this period. There he remained secluded for eighteen months, during which time he not only discovered the universal law of gravity but changed the face of science.

About these years of wonder, in a letter to Pierre Des Maizeaux, written in 1718, Newton wrote these words

> In the beginning of the year 1665 I found the method of approximating series and the rule for reducing any dignity [power] of any binomial into such a series. The same year in May I found the method of tangents of Gregory and Slusius, and in November had the direct method of fluxions and the next year [1666] in January had the theory of colours and in May following I had entrance into the inverse method of fluxions. And the same year I began to think of gravity extending to the orb of the moon ... All this was in the two plague years of 1665 and 1666, *for in those days I was in the prime of my age for invention and minded Mathematics and Philosophy more than at any time since.*

6.11 Vector Calculus

Vector calculus is the calculus of vector fields or vector-valued functions. It was developed for electromagnetism and thus it is best to be studied with electromagnetism. This tradition started with Richard Feynman in his lecture on physics and with the book *Div, Grad, Curl, and All That: An Informal Text on Vector Calculus* by Schey [58]. We also follow this approach mainly because we wanted to learn electromagnetism—a branch of physics that underlies everything in our modern world.

All of electromagnetism is contained in the Maxwell equations:

$$\nabla \cdot \boldsymbol{E} = \frac{\rho}{\epsilon_0} \tag{6.11.1a}$$

$$\nabla \times \boldsymbol{E} = -\frac{\partial \boldsymbol{B}}{\partial t} \tag{6.11.1b}$$

$$c^2 \nabla \times \boldsymbol{B} = \frac{\partial \boldsymbol{E}}{\partial t} + \frac{\boldsymbol{j}}{\epsilon_0} \tag{6.11.1c}$$

$$\nabla \cdot \boldsymbol{B} = 0 \tag{6.11.1d}$$

where ∇ is the gradient vector operator, $\nabla \cdot \boldsymbol{E}$ is the *divergence* of the electric field $\boldsymbol{E}$, $\nabla \times \boldsymbol{E}$ is the *curl* of $\boldsymbol{E}$; $\boldsymbol{B}$ is the magnetic field.

When the electric and magnetic fields do not depend on the time, i.e., the charges are permanently fixed in space or if they do more, they move as a steady flow, all of the terms in Eq. (6.11.1) which are time derivatives of the fields are zero. And we get two sets of equations. One for electrostatics:

$$\nabla \cdot \boldsymbol{E} = \frac{\rho}{\epsilon_0} \tag{6.11.2a}$$

$$\nabla \times \boldsymbol{E} = 0 \tag{6.11.2b}$$

and one for magnetostatics:

$$\nabla \times \boldsymbol{B} = \frac{\boldsymbol{j}}{c^2 \epsilon_0} \tag{6.11.3a}$$

$$\nabla \cdot \boldsymbol{B} = 0 \tag{6.11.3b}$$

Looking at these two sets of equations, we can see that electrostatics is a neat example of a vector field with zero curl and a given divergence. And magnetostatics is a neat example of a vector field with zero divergence and a given curl.

To summarize, the central object of vector calculus is vector fields $\boldsymbol{C}$. And to this object, we will of course do differentiation and integration, which leads to differential calculus of vector fields and integral calculus of vector fields, and connections between them:

- differential calculus: we have divergence $\nabla \cdot \boldsymbol{C} = \frac{\partial C_x}{\partial x} + \frac{\partial C_y}{\partial y} + \frac{\partial C_z}{\partial z}$, we have curl $\nabla \times \boldsymbol{C}$;
- integral calculus: we have line integral $\int_C \boldsymbol{C} \cdot d\boldsymbol{s}$, surface integral $\int_S \boldsymbol{C} \cdot \boldsymbol{n} dA$;
- the fundamental theorem of calculus that links line integrals to surface integrals and volume integrals: we have Green's theorem, Stokes' theorem, and Gauss's theorem. They are all generalizations of $\int_a^b {}^{df}/_{dx} dx = f(b) - f(a)$.

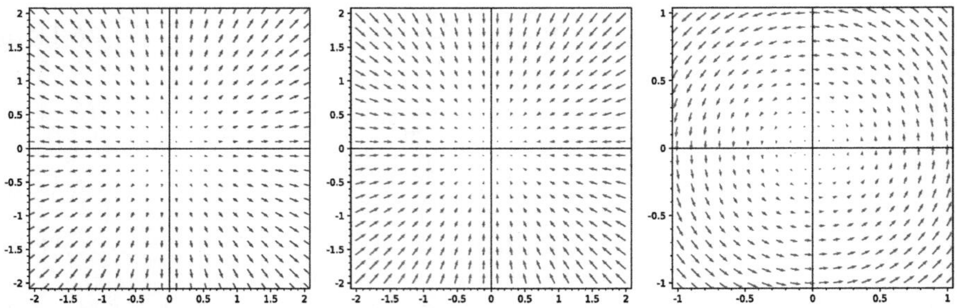

(a) $x\boldsymbol{i} + y\boldsymbol{j}$ (b) $-x\boldsymbol{i} - y\boldsymbol{j}$ (c) $-y\boldsymbol{i} + x\boldsymbol{j}$

Fig. 6.42 Some vector fields

6.11.1 Vector Fields

In vector calculus and physics, a vector field is an assignment of a vector to each point in space. A vector field in the plane (for instance), can be visualized as a collection of arrows with a given magnitude and direction, each attached to a point in the plane. Vector fields are often used to model, for example, the speed and direction of a moving fluid throughout space, or the strength and direction of some force, such as the magnetic or gravitational force, as it changes from one point to another point.

Generally a 3D vector field $\boldsymbol{F}$ can be described as:

$$\boldsymbol{F} = M(x, y, z, t)\boldsymbol{i} + N(x, y, z, t)\boldsymbol{j} + P(x, y, z, t)\boldsymbol{k} \tag{6.11.4}$$

So, a 3D vector field is similar to three ordinary functions. If the field does not depend on time t; we have a static field, then in the above equation t is omitted. And for a plane vector field, we have $\boldsymbol{F} = M(x, y, t)\boldsymbol{i} + N(x, y, t)\boldsymbol{j}$. Figure 6.42 gives some plane vector fields in which you can think of the velocity field of a fluid.

6.11.2 Central Forces and Fields

Probably the two most well-known vector fields are gravitational force and electric force. They are both central forces. The gravitational force was discovered by Newton and the electric force by Charles Coulomb.[15] In mathematical symbols, they are written as

[15] Charles-Augustin de Coulomb (1736–1806) was a French officer, engineer, and physicist. He is best known as the eponymous discoverer of what is now called Coulomb's law, the description of the electrostatic force of attraction and repulsion. He also did important work on friction. The SI unit of electric charge, the coulomb, was named in his honor in 1880.

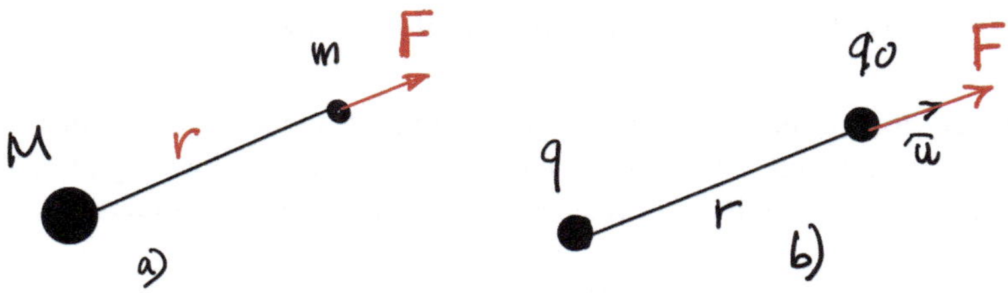

Fig. 6.43 Gravitational force between two masses M and m and electric force between two charges q_0 and q

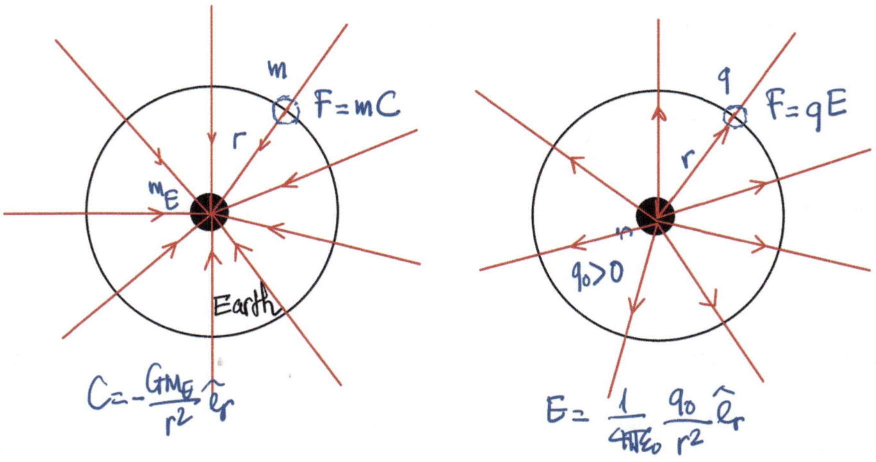

Fig. 6.44 .

$$\text{Gravitational force:} \quad \boldsymbol{F} = G\frac{Mm}{r^2}\hat{\boldsymbol{r}}$$

$$\text{Electric force:} \quad \boldsymbol{F} = \frac{1}{4\pi\epsilon_0}\frac{qq_0}{r^2}\hat{\boldsymbol{r}}$$

Remarkably these two very different forces have the same mathematical format: they are inversely proportional to the distance r between two masses M, m or two charges q and q_0, and they are proportional to the product of two masses or charges. They are known as inverse square laws. As these forces are along the line connecting the two masses (or charges), they are called central forces (Figs. 6.43 and 6.44).

$$\boldsymbol{E} = \frac{1}{4\pi\epsilon_0}\frac{q}{r^2}\hat{\boldsymbol{u}} \tag{6.11.5}$$

6.11.3 Work Done by a Force and Line Integrals

To introduce the concept of line integral, let's go back to the well-known conservation of energy principle that states, for 1D motion along a vertical line (e.g., free falling motion), that the sum of the kinetic energy and potential energy is constant

$$\underbrace{0.5mv^2}_{\text{K.E.}} + \underbrace{mgh}_{\text{P.E}} = \text{const} \tag{6.11.6}$$

And we want to verify whether this principle is correct. We use Newton's second law $F = ma = mdv/dt$, but focus on energy aspects. Let's calculate the change of the kinetic energy T:

$$\frac{dT}{dt} = \frac{d}{dt}\left(\frac{1}{2}mv^2\right) = mv\frac{dv}{dt} = \left(m\frac{dv}{dt}\right)v = Fv \tag{6.11.7}$$

Since $F = -mg$ and $v = {}^{dh}/_{dt}$, we get (assuming the mass is constant)

$$\frac{dT}{dt} = Fv = -mg\frac{dh}{dt} = -\frac{d}{dt}(mgh)$$

Hence, the change in the kinetic energy turns into potential energy, and thus Eq. (6.11.6) is indeed correct.

So, from Newton's law, we have discovered an interesting fact about energy conservation. But it was only for the simple problem of free fall. Will this energy principle work for other cases? Let's check! In 3D, the kinetic energy T for a particle of mass m traveling along a 3D curve is given by

$$T = \frac{1}{2}\left(mv_x^2 + mv_y^2 + mv_z^2\right)$$

Thus, its rate of change is

$$\frac{dT}{dt} = mv_x\frac{dv_x}{dt} + mv_y\frac{dv_y}{dt} + mv_z\frac{dv_z}{dt} = \boldsymbol{F}\cdot\boldsymbol{v} \tag{6.11.8}$$

The term $\boldsymbol{F}\cdot\boldsymbol{v}$ is called the power. Replacing $\boldsymbol{v}$ as ${}^{d\boldsymbol{s}}/_{dt}$, where $d\boldsymbol{s} = (dx, dy, dz)^\top$, we then have

$$\frac{dT}{dt} = \boldsymbol{F}\cdot\boldsymbol{v} = \boldsymbol{F}\cdot\frac{d\boldsymbol{s}}{dt} \tag{6.11.9}$$

Even though the trajectory is a 3D curve, the only non-zero force component is $F_z = -mg$, thus we have

$$\frac{dT}{dt} = (-mg)\frac{dz}{dt} = -\frac{d}{dt}(mgz)$$

And again, energy conservation works.

We have a tiny change of T w.r.t. a tiny change in time, Eq. (6.11.9). Integral calculus gives us the total change when the particle traverses the entire path, denoted by C. From Eq. (6.11.9) we obtain $dT = \boldsymbol{F} \cdot d\boldsymbol{s}$, and integrating this gives us the total change of the kinetic energy

$$\boxed{\Delta T = \int_C \boldsymbol{F} \cdot d\boldsymbol{s}} \tag{6.11.10}$$

This integral (a significant integral) is named *a line integral of a vector field.* In mechanics, this integral is called *the work done by a force.* And Eq. (6.11.10) is known as the work-kinetic energy theorem: the change in a particle's KE as it moves from point 1 to point 2 (the end points of C) is the work done by the force.

Let's say a few words about the unit of work. As work is defined as force multiplied by distance, its SI unit is Newton · meter, which is one Joule.[16]

Don't let the name line integral fool you, the integral path C is actually a curve. As $\boldsymbol{F} \cdot d\boldsymbol{s}$ is a number the line integral is simply an extension of $\int_a^b f(x)dx$. Instead of moving on the horizontal $x-$ line from $(a, 0)$ to $(b, 0)$, now we traverse a spatial curve C. Obviously when this curve happens to be the horizontal line, the line integral is reduced to the ordinary integral. So, actually nothing is too new here.

For the evaluation of a line integral, it is convenient to use a parametric representation for the curve C. That is, $C : (x(t), y(t))$ for $a \leq t \leq b$. Then, Eq. (6.11.10) becomes, for a 2D vector field $\boldsymbol{F} = M(x, y)\boldsymbol{i} + N(x, y)\boldsymbol{j}$:

$$\int_C \boldsymbol{F} \cdot d\boldsymbol{s} = \int_a^b \begin{bmatrix} M(x(t), y(t)) \\ N(x(t), y(t)) \end{bmatrix} \cdot \begin{bmatrix} x'(t) \\ y'(t) \end{bmatrix} dt \tag{6.11.11}$$

The final integral is simply an integral of the form $\int_a^b f(t)dt$, which can be evaluated using standard techniques of calculus. In what follows, we present a few examples.

Example 6.3 Let's consider this vector field $\boldsymbol{F} = -y\boldsymbol{i} + x\boldsymbol{j}$ (see Fig. 6.42c), and the path is the full unit circle centered at (2, 0), and it is traversed counterclockwise. First, we parametrize C, then just apply Eq. (6.11.11):

$$\left.\begin{aligned} x &= 2 + \cos t \\ y &= \sin t \end{aligned}\right\} \Longrightarrow \left\{\begin{aligned} dx &= -\sin t dt \\ dy &= +\cos t dt \end{aligned}\right., \quad \boldsymbol{F} \cdot d\boldsymbol{s} = (-\sin t)(-\sin t)dt + (2 + \cos t)(\cos t)dt$$

So, (the symbol $\oint$ to designate that the curve is closed)

$$\oint \boldsymbol{F} \cdot d\boldsymbol{s} = \int_0^{2\pi} (1 + 2\cos t)dt = 2\pi$$

---

[16] One joule is equal to the amount of work done when a force of 1 newton displaces a mass through a distance of 1 meter in the direction of the force applied. It is named after the English physicist James Prescott Joule (1818–1889).

The result is positive which is expected because the force and the path are both counterclockwise.

Example 6.4 Let's consider this vector field $\boldsymbol{F} = 2x\boldsymbol{i} + 2y\boldsymbol{j}$ (see Fig. 6.42a), and the path is the full unit circle centered at $(2, 0)$. Note that the vector field $\boldsymbol{F}$ is the gradient of this scalar field $\psi = x^2 + y^2$.

We have

$$\left.\begin{array}{l} x = 2 + \cos t \\ y = \sin t \end{array}\right\} \Longrightarrow \left\{\begin{array}{l} dx = -\sin t dt \\ dy = +\cos t dt \end{array}\right., \quad \boldsymbol{F} \cdot d\boldsymbol{s} = (4 + 2\cos t)(-\sin t)dt + (2\sin t)(\cos t)dt$$

Thus,

$$\oint \boldsymbol{F} \cdot d\boldsymbol{s} = -4 \int_0^{2\pi} \sin t dt = 4\cos t]_0^{2\pi} = 0 \tag{6.11.12}$$

So, *the line integral of a gradient field along a closed curve is zero*! Let's see would we also get zero if the path is not a closed curve. Assume the path is just the first quarter of the circle, and the line integral is

$$\int \boldsymbol{F} \cdot d\boldsymbol{s} = -4 \int_0^{\pi/2} \sin t dt = -4$$

which is not zero.

Now, we suspect that there is something special about the line integral of a gradient vector. But a line integral is a generalization of $\int_a^b f(x)dx$, which satisfies the fundamental theorem of calculus:

$$\int_a^b \frac{dF}{dx} dx = F(b) - F(a)$$

So, the equivalent counterpart for line integrals should look like this:

$$\underset{\text{along } C}{\int_1^2} \nabla\psi \cdot d\boldsymbol{s} = \psi(2) - \psi(1)$$

And it turns out that our guess is correct. Suppose that we have a scalar field $\psi(x, y)$ and two points 1 and 2. We denote by $\psi(1)$ the field at point 1 and by $\psi(2)$ the field at point 2. A curve C joints these two points (Fig. 6.45). We have the following theorem:

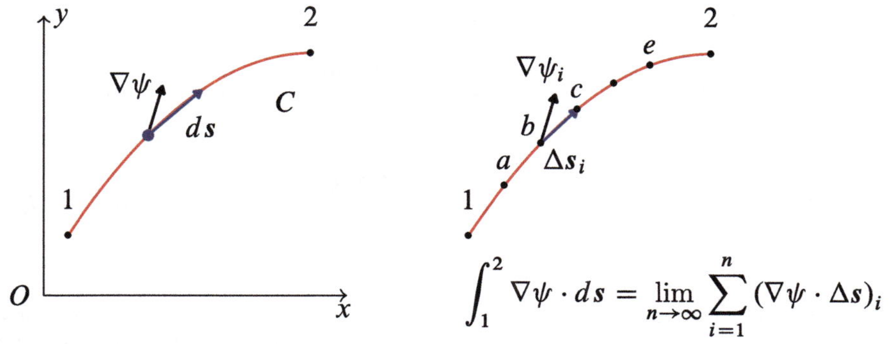

Fig. 6.45 Line integral as a Riemann sum

Theorem 6.11.1 Fundamental Theorem For Line Integrals

$$\int_{\substack{1\\ \text{along } C}}^{2} \nabla\psi \cdot d\boldsymbol{s} = \psi(2) - \psi(1) \tag{6.11.13}$$

which states that the line integral along the curve C of the dot product of a gradient $\nabla\psi$*—a vector field—with* $d\boldsymbol{s}$*—another vector which is the infinitesimal line segment—equals the difference of* ψ *evaluated at the two end points of the curve C.*

It is because of this theorem that the integral in Eq. (6.11.12) is zero, as the two end points are the same. Also because of this theorem that the line integral of a gradient vector is *path-independent*. That is, no matter how we go from point 1 to point 2, the integral is the same.

Proof [Proof of theorem 6.11.1]. We use the definition of an integral as a Riemann sum to prove the above theorem. To this end, we divide the curve C into many segments (Fig. 6.45). Then, we can write the integral as

$$\int_1^2 \nabla\psi \cdot d\boldsymbol{s} = \lim_{n\to\infty}\sum_{i=1}^{n}(\nabla\psi \cdot \Delta\boldsymbol{s})_i$$

Now, what is $\nabla\psi \cdot \Delta\boldsymbol{s}$? It is the change of ψ along $\Delta\boldsymbol{s}$. Remember the directional derivative? Here the direction is along the curve. So, we can compute this term for all n segments:

$$\begin{aligned}
(\nabla\psi \cdot \Delta\boldsymbol{s})_1 &= \psi(a) - \psi(1)\\
(\nabla\psi \cdot \Delta\boldsymbol{s})_2 &= \psi(b) - \psi(a)\\
(\nabla\psi \cdot \Delta\boldsymbol{s})_n &= \psi(2) - \psi(e)
\end{aligned}$$

If we sum up all the finite differences we get $\psi(2) - \psi(1)$. ■

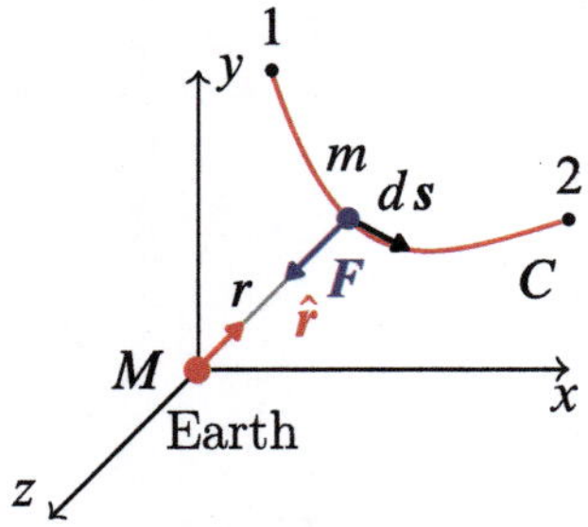

$$\boldsymbol{F} = -\frac{GMm}{r^2}\hat{\boldsymbol{r}}, \quad \hat{\boldsymbol{r}} = \frac{1}{r}(x, y, z)^\top$$

$$r^2 = x^2 + y^2 + z^2$$

$$\boldsymbol{F} \cdot \boldsymbol{ds} = -\frac{GMm}{r^2}\left(\frac{1}{r}\right)(xdx + ydy + zdz)$$

Fig. 6.46 Work of the gravitational force in moving a mass m at (x, y, z) from point 1 to point 2 along a curved path C. Note that $\boldsymbol{ds} = (dx, dy, dz)^\top$. There is a minus sign in the force because $\boldsymbol{F}$ and $\hat{\boldsymbol{r}}$ are in opposite direction

6.11.4 Work of Gravitational and Electric Forces

We have seen previously that the work done by gravity near the earth surface in moving a mass from point 1 to point 2 depends only on the vertical distance difference of these two end points. In other words, the work which is a line integral is independent of the path. No matter which path the object is moving, the work is the same! What is an interesting thing. Now we are going to examine to see if this holds for harder cases.

We compute the work done by the gravity in moving a mass m from point 1 to point 2 along a curved path C as shown in Fig. 6.46. The origin of the coordinate system is put at the earth of mass M. Refering to Fig. 6.46 for the computation of $\boldsymbol{F} \cdot \boldsymbol{ds}$, we have

$$\boldsymbol{F} \cdot \boldsymbol{ds} = \left(-\frac{GMm}{r^2}\right)\frac{1}{r}(xdx + ydy + zdz)$$

But, as $r^2 = x^2 + y^2 + z^2$, we have $rdr = xdx + ydy + zdz$, thus

$$\boldsymbol{F} \cdot \boldsymbol{ds} = \left(-\frac{GMm}{r^2}\right)dr$$

So the work is written as

$$W = \int_C \boldsymbol{F} \cdot \boldsymbol{ds} = -GMm\int \frac{dr}{r^2} = GMm\left(\frac{1}{r_2} - \frac{1}{r_1}\right) \tag{6.11.14}$$

And this work is also independent of the path! And if C is a close path, W would be zero.

We know that the work done is equal to the change in the kinetic energy (that is $W = \Delta T$). And Eq. (6.11.14) shows that work done is also a change of something: the RHS of that equation is the difference of two terms which indicates a change of something that we label as U. Our aim is now to find the expression for U. We have, $W = \Delta T$ and $W = -\Delta U$, so

$$\left.\begin{aligned} W &= +\Delta T \\ W &= -\Delta U \end{aligned}\right\} \Longrightarrow \Delta(T + U) = 0 \quad \text{(energy is conserved)} \tag{6.11.15}$$

From Eqs. (6.11.15) and (6.11.14) we can determine the expression for U:

$$GMm\left(\frac{1}{r_2} - \frac{1}{r_1}\right) = -\Delta U \Longrightarrow \boxed{U(r) = -\frac{GMm}{r}} \tag{6.11.16}$$

And $U(r)$ is called the potential of the gravitational force.

6.11.5 Fluxes and Divergence

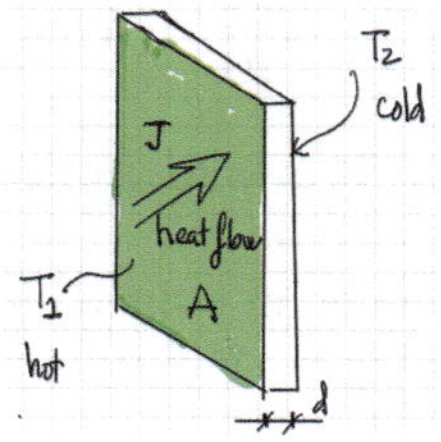

To introduce the concept of flux, let us consider the problem of heat conduction, first solved by Joseph Fourier in 1855. Assume we have a thin slab made of a certain material. One face of the slab is heated to a temperate T_1 and the other face is heated to $T_2 < T_1$. Experiments demonstrate that there is heat flow through the slab. The amount of thermal energy per unit time flows through the slab, denoted by Q, is given by (in SI system the unit of Q is W[17] or J/s)

$$Q = k\frac{(T_1 - T_2)A}{d} \quad \text{(W=J/s)} \tag{6.11.17}$$

where k is the thermal conductivity of the material (SI unit is W/(mK)). This equation was obtained based on experimental observations that the rate of heat conduction through a slab is proportional to the temperature difference across the slab $(T_1 - T_2)$ and the heat transfer area (A), but it is inversely proportional to the thickness of the slab d.

Now if we shrink the slab thickness d to zero so that we have the derivative of the temperature, and divide the above equation by A (and thus get rid of that), we get the following differential form of the one-dimensional Fourier law for heat conduction:

[17] Named after James Watt (1736–1819), a Scottish inventor, mechanical engineer, and chemist.

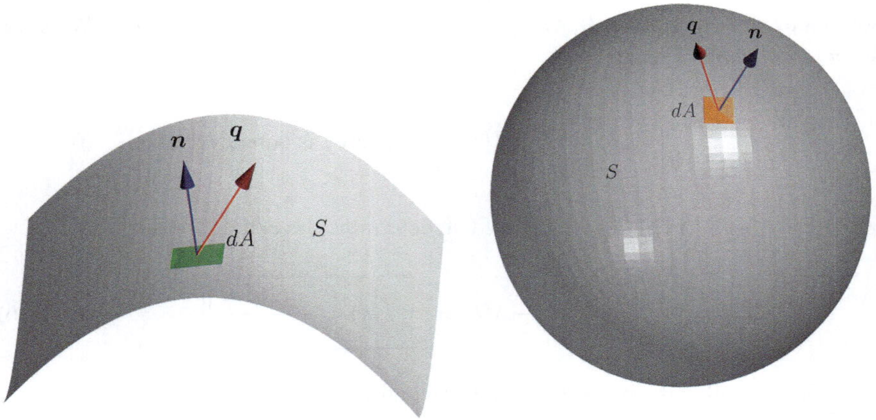

Fig. 6.47 An open or close surface with an infinitesimal surface area dA with $\boldsymbol{n}$ being its unit normal vector pointing outward. By unit normal I meant that $\|\boldsymbol{n}\| = 1$

$$q = -k\frac{dT}{dx} \quad (\text{W/m}^2) \tag{6.11.18}$$

where q is the heat flux density. The word flux comes from Latin: fluxus means "flow", and fluere is "to flow".

Now we move to heat conduction in a three-dimensional body of complicated geometry. The generalization of Eq. (6.11.18) is

$$\boldsymbol{q} = -k\nabla T \tag{6.11.19}$$

where ∇T denotes the gradient of the temperature field.

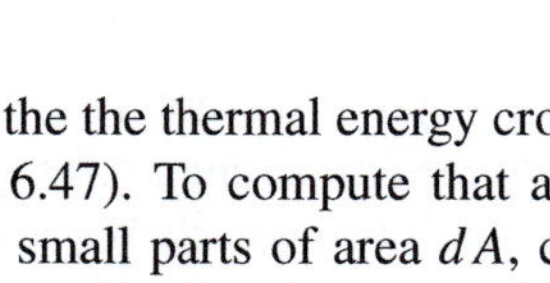

The question now is what is the the thermal energy crossing a surface? The surface can be open or closed (Fig. 6.47). To compute that amount of energy, we divide the surface into many many small parts of area dA, compute the energy crossing each part and sum all those energies. Now comes the key observation: If the surface element is not perpendicular to $\boldsymbol{q}$, the amount of thermal energy crossing it is smaller as the tangential component of the flow does not contribute to the flow across the surface. The amount of thermal energy passing though dA per unit time is then given by $\boldsymbol{q} \cdot \boldsymbol{n} dA$. The proof goes like this. Assume that the heat flux density $\boldsymbol{q}$ is perpendicular to the surface of area A_1. Thus, the amount of thermal energy crosses this surface per unit time is $Q = qA_1$. This same amount of energy is passing through the surface of area $A_2 = A_1/\cos\theta$. Thus, the heat flux through A_2 is

$$\frac{Q}{A_1/\cos\theta} = \frac{qA_1}{A_1}\cos\theta = q\cos\theta = \boldsymbol{q}\cdot\boldsymbol{n}$$

In the last step, Eq. (9.1.6) relating the dot product and the cosine of the angle was used. Noting that $\|\boldsymbol{n}\| = 1$.

And the total flux of heat through a surface S is the sum of all the fluxes through the small surface elements dA:

$$\text{heat flux} = \int_S \boldsymbol{q}\cdot\boldsymbol{n}dA$$

Now we generalize this concept of heat flux to any vector field $\boldsymbol{C}$:

$$\text{flux} = \int_S \boldsymbol{C}\cdot\boldsymbol{n}dA \tag{6.11.20}$$

So in vector calculus *a flux is a surface integral of the normal component of a vector*.

Imagine that we have a volume V with surface S (Fig. 6.48). Now we cut that volume into two volumes V_1 and V_2 by a plane S_{ab}. The first volume is enclosed by surface S_1 which consists of a part of the original surface S_a and S_{ab}. The second volume is bounded by surface S_2 which consists of the other part of the original surface S_b and S_{ab}. If we compute the flux of a vector field $\boldsymbol{C}$ through the surface S_1 and the flux through S_2, we get:

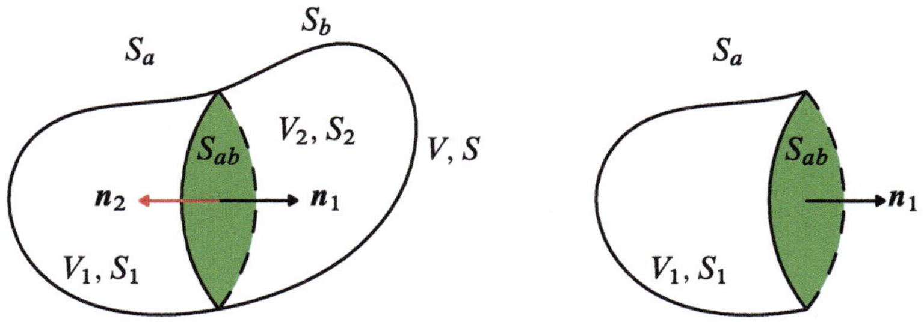

Fig. 6.48 The flux through the complete outer surface S of a volume V can be considered as the sum of the fluxes out of the two pieces into which the volume was broken

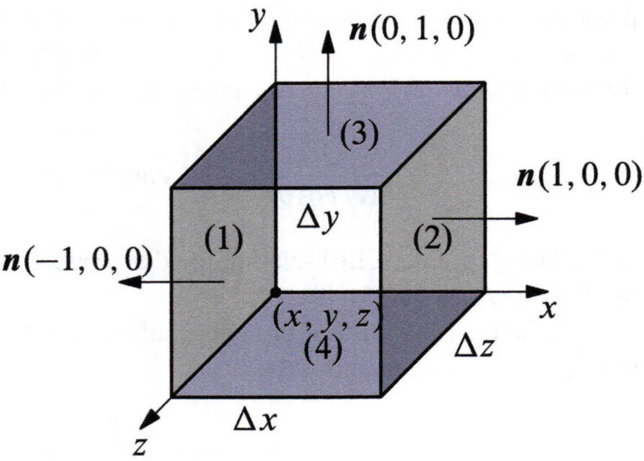

Fig. 6.49 Flux of a vector field $\boldsymbol{C}$ through the faces of an infinitesimal cube

$$\text{flux through } S_1\text{: } \int_{S_a} \boldsymbol{C}\cdot\boldsymbol{n}dA + \int_{S_{ab}} \boldsymbol{C}\cdot\boldsymbol{n}_1 dA$$
$$\text{flux through } S_2\text{: } \int_{S_b} \boldsymbol{C}\cdot\boldsymbol{n}dA + \int_{S_{ab}} \boldsymbol{C}\cdot\boldsymbol{n}_2 dA$$

Noting that $\boldsymbol{n}_2 = -\boldsymbol{n}_1$, when we sum these two fluxes, the red terms cancel out, and we obtain this interesting fact about flux: the flux through the complete outer surface S can be considered as the sum of the fluxes out of the two pieces into which the volume was broken. And nothing can stop us from dividing V_1 into two little pieces and regardless of how we divide the original volume we always get that the flux through the original outer surface S is equal to the sum of the fluxes out of all the little interior pieces.

We continue that division process until we get an infinitesimal little piece. And that is a very small cube. Now, we're going to compute the flux of a vector field $\boldsymbol{C}$ through the faces of an infinitesimal cube. And of course we choose a special cube, one that is aligned with the coordinate axes (Fig. 6.49).

The flux through faces 1 and 2, defined by $\int \boldsymbol{C}\cdot\boldsymbol{n}dA$, are given by (note that the normals of these faces are parallel to the $x-$ direction so other components of $\boldsymbol{C}$ are irrelevant)

$$\text{flux through face } 1 = -C_x(1)\Delta y\Delta z$$
$$\text{flux through face } 2 = +C_x(2)\Delta y\Delta z$$

And as the cube is tiny, the field is constant over these faces. So, for face 1, the field is $C_x(1)$ where 1 is any point on this face.

Along the $x-$direction, the field is changing, so we have

$$C_x(2) = C_x(1) + \frac{\partial C_x}{\partial x}\Delta x \tag{6.11.21}$$

which is correct as Δx is small. Thus, we can compute the flux through faces 1/2, and similarly for faces 3/4 and 5/6. They are given by

$$\text{flux through faces 1/2} = \frac{\partial C_x}{\partial x}\Delta x \Delta y \Delta z$$
$$\text{flux through faces 3/4} = \frac{\partial C_y}{\partial y}\Delta x \Delta y \Delta z$$
$$\text{flux through faces 5/6} = \frac{\partial C_z}{\partial z}\Delta x \Delta y \Delta z$$

which gives us the total flux through all the six faces of the small cube with surface S:

$$\int_S \boldsymbol{C} \cdot \boldsymbol{n} dA = \left(\frac{\partial C_x}{\partial x} + \frac{\partial C_y}{\partial y} + \frac{\partial C_z}{\partial z}\right)\Delta V \tag{6.11.22}$$

where $\Delta V = \Delta x \Delta y \Delta z$ is the volume of the cube. The red term is given a special name—the divergence of $\boldsymbol{C}$. Thus, the divergence of a 3D vector is defined as

$$\nabla \cdot \boldsymbol{C} = \text{div}\, \boldsymbol{C} = \frac{\partial C_x}{\partial x} + \frac{\partial C_y}{\partial y} + \frac{\partial C_z}{\partial z} \tag{6.11.23}$$

What does Eq. (6.11.22) mean? It tells us that, for an infinitesimal cube, the outward flux of the cube is equal to the divergence of the vector multiplied with the volume of the cube. To better understand the meaning of this new divergence concept, we consider three vector fields and compute the corresponding divergences (Fig. 6.50). Think of these vector fields as the velocities of some moving fluid. Now put a sphere at the origin and the fluid can go in and out of this sphere. In Fig. 6.50a, $\nabla \cdot \boldsymbol{C} > 0$ indicates that, due to Eq. (6.11.22), the fluid is moving out of the sphere. On the

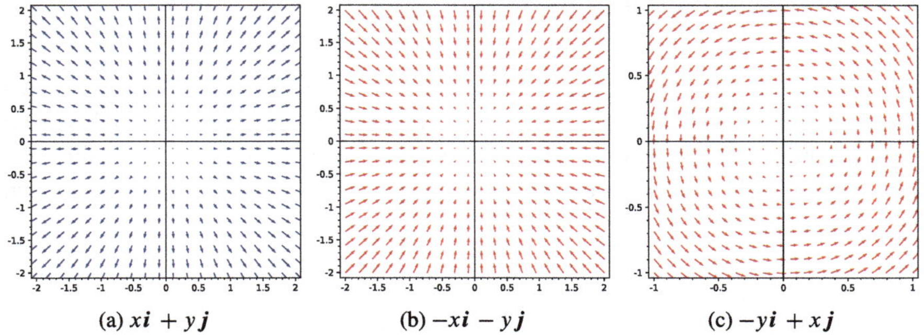

Fig. 6.50 Some 2D vector fields and their divergences: **a** $\nabla \cdot \boldsymbol{C} = 2 > 0$, **b** $\nabla \cdot \boldsymbol{C} = -2 < 0$ and **c** $\nabla \cdot \boldsymbol{C} = 0$. You're recommended to watch this amazing animation for a better understanding of the meaning of the divergence and curl

contrary, in Fig. 6.50b, the fluid is entering the sphere, thus $\nabla \cdot \boldsymbol{C} < 0$. Finally, the fluid in Fig. 6.50c is just swirling around: there is no fluid moving out of the sphere—$\nabla \cdot \boldsymbol{C} = 0$. If the divergence cannot describe a rotating fluid, then we need another concept. And indeed, the curl of the fluid velocity field does just that (Sect. 6.11.7).

6.11.6 Gauss's Theorem

Gauss's theorem is a relation between a volume integral (a triple integral) and a surface integral. Assume that we have a solid with volume V and it is enclosed by a surface S. We have a vector field $\boldsymbol{C}$ inside the volume. If we divide this solid into many many infinitesimal cubes, then for each cube, from Eq. (6.11.22), we have

$$\int_{\text{cube faces}} \boldsymbol{C} \cdot \boldsymbol{n} dA = \nabla \cdot \boldsymbol{C} \Delta V \tag{6.11.24}$$

And if we sum up all these tiny cubes, the right-hand side of Eq. (6.11.24) is the volume integral of the divergence of $\boldsymbol{C}$. How about the left-hand side? It is the flux of $\boldsymbol{C}$ through the solid surface S; see the discussion related to Fig. 6.48. And that is, Gauss's theorem or Gauss's divergence theorem[18]:

$$\text{Gauss's divergence theorem:} \quad \int_S \boldsymbol{C} \cdot \boldsymbol{n} dA = \int_V \nabla \cdot \boldsymbol{C} dV \tag{6.11.25}$$

In Sect. 7.5.2 I provide one application of Gauss's divergence theorem to derive the three-dimensional heat conduction equation.

6.11.7 Circulation of a Fluid and Curl

We have met the line integral of a force field of the form $\int_1^2 \boldsymbol{F} \cdot d\boldsymbol{s}$. This has a physical meaning of the work done by the force in moving the object from point 1 to point 2 on a curved path. Now, we study again the line integral but with two differences: instead of a force field, we consider the velocity field of a moving fluid, and the path is a close one.

---

[18] This proof is however not mathematically rigorous. It is certainly true that any domain can be cut up into cubes/boxes. But most domains have a curved boundary, so the domain is unlikely to be a union of boxes. It is not uncommon to argue that by taking the boxes to be smaller and smaller we can approximate any reasonable domain better and better, and hence taking some sort of limit, the divergence theorem follows for any such domain.

Fig. 6.51 Circulation of $\boldsymbol{C}$ around a rectangle of sides $\Delta x \times \Delta y$. The area of this small rectangle is Δa

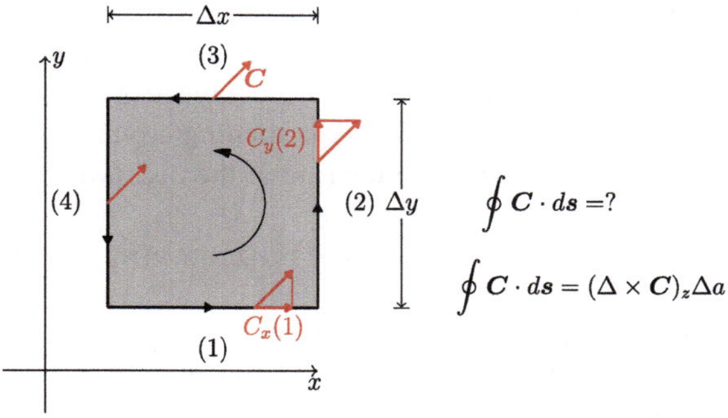

To start with we take a very special path: the boundary of a rectangle living on the xy-plane (Fig. 6.51). The rectangle is infinitesimally small, the path of the integral is its boundary with a counterclockwise orientation as indicated.

Now, the circulation of the fluid around the rectangle is the line integral along the rectangle boundary of the tangential component of the vector field or $\boldsymbol{C} \cdot d\boldsymbol{s}$. The line integral is broken into four integrals along the four sides. Take the side 1 for example, using the mean value theorem for integral, i.e., Eq. (4.11.3) we can write

$$\int_{\text{side 1}} \boldsymbol{C} \cdot d\boldsymbol{s} = C_x(1)\Delta x$$

where $C_x(1)$ is the value of C_x evaluated at some point on the side 1. It does not matter the precise location of this point. Doing similarly for other sides, the integral is given by

$$\oint \boldsymbol{C} \cdot d\boldsymbol{s} = C_x(1)\Delta x + C_y(2)\Delta y - C_x(3)\Delta x - C_y(4)\Delta y \tag{6.11.26}$$

Similarly to what we have done to get the divergence, we group the red terms and blue terms:

$$\begin{aligned}
\text{circulation along sides 1/3} &= [C_x(1) - C_x(3)]\Delta x = -\frac{\partial C_x}{\partial y}\Delta x \Delta y \\
\text{circulation along sides 2/4} &= [C_y(2) - C_y(4)]\Delta x = +\frac{\partial C_y}{\partial x}\Delta x \Delta y
\end{aligned}$$

Substitution of this into Eq. (6.11.26) we obtain:

$$\oint \boldsymbol{C} \cdot d\boldsymbol{s} = \left(\frac{\partial C_y}{\partial x} - \frac{\partial C_x}{\partial y}\right)\Delta x \Delta y \tag{6.11.27}$$

Now is the time to check if what we have obtained is really capturing the tendency of rotation. Just use the examples shown in Fig. 6.50. For the left and middle figures, the red term $C_{y,x} - C_{x,y} = 0$ and obviously these two fluids are not rotating. For the right figure, $C_{y,x} - C_{x,y} = 2$ and the fluid in that figure is counterclockwise curling.

If you're still not yet convinced, consider now the uniform circular motion discussed in Sect. 6.10.6. A disk is rotating around the z-axis with an angular velocity ω. A point $P(x, y, z)$ on the ring of the disk (with radius r) has a velocity vector

$$\boldsymbol{v} = -\omega y\boldsymbol{i} + \omega x\boldsymbol{j}, \text{or } \boldsymbol{v}(x, y, z) = (-\omega y, \omega x)$$

If we plot this velocity field it looks exactly similar to the one given in Fig. 6.50c. Then, the red term in Eq. (6.11.27) but applied to $\boldsymbol{v}$ (instead of $\boldsymbol{C}$, noting they're both vectors) is given by

$$\frac{\partial v_y}{\partial x} - \frac{\partial v_x}{\partial y} = 2\omega$$

Indeed that red term is an indication of a rotation.

Instead of considering a rectangle in the xy-plane, we can consider rectangles in the yz- and zx-planes. Altogether, the circulations are given by

$$\begin{aligned}
\text{rectangle in } xy \text{ plane:} &= \oint \boldsymbol{C} \cdot d\boldsymbol{s} = \left(\frac{\partial C_y}{\partial x} - \frac{\partial C_x}{\partial y}\right) \Delta x \Delta y \\
\text{rectangle in } yz \text{ plane:} &= \oint \boldsymbol{C} \cdot d\boldsymbol{s} = \left(\frac{\partial C_z}{\partial y} - \frac{\partial C_y}{\partial z}\right) \Delta y \Delta z \\
\text{rectangle in } zx \text{ plane:} &= \oint \boldsymbol{C} \cdot d\boldsymbol{s} = \left(\frac{\partial C_x}{\partial z} - \frac{\partial C_z}{\partial x}\right) \Delta z \Delta x
\end{aligned}$$

The three terms in the brackets are the three Cartesian components of a vector called the curl of $\boldsymbol{C}$, written as $\nabla \times \boldsymbol{C}$ (read del cross C) where $\square \times \square$ is the cross-product (see Sect. 9.1.5 for a discussion on the cross-product between two vectors). One way to memorize the formula for the curl of a vector field is to use the determinant of the following 3×3 matrix:

$$\begin{vmatrix} \boldsymbol{i} & \boldsymbol{j} & \boldsymbol{k} \\ \dfrac{\partial}{\partial x} & \dfrac{\partial}{\partial y} & \dfrac{\partial}{\partial z} \\ C_x & C_y & C_z \end{vmatrix} = \left(\frac{\partial C_z}{\partial y} - \frac{\partial C_y}{\partial z}\right)\boldsymbol{i} + \left(\frac{\partial C_x}{\partial z} - \frac{\partial C_z}{\partial x}\right)\boldsymbol{j} + \left(\frac{\partial C_y}{\partial x} - \frac{\partial C_x}{\partial y}\right)\boldsymbol{k} \tag{6.11.28}$$

Now we return to Eq. (6.11.27) and observe that the term in the brackets is just the $z-$component of $\nabla \times \boldsymbol{C}$. And $\Delta x \Delta y$ is the area of our little square Δa. Thus,

$$\oint \boldsymbol{C} \cdot d\boldsymbol{s} = (\nabla \times \boldsymbol{C}) \cdot \boldsymbol{n} \Delta a \tag{6.11.29}$$

6.11.8 Curl and Stokes' Theorem

And guess what we are going to do now? Having established the line integral around the close boundary of a flat rectangle, we now move to the harder problem: the line integral around a spatial curve Γ which is the boundary of some surface S. The idea is similar to what we have done to get the Gauss theorem. The surface S is divided into many many small parts, each part can be considered as a flat rectangle (left picture in Fig. 6.52). The line integral around Γ is then the sum of the line integral around the small rectangles Γ_i (along common side Γ_{12} the line integrals cancel each other similar to Fig. 6.48).

Thus we have

$$\oint_{\Gamma} \boldsymbol{C} \cdot d\boldsymbol{s} = \sum_i \oint_{\Gamma_i} \boldsymbol{C} \cdot d\boldsymbol{s} = \sum_i (\nabla \times \boldsymbol{C}) \cdot \boldsymbol{n} \Delta a = \int_S (\nabla \times \boldsymbol{C}) \cdot \boldsymbol{n} dA$$

which is the Stokes theorem or the Kelvin-Stokes theorem. It is named after Lord Kelvin and George Stokes.

$$\text{Stokes' theorem:} \quad \int_S (\nabla \times \boldsymbol{C}) \cdot \boldsymbol{n} dA = \oint_{\Gamma} \boldsymbol{C} \cdot d\boldsymbol{s} \qquad (6.11.30)$$

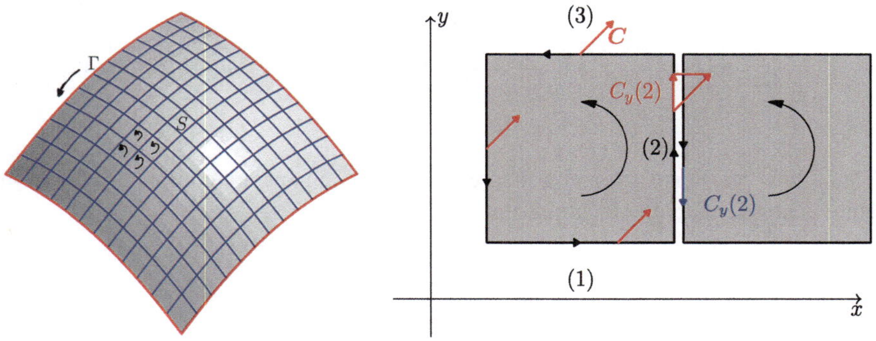

Fig. 6.52 The line integral around Γ is then the sum of the line integral around the small rectangles Γ_i. Along the common side the line integrals cancel each other: on side (2) of the left flat rectangle, the line integral is $C_y(2)\Delta y$ and on the side (3) of the right rectangle, it is $-C_y(2)\Delta y$. When sum up they cancel each other. Noting that the two rectangles are intentionally drawn apart for illustration purposes only

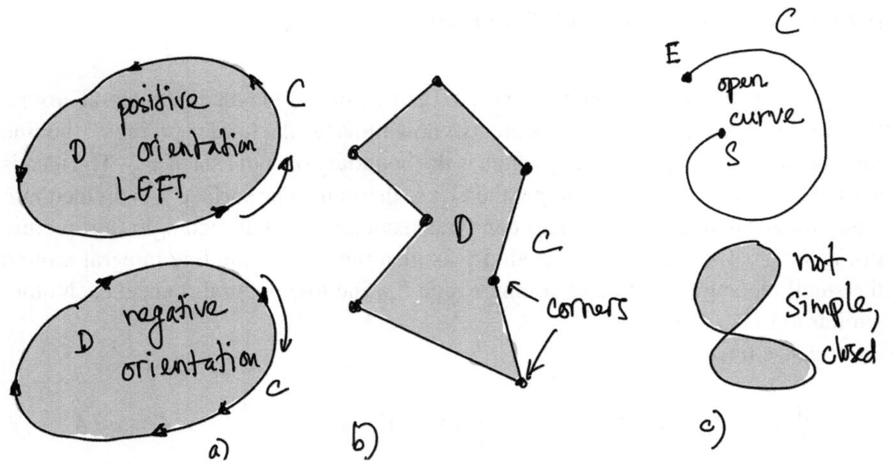

Fig. 6.53 Illustration of positively oriented, piecewise smooth, simple closed curves

6.11.9 *Green's Theorem*

If we return to 2D planes, then Stokes theorem becomes Green's theorem, which is named after the British mathematical physicist George Green. As $\boldsymbol{C}$ is now a two-dimensional vector field, the integrand in the surface integral is simply the z-component of the curl of $\boldsymbol{C}$. Thus, Green's theorem states that

$$\text{Green's theorem:} \quad \int_S \left(\frac{\partial C_y}{\partial x} - \frac{\partial C_x}{\partial y}\right) dA = \oint_\Gamma (C_x dx + C_y dy) \quad (6.11.31)$$

That's how physicists present a theorem. Mathematicians are completely different. Here is how a mathematician presents Green's theorem.

Theorem 6.11.2 Green's theorem
Let C be a positively oriented, piecewise smooth, simple closed curve in the plane and let D be the region bounded by C. If $P(x, y)$ and $Q(x, y)$ are two continuously differentiable functions on D, then

$$\int_D \left(\frac{\partial Q}{\partial x} - \frac{\partial P}{\partial y}\right) dA = \int_C (Pdx + Qdy)$$

The main content is of course the same but with rigor. To use the theorem properly we need to pay attention to the conditions mentioned in the theorem, especially about the curve C (Fig. 6.53). For example, if the curve is open, forget Green's theorem.

History note 6.3: George Green (14 July 1793–31 May 1841)

George Green (14 July 1793–31 May 1841) was a British mathematical physicist who wrote *An Essay on the Application of Mathematical Analysis to the Theories of Electricity and Magnetism* in 1828. The essay introduced several important concepts, among them a theorem similar to the modern Green's theorem, the idea of potential functions as currently used in physics, and the concept of what are now called Green's functions. Green was the first person to create a mathematical theory of electricity and magnetism and his theory formed the foundation for the work of other scientists such as James Clerk Maxwell, William Thomson, and others. His work on potential theory ran parallel to that of Carl Friedrich Gauss.
The son of a prosperous miller and a miller by trade himself, Green was *almost completely self-taught in mathematical physics*; he published his most important work 5 years before he went to the University of Cambridge at the age of 40. He graduated with a BA in 1838 as a fourth Wrangler (the fourth highest scoring student in his graduating class, coming after James Joseph Sylvester who scored second).

6.11.10 Grad, Div, Curl and Identities

While working on a scalar function of multivariable, e.g., $T(x, y, z)$ we discovered the gradient vector, denoted by ∇f or grad f. This gradient vector allows us to answer the question how much the function will change along any direction. It has a meaning that it provides the direction of maximum change.

On the other hand, while working with vector fields, we have discovered two new things: the divergence of a vector field $\boldsymbol{C}$, denoted by $\nabla \cdot \boldsymbol{C}$ or div $\boldsymbol{C}$ and the curl of a vector field $\nabla \times \boldsymbol{C}$ or curl $\boldsymbol{C}$.

For a function $f(x, y, z)$, its gradient is a vector defined as

$$\nabla f = \left(\frac{\partial f}{\partial x}, \frac{\partial f}{\partial y}, \frac{\partial f}{\partial z}\right)$$

Now we do something remarkable, we remove f from the above, and define a *gradient operator* as:

$$\nabla = \left(\frac{\partial}{\partial x}, \frac{\partial}{\partial y}, \frac{\partial}{\partial z}\right)$$

And this operator is a vector. But it is not a vector on its own. We have to attach it to something else so that it has a meaning. What can we do with this vector? Recall that we can multiply a vector with a scalar, we can do a dot product for two vectors and finally we can do a cross-product for two vectors. Now, we define all these operations for our new vector ∇ with a scalar f and a vector field $\boldsymbol{C}$:

$$\begin{aligned}
&\text{scalar multiplication: } \nabla f &&= \left(\frac{\partial f}{\partial x}, \frac{\partial f}{\partial y}, \frac{\partial f}{\partial z}\right) \\
&\text{dot product: } \nabla \cdot \boldsymbol{C} &&= \left(\frac{\partial}{\partial x}, \frac{\partial}{\partial y}, \frac{\partial}{\partial z}\right) \cdot (C_x, C_y, C_z) = \frac{\partial C_x}{\partial x} + \frac{\partial C_y}{\partial y} + \frac{\partial C_z}{\partial z} \\
&\text{cross-product: } \nabla \times \boldsymbol{C} &&= \left(\frac{\partial}{\partial x}, \frac{\partial}{\partial y}, \frac{\partial}{\partial z}\right) \times (C_x, C_y, C_z)
\end{aligned} \tag{6.11.32}$$

What we have achieved? Except for ∇f (which is where we started), we have obtained the divergence and curl of a vector field, which matches the definition discovered previously when we were doing physics!

Having now the new stuff, we're going to find the rules for them. And of course we base our thinking on the rules that we know for the differentiation of functions of a single variable. For two functions $f(x)$ and $g(x)$, we know the sum and product rule:

$$\begin{aligned}
&\text{sum rule: } \frac{d}{dx}(f+g) = \frac{df}{dx} + \frac{dg}{dx} \\
&\text{product rule: } \frac{d}{dx}(fg) = \frac{df}{dx}g + \frac{dg}{dx}f
\end{aligned}$$

From this sum rule, now considering $f(x, y, z)$, $g(x, y, z)$ and two vector fields $\boldsymbol{a}$ and $\boldsymbol{b}$, we have the sum rules

$$\begin{aligned}
&\text{sum rule 1: } \nabla(f+g) &&= \nabla f + \nabla g \\
&\text{sum rule 2: } \nabla \cdot (\boldsymbol{a}+\boldsymbol{b}) &&= \nabla \cdot \boldsymbol{a} + \nabla \cdot \boldsymbol{b} \\
&\text{sum rule 3: } \nabla \times (\boldsymbol{a}+\boldsymbol{b}) &&= \nabla \times \boldsymbol{a} + \nabla \times \boldsymbol{b}
\end{aligned} \tag{6.11.33}$$

We have not one sum rule but three because we have three combinations of ∇, f and $\boldsymbol{a}$ as shown in Eq. (6.11.32). The proof is straightforward, so we just present the proof of the second sum rule:

$$\begin{aligned}\nabla \cdot (\boldsymbol{a}+\boldsymbol{b}) &= \frac{\partial(a_x+b_x)}{\partial x} + \frac{\partial(a_y+b_y)}{\partial y} + \frac{\partial(a_z+b_z)}{\partial z} \\ &= \frac{\partial a_x}{\partial x} + \frac{\partial b_x}{\partial x} + \frac{\partial a_y}{\partial y} + \frac{\partial b_y}{\partial y} + \frac{\partial a_z}{\partial z} + \frac{\partial b_z}{\partial z} \\ &= \nabla \cdot \boldsymbol{a} + \nabla \cdot \boldsymbol{b} \;\; \text{(collecting red terms to get div of a)}\end{aligned}$$

In some books, you can see the following proof, which is similar, but adopts index notation; the vector $\boldsymbol{a} = (a_1, a_2, a_3)$ and the coordinates are x_1, x_2, x_3:

$$\nabla \cdot (\boldsymbol{a}+\boldsymbol{b}) = \sum_{i=1}^{3} \frac{\partial(a_i+b_i)}{\partial x_i} = \sum_{i=1}^{3} \left(\frac{\partial a_i}{\partial x_i} + \frac{\partial a_i}{\partial x_i} \right) = \sum_{i=1}^{3} \frac{\partial a_i}{\partial x_i} + \sum_{i=1}^{3} \frac{\partial b_i}{\partial x_i}$$

The pros of this notation is space saving, and it works for vectors in $\mathbb{R}^n$ for any n not just three.

Now comes the product rules. First, from ∇f we have $\nabla(fg)$ and $\nabla(\boldsymbol{a} \cdot \boldsymbol{b})$. Second, from $\nabla \cdot \boldsymbol{a}$ we have $\nabla \cdot (f\boldsymbol{a})$ and $\nabla \cdot (\boldsymbol{a} \times \boldsymbol{b})$. Third, from $\nabla \times \boldsymbol{a}$ we have $\nabla \times (f\boldsymbol{a})$ and $\nabla \times (\boldsymbol{a} \times \boldsymbol{b})$. Totally, we have six product rules, they are given by

$$\begin{aligned}&\text{product rule 1: } \nabla(fg) = g\nabla f + f\nabla g \\ &\text{product rule 2: } \nabla(\boldsymbol{a} \cdot \boldsymbol{b}) = ? \\ &\text{product rule 3: } \nabla \cdot (f\boldsymbol{a}) = f(\nabla \cdot \boldsymbol{a}) + \nabla f \cdot \boldsymbol{a} \\ &\text{product rule 4: } \nabla \cdot (\boldsymbol{a} \times \boldsymbol{b}) = (\nabla \times \boldsymbol{a}) \cdot \boldsymbol{b} - (\nabla \times \boldsymbol{b}) \cdot \boldsymbol{a} \\ &\text{product rule 5: } \nabla \times (f\boldsymbol{a}) = f(\nabla \times \boldsymbol{a}) + (\nabla f) \times \boldsymbol{a} \\ &\text{product rule 6: } \nabla \times (\boldsymbol{a} \times \boldsymbol{b}) = ?\end{aligned} \tag{6.11.34}$$

Proof of rules 1 and 3 is simple (and rules 1/3 have the same form). The product rule 5 can be guessed from rule 3 and can be proved straightforwardly. The proof follows the same idea as that of the proof of the sum rules. The form of rule 4 can be guessed: $\nabla \cdot (\boldsymbol{a} \times \boldsymbol{b})$ is a scalar, and if the pattern of the derivative of fg still applies $\nabla \cdot (\boldsymbol{a} \times \boldsymbol{b})$ should consist of two scalar terms: one involves the dot product of the curl of $\boldsymbol{a}$ and the other vector and the other term containing the dot product of the curl of $\boldsymbol{b}$ and the other vector. What is weird is the minus sign not plus.

Second derivatives The grad, div and curl operators involve only first derivative. How about second derivatives?

- Start with a scalar $f(x, y, z)$; we have ∇f, which is a vector. And for a vector we can do a div and a curl, so we will have $\nabla \cdot (\nabla f)$ and $\nabla \times (\nabla f)$;
- Start with a vector field $\boldsymbol{C}$; we have $\nabla \cdot \boldsymbol{C}$ which is a scalar, and for a scalar we can do a grad on it: $\nabla(\nabla \cdot \boldsymbol{C})$;
- Start with a vector field $\boldsymbol{C}$; we have $\nabla \times \boldsymbol{C}$ which is a vector, and for a vector we can do a div on it: $\nabla \cdot (\nabla \times \boldsymbol{C})$, or we can do a curl on it: $\nabla \times (\nabla \times \boldsymbol{C})$.

We now compute all these possibilities and see what we will get. Let's start with $\nabla \cdot (\nabla f)$:

$$\nabla \cdot (\nabla f) = \frac{\partial^2 f}{\partial x^2} + \frac{\partial^2 f}{\partial y^2} + \frac{\partial^2 f}{\partial z^2} = \nabla^2 f = \Delta f$$

So, $\nabla \cdot (\nabla f)$ is a scalar and called the Laplacian of f, denoted by $\nabla^2 f$. This operator appears again and again in physics (and engineering). We can define the Laplacian of a vector field $\boldsymbol{C}$ as a vector field with the components being the Laplacian of the components of the vector:

$$\nabla^2 \boldsymbol{C} = (\nabla^2 C_x, \nabla^2 C_y, \nabla^2 C_z)$$

Moving on to $\nabla \times (\nabla f)$, which is the curl of the grad of f. It is a zero vector, due to this property of partial derivative $\frac{\partial^2 f}{\partial x \partial y} = \frac{\partial^2 f}{\partial y \partial x}$. It is interesting that $\nabla \cdot (\nabla \times \boldsymbol{C})$, which is the div of a curl, is also zero.

We now summarize all the results:

$$\begin{aligned}
&\text{Laplacian:} && \nabla \cdot (\nabla f) = \nabla^2 f = \Delta f = \tfrac{\partial^2 f}{\partial x^2} + \tfrac{\partial^2 f}{\partial y^2} + \tfrac{\partial^2 f}{\partial z^2} \\
&\text{Curl of the grad:} && \nabla \times (\nabla f) = \mathbf{0} \\
&\text{Div of a curl:} && \nabla \cdot (\nabla \times \boldsymbol{C}) = 0 \\
&\text{Grad of a div:} && \nabla(\nabla \cdot \boldsymbol{C}) : \text{nothing special} \\
&\text{Curl of a curl:} && \nabla \times (\nabla \times \boldsymbol{C}) = \nabla(\nabla \cdot \boldsymbol{C}) - \nabla^2 \boldsymbol{C}
\end{aligned} \tag{6.11.35}$$

You can check the last formula by computing the components of $\nabla \times \boldsymbol{C}$, and then computing the curl of that vector, and you will see the RHS appear. The formula is not important, what is important is that the curl of a curl does not give us anything new.

6.11.11 Integration by Parts

Integration by parts is

$$\int_a^b f\left(\frac{dg}{dx}\right) dx = -\int_a^b g\left(\frac{df}{dx}\right) dx + [fg]_a^b$$

which comes from the product rule and the fundamental theorem of calculus (ordinary calculus). And of course, we're going to develop a 3D version of integration by parts. And the machinery is similar: product rule and the fundamental theorem of vector calculus.

Starting with this product rule (check Eq. (6.11.34)),

$$\nabla \cdot (f\boldsymbol{a}) = f(\nabla \cdot \boldsymbol{a}) + \nabla f \cdot \boldsymbol{a}$$

Integrating both sides of it over a volume $\mathcal{B}$ with boundary surface $\partial\mathcal{B}$, we get

$$\int_{\mathcal{B}} \nabla \cdot (f\boldsymbol{a}) dV = \int_{\mathcal{B}} f(\nabla \cdot \boldsymbol{a}) dV + \int_{\mathcal{B}} \nabla f \cdot \boldsymbol{a} dV$$

And using Gauss's divergence theorem for the LHS to convert it to a surface integral on the boundary, we obtain

$$\int_{\partial\mathcal{B}} (f\boldsymbol{a}) \cdot \boldsymbol{n} dS = \int_{\mathcal{B}} f(\nabla \cdot \boldsymbol{a}) dV + \int_{\mathcal{B}} \nabla f \cdot \boldsymbol{a} dV$$

And a bit of rearrangement gives us:

$$\int_{\mathcal{B}} f(\nabla \cdot \boldsymbol{a}) dV = -\int_{\mathcal{B}} \nabla f \cdot \boldsymbol{a} dV + \int_{\partial\mathcal{B}} (f\boldsymbol{a}) \cdot \boldsymbol{n} dS \tag{6.11.36}$$

From this result, we can obtain the gradient theorem. Let's consider a constant vector $\boldsymbol{a}$ and a smooth function u in place of f. From Eq. (6.11.36) we get ($\nabla \cdot \boldsymbol{a} = 0$)

$$\int_{\mathcal{B}} \nabla u \cdot \boldsymbol{a} dV = \int_{\partial\mathcal{B}} (u\boldsymbol{a}) \cdot \boldsymbol{n} dS$$

And since this holds for any constant vector $\boldsymbol{a}$, we get the gradient theorem:

$$\boxed{\int_V \nabla u dV = \int_S u \boldsymbol{n} dA} \tag{6.11.37}$$

6.11.12 Green's Identities

Green's identities are a set of three identities in vector calculus relating the bulk with the boundary of a region on which differential operators act. They are named after the mathematician George Green, who discovered Green's theorem.

First identity. Assume two scalar functions $u(x, y)$ and $v(x, y)$ (extension to $u(x, y, z)$ is straightforward), we then have

$$\begin{aligned}(vu_x)_x &= v_x u_x + v u_{xx} \\ (vu_y)_y &= v_y u_y + v u_{yy}\end{aligned}$$

where the notation u_x means the first derivative of u with respect to x. Adding up these identities gives

$$\nabla \cdot (v\nabla u) = \nabla v \cdot \nabla u + v\Delta u \tag{6.11.38}$$

Integrating both sides of it over a volume $\mathcal{B}$ with boundary surface $\partial\mathcal{B}$, we get

$$\int_{\mathcal{B}} \nabla \cdot (v\nabla u) dV = \int_{\mathcal{B}} \nabla v \cdot \nabla u dV + \int_{\mathcal{B}} v\Delta u dV$$

Now, using again the Gauss divergence theorem for the LHS, we have

$$\int_{\partial\mathcal{B}} (v\nabla u) \cdot \boldsymbol{n} dS = \int_{\mathcal{B}} \nabla v \cdot \nabla u dV + \int_{\mathcal{B}} v\Delta u dV$$

which is known as the first Green's identity.

Note that $\nabla u \cdot \boldsymbol{n}$ is the directional derivative of u along the direction of $\boldsymbol{n}$. Usually mathematicians define the directional derivative in the outward normal direction as:

$$\frac{\partial u}{\partial n} := \nabla u \cdot \boldsymbol{n}$$

With this new term, the first Green's identity can also be written as, for a pair of (u, v)

$$\int_{\mathcal{B}} v\Delta u dV = -\int_{\mathcal{B}} \nabla v \cdot \nabla u dV + \int_{\partial\mathcal{B}} v\frac{\partial u}{\partial n} dS$$

Second identity. Writing the first Green's identity for two pairs, (u, v) and (v, u) we get

$$\begin{aligned}
\int_{\mathcal{B}} v\Delta u dV &= -\int_{\mathcal{B}} \nabla v \cdot \nabla u dV + \int_{\partial\mathcal{B}} v\frac{\partial u}{\partial n} dS \\
\int_{\mathcal{B}} u\Delta v dV &= -\int_{\mathcal{B}} \nabla u \cdot \nabla v dV + \int_{\partial\mathcal{B}} u\frac{\partial v}{\partial n} dS
\end{aligned}$$

What we do next? We subtract the second from the first, as the red terms cancel each other:

$$\int_{\mathcal{B}} (u\Delta v - v\Delta u)\, dV = \int_{\partial\mathcal{B}} \left(u\frac{\partial v}{\partial n} - v\frac{\partial u}{\partial n}\right) dS$$

and this is the second Green's identity.

6.11.13 Kronecker and Levi-Cavita Symbols

In the proof of the sum rule 2 in Eq. (6.11.33) we have shown that using indicial notation shortens the proof. The question now is how to prove the product rule 4 in Eq. (6.11.34) in the same way? But before that, try to prove this rule the usual way and you would understand how tedious the algebra is.

Consider the three-dimensional Euclidean space $\mathbb{R}^3$ with three orthonormal vectors $\boldsymbol{e}_i (i = 1, 2, 3)$. Then any vector, say, $\boldsymbol{a}$ can be written as[19]

$$\boldsymbol{a} = a_1\boldsymbol{e}_1 + a_2\boldsymbol{e}_2 + a_3\boldsymbol{e}_2 = a_i\boldsymbol{e}_i \tag{6.11.39}$$

where we have used Einstein summation rule in the last equality. We can write the dot product of two vectors $\boldsymbol{a}$ and $\boldsymbol{b}$ as

$$\boldsymbol{a} \cdot \boldsymbol{b} = (a_i\boldsymbol{e}_i) \cdot (b_j\boldsymbol{e}_j) = a_ib_j\boldsymbol{e}_i \cdot \boldsymbol{e}_j$$

Now, we know that the three basis vectors are orthonormal, we can easily compute the dot product of any two of them, it is given by

$$\boldsymbol{e}_i \cdot \boldsymbol{e}_j = \begin{cases} 1 \text{ if } i = j \\ 0 \text{ otherwise} \end{cases} \tag{6.11.40}$$

And we introduce the Kronecker delta symbol,[20] which is defined by

$$\delta_{ij} = \begin{cases} 1 \text{ if } i = j \\ 0 \text{ otherwise} \end{cases} \tag{6.11.41}$$

Hence

$$\boldsymbol{a} \cdot \boldsymbol{b} = a_ib_j\delta_{ij} = a_ib_i = a_jb_j = a_1b_1 + a_2b_2 + a_3b_3$$

So, the dot product gave us a new symbol δ_{ij}. The cross-product should lead to a new symbol. Let's discover that. The cross-product of two vectors $\boldsymbol{a}$ and $\boldsymbol{b}$ is a vector denoted by $\boldsymbol{a} \times \boldsymbol{b}$:

$$\boldsymbol{a} \times \boldsymbol{b} = a_i\boldsymbol{e}_i \times b_j\boldsymbol{e}_j = a_ib_j\boldsymbol{e}_i \times \boldsymbol{e}_j \tag{6.11.42}$$

And of course we're going to compute $\boldsymbol{e}_i \times \boldsymbol{e}_j$ (we know how to compute the cross-product of two vectors). The results are

$$\begin{array}{lll} \boldsymbol{e}_1 \times \boldsymbol{e}_1 = \boldsymbol{0} & \boldsymbol{e}_1 \times \boldsymbol{e}_2 = \boldsymbol{e}_3 & \boldsymbol{e}_1 \times \boldsymbol{e}_3 = -\boldsymbol{e}_2 \\ \boldsymbol{e}_2 \times \boldsymbol{e}_1 = -\boldsymbol{e}_3 & \boldsymbol{e}_2 \times \boldsymbol{e}_2 = \boldsymbol{0} & \boldsymbol{e}_2 \times \boldsymbol{e}_3 = \boldsymbol{e}_1 \\ \boldsymbol{e}_3 \times \boldsymbol{e}_1 = \boldsymbol{e}_2 & \boldsymbol{e}_3 \times \boldsymbol{e}_2 = -\boldsymbol{e}_1 & \boldsymbol{e}_3 \times \boldsymbol{e}_3 = \boldsymbol{0} \end{array} \tag{6.11.43}$$

This allows us to write

$$\boldsymbol{e}_j \times \boldsymbol{e}_k = \epsilon_{ijk}\boldsymbol{e}_i \tag{6.11.44}$$

[19] We move away from $\boldsymbol{i}$, $\boldsymbol{j}$ and $\boldsymbol{k}$ and use $\boldsymbol{e}_i$ as we are now using indicial notation. It is important to remember that these objects are vectors even though they also have an index.

[20] Obviously named after Leopold Kronecker a German mathematician.

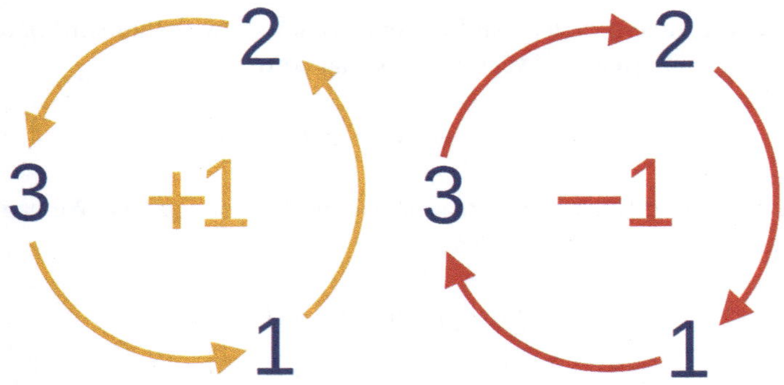

Fig. 6.54 For the indices (i, j, k) in ϵ_{ijk}, the values 1, 2, 3 occurring in the cyclic order (1, 2, 3) correspond to $\epsilon = +1$, while occurring in the reverse cyclic order correspond to $\epsilon = -1$

where ϵ_{ijk} is the permutation symbol or the Levi-Civita[21] symbol, which is defined by

$$\epsilon_{ijk} = \begin{cases} +1 \text{ if } (i, j, k) \text{ is } (1, 2, 3), (2, 3, 1), \text{ or } (3, 1, 2) \\ -1 \text{ if } (i, j, k) \text{ is } (3, 2, 1), (1, 3, 2), \text{ or } (2, 1, 3) \\ 0 \quad i = j, j = k, \text{ or } k = i \end{cases} \tag{6.11.45}$$

To ease the memorization of this symbol, I refer to Fig. 6.54.
The cross-product is now written as

$$\boldsymbol{a} \times \boldsymbol{b} = a_j b_k \boldsymbol{e}_j \times \boldsymbol{e}_k = a_j b_k \epsilon_{ijk} \boldsymbol{e}_i \tag{6.11.46}$$

Denote $\boldsymbol{c}$ as the cross-product of $\boldsymbol{a} \times \boldsymbol{b}$, then we have $\boldsymbol{c} = a_j b_k \epsilon_{ijk} \boldsymbol{e}_i$, i.e., the components of $\boldsymbol{c}$ are $c_i = a_j b_k \epsilon_{ijk}$, written explicitly

$$\begin{aligned} c_1 &= a_j b_k \epsilon_{1jk} = a_2 b_3 - a_3 b_2 \\ c_2 &= a_j b_k \epsilon_{2jk} = a_3 b_1 - a_1 b_3 \\ c_3 &= a_j b_k \epsilon_{3jk} = a_1 b_2 - a_2 b_1 \end{aligned}$$

We're now ready to prove the product rule 4 in Eq. (6.11.34) in a much elegant manner. First it is necessary to express the curl of a vector using the Levi-Civita symbol:

$$\boldsymbol{a} \times \boldsymbol{b} = a_j b_k \epsilon_{ijk} \boldsymbol{e}_i \Longrightarrow \nabla \times \boldsymbol{a} = \frac{\partial}{\partial x_j} b_k \epsilon_{ijk} \boldsymbol{e}_i = b_{k,j} \epsilon_{ijk} \boldsymbol{e}_i \tag{6.11.47}$$

---

[21] Tullio Levi-Civita (1873–1941) was an Italian mathematician, most famous for his work on absolute differential calculus (tensor calculus) and its applications to the theory of relativity, but who also made significant contributions in other areas. He was a pupil of Gregorio Ricci-Curbastro, the inventor of tensor calculus.

where the notation $b_{k,j}$ means partial derivative of b_k with respect to x_j.

$$\begin{aligned}\nabla \cdot (\boldsymbol{a} \times \boldsymbol{b}) &= \frac{\partial}{\partial x_i}(a_j b_k \epsilon_{ijk}) = (a_j b_k \epsilon_{ijk})_{,i} \\ &= \epsilon_{ijk} a_{j,i} b_k + \epsilon_{ijk} a_j b_{k,i} \\ &= \underbrace{(\epsilon_{kij} a_{j,i}) b_k}_{(\nabla \times \boldsymbol{a}) \cdot \boldsymbol{b}} - \underbrace{a_j \epsilon_{jik} b_{k,i}}_{\boldsymbol{a} \cdot (\nabla \times \boldsymbol{b})}\end{aligned}$$

where the minus comes from the fact that $\epsilon_{ijk} = -\epsilon_{jik}$, a property can be directly seen from its definition.

The Levi-Civita symbol comes back again and again whenever you have tensors, so it is a very important thing to understand. Let me just emphasize that no matter how complicated the Levi-Civita Symbol is, life would be close to unbearable if it wasn't there! In fact, it wasn't until Levi-Civita published his work on tensor analysis that Albert Einstein was able to complete his work on General Relativity. It is unbelievable that such a little permutation symbol is that useful!

6.11.14 Curvilinear Coordinate Systems

In this section we present div, grad, curl and Laplacian operators in curvilinear coordinate systems. We have seen such coordinate systems: polar coordinates, cylindrical coordinates and spherical coordinates. To illustrate the problem let's consider a scalar function $f(x, y)$ of which its gradient vector is (f_x, f_y). The question is what is the gradient vector of $f(r, \theta)$ when a polar coordinate system is used.

The first solution comes to mind is starting from $\nabla f = f_x \boldsymbol{i} + f_y \boldsymbol{j}$ we convert x, y and $\boldsymbol{i}, \boldsymbol{j}$ to r, θ and $\hat{\boldsymbol{r}}\ \hat{\boldsymbol{\theta}}$, the latter two being the unit basis vectors of a polar coordinate system. Referring to Fig. 6.55, we have

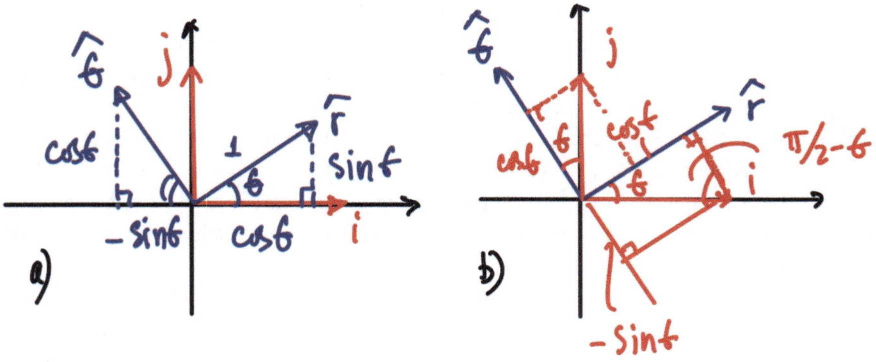

Fig. 6.55 Polar coordinates

$$\boldsymbol{i} = \cos\theta\hat{\boldsymbol{r}} - \sin\theta\hat{\boldsymbol{\theta}}, \quad \boldsymbol{j} = \sin\theta\hat{\boldsymbol{r}} + \cos\theta\hat{\boldsymbol{\theta}} \tag{6.11.48}$$

And we also have

$$r = \sqrt{x^2 + y^2}, \quad \theta = \arctan y/x \tag{6.11.49}$$

Now, we can do the variable conversion, as it is purely algebraic:

$$\begin{aligned}
\nabla f &= \frac{\partial f}{\partial x}\boldsymbol{i} + \frac{\partial f}{\partial y}\boldsymbol{j} \\
&= \left(\frac{\partial f}{\partial r}\frac{\partial r}{\partial x} + \frac{\partial f}{\partial \theta}\frac{\partial \theta}{\partial x}\right)(\cos\theta\hat{\boldsymbol{r}} - \sin\theta\hat{\boldsymbol{\theta}}) + \left(\frac{\partial f}{\partial r}\frac{\partial r}{\partial y} + \frac{\partial f}{\partial \theta}\frac{\partial \theta}{\partial y}\right)(\sin\theta\hat{\boldsymbol{r}} + \cos\theta\hat{\boldsymbol{\theta}}) \\
&= \ldots \text{(using Eq. (6.11.49))} \\
&= \frac{\partial f}{\partial r}\hat{\boldsymbol{r}} + \frac{1}{r}\frac{\partial f}{\partial \theta}\hat{\boldsymbol{\theta}}
\end{aligned} \tag{6.11.50}$$

6.12 Complex Analysis

Mathematicians were used to be skeptical about imaginary number $i^2 = -1$, but when a geometrical meaning of that was established, they embraced complex numbers and started doing wild things with them. They developed another branch of mathematics called *complex analysis*. Complex analysis refers to the calculus of complex-valued functions $f(z)$ of single complex variable z.

This section presents a brief introduction to this amazing branch of mathematics.

6.12.1 Functions of Complex Variables

In this section, we shall discuss functions of complex variables, and how to visualize them. Similar to functions of real variables which map a real number to a real number, i.e., $f : \mathbb{R} \to \mathbb{R}$, a function of one complex variable $z = x + iy$ maps z to a new complex number $w = u + iv$ according to a certain rule $f : \mathbb{C} \to \mathbb{C}$.

Let's play with some complex functions. Consider this simple function $w = f(z) = z^2$. With $z = x + iy$, we have $w = (x + iy)^2 = x^2 - y^2 + i(2xy)$. Thus, the real part of $f(z)$ is $u = x^2 - y^2$ and the imaginary part is $v = 2xy$. Let's move to another complex function. What is $f(z) = \sin z$? Using the trigonometry identity $\sin(a + b) = \sin a \cos b + \sin b \cos a$, we write

$$\sin z = \sin(x + iy) = \sin x \cos(iy) + \sin(iy)\cos x = \underbrace{\sin x \cosh y}_{u(x,y)} + i\underbrace{\sinh y \cos x}_{v(x,y)}$$

(where the identities $\cos(iy) = \cosh y$ and $\sin(iy) = i \sinh y$ were used; check Eq. (3.19.6)).

Exponential function. The exponential of a complex number is defined as

$$\boxed{e^z := 1 + \frac{z}{1!} + \frac{z^2}{2!} + \frac{z^3}{3!} + \cdots} \tag{6.12.1}$$

which is reasonable given the fact that this definition is consistent with the definition of $y = e^x$. Now, we want to check whether $e^{z_1}e^{z_2} = e^{z_1+z_2}$ using Eq. (6.12.1). Why that? Because that is the rule that the ordinary exponential function obeys. The new exponential function should obey that too! We have,

$$\begin{aligned} e^{z_1} &= 1 + \frac{z_1}{1!} + \frac{z_1^2}{2!} + \frac{z_1^3}{3!} + \cdots \\ e^{z_2} &= 1 + \frac{z_2}{1!} + \frac{z_2^2}{2!} + \frac{z_2^3}{3!} + \cdots \end{aligned}$$

And therefore, the product $e^{z_1}e^{z_2}$:

$$e^{z_1}e^{z_2} = \left(1 + \frac{z_1}{1!} + \frac{z_1^2}{2!} + \cdots\right)\left(1 + \frac{z_2}{1!} + \frac{z_2^2}{2!} + \cdots\right)$$

What we're currently dealing with is a product of two power series. It's better to develop a formula for that and we get back to $e^{z_1}e^{z_2}$ later. Considering two power series $\sum_{n=0}^{\infty} a_n x^n$, and $\sum_{m=0}^{\infty} b_m x^m$, their product is given by

$$\left(\sum_{n=0}^{\infty} a_n x^n\right)\left(\sum_{m=0}^{\infty} b_m x^m\right)$$

To get the formula, let's try the first few terms, and hope for a pattern:

$$\begin{aligned} (a_0 + a_1 x + a_2 x^2 + \cdots)(b_0 + b_1 x + b_2 x^2 + \cdots) &= (a_0 b_0)x^0 + (a_0 b_1 + a_1 b_0)x^1 + \\ &+ (a_0 b_2 + a_1 b_1 + a_2 b_0)x^2 + \cdots \end{aligned}$$

If we look at the term $(a_0 b_1 + a_1 b_0)x^1$ we can see that the sum of the indices equals the exponent of x^1 ($a_0 b_1$ has the indices sum to 1 for example). With this, we have discovered the Cauchy product formula for two power series

$$\left(\sum_{n=0}^{\infty} a_n x^n\right)\left(\sum_{m=0}^{\infty} b_m x^m\right) = \sum_{n=0}^{\infty}\left(\sum_{k=0}^{n} a_k b_{n-k}\right) x^n \tag{6.12.2}$$

With this tool, we go back to tackle the quantity $e^{z_1}e^{z_2}$, writing e^{z_1} as a power series, and using the Cauchy product formula, and the binomial theorem:

$$\begin{aligned}
e^{z_1}e^{z_2} &= \left(\sum_{n=0}^{\infty}\frac{z_1^n}{n!}\right)\left(\sum_{m=0}^{\infty}\frac{z_2^m}{m!}\right) \\
&= \sum_{n=0}^{\infty}\sum_{k=0}^{n}\frac{1}{k!(n-k)!}z_1^k z_2^{n-k} && \text{(Cauchy product)} \\
&= \sum_{n=0}^{\infty}\frac{1}{n!}\sum_{k=0}^{n}\frac{n!}{k!(n-k)!}z_1^k z_2^{n-k} && \text{(add } n!\text{)} \\
&= \sum_{n=0}^{\infty}\frac{(z_1+z_2)^n}{n!} && \text{(binomial theorem)} \\
&= e^{z_1+z_2} && \text{(def. of exponential of z)}
\end{aligned}$$

Logarithm. The task now is to define $\ln z$. We define the logarithm of a complex variable as the inverse of the exponential of a complex variable. Start with $w = u + iv \in \mathbb{C}$, compute $z = e^w$ as defined in Eq. (6.12.1). Now, the logarithm of z is defined as

$$\ln z = w$$

Writing $z = re^{i\theta}$, now we can express it another way because $z = e^w$:

$$z = e^w = e^{u+iv} = e^u e^{iv}$$

Now, we have the same complex variable written in two forms: $z = re^{i\theta}$ and $z = e^u e^{iv}$, we can deduce that

$$r = e^u (\Longrightarrow u = \ln r), \quad v = \theta + 2n\pi$$

Finally, the logarithm of a complex number is given by

$$\ln z = \ln r + i(\theta + 2n\pi), \quad r = |z|, \quad \theta = \arg z \tag{6.12.3}$$

Powers. We know how to compute $(3 + 2i)^n$, using de Moivre's formula. But we do not know what is $(3 + 2i)^{2+3i}$. Given a complex variable z and a complex constant a, we define z^a in the same manner as for real numbers:

$$z^a := e^{a \ln z}$$

Note that the RHS of this equation is completely meaningful: we know $\ln z$, thus $a \ln z$ and its exponential. Now, using Eq. (6.12.3) for $\ln z$, we obtain the expression for z^a

$$z^a = \exp\left(a[\ln r + i(\theta + 2n\pi)]\right) \tag{6.12.4}$$

As the formula involves n, z^a can be multi-valued or not, depending on a. Let's compute the nth roots of z, that is Eq. (6.12.4) with $a = 1/n$:

$$
\begin{aligned}
z^{1/n} = \sqrt[n]{z} &= \exp\left(\frac{1}{n}\ln r + \frac{1}{n}i\theta + i\frac{1}{n}2m\pi\right) \\
&= \exp\left(\frac{1}{n}\ln r\right) \times \exp\left(\frac{1}{n}i\theta\right) \times \exp\left(i\frac{1}{n}2m\pi\right) \\
&= \sqrt[n]{r}e^{i(\theta/n+2m\pi/n)}
\end{aligned}
$$

With the special case of $z = 1$ (with $r = 1, \theta = 0$), the nth root of one is thus given by

$$\sqrt[n]{1} = e^{i(2\pi/n)m}$$

which are the vertices of a regular n polygon inscribed in the unit circle.

6.12.2 Visualization of Complex Functions

Domain coloring approach. It is easy to see that we need a fourth-order dimensional space to visualize a complex function (we need x, y, u, v). As we live in a 3D space, it is difficult to visualize a 4D space. Therefore, we need a different way. One way is called *domain coloring* in which we assign a color to each point of the complex plane. By assigning points on the complex plane to different colors and brightness, domain coloring allows for a four-dimensional complex function to be easily represented and understood.

The procedure is as follows. To each point in the domain of the function, i.e., to each (x, y), do

- construct $z = f(x + iy)$;
- compute its argument $\arg z$ and its magnitude $|z|$;
- assign $\arg z$ with a hue following the color wheel, and the magnitude $|z|$ by other means, such as brightness or saturation (there are many options for this).
- convert from HSV to RGB.

The final result is a matrix of pixels of different RGB values. Figure 6.56 shows the domain coloring plots of $f(z) = \sin z^{-1}$ and $f(z) = \tan z^{-1}$. This way of visualizing complex functions was proposed by Frank Farris—an American mathematician working at Santa Clara University—possibly around 1998.

Two plane approach. Instead of using only one plane, we can visualize complex functions using two planes: the xy-plane and the uv-plane. To demonstrate the idea, consider $f(z) = z^2 + 1$, we have

$$u(x, y) = x^2 - y^2 + 1, \quad v(x, y) = 2xy$$

Now, considering the entire complex plane and two grid lines, first $x = 1$, it is mapped to

$$u(1, y) = 2 - y^2, \quad v(1, y) = 2y$$

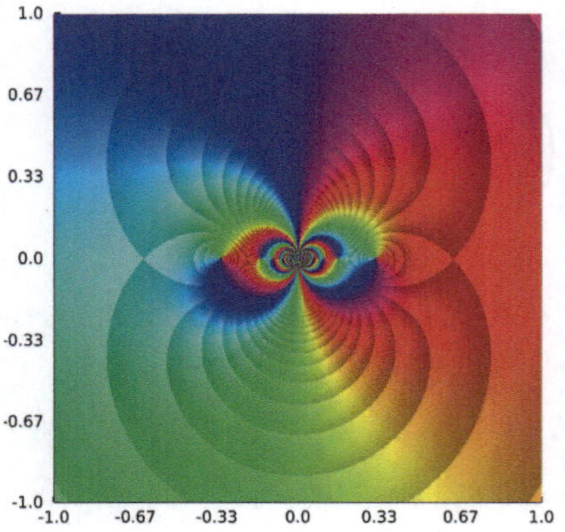

(a) $f(z) = \sin z^{-1}$ (b) $f(z) = \tan z^{-1}$

Fig. 6.56 Domain coloring based visualization of complex functions using `ComplexPortraits.jl`

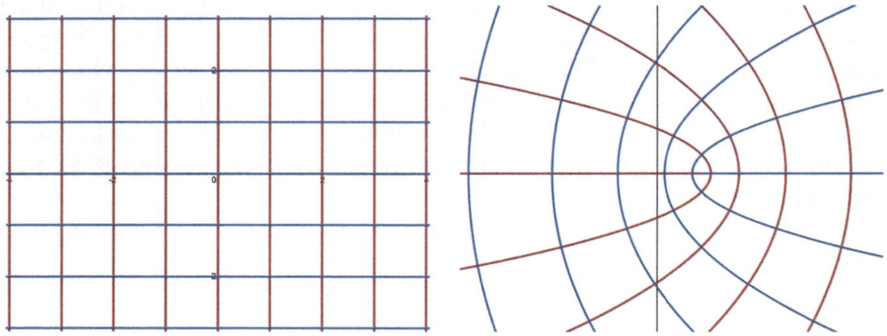

(a) (b)

Fig. 6.57 Visualization of complex functions as a mapping from the xy-plane to uv-plane (using desmos). Note that the mapping preserves the angle between the grid lines: the grid lines in the uv-plane are still perpendicular to each other. Such a mapping is called a conformal mapping

which can be combined to get $u = 2 - v^2/4$, which is a parabola. Similarly, consider the grid line $y = 1$, it is mapped to

$$u(x, 1) = x^2, \quad v(x, 1) = 2x$$

which is also a parabola. It can be shown that these two parabolas are orthogonal. We can repeat this process for other grid lines, and the result is shown in Fig. 6.57 where the grid lines $x = a$ are red colored and the lines $y = b$ are blue. The plane in Fig. 6.57a is mapped or transformed to the one in Fig. 6.57b.

6.12.3 Derivative of Complex Functions

Having now complex functions, no doubt that mathematicians are going to differentiate them. Let's do that. Recall first that for a real function $y = f(x)$, the derivative of f at x_0 is

$$f'(x_0) = \lim_{h \to 0} \frac{f(x_0 + h) - f(x_0)}{h}$$

when this limit exists. We mimic this for complex functions: the complex function $f(z) = u(x, y) + iv(x, y)$ with $z = x + iy$ has a derivative at $z_0 = x_0 + iy_0$ defined as

$$f'(z_0) = \lim_{\Delta z \to 0} \frac{f(z_0 + \Delta z) - f(z_0)}{\Delta z}$$

when this limit exists. The thing is that while for real functions there are only two ways for h to approach zero: either from the left or from the right of x_0; the only street is the number line. Now as complex numbers live on the complex plane, Δz can approach 0 from infinite number of ways. The above limit only exists (i.e., has a finite value) when this limit gets the same value no matter what direction Δz might approach 0. There are, however, two special directions:

Case 1: $\Delta z = \Delta x$.

$$\begin{aligned}
f'(z_0) &= \lim_{\Delta x \to 0} \frac{f(x_0 + \Delta x + iy_0) - f(x_0 + iy_0)}{\Delta x} \\
&= \lim_{\Delta x \to 0} \frac{u(x_0 + \Delta x, y_0) + iv(x_0 + \Delta x, y_0) - u(x_0, y_0) - iv(x_0, y_0)}{\Delta x} \\
&= \frac{\partial u}{\partial x}(x_0, y_0) + i\frac{\partial v}{\partial x}(x_0, y_0)
\end{aligned} \tag{6.12.5}$$

Case 2: $\Delta z = i\Delta y$. Following the same calculation, we get

$$f'(z_0) = \frac{\partial v}{\partial y}(x_0, y_0) - i\frac{\partial u}{\partial y}(x_0, y_0) \tag{6.12.6}$$

In order to have $f'(z_0)$, at least the two values given in Eqs. (6.12.5) and (6.12.6) must be equal because if they are not equal we definitely do not have $f'(z_0)$. And this leads to the following equations

$$\boxed{\frac{\partial u}{\partial x} = \frac{\partial v}{\partial y}, \quad \frac{\partial v}{\partial x} = -\frac{\partial u}{\partial y}} \tag{6.12.7}$$

which are now known as the Cauchy-Riemann equation.

History note 6.4: Georg Bernhard Riemann (17 September 1826–20 July 1866)

Georg Friedrich Bernhard Riemann was a German mathematician who made significant contributions to analysis, number theory, and differential geometry. In the field of real analysis, he is mostly known for the first rigorous formulation of the integral, the Riemann integral, and his work on Fourier series. His contributions to complex analysis include most notably the introduction of Riemann surfaces, breaking new ground in a natural, geometric treatment of complex analysis. His 1859 paper on the prime-counting function, containing the original statement of the Riemann hypothesis, is regarded as a foundational paper of analytic number theory. Through his pioneering contributions to differential geometry, Riemann laid the foundations of the mathematics of general relativity. He is considered by many to be one of the greatest mathematicians of all time.

GPSR Compliance

The European Union's (EU) General Product Safety Regulation (GPSR) is a set of rules that requires consumer products to be safe and our obligations to ensure this.

If you have any concerns about our products, you can contact us on ProductSafety@springernature.com

In case Publisher is established outside the EU, the EU authorized representative is:

Springer Nature Customer Service Center GmbH
Europaplatz 3
69115 Heidelberg, Germany

Batch number: 10725818

Printed by Printforce, the Netherlands

Mathematics for Engineers and Scientists

Vinh Phu Nguyen

Mathematics for Engineers and Scientists

Concepts, Applications, and History

Volume 2

Springer

Vinh Phu Nguyen
Faculty of Engineering
Department of Civil Engineering
Monash University
Melbourne, VIC, Australia

ISBN 978-981-97-4630-9 ISBN 978-981-97-4631-6 (eBook)
https://doi.org/10.1007/978-981-97-4631-6

This Springer imprint is published by the registered company Springer Nature Singapore Pte Ltd.
The registered company address is: 152 Beach Road, #21-01/04 Gateway East, Singapore 189721, Singapore

Preface

There are several issues with the traditional mathematics education. First, it focuses too much on technical details. For example, students are asked to routinely apply the formula $-b\pm\sqrt{b^2-4ac}/2a$ to solve many quadratic equations (e.g. $x^2-2x+1=0$, $x^2+5x-10=0$ etc. and the list goes on). Second, the history of mathematics is completely ignored; textbook exposition usually presents a complete reversal of the usual order of developments of mathematics. The main purpose of the textbooks is to present mathematics with its characteristic logical structure and its incomparable deductive certainty. That's why in a calculus class students are taught what is a function, then what is a limit, then what is a derivative and finally applications. The truth is the reverse: Fermat implicitly used derivative in solving maxima problems; Newton and Leibniz discovered it; Taylor, Bernoulli brothers, Euler developed it; Lagrange characterized it; and only at the end of this long period of development, that spans about 200 years, did Cauchy and Weierstrass define it. Third, there is little opportunity for students to discover (rediscover to be exact) the mathematics for themselves. Definitions, theorems are presented at the outset, the students study the proofs and do applications.

Born and grew up in Vietnam in the early 80s, I received such a mathematical education. Lack of books and guidance, I spent most of the time solving countless of mathematical exercises. Even though I remembered enjoying some of them, admittedly the goal was always to get high marks in exams and particularly pass the university entrance examination. Most of the time, it was some clever tricks that are learned, not the true meaning of the mathematical concepts or their applications. Of course why people came up with those concepts and why these concepts are so defined were not discussed by the teachers (and unfortunately I did not ask these important questions). After my bachelor, I enrolled in a master program. Again, I was on the same education route: solving as many problems as possible. And you could guess, after a master was a Ph.D. study in the Netherlands. Though I had time and freedom and resources to do whatever I felt needed, the focus was still to pass yet another form of examination—graduation. This time it is measured by a number of research papers published in a peer-reviewed journal. To pursue an academic career, I took a postdoctoral job of which the main aim is to have as many papers as possible.

As you can imagine, I became technically fluent in a narrow field but on a weak foundation.

Eventually, I got a job in a university in 2016. For the first time in my life, I did not have to "perform" but I am able to really learn things (staff in universities still need to perform to satisfy certain performance criteria which is vital for probation and promotion). This is when I started reading books not on my research field, and I found that very enjoyable.

The turning point was the book called *A Mathematician's Lament* by Paul Lockhart, a professional mathematician turned college teacher. Paul Lockhart describes how maths is incorrectly taught in schools and he provides better ways to teach maths. He continues in *Measurement* by showing us how we should learn maths by "re-discovering maths" for ourselves. That made me to decide to re-learn mathematics. But this time it must be done in a (much) more fun and efficient way. A bit of researching led me to reading the book *Learning How to Learn* by Barbara Oakley and Terry Sejnowski. The biggest lesson taken from Oakley and Sejnowski's book is that you can learn any subject if you do it properly.

So, I started learning mathematics from scratch during my free time. It started probably in 2017. I have read many books on mathematics and physics and books on the history of mathematics. I wrote some notes on my iPad recording what I have learned. Then, it came the COVID-19 pandemic, also known as the coronavirus pandemic which locked down Melbourne–the city I am living in. That was when I decided to put my iPad notes in a book format to have a coherent story which is not only beneficial to me, but it will be helpful to others, hopefully.

This book is a set of notes covering (elementary) algebra, Euclidean geometry, trigonometry, analytic geometry, calculus of functions of single variables, and probability. This covers the main content of the mathematics curriculum for high school students. These are followed by statistics, calculus of functions of more than one variable, differential equations, variational calculus, linear algebra, and numerical analysis. These topics are for undergraduate college students majoring in science, technology, engineering, and mathematics. Very few such books exist, I believe, as the two targeted audiences are too different. This one is different because it was written for me, firstly and mainly. However, I do believe that high school students can benefit from "advanced" topics by seeing what can be applications of high school mathematics and what could be extensions or better explanations thereof. On the other hand, there are college students not having a solid background in mathematics who can use the elementary parts of this book as a review.

Although it is a thick book, I have left out many topics. I do not discuss topology, graph theory, abstract algebra, differential geometry, simply because I do not know them (and plan to learn them when the time is ready). But the book goes beyond a study of mathematics just to apply it to sciences and engineering.

The style of the book, as you might guess, is informal. Mostly because I am not a mathematician and also I like a conversational tonne. This is not a traditional mathematics textbook, so it does not include many exercises. Instead it focuses on the mathematical concepts, their origin (why we need them), their definition (why they are defined like the way they're), their extension. The process leading to proofs and solutions is discussed as most often it is the first step which is hard, all the remaining is mostly labor work (involving algebra usually). And of course, history of mathematics is included by presenting major men in mathematics and their short biographies.

Of course there is no new mathematics in this book as I am not a mathematician; I do not produce new mathematics. The maths presented is standard, and thus I do not cite the exact sources. But, I do mention all the books and sources where I have learned the maths.

And finally, influenced by the fact that I am an engineer, the book introduces programming from the beginning. Thus, young students can learn mathematics and programming at the same time! For now, programming is just to automate some tedious calculations, or to compute an infinite series numerically before attacking it analytically. Or a little bit harder as to solve Newton's equations to analyze the orbit of some planets. But a soon exposure to programming is vital to their future career. Not least, coding is fun! All the code is put in github.[1] at this https://github.com/vinhphunguyen/maths_juliaaddress.

Clayton, Australia
June 2024

Vinh Phu Nguyen

[1] GitHub is a website and cloud-based service that helps developers store and manage their code, as well as track and control changes to their code.

Acknowledgments

I was lucky to get help from some people. I would like to thank "anh Bé" who tutored me, for free, on mathematics when I needed help most. To my secondary school math teacher "Thay Dieu", who refused to receive tutor fee, I want to acknowledge his generosity. To my high school math teacher "Thay Son", whose belief in me made me more confident in myself, I would like to say thank you very much. To my friend Phuong Thao, who taught me not to memorize formulas, I want to express my deepest gratitude as this simple advice has changed completely the way I have studied since. To Dr. Han Duc Tran (University of Utah) who replied "XFEM" when I asked him 20 years ago what was the hottest topic in the field of computational mechanics. His answer led me to my fateful encounter with Stephane Bordas (Prof. at Luxembourg University). And finally to Prof. Hung Nguyen-Dang, whose the EMMC master program has changed the course of my life and many other Vietnamese, "em cam on Thay rat nhieu".

In the learning process, I cannot say thank you enough to some amazing YouTube channels such as 3Blue1Brown, Mathologer, blackpenredpen, Dr. Trefor Bazett. They provide amazing, clear animation based explanation for many mathematics topics from which I have learned a lot.

I have received encouragement along this journey, and I would like to thank you Miguel Cervera at Universitat Politècnica de Catalunya whom I have never met, Laurence Brassar at University of Oxford, Haojie Lian at Taiyuan University of Technology, Stéphane Bordas at The University of Luxembourg–who was my master supervisor back in 2013. A special thank you is given to Daniel Alves Paladim–a former colleague at Cardiff University–but now working for the American self-driving car company Cruise LLC. Daniel has shared with me all the fascinating books on various topics that he has read over the years. To my close friend Chi Nguyen-Thanh (Royal HaskoningDHV Vietnam), thank you very much for your friendship and encouragement to this project.

This book was typeset with LaTeX on a MacBook. A majority of figures in the book were created using open source software `Asymptote` and `Tikz`. Other figures were generated using `geogebra, processing, Desmos, Julia, Python`. I want to say thank you to Le Huy Tien, a lecturer at Vietnam National University, Hanoi for his help and encouragement on `Asymptote` and `Tikz`.

I dedicate this book to all mathematicians who have created beautiful mathematics that helps making our world a better place to live.

Clayton, Australia
June 2024

Vinh Phu Nguyen

Contents

About the Author

Vinh Phu Nguyen born in 1980 in Hue (Vietnam). Vinh Phu Nguyen enrolled in Hai Ba Trung High School in 1995 and graduated 3 years later. He then went to Hochiminh University of Technology (known as Bach Khoa Sai gon among Vietnamese) in 1998. He graduated in 2003 with a bachelor's in civil engineering. Seeing himself unfit for the industry of civil engineering, he enrolled in the EMMC (European Master in Mechanics of Construction) program, Liège University (Belgium). Upon graduation in 2005, he got a CIFRE scholarship to go to the Ècole nationale d'ingenieurs de Saint-Etienne (ÈNISE; National Engineering School of Saint-Etienne) in France to work with Prof Jean Michel Bergheau. The Ph.D. topic seemed so difficult for him that after 1 year he left France empty-handed.

He got a second chance working with Prof. Bert Sluys at Faculty of Civil Engineering and Geosciences, Delft University of Technology (The Netherlands) in 2007. He graduated with a Ph.D. in computational mechanics in 2011 and went immediately to Johns Hopkins University (USA) for a postdoc. After a 6-month spell in the US he went back to Vietnam to get married and then in 2012 went to Cardiff University (UK) to work with Prof Stéphane Bordas, his master supervisor back in 2004, till 2014. Since 2016, he has been a lecturer at the Department of Civil Engineering, Monash University (Australia).

Dr. Nguyen has done some research on numerical methods for solving partial differential equations in solid mechanics problems (e.g. isogeometric analysis, extended finite element method, material point method) and on damage/fracture models (e.g. phase-field fracture models) for solving fracture mechanics and large deformation mechanics problems. He has published one book *Material Point Method: Theory, Implementations and Applications* with Springer Nature.

Since 2017, he became interested in the history of mathematics and science and the problems of high school education in mathematics.

Chapter 7
Differential Equations

In this chapter we discuss what probably is the most important application of calculus: differential equations. These equations are those that describe many laws of nature. In classical physics, we have to mention Newton's second law $F = m\ddot{x}$ that describes motions, Fourier's heat equation $\dot{\theta} = \kappa^2 \partial^2\theta/\partial x^2$ that describes how heat is transferred in a medium, Maxwell's equations describing electromagnetism and the Navier-Stokes equation that calculates how fluids move. In quantum mechanics, we have the Schrödinger equation. In biology, we can cite the Lotka-Volterra equations, also known as the predator-prey equations—a pair of first-order nonlinear differential equations—used to describe the dynamics of biological systems in which two species interact, one as a predator and the other as prey. In finance, there is the Black-Scholes equation.

About differential equations, the German mathematician Bernhard Riemann once said:

> ...partial differential equations are the basis of all physical theorems. In the theory of sound in gases, liquid and solids, in the investigations of elasticity, in optics, everywhere partial differential equations formulate basic laws of nature which can be checked against experiments

This chapter introduces the mathematics used to model the real world. The attention is on how to derive these equations more than on how to solve them. Yet, some exact solutions are presented. Numerical solutions to differential equations are treated later in Chap. 10. Topics which are too mathematical such as uniqueness are omitted. Also discussed is the problem of mechanical vibrations: simple harmonics and waves.

The following excellent books were consulted for the materials presented in this chapter:

V. P. Nguyen, *Mathematics for Engineers and Scientists*,
https://doi.org/10.1007/978-981-97-4631-6_7

- Partial differential equations for scientists and engineers by Stanley Farlow[1] [22];
- Classical Mechanics by John Taylor[2] [68];
- Modeling with Differential Equations by Burghes and Borrie [12].

The plan of this chapter is as follows. We start with a toy problem in Sect. 7.1 to get the feeling of what mathematical modeling looks like. Then, we become a bit more serious with a real differential equation describing the population growth (Sect. 7.2). In Sect. 7.3, we discuss ordinary differential equations. Next, we move to partial differential equations (such as the wave equation $u_{tt} = c^2 u_{xx}$). We start with Sect. 7.4 in which we get familiar with partial differential equations, discuss some terminologies. The derivation of common partial differential equations (e.g., the heat equation, the wave equation, and so on) is treated in Sect. 7.5. Section 7.7.

Harmonic oscillation is given in Sect. 7.8. How to solve the heat (diffusion) equation is presented in Sect. 7.9. Solutions of the wave equation are given in Sects. 7.10 and 7.11. It is when solving these two equations the idea of Fourier series was born.

7.1 Mathematical Models and Differential Equations

To introduce mathematical modeling and differential equations, let us consider a simple problem as follows. Assume that we want to know how long for a snow ball to completely melt. How can we start? We begin with *experiments or observations*. Assume that experiment data show us that the rate of change of the mass of the ball is proportional to the surface area of the ball. This data alone is not sufficient. We need to make some *assumptions*, i.e., we are building a simplified model of the reality. The first assumption is that the ball is always a sphere. The second assumption is that the density of snow does not change in time. These assumptions might not be enough to have a very good model, but we have to start with something anyway. In summary, we have the following set of facts to build our model:

- The rate of change of the mass of the ball is proportional to the surface area of the ball.
- At any time, the ball is a sphere.
- The density of the snow is constant.

All we have to do now is to translate the above facts (written in English) to the language of mathematics. The assumption is that *all variables are continuously smooth*. Thus, we can use differential calculus to differentiate them as we want, even though for some problems such as population growth the population is not

[1] Stanley Jerome Farlow (born 1937) is an American mathematician specializing in differential equations. For many years he has been a professor at the University of Maine. Farlow is the author of several books in mathematics.

[2] John R. Taylor (born 2 February 1939 in London). He got a BA in Mathematics, Cambridge University, 1960, and a Ph.D. in Theoretical Physics, University of California, Berkeley, 1963. Taylor is an emeritus professor of physics at the University of Colorado, Boulder.

continuous! Remember that we're building a model. As the mass is density (ρ) times volume, we can determine the mass with $r(t)$ representing the radius of the snow ball at time t.[3] And we also compute its derivative w.r.t. t (because the derivative captures changes):

$$M = \rho \frac{4}{3}\pi r^3 \Longrightarrow \frac{dM}{dt} = 4\pi\rho r^2 \frac{dr}{dt} \tag{7.1.1}$$

Using the experiment data on the rate of change of M, we can write

$$\frac{dM}{dt} = -k(4\pi r^2) \Longrightarrow 4\pi\rho r^2 \frac{dr}{dt} = -k(4\pi r^2) \Longrightarrow \boxed{\frac{dr}{dt} = -\frac{k}{\rho}} \tag{7.1.2}$$

where k is a positive constant that can only be experimentally determined. The minus sign reflects the fact that the mass is decreasing. Quantities such as ρ and k whose values do not change in time are called *parameters*.

The equation in the box is a differential equation—an equation that contains derivatives. In fact, it is an *ordinary differential equation* as there exits partial differential equations that involve partial derivatives. In this example, t is the only independent variable and $r(t)$ is the dependent variable. An ordinary differential equation expresses a relation between a dependent variable (a function), its derivatives (first, second derivatives, etc.) and the independent variable: $F(r(t), r', r'', \ldots, r^{(n)}, t) = 0$. If there are more than one independent variable, we have a partial differential equation as the derivatives are partial derivatives.

Now we have an equation. The next step is to solve it to find the solution. *A solution to a differential equation is a function that when substituted (together with all involved derivatives) into the equation results in an identity.* For example, $y = \sin x$ is a solution to the differential equation: $y' = \cos x$. For what purpose? For the prediction of the radius of the snow ball at any time instance. *It is the prediction of future events that is the ultimate goal of mathematical modeling of either natural phenomena or engineering systems.*

For this particular toy problem, it is easy to find the solution: by integrating both sides of the boxed equation in Eq. (7.1.2):

$$\frac{dr}{dt} = -c, \quad c := \frac{k}{\rho} \Longrightarrow r(t) = -ct + A \tag{7.1.3}$$

where A is a real number. But, why we get not one but many solutions? That is because the radius at time t depends, of course, on the initial radius of the ball. So, we must know this initial radius (denoted by R), then by substituting $t = 0$ in Eq. (7.1.3), we get $A = R$. Thus, $r(t) = R - ct$. Now, we can predict when the ball is completely melt, it is when $r(t_m) = 0$: $t_m = R/c$. And we need to check this against observations. If the prediction and the observation are in good agreement,

[3] The notation $r(t)$ is read "r at time t", and the parentheses tell us that our variable is a function of time.

we have discovered a law, of which the mathematical equation is the boxed equation in Eq. (7.1.2). If not, our assumptions are too strict and we need to refine them and refine our model.

7.2 Models of Population Growth

Populations are groups of organisms of the same species living in the same area at the same time. They are described by characteristics that include:

- population size: the number of individuals in the population
- population density: how many individuals are in a particular area
- population growth: how the size of the population is changing over time.

If population growth is just one of many population characteristics, what makes studying it so important? First, studying how and why populations grow (or shrink!) helps scientists make better predictions about future changes in population sizes and growth rates. This is essential for answering questions in areas such as biodiversity conservation (e.g., the polar bear population is declining, but how quickly, and when will it be so small that the population is at risk for extinction?) and human population growth (e.g., how fast will the human population grow, and what does that mean for climate change, resource use, and biodiversity?).

In what follows a simple population growth model is presented. It is based on the ideas put forward by Thomas Robert Malthus[4] in his 1798 book *An Essay on the Principle of Population*. The basic assumption of the model is that the birth rate and dead rate are proportional to the population size. Now, again, we just have to translate that assumption into mathematics. Let $N(t)$ be the population size at time t. Then, within a short time interval Δt, the births and deaths are

$$\text{births} = \alpha N(t)\Delta t, \quad \text{deaths} = \beta N(t)\Delta t$$

where α and β are real positive constants; they are similar to k in the toy model in Sect. 7.1.

With that we can determine the increase (or decrease) of the population within Δt, labeled by ΔN:

$$\text{births} - \text{deaths} = \Delta N = \gamma N(t)\Delta t, \quad \gamma := \alpha - \beta$$

You guess what we shall we do next? Dividing the above equation by Δt (so that a rate of population appears) and let $\Delta t \to 0$

[4] Thomas Robert Malthus (13/14 February 1766–23 December 1834) was an English cleric, scholar and influential economist in the fields of political economy and demography.

Table 7.1 USA population data

Year	USA population ($\times 10^6$)
1790	3.9
1800	5.3
1810	7.2

$$\frac{\Delta N}{\Delta t} = \gamma N(t) \Longrightarrow \lim_{\Delta t \to 0} \frac{\Delta N}{\Delta t} = \gamma N(t) \Longrightarrow \boxed{\dot{N} = \frac{dN}{dt} = \gamma N} \tag{7.2.1}$$

Here the overdot denotes differentiation with respect to time following Newton. Now, we have to solve the boxed ordinary differential equation. Lucky for us, we can solve this equation. The solution, i.e., $N(t)$ should involve the exponential function e^{ct} (why?). Here is how:

$$\frac{dN}{dt} = \gamma N \Longrightarrow \frac{dN}{N} = \gamma dt \Longrightarrow \int_0^t \frac{dN}{N} = \int_0^t \gamma dt \Longrightarrow \boxed{N(t) = N_0 e^{\gamma t}} \tag{7.2.2}$$

where we've assumed that the starting time is $t = 0$. Looking at the solution we can understand why this model is called an exponential growth model.

How good is this model? To answer that (pure mathematicians do not care), scientists use real data. For example, Table 7.1 is the USA population statistics taken from [12]. Of course the data is much more, but we need to use just a small portion of the data to *calibrate the model*. Calibrating a model is to find values for the parameters (or constants) in the model. In the context here, we need to find N_0 and γ using the data in Table 7.1.

We have data starting from the year of 1790, thus $t = 0$ is that year and then $N_0 = 3.9$ millions. For γ, use the data for 1800, noting that t in the model is in terms of 10 years, thus 1800 corresponds with $t = 1$:

$$5.3 = N(1) = N_0 e^{\gamma} \Longrightarrow \gamma = \ln\left(\frac{5.3}{3.9}\right) = 0.307$$

Now is time for prediction. The calibrated model is used to predict the population up to 1870. The results given in Table 7.2 indicates that the model is in good agreement until 1870, at that year the error is nearly 20%. It's time for an improved model.

When the population becomes too large, we encounter problems such as insufficient food, for example. In such cases, population growth slows down, and sometimes the death rate may exceed the birth rate. Therefore, we need a change to Malthus's model. Pierre François Verhulst did just that.

How did he do it? Quite simply, the problem with Malthus's model is that γ is a constant. Verhulst simply chose $\gamma = a - bN$. Transforming a constant function into a linear one! When the population is small (N is small), the term $-bN$ can be considered negligible, and the Verhulst model will be similar to Malthus's model.

Table 7.2 USA population data versus prediction

Year	USA population ($\times 10^6$)	Prediction ($\times 10^6$)	Error (%)
1810	7.2	7.2	0.0
1820	9.6	9.80	2.1
$\vdots$	$\vdots$	$\vdots$	$\vdots$
1870	38.6	45.47	17.8

But when N is large, $\gamma = a - bN$ will decrease significantly compared to a. Thus, Verhulst's differential equation is

$$\dot{N} = (a - bN)N, \quad a, b \in \mathbb{R}, \ a, b > 0 \tag{7.2.3}$$

And the next step is, of course, to solve it. Fortunately, we can solve this equation as follows:

$$\dot{N} = (a - bN)N \Longrightarrow \int_0^t \frac{dN}{(a - bN)N} = \int_0^t dt \tag{7.2.4}$$

Now you understand why you have to do many integration exercises in your calculus class. This is the time for you to showcase your integral skills; the red integral becomes

$$\int_0^t \frac{dN}{(a - bN)N} = \frac{1}{a}\int_0^t \frac{dN}{N} + \frac{b}{a}\int_0^t \frac{dN}{a - bN} = \frac{1}{a}\ln N\Big|_0^t - \frac{1}{a}\ln(a - bN)\Big|_0^t$$

From there, the solution $N(t)$ is

$$N(t) = \frac{aN_0}{bN_0 + (a - bN_0)e^{-at}}$$

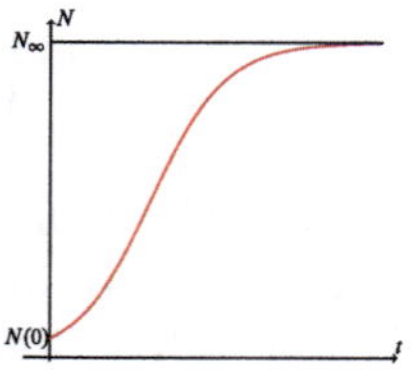

What to do with the solution? First, we need to see what happens to $N(t)$ as $t \to \infty$. Not only this is easy to do, but it also gives us long-term results (which we are usually interested in). Clearly, in that case, $N(t) \to a/b$ (because the term in blue with e^{-at} will vanish). Therefore, a/b is called the carrying capacity, denoted by N_∞. We can rewrite the solution as follows (using a, N_∞ instead of a, b now):

$$N(t) = \frac{N_\infty}{1 + \left(\frac{N_\infty}{N_0} - 1\right) e^{-at}} \tag{7.2.5}$$

The graph of $N(t)$ is presented in the figure. It accurately describes what we want: the population increases rapidly initially, then slows down, and finally approaches N_∞.

This model predicts the population of the United States very well after finding the parameters a and N_∞. I won't go into detail about this issue here.

For students who would like to become scientists trying to understand our world, no one says it best when it comes to how we—human beings—unravel the mysteries of the world, as Richard Feynman in his interesting book *The Pleasure of Finding Things Out*[5]:

> ...a fun analogy in trying to get some idea of what we're doing in trying to understand nature, is to imagine that the gods are playing some great game like chess... and you don't know the rules of the game, but you're allowed to look at the board, at least from time to time... and from these observations you try to figure out what the rules of the game are, what the rules of the pieces moving are. You might discover after a bit, for example, that when there's only one bishop around on the board that the bishop maintains its color. Later on you might discover the law for the bishop as it moves on the diagonal, which would explain the law that you understood before—that it maintained its color—and that would be analogous to discovering one law and then later finding a deeper understanding of it. Then things can happen, everything's going good, and then all of a sudden some strange phenomenon occurs in some corner, so you begin to investigate that—it's castling, something you didn't expect. We're always, by the way, in fundamental physics, always trying to investigate those things in which we don't understand the conclusions. After we've checked them enough, we're okay.

7.3 Ordinary Differential Equations

So far, we have met two ordinary differential equations and they're both of this form

$$\dot{x} = f(x, t) \tag{7.3.1}$$

In the problem of population growth, $x(t)$ is $N(t)$—the population size. As the highest derivative in the equation is one, it is called a first-order ODE. Now, we show that we can always convert a high-order ODE to a system of first-order ODEs.[6] For example, the equation for a damped harmonic oscillator is (Sect. 7.8)

$$m\ddot{x} + b\dot{x} + kx = 0 \Longleftrightarrow \ddot{x} = -\frac{b}{m}\dot{x} - \frac{k}{m}x \tag{7.3.2}$$

[5] You can watch the great man here.

[6] This is not a new idea in mathematics. We solved cubic equations by resorting to quadratic equations; quartic equations via cubic equations and so on.

Now, to remove the second derivative, we introduce a variable $x_2 = \dot{x}$; this leads to $\ddot{x} = \dot{x}_2$, and voilà we have removed the second derivative. And of course, instead of x, we use $x_1 = x$. Then, $\dot{x}_1 = \dot{x} = x_2$, and we can write $\dot{x}_2 = \ddot{x} = -(b/m)x_2 - (k/m)x_1$ from Eq. (7.3.2). Now, using matrix notation, we write

$$\left.\begin{aligned} x_1 &= x \\ x_2 &= \dot{x} \end{aligned}\right\} \Longrightarrow \begin{bmatrix} \dot{x}_1 \\ \dot{x}_2 \end{bmatrix} = \begin{bmatrix} 0 & 1 \\ -k/m & -b/m \end{bmatrix} \begin{bmatrix} x_1 \\ x_2 \end{bmatrix} \tag{7.3.3}$$

This is a system of two first-order linear ODEs with a constant coefficient matrix (the matrix does not vary with time). How about a problem with a time-dependent term like the forced oscillator of which the equation is $m\ddot{x} + b\dot{x} + kx = F \sin t$? The idea is the same, introduce another variable to get rid of t:

$$\left.\begin{aligned} x_1 &= x \\ x_2 &= \dot{x} \\ x_3 &= t \end{aligned}\right\} \Longrightarrow \begin{cases} \dot{x}_1 = x_2 \\ \dot{x}_2 = -\dfrac{k}{m}x_1 - \dfrac{b}{m}x_2 + F/m \sin(x_3) \\ \dot{x}_3 = 1 \end{cases} \tag{7.3.4}$$

So, we can now just focus on the following system of equations, which provides a general framework to study ODEs:

$$\begin{aligned} \dot{x}_1 &= f_1(x_1, \ldots, x_n) \\ &\vdots \\ \dot{x}_n &= f_n(x_1, \ldots, x_n) \end{aligned} \tag{7.3.5}$$

This general equation covers both linear systems such as the one in Eq. (7.3.3) and nonlinear ones, e.g., Eq. (7.3.4). However it is hard to solve nonlinear systems, so in the next section, we just focus on systems of linear ODEs.

7.3.1 System of Linear First-Order Equations

If we have this equation $\dot{x} = \lambda x$, we now know that (from Sect. 7.2) the solution is $x(t) = C_0 e^{\lambda t}$, where $C_0 = x(0)$. The next problem we're interested in is a system of similar equations. For example, below are two such systems

$$\begin{aligned} \dot{x}_1 &= 2x_1 \\ \dot{x}_2 &= 5x_2 \end{aligned}, \quad \begin{aligned} \dot{x}_1 &= 1x_1 + 2x_2 \\ \dot{x}_2 &= 3x_1 + 2x_2 \end{aligned}$$

We use linear algebra (matrices) to solve it, so we rewrite the above as ($\dot{\boldsymbol{x}} = (\dot{x}_1, \dot{x}_2)$)

$$\dot{\boldsymbol{x}} = \underbrace{\begin{bmatrix} 2 & 0 \\ 0 & 5 \end{bmatrix}}_{\mathbf{A}_1} \begin{bmatrix} x_1 \\ x_2 \end{bmatrix}, \quad \dot{\boldsymbol{x}} = \underbrace{\begin{bmatrix} 1 & 2 \\ 3 & 2 \end{bmatrix}}_{\mathbf{A}_2} \begin{bmatrix} x_1 \\ x_2 \end{bmatrix}$$

Before solving them, let's make one observation: it is easier to solve the first system than the second one; because the two equations in the former are *uncoupled*: we just need to solve two independent ODEs. This is reflected in the diagonal matrix $\mathbf{A}_1$ with two zeros (red terms). The solution to the first system is simply $\boldsymbol{x} = (C_1 e^{2t}, C_2 e^{5t})$. But we can also write this as[7]

$$\boldsymbol{x} = C_1 e^{2t} \begin{bmatrix} 1 \\ 0 \end{bmatrix} + C_2 e^{5t} \begin{bmatrix} 0 \\ 1 \end{bmatrix}$$

Noting that $2, 5$ are the eigenvalues of the matrix $\mathbf{A}_1$, and the two unit vectors are the eigenvectors of $\mathbf{A}_1$. Thus, the solution to a system of linear first-order differential equations can be *expressed in terms of the eigenvalues and eigenvectors of the coefficient matrix, at least when that matrix is diagonal and two eigenvalues are different.*

For the second system $\dot{\boldsymbol{x}} = \mathbf{A}_2 \boldsymbol{x}$, the matrix is not diagonal. But there is a way to diagonalize a matrix (check Sect. 9.11.4 for matrix diagonalization) using its eigenvalues λ and eigenvectors $\boldsymbol{v}$. So, we put these info below

$$\mathbf{A}_2 = \begin{bmatrix} 1 & 2 \\ 3 & 2 \end{bmatrix} : \lambda_1 = 4, \quad \lambda_2 = -1, \quad \boldsymbol{v}_1 = \begin{bmatrix} 2 \\ 3 \end{bmatrix}, \boldsymbol{v}_2 = \begin{bmatrix} -1 \\ +1 \end{bmatrix}, \quad \mathbf{P} = \begin{bmatrix} 2 & -1 \\ 3 & +1 \end{bmatrix}$$

Let $\boldsymbol{x} = \mathbf{P}\boldsymbol{y}$ and substitute that into the original system, we get (don't forget that $\dot{\boldsymbol{x}} = \mathbf{A}_2 \boldsymbol{x}$)

$$\boldsymbol{x} = \mathbf{P}\boldsymbol{y} \Longrightarrow \dot{\boldsymbol{x}} = \mathbf{P}\dot{\boldsymbol{y}} \Longrightarrow \mathbf{P}\dot{\boldsymbol{y}} = \mathbf{A}_2 \boldsymbol{x} = \mathbf{A}_2 \mathbf{P} \boldsymbol{y} \Longrightarrow \dot{\boldsymbol{y}} = \mathbf{P}^{-1} \mathbf{A}_2 \mathbf{P} \boldsymbol{y}$$

But $\mathbf{P}^{-1}\mathbf{A}_2\mathbf{P}$ is simply a diagonal matrix with the eigenvalues (of $\mathbf{A}_2$) on the diagonal, thus we can easily solve for $\boldsymbol{y}$, and from that, we obtain $\boldsymbol{x}$:

$$\dot{\boldsymbol{y}} = \begin{bmatrix} 4 & 0 \\ 0 & -1 \end{bmatrix} \boldsymbol{y} \Longrightarrow \boldsymbol{y} = \begin{bmatrix} C_1 e^{4t} \\ C_2 e^{-t} \end{bmatrix} \Longrightarrow \boxed{\boldsymbol{x} = C_1 e^{4t} \begin{bmatrix} 2 \\ 3 \end{bmatrix} + C_2 e^{-t} \begin{bmatrix} -1 \\ +1 \end{bmatrix}} \tag{7.3.6}$$

Again, we can write the solution in terms of the eigenvalues and eigenvectors of the coefficient matrix. To determine $C_{1,2}$, we need the initial condition $\boldsymbol{x}_0 = \boldsymbol{x}(0)$; substituting $t = 0$ into the boxed equation in Eq. (7.3.6), we can determine $C_{1,2}$ in terms of $\boldsymbol{x}_0$:

[7] Only if we know linear algebra we can appreciate why this form is better. So refresh your linear algebra before continuing.

$$\boldsymbol{x}_0 = C_1 \begin{bmatrix} 2 \\ 3 \end{bmatrix} + C_2 \begin{bmatrix} -1 \\ +1 \end{bmatrix} = \begin{bmatrix} 2 & -1 \\ 3 & +1 \end{bmatrix} \begin{bmatrix} C_1 \\ C_2 \end{bmatrix} \Longrightarrow \begin{bmatrix} C_1 \\ C_2 \end{bmatrix} = \begin{bmatrix} 2 & -1 \\ 3 & +1 \end{bmatrix}^{-1} \boldsymbol{x}_0$$

With a given $\boldsymbol{x}_0$, this equation gives us $C_{1,2}$ and put them in the boxed equation in Eq. (7.3.6), and we're finished. Usually, as a scientist or engineer, we stop here, but mathematicians go further. They see that

$$\boldsymbol{x} = C_1 e^{4t} \begin{bmatrix} 2 \\ 3 \end{bmatrix} + C_2 e^{-t} \begin{bmatrix} -1 \\ +1 \end{bmatrix} = \begin{bmatrix} 2 & -1 \\ 3 & +1 \end{bmatrix} \begin{bmatrix} e^{4t} & 0 \\ 0 & e^{-t} \end{bmatrix} \begin{bmatrix} 2 & -1 \\ 3 & +1 \end{bmatrix}^{-1} \boldsymbol{x}_0 \tag{7.3.7}$$

Is something useful with this new way of looking at the solution? Yes, the red matrix! It is a matrix of exponentials. What would you do next when you have seen this?

For ease of presentation, we discussed systems of only two equations, but as can be seen, the method and thus the result extends to systems of n equations (n can be 1000):

$$\begin{bmatrix} \dot{x}_1 \\ \dot{x}_2 \\ \vdots \\ \dot{x}_n \end{bmatrix} = \begin{bmatrix} A_{11} & A_{12} & \cdots & A_{1n} \\ A_{21} & A_{22} & \cdots & A_{2n} \\ \vdots & \vdots & \ddots & \vdots \\ A_{n1} & A_{n2} & \cdots & A_{nn} \end{bmatrix} \begin{bmatrix} x_1 \\ x_2 \\ \vdots \\ x_n \end{bmatrix} \Longrightarrow \boldsymbol{x} = \sum_{i=1}^{n} C_i e^{\lambda_i t} \boldsymbol{x}_i$$

where the eigenvalues of $\mathbf{A}$ are λ_i and eigenvectors are $\boldsymbol{x}_i$. Note that this solution is only possible when $\mathbf{A}$ is diagonalizable, i.e., when the eigenvectors are linear independent.

It is remarkable to look back the long journey from the simple equation $\dot{x} = \lambda x$ with the solution $x(t) = C_0 e^{\lambda t}$ to a system of as many equations as you want, and the solution is still of the same form $\sum_{i=1}^{n} C_i e^{\lambda_i t} \boldsymbol{x}_i$. It is simply remarkable!

But wait. How about non-diagonalizable matrices? The next section is answering that question.

7.3.2 Exponential of a Matrix

Now we do something extraordinary. Starting with $\dot{x} = ax$ with the solution $x(t) = ce^{at}$. Then, consider the linear system $\dot{\boldsymbol{x}} = \mathbf{A}\boldsymbol{x}$, can we write the solution as $\boldsymbol{x} = e^{\mathbf{A}t}\boldsymbol{x}_0$, with $\boldsymbol{x}_0$ being a vector? To answer that question, we need to know what exponential of a matrix means. And mathematicians define $e^{\mathbf{A}}$ by analogy to e^x:

$$e^x = 1 + x + \frac{x^2}{2!} + \frac{x^3}{3!} + \cdots \Longrightarrow \boxed{e^{\mathbf{A}} = \mathbf{I} + \mathbf{A} + \frac{\mathbf{A}^2}{2!} + \frac{\mathbf{A}^3}{3!} + \cdots} \tag{7.3.8}$$

On the RHS (of the boxed eqn), we have a sum of a bunch of matrices, thus $e^{\mathbf{A}}$ is a matrix. If we can compute the powers of a matrix (e.g., $\mathbf{A}^2$, $\mathbf{A}^3$, ...), we can compute the exponential of a matrix! Let's use the matrix $\mathbf{A}_2$ and compute $e^{\mathbf{A}t}$. For simplicity, I drop the subscript 2. The key step is to diagonalize $\mathbf{A}$[8]:

$$\mathbf{A} = \mathbf{PDP}^{-1}, \quad \mathbf{P} = \begin{bmatrix} 2 & -1 \\ 3 & +1 \end{bmatrix}, \quad \mathbf{D} = \begin{bmatrix} 4 & +0 \\ 0 & -1 \end{bmatrix}$$

Then, using the definition for $e^{\mathbf{A}}$, we can compute $e^{\mathbf{A}t}$ as follow (with $\mathbf{A}^k = \mathbf{PD}^k\mathbf{P}$, $k = 1, 2, \ldots$)

$$\begin{aligned} e^{\mathbf{A}t} &= \mathbf{I} + \mathbf{A}t + \frac{\mathbf{A}^2}{2!}t^2 + \frac{\mathbf{A}^3}{3!}t^3 + \cdots \\ &= \mathbf{PIP}^{-1} + \mathbf{PDP}^{-1}t + \frac{1}{2!}\mathbf{PD}^2\mathbf{P}^{-1}t^2 + \frac{1}{3!}\mathbf{PD}^3\mathbf{P}^{-1}t^3 + \cdots \\ &= \mathbf{P}\left(\mathbf{I} + \mathbf{D}t + \frac{1}{2!}\mathbf{D}^2t^2 + \frac{1}{3!}\mathbf{D}^3t^3 + \cdots\right)\mathbf{P}^{-1} \\ &= \mathbf{P}e^{\mathbf{D}t}\mathbf{P}^{-1} \quad \text{(the red term is } e^{\mathbf{D}t} \text{ due to (7.3.8))} \end{aligned}$$

Can we compute $e^{\mathbf{D}t}$? Because if we can then we're done. Using Eq. (7.3.8), it can be shown that

$$e^{\mathbf{D}t} = \begin{bmatrix} e^{4t} & 0 \\ 0 & e^{-t} \end{bmatrix}$$

Did we see this matrix? Yes, it is exactly the red matrix in Eq. (7.3.7)! Now we have $e^{\mathbf{A}t}$ as

$$e^{\mathbf{A}t} = \begin{bmatrix} 2 & -1 \\ 3 & +1 \end{bmatrix} \begin{bmatrix} e^{4t} & 0 \\ 0 & e^{-t} \end{bmatrix} \begin{bmatrix} 2 & -1 \\ 3 & +1 \end{bmatrix}^{-1}$$

Multiplying with $\boldsymbol{x}_0$, we get $e^{\mathbf{A}t}\boldsymbol{x}_0 = \boldsymbol{x}$—the solution we're looking for (compare with Eq. (7.3.7)). Now, we have reasons to believe that the exponential of a matrix, as we have defined it, is working.

Is there an easier way to see that $\boldsymbol{x} = e^{\mathbf{A}t}\boldsymbol{x}_0$ is the solution of $\dot{\boldsymbol{x}} = \mathbf{A}\boldsymbol{x}$? Yes, differentiate $\boldsymbol{x}$! But only if we're willing to compute the derivative of $e^{\mathbf{A}t}$. It turns out not hard at all[9]:

$$\begin{aligned} \frac{de^{\mathbf{A}t}}{dt} &= \frac{d}{dt}\left(\mathbf{I} + \mathbf{A}t + \frac{(\mathbf{A}t)^2}{2!} + \frac{(\mathbf{A}t)^3}{3!} + \cdots\right) = \mathbf{0} + \mathbf{A} + \mathbf{A}^2t + \frac{1}{2}\mathbf{A}^3t^2 + \cdots \\ &= \mathbf{A}\left(\mathbf{I} + \mathbf{A}t + \frac{1}{2}\mathbf{A}^2t^2 + \cdots\right) = \mathbf{A}e^{\mathbf{A}t} \end{aligned}$$

---

[8] Hey, isn't this section only for non-diagonalizable matrices? We're now testing the idea of $e^{\mathbf{A}}$ for the case we know the solution first. If it does not work for this case then forget the idea.

[9] So the differentiation rule: $d/dt(e^{\alpha t}) = \alpha e^{\alpha t}$ still holds if α is a matrix.

With that, we can verify whether $\boldsymbol{x} = e^{\mathbf{A}t}\boldsymbol{x}_0$ is the solution to $\dot{\boldsymbol{x}} = \mathbf{A}\boldsymbol{x}$:

$$\dot{\boldsymbol{x}} = \frac{d(e^{\mathbf{A}t}\boldsymbol{x}_0)}{dt} = \frac{de^{\mathbf{A}t}}{dt}\boldsymbol{x}_0 = \mathbf{A}(e^{\mathbf{A}t}\boldsymbol{x}_0) = \mathbf{A}\boldsymbol{x}$$

We can also prove that this is the only solution.[10]

What if the matrix is non-diagonalizable? We have computed $e^{\mathbf{A}}$ by diagonalizing the matrix and taking advantage of the fact that the exponential of a diagonal matrix is easy to get. But not all matrices are diagonalizable! For example, solving this equation: $y'' - 2y' + y = 0$. First, we convert this into a system of two first-order DEs: with $\boldsymbol{x} = (y, y')$

$$\frac{d}{dt}\begin{bmatrix} y \\ y' \end{bmatrix} = \begin{bmatrix} y' \\ 2y' - y \end{bmatrix} \Longrightarrow \dot{\boldsymbol{x}} = \mathbf{A}\boldsymbol{x}, \quad \mathbf{A} = \begin{bmatrix} +0 & 1 \\ -1 & 2 \end{bmatrix}$$

The matrix $\mathbf{A}$ is non-diagonalizable because it has repeated eigenvalues and thus linear dependent eigenvectors:

$$\lambda_1 = \lambda_2 = 1, \quad \boldsymbol{x}_1 = \begin{bmatrix} 1 \\ 1 \end{bmatrix}, \quad \boldsymbol{x}_2 = \alpha\begin{bmatrix} 1 \\ 1 \end{bmatrix}$$

We have to rely on the infinite series in Eq. (7.3.8) to compute $e^{\mathbf{A}t}$. First, massaging $\mathbf{A}$ a bit[11]:

$$\mathbf{A} = \mathbf{I} + \mathbf{A} - \mathbf{I} \Longrightarrow e^{\mathbf{A}t} = e^{(\mathbf{I}+\mathbf{A}-\mathbf{I})t} = e^{\mathbf{I}t}e^{(\mathbf{A}-\mathbf{I})t}$$

Using Eq. (7.3.8), we can compute $e^{\mathbf{I}t}$ and $e^{(\mathbf{A}-\mathbf{I})t}$, (for the latter, note that $(\mathbf{A} - \mathbf{I})^2 = \mathbf{0}$)

$$e^{\mathbf{I}t} = \mathbf{I}e^t, \quad e^{(\mathbf{A}-\mathbf{I})t} = \mathbf{I} + (\mathbf{A} - \mathbf{I})t$$

With these results, we can write $e^{\mathbf{A}t}$

$$e^{\mathbf{A}t} = \mathbf{I}e^t[\mathbf{I} + (\mathbf{A} - \mathbf{I})t] = e^t[\mathbf{I} + (\mathbf{A} - \mathbf{I})t]$$

Therefore, the solution in the form of $e^{\mathbf{A}t}\boldsymbol{x}_0$ is

$$\boldsymbol{x} = e^{\mathbf{A}t}\boldsymbol{x}_0 = e^t[\mathbf{I} + (\mathbf{A} - \mathbf{I})t]\boldsymbol{x}_0 \Longrightarrow y(t) = e^t y(0) - te^t y(0) + te^t y'(0)$$

[10] We assume that $\boldsymbol{x}(t)$ is one solution and there was another solution $\boldsymbol{y}(t)$, then we build $\boldsymbol{z} = \boldsymbol{x} - \boldsymbol{y}$. Now, letting $\boldsymbol{v}(t) = e^{-\mathbf{A}t}\boldsymbol{z}(t)$, it can be shown that $\dot{\boldsymbol{v}} = 0$: so $\boldsymbol{v}(t)$ must be constant. But $\boldsymbol{v}(0) = 0$, thus $\boldsymbol{v}(t) = \boldsymbol{z}(t) = \mathbf{0}$. Therefore, $\boldsymbol{y} = \boldsymbol{x}$: the solution is unique.

[11] We accepted that $e^{\mathbf{A}t}e^{\mathbf{B}t} = e^{(\mathbf{A}+\mathbf{B})t}$ if $\mathbf{AB} = \mathbf{BA}$, of which proof is skipped.

Is this solution correct? We can check! It is easy to see that $y = e^t$ and $y = te^t$ are two solutions to $y'' - 2y' + y = 0$. Thus, the solution is a linear combination of them. Hence, the solution obtained using the exponential of a matrix is correct.

This method was based on this trick $\mathbf{A} = \mathbf{I} + \mathbf{A} - \mathbf{I}$ and the fact that $(\mathbf{A} - \mathbf{I})^2 = \mathbf{0}$. How can we know all of this[12]? It's better to have a method that less depends on tricks.

Schur factorization. Assume a 2×2 matrix $\mathbf{A}$ with one eigenvalue λ and the associated eigenvector $\boldsymbol{v}$, i.e., $\mathbf{A}\boldsymbol{v} = \lambda\boldsymbol{v}$. Now we select a vector $\boldsymbol{w}$ such that $\boldsymbol{v}, \boldsymbol{w}$ are linear independent, thus we can write $\mathbf{A}\boldsymbol{w} = c\boldsymbol{v} + d\boldsymbol{w}$ for $c, d \in \mathbb{R}$. Now, we have

$$\begin{cases} \mathbf{A}\boldsymbol{v} = \lambda\boldsymbol{v} \\ \mathbf{A}\boldsymbol{w} = c\boldsymbol{v} + d\boldsymbol{w} \end{cases} \Longrightarrow \mathbf{A}\begin{bmatrix}\boldsymbol{v} & \boldsymbol{w}\end{bmatrix} = \begin{bmatrix}\boldsymbol{v} & \boldsymbol{w}\end{bmatrix}\begin{bmatrix}\lambda & c \\ 0 & d\end{bmatrix} \Longrightarrow \mathbf{A} = \begin{bmatrix}\boldsymbol{v} & \boldsymbol{w}\end{bmatrix}\begin{bmatrix}\lambda & c \\ 0 & d\end{bmatrix}\begin{bmatrix}\boldsymbol{v} & \boldsymbol{w}\end{bmatrix}^{-1}$$

So, we have proved that for *any* 2×2 matrix, it is always possible to *diagonalize* $\mathbf{A}$ into the form $\mathbf{PTP}^{-1}$, where $\mathbf{T}$ is an upper triangle matrix. Now, we're interested in the case $\mathbf{A}$ is defective, i.e., it has a double eigenvalue $\lambda_1 = \lambda_2 = \lambda$, thus we have[13]

$$\mathbf{A} = \begin{bmatrix}\boldsymbol{v} & \boldsymbol{w}\end{bmatrix}\begin{bmatrix}\lambda & c \\ 0 & \lambda\end{bmatrix}\begin{bmatrix}\boldsymbol{v} & \boldsymbol{w}\end{bmatrix}^{-1} \Longrightarrow \mathbf{A}^k = \begin{bmatrix}\boldsymbol{v} & \boldsymbol{w}\end{bmatrix}\begin{bmatrix}\lambda & c \\ 0 & \lambda\end{bmatrix}^k\begin{bmatrix}\boldsymbol{v} & \boldsymbol{w}\end{bmatrix}^{-1}$$

It turns out that it is easy to compute the blue term: a triangular matrix is also nice to work with. Indeed, we can decompose the blue matrix, now denoted by $\boldsymbol{\Lambda}$, into the sum of a diagonal matrix and a *nilpotent matrix*. A nilpotent matrix is a square matrix $\mathbf{N}$ such that $\mathbf{N}^p = \mathbf{0}$ for some positive integer p; the smallest such p is called the index of $\mathbf{N}$. Using the binomial theorem and the nice property of nilpotent matrices (in below the red matrix is $\mathbf{N}$ with $p = 2$), we get

$$\boldsymbol{\Lambda}^k = \left(\begin{bmatrix}\lambda & 0 \\ 0 & \lambda\end{bmatrix} + \begin{bmatrix}0 & c \\ 0 & 0\end{bmatrix}\right)^k = \begin{bmatrix}\lambda & 0 \\ 0 & \lambda\end{bmatrix}^k + k\begin{bmatrix}\lambda & 0 \\ 0 & \lambda\end{bmatrix}^{k-1}\begin{bmatrix}0 & c \\ 0 & 0\end{bmatrix} = \begin{bmatrix}\lambda^k & kc\lambda^{k-1} \\ 0 & \lambda^k\end{bmatrix}$$

Thus, using Eq. (7.3.8), we can determine $e^{\boldsymbol{\Lambda}t}$

$$e^{\boldsymbol{\Lambda}t} = \begin{bmatrix}e^{\lambda t} & cte^{\lambda t} \\ 0 & e^{\lambda t}\end{bmatrix}$$

And the solution to $\dot{\boldsymbol{x}} = \mathbf{A}\boldsymbol{x}$ is $\boldsymbol{x}(t) = e^{\mathbf{A}t}\boldsymbol{x}_0$, which can be written as

$$\boldsymbol{x}(t) = e^{\mathbf{A}t}\boldsymbol{x}_0 = \mathbf{P}e^{\boldsymbol{\Lambda}t}\mathbf{P}^{-1}\boldsymbol{x}_0 = \begin{bmatrix}\boldsymbol{v} & \boldsymbol{w}\end{bmatrix}\begin{bmatrix}e^{\lambda t} & cte^{\lambda t} \\ 0 & e^{\lambda t}\end{bmatrix}\begin{bmatrix}a \\ b\end{bmatrix} = ae^{\lambda t}\boldsymbol{v} + be^{\lambda t}(ct\boldsymbol{v} + \boldsymbol{w})$$

[12] Actually, there is a theorem called the Caley-Hamilton theorem that reveals this. The characteristic equation of $\mathbf{A}$ is $(\lambda - 1)^2 = 0$. That theorem—stating that the matrix also satisfies the characteristic equation—then gives us: $(\mathbf{A} - \mathbf{I})^2 = \mathbf{0}$.

[13] We must have $d = \lambda$ as $\mathbf{A}$ and the red matrix are similar, they have same eigenvalues.

The final step is to find $\boldsymbol{w}$ and we're done. Recall that $\mathbf{A}\boldsymbol{w} = c\boldsymbol{v} + d\boldsymbol{w}$, but $d = \lambda$, thus (redefine $\boldsymbol{w}$ as $(1/c)\boldsymbol{w}$), we obtain

$$\mathbf{A}\boldsymbol{w} = c\boldsymbol{v} + \lambda\boldsymbol{w} \Longleftrightarrow \mathbf{A}\boldsymbol{w} = \boldsymbol{v} + \lambda\boldsymbol{w} \Longleftrightarrow (\mathbf{A} - \lambda\mathbf{I})\boldsymbol{w} = \boldsymbol{v} \Longrightarrow (\mathbf{A} - \lambda\mathbf{I})^2\boldsymbol{w} = \mathbf{0}$$

We call $\boldsymbol{v}$ the eigenvector of $\mathbf{A}$, how about $\boldsymbol{w}$? Let put the equations of these two vectors together:

$$\begin{aligned} (\mathbf{A} - \lambda\mathbf{I})^1\boldsymbol{v} = \mathbf{0} \\ (\mathbf{A} - \lambda\mathbf{I})^2\boldsymbol{w} = \mathbf{0} \end{aligned} \tag{7.3.9}$$

With this, it is no surprise that mathematicians call $\boldsymbol{w}$ the *generalized eigenvectors* (of order 2) of $\mathbf{A}$. Generalized eigenvectors play a similar role for defective matrices that eigenvectors play for diagonalizable matrices. The eigenvectors of a diagonalizable matrix span the whole vector space. The eigenvectors of a defective matrix do not, but the generalized eigenvectors of that matrix do.

7.4 Partial Differential Equations: A Classification

It is more often that a quantity varies from points to points and from time to time; such quantity is a function of two variables $u(x, t)$ in the simplest setting; or it can be $u(x, y, z, t)$. Thus, we have changes of u in space when t is fixed, and we also have change of u when time is passing by at a fixed point x. Let's first introduce some short notations for all partial derivatives of $u(x, t)$:

$$u_x = \frac{\partial u}{\partial x}, \quad u_t = \frac{\partial u}{\partial t}, \quad u_{xx} = \frac{\partial^2 u}{\partial x^2}, \quad u_{tt} = \frac{\partial^2 u}{\partial t^2} \tag{7.4.1}$$

Then, a partial differential equation (PDE) in terms of $u(x, t)$ is the following equation:

$$\boxed{F(u, u_x, u_t, u_{xx}, u_{tt}) = 0} \tag{7.4.2}$$

Note that partial derivatives of order higher than 2 are not discussed. This is because in physics and engineering, we rarely see them present in differential equations.

To classify different PDEs, the concepts of order, dimension and linearity of a PDE are introduced:

Order The order of a PDE is the highest partial derivative; $u_t = u_{xx}$ is a second-order PDE;

Dimension The dimension of a PDE is the number of independent variables; $u_{tt} = u_{xx} + u_{yy}$ is a 3D PDE as it involves x, y and t;

Table 7.3 Some PDEs with associated order, dimension, and linearity

Equation	Order	Linear	Dim.
$u_t = u_{xx}$	2	✓	2
$u_{tt} = u_{xx} + u_{yy}$	2	✓	3
$xu_x + yu_y = u^2$	1	×	2

Linearity A PDE is said to be linear if the function u and all its partial derivatives appear in a linear fashion; i.e., they are not multiplied together, they are not squared, etc.

Table 7.3 presents some examples to demonstrate these concepts.

7.5 Derivation of Common PDEs

This section presents the derivation of common PDEs; those PDEs show up quite frequently in many science and engineering fields. Knowing how to get the equations is important particularly if you want to be a scientist. Mathematicians are more interested in solving the equations and determining the behavior of the solutions; for example, mathematicians are interested in questions such as whether the solutions are unique or when the solutions exist. The following quote best describes the differences between engineers, physicists, and mathematicians:

> An engineer thinks that his equations are an approximation to reality. A physicist thinks reality is an approximation to his equations. A mathematician doesn't care.

The author of this quote is commonly attributed to the British mathematician and philosopher, Alfred North Whitehead. However, there is some debate about whether he actually said it or not, as there is no clear source or reference for the quote in his writings. Nevertheless, the quote is often cited as a humorous way of highlighting the different perspectives and attitudes of engineers, physicists, and mathematicians towards the relationship between theory and reality. It suggests that engineers are more pragmatic and focused on practical solutions, physicists are more concerned with theoretical models and their correspondence to reality, while mathematicians are primarily interested in the abstract structures and relationships underlying both theory and reality.

We start with the wave equation in Sect. 7.5.1, derived centuries ago by d'Alembert in 1746. We live in a world of waves. Whenever we throw a pebble into the pond, we see the circular ripples formed on its surface which disappear gradually. The water moves up and down, and the effect, ripple, which is visible to us looks like an outwardly moving wave. When you pluck the string of a guitar, the strings move up and down, exhibiting transverse wave; the particles in the string move perpendicular to the direction of the wave propagation. The bump or rattle that we feel during an earthquake is due to seismic-S wave. It moves rock particles up and down, perpendicular to the direction of the wave propagation.

We continue in Sect. 7.5.2 with the heat equation (or diffusion equation) derived by Fourier in 1807. It was while solving this equation that Fourier developed what we now call the Fourier series.

7.5.1 Wave Equation

This section presents the derivation of the wave equation. It all started with the aim to understand the vibration of a violin string. Why this object? This is because a string can be modeled as an infinitely thin line, and its motion is constrained in a plane. So, it is simple from a mathematics point of view.

So, we consider a string fixed at two ends. At time $t = 0$, the string is horizontal and unstretched (Fig. 7.1). As the string undergoes only transverse motion, i.e., motion perpendicular to the original string, we use $u(x, t)$ to designate the transverse displacement of point x at time t. Our task is to find the equation relating $u(x, t)$ to the physics of the string.

The key idea is to use Newton's second law (what else?) for a small segment of the string. Such a segment of length Δx is shown in Fig. 7.1. What are the forces in the system? First, we have $f(x, t)$ in the vertical direction which can be gravity or any external force. This is a distributed force that is force per unit length (i.e., the

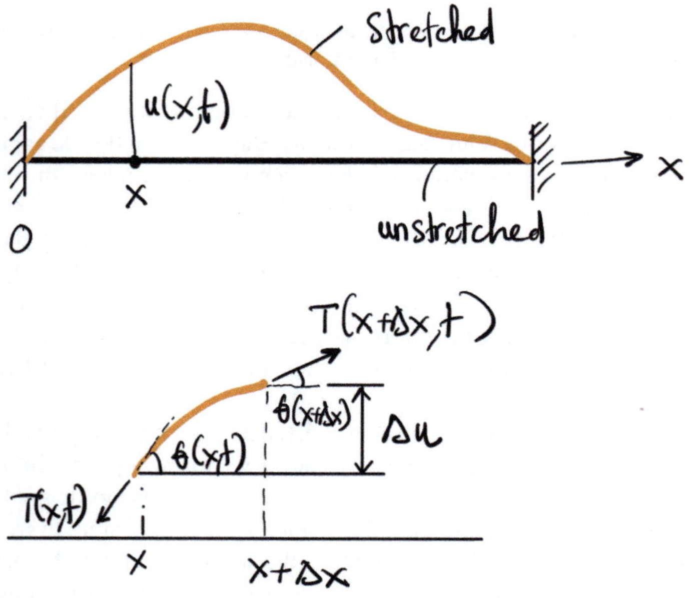

Fig. 7.1 Derivation of the wave equation: a vibrating string

total force acting on the segment is $f\Delta x$). Second, we have the tension force $T(x,t)$ inside the string. We use Newton's second law $F = ma$ in the vertical direction to write, with $a = \partial^2 u/\partial t^2$, mass is density times length, that is $m = \rho\sqrt{(\Delta x)^2 + (\Delta u)^2}$

$$\rho\sqrt{(\Delta x)^2 + (\Delta u)^2}\frac{\partial^2 u}{\partial t^2} = T(x+\Delta x, t)\sin\theta(x+\Delta x, t) - T(x,t)\sin\theta(x,t) + f(x,t)\Delta x \tag{7.5.1}$$

Dividing this equation by Δx and considering $\Delta x \to 0$, we get

$$\begin{aligned}\rho\sqrt{1+\left(\frac{\partial u}{\partial x}\right)^2}\frac{\partial^2 u}{\partial t^2} &= \frac{d}{dx}(T(x,t)\sin\theta(x,t)) + f(x,t)\\ &= \frac{\partial T}{\partial x}\sin\theta(x,t) + T(x,t)\cos\theta(x,t)\frac{\partial \theta}{\partial x} + f(x,t)\end{aligned} \tag{7.5.2}$$

We know that the derivative of $u(x,t)$ is $\tan\theta(x,t)$, so we can write

$$\tan\theta(x,t) = \frac{\partial u}{\partial x}(x,t), \quad \theta(x,t) = \arctan\left(\frac{\partial u}{\partial x}\right) \tag{7.5.3}$$

where we also need an expression for $\theta(x,t)$. From $\tan\theta(x,t)$, we can compute $\sin\theta(x,t)$, $\cos\theta(x,t)$ and from the expression for θ, we can compute the derivative of θ:

$$\sin\theta(x,t) = \sqrt{\frac{(\frac{\partial u}{\partial x})^2}{1+(\frac{\partial u}{\partial x})^2}}, \quad \cos\theta(x,t) = \sqrt{\frac{1}{1+(\frac{\partial u}{\partial x})^2}}, \quad \frac{\partial \theta}{\partial x} = \frac{\frac{\partial^2 u}{\partial x^2}}{1+(\frac{\partial u}{\partial x})^2} \tag{7.5.4}$$

Now comes the art of approximation (otherwise the problem would be too complex). We consider only small vibration, that is when $|\frac{\partial u}{\partial x}| \ll 1$,[14] and with this simplified condition the above equation becomes

$$\sin\theta(x,t) = \frac{\partial u}{\partial x}, \quad \cos\theta(x,t) = 1, \quad \frac{\partial \theta}{\partial x} = \frac{\partial^2 u}{\partial x^2} \tag{7.5.5}$$

With all this, Eq. (7.5.2) is simplified to

$$\rho\frac{\partial^2 u}{\partial t^2} = \frac{\partial T}{\partial x}\frac{\partial u}{\partial x} + T(x,t)\frac{\partial^2 u}{\partial x^2} + f(x,t) \tag{7.5.6}$$

The equation looks much simpler. But it is still unsolvable. Why? Because we have one equation but two unknowns $u(x,t)$ and $T(x,t)$. But wait, we have another Newton's second law in the horizontal direction:

$$T(x+\Delta x, t)\cos\theta(x+\Delta x, t) - T(x,t)\cos\theta(x,t) = 0 \tag{7.5.7}$$

[14] The symbol $\ll$ means much smaller than.

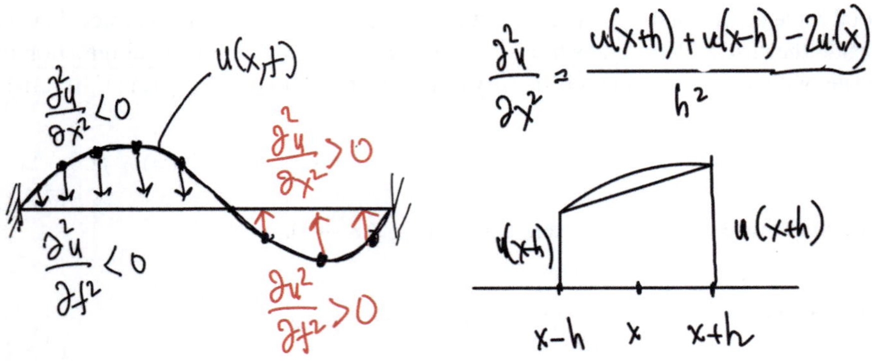

Fig. 7.2 .

Dividing it by Δx and let $\Delta x \to 0$, we get

$$\frac{\partial}{\partial x}\left[T(x,t)\cos\theta(x,t)\right] = 0 \tag{7.5.8}$$

But note that $\cos\theta(x,t) \approx 1$, thus the above equation indicates that $T(x,t)$ is constant. Let's use T to designate the tension in the string, Eq. (7.5.6) becomes

$$\rho\frac{\partial^2 u}{\partial t^2} = T\frac{\partial^2 u}{\partial x^2} + f(x,t)$$

Ignoring $f(x,t)$ and letting $c^2 = T/\rho$, we get the *wave equation*:

$$\boxed{\frac{\partial^2 u}{\partial t^2} = c^2\frac{\partial^2 u}{\partial x^2}} \tag{7.5.9}$$

What does this equation mean? On the LHS, we have the acceleration term and, on the RHS, we have the second spatial derivative of $u(x,t)$. The second spatial derivative of u measures the concavity of the curve $u(x,t)$. Thus, when the curve is concave downward, this term is negative, and thus the wave equation tells us that the acceleration is also negative and thus the string is moving downwards (Fig. 7.2).

We do not discuss the solution to the wave equation here. But even without it, we still can say something about its solutions. The first thing is that this equation is linear due to the linearity of the differentiation operator. What does this entail? Let $u(x,t)$ and $v(x,t)$ be two solutions[15] to the wave equation, that is

[15] Why the wave equation can have more than one solution? Actually, any PDE has infinitely many solutions. Think of it this way. The violin string can be bent into any shape you like before it is released and the wave equation takes over. In other words, each initial condition leads to a distinct solution.

$$\frac{\partial^2 u}{\partial t^2} = c^2 \frac{\partial^2 u}{\partial x^2}, \quad \frac{\partial^2 v}{\partial t^2} = c^2 \frac{\partial^2 v}{\partial x^2}$$

then any linear combination of these two, i.e., $\alpha u + \beta v$, where α and β are two constants, is also a solution:

$$\frac{\partial^2 (\alpha u + \beta v)}{\partial t^2} = c^2 \frac{\partial^2 (\alpha u + \beta v)}{\partial x^2}$$

3D wave equation. Having derived the 1D wave equation, the question is what is the 3D version? Let's try to guess what it would be. It should be of the same form as the 1D equation but has components relating to the other dimensions (red terms below):

$$\frac{\partial^2 u}{\partial t^2} = c^2 \left(\frac{\partial^2 u}{\partial x^2} + \frac{\partial^2 u}{\partial y^2} + \frac{\partial^2 u}{\partial z^2} \right) \tag{7.5.10}$$

It is remarkable that a model born in attempts to understand how a string vibrates now has a wide spectrum of applications. Here are some applications of the wave equation:

(a) Vibrations of a stretched membrane, such as a drumhead.
(b) Sound waves in air or other media.
(c) Waves in an incompressible fluid, such as water.
(d) Electromagnetic waves, such as light waves and radio waves.

7.5.2 *Diffusion Equation*

In the opening of his masterpiece, *The Analytical Theory of Heat* (1822), the French mathematical physicist Jean-Baptiste Joseph Fourier (1768–1830) tells us why he decided to study heat: "Heat, like gravity, penetrates every substance of the universe... The object of our work is to set forth the mathematical laws [of heat]. The theory of heat will hereafter form one of the most important branches of general physics".

And that is the beginning of what we now call the diffusion equation (or sometimes the heat equation).

If we hold a metal rod on one end and the other end of the rod is heated, after a while your hand feels the heat. This phenomenon is called *heat conduction*. This can be explained roughly as follows. Initially, before the rod is heated, the atoms (or electrons) in the rods vibrate around their equilibrium positions (you can imagine tiny particles jiggling). When one end of the rod is heated, the atoms in this part vibrate quicker and they collide with nearby atoms (that have lower temperature). Through these collisions, heat is transferred from the hotter atoms to the colder ones and eventually to the other end of the rod.

Herein, we want to make a mathematical model to describe heat conduction at a macroscopic scale (i.e., without worrying about atoms, molecules, etc.). The most

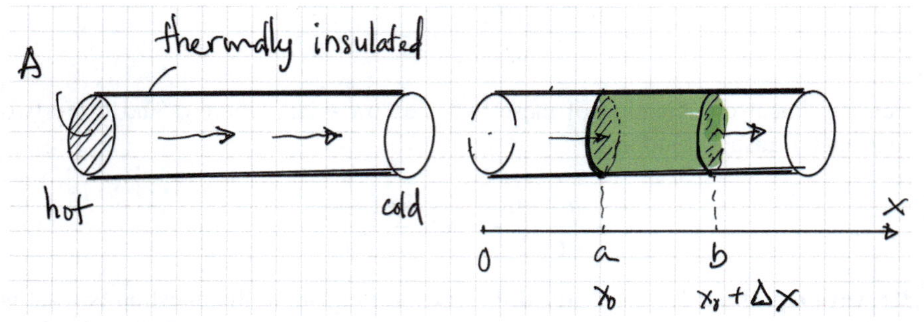

Fig. 7.3 Heat conduction in a long bar

distinct advantage of such a continuum model is that it is possible to investigate the heat conduction in a large piece of material, e.g., automotive diesel piston. Obviously, it is impossible with current computers to do so with an atomistic model.

When energy is added to a system and there is no change in the kinetic or potential energy, the temperature of the system usually rises. The quantity of energy required to raise the temperature of a given mass of a substance by some amount varies from one substance to another. To quantify this, the *specific heat* c is used, which is the amount of energy required to raise the temperature per unit mass by 1 °C. The SI unit for c is J/(kg · K).

We are deriving the equation of heat conduction inside a long thin bar of length L (to minimize the mathematical complexity). The outer surface of the bar is insulated and the left end is heated up while the right end is cooled down. Therefore, there is heat moving along the bar to the right (Fig. 7.3). Let's denote by $\theta(x, t)$ the temperature in the bar at a distance x from the left end at time t. Furthermore, let A be the cross sectional area of the bar, ρ be the density of the material making up the bar. These two quantities (A and ρ) can vary along the bar or they can be constant. For simplicity, they are considered constant subsequently though. We should always start with the simplest model that is possible.

The idea is to consider a segment of the bar, e.g., the part of the bar between $x = a$ and $x = b$, and applying the principle of *conservation of energy* to this segment. The conservation of energy is simple: the rate of change of heat inside the bar is equal to the heat flux entering the left end minus the heat flux going out the right end. The rate of change of heat is given by

$$\text{rate of change of heat} = \frac{\partial}{\partial t}\int_a^b cA\rho\theta(x, t)dx \tag{7.5.11}$$

while the heat fluxes are

$$\text{heat fluxes} = AJ(a, t) - AJ(b, t) \tag{7.5.12}$$

where J is the heat flux density. Now, we can write the equation of conservation of heat as

$$\frac{\partial}{\partial t}\int_a^b cA\rho\theta(x,t)dx = AJ(a,t) - AJ(b,t) \tag{7.5.13}$$

Using Leibniz's rule and the fundamental theorem of calculus, we can elaborate this equation as

$$\int_a^b cA\rho\frac{d\theta(x,t)}{dt}dx = -A\int_a^b \frac{dJ}{dx}dx \tag{7.5.14}$$

$$\Longrightarrow \int_a^b \left(c\rho\frac{\partial\theta(x,t)}{\partial t} + \frac{\partial J}{\partial x}\right)dx = 0 \tag{7.5.15}$$

$$\Longrightarrow c\rho\frac{\partial\theta(x,t)}{\partial t} + \frac{\partial J}{\partial x} = 0 \tag{7.5.16}$$

In the third equation, we moved from an integral equation to a partial differential equation. This is because the segment $[a, b]$ is arbitrary, so the integrand must be identically zero.

You might guess that we still miss a connection between J and $\theta(x,t)$ (one equation and two unknown variables are unsolvable). Indeed, Fourier carried out experiments to give us just that relation (known as a constitutive equation)

$$J = -k\frac{\partial\theta}{\partial x} \tag{7.5.17}$$

where k is known as the coefficient of thermal conductivity (SI unit is W/(mK)). The thermal conductivity provides an indication of the rate at which heat energy is transferred through a medium by the diffusion process.

With Eq. (7.5.17), our Eq. (7.5.16) becomes (note that k is constant):

$$c\rho\frac{\partial\theta(x,t)}{\partial t} + \frac{\partial}{\partial x}\left(-k\frac{\partial\theta}{\partial x}\right) = 0 \Longrightarrow \boxed{\frac{\partial\theta}{\partial t} = \kappa^2\frac{\partial^2\theta}{\partial x^2}}, \quad \kappa^2 = \frac{k}{c\rho} \tag{7.5.18}$$

which is a linear second-order (in space) partial differential equation. In this equation, κ^2 is called the thermal diffusivity (SI unit is m^2/s).

As it involves the second derivative of θ we need two boundary conditions on θ. They are the temperatures specified at the two ends of the bar: $\theta(0,t) = \theta_1$ and $\theta(L,t) = \theta_2$, where $\theta_{1,2}$ are real numbers representing the temperatures at the two ends (say 1 °C and 10 °C). Furthermore, we need one initial condition (as we have first derivative of θ w.r.t. time). This is the condition from which the game kicks on; it specifies the initial temperature (i.e., at time $t = 0$) along the bar (i.e., all points along the bar but the ends $0 < x < L$): $\theta(x,0) = \phi(x)$ for some function $\phi(x)$ which represents the initial temperature in the bar at $t = 0$. Altogether, the

PDE, the boundary conditions and the initial condition make an *initial-boundary value problem*:

$$\frac{\partial \theta}{\partial t} = \kappa^2 \frac{\partial^2 \theta}{\partial x^2} \quad 0 < x < L \tag{7.5.19}$$

$$\theta(x, 0) = g(x) \quad 0 \le x \le L \tag{7.5.20}$$

$$\theta(0, t) = \theta_1, \ \theta(L, t) = \theta_2 \quad t > 0 \tag{7.5.21}$$

Another derivation of Eq. (7.5.16).

We consider a segment of the bar $[x_0, x_0 + \Delta x]$, then we can write

$$\frac{d}{dt} \int_{x_0}^{x_0+\Delta x} cA\rho\theta(x, t)dx = AJ(x_0, t) - AJ(x_0 + \Delta x, t) \tag{7.5.22}$$

Intermediate value theorem of integral calculus (Eq. (4.11.3)) applied to the integral on the LHS,

$$\frac{\partial}{\partial t} cA\rho\theta(x_1, t)\Delta x = AJ(x_0, t) - AJ(x_0 + \Delta x, t) \tag{7.5.23}$$

where $x_1 \in [x_0, x_0 + \Delta x]$. Dividing both sides by Δx, we obtain

$$\frac{\partial}{\partial t} cA\rho\theta(x_1, t) = -A \left(\frac{J(x_0 + \Delta x, t) - J(x_0, t)}{\Delta x} \right) \tag{7.5.24}$$

The final step is to let Δx to go to zero, and then x_1 is x_0 and, on the RHS, we have the derivative of J evaluated at x_0.

$$c\rho \frac{\partial \theta(x_0, t)}{\partial t} = -J_x(x_0, t) \tag{7.5.25}$$

This equation holds for any x_0, we can replace x_0 by x. And we get the 1D heat diffusion equation.

3D diffusion equation. Having derived the 1D heat equation, it is not hard to derive the 3D equation. Before doing so, let's try to guess what it would be. It should be of the same form as the 1D equation but has components relating to the other dimensions (red terms below):

$$\frac{\partial \theta}{\partial t} = \kappa^2 \left(\frac{\partial^2 \theta}{\partial x^2} + \frac{\partial^2 \theta}{\partial y^2} + \frac{\partial^2 \theta}{\partial z^2} \right) \tag{7.5.26}$$

We use the Gauss's theorem, see Sect. 6.11.6, for the derivation.[16] We consider an arbitrary domain V with the surface S. The temperature is now given by $\theta(\boldsymbol{x}, t)$ where $\boldsymbol{x} = (x_1, x_2, x_3)$ is the position vector. The conservation of energy equation is

$$\begin{aligned} -\frac{\partial}{\partial t}\int_V c\rho\theta dV &= \int_S \boldsymbol{J}\cdot\boldsymbol{n} dA \\ -\frac{\partial}{\partial t}\int_V c\rho\theta dV &= \int_V \nabla\cdot\boldsymbol{J} dV \quad \text{(Gauss's theorem)} \\ \int_V \left(c\rho\frac{\partial\theta}{\partial t} + \nabla\cdot(-k\nabla\theta)\right) dV &= 0 \quad (\boldsymbol{J} = -k\nabla\theta) \end{aligned} \tag{7.5.27}$$

As the volume domain V is arbitrary, we get the well-known 3D heat equation (assuming k is constant):

$$\frac{\partial\theta(\boldsymbol{x}, t)}{\partial t} = \kappa^2 \Delta\theta(\boldsymbol{x}, t), \quad \Delta\theta := \nabla\cdot(\nabla\theta) = \sum_{i=1}^{3}\frac{\partial^2\theta}{\partial {x_i}^2} \tag{7.5.28}$$

where Δ is the Laplacian operator, named after the French mathematician Pierre-Simon Laplace (1749–1827). We see this term Δf again and again in physics. Some people say that it is the most important operator in mathematical physics.

In the above derivation, we have used the 3D version of Eq. (7.5.17):

$$\boldsymbol{J} = -k\nabla\theta \quad \text{or} \quad \begin{bmatrix} J_x \\ J_y \\ J_z \end{bmatrix} = \begin{bmatrix} k & 0 & 0 \\ 0 & k & 0 \\ 0 & 0 & k \end{bmatrix}\begin{bmatrix} \theta_{,x} \\ \theta_{,y} \\ \theta_{,z} \end{bmatrix} \tag{7.5.29}$$

The matrix form is convenient when k is not constant. In that case we say the heat conduction is not isotropic but anisotropic, and we use three different values for the diagonal terms.

Equation (7.5.26) can also be used to model other diffusion processes (that's why it is referred to as the diffusion equation rather than the restricted heat equation term). For example, if a drop of red dye is placed in a body of water, the dye will gradually spread out and permeate the entire body. If convection effects are negligible, Eq. (7.5.26) will describe the diffusion of the dye through the water; $\theta(\boldsymbol{x}, t)$ is now the concentration of dye at $\boldsymbol{x}$ and time t!

[16] Of course, it is possible to consider an infinitesimal cube and follow the same steps done for the long bar. But the divergence theorem provides a shorter way.

7.6 Linear Partial Differential Equations

We have seen a few partial differential equations, it is time to sit back and study the common features of them. That's what mathematicians do. For example, they studied quadratic equations, then cubic equations, and then n-order polynomial equations of the form $a_n x^n + a_{n-1}x^{n-1} + \cdots + a_2x^2 + a_1x + a_0 = 0$. Doing the same thing here, we can see that the wave equation, the diffusion equation, etc., can be written in this form $L(u) = F$, where L is *a linear differential operator*. Feed in a number x, the operator square root $\sqrt{\Box}$ gives another number $\sqrt{x}$. Similarly, feed in a function $u(x_1, x_2, t)$, the operator L gives another function $L(u)$. In case of the 2D wave equation, L is

$$L = \frac{\partial^2 \Box}{\partial t^2} - c^2\left(\frac{\partial^2 \Box}{\partial x_1^2} + \frac{\partial^2 \Box}{\partial x_2^2}\right)$$

Thus, feeding in a function u, we get $L(u)$:

$$L(u) = \frac{\partial^2 u}{\partial t^2} - c^2\left(\frac{\partial^2 u}{\partial x_1^2} + \frac{\partial^2 u}{\partial x_2^2}\right)$$

Now it is time to have a general expression for L, which generalizes the concrete instances we have met[17]:

$$L = a(\boldsymbol{x}) + \sum_{i=1}^{n} b_i(\boldsymbol{x})\frac{\partial}{\partial x_i} + \sum_{i,j=1}^{n} c_{ij}(\boldsymbol{x})\frac{\partial^2}{\partial x_i x_j} + \cdots \tag{7.6.1}$$

where a, b_i, c_{ij} are some coefficients, which can be constants or functions of $\boldsymbol{x}$.

But what is exactly a linear differential operator? As a linear operator is consisted of partial derivatives and the derivative of $af(x)$ is a times the derivative of f and the derivative of a sum is the sum of the derivatives, it is readily that a linear operator L satisfies the following properties for functions u, v and scalar c:

$$\begin{aligned} \text{constant taken outside the operator: } L(cu) &= cL(u) \\ \text{opetator on sum is sum of operator: } L(u+v) &= L(u) + L(v) \end{aligned} \tag{7.6.2}$$

If we combine the above two properties, we then get

$$\boxed{L(\alpha u + \beta v) = L(\alpha u) + L(\beta v) = \alpha L(u) + \beta L(v)}$$

But this is not enough for mathematicians, why just two functions u, v? Then, they go for n functions $u_1, u_2, \ldots, u_n$, and write $L(a_1u_1 + \cdots + a_nu_n) = a_1L(u_1) + \cdots + a_nL(u_n)$.

---

[17] For the wave equation, $a = 0$, $b_i = 0$, $c_{11} = c_{22} = -c^2$ and $c_{33} = 1$ noting that $x_3 = t$.

With the introduction of a linear differential operator, it is easy to define what a linear partial differential equation is. A linear partial differential equation is simply an equation of the form $L(u) = F$ where L is a linear partial differential operator and F is a function of $\boldsymbol{x}$. Such an equation is called homogeneous if $F = 0$ and inhomogeneous if $F \neq 0$.

Principle of linear superposition. Let $L(u) = F$ be a linear differential equation. The principle of linear superposition tells us that, in order to solve the equation, we break the force term F into many simple terms:

$$F = c_1 f_1 + c_2 f_2 + \cdots + c_n f_n$$

Then we solve n easier problems of the form: $L(u_i) = f_i, i = 1, 2, \ldots, n$. Finally, the solution is a linear superposition of all these u_i: $u = c_1 u_1 + \cdots + c_n u_n$.

Proof The proof is straightforward: we just need to show that $u = c_1 u_1 + \cdots + c_n u_n$ is indeed a solution of $L(u) = F$:

$$L(c_1 u_1 + \cdots + c_n u_n) = c_1 L(u_1) + c_2 L(u_2) + \cdots + c_n L(u_n) = c_1 f_1 + \cdots + c_n f_n = F$$

7.7 Dimensionless Problems

The transient heat conduction equation needs values for the heat capacity c, the density ρ, and the heat conduction coefficient k of the material. In addition, relevant values must be chosen for the initial and boundary temperatures. With a dimensionless mathematical model, as explained in this section, no physical quantities need to be assigned. Not only is this a simplification of great convenience, as one simulation is valid for any type of material, but it also actually increases the understanding of the physical problem.

This section is organized as follows. Section 7.7.1 is for a discussion on dimensions and units. If I write a physical equation with the term $\sin(t^2)$ where t denotes time, then a physicst will tell me that I made a mistake. The power law presented in Sect. 7.7.2 explains why. Section 7.7.3 talks about dimensional analysis, which can be used to check if an equation is physically correct. We shall meet the Pi theorem first formulated by American physicist and mathematician Edgar Buckingham (1867–1940) in 1914. Buckingham's work on dimensional analysis and the Pi theorem has had a significant impact on many fields of science and engineering, including fluid mechanics, thermodynamics, and electrical engineering. Finally, scaling of ordinary differential equations is treated in Sect. 7.7.4.

7.7.1 Dimensions and Units

A physical dimension is a property we associate with physical quantities for purposes of classification or differentiation. Mass, length, time, and force are examples of physical dimensions. There are fundamental and derived dimensions. Fundamental dimensions include mass, length, time, and perhaps charge and temperature.[18]

We need some suitable mathematical notation to calculate with dimensions. The dimension of length is written as $[L]$, the dimension of mass as $[M]$, the dimension of time as $[T]$. We then express the dimensions of other quantities (e.g., speed) in terms of the fundamental dimensions. For instance, the dimension of speed is $[L/T]$ or LT^{-1}. The dimension of force, another derived unit, is the same as the dimension of mass times acceleration, and hence the dimension of force is $[MLT^{-2}]$. The point is that every quantity which is not explicitly dimensionless, like a pure number, e.g., π, e, has characteristic dimensions which are not affected by the way we measure it. As we will see shortly, this provides a useful check on any calculations we do.

Units give the magnitude of some dimension relative to an arbitrary standard. For example, when we say that a person is six feet tall, we mean that person is six times as long as an object whose length is defined to be one foot. The standard size chosen is, of course, entirely arbitrary, but becomes very useful for comparing measurements made in different places and times. Several national laboratories are devoted to maintaining sets of standards, and using them to calibrate instruments.

In contrast to dimensions, of which only a few are needed, there is a multitude of units for measuring most quantities: lengths measured in inches, meters, centimeters, and kilometers. It is, therefore, always necessary to attach a unit to a number, as when giving a person's height as 175 cm or as 5 ft 9 in. *Without units, a number is at best meaningless and at worst misleading to the reader.*

The International System of Units (SI, abbreviated from the French Systéme international (d'unités)) is the modern form of the metric system. It comprises a coherent system of units of measurement starting with seven base units, which are the second (the unit of time with the symbol s), meter (length, m), kilogram (mass, kg), ampere (electric current, A), kelvin (thermodynamic temperature, K), mole (amount of substance, mol), and candela (luminous intensity, cd).

From the seven base (or fundamental) units, we can derive many more derived units. For example, what is the unit of force in SI? Using Newton's second law, we write

$$[F] = \text{kg}\frac{\text{m}}{\text{s}^2} \tag{7.7.1}$$

And to honor Newton, we invented a new unit called N, and thus $1\,\text{N} = 1\,\text{kgm/s}^2$. Similarly, we have $1\,\text{Pa} = 1\,\text{N/m}^2$ as the SI unit of pressure and stress, in honor of Blaise Pascal.

Some common consistent SI units are given in Table 7.4.

---

[18] Noting that this choice is arbitrary, it is fine to use force as a fundamental dimension instead of mass for example.

Table 7.4 Some physical quantities with corresponding dimensions and SI units

Quantity	Relation	SI (m, s, N)	Dimension
Length	–	m	$[L]$
Time	–	s	$[T]$
Mass	–	kg	$[M]$
Force	Mass × acceleration	$\text{N} = 1\,\text{kgm/s}^2$	$[MLT^{-2}]$
Pressure/stress	Force/area	$\text{Pa} = 1\,\text{N/m}^2$	$[ML^{-1}T^{-2}]$

Table 7.5 Prefixes in SI. Prefix names have been mostly chosen from Greek words (positive powers of 10) or Latin words (negative powers of 10), although recent extensions of the range of powers of 10 has resulted in the use of words from other languages. "Kilo" comes from the Greek word for 1000 (10^3), and "milli" comes from the Latin word for one thousandth (10^{-3})

Large measurements			Small measurements		
Prefix	Symbol	Multiple	Prefix	Symbol	Sub-multiple
Yotta	Y	10^{24}	Deci	d	10^{-1}
Zetta	Z	10^{21}	Centi	c	10^{-2}
Exa	E	10^{18}	Milli	m	10^{-3}
Peta	P	10^{15}	Micro	μ	10^{-6}
Tera	T	10^{12}	Nano	n	10^{-9}
Giga	G	10^{9}	Pico	p	10^{-12}
Mega	M	10^{6}	Femto	f	10^{-15}
Kilo	k	10^{3}	Atto	a	10^{-18}
Hecto	h	10^{2}	Zepto	z	10^{-21}
Deka	h	10^{1}	Yocto	y	10^{-24}

It is not the end of the story about units. Why we have meters and still need kilometers? The reason is simple: *we're unable to handle too large or too small numbers*. If we only had meter as the only unit for length, then for lengths smaller than 1 m we have to use decimals, e.g., 0.05 m. To avoid that, sub-units are developed. Instead of 0.05 m we say 5 cm. Similarly for 20 000 m, we write 20 km, which is much easier to comprehend. In conclusion, larger and smaller quantities are expressed by using appropriate *prefixes* with the base unit. Table 7.5 presents all prefixes in SI. One example is: the mass of the Earth is 5 972 Yg (yottagrams), which is 5.972×10^{24} kg.

7.7.2 Power Laws

If we stop doing calculations and make one observation about dimensions, we would see that dimensional quantities always appear in power laws. For example, we have

$[L][T]^{-1}$ (or $[L]^1[T]^{-1}$) for speed, $[MLT^{-2}]$ for force, or $[L]^3$ for volume and $[M][L]^{-3}$ for density. We can prove that as follows.

Assume that x is a quantity of dimension of length, and y is a quantity depends on x via $y = f(x)$. You can think of y as volume, for example, then $f(x) = x^3$. Suppose now that x takes two values x_1 and x_2, then we get two y's: $y_1 = f(x_1)$ and $y_2 = f(x_2)$. The crucial point is that the ratio of y_1/y_2 is a dimensionless number. Obviously, a dimensionless number exists independently of any system of units we create. That is, if we now measure x_1 and x_2 using a different system of units, the values are αx_1 and αx_2; e.g., $\alpha = 1000$ if x_1 was measured in meters and now measured in mm. Then, we have

$$\frac{y_1}{y_2} = \frac{f(x_1)}{f(x_2)} = \text{a dimensionless number} \Longrightarrow \boxed{\frac{f(x_1)}{f(x_2)} = \frac{f(\alpha x_1)}{f(\alpha x_2)}}$$

Now, our goal is to solve the boxed equation and hope that its solution is of the form $f(x) = Cx^\beta$, a power function.[19]

Now, we rearrange the boxed equation a bit and take the first derivative of two sides of the resulting equation with respect to α, we get (with the notation $f_1 = f'(x)$):

$$f(\alpha x_1) = f(\alpha x_2)\frac{f(x_1)}{f(x_2)} \Longrightarrow f_1(\alpha x_1)x_1 = f_1(\alpha x_2)x_2\frac{f(x_1)}{f(x_2)}, \quad f_1 := \frac{df}{dx}$$

The above equation holds for any value of x_1, x_2 and α. Now, setting $\alpha = 1$,

$$x_1\frac{f_1(x_1)}{f(x_1)} = x_2\frac{f_1(x_2)}{f(x_2)} \Longrightarrow x\frac{f'(x)}{f(x)} = k \Longleftrightarrow \frac{f'(x)}{f(x)} = \frac{k}{x}$$

Now, integrating both sides of the above equation, we obtain

$$\int \frac{f'(x)}{f(x)}dx = \int k\frac{1}{x}dx \Longrightarrow \ln f(x) = k\ln x + A$$

And that leads to, indeed, a power function for $f(x)$:

$$f(x) = Cx^k$$

That is good but why power functions but not other functions that we have spent a lot of time to study in calculus? The reason is simple. We can never have more complicated functions. One simple way to see this is to use Taylor series. For example, the exponential function has the following well-known Taylor series

[19] It is easy to check that if $f(x)$ is a power function then it satisfies the boxed equation. So, we're in good direction.

$$e^x = 1 + x + \frac{x^2}{2} + \cdots$$

If x was a cetain length, then e^x would require the addition of length to area (i.e., the term x^2) to volume (x^3), which is nonsense. So, if we see, *in an equation,* e^x *or* $\sin x$ *or whatever function (except* x^k*), then* x *must be a dimensionless number otherwise the equation is physically wrong.* One example: in physics we write $\sin \omega t$ all the time, the dimension of ω is $1/[T]$ and the dimension of t is $[T]$. So, ωt is dimensionless.

The next step is to consider physical quantities that depend on more than one quantities. For simplicity, I just consider a quantity z that depends on two other quantities x, y: $z = f(x, y)$. Now, doing the same thing, we will have

$$\frac{f(x_1, y_1)}{f(x_2, y_2)} = \text{dimensionless number} \Longrightarrow \frac{f(x_1, y_1)}{f(x_2, y_2)} = \frac{f(\alpha x_1, \beta y_1)}{f(\alpha x_2, \beta y_2)}$$

And we get,

$$\left.\begin{aligned} f(x, y) &= C_1 x^a \\ f(x, y) &= C_2 y^b \end{aligned}\right\} \Longrightarrow f(x, y) = C x^a y^b$$

7.7.3 Dimensional Analysis

As all students of science and engineering know, equations must be dimensionally homogeneous; that is, all terms in an equation must have the same dimensions—one cannot add apples and oranges. This simple observation forms the basis of what is called dimensional analysis.

What we are going to do is to consider some examples and for each example, we will use dimensional analysis to derive formula without actually solving the problem. While doing these examples, pay attention to any pattern. We shall then rediscover the Buckingham Pi theorem.

Example 7.1 The spring-mass system (Sect. 7.8.1) has only two quantities: the spring stiffness k with dimension $[FL^{-1}]$ and the mass m with dimension $[M]$. We know that the dimension of force is $[F] = [MLT^{-2}]$. Thus, k has a dimension of $[MT^{-2}]$. We also know that the dimension of ω_0 is $[T^{-1}]$. As this quantity is a function of m and k, we have (from the power law above)

$$\omega_0 = C m^a k^b$$

where a, b are so determined that the dimension of both sides be the same:

$$[\omega_0] = C[M^a][M^b T^{-2b}] \Longrightarrow [T^{-1}] = [M^{a+b} T^{-2b}]$$

And this gives us the following system of two linear equations to solve for a and b

$$\left.\begin{aligned} a+b&=0\\ -2b&=-1 \end{aligned}\right\} \Longrightarrow a=-1/2, \quad b=1/2$$

Thus, we obtain the formula for the angular frequency without actually solving the equation,

$$\omega_0 = C\sqrt{k/m}$$

But dimensional analysis cannot give us the value of C. For that we can either solve the problem (which is usually hard) or do an experiment. It is interesting to rewrite the above equation as

$$C = \omega_0\sqrt{m/k}$$

The number $\omega_0\sqrt{m/k}$ is called a dimensionless group. Furthermore, as it is a dimensionless number, its value is invariant under change of units. Thus, it is called a universal constant.

In summary, this example has **three** independent dimensional quantities and they need **two** fundamental dimensions ($[M]$ and $[T]$). The solution shows that there exists **one** dimensionless group.

Example 7.2 For example, suppose we want to work out how the flow Q of an ideal fluid through a hole of diameter D depends on the pressure difference Δp. It seems plausible that Q might also depend on the density of the fluid ρ, so we look for a relationship of the form:

$$Q = kD^a(\Delta p)^b\rho^c$$

Now, we write the dimensions of all quantities involved

$$[\rho] = [ML^{-3}], \ \ [D] = [L], \ \ [\Delta p] = [ML^{-1}T^{-2}], \ \ [Q] = [L^3T^{-1}]$$

Hence, the equations are

$$[L^3T^{-1}] = [L^aM^bL^{-b}T^{-2b}M^cL^{-3c}] \Longrightarrow \begin{cases} a-b-3c=3 \\ b+c=0 \\ -2b=-1 \end{cases} \Longrightarrow \begin{bmatrix} a\\ b\\ c \end{bmatrix} = \begin{bmatrix} 2\\ 0.5\\ -0.5 \end{bmatrix}$$

Thus, we obtain Q without having to actually solve the problem,

$$Q = kD^2\sqrt{\frac{\Delta p}{\rho}} \Longrightarrow \frac{Q}{D^2}\sqrt{\frac{\rho}{\Delta p}} = k$$

In summary, this example has **four** independent dimensional quantities and they need **three** fundamental dimensions ([M], [L] and [T]). The solution shows that there exists **one** dimensionless group.

Example 7.3 In the previous example, we considered only an ideal fluid, i.e., a fluid with zero viscosity. Now, suppose that we're dealing with a viscous fluid if the viscosity μ of dimension $[L^2T^{-1}]$. Now, Q is given by:

$$Q = kD^a(\Delta p)^b \rho^c \mu^d$$

Hence, the equations are

$$[L^3T^{-1}] = [L^a M^b L^{-b} T^{-2b} M^c L^{-3c} L^{2d} T^{-d}]$$

which results in the following system of linear equations (three equations for four unknowns)

$$\begin{cases} a - b - 3c + 2d = 3 \\ b + c = 0 \\ -2b - d = -1 \end{cases} \iff \begin{bmatrix} 1 & -1 & -3 & 2 \\ 0 & 1 & 1 & 0 \\ 0 & -2 & 0 & -1 \end{bmatrix} \begin{bmatrix} a \\ b \\ c \\ d \end{bmatrix} = \begin{bmatrix} 3 \\ 0 \\ -1 \end{bmatrix}$$

Using linear algebra from Chap. 9, the rank of the matrix associated to the above system is three, and the system has one free variable. Let's choose b as the free variable, we can solve for a, c, d in terms of b:

$$c = -b, \quad d = 1 - 2b, \quad a = 1 + 2b$$

Thus, Q is written as

$$Q = kD^{1+2b}(\Delta p)^b \rho^{-b} \mu^{1-2b} \tag{7.7.2}$$

If the pattern we observe from the previous two examples still works, we should have **two** dimensionless groups. This is so because there are **five** independent dimensional quantities and they need **three** fundamental dimensions ([M], [L] and [T]). Indeed, we have two dimensionless groups (highlighted red):

$$\frac{Q}{D\mu} = k\left(\frac{D^2 \Delta p}{\rho \mu^2}\right)^b \tag{7.7.3}$$

From the presented three examples, we can observe that there exists a relationship between the number of quantities, the number of fundamental dimensions and the number of dimensionless groups. Now, we need to prove it. Instead of a general proof, we consider Example 7.3 and prove that there must be one dimensionless number in this example. First, we write the dimensions of all quantities involved, but we have to explicitly write the powers of $[M]$, $[L]$ and $[T]$. For example, for $[\rho] = [M^1L^{-3}]$ (with no time), we write $[\rho] = [M^1L^{-3}T^0]$:

$$\begin{aligned}[\rho] &= [M^1L^{-3}T^0],\ [D] = [M^0L^1T^0],\ [\Delta p] = [M^1L^{-1}T^{-2}]\\ [Q] &= [M^0L^3T^{-1}],\ [\mu] = [M^0L^2T^{-1}]\end{aligned} \tag{7.7.4}$$

Now, suppose we can build a dimensionless number C of the form (power law)

$$C = \rho^{x_1} D^{x_2} \Delta p^{x_3} Q^{x_4} \mu^{x_5}$$

Noting that as C is dimensionless, its dimension is $[C] = [M^0L^0T^0]$ (or $[C] = 1$). And this leads to the following homogeneous system of equations:

$$\underbrace{\begin{bmatrix} 1 & 0 & 1 & 0 & 0\\ -3 & 1 & -1 & 3 & 2\\ 0 & 0 & -2 & -1 & -1\end{bmatrix}}_{\mathbf{A}} \begin{bmatrix} x_1\\ x_2\\ x_3\\ x_4\\ x_5\end{bmatrix} = \begin{bmatrix} 0\\ 0\\ 0\end{bmatrix} \tag{7.7.5}$$

where the first row of $\mathbf{A}$ is the powers of $[M]$ in Eq. (7.7.4). The second row is the powers of $[L]$ and so on. Thus, this matrix is called the *dimension matrix*. Are we going to solve the system in Eq. (7.7.5)? No no. That is the power of mathematics. Using the rank theorem, Theorem 9.5.4, from linear algebra which says rank($\mathbf{A}$) + nullity($\mathbf{A}$) = 5 and the fact that rank($\mathbf{A}$) = 3, we deduce that nullity($\mathbf{A}$) = 2, hence it has two solutions $\boldsymbol{x} \neq \mathbf{0}$. Therefore, we have two dimensionless numbers.

Hey, but why the rank of the dimension matrix $\mathbf{A}$ is three not less? If we use the Gauss-Jordan elimination to get the row reduced echelon form of $\mathbf{A}$, we get as the first three columns the three unit vectors of $\mathbb{R}^3$: $(1, 0, 0)$, $(0, 1, 0)$, $(0, 0, 1)$. This makes us to think of a three-dimensional space. Indeed, the three independent dimensions $[M]$, $[L]$ and $[T]$ makes a three-dimensional vector space. In this vector space, a dimensional quantity x has the coordinate vector (x_1, x_2, x_3) because we always can write

$$[x] = [M^{x_1}L^{x_2}T^{x_3}] \tag{7.7.6}$$

Now, we make another observation. In Example 7.3 the relation between the different quantities can be written in the form of Eq. (7.7.2) in terms of dimensional variables or in the equivalent form of Eq. (7.7.3) that involves only dimensionless variables. We now try to prove this. First, we need to solve Eq. (7.7.5):

$$\boldsymbol{x} = u \begin{bmatrix} 1/2 \\ -2 \\ -1/2 \\ 1 \\ 0 \end{bmatrix} + v \begin{bmatrix} 1/2 \\ -1 \\ -1/2 \\ 0 \\ 1 \end{bmatrix}$$

which allows us to find two dimensionless variables, denoted by π_1 and π_2[20]:

$$\begin{cases} \pi_1 = \rho^{1/2} D^{-2} \Delta p^{-1/2} Q \\ \pi_2 = \rho^{1/2} D^{-1} \Delta p^{-1/2} \mu \end{cases} \Longrightarrow \begin{cases} Q = \rho^{-1/2} D^2 \Delta p^{1/2} \pi_1 \\ \mu = \rho^{-1/2} D^1 \Delta p^{1/2} \pi_2 \end{cases}$$

Now, suppose that the physical law we're seeking for is given by

$$f(\rho, D, \Delta p, Q, \mu) = 0 \tag{7.7.7}$$

which can be rewritten as Q and μ can be replaced by π_1 and π_2:

$$f(\rho, D, \Delta p, Q, \mu) = f(\rho, D, \Delta p, \rho^{-1/2} D^2 \Delta p^{1/2} \pi_1, \rho^{-1/2} D^1 \Delta p^{1/2} \pi_2) = 0$$

Now we introduce a new function G depending on ρ, D, Δp, π_1, π_2:

$$G(\rho, D, \Delta p, \pi_1, \pi_2) := f(\rho, D, \Delta p, \rho^{-1/2} D^2 \Delta p^{1/2} \pi_1, \rho^{-1/2} D^1 \Delta p^{1/2} \pi_2) = 0 \tag{7.7.8}$$

Now, we can choose a particular set of units such that ρ, D, Δp have unit values, then we have $G(1, 1, 1, \pi_1, \pi_2) = 0$, which can be rewritten as $F(\pi_1, \pi_2) = 0$. Thus, we have rediscovered the following theorem.

Theorem 7.1 *The Buckingham Pi theorem*
Let

$$f(q_1, q_2, \ldots, q_m) = 0$$

be a unit free physical law that relates the dimensional quantities $q_1, \ldots, q_m$. Let $L_1, \ldots, L_n$, $n < m$, be fundamental dimensions with

$$[q_i] = [L_1^{a_{1i}} L_2^{a_{2i}} \ldots L_n^{a_{ni}}], \quad i = 1, 2, \ldots, m$$

and let $r = rank(\mathbf{A})$, where $\mathbf{A}$ is the dimension matrix given by

[20] The dimensionless combinations that we can make in a given problem are not unique: if π_1 and π_2 are both dimensionless, then so are $\pi_1 \pi_2$ and $\pi_1 + \pi_2$ and, indeed, any function that we want to make out of these two variables.

$$\mathbf{A} = \begin{bmatrix} a_{11} & a_{12} & \cdots & a_{1m} \\ a_{21} & a_{22} & \cdots & a_{2m} \\ \vdots & \vdots & \ddots & \vdots \\ a_{n1} & a_{n2} & \cdots & a_{nm} \end{bmatrix}$$

Then there exists $m - r$ independent dimensionless quantities $\pi_1, \pi_2, \ldots, \pi_{m-r}$ that can be formed from $q_1, \ldots, q_m$, and the physical law $f(q_i) = 0$ is equivalent to an equation

$$F(\pi_1, \pi_2, \ldots, \pi_{m-r}) = 0$$

expressed only in terms of the dimensionless quantities.

Note that if the chosen fundamental dimensions are independent, then r is simply the number of these fundamental dimensions.

7.7.4 Scaling of ODEs

First-order ODEs. We consider the following ODE:

$$a\dot{x} + bx = Af(t) \tag{7.7.9}$$

In the first step, we replace all dependent and independent variables by non-dimensionless counterparts. In this problem, we have two variables namely $x(t)$ and t, they are replaced by

$$\tilde{x} = \frac{x}{x_c}, \quad \tilde{t} = \frac{t}{t_c} \tag{7.7.10}$$

Of course, x_c has the same dimension as x and t_c is of the same dimension as t. That's why $\tilde{x}$ and $\tilde{t}$ are non-dimensionless variables.

Now, we write the old variables in terms of the new ones,

$$x = x_c\tilde{x}, \quad t = t_c\tilde{t} \tag{7.7.11}$$

Also the first derivative,

$$\dot{x} = \frac{dx}{dt} = \frac{dx}{d\tilde{t}}\frac{d\tilde{t}}{dt} = x_c\frac{d\tilde{x}}{d\tilde{t}}\frac{1}{t_c} \tag{7.7.12}$$

Thus, the original ODE Eq. (7.7.9) becomes

$$\frac{ax_c}{t_c}\frac{d\tilde{x}}{d\tilde{t}} + bx_c\tilde{x} = Af(t_c\tilde{t}) \tag{7.7.13}$$

Next, we divide the equation by the coefficient of the highest derivative term (red term):

$$\frac{d\tilde{x}}{d\tilde{t}} + \frac{bt_c}{a}\tilde{x} = \frac{At_c}{ax_c} f(t_c\tilde{t})$$

It is time to select x_c and t_c, so that there are less parameters possible. So, we select them so that the red/blue terms in the above equation are unity:

$$\begin{cases} \dfrac{bt_c}{a} = 1 \Longrightarrow t_c = \dfrac{a}{b} \\ \dfrac{At_c}{ax_c} = 1 \Longrightarrow x_c = \dfrac{A}{b} \end{cases} \tag{7.7.14}$$

Finally, the original ODE Eq. (7.7.9) containing three parameters, now becomes

$$\frac{d\tilde{x}}{d\tilde{t}} + \tilde{x} = F(\tilde{t}) \tag{7.7.15}$$

which is a dimensionless ODE with no parameter!

Differential operator. As a preparation for a discussion of second-order ODE in which we need to compute $\ddot{x}$, we introduce the differential operator $\frac{d}{dt}$, which we need to supply a function to compute its time derivative:

$$\frac{d}{dt} = \frac{d}{d\tilde{t}}\frac{d\tilde{t}}{dt} = \frac{1}{t_c}\frac{d}{d\tilde{t}} \tag{7.7.16}$$

The usefulness of this operator comes in when we compute the second derivative operator:

$$\begin{aligned} \frac{d^2}{dt^2} &= \frac{d}{dt}\left(\frac{d}{dt}\right) = \frac{d}{dt}\left(\frac{1}{t_c}\frac{d}{d\tilde{t}}\right) \quad \text{(use (7.7.16))} \\ &= \frac{1}{t_c}\frac{d}{d\tilde{t}}\left(\frac{1}{t_c}\frac{d}{d\tilde{t}}\right) = \frac{1}{t_c^2}\frac{d^2}{d\tilde{t}^2} \end{aligned} \tag{7.7.17}$$

and we applied Eq. (7.7.16) to the function $\frac{1}{t_c}\frac{d}{d\tilde{t}}$ in the third equality.

Second-order ODEs. Consider this second-order ODE

$$a\ddot{x} + b\dot{x} + cx = Af(t), \quad x(0) = x_0, \quad \dot{x}(0) = v_0 \tag{7.7.18}$$

Using the scaled quantities defined in Eq. (7.7.11) and the differential operators introduced previously, this equation becomes

$$\frac{ax_c}{t_c^2}\frac{d^2\tilde{x}}{d\tilde{t}^2} + \frac{bx_c}{t_c}\frac{d\tilde{x}}{d\tilde{t}} + cx_c\tilde{x} = Af(t_c\tilde{t})$$

Dividing it by the coefficient of the second derivative, we get this equation:

$$\frac{d^2\tilde{x}}{d\tilde{t}^2} + \frac{bt_c}{a}\frac{d\tilde{x}}{d\tilde{t}} + \frac{ct_c^2}{a}\tilde{x} = \frac{At_c^2}{ax_c}f(t_c\tilde{t})$$

We have two choices here: selecting t_c so that the coefficient of either the second term or the third term unity. We chose the latter[21]:

$$\frac{ct_c^2}{a} = 1 \Longrightarrow t_c = \sqrt{\frac{a}{c}} \tag{7.7.19}$$

And making $At_c^2/ax_c = 1$ gives us $x_c = A/c$.

$$\frac{d^2\tilde{x}}{d\tilde{t}^2} + \frac{b}{\sqrt{ac}}\frac{d\tilde{x}}{d\tilde{t}} + \tilde{x} = F(\tilde{t}).$$

7.8 Harmonic Oscillation

> Oscillations are ubiquitous in nature and play a crucial role in many physical phenomena, from the motion of celestial bodies to the vibrations of atoms and molecules.
>
> David Griffiths

Many kinds of motion repeat themselves over and over: the swinging pendulum of a grandfather clock and the back-and-forth motion of the pistons in a car engine. This kind of motion, called periodic motion or oscillation, is the subject of this section. Understanding periodic motion will be essential for the study of waves, sound, and light.

Observing a ball rolling back and forth in a round bowl or a pendulum that swings back and forth past its straight-down position (Fig. 7.4), we can see that a body that undergoes periodic motion always has a stable equilibrium position. When it is moved away from this position and released, a force or torque comes into play to pull it back towards equilibrium (such a force is called a *restoring force*). But by the time it gets there, it has picked up some kinetic energy, so it overshoots, stopping somewhere on the other side, and is again pulled back (by the restoring force) towards equilibrium.

When the restoring force is directly proportional to the displacement from equilibrium the oscillation is called simple harmonic motion, abbreviated SHM or simple harmonic oscillation (SHO). This section is confined to such oscillations.

---

[21] Only with hindsight we can do this. For a spring-mass system, this leads to $t_c = \sqrt{m/k}$ which is proportional to the period of the oscillation of the mass.

Fig. 7.4 A pendulum

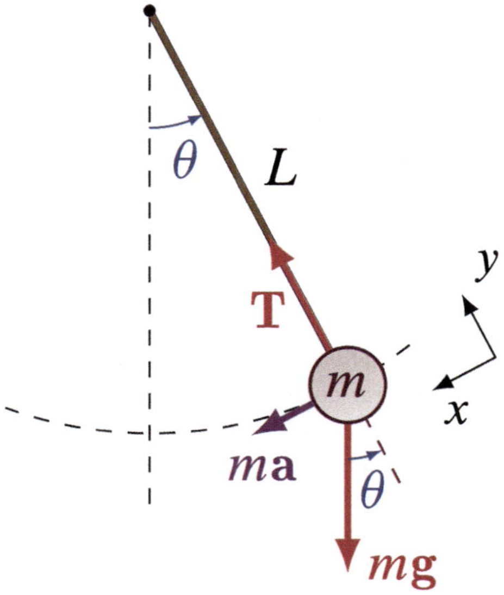

We start with the simple harmonic oscillation in Sect. 7.8.1 where we discuss the equation of motion of a spring-mass system, its solutions, and its natural frequency and period. Damped oscillations, i.e., oscillations that die out due to resistive forces are discussed in Sect. 7.8.2. Then, we present forced oscillations (those oscillations that require driving forces to maintain their motions) in Sect. 7.8.3. The discussion is confined to sinusoidal driving forces only. The phenomenon of resonance appears naturally in this context (Sect. 7.8.4). Force oscillations with any periodic driving forces are given in Sect. 7.8.5 where Fourier series are used. Sect. 7.8.6 discusses the oscillation of pendulum.

7.8.1 Simple Harmonic Oscillation

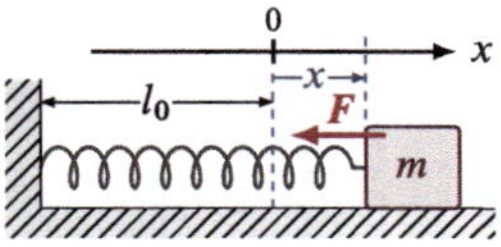

Consider a mass m attached to a weightless spring of stiffness k on a frictionless horizontal plane. Let's denote by O the equilibrium position of the mass; this is the position in which the spring is neither stretched nor compressed. Now, if we displace the mass to the right of O a distance x, the spring will try to pull it back by applying a force $-kx$ to the mass.[22] The minus sign is here to express the effect of pulling back: the force is always opposite the displacement vector. Thus, when the mass is at the left side of O the force is pointing to the right and thus the spring pushes the mass back to O. In this way we get harmonic oscillation.

Using Newton's second law we can write

$$\boxed{m\ddot{x} = -kx} \implies \ddot{x} + \omega_0^2 x = 0, \quad \omega_0^2 = \frac{k}{m} \tag{7.8.1}$$

where $\ddot{x} = d^2x/dt^2$. The notation ω_0^2 was introduced instead of ω_0 so that the maths (to be discussed) will be in a simple form. At this stage we do not know its meaning, its role is for notational convenience.

Assume that $x(t)$ is a solution of Eq. (7.8.1), then it is easy to see that $Ax(t)$ is also a solution with any $A > 0$ that is a constant. Now assume that we have two solutions to this equation, namely $x_1(t)$ and $x_2(t)$, which are independent of each other,[23] then $Ax_1(t) + Bx_2(t)$ is also a solution.[24] Actually as it contains two constants A, B it is the general solution to Eq. (7.8.1). Now we need to find two particular solutions and we are done. They are $\cos(\omega_0 t)$ and $\sin(\omega_0 t)$ which are the only functions that have the second derivatives equal minus the functions. Therefore, the general solution is[25]

$$x = A_1 \cos(\omega_0 t) + A_2 \sin(\omega_0 t) \tag{7.8.2}$$

with two constants A_1 and A_2 being real numbers. They are determined using the so-called *initial conditions*. The initial conditions specify the conditions of the system when we start the system. They include the initial position of the mass x_0 (which is $x(t)$ evaluated at $t = 0$, i.e., $x(0)$) and the initial velocity $v(0)$:

$$x(0) = A_1, \quad v(0) = \dot{x}(0) = \omega_0 A_2 \tag{7.8.3}$$

While the solution in Eq. (7.8.2) is perfectly fine, it does not immediately reveal the amplitude of the oscillation. Using the trigonometry identity $\cos(a - b) = \cos a \cos b + \sin a \sin b$, we can rewrite that equation in the following form

$$x = \sqrt{A_1^2 + A_2^2}\left(\frac{A_1}{\sqrt{A_1^2 + A_2^2}} \cos(\omega_0 t) + \frac{A_2}{\sqrt{A_1^2 + A_2^2}} \sin(\omega_0 t)\right) = A\cos(\omega_0 t - \phi) \tag{7.8.4}$$

[22] This is Hooke's law which is named after British physicist Robert Hooke (1635–1703). He first stated the law in 1676 as a Latin anagram. He published the solution of his anagram in 1678 as: ut tensio, sic vis ("as the extension, so the force" or "the extension is proportional to the force").

[23] For example $x_1(t) = \sin t$ and $x_2(t) = 5 \sin t$ are not independent. Refer to Chap. 9 for detail.

[24] You should verify this claim.

[25] We should ask why there can't be other solutions? To answer this question we need to use.

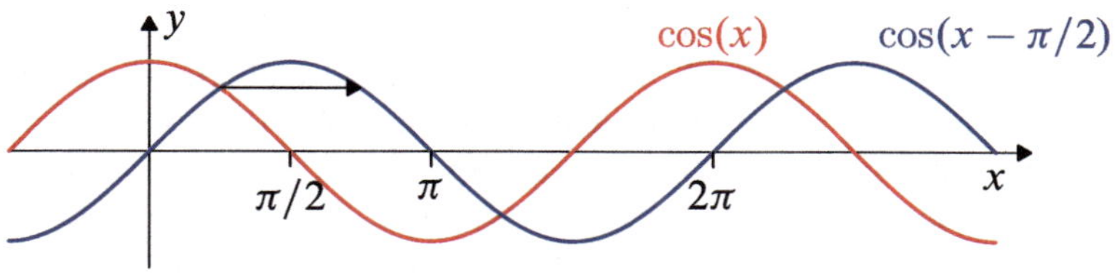

Fig. 7.5 Phase shifted angle ϕ: illustrated for $\phi = \pi/2$. The plot for $\phi = \pi/2$ is shifted to the right because it requires a larger time to achieve the same position as the $\phi = 0$ case. So a given value of y occurs later in the $\phi = \pi/2$ plot, which means that it is shifted to the right

where $A = \sqrt{A_1^2 + A_2^2}$ is the amplitude of the oscillation, i.e., the maximum displacement of the mass from equilibrium, either in the positive or negative direction. And ϕ is called *phase-shifted angle*, see Fig. 7.5. If needed, we can relate A and ϕ to A_1 and A_2: $A = \sqrt{A_1^2 + A_2^2}$ and $\cos\phi = A_1/A = x(0)/A$. Therefore, ϕ is dependent on when you pick the $t = 0$ time to be.

Simple harmonic motion is repetitive: its motion repeats after a multiple of a period. The period T is the time it takes the mass to complete one oscillation and return to the starting position. Everyone should be familiar with the period of orbit for the Earth around the Sun, which is approximately 365 d; it takes 365 d for the Earth to complete a cycle. We can find the formula for T based on this definition: the position of the mass at time t is exactly the position at time $t + T$; that is $A\cos(\omega_0(t + T) - \phi) = A\cos(\omega_0 t - \phi)$. So,

$$\omega_0(t + T) = \omega_0 t + 2\pi \Longrightarrow \boxed{T = \frac{2\pi}{\omega_0} = 2\pi\sqrt{\frac{m}{k}}} \tag{7.8.5}$$

The unit of T is second in the SI system.

Next, we mention a related quantity named frequency, usually denoted by f. Frequency helps to answer how often something happens (e.g., how many visits per day). In the case of SHO, it measures how many cycles per unit time is. There is a relation between the period T and the frequency f. To derive this relation, one example suffices. If it takes 0.1 s for one cycle (i.e., $T = 0.1$ s), there will be then 10 cycles per second. Thus,

$$f = \frac{1}{T} = \frac{\omega_0}{2\pi} \tag{7.8.6}$$

In the SI system, the unit of f is *cycles per second* or Hertz in honor of the first experimenter[26] with radio waves (which are electric vibrations). For example, if a sound wave has a frequency of 440 Hz, it means that the wave completes 440 cycles per second.

---

[26] Heinrich Rudolf Hertz (22 February 1857–1 January 1894) was a German physicist who first conclusively proved the existence of the electromagnetic waves predicted by James Clerk Maxwell's equations of electromagnetism.

While f is referred to as frequency, ω_0 is called *angular frequency*. It is such called because $\omega_0 = 2\pi f$ with the unit of radians per second. There is no circle but why angular frequency? There is a circle hidden here. Whenever we deal with sine and cosine we are dealing with the complex exponential, which in turn involves the unit circle. See Fig. 7.6 for detail. Later on, we will call ω_0 the *natural frequency of the system* when the mass is driven by a cyclic force with yet another frequency ω.

Question 15 *From Eq. (7.8.5) we can see that the period T does not depend on A: it does not matter how far the mass is from equilibrium, it will take the same time to complete one cycle. But why?*

Solution using a complex exponential. As it is more convenient to work with the exponential function than with the sine/cosine functions, we use a complex exponential to solve the SHO problem. But as $x(t)$ is real not imaginary, we use complex numbers to simplify the mathematics, and we will take $x(t)$ as the real part of the complex solution. Using complex exponential, we write $x(t)$ as[27]

$$x(t) = C_1 e^{i\omega_0 t} + C_2 e^{-i\omega_0 t}, \quad C_1, C_2 \in \mathbb{C} \tag{7.8.7}$$

Using $e^{i\theta} = \cos\theta + i\sin\theta$ in Eq. (7.8.7) and compare with Eq. (7.8.2), we can relate C_1, C_2 with $A_{1,2}$:

$$\begin{aligned} C_1 + C_2 &= A_1 \\ i(C_1 - C_2) &= A_2 \end{aligned} \tag{7.8.8}$$

Solving Eq. (7.8.8) for C_1 and C_2, we get

$$C_1 = 1/2(A_1 - iA_2), \quad C_2 = 1/2(A_1 + iA_2) \tag{7.8.9}$$

which indicates that C_2 is simply the complex conjugate of C_1: $C_2 = \bar{C}_1$. Now, we can proceed with Eq. (7.8.7) where C_2 is replaced by $\bar{C}_1$[28]:

$$\begin{aligned} x(t) &= C_1 e^{i\omega_0 t} + \bar{C}_1 e^{-i\omega_0 t} \\ &= 2\,\text{Re}[C_1 e^{i\omega_0 t}] \quad (\bar{C}_1 e^{-i\omega_0 t} \text{ is the conjugate of } C_1 e^{i\omega_0 t}) \\ &= \text{Re}[2C_1 e^{i\omega_0 t}] \quad (\text{with } 2C_1 = A_1 - iA_2 = Ae^{-i\phi}, \text{Fig. 7.6a}) \\ &= \text{Re}[Ae^{-i\phi} e^{i\omega_0 t}] = \text{Re}[Ae^{i(\omega_0 t - \phi)}] = A\cos(\omega_0 t - \phi) \end{aligned} \tag{7.8.10}$$

And we got the same result as in Eq. (7.8.4), of course.

We all know that the complex number $Ae^{i(\omega_0 t - \phi)}$ is a point on a circle of radius A on the complex plane (Fig. 7.6b). As we increase t this point travels along this

---

[27] This is so because $e^{i\omega_0 t}$ and $e^{-i\omega_0 t}$ are two solutions of Eq. (7.8.1), thus any linear combinations of them is also a solution.

[28] If not clear, check Sect. 2.25 on complex conjugate rules, particularly $\bar{u}\bar{w} = \overline{uw}$.

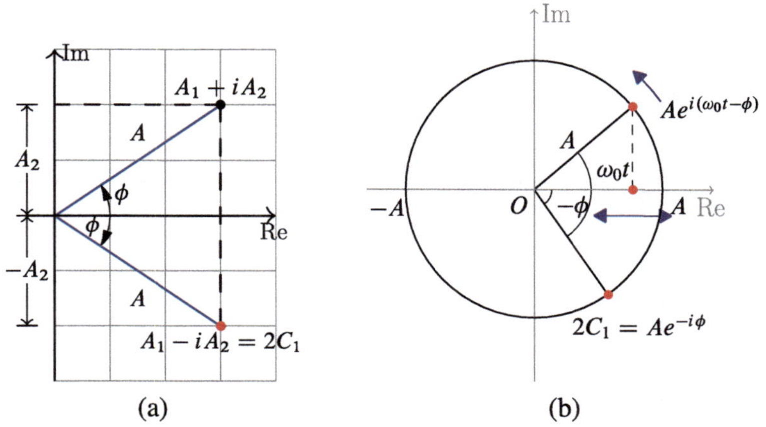

Fig. 7.6 Solving SHO using a complex exponential: the complex number $Ae^{i(\omega_0 t-\phi)}$ moves counterclockwise with angular velocity ω_0 around a circle of radius A. Its real part, $x(t)$, is the projection of the complex number onto the real axis. While the complex number goes around the circle, this projection oscillates back and forth on the x-axis

circle counterclockwise. While it is doing so, its real part, which is $A\cos(\omega_0 t - \phi)$, is moving forth and back on the real axis in the interval $[-A, A]$.

Geometric meaning of Euler's identity. Recall that we have derived Euler's identity $e^{i\pi} + 1 = 0$ in Eq. (2.25.19). Now, we can give a geometric meaning to it. Referring to Fig. 7.6 but with $A = 1$ (unit circle) and $\phi = 0$. The complex number $e^{i\omega_0 t}$ is circulating the unit circle. When $\omega_0 t = \pi$, it has traveled half of the circle and arrive at the point $(-1, 0)$ or -1. And thus $e^{i\pi} = -1$.

Plots of displacement, velocity and acceleration. To verify whether our solutions agree with our intuitive understanding of a SHO, we analyze the displacement $x(t)$, the velocity $\dot{x}$ and the acceleration $\ddot{x}$ for $A_1 = 1.0$ and $A_2 = 0.0$. That is we displace the mass (from the equilibrium) to the right a distance of A_1 and release it. The plots of x, $\dot{x}$ and $\ddot{x}$ are shown in Fig. 7.7. The expression for the velocity and acceleration are

$$x = A\cos(\omega_0 t - \phi) \Longrightarrow \dot{x} = -A\omega_0 \sin(\omega_0 t - \phi), \quad \ddot{x} = -A\omega_0^2 \cos(\omega_0 t - \phi)$$

The mass goes to the left with an increasing velocity (and acceleration). When it reaches the equilibrium point, the velocity is maximum (and so is the kinetic energy). It continues moving to the left until it reaches $-A$ at $t = 0.5$, at that point the velocity is zero (and the potential energy is maximum).

Knowing the velocity of the mass, we can now explain non-rigourously why T is independent of A. On one hand, the mass will move more slowly when it has a

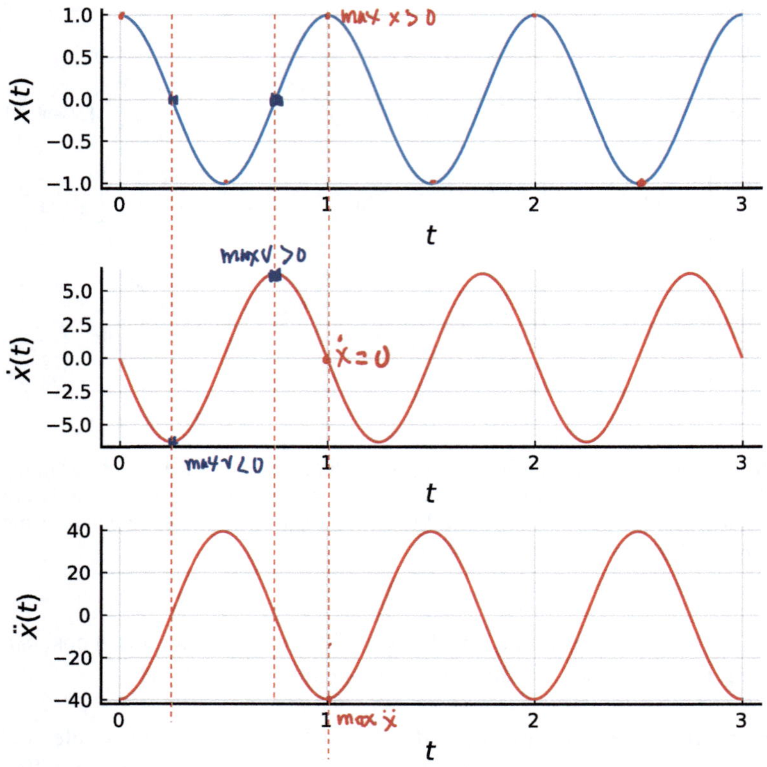

Fig. 7.7 SHO with $x = A\cos\omega t$: plots of displacement, velocity and acceleration. The frequency is $\omega_0 = 2\pi$ so that $T = 1$. The amplitude is $A = 1$

smaller amplitude. On the other hand, with a smaller amplitude the mass does not have as far to travel to complete one cycle. The two effects cancel out each other.

Energy conservation. Let's now compute the kinetic and potential energy of the SHO and see about energy conservation. The kinetic energy KE and potential energy U are given by[29]

$$\begin{aligned} KE &= \frac{1}{2}m\dot{x}^2 = \frac{1}{2}kA^2\sin^2(\omega_0 t - \phi) \\ U &= \frac{1}{2}kx^2 = \frac{1}{2}kA^2\cos^2(\omega_0 t - \phi) \end{aligned} \tag{7.8.11}$$

From that energy conservation is easily seen: $KE + U = {}^1\!/_2 kA^2$. Why the total energy is ${}^1\!/_2 kA^2$? This makes sense, because ${}^1\!/_2 kA^2$ is the potential energy of the spring when it is stretched the maximum amount (and so the mass is instantaneously at rest). It's useful to plot the evolution of the energies in time (Fig. 7.8a) to see the

[29] Don't forget that $\omega_0^2 = k/m$.

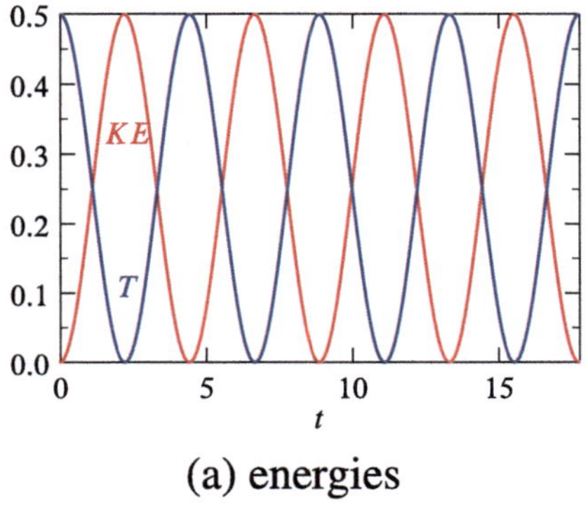

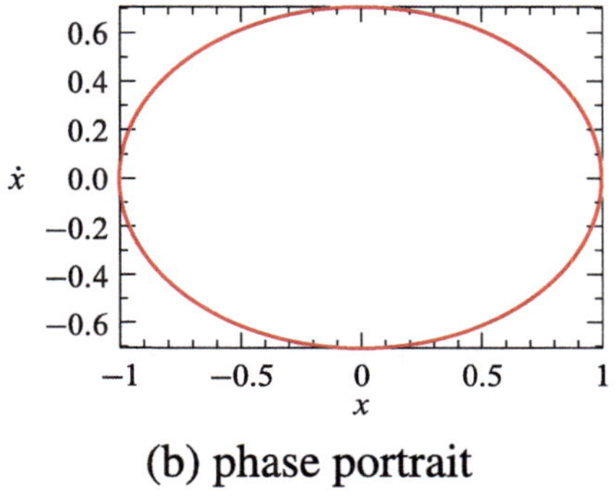

Fig. 7.8 Simple harmonic oscillator: energy conservation

exchange between kinetic and potential energies. The energy sloshes back and forth between kinetic and potential. It is all potential at the points of maximum stretching, and it is all kinetic when the mass passes through the origin. In that plot, I've used $A = 1, \phi = 0, k = 1$, and $m = 2$.

This energy conservation also gives us one more thing[30]:

$$\frac{1}{2}m\dot{x}^2 + \frac{1}{2}kx^2 = \frac{1}{2}kA^2 \Longrightarrow \boxed{\frac{x^2}{A^2} + \frac{\dot{x}^2}{(\omega_0 A)^2} = 1}$$

What is the boxed equation? It is an ellipse! So, on the $x - \dot{x}$-plane—which is called the phase plane—the trajectory of the mass is an ellipse (Fig. 7.8b). Think about it: we are dealing with a mass moving on a line, but we have a circle if we use complex numbers to study this problem, and we also met an ellipse if we use the phase plane. That's simply remarkable.

7.8.2 Damped Oscillator

We know that in reality, a spring won't oscillate forever. Frictional forces will diminish the amplitude of oscillation until eventually the system is at rest. We will now add frictional forces to the mass and spring. Imagine that the mass is put in a liquid like molasses. Only friction that is proportional to the velocity is considered. This is a pretty good approximation for a body moving at a low velocity in air, or in a liquid. So we say that the frictional force is $-b\dot{x}$. The constant $b > 0$ depends on the kind of liquid the mass is in. The negative sign, just says that the force is in the opposite direction to the body's motion. Now Newton's second law gives us the equation of motion with friction:

[30] Don't forget that $\omega_0^2 = k/m$.

$$m\ddot{x} = -b\dot{x} - kx \Longrightarrow \ddot{x} + 2\beta\dot{x} + \omega_0^2 x = 0, \quad \omega_0^2 = \frac{k}{m}, \quad 2\beta = \frac{b}{m} \tag{7.8.12}$$

We are going to use complex numbers to solve Eq. (7.8.12). Let's consider $z(t) = e^{i\omega t}$, which is the solution to the following equation

$$\ddot{z} + 2\beta\dot{z} + \omega_0^2 z = 0, \quad z = e^{i\omega t} \tag{7.8.13}$$

Now comes the reason why we used complex numbers: the derivatives of an exponential function is the product of the function and a constant! Indeed,

$$\begin{aligned} z &= e^{i\omega t} \\ \dot{z} &= i\omega e^{i\omega t} = i\omega z \\ \ddot{z} &= -\omega^2 e^{i\omega t} = -\omega^2 z \end{aligned} \tag{7.8.14}$$

Substituting z, $\dot{z}$ and $\ddot{z}$ into Eq. (7.8.13), we get the following equation

$$z(-\omega^2 + 2\beta i\omega + \omega_0^2) = 0 \tag{7.8.15}$$

which is valid for all t. Thus,

$$-\omega^2 + 2\beta i\omega + \omega_0^2 = 0 \tag{7.8.16}$$

which is a quadratic equation for ω. Solving this equation, we get:

$$\omega = i\beta \pm \sqrt{\omega_0^2 - \beta^2} \tag{7.8.17}$$

Now, we get different solutions depending on the sign of the term under the square root. In what follows, we discuss these solutions.

Weakly damped is the case when $\omega_0 > \beta$. By setting $\omega_0^d = \sqrt{\omega_0^2 - \beta^2}$, we have $\omega = i\beta \pm \omega_0^d$. So, $z = e^{i\omega t}$ is written as

$$z = e^{i\omega t} = e^{i(i\beta \pm \omega_0^d)t} = e^{(-\beta \pm i\omega_0^d)t} = e^{-\beta t} e^{\pm i\omega_0^d t} \tag{7.8.18}$$

These are two particular solutions of Eq. (7.8.13): $z_1 = e^{-\beta t} e^{i\omega_0^d t}$ and $z_2 = e^{-\beta t} e^{-i\omega_0^d t}$. Thus, the general complex solution is

$$z = C_1 e^{-\beta t} e^{i\omega_0^d t} + C_2 e^{-\beta t} e^{-i\omega_0^d t} = e^{-\beta t} \underbrace{(C_1 e^{i\omega_0^d t} + C_2 e^{-i\omega_0^d t})}_{z_0} \tag{7.8.19}$$

where C_1 and C_2 are two complex numbers. Now, we have to express z in the form $x + iy$, so that we can get the real part of it, which is the solution we are seeking of. We write z_0 as

$$
\begin{aligned}
z_0 &= [\text{Re}(C_1) + i\,\text{Im}(C_1]\left[\cos(\omega_0^d t) + i\sin(\omega_0^d t)\right] \\
&+ [\text{Re}(C_2) + i\,\text{Im}(C_2)]\left[\cos(\omega_0^d t) - i\sin(\omega_0^d t)\right] \\
&= \underbrace{(\text{Re}(C_1) + \text{Re}(C_2))}_{A}\cos(\omega_0^d t) + \underbrace{(\text{Im}(C_2) - \text{Im}(C_1))}_{B}\sin(\omega_0^d t) \\
&+ i(\ldots)
\end{aligned} \tag{7.8.20}
$$

The solution $x(t)$ is the real part of z, thus it is given by

$$
x(t) = \text{Re}\, z(t) = e^{-\beta t}\left[A\cos(\omega_0^d t) + B\sin(\omega_0^d t)\right] = e^{-\beta t} C\cos(\omega_0^d t + \theta) \tag{7.8.21}
$$

where $A, B, C \in \mathbb{R}$. Of course, we can also write the solution as $Ce^{-\beta t}\cos(\omega_0^d t - \theta)$. Is this solution correct or at least plausibly correct? To answer that question is simple: put $\beta = 0$—which is equivalent to $b = 0$—into $x(t)$ and if that $x(t)$ has the same form of the undamped solution, then $x(t)$ is ok. This can be checked to be the case, furthermore the term $e^{-\beta t}$ is indeed a decay term: the oscillation has to come to a stop due to friction.

Example 7.4 Let's consider one example with $\omega_0 = 1$, $\beta = 0.05$, $x_0 = 1.0$, $v_0 = 3.0$. We need to compute C and θ using the initial conditions. Using Eq. (7.8.21), we have

$$
\begin{aligned}
x_0 &= x(t = 0) = C\cos\theta \\
v_0 &= \dot{x}(t = 0) = -C\beta\cos(\theta) - C\omega_0^d\sin(\theta)
\end{aligned}
$$

From that, we get C and θ:

$$
\begin{aligned}
C &= \frac{x_0}{\cos\theta} \\
\theta &= \arctan\left(-\frac{v_0 + \beta x_0}{\omega_0^d x_0}\right)
\end{aligned}
$$

Now, we can plot $x(t)$ using Eq. (7.8.21) (Fig. 7.9). The code is given in Listing A.11.

Over damped is the case when $\omega_0 < \beta$. In this case, $\omega = i\beta \pm i\sqrt{\beta^2 - \omega_0^2} = i(\beta \pm \bar{\omega})$, $\bar{\omega} = \sqrt{\beta^2 - \omega_0^2}$.

$$
\left.\begin{aligned}
z_1 &= e^{i\omega_1 t} = e^{(-\beta-\bar{\omega})t} \\
z_2 &= e^{i\omega_2 t} = e^{(-\beta+\bar{\omega})t}
\end{aligned}\right\} \Longrightarrow z(t) = C_1 e^{(-\beta-\bar{\omega})t} + C_2 e^{(-\beta+\bar{\omega})t} \tag{7.8.22}
$$

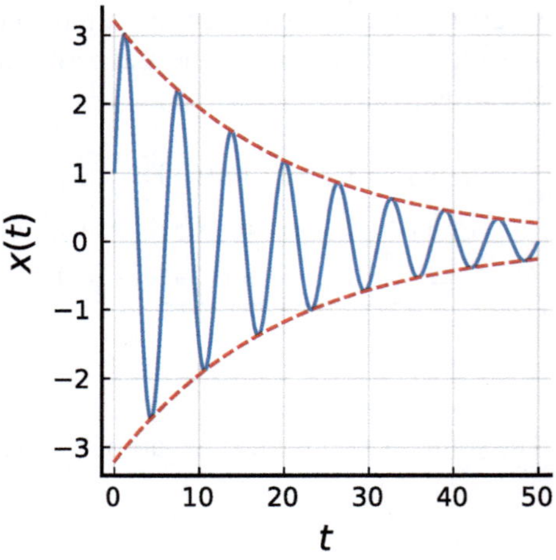

Fig. 7.9 Weakly damped oscillation can be seen as simple harmonic oscillations with an exponentially decaying amplitude $Ce^{-\beta t}$. The dashed curves are the maximum amplitudes envelop $\pm Ce^{-\beta t}$

7.8.3 Driven Damped Oscillation

Everyone knows that a swing will stop after awhile unless its motion is maintained by a parent keeps pushing it. This section studies such *forced* or *driven* oscillations. If we consider a sinusoidal driving force $f(t) = F_0 \cos(\omega t)$, the equation is given by

$$m\ddot{x} + b\dot{x} + kx = F_0 \cos(\omega t) \tag{7.8.23}$$

There are two main reasons for the importance of sinusoidal driving forces. First, there are many important systems in which the driving force is sinusoidal. The second reason is subtler. It turns out that any periodic force can be built up from sinusoidal forces using Fourier series. Therefore, before we can tackle a general driving force, we must first understand the case of sinusoidal driving forces.

Equation (7.8.23) can be rewritten as follows

$$\ddot{x} + 2\beta\dot{x} + \omega_0^2 x = f_0 \cos(\omega t), \quad \omega_0^2 = \frac{k}{m}, \quad 2\beta = \frac{b}{m}, \quad f_0 = \frac{F_0}{m} \tag{7.8.24}$$

We are going to solve this equation using a complex function $z(t) = x(t) + iy(t)$ satisfying Eq. (7.8.24):

$$\ddot{z} + 2\beta\dot{z} + \omega_0^2 z = f_0 e^{i\omega t} \tag{7.8.25}$$

It can be seen that the real part of $z(t)$, i.e., $x(t)$ is actually the solution of Eq. (7.8.24). With $z = Ce^{i\omega t}$, we compute $\dot{z}, \ddot{z}$

$$z = Ce^{i\omega t}, \quad \dot{z} = i\omega Ce^{i\omega t}, \quad \ddot{z} = -\omega^2 Ce^{i\omega t} \tag{7.8.26}$$

and substitute them into Eq. (7.8.25) to get

$$-\omega^2 C + 2\beta i\omega C + \omega_0^2 C = f_0$$

which gives us C as follows

$$\begin{aligned}
C &= \frac{f_0}{\omega_0^2 - \omega^2 + 2i\omega\beta} = \frac{f_0(\omega_0^2 - \omega^2 - 2i\omega\beta)}{(\omega_0^2 - \omega^2)^2 + 4\omega^2\beta^2} \\
&= f_0^*(\omega_0^2 - \omega^2 - 2i\omega\beta), \quad f_0^* = \frac{f_0}{(\omega_0^2 - \omega^2)^2 + 4\omega^2\beta^2}
\end{aligned}$$

Now, we write $z = Ce^{i\omega t}$ explicitly into the form $x(t) + iy(t)$ to find its real part:

$$\begin{aligned}
z &= Ce^{i\omega t} = C(\cos\omega t + i\sin\omega t) \\
&= f_0^*(\omega_0^2 - \omega^2 - 2i\omega\beta)(\cos\omega t + i\sin\omega t) \\
&= f_0^*\left[(\omega_0^2 - \omega^2)\cos\omega t + 2\omega\beta\sin\omega t\right] + if_0^*\left[(\omega_0^2 - \omega^2)\sin\omega t - 2\omega\beta\cos\omega t\right]
\end{aligned}$$

Thus, the solution to Eq. (7.8.24), which is the real part of $z(t)$ is given by

$$x(t) = \text{Re}(z) = \frac{f_0(\omega_0^2 - \omega^2)}{(\omega_0^2 - \omega^2)^2 + 4\omega^2\beta^2}\cos\omega t + \frac{2f_0\omega\beta}{(\omega_0^2 - \omega^2)^2 + 4\omega^2\beta^2}\sin\omega t \tag{7.8.27}$$

Now, we use the trigonometry identity $\cos(a - b) = \cos a\cos b + \sin a\sin b$ to rewrite $x(t)$. First, we re-arrange $x(t)$ in the form of cos cos + sin sin, then we will have a compact form for $x(t)$:

$$\begin{aligned}
x(t) &= \frac{f_0}{\sqrt{(\omega_0^2 - \omega^2)^2 + 4\omega^2\beta^2}}\left[\frac{(\omega_0^2 - \omega^2)\cos\omega t}{\sqrt{(\omega_0^2 - \omega^2)^2 + 4\omega^2\beta^2}} + \frac{2\omega\beta\sin\omega t}{\sqrt{(\omega_0^2 - \omega^2)^2 + 4\omega^2\beta^2}}\right] \\
&= A\cos(\omega t - \delta), \quad A = \frac{f_0}{\sqrt{(\omega_0^2 - \omega^2)^2 + 4\omega^2\beta^2}}, \quad \tan\delta = \frac{2\omega\beta}{\omega_0^2 - \omega^2}
\end{aligned} \tag{7.8.28}$$

We have just computed the response of the system to the driving force: a sinusoidal driving force results in a sinusoidal oscillation of the same frequency (as that of the force) and with an amplitude proportional to the amplitude of the force (that is $A \sim f_0$). All looks reasonable. Is that the final solution? No. Why? Because Eq. (7.8.28) cannot be a general solution: in it we did not see initial conditions! Where can we get a solution that involves the initial conditions? Do not forget the natural oscillation response. We're interested in the case of weakly damped only. The total solution is thus a combination of Eqs. (7.8.21) and (7.8.28). That is,

$$\boxed{x(t) = A\cos(\omega t - \delta) + Be^{-\beta t}\cos(\omega_0^d t + \theta)} \tag{7.8.29}$$

A word on the various parameters in the above equation:

- ω is the driving frequency, which can be chosen arbitrarily;
- A and δ are functions of $f_0 = F_0/m, \omega_0, 2\beta = b/m$
- $\omega_d = \sqrt{\omega_0^2 - \beta^2}$
- B and θ are determined by the initial conditions.

Example 7.5 A mass is released from rest at $t = 0$ and $x = 0$. The driven force is $f = f_0 \cos \omega t$ with $f_0 = 1000$ and $\omega = 2\pi$. Assume that the natural frequency is $\omega_0 = 5\omega = 10\pi$, and the damping is $\beta = \omega_0/20 = \pi/2$, i.e., a weakly damped oscillation.

First, we compute A and δ using Eq. (7.8.28): $A = 1.06$ and $\delta = 0.0208$. Second, we determine B and θ from the given initial conditions that $x_0 = v_0 = 0$:

$$\left.\begin{aligned} x_0 &= A\cos\delta + B\cos\theta \\ v_0 &= \omega A \sin\delta + B(-\beta\cos\theta - \omega_0^d \sin\theta) \end{aligned}\right\} \Longrightarrow \left\{\begin{aligned} B\cos\theta &= x_0 - A\cos\delta \\ \beta B\cos\theta + B\omega_0^d \sin\theta &= \omega A \sin\delta - v_0 \end{aligned}\right.$$

which yields $B = -1.056$ and $\theta = -0.054$. Using all these numbers in Eq. (7.8.29) we can plot the solution as shown in Fig. 7.10. I provide the plot of the driving force, the homogeneous solution $x_h(t)$ and the total solution $x(t)$ given in Eq. (7.8.29). Codes to produce these plots are given in Appendix A.4.

7.8.4 Resonance

By looking at the formula of the oscillation amplitude A, we can explain the phenomenon of resonance. Recall that A is given by

$$A = \frac{f_0}{\sqrt{(\omega_0^2 - \omega^2)^2 + 4\omega^2\beta^2}} \tag{7.8.30}$$

which will have a maximum value when the denominator is minimum. Note that we are not interested with using a big force to have a large amplitude. With only a relatively small force but at a correct frequency we can get a large oscillation anyway. Moreover, we are only interested in the case β is small, i.e., weakly damped. It can be seen that A is maximum when $\omega \approx \omega_0$, see Fig. 7.11a and the maximum value is

$$A_{\max} \approx \frac{f_0}{2\omega_0\beta} \tag{7.8.31}$$

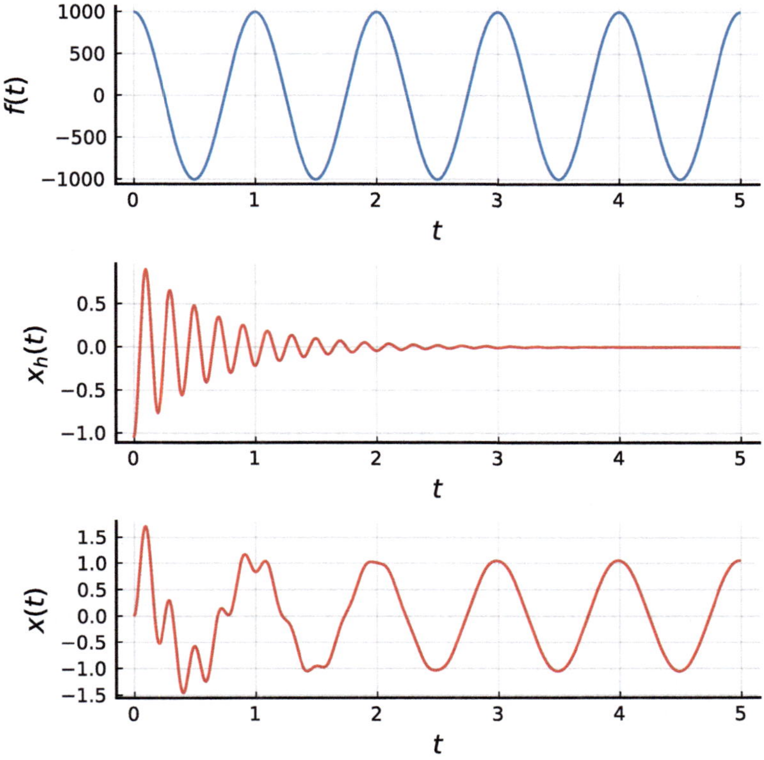

Fig. 7.10 Driven oscillation of a weakly damped spring-mass: the frequency of the force is 2π, and the natural frequency is 10π. Referring to the bottom figure, we can see that, after about 3 cycles, the motion is indistinguishable from a pure cosine, oscillating at exactly the drive frequency. The free oscillation has died out and only the long-term motion remains. In the beginning ($t \leq 3$), the effects of the transients are clearly visible: as they oscillate at a faster frequency they show up as a rapid succession of bumps and dips. In fact, you can see that there are five such bumps within the first cycle, indicating that $\omega_0 = 5\omega$

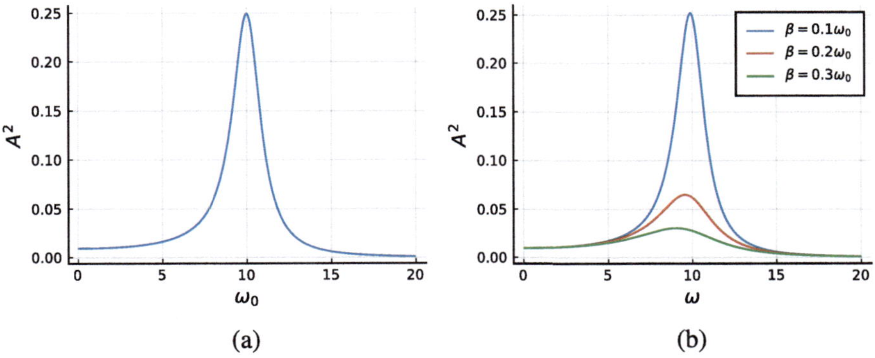

Fig. 7.11 A is maximum when $\omega \approx \omega_0$

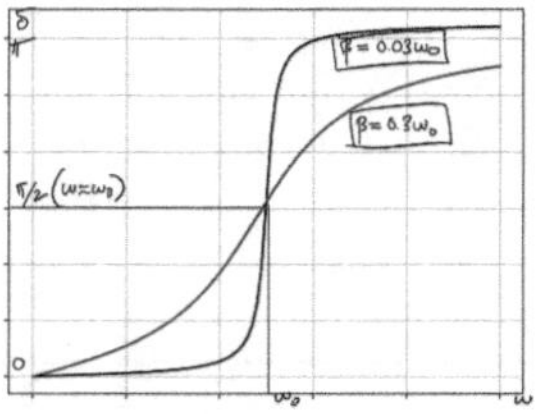

Phase at resonance. We're now interested in the phase δ when resonance occurs. Recall that the phase difference δ by which the oscillator's motion lacks behind the driving force is

$$\delta = \arctan\left(\frac{2\omega\beta}{\omega_0^2 - \omega^2}\right) \tag{7.8.32}$$

If $\omega \ll \omega_0$, $\delta \approx 0$, and thus the oscillations are almost perfectly in phase with the driving force (we can see this clearly in Fig. 7.10). At resonance $\omega = \omega_0$, $\delta = \pi/2$: the oscillations are 90° behind the driving force.

7.8.5 Driven Damped Oscillators with Any Periodic Forces

After damped oscillation with a sinusoidal driving force, it is just one more small step to tackle damped oscillation with any periodic force $f(t)$ of period τ, thanks to the genius of Fourier. The equation is now given by

$$m\ddot{x} + b\dot{x} + kx = f(t) \tag{7.8.33}$$

And we replace $f(t)$ by its Fourier series (Sect. 4.19)

$$f(t) = \sum_{n=0}^{\infty} [a_n \cos(n\omega t) + b_n \sin(n\omega t)] \tag{7.8.34}$$

with the Fourier coefficients given by (and $b_0 = 0$)

$$a_0 = \frac{1}{\tau}\int_{-\tau/2}^{\tau/2} f(t)dt, \qquad \begin{cases} a_n = \dfrac{2}{\tau}\displaystyle\int_{-\tau/2}^{\tau/2} f(t)\cos(n\omega t)dt \\ b_n = \dfrac{2}{\tau}\displaystyle\int_{-\tau/2}^{\tau/2} f(t)\sin(n\omega t)dt \end{cases} \tag{7.8.35}$$

With this replacement of $f(t)$, the equation of motion becomes

$$m\ddot{x} + b\dot{x} + kx = \sum_{n=0}^{\infty} [a_n \cos(n\omega t) + b_n \sin(n\omega t)] \tag{7.8.36}$$

What is this new form different from the original problem, Eq. (7.8.33)? Now, we have a damped SHO with infinitely many driving forces $f_0(t)$, $f_1(t)$, ... But for each of this force, we are able to solve for the solution $x_n(t)$, with $n = 0, 1, \ldots$, (I have assumed that the Fourier series contain only the cosine terms for simplicity):

$$x_n(t) = A_n \cos(n\omega t - \delta_n), \quad A_n = \frac{a_n}{\sqrt{(\omega_0^2 - n^2\omega^2)^2 + 4n^2\omega^2\beta^2}}, \quad \tan\delta_n = \frac{2n\omega\beta}{\omega_0^2 - n^2\omega^2} \tag{7.8.37}$$

And what is the final solution? It is simply the sum of all $x_n(t)$. Why that? Because our equation is linear! To see this, let's assume there are only two forces: with $f_1(t)$ we have the solution $x_1(t)$ and similarly for $f_2(t)$, so we can write:

$$\begin{aligned} m\ddot{x}_1 + b\dot{x}_1 + kx_1 &= f_1(t) \\ m\ddot{x}_2 + b\dot{x}_2 + kx_2 &= f_2(t) \end{aligned} \tag{7.8.38}$$

Adding these two equations, we get

$$m(\ddot{x}_1 + \ddot{x}_2) + b(\dot{x}_1 + \dot{x}_2) + k(x_1 + x_2) = f_1(t) + f_2(t) \tag{7.8.39}$$

which indicates that $x(t) = x_1(t) + x_2(t)$ is indeed the solution. This is known as the principle of superposition, which we discussed in Sect. 7.6. There, the discussion was abstract.

In summary, we had a hard problem (due to $f(t)$), and we replaced this $f(t)$ by many easier sinusoidal forces. For each force, we solved an easier problem and we added these solutions altogether to get the final solution. It is indeed the spirit of calculus!

7.8.6 The Pendulum

Consider a pendulum as shown in Fig. 7.4b. Newton's second law in polar coordinates in the θ direction gives us[31]:

$$F_\theta = m(2\dot{r}\dot{\theta} + r\ddot{\theta}) \tag{7.8.40}$$

And note that $F_\theta = -mg\sin\theta$, and $r = l$ is constant, thus Eq. (7.8.40) is simplified to

$$\ddot{\theta} + \frac{g}{l}\sin\theta = 0, \quad \text{or} \quad \frac{d^2\theta}{dt^2} + \frac{g}{l}\sin\theta = 0 \tag{7.8.41}$$

For small vibrations, we have $\sin\theta \approx \theta$ (remember the Taylor series for sine?). Thus, our equation is further simplified to

---

[31] Check Eq. (6.10.17) if this was not clear.

$$\ddot{\theta} + \omega^2\theta = 0, \ \omega = \sqrt{\frac{g}{l}} \Longrightarrow T = 2\pi\sqrt{\frac{l}{g}} \tag{7.8.42}$$

And voilà, we see again the simple harmonic oscillation equation! And the natural frequency (and the period) of a pendulum does not depend on the mass of the blob. And of course it does not depend on how far it swings, i.e., the initial conditions have no say on this. This fact was first observed by Galileo Galilei when he was a student of medicine watching a swinging chandelier. A historical note: it was the Dutch mathematician Christian Huygens (1629–1695) who first derived the formula for the period of a pendulum. Note that we can also use a dimensional analysis to come up with $\omega \sim \sqrt{g/l}$.

Pendulum and elliptic integral of first kind. Herein I demonstrate how an elliptic integral of the first kind shows up in the formula of the period of a pendulum when its amplitude is large. The idea is to start with Eq. (7.8.41) and massage it so that we can have dt as a function of θ. Then, integrating dt to get the period T.

We rewrite Eq. (7.8.41) using ω:

$$\frac{d^2\theta}{dt^2} + \omega^2 \sin\theta = 0 \tag{7.8.43}$$

Multiplying both sides of this equation with $\dot{\theta}$, we get

$$\frac{d^2\theta}{dt^2}\frac{d\theta}{dt} + \omega^2 \sin\theta \frac{d\theta}{dt} = 0 \tag{7.8.44}$$

Now, integrating this equation w.r.t. t, we obtain

$$\frac{1}{2}\left(\frac{d\theta}{dt}\right)^2 - \omega^2 \cos\theta = k \tag{7.8.45}$$

where k is an integration constant. To find k, let θ^* be the maximum angular displacement. At $\theta = \theta^*$ (a maxima), $\frac{d\theta}{dt}$ vanishes: $k = -\omega^2 \cos\theta^*$. Now, we can solve Eq. (7.8.45) for $\dot{\theta}$ noting that θ decreases as t increases[32]:

$$\frac{d\theta}{dt} = -\omega\sqrt{2(\cos\theta - \cos\theta^*)}, \quad \text{or} \quad \boxed{dt = -\frac{d\theta}{\omega\sqrt{2(\cos\theta - \cos\theta^*)}}} \tag{7.8.46}$$

[32] Picture moving the bob to a certain height and release it. Note that it will be easier, lot more, to derive this equation using the principle of conservation of energy: $0.5mv^2 = mg(l\cos\theta - \cos\theta^*)$, then $v = ds/dt$ with $s = l\theta$.

It's now possible to determine the period T:

$$T = 4\int_{\theta^*}^{0} dt = 4\int_0^{\theta^*} \frac{d\theta}{\omega\sqrt{2(\cos\theta - \cos\theta^*)}}$$

Now, a bit of massage using the trigonometric identity $\cos\alpha = 1 - 2\sin^2\alpha/2$ leads to

$$T = 2\sqrt{\frac{l}{g}}\int_0^{\theta^*} \frac{d\theta}{\sqrt{k^2 - \sin^2\theta/2}}, \quad k = \sin\theta^*/2 \tag{7.8.47}$$

We still need to massage this equation a bit more. Using the following change of variable[33]

$$\sin\frac{\theta}{2} = k\sin\phi \Longrightarrow \frac{1}{2}\cos\frac{\theta}{2}d\theta = k\cos\phi d\phi \Longrightarrow d\theta = \frac{2k\cos\phi d\phi}{\cos\frac{\theta}{2}} = \frac{2k\cos\phi d\phi}{\sqrt{1 - k^2\sin^2\phi}}$$

we have[34]

$$\begin{aligned} T &= 2\sqrt{\frac{l}{g}}\int_0^{\theta^*} \frac{d\theta}{\sqrt{k^2(1 - \sin^2\phi)}} \\ &= 2\sqrt{\frac{l}{g}}\int_0^{\theta^*} \frac{2k\cos\phi d\phi}{\sqrt{k^2(1 - \sin^2\phi)}\sqrt{1 - k^2\sin^2\phi}} \\ &= 4\sqrt{\frac{l}{g}}\int_0^{\pi/2} \frac{d\phi}{\sqrt{1 - k^2\sin^2\phi}} \end{aligned} \tag{7.8.48}$$

And the red integral is exactly the integral we have met in Sect. 4.9.1 when computing the length of an ellipse!

We use the generalized binomial theorem for the function under the red integral (with $x = 1$, $y = -k^2\sin^2\phi$, and $r = -0.5$):

$$\frac{1}{\sqrt{1 - k^2\sin^2\phi}} = \left(1 - k^2\sin^2\phi\right)^{-1/2} = 1 + \frac{1}{2}k^2\sin^2\phi + \frac{3}{8}k^4\sin^4\phi + \cdots$$

So, the red integral now becomes as simple as follows:

$$\int_0^{\pi/2} \frac{d\phi}{\sqrt{1 - k^2\sin^2\phi}} = \int_0^{\pi/2}\left(1 + \frac{1}{2}k^2\sin^2\phi + \frac{3}{8}k^4\sin^4\phi + \cdots\right)d\phi$$

---

33 Why this change of variable? Look at the red term in Eq. (7.8.47). This change of variable can remove the square root.

34 Note also that $k = \sin\theta^*/2$, thus when $\theta = \theta^*$, $\phi = \pi/2$.

We still have to calculate integrals like $\int \sin^2 x\, dx, \ldots$ but that's a minor detail. Finally, the period of the simple pendulum is calculated as follows:

$$T = 2\pi\sqrt{\frac{l}{g}}\left[1 + \frac{1}{4}k^2 + \left(\frac{3}{8}\right)^2 k^4 + \cdots\right]$$

Since $k = \sin(\theta^*/2)$, as we increase θ^* (the amplitude of the oscillation), the formula above shows that the period will increase accordingly.

In Sect. 10.6.6 we will return to this simple pendulum and study its motion numerically and its chaotic motion.

7.8.7 RLC Circuits

An RLC circuit is an electrical circuit consisting of a resistor (R), an inductor (L), and a capacitor (C), connected in series or in parallel. The name of the circuit is derived from the letters that are used to denote the constituent components of this circuit, where the sequence of the components may vary from RLC.

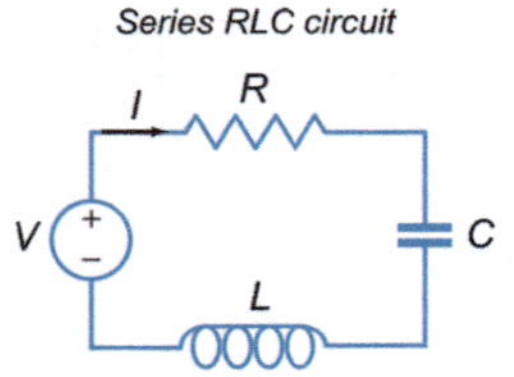

Assume that the positive direction for the current to be clockwise, and the charge $q(t)$ to be the charge on the bottom plate of the capacitor. If we follow around the circuit in the positive direction, the electric potential drops by $L\dot{I} = L\ddot{q}$ across the indcutor, by $RI = R\dot{q}$ across the resistor, and by q/C across the capacitor. Applying Kirchoff's second rule for circuits, we conclude that

$$L\ddot{q} + R\dot{q} + \frac{1}{C}q = 0 \tag{7.8.49}$$

This has exactly the form of Eq. (7.8.12) for the damped oscillator. And anything that we know about the damped oscillator will be immediately applicable to the RLC circuit. In other words, the RLC circuit is an electrical analog of a spring-mass system with damping.

Mathematicians do not care about physics or applications, what matters to them is the following *nice* equation with $a, b, c \in \mathbb{R}$

$$a\ddot{y} + b\dot{y} + cy = 0 \tag{7.8.50}$$

which they call a second-order ordinary differential equation. But now you understand why university students have to study them and similar equations. Because they describe our world quite nicely.

7.8.8 Coupled Oscillators

In the previous section we discussed the oscillation of a single body, such as a mass connected to a fixed spring. We now move to the study of the oscillation of several bodies that are coupled to each other such as the atoms making up a molecule. For this problem in which there are multiple degrees of freedom, matrices appear naturally. If you need a refresh on matrices please refer to Chap. 9.

As a simple example of coupled oscillators, consider the two carts shown in Fig. 7.12. And in this context we will meet matrices and determinants. We use Newton's second law to find the equations of motion.

Using Newton's second law we write

$$\begin{aligned} m_1\ddot{x}_1 &= -k_1x_1 + k_2(x_2 - x_1) = -(k_1 + k_2)x_1 + k_2x_2 \\ m_2\ddot{x}_2 &= -k_2(x_2 - x_1) - k_3x_2 = k_2x_1 - (k_2 + k_3)x_2 \end{aligned} \tag{7.8.51}$$

And we can write these two equations in a compact matrix form[35] as

$$\underbrace{\begin{bmatrix} m_1 & 0 \\ 0 & m_2 \end{bmatrix}}_{\mathbf{M}} \begin{bmatrix} \ddot{x}_1 \\ \ddot{x}_2 \end{bmatrix} = -\underbrace{\begin{bmatrix} k_1 + k_2 & -k_2 \\ -k_2 & k_2 + k_3 \end{bmatrix}}_{\mathbf{K}} \underbrace{\begin{bmatrix} x_1 \\ x_2 \end{bmatrix}}_{\boldsymbol{x}}, \quad \text{or} \quad \mathbf{M}\ddot{\boldsymbol{x}} = -\mathbf{K}\boldsymbol{x} \tag{7.8.52}$$

where $\mathbf{M}$ is the *mass matrix* and $\mathbf{K}$ is the spring-constant matrix or *stiffness* matrix. Note that these two matrices are symmetric. Also note that using matrix notation the equation of motion of coupled oscillators, $\mathbf{M}\ddot{\boldsymbol{x}} = -\mathbf{K}\boldsymbol{x}$, is a very natural generalization of that of a single oscillator: with just one degree of freedom, all three matrices $\boldsymbol{x}$, $\mathbf{K}$ and $\mathbf{M}$ are just ordinary numbers and we had $m\ddot{x} = -kx$.

We use complex exponentials to solve Eq. (7.8.52):

$$\boldsymbol{z} = \begin{bmatrix} z_1 \\ z_2 \end{bmatrix} = \begin{bmatrix} A_1 \\ A_2 \end{bmatrix} e^{i\omega t} = \boldsymbol{a}e^{i\omega t}, \Longrightarrow \ddot{\boldsymbol{z}} = -\omega^2\boldsymbol{a}e^{i\omega t} \tag{7.8.53}$$

---

35 Check Chap. 9 for a discussion on matrices.

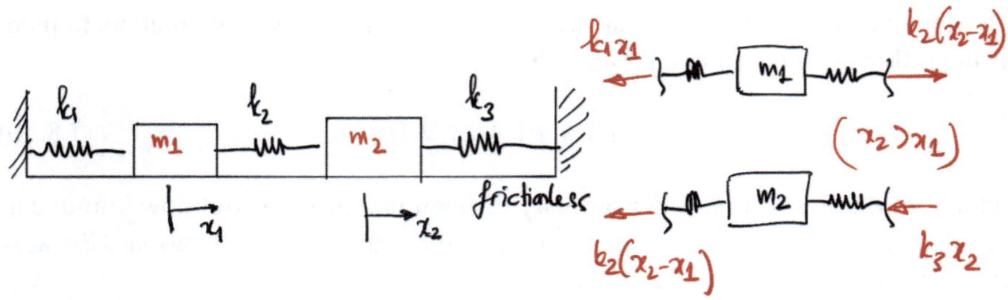

Fig. 7.12 A simple two coupled oscillators. In the absence of the spring 2, the two carts would oscillate independently of each other. It is the spring 2 that couples the two carts

Introducing $\boldsymbol{z}$ and $\ddot{\boldsymbol{z}}$ into Eq. (7.8.52) we obtain

$$(\omega^2 \mathbf{M} - \mathbf{K})\boldsymbol{a} = \mathbf{0} \tag{7.8.54}$$

which leads to the determinant of the matrix being null:

$$\det\left(\mathbf{K} - \omega^2 \mathbf{M}\right) = 0 \tag{7.8.55}$$

This is a quadratic equation for ω^2 and has two solutions for ω^2 (in general). This implies that there are two frequencies $\omega_{1,2}$ at which the carts oscillate in pure sinusoidal motion. These frequencies are called *normal frequencies*. The two sinusoidal motions associated with these normal frequencies are known as *normal modes*. The normal modes are determined by solving Eq. (7.8.54). If you know linear algebra, what we are doing here is essentially a generalized eigenvalue problem in which ω^2 play the role of eigenvalues and $\boldsymbol{a}$ play the role of eigenvectors; refer to Sect. 9.10 for more detail on eigenvalue problems.

Example 7.6 Let's consider the case of equal stiffness springs and equal masses: $k_1 = k_2 = k_3 = k$ and $m_1 = m_2 = m$. Using Eq. (7.8.55) we can determine the normal frequencies:

$$\omega_1 = \sqrt{\frac{k}{m}}, \quad \omega_2 = \sqrt{\frac{3k}{m}} \tag{7.8.56}$$

Did u notice anything special about ω_1? And we use Eq. (7.8.54) to compute $\boldsymbol{a}$:

$$\left(\begin{bmatrix} 2k & -k \\ -k & 2k \end{bmatrix} - \begin{bmatrix} \omega^2 m & 0 \\ 0 & \omega^2 m \end{bmatrix}\right)\begin{bmatrix} A_1 \\ A_2 \end{bmatrix} = \begin{bmatrix} 0 \\ 0 \end{bmatrix} \tag{7.8.57}$$

With ω_1, we solve Eq. (7.8.57) to have $A_1 = A_2 = Ae^{-i\phi_1}$. So, we have $z_1(t)$ and $z_2(t)$ and from them the real parts of the actual solutions for mode 1:

$$\left.\begin{aligned} z_1 &= Ae^{-i\phi_1}e^{i\omega_1 t} \\ z_2 &= Ae^{-i\phi_1}e^{i\omega_1 t} \end{aligned}\right\} \Longrightarrow \left\{\begin{aligned} x_1(t) &= A\cos(\omega_1 t - \phi_1) \\ x_2(t) &= A\cos(\omega_1 t - \phi_1) \end{aligned}\right. \tag{7.8.58}$$

As $x_1(t) = x_2(t)$ the two carts oscillate in a way that spring 2 is always in its unstretched configuration. In other words, spring 2 is irrelevant and thus the system oscillates with a natural frequency similar to a single oscillator (i.e., $\omega = \sqrt{k/m}$).

With ω_2, we solve Eq. (7.8.57) to have $A_1 = -A_2 = Be^{-i\phi_2}$. The mode 2 solutions are

$$\begin{aligned} x_1(t) &= +B\cos(\omega_2 t - \phi_2) \\ x_2(t) &= -B\cos(\omega_2 t - \phi_2) = B\cos(\omega_2 t - \phi_2 - \pi) \end{aligned} \tag{7.8.59}$$

These solutions tell us that when cart 1 moves to the left a distance, cart 2 moves to the right the same distance. We say that the two carts oscillate with the same amplitude but are out of phase.

Together, the general solutions are:

$$\boldsymbol{x}(t) = A\begin{bmatrix}1\\1\end{bmatrix}\cos(\omega_1 t - \phi_1) + B\begin{bmatrix}1\\-1\end{bmatrix}\cos(\omega_1 t - \phi_2) \tag{7.8.60}$$

with the four constants of integration A, B, ϕ_1, ϕ_2 to be determined from four initial conditions.

Example 7.7 This case involves equal masses, but the second spring is much less stiff: $k_1 = k_3 = k$, $k_2 \ll k$, $m_1 = m_2 = m$. The two normal frequencies are

$$\omega_1 = \sqrt{\frac{k}{m}}, \quad \omega_2 = \sqrt{\frac{k + 2k_2}{m}} \tag{7.8.61}$$

As we have discussed, spring 2 is irrelevant in mode 1, so we got the same mode 1 frequency as in Example 1. As $k_2 \ll k$, $\omega_1 \approx \omega_2$, we can write them in terms of their average ω_0 and half difference ϵ (you will see why we did this via Eq. (7.8.64); the basic idea is that we can write the solutions in two separate terms, one involves ω_0 and one involves ϵ):

$$\begin{aligned} \omega_1 &= \omega_0 - \epsilon, \\ \omega_2 &= \omega_0 + \epsilon, \end{aligned}, \quad \omega_0 = \frac{\omega_1 + \omega_2}{2}, \epsilon = \frac{\omega_2 - \omega_1}{2} \tag{7.8.62}$$

Therefore, the normal modes are

$$\left\{\begin{aligned} z_1 &= C_1 e^{i(\omega_0-\epsilon)t} = C_1 e^{i\omega_0 t}e^{-i\epsilon t} \\ z_2 &= C_1 e^{i(\omega_0-\epsilon)t} = C_1 e^{i\omega_0 t}e^{-i\epsilon t} \end{aligned}\right. \text{(mode 1)}, \left\{\begin{aligned} z_1 &= +C_2 e^{i(\omega_0+\epsilon)t} = +C_2 e^{i\omega_0 t}e^{i\epsilon t} \\ z_2 &= -C_2 e^{i(\omega_0+\epsilon)t} = -C_2 e^{i\omega_0 t}e^{i\epsilon t} \end{aligned}\right. \text{(mode 2)}$$

where $C_1, C_2 \in \mathbb{C}$. To simplify the presentation, we simply let $C_1 = C_2 = A/2$ where A is real. In this case, the general solutions are

$$z(t) = \frac{A}{2} e^{i\omega_0 t} \begin{bmatrix} e^{-i\epsilon t} + e^{i\epsilon t} \\ e^{-i\epsilon t} - e^{i\epsilon t} \end{bmatrix} = A e^{i\omega_0 t} \begin{bmatrix} \cos(\epsilon t) \\ -i \sin(\epsilon t) \end{bmatrix} \tag{7.8.63}$$

where in the last step, we have used the formula that relating sine/cosine to complex exponentials, see Sect. 2.25.5 if you do not recall this. And the real solutions are thus given by

$$\boldsymbol{x}(t) = \begin{bmatrix} A \cos(\epsilon t) \cos(\omega_0 t) \\ A \sin(\epsilon t) \sin(\omega_0 t) \end{bmatrix} \tag{7.8.64}$$

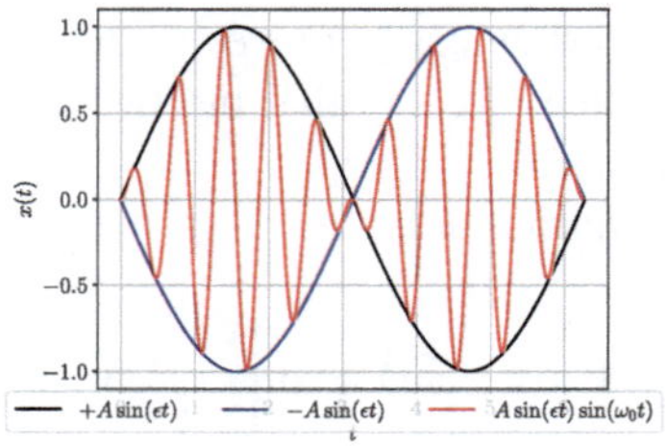

This is the solutions for the case when cart 1 is pulled a distance A to the right and released at $t = 0$ while cart 2 is stationary at its equilibrium position.[36] To illustrate this solution, we use $\omega_0 = 10$, $A = 1.0$, and $\epsilon = 1$ and consider a time duration of 2π. First, we try to understand what $A\sin(\epsilon t)\sin(\omega_0 t)$ means. As $\epsilon \ll \omega_0$ the ϵt oscillation is much slower than the $\omega_0 t$ oscillation. The former then simply acts as an envelope for the latter. In Fig. 7.13 both $x_1(t)$ and $x_2(t)$ are shown. There we can see that the motion sloshes back and forth between the two masses. At the start only the first mass is moving. But after a time of $\epsilon t = \pi/2$ (or $t = \pi/2\epsilon$), the first mass is not moving and the second mass has all the motion. Then after another time of $\pi/2\epsilon$ it switches back, and this continues forever.

Later in Sect. 7.10 we shall know that this is nothing but the beat phenomenon when two sound waves of similar frequencies meet.

7.9 Solving the Diffusion Equation

The deep study of nature is the most fertile source of mathematical discoveries.

Jean-Baptiste Joseph Fourier

We are going to discuss Fourier's solution to the following 1D diffusion equation (derived in Sect. 7.5.2). In the process, we will understand the idea of approximating

[36] This is so because at $t = 0$, $x_1 = A$ while $\dot{x}_1 = x_2 = \dot{x}_2 = 0$.

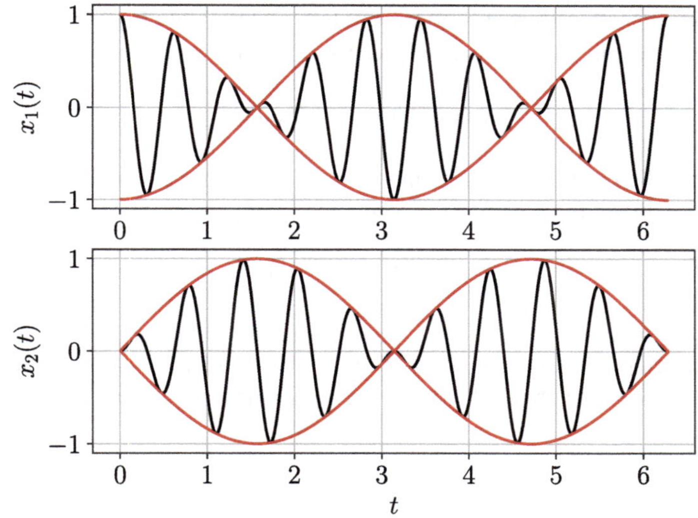

Fig. 7.13 Plot of Eq. (7.8.64) with $\omega_0 = 10$, $A = 1.0$, and $\epsilon = 1$

any periodic function by a trigonometric series now famously known as Fourier series. The equations including BCs and ICs are:

$$\frac{\partial \theta}{\partial t} = \kappa^2 \frac{\partial^2 \theta}{\partial x^2} \quad (0 < x < L,\ t > 0) \tag{7.9.1}$$

$$\theta(x, 0) = \phi(x) \quad (0 \le x \le 1) \tag{7.9.2}$$

$$\theta(0, t) = 0,\ \theta(L, t) = 0 \quad (t > 0) \tag{7.9.3}$$

Using the method of separation of variables,[37] the temperature field $\theta(x, t)$ is written as a product of two functions: one spatial function $h(x)$ and one temporal function $g(t)$:

$$\theta(x, t) = h(x)g(t) \tag{7.9.4}$$

We shall see that there exists an infinite number of solutions $\theta_n(x, t)$ of this type, which fulfill both the PDE and the BCs. They are called the *fundamental solutions*. The final solution is then found by adding up all these fundamental solutions (as the PDE is linear a combination of solutions is also a solution) so that it satisfies the initial condition. That's the plan.

---

[37] This was a clever idea of the Swiss mathematician and physicist Daniel Bernoulli (1700–1782). The first question comes to mind should be: how we know that this separation of variables would work. We do not know! Daniel probably learned this technique from his father John Bernoulli who used $y(x) = u(x)v(x)$ to solve the differential equation $y' = y(x)f(x) + y''g(x)$ for $y(x)$ [50]. Another source of motivation for the method of separation of variables is waves. If we study waves, we will observe one phenomenon called standing waves—check this youtube video out. A standing wave can be mathematically described by $g(x)h(t)$ (see also Fig. 7.17).

What we do now is, certainly, to substitute Eq. (7.9.4) into Eq. (7.9.1) to see what happens to $h(x)$ and $g(t)$. Doing so leads us to another equation:

$$h(x)g'(t) = \kappa^2 h''(x)g(t)$$

Now comes the *trick*: we separate the temporal functions (those involve t) on one side and the spatial functions on the other side:

$$\frac{g'(t)}{\kappa^2 g(t)} = \frac{h''(x)}{h(x)} \tag{7.9.5}$$

As this equation holds for any x and t, both sides must be a constant. Setting this constant by k we thus obtain the following equations (two not one)

$$\frac{g'(t)}{\kappa^2 g(t)} = k, \quad \frac{h''(x)}{h(x)} = k \tag{7.9.6}$$

Even though there are three possibilities $k = 0$, $k > 0$ and $k < 0$, the two former cases do not lead to meaningful solutions,[38] so $k < 0$, which can thus be expressed as a negative of a square: $k = -\lambda^2$. With this, our two equations become

$$g'(t) = -\lambda^2\kappa^2 g(t), \quad h''(x) + \lambda^2 h(x) = 0 \tag{7.9.7}$$

So, with the technique of separation of variables, *we have converted a single second-order PDE into a system of two first-order ODEs*. That's the key lesson! What is interesting is that it is straightforward to solve these two ODEs:

$$g(t) = A_1 e^{-\lambda^2\kappa^2 t}, \quad h(x) = A_2 \cos\lambda x + A_3 \sin\lambda x \tag{7.9.8}$$

where A_1, A_2, A_3 are arbitrary constants. With these functions substituted into Eq. (7.9.4), the temperature field is given by

$$\theta(x,t) = e^{-\lambda^2\kappa^2 t}(A\cos\lambda x + B\sin\lambda x) \tag{7.9.9}$$

where $A = A_1A_2$ and $B = A_1A_3$. We have to find A, B and λ so that $\theta(x,t)$ satisfies the BCs and IC. We handle the BCs first. A BC is of the form $\theta(x^*, t) = \theta^*$; we thus substitute x^* into Eq. (7.9.9) and equate what we obtain with θ^*. Doing that, we get

$$\begin{aligned}
\theta(0,t) &= 0 : e^{-\lambda^2\kappa^2 t} A &&= 0 \Longrightarrow A = 0\\
\theta(L,t) &= 0 : e^{-\lambda^2\kappa^2 t} B\sin(\lambda L) &&= 0 \Longrightarrow \sin\lambda = 0 \Longrightarrow \lambda = \tfrac{n\pi}{L}, \quad n = 1, 2, \ldots
\end{aligned} \tag{7.9.10}$$

[38] Why? If k is positive, then $g'(t) = \kappa^2 k g(t) > 0$, thus $g(t)$ is increasing forever. This is physically wrong as we know from daily experience that the temperature inside the bar goes to zero as time goes by.

So, we have an infinite number of solutions written as[39]

$$\theta_n(x,t) = B_n e^{-(n\pi\kappa/L)^2 t} \sin\frac{n\pi x}{L}, \quad n = 1, 2, 3, \ldots \tag{7.9.11}$$

All these $\theta_n(x, t)$ satisfy the boundary conditions (and of course the PDE). Now we tackle the initial condition. First, since the PDE is a linear equation, the sum of all the fundamental solutions is also a solution; this is known as the principle of superposition. So, we have

$$\theta(x,t) = \sum_{n=1}^{\infty} \theta_n(x,t) = \sum_{n=1}^{\infty} B_n e^{-(n\pi\kappa/L)^2 t} \sin(\frac{n\pi x}{L}) \tag{7.9.12}$$

Evaluating this solution at $t = 0$ gives us (noting that the initial condition Eq. (7.9.2) says that at $t = 0$ the temperature is $\phi(x)$):

$$\theta(x,0) = \sum_{n=1}^{\infty} B_n \sin(\frac{n\pi x}{L}) = \phi(x) \tag{7.9.13}$$

Now the problem becomes this: if we can approximate the initial temperature $\phi(x)$ as an infinite trigonometric series $\sum_{n=1}^{\infty} B_n \sin(n\pi x/L)$, then we have solved the heat equation! Now Fourier had to move away from physics to turn to mathematics: he had to find the coefficients B_n in Eq. (7.9.13). We refer to Sect. 4.19 for a discussion on how Fourier computed B_n. Then the solution to Eq. (7.9.1) is the following infinite series

$$\boxed{\theta(x,t) = \sum_{n=1}^{\infty} B_n e^{-(n\pi\kappa/L)^2 t} \sin(\frac{n\pi x}{L}), \quad B_n = \frac{2}{L}\int_0^L \phi(x)\sin(\frac{n\pi x}{L})dx} \tag{7.9.14}$$

Should we be worry about the infinity involved in this solution? No, we do not have to thanks to the term $e^{-(n\pi\kappa/L)^2 t}$ which is actually a decaying term, i.e., for large n and/or for large t, this term is small. See Fig. 7.14 for an illustration. And how fast this term is decaying depends on κ, which makes total sense.

Remark As we know from dimensional analysis, in physics the argument to the exponential function must be a dimensionless number. And $(n\pi\kappa/L)^2 t$ is indeed dimensionless: we do not worry about n and π, the SI unit for κ^2 is m^2/s, of L is m and t is s. They cancel out nicely. Similarly, the argument for sin must be dimensionless, and in $\sin n\pi x/L$, we have x in the nominator and L in the denominator. Perfect.

[39] $B = 0$ also satisfies the BCs, but it would result in a boring solution $\theta(x, t) = 0$.

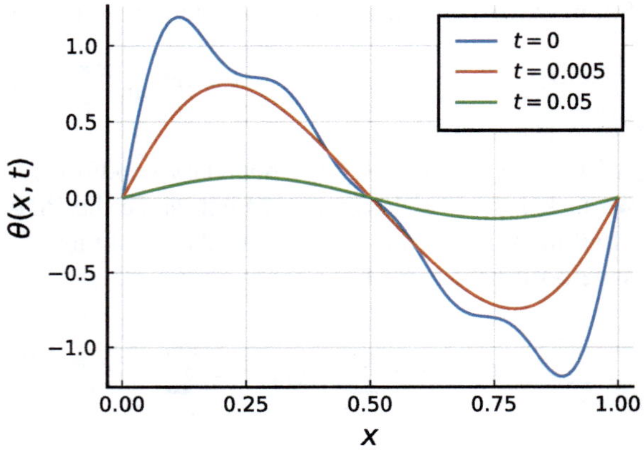

Fig. 7.14 Solution of the heat equation: high-order terms vanish first and thus the wiggles are gone first

History note 7.1: Joseph Fourier (21 March 1768–16 May 1830)

Jean-Baptiste Joseph Fourier was a French mathematician and physicist who is best known for initiating the investigation of Fourier series, which eventually developed into Fourier analysis and harmonic analysis, and their applications to problems of heat transfer and vibrations. The Fourier transform and Fourier's law of conduction are also named in his honor. Fourier is also generally credited with the discovery of the greenhouse effect.

In 1822, Fourier published his work on heat flow in *The Analytical Theory of Heat*. There were three important contributions in this work, one purely mathematical, two essentially physical. In mathematics, Fourier claimed that any function of a variable, whether continuous or discontinuous, can be expanded in a series of sines of multiples of the variable. Though this result is not correct without additional conditions, Fourier's observation that some discontinuous functions are the sum of infinite series was a breakthrough. One important physical contribution in the book was the concept of dimensional homogeneity in equations; i.e., an equation can be formally correct only if the dimensions match on either side of the equality; Fourier made important contributions to dimensional analysis. The other physical contribution was Fourier's proposal of his partial differential equation for conductive diffusion of heat. This equation is now taught to every student of mathematical physics.

7.10 Solving the Wave Equation: D'Alembert's Solution

Herein we discuss d'Alembert's solution to the wave equation. His solutions are written in terms of traveling waves. It is easy to see what is a traveling wave; it is there, in nature, waiting to be discovered by a curious mind. For example, consider a rope, which is fixed at the right end, if we hold its left end and move our hand up and down, a wave is created and travels to the right. And that's a traveling wave. Now, we need to describe it mathematically. And it turns out not so difficult.

Assume that at time $t = 0$, we have a wave of which the shape can be described by a function $y = f(x)$. Furthermore, assume that the wave travels with a constant velocity c to the right and its shape does not change in time. Then, at time t^*, the wave is given by $f(x - ct^*)$. To introduce some terminologies, let's consider the simplest traveling wave; a sine wave.

Sinusoidal waves. Now, consider a sine wave (people prefer to call it a sinusoidal wave) traveling to the right (along the x-direction) with a velocity c. As a sine wave is characterized by its height y which depends on two independent variables x (the position of a point on the wave) and time t, its equation is determined by a function $y(x, t)$, which is:

$$\boxed{y(x,t) = A\sin\left(\frac{2\pi}{\lambda}(x - ct)\right)} \tag{7.10.1}$$

The amplitude of the wave is A, the wavelength is λ. That is, the function $y(x, t)$ repeats itself each time x increases by the distance λ. Thus, the wavelength is *the spatial period of a periodic wave*.[40] It is the distance between consecutive corresponding points of the same phase on the wave, such as two adjacent crests, troughs, or zero crossings (Fig. 7.15).

So far we have focused on the shape of the entire wave at one particular time instant. Now we focus on one particular location on the wave, say x^* and let time vary. As time goes on, the wave passes by the point and makes it moves up and down. (Think of a leaf on a pond that bobs up and down with the motion of the water ripples) The motion of the point is simple harmonic. Indeed, we can show this mathematically as follows. Replacing x by x^* in Eq. (7.10.1), we have

$$y(x^*, t) = A\sin\left(\frac{2\pi}{\lambda}(x^* - ct)\right) = -A\sin\left(\frac{2\pi c}{\lambda}t - \frac{2\pi x^*}{\lambda}\right) \tag{7.10.2}$$

This is indeed the equation of a SHO (Sect. 7.8.1) with angular frequency ω and phase ϕ

$$\omega = \frac{2\pi c}{\lambda} = 2\pi f, \quad \phi = \frac{2\pi x^*}{\lambda} \tag{7.10.3}$$

[40] Note that in Sect. 7.8 we met another period, which is a temporal period. Waves are more complicated than harmonic oscillations because we have two independent variables x and t.

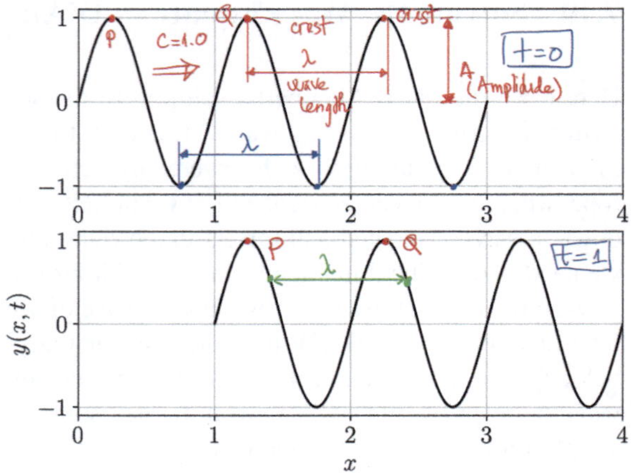

Fig. 7.15 Plots of a sine wave at two different times

where f is the frequency. Now, we can understand why the wavelength λ is defined as the distance between consecutive corresponding points of the same phase on the wave. The phase ϕ is identical for points x^* and $x^* + \lambda$. As each point in the string (e.g., x^*) oscillates back and forth in the transverse direction (not along the direction of the string), this is called a *transverse wave*.

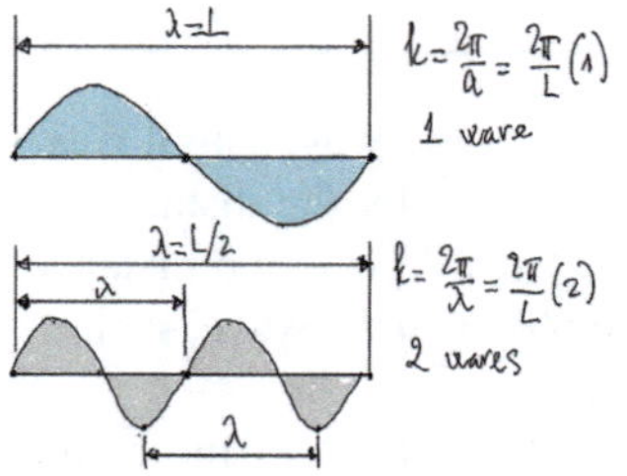

Now, I present another form of the sinusoidal wave which introduces the concept of *wavenumber*, designated by k. Obviously we can write Eq. (7.10.1) in the following form $y(x,t) = A\sin(kx - \omega t)$, with $k := {}^{2\pi}/_{\lambda}$. Referring to the figure next to the text, it is obvious that the wavenumber k tells us how many waves are there in a spatial domain of length L. More precise ${}^{k}/_{2\pi}$ is the number of waves fit inside L. We can now study what will happen if two waves of the same frequencies meet. For example if we are listening two sounds of similar frequencies, what would we hear? Writing the two sounds as

$$x_1 = \cos(\omega_1 t), \quad x_2 = \cos(\omega_2 t)$$

And what we hear is the superposition of these two sounds[41]:

[41] Note the similarity with Eq. (7.8.64).

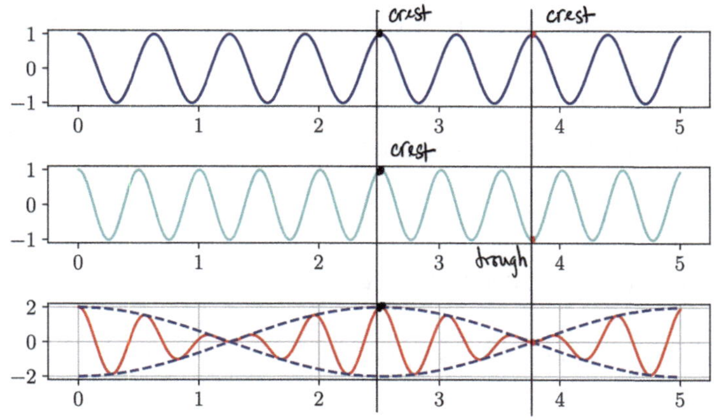

Fig. 7.16 .

$$x_1 + x_2 = \cos(\omega_1 t) + \cos(\omega_2 t) = 2\cos\left(\frac{\omega_1 + \omega_2}{2}\right) t \cos\left(\frac{\omega_1 - \omega_2}{2}\right) t$$

If we plot the waves as in Fig. 7.16 ($\omega_1/\omega_2 = 8 : 10$), we see that where the crests coincide we get a strong wave and where a trough and crest coincide we get practically zero, and then when the crests coincide again we get a strong wave again.

d'Alembert's solution. Now, we turn to d'Alembert's solutions to the wave equation. We have shown that a traveling wave (to the right) can be written as $f(x - ct)$. Thus, $f(x - ct)$, as a wave, must satisfy the wave equation. That is obvious (chain rule is what we need to verify this):

$$\frac{\partial}{\partial t^2}(f(x - ct)) = c^2 \frac{\partial}{\partial x^2}(f(x - ct))$$

And there is nothing special about a wave traveling to the right, we have another wave traveling to the left. It is given by $g(x + ct)$, and it is also a solution to the wave equation. As the wave equation is linear, $f(x - ct) + g(x + ct)$ is also a solution to the wave equation. But, we need a proof.

The equation that we want to solve is for an infinitely long string (so that we do not have to worry about what happens at the boundary):

$$u_{tt} = c^2 u_{xx} \quad -\infty < x < \infty, \quad t > 0 \tag{7.10.4}$$

$$u(x, 0) = f(x), \quad \dot{u}(x, 0) = g(x) \quad -\infty < x < \infty \tag{7.10.5}$$

where $f(x)$ is the initial shape of the string, and $g(x)$ is its initial velocity.

We introduce two new variables ξ and η as

$$\xi = x + ct, \eta = x - ct$$

which transform the PDE from $u_{tt} = c^2 u_{xx}$ to $u_{\xi\eta} = 0$, which can be solved easily:

$$u_{tt} = c^2 u_{xx} \Longrightarrow u_{\xi\eta} = 0 \Longrightarrow u(\xi, \eta) = \phi(\eta) + \psi(\xi) \Longrightarrow u(x, t) = \phi(x - ct) + \psi(x + ct)$$

Now we have to deal with the initial conditions, i.e., Eq. (7.10.5).

$$\phi(x) + \psi(x) = f(x), \quad -c\phi'(x) + c\psi'(x) = g(x)$$

And from that we obtain 'Alembert's solution:

$$u(x, t) = \frac{1}{2}(f(x - ct) + g(x + ct)) + \frac{1}{2c} \int_{x-ct}^{x+ct} g(\xi) d\xi \tag{7.10.6}$$

History note 7.2: Jean-Baptiste le Rond d'Alembert (1717–1783)

Jean-Baptiste le Rond d'Alembert (16 November 1717–29 October 1783) was a French mathematician who was a pioneer in the study of differential equations and their use of in physics. He studied the equilibrium and motion of fluids. Jean d'Alembert's father was an artillery officer, Louis-Camus Destouches and his mother was Mme de Tencin. D'Alembert was the illegitimate son from one of Mme de Tencin "amorous liaisons". His father, Louis-Camus Destouches, was out of the country at the time of d'Alembert's birth and his mother left the newly born child on the steps of the church of St Jean Le Rond. The child was quickly found and taken to a home for homeless children. He was baptized Jean Le Rond, named after the church on whose steps he had been found. When his father returned to Paris he made contact with his young son and arranged for him to be cared for by the wife of a glazier, Mme Rousseau. She would always be d'Alembert's mother in his own eyes, particularly since his real mother never recognized him as her son, and he lived in Mme Rousseau's house until he was middle-aged. Jean Le Rond d'Alembert was one of the eighteenth century's preeminent mathematicians. He was elected to the French Academy of Sciences at the age of only twenty-three. His important contributions include the d'Alembert formula, describing how strings vibrate, and the d'Alembert principle, a generalization of one of Newton's classical laws of motion.

7.11 Solving the Wave Equation

Herein we solve the problem of a finite vibrating string (of length L) which is fixed at two ends. The equations governing the motion of the string are (Sect. 7.5.1)

$$\frac{\partial^2 u}{\partial t^2} = c^2 \frac{\partial^2 u}{\partial x^2} \quad 0 < x < L, \quad t > 0 \tag{7.11.1}$$

$$u(x, 0) = f(x), \quad \dot{u}(x, 0) = g(x) \quad 0 \leq x \leq L \tag{7.11.2}$$

$$u(0, t) = 0, \quad u(L, t) = 0 \quad t > 0 \tag{7.11.3}$$

where $f(x)$ is the original position of the string and $g(x)$ is its velocity at time $t = 0$.

Using the separation of variables method, we write the solution as

$$u(x, t) = X(x)T(t) \tag{7.11.4}$$

And proceed in the same manner as for the heat equation, we now need to solve 2 equations:

$$\begin{aligned} X'' - \lambda X &= 0 \\ T'' - c^2 \lambda T &= 0 \end{aligned} \tag{7.11.5}$$

where $-\infty < \lambda < \infty$. It can be shown that only the case $\lambda < 0$ gives meaningful solution.[42] For $n = 0, 1, 2, \ldots$, we have

$$\begin{aligned} X_n &= \sin\left(\frac{n\pi x}{L}\right) \\ T_n &= A_n \cos\left(\frac{n\pi c}{L}t\right) + B_n \sin\left(\frac{n\pi c}{L}t\right) \end{aligned} \tag{7.11.6}$$

And thus the general solution is

$$u(x, t) = \sum_{n=1}^{\infty} u_n(x, t), \quad u_n(x, t) = \left[A_n \cos\left(\frac{n\pi c}{L}t\right) + B_n \sin\left(\frac{n\pi c}{L}t\right)\right] \sin\left(\frac{n\pi x}{L}\right) \tag{7.11.7}$$

where the n term $u_n(x, t)$ is called the *n-th mode of vibration* or the n-th harmonic. This solution satisfies the PDE and the BCs. If we plot these modes of vibration (Fig. 7.17), what we observe is that the wave doesn't propagate. It just sits there vibrating up and down in place. Such a wave is called a *standing wave*. Points that do not move at any time (zero amplitude of oscillation) are called *nodes*. Points where the amplitude is maximum are called *antinodes*. The simplest mode of vibration with $n = 1$ is called the fundamental, and the frequency at which it vibrates is called the fundamental frequency.

---

[42] Note that our solution must be non-zero and bounded.

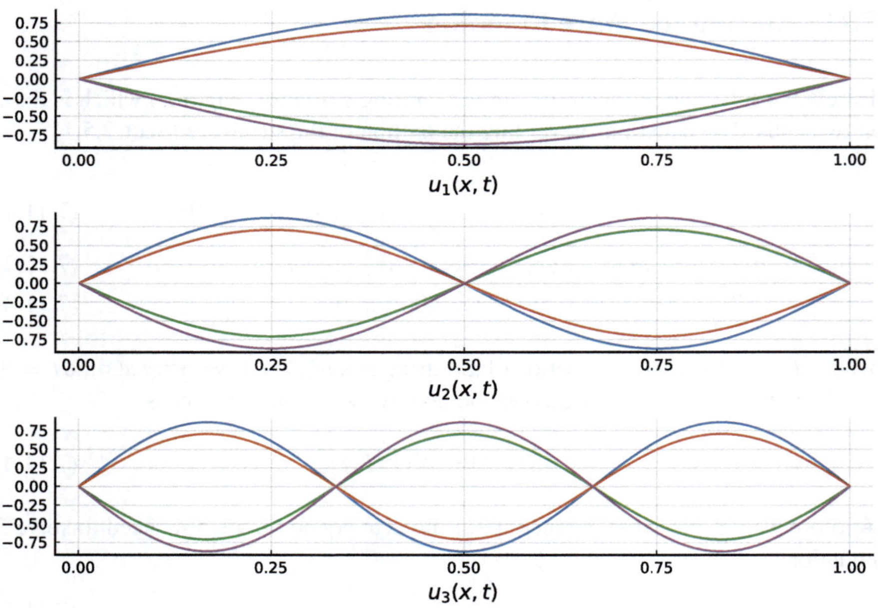

Fig. 7.17 Standing waves $u_n(x, t)$ for $n = 1, 2, 3$. Different colors are used to denote $u_n(x, t)$ for different times

But waves should be traveling, why we have standing waves here? To see why, we need to use trigonometry, particularly the product identities in Eq. (3.8.6) (e.g., $\sin\alpha\cos\beta = \sin(\alpha+\beta)+\sin(\alpha-\beta)/2$). Using these identities, we can rewrite $u_n(x, t)$ as

$$\begin{aligned} u_n(x,t) &= \frac{A_n}{2}\left(\sin\frac{n\pi}{L}(x+ct) + \sin\frac{n\pi}{L}(x-ct)\right) \\ &+ \frac{B_n}{2}\left(\cos\frac{n\pi}{L}(x-ct) - \cos\frac{n\pi}{L}(x+ct)\right) \end{aligned} \tag{7.11.8}$$

Let's now focus on the terms with A_n, we can write

$$\begin{aligned} u_n(x,t) &= \frac{A_n}{2}\sin\frac{n\pi}{L}(x-ct) + \frac{A_n}{2}\sin\frac{n\pi}{L}(x+ct) \\ &= \frac{A_n}{2}\sin\left(\frac{2\pi x}{\lambda_n} - \frac{2\pi ct}{\lambda_n}\right) + \frac{A_n}{2}\sin\left(\frac{2\pi x}{\lambda_n} + \frac{2\pi ct}{\lambda_n}\right), \quad \lambda_n = \frac{2L}{n} \end{aligned} \tag{7.11.9}$$

which is obviously the superposition of two traveling waves: the first term is a wave traveling to the right and the second travels to the left. Both waves have the same amplitude.

All points on the string oscillate at the same frequency but with different amplitudes.

Now we need to consider the initial conditions. By evaluating $u(x,t)$ and its first time derivative at $t = 0$, and using the ICs, we obtain

$$\begin{aligned} \sum_{n=1}^{\infty} A_n \sin\left(\frac{n\pi x}{L}\right) &= f(x) \\ \sum_{n=1}^{\infty} B_n \frac{n\pi c}{L} \sin\left(\frac{n\pi x}{L}\right) &= g(x) \end{aligned} \tag{7.11.10}$$

Again, we meet Fourier series.

Example 7.8 Now, assume that the initial velocity of the string is zero, thus $B_n = 0$, then the solution is

$$u(x,t) = \sum_{n=1}^{\infty} A_n \cos\left(\frac{n\pi c}{L} t\right) \sin\left(\frac{n\pi x}{L}\right), \quad A_n = \frac{2}{L}\int_0^L f(x) \sin\left(\frac{n\pi x}{L}\right) dx \tag{7.11.11}$$

What does this mean? If we break the initial shape of the string into many small components:

$$f(x) = \sum_{n=1}^{\infty} A_n \sin\left(\frac{n\pi x}{L}\right)$$

And let each component dance with $\cos\left(\frac{n\pi c}{L}t\right) A_n \sin\left(\frac{n\pi x}{L}\right)$, then adding up all these small vibrations we will get the solution to the string vibration problem. Now, we consider a plucked guitar string, it is pulled upward at the position $x = d$ so that it reaches height h. Thus, the initial position of the string is

$$f(x) = \begin{cases} \frac{hx}{d}, & \text{if } 0 \le x \le d \\ \frac{h(L-x)}{L-d}, & \text{if } d \le x \le L \end{cases}$$

Then we suddenly release the string and study its motion. As the initial velocity is zero, we just have A_n, which are computed as (Eq. (7.11.11))

$$A_n = \frac{2h}{n^2\pi^2} \frac{L^2}{d(L-d)} \sin\left(\frac{dn\pi}{L}\right)$$

7.12 Fourier Series, Fourier Transform

In this section we study Fourier series deeper than what we had done in Sect. 4.19. Even though Fourier series and related mathematics have wide applications, to ease the presentation, let's assume we have to work with functions that represent any

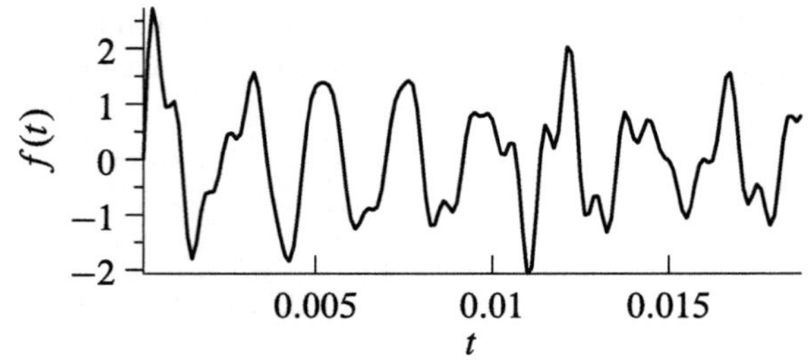

Fig. 7.18 A sound wave $f(t)$

quantity or signal that varies over time, such as the pressure of a sound wave (a radio signal), or daily temperature readings. One sound wave is given in Fig. 7.18.

7.12.1 Parseval's Theorem

To motivate the mathematics, let's assume that we have a SHO and we want to compute its average displacement $\bar{x}(t)$. One way to go is to use the root mean square (RMS). Recall that for n numbers $a_1, a_2, \ldots, a_n$, the RMS is defined as

$$RMS = \sqrt{\frac{a_1^2 + a_2^2 + \cdots + a_n^2}{n}} \tag{7.12.1}$$

Now, we extend this definition to a continuous function $f(x)$. Following the same procedure in Sect. 4.11.3 when we computed the average of a function, we get

$$\bar{f}(x) = \left(\frac{1}{L}\int_{-L}^{L} [f(x)]^2 dx\right)^{1/2} \tag{7.12.2}$$

Recall also that the Fourier series of a periodic function $f(x)$ in $[-L, L]$ is given by

$$f(x) = a_0 + \sum_{n=1}^{\infty}\left(a_n \cos\frac{n\pi x}{L} + b_n \sin\frac{n\pi x}{L}\right) \tag{7.12.3}$$

where the coefficients are

$$a_0 = \frac{1}{2L}\int_{-L}^{L} f(x)dx,\ a_n = \frac{1}{L}\int_{-L}^{L} f(x)\cos\frac{n\pi x}{L}dx,\ b_n = \frac{1}{L}\int_{-L}^{L} f(x)\sin\frac{n\pi x}{L}dx \tag{7.12.4}$$

Now, we introduce $f_N(x)$ which is a finite Fourier series of $f(x)$. That is $f_N(x)$ consists of a finite number $N \in \mathbb{N}$ of the cosine and sine terms:

$$f_N(x) = a_0 + \sum_{n=1}^{N} \left(a_n \cos \frac{n\pi x}{L} + b_n \sin \frac{n\pi x}{L} \right) \quad (7.12.5)$$

With that we compute the RMS of the difference between $f(x)$ and $f_N(x)$[43]:

$$E = \frac{1}{L} \int_{-L}^{L} (f(x) - f_N(x))^2 \, dx = \frac{1}{L} [(f, f) - 2(f, f_N) + (f_N, f_N)] \quad (7.12.6)$$

The plan is like this: if we can compute (f, f_N) and (f_N, f_N), then with the fact that $E \geq 0$, we shall get an inequality, and that inequality is the Bessel inequality. Let's start with (f_N, f_N):

$$\begin{aligned}
(f_N, f_N) &= \int_{-L}^{L} \left[a_0 + \sum_{n=1}^{N} \left(a_n \cos \frac{n\pi x}{L} + b_n \sin \frac{n\pi x}{L} \right) \right]^2 dx \\
&= \int_{-L}^{L} a_0^2 dx + \sum_{n=1}^{N} a_n^2 \int_{-L}^{L} \cos^2 \frac{n\pi x}{L} dx + \sum_{n=1}^{N} b_n^2 \int_{-L}^{L} \sin^2 \frac{n\pi x}{L} dx \\
&= L \left(2a_0^2 + \sum_{n=1}^{N} a_n^2 + \sum_{n=1}^{N} b_n^2 \right)
\end{aligned}$$

The term (f, S_N) is much easier:

$$\begin{aligned}
(f, S_N) &= \int_{-L}^{L} f(x) \left[a_0 + \sum_{n=1}^{N} \left(a_n \cos \frac{n\pi x}{L} + b_n \sin \frac{n\pi x}{L} \right) \right] dx \\
&= a_0 \int_{-L}^{L} f(x) dx + \sum_{n=1}^{N} a_n \int_{-L}^{L} f(x) \cos \frac{n\pi x}{L} dx + \sum_{n=1}^{N} b_n \int_{-L}^{L} f(x) \sin \frac{n\pi x}{L} dx \\
&= L \left(2a_0^2 + \sum_{n=1}^{N} a_n^2 + \sum_{n=1}^{N} b_n^2 \right)
\end{aligned}$$

To arrive at the final result, we just use Eq. (7.12.4) to replace the red integrals by the Fourier coefficients. Substituting all these into the second of Eq. (7.12.6), we obtain

$$E = \frac{1}{L}(f, f) - \left(2a_0^2 + \sum_{n=1}^{N} a_n^2 + \sum_{n=1}^{N} b_n^2 \right) \quad (7.12.7)$$

As $E \geq 0$, we get the following inequality:

[43] Although not necessary, I used the short notation (f, g) to denote the inner product $\int_{-L}^{L} f(x)g(x)dx$.

$$2a_0^2 + \sum_{n=1}^{N} a_n^2 + \sum_{n=1}^{N} b_n^2 \le \frac{1}{L}\int_{-L}^{L} f^2(x)dx$$

Thus, we have obtained the following result. Let $f(x)$ be a function defined on $[-L, L]$ such that $f^2(x)$ has a finite integral on $[-L, L]$, then (let N approach ∞):

$$\text{Parseval's theorem : } \boxed{2a_0^2 + \sum_{n=1}^{\infty} (a_n^2 + b_n^2) = \frac{1}{L}\int_{-L}^{L} f^2(x)dx} \tag{7.12.8}$$

Still remember the Pythagorean theorem? For a vector $\boldsymbol{a} = (a_1, a_2)$, its squared length is $||\boldsymbol{a}||^2 = a_1^2 + a_2^2$. Parseval's theorem is a generalization of the Pythagorean theorem to infinite-dimensional vector spaces. This is so for we consider the integral in Eq. (7.12.8) a squared 'length' of the function $f(x)$.

As one simple application of this theorem, we consider again the famous Basel problem. First, we note that the Fourier series for $f(x) = x$ is given by

$$f(x) = x = 2\left(\sin(x) - \frac{\sin(2x)}{2} + \frac{\sin(3x)}{3} - \frac{\sin(4x)}{4} + \dots\right)$$

The coefficients b_k of this functions are $b_k = \frac{2(-1)^{k+1}}{k}$ and Parseval's theorem then gives us:

$$\sum_{k=1}^{\infty} \frac{4}{k^2} = \frac{1}{\pi}\int_{-\pi}^{\pi} x^2 dx = \frac{2\pi^2}{3}$$

From it, we get the famous Basel problem! With powerful tools, we can solve difficult problems easily. This Basel problem has defied the likes of Leibnitz, Bernoulli brothers, and it required the genius of Euler to solve it. But now, it is just a consequence of a theorem in Fourier analysis!

7.12.2 Fourier Transforms (Fourier Integrals)

In Sect. 4.19 I have introduced the Fourier series for periodic functions. This section extends the maths to non-periodic functions. We start with the complex form of Fourier series, written as for any T−periodic functions $f(t)$:

$$f(t) = \sum_{n=-\infty}^{\infty} c_n e^{in2\pi t/T}, \quad c_n = \frac{1}{T}\int_{-T/2}^{T/2} f(t)e^{-in2\pi t/T} dt \tag{7.12.9}$$

We call the first equation in Eq. (7.12.9) a *synthesis* equation, because it tells us how to build, construct, or synthesize a periodic signal (that is how electrical engineers call $f(t)$) from other signals. In the present case, the building blocks are complex exponentials, $e^{in2\pi t/T}$ or $e^{in2\pi ft}$, and the method of constructing $f(t)$ is to add scaled or weighted versions of each complex exponential according to the set of expansion coefficients c_n. We also see that nf, the frequencies of the complex exponentials, are integer multiples of the fundamental frequency.

The key idea is: *a non-periodic function is a periodic function of an infinite period!* So, we will massage the above equations and see what happens when $T \to \infty$. First, we write $f(t)$ as

$$f(t) = \sum_{n=-\infty}^{\infty} \left(\frac{1}{T} \int_{-T/2}^{T/2} f(t) e^{-in2\pi t/T} dt \right) e^{in2\pi t/T}$$

We now introduce a new variable $k_n = {}^{n}/_{T}$ and $\Delta k = {}^{1}/_{T}$. Using these we can rewrite the above as

$$f(t) = \sum_{n=-\infty}^{\infty} \left(\int_{-T/2}^{T/2} f(t) e^{-i2\pi k_n t} dt \right) e^{i2\pi k_n t} \Delta k \tag{7.12.10}$$

We focus on the integral in the brackets, which is $C(k)$ and write $C(k)$ as

$$C(k) = \int_{-T/2}^{T/2} f(t) e^{-i2\pi kt} dt = \int_{-\infty}^{\infty} f(t) e^{-i2\pi kt} dt \quad (\text{as } T \to \infty)$$

Therefore, Eq. (7.12.10) becomes

$$f(t) = \sum_{n=-\infty}^{\infty} C(k_n) e^{i2\pi k_n t} \Delta k$$

in which the RHS is nothing but a Riemann sum, so we have

$$f(t) = \int_{-\infty}^{\infty} C(k) e^{i2\pi kt} dk$$

Therefore, we have derived the Fourier transform and its inverse for a function. Instead of $f(t)$ to denote the function, we use $g(t)$, and instead of $C(k)$, the notation $\hat{g}(f)$ is used:

$$\boxed{\hat{g}(f) = \int_{-\infty}^{\infty} g(t) e^{-2\pi i f t} dt}, \quad \boxed{g(t) = \int_{-\infty}^{\infty} \hat{g}(f) e^{2\pi i f t} df} \tag{7.12.11}$$

We call $\hat{g}(f)$ the Fourier transform of $g(t)$. And the second equation which gives us back $g(t)$ from $\hat{g}(f)$ is called the inverse Fourier transform. The two functions $g(t)$ and $\hat{g}(f)$ are termed a Fourier transform pair. We express this pair by $g(t) \Leftrightarrow \hat{g}(f)$.

What we're going to do next is to understand the Fourier transform visually. Doing so reveals that behind a complex formula is many simpler things.

7.12.3 Visualizing the Fourier Transform

Herein we're trying to understand the Fourier transform visually. To this end, I rewrite it below and highlight different terms in it:

$$\boxed{\hat{g}(f) = \int_{-\infty}^{\infty} g(t)e^{-2\pi i f t}dt} \tag{7.12.12}$$

and note that f is the frequency (number of cycles per second or Hz). This beautiful equation ranks ninth in Ian Stewart's list of the seventeen equations that changed the world.

To understand the meaning of the Fourier transform given in Eq. (7.12.12) we first focus on the integrand $g(t)e^{-2\pi i f t}$. Second, we focus on the entire integral.

Whenever we see the complex exponential we think of the unit circle on the complex plane. Thus, $e^{-2\pi i f t}$ is nothing but the graph of a point moving on the unit circle with a frequency f; that is, in one second, it moves f rounds along the unit circle clockwise. The term $g(t)e^{-2\pi i f t}$ is the product of $e^{-2\pi i f t}$ with $f(t)$. Obviously, if $f(t) = A$, which is geometrically a line, then $g(t)e^{-2\pi i f t}$ is a circle of radius A: we have wrapped a line into a circle. What interesting is when $g(t)$ is non-constant. To see what this terms is, we choose to work with a concrete sound wave of frequency f_0:

$$g(t) = 1 + \cos(2\pi f_0 t)$$

of which the graph is given in Fig. 7.19a for $f_0 = 3$ cycles per second (or Hz). Then, we fix f, let say 2 cycles per second. Then we plot $g(t)\ e^{-2\pi i f t}$. How? We use $e^{i\theta} = \cos\theta + i\sin\theta$:

$$g(t)e^{-2\pi i f t} = \underbrace{g(t)\cos(2\pi f t)}_{x(t)} - i\,\underbrace{g(t)\sin(2\pi f t)}_{y(t)}$$

Thus, we have a parametric equation $(x(t), -y(t))$, and we definitely know how to plot it. In Fig. 7.19b-d, I plot $g(t)e^{-2\pi i f t}$ for $f = 0.25, 3, 7$ Hz. We see clearly in Fig. 7.19b that the graph of $g(t)$ is wrapped around the unit circle by $g(t)e^{-2\pi i f t}$.

Now, we turn attention to the entire $\hat{g}(f)$:

$$\hat{g}(f) = \int_{-\infty}^{\infty} g(t)e^{-2\pi i f t}dt = \int_{-\infty}^{\infty} g(t)\cos(2\pi f t)dt - i\int_{-\infty}^{\infty} g(t)\sin(2\pi f t)dt \tag{7.12.13}$$

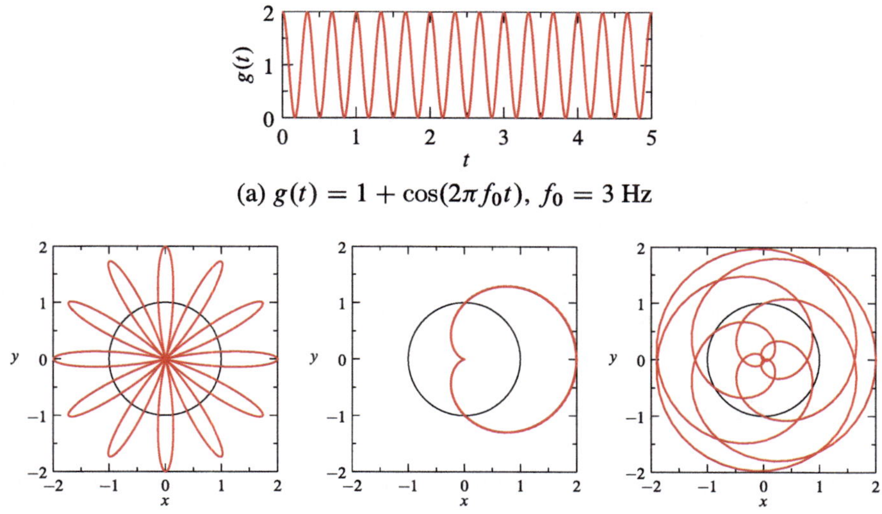

(a) $g(t) = 1 + \cos(2\pi f_0 t)$, $f_0 = 3$ Hz

(b) $g(t)e^{-2\pi i f t}$, $f = .25$ (c) $g(t)e^{-2\pi i f t}$, $f = 3$ (d) $g(t)e^{-2\pi i f t}$, $f = 7$

Fig. 7.19 Plots of $g(t)$ and $g(t)e^{-2\pi i f t}$

I introduce some terminologies now: $\int_{-\infty}^{\infty} g(t)\cos(2\pi f t)dt$ is called the *Fourier cosine transform* of $f(t)$ whereas $\int_{-\infty}^{\infty} g(t)\sin(2\pi f t)dt$ is called the *Fourier sine transform* of $f(t)$. They are the other form of the Fourier transform without using complex numbers.

Given a real frequency f, the above equation returns a complex number $\hat{g}(f)$. What we're going to do is to visualize $\hat{g}(f)$ when we vary f continuously. For simplicity, I use the plot $(f, \text{Re}\hat{g}(f))$. In Fig. 7.20b, I provide such a plot. In order to avoid evaluating the integrals in Eq. (7.12.13), I approximated it as

$$\int_{-\infty}^{\infty} g(t)\cos(2\pi f t)dt \approx \frac{1}{N}\sum_{i=1}^{N} g(t_i)\cos(2\pi f t_i)$$

The formula in the RHS reminds us of the familiar concept of center of mass in physics. So, one way to understand the integral in the Fourier transform is to see the graph in Fig. 7.19b as a bunch of points, of which the $x-$coordinate of the center of mass of them is $1/N \sum_{i=1}^{N} g(t_i)\cos(2\pi f t_i)$.

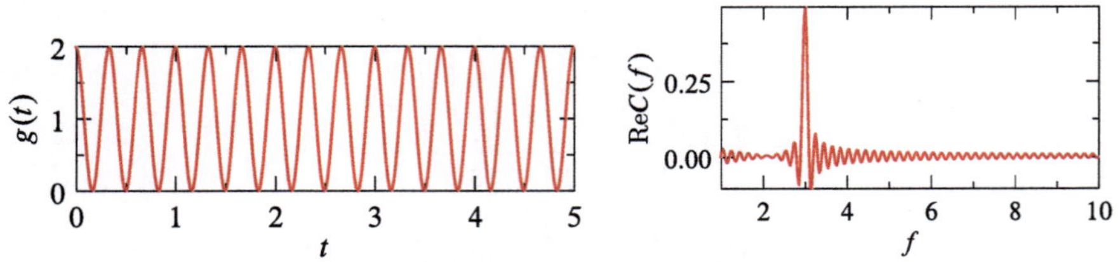

(a) $g(t) = 1 + \cos(2\pi f_0 t)$, $f_0 = 3$ Hz (b) $\frac{1}{N}\sum_{i=1}^{N} g(t_i)\cos(2\pi f t_i)$

Fig. 7.20 One interpretation of the Fourier transform: it converts a signal from the time domain, which is $g(t)$, to the frequency domain, which is $\text{Re}\hat{g}(f)$

7.12.4 The Function $\sin x/x$

We have met the function $\sin x/x$ but the origin of it was not explained. Herein, by considering a rectangular wave $g(t)$ given by

$$g(t) = \begin{cases} 1, & \text{if } |t| \le 1/2 \\ 0, & \text{if } |t| > 1/2 \end{cases} \tag{7.12.14}$$

and compute its Fourier transform $\hat{g}(f)$, we see that that function shows up. Using Eq. (7.12.11), we compute $\hat{g}(f)$ as

$$\hat{g}(f) = \int_{-\infty}^{\infty} g(t)e^{-2\pi ift}dt = \int_{-1/2}^{1/2} e^{-2\pi ift}dt = \frac{1}{-2\pi if}\left(e^{-2\pi if}\right)\Big|_{-1/2}^{1/2} = \frac{\sin \pi f}{\pi f}$$

where use was made of Eq. (2.25.21) to go from complex exponentials to the sine. And voilà the beautiful function $\sin x/x$ appears. It is said that electrical engineers see this function in their dreams.

7.12.5 The Dirac Delta Function

Assume that we want to calculate the dynamics of a billiard ball being struck by another ball at time $t = 0$. To this end, we need to model the striking force. This force is zero for $t < 0$ (the ball is not yet struck), it is something within the tiny time interval of the impact, and finally it is zero again after that. The Dirac delta function, to be introduced in this section, is suitable to model this type of force. Another example for the motivation. Suppose we have some distribution of mass on a line. If the (linear) density is $\rho(x)$ then the total amount of mass in an interval $[a, b]$ is $m = \int_a^b \rho(x)dx$. But what if we have a mass of 1, all concentrated at a single point

$x = 0$? We would need a density function that is *zero every except at* $x = 0$, and since the mass is one, we must have

$$\int_{-\infty}^{\infty} \rho(x)dx = 1$$

What kind of function is that? So, the problem is: we want to mathematically model physical effects that are *concentrated within very small volumes or times*, for which we don't actually want to worry about the microscopic details in this volume. For example, think of the concepts of a point charge, a point mass, a force plucking a string at one point, a kick that suddenly imparts some momentum to an object, and so on.

We start with a box function of unit area:

$$g_\epsilon(t) = \begin{cases} 1/2\epsilon, & \text{if } |t| < \epsilon \\ 0, & \text{if } |t| \geq \epsilon \end{cases} \tag{7.12.15}$$

For any $\epsilon > 0$ this is a perfectly ordinary, well-behaved function. But there is magic hidden in it when we consider what happens when $\epsilon \to 0$. The graph of $g_\epsilon(t)$ is given in Fig. 7.21 for two different finite values of ϵ.

Let $f(t)$ be a function which is continuous at $t = 0$, and we're interested in this integral $I = \int_{-\infty}^{\infty} g_\epsilon(t) f(t)dt$, when $\epsilon \to 0$. We can determine this integral using the mean value theorem of an integral (Eq. (4.11.3)):

$$I = \int_{-\infty}^{\infty} g_\epsilon(t) f(t)dt = \int_{-\epsilon}^{\epsilon} \frac{1}{2\epsilon} f(t)dt = 2\epsilon \left(\frac{1}{2\epsilon} f(\xi) \right) = f(\xi), \quad \xi \in (-\epsilon, \epsilon)$$

Now taking the limit when $\epsilon \to 0$: $g_\epsilon(t)$ becomes $\delta(t)$ and ξ is zero. Thus, we have

$$\lim_{\epsilon \to 0} I = \boxed{\int_{-\infty}^{\infty} \delta(t) f(t)dt = f(0)} \tag{7.12.16}$$

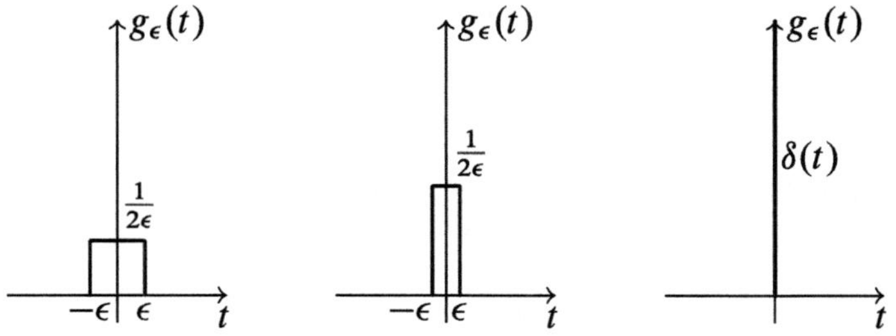

Fig. 7.21 .

The "function" $\delta(t)$ is the *Dirac delta function*, named after the British theoretical physicist Paul Dirac whom we have met in Chap. 3. Dirac introduced this function in his 1927 paper *The Physical Interpretation of the Quantum Dynamics* and in his textbook *The Principles of Quantum Mechanics*. He called it the "delta function" since he used it as a continuous analogue of the discrete Kronecker delta. Its graph is given in the right most of Fig. 7.21: it is an infinitely thin and infinitely high spike. And yet, the area under the graph of this function (whatever it means) is one: if we pick $f(t) = 1$ and plug it in the above boxed equation, we immediately obtain

$$\int_{-\infty}^{\infty} \delta(t)dt = 1.$$

7.12.6 Dirichlet Discontinuous Integral and the Sign Function

In Sect. 4.7.9 when we were talking about the Feynman trick to evaluate integrals, we derived the following result

$$\int_0^{\infty} \frac{\sin y}{y} dy = \frac{\pi}{2} \tag{7.12.17}$$

We usually work with $\sin 3y$, $\sin \pi y$ and so on; the question is: whether Eq. (7.12.17) still holds if we replace $\sin x$ by $\sin kx$? To answer this question, we do: $y = x\omega$ where x is some constant, thus $dy = xd\omega$, we then have, for $x > 0$

$$\frac{\pi}{2} = \int_0^{\infty} \frac{\sin y}{y} dy = \int_0^{\infty} \frac{\sin x\omega}{x\omega} x d\omega = \int_0^{\infty} \frac{\sin x\omega}{\omega} d\omega$$

Thus, Eq. (7.12.17) is just a special case of $k = 1$. The result holds for any $k > 0$! How about $k < 0$? The integral is simply $-\pi/2$. Thus, we obtain

$$\boxed{\int_0^{\infty} \frac{\sin \omega x}{\omega} d\omega = \begin{cases} +\frac{\pi}{2}, & \text{if } x > 0 \\ -\frac{\pi}{2}, & \text{if } x < 0 \end{cases} = \pi \left(u(x) - \frac{1}{2} \right) = \frac{\pi}{2} \text{sgn}(x)} \tag{7.12.18}$$

The integration limits are not symmetric; we need to introduce $-\infty$ to the game. Since $f(\omega) = {}^{\sin \omega x}/_{\omega}$ is an even function, we then have

$$\int_{-\infty}^{\infty} \frac{\sin \omega x}{\omega} d\omega = \pi \text{sgn}(x) \tag{7.12.19}$$

What would mathematicians do next here? They ask: how about cosine? Thus, they write

$$\int_{-\infty}^{\infty} \frac{\cos \omega x}{\omega} d\omega = 0 \tag{7.12.20}$$

Now to combine sine and cosine into one formula, we have to use $e^{ix} = \cos x + i \sin x$. Thus, we multiply Eq. (7.12.19) with i and add the result with Eq. (7.12.20). Below is what we get:

$$\int_{-\infty}^{\infty} \left(\frac{\cos \omega x}{\omega} + i \frac{\sin \omega x}{\omega} \right) d\omega = i\pi \text{sgn}(x)$$

which is nothing but

$$\int_{-\infty}^{\infty} \frac{e^{i\omega x}}{\omega} d\omega = i\pi \text{sgn}(x)$$

7.12.7 Fourier Transform from the Dirac Function

Now, I present a derivation of the Fourier transform using the Dirac delta function.[44] First, we derive yet another beautiful equation from Eq. (7.12.18). Differentiating it with respect to x, we get

$$\int_{0}^{\infty} \cos \omega x d\omega = \pi \delta(x) \Longrightarrow \int_{-\infty}^{\infty} \cos \omega x d\omega = 2\pi \delta(x)$$

Now, we play the same game: $\int_{-\infty}^{\infty} \sin \omega x d\omega = 0$, thus

$$\int_{-\infty}^{\infty} e^{i\omega x} d\omega = 2\pi \delta(x) \Longrightarrow \boxed{\delta(x) = \frac{1}{2\pi} \int_{-\infty}^{\infty} e^{i\omega x} d\omega}$$

Instead of $\delta(x)$, we use $\delta(x - y)$, the above boxed equation then gives us

$$\delta(x - y) = \frac{1}{2\pi} \int_{-\infty}^{\infty} e^{i\omega(x-y)} d\omega \tag{7.12.21}$$

Now we substitute Eq. (7.12.21) into the following,

$$h(x) = \int_{-\infty}^{\infty} \delta(x - y) h(y) dy \tag{7.12.22}$$

44 I've learned this from the book *Dr Euler's fabulous formula* of Paul Nahin.

we then get

$$
\begin{aligned}
h(x) &= \int_{-\infty}^{\infty} \frac{1}{2\pi} \int_{-\infty}^{\infty} e^{i\omega(x-y)} d\omega h(y) dy \\
&= \int_{-\infty}^{\infty} \frac{1}{2\pi} \int_{-\infty}^{\infty} e^{i\omega x} e^{-i\omega y} d\omega h(y) dy \\
&= \int_{-\infty}^{\infty} \left(\frac{1}{2\pi} \int_{-\infty}^{\infty} e^{-i\omega y} h(y) dy \right) e^{i\omega x} d\omega
\end{aligned}
$$

This is exactly the Fourier transform and its inverse; but written using ω—the angular frequency—instead of the frequency f. You can compare it with Eq. (7.12.11).

Chapter 8
Calculus of Variations

This chapter is devoted to the calculus of variations which is a branch of mathematics that allows us to find a function $y = f(x)$ that minimizes a functional—a function of function, for example, $I = \int_a^b G(y, y', y'', x)dx$ is a functional. The calculus of variations provides answers to questions like "what is the plane curve with maximum area with a given perimeter?". You might have correctly guessed the answer: in the absence of any restriction on the shape, the curve is a circle. But calculus of variation provides a proof and more.

This chapter serves only as a brief introduction to this interesting theory of mathematics. It also provides a historical account of the development of the finite element method, often regarded as one of the greatest achievements in the twentieth century.

I have used primarily the following books for the material presented herein:

- *When Least Is Best: How Mathematicians Discovered Many Clever Ways to Make Things as Small (or as Large) as Possible* by Paul Nahin[1] [48].
- *A History of the Calculus of Variations from the 17th through the 19th Century* by Herman Goldstine[2] [5].

[1] Paul Joel Nahin (born November 26, 1940) is an American electrical engineer and author who has written 20 books on topics in physics and mathematics, including biographies of Oliver Heaviside, George Boole, and Claude Shannon, books on mathematical concepts such as Euler's formula and the imaginary unit, and a number of books on the physics and philosophical puzzles of time travel. Nahin received, in 1979, the first Harry Rowe Mimno writing award from the IEEE Aerospace and Electronic Systems Society, and the 2017 Chandler Davis Prize for Excellence in Expository Writing in Mathematics.

[2] Herman Heine Goldstine (1913–2004) was a mathematician and computer scientist, who worked as the Director of the IAS machine at Princeton University's Institute for Advanced Study, and helped to develop ENIAC, the first of the modern electronic digital computers. He subsequently worked for many years at IBM as an IBM Fellow, the company's most prestigious technical position.

V. P. Nguyen, *Mathematics for Engineers and Scientists*,
https://doi.org/10.1007/978-981-97-4631-6_8

- *The Variational Principles of Mechanics* by Cornelius Lanczos[3] [38].
- *The lazy universe. An introduction to the principle of least action* by Jennifer Coopersmith[4] [14]

We start with an introduction in Sect. 8.1. Next, some elementary variational problems are given in Sect. 8.2 to illustrate the problems that this branch of mathematics has to deal with. Section 8.3 presents Lagrange's derivation of the Euler-Lagrange equation. Using this equation, Sect. 8.4 provides solutions to some elementary variational problems given in Sect. 8.2. The variational operator δy is introduced in Sect. 8.5; it is for change in a function $y(x)$ similar to dx, which is change in a number x. Two-dimensional variational problems are treated in Sect. 8.6. Boundary conditions are presented in Sect. 8.7. Section 8.8 is a brief introduction to Lagrangian mechanics—a new formulation of Newtonian mechanics using calculus of variations.

Section 8.9 discusses Ritz's direct method to solve variational problems numerically. The Ritz method begins with a functional, however, in many cases, we only have a partial differential equation, not the corresponding functional. To handle those situations, Sect. 8.10 presents the Dirichlet principle, which states that for a certain PDE we can find the associated variational principle. Then Sect. 8.11 treats what is now called the Galerkin method—a method that can solve numerically any PDE without knowing the variational principle.

The Ritz-Galerkin method is, however, limited to problems of simple geometries. Section 8.12 is devoted to a discussion on the finite element method, which can be considered as a generalization of the Ritz-Galerkin method. The finite element method can solve PDE defined on any geometry.

8.1 Introduction and Some History Comments

Calculus of variations or variational calculus is a branch of mathematics to solve the so-called *variational problems*. A variational problem is to find a function, e.g., $u(x)$ that minimizes a functional—a function of functions. A functional is mostly written as a definite integral that involves u, u' (i.e., du/dx) and x; for example, the following integral

$$I = \int_a^b F(u(x), u'(x); x)dx$$

is a functional. Briefly, if we input a function and its derivatives into a functional we get a number (i.e., I).

---

[3] Cornelius Lanczos (1893–1974) was a Hungarian-American and later Hungarian-Irish mathematician and physicist. In 1924 he discovered an exact solution of the Einstein field equation representing a cylindrically symmetric rigidly rotating configuration of dust particles. Lanczos served as assistant to Albert Einstein during the period of 1928–29.

[4] Jennifer Coopersmith (born in 1955 in Cape Town, South Africa). She obtained a BSc and a PhD in physics from King's College, University of London.

Going back in history, variational calculus started in 1696 with the famous *brachistochrone problem* stated by Johann Bernoulli.[5] In 1744, Euler gave a general solution to variational problems in the form of a differential equation—the well-known Euler-Lagrange equation. That is, the solution to a variational problem is the solution to a partial differential equation associated with the variational problem. Of course we still need to solve for this partial differential equation to get the solution to the original problem, but it was a big result. Eleven years later Euler's solution, 19-year-old Lagrange provided an elegant derivation for this equation.

There is a deep reason why variational calculus has become an important branch of mathematics. It is the fact that nature follows laws which can be expressed as variational principles. For example, Newtonian mechanics is equivalent to the least action variational principle that states that among various paths that a particle can follow, the actual path minimizes a functional. This functional is the integral of the difference between the kinetic energy and potential energy:

$$\mathcal{L}[x(t)] = \int_1^2 [\text{KE}(\dot{x}(t)) - \text{PE}(x(t))]dt$$

Yes, among infinitely many paths that a particle can choose that connect point 1 and point 2, it chooses the one that minimizes the functional $\mathcal{L}[x(t)]$. It is simply super remarkable.

Even though Euler has developed many techniques to solve the Euler-Lagrange partial differential equations, it was the physicist Walter Ritz who, in 1902, proposed a direct method to solve approximately variational problems in a systematic manner. The modifier "direct" means that one can work directly with the functional instead of first finding the associated Euler-Lagrange equation and then solving this equation; the way Euler and many other mathematicians before Ritz did.

8.2 Examples

We have seen ordinary functions such as $f(x) = x^2$ or $f(x, y) = x^2 + y^2$, but we have not seen a functional before. This section presents some examples so that we get familiar with functionals and variational problems. Note that we do not try to solve those problems in this section.

Euclidean geodesic problem is to find the shortest path joining two points (x_1, y_1) and (x_2, y_2). To this end, we are finding a curve mathematically expressed by the function $f(x)$ such that the following integral (or functional)

[5] Johann Bernoulli (1667–1748) was a Swiss mathematician and was one of the many prominent mathematicians in the Bernoulli family. He is known for his contributions to infinitesimal calculus and educating Leonhard Euler in the pupil's youth.

$$l[f(x)] = \int_{(x_1,y_1)}^{(x_2,y_2)} ds = \int_{x_1}^{x_2} \sqrt{1+(f'(x))^2}\, dx \tag{8.2.1}$$

is minimum. We use the notation $l[f(x)]$ to denote a functional l that depends on $f(x)$ (and possibly its derivatives $f'(x), f''(x), \ldots$). In this particular example, our functional depends only on the first derivative of the sought for function.

Brachistochrone problem—John and James Bernoulli 1697. Suppose a particle is allowed to slide freely and frictionlessly along a wire under gravity from a point A to point B (Fig. 8.1a). Furthermore, assume that the beads start with a zero velocity. Find the curve $y = y(x)$ that minimizes the travel time. Such a curve is called a brachistochrone curve (from Ancient Greek brákhistos khrónos "shortest time"). Surprisingly that curve is not a line.

To solve this problem we first need to compute the traveling time, then find $y(x)$ that minimizes that time. We use differential calculus to compute dt—the infinitesimal time required for the particle to travel a distance ds. We need to know the velocity of the particle for this purpose. For simplicity, we select a coordinate system as shown in Fig. 8.1b where the starting point A is at the origin and the vertical axis is pointing downward. Using the principle of conservation of energy (at time $t = 0$ and any time instance t) leads to (at $t = 0$ the total energy of the particle is zero)

$$\frac{1}{2}mv^2 - mgy = 0$$

Thus, the particle velocity, $v = {}^{ds}/_{dt}$, is given by $v = \sqrt{2gy}$. It is now possible to compute dt, and hence the total time

$$dt = \frac{ds}{v} = \frac{\sqrt{1+[y'(x)]^2}dx}{\sqrt{2gy}} \Longrightarrow t = \int_0^a \frac{\sqrt{1+[y'(x)]^2}}{\sqrt{2gy(x)}}\, dx \tag{8.2.2}$$

The shortest curve is then the one with $y(x)$ that minimizes the above integral.

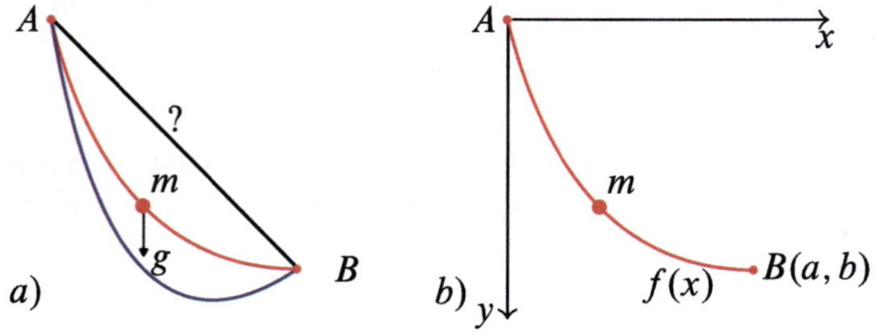

Fig. 8.1 A brachistochrone curve is a curve of shortest time or curve of fastest descent; g is the acceleration of gravity

Minimal surface of revolution. Suppose the curve $y = f(x) \geq 0$ is rotated about the x-axis. The area of the surface of the resulting solid is $2\pi \int_a^b f(x)\sqrt{1+[f'(x)]^2}dx$, check Sect. 4.9.3 for detail. Find the curve which makes this area minimal.

Galileo's hanging chain. Galileo Galilei in his Discorsi1 (1638) described a method of drawing a parabola as "Drive two nails into a wall at a convenient height and at the same level; ... Over these two nails hang a light chain ... This chain will assume the form of a parabola, ...". Unfortunately, the hanging chain does not assume the form of a parabola and Galileo's assertion became a discussion point for followers of his work. Prominent mathematicians of the time, Leibniz, Huygens and Johann Bernoulli, studied the hanging chain problem, which can be stated as: *Find the curve assumed by a loose flexible string hung freely from two fixed points*. Every person viewing power lines hanging between supporting poles is seeing Galileo's hanging chain, which is called a catenary, a name that is derived from the Latin word catena, meaning chain.

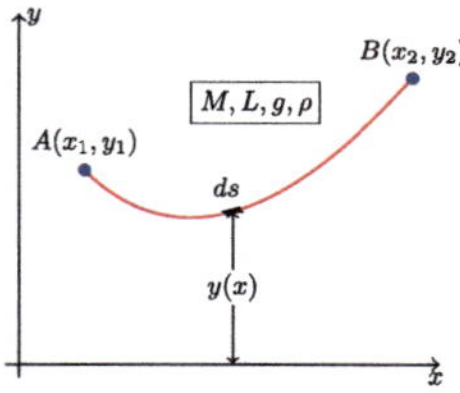

How is this problem related to the above variational problems? In other words, what quantity is to be minimized? The answer is the potential energy of the chain! Let's consider a flexible chain hung by two points A and B. The chain has a total mass M, a total length L, and thus a uniform mass per length density $\rho = M/L$. The chain is described mathematically as $y(x)$. Let's consider a (very) small segment ds of the chain locating at a distance $y(x)$ above $y = 0$, and the potential energy of this segment of mass $m = \rho ds$ is

$$mgy = \rho gyds = \rho gy\sqrt{dx^2 + dy^2}$$

Thus, the total potential energy is

$$\text{P.E} = \int_{x_1}^{x_2} \rho gy\sqrt{dx^2 + dy^2} = \int_{x_1}^{x_2} \rho gy\sqrt{1 + (y')^2}dx$$

The problem is then: find the curve $y(x)$ passing through $A(x_1, y_1)$ and $B(x_2, y_2)$ such that P.E is minimum. Not really. We forgot that not every curve is admissible; only curves of the same length L are. So, the problem must be stated like this: find the curve $y(x)$ passing through $A(x_1, y_1)$ and $B(x_2, y_2)$ such that

$$I[y, y'; x] = \int_{x_1}^{x_2} \rho gy\sqrt{1 + (y')^2}dx$$

is minimum while satisfying this constraint (check arclength in Sect. 4.9.1 if this is not clear):

$$\int_{x_1}^{x_2} \sqrt{1+(y')^2}dx = L$$

This is certainly a variational problem, but with constraints. As we have learned from calculus, we need Lagrange to handle the constraints.

Calculus-based solution of the hanging chain problem. Herein we present the calculus-based solution of the hanging chain problem. It was done by Leibniz and Johann Bernoulli before variational calculus was developed. I provide this solution to illustrate two points: (i) how calculus can be used to solve problems and (ii) how the same problem (in this context a mechanics one) can be solved by more than one way.

Considering a segment of the chain locating between x and $x+\Delta x$ as shown in Fig. 8.2; the length of this segment is Δs, there are three forces acting on this segment: the tension at the left end $T(x)$, the tension at the right end $T(x+\Delta x)$ and the gravity $\rho g \Delta s$. As this segment is stationary, i.e., not moving, the sum of total forces acting on it must be zero:

$$\begin{aligned}
&\sum F_x = 0 : T(x)\cos\alpha(x) = T(x+\Delta x)\cos\alpha(x+\Delta x) \\
&\sum F_y = 0 : T(x+\Delta x)\sin\alpha(x+\Delta x) - T(x)\sin\alpha(x) - \rho g \Delta s = 0
\end{aligned}$$

From the first equation, we deduce that the horizontal component of the tension in the chain is constant:

$$T(x)\cos\alpha(x) = T_0 = \text{constant} \Longrightarrow T(x) = \frac{T_0}{\cos\alpha(x)}$$

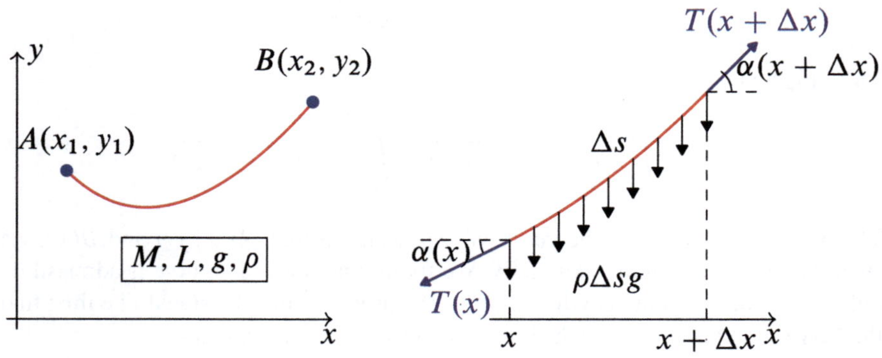

Fig. 8.2 Hanging chain problem: forces acting on a segment Δx of the chain

And from the second equation, we get

$$\Delta\left(T(x)\sin\alpha(x)\right) = \rho g \Delta s \Longrightarrow \frac{\Delta\left(T(x)\sin\alpha(x)\right)}{\Delta x} = \rho g \frac{\Delta s}{\Delta x}$$

Replacing $T(x)$ by $T_0/\cos\alpha(x)$ and considering the limit when $\Delta x \to 0$, we then have

$$\frac{d}{dx}\left(T_0 \tan\alpha(x)\right) = \rho g \sqrt{1+(y')^2}$$

And finally, we obtain the differential equation for the hanging chain (noting that T_0 is constant and $\tan\alpha(x) = y'(x)$):

$$T_0 y'' = \rho g \sqrt{1+(y')^2}$$

To solve this differential equation, we followed Vincenzo Riccati[6] with a new variable z such that $y' = z$:

$$y' = z \Longrightarrow T_0 z' = \rho g \sqrt{1+z^2} \Longleftrightarrow k \frac{dz}{\sqrt{1+z^2}} = dx, \quad k := \frac{T_0}{\rho g}$$

Now, integrating both sides we get (see Sect. 4.4.15)

$$k \int \frac{dz}{\sqrt{1+z^2}} = \int dx \Longrightarrow C_1 + k \sinh^{-1} z = x$$

where C_1 is a constant of integration, and from this we get z, and finally from $z = dy/dx$, we obtain $y(x)$:

$$z = \sinh\left(\frac{x - C_1}{k}\right) \Longrightarrow y = k\cosh\left(\frac{x - C_1}{k}\right) + C_2$$

where C_2 is yet another constant of integration. If the lowest point of the catenary is at $(0, k)$, it can be seen that $C_1 = C_2 = 0$, and the catenary has this form

$$y = k\cosh\left(\frac{x}{k}\right)$$

I hope that with this hanging chain problem, the introduction of hyperbolic functions into mathematics is easier to accept. Again, it is remarkable that mathematics, as a human invention, captures quite well natural phenomena.

[6] Vincenzo Riccati (1707–1775) was a Venetian mathematician and physicist.

8.3 Variational Problems and Euler-Lagrange Equation

The classical variational problem is finding a function $y(x)$ such that the following functional

$$I[y(x)] := \int_a^b F(y, y'; x)\, dx, \quad y(a) = A, \quad y(b) = B \tag{8.3.1}$$

is minimized. In the above A and B are two real constants and $y' = dy/dx$ is the first derivative of y. As can be seen, all the examples given in Sect. 8.2 belong to this general problem. This functional has one independent variable x and one single dependent variable $y(x)$. Thus, it is the easiest variational problem. While other mathematicians solved specific problems like those presented in Sect. 8.2, once got interested, the great Euler solved Eq. (8.3.1) once and for all. And by doing just that he pioneered a new branch of mathematics. His solution was, however, geometrical and not elegant as Lagrange's one. I refer to [38] for Euler's derivation. In what follows, we present the modern solution, which is essentially due to Lagrange when he was only 19 years old.

Before studying Lagrange's solution, let's recap how we find the minimum of $f(x)$. We denote the minimum point by x_0 and vary it a bit and saying that the corresponding change in f must be zero. Mathematically, we consider a small change in x, denoted by dx, that is, $x_0 + dx$. We compute the corresponding change in f: $df = f(x_0 + dx) - f(x_0)$. Next, we use Taylor's series for $f(x_0 + dx) = f(x_0) + f'(x_0)dx$ (higher order terms are negligible). So, $df = f'(x_0)dx$. And the condition to have a minimum at x_0 becomes $f'(x_0) = 0$.

Can we do the same thing for Eq. (8.3.1)? Lagrange did exactly just that and he invented variational calculus! Let us assume that $y(x)$ is the solution, i.e., the function that minimizes the functional. To denote a variation of $y(x)$, designated by $\overline{y(x)}$, we consider

$$\overline{y(x)} := y(x) + \epsilon\eta(x) \tag{8.3.2}$$

where $\eta(x)$ is a fixed function satisfying the conditions $\eta(a) = \eta(b) = 0$ so that $\bar{y}(a) = A$ and $\overline{\bar{y}(b)} = B$, and ϵ is a small number. See Fig. 8.3 for an illustration of $y(x)$, $\eta(x)$ and $\overline{y(x)}$. For each value of ϵ, we have a specific variation, and thus a concrete value of the functional, and among all these values the one obtained from $\epsilon = 0$ is the minimum, because we have assumed that $y(x)$ is the solution.

With the variation of the solution we proceed to the calculation of the corresponding change in the functional, denoted by dI:

Fig. 8.3 Solution function $y(x)$, $\eta(x)$ with $\eta(a) = \eta(b) = 0$ and one variation $y(x) + \epsilon_1\eta(x)$

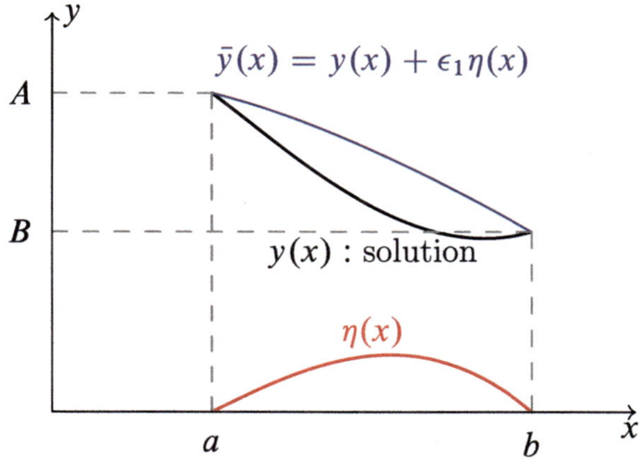

$$
\begin{aligned}
dI &= \int_a^b F(y(x) + \epsilon\eta(x), y'(x) + \epsilon\eta'(x); x)\, dx - \int_a^b F(y(x), y'(x); x)\, dx \\
&= \int_a^b \left[F(y + \epsilon\eta, y' + \epsilon\eta'; x) - F(y, y'; x)\right] dx \\
&= \int_a^b \left[\frac{\partial F}{\partial y}\eta + \frac{\partial F}{\partial y'}\eta'\right] \epsilon\, dx
\end{aligned}
\tag{8.3.3}
$$

where in the last equality we have used Taylor's series expansion for $F(y + \epsilon\eta, y' + \epsilon\eta'; x)$ around $\epsilon = 0$.

Now, as $u(x)$ is the minimal solution, one has to have $dI/\epsilon = 0$ (this is similar to $df/dx = 0$ in ordinary differential calculus). Thus, we obtain

$$
\int_a^b \left[\frac{\partial F}{\partial y}\eta + \frac{\partial F}{\partial y'}\eta'\right] dx = 0 \tag{8.3.4}
$$

In the next step, we want to get rid of η' (so that we can use a useful lemma called the fundamental lemma of variational calculus which exploits the arbitrariness of η to obtain a nice result in terms of y, no more ϵ and η), and of course the trick is integration by parts:

$$
\int_a^b \left[\frac{\partial F}{\partial y} - \frac{d}{dx}\left(\frac{\partial F}{\partial y'}\right)\right] \eta\, dx + \left[\frac{\partial F}{\partial y'}\eta\right]_a^b = 0 \tag{8.3.5}
$$

As $\eta(a) = \eta(b) = 0$, the *boundary term* (the last term in the above equation) vanishes and we get the following:

$$\int_a^b \left[\frac{\partial F}{\partial y} - \frac{d}{dx}\left(\frac{\partial F}{\partial y'}\right)\right] \eta \, dx = 0 \tag{8.3.6}$$

Using the fundamental lemma of variational calculus (which states that if $\int_a^b f(x)g(x)dx = 0$ for all $g(x)$ then $h(x) = 0$ for $x \in [a, b]$), one obtains the so-called Euler-Lagrange equation

$$\text{Euler-Lagrange equation}: \quad \frac{\partial F}{\partial y} - \frac{d}{dx}\left(\frac{\partial F}{\partial y'}\right) = 0 \tag{8.3.7}$$

Euler derived this equation before Lagrange but his derivation was not as elegant as the one presented herein which is due to Lagrange. To use Eq. (8.3.7), it should be noted that we treat y, y', x as independent variables when calculating $\frac{\partial F}{\partial y}$ and $\frac{\partial F}{\partial y'}$.

We note that the Euler-Lagrange equation in Eq. (8.3.7) is a second-order partial differential equation; this is due to the term $\frac{d}{dx}\left(\frac{\partial F}{\partial y'}\right)$ as we have the derivative of y', and thus y''.

Now to solve Eq. (8.3.1), Euler solved Eq. (8.3.7). This is known referred to as the indirect way to solving variational problems. There is a direct method to attack the variational problem Eq. (8.3.1) directly; check Sect. 8.9. However, for now we are going to use the indirect method to solve some elementary variational problems discussed in Sect. 8.2.

Stationary curves. Starting with the functional 8.3.1, we have assumed that $y(x)$ is a function that minimizes this functional, and found that it satisfies the Euler-Lagrange equation (8.3.7). Is the reverse true? That is, if $y(x)$ satisfies the Euler-Lagrange equation will it minimize the functional? The answer is, by learning from ordinary calculus, not necessarily.[7] Therefore, functions that satisfy the Euler-Lagrange equation are called stationary functions or stationary curves.

[7] For function $y = f(x)$, stationary points are those x^* such that $f'(x^*) = 0$. These points can be a maximum or a minimum or an inflection point.

History note 8.1: Joseph-Louis Lagrange (25 January 1736–10 April 1813)

Joseph-Louis Lagrange was an Italian mathematician and astronomer, later naturalized French. He made significant contributions to the fields of analysis, number theory, and both classical and celestial mechanics. As his father was a doctor in Law at the University of Torino, a career as a lawyer was planned out for him by his father, and certainly Lagrange seems to have accepted this willingly. He studied at the University of Turin and his favorite subject was classical Latin. At first he had no great enthusiasm for mathematics, *finding Greek geometry rather dull.* Lagrange's interest in mathematics began when he read a copy of Halley's 1693 work on the use of algebra in optics. In contrast to geometry, something about Halley's algebra captivated him. He devoted himself to mathematics, but largely was self taught. By age 19 he was appointed to a professorship at the Royal Artillery School in Turin. The following year, Lagrange sent Euler a better solution he had discovered for deriving the Euler-Lagrange equation in the calculus of variations. Lagrange gave us the familiar notation $f'(x)$ to represent a function's derivative, $f''(x)$ a second derivative, etc., and indeed it was he who gave us the word derivative. Mécanique analytique (1788'89) is a two volume French treatise on analytical mechanics, written by Lagrange, and published 101 years following Newton's Philosophiæ Naturalis Principia Mathematica. It consolidated into one unified and harmonious system, the scattered developments of various contributors in the historical transition from geometrical methods, as presented in Newton's Principia, to the methods of mathematical analysis. The treatise expounds *a great labor-saving and thought-saving general analytical method by which every mechanical question may be stated in a single differential equation.*

8.4 Solution of Some Elementary Variational Problems

This section presents solutions to variational problems introduced in Sect. 8.2. The idea is to use Eq. (8.3.7) to find the partial differential equation associated with a functional, and solve it. For some problems, non-variational calculus solution is also provided.

8.4.1 Euclidean Geodesic Problem

Finding the shortest path joining two points (x_1, y_1) and (x_2, y_2). For this problem F is given by (note that it does not depend on y)

$$F(y, y', x) = \sqrt{1 + (y')^2}$$

And thus

$$\frac{d}{dx}\left(\frac{\partial F}{\partial y'}\right) = \frac{d}{dx}\left[\frac{y'}{\sqrt{1 + (y')^2}}\right] = \frac{y''\sqrt{1 + (y')^2} - \dfrac{(y')^2 y''}{\sqrt{1 + (y')^2}}}{1 + (y')^2}$$

Upon substitution into the Euler-Lagrange equation in Eq. (8.3.7), one gets

$$\frac{d}{dx}\left(\frac{\partial F}{\partial y'}\right) = 0 \Longrightarrow y'' = 0 \Longrightarrow y = ax + b$$

The solution is a straight line as expected. The two coefficients a and b are determined using the boundary conditions:

$$y_1 = ax_1 + b, \quad y_2 = ax_2 + b \tag{8.4.1}$$

8.4.2 The Brachistochrone Problem

Recall that for the Brachistochrone problem, we're looking for a function $y(x)$ such that the following is minimum:

$$\sqrt{2g}t[y(x)] = \int_0^a \frac{\sqrt{1 + [y'(x)]^2}}{\sqrt{y(x)}}\, dx \tag{8.4.2}$$

So it is a classical variational calculus problem with $F = \sqrt{1+[y'(x)]^2}/\sqrt{y(x)}$. We can use the Euler-Lagrange equation with this F. But there is a better way exploiting the fact that F does not explicitly depend on x. Multiplying the Euler-Lagrange with y', we obtain

$$\left[\frac{\partial F}{\partial y} - \frac{d}{dx}\left(\frac{\partial F}{\partial y'}\right)\right] y' = 0$$

Then, a few massages to it give us:

$$\frac{\partial F}{\partial y}y' - y'\frac{d}{dx}\left(\frac{\partial F}{\partial y'}\right) = 0$$

$$\frac{dF}{dx} - \frac{\partial F}{\partial y'}y'' - y'\frac{d}{dx}\left(\frac{\partial F}{\partial y'}\right) = 0, \quad \left(\frac{dF}{dx} = \frac{\partial F}{\partial y}y' + \frac{\partial F}{\partial y}y''\right)$$

$$\frac{d}{dx}\left(F - y'\frac{\partial F}{\partial y'}\right) = 0$$

which leads to

$$F - y'\frac{\partial F}{\partial y'} = C, \quad C \text{ is a constant} \tag{8.4.3}$$

This result is known as Beltrami's identity which is the simpler version of the Euler-Lagrange equation when F does not explicitly depend on x. The identity is named after Eugenio Beltrami (1835–1900) who was an Italian mathematician notable for his work concerning differential geometry and mathematical physics.

Now we come back to the Brachistochrone problem. Using Eq. (8.4.3) for $F = \sqrt{1+[y'(x)]^2}/\sqrt{y(x)}$, we obtain

$$\frac{\sqrt{1+y'^2}}{\sqrt{y(x)}} - \frac{(y')^2}{\sqrt{y(1+y'^2)}} = C$$

And from that, we get a simpler equation (squaring both sides and some terms cancel out),

$$y(1+y'^2) = \frac{1}{C^2} \equiv A \tag{8.4.4}$$

With $y' = dy/dx$, one can solve for dx in terms of dy and y, and from that we obtain $x = \int_0^b dx$:

$$x = \int_0^b \sqrt{\frac{y}{A-y}}dy$$

Now, we're back to the old business of integral calculus: using this substitution

$$y = A\sin^2\theta/2 = A/2(1-\cos\theta)$$

we can evaluate the above integral to get

$$x = \frac{A}{2}(\theta - \sin\theta)$$

One determines A by the boundary condition that the curve passes through $B(a, b)$. The Brachistochrone curve is the one defined parametrically as

$$x = \frac{A}{2}(\theta - \sin\theta), \quad y = \frac{A}{2}(1 - \cos\theta) \tag{8.4.5}$$

And this curve is nothing but the well-known cycloid in geometry. Refer to Sect. 4.12.3 for details on this interesting curve.

8.4.3 The Brachistochrone: History and Bernoulli's Genius Solution

Johann Bernoulli posed the problem of the brachistochrone to the readers of Acta Eruditorum in June 1696. He said:

> I, Johann Bernoulli, address the most brilliant mathematicians in the world. Nothing is more attractive to intelligent people than an honest, challenging problem, whose possible solution will bestow fame and remain as a lasting monument. Following the example set by Pascal, Fermat, etc., I hope to gain the gratitude of the whole scientific community by placing before the finest mathematicians of our time a problem which will test their methods and the strength of their intellect. If someone communicates to me the solution of the proposed problem, I shall publicly declare him worthy of praise.

Bernoulli allowed six months for the solutions but none were received during this period. At the request of Leibniz, the time was publicly extended for a year and a half. At 4 p.m. on 29 January 1697 when he arrived home from the Royal Mint, Newton found the challenge in a letter from Johann Bernoulli. Newton stayed up all night to solve it and mailed the solution anonymously by the next post. Bernoulli, writing to Henri Basnage in March 1697, indicated that even though its author, "by an excess of modesty", had not revealed his name, yet even from the scant details supplied it could be recognized as Newton's work, "*as the lion by its claw*" (in Latin, tanquam ex ungue leonem). This story gives some idea of Newton's power, since Johann Bernoulli needed two weeks to solve it. Newton also wrote, "*I do not love to be dunned [pestered] and teased by foreigners about mathematical things...*", and Newton had already solved Newton's minimal resistance problem, which is considered the first of the kind in calculus of variations.

In the end, five mathematicians had provided solutions: Newton, Jakob Bernoulli, Gottfried Leibniz, Ehrenfried Walther von Tschirnhaus, and Guillaume de l'Hôpital.

Now I present the genius solution of Johann Bernoulli. He used the Snell law (see Sect. 4.5.1), re-given in Fig. 8.4a. He applied that to a medium consisting of multiple layers, Fig. 8.4b. For this medium, he got

$$\frac{\sin\alpha_1}{v_1} = \frac{\sin\alpha_2}{v_2} = \frac{\sin\alpha_3}{v_3} = \frac{\sin\alpha_4}{v_4}$$

Now, you can guess what he would do next. He imagined that if the medium has an infinite number of layers, the light would then travel in a curved path, and at any point on this path, we have

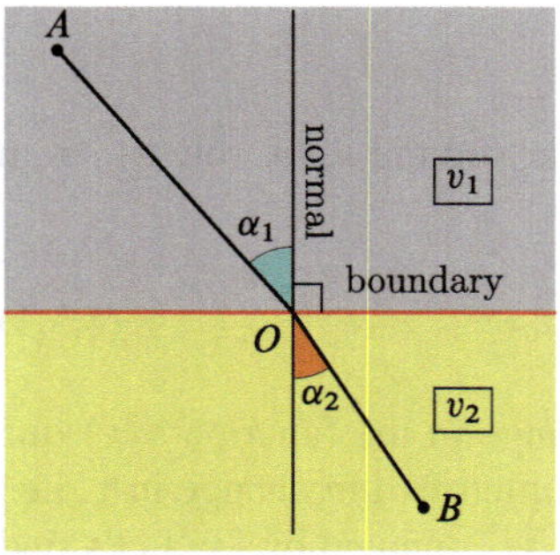

(a) $\sin\alpha_1/v_1 = \sin\alpha_2/v_2$

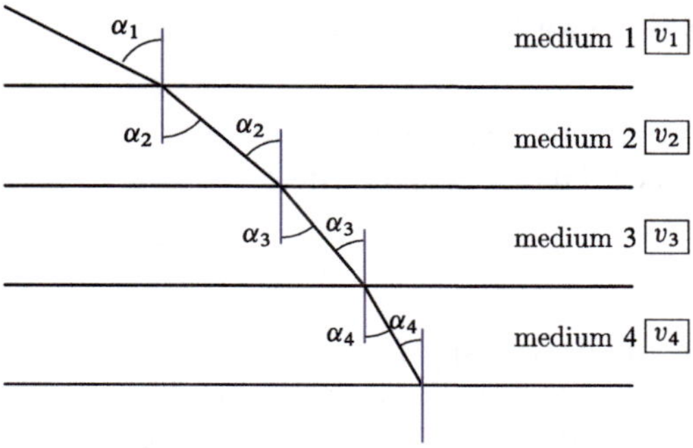

(b) a medium consisting of four layers.

Fig. 8.4 Light travels in a medium of variable density

Fig. 8.5 .

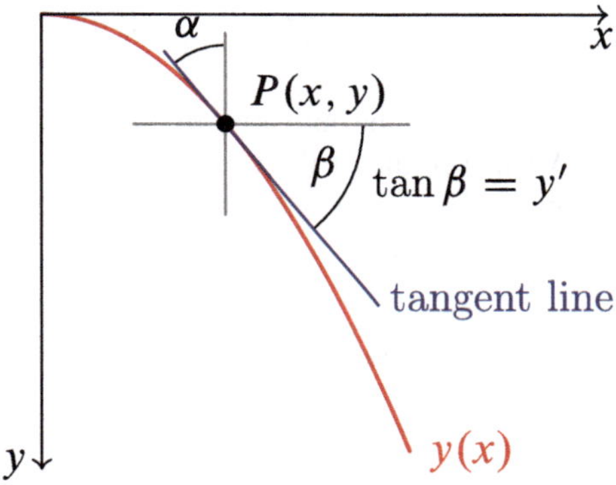

$$\frac{\sin\alpha}{v} = \text{constant} \tag{8.4.6}$$

Finally, he applied this result to the Brachistochrone problem. Referring to Fig. 8.5, and consider a point $P(x, y)$, draw a tangent line to the curve $y(x)$ at P. He computed $\sin\alpha$ in terms of y' as follows:

$$\sin\alpha = \cos\beta = \frac{1}{\sqrt{1+\tan^2\beta}} = \frac{1}{\sqrt{1+(y')^2}}$$

And the velocity $v = \sqrt{2gy}$, and thus Eq. (8.4.6) gave him:

$$\frac{1}{\sqrt{1+(y')^2}} = c\sqrt{2gy}$$

which is equivalent to Eq. (8.4.4)—the solution obtained using variational calculus.

8.5 The Variational δ Operator

We are now ready to define the variation of $y(x)$ (playing the same role as dx in standard minimum problems):

$$\delta y(x) := \overline{y(x)} - y(x) = \epsilon\eta(x) \tag{8.5.1}$$

We might ask why δy not dy? Note that dy is a change in the function $y(x)$ due to a change in x. For variational problems, we're not interested in change in x, i.e., $\delta x = 0$. Instead we need change in the function, and it is denoted by δy. Let's find what properties this δ operator possesses.

First, a variation δy is a function of x, so we can take its derivative:

$$\frac{d}{dx}\delta y = \frac{d}{dx}[\epsilon\eta(x)] = \epsilon\eta'(x) = \delta\left(\frac{dy}{dx}\right)$$

This shows that the derivative of the variation is equal to the variation of the derivative.

We also define δI the variation of the functional and it can be shown that the variation of a functional is the integral of the variation of its integrand (as the integration limits are fixed):

$$\begin{aligned}\delta\int_a^b F(y, y'; x)dx &:= \int_a^b F(y+\delta y, y'+\delta y'; x)dx - \int_a^b F(y, y'; x)dx\\ &= \int_a^b [F(y+\delta y, y'+\delta y'; x) - F(y, y'; x)]dx = \int_a^b \delta F dx\end{aligned}$$

From Eq. (8.3.3), we can compute δF as easily as (recall that $F = F(y, y'; x)$)

$$\delta F = \left[\frac{\partial F}{\partial y}\eta + \frac{\partial F}{\partial y'}\eta'\right]\epsilon = \frac{\partial F}{\partial y}\delta y + \frac{\partial F}{\partial y'}\delta y' \tag{8.5.2}$$

Observing the similarity to the total differential df of a function of two variables $f(x, y)$: $df = f_x dx + f_y dy$ when its variables change by dx and dy. We put these two side by side:

$$\begin{aligned} df &= \frac{\partial f}{\partial x}dx + \frac{\partial f}{\partial y}dy\\ \delta F &= \frac{\partial F}{\partial y}\delta y + \frac{\partial F}{\partial y'}\delta y'\end{aligned} \tag{8.5.3}$$

To summarize, here are some important properties of the δ operator:

variation/differentiation are permutable: $\frac{d}{dx}\delta y = \delta\left(\frac{dy}{dx}\right)$

variation/integration are permutable: $\delta\int_a^b F(y, y'; x)dx = \int_a^b \delta F dx$

Finally, we can see that δy is similar to the differential operator df in differential calculus; Eq. (8.5.3) is one example. That is why Lagrange selected the symbol δ. We know that $d(f+g) = df + dg$ and $d(x^2) = 2xdx$. We have counterparts for δ: for u, v are some functions

$$\begin{aligned} \delta(\alpha u + \beta v) &= \alpha\delta u + \beta\delta v \\ \delta(u^2) &= 2u\delta u \end{aligned} \tag{8.5.4}$$

Now we can use δ in the same manner we do with d. The proof is easy. For example, consider $F(u) = u^2$, when we vary the function u by δu, we get a new functional $\bar{F} = (u+\delta u)^2$. Thus, the variation in the functional is $\delta F = (u+\delta u)^2 - u^2 = 2u\delta u$.

One-dimensional variational problem with second derivatives. Find the function $y(x)$ that makes the following functional

$$J[y] := \int_a^b F(y, y', y'', x)\,dx \tag{8.5.5}$$

stationary and subjects to boundary conditions that $y(a)$, $y(b)$, $y'(a)$, $y'(b)$ fixed.

We compute the first variation δJ due to the variation in $y(x)$, δy (recall that $\delta y' = d/dx(\delta y)$ and $\delta y'' = d^2/dx^2(\delta y)$):

$$\delta J = \int_a^b \delta F\,dx = \int_a^b \left[\frac{\partial F}{\partial y}\delta y + \frac{\partial F}{\partial y'}\delta y' + \frac{\partial F}{\partial y''}\delta y''\right] dx$$

Now comes the usual integration by parts. For the term with $\delta y'$:

$$\frac{d}{dx}\left[\frac{\partial F}{\partial y'}\delta y\right] = \frac{d}{dx}\left[\frac{\partial F}{\partial y'}\right]\delta y + \frac{\partial F}{\partial y'}\delta y' \Rightarrow \int_a^b \frac{\partial F}{\partial y'}\delta y' dx = -\int_a^b \frac{d}{dx}\left[\frac{\partial F}{\partial y'}\right]\delta y dx$$

Now for the term with $\delta y''$:

$$\frac{d}{dx}\left[\frac{\partial F}{\partial y''}\delta y'\right] = \frac{d}{dx}\left[\frac{\partial F}{\partial y''}\right]\delta y' + \frac{\partial F}{\partial y''}\delta y'' \Rightarrow \int_a^b \frac{\partial F}{\partial y''}\delta y'' dx = -\int_a^b \frac{d}{dx}\left[\frac{\partial F}{\partial y''}\right]\delta y' dx$$

And still having $\delta y'$, we have to do integration by parts again:

$$\int_a^b \left[\frac{d}{dx}\left(\frac{\partial F}{\partial y''}\right)\delta y'\right] dx = -\int_a^b \frac{d^2}{dx^2}\left(\frac{\partial F}{\partial y''}\right)\delta y dx$$

Finally, the first variation δJ is given by

$$\delta J = \int_a^b \left[\frac{\partial F}{\partial y} - \frac{d}{dx}\left(\frac{\partial F}{\partial y'}\right) + \frac{d^2}{dx^2}\left(\frac{\partial F}{\partial y''}\right)\right]\delta y dx$$

which yields the following Euler-Lagrange equation:

$$\frac{\partial F}{\partial y} - \frac{d}{dx}\left(\frac{\partial F}{\partial y'}\right) + \frac{d^2}{dx^2}\left(\frac{\partial F}{\partial y''}\right) = 0$$

8.6 Multi-dimensional Variational Problems

We now extend what we have done to higher dimensions. We consider a two-dimensional space with two Cartesian coordinates x, y serving as the independent variables. We have two dependent variables $u(x, y)$ and $v(x, y)$. Now let's consider the following functional:

$$J[u(x, y), v(x, y)] := \int_{\mathcal{B}} F(u, v, u_x, v_x, u_y, v_y; x, y)\, dxdy \tag{8.6.1}$$

And we want to find functions $u(x, y)$ and $v(x, y)$ defined on a domain $\mathcal{B}$ such that J is minimum. On the boundary $\partial\mathcal{B}$ the functions are prescribed, i.e., $u = g$ and $v = h$, where g, h are known functions of (x, y).

The first variation of J, δJ, is given by

$$\delta J = \int_{\mathcal{B}} \left[\frac{\partial F}{\partial u}\delta u + \frac{\partial F}{\partial u_x}\delta u_x + \frac{\partial F}{\partial u_y}\delta u_y + \frac{\partial F}{\partial v}\delta v + \frac{\partial F}{\partial v_x}\delta v_x + \frac{\partial F}{\partial v_y}\delta v_y\right] dxdy$$

The next step is certainly to integrate by parts the second, third, fifth and sixth terms. We demonstrate the steps only for the second term, starting with:

$$\frac{\partial}{\partial x}\left(\frac{\partial F}{\partial u_x}\delta u\right) = \frac{\partial}{\partial x}\left(\frac{\partial F}{\partial u_x}\right)\delta u + \frac{\partial F}{\partial u_x}\delta u_x$$

And thus,

$$\int_{\mathcal{B}} \frac{\partial F}{\partial u_x}\delta u_x dV = \int_{\mathcal{B}} \frac{\partial}{\partial x}\left(\frac{\partial F}{\partial u_x}\delta u\right) dV - \int_{\mathcal{B}} \frac{\partial}{\partial x}\left(\frac{\partial F}{\partial u_x}\right)\delta u dV$$

Using the gradient theorem, Eq. (6.11.37), for the second term—the red term in the above equation, we obtain

$$\int_{\mathcal{B}} \frac{\partial F}{\partial u_x}\delta u_x dV = \int_{\partial\mathcal{B}} \frac{\partial F}{\partial u_x} n_x \delta u ds - \int_{\mathcal{B}} \frac{\partial}{\partial x}\left(\frac{\partial F}{\partial u_x}\right)\delta u dV$$

Repeating the same calculations for the third, fifth and sixth terms, the variation of J is eventually written as

$$\delta J = \int_{\mathcal{B}} \left[\left\{ \frac{\partial F}{\partial u} - \frac{\partial}{\partial x}\left(\frac{\partial F}{\partial u_x}\right) - \frac{\partial}{\partial y}\left(\frac{\partial F}{\partial u_y}\right) \right\} \delta u + \left\{ \frac{\partial F}{\partial v} - \frac{\partial}{\partial x}\left(\frac{\partial F}{\partial v_x}\right) - \frac{\partial}{\partial y}\left(\frac{\partial F}{\partial v_y}\right) \right\} \delta v \right] dxdy$$
$$+ \int_{\partial\mathcal{B}} \left[\left(\frac{\partial F}{\partial u_x} n_x + \frac{\partial F}{\partial u_y} n_y \right) \delta u \right] ds + \int_{\partial\mathcal{B}} \left[\left(\frac{\partial F}{\partial v_x} n_x + \frac{\partial F}{\partial v_y} n_y \right) \delta v \right] ds$$

As u, v are specified on the boundary $\partial\mathcal{B}$, $\delta u = \delta v = 0$ there. Using the fundamental lemma of variational calculus, we obtain the following Euler-Lagrange equations:

$$\begin{aligned} \frac{\partial F}{\partial u} - \frac{\partial}{\partial x}\left(\frac{\partial F}{\partial u_x}\right) - \frac{\partial}{\partial y}\left(\frac{\partial F}{\partial u_y}\right) &= 0 \\ \frac{\partial F}{\partial v} - \frac{\partial}{\partial x}\left(\frac{\partial F}{\partial v_x}\right) - \frac{\partial}{\partial y}\left(\frac{\partial F}{\partial v_y}\right) &= 0 \end{aligned} \tag{8.6.2}$$

Example 8.1 For example, if J is

$$J[u(x, y)] := \int_{\mathcal{B}} (u_x^2 + u_y^2)\, dV = \int_{\mathcal{B}} |\nabla u|^2\, dV = \int_{\mathcal{B}} \nabla u \cdot \nabla u\, dV \tag{8.6.3}$$

then Eq. (8.6.2) yields (we need to use the first equation only as there is no v function in our functional)

$$u_{xx} + u_{yy} = 0 \quad \text{or } \Delta u = 0 \quad \text{in } \mathcal{B} \tag{8.6.4}$$

Example 8.2 In the field of fracture mechanics, we have the following functional concerning a scalar field $\phi(x, y)$, where G_c, b, c_0 are real numbers and α is a function depending on ϕ:

$$J[\phi(x, y)] = \int_{\mathcal{B}} \frac{G_c}{c_0} \left(\frac{1}{b}\alpha(\phi) + b\nabla\phi \cdot \nabla\phi \right) dV \tag{8.6.5}$$

then Eq. (8.6.2) yields (we need to use the first equation only as there is no v function in our functional, and $\phi(x, y)$ plays the role of $u(x, y)$)

$$\frac{G_c}{c_0}\frac{1}{b}\alpha'(\phi) - \frac{2G_c b}{c_0}\Delta\phi = 0 \quad \text{in } \mathcal{B} \tag{8.6.6}$$

8.7 Boundary Conditions

We have seen that the first variation of any functional has two terms: one term defined inside the problem domain and one term defined on the problem boundary. For example, for $I[y] = \int_a^b F(y, y'; x)dx$, the first variation reads

$$\delta I = \int_a^b \left[\frac{\partial F}{\partial y} - \frac{d}{dx}\left(\frac{\partial F}{\partial y'}\right) \right] \delta y\, dx + \frac{\partial F}{\partial y'}(b)\delta y(b) - \frac{\partial F}{\partial y'}(a)\delta y(a) \tag{8.7.1}$$

where the red terms are the boundary terms. The Euler-Lagrange equation associated with this functional is a *second-order* partial differential equation. Thus it requires two boundary conditions (BCs) to have a unique solution. In many cases, it is easy to determine these boundary conditions. But there are also cases where it is very difficult to know the boundary conditions. This is particularly true for fourth-order PDEs.

It is a particularly beautiful feature of variational problems that *they always furnish automatically the right number of boundary conditions and the form.* All comes from the first variation of the functional. Getting back to the already mentioned δI, we have the following cases:

- **Case 1**: We *impose* the boundary conditions of this form: $y(a) = A$ and $y(b) = B$. In other words, we fix the two ends of the curve $y(x)$, and the corresponding variational problems are called *fixed end variational problems*. As fixed quantities do not vary, we have $\delta y(a) = \delta y(b) = 0$, and the boundary terms—red terms in Eq. (8.7.1)—vanish. This type of boundary condition is called *imposed boundary conditions* or *essential boundary conditions*.
- **Case 2**: We fix one end (for example, $y(a) = A$, and thus $\delta y(a) = 0$), and allow the other end to be free. As $y(b)$ can be anything, we have $\delta y(b) \neq 0$, so to have $\delta I = 0$, we need $\frac{\partial F}{\partial y'}(b) = 0$. And this is the second BC that the Euler-Lagrange equation has to satisfy. Since this BC is provided by the variational problem, it is called *natural boundary condition*. In case of the brachistochrone, this BC is translated to $y'(b) = 0$ which indicates that the tangent to the curve at $x = b$ is horizontal.

Example 8.3 an elastic bar Consider an elastic bar of length L, modulus of elasticity E, and cross-sectional area A. We denote by x the independent variable which runs from 0 to L, characterizing the position of a point of the bar. Assume that the bar is fixed at the left end ($x = 0$) and subjected to a distributed axial load $f(x)$ (per unit length) and a point load P at its right end ($x = L$). The axial displacement of the bar $u(x)$ is the function that minimizes the following potential energy:

$$\Pi[u(x)] = \int_0^L \left[\frac{EA}{2} \left(\frac{du}{dx} \right)^2 - fu \right] dx - Pu(L) \tag{8.7.2}$$

where the first term is the strain energy stored in the bar and the second and third terms denote the work done on the bar by the force f and P, respectively.

To find the Euler-Lagrange equation for this problem, we compute the first variation of the energy functional and set it to zero. The variation is given by

$$\delta\Pi = \int_0^L \left[EA \frac{du}{dx} \frac{d(\delta u)}{dx} - f\delta u \right] dx - P\delta u(L) \tag{8.7.3}$$

We need to remove $\delta u' = d/dx(\delta u)$; for this we use integration by parts. Noting that

$$\frac{d}{dx}\left(\frac{du}{dx}\delta u\right) = \frac{d^2u}{dx^2}\delta u + \frac{du}{dx}\frac{d(\delta u)}{dx}$$

Thus, we have

$$\int_0^L \frac{du}{dx}\frac{d(\delta u)}{dx}dx = \left[\frac{du}{dx}\delta u\right]_0^L - \int_0^L \frac{d^2u}{dx^2}\delta u dx$$

Equation (8.7.3) becomes

$$\begin{aligned}\delta\Pi &= -\int_0^L \left[EA\frac{d^2u}{dx^2} + f\right]\delta u dx + \left[EA\frac{du}{dx}\delta u\right]_0^L - P\delta u(L) \\ &= -\int_0^L \left[EA\frac{d^2u}{dx^2} + f\right]\delta u dx + \left[\left(EA\frac{du}{dx}\right)_{x=L} - P\right]\delta u(L)\end{aligned} \tag{8.7.4}$$

which gives the Euler-Lagrange equation

$$EA\frac{d^2u}{dx^2} + f = 0, \quad 0 < x < L$$

which requires 2 BCs: one is $u(0) = 0$—the BC that we impose upon the bar, and the other is

$$\left(EA\frac{du}{dx}\right)_{x=L} - P = 0$$

provided by the variational formulation.

Example 8.4 Consider an elastic beam of length L, modulus of elasticity E, and second moment of area I. The vertical displacement of the beam $y(x)$ is the function that minimizes the following potential energy:

$$\Pi[u(x)] = \int_0^L \left[\frac{k}{2}(y'')^2 - \rho y\right]dx, \quad k := EI \tag{8.7.5}$$

where the first term is the strain energy stored in the bar and the second term denotes the work done on the beam by the force per unit length $\rho(x)$.

We use the results developed for the functional given in Eq. (8.5.5),

$$\delta\Pi = \int_a^b \left[\frac{\partial F}{\partial y} - \frac{d}{dx}\left(\frac{\partial F}{\partial y'}\right) + \frac{d^2}{dx^2}\left(\frac{\partial F}{\partial y''}\right)\right]\delta y dx + \left[\left(\frac{\partial F}{\partial y'} - \frac{d}{dx}\frac{\partial F}{\partial y''}\right)\delta y + \frac{\partial F}{\partial y''}\delta y'\right]_0^L \tag{8.7.6}$$

With $F = {}^{k}/_{2}(y'')^2 - \rho y$, we get the Euler-Lagrange equation from the first term in $\delta\Pi = 0$,

$$ky'''' = \rho(x), \quad 0 < x < L \tag{8.7.7}$$

which is a fourth-order different equation; it requires four boundary conditions. We are demonstrating that the variational character yields all these required BCs. We note that solving this equation yields the so-called elastic curve, which is the deflected shape of a bending beam.

With Eq. (8.7.6) and $F = {}^{k}/_{2}(y'')^2 - \rho y$, the boundary term of the first variation of the functional is given by

$$\delta\Pi = k\left[-y'''(L)\delta y(L) + y'''(0)\delta y(0) + y''(L)\delta y'(L) - y''(0)\delta y'(0)\right] \tag{8.7.8}$$

And $\delta\Pi = 0$ provides all BCs that the Euler-Lagrange equation of the beam requires. There are the following cases:

- **Clamped ends**: The BCs are

$$\begin{aligned} y(0) = 0,\ \ y(L) = 0 \\ y'(0) = 0,\ \ y'(L) = 0 \end{aligned} \tag{8.7.9}$$

  That is, we fix the displacement and the rotation at both ends of the beam. As the variations of fixed quantities are zero, all the terms in Eq. (8.7.8) vanish. No natural BCs have to be added.
- **Supported ends**: in this case, the BCs are simple as

$$y(0) = 0,\ y(L) = 0 \tag{8.7.10}$$

  That is, we fix only the displacement of the two ends. Equation (8.7.8) provides two more natural BCs:

$$y''(0) = 0, \quad y''(L) = 0$$

  which indicate that the bending moments are zero at both ends.
- **One end clamped, one end free**:

$$y(0) = 0,\ y'(0) = 0 \tag{8.7.11}$$

  That is, we fix both the displacement/rotation of the left end, but leave the right end free. Equation (8.7.8) yields the remaining two BCs:

$$y''(L) = 0, \quad y'''(L) = 0 \tag{8.7.12}$$

which means that the bending moment at the right end is zero and so is the shear force there.

8.8 Lagrangian Mechanics

With the new mathematics (i.e., variational calculus) developed by himself, Lagrange created a new formulation of Newtonian mechanics. That formulation is now called the Lagrangian mechanics. To study motions, we can either use Newtonian mechanics or Lagrangian mechanics, they're giving the same result. To motivate the need of Lagrangian mechanics, we can cite one weakness of Newtonian mechanics: because Newton's 2nd law is vectorial in nature, the equations change when the coordinate system changes (Sects. 6.10.6 and 6.10.7). Lagrange gave us another way to view motion.

8.8.1 The Lagrangian, the Action, and the EL Equations

Let's consider a particle moving from point 1 to point 2; its trajectory is given by $(x(t), y(t), z(t))$. Now I present a result from variational calculus: define the following functional:

$$J[x(t), y(t), z(t)] := \int_1^2 F(x, y, z, \dot{x}, \dot{y}, \dot{z}; t)\, dt \tag{8.8.1}$$

Then, the associated Euler-Lagrange equations are (there are three equations, one for each component of the particle position vector)

$$\frac{\partial F}{\partial x} = \frac{d}{dt}\left(\frac{\partial F}{\partial \dot{x}}\right), \quad \frac{\partial F}{\partial y} = \frac{d}{dt}\left(\frac{\partial F}{\partial \dot{y}}\right), \quad \frac{\partial F}{\partial z} = \frac{d}{dt}\left(\frac{\partial F}{\partial \dot{z}}\right) \tag{8.8.2}$$

We need this result in the following discussion on Lagrangian mechanics.

The central object in Lagrangian mechanics is the Lagrangian $\mathcal{L}$ defined as the difference between the kinetic energy T and the potential energy U. And from that he computed a term called action, labeled by S, defined as

$$S = \int_{t_1}^{t_2} \mathcal{L} dt, \quad \mathcal{L} = T - U \tag{8.8.3}$$

which is an integral of the Lagrangian, from t_1 to t_2.

Now comes the magic: the particle actual trajectory is $(x(t), y(t), z(t))$ that renders the action stationary. To show that, we just need to work out the mathematics:

$$S[\ast x(t)] = \int_{t_1}^{t_2} \mathcal{L}(\ast x, \ast\dot{x}) dt, \quad T = \frac{1}{2} m(\dot{x}^2 + \dot{y}^2 + \dot{z}^2), \quad U = U(x, y, z) \tag{8.8.4}$$

This action is of the form of Eq. (8.8.1), thus its Euler-Lagrange equations can be obtained from Eq. (8.8.2): (replace F by $\mathcal{L}$)

$$\frac{\partial \mathcal{L}}{\partial x} = \frac{d}{dt}\frac{\partial \mathcal{L}}{\partial \dot{x}}, \quad \frac{\partial \mathcal{L}}{\partial y} = \frac{d}{dt}\frac{\partial \mathcal{L}}{\partial \dot{y}}, \quad \frac{\partial \mathcal{L}}{\partial z} = \frac{d}{dt}\frac{\partial \mathcal{L}}{\partial \dot{z}} \tag{8.8.5}$$

Lagrange thus obtained three equations and recall that Newton also had three equations. If Lagrange could show that his three equations are exactly Newton's equations, then he has created a new formulation of (classical) mechanics. That part is easy, we first need to compute $\frac{\partial \mathcal{L}}{\partial x}$ and $\frac{\partial \mathcal{L}}{\partial \dot{x}}$:

$$\frac{\partial \mathcal{L}}{\partial x} = -\frac{\partial U}{\partial x} = F_x; \quad \frac{\partial \mathcal{L}}{\partial \dot{x}} = \frac{\partial T}{\partial \dot{x}} = m\dot{x} \tag{8.8.6}$$

Substituting these into the first of Eq. (8.8.5), we get $F_x = m\ddot{x}$, which is nothing but Newton's 2nd law.

8.8.2 Generalized Coordinates

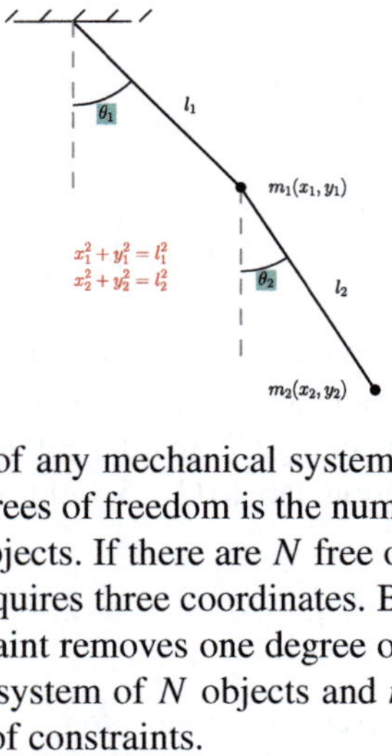

An important characteristic of any mechanical system is the *number of degrees of freedom*. The number of degrees of freedom is the number of coordinates needed to specify the location of the objects. If there are N free objects, there are $3N$ degrees of freedom as each object requires three coordinates. But if there are constraints on the objects, then each constraint removes one degree of freedom. The total number of degrees of freedom for a system of N objects and n constraints is $3N - n$. See the figure for an illustration of constraints.

To describe a system, we can use any set of parameters that unambiguously represents it. These parameters do not need to have dimensions of length (e.g., the usual Cartesian coordinates x, y, and z). They are referred to as *generalized coordinates*. In the next figure, to describe the system the two angles θ_1 and θ_2 are sufficient.

Now, we are going to prove one important result in Lagrangian mechanics. The result is that: if the EL equations hold for one set of generalized coordinates, they hold for other generalized coordinates. Assuming that we have $x_1, x_2, \ldots, x_N$ as the first set of generalized coordinates. And the EL equations hold, that is,

$$\frac{\partial \mathcal{L}}{\partial x_i} = \frac{d}{dt}\frac{\partial \mathcal{L}}{\partial \dot{x}_i}, \quad i = 1, 2, \ldots, N \tag{8.8.7}$$

Now, we have another set of generalized coordinates $q_1, q_2, \ldots, q_N$. We assume that it's always possible to go back and forth between the two coordinate systems. That is,

$$\begin{aligned} x_i &= x_i(q_1, q_2, \ldots, q_N, t) \\ q_i &= q_i(x_1, x_2, \ldots, x_N, t) \end{aligned} \tag{8.8.8}$$

What we need to prove is: the EL equations hold for q_i:

$$\frac{\partial \mathcal{L}}{\partial q_i} = \frac{d}{dt}\frac{\partial \mathcal{L}}{\partial \dot{q}_i}, \quad i = 1, 2, \ldots, N \tag{8.8.9}$$

Proof We start from the RHS of Eq. (8.8.9) with

$$\frac{\partial \mathcal{L}}{\partial \dot{q}_m} = \sum_{i=1}^{N} \frac{\partial \mathcal{L}}{\partial \dot{x}_i}\frac{\partial \dot{x}_i}{\partial \dot{q}_m}, \quad m = 1, 2, \ldots, N \tag{8.8.10}$$

From the first in Eq. (8.8.8), we have

$$\dot{x}_i = \sum_{k=1}^{N} \frac{\partial x_i}{\partial q_k}\frac{\partial q_k}{\partial t} + \frac{\partial x_i}{\partial t} \Longrightarrow \frac{\partial \dot{x}_i}{\partial \dot{q}_m} = \frac{\partial x_i}{\partial q_m} \tag{8.8.11}$$

Thus, we can write

$$\frac{\partial \mathcal{L}}{\partial \dot{q}_m} = \sum_{i=1}^{N} \frac{\partial \mathcal{L}}{\partial \dot{x}_i}\frac{\partial x_i}{\partial q_m} \tag{8.8.12}$$

From that, its time derivative is computed:

$$\begin{aligned} \frac{d}{dt}\frac{\partial \mathcal{L}}{\partial \dot{q}_m} &= \frac{d}{dt}\sum_{i=1}^{N} \frac{\partial \mathcal{L}}{\partial \dot{x}_i}\frac{\partial x_i}{\partial q_m} \\ &= \sum_{i=1}^{N} \frac{d}{dt}\frac{\partial \mathcal{L}}{\partial \dot{x}_i}\frac{\partial x_i}{\partial q_m} + \sum_{i=1}^{N} \frac{\partial \mathcal{L}}{\partial \dot{x}_i}\frac{d}{dt}\frac{\partial x_i}{\partial q_m} \\ &= \sum_{i=1}^{N} \frac{\partial \mathcal{L}}{\partial x_i}\frac{\partial x_i}{\partial q_m} + \sum_{i=1}^{N} \frac{\partial \mathcal{L}}{\partial \dot{x}_i}\frac{\partial \dot{x}_i}{\partial q_m} \\ &= \frac{\partial \mathcal{L}}{\partial q_m} \end{aligned} \tag{8.8.13}$$

where in the third equality, Eq. (8.8.7) was used for the red term and for the blue term, the order of d/dt and d/dx was switched.[8] ■

8.8.3 Examples

A bead is free to slide along a friction-less hoop of radius R. The hoop rotates with constant angular speed ω around a vertical diameter (Fig. 8.6a). Find the equation of motion for the angle θ shown.

From Fig. 8.6, we can determine the speed in the hoop direction and the direction perpendicular to the hoop. From that, the kinetic and potential energies are written as

$$T = \frac{1}{2}m\left(R^2\dot{\theta}^2 + R^2\sin^2\theta\omega^2\right), \quad U = mgR(1-\cos\theta) \tag{8.8.14}$$

Now, we compute the terms in the EL equation:

$$\begin{aligned} \frac{\partial L}{\partial \theta} &= mR^2\omega^2\sin\theta\cos\theta - mgR\sin\theta \\ \frac{\partial L}{\partial \dot{\theta}} &= mR^2\dot{\theta} \Longrightarrow \frac{d}{dt}\frac{\partial L}{\partial \dot{\theta}} = mR^2\ddot{\theta} \end{aligned} \tag{8.8.15}$$

And thus the EL equation yields the following equation of motion:

$$\frac{d}{dt}\frac{\partial \mathcal{L}}{\partial \dot{\theta}} = \frac{\partial \mathcal{L}}{\partial \theta} \Longrightarrow \boxed{\ddot{\theta} = \left(\omega^2\cos\theta - \frac{g}{R}\right)\sin\theta} \tag{8.8.16}$$

It is hard to solve this equation exactly. Still, we can get something out of Eq. (8.8.16). One thing it can tells us is equilibrium points. If we place the bead at rest (i.e., $\dot{\theta} = 0$) at an equilibrium point θ_0, it remains there. Since the bead remains at θ_0, its velocity must be constant, and thus its acceleration must be zero. So, to find equilibrium points, solve $\ddot{\theta} = 0$, which is

$$\left(\omega^2\cos\theta - \frac{g}{R}\right)\sin\theta = 0$$

[8] If it was not clear, here are the details:

$$\frac{d}{dt}\frac{\partial x_i}{\partial q_m} = \sum_{k=1}^{N}\frac{\partial}{\partial q_k}\left(\frac{\partial x_i}{\partial q_m}\right)\dot{q}_k + \frac{\partial}{\partial t}\left(\frac{\partial x_i}{\partial q_m}\right) = \sum_{k=1}^{N}\frac{\partial}{\partial q_m}\left(\frac{\partial x_i}{\partial q_k}\right)\dot{q}_k + \frac{\partial}{\partial q_m}\left(\frac{\partial x_i}{\partial t}\right)$$

Thus,

$$\frac{d}{dt}\frac{\partial x_i}{\partial q_m} = \frac{\partial}{\partial q_m}\left[\sum_{k=1}^{N}\left(\frac{\partial x_i}{\partial q_k}\right)\dot{q}_k + \frac{\partial x_i}{\partial t}\right] = \frac{\partial \dot{x}_i}{\partial q_m}.$$

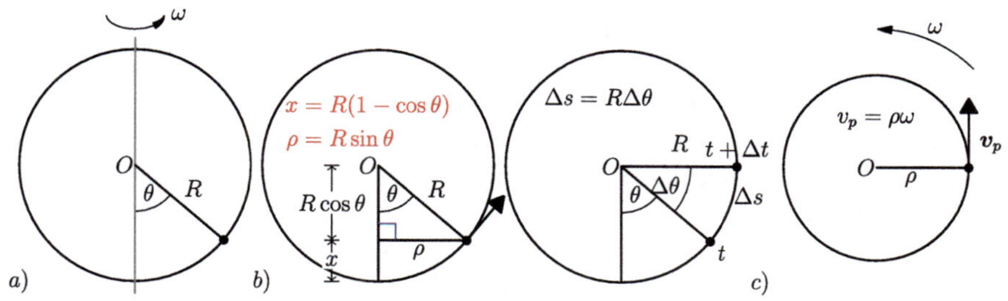

Fig. 8.6 A bead is free to slide along a friction-less hoop of radius R. The hoop rotates with constant angular speed ω. The bead position is specified by $\theta \in [0, \pi]$. The bead has two velocities: one is when the hoop is not spinning, this velocity is of magnitude of $\Delta s/\Delta t$ which is $R\dot{\theta}$ (b), and the other is due to the rotation of the hoop (c): $\rho\omega = R\sin\theta\omega$

A trigonometric equation! But this one is easy, it has totally four solutions:

$$\theta_0^1 = 0, \quad \theta_0^2 = \pi, \quad \theta_0^{3,4} = \pm\arccos\left(\frac{g}{R\omega^2}\right) \quad (\text{if } \omega^2 \geq g/R)$$

So, there are four equilibrium points if the hoop spins fast, i.e., $\omega^2 \geq g/R$. Otherwise, there are two equilibrium points $\theta_0^{1,2}$; they are the bottom and top of the hoop as you can predict. But equilibrium points can be stable or unstable. An equilibrium point is said to be stable if when the bead is at that position θ_0 and if it is given a small disturbance, it moves back to θ_0. So, our question now is among these four equilibrium points, which ones are stable.

- First case: $\omega^2 < g/R$. There are only two equilibrium points: $\theta_0^1 = 0$ and $\theta_0^2 = \pi$. Consider first $\theta_0^1 = 0$ (that is the bottom of the hoop). Close to 0, we have $\sin\theta \approx \theta$ and $\cos\theta \approx 1$, thus Eq. (8.8.16) becomes

$$\ddot{\theta} = \left(\omega^2 - \frac{g}{R}\right)\theta = -k\theta, \quad k := \frac{g}{R} - \omega^2$$

Now if the hoop spins at a small speed that $\omega^2 < g/R$, then $k > 0$. The above equation is identical to the one describing simple harmonic oscillations (discussed in Sect. 7.8). From the study of these oscillations, we know that the bead will oscillate around the bottom of the hoop. Therefore, the bottom of the hoop is a stable equilibrium point when $\omega^2 < g/R$. However, if $\omega^2 \geq g/R$, then that position is unstable.

Now we consider $\theta_0^2 = \pi$ (that is the top of the hoop). Intuitively, this must be an unstable equilibrium. We adopt a change of variable $\theta = \pi + \epsilon$ where ϵ is tiny. Equation (8.8.16) then becomes

$$\ddot{\epsilon} = \left(\omega^2 + \frac{g}{R}\right)\epsilon$$

So, this point is an unstable equilibrium point.

- Second case: $\omega^2 \geq g/R$. In this case, both $\theta_0^1 = 0$ and $\theta_0^2 = \pi$ are unstable (see the above analysis). The only two stable equilibrium points are $\theta_0^{3,4} = \pm\arccos(g/R\omega^2)$. Why they are stable? We just need to consider θ_0^3 due to symmetry. Noting that $\theta_0^3 \in [0, \pi/2]$ and in this interval the sine function is positive and the cosine is decreasing. Starting with θ_0^3 and move the bead a little bit up, we have $\ddot{\theta} < 0$ because:

$$\ddot{\theta} = \underbrace{\left(\omega^2 \cos\theta - \frac{g}{R}\right)}_{<0} \underbrace{\sin\theta}_{>0}$$

Thus, the bead is accelerating back to θ_0^3. Doing the same analysis by moving the bead a little bit down from θ_0^3 and we have $\ddot{\theta} > 0$: the bead accelerates to θ_0^3.

So, we have an interesting phenomenon. When the hoop is rotating slowly (i.e., $\omega^2 < g/R$), there is just one stable equilibrium at $\theta = 0$. If we speed up the rotation, as ω passes the critical value of $\omega^2 = g/R$, this original equilibrium becomes unstable. However, two new stable equilibrium points appear. This phenomenon—the disappearance of one stable equilibrium and appearance of other stable equilibrium points—is called a bifurcation.

8.9 Ritz' Direct Method

To introduce Ritz's method, we use the following example: finding $y(x)$ that minimizes the following functional:

$$I[y(x)] = \int_0^1 [y^2 + (y')^2]dx; \quad y(0) = y(1) = 1 \tag{8.9.1}$$

Ritz did not follow Euler, he thus did not derive the Euler-Lagrange equation associated with Eq. (8.9.1). Instead he attacks the functional directly, but he looks only for an *approximate solution* of the following form:

$$\boxed{\bar{y}(x) = \alpha + \beta x + \gamma x^2} \tag{8.9.2}$$

We should be aware that even if we can derive the Euler-Lagrange equation, it is quite often that we cannot solve it. Or it does not have solutions expressible in terms of elementary functions. Still physicists (or engineers) need a solution even not in a nice analytical expression, but in the form of a list of numbers.

If you ask why the form in Eq. (8.9.2)? Note that it is easy to work with polynomials (easy to differentiate, to integrate for example). And the first curve we normally think of is a parabola. So, it is natural to start with this polynomial form.

Because of the boundary conditions $y(0) = y(1) = 1$, $\bar{y}(x)$ has to be of the following form:

$$\bar{y}(x) = 1 + \beta x - \beta x^2 \tag{8.9.3}$$

(Use Eq. (8.9.2) for $x = 0$ and $x = 1$ with the given boundary conditions led to two equations for α, β, and γ). We can proceed with this form of $\bar{y}(x)$. But we pause here a bit to study the form of Eq. (8.9.3) carefully:

$$\bar{y}(x) = 1 + \beta x - \beta x^2 = 1 + \beta x(1 - x) \tag{8.9.4}$$

It can be seen that the red function $x(1 - x)$ is vanished at both $x = 0$ and $x = 1$; the boundary points! And the constant 1 is exactly the value of $y(x)$ at the boundary. Based on this analysis, we can, in general, seek for $\bar{y}(x)$ in the following general form:

$$\bar{y}(x) = \alpha_0(x) + \sum_{i=1}^{n} c_i \alpha_i(x) \tag{8.9.5}$$

where $\alpha_i(x)$ must be zero at the boundary points, and $\alpha_0(x)$ chosen to satisfy the non-zero boundary conditions. Note that the α_i's were called Ritz parameters.

And from $\bar{y}(x)$ in Eq. (8.9.4), we can determine its first derivative:

$$\bar{y}'(x) = \beta - 2\beta x \tag{8.9.6}$$

Introducing $\bar{y}(x)$ and $\bar{y}'(x)$ into Eq. (8.9.7), we get (obtained using a CAS as I was lazy, in the next example I will show the code):

$$I(\beta) = \frac{11}{30}\beta^2 + \frac{1}{3}\beta + 1 \tag{8.9.7}$$

which is simply an ordinary function of β, and we want to minimize I, right? That's easy now:

$$\frac{dI}{d\beta} = 0 : \frac{11}{15}\beta + \frac{1}{3} = 0 \Longrightarrow \beta = -\frac{5}{11} \tag{8.9.8}$$

Now that β has been determined, we have found the approximate solution:

$$\bar{y}(x) = 1 - \frac{5}{11}x + \frac{5}{11}x^2$$

How accurate is this solution? We can compare it with the exact $y^e(x)$ solution, which is given by

$$y^e(x) = \frac{\sinh(x) + \sinh(1 - x)}{\sinh(1)}$$

One way to check the accuracy of an approximate solution is to plot both solutions together as in Fig. 8.7a. The Ritz solution is quite good; however to have a better appreciation of the accuracy, we can plot the error function defined as the relative difference of the Ritz solution with respect to the exact one:

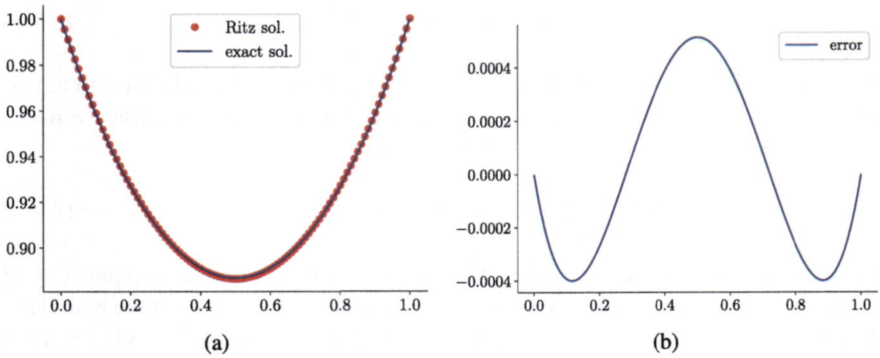

Fig. 8.7 Ritz solution vs. exact solution to the variational problem $I[y(x)] = \int_0^1 [y^2 + (y')^2]dx; \quad y(0) = y(1) = 1$

$$\text{error}(x) := \frac{y^e(x) - \bar{y}(x)}{y^e(x)}$$

Figure 8.7b shows the plot of this error.

Let's solve another problem with the Ritz method. Consider a simply supported beam of length L. Find the deflection of the beam under uniformly distributed transverse load q_0. Recall from Eq. (8.7.5) that the deflection $y(x)$ minimizes the following energy functional:

$$\Pi[u(x)] = \int_0^L \left[\frac{k}{2}(y'')^2 - q_0 y\right] dx, \quad k := EI \tag{8.9.9}$$

What are the boundary conditions? Because the beam is simply supported, its two ends cannot move down, thus $y(0) = y(L) = 0$.

Before using the Ritz method, note that the exact solution is a fourth-order polynomial:

$$y^e(x) = \frac{q_0 L^4}{24EI}\left(\frac{x}{L} - 2\frac{x^3}{L^3} + \frac{x^4}{L^4}\right) \tag{8.9.10}$$

Thus, as a first approximate solution, we seek for the following solution (what if we do not have the exact solution at hand? Then, we have to rely on the functional (8.9.9))

$$\bar{y}(x) = c_1 x(x - L) + c_2 x^2(x - L) \tag{8.9.11}$$

This form is chosen due to the fact that $\alpha_1(x) = x(x - L)$ and $\alpha_2(x) = x^2(x - L)$ vanish at $x = 0$ and $x = L$. With this $\bar{y}(x)$, I used SymPy to do everything for me, as shown in Listing 8.1.

Listing 8.1 Ritz's solution for the simply supported beam with Eq. (8.9.11).

```
using SymPy
@vars x k L q0 c1 c2
y   = c1*x*(x-L) + c2*x*x*(x-L) # approximate solution yh
ypp = diff(y,x,2)               # its 2nd derivative
F   = 0.5*k*ypp^2-q0*y          # the integrand in the functional
J   = integrate(F, (x, 0, L))   # the functional J
J1  = diff(J,c1)                # derivative of J wrt c1
J2  = diff(J,c2)                # derivative of J wrt c2
solve([J1, J2], [c1,c2])        # solve for c1 and c2
```

It is found that c_1 and c_2 are

$$c_1 = -\frac{q_0 L^2}{24EI}, \quad c_2 = 0$$

Thus, the two-parameter Ritz solution is given by

$$\bar{y}(x) = -\frac{q_0 L^2}{24EI} x(x - L) = \frac{q_0 L^4}{24EI} \left(\frac{x}{L} - \frac{x^2}{L^2} \right)$$

We now can check the accuracy. It can show that the Ritz maximum deflection, at the middle of the beam $x = L/2$, is off 20% of the exact deflection.

Even though programming gave us quickly the solution (Listing 8.1), it did not tell us everything. So, it is always a good idea to develop everything manually. Upon introduction of Eq. (8.9.11) into Eq. (8.9.9), we obtained a functional Π which is a function of c_1 and c_2. To minimize it, we set $d\Pi/dc_1 = 0$ and $d\Pi/dc_2 = 0$. Here is what we get from these two equations:

$$\begin{bmatrix} A_{11} & A_{12} \\ A_{21} & A_{22} \end{bmatrix} \begin{bmatrix} c_1 \\ c_2 \end{bmatrix} = \begin{bmatrix} b_1 \\ b_2 \end{bmatrix} \tag{8.9.12}$$

with

$$A_{ij} = \int_0^L k\alpha_i''(x)\alpha_j''(x)dx, \quad b_j = \int_0^L q_0\alpha_j(x)dx \tag{8.9.13}$$

Thus, Ritz converted a problem of solving a PDE (or minimizing a functional) to a linear algebra problem of finding the solutions to $\boldsymbol{A}\boldsymbol{c} = \boldsymbol{b}$. And the matrix is of size $n \times n$, where n is the number of terms in the Ritz approximation; furthermore, the matrix is symmetric. What is nice about Eq. (8.9.12) is that it has a pattern: the row ith can be written in this form

$$A_{ij}c_j = b_i$$

which works for any value of n. Thus, we have a recipe to build up our system, e.g., $\boldsymbol{A}$ and $\boldsymbol{b}$ to solve for c_i's.

To improve the Ritz solution, what should we do? We use a better approximation! A better approximation can be obtained if we add more terms to $\bar{u}(x)$; we add a new term $c_3x^3(x - L)$ to the two-parameter approximate $\bar{y}(x)$:

$$\bar{y}(x) = c_1x(x - L) + c_2x^2(x - L) + c_3x^3(x - L)$$

Repeat the same procedure by modifying the code in Listing 8.1, we get[9]

$$c_1 = -\frac{q_0L^2}{24EI}, \quad c_2 = -\frac{q_0L}{24EI}, \quad c_3 = \frac{q_0}{24EI}$$

Thus, the three-parameter Ritz solution is given by

$$\bar{y}(x) = \frac{q_0L^4}{24EI}\left[\frac{x}{L} - 2\frac{x^3}{L^3} + \frac{x^4}{L^4}\right]$$

which is exactly the exact solution!

8.10 What if There is No Functional to Start With?

At this stage, we should pause and think about what we have covered. The story is like this. Some separate minimization problems were proposed by mathematicians like the Bernoulli; they all have the same form of seeking a function to minimize an integral that depends on the function and its derivatives. While eminent mathematicians in the eighteenth century had solved these so-called variational problems, they did not develop a systematic approach. Then came Euler and Lagrange. Euler and Lagrange proceeded on what Gander and Wanner in their interesting article *From Euler, Ritz, and Galerkin to modern computing* [25] call the Euler-Lagrange highway, see the left branch of Fig. 8.8. They took the variation of the functional, then integrated by parts, and used the fundamental lemma of variational calculus to obtain a partial differential equation—now bears their name. Thus, the solution to the original variational problem is now the solution to the Euler-Lagrange equation. Then, Euler had spent time to develop techniques to solve his PDE including the first version of the finite difference method.[10]

On the other hand, in physics quite often we need to solve a partial differential equation. On such example is Laplace's equation. In mathematics and physics, Laplace's equation is a second-order partial differential equation named after Pierre-Simon Laplace, who first studied its properties. One example is: we have a thin plate and its edge is heated up to a certain degree, then we ask this question:

[9] If you want to do it manually, then use Eq. (8.9.13) to compute the members A_{ij} of the 3×3 matrix $\boldsymbol{A}$, and 3×1 vector $\boldsymbol{b}$. Solving the equation $\boldsymbol{Ac} = \boldsymbol{b}$ gives you exactly the same c_i's.

[10] The finite difference method is discussed in Sect. 10.7.

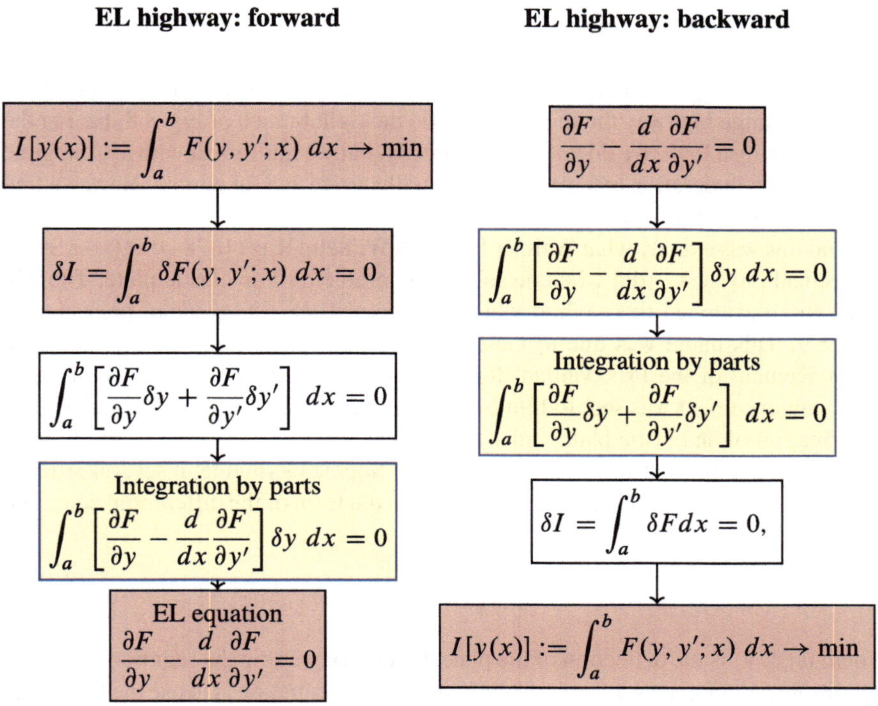

Fig. 8.8 The Euler-Lagrange highway of variational calculus: forward direction from a functional to the Euler-Lagrange PDE and the backward direction from a PDE to a functional

what is the temperature inside the plate? That temperature is the solution to Laplace's equation, if $u(x, y)$ denotes the temperature in the plate:

$$\Delta u = 0 \text{ in } \mathcal{B}, \quad \Delta u = \frac{\partial^2 u}{\partial x^2} + \frac{\partial^2 u}{\partial y^2} \tag{8.10.1}$$

where Δ is the Laplacian operator, see Eq. (6.11.35). Equation (8.10.1) means that $u(x, y)$ is a function such that $\Delta u = 0$ for all points in the plate or $(x, y) \in \mathcal{B}$.

Now, we start with a partial differential equation, and some mathematicians asked the question: whether there exists a functional associated with this equation? And the answer to this question in the case of Laplace's equation is yes; a result which is now known as Dirichlet's principle. Dirichlet's principle states that (a proof of which will be presented shortly), if the function u is the solution to Laplace's equation, Eq. (8.10.1), with boundary condition $u = g$ on the boundary $\partial\mathcal{B}$, then u can be obtained as the minimizer of the Dirichlet energy functional

$$E[v] = \int_{\mathcal{B}} \frac{1}{2}\|\nabla v\|^2 dV \tag{8.10.2}$$

The name "Dirichlet's principle" is due to Riemann, who applied it in the study of complex analytic functions.

What is the significance of Dirichlet's principle? It tells us that we can go the Euler-Lagrange highway the inverse way, see the right branch of Fig. 8.8. Facing the task of solving a PDE, we do not solve it directly, but we multiply it with δy, integrate the result, do integration by parts, and eventually arrive at a functional. Now, we find the minimizer of this functional.

And this was exactly what Walther Heinrich Wilhelm Ritz (1878–1909)—a Swiss theoretical physicist—did when he solved the problem of an elastic plate. Thus, in 1915 Ritz developed the method which was coined *the Ritz method*, presented in Sect. 8.9. This name was due to Galerkin. The main motivation for Ritz was the announcement of the Prix Vaillant for 1907 by the Academy of Science in Paris. This announcement was sent to him by his friend Paul Ehrenfest on a postcard. The deformation of an elastic plate under an external force $f(x, y)$ was a very difficult problem at that time; it was first considered by Sophie Germaine in several articles. The breakthrough was achieved by Kirchhoff in the form of the differential equation:

$$\frac{\partial^4 w}{\partial x^4} + 2\frac{\partial^4 w}{\partial x^2 y^2} + \frac{\partial^4 w}{\partial y^4} = f(x, y) \tag{8.10.3}$$

where $w(x, y)$ is the deflection of the plate. Of course we skip the required boundary conditions. A compact way to write the bending plate equation is to use the Laplacian operator Δ:

$$\Delta\Delta w = f(x, y) \tag{8.10.4}$$

Ritz went the Euler-Lagrange highway backwards, and came up with the following functional:

$$J[w(x, y)] = \int_{\mathcal{B}} \left[\frac{1}{2}(\Delta w)^2 - fw\right] dV \to \min \tag{8.10.5}$$

Then, he introduced his approximation for the solution function $w(x, y)$, assuming that the boundary condition is zero deflection on the plate edges:

$$\bar{w}(x, y) = c_1\psi_1(x, y) + c_2\psi_2(x, y) + \cdots + c_n\psi_n(x, y) \tag{8.10.6}$$

Substitution of this into Eq. (8.10.5), we have $J(c_1, c_2, \ldots)$, and minimizing it gives us a system of linear equations to solve for the Ritz parameters c_i's. The effort was high as Ritz did not have computer to help him, but of course he managed to get good results.

Because we need the functions $\psi_i(x, y)$ to be zero on the plate boundary, Ritz selected the easiest plate problem: a square plate of size 2×2. Thus, $\psi_1(x, y) =$

$(1-x^2)^2(1-y^2)^2$, with the origin of the coordinate system at the plate center, and so on.[11]

Proof of Dirichlet's principle. Assume that u is the solution to Laplace's equation, thus $\Delta u = 0$ in $\mathcal{B}$. Furthermore, we have $u = g$ on $\partial\mathcal{B}$. We have to show that

$$E[u] \leq E[w] \quad \text{for all } w \text{ such that } w = g \text{ on } \mathcal{B}$$

Now consider this function $v = u - w$, we have $v = 0$ on $\partial\mathcal{B}$. We now write $w = u - v$, and compute $E[w]$ using Eq. (8.10.2); if the final step was not clear, note that ∇u is a vector and check Box 9.2 for rules of the dot product:

$$\begin{aligned} E[w] &= \int_{\mathcal{B}} \frac{1}{2}\|\nabla(u-v)\|^2 dV = \frac{1}{2}\int_{\mathcal{B}} \nabla(u-v)\cdot\nabla(u-v)dV \\ &= \frac{1}{2}\int_{\mathcal{B}} \left[\|\nabla u\|^2 + \|\nabla v\|^2 + 2\nabla u\cdot\nabla v\right]dV \end{aligned}$$

Note that $\int_{\mathcal{B}} 2\nabla u\cdot\nabla v dV = 0$, thanks to the first Green's identity, see Sect. 6.11.12:

$$\int_{\mathcal{B}} \nabla u\cdot\nabla v dV = \int_{\partial\mathcal{B}} (v\nabla u)\cdot\boldsymbol{n}dS - \int_{\mathcal{B}} v\Delta u dV = 0$$

In the above integrals on the RHS, we have $\Delta u = 0$ in $\mathcal{B}$ and $v = 0$ on $\partial\mathcal{B}$; that's why $\int_{\mathcal{B}} \nabla u\cdot\nabla v dV = 0$. Thus,

$$E[w] = E[u] + E[v] \geq E[u], \quad (\text{because } E[v] \geq 0\,)$$

■

8.11 Galerkin Methods

The Ritz method was picked up by Russian mathematicians and engineers. For example, Ivan Grigoryevich Bubnov (1872–1919), a Russian marine engineer and designer of submarines for the Imperial Russian Navy, and Boris Galerkin (1871–1945), a Soviet mathematician and an engineer, used the method for practical problems and also developed new developments.

To solve the beam problem in Eq. (8.9.9), Bubnov used trigonometric functions instead of polynomials. For example, his two-parameter approximation reads

$$\bar{y}(x) = c_1 \sin\frac{\pi x}{L} + c_2 \sin\frac{3\pi x}{L} \tag{8.11.1}$$

[11] What if we have to deal with an L-shaped plate? Or even worse an arbitrary three-dimensional shape? To that we need an extended version of the Ritz method known as the finite element method.

And the corresponding solution is

$$\bar{y}(x) = \frac{4q_0L^4}{EI\pi^5}\sin\frac{\pi x}{L} + \frac{4q_0L^4}{243EI\pi^5}\sin\frac{3\pi x}{L} \tag{8.11.2}$$

You can plot the exact solution in Eq. (8.9.10), the two-parameter Ritz solution using polynomials in Eq. (8.9.2), and Bubnov's solution in Eq. (8.11.2), and you will see that using two trigonometric functions yield a better solution than using two polynomials. But, what is more is that, during the computation, you see that $A_{12} = A_{21} = 0$. Actually it was not a surprise to Bubnov. He knew the orthogonality of trigonometric functions (Fourier's work) and took advantage of it to simplify the computations. Note that because the diagonal terms in matrix $\boldsymbol{A}$ are zero, solving $\boldsymbol{Ac} = \boldsymbol{b}$ is super easy in this case.

Another contribution that Bubnov and Galerkin made is their observation that it is not necessary to know the functional associated with the given PDE to find the Ritz solution. So, they also followed the Euler-Lagrange highway backwards, but stopped before the final destination, Fig. 8.8:

$$\frac{\partial F}{\partial y} - \frac{d}{dx}\frac{\partial F}{\partial y'} = 0 \Longrightarrow \int_a^b \left[\frac{\partial F}{\partial y} - \frac{d}{dx}\frac{\partial F}{\partial y'}\right]\delta y\, dx = 0 \Longrightarrow \boxed{\int_a^b \left[\frac{\partial F}{\partial y}\delta y + \frac{\partial F}{\partial y'}\delta y'\right] dx = 0}$$

With the boxed equation, they introduced the usual Ritz approximations for y and δy to obtain a system of linear equations. To demonstrate their method, we solve the bending beam problem again, starting with the PDE:

$$ky'''' = q_0 \quad 0 < x < L, \quad y(0) = y(L) = 0;\ y''(0) = y''(L) = 0 \tag{8.11.3}$$

We first put the PDE in the form $ky'''' - q_0 = 0$, multiply that with δy, and integrate over the problem domain:

$$\int_0^L (ky'''' - q_0)\delta y dx = 0 \tag{8.11.4}$$

Then, integrating by parts twice to get

$$\boxed{\int_0^L (ky''\delta y'' - q_0\delta y)dx = 0} \tag{8.11.5}$$

Of course, this equation is nothing but the variation of a functional being set to zero. But we do not need to know the form of that functional, if our aim is primarily to find the solution $y(x)$.

Why integration by parts? In theory, we can stop at Eq. (8.11.4), and introduce the Ritz approximation into it to get a system of equations to solve for the Ritz parameters. However, it involves y'''', thus the Ritz approximation for y must use at least a third-order polynomial. Furthermore, we have asymmetry in the formulation: there is y'''' and only δy. A simple integration by parts solves these two issues! Just one simple integration by parts, and we get Eq. (8.11.5) in which the derivative of $y(x)$ has been lowered from four to two, and that is passed to $\delta y''$. Thus, we have a symmetric formulation. Thanks to this, the resulting matrix $\boldsymbol{A}$ will be symmetric, i.e., $A_{ij} = A_{ji}$.

Now, Galerkin used the Ritz approximation for $y(x)$. For illustration, only two terms are used,

$$y = c_1\phi_1(x) + c_2\phi_2(x) \Longrightarrow y'' = c_1\phi_1''(x) + c_2\phi_2''(x) \tag{8.11.6}$$

What about the variation δy? What should be its approximation? As a variation is a small perturbation to the actual solution $y(x)$, if $y(x)$ is of the form $c_i\phi_i(x)$, then its variation is of the same form[12]:

$$\delta y = d_1\phi_1(x) + d_2\phi_2(x) \Longrightarrow \delta y'' = d_1\phi_1''(x) + d_2\phi_2''(x) \tag{8.11.7}$$

What are the d_i's? They are real numbers which can be of any value, because a variation is anything that is zero at the boundary.

With these approximations of y and δy introduced into Eq. (8.11.5), we get

$$\int_0^L [k(c_1\phi_1'' + c_2\phi_2'')(d_1\phi_1'' + d_2\phi_2'') - q_0(d_1\phi_1 + d_2\phi_2]dx = 0$$

which is re-arranged in the form of $()d_1 + ()d_2 = 0$:

$$\begin{gathered}\left[\left(\int_0^L k\phi_1''\phi_1''dx\right)c_1 + \left(\int_0^L k\phi_1''\phi_2''dx\right)c_2 - \int_0^L q_0\phi_1 dx\right]d_1 + \\ \left[\left(\int_0^L k\phi_1''\phi_2''dx\right)c_1 + \left(\int_0^L k\phi_2''\phi_2''dx\right)c_2 - \int_0^L q_0\phi_2 dx\right]d_2 = 0\end{gathered} \tag{8.11.8}$$

Now, because d_1 and d_2 are arbitrary, we conclude that the two bracket terms must be zeroes:

$$\begin{aligned}\left(\int_0^L k\phi_1''\phi_1''dx\right)c_1 + \left(\int_0^L k\phi_1''\phi_2''dx\right)c_2 &= \int_0^L q_0\phi_1 dx \\ \left(\int_0^L k\phi_1''\phi_2''dx\right)c_1 + \left(\int_0^L k\phi_2''\phi_2''dx\right)c_2 &= \int_0^L q_0\phi_2 dx\end{aligned}$$

[12] In theory, the only requirement is that $\delta y(0) = \delta y(L) = 0$. Thus, it is possible to use another approximation for it, for example, $\delta y = d_i\psi_i(x)$. But that would be some years later after Galerkin's work. Advancements are made in small steps.

Look at what we have obtained? A system of equations to determine the Ritz coefficients, and the system is identical to the one got from the Ritz method, see Eqs. (8.9.12) and (8.9.13). That's probably why Galerkin called his method the Ritz method, and nowadays we call what Galerkin did the Galerkin method!

Let's summarize the steps of the method, which I refer to as the Bubnov-Galerkin method—a common term nowadays—in Box 8.1, even though a better term should have been Ritz-Bubnov-Galerkin method. What more this method gives us compared with its predecessor that Ritz developed? It has a wider range of applications as there are many partial differential equations that are not Euler-Lagrange equations of any variational problem.

Box 8.1: Bubnov-Galerkin method to solve numerically any PDE.

- Starting point: the PDE

$$ky'''' = q_0 \quad 0 < x < L$$

- Derive the weak form (multiply the PDE with δy, integrate over the domain, integrating by parts)

$$\int_0^L (ky''\delta y'' - q_0\delta y)dx = 0$$

- Approximating the function $y(x)$ using $\phi_i(x)$

$$y = \phi_0(x) + \sum_{i=1}^{n} c_i\phi_i(x)$$

- Approximating the variation $\delta y(x)$ also using $\phi_i(x)$

$$\delta y = \sum_{i=1}^{n} d_i\phi_i(x)$$

- Obtain a system of linear equations to solve for c_i's

$$A_{ij}c_j = b_j$$

8.12 The Finite Element Method

8.12.1 Basic Idea

The Ritz-Galerkin method has one serious limitation: it is difficult to apply the method to PDEs with complex geometry. For one-dimensional problems, this limitation does not present to us. Only in two dimensions it shows up. In 1942, Richard Courant, in his classic paper entitled "Variational methods for the solution of problems of equilibrium and vibrations", presented the first appearance of what we now call the finite element method. Unfortunately, the relevance of this article was not recognized at the time and the idea was forgotten. In the early 1950s the method was rediscovered by aerospace engineers at Boeing (MJ Turner) and structural engineers (J. H. Argyris). The term "finite elements" was coined by Ray Clough[13] in his classic paper "The Finite Element Method in Plane Stress Analysis" in 1960. The mathematical analysis of finite element approximations began much later, in the 1960s, the first important results being due to Milos Zlamal in 1968. Since then finite element methods have been developed into one of the most general and powerful classes of techniques for the numerical solution of partial differential equations and are widely used in engineering design and analysis.

The finite element method is a Ritz-Galerkin method but with one vital difference (Fig. 8.9) regarding the construction of the approximate solution:

- The domain is divided (or partitioned) into a number of sub-domains called *elements*. These elements are of simple shapes: in 2D the elements are triangles and quadrilaterals and in 3D they are tetrahedrals and hexahedrals. The vertices of the elements are called the *nodes*. The elements, the nodes, and the relation between elements/nodes altogether make a mesh. Let n be the total number of nodes in the mesh.
- The theory of interpolation is used to build the approximate solution. Assuming that $u(x)$ is the function we're trying to find, and let's denote by u_I the value of $u(x)$ at node I, then the approximate solution is written as

$$u(x) = \sum_{I}^{n} N_I(x) u_I \tag{8.12.1}$$

where $N_I(x)$ are called the shape functions. The shape functions are constructed such that they satisfy the Kronecker delta property:

$$N_I(x_J) = \delta_{IJ}, \quad \delta_{IJ} = \begin{cases} 1 & \text{if } I = J \\ 0 & \text{otherwise} \end{cases} \tag{8.12.2}$$

[13] Ray William Clough (1920–2016) was Byron L. and Elvira E. Nishkian Professor of Structural Engineering in the Department of Civil Engineering at the University of California, Berkeley and one of the founders of the finite element method.

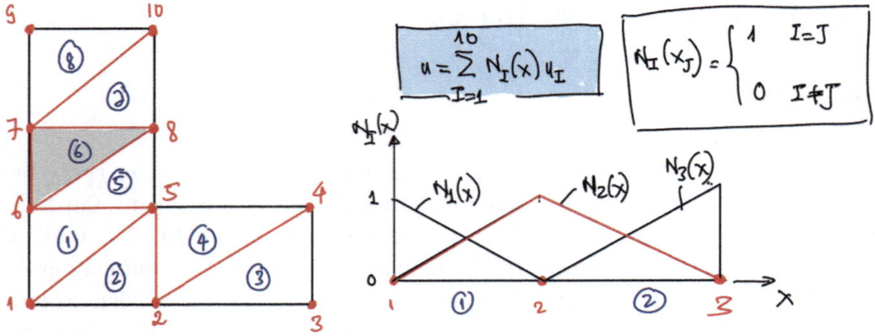

Fig. 8.9 Basic ideas of the finite element method: (1) domain division into triangular elements connected via the notes and (2) FE approximation using local shape functions

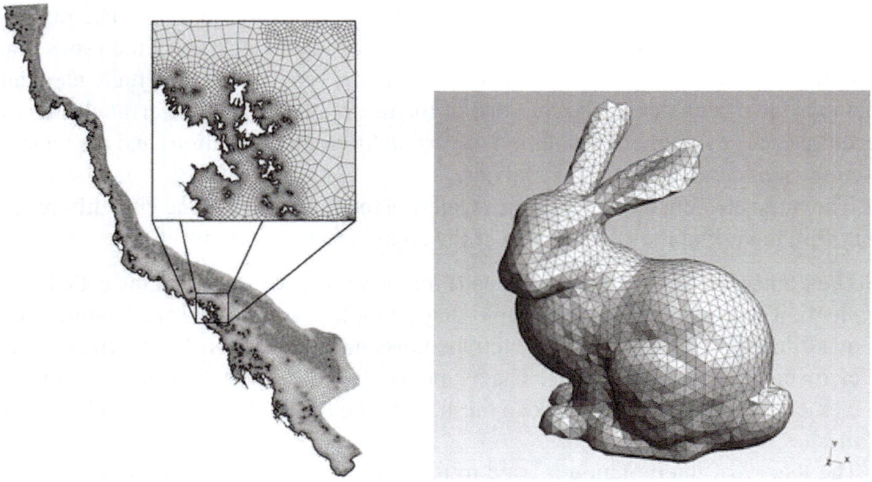

(a) 2D (b) 3D

Fig. 8.10 The finite element method enjoys a geometry flexibility: it can handle any geometry. These images are obtained using `Gmsh` [26]

Therefore, $u(x_J) = \sum_I N_I(x_J)u_I = u_J$. The Ritz parameters u_I now have a meaning: it is the value of the function evaluated at the nodes. Furthermore, the shape functions have local support, i.e., $N_I(x)$ is non-zero only over few elements connecting node I; see the right figure (bottom) of Fig. 8.9.

The finite element method is extremely flexible about geometry. It can solve PDEs on arbitrary 3D domains (Fig. 8.10). Furthermore, because the approximation is local, the construction of $N_I(x)$ is easier (than to build shape functions over the entire domain).

8.12.2 FEM for 1D Wave Equation

For a simplest introduction to FEM, let's consider the one-dimensional momentum equation that governs the deformation of a bar due to applied external forces:

$$\rho\frac{\partial^2 u}{\partial t^2} = E\frac{\partial^2 u}{\partial x^2} + \rho b \tag{8.12.3}$$

where $u(x, t)$ is the displacement field, E is Young's modulus of the material, ρ is the density, and b is the body force. The spatial domain is $0 \leq x \leq L$, L is the length of the bar and the time domain is $0 \leq t \leq T$.

For the case of zero body force (i.e., $b = 0$) the above equation becomes the well-known one-dimensional wave equation written as

$$\frac{\partial^2 u}{\partial t^2} = c^2\frac{\partial^2 u}{\partial x^2}, \quad c = \sqrt{\frac{E}{\rho}} \tag{8.12.4}$$

In order for a PDE to have unique solutions, initial and boundary conditions have to be provided. For example, the so-called Dirichlet boundary conditions read

$$u(0, t) = a, \quad u(L, t) = b, \quad t > 0 \tag{8.12.5}$$

where a, b are some constants. As Eq. (8.12.4) involves second derivative with respect to t, two initial conditions are required which are given by

$$u(x, 0) = f(x), \quad \dot{u}(x, 0) = g(x) \tag{8.12.6}$$

where $\dot{u} := du/dt$ and f, g are some known functions (i.e., data of the problem).

Putting all the above together we come up with the following initial-boundary value problem:

$$\begin{aligned} &\frac{\partial^2 u}{\partial t^2} = c^2\frac{\partial^2 u}{\partial x^2} && \text{(wave equation)} \\ &u(0, t) = a, \quad u(L, t) = b, \quad t > 0 && \text{(boundary conditions)} \\ &u(x, 0) = f(x), \quad \dot{u}(x, 0) = g(x) && \text{(initial conditions)} \end{aligned} \tag{8.12.7}$$

Equation (8.12.7) is called a *strong form* of the wave equation. The finite element methods (or generally Galerkin-based methods) adopt a weak formulation where the partial differential equations are restated in an integral form called the *weak form*. A weak form of the differential equations is equivalent to the strong form. In many disciplines, the weak form has a physical meaning; for example, the weak form of the momentum equation is called the principle of virtual work in solid/structural mechanics.

To obtain the weak form, one multiplies the PDE, i.e., the wave equation in this particular context, with an arbitrary function $w(x)$, called the *weight function*, and integrate the resulting equation over the entire domain. That is,

$$\int_0^L \left[\frac{\partial^2 u}{\partial t^2} - c^2 \frac{\partial^2 u}{\partial x^2}\right] w(x)dx = 0, \quad \forall w(x) \text{ with } w(0) = w(L) = 0 \qquad (8.12.8)$$

The arbitrariness of the weight function is crucial, as otherwise a weak form is not equivalent to the strong form. In this way, the weight function can be thought of as an enforcer: whatever it multiplies is enforced to be zero by its arbitrariness.

Using the integration by parts for the second term, the above equation becomes

$$\int_0^L \frac{\partial^2 u}{\partial t^2} w(x)dx + c^2 \int_0^L \frac{\partial u}{\partial x}\frac{\partial w}{\partial x} dx = 0 \qquad (8.12.9)$$

where the spatial derivative of the unknown field, $u(x, t)$, was lowered from two to one.

The weak form of the wave equation is thus given by: find the smooth function $u(x, t)$ such that

$$\begin{aligned} \int_0^L \frac{\partial^2 u}{\partial t^2} w(x)dx + c^2 \int_0^L \frac{\partial u}{\partial x}\frac{\partial w}{\partial x} dx = 0 \\ u(0, t) = a, \quad u(L, t) = b \\ u(x, 0) = f(x), \quad u(x, 0) = g(x) \end{aligned} \qquad (8.12.10)$$

for all $w(x)$ with $w(0) = w(L) = 0$.

Our weak form has both spatial and temporal variables. One simple method to deal with them is the method of lines. The method of lines proceeds by first discretizing the spatial derivatives only and leaving the time variable continuous. Therefore, the approximation of the unknown field $u(x, t)$ is written as

$$u(x, t) \approx u^h(x, t) = \sum_I^n N_I(x) u_I(t) \qquad (8.12.11)$$

where $N_I(x)$ are the shape functions and $u_I(t)$ denotes the value of u at point I at time instant t and constitutes the unknowns to be solved. The weak form (8.12.10) requires the acceleration and the first spatial derivative of $u(x, t)$, they are given by

$$\frac{\partial^2 u}{\partial t^2} = \sum_I^n N_I(x)\ddot{u}_I(t), \quad \frac{\partial u}{\partial x} = \sum_I^n \frac{dN_I(x)}{dx} u_I(t)$$

Even though there are many choices for the weight functions w, in the Bubnov-Galerkin method, which is the most commonly used method at least for solid mechanics applications, the weight function is approximated using the same shape functions as u. That is,

$$w(x,t) = \sum_{I}^{n} N_I(x) w_I \tag{8.12.12}$$

where w_I are the nodal values of the weight function; they are not functions of time. It is straightforward to compute w' required in Eq. (8.12.10).

With these approximations, the weak form of the wave equation, i.e., Eq. (8.12.10), becomes: find u_J such that

$$\int_0^L (N_I(x)\ddot{u}_I)(N_J(x)w_J)\,dx + c^2 \int_0^L \left(\frac{dN_I}{dx}u_I\right)\left(\frac{dN_J}{dx}w_J\right) dx = 0 \tag{8.12.13}$$

for all w_J. Note that the Einstein summation rule was adopted: indices which are repeated twice in a term are summed.

The arbitrariness of w_J results in the following system of ordinary differential equations[14]

$$\begin{bmatrix} \int_0^L N_1N_1dx & \int_0^L N_1N_2dx & \dots & \int_0^L N_1N_ndx \\ \int_0^L N_2N_1dx & \int_0^L N_2N_2dx & \dots & \int_0^L N_2N_ndx \\ \vdots & \vdots & \vdots & \vdots \\ \int_0^L N_nN_1dx & \int_0^L N_nN_2dx & \dots & \int_0^L N_nN_ndx \end{bmatrix} \begin{bmatrix} \ddot{u}_1 \\ \ddot{u}_2 \\ \vdots \\ \ddot{u}_n \end{bmatrix}$$

$$+c^2 \begin{bmatrix} \int_0^L dN_1dN_1dx & \int_0^L dN_1dN_2dx & \dots & \int_0^L dN_1dN_ndx \\ \int_0^L dN_2dN_1dx & \int_0^L dN_2dN_2dx & \dots & \int_0^L dN_2dN_ndx \\ \vdots & \vdots & \vdots & \vdots \\ \int_0^L dN_ndN_1dx & \int_0^L dN_ndN_2dx & \dots & \int_0^L dN_ndN_ndx \end{bmatrix} \begin{bmatrix} u_1 \\ u_2 \\ \vdots \\ u_n \end{bmatrix} = \begin{bmatrix} 0 \\ 0 \\ \vdots \\ 0 \end{bmatrix} \tag{8.12.14}$$

where the short notation dN_I is the first spatial derivative of the shape function N_I, $dN_I = dN_I/dx$. The integrals in the above equation are called weak form integrals.

[14] This is exactly identical to what we have done in the Ritz-Galerkin method, see, for example, Eq. (8.11.8).

For this simple 1D problem, they can be exactly computed, but generally, numerical integration is used to evaluate these integrals.

And Eq. (8.12.14) can be cast in the following compact equation using a matrix notation:

$$\mathbf{M}\ddot{\boldsymbol{u}} + \mathbf{K}\boldsymbol{u} = \mathbf{0}, \quad M_{IJ} = \int_0^L N_I N_J dx, \quad K_{IJ} = c^2 \int_0^L dN_I dN_J dx \tag{8.12.15}$$

where $\boldsymbol{u}$ and $\ddot{\boldsymbol{u}}$ are the vectors of displacements and accelerations of the whole problem, respectively; they are one-dimensional arrays of length n. $\mathbf{M}$ and $\mathbf{K}$ are the mass and stiffness matrix—matrices of dimension $n \times n$. Furthermore, these matrices are symmetric.

Equation (8.12.15) is referred to as the *semi-discrete equation* as the time has not been yet discretized. Any time integration methods for ODEs can be used to discretize Eq. (8.12.15) in time. Refer to Sect. 10.6 for detail. After having obtained $u_I(t)$, Eq. (8.12.11) is used to compute the function at any other points.

8.12.3 Role of FEM in Computational Sciences and Engineering

The finite element method is an important tool in computational sciences and engineering (CSE). CSE is a relatively new discipline that deals with the development and application of computational models, often coupled with high-performance computing, to solve complex problems arising in engineering analysis and design (computational engineering) as well as in natural phenomena (computational science). CSE has been described as the "third mode of discovery" next to theory and experimentation.

Within the realm of CSE these are steps to solve a problem:

- First, a mathematical model that best describes the problem is selected or developed. This step of model development is done manually by people with sufficient mathematical skills. A majority of mathematical model is developed using calculus and thus they are continuous models not suitable for digital computers.
- Second, a computational model of this mathematical model is derived. A computational model is an approximation to the mathematical model and is in a discrete form which can be solved using computers.
- Third, this discrete model is implemented in a programming language (Fortran in the past and C++ and Python nowadays) to have a computational code or platform.

Computer simulations are not only useful to solve problems too complex to be resolved analytically, but are also increasingly replacing costly and time-consuming experiments. Furthermore, they can provide tremendous information at scales of space and time where experimental visualization is difficult or impossible. And finally, simulations also have a value in their ability to predict the behavior of materials and structures that are yet to be created; experiments are limited to materials and structures that have already been created.

Chapter 9
Linear Algebra

This chapter is about linear algebra. Linear algebra is central to almost all areas of mathematics. Linear algebra is also used in most sciences and fields of engineering. Thus, it occupies a vital part in the university curriculum. Linear algebra is all about matrices, vector spaces, systems of linear equations, eigenvectors, you name it. It is common that a student of linear algebra can do the computations (e.g., compute the determinant of a matrix, or the eigenvector), but he/she usually does not know the why and the what—the theoretical essence of the subject. This chapter hopefully provides some answers to these questions.

There is one more strong motivation to learn linear algebra: it plays a vital part in machine learning, which is basically ubiquitous in our modern lives.

The following books were consulted for the materials presented in this chapter:

- *Introduction to Applied Linear Algebra: Vectors, Matrices, and Least Squares* by Stephen Boyd[1] and Lieven Vandenberghe[2].
- *Introduction to Linear Algebra* by the famous maths teacher Gilbert Strang[3] [24].
- *Linear Algebra: A Modern Introduction* by David Poole[4] [54].
- *Linear Algebra And Learning from Data* by Strang [66].

---

[1] Stephen Boyd is the Samsung Professor of Engineering, and Professor of Electrical Engineering in the Information Systems Laboratory at Stanford University. His current research focus is on convex optimization applications in control, signal processing, machine learning, and finance.

[2] Lieven Vandenberghe is a Professor of Electrical Engineering at the University of California, Los Angeles.

[3] His lectures are available at https://www.youtube.com/watchv=ZK3O402wf1c&list=PL49CF3715CB9EF31D&index=1.

[4] David Poole is a Professor of Mathematics at Trent University. He has been recognized with a number of awards for his inspirational teaching. His research interests are algebra, discrete mathematics, ring theory, and mathematics education.

V. P. Nguyen, *Mathematics for Engineers and Scientists*,
https://doi.org/10.1007/978-981-97-4631-6_9

I follow David Poole's organization for the subject to a great extent. Sometimes I felt lost reading Strang's [24]. With Poole, I could read from the beginning to the end of his book. Even though I understand that my linear algebra is still shaky (it is a big field and I rarely did exercises), thus reading Strang's [66] was useful. That book gave a concise review of linear algebra required to be used in applications. If I could understand Strang this time, I can say that I understand linear algebra.

The chapter starts with the familiar physical vectors in the 2D plane and in the 3D space we are living in (Sect. 9.1). Nothing is abstract and it is straightforward to introduce vector-vector addition and scalar-vector multiplication—the two most important vector operations in linear algebra. For use in vector calculus (and applications in physics), the cross product of two 3D vectors is also presented. But keep in mind that this product (with a weird definition and we can define a cross product of two 3D vectors only) is not used in linear algebra. The description, using vectors, of lines and planes is discussed, which plays an important role later.

Section 9.2 then presents a generalization of 2D and 3D vectors to vectors in $\mathbb{R}^n$—the n-dimensional space, whatever it is geometrically. The section introduces the important concept of linear combinations of a set of vectors, which plays a vital role in the treatment of systems of linear equations.

Systems of linear equations, those of the form $\mathbf{A}\boldsymbol{x} = \boldsymbol{b}$, are the subject of Sect. 9.3. More than 2000 years ago Chinese mathematicians already knew how to solve these systems. Due to its linearity solving a system of linear equations is not hard. But we introduce the new concept of matrix to the subject, and of course the Gaussian elimination method to take a matrix associated to $\mathbf{A}\boldsymbol{x} = \boldsymbol{b}$ to a row (reduced) echelon form.

And with that we study the algebraic rules of matrices; how we can add two matrices, multiplying a matrix with a vector and so on. The subject is known as matrix algebra (Sect. 9.4). Also discussed are the transpose of a matrix, the inverse of a matrix, the LU factorization of a matrix. Subspaces, basis, and dimension are treated in Sect. 9.5. A brief introduction to linear transformation is given to illustrate the geometric meaning of matrix-vector multiplication as well as the geometric meaning of the determinant of a matrix (Sect. 9.6).

In Sect. 9.7, the use of `Julia` to do some linear algebra calculations is presented. The purpose is to use computers to do boring/tedious computations so that we can focus on the meaning underlying those computations. There is no gain if one can compute the determinant of a 10×10 matrix manually but does not understand what a determinant is!

Section 9.8 is all about orthogonality: one vector orthogonal to another vector, vector orthogonal to a set of vectors, vectors orthogonal to a subspace, and so on. Orthogonal projection is presented together with the Gram-Schmidt orthogonalization process to build an orthogonal basis of a subspace. And that leads to another matrix factorization—the QR factorization.

Section 9.9 is about determinants of matrices. With the introduction of linear transformation, it is clear to get the formula for the determinant of 2×2 matrices and of 3×3 matrices. The derivation is purely geometric. Based on this geometric meaning of the determinant of 3×3 matrices, some properties of the determinant are observed. And we then derive a general formula for the determinant of an $n \times n$ matrix based on these properties.

Section 9.10 is about eigenvalue problem. An eigenvector $\boldsymbol{x}$ of a matrix $\mathbf{A}$ is a vector such that $\mathbf{A}\boldsymbol{x}$ is in the *same direction as the vector* $\boldsymbol{x}$: $\mathbf{A}\boldsymbol{x} = \lambda\boldsymbol{x}$. In other words, multiplying matrix $\boldsymbol{A}$ with an eigenvector gives a new vector $\lambda\boldsymbol{x}$ where λ—an *eigenvalue*—is a stretching/shrinking factor.

The cultivation of linear algebra is vector spaces (Sect. 9.11). Vector spaces not only include the space of n-dimensional vectors but also functions, matrices. Section 9.12.1 is about a factorization for rectangle matrices—the so-called singular value decomposition. Complex matrices are briefly treated in Sect. 9.13. Finally, two applications of linear algebra: least square problems (Sect. 9.14) and principal component analysis (Sect. 9.15) end the chapter.

9.1 Vector in $\mathbb{R}^3$

To begin our journey about vector algebra let's do some observation about various concepts we use daily. For example, consider a cube of side 2 cm; its volume is 8 cm^3. Now if we rotate this cube, whatever the rotation angle is, its volume is always 8 cm^3. We say that volume is a *direction-independent quantity*. Mass, volume, density, and temperature are such quantities. The formal term for them is *scalar quantities*. To specify a scalar quantity, we need only to provide its *magnitude* (8 cm^3, for example). And we know how to do mathematics with these scalars: we can add, subtract, multiply, take roots, etc. Furthermore, we know the rules of these operations, see, e.g., Eq. (2.1.2).

On the other hand, there are quantities that are *direction-dependent*. It is not hard to see that velocity is such a quantity. We need to specify the magnitude (or speed) and a direction when speaking of a velocity. After all, your car is running at 50 km/h north-west is completely different from 50 km/h south-east. Quantities such as velocity, force, acceleration, and (linear and angular) momentum are called *vectorial quantities*; they need a magnitude and a direction.

Geometrically, we use arrows to represent vectors (Fig. 9.1). Symbolically, we can write $\overrightarrow{AB}$ or a boldface $\boldsymbol{a}$—a notation introduced by Josiah Willard Gibbs (1839–1903), an American scientist. We employ Gibbs' notation in this book. So, in what follows $\boldsymbol{a}$ (and similar symbols such as $\boldsymbol{b}$) are vectors. However, in some figures, the old $\overrightarrow{AB}$ still exist as it's easier to draw an arrow.

Now, we need to define some operations for vectors similar to what we have done for numbers. It turns out there are only a few: addition of vectors (two or more), multiplication of a vector with a scalar, dot product of two vectors (yielding a scalar), and cross product of two vectors giving a vector (remember the torque in physics?).

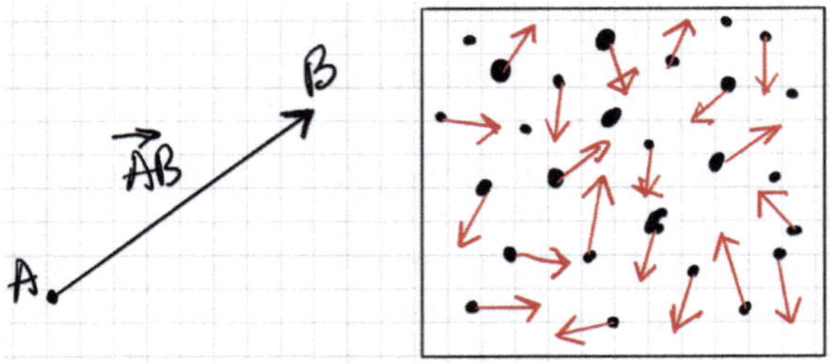

Fig. 9.1 Vectors are geometrically represented by arrows. A is the head of the vector and B is its tail

9.1.1 Addition and Scalar Multiplication

Addition of two vectors is simple: if we walk from A to B, then from B to C it is equivalent to walking from A to C directly, see Fig. 9.2a. So, to compute the sum of $\boldsymbol{a}$ and $\boldsymbol{b}$, we move the head of $\boldsymbol{b}$ to the tail of $\boldsymbol{a}$. Doing so does not change $\boldsymbol{b}$ as *two vectors are the same if they have identical lengths and directions*. And that is the well-known parallelogram rule.

Having defined the addition operation, we need to find the properties that vector addition obeys. From Fig. 9.2a, we can see immediately that $\boldsymbol{a}+\boldsymbol{b}=\boldsymbol{b}+\boldsymbol{a}$. Furthermore, it can be seen that $(\boldsymbol{a}+\boldsymbol{b})+\boldsymbol{c}=\boldsymbol{a}+(\boldsymbol{b}+\boldsymbol{c})$. That is, addition of vectors follows the commutative and associative rules similar to numbers. Why $(\boldsymbol{a}+\boldsymbol{b})+\boldsymbol{c}=\boldsymbol{a}+(\boldsymbol{b}+\boldsymbol{c})$ useful? Because it allows us not to worry about the order and thus we can remove the brackets unambiguously.

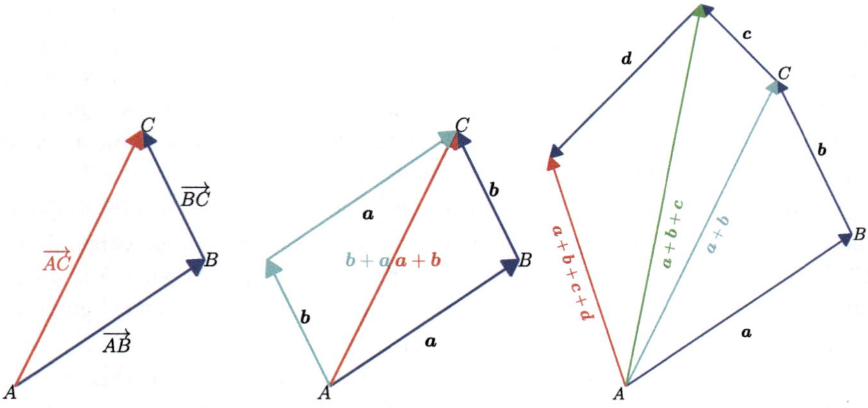

(a) Addition of 2 vectors (b) Addition of more than 2 vecs

Fig. 9.2 Addition of vectors: **a** addition of two vectors: the parallelogram rule and **b** addition of more than two vectors

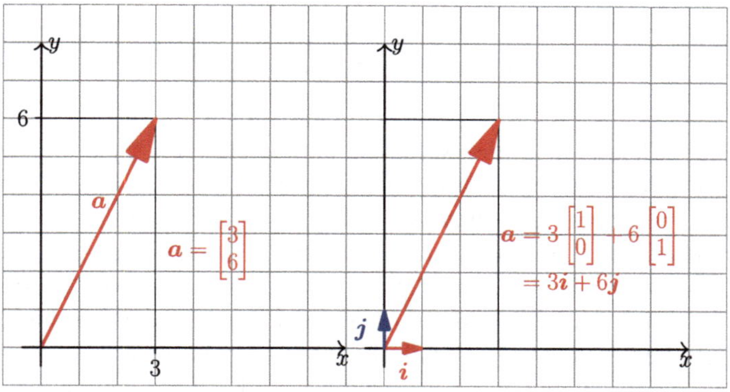

Fig. 9.3 With the introduction of a coordinate system, any vector is represented by an ordered pair of numbers (x, y) in 2D, written as a column vector, or an ordered triplet of numbers (x, y, z) in 3D. To save space, in text we write $\boldsymbol{a} = (a_1, a_2)^\top$ instead of $\boldsymbol{a} = \begin{bmatrix} a_1 \\ a_2 \end{bmatrix}$. There is more to say about the transpose operator $\Box^\top$. Noting that a vector is a geometrical object, not a list of numbers. That list (x, y, z) is just a representation of a vector in a chosen coordinate system

Repeated addition leads to multiplication. If we add a vector $\boldsymbol{a}$ to itself we get $2\boldsymbol{a}$, a vector that has the same direction with $\boldsymbol{a}$ but twice the length. We can generalize this by defining a scalar multiplication for vectors. Given $\alpha \in \mathbb{R}$, $\alpha\boldsymbol{a}$ is the scaled vector that has a new length being the length of the original vector multiplied by α, but maintains the direction of $\boldsymbol{a}$. From first principles of Euclidean geometry (e.g., similar triangles), we can see that $\alpha(\boldsymbol{a} + \boldsymbol{b}) = \alpha\boldsymbol{a} + \alpha\boldsymbol{b}$.

Up to this point we have considered vectors as purely geometrical objects. To simplify the computations, we adopt the approach of analytic geometry—use algebra to describe geometrical objects. To this end, we use a Cartesian coordinate system, where each point is described by an ordered pair[5] of numbers (x, y) in 2D or an ordered triplet of numbers (x, y, z) or (x_1, x_2, x_3) in 3D. A vector is then a directed line segment from the origin to any point in space (Fig. 9.3). The list (x_1, x_2, x_3) are called the coordinates or the components of the vector.

On this plane we see a remarkable thing: *any vector*, say $\boldsymbol{a} = (a_1, a_2)^\top$, is obtained by going to the right (from the origin) a distance a_1 and then going vertically a distance a_2, see the right figure in Fig. 9.3. We can write this down as

$$\boldsymbol{a} = a_1 \begin{bmatrix} 1 \\ 0 \end{bmatrix} + a_2 \begin{bmatrix} 0 \\ 1 \end{bmatrix} \tag{9.1.1}$$

or, with the introduction of two new vectors $\boldsymbol{i}$ and $\boldsymbol{j}$ called *the unit coordinate vectors*[6]:

[5] The word ordered is used because (x, y) is totally different from (y, x).

[6] Now imagine that we scale these unit vectors to $2\boldsymbol{i}$ and $2\boldsymbol{j}$ and use the scaled vectors as the new basis vectors. What will happen to the components of our vector $\boldsymbol{a}$? Apparently, its components will

$$\boldsymbol{a} = a_1\boldsymbol{i} + a_2\boldsymbol{j}, \quad \boldsymbol{i} := \begin{bmatrix} 1 \\ 0 \end{bmatrix}, \ \boldsymbol{j} := \begin{bmatrix} 0 \\ 1 \end{bmatrix} \tag{9.1.2}$$

Of course for 3D, we have three such vectors $\boldsymbol{i} = (1, 0, 0)^\top$, $\boldsymbol{j} = (0, 1, 0)^\top$, and $\boldsymbol{k} = (0, 0, 1)^\top$. Why writing such trivial equation such as Eq. (9.1.2)? Because it says that any vector can be written as a *linear combination of the unit coordinate vectors*. In other words, we say that the two unit coordinate vectors *span* the 2D space. This is how mathematicians express the idea that "the two directions—east and north—are sufficient to get us anywhere on a plane". Note that this geometric view does not, however, exist if we talk about high-dimensional spaces.

Vector addition is simple with components: to add vectors, add the components. The proof is straightforward as follows, where $\boldsymbol{a} = (a_1, a_2, a_3)^\top$

$$\begin{aligned}
\boldsymbol{a} + \boldsymbol{b} &= (a_1\boldsymbol{i} + a_2\boldsymbol{j} + a_3\boldsymbol{k}) + (b_1\boldsymbol{i} + b_2\boldsymbol{j} + b_3\boldsymbol{k}) \\
&= (a_1 + b_1)\boldsymbol{i} + (a_2 + b_2)\boldsymbol{j} + (a_3 + b_3)\boldsymbol{k}
\end{aligned}$$

Similarly, to scale a vector, scale its components: for a vector in 2D, $\alpha\boldsymbol{a} = (\alpha a_1, \alpha a_2)$. Do we have to define vector subtraction? No! This is because $\boldsymbol{a} - \boldsymbol{b} = \boldsymbol{a} + (-1)\boldsymbol{b}$. Scaling a vector with a negative number changes its length and flips its direction.

Being able to be added, and scaled by a number, it is natural to compute a vector given by $\alpha_1\boldsymbol{a}_1 + \alpha_2\boldsymbol{a}_2 + \cdots + \alpha_n\boldsymbol{a}_n$—a linear combination of n vectors $\boldsymbol{a}_i$. We have seen such combination in Eq. (9.1.2).

With components, it is easy to prove $\alpha(\boldsymbol{a} + \boldsymbol{b}) = \alpha\boldsymbol{a} + \alpha\boldsymbol{b}$. Indeed, $\alpha(a_i + b_i) = \alpha a_i + \alpha b_i$. Similar trivial proofs show up frequently in linear algebra.

Box 9.1 summarizes the laws of vector addition and scalar multiplication. Note that $\boldsymbol{0}$ is the zero vector, i.e., $\boldsymbol{0} = (0, 0, 0)^\top$ for 3D vectors.

Box 9.1: The laws of vector addition and scalar multiplication.

(a): commutative law	$\boldsymbol{a} + \boldsymbol{b}$	$= \boldsymbol{b} + \boldsymbol{a}$
(b): associative law	$\boldsymbol{a} + (\boldsymbol{b} + \boldsymbol{c})$	$= (\boldsymbol{a} + \boldsymbol{b}) + \boldsymbol{c}$
(c): zero vector	$\boldsymbol{a} + \boldsymbol{0}$	$= \boldsymbol{a}$
(d): distributive law	$\alpha(\boldsymbol{a} + \boldsymbol{b})$	$= \alpha\boldsymbol{a} + \alpha\boldsymbol{b}$
(e): distributive law	$(\alpha + \beta)\boldsymbol{a}$	$= \alpha\boldsymbol{a} + \beta\boldsymbol{a}$
(f):	$1\boldsymbol{a}$	$= \boldsymbol{a}$
(g):	$\alpha(\beta\boldsymbol{a})$	$= (\alpha\beta)\boldsymbol{a}$

be $0.5a_1, 0.5a_2$. As the basis vectors get bigger the components of the vector get smaller. That's why the better name for them is: contravariant components. This is important when we study tensors.

9.1.2 Dot Product

While vector addition and scalar-vector multiplication are quite natural, it is hard to immediately grasp the dot product of two vectors. We give the definition first and from there we deduce the meaning of the dot product. At the end of the section, we provide a discussion that leads to the definition of the dot product. The discussion is based on the observation that the *length of a vector does not change whatever transformation we apply to the vector*.

Definition 9.1.1 The dot product of two 3D vectors $\boldsymbol{a} = (a_1, a_2, a_3)^\top$ and $\boldsymbol{b} = (b_1, b_2, b_3)^\top$ is a number defined as

$$\boldsymbol{a} \cdot \boldsymbol{b} = a_1 b_1 + a_2 b_2 + a_3 b_3 \tag{9.1.3}$$

Why this definition? One way to understand is to consider the special case that the two vectors are the same. When $\boldsymbol{b} = \boldsymbol{a}$, we have $\boldsymbol{a} \cdot \boldsymbol{a} = a_1^2 + a_2^2 + a_3^2$, which is the square of the length of $\boldsymbol{a}$, see Fig. 9.4. So, the dot product gives us the length of a vector, defined by $\|\boldsymbol{a}\| := \sqrt{\boldsymbol{a} \cdot \boldsymbol{a}}$. We recall that the notation $|x|$ gives the distance from x to 0. Note the similarity in the notations.

The dot product has many applications. For example, the kinetic energy of a 1D point mass m with speed v is $0.5mv^2$ and its extension to 3D is $0.5m\boldsymbol{v} \cdot \boldsymbol{v}$. The work done by a force $\boldsymbol{F}$ is $\int_1^2 \boldsymbol{F} \cdot d\boldsymbol{s}$. And the list goes on.

There is a geometric meaning of this dot product: $\boldsymbol{a} \cdot \boldsymbol{b} = \|\boldsymbol{a}\| \|\boldsymbol{b}\| \cos(\boldsymbol{a}, \boldsymbol{b})$. The notation $(\boldsymbol{a}, \boldsymbol{b})$ means the angle between the two vectors $\boldsymbol{a}$ and $\boldsymbol{b}$. The proof is based

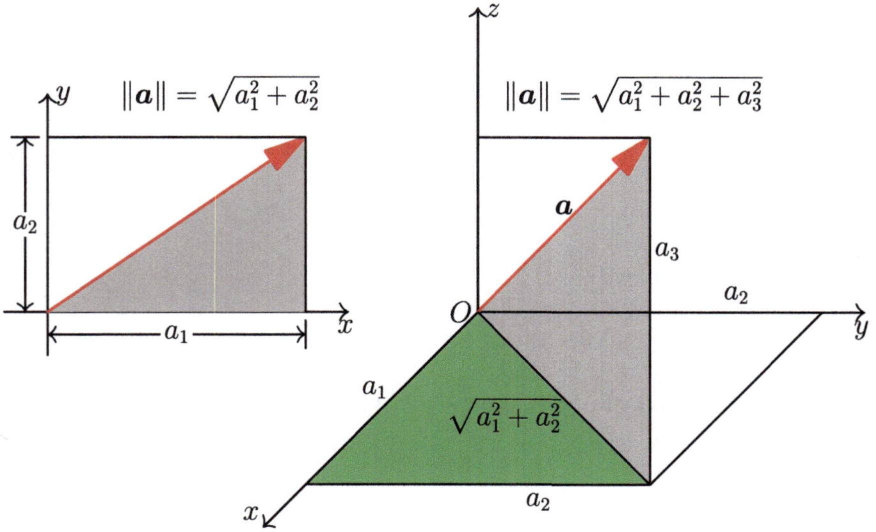

Fig. 9.4 Length of a 2D and 3D vector: $\|\boldsymbol{a}\| = \sqrt{a_1^2 + a_2^2 + a_3^2}$ from the Pythagorean theorem

on the generalized Pythagorean theorem $c^2 = a^2 + b^2 - 2ab\cos C$ (Sect. 3.17). We need a triangle here: two edges are vectors $\boldsymbol{a}$ and $\boldsymbol{b}$, and the remaining edge is $\boldsymbol{c} = \boldsymbol{b} - \boldsymbol{a}$. To this triangle, we can write (using the generalized Pythagorean theorem)

$$\|\boldsymbol{b} - \boldsymbol{a}\|^2 = \|\boldsymbol{a}\|^2 + \|\boldsymbol{b}\|^2 - 2\|\boldsymbol{a}\|\|\boldsymbol{b}\|\cos\theta \tag{9.1.4}$$

On the other hand, the squared length of vector $\boldsymbol{b} - \boldsymbol{a}$ can also be written as using the dot product and its property $\boldsymbol{a} \cdot (\boldsymbol{b} \pm \boldsymbol{c}) = \boldsymbol{a} \cdot \boldsymbol{b} \pm \boldsymbol{a} \cdot \boldsymbol{c}$ (known as the distributive law, see Box 9.2):

$$\begin{aligned}\|\boldsymbol{b} - \boldsymbol{a}\|^2 &= (\boldsymbol{b} - \boldsymbol{a}) \cdot (\boldsymbol{b} - \boldsymbol{a}) \\ &= \boldsymbol{b} \cdot \boldsymbol{b} + \boldsymbol{a} \cdot \boldsymbol{a} - 2\boldsymbol{a} \cdot \boldsymbol{b} = \|\boldsymbol{a}\|^2 + \|\boldsymbol{b}\|^2 - 2\boldsymbol{a} \cdot \boldsymbol{b}\end{aligned} \tag{9.1.5}$$

From Eqs. 9.1.4, 9.1.5, we get

$$\boxed{\boldsymbol{a} \cdot \boldsymbol{b} = \|\boldsymbol{a}\|\|\boldsymbol{b}\|\cos\theta} \tag{9.1.6}$$

And this formula reveals one nice geometric property. As $\cos\theta = 0$ when $\theta = \pi/2$, *two vectors are perpendicular/orthogonal to each other if their dot product is zero.* We can now see that the unit vectors $\boldsymbol{i}, \boldsymbol{j}, \boldsymbol{k}$ are mutually perpendicular: $\boldsymbol{i} \cdot \boldsymbol{j} = 0$, $\boldsymbol{i} \cdot \boldsymbol{k} = 0$, $\boldsymbol{j} \cdot \boldsymbol{k} = 0$. Why call them unit vectors? Because their lengths are 1. We can always make a non-unit vector a unit vector simply by dividing it by its length, a process known as normalizing a vector:

$$\text{normalizing a vector:} \quad \hat{\boldsymbol{v}} = \frac{\boldsymbol{v}}{\|\boldsymbol{v}\|} \tag{9.1.7}$$

When we need just the direction of a vector, $\frac{\boldsymbol{v}}{\|\boldsymbol{v}\|}$ is the answer.

Again, we observe some properties or laws governing the behavior of the dot product. We summarize them in Box 9.2. The proofs are quite straightforward and thus skipped. From (a) and (b), we are going to derive another rule with $\boldsymbol{a} = \boldsymbol{e} + \boldsymbol{f}$

$$\boldsymbol{a} \cdot (\boldsymbol{b} + \boldsymbol{c}) = \boldsymbol{a} \cdot \boldsymbol{b} + \boldsymbol{a} \cdot \boldsymbol{c} \Longleftrightarrow (\boldsymbol{e} + \boldsymbol{f}) \cdot (\boldsymbol{b} + \boldsymbol{c}) = (\boldsymbol{e} + \boldsymbol{f}) \cdot \boldsymbol{b} + (\boldsymbol{e} + \boldsymbol{f}) \cdot \boldsymbol{c}$$

And using (a, b) again, we have

$$(\boldsymbol{e} + \boldsymbol{f}) \cdot (\boldsymbol{b} + \boldsymbol{c}) = \boldsymbol{e} \cdot \boldsymbol{b} + \boldsymbol{e} \cdot \boldsymbol{c} + \boldsymbol{f} \cdot \boldsymbol{b} + + \boldsymbol{f} \cdot \boldsymbol{c}$$

And what is this? This is the FOIL (First-Outer-Inner-Last) rule of algebra discussed in Sect. 2.1!

Box 9.2: The laws of the dot product.

$$\begin{array}{lll} \text{(a): commutative law} & \boldsymbol{a}\cdot\boldsymbol{b} & = \boldsymbol{b}\cdot\boldsymbol{a} \\ \text{(b): distributive law} & \boldsymbol{a}\cdot(\boldsymbol{b}+\boldsymbol{c}) & = \boldsymbol{a}\cdot\boldsymbol{b}+\boldsymbol{a}\cdot\boldsymbol{c} \\ \text{(c):} & (\alpha\boldsymbol{a})\cdot\boldsymbol{b} & = \alpha(\boldsymbol{a}\cdot\boldsymbol{b}) = \boldsymbol{a}\cdot(\alpha\boldsymbol{b}) \\ \text{(d):} & \boldsymbol{a}\cdot\boldsymbol{a} & \geq 0 \quad \text{(equality holds iff } \boldsymbol{a}=\boldsymbol{0}) \end{array}$$

One application of (b): if $\boldsymbol{a}$ is perpendicular to both $\boldsymbol{b}$ and $\boldsymbol{c}$ (i.e., $\boldsymbol{a}\cdot\boldsymbol{b} = \boldsymbol{a}\cdot\boldsymbol{c} = 0$), then it is perpendicular to $\boldsymbol{b}+\boldsymbol{c}$ and in fact it is perpendicular to all linear combinations of $\boldsymbol{b}$ and $\boldsymbol{c}$ (or it is perpendicular to the plane spanned by $\boldsymbol{b}$ and $\boldsymbol{c}$):

$$\boldsymbol{a}\cdot(\alpha\boldsymbol{b}+\beta\boldsymbol{c}) = \alpha(\boldsymbol{a}\cdot\boldsymbol{b}) + \beta(\boldsymbol{a}\cdot\boldsymbol{c}) = \alpha 0 + \beta 0 = 0$$

The triangle inequality. Consider two vectors $\boldsymbol{a}$ and $\boldsymbol{b}$, they make two edges of a triangle, the remaining edge is its sum $\boldsymbol{a}+\boldsymbol{b}$. From the property of triangle, we then have the following inequality:

$$||\boldsymbol{a}+\boldsymbol{b}|| \leq ||\boldsymbol{a}|| + ||\boldsymbol{b}|| \tag{9.1.8}$$

Proof We need to use the Cauchy-Schwarz inequality proved in Sect. 2.20.3. Note that Eq. (9.1.6) also provides a geometric proof for the Cauchy-Schwarz inequality at least for 2D/3D cases.[7] Now, we can write[8]:

$$\begin{aligned} (\boldsymbol{a}+\boldsymbol{b})\cdot(\boldsymbol{a}+\boldsymbol{b}) &= \boldsymbol{a}\cdot\boldsymbol{a} + 2\boldsymbol{a}\cdot\boldsymbol{b} + \boldsymbol{b}\cdot\boldsymbol{b} \\ &\leq ||\boldsymbol{a}||^2 + 2||\boldsymbol{a}||||\boldsymbol{b}|| + ||\boldsymbol{b}||^2 \quad \text{(Cauchy-Schwarz inequality)} \\ &\leq (||\boldsymbol{a}|| + ||\boldsymbol{b}||)^2 \end{aligned}$$

■

And if we have something for two vectors, we should extend that to n vectors. First, it's easy to see that, for three vectors we have

$$||\boldsymbol{a}+\boldsymbol{b}+\boldsymbol{c}|| \leq ||\boldsymbol{a}|| + ||\boldsymbol{b}|| + ||\boldsymbol{c}||$$

Proof goes as: using Eq. (9.1.8) with two vectors $\boldsymbol{a}$ and $\boldsymbol{d} = \boldsymbol{b}+\boldsymbol{c}$, then Eq. (9.1.8) gain for $\boldsymbol{b}$ and $\boldsymbol{c}$. You see the pattern to go to n vectors. And to practice proof by induction you can prove the general case.

---

[7] As $\boldsymbol{a}\cdot\boldsymbol{b} = ||\boldsymbol{a}||||\boldsymbol{b}||\cos\theta$, we have $\boldsymbol{a}\cdot\boldsymbol{b} \leq ||\boldsymbol{a}||||\boldsymbol{b}||$.

[8] We can use this to prove Pythagoras theorem: if $\boldsymbol{a}$ is orthogonal to $\boldsymbol{b}$ then $\boldsymbol{a}\cdot\boldsymbol{b} = 0$, thus we have $(\boldsymbol{a}+\boldsymbol{b})\cdot(\boldsymbol{a}+\boldsymbol{b}) = \boldsymbol{a}\cdot\boldsymbol{a} + \boldsymbol{b}\cdot\boldsymbol{b}$. which is nothing than $||\boldsymbol{a}+\boldsymbol{b}||^2 = ||\boldsymbol{a}||^2 + ||\boldsymbol{b}||^2$. And this vector-based proof of the Pythagoras theorem works for 2D and 3D and actually nD.

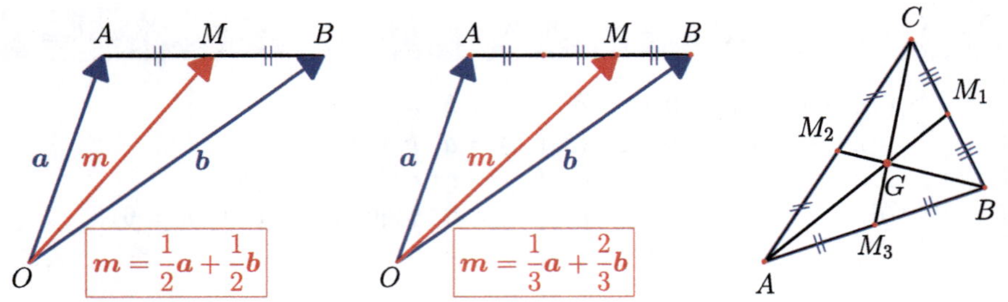

Fig. 9.5 Solving plane geometry using vectors. A median of a triangle is the line segment from a vertex to the mid-point of the opposite side. So, AM_1 is a median of the triangle ABC. We want to prove this fact: in any triangle, the three medians intersect at a common point (G) which is 2/3 of the way along each median. In other words, for each median, the distance from a vertex to the G is twice that of the distance from G to the mid-point of the side opposite that vertex. And that common point is called the centroid of the triangle

Solving plane geometry problems using vectors. Vectors can be used to solve easily many plane geometry problem (algebraic manipulations of some vectors only), see Fig. 9.5 for some examples. First, we consider a segment AB with M being its mid-point. Let's denote by $\boldsymbol{a}$ and $\boldsymbol{b}$ the vectors from the origin to A and B, and $\boldsymbol{m}$ for point M. And we would like to express $\boldsymbol{m}$ in terms of $\boldsymbol{a}, \boldsymbol{b}$.

It is not hard to derive the result shown in the left of Fig. 9.5:

$$\begin{cases} \overrightarrow{AM} = \boldsymbol{m} - \boldsymbol{a} \\ \overrightarrow{AM} = \frac{1}{2}\overrightarrow{AB} = \frac{1}{2}(\boldsymbol{b} - \boldsymbol{a}) \end{cases} \Longrightarrow \boldsymbol{m} - \boldsymbol{a} = \frac{1}{2}(\boldsymbol{b} - \boldsymbol{a}) \Longrightarrow \boldsymbol{m} = \frac{1}{2}(\boldsymbol{a} + \boldsymbol{b})$$

And in the same manner, we get the result in the middle picture of the mentioned figure. Now, we're ready to prove the theorem about the centroid of a triangle.

We consider the median CM_3, and point G such that $GM_3 = {}^1\!/_3 CM_3$. Using the results of Fig. 9.5, we have

$$\begin{cases} \overrightarrow{OM_3} = \frac{1}{2}(\boldsymbol{a} + \boldsymbol{b}) \\ \overrightarrow{OG} = \frac{1}{3}\boldsymbol{c} + \frac{2}{3}\left(\frac{1}{2}(\boldsymbol{a} + \boldsymbol{b})\right) = \frac{1}{3}(\boldsymbol{a} + \boldsymbol{b} + \boldsymbol{c}) \end{cases}$$

Of course, we next consider the median AM_1, and a point G' such that $G'M_1 = {}^1\!/_3 AM_1$. It can be shown that $\overrightarrow{OG'} = {}^1\!/_3(\boldsymbol{a} + \boldsymbol{b} + \boldsymbol{c})$. Thus, G' is nothing but G. And finally, considering the median BM_2 and we're done.

Another way to come up with the dot product.

It is obvious that the length of a vector, which is a scalar quantity, is invariant under translation and rotation. That is, if we rotate a vector, its length does not change. So, we can define a "dot product" that applies to a single vector only, i.e., $\boldsymbol{a} \cdot \boldsymbol{a} = a_1^2 + a_2^2 + a_3^2$. We can thus write

$$\begin{aligned} \boldsymbol{a} \cdot \boldsymbol{a} &= \|\boldsymbol{a}\|^2 = \text{constant} \\ \boldsymbol{b} \cdot \boldsymbol{b} &= \|\boldsymbol{b}\|^2 = \text{constant} \\ (\boldsymbol{a}+\boldsymbol{b}) \cdot (\boldsymbol{a}+\boldsymbol{b}) &= \|\boldsymbol{a}+\boldsymbol{b}\|^2 = \text{constant} \end{aligned}$$

The length of vector $\boldsymbol{a} + \boldsymbol{b}$ can be evaluated using our dot product definition:

$$\begin{aligned} (\boldsymbol{a}+\boldsymbol{b}) \cdot (\boldsymbol{a}+\boldsymbol{b}) &= (a_1+b_1)^2 + (a_2+b_2)^2 + (a_3+b_3)^2 \\ &= \underbrace{(a_1^2+a_2^2+a_3^2)}_{\text{constant}} + \underbrace{(b_1^2+b_2^2+b_3^2)}_{\text{constant}} + 2(a_1b_1+a_2b_2+a_3b_3) \end{aligned}$$

So, we come up with the fact that $a_1b_1 + a_2b_2 + a_3b_3$ is also constant. That is why people came up with this dot product between two vectors. It preserves lengths and angle.

9.1.3 Lines and Planes

Using vectors we can write equation of lines in 2D and 3D uniformly. There are two ways. In the first way, one uses a normal vector to the line (let's denote that normal by $\boldsymbol{n} = (a, b)$) and a point $P_0(x_0, y_0)$ on the line, see Fig. 9.6. Then consider an arbitrary point $P(x, y)$ on the line. The fact that PP_0 is perpendicular to the normal gives us the equation of the line. Of course, perpendicularity is expressed using the dot product. So:

$$(a, b) \cdot (x - x_0, y - y_0) = 0 \Longrightarrow a(x - x_0) + b(y - y_0) = 0, \quad \text{or} \quad ax + by = c$$

Noting that scaling $\boldsymbol{n}$ does not change the equation, so we usually just need to use a unit vector $\boldsymbol{n}$. Geometry becomes easy with numbers!

In the second way, one uses a vector tangent to the line called a *direction vector* $\boldsymbol{d}$. Now, any point P on the line is just a step from $P_0(x_0, y_0)$ along $\boldsymbol{d}$. Thus, the equation of the line is: $\boldsymbol{r}_0 + t\boldsymbol{d}$ where $t \in \mathbb{R}$ denotes the step size; the resulting equation has a vectorial form, see Fig. 9.6. Later on for linear algebra, the vector form is helpful,

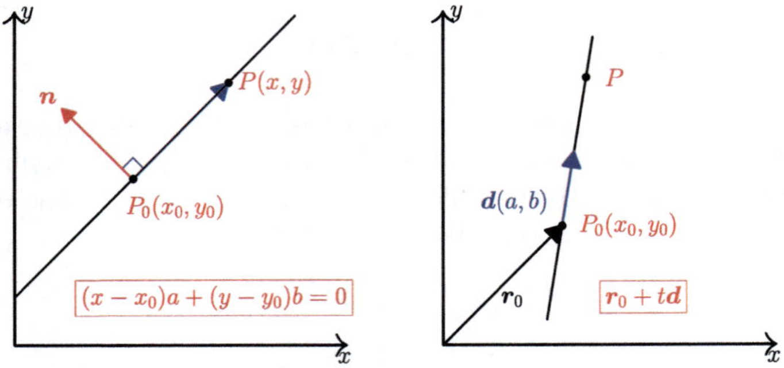

Fig. 9.6 Equations of a line using vectors

as it shows that a line passing through the origin (with $\boldsymbol{r}_0 = (0, 0)$) can be expressed as a scalar of a direction vector.

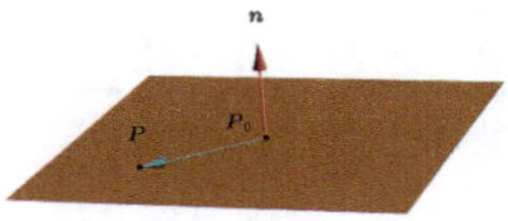

With the dot product we can now write the equation for a plane in 3D. In 2D, a line needs a point (x_0, y_0) and a slope. For a plane, we need also a point $P_0 = (x_0, y_0, z_0)$ and a normal $\boldsymbol{n} = (a, b, c)$ (not a slope as there are infinitely many tangents to a plane). For a point $P = (x, y, z)$ on the plane, the vector from P_0 to P is perpendicular to the normal. And of course perpendicularity is expressed by the dot product of these two vectors:

$$(x - x_0)a + (y - y_0)b + (z - z_0)c = 0, \quad \text{or} \quad ax + by + cz = d \tag{9.1.9}$$

with $d = ax_0 + by_0 + cz_0$. And that is the equation of a plane: $ax + by + cz = d$. Noting the similarity with the equation $ax + by = c$ of a line.

Using two direction vectors $\boldsymbol{u} = (u_1, u_2, u_3)$, $\boldsymbol{v} = (v_1, v_2, v_3)$ living on the plane, which are not parallel,[9] we can write the equation for a plane in 3D passing through the point $\boldsymbol{P}_0$. To get this equation, consider a point $P = (x, y, z)^\top$ on the plane, then the vector P_0P is a vector on the plane. And that vector can be written as a linear combination of $\boldsymbol{u}$ and $\boldsymbol{v}$:

$$\boldsymbol{x} = \boldsymbol{P}_0 + \boldsymbol{u}s + t\boldsymbol{v}, \quad \text{or} \quad \begin{bmatrix} x \\ y \\ z \end{bmatrix} = \begin{bmatrix} x_0 \\ y_0 \\ z_0 \end{bmatrix} + s \begin{bmatrix} u_1 \\ u_2 \\ u_3 \end{bmatrix} + t \begin{bmatrix} v_1 \\ v_2 \\ v_3 \end{bmatrix} \tag{9.1.10}$$

Again, a plane passing through the origin can be expressed as a linear combination of two (direction) vectors:

[9] We need two directions, if $\boldsymbol{u}$ is parallel to $\boldsymbol{v}$, we would have only one direction.

Table 9.1 Lines and planes in $\mathbb{R}^2$ and $\mathbb{R}^3$: a summary

Objects	Dim. of objects	General form	Vector form
Lines in 2D	1	$ax + by = c$	$\boldsymbol{x} = \boldsymbol{p} + s\boldsymbol{u}$
Lines in 3D	1	$\begin{cases} a_1x + b_1y + c_1z = d_1 \\ a_2x + b_2y + c_2z = d_2 \end{cases}$	$\boldsymbol{x} = \boldsymbol{p} + s\boldsymbol{u}$
Planes in 3D	2	$ax + by + cz = d$	$\boldsymbol{x} = \boldsymbol{p} + s\boldsymbol{u} + t\boldsymbol{v}$

$$\boxed{\text{Plane through}(0, 0, 0)\text{with two direction vectors}\boldsymbol{u}, \boldsymbol{v} : \quad \boldsymbol{x} = \boldsymbol{u}s + t\boldsymbol{v}} \tag{9.1.11}$$

When u and t take all the values in $\mathbb{R}$, $\boldsymbol{x} = \boldsymbol{u}s + t\boldsymbol{v}$ generates all the vectors (infinitely many of them) lying on this plane. We can see that this plane in $\mathbb{R}^3$ is similar to the plane $\mathbb{R}^2$: if we take a linear combination of $\boldsymbol{u}, \boldsymbol{v}$ we can never escape the plane. It is a space of itself and later on it leads to the important concept of subspace (Table 9.1).

9.1.4 Projections

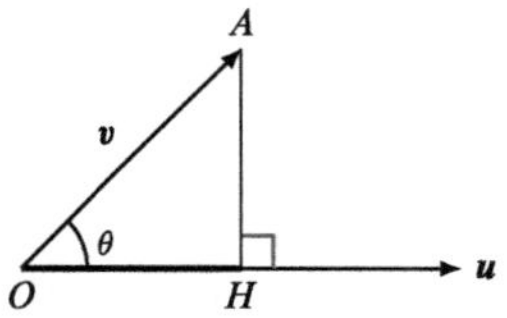

Considering two vectors $\boldsymbol{u}$ and $\boldsymbol{v}$ that make an angle θ. In case that $\boldsymbol{u}$ is short we can always scale it $\alpha\boldsymbol{u}$ and get a line of which direction is determined by $\boldsymbol{u}$. Let's denote by $\boldsymbol{p}$ the projection of $\boldsymbol{v}$ on $\boldsymbol{u}$ (or on the line $\alpha\boldsymbol{u}$). We obtain this projection by dropping a line from A perpendicular to $\boldsymbol{u}$. Then, $\boldsymbol{p} = \overrightarrow{OH}$. The idea of a vector projection in its simplest form is just the question of how much one vector goes in the direction of another. We have $\boldsymbol{p} = OH\boldsymbol{u}/||\boldsymbol{u}||$. And consider the right triangle OHA, we also have $OH = ||\boldsymbol{v}|| \cos\theta$, now relating $\cos\theta$ to the dot product of $\boldsymbol{u}, \boldsymbol{v}$, we can write $\boldsymbol{p}$ as (another common notation for vector projection $\text{proj}_{\boldsymbol{v}}(\boldsymbol{u})$ is also introduced):

$$\boldsymbol{p} = \text{proj}_{\boldsymbol{u}}(\boldsymbol{v}) = ||\boldsymbol{v}|| \cos\theta \frac{\boldsymbol{u}}{||\boldsymbol{u}||} = ||\boldsymbol{v}|| \frac{\boldsymbol{u} \cdot \boldsymbol{v}}{||\boldsymbol{u}|| ||\boldsymbol{v}||} \frac{\boldsymbol{u}}{||\boldsymbol{u}||} = \left(\frac{\boldsymbol{u} \cdot \boldsymbol{v}}{\boldsymbol{u} \cdot \boldsymbol{u}}\right) \boldsymbol{u} \tag{9.1.12}$$

Finding a projection of a vector onto another one has many applications. For example, calculation of the distance from a point to a line in space is one of them, but not an important one. As can be seen, while finding the projection of $\boldsymbol{v}$ on $\boldsymbol{u}$, we also get the vector perpendicular to $\boldsymbol{u}$ (vector $\overrightarrow{AB}$). This is very useful later on (Sect. 9.8.6). But I want to show you what will come next. The vector $\boldsymbol{p}$ is, among all vectors

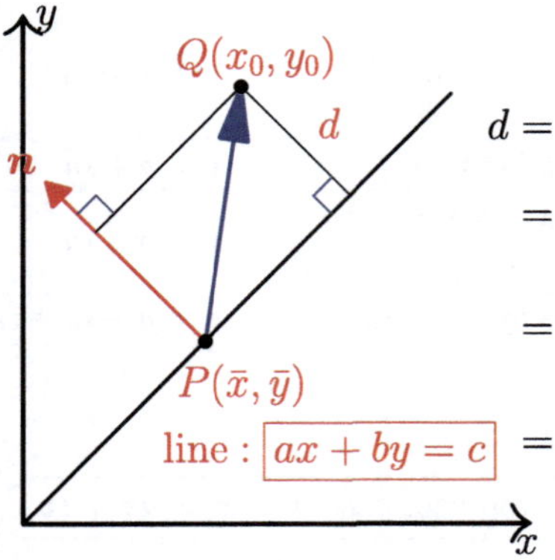

$$
\begin{aligned}
d &= \|\text{proj of } PQ \text{ on } \boldsymbol{n}\| \\
&= \frac{|\boldsymbol{n} \cdot PQ|}{\|\boldsymbol{n}\|} \\
&= \frac{|(a, b) \cdot (x_0 - \bar{x}, y_0 - \bar{y})|}{\sqrt{a^2 + b^2}} \\
&= \frac{|ax_0 + bx_0 - c|}{\sqrt{a^2 + b^2}}
\end{aligned}
$$

Fig. 9.7 Distance from a point $Q(x_0, y_0)$ to a 2D line $ax + by = c$. Note that $a\bar{x} + b\bar{y} = c$ as $(\bar{x}, \bar{y})$ is a point on the line

along the line defined by $\boldsymbol{u}$, the closest vector to $\boldsymbol{v}$. This will be generalized to the best approximation theorem when we extend our 3D space to n-dimensional space (Sect. 9.11.9).

The length of the projected vector can be computed as

$$
||\boldsymbol{p}|| = \left|\left|\left(\frac{\boldsymbol{u} \cdot \boldsymbol{v}}{\boldsymbol{u} \cdot \boldsymbol{u}}\right)\boldsymbol{u}\right|\right| = \left|\left(\frac{\boldsymbol{u} \cdot \boldsymbol{v}}{\boldsymbol{u} \cdot \boldsymbol{u}}\right)\right| \, ||\boldsymbol{u}|| = \frac{|\boldsymbol{u} \cdot \boldsymbol{v}|}{||\boldsymbol{u}||} \tag{9.1.13}
$$

One application of this formula is to compute the distance from a point $B(x_0, y_0, z_0)$ to a plane $P : ax + by + cz = d$. To derive the formula for this distance, first we consider a simpler problem: distance from a point to a 2D line (Fig. 9.7). Then, it is a simple generalization to 3D:

$$
d(B, P) = \frac{|ax_0 + by_0 + cz_0 - d|}{\sqrt{a^2 + b^2 + c^2}}
$$

Projection of a point on a line. Consider a point P_0 and a line passing through two points A and B. Now, we want to find the projection of P_0 on that line. We first project vector AP_0 onto vector $\boldsymbol{B} - \boldsymbol{A}$ using Eq. (9.1.12), then we add $\boldsymbol{A}$ to that result:

$$
\boldsymbol{P} = \boldsymbol{A} + \frac{(\boldsymbol{P_0} - \boldsymbol{A}) \cdot (\boldsymbol{B} - \boldsymbol{A})}{||\boldsymbol{B} - \boldsymbol{A}||^2}(\boldsymbol{B} - \boldsymbol{A})
$$

9.1.5 Cross Product

There are different ways to introduce the cross product. I prefer the way in which the cross product would appear naturally when we talk about rotational motions. I refer

Table 9.2 Analogy between linear motion and angular motion

Linear motion	Angular motion
Linear displacement Δx	Angular displacement $\Delta\theta$
Linear velocity $\Delta x/\Delta t$	Angular velocity $\Delta\theta/\Delta t$
Linear acceleration $\Delta^2 x/\Delta t^2$	Angular acceleration $\Delta\omega/\Delta t$
Work done $\Delta W = \boldsymbol{F} \cdot \Delta\boldsymbol{x}$	Work done $\Delta W = ? \times \Delta\theta$

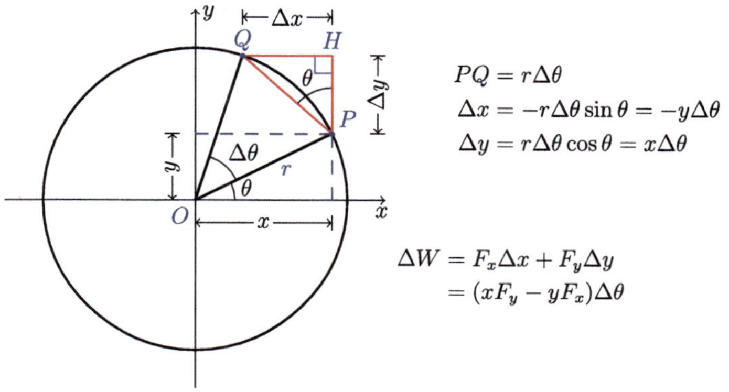

Fig. 9.8 Work in terms of $\Delta\theta$. Noting that $\Delta\theta$ is tiny, that's why we have $PQ = r\theta$ and PQ is perpendicular to OP. There is a minus in Δx because x is decreasing

to Table 9.2 for analogies between linear (or translational) and rotational motions. For rotational motions, we use a rotation angle to measure the movement. This way of presenting the cross product is due to Feynman in his celebrated lectures on physics, volume I.

First, we consider a two-dimensional rotation, i.e., an object is circulating around in the xy-plane (Fig. 9.8). Our analysis is guided by the last row in Table 9.2. That is we are going to write the work $\boldsymbol{F} \cdot \Delta\boldsymbol{x}$ in terms of $\Delta\theta$. If force times distance is work, then we define torque such that work is torque times angle. That is our plan.

Assume that at a given time instant, the object is located at point P, which is specified by (x, y) using the Cartesian coordinates or (r, θ) using polar coordinates. A moment later, under the influence of a force $\boldsymbol{F}$ it moves to point Q by rotating a *tiny angle* of $\Delta\theta$. We compute the change in positions Δx and Δy in terms of $\Delta\theta$. Then, we compute the work $\Delta W = F_x\Delta_x + F_y\Delta_y = (xF_y - yF_x)\Delta\theta$. So this term $(xF_y - yF_x)$—a strange-looking combination of the force and the distance—should be defined as torque which is a kind of force that makes objects turn.

Yes, we have obtained one formula for the torque. But we can also obtain another formula for it to reveal the geometry behind the algebraic expression. To this end, we recall that work is tangential force multiplied with displacement. As seen from Fig. 9.9, torque can also be defined as the magnitude of the force times the length

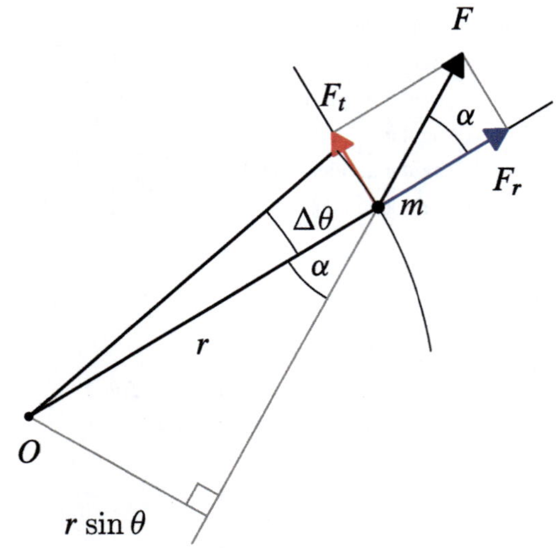

$$\begin{aligned}\Delta W &= F_t \times r\Delta\theta \\ &= (F\sin\alpha) \times r\Delta\theta \\ &= (Fr\sin\alpha) \times \Delta\theta\end{aligned}$$

Fig. 9.9 Torque is defined as the magnitude of the force times the length of the level arm: $\Gamma = Fr\sin\theta$. A force $\boldsymbol{F}$ is acting at an object m locating at $\boldsymbol{r}$. When the object has rotated a small angle $\Delta\theta$, the object has moved tangentially a distance of $r\Delta\theta$. Thus the work is $\Delta W = (F\sin\alpha) \times (r\Delta\theta)$

of the level arm. And this formula agrees with our experiences with torques: if the force is radial, i.e., $\alpha = 0$ the torque is zero, or for zero length of level arm the torque is also zero.

With forces, we have linear momentum $\boldsymbol{p} = m\boldsymbol{v}$ and Newton's second law saying that the external force is equal to the time derivative of the linear momentum: $\boldsymbol{F}^{\text{ext}} = \dot{\boldsymbol{p}}$. A question arises, with torques, do we have another kind of momentum in the sense that $\boldsymbol{\Gamma}^{\text{ext}} = \dot{\square}$. Let's do the analysis. We start with the formula for the torque, $\Gamma = xF_y - yF_x$, then we replace F_x and F_y using Newton's second law so that derivative with time appears:

$$\Gamma = xF_y - yF_x = xm\frac{dv_y}{dt} - ym\frac{dv_x}{dt} = \frac{d}{dt}(xmv_y - ymv_x) = \frac{d}{dt}(xp_y - yp_x) \tag{9.1.14}$$

Indeed, the torque is the time rate of change of something. And that something $xp_y - yp_x$ is what we now call the *angular momentum*, denoted by L. And by doing the same analysis as done in Fig. 9.9 for the torque, we can see that the angular momentum is the magnitude of the linear momentum times the length of the level arm.

We have conservation of linear momentum when the total external force in a system is zero. Do we have the same principle for angular momentum? As can be seen from Fig. 9.10 for a system of two particles, the torque due to $\boldsymbol{F}_{12}$ cancels the torque due to $\boldsymbol{F}_{21}$. Thus, the rate of change of the total momenta depends only on the external torques:

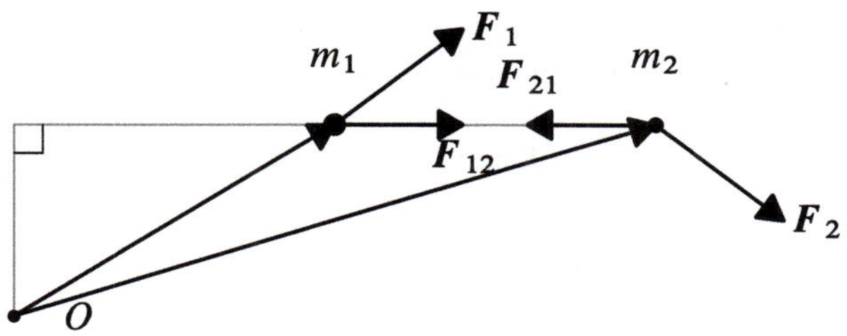

Fig. 9.10 The torque due to $\boldsymbol{F}_{12}$ cancels the torque due to $\boldsymbol{F}_{21}$ due to Newton's third law of action and reaction $\boldsymbol{F}_{12} = -\boldsymbol{F}_{21}$ and the level arms are the same

$$\left.\begin{aligned}\frac{dL_1}{dt} &= \Gamma_1^{\text{ext}} + \Gamma_{12}\\ \frac{dL_2}{dt} &= \Gamma_2^{\text{ext}} + \Gamma_{21}\end{aligned}\right\} \Longrightarrow \frac{dL}{dt} = \Gamma_1^{\text{ext}} + \Gamma_2^{\text{ext}} \tag{9.1.15}$$

Thus, if the net torque is zero, the angular momentum is conserved. Indeed, we also have an analog for the principle of conservation of linear momentum. This encourages us to keep moving on. We have kinetic energy for translational motions, what it will look like for rotational motions?

Kinetic energy is $T = 0.5mv^2$: mass time velocity squared. So we anticipate that for rotations, it should be $T = 0.5f(m)\omega^2$. Let's do the maths (note that $v = r\omega$ see Fig. 6.39):

$$T = \frac{1}{2}mv^2 = \frac{1}{2}mr^2\omega^2 \Longrightarrow I = mr^2 \tag{9.1.16}$$

The quantity $I = mr^2$ is called *moment of inertia* by Leonhard Euler. It is a function of mass (of course) but it depends also on r, i.e., how far the mass is away from the rotation axis, see for an application in Fig. 9.11.

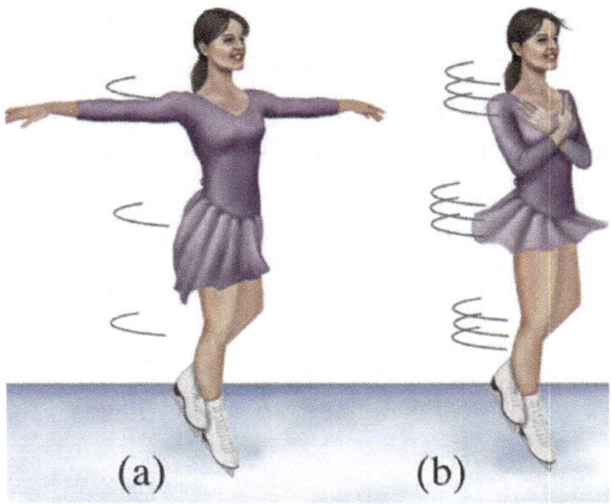

Fig. 9.11 Moment of inertia in rotations: it is a function of mass (of course) but it depends also on r, i.e., how far the mass is away from the rotation axis. A spinning figure skater pull in her outstretched arms to spin faster. This is because the angular momentum $l = I\omega$ is conserved, when I is decreased, ω is increased, i.e., spinning faster

Now, if we repeat the analysis that we have just done in the xy-plane but now for the yz-plane and $zx-$plane, we obtain three terms:

$$\begin{aligned} xy \text{ plane}: &\quad xF_y - yF_x \\ yz \text{ plane}: &\quad yF_z - zF_y \\ zx \text{ plane}: &\quad yF_z - zF_y \end{aligned} \tag{9.1.17}$$

And that is the torque which is defined from two vectors $\boldsymbol{r} = (x, y, z)$ and $\boldsymbol{F}$; $xF_y - yF_x$ is just the $z-$component of this torque. Now we generalize that to any two 3D vectors $\boldsymbol{a}$ and $\boldsymbol{b}$:

$$\boxed{\boldsymbol{c} := \boldsymbol{a} \times \boldsymbol{b} \Longrightarrow \boldsymbol{c} = \begin{bmatrix} a_2b_3 - a_3b_2 \\ a_3b_1 - a_1b_3 \\ a_1b_2 - a_2b_1 \end{bmatrix}} \tag{9.1.18}$$

From this definition, it can be seen that $\boldsymbol{b} \times \boldsymbol{a} = -\boldsymbol{a} \times \boldsymbol{b}$:

$$\boldsymbol{b} \times \boldsymbol{a} = \begin{bmatrix} b_2a_3 - b_3a_2 \\ b_3a_1 - b_1a_3 \\ b_1a_2 - b_2a_1 \end{bmatrix} = -\boldsymbol{a} \times \boldsymbol{b} \tag{9.1.19}$$

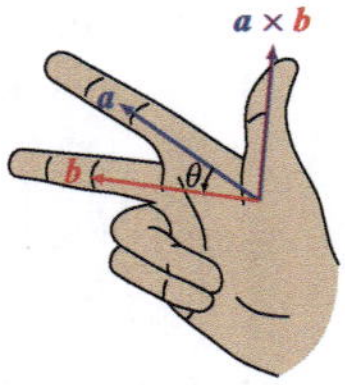

The vector product is not commutative! One consequence is that $\boldsymbol{a} \times \boldsymbol{a} = \boldsymbol{0}$. Now, we need to know the direction of $\boldsymbol{a} \times \boldsymbol{b}$. Just apply Eq. (9.1.18) to two special vectors $(1, 0, 0)$ and $(0, 1, 0)$, and we get the cross product of them is $(0, 0, 1)$, which is perpendicular to $(1, 0, 0)$ and $(0, 1, 0)$. The rule is: $\boldsymbol{c}$ is perpendicular to both $\boldsymbol{a}, \boldsymbol{b}$. This can be proved simply by just calculating the dot product of $\boldsymbol{a} \times \boldsymbol{b}$ with $\boldsymbol{a}$, and you will see it is zero. But $\boldsymbol{c}$ points up or down? The right-hand rule tells us which exact direction it follows (see figure next to the text).

We now know the direction of the cross product, how about its length? Let's compute it and see what we shall get:

$$\begin{aligned} \|\boldsymbol{a} \times \boldsymbol{b}\|^2 &= (b_2a_3 - b_3a_2)^2 + (b_3a_1 - b_1a_3)^2 + (b_1a_2 - b_2a_1)^2 \\ &= (a_1^2 + a_2^2 + a_3^2)(b_1^2 + b_2^2 + b_3^2) - (a_1b_1 + a_2b_2 + a_bb_3)^2 \\ &= \|\boldsymbol{a}\|^2\|\boldsymbol{b}\|^2 - \|\boldsymbol{a}\|^2\|\boldsymbol{b}\|^2\cos^2\theta, \quad \text{(used dot product formula)} \\ &= \|\boldsymbol{a}\|^2\|\boldsymbol{b}\|^2\sin^2\theta \end{aligned}$$

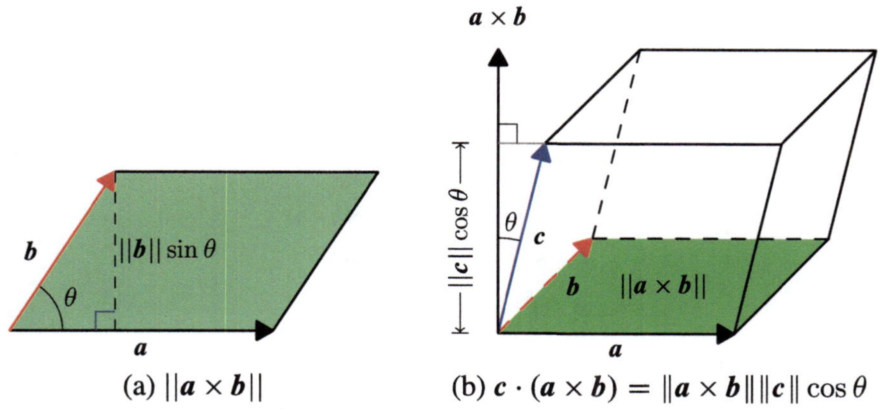

Fig. 9.12 A geometric interpretation of the cross product of two vectors: the length of the cross product of $\boldsymbol{a}$ and $\boldsymbol{b}$ is the area of the parallelogram formed by $\boldsymbol{a}$ and $\boldsymbol{b}$ (**a**); $\boldsymbol{c} \cdot (\boldsymbol{a} \times \boldsymbol{b})$ is the volume of a parallelepiped with three sides being our three vectors $\boldsymbol{a}, \boldsymbol{b}, \boldsymbol{c}$ (**b**)

We get a nice formula for the length of the cross product of two 3D vectors $\boldsymbol{a}$ and $\boldsymbol{b}$ in terms of the length of the vectors and the angle between them:

$$\|\boldsymbol{a} \times \boldsymbol{b}\| = \|\boldsymbol{a}\| \|\boldsymbol{b}\| \sin\theta \tag{9.1.20}$$

Note the striking similarity with Eq. (9.1.6) about the dot product! With the dot product we have $\cos\theta$, and now with the cross product we have $\sin\theta$. The dot product tells us when two vectors are perpendicular and the cross product tells us when they are parallel. Perfect duo. A geometric interpretation of this formula is that the length of the cross product of $\boldsymbol{a}$ and $\boldsymbol{b}$ is the area of the parallelogram formed by $\boldsymbol{a}$ and $\boldsymbol{b}$. We also get a result: *the area of a triangle formed by* $\boldsymbol{a}$ and $\boldsymbol{b}$ is $0.5\|\boldsymbol{a} \times \boldsymbol{b}\|$. See Fig. 9.12a.

As the area of a triangle formed by $\boldsymbol{a}$ and $\boldsymbol{b}$ is $0.5\|\boldsymbol{a} \times \boldsymbol{b}\|$, if the three vertices are (x_1, y_1), (x_2, y_2), and (x_3, y_3), the area of the triangle explicitly expressed in terms of the coordinates of its vertices is given by

$$A = \frac{1}{2} \det \begin{bmatrix} 1 & 1 & 1 \\ x_1 & x_2 & x_3 \\ y_1 & y_2 & y_3 \end{bmatrix} \tag{9.1.21}$$

Here are some rules regarding the cross product:

$$
\begin{aligned}
\boldsymbol{a} \times \boldsymbol{b} &= -\boldsymbol{b} \times \boldsymbol{a} \\
\boldsymbol{a} \times \boldsymbol{a} &= \boldsymbol{0} \\
(\alpha \boldsymbol{a}) \times \boldsymbol{b} &= \alpha(\boldsymbol{a} \times \boldsymbol{b}) = \boldsymbol{a} \times (\alpha \boldsymbol{b}) \\
\boldsymbol{a} \times (\boldsymbol{b} + \boldsymbol{c}) &= \boldsymbol{a} \times \boldsymbol{b} + \boldsymbol{a} \times \boldsymbol{c} \\
(\boldsymbol{a} + \boldsymbol{b}) \times \boldsymbol{c} &= \boldsymbol{a} \times \boldsymbol{c} + \boldsymbol{b} \times \boldsymbol{c} \\
\boldsymbol{a} \times (\boldsymbol{b} \times \boldsymbol{c}) &= \boldsymbol{b}(\boldsymbol{a} \cdot \boldsymbol{c}) - \boldsymbol{c}(\boldsymbol{a} \cdot \boldsymbol{b}) \\
(\boldsymbol{a} \times \boldsymbol{b})^2 &= \boldsymbol{a}^2 \boldsymbol{b}^2 - (\boldsymbol{a} \cdot \boldsymbol{b})^2 \\
\boldsymbol{c} \cdot (\boldsymbol{a} \times \boldsymbol{b}) &= (\boldsymbol{c} \times \boldsymbol{a}) \cdot \boldsymbol{b}
\end{aligned}
\tag{9.1.22}
$$

The first three rules are straightforward. How others have been discovered? Herein, we prove the last rule, known as the *scalar triple product of three vectors*. As two vectors give us an area so three vectors could give us a volume. So, let's build a box with three sides being our three vectors $\boldsymbol{a}, \boldsymbol{b}, \boldsymbol{c}$ (see Fig. 9.12); this box is called a parallelepiped.[10] It is seen that the volume of this box is $\boldsymbol{c} \cdot (\boldsymbol{a} \times \boldsymbol{b})$: consider the base with two sides $\boldsymbol{a}, \boldsymbol{b}$, its area is $\|\boldsymbol{a} \times \boldsymbol{b}\|$; the volume is: base area times the height; that is, $\|\boldsymbol{a} \times \boldsymbol{b}\| \|\boldsymbol{c}\| \cos\theta$. As the volume does not change if we consider a different base, the rule of the scalar triple product of three vectors is proved. Of course, a proof using pure algebra exists:

$$
\begin{aligned}
\boldsymbol{c} \cdot (\boldsymbol{a} \times \boldsymbol{b}) &= c_1(a_2b_3 - a_3b_2) + c_2(a_3b_1 - a_1b_3) + c_3(a_1b_2 - a_2b_1) \\
&= b_1(c_2a_3 - c_3a_2) + b_2(c_3a_1 - c_1a_3) + b_3(c_1a_2 - c_2a_1) = \boldsymbol{b} \cdot (\boldsymbol{c} \times \boldsymbol{a})
\end{aligned}
$$

The rule $\boldsymbol{a} \times (\boldsymbol{b} \times \boldsymbol{c}) = \boldsymbol{b}(\boldsymbol{a} \cdot \boldsymbol{c}) - \boldsymbol{c}(\boldsymbol{a} \cdot \boldsymbol{b})$ is known as the triple product. You're encouraged to prove it using of course the definition of cross product. You would realize that the process is tedious and boring (lengthy algebraic expressions). Refer to Sect. 6.11.13 for a more elegant proof when we're equipped with more mathematics tools.

9.1.6 Hamilton and Quartenions

> The essence of mathematics lies in its freedom. (George Cantor)

This section is about the story of how Hamilton discovered quartenions in 1843. The story started with complex numbers (Sect. 2.25). Let's consider two complex numbers $z_1 = a + bi$ and $z_2 = c + di$, where $a, b, c, d \in \mathbb{R}$ and $i^2 = -1$. Addition/subtraction of complex numbers is straightforward, but multiplication is much harder. So, we focus on the product of z_1 and z_2:

---

[10] Parallelepiped is a 3D shape whose faces are all parallelograms. It is obtained from a Greek word which means "an object having parallel plane".

$$z_1 z_2 = (ac - bd) + (ad + bc)i$$

Note that to get this result we only needed to use high school algebra and $i^2 = -1$. Thus, the modulus (or length) of $z_1 z_2$ is $|z_1 z_2| = \sqrt{(ac - bd)^2 + (ad + bc)^2}$. Next, we're trying to find the relation between $|z_1 z_2|$ and $|z_1|$ and $|z_2|$. To this end, we square $|z_1 z_2|$ and obtain:

$$|z_1 z_2|^2 = (ac - bd)^2 + (ad + bc)^2 = (a^2 + b^2)(c^2 + d^2) = |z_1|^2 |z_2|^2 \quad (9.1.23)$$

or

$$\boxed{|z_1 z_2| = |z_1||z_2|} \quad (9.1.24)$$

And this result is called the law of the moduli by Hamilton: it states that the modulus of the product of two complex numbers is equal to the product of the modulus of the two numbers.

Hamilton wanted to extend complex numbers—which he called couples as each complex number contains two real numbers—to triplets. Thus, he considered a triplet of the following form:

$$z = a + bi + cj, \quad \text{with}\, i^2 = j^2 = -1 \text{and}\, ij = ji$$

Hamilton considered $ij = ji$ because at that time Hamilton still insisted on the commutativity of multiplication. Although it is straightforward to add two triplets, multiplication was, however, not easy to even a mathematician of high caliber such as Hamilton. He wrote to his son Archibald shortly before his death:

> Every morning in the early part of the above-cited month, on my coming down to breakfast, your brother William Edwin and yourself used to ask me, 'Well, Papa, can you multiply triplets?' Whereto I was obliged to reply, with a sad shake of the head, 'No, I can only add and subtract them.'

He started with zz or z^2, and he obtained:

$$z^2 = (a + bi + cj)(a + bi + cj) = (a^2 - b^2 - c^2) + 2abi + 2acj + 2bcij \quad (9.1.25)$$

The red term troubled him. To get a triplet from z^2, he needed to have $ij = a_1 + a_2 i + a_3 j$ with $a_i \in \mathbb{R}$. But this is impossible:

$$\begin{aligned}
ij &= a_1 + a_2 i + a_3 j \\
i^2 j &= a_1 i + a_2 i^2 + a_3 ij && \text{(multiplying the above by} i) \\
-j &= a_1 i - a_2 + a_3 ij && (i^2 = -1) \\
-j &= a_1 i - a_2 + a_3(a_1 + a_2 i + a_3 j) && \text{(replacing } ij \text{using 1st eq.)} \\
-j &= a_1 a_3 - a_2 + (a_1 + a_2 a_3)i + a_3^2 j
\end{aligned}$$

The last equation holds only when $a_3^2 = -1$, which is impossible as a_3 is a real number. So, ij cannot be a triplet.

But if this troubling term $2bcij$ is zero, then it is simple to see that $|z^2| = (a^2 + b^2 + c^2)$, which is $|z||z|$. The law of the moduli, Eq. (9.1.24), works! But when $2bcij$ is zero? It is absurd to think that $ij = 0$. So, Hamilton thought that if $ij \neq ji$, then it is possible for the red term to vanish. So, with $ij \neq ji$, he computed z^2:

$$(a + bi + cj)(a + bi + cj) = (a^2 - b^2 - c^2) + 2abi + 2acj + bc(ij + ji) \tag{9.1.26}$$

If $ij = -ji$, then the red term in the above expression is zero, and the law of the moduli holds. At this time, due to the red term, Hamilton decided that he had to consider not triplets but quadruplets of the form $z = a + bi + cj + dk$. This k is for $ij = -ji = k$! He called such number z a quartenion. He defined the modulus of a quartenion is $\sqrt{a^2 + b^2 + c^2 + d^2}$, which is reasonable.

What should be the rules of i, j, k? We have $i^2 = -1$, thus we should have $j^2 = k^2 = -1$. After all, there is no reason that i is more special than j and k. And we need $ij = -ji$, and Hamilton considered $ij = -ji = k$. Thus, his i, j, k must satisfy the following[11]:

$$\begin{aligned} i^2 &= j^2 = k^2 = -1 \\ ij &= -ji = k \\ jk &= -kj = i \\ ki &= -i = j \end{aligned} \tag{9.1.27}$$

Hamilton now needed to verify that his quartenions satisfy the rule of modulus (Eq. (9.1.24)). He computed z^2 and with Eq. (9.1.27), he got:

$$\begin{aligned} (a + bi + cj + dk)(a + bi + cj + dk) &= a^2 + abi + acj + adk + \\ &+ abi - b^2 + bcij + bdik \\ &+ acj + bcji - c^2 + cdjk \\ &+ adk + bdki + dckj - d^2 \\ &= (a^2 - b^2 - c^2 - d^2) + 2abi + 2acj + 2adk \end{aligned}$$

Thus, the modulus of zz is

$$|zz| = \sqrt{(a^2 - b^2 - c^2 - d^2)^2 + (2ab)^2 + (2ac)^2 + (2ad)^2} = a^2 + b^2 + c^2 + d^2 = |z||z|$$

Therefore, we have again the old rule about modulus that $|zz| = |z||z|$.

[11] Which can also be compactly written as $i^2 = j^2 = k^2 = ijk = -1$.

Hamilton's discovery of the quartenions was one of those very rare incidents in science where a breakthrough was captured in real time. Hamilton had been working on this problem for over 10 years, and finally had a breakthrough on 16 October 1843 while on a walk along the Royal Canal in Dublin towards the Royal Irish Academy with his wife, Lady Hamilton. And when this exciting idea took hold, he couldn't resist the urge to etch his new equation into the stone of Broom Bridge and give life to a new system of four-dimensional numbers.

Hamilton described the "eureka" moment in a letter to his son some years later:

> Although your mother talked with me now and then, yet an undercurrent of thought was going on in my mind, which gave at last a result, whereof it is not too much to say that I felt at once an importance. An electric current seemed to close; and a spark flashed forth, the herald (as I foresaw, immediately) of many long years to come of definitely directed thought and work . . . Nor could I resist the impulse-unphilosophical as it may have been-to cut with a knife on a stone of Brougham Bridge as we passed it, the fundamental formula ...

Hamilton had created a completely new structure in mathematics. What is interesting is that the quartenions did not satisfy the commutative rule $ab = ba$ (note that complex numbers still follow this rule). This did not bother Hamilton because *this is what usually happens in nature*. For example, consider an empty swimming pool and the two operations of diving into the pool head first and turning the water on. The order in which the operations take place is important! The set of all quartenions is now denoted by $\mathbb{H}$ to honor Hamilton.

It was Hamilton who gave us the terms scalar and vector for he considered the quartenion $a + bi + cj + dk$ as consisted of a scalar part (a) and a vector part $bi + cj + dk$. Considering two quartenions with zero scalar parts $\alpha = xi + yj + zk$ and $\alpha' = x'i + y'j + z'k$, he computed their product using Eq. (9.1.27):

$$\begin{aligned}\alpha\alpha' &= (xi + yj + zk)(x'i + y'j + z'k)\\ &= -(xx' + yy' + zz') + (yz' - zy')i + (zx' - xz')j + (xy' - x'y)k\end{aligned}$$

What is the red term? It is nothing but the dot product of two 3D vectors. And the blue term is nothing but the cross product. Gibbs gave us the dot product and the cross product. But it was Hamilton who was the first to write down these products.

9.2 Vectors in $\mathbb{R}^n$

So we have seen 2D and 3D vectors. They are easy to grasp as we have counterparts in real life. But mathematicians do not stop there. Or actually they encounter problems in which they have to stretch their imaginations. One such problem is solving a system of large simultaneous equations, like the following one:

$$\begin{aligned} 2x_1 + 3x_2 + 4x_3 + x_4 &= 5 \\ x_1 + 2x_2 + 3x_3 - x_4 &= 1 \\ 2x_1 + 3x_2 + x_3 - x_4 &= 7 \\ 3x_1 + 2x_2 + 2x_3 - 3x_4 &= 2 \end{aligned}$$

which they simply write $\mathbf{A}\boldsymbol{x} = \boldsymbol{b}$ where the vector $\boldsymbol{x} = (x_1, x_2, x_3, x_4)$ is a vector in a four-dimensional space. And if we have a system of 1000 equations for 1000 unknowns, we are talking about its solution as a vector living in a 1000-dimensional space! Obviously it is impossible to visualize spaces of dimensions higher than 3, the study of vectors in higher dimensional spaces must proceed entirely by analytic means.

In this section, we move to spaces of n dimensions where n is most of the time (much) larger than three. We use the symbol $\boldsymbol{x} \in \mathbb{R}^n$ to denote such a vector, and we write (with respect to a chosen basis which is usually the standard basis)

$$\boldsymbol{x} = \begin{bmatrix} x_1 \\ x_2 \\ \vdots \\ x_n \end{bmatrix}, \quad \text{or } \boldsymbol{x} = (x_1, x_2, \ldots, x_n)$$

where the second notation is to save space. When we say a vector we mean a *column vector*. For 2D/3D vectors, we called x_i the ith coordinate. However, for n-dimensional vector we call it the ith component, as $\boldsymbol{x}$ is no longer representing a positional vector. Actually, x_i can be anything: price of a product, deflection of a point in a beam, etc. *It should be emphasized that a vector exists independent of a coordinate system. So, when we write (or see) $\boldsymbol{x} = (x_1, x_2, \ldots, x_n)$, we should be aware that a certain choice of a coordinate system was made.*

For vectors $\boldsymbol{a}$ and $\boldsymbol{b}$ in a n-dimensional space and a scalar α, we have the following definitions for vector addition, scalar vector multiplication, and dot product of two vectors, which are merely extensions of what we know for 3D vectors:

$$\text{addition:} \quad \boldsymbol{a} + \boldsymbol{b} = \sum_{i=1}^{n} (a_i + b_i)$$

$$\text{scalar multiplication:} \quad \alpha\boldsymbol{a} = (\alpha a_1, \alpha a_2, \ldots, \alpha a_n)$$

$$\text{dot product:} \quad \boldsymbol{a} \cdot \boldsymbol{b} = \sum_{i=1}^{n} a_i b_i = a_i b_i$$

$$\text{length (norm):} \quad ||\boldsymbol{a}|| = \sqrt{\boldsymbol{a} \cdot \boldsymbol{a}} = \left(\sum_i a_i^2\right)^{1/2}$$

where we have used Einstein summation rule in $\sum_{i=1}^n a_i b_i = a_i b_i$. According to this rule, when an index variable (i in this example) appears twice in a single term, it implies summation of that term over all the values of the index. The index i is thus named summation index or dummy index. The dummy word is used because we can replace it by any other symbol: $\sum_{i=1}^n a_i b_i = \sum_{j=1}^n a_j b_j = a_j b_j$.[12]

Remark All the rules about vector addition and scalar vector multiplication in Box 9.1 still apply for vectors in $\mathbb{R}^n$. And note that we did not define the cross product for vectors living in a space with dimensions larger than three! Lucky for us that in the world of linear algebra we do not need the cross product.

Notation $\mathbb{R}^n$. Let's discuss how mathematicians say about 1D, 2D, 3D, and nD spaces. When x is a number living on the number line, they write $x \in \mathbb{R}$. When a point $\boldsymbol{x} = (x, y)$ lives on a plane, they write $\boldsymbol{x} \in \mathbb{R}^2$; this is because $x \in \mathbb{R}$ and $y \in \mathbb{R}$. Similarly, they write $\boldsymbol{x} \in \mathbb{R}^3$ and $\boldsymbol{x} \in \mathbb{R}^n$. This notation follows the Cartesian product of two sets discussed in Sect. 5.5.

We have special numbers: 0 and 1, and we also have special vectors. The zero vector $\mathbf{0}$, note the bold font for 0, has all components being zeros, and the one vector $\mathbf{1}$ has all components equal to one. And the unit vectors (remember $\boldsymbol{i}, \boldsymbol{j}, \boldsymbol{k}$ of the 3D space?):

$$\boldsymbol{e}_1 = \begin{bmatrix} 1 \\ 0 \\ 0 \\ \vdots \\ 0 \end{bmatrix}, \quad \boldsymbol{e}_2 = \begin{bmatrix} 0 \\ 1 \\ 0 \\ \vdots \\ 0 \end{bmatrix}, \quad \ldots, \boldsymbol{e}_n = \begin{bmatrix} 0 \\ 0 \\ 0 \\ \vdots \\ 1 \end{bmatrix} \tag{9.2.1}$$

That is vector $\boldsymbol{e}_i$ has all component vanished except the ith component which is one.

Linear combination. If $\boldsymbol{u}_1, \ldots, \boldsymbol{u}_m$ are m vectors in $\mathbb{R}^n$ and $\alpha_1, \ldots, \alpha_m$ are m real numbers, then the vector

$$\alpha_1 \boldsymbol{u}_1 + \alpha_2 \boldsymbol{u}_2 + \cdots + \alpha_m \boldsymbol{u}_m \tag{9.2.2}$$

is called a linear combination of the vectors $\boldsymbol{u}_1, \ldots, \boldsymbol{u}_m$. The scalars $\alpha_1, \ldots, \alpha_m$ are the coefficients of the combination.

[12] Of course it is not a requirement to use Einstein notation in linear algebra, but it can be very useful elsewhere.

For some special values for α_i, we obtain some special combinations:

$$
\begin{aligned}
\text{sum } (\alpha_i = 1)\text{:} &\quad \boldsymbol{u}_1 + \boldsymbol{u}_2 + \cdots + \boldsymbol{u}_m \\
\text{average } (\alpha_i = 1/m)\text{:} &\quad \frac{1}{m}\left[\boldsymbol{u}_1 + \boldsymbol{u}_2 + \cdots + \boldsymbol{u}_m\right] \\
\text{center of mass } \boldsymbol{R}_{\text{CM}}\text{:} &\quad \frac{m_1\boldsymbol{r}_1 + m_2\boldsymbol{r}_2 + \cdots + m_n\boldsymbol{r}_n}{m_1 + m_2 + \cdots + m_n}
\end{aligned}
$$

And we shall see more and more linear combinations of vectors in coming sections. *The key operation in linear algebra is taking a (linear) combination of some vectors.* One special linear combination is that any vector $\in \mathbb{R}^n$ can be written as a linear combination of the unit vectors with its components being the coefficients:

$$
\begin{bmatrix} a_1 \\ a_2 \\ \vdots \\ a_n \end{bmatrix} = a_1 \begin{bmatrix} 1 \\ 0 \\ \vdots \\ 0 \end{bmatrix} + a_2 \begin{bmatrix} 0 \\ 1 \\ \vdots \\ 0 \end{bmatrix} + \cdots + a_n \begin{bmatrix} 0 \\ 0 \\ \vdots \\ 1 \end{bmatrix} \tag{9.2.3}
$$

9.3 System of Linear Equations

The central problem of linear algebra is to solve a system of linear equations. In these linear equations, we never meet xy or $\sin x$, we just have unknowns multiplied by constants, e.g., $2x + 3y = 7$. If you have not see any system of linear equations before, check Sect. 2.16 out first. Let's start humbly with a system of two equations for two unknowns:

$$
\begin{aligned}
2x - y &= 1 \\
x + y &= 5
\end{aligned} \tag{9.3.1}
$$

All of us know the technique to solve it: *elimination method*. We keep the first equation, but replace the second by the sum of the second equation and the first (to remove y):

$$
\begin{aligned}
2x - 1y &= 1 \\
3x + 0y &= 6
\end{aligned} \tag{9.3.2}
$$

Then, we have $x = 2$ from the second equation, and back substituting $x = 2$ into the first equation gives us $y = 3$. This is pretty easy. What is interesting is the fact that we write the second equation $3x = 6$ as $3x + 0y = 6$. Furthermore, we can work on the two equations without referring to x, y (after all, instead of x, y we can equally use u, v or whatever pleases us); we just need to focus on the numbers $2, -1, 1, 1, 1, 5$. So, we put the numbers appearing in the LHS in a rectangular array with 2 rows and 2 columns, denoted by a capital boldface symbol $\mathbf{A}$, the numbers in the RHS in a vector ($\boldsymbol{b}$), and the unknowns in another vector ($\boldsymbol{x}$):

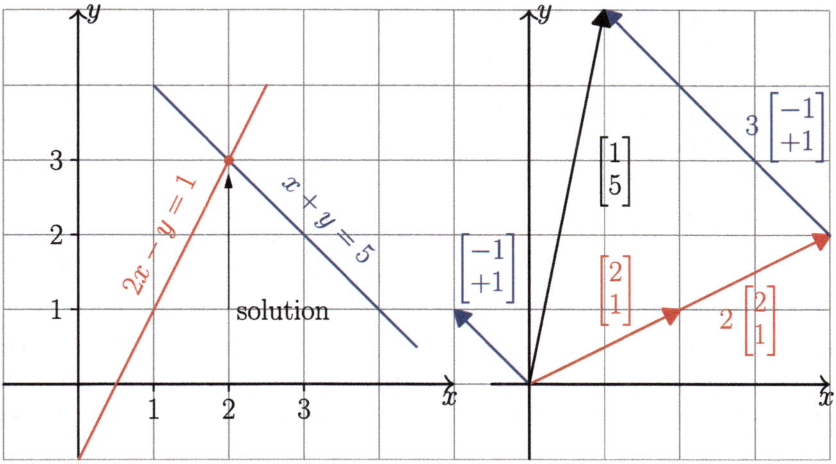

Fig. 9.13 System of linear equations: row view (left) and column view (right)

$$\begin{bmatrix}2 & -1\\1 & 1\end{bmatrix}\begin{bmatrix}x\\y\end{bmatrix}=\begin{bmatrix}1\\5\end{bmatrix}, \quad \text{or } \mathbf{A}\boldsymbol{x}=\boldsymbol{b} \tag{9.3.3}$$

and this 2 row and 2 col array is called *the coefficient matrix*[13] and the vector on the RHS is called the RHS vector. Note that this is not simply a notation. Equation (9.3.3) says that the matrix $\mathbf{A}$ acts on the vector $\boldsymbol{x}$ to produce the vector $\boldsymbol{b}$. Matrices do something as they are associated with linear transformations. More about this later in Section 9.11.3.

In a matrix there are rows and columns, thus we can view Eq. (9.3.3) from the row picture or the column picture. In the row picture, each row is an equation, which is geometrically a line in a 2D plane. There are two lines, Fig. 9.13—left, and they intersect at (2, 3), which is the solution of the system. And this solution is unique, as there is no other solutions.

In the column picture, we do not see two equations with scalar unknowns x and y, but we see only one vector equation:

$$x\begin{bmatrix}2\\1\end{bmatrix}+y\begin{bmatrix}-1\\1\end{bmatrix}=\begin{bmatrix}1\\5\end{bmatrix} \tag{9.3.4}$$

And we are seeking for the right linear combination of the columns of the coefficient matrix to get the RHS vector. In Fig. 9.13—right, we see that if we go along the first column two times its length and then follow the second column three times the length, then we reach the RHS (1, 5).

[13] Historically it was the nineteenth-century English mathematician James Sylvester (1814–1897) who first coined the term matrix, even though Chinese mathematicians knew about matrices from the tenth to second century BCE, written in *The Nine Chapters on the Mathematical Art*.

For this simple example in 2D, the row picture is easier to work with. However, for a system of more than three unknowns such a geometric view does not exist.

No solution and many solutions. Using the row picture it is easy to see that $\mathbf{A}\boldsymbol{x} = \boldsymbol{b}$ either: (i) has a unique solution, (ii) has no solution, and (iii) has many solutions. The following systems have no solution and many solutions:

$$\begin{cases} 2x - y = 1 \\ 2x - y = -2 \end{cases}, \quad \begin{cases} 2x - y = 1 \\ 4x - 2y = 2 \end{cases} \tag{9.3.5}$$

In the first system, the two lines are parallel and thus do not intersect. In the second system, the second equation is just a multiple of the first; we have then just one equation and all the points on the line of the first equation are the solutions, see Fig. 9.14.

Underdetermined versus overdetermined systems. A system of linear equations is considered underdetermined if there are fewer equations than unknowns. On the other hand, in an overdetermined system, there are more equations than unknowns. For example,

$$\begin{bmatrix} 1 & 2 & 2 & 2 \\ 2 & 4 & 6 & 8 \\ 3 & 6 & 8 & 10 \end{bmatrix} \begin{bmatrix} x_1 \\ x_2 \\ x_3 \\ x_4 \end{bmatrix} = \begin{bmatrix} 1 \\ 2 \\ 5 \\ 7 \end{bmatrix} \text{ (underdetermined)}, \quad \begin{bmatrix} 1 & 2 & 3 \\ 2 & 4 & 6 \\ 2 & 6 & 8 \\ 2 & 8 & 10 \end{bmatrix} \begin{bmatrix} x_1 \\ x_2 \\ x_3 \end{bmatrix} = \begin{bmatrix} 3 \\ 4 \\ 5 \end{bmatrix}$$

The first matrix has more columns than rows—it is short and wide. The second matrix has more rows than columns—it is thin and tall.

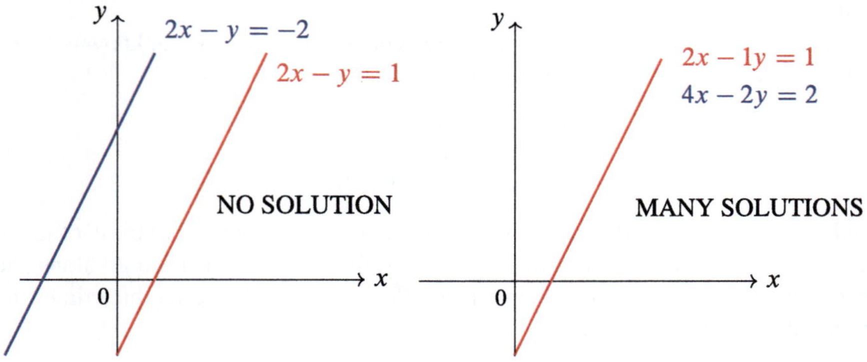

Fig. 9.14 Linear systems of equations: no solution versus infinitely many solutions

Elementary row operations. It is clear that we can perform some massages to a system of linear equations without altering the solutions. For example, the system in Eq. (9.3.1) is equivalent to the following ones:

$$\begin{cases} x + y = 5 \\ 2x - y = 1 \end{cases} \Longleftrightarrow \begin{cases} 2(x + y) = 10 \\ 2x - y = 1 \end{cases} \Longleftrightarrow \begin{cases} x + y = 5 \\ 3x = 6 \end{cases}$$

in which the first system was obtained by swapping the two original equations; from it, the second system obtained by multiplying the first equation by two and the third system by adding the first equation to the second equation. Using the row picture, what we have done is called *elementary row operations*.[14] This is because the coefficients of the system are stored in the coefficient matrix, and thus what done to the equations are done to the rows of this matrix. There are only three types of elementary row operations:

- (Row Swap) Exchange any two rows.
- (Scalar Multiplication) Multiply any row by a constant.
- (Row Sum) Add a multiple of one row to another row.

The Gaussian elimination method, discussed in the next section, uses the elementary row operations to transform the system into a simpler form.

9.3.1 Gaussian Elimination Method

To demonstrate the Gaussian elimination method, let's consider the following system of three unknowns and three equations:

$$\begin{aligned} 2x_1 + 4x_2 - 2x_3 &= 2 \\ 4x_1 + 9x_2 - 3x_3 &= 8 \\ -2x_1 - 3x_2 + 7x_3 &= 10 \end{aligned}$$

which is rewritten in this matrix form:

$$\mathbf{A}\boldsymbol{x} = \boldsymbol{b}, \quad \mathbf{A} = \begin{bmatrix} 2 & 4 & -2 \\ 4 & 9 & -3 \\ -2 & -3 & 7 \end{bmatrix}, \quad \boldsymbol{x} = \begin{bmatrix} x_1 \\ x_2 \\ x_3 \end{bmatrix}, \quad \boldsymbol{b} = \begin{bmatrix} 2 \\ 8 \\ 10 \end{bmatrix}$$

Once the elimination process—to be discussed shortly—has been done, we get a new form $\mathbf{U}\boldsymbol{x} = \boldsymbol{c}$:

---

[14] We only mentioned about multiplying a row by a constant, but if the constant is $1/c$, where $c \neq 0$, then we also cover division. Similarly, by adding a negative multiple of one row to another, we're actually subtracting.

$$\mathbf{U} = \begin{bmatrix} 2 & 4 & -2 \\ 0 & 1 & 1 \\ 0 & 0 & 4 \end{bmatrix}, \quad \boldsymbol{c} = \begin{bmatrix} 2 \\ 4 \\ 8 \end{bmatrix}$$

where $\mathbf{U}$, a matrix of which all elements below the main diagonal are zeros, is called an *upper triangular matrix*; the non-zero red terms form a triangle. All the pivots of this upper triangular matrix are on the diagonal. Obviously solving $\mathbf{U}\boldsymbol{x} = \boldsymbol{c}$ is super easy: back substitution. The last row gives us $4x_3 = 8$ or $x_3 = 2$, substituting that x_3 into the second row: $x_2 + x_3 = 4$ we get $x_2 = 2$. Finally substituting x_3, x_2 into the first row we get $x_1 = -1$.

The elimination process brings $\mathbf{A}$ to $\mathbf{U}$ which is in a row echelon form (REF). A matrix is said to be in row echelon form if all entries below the pivots are zero.

Now, I present the elimination process. We start with the elimination of x_1 in the second row (or equivalently the blue number 4); this is obtained by subtracting two times the first row from the second row (the red number 2 is the first non-zero in the row that does the elimination, it is called a pivot):

$$\begin{array}{r} 2x_1 + 4x_2 - 2x_3 = 2 \\ 4x_1 + 9x_2 - 3x_3 = 8 \\ -2x_1 - 3x_2 + 7x_3 = 10 \end{array} \quad \overset{R2 \leftrightarrow R2-2R1}{\Longrightarrow} \quad \begin{array}{r} 2x_1 + 4x_2 - 2x_3 = 2 \\ 0x_1 + 1x_2 + 1x_3 = 4 \\ -2x_1 - 3x_2 + 7x_3 = 10 \end{array}$$

We observe that after this elimination step, only the second equation changes, highlighted by the red terms. We continue to remove x_1 in the third equation, or, in other words, remove -2 below the first zero in the second equation:

$$\begin{array}{r} 2x_1 + 4x_2 - 2x_3 = 2 \\ 0x_1 + 1x_2 + 1x_3 = 4 \\ -2x_1 - 3x_2 + 7x_3 = 10 \end{array} \quad \overset{R3 \leftrightarrow R3+R1}{\Longrightarrow} \quad \begin{array}{r} 2x_1 + 4x_2 - 2x_3 = 2 \\ 0x_1 + \boxed{1}x_2 + 1x_3 = 4 \\ 0x_1 + 1x_2 + 5x_3 = 12 \end{array}$$

Now, the first column has been finished, we move to the second column, and we want to remove x_2 in the third equation, i.e., the one below the pivot on row 2:

$$\begin{array}{r} 2x_1 + 4x_2 - 2x_3 = 2 \\ 0x_1 + 1x_2 + 1x_3 = 4 \\ 0x_1 + 1x_2 + 5x_3 = 12 \end{array} \quad \overset{R3 \leftrightarrow R3-R2}{\Longrightarrow} \quad \begin{array}{r} 2x_1 + 4x_2 - 2x_3 = 2 \\ 0x_1 + \boxed{1}x_2 + 1x_3 = 4 \\ 0x_1 + 0x_2 + 4x_3 = 8 \end{array}$$

9.3.2 *The Gauss-Jordan Elimination Method*

As the Gaussian elimination applies to both the coefficient matrix and the RHS vector, it is more efficient to put the matrix and the vector into a so-called augmented matrix and carry out the elimination:

$$[\mathbf{A}|\boldsymbol{b}] = \left[\begin{array}{ccc|c} 2 & 4 & -2 & 2 \\ 4 & 9 & -3 & 8 \\ -2 & -3 & 7 & 10 \end{array}\right] \Longrightarrow \left[\begin{array}{ccc|c} 2 & 4 & -2 & 2 \\ 0 & 1 & 1 & 4 \\ 0 & 0 & 4 & 8 \end{array}\right]$$

Gauss would finish here and do back substitution. Jordan[15] continued with elimination until the left block is the unit matrix: **A** becomes **I**. And the obtained form is called the reduced row echelon form; it makes the back substitution super easy. A matrix is said to be in reduced row echelon form (RREF) if all the entries below and above the pivots are zero. What we have to do is to remove the red terms—making zeros above the pivots and making the pivots ones:

$$\left[\begin{array}{ccc|c} 2 & 4 & -2 & 2 \\ 0 & \boxed{1} & 1 & 4 \\ 0 & 0 & \boxed{4} & 8 \end{array}\right] \Longrightarrow \left[\begin{array}{ccc|c} 2 & 0 & -6 & -14 \\ 0 & 1 & 1 & 4 \\ 0 & 0 & 4 & 8 \end{array}\right] \Longrightarrow \left[\begin{array}{ccc|c} 2 & 0 & 0 & -2 \\ 0 & 1 & 0 & 2 \\ 0 & 0 & 1 & 2 \end{array}\right] \Longrightarrow \left[\begin{array}{ccc|c} 1 & 0 & 0 & -1 \\ 0 & 1 & 0 & 2 \\ 0 & 0 & 1 & 2 \end{array}\right]$$

The solution is now simply the right block, which is $(-1, 2, 2)$. Note that the columns in **A** transformed to the three unit vectors $(1, 0, 0)$, $(0, 1, 0)$, and $(0, 0, 1)$ of $\mathbb{R}^3$ in the reduced row echelon form.

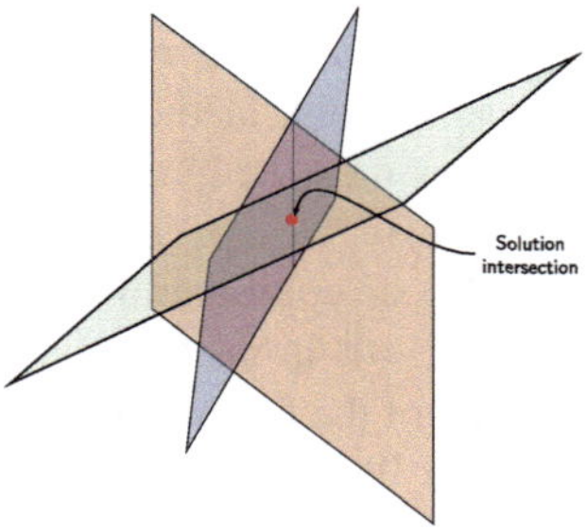

Is this solution making sense? We have three unknowns and three equations; each equation is then a plane in $\mathbb{R}^3$. The intersection of two such planes gives a line, and a line intersects the remaining plane at a single point (if it is not parallel to the plane): that is the solution to the system. This system is similar to Fig. 9.13—left; it is harder to plot three planes and show their intersection. See the figure for an illustration (noting that this is not for the system we're solving).

---

[15] Wilhelm Jordan (1842–1899) was a German geodesist who conducted surveys in Germany and Africa. He is remembered among mathematicians for the Gauss-Jordan elimination algorithm, with Jordan improving the stability of the algorithm so it could be applied to minimizing the squared error in the sum of a series of surveying observations. This algebraic technique appeared in the third edition (1888) of his Textbook of Geodesy.Wilhelm Jordan is not to be confused with the French mathematician Camille Jordan (Jordan curve theorem), nor with the German physicist Pascual Jordan (Jordan algebras).

Many solutions: underdetermined systems. We consider the following underdetermined system where there are more unknowns than equations:

$$\begin{aligned} x_1 - x_2 - x_3 + 2x_4 &= 1 \\ 2x_1 - 2x_2 - x_3 + 3x_4 &= 3 \\ -1x_1 + 1x_2 - x_3 + 0x_4 &= -3 \end{aligned} \implies \left[\begin{array}{cccc|c} 1 & -1 & -1 & 2 & 1 \\ 2 & -2 & -1 & 3 & 3 \\ -1 & 1 & -1 & 0 & -3 \end{array}\right] \implies \left[\begin{array}{cccc|c} \boxed{1} & -1 & 0 & 1 & 2 \\ 0 & 0 & \boxed{1} & -1 & 1 \\ 0 & 0 & 0 & 0 & 0 \end{array}\right]$$

where to save space we have carried out the Gauss-Jordan elimination process in the final step.[16] Looking at the RREF, we have the third row full of zeros: it is meaningless because it is equivalent to the equation $0 = 0$. This indicates that the hyperplane $-1x_1 + 1x_2 - x_3 + 0x_4 = -3$ is just a linear combination of the other hyperplanes. Indeed, the third row of $\mathbf{A}$ is equal to three times the first row minus two times the second one.

Now, we have four unknowns but only two equations; there are so many freedom here. We say that there are $4 - 2 = 2$ free variables. And we also have two pivots (indicated by boxes in the above equation). The columns containing the pivots are called the *pivot columns*; in this example, they are the first and third columns. They are of course the unit vectors $(1, 0, 0)$ and $(0, 1, 0)$ of $\mathbb{R}^3$. The other columns are called the non-pivot columns; they are the second and fourth columns.

Now comes an important fact: *the non-pivot columns can be written as linear combinations of the pivot columns*. Look at the first non-pivot column, it is the second column. Its non-zero entries must be in the first entry (if not the case, then it would be a pivot column). Obviously, we can write $(-1, 0, 0) = (-1) \times (1, 0, 0)$. The first non-pivot column is a linear combination of the first pivot column. The second non-pivot column is $(1, -1, 0)$: it has the non-zero entries at the first two slots, thus it is a linear combination of the first two unit vectors (or the first two pivot columns): $(1, -1, 0) = (1) \times (1, 0, 0) + (-1) \times (0, 1, 0)$. To illustrate this point, let's consider a RREF for a 4×6 matrix with three pivots:

$$\mathbf{R} = \begin{bmatrix} \boxed{1} & b_{12} & 0 & b_{14} & 0 & b_{16} \\ 0 & 0 & \boxed{1} & b_{24} & 0 & b_{26} \\ 0 & 0 & 0 & 0 & \boxed{1} & b_{36} \\ 0 & 0 & 0 & 0 & 0 & 0 \end{bmatrix}$$

Another important fact: in the RREF the 4th col is the 1st col minus the third col, if not clear check again Eq. (9.2.3). And we also have the same relation in $\mathbf{A}$: check that the 4th col of $\mathbf{A}$ is exactly the 1st col minus the third one. To explain why we need to consider $\mathbf{A}\boldsymbol{x} = \mathbf{0}$ discussed in Sect. 9.3.3.[17]

It is a choice we made to select the variables associated with the non-pivot columns as the free variables, and compute other variables, called the pivot variables, in terms

[16] As I did not aim to practice the Gauss-Jordan method, I used `Julia` to do this for me. The aim was to see the solution of the system.

[17] The short answer is that $\mathbf{A}\boldsymbol{x} = \mathbf{0}$ is equivalent to $\mathbf{R}\boldsymbol{x} = \mathbf{0}$.

of the free ones. Thus, x_2, x_4 are the free variables and x_1, x_3 are the pivot variables. For the free variables we can assign $x_2 = s$ and $x_4 = t$, then

$$\begin{aligned} x_1 - x_2 + x_4 &= 2 \\ x_3 - x_4 &= 1 \end{aligned} \Longrightarrow \begin{aligned} x_1 &= 2 + s - t \\ x_3 &= 1 + t \end{aligned} \Longrightarrow \boldsymbol{x} = \begin{bmatrix} 2+s-t \\ s \\ 1+t \\ t \end{bmatrix} = \begin{bmatrix} 2 \\ 0 \\ 1 \\ 0 \end{bmatrix} + s \begin{bmatrix} 1 \\ 1 \\ 0 \\ 0 \end{bmatrix} + t \begin{bmatrix} -1 \\ 0 \\ 1 \\ 1 \end{bmatrix} \tag{9.3.6}$$

This specific example tells us that the number of free variables equals the number of unknowns minus the number of non-zero rows in the echelon form of **A**. Thus, we need to introduce another number that characterizes the matrix better (for a matrix we have already two numbers: the number of rows and cols): that is the concept of the rank of the matrix.

Definition 9.3.1 The rank of a matrix is the number of non-zero rows in its row echelon form (or its reduced REF). It is also the number of pivots.

Theorem 9.3.1 (The rank theorem) *Let* **A** *be the coefficient matrix of a system of linear equations with n variables. If the system is solvable (or consistent), then*

$$\textit{number of free variables} = n - \text{rank}(\mathbf{A})$$

9.3.3 *Homogeneous Linear Systems*

Now the focus is on the solutions to $\mathbf{A}\boldsymbol{x} = \mathbf{0}$. Such a system is called a homogeneous system. The coefficient matrix **A** is of shape $m \times n$ which can be either rectangular or square. There should be three questions to ask now

- Why $\mathbf{A}\boldsymbol{x} = \mathbf{0}$ called a homogeneous system?
- And why we care about such systems?
- What can be the solutions of such systems?

The answer to the first question is simple: if $\boldsymbol{x}^*$ is a solution we have $\mathbf{A}\boldsymbol{x}^* = \mathbf{0}$, and thus $\mathbf{A}(c\boldsymbol{x}^*) = \mathbf{0}$ with $c \in \mathbb{R}$; in other words, $c\boldsymbol{x}^*$ is also a solution. And that's why mathematicians call $\mathbf{A}\boldsymbol{x} = \mathbf{0}$ a homogeneous equation. If the RHS is not $\mathbf{0}$, then we get an inhomogeneous system.

We focus on the third question for now. It is obvious that one possible solution is the zero vector, which is called understandably *the trivial solution*. This is similar to the equation $5x = 0$. But for the equation $0x = 0$, then there are infinitely many solutions. So, $\mathbf{A}\boldsymbol{x} = \mathbf{0}$ either has one unique solution which is the zero vector or has infinitely many solutions. From the previous section, we know that only when we have free variables we have infinitely many solutions.

Theorem 9.3.2 *If* $[\mathbf{A}|\mathbf{0}]$ *is a homogeneous system of m linear equations with n unknowns, where* $m < n$*, then the system has infinitely many solutions.*

Proof Note that the system is solvable, then we use the rank theorem to have

$$\text{number of free variables} = n - \text{rank}(\mathbf{A})$$

Now comes the fact that $\text{rank}(\mathbf{A}) \leq m$,[18] thus

$$\text{number of free variables} = n - \text{rank}(\mathbf{A}) \geq n - m > 0$$

which indicates that there is at least one free variable, and hence infinitely many solutions. ■

9.3.4 Spanning Sets of Vectors and Linear Independence

One important operation in linear algebra is to consider a linear combination of a given set of vectors. If $S = \{\boldsymbol{v}_1, \boldsymbol{v}_2, ..., \boldsymbol{v}_k\}$, then *one* linear combination of $\boldsymbol{v}_1, \boldsymbol{v}_2, ..., \boldsymbol{v}_k$ is $\sum_{i=1}^{k} \alpha_i \boldsymbol{v}_i$. More often that not, we're interested in *ALL* the linear combinations of $\boldsymbol{v}_1, \boldsymbol{v}_2, ..., \boldsymbol{v}_k$. To this end, the concept of a spanning set of vectors is introduced.

Definition 9.3.2 If $S = \{\boldsymbol{v}_1, \boldsymbol{v}_2, ..., \boldsymbol{v}_k\}$ is a set of vectors in $\mathbb{R}^n$, then the set of ALL linear combination of $\boldsymbol{v}_1, \boldsymbol{v}_2, ..., \boldsymbol{v}_k$ is called the span of $\boldsymbol{v}_1, \boldsymbol{v}_2, \ldots, \boldsymbol{v}_k$, and is denoted by $\text{span}(\boldsymbol{v}_1, \boldsymbol{v}_2, \ldots, \boldsymbol{v}_k)$ or $\text{span}(S)$. If $\text{span}(S) = \mathbb{R}^n$, then S is called a spanning set for $\mathbb{R}^n$.

Example 9.1 Show that $\mathbb{R}^2 = \text{span}(\{(2, -1), (1, 3)\})$. What we need to prove is that, for an arbitrary vector in $\mathbb{R}^2$, namely, (a, b) it is possible to write it as a linear combination of $\{(2, -1), (1, 3)\}$. That is, the following system

$$\begin{aligned} 2x + y &= a \\ -x + 3y &= b \end{aligned}$$

always has solution for all a, b. We can use the Gaussian elimination to solve this system and see that it always has solution.

Example 9.2 Find the $\text{span}(\{(1, 0), (0, 1), (2, 3)\})$. We simply use the definition to compute the span:

$$\text{span}(\{(1, 0), (0, 1), (2, 3)\}) = c_1 \begin{bmatrix} 1 \\ 0 \end{bmatrix} + c_2 \begin{bmatrix} 0 \\ 1 \end{bmatrix} + c_3 \begin{bmatrix} 2 \\ 3 \end{bmatrix}$$

[18] Rank of $\mathbf{A}$ is the number of non-zero rows and we have maximum m rows.

What is interesting is that the third vector (2, 3) is nothing new, it is a linear combination of the first two, so the span can be written in terms of only the first two vectors:

$$\text{span}(\{(1,0),(0,1),(2,3)\}) = c_1 \begin{bmatrix}1\\0\end{bmatrix} + c_2 \begin{bmatrix}0\\1\end{bmatrix} + c_3 \left(2\begin{bmatrix}1\\0\end{bmatrix} + 3\begin{bmatrix}0\\1\end{bmatrix}\right) = \alpha \begin{bmatrix}1\\0\end{bmatrix} + \beta \begin{bmatrix}0\\1\end{bmatrix}$$

Linear independence. We have seen that in matrices, it is possible that some columns can be written in terms of others. For example, we can have

$$\boldsymbol{a}_3 = -2\boldsymbol{a}_1 - 3\boldsymbol{a}_1$$

In this case, we say that the three columns or vectors are linear dependent. Noting that the above writing is not symmetric, as $\boldsymbol{a}_3$ was received special treatment. Thus, mathematicians will rewrite the above relation as

$$-2\boldsymbol{a}_1 - 3\boldsymbol{a}_1 - \boldsymbol{a}_3 = \boldsymbol{0}$$

And with that we have the following definitions about linear independence/dependence of a set of vectors.

Definition 9.3.3 A collection of k vectors $\boldsymbol{u}_1, \ldots, \boldsymbol{u}_k$ is *linear dependent* if

$$\alpha_1 \boldsymbol{u}_1 + \alpha_2 \boldsymbol{u}_2 + \cdots + \alpha_k \boldsymbol{u}_k = \boldsymbol{0}$$

holds with at least one $\alpha_k \neq 0$.

Definition 9.3.4 A collection of k vectors $\boldsymbol{u}_1, \ldots, \boldsymbol{u}_k$ is *linear independent* if it is not linear dependent. That is,

$$\alpha_1 \boldsymbol{u}_1 + \alpha_2 \boldsymbol{u}_2 + \cdots + \alpha_k \boldsymbol{u}_k = \boldsymbol{0} \Longrightarrow \alpha_i = 0 \ (i = 1, 2, \ldots, k)$$

In summary, a collection of n vectors is said to be linear dependent if we can express one vector in terms of the $n-1$ remaining vectors. On the other hand, it is linear independent if none of them can be written as a linear combination of others: they are independent.

Here is an important fact: *Any set of vectors containing the zero vector is linearly dependent*. This is because we can always write: $1\boldsymbol{0} + 0\boldsymbol{a}_2 + \cdots + 0\boldsymbol{a}_k = \boldsymbol{0}$. Thus, according to our definition, the set $\{\boldsymbol{0}, \boldsymbol{v}_1, \boldsymbol{v}_2, \ldots, \boldsymbol{v}_k\}$ is linear dependent.

Example 9.3 Determine whether the vectors $\{(1,2,0),(1,1,-1),(1,4,2)\}$ are linear independent. This is equivalent to see if the following system

$$\begin{bmatrix}1 & 1 & 1\\2 & 1 & 4\\0 & -1 & 2\end{bmatrix}\begin{bmatrix}\alpha_1\\\alpha_2\\\alpha_3\end{bmatrix} = \begin{bmatrix}0\\0\\0\end{bmatrix}$$

has trivial solution (zero vector) or not. Using the Gauss elimination method, we get one zero row, thus this system has infinitely many solutions, and one solution is not the zero vector. Thus, the vectors are linear dependent.

It can be seen then that in a 2D plane, 3 (or more) vectors are surely linear dependent. This can be intuitively explained: on a 2D plane, two directions (two vectors which are not parallel) are sufficient to get us anywhere, so the third vector can be nothing new: it must be a combination of the first two directions. Similarly, in a 3D space, any four vectors are linearly dependent. We can state this fact as the following theorem:

Theorem 9.3.3 *Any set of m vectors in $\mathbb{R}^n$ is linearly dependent if $m > n$.*

Proof The proof is based on Theorem 9.3.2, which tells us that a system of equation $\mathbf{A}\boldsymbol{x} = \mathbf{0}$, where $\mathbf{A}$ is a $n \times m$ matrix, has a nontrivial solution whenever $n < m$. Thus, we build $\mathbf{A}$ with its columns are the set of m vectors in $\mathbb{R}^n$. Because $\boldsymbol{x} \neq \mathbf{0}$, the columns of $\mathbf{A}$ are linearly dependent.[19] ■

9.4 Matrix Algebra

This section is about operations that can be performed on matrices and the rules that govern these operations. Many rules are similar to the rules of vectors (which are similar to the arithmetic rules of numbers). This is so because a vector is a list of numbers and a matrix is a list of vectors. First, we start with a formal definition of what is a matrix.

Definition 9.4.1 A matrix is a rectangular array of numbers called entries or elements, of the matrix.

The size of a matrix gives the number of rows and columns it has. An $m \times n$ matrix[20] has m rows and n columns:

$$\mathbf{A} = \begin{bmatrix} A_{11} & A_{12} & \cdots & A_{1n} \\ A_{21} & A_{22} & \cdots & A_{2n} \\ \vdots & \vdots & \ddots & A_{1n} \\ A_{m1} & A_{m2} & \cdots & A_{mn} \end{bmatrix}, \quad \text{or } \mathbf{A} = \begin{bmatrix} \boldsymbol{A}_1 & \boldsymbol{A}_2 & \cdots & \boldsymbol{A}_n \end{bmatrix}$$

and we denote by A_{ij} the entry at row i and column j of $\mathbf{A}$. The columns of $\mathbf{A}$ are vectors in $\mathbb{R}^m$ (i.e., they have m components) and the rows of $\mathbf{A}$ are vectors in $\mathbb{R}^n$. In the above, the columns of $\mathbf{A}$ are $\boldsymbol{A}_i$, $i = 1, 2, \ldots, n$. When $m = n$ we have a square

---

[19] Do not forget the column picture of $\mathbf{A}\boldsymbol{x} = \boldsymbol{b}$ that $\boldsymbol{x}$ is the coefficients of the linear combination of $\mathbf{A}$'s columns.

[20] Pronounced m by n matrix.

matrix. The most special square matrix is the identity matrix **I**, or $\mathbf{I}_n$ to explicitly reveal the size, where all the entries on the diagonal are 1: $I_{ii} = 1$:

$$\mathbf{I} = \mathbf{I}_n := \begin{bmatrix} 1 & 0 & \cdots & 0 \\ 0 & 1 & \cdots & 0 \\ \vdots & \vdots & \ddots & 0 \\ 0 & 0 & \cdots & 1 \end{bmatrix} = \begin{bmatrix} \boldsymbol{e}_1 & \boldsymbol{e}_2 & \cdots & \boldsymbol{e}_n \end{bmatrix} \tag{9.4.1}$$

This matrix is called the identity matrix because $\mathbf{I}\boldsymbol{x} = \boldsymbol{x}$ for all $\boldsymbol{x}$, it is the counterpart of number one. As can be seen, **I** consists of all unit vectors in $\mathbb{R}^n$.

9.4.1 Matrix Operations

We can do things with numbers and vectors; things such as addition and multiplication. It is no surprise that we can add (and subtract) two matrices, we can multiply a scalar with a matrix, we can multiply a matrix with a vector and finally we can multiply (and divide) two matrices.

Considering two $m \times n$ matrices **A** and **B**, the sum of the two matrices is an $m \times n$ matrix defined as

$$\boxed{(\mathbf{A} + \mathbf{B})_{ij} = A_{ij} + B_{ij}} \tag{9.4.2}$$

So, to add two matrices, add the entries. This is similar to adding vectors! Similarly, we can scale a matrix by a factor c as

$$\boxed{(c\mathbf{A})_{ij} = cA_{ij}} \tag{9.4.3}$$

which is similar to scaling a vector.

The next operation is multiplication of a $m \times n$ matrix **A** with a n-vector $\boldsymbol{x}$ (a vector in $\mathbb{R}^n$ is referred to as n-vector). The result is a vector of length m of which the ith entry is given by

$$\boxed{(\mathbf{A}\boldsymbol{x})_i = \sum_{k=1}^{n} A_{ik}x_k} \tag{9.4.4}$$

That is, the ith entry of the result vector is *the dot product of the ith row of* **A** *and* $\boldsymbol{x}$. This definition comes directly from the system $\mathbf{A}\boldsymbol{x} = \boldsymbol{b}$. Because the dot product has the distributive property that $\boldsymbol{a} \cdot (\boldsymbol{b} + \boldsymbol{c}) = \boldsymbol{a} \cdot \boldsymbol{b} + \boldsymbol{a} \cdot \boldsymbol{c}$, the matrix-vector multiplication also has the same property:

$$\mathbf{A}(\boldsymbol{a}+\boldsymbol{b}) \stackrel{\text{def}}{=} \begin{bmatrix} \text{row 1 of A} \cdot (\boldsymbol{a}+\boldsymbol{b}) \\ \text{row 2 of A} \cdot (\boldsymbol{a}+\boldsymbol{b}) \\ \vdots \\ \text{row m of A} \cdot (\boldsymbol{a}+\boldsymbol{b}) \end{bmatrix} = \begin{bmatrix} \text{row 1 of A} \cdot \boldsymbol{a} + \text{row 1 of A} \cdot \boldsymbol{b} \\ \text{row 2 of A} \cdot \boldsymbol{a} + \text{row 1 of A} \cdot \boldsymbol{b} \\ \vdots \\ \text{row m of A} \cdot \boldsymbol{a} + \text{row 1 of A} \cdot \boldsymbol{b} \end{bmatrix} = \mathbf{A}\boldsymbol{a} + \mathbf{A}\boldsymbol{b}$$

Now comes the harder matrix-matrix multiplication. One simple example for the motivation: considering the following two linear systems:

$$\begin{aligned} x_1 + 2x_2 &= y_1 \\ 0x_1 + 3x_2 &= y_2 \end{aligned}, \quad \begin{aligned} y_1 - y_2 &= z_1 \\ -2y_1 + 0y_2 &= z_2 \end{aligned}$$

Now, we want to eliminate y_1, y_2 to have a system with unknowns x_1, x_2. This is simple, we can just substitute y_1, y_2 in the second system by the first system. The result is

$$\begin{aligned} x_1 - x_2 &= z_1 \\ -2x_1 - 4x_2 &= z_2 \end{aligned} \tag{9.4.5}$$

Now, we do the same but using matrices:

$$\begin{bmatrix} 1 & -1 \\ -2 & 0 \end{bmatrix} \begin{bmatrix} 1 & 2 \\ 0 & 3 \end{bmatrix} \begin{bmatrix} x_1 \\ x_2 \end{bmatrix} = \begin{bmatrix} z_1 \\ z_2 \end{bmatrix}$$

Thus, the product of the two matrices in this equation must be another 2×2 matrix, and this matrix must be, because we know the result from Eq. (9.4.5),

$$\begin{bmatrix} 1 & -1 \\ -2 & 0 \end{bmatrix} \begin{bmatrix} 1 & 2 \\ 0 & 3 \end{bmatrix} = \begin{bmatrix} 1 & -1 \\ -2 & -4 \end{bmatrix}$$

This result can be obtained if we first multiply the left matrix on the LHS with the first column of the right matrix (red colored), and we get the first column of the RHS matrix. Doing the same we get the second column. And with that, we now can define the rule for matrix-matrix multiplication. Assume that **A** is an $m \times n$ matrix and **B** is an $n \times p$ matrix, then the product **AB** is an $m \times p$ matrix of which the ij entry is

$$\boxed{(\mathbf{AB})_{ij} = \sum_{k=1}^{n} \mathbf{A}_{ik}\mathbf{B}_{kj}} \tag{9.4.6}$$

In words: *the entry at row i and column j of the product* **AB** *is the dot product of row i of* **A** *and column j of* **B**.[21] And we understand why for matrix-matrix multiplication the number of columns in the first matrix must be equal to the number of rows in the second matrix.

[21] Thus matrix-matrix multiplication is not actually something entirely new.

It must be emphasized that the above definition of matrix-matrix multiplication is not the only way to look at this multiplication. In Sect. 9.4.4, other ways are discussed. This definition is used for the actual computation of the matrix-matrix product, but it does not tell much what is going on.

Remark Of course you can define matrix-matrix multiplication in a different way, and in the process you would create another branch of algebra. However, the presented definition is compatible with matrix-vector multiplication. Thus, it inherits many nice properties as we shall discuss shortly.

9.4.2 The Laws for Matrix Operations

Let's denote by **A**, **B**, and **C** three matrices (of appropriate shapes) and a real number α, we obtain the following laws for matrix operations, which are exactly identical to the arithmetic rules of real numbers (except the broken $\mathbf{AB} \neq \mathbf{BA}$):

$$
\begin{array}{lll}
\text{(a): commutative law} & \mathbf{A}+\mathbf{B} & =\mathbf{B}+\mathbf{A} \\
\text{(b): distributive law} & \alpha(\mathbf{A}+\mathbf{B}) & =\alpha\mathbf{A}+\alpha\mathbf{B} \\
\text{(c): associative law} & \mathbf{A}+(\mathbf{B}+\mathbf{C}) & =(\mathbf{A}+\mathbf{B})+\mathbf{C} \\
\text{(d): associative law for ABC} & \mathbf{A}(\mathbf{BC}) & =(\mathbf{AB})\mathbf{C} \\
\text{(e): distributive law (left)} & \mathbf{A}(\mathbf{B}+\mathbf{C}) & =\mathbf{AB}+\mathbf{AC} \\
\text{(f): distributive law (right)} & (\mathbf{A}+\mathbf{B})\mathbf{C} & =\mathbf{AC}+\mathbf{BC} \\
\text{(f): broken commutative law (multiplication)} & \mathbf{AB} & \neq\mathbf{BA}
\end{array}
\tag{9.4.7}
$$

Certainly mathematicians ask for proofs. Proving the first three laws is straightforward. This is not unexpected as these laws are exactly identical to the laws for vector addition and scalar multiplication. If we want we can think of a matrix as a "long" vector (and this is actually how computers store matrices).

For the distributive law from the left: we consider one column of $\mathbf{A}(\mathbf{B}+\mathbf{C})$, it is $\mathbf{A}(\boldsymbol{b}_i+\boldsymbol{c}_i)$, which is $\mathbf{A}\boldsymbol{b}_i+\mathbf{A}\boldsymbol{c}_i$ (due to the linearity of matrix-vector multiplication).

After multiplication is powers, so we now define powers of a matrix. With p is a positive integer, the pth power of a square matrix **A** is defined as

$$\mathbf{A}^p := \mathbf{AAA}\cdots\mathbf{A} \;\; (p \text{ factors})$$

And the usual laws of exponents e.g., $2^m \times 2^n = 2^{n+m}$ hold for matrix powers:

$$(\mathbf{A}^p)(\mathbf{A}^q) = \mathbf{A}^{p+q}, \quad (\mathbf{A}^p)^q = \mathbf{A}^{pq}$$

Similar to $2^0 = 1$, when $p = 0$ we have $\mathbf{A}^0 = \mathbf{I}$—the identity matrix.

9.4.3 Transpose of a Matrix

Considering two (column) vectors $\boldsymbol{x} = (x, y)$ and $\boldsymbol{a} = (a, b)$ in $\mathbb{R}^2$. Our problem is how to write the dot product $\boldsymbol{x} \cdot \boldsymbol{a} = ax + by$ using matrix notation. We cannot use $\boldsymbol{xa}$ because matrix multiplication requires that the shapes of the matrices must be compatible. We can solve the problem if we turn the column vector $\boldsymbol{x}$ into a row vector, then we're done:

$$\boldsymbol{x} \cdot \boldsymbol{a} = ax + by = \begin{bmatrix} x & y \end{bmatrix} \begin{bmatrix} a \\ b \end{bmatrix} = \boldsymbol{x}^\top \boldsymbol{a} \quad (= \boldsymbol{a}^\top \boldsymbol{x}) \tag{9.4.8}$$

where the notation $\boldsymbol{x}^\top$ is to denote the transpose of $\boldsymbol{x}$. It turns a column vector into a row vector. As a matrix can be seen as a collection of some column vectors, we can also transpose a matrix.

With two vectors we can multiply them to get a number with the above dot product. A question should arise: is this possible to get a matrix from the product of two vectors? The answer is yes:

$$\boldsymbol{a} = \begin{bmatrix} 1 \\ 2 \end{bmatrix}, \quad \boldsymbol{b} = \begin{bmatrix} 3 \\ 4 \end{bmatrix} \Longrightarrow \boldsymbol{ab}^\top = \begin{bmatrix} 1 \\ 2 \end{bmatrix} \begin{bmatrix} 3 & 4 \end{bmatrix} = \begin{bmatrix} 3 & 4 \\ 6 & 8 \end{bmatrix}$$

So, a vector $\boldsymbol{a}$ of length m with a vector $\boldsymbol{b}$ of length n via the outer product $\boldsymbol{ab}^\top$ yields an $m \times n$ matrix.

Definition 9.4.2 The transpose of an $m \times n$ matrix $\mathbf{A}$ is the $n \times m$ matrix $\mathbf{A}^\top$ obtained by interchanging the rows and columns of $\mathbf{A}$. That is the ith column of $\mathbf{A}^\top$ is the ith row of $\mathbf{A}$.

One example suffices to clarify the definition:

$$\mathbf{A} = \begin{bmatrix} 1 & 2 & 3 \\ 2 & 4 & 6 \end{bmatrix} \Longrightarrow \mathbf{A}^\top = \begin{bmatrix} 1 & 2 \\ 2 & 4 \\ 3 & 6 \end{bmatrix}$$

With the introduction of the transpose operator, we can define a symmetric matrix as

Definition 9.4.3 A square matrix $\mathbf{A}$ of size $n \times n$ is symmetric if it is equal to its transpose, or $A_{ij} = Aji$.

Obviously transpose is an operator or a function, and thus it obeys certain rules. Here are some basic rules regarding the transpose operator for matrices:

$$\begin{aligned} &(a)\!: \ (\mathbf{A}^\top)^\top &&= \mathbf{A} \\ &(b)\!: \ (\mathbf{A}+\mathbf{B})^\top &&= \mathbf{A}^\top + \mathbf{B}^\top \\ &(c)\!: \ (k\mathbf{A})^\top &&= k\mathbf{A}^\top \\ &(d)\!: \ (\mathbf{AB})^\top &&= \mathbf{B}^\top\mathbf{A}^\top \end{aligned} \tag{9.4.9}$$

Recall that in Sect. 4.2.1, we have seen that it is possible to decompose any function into an even function and an odd function:

$$f(x) = \frac{1}{2}\left[f(x) + f(-x)\right] + \frac{1}{2}\left[f(x) - f(-x)\right]$$

in which the first term is an even function, i.e., $g(-x) = g(x)$ and the second is an odd function, i.e., $g(-x) = -g(x)$ (see Sect. 4.2.1). And we have applied this decomposition to the exponential function $y = e^x$, we had $e^x = \frac{1}{2}\left[e^x + e^{-x}\right] + \frac{1}{2}\left[e^x - e^{-x}\right]$, which led to the definition of the hyperbolic cosine and sine functions. Now, we do the same thing for square matrices. Given a square matrix **A**, we can write

$$\mathbf{A} = \frac{1}{2}(\mathbf{A} + \mathbf{A}^\top) + \frac{1}{2}(\mathbf{A} - \mathbf{A}^\top)$$

and applying that decomposition to the following matrix:

$$\mathbf{A} = \begin{bmatrix} 1 & 2 & 3 \\ 4 & 5 & 6 \\ 7 & 8 & 9 \end{bmatrix} = \frac{1}{2}\begin{bmatrix} 2 & 6 & 10 \\ 6 & 10 & \mathbf{14} \\ 10 & \mathbf{14} & 18 \end{bmatrix} + \frac{1}{2}\begin{bmatrix} 0 & -2 & -4 \\ 2 & 0 & -2 \\ 4 & 2 & 0 \end{bmatrix}$$

we get a symmetric matrix, and a *skew-symmetric* matrix (the second matrix). A skew-symmetric matrix **A** is a square matrix with the property $\mathbf{A}^\top = -\mathbf{A}$.

9.4.4 Partitioned Matrices

Considering two 3 × 3 matrices **A** and **B**, its product **AB** is given by

$$\mathbf{AB} = \begin{bmatrix} \text{(row 1 of A)} \cdot \text{(col 1 of B)} & \text{(row 1 of A)} \cdot \text{(col 2 of B)} & \text{(row 1 of A)} \cdot \text{(col 3 of B)} \\ \text{(row 2 of A)} \cdot \text{(col 1 of B)} & \text{(row 2 of A)} \cdot \text{(col 2 of B)} & \text{(row 2 of A)} \cdot \text{(col 3 of B)} \\ \text{(row 3 of A)} \cdot \text{(col 1 of B)} & \text{(row 3 of A)} \cdot \text{(col 2 of B)} & \text{(row 3 of A)} \cdot \text{(col 3 of B)} \end{bmatrix}$$

Thus, we can split **B** into three columns $\boldsymbol{B}_1$, $\boldsymbol{B}_2$, $\boldsymbol{B}_3$, and **AB** is equal to the product of **A** with each column, and the results put together:

$$\mathbf{AB} = \mathbf{A}\left[\,\boldsymbol{B}_1 | \boldsymbol{B}_2 | \boldsymbol{B}_3\,\right] = \left[\,\mathbf{A}\boldsymbol{B}_1 | \mathbf{A}\boldsymbol{B}_2 | \mathbf{A}\boldsymbol{B}_3\,\right]$$

The form on the right is called the matrix-column representation of the product. What does this representation tell us? It tells us that the columns of **AB** are the linear combinations of the columns of **A** (e.g., $\mathbf{A}\boldsymbol{B}_1$ is a linear combination of the cols of **A** from the definition of matrix-vector multiplication). And that leads to the linear combination of all columns of **AB** is just a linear combination of the columns of **A**. Later on, this results in $\text{rank}(\mathbf{AB}) \leq \text{rank}(\mathbf{A})$.

And nothing stops us to partition matrix **A**, but we have to split it by rows:

$$\mathbf{AB} = \left[\begin{array}{c} \boldsymbol{A}_1 \\ \hline \boldsymbol{A}_2 \\ \hline \boldsymbol{A}_3 \end{array}\right] \mathbf{B} = \left[\begin{array}{c} \boldsymbol{A}_1\mathbf{B} \\ \hline \boldsymbol{A}_2\mathbf{B} \\ \hline \boldsymbol{A}_3\mathbf{B} \end{array}\right]$$

And this is called the row-matrix representation of the product.

It is also possible to partition both matrices, and we obtain the column-row representation of the product:

$$\mathbf{AB} = \left[\begin{array}{c|c|c} \boldsymbol{A}_1 & \boldsymbol{A}_2 & \boldsymbol{A}_3 \end{array}\right] \left[\begin{array}{c} \boldsymbol{B}_1 \\ \hline \boldsymbol{B}_2 \\ \hline \boldsymbol{B}_3 \end{array}\right] = \underbrace{\boldsymbol{A}_1\boldsymbol{B}_1 + \boldsymbol{A}_2\boldsymbol{B}_2 + \boldsymbol{A}_3\boldsymbol{B}_3}_{\text{sum of rank 1 matrices}}$$

This reminds us of the dot product, but the individual terms are matrices not scalars because $\boldsymbol{A}_1\boldsymbol{B}_1$ is the outer product. For example, $\boldsymbol{A}_1\boldsymbol{B}_1$ is a 3×3 matrix as $\boldsymbol{A}_1$ is a 3×1 matrix and $\boldsymbol{B}_1$ is a 1×3 matrix:

$$\boldsymbol{A}_1\boldsymbol{B}_1 = \begin{bmatrix} A_{11} \\ A_{21} \\ A_{31} \end{bmatrix} \begin{bmatrix} B_{11} & B_{12} & B_{13} \end{bmatrix} = \begin{bmatrix} A_{11}B_{11} & A_{11}B_{12} & A_{11}B_{13} \\ A_{21}B_{11} & A_{21}B_{12} & A_{21}B_{13} \\ A_{31}B_{11} & A_{31}B_{12} & A_{31}B_{13} \end{bmatrix}$$

Matrices like $\boldsymbol{A}_1\boldsymbol{B}_1$ are called rank-1 matrices because their rank is one; this is because the rank of either $\boldsymbol{A}_1$ or $\boldsymbol{B}_1$ is one.[22]

Each of the forgoing partitions is a special case of partitioning a matrix in general. A matrix is said to be partitioned if horizontal and vertical lines are introduced, subdividing it into submatrices or blocks. Partitioning a matrix allows it to be written as a matrix whose entries are its blocks (which are matrices of themselves). For example,

$$\mathbf{A} = \left[\begin{array}{ccc|cc} 1 & 0 & 0 & 2 & -1 \\ 0 & 1 & 0 & 1 & 3 \\ 0 & 0 & 1 & 4 & 0 \\ \hline 0 & 0 & 0 & 1 & 6 \\ 0 & 0 & 0 & 7 & 1 \end{array}\right] = \left[\begin{array}{c|c} \mathbf{I} & \mathbf{A}_{12} \\ \hline \mathbf{0} & \mathbf{A}_{22} \end{array}\right], \quad \mathbf{B} = \left[\begin{array}{cc|cc|c} 4 & 3 & 1 & 2 & 1 \\ -1 & 2 & 2 & 1 & 1 \\ 1 & -5 & 3 & 3 & 1 \\ \hline 1 & 0 & 0 & 0 & 2 \\ 0 & 1 & 0 & 0 & 3 \end{array}\right] = \left[\begin{array}{c|c|c} \mathbf{B}_{11} & \mathbf{B}_{12} & \mathbf{B}_{13} \\ \hline \mathbf{I} & \mathbf{0} & \mathbf{B}_{23} \end{array}\right]$$

where **A** has been partitioned into a 2×2 matrix and **B** as a 2×3 matrix. (Note that **I** was used to denote the identity matrix but the size of **I** varies; similarly for **0**.) With these partitions, the product **AB** can be computed blockwise as if the entries are numbers:

[22] This comes from the fact that $\text{rank}(\mathbf{AB}) \leq \min(\text{rank}(\mathbf{A}), \text{rank}(\mathbf{B}))$. Another way to see this is: $\boldsymbol{A}_1\boldsymbol{B}_1$ is a linear combination of $\boldsymbol{A}_1$, thus it is just a line in the direction of $\boldsymbol{A}_1$ and a line has rank 1.

$$\mathbf{AB} = \begin{bmatrix} \mathbf{I} & \mathbf{A}_{12} \\ \mathbf{0} & \mathbf{A}_{22} \end{bmatrix} \begin{bmatrix} \mathbf{B}_{11} & \mathbf{B}_{12} & \mathbf{B}_{13} \\ \mathbf{I} & \mathbf{0} & \mathbf{B}_{23} \end{bmatrix} = \begin{bmatrix} \mathbf{B}_{11} + \mathbf{A}_{12} & \mathbf{B}_{12} & \mathbf{B}_{13} + \mathbf{A}_{12}\mathbf{B}_{23} \\ \mathbf{A}_{22} & \mathbf{0} & \mathbf{A}_{22}\mathbf{B}_{23} \end{bmatrix}$$

Using `Julia` it is quick to check that the usual way to compute **AB** gives the same result as the way using partitioned matrices.

9.4.5 *Inverse of a Matrix*

Inverse matrices are related to inverse functions. Recall that start with an angle x and press the sin button on a calculation we get $y = \sin x$. Now, to get back to where we have started, i.e., x, press the inverse function arcsin and we have $\arcsin y = x$. Now, we have a square matrix $\mathbf{A}$, a vector $\boldsymbol{x}$, when $\mathbf{A}$ acts on $\boldsymbol{x}$ we get a new vector $\boldsymbol{b}$. To get back to $\boldsymbol{x}$, do:

$$\mathbf{A}\boldsymbol{x} = \boldsymbol{b} \Longrightarrow \underbrace{\mathbf{A}^{-1}\mathbf{A}}_{\mathbf{I}}\boldsymbol{x} = \mathbf{A}^{-1}\boldsymbol{b} \Longrightarrow \mathbf{A}^{-1}\boldsymbol{b} = \boldsymbol{x} \tag{9.4.10}$$

The matrix $\mathbf{A}^{-1}$ is called the *left inverse matrix* of $\mathbf{A}$. There exists the right inverse matrix of $\mathbf{A}$ as well: it is defined by $\mathbf{A}\mathbf{A}^{-1} = \mathbf{I}$. If a matrix is invertible, then its inverse, $\mathbf{A}^{-1}$, is the matrix that inverts $\mathbf{A}$:

$$\mathbf{A}^{-1}\mathbf{A} = \mathbf{I} \quad \text{and} \quad \mathbf{A}\mathbf{A}^{-1} = \mathbf{I} \tag{9.4.11}$$

Property 1. If a matrix is invertible, its inverse is unique.

Property 2. The inverse of the product **AB** is the product of the inverses, but in reverse order:

$$(\mathbf{AB})^{-1} = \mathbf{B}^{-1}\mathbf{A}^{-1}$$

Even though this is natural,[23] an algebraic proof goes:

$$(\mathbf{AB})^{-1}(\mathbf{AB}) = (\mathbf{B}^{-1}\mathbf{A}^{-1})(\mathbf{AB}) = \mathbf{B}^{-1}(\mathbf{A}^{-1}\mathbf{A})\mathbf{B} = \mathbf{B}^{-1}\mathbf{IB} = \mathbf{B}^{-1}\mathbf{B} = \mathbf{I}$$

And of course starting with this property for two matrices, we can develop the rule for three matrices,

$$(\mathbf{ABC})^{-1} = \mathbf{C}^{-1}\mathbf{B}^{-1}\mathbf{A}^{-1}$$

and then for any number of matrices that we want.

Property 3. If $\mathbf{A}$ is invertible then $\mathbf{A}^{-1}$ is invertible and its inverse is $\mathbf{A}$. That is $(\mathbf{A}^{-1})^{-1} = \mathbf{A}$.

[23] This property is sometimes called the socks-and-shoe rule: you put in the socks and then the shoe. Now, you take off the shoe first, then remove the socks.

Property 4. If $\mathbf{A}$ is invertible then $\mathbf{A}^\top$ is invertible and $(\mathbf{A}^\top)^{-1} = (\mathbf{A}^{-1})^\top$[24].

Property 5. If $\mathbf{A}$ is invertible then $\mathbf{A}^n$ is invertible for all non-negative integer n and $(\mathbf{A}^n)^{-1} = (\mathbf{A}^{-1})^n$.[25]

Elementary matrices. We are going to use matrix multiplication to describe the Gaussian elimination method used in solving $\mathbf{A}\boldsymbol{x} = \boldsymbol{b}$. The key idea is that each elimination step is corresponding with the multiplication of an elimination matrix $\mathbf{E}$ with the augmented matrix.

We reuse the example in Sect. 9.3.1. We're seeking for a matrix $\mathbf{E}$ that expresses the process of subtracting two times the first equation from the second equation. To find that matrix, look at the RHS vector: we start with (2, 8, 10) and we get (2, 4, 10) after the elimination step; this can be nearly achieved with:

$$\begin{bmatrix} 1 & 0 & 0 \\ 0 & 1 & 0 \\ 0 & 0 & 1 \end{bmatrix} \begin{bmatrix} 2 \\ 8 \\ 10 \end{bmatrix} \approx \begin{bmatrix} 2 \\ 4 \\ 10 \end{bmatrix}$$

We need to change this matrix slightly as follows, and we get what we have wanted for:

$$\begin{bmatrix} 1 & 0 & 0 \\ -2 & 1 & 0 \\ 0 & 0 & 1 \end{bmatrix} \begin{bmatrix} 2 \\ 8 \\ 10 \end{bmatrix} = \begin{bmatrix} 2 \\ 4 \\ 10 \end{bmatrix}$$

Thus, starting from the identity matrix $\mathbf{I}$: $\mathbf{I}\boldsymbol{b} = \boldsymbol{b}$, the elimination matrix $\mathbf{E}_{21}$ is $\mathbf{I}$ with the extra non-zero entry -2 in the (2, 1) position. How to get that -2 from $\mathbf{I}$? Replacing the second row (of $\mathbf{I}$) by subtracting two times the first row from the second row. But that is exactly what we wanted for $\boldsymbol{b}$!

Multiplying $\mathbf{E}_{21}$ with $\mathbf{A}$ has the same effect:

$$\begin{bmatrix} 1 & 0 & 0 \\ -2 & 1 & 0 \\ 0 & 0 & 1 \end{bmatrix} \begin{bmatrix} 2 & 4 & -2 \\ 4 & 9 & -3 \\ -2 & -3 & 7 \end{bmatrix} = \begin{bmatrix} 2 & 4 & -2 \\ 0 & 1 & 1 \\ -2 & -3 & 7 \end{bmatrix}$$

Definition 9.4.4 An elementary matrix is a matrix that can be obtained from the identity matrix by *one single elementary row operation*. Multiplying a matrix $\mathbf{A}$ by an elementary matrix $\mathbf{E}$ (on the left) causes $\mathbf{A}$ to undergo the elementary row

---

[24] Proof: $\mathbf{A}^{-1}\mathbf{A} = \mathbf{I}$, thus $\mathbf{A}^\top(\mathbf{A}^{-1})^\top = \mathbf{I}$

[25] Proof for $n = 2$: $\mathbf{A}^2(\mathbf{A}^{-1})^2 = \mathbf{A}\mathbf{A}\mathbf{A}^{-1}\mathbf{A}^{-1} = \mathbf{A}\mathbf{I}\mathbf{A}^{-1} = \mathbf{A}\mathbf{A}^{-1} = \mathbf{I}$.

operation represented by **E**. This can be expressed by symbols, where $\mathcal{R}$ denotes a row operation:

$$\mathbf{A}' = \mathcal{R}(\mathbf{A}) \Longleftrightarrow \mathbf{A}' = \mathbf{E}_{\mathcal{R}}\mathbf{A} \tag{9.4.12}$$

Now, as the row operation affects the matrix **A** and the RHS vector $\boldsymbol{b}$ altogether, we can put the coefficient matrix **A** and the RHS vector $\boldsymbol{b}$ side by side to get the so-called augmented matrix, and we apply the elimination operation to this augmented matrix by left multiplying it with $\mathbf{E}_{21}$:

$$\mathbf{E}_{21}\left[\mathbf{A}\ \boldsymbol{b}\right] = \left[\mathbf{E}_{21}\mathbf{A}\ \mathbf{E}_{21}\boldsymbol{b}\right] = \left[\begin{array}{ccc|c} 2 & 4 & -2 & 2 \\ 0 & 1 & 1 & 4 \\ -2 & -3 & 7 & 10 \end{array}\right] \tag{9.4.13}$$

To proceed, we want to eliminate –2 using the pivot 2 (red). The row operation is: replacing row 3 by row 3 + row 1, and that can be achieved with matrix $\mathbf{E}_{31}$ as follows (obtained from **I** by replacing its row 3 by row 3 + row 1):

$$\mathbf{E}_{31} = \begin{bmatrix} 1 & 0 & 0 \\ 0 & 1 & 0 \\ 1 & 0 & 1 \end{bmatrix}$$

to remove x_1 in the third equation. Together, the two elimination steps can be expressed as

$$\mathbf{E}_{31}\mathbf{E}_{21}\left[\mathbf{A}\ \boldsymbol{b}\right] = \left[\mathbf{E}_{31}\mathbf{E}_{21}\mathbf{A}\ \mathbf{E}_{31}\mathbf{E}_{21}\boldsymbol{b}\right] = \left[\begin{array}{ccc|c} 2 & 4 & -2 & 2 \\ 0 & 1 & 1 & 4 \\ 0 & 1 & 5 & 12 \end{array}\right]$$

Finally, we use $\mathbf{E}_{32}$ as follows (we want to remove the blue 1, or x_2 in the row 3, and that is obtained by replacing row 3 with row 3 minus row 2)

$$\mathbf{E}_{32} = \begin{bmatrix} 1 & 0 & 0 \\ 0 & 1 & 0 \\ 0 & -1 & 1 \end{bmatrix}$$

Altogether, the three elimination steps can be expressed as

$$\mathbf{E}_{32}\mathbf{E}_{31}\mathbf{E}_{21}\left[\mathbf{A}\ \boldsymbol{b}\right] = \left[\mathbf{E}_{32}\mathbf{E}_{31}\mathbf{E}_{21}\mathbf{A}\ \mathbf{E}_{32}\mathbf{E}_{31}\mathbf{E}_{21}\boldsymbol{b}\right] = \left[\begin{array}{ccc|c} 2 & 4 & -2 & 2 \\ 0 & 1 & 1 & 4 \\ 0 & 0 & 4 & 8 \end{array}\right] \tag{9.4.14}$$

And we have obtained the same matrix **U** that we got before. Notice the pivots along the diagonal.

The inverse of an elementary matrix. The inverse of an elementary matrix **E** is also an elementary matrix that undoes the row operation that **E** has done. For example,

$$\mathbf{E}_{32} = \begin{bmatrix} 1 & 0 & 0 \\ 0 & 1 & 0 \\ 0 & -1 & 1 \end{bmatrix} \Longrightarrow (\mathbf{E}_{32})^{-1} = \begin{bmatrix} 1 & 0 & 0 \\ 0 & 1 & 0 \\ 0 & 1 & 1 \end{bmatrix}$$

Finding the inverse: Gauss-Jordan elimination method. We have an invertible matrix and we want to find its inverse. To illustrate the method, let's consider a 3×3 matrix **A**. We know that its inverse $\mathbf{A}^{-1}$ is a 3×3 matrix such that $\mathbf{A}\mathbf{A}^{-1} = \mathbf{I}$. Let's denote by $\boldsymbol{x}_1, \boldsymbol{x}_2, \boldsymbol{x}_3$ the three columns of $\mathbf{A}^{-1}$. We're looking for these columns. The equation $\mathbf{A}\mathbf{A}^{-1} = \mathbf{I}$ is equivalent to three systems of linear equations, one for each column:

$$\mathbf{A}\boldsymbol{x}_i = \boldsymbol{e}_i, \quad i = 1, 2, 3 \tag{9.4.15}$$

where $\boldsymbol{e}_i$ are the unit vectors.

We know how to solve a system of linear equations, using the Gaussian elimination method. The idea of the Gauss-Jordan elimination method is to solve Eq. (9.4.15) altogether. So, the augmented matrix is $[\mathbf{A}|\ \boldsymbol{e}_1\ \boldsymbol{e}_2\ \boldsymbol{e}_3]$ or $[\mathbf{A}\ \mathbf{I}]$, and we perform the usual row operations on it. Let's consider a concrete matrix with 2's on the diagonal and -1's next to the 2's, then the augmented matrix is

$$\left[\begin{array}{ccc|ccc} 2 & -1 & 0 & 1 & 0 & 0 \\ -1 & 2 & -1 & 0 & 1 & 0 \\ 0 & -1 & 2 & 0 & 0 & 1 \end{array}\right]$$

The Gaussian elimination steps are

$$\Longrightarrow \left[\begin{array}{ccc|ccc} 2 & -1 & 0 & 1 & 0 & 0 \\ 0 & 3/2 & -1 & 1/2 & 1 & 0 \\ 0 & -1 & 2 & 0 & 0 & 1 \end{array}\right] \qquad (1/2 \text{ row } 1 + \text{row } 2)$$

$$\Longrightarrow \left[\begin{array}{ccc|ccc} 2 & -1 & 0 & 1 & 0 & 0 \\ 0 & 3/2 & -1 & 1/2 & 1 & 0 \\ 0 & 0 & 4/3 & 1/3 & 2/3 & 1 \end{array}\right] \qquad (2/3 \text{ row } 2 + \text{row } 3)$$

What we have to do is to remove the red terms—making zeros above the pivots:

$$\Longrightarrow \left[\begin{array}{ccc|ccc} 2 & -1 & 0 & 1 & 0 & 0 \\ 0 & 3/2 & 0 & 3/4 & 3/2 & 3/4 \\ 0 & -1 & 2 & 0 & 0 & 1 \end{array}\right] \quad \text{(3/4 row 3 + row 2)}$$

$$\Longrightarrow \left[\begin{array}{ccc|ccc} 2 & 0 & 0 & 3/2 & 1 & 1/2 \\ 0 & 3/2 & 0 & 3/4 & 3/2 & 3/4 \\ 0 & 0 & 4/3 & 1/3 & 2/3 & 1 \end{array}\right] \quad \text{(2/3 row 2 + row 1)}$$

$$\Longrightarrow \left[\begin{array}{ccc|ccc} 1 & 0 & 0 & 3/4 & 1/2 & 1/4 \\ 0 & 1 & 0 & 1/2 & 1 & 1/2 \\ 0 & 0 & 1 & 1/4 & 1/2 & 3/4 \end{array}\right] \quad \text{(making the pivots of each row equal 1)}$$

Now, the three columns of $\mathbf{A}^{-1}$ are in the second half of the above matrix.[26] Thus,

$$\left[\mathbf{A}|\mathbf{I}\right] \Longrightarrow \left[\mathbf{I}|\mathbf{A}^{-1}\right]$$

which can be written as

$$\mathcal{R}_k\left(\ldots\mathcal{R}_2\left(\mathcal{R}_1\left(\left[\mathbf{A}|\mathbf{I}\right]\right)\right)\right) = \left[\mathbf{I}|\mathbf{A}^{-1}\right]$$

And each row operation corresponds with an elementary matrix, so the above can be also written as

$$\mathbf{E}_k\ldots\mathbf{E}_2\mathbf{E}_1\left[\mathbf{A}|\mathbf{I}\right] = \left[\mathbf{I}|\mathbf{A}^{-1}\right]$$

From that, we obtain

$$\begin{aligned} \mathbf{E}_k\ldots\mathbf{E}_2\mathbf{E}_1\mathbf{A} &= \mathbf{I} \\ \mathbf{E}_k\ldots\mathbf{E}_2\mathbf{E}_1 &= \mathbf{A}^{-1} \end{aligned}$$

Taking the inverse of the second equation and using the rule $(\mathbf{AB}^{-1} = \mathbf{B}^{-1}\mathbf{A}^{-1}$, we can express $\mathbf{A}$ as a product of the inverses of $\mathbf{E}_i$:

$$\mathbf{A} = \mathbf{E}_1^{-1}\mathbf{E}_2^{-1}\cdots\mathbf{E}_k^{-1} \tag{9.4.16}$$

As the inverse of an elementary matrix is also an elementary matrix, this tells us that every invertible matrix can be decomposed as the product of elementary matrices.

9.4.6 LU Decomposition/Factorization

This section shows that the Gaussian elimination process results in a factorization of the matrix $\mathbf{A}$ into two matrices: one lower triangular matrix $\mathbf{L}$ and the familiar upper triangular matrix $\mathbf{U}$ that we have met. Recall Eq. (9.4.14) that

[26] This is because the first column after the vertical bar is $\boldsymbol{x}_1$, the first column of the inverse of $\mathbf{A}$.

$$\mathbf{E}_{32}\mathbf{E}_{31}\mathbf{E}_{21}\left[\mathbf{A}\ \boldsymbol{b}\right] = \left[\mathbf{E}_{32}\mathbf{E}_{31}\mathbf{E}_{21}\mathbf{A}\ \mathbf{E}_{32}\mathbf{E}_{31}\mathbf{E}_{21}\boldsymbol{b}\right] = \left[\begin{array}{ccc|c} 2 & 4 & -2 & 2 \\ 0 & 1 & 1 & 4 \\ 0 & 0 & 4 & 8 \end{array}\right]$$

From which we can write

$$\mathbf{E}_{32}\mathbf{E}_{31}\mathbf{E}_{21}\mathbf{A} = \mathbf{U} \Longrightarrow \mathbf{A} = (\mathbf{E}_{32}\mathbf{E}_{31}\mathbf{E}_{21})^{-1}\mathbf{U} = \mathbf{E}_{21}^{-1}\mathbf{E}_{31}^{-1}\mathbf{E}_{32}^{-1}\mathbf{U}$$

From Property 3 of matrix inverse, we know the inverse matrices $\mathbf{E}_{21}^{-1}, \mathbf{E}_{31}^{-1}, \mathbf{E}_{32}^{-1}$: they are all lower triangular matrices with 1s on the diagonal. Therefore, we get a lower triangular matrix as their product. Thus, we have decomposed $\mathbf{A}$ into two matrices:

$$\mathbf{A} = \begin{bmatrix} 1 & 0 & 0 \\ 2 & 1 & 0 \\ -1 & 1 & 1 \end{bmatrix} \begin{bmatrix} 2 & 4 & -2 \\ 0 & 1 & 1 \\ 0 & 0 & 4 \end{bmatrix}$$

just similar to how we can decompose a number, e.g., $12 = 2 \times 6$. And this is always a good thing: dealing with 2 and 6 is much easier than with 12. $\mathbf{L}$ and $\mathbf{U}$ contain many zeros.

What is the benefits of this decomposition? It is useful because we replace $\mathbf{A}\boldsymbol{x} = \boldsymbol{b}$ into two problems with triangular matrices:

$$\mathbf{A}\boldsymbol{x} = \boldsymbol{b} \Longleftrightarrow \mathbf{LU}\boldsymbol{x} = \boldsymbol{b} \Longleftrightarrow \begin{cases} \mathbf{L}\boldsymbol{y} = \boldsymbol{b} \\ \mathbf{U}\boldsymbol{x} = \boldsymbol{y} \end{cases} \tag{9.4.17}$$

in which we first solve for $\boldsymbol{y}$, then solve for $\boldsymbol{x}$. Using the LU decomposition method to solve $\mathbf{A}\boldsymbol{x} = \boldsymbol{b}$ is faster than the Gaussian elimination method when we have a constant matrix $\mathbf{A}$ but many different RHS vectors $\boldsymbol{b}_1, \boldsymbol{b}_2, \ldots$ This is because we just need to factor $\mathbf{A}$ into $\mathbf{LU}$ once.

Another benefit of the LU decomposition is that it allows us to compute the determinant of a matrix as the product of the pivots of the $\mathbf{U}$ matrix:

$$\det(\mathbf{A}) = \det(\mathbf{LU}) = \det(\mathbf{L})\det(\mathbf{U}) = 1 \times \prod_i u_i \tag{9.4.18}$$

where u_i are the entries on the diagonal of $\mathbf{U}$ (pivots). There are more to say about determinants in Sect. 9.9.

9.5 Subspaces, Basis, Dimension, and Rank

Subspaces. Inside a vector space there might be a subspace, that is, a smaller set of vectors but big enough to be a space of itself. One example to demonstrate the idea. Note that a plane passing through the origin can be expressed as a linear combination of two (direction) vectors: $P : \boldsymbol{x} = u\boldsymbol{s} + t\boldsymbol{v}$, where $\boldsymbol{s}, \boldsymbol{v} \in \mathbb{R}^3$ and $u, t \in \mathbb{R}$. Now, considering two vectors $\boldsymbol{x}_1$ and $\boldsymbol{x}_2$ lying on this plane, we can write

$$\begin{aligned} \boldsymbol{x}_1 &= u_1\boldsymbol{s} + t_1\boldsymbol{v} \\ \boldsymbol{x}_2 &= u_2\boldsymbol{s} + t_2\boldsymbol{v} \end{aligned} \implies \begin{aligned} \boldsymbol{x}_1 + \boldsymbol{x}_2 &= (u_1 + u_2)\boldsymbol{s} + (t_1 + t_2)\boldsymbol{v} \in P, \\ \alpha\boldsymbol{x}_1 &= \alpha u_1\boldsymbol{s} + \alpha t_1\boldsymbol{v} \in P \end{aligned}$$

This indicates that if we take two vectors on this plane, their sum is also on this plane and the product of one vector with a real number is also on the plane. We say that: The plane going through the origin (0, 0, 0) is a subspace of $\mathbb{R}^3$. And this example leads to the following definition of a subspace.

Definition 9.5.1 A subspace of $\mathbb{R}^n$ is a set of vectors in $\mathbb{R}^n$ that satisfies two requirements: if $\boldsymbol{u}$ and $\boldsymbol{v}$ are two vectors in the subspace and α is a scalar, then

$$(i)\ \boldsymbol{u} + \boldsymbol{v} \text{ is in the subspace} \quad (ii)\ \alpha\boldsymbol{u} \text{ is in the subspace}$$

We can combine the two requirements into one: $\alpha\boldsymbol{u}$ is in the subspace and $\beta\boldsymbol{v}$ is in the subspace (from requirement 2), then $\alpha\boldsymbol{u} + \beta\boldsymbol{v}$ is also in the subspace (requirement 1). And that means that the linear combination of $\boldsymbol{u}$ and $\boldsymbol{v}$ is in the subspace:

$$\boxed{\text{if } \boldsymbol{u} \text{and} \boldsymbol{v} \text{in the subspace then } \alpha\boldsymbol{u} + \beta\boldsymbol{v} \text{ is in the subspace}} \tag{9.5.1}$$

If we take $\alpha = 0$, then $\alpha\boldsymbol{u} = \boldsymbol{0}$ which is in a subspace. *A subspace must contain the zero vector*. Back to the example of a plane in $\mathbb{R}^3$, the plane went through (0, 0, 0).

The plane $P : \boldsymbol{x} = u\boldsymbol{s} + t\boldsymbol{v}$ is a subspace. This leads us to think that given a set of vectors $\boldsymbol{v}_1, \boldsymbol{v}_2, \ldots, \boldsymbol{v}_k$ in $\mathbb{R}^n$, all the linear combinations of them are a subspace of $\mathbb{R}^n$. And that is true:

$$\begin{cases} \boldsymbol{u} = \alpha_i\boldsymbol{v}_i \\ \boldsymbol{v} = \beta_i\boldsymbol{v}_i \end{cases} \implies \begin{cases} \boldsymbol{u} + \boldsymbol{v} = (\alpha_i + \beta_i)\boldsymbol{v}_i, \\ \gamma\boldsymbol{u} = \gamma\alpha_i\boldsymbol{v}_i \end{cases}$$

This gives us the following theorem (check definition 9.3.2 for what a span is).

Theorem 9.5.1 (Span is a subspace) *Let $\boldsymbol{v}_1, \boldsymbol{v}_2, \ldots, \boldsymbol{v}_k$ be vectors in $\mathbb{R}^n$. Then span$(\boldsymbol{v}_1, \boldsymbol{v}_2, \ldots, \boldsymbol{v}_k)$ is a subspace of $\mathbb{R}^n$.*

And this theorem leads to the following subspaces of matrices: column space, row space, nullspace.

Subspaces associated with matrices. We know that solving $\mathbf{A}\boldsymbol{x} = \boldsymbol{b}$ is to find the linear combination of the columns of $\mathbf{A}$ with the coefficients being the components of vector $\boldsymbol{x}$ so that this combination is exactly $\boldsymbol{b}$. And this leads naturally to the concept of the *column space of a matrix*. And why not row space. And there are more. We put all these subspaces related to a matrix in the following definition.

Definition 9.5.2 Let **A** be an $m \times n$ matrix.

(a) The row space of **A** is the subspace $R(\mathbf{A})$ of $\mathbb{R}^n$ spanned by the rows of **A**.
(b) The column space of **A** is the subspace $C(\mathbf{A})$ of $\mathbb{R}^m$ spanned by the columns of **A**.
(c) The null space of **A** is the subspace $N(\mathbf{A})$ of $\mathbb{R}^n$ that contains all the solutions to $\mathbf{A}\boldsymbol{x} = \mathbf{0}$.

With this definition, we can deduce that $\mathbf{A}\boldsymbol{x} = \boldsymbol{b}$ is solvable if and only if $\boldsymbol{b}$ is in the column space of **A**. Therefore, $C(\mathbf{A})$ describes all the attainable right-hand side vectors $\boldsymbol{b}$.

Basis. A plane through (0, 0, 0) in $\mathbb{R}^3$ is spanned by two linear independent vectors. Fewer than two independent vectors will not work; more than two is not necessary (e.g., three vectors in $\mathbb{R}^3$, assuming that the third vector is a combination of the first two, then a linear combination of these three vectors is essentially a combination of the first two vectors). We just need a smallest number of independent vectors.

Definition 9.5.3 A basis for a subspace S of $\mathbb{R}^n$ is a set of vectors in S that

(a) spans S and
(b) is linear independent.

The first requirement makes sure that a sufficient number of vectors is included in a basis, and the second requirement ensures that a basis contains a minimum number of vectors that spans the subspace. We do not need more than that.

It is easy to see that the following sets of vectors are the bases for $\mathbb{R}^2$ (because they span $\mathbb{R}^2$ and they are linear independent):

$$\left(\begin{bmatrix}1\\0\end{bmatrix}, \begin{bmatrix}0\\1\end{bmatrix}\right), \quad \left(\begin{bmatrix}1\\0\end{bmatrix}, \begin{bmatrix}1\\1\end{bmatrix}\right)$$

Even though $\mathbb{R}^2$ has many bases, these bases all have the same number of vectors (2). And this is true for any subspace by the following theorem.

Theorem 9.5.2 (The basis theorem) *Let S be a subspace of $\mathbb{R}^n$. Then any two bases of S have the same number of vectors.*

Proof Let $\mathcal{B} = \{\boldsymbol{u}_1, \boldsymbol{u}_2, \ldots, \boldsymbol{u}_s\}$ and $\mathcal{C} = \{\boldsymbol{v}_1, \boldsymbol{v}_2, \ldots, \boldsymbol{v}_r\}$ be two bases of $\mathbb{R}^n$. We want to prove that $s = r$. As Sherlock Holmes noted, "When you have eliminated the impossible, whatever remains, however improbable, must be the truth" (from *The Sign of Four* by Sir Arthur Conan Doyle), we will prove that neither $s > r$ nor $s < r$ is possible, and thus we're left with $r = s$. Assuming first that $s < r$, we then prove that $\mathcal{C}$ is linear dependent,[27] which contradicts the fact that it is a basis. ■

[27] Express each $\boldsymbol{v}_i$ in terms of $\boldsymbol{u}_1, \boldsymbol{u}_2, \ldots$. Then build $c_1\boldsymbol{v}_1 + \cdots = \mathbf{0}$, which in turn is in terms of $()\boldsymbol{u}_1 + ()\boldsymbol{u}_2 + \cdots = \mathbf{0}$. As $\mathcal{B}$ is a basis all the terms in the brackets must be zero. This is equivalent to a linear system $\mathbf{A}\boldsymbol{c} = \mathbf{0}$ with $\mathbf{A} \in \mathbb{R}^{s \times r}$. This system has a nontrivial solution $\boldsymbol{c}$ due to Theorem 9.3.2.

Any two bases of a subspace of $\mathbb{R}^n$ have the same number of vectors. That number should be special. Indeed, it is the dimension of the subspace. So, we have the following definition for it.

Definition 9.5.4 Let S be a subspace of $\mathbb{R}^n$, then the number of vectors in a basis for S is called the dimension of S, denoted by $\dim(S)$. Using the language of set theory, the dimension of S is the cardinality of one basis of S.

Example 9.4 Find a basis for the row space of

$$\mathbf{A} = \begin{bmatrix} 1 & 13 & 1 & 6 \\ 2 & -10 & 1 & -1 \\ -3 & 21 & -2 & 1 \\ 4 & 16 & 1 & 3 \end{bmatrix}$$

The way to do is the observation that if we perform a number of row elementary operations on $\mathbf{A}$ to get another matrix $\mathbf{B}$, then $R(\mathbf{A}) = R(\mathbf{B})$.[28] So, the same old tool of Gauss-Jordan elimination gives us:

$$\mathbf{R} = \begin{bmatrix} 1\,0\,1\,0 & -1 \\ 0\,1\,2\,0 & 3 \\ 0\,0\,0\,1 & 4 \\ 0\,0\,0\,0 & 0 \end{bmatrix}$$

Now, the final row consists of all zeros is useless; thus the first three non-zero rows form a basis for $R(\mathbf{A})$.[29] And we also get $\dim(R(\mathbf{A})) = 3$.

Example 9.5 Find a basis for the column space of $\mathbf{A}$ given in Example 9.4. We have row operations not column operations. So, one solution is to transpose the matrix to get $\mathbf{A}^\top$ in which the rows are the columns of $\mathbf{A}$. With $\mathbf{A}^\top$, we can proceed as in the previous example. The second way is better as we just work with $\mathbf{A}$. Noting that basis is about the linear independence of the columns of $\mathbf{A}$. That is to see $\mathbf{A}\boldsymbol{x} = \mathbf{0}$ has a zero vector as a solution or not. With this view, we can study $\mathbf{R}\boldsymbol{x} = \mathbf{0}$ instead where $\mathbf{R}$ is the RREF of $\mathbf{A}$.

There are three pivot columns in $\mathbf{R}$: the first, second, and fourth columns. These pivot columns are the standard unit vectors $\boldsymbol{e}_i$, so they are linear independent. The pivot columns also span the column space of $\mathbf{R}$.[30] Now, we know that the pivot

[28] The rows of $\mathbf{B}$ are simply linear combinations of the rows of $\mathbf{A}$, thus the linear combination of the rows of $\mathbf{B}$ is a linear combination of all rows of $\mathbf{A}$. This leads to $R(\mathbf{B}) \subset R(\mathbf{A})$. But the row operations can be reversed to go from $\mathbf{B}$ to $\mathbf{A}$, so we also have $R(\mathbf{A}) \subset R(\mathbf{B})$.

[29] Why? Because the non-zero rows are independent.

[30] This is because the non-pivot columns are linear combinations of the pivot ones, they do not add new thing to the span.

columns of **R** are a basis for the column space of **R**. And this means that the pivot columns of **A** are a basis for the column space of **A**. And we also obtain $\dim(C(\mathbf{A})) = 3)$[31].

From the previous examples, we see that the column and row space of that specific matrix have the same dimension. And in fact it is true for any matrix. So, we have the following theorem.

Theorem 9.5.3 *The row and column spaces of a matrix have the same dimension.*

A nice thing with this theorem is that it allows us to have a better definition for the rank of a matrix. *The rank of a matrix is the dimension of its row and column spaces.* Compared with the definition of the rank as the number of non-zero rows, this definition is symmetric with both rows and columns. And it should be. With this row-column symmetry, it is no surprise that $\text{rank}(\mathbf{A}) = \text{rank}(\mathbf{A}^\top)$.

Suppose that **A** and **B** are two matrices such that **AB** makes sense, from the definition of matrix-matrix product, we know that the columns of **AB** are linear combinations of the columns of **A**. Thus $C(\mathbf{AB}) \subseteq C(\mathbf{A})$. Therefore, $\text{rank}(\mathbf{AB}) \leq \text{rank}(\mathbf{A})$. Similarly, we have $R(\mathbf{AB}) \subseteq R(\mathbf{B})$. Then, $\text{rank}(\mathbf{AB}) \leq \text{rank}(\mathbf{B})$. Finally, $\text{rank}(\mathbf{AB}) \leq \min(\text{rank}(\mathbf{A}), \text{rank}(\mathbf{B}))$.

Proof (Proof of Theorem 9.5.3) Consider a matrix **A**, and we need to prove that $\dim(R(\mathbf{A})) = \dim(C(\mathbf{A}))$. We start with the row space with the fact that $R(\mathbf{A}) = R(\mathbf{R})$ where **R** is the RREF of **A**. Thus, $\dim(R(\mathbf{A})) = \dim(R(\mathbf{R}))$. But $\dim(R(\mathbf{R}))$ is equal to the number of unit pivots, which equals to the number of pivot columns of **A**. And we know that the pivot columns of **A** are $C(\mathbf{A})$. ■

We have the dimension for the row space and column space. What about the null space?

Definition 9.5.5 The nullity of a matrix **A** is the dimension of its null space and is denoted by nullity(**A**).

Example 9.6 Find a basis for the null space of **A** given in Example 9.4. This is equivalent to solving the homogeneous system $\mathbf{A}\boldsymbol{x} = \mathbf{0}$. We get the RREF as

$$\mathbf{A} = \left[\begin{array}{rrrrr|r} 1 & 1 & 3 & 1 & 6 & 0 \\ 2 & -1 & 0 & 1 & -1 & 0 \\ -3 & 2 & 1 & -2 & 1 & 0 \\ 4 & 1 & 6 & 1 & 3 & 0 \end{array}\right] \Longrightarrow \left[\begin{array}{rrrrr|r} 1 & 0 & 1 & 0 & -1 & 0 \\ 0 & 1 & 2 & 0 & 3 & 0 \\ 0 & 0 & 0 & 1 & 4 & 0 \\ 0 & 0 & 0 & 0 & 0 & 0 \end{array}\right]$$

Looking at the matrix **R**, we know that there are two free variables x_3, x_5. We then solve for the pivot variables in terms of the free ones with $x_3 = s$ and $x_5 = t$:

[31] Be careful that $C(\mathbf{A}) \neq C(\mathbf{R})$

$$\begin{bmatrix} x_1 \\ x_2 \\ x_3 \\ x_4 \\ x_5 \end{bmatrix} = \begin{bmatrix} -s+t \\ -2s-3t \\ s \\ -4t \\ t \end{bmatrix} = s\begin{bmatrix} -1 \\ -2 \\ 1 \\ 0 \\ 0 \end{bmatrix} + t\begin{bmatrix} 1 \\ -3 \\ 0 \\ -4 \\ 1 \end{bmatrix}$$

Therefore, the null space of $\mathbf{A}$ has a basis of the two red vectors. And the nullity of $\mathbf{A}$ is 2.

Theorem 9.5.4 (Rank theorem) *Let* $\mathbf{A}$ *be an* $m \times n$ *matrix, then*

$$rank(\mathbf{A}) + nullity(\mathbf{A}) = n$$

Theorem 9.5.5 *Let* $\mathbf{A}$ *be an* $m \times n$ *matrix, then*

(a) $rank(\mathbf{A}^\top\mathbf{A}) = rank(\mathbf{A})$.
(b) The matrix $\mathbf{A}^\top\mathbf{A}$ *is invertible if and only if* $rank(\mathbf{A}) = n$.

Proof Using Theorem 9.5.4 for matrices $\mathbf{A}$ and $\mathbf{A}^\top\mathbf{A}$ (both have the same number of cols n), we have

$$\text{rank}(\mathbf{A}) + \text{nullity}(\mathbf{A}) = n, \quad \text{rank}(\mathbf{A}^\top\mathbf{A}) + \text{nullity}(\mathbf{A}^\top\mathbf{A}) = n$$

Thus, we only need to show that $\text{nullity}(\mathbf{A}) = \text{nullity}(\mathbf{A}^\top\mathbf{A})$. In other words, we have to show that if $\boldsymbol{x}$ is a solution to $\mathbf{A}\boldsymbol{x} = \mathbf{0}$, then it is also a solution to $\mathbf{A}^\top\mathbf{A}\boldsymbol{x} = \mathbf{0}$ and vice versa. I present only the way from $\mathbf{A}^\top\mathbf{A}\boldsymbol{x} = \mathbf{0}$ to $\mathbf{A}\boldsymbol{x} = \mathbf{0}$:

$$\boldsymbol{x}^\top\mathbf{A}^\top\mathbf{A}\boldsymbol{x} = 0 \Longleftrightarrow (\mathbf{A}\boldsymbol{x}) \cdot (\mathbf{A}\boldsymbol{x}) = 0 \Longrightarrow \mathbf{A}\boldsymbol{x} = 0$$

The last step is due to the property of the dot product, see Box 9.2, property (d). ■

Expansion in a basis or coordinates. Let S be a subspace of $\mathbb{R}^n$ and let $\mathcal{B} = \{\boldsymbol{u}_1, \ldots, \boldsymbol{u}_k\}$ be a basis for S, then *any vector* $\boldsymbol{b} \in S$ can be written as a linear combination of the basis in a unique way.

Proof The proof is as follows (where we write two linear combinations for $\boldsymbol{b}$ and subtract them and use the definition of linear independent vectors to show that the two set of coefficients are identical):

$$\begin{aligned} \boldsymbol{b} &= \alpha_1\boldsymbol{u}_1 + \alpha_2\boldsymbol{u}_2 + \cdots + \alpha_k\boldsymbol{u}_k \\ \boldsymbol{b} &= \beta_1\boldsymbol{u}_1 + \beta_2\boldsymbol{u}_2 + \cdots + \beta_k\boldsymbol{u}_k \\ \Longrightarrow \mathbf{0} &= (\alpha_1 - \beta_1)\boldsymbol{u}_1 + (\alpha_2 - \beta_2)\boldsymbol{u}_2 + \cdots + (\alpha_k - \beta_k)\boldsymbol{u}_k \end{aligned}$$

As $\boldsymbol{u}_1, \ldots, \boldsymbol{u}_k$ are linear independent, it must follow that $\beta_i = \alpha_i$ for $i = 1, 2, \ldots, k$. This is one common way to prove something is unique: we assume this something can be written in two ways and prove that two ways are identical. ■

If S is a subspace of $\mathbb{R}^n$ and $\mathcal{B} = \{\boldsymbol{u}_1, \ldots, \boldsymbol{u}_k\}$ a basis for S, then $\boldsymbol{b} \in S$ can be written as $c_i \boldsymbol{u}_i$. There is a special name to these c_i's. So, we have the following definition for them.

Definition 9.5.6 Let S be a subspace of $\mathbb{R}^n$ and let $\mathcal{B} = \{\boldsymbol{u}_1, \ldots, \boldsymbol{u}_k\}$ be a basis for S, and $\boldsymbol{b} \in S$ is a vector in S. We can then write $\boldsymbol{b} = c_i \boldsymbol{u}_i$. Then, $(c_1, c_2, \ldots, c_k)$ are called the coordinates[32] of $\boldsymbol{b}$ with respect to $\mathcal{B}$. And the vector making of c's is called the coordinate vector of $\boldsymbol{b}$ with respect to $\mathcal{B}$.

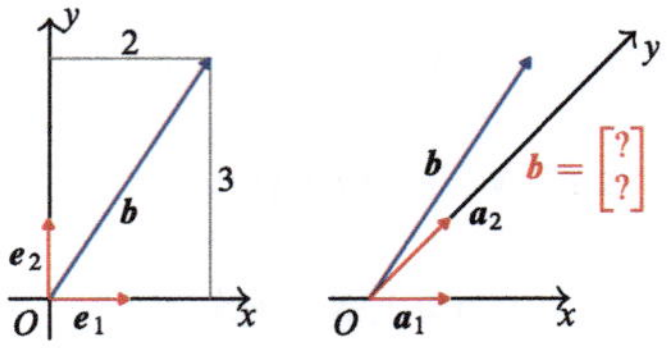

Let's demonstrate the fact that the same vector will have different coordinates in different bases. For example, in 2D, consider two bases: the first one is the traditional one that Descartes gave us: $\boldsymbol{e}_1 = (1, 0)$ and $\boldsymbol{e}_2 = (0, 1)$. The second basis is $\boldsymbol{a}_1 = (1, 0)$ and $\boldsymbol{a}_2 = (1, 1)$. Now, consider a fixed point $\boldsymbol{p} = (2, 3)$ in the first base. In the second basis, we write

$$\begin{bmatrix} 2 \\ 3 \end{bmatrix} = (-1) \begin{bmatrix} 1 \\ 0 \end{bmatrix} + (3) \begin{bmatrix} 1 \\ 1 \end{bmatrix}$$

That is, in the second basis, the coordinates of $\boldsymbol{p}$ are $(-1, 3)$. How did we find out the coordinates? We had to solve the following system of equations:

$$\alpha \begin{bmatrix} 1 \\ 0 \end{bmatrix} + \beta \begin{bmatrix} 1 \\ 1 \end{bmatrix} = \begin{bmatrix} 2 \\ 3 \end{bmatrix}$$

which is easy but nevertheless taking time. Imagine that if our space is $\mathbb{R}^n$, then we would need to solve a system of n linear equations for n unknowns. A time-consuming part! Why things are easy for $\boldsymbol{e}_1 = (1, 0)$ and $\boldsymbol{e}_2 = (0, 1)$? They are orthogonal to each other. We shall discuss orthogonal vectors in Sect. 9.8.

9.6 Introduction to Linear Transformation

Let's consider this function of one variable $y = f(x) = ax$ where $a, x \in \mathbb{R}$. This function has the following two properties

$$\begin{aligned} \text{linearity:} \quad & f(x_1 + x_2) = f(x_1) + f(x_2) \\ \text{homogeneity:} \quad & f(\alpha x_1) = \alpha f(x_1) \end{aligned}$$

[32] Some authors use expansion coefficients instead of coordinates.

which also means that $f(\alpha x_1 + \beta x_2) = \alpha f(x_1) + \beta f(x_2)$. The function $y = g(x) = ax + b$, albeit also a linear function, does not satisfy these two properties: it is not a linear function. But $y = g(x) = ax + b$ is an *affine function*.

Any function possesses the linearity property of $f(\alpha x_1 + \beta x_2) = \alpha f(x_1) + \beta f(x_2)$ is called a linear function. And there exists lots of such functions. But we need to generalize our concept of function. A function $f : \mathcal{D} \to \mathcal{R}$ maps an object of D to an object of $\mathcal{R}$. By objects, we mean anything: a number x, a point in 3D space $\boldsymbol{x} = (x, y, z)$, a point in a n-dimensional space, a function, a matrix, etc.

Of course linear algebra studies vectors and functions that take a vector and return another vector. However, a new term is used: instead of functions, mathematicians use transformations. For a vector $\boldsymbol{u} \in \mathbb{R}^n$, a transformation T turns it into a new vector $\boldsymbol{v} \in \mathbb{R}^m$. For example, we can define $T : \mathbb{R}^2 \to \mathbb{R}^3$ as

$$T\left(\begin{bmatrix} x_1 \\ x_2 \end{bmatrix}\right) = \begin{bmatrix} x_1 + x_2 \\ x_1 - x_2 \\ x_1 x_2 \end{bmatrix}$$

However, among many types of transformation, linear algebra focuses on one special transformation: linear transformation. This is similar to ordinary calculus focus on functions that are differentiable.

Definition 9.6.1 A linear transformation is the transformation $T : \mathbb{R}^n \to \mathbb{R}^m$ satisfying the following two properties:

$$\begin{aligned} \text{linearity:} \quad & T(\boldsymbol{u} + \boldsymbol{v}) = T(\boldsymbol{u}) + T(\boldsymbol{v}) \\ \text{homogeneity:} \quad & T(\alpha \boldsymbol{u}) = \alpha T(\boldsymbol{u}) \end{aligned}$$

for all $\boldsymbol{u}, \boldsymbol{v} \in \mathbb{R}^n$ and $\alpha \in \mathbb{R}$. The domain of T is $\mathbb{R}^n$ and the codomain of T is $\mathbb{R}^m$. For a vector $\boldsymbol{v}$ in the domain of T, the vector $T(\boldsymbol{v})$ in the codomain is called the image of $\boldsymbol{v}$ under the action of T. The set of all possible images $T(\boldsymbol{v})$ is called the range of T.

For abstract concepts (concepts for objects do not exist in real life) we need to think about some examples to understand more about them. So, in what follows we present some linear transformations.

Some 2D linear transformations. Figure 9.15 shows a shear transformation. The equation for a 2D shear transformation is

$$T\left(\begin{bmatrix} x_1 \\ x_2 \end{bmatrix}\right) = \begin{bmatrix} x_1 + \lambda x_2 \\ x_2 \end{bmatrix} \tag{9.6.1}$$

If we apply this transformation to the two unit vectors $\boldsymbol{i}$ and $\boldsymbol{j}$, $\boldsymbol{i}$ is not affected but $\boldsymbol{j}$ is sheared to the right ($\lambda = 1$ in the figure). So the unit square made by $\boldsymbol{i}$ and $\boldsymbol{j}$ was transformed to a parallelogram.

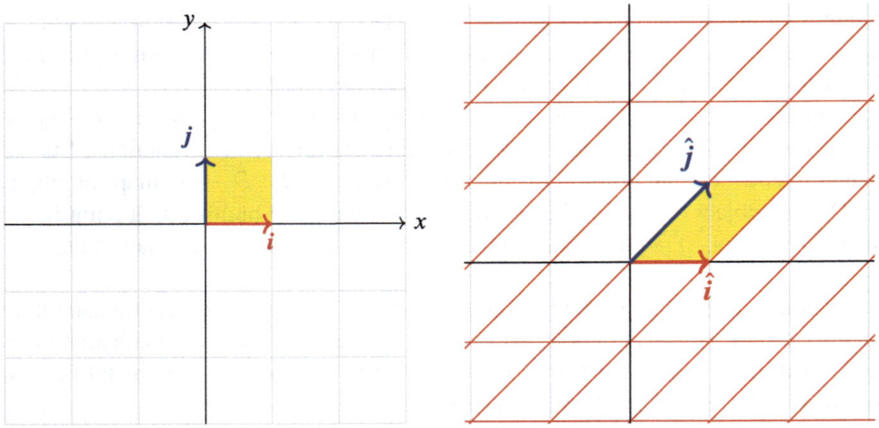

(a) A square before transformation (b) A parallelogram after transformation

Fig. 9.15 Shear transformation is a linear transformation from plane to plane. Side note: a shear transformation does not change the area. That's why a parallelogram has the same area as the rectangle of same base and height

In Fig. 9.15, we applied the transformation T to all the grid lines of the 2D space. You can see that grid lines (which are gray lines in Fig. 9.15a) are transformed to lines (red lines in Fig. 9.15b), the origin is kept fixed and equally spaced points transformed to equally spaced points. These are the consequence of the following properties of any linear transformation.

Let $T : \mathbb{R}^n \to \mathbb{R}^m$ be a linear transformation, then

(a) $T(\mathbf{0}) = \mathbf{0}$.[33]
(b) For a set of vectors $\boldsymbol{v}_1, \boldsymbol{v}_2, \ldots, \boldsymbol{v}_k$ and set of scalars $c_1, c_2, \ldots, c_k$, we have[34]

$$T(c_1\boldsymbol{v}_1 + c_2\boldsymbol{v}_2 + \cdots + c_k\boldsymbol{v}_k) = c_1T(\boldsymbol{v}_1) + c_2T(\boldsymbol{v}_2) + \cdots + c_kT(\boldsymbol{v}_k)$$

The second property is the mathematical expression of the fact that *linear transformations preserve linear combinations*. For example, if $\boldsymbol{v}$ is a certain linear combination of other vectors $\boldsymbol{s}$, $\boldsymbol{t}$, and $\boldsymbol{u}$, say $\boldsymbol{v} = 3\boldsymbol{s} + 5\boldsymbol{t} - 2\boldsymbol{u}$, then $T(\boldsymbol{v})$ is the same linear combination of the images of those vectors, that is, $T(\boldsymbol{v}) = 3T(\boldsymbol{s}) + 5T(\boldsymbol{t}) - 2T(\boldsymbol{u})$.

The standard matrix associated with a linear transformation. Consider again the linear transformation in Eq. (9.6.1). Now, we choose three vectors: the first two are very specials—they are the unit vectors $\boldsymbol{e}_1 = (1, 0)$ and $\boldsymbol{e}_2 = (0, 1)$; the third vector is arbitrary $\boldsymbol{a} = (1, 2)$. After the transformation T, we get three new vectors:

$$T(\boldsymbol{e}_1) = (1, 0), \quad T(\boldsymbol{e}_2) = (1, 1), \quad T(\boldsymbol{a}) = (3, 2)$$

[33] Proof: $T(\boldsymbol{v}) = T(\mathbf{0} + \boldsymbol{v}) = T(\mathbf{0}) + T(\boldsymbol{v})$.

[34] Proof for $k = 2$: $T(c_1\boldsymbol{v}_1 + c_2\boldsymbol{v}_2) = T(c_1\boldsymbol{v}_1) + T(c_2\boldsymbol{v}_2) = c_1T(\boldsymbol{v}_1) + c_2T(\boldsymbol{v}_2)$.

As $\boldsymbol{a} = \boldsymbol{e}_1 + 2\boldsymbol{e}_2$ and a linear transformation preserves the linear combination, we have

$$T(\boldsymbol{a}) = 1T(\boldsymbol{e}_1) + 2T(\boldsymbol{e}_2)$$

Knowing matrix-vector multiplication as a linear combination of some columns, we can write $T(\boldsymbol{a})$ as a matrix-vector multiplication:

$$T(\boldsymbol{a}) = \begin{bmatrix} 1 & 1 \\ 0 & 1 \end{bmatrix} \begin{bmatrix} 1 \\ 2 \end{bmatrix}$$

Of course carrying out this matrix-vector multiplication will give us the same result as of direct use of Eq. (9.6.1). It is even slower. Why bother then? Because, a linear transformation $T : \mathbb{R}^n \to \mathbb{R}^m$ determines an $m \times n$ matrix $\mathbf{A}$, and conversely an $m \times n$ matrix $\mathbf{A}$ determines a linear transformation $T : \mathbb{R}^n \to \mathbb{R}^m$. This is important as from now on when we see $\mathbf{A}\boldsymbol{x} = \boldsymbol{b}$, we do not see a bunch of meaningless numbers, but we see it as a linear transformation that $\mathbf{A}$ acts on $\boldsymbol{x}$ to bring it to $\boldsymbol{b}$.

We now just need to generalize what we have done. Let's consider a linear transformation $T : \mathbb{R}^n \to \mathbb{R}^m$. Now, for a vector $\boldsymbol{u} = (u_1, u_2, \ldots, u_n)$ in $\mathbb{R}^n$, we can always write $\boldsymbol{u}$ as a linear combination of the *standard* basis vectors $\boldsymbol{e}_i$ (we can use a different basis, but that leads to a different matrix):

$$\boldsymbol{u} = u_1\boldsymbol{e}_1 + u_2\boldsymbol{e}_2 + \cdots + u_n\boldsymbol{e}_n$$

So, the linear transformation applied to $\boldsymbol{u}$ can be written as

$$T(\boldsymbol{u}) = T(u_1\boldsymbol{e}_1 + u_2\boldsymbol{e}_2 + \cdots + u_n\boldsymbol{e}_n) = u_1T(\boldsymbol{e}_1) + u_2T(\boldsymbol{e}_2) + \cdots + u_nT(\boldsymbol{e}_n) \tag{9.6.2}$$

which indicates that the transformed vector $T(\boldsymbol{u})$ is a linear combination of the transformed basis vectors, i.e., $T(\boldsymbol{e}_i)$, in which the coefficients are the coordinates of the vector. In other words, if we know where the basis vectors land after the transformation, we can determine where any vector $\boldsymbol{u}$ lands in the transformed space.

Now, assume that the n basis vectors in $\mathbb{R}^n$ are transformed to n vectors in $\mathbb{R}^m$ with coordinates (implicitly assumed that the standard basis for $\mathbb{R}^m$ was used)

$$\begin{aligned} T(\boldsymbol{e}_1) &= (a_{11}, a_{21}, \ldots, a_{m1}) \\ T(\boldsymbol{e}_2) &= (a_{12}, a_{22}, \ldots, a_{m2}) \\ \ldots &= \ldots \\ T(\boldsymbol{e}_n) &= (a_{1n}, a_{2n}, \ldots, a_{mn}) \end{aligned}$$

So we can characterize a linear transformation by storing $T(\boldsymbol{e}_i)$, $i = 1, 2, \ldots, n$ in an $m \times n$ matrix like this

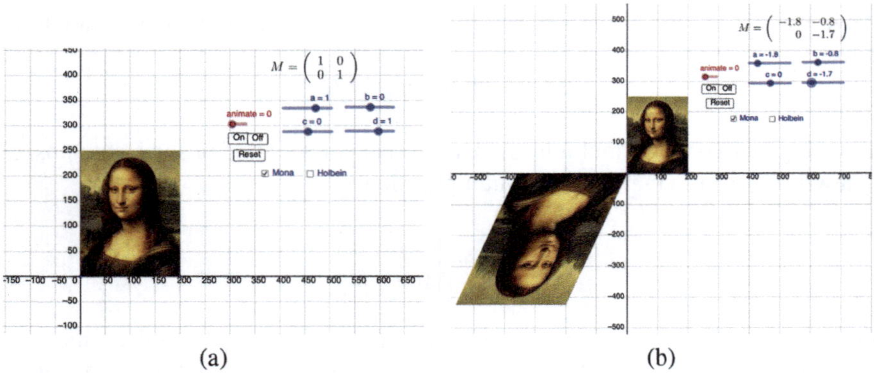

Fig. 9.16 Transformation in a plane: `geogebra` applet

$$\mathbf{A} := \begin{bmatrix} | & | & | & | \\ T(\boldsymbol{e}_1) & T(\boldsymbol{e}_2) & \cdots & T(\boldsymbol{e}_n) \\ | & | & | & | \end{bmatrix} = \begin{bmatrix} a_{11} & a_{12} & \cdots & a_{1n} \\ a_{21} & a_{22} & \cdots & a_{2n} \\ \vdots & \vdots & \ddots & a_{1n} \\ a_{m1} & a_{m2} & \cdots & a_{mn} \end{bmatrix} \tag{9.6.3}$$

That is, each column of this matrix is $T(\boldsymbol{e}_i)$, which is a vector of length m. This matrix is called the standard matrix representing the linear transformation T. Why standard? Because we have used one standard basis for $\mathbb{R}^n$ and another standard basis for $\mathbb{R}^m$.

With this introduction of **A**, the linear transformation in Sect. 9.6.2 can be rewritten as a matrix-vector product:

$$T(\boldsymbol{u}) := \mathbf{A}\boldsymbol{u} \tag{9.6.4}$$

A visual way to understand linear transformations is to use a `geogebra` applet[35] and play with it. In Fig. 9.16, we present some transformations of a small image of Mona Lisa. By changing the transformation matrix M, we can see the effect of the transformation immediately.

Proof Now we can prove that a linear transformation maps a line to a line. Considering a line l described by $\boldsymbol{p} + t\boldsymbol{d}$ for $t \in \mathbb{R}$. Then, for every $t \in \mathbb{R}$, we have

$$T(\boldsymbol{p} + t\boldsymbol{d}) = \mathbf{A}(\boldsymbol{p} + t\boldsymbol{d}) = \mathbf{A}\boldsymbol{p} + t\mathbf{A}\boldsymbol{d} = \boldsymbol{p}_1 + t\boldsymbol{d}_1$$

Hence $T(l) = l_1$, where $l_1 : \boldsymbol{p}_1 + t\boldsymbol{d}_1$ is obviously a line. Using similar arguments, it can be shown that a linear transformation preserves the property of parallelism among lines and line segments (i.e., parallel lines are mapped to parallel lines).

To prove equally spaced points are mapped to equally spaced points, we consider points A and B such that A is specified by vector $\boldsymbol{u}$ and B by $\boldsymbol{v}$, and $\boldsymbol{v} = 2\boldsymbol{u}$ (or

[35] It can be found easily using Google https://www.geogebra.org/m/pDU4peV5.

$OA = AB$). After a linear transformation, A becomes A' and B becomes B' specified by vectors $\mathbf{A}\boldsymbol{u}$ and $\mathbf{A}(2\boldsymbol{u}) = 2(\mathbf{A}\boldsymbol{u})$. Hence, $OA' = A'B'$. ■

Determinants. While playing with the `geogebra` applet we can see that sometimes a transformation enlarges the image and sometimes it shrinks the image. Can we quantify this effect of a linear transformation? Let's do it, but in a plane only. We consider a general transformation matrix

$$\mathbf{A} = \begin{bmatrix} a & b \\ c & d \end{bmatrix}$$

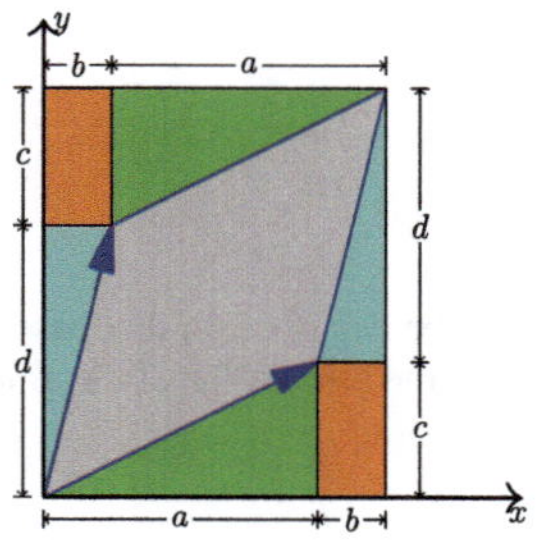

which tells us that the unit vector $\boldsymbol{i}$ is now at (a, c) and $\boldsymbol{j}$ is at (b, d). We are going to compute the area of the parallelogram made up by these two vectors. This parallelogram is what the unit square (which has an area of 1) has been transformed to. Based on the next figure, this area is $ad - bc$: the area of the big square minus the total area of those simple shapes. So, any unit square in the plane is transformed to a parallelogram with an area of $ad - bc$. What about a square of 2×2? It is transformed to a parallelogram of area $4(ad - bc)$. So, $ad - bc$ is the scaling of the transformation. But how about a curvy domain in the plane? Is it still true that its area is scaled up by the same amount? We're in linear algebra, but do not forget calculus! The area of any shape is equal to the sum of the area of infinitely many unit squares, and each small unit square is scaled by $ad - bc$ and thus the area of any shape is scaled by $ad - bc$. Mathematicians call this scaling *the determinant of the transformation matrix*. They use either $\det\mathbf{A}$ or $|\mathbf{A}|$ to denote the determinant of matrix $\mathbf{A}$.

Looking at the determinant (i.e., $ad - bc$) of a 2×2 matrix it can be suspected that this determinant can be positive or negative. See Fig. 9.17 for an illustration of what a negative determinant means.

It is obvious that the next move is to repeat the same analysis but in 3D. The three unit vectors make a cube of which volume is one, see Fig. 9.18a. After the linear transformation, these vectors are transformed to $\boldsymbol{a} = (a, d, g)$, $\boldsymbol{b} = (b, e, h)$, and $\boldsymbol{c} = (c, f, i)$. Thus, we consider the following 3×3 matrix:

$$\mathbf{A} = \begin{bmatrix} a & b & c \\ d & e & f \\ g & h & i \end{bmatrix}$$

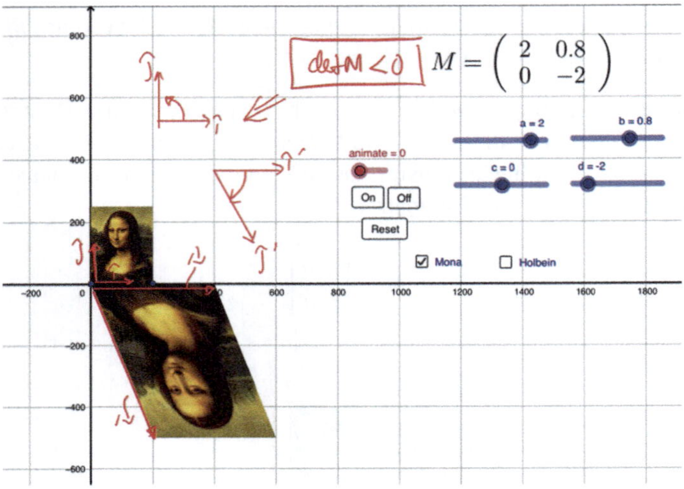

Fig. 9.17 Determinant of a matrix can be negative. In that case the linear transformation flips the space or changes the orientation. Look at the orientation of the unit vectors before the transformation and after

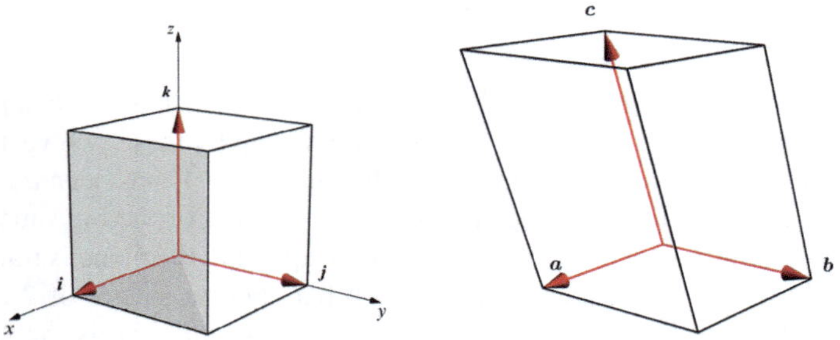

(a) before transformation (b) after transformation

Fig. 9.18 The determinant of a 3×3 matrix is the volume of the parallelepiped formed by the columns of the matrix

which is the matrix of a 3D linear transformation. The determinant of this matrix is the volume of the parallelepiped formed by these three vectors (Fig. 9.18b). We know how to compute such a volume using the scalar triple product in Sect. 9.1.5 (particularly Fig. 9.12b):

$$\text{volume} = \boldsymbol{c} \cdot (\boldsymbol{a} \times \boldsymbol{b}) = (aei + dhc + gbf) - (ceg + ahf + dbi)$$

and this is the determinant of our 3×3 matrix **A**.

That's the most we can do about determinant using geometry. We cannot find out the formula for the determinant of a 4×4 matrix. How did mathematicians proceed then? We refer to Sect. 9.9 for more on the determinant of a square matrix.

Matrix-matrix product as composition of transformations. Still remember function composition, like $\sin(x^3)$, which is composed of two functions: $y = \sin t$ and $t = x^3$? Now we apply the same concept but our functions are linear transformations. And doing it will reveal the rule for matrix-matrix multiplication.

Assume we have a linear transformation $T : \mathbb{R}^n \to \mathbb{R}^m$ and a second linear transformation $S : \mathbb{R}^m \to \mathbb{R}^p$. From the previous sub-section, we know that there exists a matrix $\mathbf{A}$, of size $m \times n$, associated with the transformation T and another matrix $\mathbf{B}$ (of size $p \times m$) associated with S. Now, consider a composite transformation of first applying $T(\boldsymbol{u})$ and second applying S to the outcome of the first transformation. Mathematically, we write $(S \circ T)(\boldsymbol{u}) = S(T(\boldsymbol{u}))$ which transform $\boldsymbol{u} \in \mathbb{R}^n$ to $\mathbb{R}^p$.

Assume that there exists a matrix $\mathbf{C}$[36] associated with $(S \circ T)(\boldsymbol{u})$. Then the jth column of $\mathbf{C}$ is $(S \circ T)(\boldsymbol{e}_j)$:

$$\mathbf{C}[:, j] = (S \circ T)(\boldsymbol{e}_j) = S(T(\boldsymbol{e}_j)) = S(\mathbf{A}[:, j]) = \mathbf{BA}[:, j]$$

Therefore,

$$\boxed{\mathbf{BA} = \begin{bmatrix}\mathbf{BA}_1 & \mathbf{BA}_2 & \cdots & \mathbf{BA}_n\end{bmatrix}} \tag{9.6.5}$$

So, if we denote by $\mathbf{C} = \mathbf{BA}$ the matrix-matrix multiplication, then the jth column of $\mathbf{C}$ is the product of matrix $\mathbf{B}$ and the jth column of $\mathbf{A}$. Using Eq. (9.4.4), we can write the entry C_{ij} as $C_{ij} = \sum_{k=1}^{m} B_{ik} A_{kj}$ which is the familiar matrix-matrix multiplication rule: the entry at row i and column j of $\mathbf{BA}$ is the dot product of row i of $\mathbf{B}$ and column j of $\mathbf{A}$.

We have $ab = ba$ for $a, b \in \mathbb{R}$. Do we have the same for matrix-matrix product: $\mathbf{AB} = \mathbf{BA}$ (of course if their sizes are consistent for matrix-matrix multiplication). The answer is no. And the proof is simple: $f(g(x)) \neq g(f(x))$, for example, $\sin(x^3) \neq \sin^3 x$.

How about $\mathbf{ABC}$? From function composition discussed in Sect. 4.2.3, we know that it is associative, so $(\mathbf{AB})\mathbf{C} = \mathbf{A}(\mathbf{BC})$. This is a nice proof, much better than the proof that is based on the definition of matrix-matrix multiplication (you can try it to see my point).

With the geometric meaning of determinant and matrix-matrix product, it is easy to see that the determinant of the product of two matrices is the product of the determinants of each matrix:

$$|\mathbf{AB}| = |\mathbf{A}||\mathbf{B}| \tag{9.6.6}$$

This is because $\mathbf{AB}$ is associated with first a linear transformation which area scaling of $|\mathbf{B}|$, followed by another transformation which area scaling of $|\mathbf{A}|$. Thus, in total, the area scaling should be $|\mathbf{A}||\mathbf{B}|$.

---

[36] We can see this by $(S \circ T)(\boldsymbol{u}) = S(\mathbf{A}\boldsymbol{u}) = \mathbf{B}(\mathbf{A}\boldsymbol{u}) = (\mathbf{BA})\boldsymbol{u}$.

9.7 Linear Algebra with `Julia`

Let the computer do the computation so that we can focus on the theory of linear algebra. This would not serve you well in exams, but I am not fan of exams. To serve this purpose, in Listing 9.1 I summarize some common matrix operations using `Julia`.

Listing 9.1 Basic linear algebra in `Julia`.

```
using LinearAlgebra                  # you have to install this package first
using RowEchelon
A   = [2 4 -2;4 9 -3;-2 -3 7]        # create a 3x3 matrix
b   = [2,8,10]                       # create a vector of length 3
x   = A\b                            # solving Ax=b
E21 = [1 0 0;-2 0 1;0 0 1]
E21*A                                # matrix matrix multiplication
detA = det(A)                        # determinant of A
invE = inv(E)                        # inverse of E
rref(A)                              # get reduced row echelon form
A'                                   # get transpose of A
x     = A[1,:]                       # get 1st row
y     = A[:,1]                       # get 1st col
AA    = zeros(3,3)                   # 3x3 zero matrix
BB    = ones(3,3)                    # 3x3 one matrix
dot(x,y)                             # dot product of two vecs
V     = eigvecs(A)                   # cols of matrix V  = eigenvectors of A
v     = eigvals(A)                   # vector v contains eigenvalues of A
svd(A)                               # SVD of A
norm(b, Inf)                         # max norm of b
norm(b, 1)                           # sum norm of b, or 1-norm of b, see Norms
```

9.8 Orthogonality

We begin with orthogonal vectors in Sect. 9.8.1. Orthogonality of two n-vectors is a generalization of the notion of perpendicularity of two vectors in $\mathbb{R}^3$. Orthogonal vectors are always linear independent and thus make a good base for a subspace. When these vectors are normalized, i.e., having unit lengths, they make orthonormal basis vectors (Sect. 9.8.2). Stacking orthonormal vectors column by column and we obtain an orthogonal matrix (Sect. 9.8.3).

Considering a plane in $\mathbb{R}^3$ and a vector $\boldsymbol{n}$ normal to the plane, then $\boldsymbol{n}$ is orthogonal to all vectors in the plane. If we extend this to any subspace of $\mathbb{R}^n$, we meet the concept of orthogonal complements and we discuss it in Sect. 9.8.4.

We are familiar with the orthogonal projection of a vector on another vector, and now extend this to the projection of a vector on a subspace (Sect. 9.8.5). The Gram-Schmidt algorithm takes a set of linear independent vectors $\boldsymbol{v}_1, \boldsymbol{v}_2, \ldots, \boldsymbol{v}_k$ and generates an orthogonal linear independent set of vectors $\boldsymbol{u}_1, \boldsymbol{u}_2, \ldots, \boldsymbol{u}_k$ (Sect. 9.8.6).

The Gauss elimination process of $\mathbf{A}\boldsymbol{x} = \boldsymbol{b}$ results in the LU factorization: $\mathbf{A} = \mathbf{LU}$. Now, the Gram-Schmidt orthogonalization process applied to the linear independent columns of a matrix $\mathbf{A}$ results in another factorization—known as the QR factorization: $\mathbf{A} = \mathbf{QR}$ (Sect. 9.8.7).

9.8.1 Orthogonal Vectors and Orthogonal Bases

We know that two vectors in $\mathbb{R}^2$ or $\mathbb{R}^3$, $\boldsymbol{a}$ and $\boldsymbol{b}$, are called orthogonal when their dot product is zero, i.e., $\boldsymbol{a} \cdot \boldsymbol{b} = 0$. We now extend this to vectors in $\mathbb{R}^n$. So, vectors $\boldsymbol{x}, \boldsymbol{y}$ in $\mathbb{R}^n$ are said to be orthogonal (denoted by $\boldsymbol{x} \perp \boldsymbol{y}$) if $\boldsymbol{x} \cdot \boldsymbol{y} = 0$ or $\boldsymbol{x}^\top \boldsymbol{y} = 0$. Usually, we are interested in a bunch of vectors that are orthogonal to each other as the following definition.

Definition 9.8.1 A set of vectors $\boldsymbol{a}_1, \ldots, \boldsymbol{a}_k$ in $\mathbb{R}^n$ is an orthogonal set if all pairs of distinct vectors in the set are orthogonal. That is, if

$$\boldsymbol{a}_i \cdot \boldsymbol{a}_j = 0 \text{ for any} i, j \text{with} i \neq j, \quad i, j = 1, 2, \ldots, k$$

The most famous example of an orthogonal set of vectors is the standard basis $\{\boldsymbol{e}_1, \boldsymbol{e}_2, \ldots, \boldsymbol{e}_n\}$ of $\mathbb{R}^n$. And we know that these basic vectors are linear independent. Therefore, we guess that orthogonal vectors are linear independent. And that guess is correct as stated by the following theorem.

Theorem 9.8.1 (Orthogonality-Independence) *Given a set of non-zero orthogonal vectors $\boldsymbol{a}_1, \ldots, \boldsymbol{a}_k$ in $\mathbb{R}^n$, then they are linear independent.*

Proof The idea is to assume that there is a zero vector expressed as a linear combination of these orthogonal vectors. Then take the dot product of two sides with $\boldsymbol{a}_i$ and use the orthogonality to obtain $\alpha_i = 0$ for $i = 1, 2, \ldots$[37]:

$$\begin{aligned}
&\alpha_1\boldsymbol{a}_1 + \alpha_2\boldsymbol{a}_2 + \cdots + \alpha_k\boldsymbol{a}_k = \boldsymbol{0} \\
\Longrightarrow\ &\boldsymbol{a}_i \cdot (\alpha_1\boldsymbol{a}_1 + \alpha_2\boldsymbol{a}_2 + \cdots + \alpha_k\boldsymbol{a}_k) = 0 \\
\Longrightarrow\ &\alpha_i(\boldsymbol{a}_i \cdot \boldsymbol{a}_i) = 0 \\
\Longrightarrow\ &\alpha_i = 0
\end{aligned} \tag{9.8.1}$$

■

Example 9.7 Considering these three vectors in $\mathbb{R}^3$: $\boldsymbol{v}_1 = (2, 1, -1)$, $\boldsymbol{v}_2 = (0, 1, 1)$, and $\boldsymbol{v}_3 = (1, -1, 1)$. We can see that: (i) they form an orthogonal set of vectors, then

[37] If the last step was not clear, just use a specific $\boldsymbol{a}_1$, and assuming there are only three vectors $\boldsymbol{a}_1, \boldsymbol{a}_2, \boldsymbol{a}_3$. Then, the LHS of the second line in Eq. (9.8.1) is: $\boldsymbol{a}_1 \cdot (\alpha_1\boldsymbol{a}_1 + \alpha_2\boldsymbol{a}_2 + \alpha_3\boldsymbol{a}_3)$, which is $\alpha_1\boldsymbol{a}_1 \cdot \boldsymbol{a}_1 + \alpha_2\boldsymbol{a}_1 \cdot \boldsymbol{a}_2 + \alpha_3\boldsymbol{a}_1 \cdot \boldsymbol{a}_3 = \alpha_1||\boldsymbol{a}_1|| + 0 + 0$. And thus, we get $\alpha_1 = 0$. Similarly, we get $\alpha_2 = 0$ if we started with $\boldsymbol{a}_2$ and so on.

(ii) from Theorem 9.8.1, they are linear independent, then (iii) three independent vectors in $\mathbb{R}^3$ form a basis for $\mathbb{R}^3$. If these vectors form a basis, then we can find the coordinates of any vector in $\mathbb{R}^3$ with respect to this basis. Find the coords of $\boldsymbol{v} = (1, 2, 3)$.

We simply have to solve the following system to find the coordinates of $\boldsymbol{v} = (1, 2, 3)$:

$$\begin{bmatrix} 2 & 0 & 1 \\ 1 & 1 & -1 \\ -1 & 1 & 1 \end{bmatrix} \begin{bmatrix} c_1 \\ c_2 \\ c_3 \end{bmatrix} = \begin{bmatrix} 1 \\ 2 \\ 3 \end{bmatrix} \Longrightarrow \boldsymbol{c} = \begin{bmatrix} 1/6 \\ 5/2 \\ 2/3 \end{bmatrix}$$

Solving a 3×3 system is not hard, but what if the question is for a vector in $\mathbb{R}^{100}$? Is there any better way? The answer is yes, and thus orthogonal bases are very nice to work with. We need to define first what an orthogonal basis is.

Definition 9.8.2 An orthogonal basis for a subspace S of $\mathbb{R}^n$ is a basis of S that is an orthogonal set.

Now, we are going to find out the coordinates of $\boldsymbol{v} = (1, 2, 3)$, considered in Example 9.7, using an easier way. We write $\boldsymbol{v}$ in terms of the orthogonal basis vectors $(\boldsymbol{v}_1, \boldsymbol{v}_2, \boldsymbol{v}_3)$, and we take the dot product of both sides with $\boldsymbol{v}_1$, it is then due to the orthogonality that all terms vanish, and we're left with c_1:

$$\begin{aligned} \boldsymbol{v} &= c_1\boldsymbol{v}_1 + c_2\boldsymbol{v}_2 + c_3\boldsymbol{v}_3 \\ \Longrightarrow \boldsymbol{v} \cdot \boldsymbol{v}_1 &= (c_1\boldsymbol{v}_1 + c_2\boldsymbol{v}_2 + c_3\boldsymbol{v}_3) \cdot \boldsymbol{v}_1 \\ \Longrightarrow \boldsymbol{v} \cdot \boldsymbol{v}_1 &= c_1(\boldsymbol{v}_1 \cdot \boldsymbol{v}_1) \Longrightarrow \boxed{c_1 = \frac{\boldsymbol{v} \cdot \boldsymbol{v}_1}{\boldsymbol{v}_1 \cdot \boldsymbol{v}_1} = \frac{\boldsymbol{v} \cdot \boldsymbol{v}_1}{\|\boldsymbol{v}_1\|^2}} \end{aligned}$$

What does this formula tell us? That is, to find c_1, just compute two dot products: one of $\boldsymbol{v}$ with the first basis vector, and the other is the squared length of this basis vector. The ratio of these two products is c_1.

Nothing can be simpler. Wait, I wish we did not have to do the division with the squared length of $\boldsymbol{v}_1$. It is possible if that vector has a unit length. And we know that we can always make a non-unit vector a unit vector simply by dividing it by its length, a process known as normalizing a vector, see Eq. (9.1.7). Thus, we now move from orthogonal bases to orthonormal bases.

9.8.2 Orthonormal Vectors and Orthonormal Bases

First, a definition is given for orthonormal vectors and orthonormal basis.

Definition 9.8.3 A set of vectors in $\mathbb{R}^n$ is an orthonormal set if it is an orthogonal set of unit vectors. An orthonormal basis for a subspace S of $\mathbb{R}^n$ is a basis of S that is an orthonormal set.

If a collection of vectors $\boldsymbol{a}_1, \ldots, \boldsymbol{a}_k$ is mutually orthogonal and $||\boldsymbol{a}_i|| = 1$, $i = 1, 2, \ldots, k$ (i.e., having unit length), then it is *orthonormal*. We can combine the two conditions of orthogonality and normality to have:

$$\text{Orthonormal vectors } \boldsymbol{a}_1, \boldsymbol{a}_2, \ldots: \quad \boldsymbol{a}_i \cdot \boldsymbol{a}_j = \begin{cases} 1 & i = j \\ 0 & i \neq j \end{cases} = \delta_{ij}$$

where we have introduced the Kronecker delta notation (named after Leopold Kronecker[38]) δ_{ij}.

A vector $\boldsymbol{b}$ in a subspace S with an orthonormal basis $\boldsymbol{v}_1, \boldsymbol{v}_2, \ldots, \boldsymbol{v}_k$ has coordinates with respect to the basis given by

$$\boldsymbol{b} = \alpha_1 \boldsymbol{v}_1 + \alpha_2 \boldsymbol{v}_2 + \cdots + \alpha_k \boldsymbol{v}_k, \quad \alpha_i = \boldsymbol{b} \cdot \boldsymbol{v}_i \tag{9.8.2}$$

Did we see this before? Still remember Monsieur Fourier? What he did was to write a periodic function $f(x)$ as a linear combination of the sine/cosine functions:

$$f(x) = a_0 + \sum_{n=1}^{\infty} \left(a_n \cos \frac{n\pi x}{L} + b_n \sin \frac{n\pi x}{L} \right)$$

And below is how he had obtained the coefficients in this linear combination[39]:

$$\begin{aligned} a_n &= \frac{1}{L} \int_{-L}^{L} f(x) \cos \frac{n\pi x}{L} dx \\ b_n &= \frac{1}{L} \int_{-L}^{L} f(x) \sin \frac{n\pi x}{L} dx \end{aligned}$$

What he was doing? To find the coefficients a_n, he multiplied the function $f(x)$ with $\cos {n\pi x}/{L}$ and integrated. This is similar to $\boldsymbol{b} \cdot \boldsymbol{v}_i$ and herein the basis vectors are the functions $\sin x, \cos x, \sin 2x, \cos 2x, \ldots$ They are orthonormal to each other. Of course we need to define what the dot product of two functions is. See Eq. (9.11.9) for the definition of the dot product of two functions.

We have gone a long way: two vectors in $\mathbb{R}^2$ can be orthogonal to each other, then two n-vectors can also be orthogonal to each other. We even have two functions orthogonal to each other. Why not two orthogonal matrices?

---

[38] Leopold Kronecker (7 December 1823–29 December 1891) was a German mathematician who worked on number theory, algebra, and logic. He criticized Georg Cantor's work on set theory, and was quoted by Weber (1893) as having said, "God made the integers, all else is the work of man".

[39] To be historically precise, Euler did this before Fourier, even though Euler doubted the idea of trigonometric expansion of a periodic function.

9.8.3 Orthogonal Matrices

Orthonormal vectors are special because of $\boldsymbol{v}_i \cdot \boldsymbol{v}_j = \delta_{ij}$. If we put some orthonormal vectors in a matrix, we should get a special matrix. Let's do that with three orthonormal vectors $\boldsymbol{v}_1, \boldsymbol{v}_2, \boldsymbol{v}_3 \in \mathbb{R}^3$. They make a 3×3 matrix $\mathbf{A}$. Now, consider the product $\mathbf{A}^\top\mathbf{A}$, and see what matrix we shall get:

$$\begin{bmatrix} - \boldsymbol{v}_1 - \\ - \boldsymbol{v}_2 - \\ - \boldsymbol{v}_3 - \end{bmatrix} \begin{bmatrix} | & | & | \\ \boldsymbol{v}_1 & \boldsymbol{v}_2 & \boldsymbol{v}_3 \\ | & | & | \end{bmatrix} = \begin{bmatrix} 1 & 0 & 0 \\ 0 & 1 & 0 \\ 0 & 0 & 1 \end{bmatrix}$$

We got an identity matrix, which is special. This reminds us of the inverse, and indeed we have some special matrix—a matrix of which *the inverse is equal to the transpose*:

$$\mathbf{A}^\top\mathbf{A} = \mathbf{I} \Longrightarrow \mathbf{A}^\top = \mathbf{A}^{-1}$$

And this leads to the following special matrix whose inverse is simply its transpose. The notation $\mathbf{Q}$ is reserved for such matrices.

Definition 9.8.4 An $n \times n$ matrix $\mathbf{Q}$ whose columns form an orthonormal set is called an orthogonal matrix.

We now present an example of an orthogonal matrix. Assume that we want to rotate a point P with coordinates $\boldsymbol{x}$ to P' an angle β as shown in Fig. 9.19. The coordinates of P', $\boldsymbol{x}'$, are given by

$$\boldsymbol{x}' = \mathbf{R}\boldsymbol{x}, \quad \mathbf{R} = \begin{bmatrix} \cos\beta & -\sin\beta \\ \sin\beta & +\cos\beta \end{bmatrix}$$

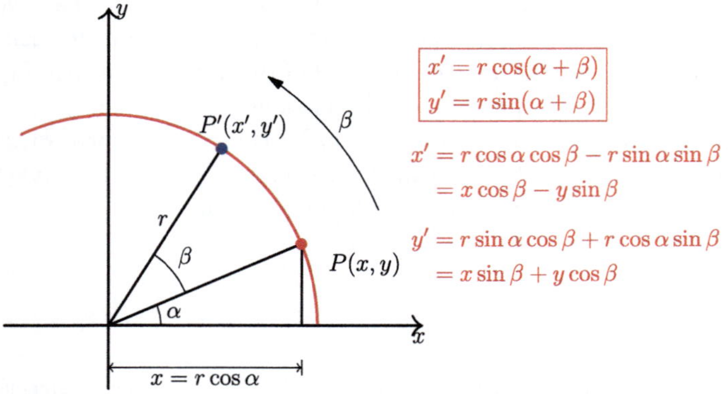

Fig. 9.19 Rotation in a plane is a matrix transformation that preserves length. The matrix of the rotation is an orthogonal matrix

It is easy to check that the columns of $\mathbf{R}$ are orthonormal vectors. Therefore, $\mathbf{R}^\top\mathbf{R} = \mathbf{I}$, which can be checked directly. We know that any *rotation preserves length* (that is, $||\boldsymbol{x}'|| = ||\boldsymbol{x}||$ or $||\mathbf{R}\boldsymbol{x}|| = ||\boldsymbol{x}||$), which is known as isometry in geometry. It turns out that every orthogonal matrix transformation is an isometry. Note also that $\det \mathbf{R} = 1$. It is not a coincidence. Indeed, from the property $\mathbf{A}^\top\mathbf{A} = \mathbf{I}$, we can deduce the determinant of $\mathbf{A}$:

$$\mathbf{A}^\top\mathbf{A} = \mathbf{I} \Longrightarrow \det(\mathbf{A}^\top\mathbf{A}) = 1 \Longrightarrow (\det(\mathbf{A}))^2 = 1 \Longrightarrow \det(\mathbf{A}) = \pm 1$$

I have used $\det(\mathbf{AB}) = \det(\mathbf{A})\det(\mathbf{B})$ and $\det(\mathbf{A}^\top) = \det(\mathbf{A})$ in the above calculations. With this special example of an orthogonal matrix (and its properties), we now have a theorem on orthogonal matrices.

Theorem 9.8.2 *Let $\mathbf{Q}$ be an $n \times n$ matrix. The following statements are equivalent:*

(a) *$\mathbf{Q}$ is orthogonal.*
(b) *$\mathbf{Q}\boldsymbol{x} \cdot \mathbf{Q}\boldsymbol{y} = \boldsymbol{x} \cdot \boldsymbol{y}$ for all $\boldsymbol{x}, \boldsymbol{y} \in \mathbb{R}^n$.*
(c) *$||\mathbf{Q}\boldsymbol{x}|| = ||\boldsymbol{x}||$ for all $\boldsymbol{x} \in \mathbb{R}^n$.*

Proof We prove (a) to (b) first:

$$\mathbf{Q}\boldsymbol{x} \cdot \mathbf{Q}\boldsymbol{y} = (\mathbf{Q}\boldsymbol{x})^\top(\mathbf{Q}\boldsymbol{y}) = (\boldsymbol{x}^\top\mathbf{Q}^\top)\mathbf{Q}\boldsymbol{y} = \boldsymbol{x}^\top(\mathbf{Q}^\top\mathbf{Q})\boldsymbol{y} = \boldsymbol{x}^\top\mathbf{I}\boldsymbol{y} = \boldsymbol{x}^\top\boldsymbol{y} = \boldsymbol{x} \cdot \boldsymbol{y}$$

Going from (b) to (c) is easy: use (b) with $\boldsymbol{y} = \boldsymbol{x}$. We need to go backwards: (c) to (b) to (a), which is left as an exercise. Check Poole's book if stuck. ■

9.8.4 Orthogonal Complements

Considering a plane in $\mathbb{R}^3$ and a vector $\boldsymbol{n}$ normal to the plane, then $\boldsymbol{n}$ is orthogonal to all vectors in the plane. We now extend this to any subspace of $\mathbb{R}^n$.

Definition 9.8.5 Let W be a subspace of $\mathbb{R}^n$.

(a) We say that a vector $\boldsymbol{v}$ is orthogonal to W if it is orthogonal to every vector in W.[40]
(b) The set of all vectors that are orthogonal to W is called the orthogonal complement of W, denoted by $W^\perp$. That is,

$$W^\perp = \{\boldsymbol{v} \in \mathbb{R}^n : \boldsymbol{v} \cdot \boldsymbol{w} = 0 \ \text{ for all } \boldsymbol{w} \in W\}$$

(c) Two subspaces S and W are said to be orthogonal, i.e., $S \perp W$ if and only if $\boldsymbol{x} \perp \boldsymbol{y}$, or, $\boldsymbol{x}^\top\boldsymbol{y} = 0$ for all $\boldsymbol{x} \in S$ and for all $\boldsymbol{y} \in W$.

[40] For a vector to be orthogonal to a subspace, it just needs to be orthogonal to the span of that subspace.

This definition actually consists of three definitions. The first one extend the idea that we discussed in the beginning of this section. Why we need $W^{\perp}$? Because it is a subspace. We know how to prove whether something is a subspace: Assume that $\boldsymbol{v}_1, \boldsymbol{v}_2 \in W^{\perp}$, and we need to show that $c_1\boldsymbol{v}_1 + c_2\boldsymbol{v}_2$ is also in $W^{\perp}$:

$$\begin{cases} \boldsymbol{v}_1 \cdot \boldsymbol{w} = 0 \\ \boldsymbol{v}_2 \cdot \boldsymbol{w} = 0 \end{cases} \Longrightarrow (c_1\boldsymbol{v}_1 + c_2\boldsymbol{v}_2) \cdot \boldsymbol{w} = 0$$

And the third definition is about orthogonality of two subspaces. We has gone a long way from orthogonality of two vectors in $\mathbb{R}^2$ to that of two vectors in $\mathbb{R}^n$, then to the orthogonality of one vector with a subspace and finally to the orthogonality of two subspaces.

Fundamental subspaces of a matrix. With the concept of orthogonal complements, we can understand more about the row space, column space, and null space of a matrix. Furthermore, we will see that there is another subspace associated with a matrix. The results are summarized in the following theorem.

Theorem 9.8.3 *Let* $\mathbf{A}$ *be an* $m \times n$ *matrix. Then the orthogonal complement of the row space of* $\mathbf{A}$ *is the null space of* $\mathbf{A}$*, and the orthogonal complement of the column space of* $\mathbf{A}$ *is the null space of* $\mathbf{A}^{\top}$*:*

$$(R(\mathbf{A}))^{\perp} = N(\mathbf{A}), \quad (C(\mathbf{A}))^{\perp} = N(\mathbf{A}^{\top})$$

The proof is straightforward. The null space of $\mathbf{A}$ is all vector $\boldsymbol{x}$ such that $\mathbf{A}\boldsymbol{x} = \mathbf{0}$, and from matrix-vector multiplication, this is equivalent to saying that $\boldsymbol{x}$ is orthogonal to the rows of $\mathbf{A}$. Now, replace $\mathbf{A}$ by its transpose, then we have the second result in the theorem above.

To conclude, an $m \times n$ matrix $\mathbf{A}$ has four subspaces, namely, $R(\mathbf{A})$, $N(A)$, $C(\mathbf{A})$, $N(\mathbf{A}^{\top})$. But they go in pairs: the first two are orthogonal complements in $\mathbb{R}^n$, and the last two are orthogonal in $\mathbb{R}^m$.

9.8.5 Orthogonal Projections

We recall the orthogonal projection of a vector $\boldsymbol{u}$ onto another vector $\boldsymbol{v}$, defined by the projection operator $\text{proj}_{\boldsymbol{v}}(\boldsymbol{u})$, check Sect. 9.1.4 if needed:

$$\text{proj}_{\boldsymbol{v}}(\boldsymbol{u}) := \frac{\boldsymbol{u} \cdot \boldsymbol{v}}{\boldsymbol{v} \cdot \boldsymbol{v}}\boldsymbol{v} \tag{9.8.3}$$

While projecting $\boldsymbol{u}$ onto vector $\boldsymbol{v}$, we also get $\text{perp}_{\boldsymbol{v}}(\boldsymbol{u}) := \boldsymbol{u} - \text{proj}_{\boldsymbol{v}}(\boldsymbol{u})$, which is orthogonal to $\boldsymbol{v}$, see Fig. 9.20a.[41] This indicates that we can decompose a vector $\boldsymbol{u}$ into two vectors,

$$\boldsymbol{u} = \text{proj}_{\boldsymbol{v}}(\boldsymbol{u}) + \text{perp}_{\boldsymbol{v}}(\boldsymbol{u})$$

of which one is in $\text{span}(\boldsymbol{v})$ and the other is in $\text{span}(\boldsymbol{v})^{\perp}$.

The next step is, of course, to project a vector $\boldsymbol{u}$ on a plane in $\mathbb{R}^3$. Suppose that this plane (of dim 2) has a basis $(\boldsymbol{i}, \boldsymbol{j})$. We can project $\boldsymbol{u}$ onto the first basis vector $\boldsymbol{i}$, then project it onto the second basis and sum the two, see Fig. 9.20b,

$$\text{proj}_{i,j}(\boldsymbol{u}) := \text{proj}_{i}(\boldsymbol{u}) + \text{proj}_{j}(\boldsymbol{u}) \tag{9.8.4}$$

Is this still an orthogonal projection? We just need to check whether $\text{proj}_{i,j}(\boldsymbol{u}) \cdot \boldsymbol{i} = 0$ and $\text{proj}_{i,j}(\boldsymbol{u}) \cdot \boldsymbol{j} = 0$. The answer is yes, and due to the fact that $\boldsymbol{i} \perp \boldsymbol{j}$.

Definition 9.8.6 Let W be a subspace of $\mathbb{R}^n$ and let $\{\boldsymbol{v}_1, \boldsymbol{v}_2, \ldots, \boldsymbol{v}_k\}$ be an orthogonal basis for W. For any vector $\boldsymbol{v} \in \mathbb{R}^n$, the orthogonal projection of $\boldsymbol{v}$ onto W is defined as

$$\text{proj}_W(\boldsymbol{v}) = \left(\frac{\boldsymbol{v}_1 \cdot \boldsymbol{v}}{\boldsymbol{v}_1 \cdot \boldsymbol{v}_1}\right)\boldsymbol{v}_1 + \left(\frac{\boldsymbol{v}_2 \cdot \boldsymbol{v}}{\boldsymbol{v}_2 \cdot \boldsymbol{v}_2}\right)\boldsymbol{v}_2 + \cdots + \left(\frac{\boldsymbol{v}_k \cdot \boldsymbol{v}}{\boldsymbol{v}_k \cdot \boldsymbol{v}_k}\right)\boldsymbol{v}_k$$

The component of $\boldsymbol{v}$ orthogonal to W is the vector

$$\text{perp}_W(\boldsymbol{v}) = \boldsymbol{v} - \text{proj}_W(\boldsymbol{v})$$

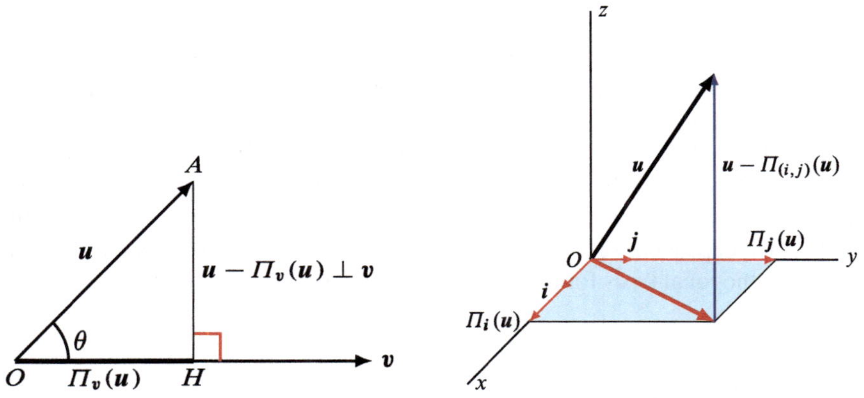

(a) Projection on a line (b) Projection on a plane

Fig. 9.20 Orthogonal projection of a vector onto another vector (or a line) and onto a plane

[41] Proof: $\boldsymbol{v} \cdot \text{perp}_{\boldsymbol{v}}(\boldsymbol{u}) = \boldsymbol{v} \cdot (\boldsymbol{u} - {}^{\boldsymbol{u}\cdot\boldsymbol{v}}/_{\boldsymbol{v}\cdot\boldsymbol{v}}\boldsymbol{v}) = \boldsymbol{v} \cdot \boldsymbol{u} - \boldsymbol{u} \cdot \boldsymbol{v} = 0$.

9.8.6 Gram-Schmidt Orthogonalization Process

The Gram-Schmidt algorithm takes a set of linear independent vectors $\boldsymbol{v}_1, \boldsymbol{v}_2, \ldots, \boldsymbol{v}_k$ and generates an orthogonal linear independent set of vectors $\boldsymbol{u}_1, \boldsymbol{u}_2, \ldots, \boldsymbol{u}_k$. In the process, if needed, these vectors can be normalized to get an orthonormal set $\boldsymbol{e}_1, \boldsymbol{e}_2, \ldots, \boldsymbol{e}_k$.

The method is named after the Danish actuary and mathematician Jørgen Pedersen Gram (1850–1916) and the Baltic German mathematician Erhard Schmidt (1876–1959), but Pierre-Simon Laplace had been familiar with it before Gram and Schmidt.

The idea is to start with the first vector $\boldsymbol{v}_1$, nothing to do here so we take $\boldsymbol{u}_1 = \boldsymbol{v}_1$ and normalize it to get $\boldsymbol{e}_1$. Next, move to the second vector $\boldsymbol{v}_2$, we make it orthogonal to $\boldsymbol{u}_1$ by $\boldsymbol{u}_2 = \boldsymbol{v}_2 - \text{proj}_{\boldsymbol{u}_1}(\boldsymbol{v}_2)$. Then, we normalize $\boldsymbol{u}_2$. Now, move to the third vector $\boldsymbol{v}_3$. We make it orthogonal to the hyperplane spanned by $\boldsymbol{u}_1$ and $\boldsymbol{u}_2$. And the process keeps going until the last vector:

$$
\begin{aligned}
\boldsymbol{u}_1 &= \boldsymbol{v}_1, & \boldsymbol{e}_1 &= \frac{\boldsymbol{u}_1}{||\boldsymbol{u}_1||}\\
\boldsymbol{u}_2 &= \boldsymbol{v}_2 - \text{proj}_{\boldsymbol{u}_1}(\boldsymbol{v}_2), & \boldsymbol{e}_2 &= \frac{\boldsymbol{u}_2}{||\boldsymbol{u}_2||}\\
\boldsymbol{u}_3 &= \boldsymbol{v}_3 - \text{proj}_{\boldsymbol{u}_1}(\boldsymbol{v}_3) - \text{proj}_{\boldsymbol{u}_2}(\boldsymbol{v}_3), & \boldsymbol{e}_3 &= \frac{\boldsymbol{u}_3}{||\boldsymbol{u}_3||}\\
&\vdots\\
\boldsymbol{u}_k &= \boldsymbol{v}_k - \sum_{i}^{k-1} \text{proj}_{\boldsymbol{u}_i}(\boldsymbol{v}_k), & \boldsymbol{e}_k &= \frac{\boldsymbol{u}_k}{||\boldsymbol{u}_k||}
\end{aligned}
$$

The calculation of the sequence $\boldsymbol{u}_1, \ldots, \boldsymbol{u}_k$ is known as Gram-Schmidt orthogonalization, while the calculation of the sequence $\boldsymbol{e}_1, \ldots, \boldsymbol{e}_k$ is known as Gram-Schmidt orthonormalization as the vectors are normalized.

If we denote by $W_1 = \text{span}(\boldsymbol{v}_1)$, then $\boldsymbol{u}_1$ is the basis of W_1. Moving on, let $W_2 = \text{span}(\boldsymbol{v}_1, \boldsymbol{v}_2)$, we guess that $\{\boldsymbol{u}_1, \boldsymbol{u}_2\}$ are the basis vectors for W_2. Why? First, by definition $\boldsymbol{u}_1$ and $\boldsymbol{u}_2$ are linear combinations of $\boldsymbol{v}_1$ and $\boldsymbol{v}_2$, thus they are in W_2. Second, they are linear independent (because they're orthogonal). And finally, $\{\boldsymbol{u}_1, \boldsymbol{u}_2, \ldots, \boldsymbol{u}_k\}$ form an orthogonal basis for the subspace $W_k = \text{span}(\boldsymbol{v}_1, \boldsymbol{v}_2, \ldots, \boldsymbol{v}_k)$.

9.8.7 QR Factorization

The Gauss elimination process of $\mathbf{A}\boldsymbol{x} = \boldsymbol{b}$ results in the LU factorization: $\mathbf{A} = \mathbf{LU}$. Now, the Gram-Schmidt orthogonalization process applied to the linear independent columns of a matrix $\mathbf{A}$ results in another factorization—known as the QR factorization: $\mathbf{A} = \mathbf{QR}$. To demonstrate this factorization, consider a matrix with three independent columns $\mathbf{A} = [\boldsymbol{a}_1|\boldsymbol{a}_2|\boldsymbol{a}_3]$. Applying the Gram-Schmidt orthonormalization to these three vectors we obtain $\boldsymbol{e}_1, \boldsymbol{e}_2, \boldsymbol{e}_3$. We can write then

$$\begin{aligned}
\boldsymbol{a}_1 &= (\boldsymbol{e}_1, \boldsymbol{a}_1)\boldsymbol{e}_1 \\
\boldsymbol{a}_2 &= (\boldsymbol{e}_1, \boldsymbol{a}_2)\boldsymbol{e}_1 + (\boldsymbol{e}_2, \boldsymbol{a}_2)\boldsymbol{e}_2 \\
\boldsymbol{a}_3 &= (\boldsymbol{e}_1, \boldsymbol{a}_3)\boldsymbol{e}_1 + (\boldsymbol{e}_2, \boldsymbol{a}_3)\boldsymbol{e}_2 + (\boldsymbol{e}_3, \boldsymbol{a}_3)\boldsymbol{e}_3
\end{aligned}$$

which can be written as (using block matrices multiplication)

$$\mathbf{A} = \begin{bmatrix}\boldsymbol{a}_1 & \boldsymbol{a}_2 & \boldsymbol{a}_3\end{bmatrix} = \begin{bmatrix}\boldsymbol{e}_1 & \boldsymbol{e}_2 & \boldsymbol{e}_3\end{bmatrix}\begin{bmatrix}(\boldsymbol{e}_1, \boldsymbol{a}_1) & (\boldsymbol{e}_1, \boldsymbol{a}_2) & (\boldsymbol{e}_1, \boldsymbol{a}_3) \\ 0 & (\boldsymbol{e}_2, \boldsymbol{a}_2) & (\boldsymbol{e}_2, \boldsymbol{a}_3) \\ 0 & 0 & (\boldsymbol{e}_3, \boldsymbol{a}_3)\end{bmatrix} = \mathbf{QR}$$

The matrix $\mathbf{Q}$ consists of orthonormal columns and thus is an orthogonal matrix (that explains why the notation $\mathbf{Q}$ was used). The matrix $\mathbf{R}$ is an upper triangular matrix.

History note 9.1: James Joseph Sylvester (1814–1897)

James Joseph Sylvester was an English mathematician. He made fundamental contributions to matrix theory, invariant theory, number theory, partition theory, and combinatorics. He played a leadership role in American mathematics in the later half of the 19th century as a professor at the Johns Hopkins University and as founder of the American Journal of Mathematics. At his death, he was a professor at Oxford University.

James Joseph was born in London on 3 September 1814, the son of Abraham Joseph, a Jewish merchant. James later adopted the surname Sylvester when his older brother did so upon emigration to the United States-a country which at that time required all immigrants to have a given name, a middle name, and a surname. Sylvester began his study of mathematics at St John's College, Cambridge in 1831, where his tutor was John Hymers. Although his studies were interrupted for almost two years due to a prolonged illness, he nevertheless ranked second in Cambridge's famous mathematical examination, the tripos, for which he sat in 1837.

9.9 Determinant

To derive the formula for the determinant of a square matrix $n \times n$ when $n > 3$, we cannot rely on geometry. To proceed, it is better to deduce the properties of the determinant from the special cases of 2×2 and 3×3 matrices. From those properties, we can define what should be a determinant. It is not so hard to observe the following properties of the determinant of a 2×2 matrix (they also apply for 3×3 matrices):

- The determinant of the 2×2 unit matrix is one; this is obvious because this matrix does not change the unit square at all.
- If the two columns of a 2×2 matrix are the same, its determinant is zero; this is obvious either from the formula or from the fact that the two transformed basic vectors collapse onto each other, a domain transforms to a line with zero area.
- If one column is a multiple of the other column, the determinant is also zero. The explanation is similar to the previous property; this one is a generalization of the previous property.
- If one column of a 2×2 matrix is scaled by a factor α, the determinant is scaled by the same factor:

$$\mathbf{B} = \begin{bmatrix} a & \alpha b \\ c & \alpha d \end{bmatrix} \implies |\mathbf{B}| = \alpha ad - \alpha cd = \alpha|\mathbf{A}|$$

- Additive property:

$$|[\boldsymbol{u} \quad \boldsymbol{v} + \boldsymbol{w}]| = |[\boldsymbol{u} \quad \boldsymbol{v}]| + |[\boldsymbol{u} \quad \boldsymbol{w}]|$$

  This is a consequence of the fact that we can decompose the area into two areas, see Fig. 9.21.
- If we interchange the columns of **A**, the determinant changes sign (changes by a factor of -1):

$$\det \begin{bmatrix} b & a \\ d & c \end{bmatrix} = bc - da = -(ad - bc) = -\det \begin{bmatrix} a & b \\ c & d \end{bmatrix}$$

9.9.1 Defining the Determinant in Terms of Its Properties

Up to now we know that a matrix 2×2 or 3×3 has a number associated with it, which is called the determinant of the matrix. We can see it as a function $D : \mathbb{R}^{n \times n} \to \mathbb{R}$ which assigns to each $n \times n$ matrix a single real number. We write $D(\mathbf{A})$ to label this number, and we also write D in terms of the columns of **A**: $D(\boldsymbol{a}_1, \boldsymbol{a}_2, \ldots, \boldsymbol{a}_n)$

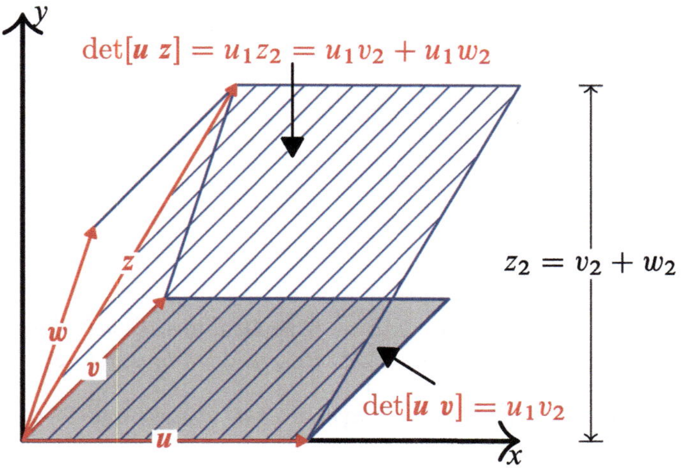

Fig. 9.21 Additive area property: to ease the proof, vector $\boldsymbol{u}$ is aligned to the x-axis

where $\boldsymbol{a}_1, \boldsymbol{a}_2, \ldots, \boldsymbol{a}_n$ are the columns of the matrix $\mathbf{A}$. We did this because from the previous discussion we know that the determinant depends heavily on the columns of the matrix.

Now, we propose the following properties for D inspired from the properties of the determinants of 3×3 matrices.

Property 1. $D(\mathbf{I}) = 1$.

Property 2. $D(\boldsymbol{a}_1, \boldsymbol{a}_2, \ldots, \boldsymbol{a}_n) = 0$ if $\boldsymbol{a}_i = \boldsymbol{a}_j$ for $i \neq j$.

Property 3. If $n - 1$ columns of $\mathbf{A}$ held fixed, then $D(\mathbf{A})$ is a linear function of the remaining column. Stated in terms of the jth column, this property says that:

$$D(\boldsymbol{a}_1, \ldots, \boldsymbol{a}_{j-1}, \boldsymbol{u} + \alpha \boldsymbol{v}, \ldots, \boldsymbol{a}_n) = D(\boldsymbol{a}_1, \ldots, \boldsymbol{a}_{j-1}, \boldsymbol{u}, \ldots, \boldsymbol{a}_n) + \alpha D(\boldsymbol{a}_1, \ldots, \boldsymbol{a}_{j-1}, \boldsymbol{v}, \ldots, \boldsymbol{a}_n)$$

This comes from the additive area property and the fact that if we scale one column by α, the determinant is scaled by the same factor.

Property 4. D is an alternating function of the columns, i.e., if two columns are interchanged, the value of D changes by a factor of -1. Let's focus on columns ith and jth, so we write $D(\boldsymbol{a}_i, \boldsymbol{a}_j)$ leaving other columns untouched and left behind the scene. What we need to show is that $D(\boldsymbol{a}_j, \boldsymbol{a}_i) = -D(\boldsymbol{a}_i, \boldsymbol{a}_j)$.

Proof The proof is based on Property 2 and Property 3. The trick of using Property 2 is to add zero or subtract zero to a quantity.

$$
\begin{aligned}
D(\boldsymbol{a}_j, \boldsymbol{a}_i) &= D(\boldsymbol{a}_j, \boldsymbol{a}_i) + \underbrace{D(\boldsymbol{a}_i, \boldsymbol{a}_i)}_{0} && \text{(added 0 due to Property 2)} \\
&= D(\boldsymbol{a}_i + \boldsymbol{a}_j, \boldsymbol{a}_i) && \text{(due to Property 3)} \\
&= D(\boldsymbol{a}_i + \boldsymbol{a}_j, \boldsymbol{a}_i) - \underbrace{D(\boldsymbol{a}_i + \boldsymbol{a}_j, \boldsymbol{a}_i + \boldsymbol{a}_j)}_{0} && \text{(minus 0 due to Property 2)} \\
&= D(\boldsymbol{a}_i + \boldsymbol{a}_j, -\boldsymbol{a}_j) && \text{(due to Property 3)} \\
&= D(\boldsymbol{a}_i, -\boldsymbol{a}_j) - \underbrace{D(\boldsymbol{a}_j, \boldsymbol{a}_j)}_{0} && \text{(due to Property 3)} \\
&= -D(\boldsymbol{a}_i, \boldsymbol{a}_j) && \text{(due to Property 3 with} \alpha = -1)
\end{aligned}
$$

■

Property 5. If the columns of **A** are linear dependent then $D = 0$. One interesting case is that if **A** has at least one row of all zeros, its determinant is zero.[42]

Proof Without loss of generality, we can express $\boldsymbol{a}_1$ as $\boldsymbol{a}_1 = \alpha_2\boldsymbol{a}_2 + \alpha_3\boldsymbol{a}_3 + \cdots + \alpha_n\boldsymbol{a}_n$. Now, $D(\mathbf{A})$ is computed as

$$
\begin{aligned}
D &= D(\boldsymbol{a}_1, \boldsymbol{a}_2, \ldots, \boldsymbol{a}_n) \\
&= D(\alpha_2\boldsymbol{a}_2 + \alpha_3\boldsymbol{a}_3 + \cdots + \alpha_n\boldsymbol{a}_n, \boldsymbol{a}_2, \ldots, \boldsymbol{a}_n) \\
&= D(\alpha_2\boldsymbol{a}_2, \boldsymbol{a}_2, \ldots, \boldsymbol{a}_n) + D(\alpha_3\boldsymbol{a}_3, \boldsymbol{a}_2, \ldots, \boldsymbol{a}_n) + \cdots + D(\alpha_n\boldsymbol{a}_n, \boldsymbol{a}_2, \ldots, \boldsymbol{a}_n) \\
&= \alpha_2 D(\boldsymbol{a}_2, \boldsymbol{a}_2, \ldots, \boldsymbol{a}_n) + \alpha_3 D(\boldsymbol{a}_3, \boldsymbol{a}_2, \boldsymbol{a}_3, \ldots, \boldsymbol{a}_n) + \cdots + \alpha_n D(\boldsymbol{a}_n, \boldsymbol{a}_2, \ldots, \boldsymbol{a}_n) \\
&= 0 + 0 + \cdots + 0 \quad \text{(Property 2)}
\end{aligned}
$$

where Property 3 was used in the third equality, Property 3 again in the fourth equality (with $\alpha = 0$). ■

Property 6. Adding a multiple of one column to another one does not change the determinant.

Proof Suppose we obtain matrix **B** from **A** by adding α times column j to column i. Then,

$$
\begin{aligned}
D(\mathbf{B}) &= D(\boldsymbol{a}_1, \ldots, \boldsymbol{a}_{i-1}, \boldsymbol{a}_i + \alpha\boldsymbol{a}_j, \ldots, \boldsymbol{a}_n) \\
&= D(\boldsymbol{a}_1, \ldots, \boldsymbol{a}_{i-1}, \boldsymbol{a}_i, \ldots, \boldsymbol{a}_n) + \alpha D(\boldsymbol{a}_1, \ldots, \boldsymbol{a}_{i-1}, \boldsymbol{a}_j, \ldots, \boldsymbol{a}_n) \text{ (Property 3)} \\
&= D(\boldsymbol{a}_1, \ldots, \boldsymbol{a}_{i-1}, \boldsymbol{a}_i, \ldots, \boldsymbol{a}_n) \quad \text{(second red term is zero of Property 2)} \\
&= D(\mathbf{A})
\end{aligned}
$$

■

[42] In case it is not clear. Any set of vectors containing the zero vector is linearly dependent: $\mathbf{10} + 0\boldsymbol{a}_2 + \cdots + 0\boldsymbol{a}_k = \mathbf{0}$.

9.9.2 Determinant of Elementary Matrices

It is obvious that we have

$$\det\begin{bmatrix} a & 0 \\ 0 & b \end{bmatrix} = ab, \quad \det\begin{bmatrix} a & 0 & 0 \\ 0 & b & 0 \\ 0 & 0 & c \end{bmatrix} = abc$$

which can be verified using the formula, or from the geometric meaning of the determinant.

What is more interesting is the following results:

$$\det\begin{bmatrix} a & c \\ 0 & b \end{bmatrix} = ab, \quad \det\begin{bmatrix} a & d & e \\ 0 & b & f \\ 0 & 0 & c \end{bmatrix} = abc$$

A geometry explanation for these results is that for the 2D matrix, shearing a rectangle does not change its area, and for the 3D matrix, shearing a cube also does not change its volume. Still we need an algebraic proof so that it can be extended to larger matrices. For the 3×3 matrix, the second column can be decomposed as

$$\begin{bmatrix} d \\ b \\ 0 \end{bmatrix} = \begin{bmatrix} d \\ 0 \\ 0 \end{bmatrix} + \begin{bmatrix} 0 \\ b \\ 0 \end{bmatrix}$$

Then, using Property 3, its determinant is given by

$$\begin{vmatrix} a & d & e \\ 0 & b & f \\ 0 & 0 & c \end{vmatrix} = \begin{vmatrix} a & d & e \\ 0 & 0 & f \\ 0 & 0 & c \end{vmatrix} + \begin{vmatrix} a & 0 & e \\ 0 & b & f \\ 0 & 0 & c \end{vmatrix} = \begin{vmatrix} a & 0 & e \\ 0 & b & f \\ 0 & 0 & c \end{vmatrix}$$

The red determinant is zero because of Property 5: the first and second columns are linear dependent. Now, we do the same thing for the determinant left by decomposing column 3:

$$\begin{vmatrix} a & 0 & e \\ 0 & b & f \\ 0 & 0 & c \end{vmatrix} = \begin{vmatrix} a & 0 & e \\ 0 & b & 0 \\ 0 & 0 & 0 \end{vmatrix} + \begin{vmatrix} a & 0 & 0 \\ 0 & b & f \\ 0 & 0 & 0 \end{vmatrix} + \begin{vmatrix} a & 0 & 0 \\ 0 & b & 0 \\ 0 & 0 & c \end{vmatrix} = \begin{vmatrix} a & 0 & 0 \\ 0 & b & 0 \\ 0 & 0 & c \end{vmatrix} = abc$$

Property 7. The determinant of a triangular matrix is the product of its diagonal entries. This property results in another fact that if $\mathbf{A}$ is a triangular matrix, its transpose is also a triangular matrix with the same entries on the diagonal, thus $D(\mathbf{A}) = D(\mathbf{A}^\top)$. This holds for any square matrix, not just for triangular matrix.

Property 8. $D(\mathbf{A}^\top) = D(\mathbf{A})$. The proof goes as: If $\mathbf{A}$ is invertible, it can be written as a product of some elementary matrices:

$$\mathbf{A} = \mathbf{E}_1\mathbf{E}_2\cdots\mathbf{E}_k$$

Thus, with $D(\mathbf{EF}) = D(\mathbf{E})D(\mathbf{F})$, we can write

$$\begin{aligned}
D(\mathbf{A}) &= D(\mathbf{E}_1\mathbf{E}_2\cdots\mathbf{E}_k) = D(\mathbf{E}_1)D(\mathbf{E}_2)\cdots D(\mathbf{E}_k)\\
&= D(\mathbf{E}_1^\top)D(\mathbf{E}_2^\top)\cdots D(\mathbf{E}_k^\top) = D(\mathbf{E}_k^\top)\cdots D(\mathbf{E}_2^\top)D(\mathbf{E}_1^\top)\\
&= D(\mathbf{E}_k^\top\cdots\mathbf{E}_2^\top\mathbf{E}_1^\top) = D((\mathbf{E}_1\mathbf{E}_2\cdots\mathbf{E}_k)^\top) = D(\mathbf{A}^\top)
\end{aligned}$$

where the fact that for an elementary matrix $\mathbf{E}$, $D(\mathbf{E}^\top) = D(\mathbf{E})$ was used. The importance of Property 7 is that it allows us to conclude that all the properties of the determinant that we have stated concerning the columns also work for rows, e.g., if two rows of a matrix are the same its determinant is zero. This is so because the columns of $\mathbf{A}^\top$ are the rows of $\mathbf{A}$.

Property 9. If $\mathbf{A}$ is invertible then we have $\det(\mathbf{A}^{-1}) = 1/\det(\mathbf{A})$.[43] So, we do not need to know what $\mathbf{A}^{-1}$ is, still we can compute its determinant.

9.9.3 A Formula for the Determinant

Let's start with a 3×3 matrix. We can compute its determinant as follows (decomposing the first column[44] as the sum of three vectors and use Property 3):

$$\begin{vmatrix} a_{11} & a_{12} & a_{13}\\ a_{21} & a_{22} & a_{23}\\ a_{31} & a_{32} & a_{33}\end{vmatrix} = \begin{vmatrix} a_{11} & a_{12} & a_{13}\\ 0 & a_{22} & a_{23}\\ 0 & a_{32} & a_{33}\end{vmatrix} + \begin{vmatrix} 0 & a_{12} & a_{13}\\ a_{21} & a_{22} & a_{23}\\ 0 & a_{32} & a_{33}\end{vmatrix} + \begin{vmatrix} 0 & a_{12} & a_{13}\\ 0 & a_{22} & a_{23}\\ a_{31} & a_{32} & a_{33}\end{vmatrix}$$

Next, for the second determinant in the RHS, we exchange row 1 and row 2 (Property 4), we get a minus. Then, for the third determinant in the RHS, we exchange rows 1/3 and another exchange between rows 3/2 (Property 4 with two minuses we get a plus):

$$\begin{vmatrix} a_{11} & a_{12} & a_{13}\\ a_{21} & a_{22} & a_{23}\\ a_{31} & a_{32} & a_{33}\end{vmatrix} = \begin{vmatrix} a_{11} & a_{12} & a_{13}\\ 0 & a_{22} & a_{23}\\ 0 & a_{32} & a_{33}\end{vmatrix} - \begin{vmatrix} a_{21} & a_{22} & a_{23}\\ 0 & a_{12} & a_{13}\\ 0 & a_{32} & a_{33}\end{vmatrix} + \begin{vmatrix} a_{31} & a_{32} & a_{33}\\ 0 & a_{12} & a_{13}\\ 0 & a_{22} & a_{23}\end{vmatrix} \tag{9.9.1}$$

---

[43] We start with $\mathbf{AA}^{-1} = \mathbf{I}$, which leads to $\det(\mathbf{A}^{-1})\det(\mathbf{A}) = \det(\mathbf{I}) = 1$.

[44] There is nothing special about the first column; this is just one way to go.

The nice thing we get is that all the three determinants in the RHS are of this form:

$$\begin{vmatrix} a_{11} & d_{12} & d_{13} \\ 0 & b_{22} & b_{23} \\ 0 & b_{32} & b_{33} \end{vmatrix} = \det\left(\begin{bmatrix} a_{11} & d_{12} & d_{13} \\ 0 & & \mathbf{B} \\ 0 & & \end{bmatrix}\right)$$

which can be rewritten as (to get a lower triangular matrix)

$$\begin{vmatrix} a_{11} & d_{12} & d_{13} \\ 0 & b_{22} & b_{23} \\ 0 & b_{32} & b_{33} \end{vmatrix} = \begin{vmatrix} a_{11} & d_{12} & d_{13} \\ 0 & b_{22} & b_{23} \\ 0 & 0 & b_{33} - b_{23}b_{32}/b_{22} \end{vmatrix} = a_{11}b_{22}\left(b_{33} - \frac{b_{23}b_{32}}{b_{22}}\right)$$
$$= a_{11}(b_{22}b_{33} - b_{32}b_{23})$$

where in the first equality Property 6 (for row) was used and in the second equality, Property 7 was used. The red term is called a *cofactor*, and it is exactly the determinant of **B**. With this result, Eq. (9.9.1) can be rewritten as

$$\begin{vmatrix} a_{11} & a_{12} & a_{13} \\ a_{21} & a_{22} & a_{23} \\ a_{31} & a_{32} & a_{33} \end{vmatrix} = a_{11}\begin{vmatrix} a_{22} & a_{23} \\ a_{32} & a_{33} \end{vmatrix} - a_{21}\begin{vmatrix} a_{12} & a_{13} \\ a_{32} & a_{33} \end{vmatrix} + a_{31}\begin{vmatrix} a_{12} & a_{13} \\ a_{22} & a_{23} \end{vmatrix} \tag{9.9.2}$$

Finally, noting that the matrix **B** is obtained by deleting a certain row and column of **A**. So, we define $\mathbf{A}_{ij}$ the matrix obtained by deleting row ith and column jth of **A**. With this definition, the determinant of **A** can be expressed as

$$\begin{vmatrix} a_{11} & a_{12} & a_{13} \\ a_{21} & a_{22} & a_{23} \\ a_{31} & a_{32} & a_{33} \end{vmatrix} = a_{11}|\mathbf{A}_{11}| - a_{21}|\mathbf{A}_{21}| + a_{31}|\mathbf{A}_{31}| \tag{9.9.3}$$

There is a pattern in this formula. Also this formula works for 2×2 matrix (you can check it). So, for a $n \times n$ matrix **A**, its determinant is given by

$$|\mathbf{A}| = a_{11}|\mathbf{A}_{11}| - a_{21}|\mathbf{A}_{21}| + a_{31}|\mathbf{A}_{31}| - \cdots + a_{n1}|\mathbf{A}_{n1}| = \sum_{i=1}^{n}(-1)^{i-1}a_{i1}|\mathbf{A}_{i1}| \tag{9.9.4}$$

This is called *the cofactor expansion along the first column of* **A**.

But why only column 1? We can choose any column to start and for column jth, we have the following definition of the determinant:

$$\boxed{|\mathbf{A}| = \sum_{i=1}^{n}(-1)^{i+j}a_{ij}|\mathbf{A}_{ij}|} \tag{9.9.5}$$

Why this definition works? Because it allows us to define the determinant of a matrix inductively: we define the determinant of an $n \times n$ matrix in terms of the determinants of $(n-1) \times (n-1)$ matrices. We begin by defining the determinant of a 1×1 matrix $\mathbf{A} = [a]$ by $\det(\mathbf{A}) = a$. Then we proceed to 2×2 matrices, then 3×3 and so on. This is similar to how the factorial was defined: $n! = n(n-1)!$. Note that a definition is not the best way to compute the factorial and also the determinant.

9.9.4 Cramer's Rule

Cramer's rule solves $\mathbf{A}\boldsymbol{x} = \boldsymbol{b}$ using determinants. Its unique feature is that it provides an explicit formula for $\boldsymbol{x}$. The key idea is (assuming a system of three equations and three unknowns) the following identity:

$$\begin{bmatrix} & & \\ & \mathbf{A} & \\ & & \end{bmatrix} \begin{bmatrix} x_1 & 0 & 0 \\ x_2 & 1 & 0 \\ x_3 & 0 & 1 \end{bmatrix} = \begin{bmatrix} b_1 & a_{12} & a_{13} \\ b_2 & a_{22} & a_{23} \\ b_3 & a_{32} & a_{33} \end{bmatrix} := \mathbf{B}_1$$

where $\mathbf{B}_1$ is the matrix $\mathbf{A}$ with its first col replaced by vector $\boldsymbol{b} = (b_1, b_2, b_3)^\top$. Now, taking the determinant of both sides, noting that the determinant of the red matrix (a nice triangular matrix) is x_1 and the determinant of the product is equal to the product of the determinants:

$$x_1|\mathbf{A}| = |\mathbf{B}_1|$$

which gives us x_1:

$$x_1 = \frac{|\mathbf{B}_1|}{|\mathbf{A}|}$$

which is strikingly similar to $x = b/a$ for the linear equation $ax = b$. But now, we have to live with determinants. Similarly, we have $x_2 = {}^{|\mathbf{B}_2|}/_{|\mathbf{A}|}$. The geometric meaning of Cramer's rule is given in Fig. 9.22 for the case of 2×2 matrices for $y = {}^{|\mathbf{B}_2|}/_{|\mathbf{A}|}$ (noting that y is x_2). The area of the parallelogram formed by $\boldsymbol{e}_1$ and $\boldsymbol{x}$ is y (or x_2). After the transformation by $\mathbf{A}$, $\boldsymbol{e}_1$ becomes $\boldsymbol{a}_1 = (a_{11}, a_{21})$ and $\boldsymbol{x}$ becomes $\boldsymbol{b}$. The transformed parallelogram's area is thus $\det([\boldsymbol{a}_1\ \boldsymbol{b}])$. But we know that this new area is the original area scaled by the determinant of $\mathbf{A}$. The Cramer rule follows.

It is now possible to have Cramer's rule for a system of n equations for n unknowns, if $|\mathbf{A}| \neq 0$

$$\boxed{x_1 = \frac{|\mathbf{B}_1|}{|\mathbf{A}|},\quad x_2 = \frac{|\mathbf{B}_2|}{|\mathbf{A}|}, \ldots, \mathbf{B}_j \text{is matrix} \mathbf{A} \text{with} j\text{th col replaced by} \boldsymbol{b}} \tag{9.9.6}$$

This rule is named after the Genevan mathematician Gabriel Cramer (1704–1752), who published the rule for an arbitrary number of unknowns in 1750, although Colin

Maclaurin also published special cases of the rule in 1748 (and possibly knew of it as early as 1729).

Cramer's rule is of theoretical value than practical as it is not efficient to solve $\mathbf{A}\boldsymbol{x} = \boldsymbol{b}$ using Cramer's rule; use Gaussian elimination instead. However, it leads to a formula of the inverse of a matrix in terms of the determinant of the matrix. We discuss this now.

Cramer's rule and inverse of a matrix. Suppose that we want to find the inverse of 2×2 matrix $\mathbf{A}$. Let's denote

$$\mathbf{A}^{-1} = \begin{bmatrix} x_1 & y_1 \\ x_2 & y_2 \end{bmatrix}$$

We then solve for x_1, x_2, y_1, y_2 such that $\mathbf{A}\mathbf{A}^{-1} = \mathbf{I}$, or two systems of linear equations:

$$\mathbf{A}\begin{bmatrix} x_1 \\ x_2 \end{bmatrix} = \begin{bmatrix} 1 \\ 0 \end{bmatrix}, \quad \mathbf{A}\begin{bmatrix} y_1 \\ y_2 \end{bmatrix} = \begin{bmatrix} 0 \\ 1 \end{bmatrix}$$

And Cramer's rule is used for this, so we have

$$x_1 = \frac{\det \begin{bmatrix} 1 & a_{12} \\ 0 & a_{22} \end{bmatrix}}{\det \mathbf{A}}, \quad x_2 = \frac{\det \begin{bmatrix} a_{11} & 1 \\ a_{21} & 0 \end{bmatrix}}{\det \mathbf{A}}, \quad y_1 = \frac{\det \begin{bmatrix} 0 & a_{12} \\ 1 & a_{22} \end{bmatrix}}{\det \mathbf{A}}, \quad y_2 = \frac{\det \begin{bmatrix} a_{11} & 0 \\ a_{21} & 1 \end{bmatrix}}{\det \mathbf{A}}$$

Thus, we obtain the explicit formula for the inverse of a 2×2 matrix: (And also understand why for an invertible matrix, the determinant must be non-zero.)

$$\mathbf{A} = \begin{bmatrix} a & b \\ c & d \end{bmatrix} \Longrightarrow \mathbf{A}^{-1} = \frac{1}{ad - bc}\begin{bmatrix} d & -b \\ -c & a \end{bmatrix} \tag{9.9.7}$$

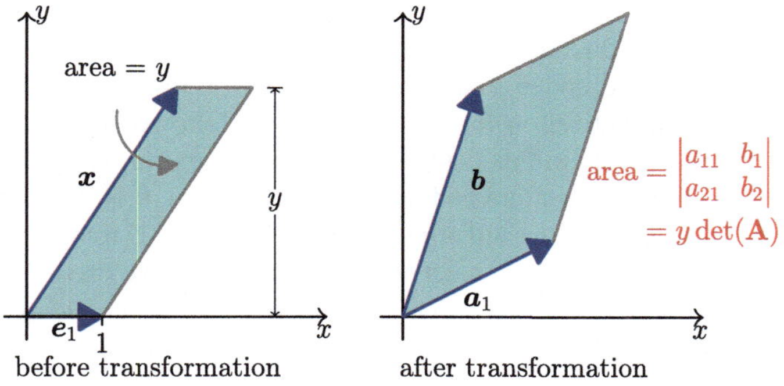

Fig. 9.22 Geometric meaning of Cramer's rule illustrated for 2×2 matrices. Considering $\mathbf{A}$ as a linear transformation which transforms $\boldsymbol{e}_1$ into $\boldsymbol{a}_1$—the first col of $\mathbf{A}$, and $\boldsymbol{x}$ into $\boldsymbol{b}$

What next? Many people would go for a general $n \times n$ matrix, but I am slow, so I do the same for a 3×3 matrix. However, I just need to compute the (3, 2) entry of the inverse:

$$(\mathbf{A}^{-1})_{32} = \frac{\det\left(\begin{bmatrix} a_{11} & a_{21} & 0 \\ a_{21} & a_{22} & 1 \\ a_{31} & a_{32} & 0 \end{bmatrix}\right)}{|\mathbf{A}|} = \frac{(-1)\det\left(\begin{bmatrix} a_{11} & a_{21} \\ a_{31} & a_{32} \end{bmatrix}\right)}{|\mathbf{A}|} = \frac{|\mathbf{A}_{23}|}{|\mathbf{A}|}$$

Notice that the nominator of the (3, 2) entry of the inverse matrix is the cofactor $|\mathbf{A}_{23}|$. Now, we have the formula for the inverse of a $n \times n$ matrix:

$$\left(\mathbf{A}^{-1}\right)_{ij} = \frac{\det \mathbf{A}_{ji}}{\det \mathbf{A}}, \quad \mathbf{A}^{-1} = \frac{1}{\det \mathbf{A}} \text{adj } \mathbf{A}, \quad \text{adj } \mathbf{A} = \begin{bmatrix} |A_{11}| & |A_{21}| & \cdots & |A_{n1}| \\ |A_{12}| & |A_{22}| & \cdots & |A_{n2}| \\ \vdots & \vdots & \ddots & \vdots \\ |A_{1n}| & |A_{2n}| & \cdots & |A_{nn}| \end{bmatrix} \tag{9.9.8}$$

where two formulae are presented: the first one is for the ij-entry of the $\mathbf{A}^{-1}$, and the second one is for the entire matrix $\mathbf{A}^{-1}$ with the introduction of the so-called adjoint (or adjugate) matrix of $\mathbf{A}$. This matrix is the transpose of the matrix of cofactors of $\mathbf{A}$.

9.10 Eigenvectors and Eigenvalues

The definition of eigenvectors and eigenvalues is actually simple: an eigenvector $\boldsymbol{x}$ of a matrix $\mathbf{A}$ is a vector such that $\mathbf{A}\boldsymbol{x}$ is in the *same direction as the vector* $\boldsymbol{x}$: $\mathbf{A}\boldsymbol{x} = \lambda\boldsymbol{x}$. In other words, multiplying matrix $\boldsymbol{A}$ with an eigenvector gives a new vector $\lambda\boldsymbol{x}$ where λ—an *eigenvalue*—is a stretching/shrinking factor. The computation of eigenvectors and eigenvalues for small matrices of sizes 2×2, 3×3 can be done manually and in a quite straightforward manner.

However, it is hard to understand why people came up with the idea of eigenvectors. To present a motivation for eigenvectors, we followed Euler in his study of rotation of rigid bodies. In this context, eigenvectors appear naturally.[45] So, we discuss briefly angular momentum and inertia tensor in Sect. 9.10.1. Then, in Sect. 9.10.2 we discuss principal axes and principal moments for a 3D rotating rigid body. From this starting point, we leave mechanics behind, and move on to the maths of eigenvectors. I have read *An Introduction To Mechanics* by Daniel Kleppner, Robert Kolenkow [34] and *Classical Mechanics* by John Taylor [68] for the materials in this section.

[45] Eigenvectors appear in many fields and thus I do not know exactly in what context eigenvalues first appeared. My decision to use the rotation of rigid bodies as a natural context for eigenvalues is that the maths is not hard.

9.10.1 Angular Momentum and Inertia Tensor

Let's consider a rigid body which is divided into many small pieces with masses m_α ($\alpha = 1, 2, 3, \ldots$). Now assume that this rigid body is rotating about an arbitrary axis with an angular velocity $\boldsymbol{\omega}$. The total angular momentum of that body is given by

$$\boldsymbol{l} = \sum_\alpha \boldsymbol{r}_\alpha \times \boldsymbol{p}_\alpha \tag{9.10.1}$$

where $\boldsymbol{p}_\alpha = m_\alpha \boldsymbol{\omega} \times \boldsymbol{r}_\alpha$, and $\boldsymbol{r}_\alpha$ denotes the position vector of mass α.[46] With the vector identity $\boldsymbol{a} \times (\boldsymbol{b} \times \boldsymbol{c}) = \boldsymbol{b}(\boldsymbol{a} \cdot \boldsymbol{c}) - \boldsymbol{c}(\boldsymbol{a} \cdot \boldsymbol{b})$, we can elaborate the angular momentum $\boldsymbol{l}$ further as

$$\boldsymbol{l} = \sum_\alpha m_\alpha \boldsymbol{r}_\alpha \times (\boldsymbol{\omega} \times \boldsymbol{r}_\alpha) = \sum_\alpha \left(m_\alpha r_\alpha^2 \boldsymbol{\omega} - m_\alpha \boldsymbol{r}_\alpha (\boldsymbol{r}_\alpha \cdot \boldsymbol{\omega})\right) \tag{9.10.2}$$

With a coordinate system, the angular velocity and position vector are written as

$$\boldsymbol{\omega} = \begin{bmatrix} \omega_x \\ \omega_y \\ \omega_z \end{bmatrix}, \quad \boldsymbol{r}_\alpha = \begin{bmatrix} x_\alpha \\ y_\alpha \\ z_\alpha \end{bmatrix}$$

Thus, we can work out explicitly the components of the angular momentum in Eq. (9.10.2) as

$$\begin{bmatrix} l_x \\ l_y \\ l_z \end{bmatrix} = \sum_\alpha m_\alpha \begin{bmatrix} (y_\alpha^2 + z_\alpha^2)\omega_x - x_\alpha y_\alpha \omega_y - x_\alpha z_\alpha \omega_z \\ -y_\alpha x_\alpha \omega_x + (x_\alpha^2 + z_\alpha^2)\omega_y - y_\alpha z_\alpha \omega_z \\ -z_\alpha x_\alpha \omega_x - z_\alpha y_\alpha \omega_y + (x_\alpha^2 + y_\alpha^2)\omega_z \end{bmatrix}$$

which can be rewritten in matrix-vector notation as

$$\begin{bmatrix} l_x \\ l_y \\ l_z \end{bmatrix} = \underbrace{\begin{bmatrix} \sum m_\alpha (y_\alpha^2 + z_\alpha^2) & -\sum m_\alpha x_\alpha y_\alpha & -\sum m_\alpha x_\alpha z_\alpha \\ -\sum m_\alpha y_\alpha x_\alpha & \sum m_\alpha (x_\alpha^2 + z_\alpha^2) & -\sum m_\alpha y_\alpha z_\alpha \\ -\sum m_\alpha z_\alpha x_\alpha & -\sum m_\alpha z_\alpha y_\alpha & \sum m_\alpha (x_\alpha^2 + y_\alpha^2) \end{bmatrix}}_{\mathbf{I}_\omega} \begin{bmatrix} \omega_x \\ \omega_y \\ \omega_z \end{bmatrix} \tag{9.10.3}$$

The matrix $\mathbf{I}_\omega$ is called the *moment of inertia matrix*; it is a symmetric matrix.[47]

Next, we show that by calculating the kinetic energy of a 3D rotating body, the matrix $\mathbf{I}_\omega$ shows up again. The kinetic energy is given by

---

[46] The length of this position vector is denoted by r_α.

[47] To be precise, $\mathbf{I}_\omega$ is a second-order tensor, and its representation in a coordinate system is a matrix. However, for the discussion herein, the fact that $\mathbf{I}_\omega$ is a tensor is not important.

$$K = \sum_\alpha \frac{m_\alpha \boldsymbol{v}_\alpha \cdot \boldsymbol{v}_\alpha}{2} = \sum_\alpha \frac{m_\alpha (\boldsymbol{r}_\alpha \times \boldsymbol{\omega}) \cdot (\boldsymbol{r}_\alpha \times \boldsymbol{\omega})}{2} \tag{9.10.4}$$

which in conjunction with this vector identity $||\boldsymbol{a} \times \boldsymbol{b}|| = ||\boldsymbol{a}||^2||\boldsymbol{b}||^2 - (\boldsymbol{a} \cdot \boldsymbol{b})^2$[48] becomes

$$K = \sum_\alpha \frac{m_\alpha (r_\alpha^2 \omega^2 - (\boldsymbol{r}_\alpha \cdot \boldsymbol{\omega})^2)}{2} \tag{9.10.5}$$

Using the components of $\boldsymbol{r}_\alpha$ and $\boldsymbol{\omega}$, K is written as

$$\begin{aligned} K = \frac{1}{2}\Big(\sum_\alpha m_\alpha (y_\alpha^2 + z_\alpha^2)\omega_x^2 + m_\alpha (x_\alpha^2 + z_\alpha^2)\omega_y^2 + m_\alpha (x_\alpha^2 + y_\alpha^2)\omega_z^2 - \\ - 2m_\alpha x_\alpha y_\alpha \omega_x \omega_y - 2m_\alpha y_\alpha z_\alpha \omega_y \omega_z - 2m_\alpha x_\alpha z_\alpha \omega_x \omega_z \Big) \end{aligned} \tag{9.10.6}$$

which is a quadratic form; check Sect. 6.7.4 for a refresh. So, we can rewrite it in this familiar vector-matrix-vector product, and of course the matrix is **I**:

$$K = \frac{1}{2}\boldsymbol{\omega}^\top \mathbf{I}_\omega \boldsymbol{\omega} \tag{9.10.7}$$

Moment of inertia for continuous bodies. For a continuous body $\mathcal{B}$, its matrix of moment of inertia is given by (sum is replaced by integral and mass is replaced by ρdV)

$$\begin{aligned} I_{xx} &= \int_\mathcal{B} \rho(y^2 + z^2)dV, \quad I_{yy} = \int_\mathcal{B} \rho(x^2 + z^2)dV, \quad I_{zz} = \int_\mathcal{B} \rho(x^2 + y^2)dV \\ I_{xy} &= -\int_\mathcal{B} \rho xydV, \qquad I_{xz} = -\int_\mathcal{B} \rho xzdV, \qquad I_{yz} = -\int_\mathcal{B} \rho yzdV \end{aligned} \tag{9.10.8}$$

Example 9.8 As the first example, compute the matrix of inertia for a cube of side a and mass M (the mass is uniformly distributed, i.e., the density is constant) for two cases: (a) for a rotation w.r.t. to one corner and (b) w.r.t. to the center of the cube. The coordinate system axes are parallel to the sides.

For case (a), we have

$$\begin{aligned} I_{xx} &= I_{yy} = I_{zz} = \int \rho y^2 dV + \int \rho z^2 dV = 2\rho \int_0^a dx \int_0^a y^2 dy \int_0^a dz = \frac{2Ma^2}{3} \\ I_{xy} &= I_{xz} = I_{yz} = -\rho \int_0^a xdx \int_0^a ydy \int_0^a dz = -\frac{Ma^2}{4} \end{aligned}$$

[48] Check the discussion around Eq. (9.1.20) if this identity is not clear.

where $M = \rho a^3$. Thus, the inertia matrix is given by (this matrix has a determinant of $242^{Ma^2}/_{12}$)

$$\mathbf{I}_\omega = \frac{Ma^2}{12}\begin{bmatrix} 8 & -3 & -3 \\ -3 & 8 & -3 \\ -3 & -3 & 8 \end{bmatrix} \tag{9.10.9}$$

Now, we will compute the angular momentum if the cube is rotated around the x-axis (due to symmetry it does not matter which axis is chosen) with an angular velocity $\boldsymbol{\omega} = (\omega, 0, 0)$. The angular velocity in this case is

$$\boldsymbol{l} = \frac{Ma^2}{12}\begin{bmatrix} 8 & -3 & -3 \\ -3 & 8 & -3 \\ -3 & -3 & 8 \end{bmatrix}\begin{bmatrix} \omega \\ 0 \\ 0 \end{bmatrix} = \frac{Ma^2}{12}\begin{bmatrix} 8\omega \\ -3\omega \\ -3\omega \end{bmatrix}$$

What we learn from this? Two things: first the inertia matrix is full and the angular momentum is not parallel to the angular velocity. That is, $\mathbf{I}\boldsymbol{\omega}$ is in different direction than $\boldsymbol{\omega}$. Let's see what we get if the angular velocity is along the diagonal of the cube, i.e., $\boldsymbol{\omega} = {}^{\omega}/_{\sqrt{3}}(1, 1, 1)$:

$$\boldsymbol{l} = \frac{Ma^2}{12}\frac{\omega}{\sqrt{3}}\begin{bmatrix} 8 & -3 & -3 \\ -3 & 8 & -3 \\ -3 & -3 & 8 \end{bmatrix}\begin{bmatrix} 1 \\ 1 \\ 1 \end{bmatrix} = \frac{Ma^2}{12}\frac{\omega}{\sqrt{3}}\begin{bmatrix} 2 \\ 2 \\ 2 \end{bmatrix} = \frac{Ma^2}{6}\boldsymbol{\omega}$$

In this case, the angular momentum is parallel to the angular velocity. In other words, $\mathbf{I}_\omega\boldsymbol{\omega} = \lambda\boldsymbol{\omega}$, $\lambda = Ma^2/6$.

For case (b), we have (same calculations with different integration limits from $-a/2$ to $a/2$ instead)

$$I_{xx} = I_{yy} = I_{zz} = \int \rho y^2 dV + \int \rho z^2 dV = 2\rho\int_{-a/2}^{a/2} dx \int_{-a/2}^{a/2} y^2 dy \int_{-a/2}^{a/2} dz = \frac{Ma^2}{6}$$

$$I_{xy} = I_{xz} = I_{yz} = -\rho\int_{-a/2}^{a/2} x dx \int_{-a/2}^{a/2} y dy \int_{-a/2}^{a/2} dz = 0$$

Actually I_{xy} is zero because the integrand is an odd function xy. Another explanation is, by looking at Fig. 9.23, we see that the material on the side above the plane $y = 0$ cancels the contribution of the material below this plane (so $I_{xy} = I_{yz} = 0$). Thus, the inertia matrix is given by

$$\mathbf{I}_\omega = \frac{Ma^2}{6}\begin{bmatrix} 1 & 0 & 0 \\ 0 & 1 & 0 \\ 0 & 0 & 1 \end{bmatrix}$$

If we compute now the angular momentum for any angular velocity $\boldsymbol{\omega}$, we get $\boldsymbol{l} = ({}^{Ma^2}/_6)\boldsymbol{\omega}$ because $\mathbf{I}_\omega$ is a multiple of the identity matrix (the red matrix). So, we

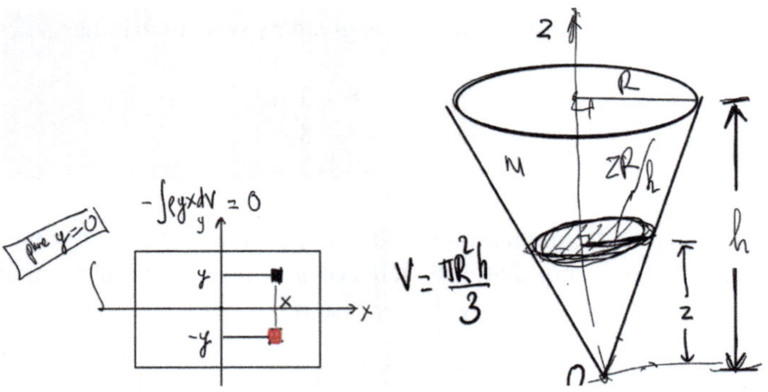

Fig. 9.23 .

see two things: (1) the inertia matrix is diagonal (entries not in the diagonal are all zeros) and (2) the angular momentum is parallel to the angular velocity, or $\mathbf{I}\boldsymbol{\omega} = \lambda\boldsymbol{\omega}$, $\lambda = Ma^2/6$. And this holds for any $\boldsymbol{\omega}$ because of the infinite symmetry of a cube w.r.t. to its center.

Example 9.9 The second example is finding the inertia matrix for a spinning top that is a uniform solid cone (mass M, height h, and base radius R) spinning about its tip O; cf. Fig. 9.23. The z-axis is chosen along the axis of symmetry of the cone.

All the integrals in the inertia matrix are computed using cylindrical coordinates. Due to symmetry, all the non-diagonal terms are zero, and $I_{xx} = I_{yy}$. So, we just need to compute three diagonal terms. Let's start with I_{zz}, but not I_{xx} (we will see why this saves us some calculations):

$$\begin{aligned} I_{zz} &= \int \rho(x^2+y^2)dV = \rho \int r^3 dr d\theta dz \\ &= \rho \int_0^h \left[\int_0^{zR/h} r^3 dr \int_0^{2\pi} d\theta \right] dz = \frac{3M}{10}R^2 \end{aligned}$$

From this we also get $\int \rho y^2 dV = I_{zz}/2 = (3M/20)R^2$. And this saves us a bit of work when calculating I_{xx}:

$$\begin{aligned} I_{xx} &= \int \rho(y^2+z^2)dV = \int \rho y^2 dV + \int \rho z^2 dV \\ &= (3M/20)R^2 + \rho \int_0^h \left[\int_0^{zR/h} r dr \int_0^{2\pi} d\theta \right] z^2 dz = \frac{3M}{20}(R^2+4h^2) \end{aligned}$$

So, the inertia matrix for this cone is a diagonal matrix:

$$\mathbf{I}_\omega = \begin{bmatrix} \lambda_1 & 0 & 0 \\ 0 & \lambda_1 & 0 \\ 0 & 0 & \lambda_2 \end{bmatrix}, \quad \lambda_1 = \frac{3M}{20}(R^2+4h^2), \quad \lambda_2 = \frac{3M}{20}R^2$$

We get a diagonal matrix. For an angular velocity $(\omega_x, \omega_y, \omega_z)$, the corresponding angular momentum is $(\lambda_1\omega_x, \lambda_1\omega_y, \lambda_2\omega_z)$. To get something interesting, consider this angular velocity $\boldsymbol{\omega} = (\omega, 0, 0)$ (that is, rotation about the x-axis), then the angular momentum is $(\lambda_1\omega, 0, 0)$ or $\lambda_1\boldsymbol{\omega}$.

9.10.2 Principal Axes and Eigenvalue Problems

We have studied some inertia matrices and we have observed that for the same solid, depending on the chosen axes, its inertia matrices are either full or diagonal, and when the inertia matrix is diagonal, if the rotation axis is one of the axes, then the angular momentum is parallel to the rotation axis. We also observe that by exploiting the symmetry of the solid, we can select axes so that the inertia matrix is diagonal. A question naturally arise: *does a non-symmetric solid have axes such that its matrix of inertia is diagonal?* The answer is yes (probably due to Euler) and such axes are called *principal axes* by the great man. The diagonal inertia matrix has this form:

$$\mathbf{I}_\omega = \begin{bmatrix} \lambda_1 & 0 & 0 \\ 0 & \lambda_2 & 0 \\ 0 & 0 & \lambda_3 \end{bmatrix}$$

where λ_i are called *principal moments*.

Ok, now we have two problems. The first problem is how to prove that any non-symmetric solid has principal axes and the second problem is how to find the principal axes. Herein, we focus on the second problem, being pragmatic. But wait, why the angular momentum being parallel to the rotation axis is important? Otherwise, people did not spend time studying this case.

To find the principal axes, we use the fact that for a principal axis through a certain origin O, if the angular velocity points along this axis, then the angular momentum is parallel to $\boldsymbol{\omega}$, that is:

$$\mathbf{I}_\omega\boldsymbol{\omega} = \lambda\boldsymbol{\omega} \tag{9.10.10}$$

And this is an *eigenvalue equation*. A vector $\boldsymbol{\omega}$ satisfying Eq. (9.10.10) is called an *eigenvector*, and the corresponding number λ, the corresponding *eigenvalue*.[49] To solve the eigenvalue equation, we rewrite it in this form $(\mathbf{I}_\omega - \lambda\mathbf{I})\boldsymbol{\omega} = \mathbf{0}$. This equation only has non-zero solution (i.e., $\boldsymbol{\omega} \neq \mathbf{0}$) only when the determinant of the coefficient matrix is zero (if the determinant is not zero, then the only solution is $\boldsymbol{\omega} = \mathbf{0}$, similar to equation $2x = 0$). That is,

$$\det(\mathbf{I}_\omega - \lambda\mathbf{I}) = 0 \tag{9.10.11}$$

[49] The German adjective eigen means "own" or "characteristic of". Eigenvalues and eigenvectors are characteristic of a matrix in the sense that they contain important information about the nature of the matrix.

This is called the *characteristic equation* which is a cubic equation in terms of λ. Solving this for λ and substituting λ into Eq. (9.10.10), we get a system of linear equations for three unknowns $\boldsymbol{\omega}$ of which solutions are the eigenvectors (or principal axes).

We consider the cube example again (case a). The characteristic equation is, see Eq. (9.10.9)

$$\begin{vmatrix} 8\mu-\lambda & -3\mu & -3\mu \\ -3\mu & 8\mu-\lambda & -3\mu \\ -3\mu & -3 & 8\mu-\lambda \end{vmatrix} = 0 \Longrightarrow (2\mu-\lambda)(11\mu-\lambda)^2 = 0 \Longrightarrow \begin{cases} \lambda_1 = 2\mu \\ \lambda_2 = 11\mu \\ \lambda_3 = 11\mu \end{cases}$$

with $\mu = Ma^2/12$. First observation: $\lambda_1 + \lambda_2 + \lambda_3 = 24\mu$ and is equal to $I_{11} + I_{22} + I_{33}$. Second observation $\lambda_1\lambda_2\lambda_3 = 242\mu^3$, which is $\det \mathbf{I}_\omega$. So, at least for this example, the sum of the eigenvalues is equal to the trace of the matrix, and the product of the eigenvalues is equal to the determinant of the matrix.

For the first eigenvalue $\lambda = 2\mu$, we have this system of equations:

$$\begin{bmatrix} 6\mu & -3\mu & -3\mu \\ -3\mu & 6\mu & -3\mu \\ -3\mu & -3\mu & 6\mu \end{bmatrix} \begin{bmatrix} \omega_1 \\ \omega_2 \\ \omega_3 \end{bmatrix} = \begin{bmatrix} 0 \\ 0 \\ 0 \end{bmatrix}$$

of which the solution is $\omega_1 = \omega_2 = \omega_3$. So, the first principal axis is $\boldsymbol{a}_1 = (1/\sqrt{3})(1, 1, 1)$.

For the second and third eigenvalues $\lambda = 11\mu$, we have this system of equations:

$$\begin{bmatrix} -3\mu & -3\mu & -3\mu \\ -3\mu & -3\mu & -3\mu \\ -3\mu & -3\mu & -3\mu \end{bmatrix} \begin{bmatrix} \omega_1 \\ \omega_2 \\ \omega_3 \end{bmatrix} = \begin{bmatrix} 0 \\ 0 \\ 0 \end{bmatrix}$$

of which the solution is $\omega_1 + \omega_2 + \omega_3 = 0$. We are looking for the other two axes, so we think of vectors perpendicular to the first principal axis, i.e., $\boldsymbol{a}_1$. So, we write $\omega_1 + \omega_2 + \omega_3 = 0$ as $\boldsymbol{\omega} \cdot \boldsymbol{a}_1 = 0$. *This indicates that the other two axes are perpendicular to the first axis.* Later on we shall prove that the eigenvectors corresponding to distinct eigenvalues are orthogonal if the matrix is symmetric.

Principal stresses and principal planes. It is a fact that the same thing happens again and again in many different fields. Herein, I demonstrate this by presenting principal stresses and principal planes from a field called solid mechanics or mechanics of materials. This field is studied by civil engineers, mechanical engineers, aerospace engineers, and those people who want to design structures and machines (i.e., those made of solid materials).

Similar to $\mathbf{I}_\omega$, $\boldsymbol{\omega}$, and $\boldsymbol{l}$, in solid mechanics there are the (second order) stress tensor $\boldsymbol{\sigma}$, the normal vector $\boldsymbol{n}$, and the traction vector $\boldsymbol{t}$. And we also have a relation between them discovered by Cauchy:

$$\boldsymbol{t} = \boldsymbol{\sigma n} \tag{9.10.12}$$

Again $\boldsymbol{t}$ is in general not in the same direction as $\boldsymbol{n}$. So, principal planes are those with normal vectors $\boldsymbol{n}$ such that $\boldsymbol{\sigma n} = \sigma \boldsymbol{n}$, with σ being called the principal stresses (there are three principal stresses).

9.10.3 Eigenvalues and Eigenvectors

We now provide a formal definition of eigenvectors and eigenvalues of a square matrix.

Definition 9.10.1 Let $\mathbf{A}$ be an $n \times n$ matrix. A scalar λ is called an eigenvalue of $\mathbf{A}$ if there is a non-zero vector $\boldsymbol{x}$ such that $\mathbf{A}\boldsymbol{x} = \lambda\boldsymbol{x}$. Such a vector is called an eigenvector of $\mathbf{A}$ corresponding to λ.

If $\boldsymbol{x}$ is an eigenvector of $\mathbf{A}$ with the corresponding eigenvalue λ, then $\mathbf{A}\boldsymbol{x} = \lambda x$, which leads to $\mathbf{A}(c\boldsymbol{x}) = c(\mathbf{A}\boldsymbol{x}) = c\lambda\boldsymbol{x} = \lambda(c\boldsymbol{x})$. This means that any non-zero multiple of $\boldsymbol{x}$ (that is, $c\boldsymbol{x}$) is also an eigenvector. Thus, if we want to search for eigenvectors geometrically, we need only to consider the effect of $\mathbf{A}$ on unit vectors. Figure 9.24 shows what happens when we transform unit vectors with matrices. All transformed vectors lie on the surface of an ellipse; refer to Sect. 9.12.2 for an explanation.

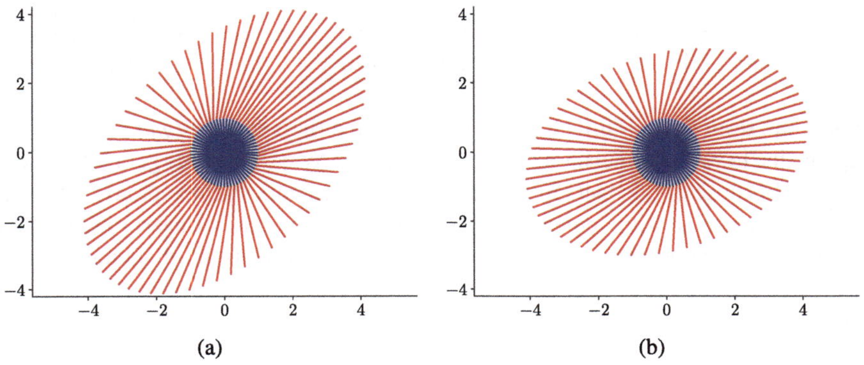

Fig. 9.24 Eigenpicture: $\boldsymbol{x}$ are points on the unit circle (highlighted by blue) and the transformed vectors $\mathbf{A}\boldsymbol{x}$, highlighted by red, are plotted head to tail with $\boldsymbol{x}$. The eigenvector is the one in which the blue and red vectors are aligned

Example 9.10 Find the eigenvalues and the eigenspaces of

$$\mathbf{A} = \begin{bmatrix} 0 & +1 & 0 \\ 0 & +0 & 1 \\ 2 & -5 & 4 \end{bmatrix}$$

The characteristic polynomial is

$$\det(\mathbf{A} - \lambda\mathbf{I}) = -\lambda^3 + 4\lambda^2 - 5\lambda + 2 = -(\lambda - 1)(\lambda - 1)(\lambda - 2)$$

Thus, the characteristic equation is $(\lambda - 1)(\lambda - 1)(\lambda - 2) = 0$, which has solutions $\lambda_1 = \lambda_2 = 1$ and $\lambda_3 = 2$. Note that even though $\mathbf{A}$ is a 3×3 matrix, it has only two distinct eigenvalues. But if we count multiplicities (repeated roots), then $\mathbf{A}$ has exactly three eigenvalues. The *algebraic multiplicity* of an eigenvalue is its multiplicity as a root of the characteristic equation. Thus, $\lambda = 1$ has algebraic multiplicity 2 and $\lambda = 2$ has algebra multiplicity 1.

Now, to find the eigenvectors for a certain λ, we such for $\boldsymbol{x}$ such that

$$(\mathbf{A} - \lambda\mathbf{I})\boldsymbol{x} = 0$$

Thus, the eigenvector $\boldsymbol{x}$ is in the null space of $\mathbf{A} - \lambda\mathbf{I}$. The set of all eigenvectors and the zero vector forms a subspace known as an eigenspace and denoted by E_λ. Now for $\lambda_1 = \lambda_2 = 1$ we need to find the null space of $\mathbf{A} - \mathbf{I}$ (using Gauss elimination)[50]:

$$\mathbf{A} - \mathbf{I} = \begin{bmatrix} -1 & 1 & 0 \\ 0 & -1 & 1 \\ 2 & -5 & 3 \end{bmatrix} \Longrightarrow \left[\begin{array}{ccc|c} -1 & 0 & -1 & 0 \\ 0 & 1 & -1 & 0 \\ 0 & 0 & 0 & 0 \end{array}\right] \Longrightarrow E_1 = \text{span}\left(\begin{bmatrix} 1 \\ 1 \\ 1 \end{bmatrix}\right)$$

Similarly, to find the eigenvectors for $\lambda_3 = 2$, we look for the null space of $\mathbf{A} - 2\mathbf{I}$:

$$\mathbf{A} - 2\mathbf{I} = \begin{bmatrix} -2 & 1 & 0 \\ 0 & -2 & 1 \\ 2 & -5 & 2 \end{bmatrix} \Longrightarrow \left[\begin{array}{ccc|c} 1 & 0 & -1/4 & 0 \\ 0 & 1 & -1/2 & 0 \\ 0 & 0 & 0 & 0 \end{array}\right] \Longrightarrow E_2 = \text{span}\left(\begin{bmatrix} 1 \\ 2 \\ 4 \end{bmatrix}\right)$$

Note that, $\dim(E_1) = \dim(E_2) = 1$. Let us define the *geometric multiplicity* of an eigenvalue to be the dimension of its eigenspace. Why we need this geometric multiplicity? Because of this fact: an $n \times n$ matrix is diagonalizable if and only if the sum of the dimensions of the eigenspaces is n or the matrix has n linearly independent eigenvectors. (Thus, the matrix considered in this example is not diagonalizable.)

[50] Why we see a row full of zeros? This is because $\mathbf{A} - \mathbf{I}$ is singular by definition of eigenvectors.

9.10.4 More on Eigenvectors/Eigenvalues

If we take a 3×3 lower (or upper) triangular matrix $\mathbf{A}$ and compute its eigenvalues, we shall find that the process is easy. This is because when $\mathbf{A}$ is a triangular matrix, so is $\mathbf{A} - \lambda\mathbf{I}$: this is a triangular matrix with diagonal entries $a_{ii} - \lambda$. And we know that the determinant of a triangular matrix is the product of the diagonal terms, thus the characteristic equation is simply the following factored cubic equation:

$$(a_{11} - \lambda)(a_{22} - \lambda)(a_{33} - \lambda) = 0$$

From that, the eigenvalues are simply the diagonal entries of the triangular matrix. This fact holds for any $n \times n$ triangular matrix, including diagonal matrices.

Below are some properties of eigenvectors and eigenvalues (Properties 1 and 3 look similar, but pay attention to the highlighted words):

1. If $\mathbf{A}\boldsymbol{x} = \lambda\boldsymbol{x}$, then $\mathbf{A}^2\boldsymbol{x} = \lambda^2\boldsymbol{x}$ and $\mathbf{A}^n\boldsymbol{x} = \lambda^n\boldsymbol{x}$, for a **positive integer** n.[51]
2. If $\mathbf{A}\boldsymbol{x} = \lambda\boldsymbol{x}$, then $\mathbf{A}^{-1}\boldsymbol{x} = \lambda^{-1}\boldsymbol{x}$.
3. If $\mathbf{A}\boldsymbol{x} = \lambda\boldsymbol{x}$, then $\mathbf{A}^n\boldsymbol{x} = \lambda^n\boldsymbol{x}$, for **any integer** n.[52]
4. A square matrix $\mathbf{A}$ is invertible if and only if 0 is not an eigenvalue of $\mathbf{A}$.[53]
5. Let $\mathbf{A}$ be an $n \times n$ matrix and let $\lambda_1, \lambda_2, \ldots, \lambda_m$ be distinct eigenvalues of $\mathbf{A}$ with corresponding eigenvectors $\boldsymbol{v}_1, \boldsymbol{v}_2, \ldots, \boldsymbol{v}_m$. Then, these eigenvectors are linear independent.
6. Let $\lambda_1, \lambda_2, \ldots, \lambda_n$ be a complete set of eigenvalues of an $n \times n$ matrix $\mathbf{A}$, then $\det(\mathbf{A}) = \prod_i \lambda_i$, and $\mathrm{tr}(\mathbf{A}) = \sum_i \lambda_i$.

Proof [Proof of 5] For simplicity the proof is for a 2×2 matrix only. The two eigenvectors of $\mathbf{A}$ are $\boldsymbol{x}_1, \boldsymbol{x}_2$. Suppose that $c_1\boldsymbol{x}_1 + c_2\boldsymbol{x}_2 = \mathbf{0}$. Multiplying it with $\mathbf{A}$ yields $c_1\lambda_1\boldsymbol{x}_1 + c_2\lambda_2\boldsymbol{x}_2 = \mathbf{0}$ and multiplying it with λ_2 gives $c_1\lambda_2\boldsymbol{x}_1 + c_2\lambda_2\boldsymbol{x}_2 = \mathbf{0}$. Subtracting the obtained two equations yields

$$(\lambda_1 - \lambda_2)c_1\boldsymbol{x}_1 = \mathbf{0}$$

Now that $\lambda_1 \neq \lambda_2$ and $\boldsymbol{x}_1 \neq \mathbf{0}$ (the premise of the problem), thus we must have $c_1 = 0$. Doing the same thing we also get $c_2 = 0$. Thus, the eigenvectors are linear independent. ■

---

[51] We write $\mathbf{A}^2\boldsymbol{x} = \mathbf{A}\mathbf{A}\boldsymbol{x} = \mathbf{A}(\mathbf{A}\boldsymbol{x}) = \mathbf{A}(\lambda\boldsymbol{x}) = \lambda(\mathbf{A}\boldsymbol{x}) = \lambda(\lambda\boldsymbol{x}) = \lambda^2\boldsymbol{x}$.

[52] This holds because $(\mathbf{A}^n)^{-1} = (\mathbf{A}^{-1})^n$ for positive integer n.

[53] Since $\mathbf{A}$ is only invertible when $\det\mathbf{A} \neq 0$, which is equivalent to $\det(\mathbf{A} - 0\mathbf{I}) \neq 0$. Thus 0 is not an eigenvalue of $\mathbf{A}$ when it is invertible.

9.10.5 Symmetric Matrices

The question addressed in this section is: what is special about $\mathbf{A}\boldsymbol{x} = \lambda\boldsymbol{x}$ when $\mathbf{A}$ is symmetric? As we shall see, there exist many nice results about eigenvectors/eigenvalues when the matrix is symmetric. Strang wrote "*It is no exaggeration to say that symmetric matrices are the most important matrices the world will ever see*".

The first nice result is stated in the following theorem.

Theorem 9.10.1 *If* $\mathbf{A}$ *is a symmetric real matrix, then its eigenvalues are real.*

Proof How we're going to prove this theorem? Let us denote by $\boldsymbol{x}$ and λ the eigenvector and eigenvalue of $\mathbf{A}$; λ might be a complex number of the form $a + bi$ and the components of $\boldsymbol{x}$ may be complex numbers. Our task is now to prove that λ is real. One way is to prove that the complex conjugate of λ, which is $\overline{\lambda} = a - bi$, is equal to λ. That is, prove $\overline{\lambda} = \lambda$. To this end, we need to extend the notion of complex conjugate to vectors and matrices. It turns out to be easy: just replace the entries of vectors/matrices by the conjugates. That is, if $\mathbf{A} = [a_{ij}]$, then its conjugate $\overline{\mathbf{A}} = [\overline{a}_{ij}]$. Properties of complex conjugates as discussed in Sect. 2.25 still apply for matrices/vectors, e.g., $\overline{\mathbf{AB}} = \bar{\mathbf{A}}\bar{\mathbf{B}}$.

We start with $\mathbf{A}\boldsymbol{x} = \lambda\boldsymbol{x}$, and to make $\overline{\lambda}$ appear, take the conjugate of this equation to get $\overline{\mathbf{A}\boldsymbol{x}} = \overline{\lambda\boldsymbol{x}}$. But $\overline{\lambda\boldsymbol{x}} = \overline{\lambda}\overline{\boldsymbol{x}}$ and $\overline{\mathbf{A}\boldsymbol{x}} = \overline{\mathbf{A}}\overline{\boldsymbol{x}} = \mathbf{A}\overline{\boldsymbol{x}}$ (this is so for $\mathbf{A}$ is real, which means that $\overline{\mathbf{A}} = \mathbf{A}$). Thus, we have

$$\mathbf{A}\overline{\boldsymbol{x}} = \overline{\lambda}\overline{\boldsymbol{x}}$$

Now, to use the information that $\mathbf{A}$ is symmetric (which means that $\mathbf{A}^\top = \mathbf{A}$), we transpose the above equation ($\mathbf{A}\overline{\boldsymbol{x}} = \overline{\lambda}\overline{\boldsymbol{x}}$):

$$\overline{\boldsymbol{x}}^\top\mathbf{A} = \overline{\lambda}\overline{\boldsymbol{x}}^\top$$

Now we have two equations:

$$\mathbf{A}\boldsymbol{x} = \lambda\boldsymbol{x}, \quad \overline{\boldsymbol{x}}^\top\mathbf{A} = \overline{\lambda}\overline{\boldsymbol{x}}^\top$$

Next, we compute the dot product of the first equation with $\overline{\boldsymbol{x}}^\top$, and the dot product of the second equation with $\boldsymbol{x}$, we then obtain

$$\overline{\boldsymbol{x}}^\top\mathbf{A}\boldsymbol{x} = \overline{\boldsymbol{x}}^\top\lambda\boldsymbol{x}, \quad \overline{\boldsymbol{x}}^\top\mathbf{A}\boldsymbol{x} = \overline{\lambda}\overline{\boldsymbol{x}}^\top\boldsymbol{x}, \quad \Longrightarrow \lambda\overline{\boldsymbol{x}}^\top\boldsymbol{x} = \overline{\lambda}\overline{\boldsymbol{x}}^\top\boldsymbol{x} \Longrightarrow (\lambda - \overline{\lambda})\overline{\boldsymbol{x}}^\top\boldsymbol{x} = 0$$

But, $\overline{\boldsymbol{x}}^\top\boldsymbol{x} \neq 0$ as $\boldsymbol{x}$ is not a zero vector (it is an eigenvector). Thus, we must have $\lambda = \overline{\lambda}$ or $a + bi = a - bi$ which leads to $b = 0$. Hence, the eigenvalues are real. ■

We know that for any square matrix, eigenvectors corresponding to distinct eigenvalues are linear independent. For symmetric matrices, something stronger is true: such eigenvectors are orthogonal.[54] So, we have the following theorem.

Theorem 9.10.2 *If* $\mathbf{A}$ *is a symmetric matrix, then any two eigenvectors corresponding to distinct eigenvalues of* $\mathbf{A}$ *are orthogonal.*

The proof of this theorem is not hard, but why we know this result? In Sect. 9.11.4 on matrix diagonalization, we know that we can decompose $\mathbf{A}$ as $\mathbf{A} = \mathbf{V}\boldsymbol{\Lambda}\mathbf{V}^{-1}$. Transposing it gives us $\mathbf{A}^\top = (\mathbf{V}^{-1})^\top\boldsymbol{\Lambda}\mathbf{V}^\top$. As $\mathbf{A}$ is symmetric, we then have $\mathbf{V}\boldsymbol{\Lambda}\mathbf{V}^{-1} = (\mathbf{V}^{-1})^\top\boldsymbol{\Lambda}\mathbf{V}^\top$. We then guess that $\mathbf{V}^\top = \mathbf{V}^{-1}$. Or, $\mathbf{V}^\top\mathbf{V} = \mathbf{I}$: $\mathbf{V}$ is an orthogonal matrix!

Theorem 9.10.3 (Spectral theorem) *Let* $\mathbf{A}$ *be an* $n \times n$ *real symmetric matrix, then it has the factorization* $\mathbf{A} = \mathbf{Q}\boldsymbol{\Lambda}\mathbf{Q}^\top$ *with real eigenvalues in* $\boldsymbol{\Lambda}$ *and orthonormal eigenvectors in the columns of* $\mathbf{Q}$*:*

$$\mathbf{A} = \mathbf{Q}\boldsymbol{\Lambda}\mathbf{Q}^{-1} = \mathbf{Q}\boldsymbol{\Lambda}\mathbf{Q}^\top \quad \textit{with} \quad \mathbf{Q}^\top = \mathbf{Q}^{-1}$$

Next, we derive the so-called spectral decomposition of $\mathbf{A}$. To see the point, assume that $\mathbf{A}$ is a 2×2 matrix, we can then write (from the Spectral theorem)

$$\mathbf{A} = \mathbf{Q}\boldsymbol{\Lambda}\mathbf{Q}^\top = \begin{bmatrix}\mathbf{q}_1 & \mathbf{q}_2\end{bmatrix}\begin{bmatrix}\lambda_1 & 0\\ 0 & \lambda_2\end{bmatrix}\begin{bmatrix}\mathbf{q}_1^\top\\ \mathbf{q}_2^\top\end{bmatrix} = \begin{bmatrix}\mathbf{q}_1 & \mathbf{q}_2\end{bmatrix}\begin{bmatrix}\lambda_1\mathbf{q}_1^\top\\ \lambda_2\mathbf{q}_2^\top\end{bmatrix} = \sum_{i=1}^{2}\lambda_i\mathbf{q}_i\mathbf{q}_i^\top \tag{9.10.13}$$

9.10.6 Quadratic Forms and Positive Definite Matrices

We have seen quadratic forms (e.g., $ax^2 + bxy + cy^2$) when discussing the extrema of functions of two variables (Sect. 6.7) and when talking about the kinetic energy of a 3D rotating body (Sect. 9.10.1). Now is the time for a formal definition of quadratic forms:

Definition 9.10.2 A quadratic form in n variables is a function $f : \mathbb{R}^n \to \mathbb{R}$ of the form:

$$f(\boldsymbol{x}) = \boldsymbol{x}^\top\mathbf{A}\boldsymbol{x}$$

where $\mathbf{A}$ is a symmetric $n \times n$ matrix and $\boldsymbol{x} \in \mathbb{R}^n$. We refer to $\mathbf{A}$ as the matrix associated with the quadratic form f.

---

[54] The proof goes as $\lambda_1\boldsymbol{x}_1 \cdot \boldsymbol{x}_2 = \cdots = \lambda_2\boldsymbol{x}_1 \cdot \boldsymbol{x}_2$, thus $(\lambda_1 - \lambda_2)\boldsymbol{x}_1 \cdot \boldsymbol{x}_2 = 0$. But $\lambda_1 \neq \lambda_2$.

Definiteness of quadratic forms. Let $\boldsymbol{x}$ be a vector in $\mathbb{R}^n$ and $f(\boldsymbol{x}) = \boldsymbol{x}^\top \mathbf{A}\boldsymbol{x}$ is a quadratic form. An important property of $f(\boldsymbol{x})$ is its definiteness, defined as

$$\begin{array}{ll} f(\boldsymbol{x}) \text{ is} \textbf{positive definite} & f(\boldsymbol{x}) > 0 \;\; \forall \boldsymbol{x} \neq \mathbf{0} \\ f(\boldsymbol{x}) \text{ is} \textbf{positive semi-definite} & f(\boldsymbol{x}) \geq 0 \;\; \forall \boldsymbol{x} \\ f(\boldsymbol{x}) \text{ is} \textbf{negative definite} & f(\boldsymbol{x}) < 0 \;\; \forall \boldsymbol{x} \neq \mathbf{0} \\ f(\boldsymbol{x}) \text{ is} \textbf{negative semi-definite} & f(\boldsymbol{x}) \leq 0 \;\; \forall \boldsymbol{x} \\ f(\boldsymbol{x}) \text{ is} \textbf{indefinite} & f(\boldsymbol{x}) \text{ takes both +/- values} \end{array} \tag{9.10.14}$$

If $f(\boldsymbol{x})$ is positive definite, then its associated matrix $\mathbf{A}$ is said to be a *positive definite matrix*.

The next problem we have to solve is: when a quadratic form is positive definite? What is then the properties of $\mathbf{A}$? To answer this question, one observation is that, if there is no cross term in $f(\boldsymbol{x})$, then it is easy to determine its positive definiteness. One example is enough to convince us: $f(\boldsymbol{x}) = 2x^2 + 4y^2$ is positive semi-definite (PSD). Furthermore, without the cross term, the associated matrix is diagonal:

$$f(\boldsymbol{x}) = 2x^2 + 4y^2 = \begin{bmatrix} x & y \end{bmatrix} \begin{bmatrix} 2 & 0 \\ 0 & 4 \end{bmatrix} \begin{bmatrix} x \\ y \end{bmatrix} \tag{9.10.15}$$

Diagonal matrices? We need the spectral theorem (Theorem 9.10.3) that states that an $n \times n$ real symmetric matrix has the factorization $\mathbf{A} = \mathbf{Q}\boldsymbol{\Lambda}\mathbf{Q}^\top$ with real eigenvalues in the diagonal matrix $\boldsymbol{\Lambda}$ and orthonormal eigenvectors in the columns of $\mathbf{Q}$. Thus, we do a change of variable $\boldsymbol{x} = \mathbf{Q}\boldsymbol{y}$, and compute the quadratic form with this new variable $\boldsymbol{y}$, magic will happen[55]:

$$f(\boldsymbol{x}) = \boldsymbol{x}^\top \mathbf{A}\boldsymbol{x} = (\mathbf{Q}\boldsymbol{y})^\top \mathbf{A}(\mathbf{Q}\boldsymbol{y}) = \boldsymbol{y}^\top \underbrace{\mathbf{Q}^\top \mathbf{A}\mathbf{Q}}_{\boldsymbol{\Lambda}} \boldsymbol{y} = \boldsymbol{y}^\top \boldsymbol{\Lambda} \boldsymbol{y} = \sum_{i=1}^{n} \lambda_i y_i^2 \tag{9.10.16}$$

Obviously $\boldsymbol{y}^\top \boldsymbol{\Lambda} \boldsymbol{y}$ is of the form in the RHS of Eq. (9.10.15).

With this result, i.e., Eq. (9.10.16), it is now straightforward to say about the definiteness of a quadratic form: *it all depends on the eigenvalues* of $\mathbf{A}$:

$$\begin{array}{ll} f(\boldsymbol{x}) \text{ is} \textbf{positive definite} & \mathbf{A} \text{has all positive eigenvalues} \\ f(\boldsymbol{x}) \text{ is} \textbf{positive semi-definite} & \mathbf{A} \text{has all non-negative eigenvalues} \\ f(\boldsymbol{x}) \text{ is} \textbf{negative definite} & \mathbf{A} \text{has all negative eigenvalues} \\ f(\boldsymbol{x}) \text{ is} \textbf{negative semi-definite} & \mathbf{A} \text{has all non-positive eigenvalues} \\ f(\boldsymbol{x}) \text{ is} \textbf{indefinite} & \mathbf{A} \text{has both positive/negative eigenvalues} \end{array} \tag{9.10.17}$$

Principal axes theorem and ellipses. Equation (9.10.16) is the theorem of principal axes in mechanics. This theorem tells us that any quadratic form can be written in a form without the cross terms. This is achieved by using a change of variable

---

55 We cannot know this will work, but we have to try and usually pieces of mathematics fit nicely together.

$\boldsymbol{x} = \mathbf{Q}\boldsymbol{y}$. Now, we explain the name of the theorem. Consider the following conic section (Sect. 4.1.5):

$$5x^2 + 8xy + 5y^2 = 1 \Longleftrightarrow \boldsymbol{x}^\top \mathbf{A}\boldsymbol{x} = 1, \quad \boldsymbol{x} = (x, y), \quad \mathbf{A} = \begin{bmatrix} 5 & 4 \\ 4 & 5 \end{bmatrix}$$

First, the eigenvectors and eigenvalues of $\mathbf{A}$[56]:

$$\lambda_1 = 1, \lambda_2 = 9; \quad \boldsymbol{v}_1 = (1/\sqrt{2}, -1/\sqrt{2}), \ \boldsymbol{v}_2 = (1/\sqrt{2}, 1/\sqrt{2})$$

Then, the following change of variable (If you check Sect. 4.1.5 again you would see that this change of variable is exactly the rotation mentioned in that section. Here, we have $A = C = 5$, thus the rotation angle is $-\pi/4$.)

$$\boldsymbol{x} = \mathbf{Q}\boldsymbol{x}', \quad \mathbf{Q} = \begin{bmatrix} +1/\sqrt{2} & 1/\sqrt{2} \\ -1/\sqrt{2} & 1/\sqrt{2} \end{bmatrix}, \quad \boldsymbol{x}' = (y_1, y_2)$$

results in (see Eq. (9.10.16))

$$1y_1^2 + 9y_2^2 = 1, \quad \text{or} \quad \left(\frac{y_1}{1/\sqrt{1}}\right)^2 + \left(\frac{y_1}{1/\sqrt{9}}\right)^2 = 1$$

Thus, our conic is an ellipse. The second expression gives us the length of the ellipse axes. Now, to graph this ellipse we need to know its axes. To this end, we need to know where is the unit vector in the (y_1, y_2) coordinate systems: $\boldsymbol{e}'_1 = (1, 0)$. Using $\boldsymbol{x} = \mathbf{Q}\boldsymbol{y}$, we have

$$\mathbf{Q}\boldsymbol{e}'_1 = \begin{bmatrix} +1/\sqrt{2} & 1/\sqrt{2} \\ -1/\sqrt{2} & 1/\sqrt{2} \end{bmatrix} \begin{bmatrix} 1 \\ 0 \end{bmatrix} = \begin{bmatrix} +1/\sqrt{2} \\ -1/\sqrt{2} \end{bmatrix}$$

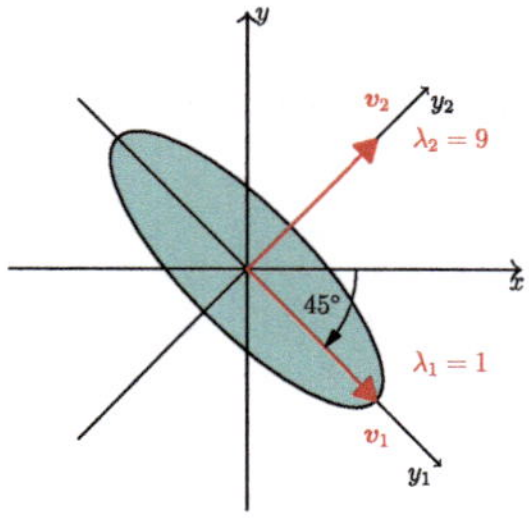

[56] You can reverse the direction of $\boldsymbol{v}_1$.

which is the first eigenvector of $\mathbf{A}$. Similarly, $\boldsymbol{e}_2' = (0, 1)$ is the second eigenvector. Thus, the eigenvectors of $\mathbf{A}$—the matrix associated with a quadratic form—give the directions of the principal axes of the corresponding graph of the quadratic form. This explains why $\mathbf{A} = \mathbf{Q}\boldsymbol{\Lambda}\mathbf{Q}^\top$ is called the principal axis theorem—it displays the axes. What is more, the eigenvalues of $\mathbf{A}$ give us the lengths of the axes. The smaller eigenvalue (1) gives the length of semi-major axis ($1/\sqrt{1}$) and the larger eigenvalue (9) gives the shorter axis (of half length $1/\sqrt{9}$). This geometry will help us to solve constrained optimization problems relating to quadratic forms as explained in what follows.

Constrained optimization problems. I present herein now one application about the definiteness of a quadratic form. Assume that a quadratic form $f(\boldsymbol{x}) = \boldsymbol{x}^\top \mathbf{A}\boldsymbol{x}$ is positive semi-definite, then since $f(\mathbf{0}) = 0$, the minimum value of $f(\boldsymbol{x})$ is zero, without calculus. It is more often that we have to find the maximum/minimum of $f(\boldsymbol{x})$ with $\boldsymbol{x}$ subjected to the constraint $\|\boldsymbol{x}\| = 1$. Thus, we pose the following constrained optimization problem[57]

$$\max_{\boldsymbol{x}\neq\mathbf{0}} \frac{\boldsymbol{x}^\top \mathbf{A}\boldsymbol{x}}{\boldsymbol{x}^\top \boldsymbol{x}} \quad \text{or} \quad \max_{\|\boldsymbol{x}\|=1} \boldsymbol{x}^\top \mathbf{A}\boldsymbol{x}$$

The term $(\boldsymbol{x}^\top \mathbf{A}\boldsymbol{x})/\boldsymbol{x}^\top \boldsymbol{x}$ is called a Rayleigh quotient $R(\boldsymbol{x})$. The solution to this problem actually lies in Eq. (9.10.16): to see that just look at $f = 1y_1^2 + 9y_2^2 = 1$ with the constraint $y_1^2 + y_2^2 = 1$, the maximum is $f = 9$, which is the maximum eigenvalue of the matrix associated with the quadratic form. Thus, we sort the eigenvalues of $\mathbf{A}$ in a descending order $\lambda_1 \geq \lambda_2 \geq \cdots \geq \lambda_n$, then

$$\begin{aligned}
f(\boldsymbol{x}) &= \sum_{i=1}^{n} \lambda_i y_i^2 = \lambda_1 y_1^2 + \lambda_2 y_2^2 + \cdots + \lambda_n y_n^2 \\
&\leq \lambda_1 y_1^2 + \lambda_1 y_2^2 + \cdots + \lambda_1 y_n^2 \\
&\leq \lambda_1 \left(y_1^2 + y_2^2 + \cdots + y_n^2\right) = \lambda_1
\end{aligned}$$

Noting that $\|\boldsymbol{y}\| = 1$; that's why we got the last result.

[57] To see that the two forms are equivalent, we can do this massage:

$$\frac{\boldsymbol{x}^\top \mathbf{A}\boldsymbol{x}}{\boldsymbol{x}^\top \boldsymbol{x}} = \frac{\boldsymbol{x}^\top \mathbf{A}\boldsymbol{x}}{\|\boldsymbol{x}\|^2} = \frac{\boldsymbol{x}^\top}{\|\boldsymbol{x}\|}\mathbf{A}\frac{\boldsymbol{x}}{\|\boldsymbol{x}\|} = \boldsymbol{y}^\top \mathbf{A}\boldsymbol{y} \quad (\|\boldsymbol{y}\| = 1)$$

Another derivation.

A symmetric matrix $\mathbf{A}$ has orthonormal eigenvectors $\boldsymbol{v}_1, \ldots, \boldsymbol{v}_n$. Thus, we can write $\boldsymbol{x} = c_1\boldsymbol{v}_1 + \cdots + c_n\boldsymbol{v}_n$, hence

$$\|\boldsymbol{x}\|^2 = \boldsymbol{x}^\top\boldsymbol{x} = c_1^2 + \cdots + c_n^2, \quad \boldsymbol{x}^\top\mathbf{A}\boldsymbol{x} = \lambda_1 c_1^2 + \cdots + \lambda_n c_n^2$$

Therefore, the Rayleigh quotient $R(\boldsymbol{x})$ is now written as

$$\frac{\boldsymbol{x}^\top\mathbf{A}\boldsymbol{x}}{\boldsymbol{x}^\top\boldsymbol{x}} = \frac{\lambda_1 c_1^2 + \cdots + \lambda_n c_n^2}{c_1^2 + \cdots + c_n^2}$$

From that it can be seen that the maximum of $R(\boldsymbol{x})$ is λ_1. What is nice with this form of $R(\boldsymbol{x})$ is the ease to find the maximum of $R(\boldsymbol{x})$ when the constraint is that $\boldsymbol{x}$ is perpendicular to $\boldsymbol{v}_1$. This constraint means that $c_1 = 0$, thus

$$\max \frac{\boldsymbol{x}^\top\mathbf{A}\boldsymbol{x}}{\boldsymbol{x}^\top\boldsymbol{x}} = \max \frac{\lambda_2 c_2^2 + \cdots + \lambda_n c_n^2}{c_2^2 + \cdots + c_n^2} = \lambda_2$$

9.11 Vector Spaces

Up to this point we have seen many mathematical objects: numbers, vectors, matrices, and functions. Do these different objects share any common thing? Many, actually. First, we can add two numbers, we can add two vectors, we can add two matrices, and of course we can add two functions. Second, we can multiply a vector by a scalar, a matrix by a scalar, and a function by a scalar. Third, adding two vectors gives us a new vector, adding two matrices returns a matrix, and adding two functions gives us a function (not anything else).

We believe the following equation showing a vector in $\mathbb{R}^4$, a polynomial of degree less than or equal 3, and a 2×2 matrix

$$\boldsymbol{u} = \begin{bmatrix} a \\ b \\ c \\ d \end{bmatrix}, \quad p(x) = a + bx + cx^2 + dx^3, \quad \mathbf{A} = \begin{bmatrix} a & b \\ c & d \end{bmatrix}$$

is a good illustration that all these objects are related. After all, they are represented by four numbers a, b, c, d.[58]

It seems reasonable and logical now for mathematicians to *unify all these seemingly different but similar objects*. Here comes *vector spaces*, which constitute the most abstract part of linear algebra. The term vector spaces is a bit confusing because not all objects in vector spaces are vectors, e.g., matrices are not vectors. A better name would probably be linear spaces. About the power of algebra, Jean le Rond d'Alambert wrote: "*Algebra is generous; she often gives more than is asked of her*".

9.11.1 Vector Spaces

To define a vector space, let V be a set of objects $\boldsymbol{u}$, $\boldsymbol{v}$, $\boldsymbol{w}$, ... on which two operations, called addition and scalar multiplication, are defined: the sum of $\boldsymbol{u}$ and $\boldsymbol{v}$ is denoted by $\boldsymbol{u}+\boldsymbol{v}$, and if α is a scalar, the scalar multiple of $\boldsymbol{v}$ is denoted by $\alpha\boldsymbol{v}$. Then, V is defined as a vector space (sometimes also referred to as linear space) if the following ten axioms are satisfied (α, β are scalars):

$$\begin{array}{ll}
\text{(1) commutativity of addition:} & \boldsymbol{u}+\boldsymbol{v}=\boldsymbol{v}+\boldsymbol{u} \\
\text{(2) associativity of addition:} & (\boldsymbol{u}+\boldsymbol{v})+\boldsymbol{w}=\boldsymbol{u}+(\boldsymbol{v}+\boldsymbol{w}) \\
\text{(3) identity element of addition:} & \boldsymbol{u}+\boldsymbol{0}=\boldsymbol{u} \\
\text{(4) inverse element of addition:} & \boldsymbol{u}+(-\boldsymbol{u})=\boldsymbol{0} \\
\text{(5) distributivity w.r.t. vector addition:} & \alpha(\boldsymbol{u}+\boldsymbol{v})=\alpha\boldsymbol{u}+\alpha\boldsymbol{v} \\
\text{(6) distributivity w.r.t field addition:} & (\alpha+\beta)\boldsymbol{u}=\alpha\boldsymbol{u}+\beta\boldsymbol{v} \\
\text{(7) distributivity:} & \alpha(\beta\boldsymbol{u})=(\alpha\beta)\boldsymbol{u} \\
\text{(8) identity element of multiplication:} & 1\boldsymbol{u}=\boldsymbol{u} \\
\text{(9) closure under addition:} & \boldsymbol{u}+\boldsymbol{v}\in V \\
\text{(10) closure under multiplication:} & \beta\boldsymbol{u}\in V
\end{array} \tag{9.11.1}$$

So, a vector space is a set of objects called vectors, which may be added together and multiplied ("scaled") by numbers, called scalars and these vectors satisfy the above ten axioms. Sometimes we see this notation $(V, \mathbb{R}, +, \cdot)$ to denote a vector space V over $\mathbb{R}$ with the two operations of addition and multiplication.

Example 1 Of course $\mathbb{R}^n$ with $n \geq 1$ is a vector space. All the ten axioms of a vector space can be verified easily.

Example 2 Let $\mathcal{P}_2$ be the set of all polynomials of degree less than or equal 2 with real coefficients. To see if $\mathcal{P}_2$ is a vector space, we first need to define the two basic operations of addition and scalar multiplication. If $p(x)$, $q(x)$ are two objects in

---

[58] We can view $p(x) = a + bx + cx^2 + dx^3$ as a space—similar to $\mathbb{R}^n$—with a basis of $\{1, x, x^2, x^3\}$. Thus, (a, b, c, d) are the coordinates of $p(x)$ with respect to that basis. And (a, b, c, d) can also be seen as the coordinates of a point in $\mathbb{R}^4$!

$\mathcal{P}_2$, then $p(x) = a_0 + a_1x + a_2x^2$ and $q(x) = b_0 + b_1x + b_2x^2$. Addition and scalar multiplication are defined as

$$p(x) + q(x) = (a_0 + b_0) + (a_1 + b_1)x + (a_2 + b_2)x^2, \quad \alpha p(x) = \alpha a_0 + \alpha a_1 x + \alpha a_2 x^2$$

This verifies the last two axioms on closure. The identity element for addition is the polynomial with all coefficients being zero. The inverse element of addition of $p(x)$ is $-p(x) = -a_0 - a_1x - a_2x^2$. Verification of other axioms is straightforward as they come from the arithmetic rules of real numbers.

Example 3 Let us denote by $\mathcal{F}$ the set of all real-valued functions defined on the real line. If $f(x)$ and $g(x)$ are such two functions and α is a scalar, then we define $(f + g)(x)$ and $\alpha f(x)$ as

$$(f + g)(x) := f(x) + g(x), \quad (\alpha f)(x) := \alpha f(x) \tag{9.11.2}$$

The zero function is $f(x) = 0$ for all x. The negative function $(-f)(x)$ is $-f(x)$. It can then be seen that $\mathcal{F}$ is a vector space, but *a vector space of infinite dimension*. Usually linear algebra deals with finite-dimensional vector spaces and functional analysis concerns infinite-dimensional vector space. But we do not follow this convention and cover both spaces in this chapter. Similarly, we have another vector space $\mathcal{F}[a, b]$ that contains all real-valued functions defined on the interval $[a, b]$.

Example 4 All rectangular matrices of shape $m \times n$ belong to a vector space $\mathbb{R}^{m \times n}$. From Sect. 9.4.2, we can verify that matrices obey the ten axioms of linear spaces. And the columns of a $m \times n$ matrix are also vector spaces because a column is a $\mathbb{R}^m$ vector.

If matrices are vectors, then we can do a linear combination of matrices, we can talk about linearly independent matrices. For example, consider the space of all 2×2 matrices M. It is obvious that we can write any such matrix as

$$\begin{bmatrix} a & b \\ c & d \end{bmatrix} = a\begin{bmatrix} 1 & 0 \\ 0 & 0 \end{bmatrix} + b\begin{bmatrix} 0 & 1 \\ 0 & 0 \end{bmatrix} + c\begin{bmatrix} 0 & 0 \\ 1 & 0 \end{bmatrix} + d\begin{bmatrix} 0 & 0 \\ 0 & 1 \end{bmatrix}$$

The red matrices are linear independent, and they are the basis vectors of M; they play the same roles of the unit vectors $\boldsymbol{e}_i$ that we're familiar with.

If $\boldsymbol{a} + \boldsymbol{c} = \boldsymbol{b} + \boldsymbol{c}$ then $\boldsymbol{a} = \boldsymbol{b}$ for $\boldsymbol{a}, \boldsymbol{b}, \boldsymbol{c}$ being scalars, $n-$ vectors or matrices. Thus, we guess that this holds for any vector in a vector space. The following theorem is a summary of some properties that vectors in a vector space satisfy. These properties are called the trivial consequences of the axioms as they look obvious.

Theorem 9.11.1 *Let V be a vector space, and $\boldsymbol{a}, \boldsymbol{b}, \boldsymbol{c}$ be vectors in V and c is a scalar. Then, we have*

(a) If $\boldsymbol{a} + \boldsymbol{c} = \boldsymbol{b} + \boldsymbol{c}$ then $\boldsymbol{a} = \boldsymbol{b}$.
(b) If $\boldsymbol{a} + \boldsymbol{b} = \boldsymbol{b}$ then $\boldsymbol{a} = \boldsymbol{0}$.

(c) $\lambda\mathbf{0} = \mathbf{0}$, $0\boldsymbol{v} = \mathbf{0}$.
(d) $(-1)\boldsymbol{v} = -\boldsymbol{v}$.
(e) *If* $c\boldsymbol{v} = \mathbf{0}$, *then either* $c = 0$ *or* $\boldsymbol{v} = \mathbf{0}$.

Proof Proof of (a) goes as follows:

$$\begin{aligned}
\boldsymbol{a} &= \boldsymbol{a} + \mathbf{0} && \text{(axiom 3)} \\
&= \boldsymbol{a} + (\boldsymbol{c} + \boldsymbol{x}) && (\boldsymbol{x}\text{is the identity element for addition of } \boldsymbol{c}) \\
&= (\boldsymbol{a} + \boldsymbol{c}) + \boldsymbol{x} && \text{(axiom 2)} \\
&= (\boldsymbol{b} + \boldsymbol{c}) + \boldsymbol{x} && \text{(given)} \\
&= \boldsymbol{b} + (\boldsymbol{c} + \boldsymbol{x}) && \text{(axiom 2)} \\
&= \boldsymbol{b} + \mathbf{0} && (\boldsymbol{x}\text{is the identity element for addition of } \boldsymbol{c}) \\
&= \boldsymbol{b} && \text{(axiom 3)}
\end{aligned}$$

Proof of (b) is based on (a):

$$\boldsymbol{a} + \boldsymbol{b} = \boldsymbol{b} = \boldsymbol{b} + \mathbf{0} \Longrightarrow \boldsymbol{a} = \mathbf{0} \quad \text{(using (a))}$$

Proof of (c) is based on axioms 5/6[59]:

$$\lambda\mathbf{0} = \lambda(\mathbf{0} + \mathbf{0}) \overset{ax.5}{=} \lambda\mathbf{0} + \lambda\mathbf{0} \overset{(b)}{\Longrightarrow} \lambda\mathbf{0} = \mathbf{0}$$

$$0\boldsymbol{v} = (0 + 0)\boldsymbol{v} \overset{ax.5}{=} 0\boldsymbol{v} + 0\boldsymbol{v} \overset{(b)}{\Longrightarrow} 0\boldsymbol{v} = \mathbf{0}$$

Proof of (d) is

$$\boldsymbol{v} + (-1)\boldsymbol{v} = 1\boldsymbol{v} + (-1)\boldsymbol{v} \overset{ax.6}{=} (1 + (-1))\boldsymbol{v} = 0\boldsymbol{v} \overset{(c)}{=} \mathbf{0}$$

But we know that $\boldsymbol{v} + (-\boldsymbol{v}) = \mathbf{0}$, thus $(-1)\boldsymbol{v} = -\boldsymbol{v}$. Proof of (e) is (we're interested in the case $c \neq 0$ only, otherwise (e) is simply (c))

$$\boldsymbol{v} \overset{ax.8}{=} 1\boldsymbol{v} = \left(c\frac{1}{c}\right)\boldsymbol{v} \overset{ax.7}{=} \frac{1}{c}(c\boldsymbol{v}) = \frac{1}{c}\mathbf{0} \overset{(c)}{=} \mathbf{0}$$

■

9.11.2 Change of Basis

Similar to $\mathbb{R}^n$—which is a vector space—there exists subspaces inside a vector space V, each subspace has a basis that is a set of vectors that are linear independent and

[59] Why? Because these axioms involve the scalar multiplication of vectors.

span the subspace. I do not state the definitions of those concepts here for sake of brevity. Below are some examples.

Example 1. In $\mathcal{P}_2$, determine whether the set $\{1+x, x+x^2, 1+x^2\}$ is linearly independent. Let c_1, c_2, c_3 be scalars such that

$$c_1(1+x) + c_2(x+x^2) + c_3(1+x^2) = 0$$

This is equivalent to

$$(c_1+c_3) + (c_1+c_2)x + (c_2+c_3)x^2 = 0 \times 1 + 0 \times x + 0 \times x^2$$

Equating the coefficients of $1, x, x^2$ results in

$$c_1 + c_3 = 0 \quad c_1 + c_2 = 0, \quad c_2 + c_3 = 0 \Longrightarrow c_1 = c_2 = c_3 = 0$$

It follows that $\{1+x, x+x^2, 1+x^2\}$ is linearly independent.

Example 2. Show that the set $\{1+x, x+x^2, 1+x^2\}$ is a basis for $\mathcal{P}_2$. In the previous example, we have shown that this set is linearly independent. Now to prove that it is the basis for $\mathcal{P}_2$, we just need to show that it spans $\mathcal{P}_2$. In other words, we need to show that we can always find c_1, c_2, c_3 such that

$$c_1(1+x) + c_2(x+x^2) + c_3(1+x^2) = a + bx + cx^2$$

holds for all a, b, c. This is equivalent to

$$(c_1+c_3) + (c_1+c_2)x + (c_2+c_3)x^2 = a + bx + cx^2$$

And we then get again a linear system:

$$c_1 + c_3 = a \quad c_1 + c_2 = b, \quad c_2 + c_3 = c$$

The coefficient matrix of this system is invertible, thus it has a solution. As $\{1+x, x+x^2, 1+x^2\}$ is a basis for $\mathcal{P}_2$, we deduce that $\dim(\mathcal{P}_2) = 2$. And it is a finite-dimensional subspace. The following definition aims to make this precise.

Definition 9.11.1 A vector space V is called finite dimensional if it has a basis consisting of finitely many vectors. The dimension of V, denoted by $\dim V$, is the number of vectors in a basis for V. The dimension of the zero vector space $\{\mathbf{0}\}$ is defined to be zero. A vector space that has no finite basis is called infinite dimensional.

Coordinates. Consider a vector space V with a basis $\mathcal{B} = \{\boldsymbol{v}_1, \boldsymbol{v}_2, \ldots, \boldsymbol{v}_n\}$, any vector $\boldsymbol{v} \in V$ can be written as a unique linear combination of $\boldsymbol{v}_i$'s: $\boldsymbol{v} = c_1\boldsymbol{v}_1 + c_2\boldsymbol{v}_2 + \cdots + c_n\boldsymbol{v}_n$. The vector $[\boldsymbol{v}]_{\mathcal{B}} = (c_1, c_2, \ldots, c_n)$ is called *the coordinate vector of* $\boldsymbol{v}$

with respect to the basis $\mathcal{B}$, or the $\mathcal{B}$-coordinates of $\boldsymbol{v}$. Whatever $\boldsymbol{v}$ is, its $\mathcal{B}-$coordinates is a vector in the familiar $\mathbb{R}^n$. Now with the new object $[\boldsymbol{v}]_{\mathcal{B}}$, certainly we have some rules that this object obey, as stated by the following theorem.[60]

Theorem 9.11.2 *Consider a vector space V with a basis* $\mathcal{B} = \{\boldsymbol{v}_1, \boldsymbol{v}_2, \ldots, \boldsymbol{v}_n\}$. *If we have two vectors* $\boldsymbol{u}$ *and* $\boldsymbol{v}$ *in V and we know their coordinates* $[\boldsymbol{u}]_{\mathcal{B}}$ *and* $[\boldsymbol{v}]_{\mathcal{B}}$, *then we can determine the coordinates of their sum and the coordinates of* $\alpha\boldsymbol{v}$

(a) $[\boldsymbol{u} + \boldsymbol{v}]_{\mathcal{B}} = [\boldsymbol{u}]_{\mathcal{B}} + [\boldsymbol{v}]_{\mathcal{B}}$.
(b) $[\alpha\boldsymbol{v}]_{\mathcal{B}} = \alpha[\boldsymbol{u}]_{\mathcal{B}}$.

Example 9.11 Consider a vector in $\mathcal{P}_2$: $p(x) = a + bx + cx^2$. If we use the standard basis $\mathcal{B} = \{1, x, x^2\}$ for $\mathcal{P}_2$, then it is easy to see that the coordinate vectors of $p(x)$ w.r.t. $\mathcal{B}$ are

$$[p(x)]_{\mathcal{B}} = \begin{bmatrix} a & b & c \end{bmatrix}^\top$$

which is simply a vector in $\mathbb{R}^3$. Thus, $[p(x)]_{\mathcal{B}}$ connects the possibly unfamiliar space $\mathcal{P}_2$ with the familiar space $\mathbb{R}^3$. Points in $\mathcal{P}_2$ can now be identified by their coordinates in $\mathbb{R}^3$, and every vector-space calculation in $\mathcal{P}_2$ is accurately reproduced in $\mathbb{R}^3$ (and vice versa). Note that $\mathcal{P}_2$ is not $\mathbb{R}^3$ but it does look like $\mathbb{R}^3$ as a vector space.

What is $[1]_{\mathcal{B}}$? As we can write $1 = (1)(1) + 0(x) + 0(x^2)$, therefore $[1]_{\mathcal{B}} = (1, 0, 0) = \boldsymbol{e}_1$. Similarly, $[x]_{\mathcal{B}} = (0, 1, 0) = \boldsymbol{e}_2$. So, if $\mathcal{B} = \{\boldsymbol{v}_1, \boldsymbol{v}_2, \ldots, \boldsymbol{v}_n\}$ is a basis for a vector space, then $[\boldsymbol{v}_i]_{\mathcal{B}} = \boldsymbol{e}_i$.

The above example demonstrates that there is a connection between a vector space V and $\mathbb{R}^n$, and the following theorem is one of such connection. We shall use this theorem in Definition 9.11.2 when we discuss the change-of-basis matrix and use it to show that this matrix is invertible.

Theorem 9.11.3 *Let* $\mathcal{B} = \{\boldsymbol{v}_1, \boldsymbol{v}_2, \ldots, \boldsymbol{v}_n\}$ *be a basis for a vector space V and let* $\boldsymbol{u}_1, \boldsymbol{u}_2, \ldots, \boldsymbol{u}_k$ *be vectors in V, then* $\{\boldsymbol{u}_1, \boldsymbol{u}_2, \ldots, \boldsymbol{u}_k\}$ *is linear independent in V if and only if* $\{[\boldsymbol{u}_1]_{\mathcal{B}}, [\boldsymbol{u}_2]_{\mathcal{B}}, \ldots, [\boldsymbol{u}_k]_{\mathcal{B}}\}$ *is linear independent in* $\mathbb{R}^n$.

Proof First, we prove that if $\{\boldsymbol{u}_1, \boldsymbol{u}_2, \ldots, \boldsymbol{u}_k\}$ is linear independent in V then $\{[\boldsymbol{u}_1]_{\mathcal{B}}, [\boldsymbol{u}_2]_{\mathcal{B}}, \ldots, [\boldsymbol{u}_k]_{\mathcal{B}}\}$ is linear independent in $\mathbb{R}^n$. To this end, we consider a linear combination of $\{[\boldsymbol{u}_1]_{\mathcal{B}}, [\boldsymbol{u}_2]_{\mathcal{B}}, \ldots, [\boldsymbol{u}_k]_{\mathcal{B}}\}$ and set it to zero

$$c_1[\boldsymbol{u}_1]_{\mathcal{B}} + c_2[\boldsymbol{u}_2]_{\mathcal{B}} + \cdots + c_k[\boldsymbol{u}_k]_{\mathcal{B}} = \boldsymbol{0}$$

Our task is now to show that $c_1 = c_2 = \cdots = c_k = 0$. Theorem 9.11.2 allows us to rewrite the above as

$$[c_1\boldsymbol{u}_1 + c_2\boldsymbol{u}_2 + \cdots + c_k\boldsymbol{u}_k]_{\mathcal{B}} = \boldsymbol{0}$$

which means that the coordinate vector of $c_1\boldsymbol{u}_1 + c_2\boldsymbol{u}_2 + \cdots + c_k\boldsymbol{u}_k$ w.r.t. $\mathcal{B}$ is the zero vector. Therefore, we can write

---

[60] The proof is straightforward and uses the definition of $[\boldsymbol{v}]_{\mathcal{B}}$. If you're stuck check [54].

$$c_1\boldsymbol{u}_1 + c_2\boldsymbol{u}_2 + \cdots + c_k\boldsymbol{u}_k = 0\boldsymbol{v}_1 + 0\boldsymbol{v}_2 + \cdots + 0\boldsymbol{v}_n = \boldsymbol{0}$$

Since $\{\boldsymbol{u}_1, \boldsymbol{u}_2, \ldots, \boldsymbol{u}_k\}$ is linear independent the above equation forces c_i's to be all zero. ■

Change of basis. Now, we discuss the topic of change of bases. The reason is simple: it is convenient to work with some bases than others. We study how to do a change of bases herein. Consider the easy $\mathbb{R}^2$ plane with two nonstandard bases: $\mathcal{B}$ with $\boldsymbol{u}_1 = (-1, 2)$ and $\boldsymbol{u}_2 = (2, -1)$ and $\mathcal{C}$ with $\boldsymbol{v}_1 = (1, 0)$ and $\boldsymbol{v}_2 = (1, 1)$. Certainly, all these vectors (e.g., $\boldsymbol{u}_1$) are written with respect to the standard basis $(1, 0)$ and $(0, 1)$. The question is: given a vector $\boldsymbol{x}$ with $[\boldsymbol{x}]_{\mathcal{B}} = (1, 3)$, what is $[\boldsymbol{x}]_{\mathcal{C}}$?

The first thing we need to do is to write the basis vectors of $\mathcal{B}$ in terms of those of $\mathcal{C}$[61]:

$$\begin{bmatrix}-1\\+2\end{bmatrix} = -3\begin{bmatrix}1\\0\end{bmatrix} + 2\begin{bmatrix}1\\1\end{bmatrix} \Longrightarrow [\boldsymbol{u}_1]_{\mathcal{C}} = \begin{bmatrix}-3\\+2\end{bmatrix}$$
$$\begin{bmatrix}+2\\-1\end{bmatrix} = +3\begin{bmatrix}1\\0\end{bmatrix} - 1\begin{bmatrix}1\\1\end{bmatrix} \Longrightarrow [\boldsymbol{u}_2]_{\mathcal{C}} = \begin{bmatrix}+3\\-1\end{bmatrix}$$

Now, the vector $\boldsymbol{x}$ with $[\boldsymbol{x}]_{\mathcal{B}} = (1, 3)$ is $\boldsymbol{x} = 1\boldsymbol{u}_1 + 3\boldsymbol{u}_2$. Thus,

$$[\boldsymbol{x}]_{\mathcal{C}} = [1\boldsymbol{u}_1 + 3\boldsymbol{u}_2]_{\mathcal{C}} = 1[\boldsymbol{u}_1]_{\mathcal{C}} + 3[\boldsymbol{u}_2]_{\mathcal{C}} = \begin{bmatrix}[\boldsymbol{u}_1]_{\mathcal{C}} & [\boldsymbol{u}_2]_{\mathcal{C}}\end{bmatrix}\begin{bmatrix}1\\3\end{bmatrix} = \begin{bmatrix}-3 & 3\\2 & -1\end{bmatrix}\begin{bmatrix}1\\3\end{bmatrix} = \begin{bmatrix}6\\-1\end{bmatrix}$$

where Theorem 9.11.2 was used in the second step. And with the red matrix, denoted for now by $\mathbf{P}$, whose columns are the coordinate vectors of the basis vectors in $\mathcal{B}$ w.r.t. $\mathcal{C}$, the calculation of the coordinates of any vector in $\mathcal{C}$ is easy: $[\boldsymbol{x}]_{\mathcal{C}} = \mathbf{P}[\boldsymbol{x}]_{\mathcal{B}}$.

Thus, we have the following definition of this important matrix.

Definition 9.11.2 Let $\mathcal{B} = \{\boldsymbol{u}_1, \boldsymbol{u}_2, \ldots, \boldsymbol{u}_n\}$ and $\mathcal{C} = \{\boldsymbol{v}_1, \boldsymbol{v}_2, \ldots, \boldsymbol{v}_n\}$ be bases for a vector space V. The $n \times n$ matrix whose columns are the coordinate vectors $[\boldsymbol{u}_1]_{\mathcal{C}}, \ldots, [\boldsymbol{u}_n]_{\mathcal{C}}$ of the vectors in the old basis $\mathcal{B}$ with respect to the new basis $\mathcal{C}$ is denoted by $\mathbf{P}_{\mathcal{C}\leftarrow\mathcal{B}}$ and is called the change-of-basis matrix from $\mathcal{B}$ to $\mathcal{C}$.

That matrix allows us to compute the coordinates of a vector in the new base:

$$[\boldsymbol{x}]_{\mathcal{C}} = \mathbf{P}_{\mathcal{C}\leftarrow\mathcal{B}}[\boldsymbol{x}]_{\mathcal{B}}$$

Change-of-basis formula relates the coordinates of one and the same vector in two different bases, whereas a linear transformation relates coordinates of two different vectors in the same basis. One more thing is that $\mathbf{P}_{\mathcal{C}\leftarrow\mathcal{B}}$ is invertible, thus we can always go forth and back between the bases:

$$[\boldsymbol{x}]_{\mathcal{C}} = \mathbf{P}_{\mathcal{C}\leftarrow\mathcal{B}}[\boldsymbol{x}]_{\mathcal{B}} \Longrightarrow [\boldsymbol{x}]_{\mathcal{B}} = \mathbf{P}^{-1}_{\mathcal{C}\leftarrow\mathcal{B}}[\boldsymbol{x}]_{\mathcal{C}}$$

---

[61] You can either draw these vectors and see this or you can simply solve two 2-by-2 systems.

Why the change-of-basis matrix is invertible? This is thanks to theorem 9.11.3: the vectors $\{\boldsymbol{u}_1, \boldsymbol{u}_2, \ldots, \boldsymbol{u}_n\}$ are linear independent in V, thus the vectors $\{[\boldsymbol{u}_1]_C, \ldots, [\boldsymbol{u}_n]_C\}$ are linear independent in $\mathbb{R}^n$: the columns of the change-of-basis matrix are thus linear independent. Hence, it is invertible.

9.11.3 Linear Transformations

In Sect. 9.6, a brief introduction to linear transformations from $\mathbb{R}^n$ to $\mathbb{R}^m$ was presented. In this section, we present linear transformations from a vector space V to another vector space W. I do not repeat the definition because we just need to replace $\mathbb{R}^n$ with V and $\mathbb{R}^m$ with W. Instead, "new" linear transformations are shown in what follows.

Example 1. The differential operator, $D(f) = df/dx$, is a linear transformation because

$$\frac{d(f+g)}{dx} = \frac{df}{dx} + \frac{dg}{dx}, \quad \frac{d(cf)}{dx} = c\frac{df}{dx}$$

Example 2. Let $\mathcal{F}[a, b]$ be a vector space of all real-valued functions defined on the interval $[a, b]$. The integration operator $S : \mathcal{F}[a, b] \to R$ by $S(f) = \int_a^b f(x)dx$ is a linear transformation.

Linear transformation is a fancy term and thus seems scary. Let's get back to the friendly $y = f(x)$: pop in a number x and it is transformed to a new number $f(x)$. Thus, a linear transformation is simply a generalization of the concept of function, instead of taking a single number now it takes in a vector and gives another vector. The key difference is that linear transformations are similar to $y = ax$ not $y = \sin x$: the transformation is linear only. In Sect. 4.2.4, we have discussed the concept of range of a function. We extend that to linear transformation and introduce a new concept: kernel of the transformation. For $y = f(x)$, the roots of this function are all x^* such that $f(x^*) = 0$. The kernel of a linear transformation is exactly this.

Definition 9.11.3 Let $T : V \to W$ be a linear transformation.

(a) The kernel of T, denoted by $\ker(T)$, is the set of all vectors in V that are mapped by T to $\mathbf{0}$ in W. That is,

$$\ker(T) = \{\boldsymbol{v} \in V : T(\boldsymbol{v}) = \mathbf{0}\}$$

(b) The range of T, denoted by range(T), is the set of all vectors in W that are images of vectors in V under T. That is,

$$\text{range}(T) = \{\boldsymbol{w} \in W : \boldsymbol{w} = T(\boldsymbol{v}) \text{ for some} \boldsymbol{v} \in V\}$$

Definition 9.11.4 Rank and nullity of a transformation Let $T : V \to W$ be be a linear transformation. The rank of T is the dimension of the range of T and is denoted by rank(T). The nullity of T is the dimension of the kernel of T and is denoted by nullity(T).

One-to-one and onto linear transformation. If we consider the function $y = x^2$ we have $y(-x) = y(x) = x^2$; that is, the horizontal line $y = x_0^2$ cuts the curve $y = x^2$ at two points. In that case it is impossible to go inverse: from x^2 to $-x$ or x? We say that the function $y = x^2$ is not *one to one*. Functions such as $y = e^x$ or $y = x^3$ are one-to-one functions.

A function is called onto if its range is equal to its codomain. The function $\sin : \mathbb{R} \to \mathbb{R}$ is not onto. Indeed, taking $b = 2$, the equation $\sin(x) = 2$ has no solution. The range of the sine function is the closed interval $[-1, 1]$, which is smaller than the codomain $\mathbb{R}$. The function $y = e^x$ is not onto: the range of $y = e^x$ is $(0, \infty)$ which is not $\mathbb{R}$. Functions such as $y = x^3$ are onto functions.

A function such as $y = x^3$, which is *both one to one and onto*, is special: we can *always* perform an inverse: $x \to x^3 \to x$. Now, we generalize all this to linear transformations.

Definition 9.11.5 Consider a linear transformation $T : V \to W$.

(a) T is called one to one if it maps distinct vectors in V to distinct vectors in W. That is, for all $\boldsymbol{u}$ and $\boldsymbol{v}$ in V, then $\boldsymbol{u} \neq \boldsymbol{v}$ implies that $T(\boldsymbol{u}) \neq T(\boldsymbol{v})$.
(b) T is called onto if range(T) $= W$. In other words, the range of T is equal to the codomain of T. Or, every vector in the codomain is the output of some input vector. That is, for all $\boldsymbol{w} \in W$, there is at least one $\boldsymbol{v} \in V$ such that $T(\boldsymbol{v}) = \boldsymbol{w}$.

Again the definition above is not useful to check whether a transformation is one to one or onto. There exists theorems which provide simpler ways to do that. Below is such a theorem:

(a) A linear transformation $T : V \to W$ is one to one if $\ker(T) = \{\mathbf{0}\}$.
(b) Let $T : V \to W$ be a one-to-one linear transformation. If $S = \{\boldsymbol{v}_1, \boldsymbol{v}_2, \ldots, \boldsymbol{v}_k\}$ is a linear independent set in V, then $T(S) = \{T(\boldsymbol{v}_1), T(\boldsymbol{v}_2), \ldots, T(\boldsymbol{v}_k)\}$ is a linear independent set in W.
(c) A linear transformation $T : V \to W$ is invertible if it is one to one and onto.

Isomorphism of vector spaces. In Example 9.11, we have seen that points in $\mathcal{P}_2$ can now be identified by their coordinates in $\mathbb{R}^3$, and every vector-space calculation in $\mathcal{P}_2$ is accurately reproduced in $\mathbb{R}^3$ (and vice versa). The concept of isomorphism is the generalization of this idea.

Definition 9.11.6 A linear transformation $T : V \to W$ is called an isomorphism if it is one to one and onto. If V and W are two vector spaces such that there is an isomorphism form V to W, then we say that V is isomorphic to W and write $V \cong W$.

The idea is that an isomorphism $T : V \to W$ means that W is "just like" V in the context of any question involving addition and scalar multiplication. The word isomorphism and isomorphic are derived from the Greek words *isos*, meaning "equal" and *morph*, meaning "shape".

Example 9.12 Show that $\mathcal{P}_{n-1}$ and $\mathbb{R}^n$ are isomorphic. To this end, we need to prove that there exists a linear transformation $T : \mathcal{P}_{n-1} \to \mathbb{R}^n$ that is one to one and onto. Actually, we already knew such transformation: the one that gives us the coordinates of a vector in $\mathcal{P}_{n-1}$ with respect to a basis of $\mathcal{P}_{n-1}$.

Let $\mathcal{E} = \{1, x, ..., x^{n-1}\}$ be a basis for $\mathcal{P}_{n-1}$. Then, any vector $p(x)$ in $\mathcal{P}_{n-1}$ can be written as

$$p(x) = a_0(1) + a_1(x) + \cdots + a_{n-1}(x^{n-1}) \Longrightarrow [p(x)]_{\mathcal{E}} = (a_0, a_1, \ldots, a_{n-1})$$

Now, we define the following transformation $T : \mathcal{P}_{n-1} \to \mathbb{R}^n$ (this transformation is known as a coordinate map)

$$T(p(x)) := [p(x)]_{\mathcal{E}}$$

Is this a linear transformation? Yes, thanks to Theorem 9.11.2. What left is to prove that T is one to one and onto. For the former, just need to show that $\ker(T) = \{0\}$. For the latter, $\dim \mathcal{P}_{n-1} = \dim \mathbb{R}^n = n$. I refer to Table 9.3 for an example of this isomorphism.

Matrix associated with a linear transformation. Let V and W be two finite-dimensional vector spaces with bases $\mathcal{B}$ and $\mathcal{C}$, respectively, where $\mathcal{B} = \{\boldsymbol{v}_1, \boldsymbol{v}_2, \ldots, \boldsymbol{v}_n\}$. Now consider a linear transformation $T : V \to W$. Our task is to find the matrix associated with T. To this end, consider a vector $\boldsymbol{u} \in V$, we can write it as

$$\boldsymbol{u} = u_1\boldsymbol{v}_1 + u_2\boldsymbol{v}_2 + \cdots + u_n\boldsymbol{v}_n$$

Table 9.3 The parallel universes of $\mathcal{P}_2$ and $\mathbb{R}^3$: $\mathcal{P}_2$ is isomorphic to $\mathbb{R}^3$ by the coordinate map $T(p(x)) := [p(x)]_{\mathcal{E}}$ where $\mathcal{E} = \{1, t, t^2\}$ is the standard basis in $\mathcal{P}_2$

$\mathcal{P}_2$	$\mathbb{R}^3$
$p(t) = a + bt + ct^2$	$\begin{bmatrix} a \\ b \\ c \end{bmatrix}$
$(-1 + 2t + 3t^2) + (2 + 4t + 3t^2) = 1 + 6t + 6t^2$	$\begin{bmatrix} -1 \\ 2 \\ 3 \end{bmatrix} + \begin{bmatrix} 2 \\ 4 \\ 3 \end{bmatrix} = \begin{bmatrix} 1 \\ 6 \\ 6 \end{bmatrix}$
$3(2 + t + 3t^2) = 6 + 3t + 9t^2$	$3\begin{bmatrix} 2 \\ 1 \\ 3 \end{bmatrix} = \begin{bmatrix} 6 \\ 3 \\ 9 \end{bmatrix}$

So, the linear transformation T applied to $\boldsymbol{u}$ can be written as

$$T(\boldsymbol{u}) = T(u_1\boldsymbol{v}_1 + u_2\boldsymbol{v}_2 + \cdots + u_n\boldsymbol{v}_n) = u_1T(\boldsymbol{v}_1) + u_2T(\boldsymbol{v}_2) + \cdots + u_nT(\boldsymbol{v}_n)$$

Now $T(\boldsymbol{u})$ is a vector in W, and with respect to the basis $\mathcal{C}$, its coordinates are

$$\begin{aligned} [T(\boldsymbol{u})]_{\mathcal{C}} &= [u_1T(\boldsymbol{v}_1) + u_2T(\boldsymbol{v}_2) + \cdots + u_nT(\boldsymbol{v}_n)]_{\mathcal{C}} \\ &= u_1[T(\boldsymbol{v}_1)]_{\mathcal{C}} + u_2[T(\boldsymbol{v}_2)]_{\mathcal{C}} + \cdots + u_n[T(\boldsymbol{v}_n)]_{\mathcal{C}} \end{aligned}$$

So we can characterize a linear transformation by storing $[T(\boldsymbol{v}_i)]_{\mathcal{C}}$, $i = 1, 2, \ldots, n$ in an $m \times n$ matrix like this

$$[T]_{\mathcal{C}\leftarrow\mathcal{B}} := \begin{bmatrix} | & | & | & | \\ [T(\boldsymbol{v}_1)]_{\mathcal{C}} & [T(\boldsymbol{v}_2)]_{\mathcal{C}} & \cdots & [T(\boldsymbol{v}_n)]_{\mathcal{C}} \\ | & | & | & | \end{bmatrix} \tag{9.11.3}$$

This matrix is called the matrix of T with respect to the bases $\mathcal{B}$ and $\mathcal{C}$. Then, any vector $\boldsymbol{x} \in V$ with $\mathcal{B}-$coordinate vector $[\boldsymbol{x}]_{\mathcal{B}}$ is transformed to vector $T(\boldsymbol{x}) \in W$ with $\mathcal{C}-$coordinate vector $[T(\boldsymbol{x})]_{\mathcal{C}}$:

$$\boxed{[T(\boldsymbol{x})]_{\mathcal{C}} = [T]_{\mathcal{C}\leftarrow\mathcal{B}}[\boldsymbol{x}]_{\mathcal{B}}} \tag{9.11.4}$$

Thus, we have shown that any linear transformation can be described by a matrix. In the special case where $V = W$ and $\mathcal{B} = \mathcal{C}$, we have

$$\boxed{[T(\boldsymbol{x})]_{\mathcal{B}} = [T]_{\mathcal{B}}[\boldsymbol{x}]_{\mathcal{B}}} \tag{9.11.5}$$

Matrices of a transformation in different bases. Consider a linear transformation $T : V \to V$. The choice of a basis for V identifies this transformation with a matrix multiplication. Now, we consider two bases, then we will have two matrices:

$$\begin{aligned} &\text{basis } \mathcal{B} : \mathbb{R}^n \to \mathbb{R}^n : [\boldsymbol{x}]_{\mathcal{B}} \to [T]_{\mathcal{B}}[\boldsymbol{x}]_{\mathcal{B}} \\ &\text{basis } \mathcal{C} : \mathbb{R}^n \to \mathbb{R}^n : [\boldsymbol{x}]_{\mathcal{C}} \to [T]_{\mathcal{C}}[\boldsymbol{x}]_{\mathcal{C}} \end{aligned}$$

Our problem is now to relate $[T]_{\mathcal{B}}$ to $[T]_{\mathcal{C}}$ or vice versa. First, we consider an arbitrary vector $\boldsymbol{x} \in V$ and the basis $\mathcal{B}$, we can write the transformation T on $\boldsymbol{x}$ as

$$[T(\boldsymbol{x})]_{\mathcal{B}} = [T]_{\mathcal{B}}[\boldsymbol{x}]_{\mathcal{B}}$$

Now, we look at the transformed vector $T(\boldsymbol{x})$ but in the basis $\mathcal{C}$, by multiplying $[T(\boldsymbol{x})]_{\mathcal{B}}$ with the change-of-basis matrix $\mathbf{P}_{\mathcal{C}\leftarrow\mathcal{B}}$:

$$\underbrace{\mathbf{P}_{\mathcal{C}\leftarrow\mathcal{B}}[T(\boldsymbol{x})]_{\mathcal{B}}}_{[T(\boldsymbol{x})]_{\mathcal{C}}} = \mathbf{P}_{\mathcal{C}\leftarrow\mathcal{B}}[T]_{\mathcal{B}}[\boldsymbol{x}]_{\mathcal{B}}$$

Of course we have $[T(\boldsymbol{x})]_{\mathcal{C}} = [T]_{\mathcal{C}}[\boldsymbol{x}]_{\mathcal{C}}$, thus

$$[T]_{\mathcal{C}}[\boldsymbol{x}]_{\mathcal{C}} = \mathbf{P}_{\mathcal{C}\leftarrow\mathcal{B}}[T]_{\mathcal{B}}[\boldsymbol{x}]_{\mathcal{B}}$$

Now, to get rid of $[\boldsymbol{x}]_{\mathcal{C}}$, we replace it by $\mathbf{P}_{\mathcal{C}\leftarrow\mathcal{B}}[\boldsymbol{x}]_{\mathcal{B}}$, and obtain

$$[T]_{\mathcal{C}}\mathbf{P}_{\mathcal{C}\leftarrow\mathcal{B}}[\boldsymbol{x}]_{\mathcal{B}} = \mathbf{P}_{\mathcal{C}\leftarrow\mathcal{B}}[T]_{\mathcal{B}}[\boldsymbol{x}]_{\mathcal{B}}$$

This equation holds for any $[\boldsymbol{x}]_{\mathcal{B}}$, thus we get the following identity $[T]_{\mathcal{C}}\mathbf{P}_{\mathcal{C}\leftarrow\mathcal{B}} = \mathbf{P}_{\mathcal{C}\leftarrow\mathcal{B}}[T]_{\mathcal{B}}$ from which we obtain

$$\boxed{[T]_{\mathcal{B}} = \mathbf{P}^{-1}_{\mathcal{C}\leftarrow\mathcal{B}}[T]_{\mathcal{C}}\mathbf{P}_{\mathcal{C}\leftarrow\mathcal{B}}} \tag{9.11.6}$$

This is often used when we are trying to find a good basis with respect to which the matrix of a linear transformation is particularly simple (e.g., diagonal). For example, we can ask whether there is a basis $\mathcal{B}$ such that the matrix $[T]_{\mathcal{B}}$ of $T : V \to V$ is a diagonal matrix. The next section is answering this question.

9.11.4 Diagonalizing a Matrix

A diagonal matrix is so nice to work with. For example, the eigenvalues can be read off immediately—the entries on the diagonal. It turns out that we can always transform a full matrix to a diagonal one using eigenvalues and eigenvectors. This is not so surprising if we already know principal axes of rotating rigid bodies. Let's start with an example.

Example 9.13 Let's consider the following matrix, which is associated to a linear transformation T, with its eigenvalues and eigenvectors:

$$\mathbf{A} = \begin{bmatrix} 3 & 1 \\ 0 & 2 \end{bmatrix}, \quad \lambda_1 = 3, \quad \lambda_2 = 2, \quad \boldsymbol{v}_1 = \begin{bmatrix} 1 \\ 0 \end{bmatrix}, \quad \boldsymbol{v}_2 = \begin{bmatrix} -1 \\ +1 \end{bmatrix}$$

Now, we consider two bases: the first basis $\mathcal{C}$ is the standard basis with $(1, 0)$ and $(0, 1)$ as the basis vectors, and the second basis $\mathcal{B}$ with the basis vectors being the eigenvectors $\boldsymbol{v}_1, \boldsymbol{v}_2$. Now, we have

$$[T]_{\mathcal{C}} = \mathbf{A}, \quad \mathbf{P}_{\mathcal{C}\leftarrow\mathcal{B}} = \begin{bmatrix} \boldsymbol{v}_1 & \boldsymbol{v}_2 \end{bmatrix} = \begin{bmatrix} 1 & -1 \\ 0 & 1 \end{bmatrix}$$

Now using Eq. (9.11.6) the transformation T—that is associated with $\mathbf{A}$ with respect to $\mathcal{C}$—is now given with respect to the eigenbasis $\mathcal{B}$:

$$[T]_{\mathcal{B}} = \mathbf{P}_{\mathcal{C}\leftarrow\mathcal{B}}^{-1}[T]_{\mathcal{C}}\mathbf{P}_{\mathcal{C}\leftarrow\mathcal{B}} = \begin{bmatrix}1 & -1\\ 0 & 1\end{bmatrix}^{-1}\begin{bmatrix}3 & 1\\ 0 & 2\end{bmatrix}\begin{bmatrix}1 & -1\\ 0 & 1\end{bmatrix} = \begin{bmatrix}3 & 0\\ 0 & 2\end{bmatrix}$$

Look at what we have obtained: a diagonal matrix with the eigenvalues on the diagonal! In other words, we have diagonalized the matrix **A**.

Suppose the $n \times n$ matrix **A** has n linearly independent eigenvectors $\boldsymbol{v}_1, \boldsymbol{v}_2, \ldots, \boldsymbol{v}_n$ (and eigenvalues $\lambda_1, \lambda_2, \ldots, \lambda_n$). Put these eigenvectors into the columns of an *eigenvector matrix* **V**. We now compute **A** times **V**:

$$\mathbf{AV} := \mathbf{A}\begin{bmatrix}\boldsymbol{v}_1 & \boldsymbol{v}_2 & \cdots & \boldsymbol{v}_n\end{bmatrix} = \begin{bmatrix}\mathbf{A}\boldsymbol{v}_1 & \mathbf{A}\boldsymbol{v}_2 & \cdots & \mathbf{A}\boldsymbol{v}_n\end{bmatrix} = \begin{bmatrix}\lambda_1\boldsymbol{v}_1 & \lambda_2\boldsymbol{v}_2 & \cdots & \lambda_n\boldsymbol{v}_n\end{bmatrix}$$

Now, the trick is to split the matrix $\begin{bmatrix}\lambda_1\boldsymbol{v}_1 & \lambda_2\boldsymbol{v}_2 & \cdots & \lambda_n\boldsymbol{v}_n\end{bmatrix}$ into **V** times a diagonal matrix $\boldsymbol{\Lambda}$ with λ_i's on the diagonal:

$$\mathbf{AV} = \begin{bmatrix}\lambda_1\boldsymbol{v}_1 & \lambda_2\boldsymbol{v}_2 & \cdots & \lambda_n\boldsymbol{v}_n\end{bmatrix} = \begin{bmatrix}\boldsymbol{v}_1 & \boldsymbol{v}_2 & \cdots & \boldsymbol{v}_n\end{bmatrix}\begin{bmatrix}\lambda_1 & 0 & \cdots & 0\\ 0 & \lambda_2 & \cdots & 0\\ \vdots & \vdots & \ddots & \vdots\\ 0 & 0 & \cdots & \lambda_n\end{bmatrix} = \mathbf{V}\boldsymbol{\Lambda}$$

If this is not clear, check Sect. 9.4.4 on the matrix-column representation of the product **AB**: $\mathbf{AB} = \begin{bmatrix}\mathbf{A}\boldsymbol{B}_1 & \mathbf{A}\boldsymbol{B}_2 & \mathbf{A}\boldsymbol{B}_3\end{bmatrix}$. And $\mathbf{A}\boldsymbol{B}_1$ is a linear combination of the cols of **A** with the coefficients being the components of $\boldsymbol{B}_1$. Here, **A** is **V** and $\boldsymbol{B}_1 = (\lambda_1, 0, \ldots)$.

Thus we have obtained $\mathbf{AV} = \mathbf{V}\boldsymbol{\Lambda}$ and since **V** has linear independent columns, it can be inverted, so we can diagonalize **A**:

$$\mathbf{AV} = \mathbf{V}\boldsymbol{\Lambda} \Longrightarrow \boxed{\mathbf{A} = \mathbf{V}\boldsymbol{\Lambda}\mathbf{V}^{-1}}$$

With this form, it is super easy to compute powers of **A**. For example,

$$\mathbf{A}^3 = (\mathbf{V}\boldsymbol{\Lambda}\mathbf{V}^{-1})(\mathbf{V}\boldsymbol{\Lambda}\mathbf{V}^{-1})(\mathbf{V}\boldsymbol{\Lambda}\mathbf{V}^{-1}) = \mathbf{V}\boldsymbol{\Lambda}(\mathbf{V}^{-1}\mathbf{V})\boldsymbol{\Lambda}(\mathbf{V}^{-1}\mathbf{V})\boldsymbol{\Lambda}\mathbf{V}^{-1}) = \mathbf{V}\boldsymbol{\Lambda}^3\mathbf{V}^{-1}$$

And nothing can stop us from going to $\mathbf{A}^k = \mathbf{V}\boldsymbol{\Lambda}^k\mathbf{V}^{-1}$ whatever k might be: 1000 or 10 000. This equation tells us that the eigenvalues of $\mathbf{A}^k$ are $\lambda_1^k, \ldots, \lambda_n^k$, and the eigenvectors of $\mathbf{A}^k$ are the same as the eigenvectors of **A**.

9.11.5 Inner Product and Inner Product Spaces

In this section, we present the inner product which is an extension of the familiar dot product between two vectors in $\mathbb{R}^n$. First, let's recall the dot product of two n-vectors:

$$\boldsymbol{a} \cdot \boldsymbol{b} = \sum_{i=1}^{n} a_i b_i \tag{9.11.7}$$

This dot product has these properties: $\boldsymbol{a} \cdot \boldsymbol{b} = \boldsymbol{b} \cdot \boldsymbol{a}$, $\boldsymbol{a} \cdot \boldsymbol{a} \geq 0$, and $(\alpha \boldsymbol{a} + \beta \boldsymbol{b}) \cdot \boldsymbol{c} = \alpha(\boldsymbol{a} \cdot \boldsymbol{c}) + \beta(\boldsymbol{b} \cdot \boldsymbol{c})$. Now, we define an inner product between two vectors $\boldsymbol{a}, \boldsymbol{b}$ in a vector space V, denoted by $\langle \boldsymbol{a}, \boldsymbol{b} \rangle$[62], which is an operation that assigns these two vectors a real number such that this product has properties identical to those of the dot product:

$$\begin{aligned} \text{symmetry:} \quad & \langle \boldsymbol{a}, \boldsymbol{b} \rangle = \langle \boldsymbol{b}, \boldsymbol{a} \rangle \\ \text{positivity:} \quad & \langle \boldsymbol{a}, \boldsymbol{a} \rangle \geq 0 \\ \text{positivity:} \quad & \langle \boldsymbol{a}, \boldsymbol{a} \rangle = 0 \;\; \text{if and only if} \;\; \boldsymbol{a} = \boldsymbol{0} \\ \text{linearity:} \quad & \langle \alpha \boldsymbol{a} + \beta \boldsymbol{b}, \boldsymbol{c} \rangle = \alpha \langle \boldsymbol{a}, \boldsymbol{c} \rangle + \beta \langle \boldsymbol{b}, \boldsymbol{c} \rangle \end{aligned} \tag{9.11.8}$$

From the linearity property, we can show that the inner product has the bilinearity property that reads[63]

$$\langle a\boldsymbol{x} + b\boldsymbol{y}, c\boldsymbol{u} + d\boldsymbol{v} \rangle = ac\langle \boldsymbol{x}, \boldsymbol{u} \rangle + ad\langle \boldsymbol{x}, \boldsymbol{v} \rangle + bc\langle \boldsymbol{y}, \boldsymbol{u} \rangle + bd\langle \boldsymbol{y}, \boldsymbol{v} \rangle$$

The word bilinearity is used to indicate that the inner product is linear with respect to both input vectors.

Proof

$$\begin{aligned} \langle a\boldsymbol{x} + b\boldsymbol{y}, c\boldsymbol{u} + d\boldsymbol{v} \rangle &= \langle a\boldsymbol{x}, c\boldsymbol{u} + d\boldsymbol{v} \rangle + \langle b\boldsymbol{y}, c\boldsymbol{u} + d\boldsymbol{v} \rangle \quad \text{(linearity prop.)} \\ &= \langle c\boldsymbol{u} + d\boldsymbol{v}, a\boldsymbol{x} \rangle + \langle c\boldsymbol{u} + d\boldsymbol{v}, b\boldsymbol{y} \rangle \quad \text{(symmetry prop.)} \\ &= \langle c\boldsymbol{u}, a\boldsymbol{x} \rangle + \langle d\boldsymbol{v}, a\boldsymbol{x} \rangle + \langle c\boldsymbol{u}, b\boldsymbol{y} \rangle + \langle d\boldsymbol{v}, b\boldsymbol{y} \rangle \quad \text{(linearity prop.)} \\ &= ca\langle \boldsymbol{u}, \boldsymbol{x} \rangle + da\langle \boldsymbol{v}, \boldsymbol{x} \rangle + cb\langle \boldsymbol{u}, \boldsymbol{y} \rangle + db\langle \boldsymbol{v}, \boldsymbol{y} \rangle \quad \text{(linearity prop.)} \end{aligned}$$

■

Example 9.14 Let $\boldsymbol{u} = (u_1, u_2)$ and $\boldsymbol{v} = (v_1, v_2)$ be two vectors in $\mathbb{R}^2$. Then, the following

$$\langle \boldsymbol{u}, \boldsymbol{v} \rangle = 5u_1 v_1 + 7u_2 v_2$$

defines an inner product. It's not hard to check that this really satisfies all the properties in Eq. (9.11.8). Now, we generalize it to $\mathbb{R}^n$. Let $\boldsymbol{u} = (u_1, u_2, \ldots, u_n)$ and $\boldsymbol{v} = (v_1, v_2, \ldots, v_n)$ be two vectors in $\mathbb{R}^n$ and $w_1, w_2, \ldots, w_n$ are n positive real numbers which are called the weights, then

[62] Other notation for the inner product is $(\boldsymbol{a}, \boldsymbol{b})$.

[63] With $\beta = 0$, the linearity property gives us $\langle \alpha \boldsymbol{a}, \boldsymbol{c} \rangle = \alpha \langle \boldsymbol{a}, \boldsymbol{c} \rangle$. And from that we also have $\langle \alpha \boldsymbol{a}, \gamma \boldsymbol{c} \rangle = \alpha\gamma \langle \boldsymbol{a}, \boldsymbol{c} \rangle$.

$$\langle \boldsymbol{u}, \boldsymbol{v} \rangle = w_1 u_1 v_1 + w_2 u_2 v_2 + \cdots + w_n u_n v_n = \boldsymbol{u}^\top \mathbf{W} \boldsymbol{v}, \quad \mathbf{W} = \begin{bmatrix} w_1 & \cdots & 0 \\ \vdots & \ddots & \vdots \\ 0 & \cdots & w_n \end{bmatrix}$$

defines an inner product called a weighted dot product.

A vector space equipped with an inner product is called an *inner product space*. Don't be scared as the space $\mathbb{R}^n$ is an inner product space! It must be as it was the inspiration for mathematicians to generalize it to inner product spaces. We shall meet other inner product spaces when we define concrete inner product. Up to now we do not yet define what $\langle \boldsymbol{a}, \boldsymbol{b} \rangle$ is! Similar to how the dot product defines length, distance, and orthogonality, the inner product enables us to define these concepts for vectors in an inner product space.

Definition 9.11.7 Let $\boldsymbol{u}$ and $\boldsymbol{v}$ be two vectors in an inner product space V.

(a) The length (or norm) of $\boldsymbol{v}$ is $||\boldsymbol{v}|| = \sqrt{\langle \boldsymbol{v}, \boldsymbol{v} \rangle}$.
(b) The distance between $\boldsymbol{u}$ and $\boldsymbol{v}$ is $d(\boldsymbol{u}, \boldsymbol{v}) = ||\boldsymbol{u} - \boldsymbol{v}||$.
(c) $\boldsymbol{u}$ and $\boldsymbol{v}$ are orthogonal if $\langle \boldsymbol{u}, \boldsymbol{v} \rangle = 0$.

Example 5 If we consider two functions f and g in $C[a, b]$—the vector space of continuous functions in $[a, b]$, show that

$$\langle f, g \rangle = \int_a^b f(x)g(x)dx \tag{9.11.9}$$

defines an inner product on $C[a, b]$.

We need, of course, to verify that this definition satisfies all four conditions in Eq. (9.11.8). This satisfaction comes from the properties of definite integrals.

Gram-Schmidt orthogonalization and Legendre polynomials. If we apply the Gram-Schmidt orthogonalization to $1, x, x^2, \ldots$, but with Eq. (9.11.9) in place of the dot product, we will obtain the so-called Legendre polynomials.

Applying the Gram-Schmidt orthogonalization to $1, x, x^2, x^3$ we obtain the first four Legendre polynomials (note that Legendre polynomials are defined on the interval $[-1, 1]$):

$$\begin{aligned}
L_0(x) &= 1 \\
L_1(x) &= x - \frac{\langle 1, x \rangle}{\langle 1, 1 \rangle} 1 = x - \frac{1}{2} \int_{-1}^{1} x dx = x \\
L_2(x) &= x^2 - \frac{\langle 1, x^2 \rangle}{\langle 1, 1 \rangle} 1 - \frac{\langle x, x^2 \rangle}{\langle x, x \rangle} x = x^2 - \frac{1}{3} \\
L_3(x) &= x^3 - \frac{\langle 1, x^3 \rangle}{\langle 1, 1 \rangle} 1 - \frac{\langle x, x^3 \rangle}{\langle x, x \rangle} x - \frac{\langle x^2, x^3 \rangle}{\langle x^2, x^2 \rangle} x^2 = x^3 - \frac{3}{5} x
\end{aligned} \tag{9.11.10}$$

Table 9.4 The first six Legendre polynomials

n	$L_n(x)$
0	1
1	x
2	$\frac{1}{2}(3x^2 - 1)$
3	$\frac{1}{2}(5x^3 - 3x)$
4	$\frac{1}{8}(35x^4 - 30x^2 + 3)$
5	$\frac{1}{8}(63x^5 - 70x^3 + 15x)$

Actually, we need to scale these polynomials so that $L_n(1) = 1$, then we have the so-called standard Legendre polynomials as shown in Table 9.4. One surprising fact about Legendre polynomials is that, their roots are symmetrical with respect to $x = 0$, and $L_n(x)$ has n real roots within $[-1, 1]$, see Fig. 9.25. And these roots define the quadrature points in Gauss' rule—a well-known numerical integration rule (Sect. 10.4.3).

Adrien-Marie Legendre (1752–1833) was a French mathematician who made numerous contributions to mathematics. Well-known and important concepts such as the Legendre polynomials and Legendre transformation are named after him.

Now, we focus on the inner product space of polynomials. Because Legendre polynomials are orthogonal to each other, they can be the basis for the inner product space of polynomials. For example, any 2nd-degree polynomial can be uniquely written as

$$p_2(x) = c_0L_0(x) + c_1L_1(x) + c_2L_2(x)$$

where $L_i(x)$ are the orthogonal Legendre polynomials, see Table 9.4. Next, we compute the inner product of $p_2(x)$ with $L_3(x)$, because the result is beautiful:

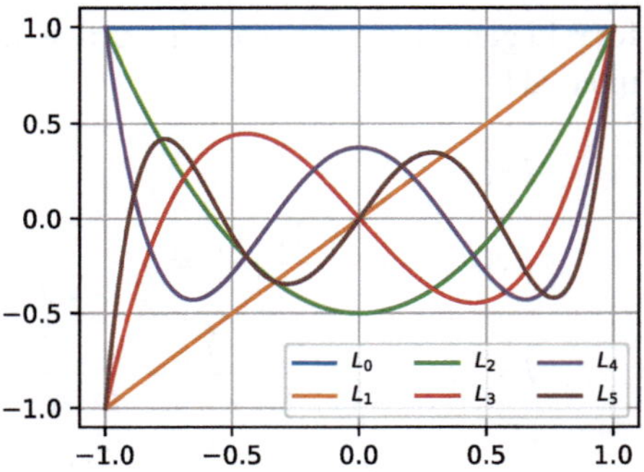

Fig. 9.25 Plots of some Legendre polynomials

$$\begin{aligned}\int_{-1}^{1} L_3(x)p_2(x)dx &= \int_{-1}^{1} [c_0L_0(x) + c_1L_1(x) + c_2L_2(x)]\, L_3(x)dx \\ &= c_0\int_{-1}^{1} L_0(x)L_3(x)dx + c_1\int_{-1}^{1} L_1(x)L_3(x)dx + c_3\int_{-1}^{1} L_2(x)L_3(x)dx \\ &= 0\end{aligned}$$

This is due to the orthogonality of Legendre polynomials. We will use this in Sect. 10.4.3 to derive the famous Gauss-Legendre quadrature rule.

The Cauchy-Schwarz inequality. In Sect. 2.20.3, we have met the Cauchy-Schwarz inequality. At that time, we did not know of $\mathbb{R}^n$. But now, we can see that this inequality can be, for two vectors $\boldsymbol{u}$ and $\boldsymbol{v}$ in $\mathbb{R}^n$, written as

$$|\boldsymbol{u}\cdot\boldsymbol{v}| \le ||\boldsymbol{u}||||\boldsymbol{v}||$$

The nice thing of mathematics is that the same inequality holds for two vectors in an inner product space. We just replace the dot product by the more general inner product.

Proof The proof is pretty similar to the one given in Sect. 2.20.3. We construct the following function:

$$f(t) = \langle \boldsymbol{u} + t\boldsymbol{v}, \boldsymbol{u} + t\boldsymbol{v}\rangle$$

which is always non-negative, i.e., $f(t) \ge 0\, \forall t$ (due to Eq. (9.11.8)). We can rewrite $f(t)$ as

$$\begin{aligned}f(t) &= \langle \boldsymbol{u} + t\boldsymbol{v}, \boldsymbol{u} + t\boldsymbol{v}\rangle \\ &= \langle \boldsymbol{v}, \boldsymbol{v}\rangle t^2 + 2\langle \boldsymbol{u}, \boldsymbol{v}\rangle t + \langle \boldsymbol{u}, \boldsymbol{u}\rangle \ge 0 \quad \text{for all } t\end{aligned}$$

So, $f(t)$ is a quadratic function in t, we hence compute the discriminant Δ and it has to be less than or equal to 0:

$$\Delta = 4\langle \boldsymbol{u}, \boldsymbol{v}\rangle^2 - 4\langle \boldsymbol{v}, \boldsymbol{v}\rangle\langle \boldsymbol{u}, \boldsymbol{u}\rangle \le 0$$

■

And with this, we also get the triangle inequality for vectors in an inner product space:

$$||\boldsymbol{a} + \boldsymbol{b}|| \le ||\boldsymbol{a}|| + ||\boldsymbol{b}|| \tag{9.11.11}$$

(If you forgot how to prove this, check Eq. (9.1.8).)

9.11.6 Norm, Distance, and Normed Vector Spaces

We live in a 3D Euclidean world, and therefore concepts from Euclidean geometry govern our way of looking at the world. For example, when thinking about the distance between two points A and B, we thought of the shortest distance as $\sqrt{(x_B - x_A)^2 + (y_B - y_A)^2}$. Thus, we're using the length defined as the square root of the inner product of $\boldsymbol{x}_B - \boldsymbol{x}_A$ with itself. However, there exists different types of distance. For example, suppose we're at an intersection in a city, trying to get to another intersection. In this case, we do not use the Euclidean distance, instead we use the so-called taxicab distance. This is so because that is how taxicab drivers measure distance.

This section presents new ways to measure distance. The starting point cannot be the inner product (from which we can only define the Euclidean length or norm). Instead, we start directly with the concept of a norm with certain properties that we want it to have.

Definition 9.11.8 A norm on a vector space V is a mapping that associated with each vector $\boldsymbol{v}$ a real number $||\boldsymbol{v}||$, called the norm of $\boldsymbol{v}$, such that the following properties are satisfied for all vectors $\boldsymbol{u}$ and $\boldsymbol{v}$ and all scalars c:

(a) (non-negativity) $||\boldsymbol{v}|| \geq 0$, and $||\boldsymbol{v}|| = 0$ if and only if $\boldsymbol{v} = \boldsymbol{0}$.
(b) (scaling) $||c\boldsymbol{v}|| = |c|||\boldsymbol{v}||$.
(c) (triangle inequality) $||\boldsymbol{u} + \boldsymbol{v}|| \leq ||\boldsymbol{u}|| + ||\boldsymbol{v}||$.

A vector space equipped with a norm is called a normed vector space.

In the following example, we consider the vector space $\mathbb{R}^n$ and show that there are many norms rather than the usual Euclidean norm.

Example 9.15 Consider $\boldsymbol{v} = (v_1, v_2, \ldots, v_n)$, the following are common norms for $\boldsymbol{v}$:

(a) (l^1) $||\boldsymbol{v}||_1 = |v_1| + |v_2| + \cdots + |v_n|$
(b) (l^2) $||\boldsymbol{v}||_2 = \left(|v_1|^2 + |v_2|^2 + \cdots + |v_n|^2\right)^{1/2}$
(c) (l^∞) $||\boldsymbol{v}||_\infty = \max\{|v_1|, |v_2|, \ldots, |v_n|\} = \max_i |v_i|$
(d) (l^p) $||\boldsymbol{v}||_p = (|v_1|^p + |v_2|^p + \cdots + |v_n|^p)^{1/p}, \quad 1 \leq p < \infty$

where $||\boldsymbol{v}||_2$ is the usual Euclidean norm. It is not hard to prove that l^1, l^2, and l^∞ are indeed norms (we just need to verify the three properties stated in the definition of a norm). For l^p, the proof is harder and thus skipped. Note that I wrote $|v_1|^2$ instead

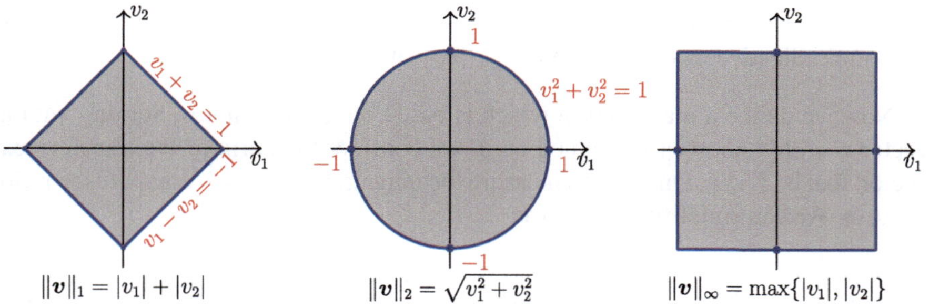

Fig. 9.26 Geometry of different norms: illustrated in $\mathbb{R}^2$ with $\boldsymbol{v} = (v_1, v_2)$

of v_1^2 because the discussion covers complex vectors as well. Thus, the symbol $|\square|$ indicates the modulus.

Figure 9.26 presents the geometry of these norms in $\mathbb{R}^2$. Is this just for fun? Maybe, but it reveals that the different norms are close to each other. Precisely, the norms are all equivalent on $\mathbb{R}^n$ in the sense that[64]

$$\|\boldsymbol{v}\|_2 \leq \|\boldsymbol{v}\|_1 \leq \sqrt{n}\|\boldsymbol{v}\|_2$$

9.11.7 Matrix Norms

To tell when a matrix is large or small or to know when the two matrices are close to each other, we need to define the norm of a matrix. If we can accept that a function has a length, then it is reasonable that a matrix has a norm (kind of length). We do not know the definition of that, but we know what properties a matrix norm should have. So, we define a matrix norm based on these properties. Later on, once we have found the formula for the norm, we check whether it satisfies all these properties. This is similar to how we defined the determinant of a matrix. You will see that a matrix norm has many properties similar to those of a vector norm. But with one additional property. Why? Because we can multiply two matrices, but we cannot multiply two vectors!

Definition 9.11.9 A norm on a matrix space M_{nn} is a mapping that associated with each matrix $\mathbf{A}$ a real number $||\mathbf{A}||$, called the norm of $\mathbf{A}$, such that the following properties are satisfied for all matrices $\mathbf{A}$ and $\mathbf{B}$ and all scalars c:

(a) (non-negativity) $||\mathbf{A}|| \geq 0$, and $||\mathbf{A}|| = 0$ if and only if $\mathbf{A} = \mathbf{0}$.
(b) (scaling) $||c\mathbf{A}|| = |c|||\mathbf{A}||$.

---

[64] One proof is $\|\boldsymbol{v}\|_1 = \sum_i |v_i| = \sum_i |v_i|^2 1 \leq \sqrt{\sum_i |v_i|}\sqrt{1^2 + \cdots + 1^2}$ using the Cauchy-Schwarz inequality.

(c) (triangle inequality) $||\mathbf{A}+\mathbf{B}|| \le ||\mathbf{A}|| + ||\mathbf{B}||$.
(d) (multiplicative inequality) $||\mathbf{AB}|| \le ||\mathbf{A}||||\mathbf{B}||$.

Now we define a matrix norm which is based on a vector norm. Starting with a vector $\boldsymbol{x}$ with a norm $||\square||$ defined on it, we consider the norm of the transformed vector, that is, $\|\mathbf{A}\boldsymbol{x}\|$. One way to measure the magnitude of $\mathbf{A}$ is to compute the ratio $\|\mathbf{A}\boldsymbol{x}\|/\|\boldsymbol{x}\|$. We can simplify this ratio as

$$\frac{||\mathbf{A}\boldsymbol{x}||}{||\boldsymbol{x}||} = \frac{1}{\|\boldsymbol{x}\|}\|\mathbf{A}\boldsymbol{x}\| \overset{(a)}{=} \|[\|\Big]\frac{1}{||\boldsymbol{x}||}\mathbf{A}\boldsymbol{x} = \|[\|\Big]\mathbf{A}\frac{\boldsymbol{x}}{||\boldsymbol{x}||} = \|\mathbf{A}\boldsymbol{x}^*\|, \quad \left(\boldsymbol{x}^* = \frac{\boldsymbol{x}}{||\boldsymbol{x}||}\right)$$

where the scaling property of a vector norm $\|c\boldsymbol{x}\| = |c|\|\boldsymbol{x}\|$ was used in (a) with $c = 1/\|\boldsymbol{x}\|$. As a norm is just one single number, so we are interested only in the maximum of the ratio $\|\mathbf{A}\boldsymbol{x}\|/\|\boldsymbol{x}\|$:

$$\max_{\|\boldsymbol{x}\|\neq 0} \frac{||\mathbf{A}\boldsymbol{x}||}{||\boldsymbol{x}||} = \max_{\|\boldsymbol{x}^*\|=1} \|\mathbf{A}\boldsymbol{x}^*\|$$

Mathematicians then define the *operator norm*, of a matrix, induced by the vector norm $\|\boldsymbol{x}\|$ as[65]:

$$\boxed{\|\mathbf{A}\| = \max_{\|\boldsymbol{x}\|=1} \|\mathbf{A}\boldsymbol{x}\|}$$

We think of $\|\boldsymbol{x}\|_1$, $\|\boldsymbol{x}\|_2$, and $\|\boldsymbol{x}\|_\infty$ as the important vector norms. Then, we have three corresponding matrix norms:

$$||\mathbf{A}||_1 = \max_{\|\boldsymbol{x}\|_1=1} \|\mathbf{A}\boldsymbol{x}\|_1, \quad ||\mathbf{A}||_2 = \max_{\|\boldsymbol{x}\|_2=1} \|\mathbf{A}\boldsymbol{x}\|_2, \quad ||\mathbf{A}||_\infty = \max_{\|\boldsymbol{x}\|_\infty=1} \|\mathbf{A}\boldsymbol{x}\|_\infty$$

The definition looks scary but it turns out that we can actually compute the norms quite straightforwardly at least for the 1-norm and the $\infty-$norm. For $||\mathbf{A}||_2$ we need the singular value decomposition, so the discussion of that norm is postponed to Sect. 9.12.3. I want to start with $||\mathbf{A}||_1$ for simple 2×2 matrices:

$$\mathbf{A} = \begin{bmatrix} a & b \\ c & d \end{bmatrix} \Longrightarrow \boldsymbol{y} := \mathbf{A}\boldsymbol{x} = \begin{bmatrix} ax_1 + bx_2 \\ cx_1 + dx_2 \end{bmatrix} \Longrightarrow \|\boldsymbol{y}\|_1 = |ax_1 + bx_2| + |cx_1 + dx_2|$$

Now, to find $||\mathbf{A}||_1$, we just need to find the maximum of $|ax_1 + bx_2| + |cx_1 + dx_2|$ subjecting to the constraint $|x_1| + |x_2| = 1$:

$$\begin{aligned} \|\boldsymbol{y}\|_1 &\le |x_1||a| + |x_2||b| + |x_1||c| + |x_2||d| \\ &\le |x_1|(|a| + |c|) + |x_2|(|b| + |d|) \\ &\le (|x_1| + |x_2|)M = M, \quad M = \max\{|a| + |c|, |b| + |d|\} \end{aligned}$$

---

[65] Of course we have to check the conditions in 9.11.9. I skipped that part. Check [54].

Thus, $||\mathbf{A}||_1$ is simply the largest absolute column sum of the matrix. Not satisfied with this simple English, mathematicians write

$$||\mathbf{A}||_1 = \max_{j=1,\ldots,n} \sum_{i=1}^{n} |A_{ij}|$$

The red term is the absolute column sum and the blue operator is to get the largest of these sums. As simple as that.

Follow the same steps, it can be shown that $||\mathbf{A}||_\infty$ is the largest absolute row sum of the matrix. The proof for an $n \times n$ matrix for $||\mathbf{A}||_1$ is not hard but for $||\mathbf{A}||_\infty$ it is harder.

9.11.8 *The Condition Number of a Matrix*

When we solve some systems of linear equations $\mathbf{A}\boldsymbol{x} = \boldsymbol{b}$, we often see that small changes in the entries of $\mathbf{A}$ or $\boldsymbol{b}$ can produce large changes in the solutions $\boldsymbol{x}$. For example, considering the following system of equations:

$$\begin{bmatrix} 1 & 1 \\ 1 & 1.0005 \end{bmatrix} \begin{bmatrix} x_1 \\ x_2 \end{bmatrix} = \begin{bmatrix} 3 \\ 3.0010 \end{bmatrix} \Longrightarrow \begin{bmatrix} x_1 \\ x_2 \end{bmatrix} = \begin{bmatrix} 1 \\ 2 \end{bmatrix}$$

If the matrix is slightly changed, we obtain a completely different solution:

$$\begin{bmatrix} 1 & 1 \\ 1 & 1.0010 \end{bmatrix} \begin{bmatrix} x_1 \\ x_2 \end{bmatrix} = \begin{bmatrix} 3 \\ 3.0010 \end{bmatrix} \Longrightarrow \begin{bmatrix} x_1 \\ x_2 \end{bmatrix} = \begin{bmatrix} 2 \\ 1 \end{bmatrix}$$

Now the problem is to study when this happens or, in other words, there is any measure of $\mathbf{A}$ that can quantify this behavior? The answer is yes and that measure is what we call the condition (or conditioning) number of the matrix. To work out this number, we consider a general $\mathbf{A}\boldsymbol{x} = \boldsymbol{b}$ and $\mathbf{A}'\boldsymbol{x}' = \boldsymbol{b}$ where $\mathbf{A}'$ is slightly different from $\mathbf{A}$. As $\mathbf{A}'$ is slightly different from $\mathbf{A}$, we can write it as $\mathbf{A}' = \mathbf{A} + \Delta\mathbf{A}$. Similarly, we write $\boldsymbol{x}' = \boldsymbol{x} + \Delta x$. If we can compute the norm of Δx we will know when this change in the solution is large or small. That's the plan and it is natural.

Starting with $\mathbf{A}'\boldsymbol{x}' = \boldsymbol{b}$, and replacing $\mathbf{A}' = \mathbf{A} + \Delta\mathbf{A}$ and $\boldsymbol{x}' = \boldsymbol{x} + \Delta x$, we have

$$\mathbf{A}'\boldsymbol{x}' = \boldsymbol{b} \Longleftrightarrow (\mathbf{A} + \Delta\mathbf{A})(\boldsymbol{x} + \Delta x) = \boldsymbol{b} \Longleftrightarrow \Delta\boldsymbol{x} = -\mathbf{A}^{-1}\Delta\mathbf{A}\boldsymbol{x}'$$

Now, we can compute the norm of $\Delta\boldsymbol{x}$

$$\|\Delta\boldsymbol{x}\| = \|[\|]-\mathbf{A}^{-1}\Delta\mathbf{A}\boldsymbol{x}' = \|\mathbf{A}^{-1}\Delta\mathbf{A}\boldsymbol{x}'\| \leq \|\mathbf{A}^{-1}\|\|\Delta\mathbf{A}\boldsymbol{x}'\| \leq \|\mathbf{A}^{-1}\|\|\Delta\mathbf{A}\|\|\boldsymbol{x}'\|$$

(In the first inequality I have used the multiplicative inequality stated in 9.11.9.) Thus,

$$\frac{\|\Delta x\|}{\|x'\|} \leq \|\mathbf{A}^{-1}\|\|\Delta\mathbf{A}\| = \|\mathbf{A}^{-1}\|\|\mathbf{A}\|\frac{\|\Delta\mathbf{A}\|}{\|\mathbf{A}\|}$$

And the red term is defined as the condition number of **A**, denoted by cond(**A**). Why we had to make $\|\mathbf{A}\|$ appear in the above? Because only the relative change in the matrix (e.g., $\|\Delta\mathbf{A}\|/\|\mathbf{A}\|$) makes sense. Thus, the conditioning number gives an upper bound on the relative change in the solution:

$$\frac{\|\Delta x\|}{\|x'\|} \leq \text{cond}(\mathbf{A})\frac{\|\Delta\mathbf{A}\|}{\|\mathbf{A}\|}$$

It is certain that the conditioning number of a matrix depends on the choice of the matrix norm used. The most commonly used norms are $\|\mathbf{A}\|_1$ and $\|\mathbf{A}\|_\infty$. Below is one example.

Example 9.16 Find the conditioning number of the matrix **A** given in the beginning of this section. We need to compute $\mathbf{A}^{-1}$:

$$\mathbf{A} = \begin{bmatrix} 1 & 1 \\ 1 & 1.0005 \end{bmatrix}, \quad \mathbf{A}^{-1} = \begin{bmatrix} +2001 & -2000 \\ -2000 & +2000 \end{bmatrix}$$

Then, the norms of **A** and its inverse, and the condition number are

$$\begin{aligned} \|\mathbf{A}\|_1 &= 2.0005, \ \|\mathbf{A}^{-1}\|_1 = 4001 \Longrightarrow \text{cond}_1(\mathbf{A}) \approx 8004 \\ \|\mathbf{A}\|_\infty &= 2.0005, \ \|\mathbf{A}^{-1}\|_\infty = 4001 \Longrightarrow \text{cond}_\infty(\mathbf{A}) \approx 8004 \end{aligned}$$

If we compute $\text{cond}_2(\mathbf{A})$ it is about 8 002. Thus, when the condition number of a matrix is large for a compatible matrix norm, it will be large for other norms. And that saves us from having to compute different condition numbers! To appreciate that this matrix **A** has a large condition number, consider now the well-behaved matrix in Eq. (9.3.1), its condition number is just three. Matrices such as **A** with large condition numbers are called *ill-conditioned* matrices.

9.11.9 The Best Approximation Theorem

A common problem in mathematics is finding the vector $\boldsymbol{w}$ in a subspace W of a vector space V that best approximates (i.e., is closest to) a given vector $\boldsymbol{v}$ in V. For example, given vector $\boldsymbol{v}$ in $\mathbb{R}^2$, find the vector $\boldsymbol{w}$ living in the span of vector $\boldsymbol{u}$ that is closest to it (Fig. 9.27a). This problem gives rise to the following definition of what is the best approximation.

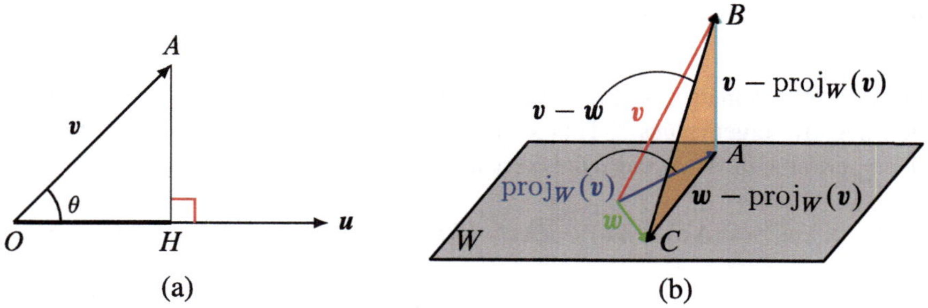

Fig. 9.27 Best approximation theorem: If W is a subspace of a normed linear space V and if $\boldsymbol{v}$ is a vector in V, then the best approximation to $\boldsymbol{v}$ in W is the vector $\text{proj}_W(\boldsymbol{v})$

Definition 9.11.10 If W is a subspace of a normed linear space V and if $\boldsymbol{v}$ is a vector in V, then the best approximation to $\boldsymbol{v}$ in W is the vector $\boldsymbol{v}^*$ in W such that

$$\|\boldsymbol{v} - \boldsymbol{v}^*\| \leq \|\boldsymbol{v} - \boldsymbol{w}\|$$

for every vector $\boldsymbol{w}$ in W.

Now, the question is what is $\boldsymbol{v}^*$? Actually we already met the answer: the projection of $\boldsymbol{v}$ onto W is the answer. This guess is based on the standard geometry of $\mathbb{R}^3$. Refer to Fig. 9.27b, and consider the right triangle ABC, the Pythagorean theorem gives us

$$\|\boldsymbol{w} - \text{proj}_W(\boldsymbol{v})\|^2 + \|\boldsymbol{v} - \text{proj}_W(\boldsymbol{v})\|^2 = \|\boldsymbol{v} - \boldsymbol{w}\|^2$$

which immediately results in

$$\|\boldsymbol{v} - \text{proj}_W(\boldsymbol{v})\|^2 \leq \|\boldsymbol{v} - \boldsymbol{w}\|^2 \Longrightarrow \|\boldsymbol{v} - \text{proj}_W(\boldsymbol{v})\| \leq \|\boldsymbol{v} - \boldsymbol{w}\|$$

And we have just proved the best approximation theorem.

9.12 Singular Value Decomposition

For a square matrix $\mathbf{A}$, which might be not symmetric, we have the factorization $\mathbf{A} = \mathbf{V\Lambda V}^{-1}$. When the matrix is symmetric, we have another factorization $\mathbf{A} = \mathbf{Q\Lambda Q}^\top$. All these decompositions are based on eigenvalues/eigenvectors $\mathbf{A}\boldsymbol{x} = \lambda\boldsymbol{x}$. If $\mathbf{A}$ is non-square, i.e., $\mathbf{A}$ is an $m \times n$ matrix, we cannot have $\mathbf{A}\boldsymbol{x} = \lambda\boldsymbol{x}$ as the left side is in $\mathbb{R}^m$ and the right side is in $\mathbb{R}^n$. The singular value decomposition fills this gap in a perfect way. Now, this section presents this invaluable decomposition.

9.12.1 Singular Values

For any $m \times n$ matrix $\mathbf{A}$, the matrix $\mathbf{A}^\top\mathbf{A}$ is a symmetric $n \times n$ matrix with real non-negative eigenvalues.[66] That's one thing special about $\mathbf{A}^\top\mathbf{A}$. Let's denote by λ the eigenvalue of $\mathbf{A}^\top\mathbf{A}$ and $\boldsymbol{v}$ the corresponding unit eigenvector, then we have

$$0 \le ||\mathbf{A}\boldsymbol{v}||^2 = (\mathbf{A}\boldsymbol{v}) \cdot (\mathbf{A}\boldsymbol{v}) = (\mathbf{A}\boldsymbol{v})^\top(\mathbf{A}\boldsymbol{v}) = \boldsymbol{v}^\top\mathbf{A}^\top\mathbf{A}\boldsymbol{v} = \boldsymbol{v}^\top\lambda\boldsymbol{v} = \lambda||\boldsymbol{v}||^2 = \lambda$$

Thus, λ is the squared length of the vector $\mathbf{A}\boldsymbol{v}$. So, for a rectangular matrix, we do not have eigenvalues but we have singular values, which are the eigenvalues of $\mathbf{A}^\top\mathbf{A}$:

Definition 9.12.1 If $\mathbf{A}$ is an $m \times n$ matrix, the singular values of $\mathbf{A}$ are the square roots of the eigenvalues of $\mathbf{A}^\top\mathbf{A}$ and are denoted by $\sigma_1, \sigma_2, \ldots, \sigma_n$. It is conventional to arrange the singular values in a descending order: $\sigma_1 \ge \sigma_2 \ge \cdots \ge \sigma_n$.

We can find the rank of $\mathbf{A}$ by counting the number of non-zero singular values. Indeed, from Theorem 9.5.5 we have $\text{rank}(\mathbf{A}) = \text{rank}(\mathbf{A}^\top\mathbf{A})$. But,

$$\text{rank}(\mathbf{A}^\top\mathbf{A}) = \text{rank}(\mathbf{Q}\boldsymbol{\Lambda}\mathbf{Q}^\top) \le \min(\text{rank}(\mathbf{Q}), \text{rank}(\boldsymbol{\Lambda})) = r$$

9.12.2 Singular Value Decomposition

Now the problem we want to solve is: starting with two orthogonal vectors $\boldsymbol{v}_1$ and $\boldsymbol{v}_2$. They are transformed by a matrix $\mathbf{A}$ to $\boldsymbol{y}_1$ and $\boldsymbol{y}_2$: $\mathbf{A}\boldsymbol{v}_1 = \boldsymbol{y}_1$ and $\mathbf{A}\boldsymbol{v}_2 = \boldsymbol{y}_2$. We want those transformed vectors to be orthogonal (to another) too. Recall from Sect. 9.12.1, the length of $\boldsymbol{y}_1$ is the singular value σ_1 of $\mathbf{A}^\top\mathbf{A}$. Therefore, we can write $\boldsymbol{y}_1 = \sigma_1\boldsymbol{u}_1$ with $\boldsymbol{u}_1$ is a unit vector. Now we write

$$\mathbf{A}\boldsymbol{v}_1 = \sigma_1\boldsymbol{u}_1, \quad \mathbf{A}\boldsymbol{v}_2 = \sigma_2\boldsymbol{u}_2 \Longrightarrow \mathbf{A}\begin{bmatrix}\boldsymbol{v}_1 & \boldsymbol{v}_2\end{bmatrix} = \begin{bmatrix}\sigma_1\boldsymbol{u}_1 & \sigma_2\boldsymbol{u}_2\end{bmatrix} = \begin{bmatrix}\boldsymbol{u}_1 & \boldsymbol{u}_2\end{bmatrix}\begin{bmatrix}\sigma_1 & 0 \\ 0 & \sigma_2\end{bmatrix}$$

Now, we introduce the matrix $\mathbf{V} = [\boldsymbol{v}_1 \;\; \boldsymbol{v}_2]$, matrix $\mathbf{U} = [\boldsymbol{u}_1 \;\; \boldsymbol{u}_2]$ and Σ is the diagonal matrix containing $\sigma_{1,2}$. The above equation then becomes

$$\mathbf{AV} = \mathbf{U}\Sigma \Longrightarrow \boxed{\mathbf{A} = \mathbf{U}\Sigma\mathbf{V}^\top}$$

[66] But if you're wondering why we know to consider $\mathbf{A}^\top\mathbf{A}$ in the first place, I do not have a correct historical answer. However this might help. Start with a rectangular matrix $\mathbf{A}$, it transforms a vector $\boldsymbol{x}$ into $\boldsymbol{y} = \mathbf{A}\boldsymbol{x}$. If we ask what is the length of $\boldsymbol{y}$, then $\mathbf{A}^\top\mathbf{A}$ appears. Indeed, $||\boldsymbol{y}||^2 = (\mathbf{A}\boldsymbol{x})^\top(\mathbf{A}\boldsymbol{x}) = \boldsymbol{x}^\top\mathbf{A}^\top\mathbf{A}\boldsymbol{x}$.

And the decomposition in the box is the *singular value decomposition* of $\mathbf{A}$. Why $\boldsymbol{y}_1$ is orthogonal to $\boldsymbol{y}_2$? To see this, suppose $\boldsymbol{v}_i$ is the eigenvector of $\mathbf{A}^\top\mathbf{A}$ corresponding to the eigenvalue λ_i. Then, for $i \neq j$, we have

$$(\mathbf{A}\boldsymbol{v}_i) \cdot (\mathbf{A}\boldsymbol{v}_j) = (\mathbf{A}\boldsymbol{v}_i)^\top(\mathbf{A}\boldsymbol{v}_j) = \boldsymbol{v}_i^\top\mathbf{A}^\top\mathbf{A}\boldsymbol{v}_j = \lambda_j\boldsymbol{v}_i^\top\boldsymbol{v}_j = 0$$

The final equality is due to the fact that the eigenvectors of the symmetric matrix $\mathbf{A}^\top\mathbf{A}$ are orthogonal.

Example 9.17 Find a singular value decomposition for the following matrix:

$$\mathbf{A} = \begin{bmatrix} 1 & 1 & 0 \\ 0 & 0 & 1 \end{bmatrix}$$

The first step is to consider the matrix $\mathbf{A}^\top\mathbf{A}$ and find its eigenvalues/eigenvectors:

$$\mathbf{A}^\top\mathbf{A} = \begin{bmatrix} 1 & 1 & 0 \\ 1 & 1 & 0 \\ 0 & 0 & 1 \end{bmatrix} \Longrightarrow \boldsymbol{v}_1 = \begin{bmatrix} 1/\sqrt{2} \\ 1/\sqrt{2} \\ 0 \end{bmatrix}, \boldsymbol{v}_2 = \begin{bmatrix} 0 \\ 0 \\ 1 \end{bmatrix}, \boldsymbol{v}_3 = \begin{bmatrix} -1/\sqrt{2} \\ 1/\sqrt{2} \\ 0 \end{bmatrix}$$

with corresponding eigenvalues $\lambda_1 = 2, \lambda_2 = 1, \lambda_3 = 0$. Note that as $\text{rank}(\mathbf{A}) = 2$, we have $\text{rank}(\mathbf{A}^\top\mathbf{A}) = 2$, thus one eigenvalue must be zero. Note also that as $\mathbf{A}^\top\mathbf{A}$ is symmetric, $\{\boldsymbol{v}_i\}$ is an orthogonal set. Thus, $\mathbf{V}$ and Σ are given by ($\sigma_i = \sqrt{\lambda_i}$)

$$\mathbf{V} = \begin{bmatrix} 1/\sqrt{2} & 0 & -1/\sqrt{2} \\ 1/\sqrt{2} & 0 & +1/\sqrt{2} \\ 0 & 1 & 0 \end{bmatrix}, \quad \Sigma = \begin{bmatrix} \sqrt{2} & 0 & 0 \\ 0 & 1 & 0 \end{bmatrix}$$

To find $\mathbf{U}$ find $\boldsymbol{u}_i$:

$$\boldsymbol{u}_1 = \frac{1}{\sigma_1}\mathbf{A}\boldsymbol{v}_1 = (1, 0), \quad \boldsymbol{u}_2 = \frac{1}{\sigma_2}\mathbf{A}\boldsymbol{v}_2 = (0, 1)$$

These two vectors are already an orthonormal basis. Now, we have $\mathbf{U}$, $\mathbf{V}$ and Σ, then the SVD of $\mathbf{A}$ is

$$\underbrace{\begin{bmatrix} 1 & 1 & 0 \\ 0 & 0 & 1 \end{bmatrix}}_{\mathbf{A}} = \underbrace{\begin{bmatrix} 1 & 0 \\ 0 & 1 \end{bmatrix}}_{\mathbf{U}} \underbrace{\begin{bmatrix} \sqrt{2} & 0 & 0 \\ 0 & 1 & 0 \end{bmatrix}}_{\Sigma} \underbrace{\begin{bmatrix} 1/\sqrt{2} & 1/\sqrt{2} & 0 \\ 0 & 0 & 1 \\ -1/\sqrt{2} & 1/\sqrt{2} & 0 \end{bmatrix}}_{\mathbf{V}^\top}$$

Using `Julia` we can easily verify that the above is correct. Thus, we have singular value decomposed a rectangular matrix!

Hope that this example demonstrates what a SVD is. Now, we give the formal definition of it and then we need to prove that it is always possible to do a SVD for any matrix.

Definition 9.12.2 Let $\mathbf{A}$ be an $m \times n$ matrix with singular values $\sigma_1 \geq \sigma_2 \geq \cdots \geq \sigma_n \geq 0$. Let r denote the number of non-zero singular values of $\mathbf{A}$. A singular value decomposition of $\mathbf{A}$ is the following factorization $\mathbf{A} = \mathbf{U}\Sigma\mathbf{V}^\top$, where $\mathbf{U}$ is an $m \times m$ orthogonal matrix, $\mathbf{V}$ is an $n \times n$ orthogonal matrix, and Σ is an $m \times n$ diagonal matrix whose ith diagonal entry is the ith singular value σ_i for $i = 1, 2, ..., r$. All other entries of Σ are zero.

Proof We now prove that we can always do a SVD for $\mathbf{A}$. The idea of the proof is to show that for any vector $\boldsymbol{x} \in \mathbb{R}^n$, we have $\mathbf{A}\boldsymbol{x} = \mathbf{U}\Sigma\mathbf{V}^\top\boldsymbol{x}$. If so, then of course $\mathbf{A} = \mathbf{U}\Sigma\mathbf{V}^\top$. To this end, we start with $\mathbf{V}^\top\boldsymbol{x}$, then $\Sigma\mathbf{V}^\top\boldsymbol{x}$:

$$\mathbf{V}^\top\boldsymbol{x} = \begin{bmatrix} \boldsymbol{v}_1 \cdot \boldsymbol{x} \\ \boldsymbol{v}_2 \cdot \boldsymbol{x} \\ \vdots \\ \boldsymbol{v}_n \cdot \boldsymbol{x} \end{bmatrix} \Longrightarrow \Sigma\mathbf{V}^\top\boldsymbol{x} = \begin{bmatrix} \sigma_1\boldsymbol{v}_1 \cdot \boldsymbol{x} \\ \sigma_2\boldsymbol{v}_2 \cdot \boldsymbol{x} \\ \vdots \\ \sigma_r\boldsymbol{v}_r \cdot \boldsymbol{x} \\ 0 \end{bmatrix} = \begin{bmatrix} \sigma_1\boldsymbol{v}_1^\top\boldsymbol{x} \\ \sigma_2\boldsymbol{v}_2^\top\boldsymbol{x} \\ \vdots \\ \sigma_r\boldsymbol{v}_r^\top\boldsymbol{x} \\ 0 \end{bmatrix}$$

Now, we consider $\mathbf{U}\Sigma\mathbf{V}^\top\boldsymbol{x}$, noting that $\mathbf{U}$ contains $\boldsymbol{u}_i = \sigma_i^{-1}\mathbf{A}\boldsymbol{v}_i$:

$$\begin{aligned}
\mathbf{U}\Sigma\mathbf{V}^\top\boldsymbol{x} &= \boldsymbol{u}_1\sigma_1\boldsymbol{v}_1^\top\boldsymbol{x} + \cdots + \boldsymbol{u}_r\sigma_r\boldsymbol{v}_r^\top\boldsymbol{x} \quad (\mathbf{A}\boldsymbol{x}\text{is a linear combination of the cols of}\mathbf{A}) \\
&= \sigma_1^{-1}\mathbf{A}\boldsymbol{v}_1\sigma_1\boldsymbol{v}_1^\top\boldsymbol{x} + \cdots + \sigma_r^{-1}\mathbf{A}\boldsymbol{v}_r\sigma_r\boldsymbol{v}_n^\top\boldsymbol{x} \quad (\text{use } \boldsymbol{u}_i = \sigma_i^{-1}\mathbf{A}\boldsymbol{v}_i) \\
&= \mathbf{A}\boldsymbol{v}_1\boldsymbol{v}_1^\top\boldsymbol{x} + \cdots + \mathbf{A}\boldsymbol{v}_n\boldsymbol{v}_n^\top\boldsymbol{x} \quad (\mathbf{A}\boldsymbol{v}_i = \mathbf{0}, i > r) \\
&= \mathbf{A}\underbrace{(\boldsymbol{v}_1\boldsymbol{v}_1^\top + \cdots + \boldsymbol{v}_n\boldsymbol{v}_n^\top)}_{\mathbf{I}}\boldsymbol{x} = \mathbf{A}\boldsymbol{x}
\end{aligned}$$

So, in the third equality, we just added a bunch of zero vectors. Note that $\mathbf{A}\boldsymbol{v}_i = \mathbf{0}$, $i > r$ because we have only r non-zero singular values. The final equality comes from the fact that if $\{\boldsymbol{v}_1, \ldots, \boldsymbol{v}_n\}$ is an orthonormal set then $\boldsymbol{v}_1\boldsymbol{v}_1^\top + \cdots + \boldsymbol{v}_n\boldsymbol{v}_n^\top = \mathbf{I}$. ■

Left and right singular vectors. We have $\mathbf{A}^\top\mathbf{A}$ with the eigenvectors $\boldsymbol{v}_k$. How about $\boldsymbol{u}_k$? Are they the eigenvectors of some matrix? The answer is yes: it is the eigenvector of $\mathbf{A}\mathbf{A}^\top$. Maths is really nice, isn't it. The proof goes as

$$\mathbf{A}\mathbf{A}^\top\boldsymbol{u}_k = \mathbf{A}\mathbf{A}^\top\frac{\mathbf{A}\boldsymbol{v}_k}{\sigma_k} = \frac{\mathbf{A}[(\mathbf{A}^\top\mathbf{A})\boldsymbol{v}_k]}{\sigma_k} = \frac{\mathbf{A}\sigma_k^2\boldsymbol{v}_k}{\sigma_k} = \sigma_k^2\boldsymbol{u}_k$$

The key to the proof was the fact that $(\mathbf{A}\mathbf{A}^\top)\mathbf{A} = \mathbf{A}(\mathbf{A}^\top\mathbf{A})$. Some new terms: the $\boldsymbol{v}_k$ are called the right singular vectors and the $\boldsymbol{u}_k$ are called the left singular vectors.

Geometry of the SVD. We have seen in Fig. 9.24 that the linear transformation $\mathbf{A}\boldsymbol{x}$ transforms a circle in $\mathbb{R}^2$ into an ellipse in $\mathbb{R}^2$. With the SVD, it can be proved that an $m \times n$ matrix $\mathbf{A}$ maps a unit sphere in $\mathbb{R}^n$ into an ellipsoid in $\mathbb{R}^m$. Here is the proof. Consider a unit vector $\boldsymbol{x} \in \mathbb{R}^n$, and its image $\boldsymbol{y} = \mathbf{A}\boldsymbol{x} \in \mathbb{R}^m$:

$$\boldsymbol{x} = x_1\boldsymbol{v}_1 + x_2\boldsymbol{v}_2 + \cdots + x_n\boldsymbol{v}_n \Longrightarrow \boldsymbol{y} = \mathbf{A}\boldsymbol{x} = x_1\sigma_1\boldsymbol{u}_1 + \cdots + x_r\sigma_r\boldsymbol{u}_r$$

The image vector $\boldsymbol{y}$ has coordinates $(y_1, y_2, \ldots, y_r) = (x_1\sigma_1, \ldots, x_r\sigma_r)$, where

$$\left(\frac{y_1}{\sigma_1}\right)^2 + \left(\frac{y_2}{\sigma_2}\right)^2 + \cdots + \left(\frac{y_r}{\sigma_r}\right)^2 = \sum_{i=1}^{r} x_i^2 \le 1$$

The last inequality comes from the unit vector $\boldsymbol{x}$. Now, if $r = n$ (i.e., the matrix $\mathbf{A}$ is a full column rank matrix), then in the above inequality we have equal sign, and thus the image $\mathbf{A}\boldsymbol{x}$ is the surface of the ellipsoid. On the other hand, if $r < n$, then the image is a solid ellipsoid in $\mathbb{R}^m$.

We can even have a geometry interpretation of the different matrices in an SVD. For that we have to restrict to a plane. Start with a unit vector $\boldsymbol{x} \in \mathbb{R}^2$. Now the transformation $\mathbf{A}\boldsymbol{x}$ is $\left(\mathbf{U}\Sigma\mathbf{V}^\top\right)\boldsymbol{x}$. From Sect. 9.6 on linear transformation, we know that we're dealing with a composite transformation. And we handle it from right to left. So, we start with $\mathbf{V}^\top\boldsymbol{x}$, which is a rotation, thus we get a circle from a circle. But now we see the transformed circle in the plane in which the axes are $\boldsymbol{v}_1$ and $\boldsymbol{v}_2$ (Fig. 9.28). Then comes $\Sigma\left(\mathbf{V}^\top\boldsymbol{x}\right)$ which simply stretches (sometimes shrinks) our circle (the second circle from the left) to an ellipse. Finally, $\mathbf{U}$ is a rotation and we got an oblique ellipse as the final $\mathbf{A}\boldsymbol{x}$.

A byproduct of this is that we are now able to compute $||\mathbf{A}||_2$, it is simply σ_1: $||\mathbf{A}||_2 = \sigma_1$.

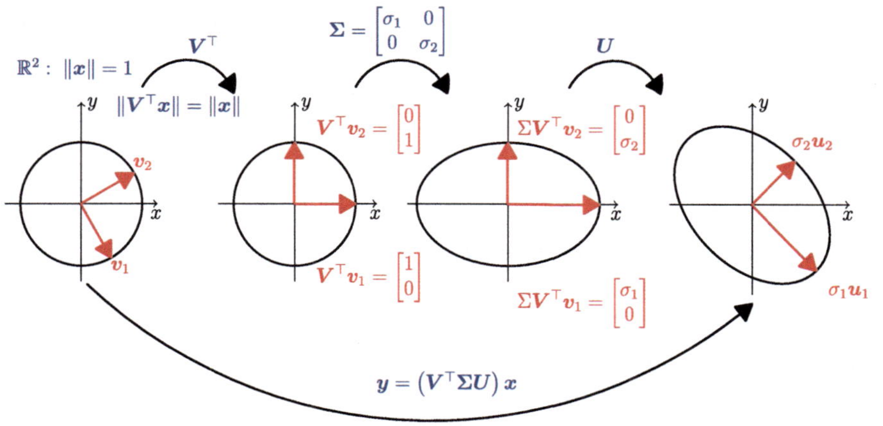

Fig. 9.28 The geometry of the singular value decomposition

9.12.3 Matrix Norms and the Condition Number

Recall that the 2-condition number of matrix **A** is defined as

$$\text{cond}_2(\mathbf{A}) = \|\mathbf{A}^{-1}\|_2 \|\mathbf{A}\|_2$$

Now, with the SVD of **A**, we can compute these norms and thus the 2-condition number.

As shown in Fig. 9.28, the norm of **A** is simply its largest singular value:

$$\|\mathbf{A}\|_2 = \max_{\|\boldsymbol{x}\|_2=1} \|\mathbf{A}\boldsymbol{x}\|_2 = \sigma_1$$

The inverse of **A** (if it exists) can be determined easily from the SVD $\mathbf{A} = \mathbf{U}\Sigma\mathbf{V}^\top$, namely,

$$\mathbf{A}^{-1} = \mathbf{V}\Sigma^{-1}\mathbf{U}^\top \tag{9.12.1}$$

where Σ^{-1} is a diagonal matrix with $1/\sigma_i$ on the diagonal. The reason is simple using the idea of inverse mapping by undoing each of the three operations shown in Fig. 9.28. First, undo the last rotation by multiplying with $\mathbf{U}^\top$; second un-stretch by multiplying by $1/\sigma_i$ along each axis; third, un-rotate by multiplying by **V**. If you need to see an algebra proof, here it is:

$$\mathbf{A}^{-1}\mathbf{A} = \left(\mathbf{V}\Sigma^{-1}\mathbf{U}^\top\right)\left(\mathbf{U}\Sigma\mathbf{V}^\top\right) = \mathbf{V}\Sigma^{-1}\left(\mathbf{U}^\top\mathbf{U}\right)\Sigma\mathbf{V}^\top = \mathbf{V}\left(\Sigma^{-1}\Sigma\right)\mathbf{V}^\top = \mathbf{V}\mathbf{V}^\top = \mathbf{I}$$

The 2-norm of $\mathbf{A}^{-1}$ is its maximum singular value which is $1/\sigma_n$:

$$\|\mathbf{A}^{-1}\|_2 = \max_{\|\boldsymbol{x}\|_2=1} \|\mathbf{A}^{-1}\boldsymbol{x}\|_2 = \frac{1}{\sigma_n}$$

Now, the 2-condition number of **A** is simply the ratio of the maximum singular value and minimum singular value, or

$$\text{cond}_2(\mathbf{A}) = \frac{\sigma_1}{\sigma_n} \geq 1 \tag{9.12.2}$$

9.12.4 Low Rank Approximations

We have seen Taylor series and Fourier series of which the main idea is to expand or decompose a function into a sum of many pieces. Thus, it is not a surprise when we can do the same thing with a matrix:

$$\mathbf{U}\Sigma\mathbf{V}^\top\boldsymbol{x} = \sigma_1\boldsymbol{u}_1\boldsymbol{v}_1^\top\boldsymbol{x} + \cdots + \sigma_r\boldsymbol{u}_r\boldsymbol{v}_r^\top\boldsymbol{x} \Longrightarrow \mathbf{A} = \sigma_1\boldsymbol{u}_1\boldsymbol{v}_1^\top + \cdots + \sigma_r\boldsymbol{u}_r\boldsymbol{v}_r^\top$$

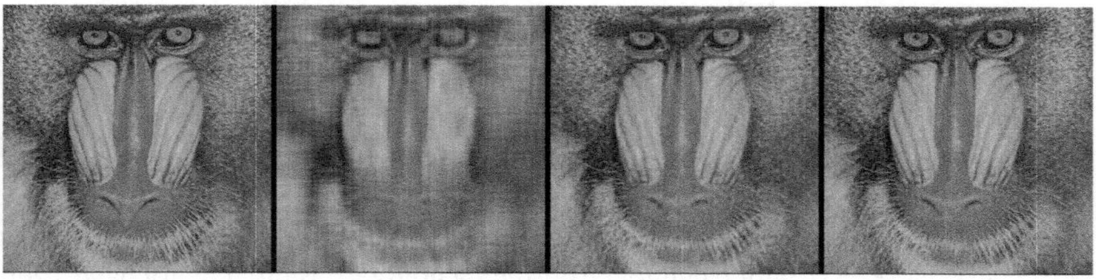

Fig. 9.29 From left to right: original image, 10, 50, and 100 singular values

Similar to what we have done to Taylor series, we truncate the sum on the RHS of $\mathbf{A}$ to get $\mathbf{A}_k$—a rank k matrix:

$$\mathbf{A}_k = \sigma_1 \boldsymbol{u}_1 \boldsymbol{v}_1^\top + \cdots + \sigma_k \boldsymbol{u}_k \boldsymbol{v}_k^\top$$

And we expect there exists a truth between $\mathbf{A}$ and $\mathbf{A}_k$. And this truth was discovered by Schmidt in 1907, which was later proved by Eckart and Young in 1936 and by Mirsky in 1955. The theorem is now called the Eckart-Young-Mirsky theorem stating that $\mathbf{A}_k$ *is the closet rank k matrix to* $\mathbf{A}$. Obviously we need to use matrix norms to express this theorem:

Theorem 9.12.1 (The Eckart-Young-Mirsky theorem) *If* $\mathbf{B}$ *has rank k then*

$$\|\mathbf{A} - \mathbf{B}\| \geq \|\mathbf{A} - \mathbf{A}_k\|, \quad \mathbf{A}_k = \sigma_1 \boldsymbol{u}_1 \boldsymbol{v}_1^\top + \cdots + \sigma_k \boldsymbol{u}_k \boldsymbol{v}_k^\top$$

In what follows we give an application of this theorem in image compression.

SVD in image compression. Suppose that the original image is a gray image of size $(512, 512)$, and we rebuild the image with 50 singular values, then we only need to save $2 \times 512 \times 50 + 50$ numbers to rebuild the image, while original image has 512×512 numbers. Hence, this gives us a compression ratio 19.55% if we don't consider the storage type. Figure 9.29 presents one example and the code to produce it is given in Listing A.23.

9.13 Complex Vectors and Complex Matrices

A complex vector is a vector whose components are complex numbers. For example, $z = (1 + 2i, 3 - 4i)$ is a complex vector, we use the notation $z \in \mathbb{C}^2$ for this. A general n-complex vector is given by

$$z = \begin{bmatrix} a_1 + ib_1 & a_2 + ib_2 & \cdots & a_n + ib_n \end{bmatrix}^\top$$

9.13.1 Complex Dot Product

The first question we have to ask is: how we compute the length of a complex vector? To find the answer we look at the old friend real vectors. If $\boldsymbol{a}$ is a real n-vector, then its length is $(a_1^2 + \cdots + a_n^2)^{1/2}$. Can we use this for complex vectors? Just try for a vector and see what we shall get. As simple as that. For example, try $\boldsymbol{z} = (1, i)$, then $||z|| = \sqrt{1^2 + i^2} = 0$, which cannot be correct: a non-zero vector cannot have a zero length!

To find the formula, let's try another vector that we know its length for sure. It is $z = (a + bi)$! A vector with only one component. Its length is $||z|| = \sqrt{a^2 + b^2}$. This does not come from $(a + bi)(a + bi)$ but from $(a + bi)(a - bi)$. And this leads to the following definition of the complex dot product.

Definition 9.13.1 If $\boldsymbol{u} = (u_1, u_2, \ldots, u_n)$ and $\boldsymbol{v} = (v_1, v_2, \ldots, v_n)$ are vectors in $\mathbb{C}^n$, then the complex dot product of them is defined by

$$\boldsymbol{u} \cdot \boldsymbol{v} = \bar{\boldsymbol{u}}^\top \boldsymbol{v} = \boldsymbol{u}^{\mathrm{H}} \boldsymbol{v} = \bar{u}_1 v_1 + \bar{u}_2 v_2 + \cdots + \bar{u}_n v_n$$

where $\bar{u}_i$ is the complex conjugate of u_i. Recall that $z = a + bi$, then $\bar{z} = a - bi$.

So, to compute the dot product of two complex vectors, get the conjugate of the first vector, transpose it, and multiply that with the second vector. Instead of writing $\bar{\boldsymbol{u}}^\top$, which requires two symbols, we write $\boldsymbol{u}^{\mathrm{H}}$. This is $\boldsymbol{u}$ Hermitian, the conjugate (of the) transpose of $\boldsymbol{u}$. The new word is pronounced "Hermeeshan".[67]

Now, using this definition of the complex dot product we can get the correct length of a complex vector.

$$||z|| = \left(\boldsymbol{z}^{\mathrm{H}} \boldsymbol{z}\right)^{1/2} = (|z_1|^2 + \cdots + |z_n|^2)^{1/2}$$

Recall that absolute value of a complex number $z = a + bi$ is $|z| = r = \sqrt{a^2 + b^2}$. Now, we return to the vector $\boldsymbol{z} = (1, i)$, then $||z|| = \sqrt{1^2 + (-i)(i)} = \sqrt{2}$.

9.13.2 Hermitian and Unitary Matrices

If $\mathbf{A}$ is a complex matrix, then its Hermitian is $\mathbf{A}^{\mathrm{H}} = \bar{\mathbf{A}}^\top$. Just like complex vectors. For instance,

$$\text{if } \mathbf{A} = \begin{bmatrix} 1 & i \\ 0 & 1+i \end{bmatrix} \Longrightarrow \mathbf{A}^{\mathrm{H}} = \begin{bmatrix} 1 & 0 \\ -i & 1-i \end{bmatrix}$$

Among real matrices, symmetric matrices form the most important special class: $\mathbf{S} = \mathbf{S}^\top$. Among complex matrices, the special class contains the Hermitian matrices:

[67] Named after the French mathematician Charles Hermite (1822–1901).

those that are equal to their Hermitian. That is, $\mathbf{A} = \mathbf{A}^{\mathrm{H}}$. Since taking the conjugate of a real matrix has no effect on it, every real symmetric matrix is Hermitian.

9.14 Least Squares Problems

Astronomy is the first scientific discipline to demand precise measurements. Therefore, it is also the first scientific field to grapple with measurement errors and face questions about how to approach situations where multiple observations yield different values for the same quantity. For example, consider the position of a certain star at a certain time, with five individuals providing five different numbers. It wasn't until the eighteenth century that scientists unified, at that time, the average value of all measurements as the representative figure for the quantity under consideration. Why is the average value chosen? The method of least squares, developed by the French mathematician Adrien-Marie Legendre, provides the answer.

The method is as follows: suppose we have a physical quantity, and its true value is x^* (which we want to find the value but we just don't know that yet). To determine it, we conduct measurements and obtain n data points $x_1, x_2, ..., x_n$. Legendre stated that x^* we're looking for will minimize the sum of squared errors d:

$$d = (x_1 - x^*)^2 + (x_2 - x^*)^2 + \cdots + (x_n - x^*)^2 \tag{9.14.1}$$

Why Legendre chose d like that is a difficult question. However, once Legendre pointed it out, the rest is just a calculus exercise. Minimum? Take the derivative of d with respect to x^* and set it equal to 0:

$$-2(x_1 - x^*) - 2(x_2 - x^*) - \cdots - 2(x_n - x^*) = 0 \Longrightarrow x^* = \frac{x_1 + x_2 + \cdots + x_n}{n} \tag{9.14.2}$$

Indeed, the average minimizes the error according to Legendre's chosen minimum criterion! But doesn't taking the derivative only give us a minimum value? How can we be sure it's a minimum and not a maximum? I'll leave that for you to answer.

After that humbling introduction to the least squares method, we start in Sect. 9.14.1 a statement of a general least squares problem and we discuss its solution in Sect. 9.14.2.

9.14.1 Problem Statement

In many scientific problems, experimental data are used to infer a mathematical relationship among the variables being measured. In the simplest case, there is one single independent variable and one dependent variable. Then, the data come in the form of two measurements: one for the independent variables and one for the

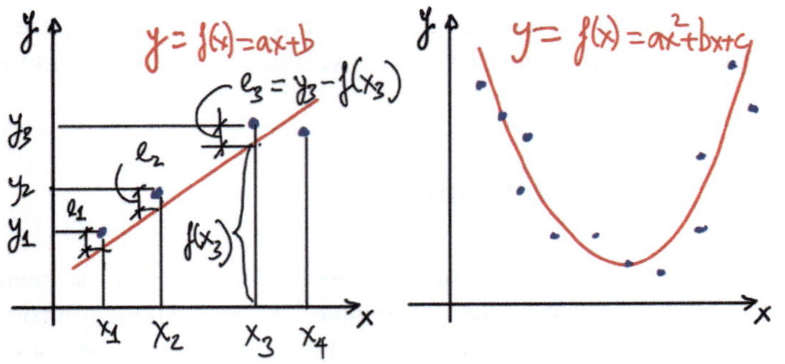

Fig. 9.30 Fitting a curve through a cloud of points

dependent variable. Thus, we have a set of points (x_i, y_i), and we are looking for a function that best approximates the relationship between the independent variable x and the dependent variable y. Once we have found that function (e.g., $y = f(x)$), then we can *make predictions*: given any x^* (not in the experimental data), we can determine the corresponding $y = f(x^*)$. Figure 9.30 gives two examples.

Suppose that the function relating x and y is a linear function $y = f(x) = ax + b$ or a quadratic function $y = ax^2 + bx + c$. The problem is then to determine the parameters a, b or a, b, c (for the quadratic case) so that the model best fits the data. It was found[68] that to get the best fit, we need to minimize the sum of the squares of the error, where the error is the difference between the data and the model evaluated at x_i:

$$\text{minimize}: \quad S = \sum_{i=1}^{n} e_i^2, \quad e_i = y_i - f(x_i)$$

Even though this problem can be solved by calculus (i.e., setting the derivative of S w.r.t a and b to zero), I prefer to use linear algebra to solve it. Why? To understand more about linear algebra! To this end, we introduce the error vector $\boldsymbol{e} = (e_1, e_2, \ldots, e_n)$ where $e_i = y_i - f(x_i)$. Let's start with the simplest case where $f(x) = \alpha x + \beta$, then we can write the error function as

$$\boldsymbol{e} = \begin{bmatrix} e_1 \\ e_2 \\ \vdots \\ e_n \end{bmatrix} = \underbrace{\begin{bmatrix} y_1 \\ y_2 \\ \vdots \\ y_n \end{bmatrix}}_{\boldsymbol{b}} - \underbrace{\begin{bmatrix} x_1 & 1 \\ x_2 & 1 \\ \vdots & \vdots \\ x_n & 1 \end{bmatrix}}_{\mathbf{A}} \underbrace{\begin{bmatrix} \alpha \\ \beta \end{bmatrix}}_{\boldsymbol{x}} \tag{9.14.3}$$

[68] By Roger Cotes, Legendre, and Gauss. In 1809, Carl Friedrich Gauss published his method (of least squares) of calculating the orbits of celestial bodies.

In statistics, the matrix $\mathbf{A}$ is called *design matrix*. Usually we have lots of data this matrix is, therefore, skinny meaning that it has more rows than columns. Now the problem is: to find $\boldsymbol{x} = (\alpha, \beta)$ to minimize S, which is equivalent to minimize $\|\boldsymbol{e}\|$ (where $||\boldsymbol{v}||$ is the Euclidean norm), which in turn is equivalent to minimize $\|\boldsymbol{b} - \mathbf{A}\boldsymbol{x}\|$. What we have just achieved? We have converted the least squares problem to a linear algebra problem of solving $\mathbf{A}\boldsymbol{x} = \boldsymbol{b}$, but with a rectangular matrix. This overdetermined system is unsolvable in the traditional sense that no $\boldsymbol{x}^*$ would make $\mathbf{A}\boldsymbol{x}^*$ equals $\boldsymbol{b}$. Thus, we ask for a vector $\boldsymbol{x}^*$ that minimizes $\|\boldsymbol{b} - \mathbf{A}\boldsymbol{x}\|$, such a vector is called *the least square solution to* $\mathbf{A}\boldsymbol{x} = \boldsymbol{b}$. So, we have the following definition:

Definition 9.14.1 Least squares problem If $\mathbf{A}$ is an $m \times n$ matrix and $\boldsymbol{b}$ is in $\mathbb{R}^n$, a least squares solution of $\mathbf{A}\boldsymbol{x} = \boldsymbol{b}$ is a vector $\boldsymbol{x}^*$ such that

$$\|\boldsymbol{b} - \mathbf{A}\boldsymbol{x}^*\| \leq \|\boldsymbol{b} - \mathbf{A}\boldsymbol{x}\|$$

for all $\boldsymbol{x}$ in $\mathbb{R}^n$.

9.14.2 Solution of the Least Squares Problem

We are seeking for a vector $\boldsymbol{x}$ that minimizes $\|\boldsymbol{b} - \mathbf{A}\boldsymbol{x}\|$. Noting that $\mathbf{A}\boldsymbol{x}$ is a vector living in the column space of $\mathbf{A}$. So now, the problem is: find a vector $\boldsymbol{y}$ in $C(\mathbf{A})$ that is closest to $\boldsymbol{b}$. According to the best approximation theorem (Sect. 9.11.9), the solution is then the projection of $\boldsymbol{b}$ onto $C(\mathbf{A})$:

$$\mathbf{A}\boldsymbol{x}^* = \text{proj}_{C(\mathbf{A})}(\boldsymbol{b})$$

We do not have to solve this system to get $\boldsymbol{x}^*$, a bit of algebra leads to

$$\boldsymbol{b} - \mathbf{A}\boldsymbol{x}^* = \boldsymbol{b} - \text{proj}_{C(\mathbf{A})}(\boldsymbol{b}) = \text{perp}_{C(\mathbf{A})}(\boldsymbol{b})$$

which means that $\boldsymbol{b} - \mathbf{A}\boldsymbol{x}^*$ is perpendicular to the columns of $\mathbf{A}$:

$$\boldsymbol{a}_i \cdot (\boldsymbol{b} - \mathbf{A}\boldsymbol{x}^*) = 0, \quad i = 1, 2, \ldots, n$$

Thus, we obtain the following equation known as the normal equation, and from it we can find $\boldsymbol{x}^*$

$$\mathbf{A}^\top\mathbf{A}\boldsymbol{x}^* = \mathbf{A}^\top\boldsymbol{b} \Longrightarrow \boxed{\boldsymbol{x}^* = (\mathbf{A}^\top\mathbf{A})^{-1}\mathbf{A}^\top\boldsymbol{b}} \tag{9.14.4}$$

The solution given in the box holds only when $\text{rank}(\mathbf{A}) = n$; that is, when all the columns of $\mathbf{A}$ are linear independent.[69] In that case, due to Theorem 9.5.5 which

[69] When that happens? Take Eq. (9.14.3) as an example, for this design matrix to have a rank of 2, at least there must be two different x_i's.

states that $\text{rank}(\mathbf{A}^\top\mathbf{A}) = \text{rank}(\mathbf{A}) = n$. An $n \times n$ matrix has a rank of n, it is invertible. That's why the normal equation has the unique solution expressed in terms of the inverse of $\mathbf{A}^\top\mathbf{A}$.

Calculus-based solution. We just need to rewrite S as

$$\text{minimize}: \quad S = \frac{1}{2}\|\boldsymbol{e}\|^2 = \frac{1}{2}\boldsymbol{e}^\top\boldsymbol{e}$$

and with $\boldsymbol{e} = \boldsymbol{b} - \mathbf{A}\boldsymbol{x}$, S becomes

$$S = \frac{1}{2}(\boldsymbol{b} - \mathbf{A}\boldsymbol{x})^\top(\boldsymbol{b} - \mathbf{A}\boldsymbol{x}) = \frac{1}{2}\boldsymbol{x}^\top\mathbf{A}^\top\mathbf{A}\boldsymbol{x} - \boldsymbol{b}^\top\mathbf{A}\boldsymbol{x} + \frac{1}{2}\boldsymbol{b}^\top\boldsymbol{b}$$

Now, we use $dS/d\boldsymbol{x} = 0$ (check Sect. 10.9.2 to know how to do differentiation with matrices):

$$\frac{dS}{d\boldsymbol{x}} = \mathbf{A}^\top\mathbf{A}\boldsymbol{x} - \mathbf{A}^\top\boldsymbol{b} \Longrightarrow \mathbf{A}^\top\mathbf{A}\boldsymbol{x} - \mathbf{A}^\top\boldsymbol{b} = \mathbf{0} \Longrightarrow \boldsymbol{x} = (\mathbf{A}^\top\mathbf{A})^{-1}\mathbf{A}^\top\boldsymbol{b}$$

And we have obtained the same result.

Pseudoinverse. For the square matrix $\mathbf{A}$, the solution to $\mathbf{A}\boldsymbol{x} = \boldsymbol{b}$ is written in terms of its inverse matrix: $\boldsymbol{x} = \mathbf{A}^{-1}\boldsymbol{b}$. We should do the same thing for rectangular matrices! And that leads to the concept of pseudoinverse matrix of which definition comes from $\boldsymbol{x}^* = (\mathbf{A}^\top\mathbf{A})^{-1}\mathbf{A}^\top\boldsymbol{b}$.

Definition 9.14.2 If $\mathbf{A}$ is a matrix with linearly independent columns, then the pseudoinverse of $\mathbf{A}$ is the matrix $\mathbf{A}^+$ defined by

$$\mathbf{A}^+ = (\mathbf{A}^\top\mathbf{A})^{-1}\mathbf{A}^\top$$

Fitting a cloud of points with a parabola. The least squares method just works when $f(x) = \alpha x^2 + \beta x + \gamma$. Everything is the same, except that we have a bigger design matrix and we have three unknowns to solve for:

$$\mathbf{A} = \begin{bmatrix} x_1^2 & x_1 & 1 \\ x_2^2 & x_2 & 1 \\ \vdots & \vdots & \vdots \\ x_n^2 & x_n & 1 \end{bmatrix}, \quad \boldsymbol{x} = \begin{bmatrix} \alpha \\ \beta \\ \gamma \end{bmatrix}, \quad \boldsymbol{b} = \begin{bmatrix} y_1 \\ y_2 \\ \vdots \\ y_n \end{bmatrix}$$

Fitting a cloud of 3D points with a plane. So far we just dealt with $y = f(x)$. How about $z = f(x, y)$? No problem, the exact same method works too. Assume that we want to find the best plane $z = \alpha x + \beta y + \gamma$, then $\mathbf{A}$, $\boldsymbol{x}$ and $\boldsymbol{b}$ are given by

$$\mathbf{A} = \begin{bmatrix} x_1 & y_1 & 1 \\ x_2 & y_2 & 1 \\ \vdots & \vdots & \vdots \\ x_n & y_n & 1 \end{bmatrix}, \quad \boldsymbol{x} = \begin{bmatrix} \alpha \\ \beta \\ \gamma \end{bmatrix}, \quad \boldsymbol{b} = \begin{bmatrix} z_1 \\ z_2 \\ \vdots \\ z_n \end{bmatrix}$$

9.15 Principal Component Analysis (PCA)

In many problems, we have a matrix of data (coming from certain measurements). For example, there are n samples and for each sample we are measuring m variables. Thus the data matrix $\mathbf{A}$ has n columns and m rows. Geometrically we have n points in the $\mathbb{R}^m$ space. Most often $m > 3$ which makes visualization and understanding of this data very hard. Principal component analysis (with abbreviation of PCA) provides a way to understand this data.

As one example, suppose we have a bunch of points in a 2D space generated using a bivariate normal distribution (Sect. 5.18.6) as shown in Fig. 9.31a. And an application of the PCA can give us the two lines shown there; these lines are the first two principal components of the data points. If we project the data to the space spanned by these two components we get the plot as shown in Fig. 9.31b. This sounds similar to something we know, doesn't it? Yes. If we consider the points as belong to a solid, then these two lines are indeed the principal axes of it. So, you might correctly guess that in PCA we shall see eigenvectors and eigenvalues. What else?

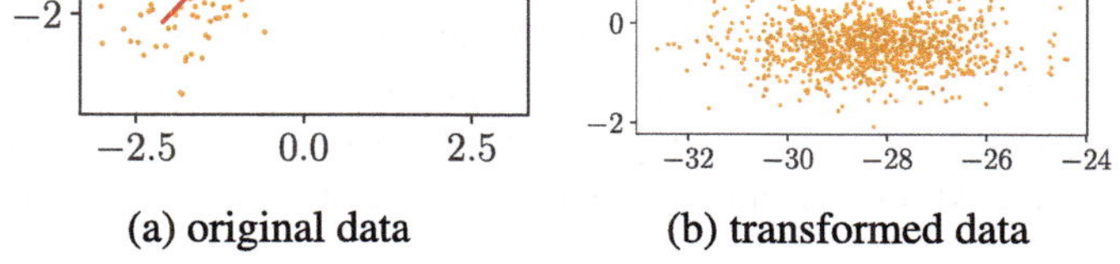

Fig. 9.31 Demonstration of the PCA for 2D data. The job of PCA is to give us the principal components—the two lines shown here

The starting point of the PCA for the data matrix **A** is the covariance matrix **S** of the data (Sect. 5.12.5). This is a symmetric positive semi-definite matrix of dimension $m \times m$. According to the spectral theorem (Theorem 9.10.3), **S** has a spectral decomposition $\mathbf{S} = \mathbf{Q}\boldsymbol{\Lambda}\mathbf{Q}^\top$ with real eigenvalues in $\boldsymbol{\Lambda}$ and orthonormal eigenvectors in the columns of **Q**:

$$\mathbf{S} = \mathbf{Q}\boldsymbol{\Lambda}\mathbf{Q}^{-1} = \mathbf{Q}\boldsymbol{\Lambda}\mathbf{Q}^\top \quad \text{with } \mathbf{Q}^\top = \mathbf{Q}^{-1}$$

Now, if we label by λ_1 the maximum of all the eigenvalues of **S**, then from Sect. 9.10.6, we know that

$$\lambda_1 = \max_{\|\boldsymbol{u}=1\|} \boldsymbol{u}^\top \mathbf{S}\boldsymbol{u}$$

And this happens when $\boldsymbol{u} = \boldsymbol{u}_1$ where $\boldsymbol{u}_1$ is the eigenvector corresponding to λ_1.[70] To understand the geometry of the PCA (which is important), we now try to understand the geometric meaning of the term $\boldsymbol{u}_1^\top \mathbf{S}\boldsymbol{u}_1$. To this end, we confine to the 2D plane, i.e., $m = 2$, and we can write $\boldsymbol{u}_1^\top \mathbf{S}\boldsymbol{u}_1$ as (see Eq. (5.12.21) for $\mathbf{S}$[71])

$$\begin{aligned}
\boldsymbol{u}_1^\top \mathbf{S}\boldsymbol{u}_1 &= \frac{1}{n-1}\boldsymbol{u}_1^\top \begin{bmatrix} \sum_i x_i^2 & \sum_i x_i y_i \\ \sum_i x_i y_i & \sum_i y_i^2 \end{bmatrix} \boldsymbol{u}_1 = \frac{1}{n-1}\sum_i \boldsymbol{u}_1^\top \begin{bmatrix} x_i^2 & x_i y_i \\ x_i y_i & y_i^2 \end{bmatrix} \boldsymbol{u}_1 \\
&= \frac{1}{n-1}\sum_{i=1}^{n} \boldsymbol{u}_1^\top \boldsymbol{x}_i \boldsymbol{x}_i^\top \boldsymbol{u}_1 = \frac{1}{n-1}\sum_{i=1}^{n} (\boldsymbol{x}_i^\top \boldsymbol{u}_1)^2, \quad \boldsymbol{x}_i = (x_i, y_i)
\end{aligned}$$

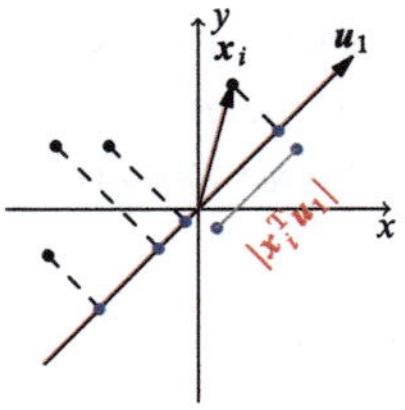

Recognizing that $|\boldsymbol{x}_i^\top \boldsymbol{u}_1|$ is the length of the projected vector of $\boldsymbol{x}_i$ on $\boldsymbol{u}_1$,[72] the term $\boldsymbol{u}_1^\top \mathbf{S}\boldsymbol{u}_1$ is then the sum of squares of the projection of all data points on the line with direction given by $\boldsymbol{u}_1$ (referring to the figure). In summary, we have found the axis $\boldsymbol{u}_1$ which gives the maximum variance of the data (in the parlance of PCA, this axis is the first principal component). Which also means that this axis yields the minimum of the squared distances from data points to the line. If we wish we can find the second axis given by, what else, the second eigenvector $\boldsymbol{u}_2$ (corresponding with the second largest eigenvalue λ_2). Along this axis the variance is also maximum. And we can continue with other eigenvectors, thus we can project our data points to a

[70] Note that the eigenvectors are normalized, i.e., having unit lengths.

[71] Recall that $\mathbf{S} = \text{Cov}(X, Y) = \frac{1}{n-1}\sum_{i=1}^{n}(x_i - \bar{x})(y_i - \bar{y})$

[72] If not clear, check Sect. 9.1.4, in particular, Eq. (9.1.13).

k-dimensional space spanned by $\boldsymbol{u}_1, \ldots, \boldsymbol{u}_k$. Indeed, if we put these eigenvectors in matrix $\mathbf{Q}_k$—an $m \times k$ matrix, then $\boldsymbol{Y} = \mathbf{Q}_k^\top \mathbf{A}$ is the transformed data points living in a k-dimensional space where $k \ll m$.

To help beginners to PCA, the code used to generate Fig. 9.31 is given in Listing 9.2.

Listing 9.2 PCA from scratch using `Julia`.

```
using MultivariateStats, Plots, Distributions, PyCall, Statistics
function manualPCA(xo)
  xs = xo[1, :]; ys = xo[2, :]
  xs_centered = xs .- mean(xs); ys_centered = ys .- mean(ys)
  x         = hcat(xs_centered, ys_centered)
  x         = x'
  covMat    = cov(x')           # covariance matrix S of the data
  ev        = eigvals(covMat) # compute eigenvalues of S
  eigOrder = sortperm(ev,rev=true)# indices  sorting input array in descending order
  vecs      = eigvecs(covMat)[:, eigOrder] # sorted eigenvectors
  return  (ev, vecs, x)
end
meanVect = [20, 20]
covMat   = [1 .8; .8 1]
N        = 10^3
points   = rand(MvNormal(meanVect, covMat), N)
ev, vecs, pts = manualPCA(points)
u1 = 3*vecs[:, 1]
u2 = vecs[:, 2]
trans_pts1 = vecs[:,1]' * points # transformed data Y
trans_pts2 = vecs[:,2]' * points
fig, (ax) = plt.subplots(1, figsize=set_size())
plt.gca().set_aspect("equal") # without it, two components are not shown perpen.
ax.scatter(pts[1, :], pts[2, :], color="orange", marker="o", s=1)
ax.plot([-u1[1], u1[1]], [-u1[2], u1[2]], color="red")
ax.plot([-u2[1], u2[1]], [-u2[2], u2[2]], color="purple")
```

Chapter 10
Numerical Analysis

Numerical analysis is an area of mathematics that creates, analyzes, and implements algorithms for obtaining *numerical solutions* to problems involving continuous variables. The Newton-Raphson method to solve numerically the equation $\tan x = x$ is one example. The Gauss quadrature method to numerically evaluate any definite integral $\int_a^b f(x)dx$ is also one example. The finite difference method to solve ordinary and partial differential equations is yet another example.

Numerical solutions are numbers not closed-form expressions. For example, it is possible to solve the quadratic equation $ax^2 + bx + c = 0$ exactly to get the well-known closed-form solutions $x_{1,2} = {}^{-b\pm\sqrt{b^2-4ac}}/_{2a}$. Such solutions do not exist for polynomial equations of fifth order or higher and for transcendental equations such as $\tan x = x$. However, the Newton-Raphson method can solve all the equations efficiently, but it only gives us numerical solutions. For example, applying to $\tan x = x$, it gives us 4.49340946. Once we have an algorithm the procedure to the solution becomes mechanical: we simply follow the algorithm step by step and voilà, there comes the solution. No genius is needed, and that's why scientists and engineers love numerical methods.

The following books were consulted for the majority of the material presented in this chapter:

- *Approximation Theory and Approximation Practice* by Lloyd Trefethen[1] [69].
- *Numerical methods for scientists and engineers* by Richard Hamming[2] [29].

---

[1] Lloyd Nicholas Trefethen (born 30 August 1955) is an American mathematician, Professor of Numerical Analysis and Head of the Numerical Analysis Group at the Mathematical Institute, University of Oxford. He is perhaps best known for his work on pseudospectra of non-normal matrices and operators.

[2] Richard Wesley Hamming (1915–1998) was an American mathematician whose work had many implications for computer engineering and telecommunications. His contributions include the Hamming code (which makes use of a Hamming matrix), the Hamming window, Hamming numbers, sphere packing (or Hamming bound), and the Hamming distance.

V. P. Nguyen, *Mathematics for Engineers and Scientists*,
https://doi.org/10.1007/978-981-97-4631-6_10

- *Analysis of numerical methods*, by Eugene Isaacson[3] and Herbert Keller[4] [32].
- *Finite Difference Computing with PDEs: A Modern Software Approach* by Hans Petter Langtangen[5] and Svein Linge [40].
- *Computational Fluid Dynamics the basic and applications* by John Anderson[6] [4]
- *Computational Physics* by Nicholas J. Giordano and Hisao Nakanishi [27].

I strongly recommend the book of Anderson; it is so well written and a joy to read. Even though it addresses numerical methods to solve the Navier-Stokes equations (which are of interest only to people fascinated by the behavior of fluids), he explains things so clearly.

10.1 Introduction

Suppose we have to compute the following sum, for many values of α:

$$f(\alpha) = \sum_{k=0}^{3} a_k \cos(k\alpha) \tag{10.1.1}$$

where a_k are known constants. The first solution is, of course, to compute term by term and add them up. What do you think if someone tell you that there is a much better method to compute $f(\alpha)$? The secret is that $\cos\alpha$, $\cos 2\alpha$ and so on, they are all related. Recall that, we have derived such a relation in Eq. (3.12.5), re-given here

$$\cos(k\alpha) = 2\cos\alpha\cos(k-1)\alpha - \cos(k-2)\alpha, \quad \text{if } k \geq 2 \tag{10.1.2}$$

A name was given to such formula as it occurs quite often in mathematics. This is known as the *three-term recurrence relation* because it involves three terms. Even with the hint that this recurrence is the key to an efficient computation of the mentioned sum, it is really hard to know where to start. Unless you know where to look for inspiration, and it comes in the name of the Horner method in polynomial evaluation.

---

[3] Eugene Isaacson (1919–2008), was a US mathematician who pioneered modern numerical analysis. He was a mathematics and physics graduate of City College in New York, he then entered the graduate program in mathematics at New York University gaining a PhD on water waves on sloping beaches in 1949. His academic career was then spent at the Courant Institute until his retirement.

[4] Herbert Bishop Keller (1925–2008) was an American applied mathematician and numerical analyst. He was Professor of Applied Mathematics, Emeritus, at the California Institute of Technology.

[5] Hans Petter Langtangen (1962–2016) was a Norwegian Scientist trained in mechanics and scientific computing. Langtangen was the Director of the Centre for Biomedical Computing, a Norwegian Center of Excellence hosted by Simula Research Laboratory. Langtangen promoted the use of Python for scientific computing through numerous journal papers and conference talks.

[6] John D. Anderson Jr. (born 1 October 1937) is the Curator of Aerodynamics at the National Air and Space Museum at the Smithsonian Institution in Washington, DC, Professor Emeritus in the Department of Aerospace Engineering at the University of Maryland, College Park.

Horner's method. In Sect. 2.29.6, the Horner method was presented as an efficient way to evaluate any polynomial at a point x_0. As a recap, let's consider a specific cubic polynomial $p(x) = 2x^3 - 6x^2 + 2x + 1$. In Horner's method, we massage $p(x_0)$ a bit as

$$p(x_0) = 2x_0^3 - 6x_0^2 + 2x_0 + 1 = x_0[2x_0^2 - 6x_0 + 2] + 1 = x_0[x_0(2x_0 - 6) + 2] + 1$$

To implement Horner's method, a new sequence of constants is defined recursively as follows:

$$\begin{aligned} b_3 &= a_3 & b_3 &= 2 \\ b_2 &= x_0 b_3 + a_2 & b_2 &= 2x_0 - 6 \\ b_1 &= x_0 b_2 + a_1 & b_1 &= x_0(2x_0 - 6) + 2 \\ b_0 &= x_0 b_1 + a_0 & b_0 &= x_0(x_0(2x_0 - 6) + 2) + 1 \end{aligned}$$

where the left column is for a general cubic polynomial whereas the right column is for the specific $p(x) = 2x^3 - 6x^2 + 2x + 1$. Then, $p(x_0) = b_0$. As to finding the consecutive b-values, we start with determining b_3, which is simply equal to a_3. We then work our way down to the other b's, using the recursive formula:

$$b_{k-1} = a_{k-1} + b_k x_0$$

until we arrive at b_0. This relation can also be written as

$$\boxed{b_k = a_k + b_{k+1} x_0} \tag{10.1.3}$$

But what is the relation between the sum in Eq. (10.1.1) and a polynomial? To see that relation, we need to write an n-order polynomial using the sum notation:

$$p_n(x_0) = \sum_{k=0}^{n} a_k x^k, \quad x^k = x x^{k-1}$$

Now, we can see that the sum in Eq. (10.1.1) and a polynomial are of the same form:

$$f(x) = \sum_{k=0}^{n} a_k \psi_k(x) \tag{10.1.4}$$

where $\psi_k(x)$ has either a three-term recurrence relation or a two-term recurrence relation (in the case $\psi_k(x) = x^k$).

Inspired by Eq. (10.1.3), we define the sequence of b_k's as, where the only difference is the red term which is related to $\cos(k-2)\alpha$ in the three-term recurrence relation (and of course $2\cos\alpha$ replaced x):

$$b_k = a_k + (2\cos\alpha) b_{k+1} - b_{k+2} \Longrightarrow a_k = b_k + b_{k+2} - 2\cos\alpha b_{k+1} \tag{10.1.5}$$

To compute the sum in Eq. (10.1.1), we compute a_0, a_1, a_2, a_3 in terms of b_i's:

$$\begin{aligned} a_3 &= b_3 \\ a_2 &= b_2 - 2\cos\alpha b_3 \\ a_1 &= b_1 + b_3 - 2\cos\alpha b_2 \\ a_0 &= b_0 + b_2 - 2\cos\alpha b_1 \end{aligned}$$

Substitution of a_i's into Eq. (10.1.1), and re-arranging the terms in this form $b_0 + b_1(\cdots) + b_2(\cdots) + b_3(\cdots)$:

$$\begin{aligned} f(\alpha) &= \sum_{k=0}^{3} a_k \cos(k\alpha) \\ &= (b_0 + b_2 - 2\cos\alpha b_1) + (b_1 + b_3 - 2\cos\alpha b_2)\cos\alpha \\ &\quad + (b_2 - 2\cos\alpha b_3)\cos 2\alpha + b_3 \cos 3\alpha \\ &= b_3(\cos 3\alpha + \cos\alpha - 2\cos\alpha\cos 2\alpha) + b_2(\cos 2\alpha + 1 - 2\cos^2\alpha) \\ &\quad + b_1(-\cos\alpha) + b_0 \end{aligned}$$

Amazingly, all the red terms are zeros because of (10.1.2), thus the scary sum is just the following simple formula:

$$\sum_{k=0}^{3} a_k \cos(k\alpha) = b_0 - b_1 \cos\alpha \tag{10.1.6}$$

This is Clenshaw's algorithm, named after the English mathematician Charles William Clenshaw (1926–2004) who published this method in 1955.

10.2 Numerical Differentiation

Numerical differentiation deals with numerical approximations of derivatives. The first questions that come up to mind is: why do we need to approximate derivatives at all? After all, we do know how to analytically differentiate every function. Nevertheless, there are several reasons as of why we still need to approximate derivatives. The most important application of numerical differentiation is in numerically solving ordinary and partial differential equations (Sect. 10.6). When approximating solutions to ordinary (or partial) differential equations, we typically represent the solution as a discrete approximation that is defined on a grid. Since we then have to evaluate derivatives at the grid points, we need to be able to come up with methods for approximating the derivatives at these points, and this will typically be done using only values that are defined on that grid.

10.2.1 First-Order Derivatives

The starting point is Taylor's theorem (Sect. 4.15.10):

$$\begin{aligned} f(x+h) &= f(x) + f'(x)h + \frac{f''(\xi_1)}{2!}h^2, \quad \xi_1 \in (x, x+h) \\ f(x-h) &= f(x) - f'(x)h + \frac{f''(\xi_2)}{2!}h^2, \quad \xi_2 \in (x-h, x) \end{aligned} \tag{10.2.1}$$

From that, by ignoring the red terms, we obtain different ways to approximate the first derivative of $f(x)$:

$$\begin{aligned} \text{forward difference:} \quad f'(x) &\approx \frac{f(x+h) - f(x)}{h}, \quad \text{error} \sim \frac{f''(\xi)}{2}h \\ \text{backward difference:} \quad f'(x) &\approx \frac{f(x) - f(x-h)}{h} \end{aligned}$$

Since the approximations are obtained by truncating the term $(f''(\xi)/2!)h^2$ from the exact formula (10.2.1), this term is the error in our approximations, and is called the *truncation error*. When the truncation error is of the order of $\mathcal{O}(h)$, we say that the method is a first-order method. We refer to a method as a pth-order method if the truncation error is of the order of $\mathcal{O}(h^p)$. The forward difference was used to develop the famous Euler's method which is commonly used to solve ordinary differential equations.

To develop a second-order method we use more terms in the Taylor series including $f''(x)$:

$$\begin{aligned} f(x+h) &= f(x) + f'(x)h + \frac{f''(x)}{2!}h^2 + \frac{f'''(\xi_1)}{3!}h^3, \quad \xi_1 \in (x, x+h) \\ f(x-h) &= f(x) - f'(x)h + \frac{f''(x)}{2!}h^2 - \frac{f'''(\xi_2)}{3!}h^3, \quad \xi_2 \in (x-h, x) \end{aligned} \tag{10.2.2}$$

And subtracting the first from the second, we arrive at

$$f'(x) = \frac{f(x+h) - f(x-h)}{2h} - \left[\frac{f'''(\xi_1) + f'''(\xi_2)}{12}\right]h^2$$

which yields the so-called centered difference for the first derivative:

$$f'(x) \approx \frac{f(x+h) - f(x-h)}{2h} \tag{10.2.3}$$

This approximation is a second-order method by construction as the error is $\sim h^2$. To demonstrate the performance of these approximations, let's consider the function

Table 10.1 Finite difference approximations of $f'(x)$ for $f(x) = \sin x + \cos x$. Errors of one-sided difference (forward/backward) versus two-sided centered difference

h	Forward diff.	Backward diff.	Centered diff.
0.2500	0.1347	0.1140	0.0104
0.1250	0.0650	0.0598	0.0026
0.0625	0.0319	0.0306	0.0007
0.0313	0.0158	0.0155	0.0002

$f(x) = \sin x + \cos x$ and compute $f'(0)$ and the errors (noting that the exact value is 1). The results are shown in Table 10.1.

The result clearly indicates that as h is halved, the error of one-sided differences is only halved (in Table 10.1, starting from the first row and going down, each time h is half of the previous row), but the error of centered difference is decreased four times.

10.2.2 Second-Order Derivatives

To get a formula for $f''(x)$, we also start with Eq. (10.2.2)

$$\begin{aligned} f(x+h) &= f(x) + f'(x)h + \frac{f''(x)}{2!}h^2 + \frac{f'''(x)}{3!}h^3 + \frac{f^{(4)}(\xi_1)}{4!}h^4 \\ f(x-h) &= f(x) - f'(x)h + \frac{f''(x)}{2!}h^2 - \frac{f'''(x)}{3!}h^3 + \frac{f^{(4)}(\xi_2)}{4!}h^4 \end{aligned} \tag{10.2.4}$$

However, now we add them up to get

$$f''(x) = \frac{f(x+h) - 2f(x) + f(x-h)}{h^2} + \mathcal{O}(h^2) \tag{10.2.5}$$

This approximation was used to develop the famous Verlet's method which is commonly used to solve Newton's equations of motions $F = ma$.

10.3 Interpolation

Assume that we are back in time to the period of no calculators and no formula for calculating sine. Luckily, some people made up a table of sine of 1°, 5°, 10°, 15°,... But we need sin 2°. What are we going to do? We will use a method that has become to what we know today as *interpolation*.[7] In the first attempt, we assume that the two data points $(1°, \sin 1°)$, $(5°, \sin 5°)$ are connected by a line. We can determine the equation, let's call it $f(x)$, for this line (because it is straightforward). Having such an equation, it is a simple task to compute $\sin 2°$, it is $f(2°)$.

Then, we realize that our assumption was too crude. In need of higher accuracy, instead of a line joining the two points, we assume a parabola joining three data points. Generally, interpolation is where an approximating function is constructed in such a way as to *agree perfectly* with the usually unknown original function *at the given measurement/data points*.

There exits another situation where we need to do interpolation. Suppose that we have a very complex function $y = f(x)$ that we do not want to work with it directly. So, we can generate some data points $(x_i, f(x_i))$ and use them to generate an interpolating function that matches $f(x)$ only at x_i. The thing is usually the interpolating function is simple to work with, e.g., it is a polynomial.

10.3.1 Polynomial Interpolations

Given two points (x_1, y_1) and (x_2, y_2), where $x_1 \neq x_2$, there is one and only one line joining them. And its equation is

$$y = \left(\frac{y_1 - y_2}{x_1 - x_2}\right) x + \left(\frac{y_2 x_1 - y_1 x_2}{x_1 - x_2}\right) \tag{10.3.1}$$

This is straightforward as it should be. How about finding the parabola passing through three points (x_1, y_1), (x_2, y_2), and (x_3, y_3)? The same approach of $y = ax^2 + bx + c$ and 3 equations for 3 unknowns to determine a, b, c would work, but it is laborious. The situation is even worse if we have to find the curve going through 10 points. There is a much better way and it is hidden in Eq. (10.3.1).

The idea is to rewrite Eq. (10.3.1) as

$$y = \left(\frac{x - x_2}{x_1 - x_2}\right) y_1 + \left(\frac{x_1 - x}{x_1 - x_2}\right) y_2 \tag{10.3.2}$$

Why this form is better than the previous one? It is because with this form, it is immediately clear that when $x = x_1$, $y = y_1$ because the blue term is zero or when $x = x_2$, $y = y_2$ as the red term vanishes. Thus, y has this form $y = u(x)y_1 + v(x)y_2$ with $u(x_1) = 1$, $u(x_2) = 0$ and $v(x_1) = 0$, $v(x_2) = 1$. Note that $u(x) + v(x) = 1$.

With this, we suspect that for a parabola its equation should have this form:

---

[7] The word "interpolation" originates from the Latin verb interpolare, a contraction of "inter", meaning "between", and "polare", meaning "to polish". That is to say, to smooth in between given pieces of information.

$$y = u(x)y_1 + v(x)y_2 + w(x)y_3$$

where $u(x_1) = 1$, $u(x_2) = 0$, and $u(x_3) = 0$. The following form satisfies the last two conditions:

$$u(x) = k(x - x_2)(x - x_3)$$

And the first condition gives us $k = 1/(x_1 - x_2)(x_1 - x_3)$, thus

$$u(x) = \frac{(x - x_2)(x - x_3)}{(x_1 - x_2)(x_1 - x_3)} \tag{10.3.3}$$

Similarly, we get the expressions for $v(x)$ and $w(x)$

$$v(x) = \frac{(x - x_1)(x - x_3)}{(x_2 - x_1)(x_2 - x_3)}, \quad w(x) = \frac{(x - x_1)(x - x_2)}{(x_3 - x_1)(x_3 - x_2)}$$

and then the quadratic interpolation is

$$y = \frac{(x - x_2)(x - x_3)}{(x_1 - x_2)(x_1 - x_3)}y_1 + \frac{(x - x_1)(x - x_3)}{(x_2 - x_1)(x_2 - x_3)}y_2 + \frac{(x - x_1)(x - x_2)}{(x_3 - x_1)(x_3 - x_2)}y_3 \tag{10.3.4}$$

At this point, we should check whether what we have observed, $u(x) + v(x) = 1$, continues holding. That is, $u(x) + v(x) + w(x) \stackrel{?}{=} 1$. The algebra might be messy, but the identity holds.

Now, we can write the equation for a 17th-degree polynomial passing through 18 points. But the equation would be so lengthy. We need to introduce some short notations. First, for $n + 1$ points $(x_0, y_0), \ldots, (x_j, y_j), \ldots, (x_n, y_n)$ the interpolating polynomial is now given by

$$\boxed{y(x) = \sum_{i=0}^{n} l_i(x)y_i} \tag{10.3.5}$$

What is this? It is (AGAIN!) a linear combination of some functions $l_i(x)$ with coefficients being y_i. In this equation, $l_i(x)$ is written as (after examining the form of u, v, w, see again Eq. (10.3.3))

$$\boxed{l_i(x) = \prod_{\substack{j=0 \\ j \neq i}}^{n} \frac{x - x_j}{x_i - x_j}} \tag{10.3.6}$$

and are the so-called Lagrange basis polynomials. Plots of linear and quadratic Lagrange polynomials are given in Fig. 10.1. Although named after Joseph-Louis Lagrange, who published it in 1795, the method was first discovered in 1779 by

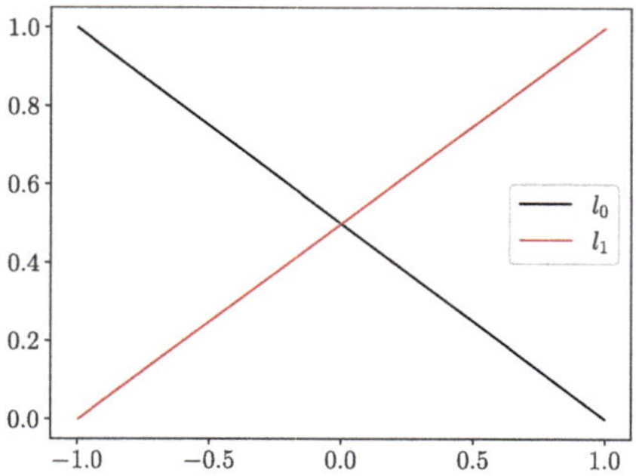

(a) linear l_0 and l_1

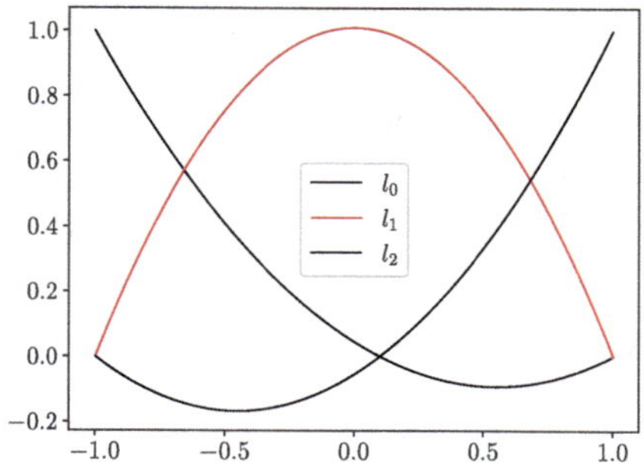

(b) quadratic $l_i(x), i = 0, 1, 2$

Fig. 10.1 Plots of linear and quadratic Lagrange basis functions in $[-1, 1]$. It is clear that $l_i(x_j) = \delta_{ij}$

Table 10.2 Data points

x	$f(x)$
0	0
1	0.8415
2	0.9093
3	0.1411
4	−0.7568
5	−0.9589
6	−0.2794

the English mathematician Edward Waring (1736–1798). We mentioned this not to imply that Lagrange is not great. He is one of the greatest of all time. Just to mention that sometimes credit was not given to the first discoverer. About this topic, some more examples are: the Lagrange interpolation formula was discovered by Waring, the Gibbs phenomenon was discovered by Wilbraham, and the Hermite integral formula is due to Cauchy. These are just some of the instances of Stigler's law[8] in approximation theory.

Example. There are 7 data points given in Table 10.2. And we use Lagrange interpolation to find the 6th-degree polynomial passing through all these points. As I am lazy (already in the early 40 s when doing this), I did not explicitly compute $l_i(x)$.

[8] Stigler's law of eponymy, proposed by statistician Stephen Stigler in his 1980 publication Stigler's law of eponymy, states that no scientific discovery is named after its original discoverer. Examples include Hubble's law, which was derived by Georges Lemaître 2 years before Edwin Hubble, the Pythagorean theorem, which was known to Babylonian mathematicians before Pythagoras, and Halley's Comet, which was observed by astronomers since at least 240 BC. Stigler himself named the sociologist Robert K. Merton as the discoverer of "Stigler's law" to show that it follows its own decree, though the phenomenon had previously been noted by others.

Instead I wrote a Julia code (Listing A.12) and with it got Fig. 10.2: a nice curve joining all the points.

Runge's phenomenon. Consider the Runge function

$$f(x) = \frac{1}{1+25x^2} \tag{10.3.7}$$

And we want to use equidistant points x_i between -1 and 1 such that:

$$x_i = -1 + \frac{2i}{n}, \quad i = \{0, 1, 2, \ldots, n\}$$

to construct Lagrange polynomials that can capture this function. Then, we hope that a 5th-degree Lagrange polynomial can fit Runge's function. But it does not do a good job. Well, after all just 6 points were used. Then, we used 10 points to have a 9th-degree Lagrange polynomial, and this is even worse: there is oscillation at the edges of the interval, even though far from the edges, the approximation is quite good (Fig. 10.3). This is known as Runge's phenomena.

Interpolation error. Let f be an m times continuously differentiable function on $[a, b]$.[9] Suppose we have m sampling points $x_1, x_2, \ldots, x_m$ and we construct a $m-1$ degree polynomial $p(x)$ going through these points. Then for any $x \in [a, b]$, we have

$$R(x) := f(x) - p(x) = \frac{f^{(m)}(\xi)}{m!}(x - x_1)(x - x_2)\cdots(x - x_m) \tag{10.3.8}$$

for some $\xi \in [a, b]$. It follows that

$$|f(x) - p(x)| \le \frac{\phi(x)}{m!} \max_{y\in[a,b]} |f^{(m)}(y)|, \quad \phi(x) = \prod_{i=1}^{m}(x - x_i) \tag{10.3.9}$$

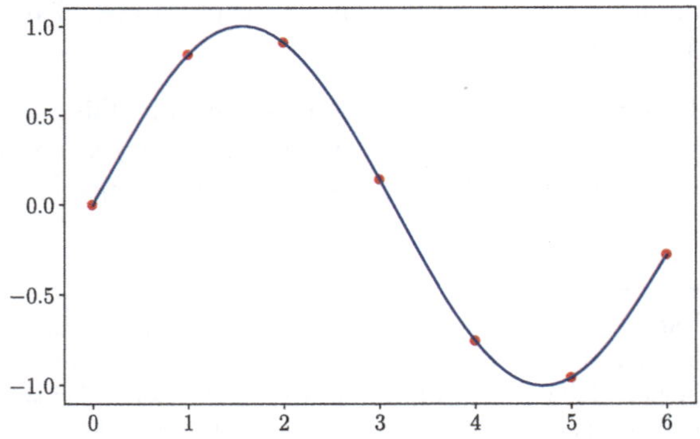

Fig. 10.2 Lagrange interpolating function

[9] Which can be compactly written as f is $C^m([a, b])$.

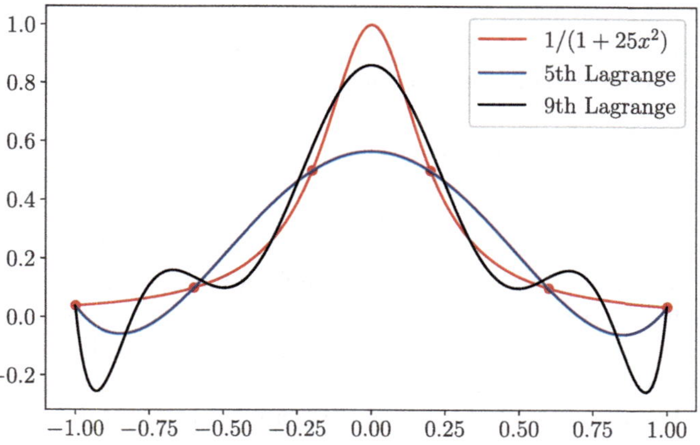

Fig. 10.3 Runge's phenomena. This happens only for *high-order polynomials and equi-spaced points*. This is named Runge's phenomenon as it was discovered by the German mathematician Carl David Tolmé Runge (1856–1927) in 1901 when exploring the behavior of errors when using polynomial interpolation to approximate certain functions. The discovery was important because it shows that going to higher degrees does not always improve accuracy. Note that this phenomenon is similar to the Gibbs phenomenon in Fourier series (Sect. 4.19)

And this theorem explains the Runge phenomenon in which the derivatives blow up, Fig. 10.4b. Note that $\phi(x)$ is a monic polynomial, which is a single-variable polynomial in which the leading coefficient (the non-zero coefficient of highest degree) is equal to 1. An n degree monic is of this form $x^n + c_{n-1}x^{n-1} + \cdots + c_2x^2 + c_1x + c_0$.

Properties. If $f(x)$ is a polynomial of degree less than or equal n, and we use $n+1$ points $(x_i, f(x_i))$ to construct a Lagrange interpolating function $y(x)$. Then, $y(x) \equiv f(x)$ or, in other words, the Lagrange interpolation is exact. Another property

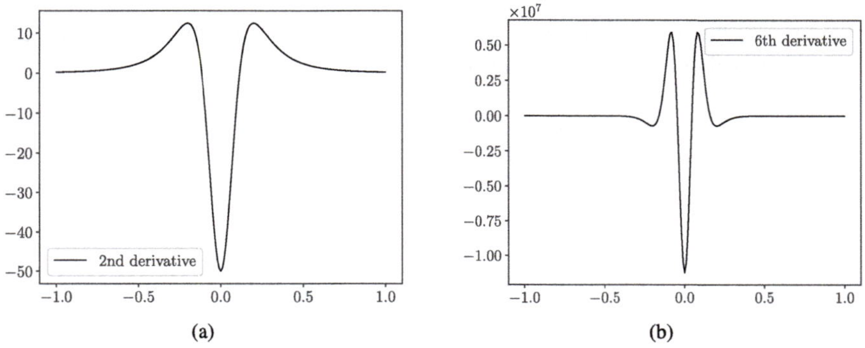

Fig. 10.4 Derivatives of the Runge function $f(x) = 1/1+25x^2$. Note that, I used SymPy to automatically compute $f^{(m)}(x)$ and evaluate the resulting expression at sampling points in $[-1, 1]$ to generate these plots. We should take advantage of CAS to focus on other things

is that the polynomial interpolant is unique.[10] And this uniqueness allows us to state that $\sum_i l_i(x) = 1$ for all x—a fact that we have observed for $n = 2$ and $n = 3$.[11]

Now, the Lagrange basis functions have two properties, as stated below:

$$\begin{aligned} &\text{Kronecker delta property} \quad l_i(x_j) = \delta_{ij} \\ &\text{Partition of unity property} \sum_{i=0}^{n} l_i(x) = 1 \quad \text{for all} x \end{aligned} \tag{10.3.10}$$

Applications. Lagrange polynomials are used to derive Newton-Cotes numerical integration rule. But its most well-known application is as the basis for the finite element method—a very powerful numerical method for solving complex 1D/2D/3D partial differential equations. Civil engineers use this method in the design process of buildings, bridges. Mechanical engineers use it to design cars. Aerospace engineers also use it. Just to name a few. Section 8.12 briefly presented this method and its early history.

Motivation. If you're wondering what ensures that there exists a polynomial that can interpolate a given function, rest assured, it is the Weierstrass approximation theorem.

Theorem 10.3.1 Weierstrass approximation theorem *Let f be a real-valued function defined on an interval $[a, b]$ of $\mathbb{R}$. Then, for any $\epsilon > 0$, there exists a polynomial $p(x)$ such that*

$$|f(x) - p(x)| < \epsilon \quad \textit{for all} x \in [a, b]$$

This theorem does not tell us what is the expression of $p(x)$; you have to find it for yourself! But it motivates mathematicians: if you work hard, you can find a polynomial that can approximate well any function.

Vandermonde matrix. Let's attack the interpolation problem directly and the Vandermonde matrix will show up. We use an n degree polynomial of this form

$$P_n(x) = a_0 + a_1x + a_2x^2 + \cdots + a_nx^n$$

to interpolate the $n + 1$ points (x_i, y_i), $i = 0, 1, 2, \ldots$ We have this system of linear equations to solve for the coefficients a_i:

$$\begin{aligned} a_0 + a_1x_0 + a_2x_0^2 + \cdots + a_nx_0^n &= y_0 \\ a_0 + a_1x_1 + a_2x_1^2 + \cdots + a_nx_1^n &= y_1 \\ \vdots \quad &= \vdots \\ a_0 + a_1x_n + a_2x_n^2 + \cdots + a_nx_n^n &= y_n \end{aligned}$$

[10] How to prove this? In maths, to prove something is unique, we can assume that there are two versions of it, and prove that the two are the same.

[11] Consider the case where $f_i = 1$ for x_i, $i = 0, 1, ..., n$. Through these points, there is the horizontal line $y(x) = 1$, and this line is the only polynomial that interpolates the points. Thus, Eq. (10.3.5) leads to $\sum_i l_i(x) = 1$.

which can be rewritten in a matrix notation as

$$\begin{bmatrix} 1 & x_0 & x_0^2 & \cdots & x_0^n \\ 1 & x_1 & x_1^2 & \cdots & x_1^n \\ \vdots & \vdots & \vdots & \ddots & \vdots \\ 1 & x_n & x_n^2 & \cdots & x_n^n \end{bmatrix} \begin{bmatrix} a_0 \\ a_1 \\ \vdots \\ a_n \end{bmatrix} = \begin{bmatrix} y_0 \\ y_1 \\ \vdots \\ y_n \end{bmatrix} \tag{10.3.11}$$

The red beautiful matrix is the Vandermonde matrix, named after Alexandre-Théophile Vandermonde (1735–1796)—a French mathematician, musician, and chemist. Now, as an exercise to determinant, we're going to compute the determinant of the Vandermonde matrix. And from that determinant we can also prove that the interpolating polynomial is unique.

It's easier to deal with the transpose of the Vandermonde matrix, so we consider the transpose:

$$\mathbf{V} = \begin{bmatrix} 1 & x_0 & x_0^2 & \cdots & x_0^n \\ 1 & x_1 & x_1^2 & \cdots & x_1^n \\ \vdots & \vdots & \vdots & \ddots & \vdots \\ 1 & x_n & x_n^2 & \cdots & x_n^n \end{bmatrix} \Longrightarrow \mathbf{V}^\top = \begin{bmatrix} 1 & 1 & \cdots & 1 \\ x_0 & x_1 & \cdots & x_n \\ x_0^2 & x_1^2 & \cdots & x_n^2 \\ \vdots & \vdots & \ddots & \vdots \\ x_0^n & x_1^n & \cdots & x_n^n \end{bmatrix}$$

Now, we consider a simpler problem with only 4 points:

$$\det(\mathbf{V}^\top) = \begin{vmatrix} 1 & 1 & 1 & 1 \\ x_0 & x_1 & x_2 & x_3 \\ x_0^2 & x_1^2 & x_2^2 & x_3^2 \\ x_0^3 & x_1^3 & x_2^3 & x_3^3 \end{vmatrix} = \begin{vmatrix} 1 & 1 & 1 & 1 \\ 0 & x_1 - x_0 & x_2 - x_0 & x_3 - x_0 \\ 0 & x_1^2 - x_1x_0 & x_2^2 - x_2x_0 & x_3^2 - x_3x_0 \\ 0 & x_1^3 - x_1^2x_0 & x_2^3 - x_2^2x_0 & x_3^3 - x_3^2x_0 \end{vmatrix}$$

where the second row was replaced by second row minus x_0 times the first row; the third row by the third row minus x_0 times the second row, and so on. Now, of course we expand by the first column and do some factorizations to get, check Sect. 9.9.3 if something was not clear:

$$\begin{aligned} \det(\mathbf{V}^\top) &= \begin{vmatrix} x_1 - x_0 & x_2 - x_0 & x_3 - x_0 \\ x_1^2 - x_1x_0 & x_2^2 - x_2x_0 & x_3^2 - x_3x_0 \\ x_1^3 - x_1^2x_0 & x_2^3 - x_2^2x_0 & x_3^3 - x_3^2x_0 \end{vmatrix} \\ &= (x_1 - x_0)(x_2 - x_0)(x_3 - x_0) \begin{vmatrix} 1 & 1 & 1 \\ x_1 & x_2 & x_3 \\ x_1^2 & x_2^2 & x_3^2 \end{vmatrix} \end{aligned}$$

Now the red determinant should not be a problem for us, we can write immediately the answer

$$\begin{vmatrix} 1 & 1 & 1 \\ x_1 & x_2 & x_3 \\ x_1^2 & x_2^2 & x_3^2 \end{vmatrix} = (x_2 - x_1)(x_3 - x_1) \begin{vmatrix} 1 & 1 \\ x_2 & x_3 \end{vmatrix} = (x_2 - x_1)(x_3 - x_1)(x_3 - x_2)$$

The final result is then given by

$$\begin{vmatrix} 1 & 1 & 1 & 1 \\ x_0 & x_1 & x_2 & x_3 \\ x_0^2 & x_1^2 & x_2^2 & x_3^2 \\ x_0^3 & x_1^3 & x_2^3 & x_3^3 \end{vmatrix} = (x_1 - x_0)(x_2 - x_0)(x_3 - x_0)(x_2 - x_1)(x_3 - x_1)(x_3 - x_2)$$

$$= \prod_{i=0}^{2} \prod_{j=i+1}^{3} (x_j - x_i) = \prod_{0 \le i < j \le 2} (x_j - x_i)$$

As x_i's are distinct, the determinant is different from zero, thus the Vandermonde matrix is invertible. Thus, Eq. (10.3.11) has a unique solution. In other words, there is only one single polynomial passing through all data points.

10.3.2 Chebyshev Polynomials

Recall that in Eq. (3.12.5) we have derived a recursive formula for $\cos n\alpha$:

$$\cos(n\alpha) = \begin{cases} 1, & \text{if } n = 0 \\ \cos\alpha, & \text{if } n = 1 \\ 2\cos\alpha\cos(n-1)\alpha - \cos(n-2)\alpha, & \text{if } n \ge 2 \end{cases} \tag{10.3.12}$$

The Chebyshev polynomials are two sequences of polynomials related to the cosine and sine functions, notated as $T_n(x)$ and $U_n(x)$. They can be defined in several equivalent ways; in this section, the polynomials are defined by starting with trigonometric functions. The Chebyshev polynomials of the first kind $T_n(x)$ are defined in this way. Note that from the above equation, $\cos(n\alpha)$ is a polynomial in terms of $\cos\alpha$, e.g., $\cos 3\alpha = 4(\cos\alpha)^3 - 3\cos(\alpha)$. For n being a fixed counting number, the Chebyshev polynomial is defined to be that polynomial of cosine:

$$T_n(\cos\alpha) = \cos(n\alpha)$$

Change of variable $x = \cos\alpha$, and we get

$$T_n(x) := \cos(n \arccos x) \tag{10.3.13}$$

These polynomials were named after Pafnuty Chebyshev. The letter T is used, by Berstein, because of the alternative transliterations of the name Chebyshev as Tchebycheff, Tchebyshev (French) or Tschebyschow (German). Pafnuty Lvovich Chebyshev (1821–1894) was a Russian mathematician and considered to be the founding father of Russian mathematics.

The recursive definition of $T_n(x)$ follows from the recursive formula for $\cos n\alpha$:

$$T_n(x) = \begin{cases} 1, & \text{if } n = 0 \\ x, & \text{if } n = 1 \\ 2xT_{n-1}(x) - T_{n-2}(x), & \text{if } n \geq 2 \end{cases} \tag{10.3.14}$$

The first four Chebyshev polynomials are obtained using (10.3.14)

$$\begin{aligned} T_0(x) &= 1 &&= 1 \\ T_1(x) &= x &&= 2^0x^1 \\ T_2(x) &= 2x^2 - 1 &&= 2^1x^2 - 1 \\ T_3(x) &= 4x^3 - 3x &&= 2^2x^3 - 3x \\ T_4(x) &= 8x^4 - 8x^2 + 1 &&= 2^3x^4 - 8x^2 + 1 \end{aligned} \tag{10.3.15}$$

From this, we can see that $T_n(x)$ is an n-degree polynomial. Furthermore, the leading coefficient of $T_n(x)$ is 2^{n-1}. Plots of the first four $T_n(x)$ are given in Fig. 10.5. We can see that $|T_n(x)| \leq 1$, which is expected as $T_n(\cos\alpha) = \cos(n\alpha)$. And $T_n(x)$ has n real roots which lead to the following concept.

Chebyshev nodes are the roots of the Chebyshev polynomial of the first kind of degree n. To find the roots, just use (10.3.13):

$$T_n(x) = 0 \iff \cos(n \arccos x) = 0 \iff n \arccos x = \frac{\pi}{2} + k\pi$$

Therefore, for a given positive integer n the Chebyshev nodes in the interval $(-1, 1)$ are

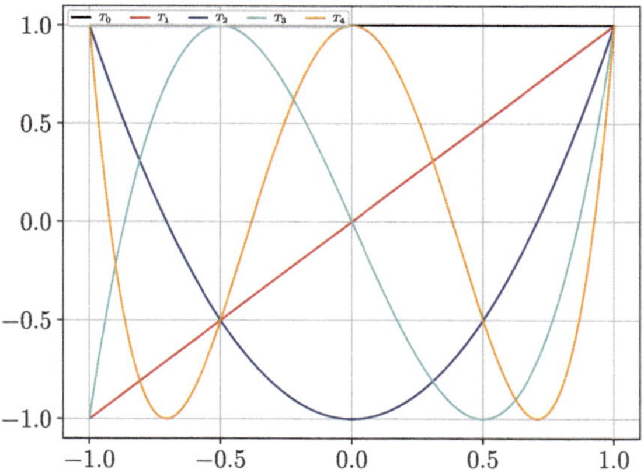

Fig. 10.5 Plots of the first four Chebyshev polynomials $T_n(x)$. Check source file `lagrange-interpolation`

$$x_k = \cos\left[\frac{\pi}{n}\left(k + \frac{1}{2}\right)\right], \quad k = 0, 1, \ldots, n-1 \tag{10.3.16}$$

It's a good habit to plot the nodes to see how they distribute in $[-1, 1]$. Figure 10.6 is such a plot. Also plotted are the angles $\frac{\pi}{n}\left(k + \frac{1}{2}\right)$ which correspond to equally spaced points on the upper half of the unit circle.

As $T_n(x)$ is an n-degree polynomial with leading coefficient of 2^{n-1} and it has n roots $x_1, x_2, \ldots, x_n$, which are the Chebyshev nodes, we can write $T_n(x)$ in this factor form:

$$T_n(x) = 2^{n-1}(x - x_1)(x - x_2)\cdots(x - x_n)$$

Then, due to the fact that $|T_n(x)| \le 1$, we have

$$2^{n-1}(x - x_1)(x - x_2)\cdots(x - x_n) \le 1 \Longrightarrow \prod_{i=1}^{n}(x - x_i) \le \frac{1}{2^{n-1}}$$

If we use the Chebyshev nodes in a polynomial approximation, then Eq. (10.3.9) gives us

$$|f(x) - p(x)| \le \frac{1}{n!2^{n-1}} \max_{y \in [a,b]} |f^{(n)}(y)| \tag{10.3.17}$$

And we hope that the denominator with $n!$ and 2^{n-1} will dominate when n is large (compared with $|f^{(n)}(y)|$), and thus the error $|f(x) - p(x)|$ will decrease to zero. And we have a better approximation. Of course we verify our guess with the Runge function (that troubled Lagrange polynomial with equally spaced points), and the result shown in Fig. 10.7 confirms our analysis.

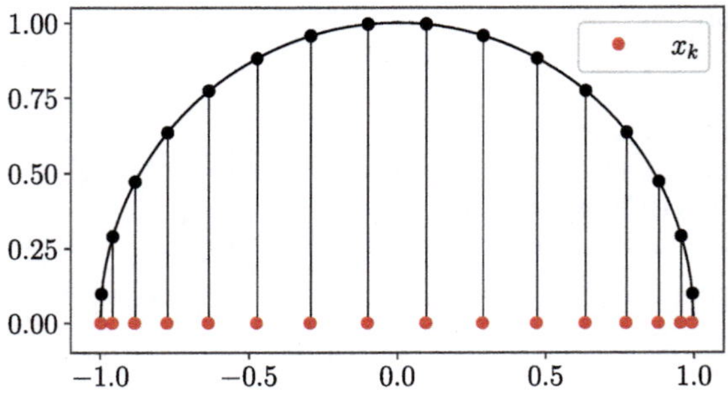

Fig. 10.6 Distribution of Chebyshev nodes for $n = 16$

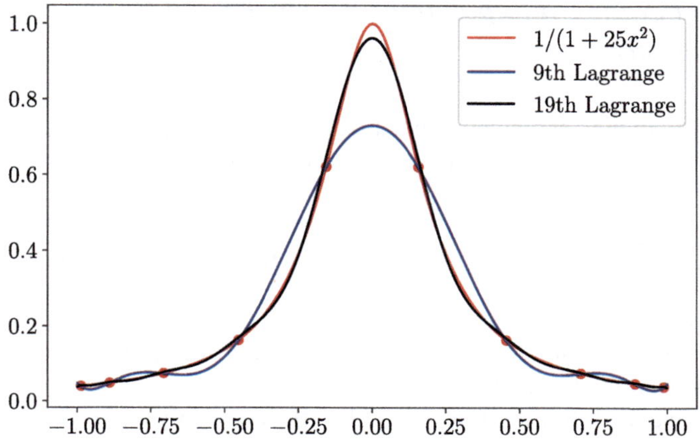

Fig. 10.7 Approximation of Runge's function using Chebyshev nodes: 10 nodes (red points) and 20 nodes. No more oscillations near −1 and 1

Now we discuss the orthogonality of Chebyshev functions. Recall that

$$I = \int_0^\pi \cos n\alpha \cos m\alpha d\alpha = 0 \quad (m \neq n) \tag{10.3.18}$$

A change of variable from α to x

$$x = \cos\alpha \Longrightarrow dx = -\sin\alpha d\alpha = -\sqrt{1-x^2}d\alpha$$

reveals that I is given by

$$I = \int_{-1}^{1} \cos(n \arccos x)\cos(m \arccos x)\frac{dx}{\sqrt{1-x^2}} = 0$$

And that is the orthogonality of Chebyshev polynomials:

$$\boxed{\int_{-1}^{1} T_n(x)T_m(x)\frac{1}{\sqrt{1-x^2}}dx = 0 \quad (m \neq n)} \tag{10.3.19}$$

10.3.3 Lagrange Interpolation: Efficiency and Barycentric Forms

The efficiency of the Lagrange interpolation is not good. For example, consider again the case of a quadratic interpolation going through three points:

$$y = \frac{(x-x_2)(x-x_3)}{(x_1-x_2)(x_1-x_3)}y_1 + \frac{(x-x_1)(x-x_3)}{(x_2-x_1)(x_2-x_3)}y_2 + \frac{(x-x_1)(x-x_2)}{(x_3-x_1)(x_3-x_2)}y_3 \tag{10.3.20}$$

For each value of x, to evaluate $y(x)$ one needs 18 multiplications/divisions and 15 additions/subtractions. The efficiency can be improved just by simple algebraic manipulations of the formula.

First, define the following function (called the node polynomial):

$$l(x) = (x-x_1)(x-x_2)(x-x_3)$$

and the following numbers, which are independent of x, and thus can be computed once and for all x, i.e., out of the loop when computing $y(x)$ (note that λ_i are also independent of f_i, so the same calculation can be used to interpolate different data!)

$$\lambda_1 = \frac{1}{(x_1-x_2)(x_1-x_3)}, \quad \lambda_2 = \frac{1}{(x_2-x_1)(x_2-x_3)}, \quad \lambda_3 = \frac{1}{(x_3-x_1)(x_3-x_2)}$$

then, Eq. (10.3.20) can be rewritten as

$$y = l(x)\left(\frac{\lambda_1}{x-x_1}y_1 + \frac{\lambda_2}{x-x_2}y_2 + \frac{\lambda_3}{x-x_3}y_3\right)$$

And thus, for the general case, the new form of the Lagrange interpolation is given by (first done by Jacobi in his PhD thesis)

$$\boxed{y(x) = l(x)\sum_{i=0}^{n}\frac{\lambda_i}{x-x_i}y_i, \quad l(x) = \prod_{i=0}^{n}(x-x_i), \quad \lambda_i = \frac{1}{\prod_{j\neq i} x_i - x_j}} \tag{10.3.21}$$

It can be seen that, in this form, the Lagrange basis $l_i(x)$ is written as

$$l_i(x) = l(x)\frac{\lambda_i}{x-x_i} \tag{10.3.22}$$

To test the efficiency of this new form, one can try to use random data. For example, in Fig. 10.8, 80 random y_i in $[-1, 1]$ are generated corresponding to 80 Chebyshev nodes. Then, Eq. (10.3.21) was used to compute $y(x)$ at 2001 drawing points to get the interpolating polynomial (the blue curve in the figure). The new form is about 1.5 times faster than the original form.

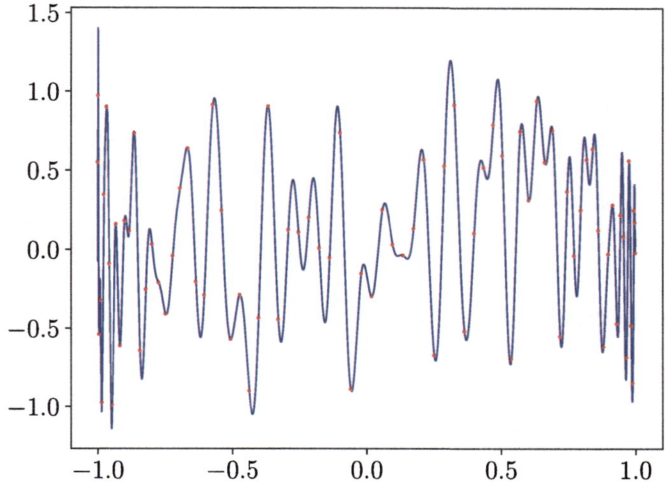

Fig. 10.8 A Lagrange interpolating polynomial through 80 random values at 80 Chebyshev nodes. The solid red dots are the data points

But that's not the end of the story. We can massage the formula to get more of it. Using the PoU property of $l_i(x)$, we can find a formula of $l(x)$ as

$$\sum_i l_i(x) = 1 \Longrightarrow \sum_{i=1}^{n} l(x)\frac{\lambda_i}{x - x_i} = 1 \Longrightarrow l(x) = \frac{1}{\sum_{i=1}^{n} \frac{\lambda_i}{x-x_i}} \tag{10.3.23}$$

With this new form of $l(x)$, Eq. (10.3.21) becomes

$$\boxed{y(x) = \sum_{i=0}^{n} \frac{\lambda_i y_i}{x - x_i} \Bigg/ \sum_{i=1}^{n} \frac{\lambda_i}{x - x_i}, \quad \lambda_i = \frac{1}{\prod_{j\neq i} x_i - x_j}} \tag{10.3.24}$$

What's special about this form, beside the fact that it is more efficient than the previous forms[12]? Actually, this formula has a form that most of us are familiar with. To show that, let's introduce this symbol

$$w_i = \frac{\lambda_i}{x - x_i} \tag{10.3.25}$$

Equation (10.3.24) then becomes

[12] You can check this by implementing this form and comparing with the others. In `Julia` you can use the package `BenchmarkTools` for measuring the running time of a program.

$$y(x) = \frac{\sum_{i=0}^{n} w_i y_i}{\sum_{i=1}^{n} w_i} \tag{10.3.26}$$

This has the exactly same form of the center of mass in physics, see Eq. (6.8.17), if we think of w_i as the masses of particles. Barycenter is the term used in astrophysics for the center of mass of two or more bodies orbiting each other. Therefore, Eq. (10.3.24) is called the *barycentric form*.

10.4 Numerical Integration

This section is about the computation of definite integrals (e.g., $I = \int_a^b f(x)dx$). The idea is to compute these integrals without finding anti-derivatives. Note that we are in the field of applied mathematics, no more exact solutions. It is obvious that we have to use the old idea of chopping the plane into many small pieces (for instance, thin rectangles). But we cannot go to infinity. Not only because no computer is powerful enough for that, but also that we do not need that extremely high accuracy for applications. Instead, what we need to do is to find a smart way of chopping so that we have a finite sum and still an accurate evaluation of the integrals.

Let's start with the simple linear function $y = x$. The integral $I = \int_0^1 xdx$ has a value of 0.5. And as the first option, we divide the region by n equal rectangles (Fig. 10.9). Using Eq. (4.3.3), the numerical value of this integral is given by (noting that $\Delta = 1/n$)

$$I(n) = \sum_{i=1}^{n} (i\Delta)\Delta = \frac{1}{n^2}\sum_i i = \frac{n(n+1)}{2n^2} \tag{10.4.1}$$

where we have used the formula of the sum of the first n positive integers, see Eq. (2.6.2). For various values of n, the corresponding values of $I(n)$ are given in Table 10.3. We can observe a few things from this table. First, $I(n)$ always overestimates I—this should be obvious by looking at Fig. 10.9. Second, we need 500000 intervals to get an accuracy of 6 decimals. This is not practically useful. Note that for a general function it is impossible to have a final formula for $I(n)$ as in Eq. (10.4.1); instead we have to compute $I(n)$ as a sum, i.e., $\sum f(x_i)\Delta$. With $n = 500000$ we need such number of function evaluation $f(x_i)$ and such a number of multiplications. That's a lot of work for a simple function!

Table 10.3 Numerical integration of $\int_0^1 xdx$. Exact value is 0.5

n	1	10	100	1000	...	500000
Δ	1	0.1	0.01	0.001	...	$2e-6$
$I(n)$	1	0.55	0.505	0.5005	...	0.500001
$I(n) - I$	0.5	0.05	0.005	0.0005	...	0.500001

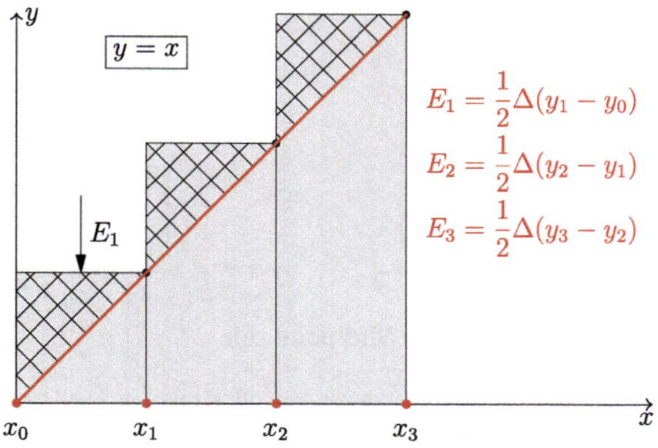

Fig. 10.9 Numerical integration of $y = x$ from 0 to 1 using three ($n = 3$) equal sub-intervals and the right points. Obviously the area is overestimated by an amount of $E_1 + E_2 + E_3$, which is the area of the crossed triangles above the curve $y = x$

As for any approximation we need to know the associated error with our numerical integral. Looking at Fig. 10.9, the error, denoted by $E(3)$, is obviously:

$$E(3) = E_1 + E_2 + E_3 = \frac{1}{2}\Delta[(y_1 - y_0) + (y_2 - y_1) + (y_3 - y_2)] = \frac{1}{2}\Delta(y_3 - y_0) = \frac{1}{2}\Delta \tag{10.4.2}$$

where $y_3 = f(1) = 1$ and $y_0 = f(0) = 0$, and this error can be generalized to $E(n) = 0.5\Delta$. The data (last row in Table 10.3) confirms this. Now, we can understand why the sequence $(E(n))$ converges slowly to 0.5. This is because the error is proportional only to Δ. We desperately need better methods, those for which the error is proportional to Δ^2 or higher powers of Δ.

10.4.1 Trapezoidal and Mid-Point Rule

The poor performance of a mere application of the definition of a definite integral but with finite terms is due to the fact that the thin rectangles do not faithfully align with the curve. A better approximation, known as the mid-point rule, is obtained by also dividing the interval $[a, b]$ into n equal sub-intervals as before; however, the height of a slice is computed at the mid-point of a sub-interval (Fig. 10.10a). The corresponding integral is thus given by

$$M(n) = \sum_{i=0}^{n-1} \Delta \times f\left((2i + 1)\frac{\Delta}{2}\right) \tag{10.4.3}$$

We use the symbol $M(n)$ to remind us it is a mid-point rule. It can be seen from Fig. 10.10a that this mid-point rule gives exact value of $\int_0^1 x dx$. We can also get the same value algebraically.

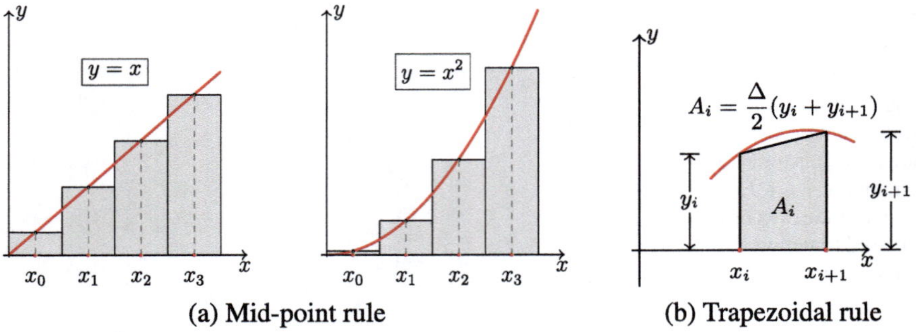

Fig. 10.10 Second-order quadrature rules: mid-point and trapezoidal rule. In a mid-point rule, each slice is still a rectangle of which the height is evaluated at the mid-point, whereas in the trapezoidal rule each slice is now a trapezoidal

Table 10.4 Performance of the mid-point rule versus the for $\int_0^1 x^2 dx$ (exact value is 1/3)

n	Δ	$S_1(n)$	$S_1(n) - 1/3$	$M(n)$	$M(n) - 1/3$
1	1.0	1.00000000	6.666667e-01	0.25000000	−8.333333e-02
10	0.1	0.38500000	5.166667e-02	0.33250000	−8.333333e-04
100	0.01	0.33835000	5.016670e-03	0.33332500	−8.333333e-06
1000	0.001	0.33383350	5.001700e-04	0.33333325	−8.333333e-08

Let's see the performance of the mid-point rule for a harder function $y = x^2$. That is, we're computing $\int_0^1 x^2 dx = 1/3$. The results, given in Table 10.4, indicate that it is a second-order method (look at the last column).

In a similar manner, one can develop a trapezoidal rule where each slice is now a trapezoidal, because we know how to compute the area of a trapezoidal. Thus, the integral is given by (T in $T(n)$ is for trapezoidal):

$$\begin{aligned} T(n) &= \frac{\Delta}{2}[(y_0 + y_1) + (y_1 + y_2) + \cdots + (y_{n-1} + y_n)] \\ &= \frac{\Delta}{2}[y_0 + 2y_1 + 2y_2 + \cdots + y_n] \end{aligned} \tag{10.4.4}$$

In Table 10.5, we compare the mid-point rule and the trapezoidal rule for $\int_0^1 x^2 dx$. Both are second-order methods, but still not efficient as we need 100 intervals just for an accuracy of 6 decimals. We need better methods. To have better methods, we need to change point of view. All the methods discussed so far focus on the way the area of each thin slice is computed; the integrand $y = f(x)$ was not touched!

Table 10.5 Performance of the mid-point rule versus the trapezoidal rule for $\int_0^1 x^2 dx$

n	Δ	$T(n)$	$T(n) - 1/3$	$M(n)$	$M(n) - 1/3$
1	1.0	0.50000000	1.6666667e-1	0.25000000	−8.333333e-02
10	0.1	0.33450000	1.166670e-03	0.33250000	−8.333333e-04
100	0.01	0.33334950	1.617000e-05	0.33332500	−8.333333e-06
1000	0.001	0.33333350	1.700000e-07	0.33333325	−8.333333e-08

10.4.2 Simpson's Rule

The Simpson's rule is based on the approximation of the function $f(x)$ by a quadratic function $y = a_2x^2 + a_1x + a_0$. The idea is similar to the other rules we have discussed: dividing the interval into a number of sub-intervals, but instead of approximating the curve in any sub-interval by a line as in the trapezoidal rule, *Simpson approximated that segment by a parabola.* And we know how to integrate a parabola. Thomas Simpson (1710–1761) was a British mathematician and inventor known for the eponymous Simpson's rule to approximate definite integrals. The attribution, as often in mathematics, can be debated: this rule had been found 100 years earlier by Johannes Kepler, and in German it is called Keplersche Fassregel.

Assume that we consider the bi-unit interval $[-1, 1]$ and we consider three points on it: $(-1, f(-1))$, $(0, f(0))$ and $(1, f(1))$, and we find a parabola of the form $g(x) = a_2x^2 + a_1x + a_0$ passing through these three points. It is straightforward to find these coefficients a_i[13]:

$$a_2 = \frac{1}{2}[f(-1) + f(1)] - f(0), \quad a_1 = \frac{1}{2}[f(1) - f(-1)], \quad a_0 = 2f(0) \tag{10.4.5}$$

Now, we can approximate the integral in $[-1, 1]$ by replacing the function by $g(x)$

$$\begin{aligned}\int_{-1}^{1} f(x)dx &\approx \int_{-1}^{1} g(x)dx \\ &= \int_{-1}^{1} (a_2x^2 + a_1x + a_0)dx \\ &= \frac{2a_2}{3} + 2a_0 = \frac{1}{3}\Big[f(-1) + 4f(0) + f(1)\Big] \quad (Eq.(10.4.5))\end{aligned} \tag{10.4.6}$$

[13] Using the method of undetermined coefficients to get three equations for three unknowns a_0, a_1, a_2.

Table 10.6 Performance of Simpson's rule for integral of x^2, x^3, x^4 from 0 to 1

n	1	10	100
Δ	1.00e+00	1.00e-01	1.00e-02
Error for $y = x^2$	0.00e+00	0.00e+00	0.00e+00
Error for $y = x^3$	0.00e+00	0.00e+00	0.00e+00
Error for $y = x^4$	8.33e-03	8.33e-07	8.33e-11

Going from $[-1, 1]$ to $[c, d]$ is easy. By using a change of variable,[14] we thus obtain the well-known Simpson's rule:

$$\int_c^d f(x)dx \approx \frac{d-c}{6}\left[f(c) + 4f\left(\frac{c+d}{2}\right) + f(d)\right] \tag{10.4.7}$$

More often we need to break the interval $[a, b]$ into n equal sub-intervals of length $\Delta = (b-a)/n$ and apply the Simpson rule for each interval:

$$\begin{aligned}\int_a^b f(x)dx &\approx \sum_{i=1}^{n} \int_{a+(i-1)\Delta}^{a+i\Delta} f(x)dx \\ &= \sum_{i=1}^{n} \frac{\Delta}{6}\left[f(a+(i-1)\Delta) + 4f(a+i\Delta-\Delta/2) + f(a+i\Delta)\right]\end{aligned} \tag{10.4.8}$$

We test the performance of Simpson's rule for x^2, x^3 and x^4. The Julia code is given in Listing A.10 which is based on Eq. (10.4.8). The error for $y = x^2$ is zero which is expected. The error is also zero for $y = x^3$, which is a surprise. And the error for $y = x^4$ is proportional to Δ^4; Simpson's rule is a fourth-order method, which explains its popularity in calculators and codes (Table 10.6).

Another derivation. By now we can see that all quadrature rules have this common form

$$\boxed{\int_a^b f(x)dx = \sum_i w_i f(x_i)} \tag{10.4.9}$$

that is, the sum of $f(x)$ evaluated at some points x_i multiplied with a weight w_i. In other words, the integral is a *weighted sum of function values at specially selected locations*. So, we can select *a prior* x_i's—the quadrature points—and determine the corresponding weights w_i. The first choice is to use equally spaced quadrature

[14] If you're not clear of this change of variable, check Sect. 10.4.3.

points. For example, $\int_{-1}^{1} f(x)dx$ can be computed as with 3 equally spaced points at $-1, 0, 1$:

$$\int_{-1}^{1} f(x)dx = w_1 f(-1) + w_2 f(0) + w_3 f(1) \tag{10.4.10}$$

The problem is now how to determine the weights w_i. We use Simpson's idea of parabolic approximation to replace $f(x)$ by $ax^2 + bx + c$. With this $f(x)$, Eq. (10.4.10) becomes

$$\begin{aligned}\frac{2}{3}a &= w_1(a - b + c) + w_2(c) + w_3(a + b + c)\\ &= a(w_1 + w_3) + b(w_3 - w_1) + c(w_1 + w_2 + w_3)\end{aligned}$$

So we have two expressions supposed to be identical for all values of a, b, c. This can happen only when:

$$\left.\begin{aligned} w_1 + w_3 &= 2/3\\ w_1 - w_3 &= 0\\ w_1 + w_2 + w_3 &= 0 \end{aligned}\right\} \Longrightarrow w_1 = w_3 = \frac{1}{3}, \quad w_2 = \frac{4}{3}$$

which is the same result we have obtained in Eq. (10.4.6).

Newton-Cotes rule. It can be seen that the mid-point rule can be derived similarly to the Simpson rule by approximating the function $f(x)$ with a constant function within each slice. And the trapezoidal rule is where a linear approximation to the function was used. Actually these rules are special cases of the so-called Newton-Cotes rules. Note that, in Newton-Cotes rules, the quadrature points are evenly spaced along the interval and thus known. We just need to find the quadrature weights w_i.

10.4.3 Gauss's Rule

Gauss also considered this integral $\int_{-1}^{1} f(x)dx$. But he wanted to beat Newton-Cotes by having less quadrature points. He also used $\int_{-1}^{1} f(x)dx = \sum_i w_i f(x_i)$, but the quadrature points x_i are not selected *a priori*, they are also unknowns to be determined together with the weights w_i.

Two-point Gauss rule. In the two-point Gauss rule, two quadrature points are used, thus we write

$$\int_{-1}^{1} f(x)dx = w_1 f(x_1) + w_2 f(x_2) \tag{10.4.11}$$

To determine the four unknowns (i.e., (x_1, x_2, w_1, w_2)), we need four equations. Gauss's idea is to exactly integrate these functions $1, x, x^2, x^3$. Using Eq. (10.4.11) for these 4 functions, we have

$$\begin{aligned}
f(x) = 1: &\quad 2 = w_1 + w_2 \\
f(x) = x: &\quad 0 = w_1 x_1 + w_2 x_2 \\
f(x) = x^2: &\quad \frac{2}{3} = w_1 x_1^2 + w_2 x_2^2 \\
f(x) = x^3: &\quad 0 = w_1 x_1^3 + w_2 x_2^3
\end{aligned}$$

Four equations and four unknowns should be ok. But the equations are nonlinear. How to solve them? Lucky for us, the equations are symmetric: changing w_1 with w_2 does not change the equations! So we know $w_1 = w_2$ and thus from the first equation they are both equal to one. Symmetry demands that $x_1 = -x_2$. Then, it is straightforward to get $x_1 = -1/\sqrt{3}$ and $x_2 = 1/\sqrt{3}$. The two-point Gauss rule is thus given by

$$\int_{-1}^{1} f(x)dx \approx 1 \times f\left(-\frac{1}{\sqrt{3}}\right) + 1 \times f\left(\frac{1}{\sqrt{3}}\right)$$

So, with two quadrature points (now referred to as Gauss points) Gauss quadrature can integrate exactly cubic polynomials, by its very definition.

Three-point Gauss rule. In the same manner, we can develop the three-point Gauss rule:

$$\int_{-1}^{1} f(x)dx = w_1 f(x_1) + w_2 f(x_2) + w_3 f(x_3) \tag{10.4.12}$$

To determine the six unknowns, we need six equations. So, the idea is to exactly integrate these six functions $1, x, x^2, x^3, x^4, x^5$. Using Eq. (10.4.12) for these six functions, we have

$$\begin{aligned}
f(x) = 1: &\quad 2 = w_1 + w_2 + w_3 \\
f(x) = x: &\quad 0 = w_1 x_1 + w_2 x_2 + w_3 x_3 \\
f(x) = x^2: &\quad \frac{2}{3} = w_1 x_1^2 + w_2 x_2^2 + w_3 x_3^2 \\
f(x) = x^3: &\quad 0 = w_1 x_1^3 + w_2 x_2^3 + w_3 x_3^3 \\
f(x) = x^4: &\quad \frac{2}{5} = w_1 x_1^4 + w_2 x_2^4 + w_3 x_3^4 \\
f(x) = x^5: &\quad 0 = w_1 x_1^5 + w_2 x_2^5 + w_3 x_3^5
\end{aligned}$$

Again, symmetry will help us to solve this scary-looking equations:

Table 10.7 Gauss-Legendre quadrature formulas on $[-1, 1]$

n	ξ_i	w_i
1	0.	2.0000000000
2	±0.5773502692	1.0000000000
3	±0.7745966692	0.5555555556
	0.	0.8888888889
4	±0.8611363116	0.3478548451
	±0.3399810436	0.6521451549

$$\begin{aligned} x_1 &= -x, \quad w_1 = w \\ x_2 &= \ 0, \quad w_2 = w_2 \\ x_3 &= \ x, \quad w_3 = w \end{aligned}$$

where $x > 0$. Now, the system is simplified to

$$\left.\begin{aligned} 2w + w_2 &= 2 \\ 2wx^2 &= \frac{2}{3} \\ 2wx^4 &= \frac{2}{5} \end{aligned}\right\} \Longrightarrow x = \sqrt{\frac{3}{5}}, \quad w = \frac{5}{9}, \quad w_2 = \frac{8}{9}$$

The three-point Gauss rule is thus given by

$$\int_{-1}^{1} f(x)dx \approx \frac{5}{9} f(-\sqrt{3/5}) + \frac{8}{9} f(0) + \frac{5}{9} f(\sqrt{3/5}) \tag{10.4.13}$$

So, with three Gauss points Gauss quadrature can integrate exactly quintic polynomials. We can generalize this to: using n Gauss points, Gauss' rule can integrate exactly polynomials of degree equal or less than $2n - 1$.

How we are going to develop four-point Gaussian quadrature and higher order versions? The way we just used would become tedious. But wait. The quadrature points x_i are special. Can you say what they are? Yes, *they are the roots of Legendre polynomials*, see Table 9.4. That's why Gaussian quadrature is also referred to as Gauss-Legendre (GL) quadrature. While this is a pleasant surprise, we need to be able to explain why Legendre polynomials appear here. Then, nice formula will appear and derivation of GL quadrature of any points will be a breeze. Table 10.7 presents values for some GL rules.

Arbitrary interval. We need $\int_a^b f(x)dx$ not $\int_{-1}^1 f(\xi)d\xi$. A simple change of variable is needed: $x = 0.5(1 - \xi)a + 0.5(1 + \xi)b$. So, the n point GL quadrature is given by

$$\int_a^b f(x)dx = \frac{b-a}{2}\int_{-1}^{1} f(x(\xi))d\xi \approx \frac{b-a}{2}\left[\sum_i w_i f\left(\frac{a+b}{2}+\frac{b-a}{2}\xi_i\right)\right] \tag{10.4.14}$$

which can accurately integrate any polynomial of degree less than or equal to $2n-1$.

Derivation of Gauss-Legendre quadrature rule. Herein we present the derivation of the Gauss-Legendre quadrature rule using orthogonal Legendre polynomials. Assume that we want to compute the following integral where $p_5(x)$ is any 5th-degree polynomial:

$$I = \int_{-1}^{1} p_5(x)dx \tag{10.4.15}$$

We do not compute this integral directly, but we massage $p_5(x)$ a bit: we divide it by the Legendre polynomial $L_3(x)$:

$$p_5(x) = Q_2(x)L_3(x) + R_2(x) \tag{10.4.16}$$

where $Q_2(x)$ and $R_2(x)$ are polynomials of degree 2 at most. Now, the integral becomes

$$I = \int_{-1}^{1} [Q_2(x)L_3(x) + R_2(x)]dx = \int_{-1}^{1} R_2(x)dx \tag{10.4.17}$$

We converted an integral of a 5th-degree polynomial to the integral of a 2nd-degree polynomial! (This is so because $Q_2(x)$ and $L_3(x)$ are orthogonal, i.e., $\int_{-1}^{1} Q_2(x)L_3(x)dx = 0$, check Sect. 9.11.5 if this is not clear). Now, we tackle the problem of how to compute the integral of $R_2(x)$ without knowing its expression. But, we do know $p_5(x)$. So, if we use the roots of $L_3(x)$, denoted by x_0, x_1, x_2, we have this, from Eq. (10.4.16)

$$p_5(x_i) = R_2(x_i), \quad \text{or} \quad R_2(x_i) = p_5(x_i) \tag{10.4.18}$$

Now, the problem is easier. We build a Lagrange polynomial interpolating the points $(x_i, R_2(x_i))$ or $(x_i, p_5(x_i))$. But this polynomial is exactly $R_2(x)$, so we have

$$R_2(x) = \sum_{i=0}^{2} l_i(x)p_5(x_i) \tag{10.4.19}$$

With all these results, the original integral can be computed as

$$I = \int_{-1}^{1} p_5(x)dx = \int_{-1}^{1} R_2(x)dx = \int_{-1}^{1} \sum_{i=0}^{2} l_i(x) p_5(x_i)dx = \sum_{i=0}^{2} p_5(x_i) \int_{-1}^{1} l_i(x)dx$$

$$= \sum_{i=0}^{2} p_5(x_i)w_i, \quad w_i := \int_{-1}^{1} l_i(x)dx, \quad x_i \text{ are roots of } L_3(x) = 0$$

Now, we understand why GL points are the roots of Legendre polynomials. You should double check the values in Table 10.7 using this.

10.4.4 Two- and Three-Dimensional Integrals

It is a straightforward step to go from one-dimensional integrals to two-dimensional integrals. We just keep one variable fixed, and integrate the integral using any known 1D quadrature rule. Then, we apply again a 1D rule for the remaining integral:

$$\int_{-1}^{1}\int_{-1}^{1} f(\xi, \eta)d\xi d\eta = \int_{-1}^{1} \left[\int_{-1}^{1} f(\xi, \eta)d\xi \right] d\eta = \int_{-1}^{1} \left[\sum_{i=1}^{n} w_i f(\xi_i, \eta) \right] d\eta$$

$$= \sum_{i=1}^{n} \int_{-1}^{1} w_i f(\xi_i, \eta)d\eta = \sum_{i=1}^{n}\sum_{j=1}^{n} w_i w_j f(\xi_i, \eta_j)$$

Sometimes a short notation is used, and we write

$$\int_{-1}^{1}\int_{-1}^{1} f(\xi, \eta)d\xi d\eta = \sum_{k=1}^{n\times n} w_k f(\boldsymbol{\xi}_k)$$

10.4.5 Monte Carlo Integration

This section presents the Monte Carlo method for calculating the definite integral $F = \int_a^b f(x)dx$. We already know the concept of the average value of the function $f(x)$, denoted as $\langle f \rangle$, and we have the following formula (check Sect. 4.11.3 if not clear):

$$F = \int_{a}^{b} f(x)dx = (b-a)\langle f \rangle \tag{10.4.20}$$

Looking at this formula, we can compute $\int_a^b f(x)dx$ if we know $\langle f\rangle$. Of course, we cannot compute $\langle f\rangle$ using integration! Therefore, we generate N random numbers X_i uniformly distributed over $[a,b]$. Then, we calculate the average value of $f(x)$ over $[a,b]$ as follows:

$$\langle f\rangle = \frac{\sum_{i=1}^{N} f(X_i)}{N} \tag{10.4.21}$$

And like that, the definite integral of any function becomes as easy as eating candy. First, let me introduce some terms; we call $\langle F^N\rangle$ the Monte Carlo estimate of F, and $\langle F^N\rangle$ is calculated as follows:

$$\langle F^N\rangle = (b-a)\frac{1}{N}\sum_{i=1}^{N} f(X_i) \tag{10.4.22}$$

Clearly, the above formula can be written as $\sum_i w_i f(x_i)$, similar to the traditional way of approximating definite integrals (see ??). The difference is that while Simpson and Gauss had to rack their brains to find w_i, x_i, with Monte Carlo, we only need to generate N random numbers X_i uniformly distributed over $[a,b]$, with the same weight $w_i = (b-a)/N$.

What ensures that Eq. (10.4.22) is a good approximation of $\int_a^b f(x)dx$? It's the law of large numbers. Implementing this method is also simple: just two lines of code (Listing 10.1). (Of course, you can write longer, and less concise code. In fact, this code was written by the AI GitHub Copilot for me. I just had to type a comment saying what I wanted!)

Listing 10.1 Monte Carlo integration for $\int_a^b f(x)dx$.

```
function MC_definite_integral(f,a,b,N)
  x = a .+ (b-a)*rand(N)
  return (b-a)*sum(f.(x))/N
end
f(x)  = sqrt(1-x^2)
a     = 0.0; b = 1.0; N = 10^6
pii   = 4. * MC_definite_integral(f, a, b, N)
```

In the previous section, we started with the concept of the average value of a function and developed the Monte Carlo estimate formula for a definite integral. Based on that, we will now refine this formula using the concept of the expectation of a random variable. We will prove that the definite integral can be seen as the expectation of a random variable, and we know how to calculate the average of a random variable!

Let's assume we have a random variable X with probability density function $f_X(x)$, and X takes values in $[a,b]$. We consider the random variable Y given by $Y = g(X)$ where g is some function. In that case, we have

$$\mathbb{E}[Y] = \mathbb{E}[g(X)] = \int_a^b g(x) f_X(x) dx \tag{10.4.23}$$

Now let's consider the case where X has a uniform distribution on $[a, b]$. In this case, the probability density function of X is $f_X(x) = 1/b-a$ (this function is a constant, hence the name uniform distribution). We can write $\int_a^b g(x)$ as follows (so that $\mathbb{E}[Y]$ appears according to Eq. (10.4.23)):

$$\int_a^b g(x)dx = (b-a)\int_a^b g(x)\frac{1}{b-a}dx = (b-a)\int_a^b g(x) f_X(x) dx = (b-a)\mathbb{E}[Y] \tag{10.4.24}$$

Thus, we have transformed the problem of calculating the integral $\int_a^b g(x)dx$ into the problem of calculating the expectation $\mathbb{E}[Y]$. Of course, X does not necessarily have to follow a uniform distribution. In the general case, we have (always remember the form of Eq. (10.4.23))

$$\int_a^b g(x)dx = \int_a^b \frac{g(x)}{f_X(x)} f_X(x) dx = \mathbb{E}[Y], \quad Y = \frac{g(x)}{f_X(x)} \tag{10.4.25}$$

Let's consider an illustrative example. For instance, to compute the following integral (which is just $\Phi(z)$ that we are familiar with) using the Monte Carlo method:

$$I = \frac{1}{\sqrt{2\pi}} \int_{-\infty}^{z} \exp\left(-\frac{u^2}{2}\right) du$$

What we need to do is to rewrite this integral so that $\mathbb{E}[Y]$ appears:

$$I = \int_{-\infty}^{z} \frac{1}{\sqrt{2\pi}} \exp\left(-\frac{u^2}{2}\right) du = \int_{-\infty}^{z} 1 \times \frac{1}{\sqrt{2\pi}} \exp\left(-\frac{u^2}{2}\right) du + \int_{z}^{\infty} 0 \times \frac{1}{\sqrt{2\pi}} \exp\left(-\frac{u^2}{2}\right) du$$

Thus, I takes the form:

$$I = \int_{-\infty}^{\infty} g(x) f_X(x) dx = \mathbb{E}[Y], \quad Y = g(X) = 1 \ \text{for} \, x \leq z, \quad f_X(x) = \frac{1}{\sqrt{2\pi}} \exp\left(-\frac{x^2}{2}\right)$$

So I is computed as the average value of the random variable Y. The code is given in Listing 10.2. In the first two lines, we generate N standard normally distributed random variables $X_1, X_2, ..., X_N$ (due to the form of $f_X(x)$). Then we calculate Y using the function $g(x)$. Finally, we calculate the average value of Y to find I.

Listing 10.2 Monte Carlo method for $I = \int_{-\infty}^{z} {}^{1}/\sqrt{2\pi}\, \exp\left(-{}^{u^2}/2\right) du$.

```
using Distributions
std_var = Normal(0,1)           # this is X
sample  = rand(std_var, 10^6)   # this is X_1,X_2,...
Y       = Int.(sample .<= z)    # this is Y, convert from bool to Int
println(mean(Y))                # this is I
```

When to use a uniform distribution and when to use another distribution depends on the specific function $g(x)$ and its behavior over the given interval $[a, b]$. An example will clarify this issue. Suppose we want to compute the following integral using the Monte Carlo method:

$$I = \int_{-4}^{4} (1 + x^2)e^{-x^4/4}\, dx$$

We use `QuadGK` and obtain the result $I = 4.29669427$, which we consider accurate and use as a benchmark for comparison. Now, we will use Monte Carlo with two different distributions. First, we use a uniform distribution, meaning we employ Eq. (10.4.22). Second, we use a standard normal distribution, meaning we use Eq. (10.4.25) with $f_X(x) = {}^{1}/\sqrt{2\pi}\, \exp\left(-{}^{x^2}/2\right)$. Of course, $g(x) = (1 + x^2)e^{-x^4/4}$.

To understand the difference between the two distributions, we plot Fig. 10.11. By looking at it, can you guess which distribution yields more accurate results for the same N? The one that used the normal distribution is called importance sampling.

The real power of MC method is for integrals in multiple dimension. Assume that we need to compute the following triple integral:

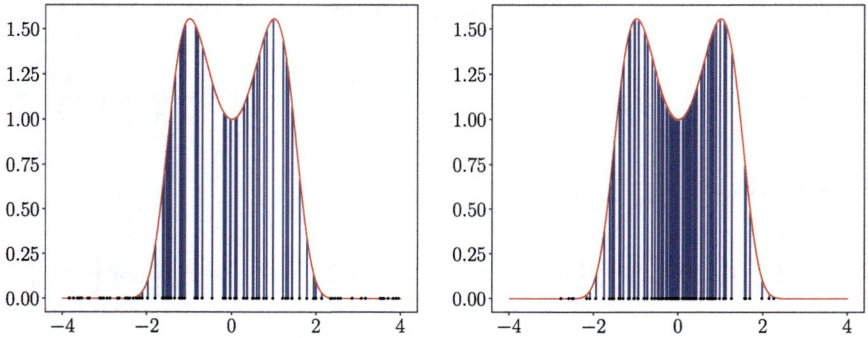

Fig. 10.11 Graph of the function $g(x) = (1 + x^2)e^{-x^4/4}$ (in red). On the left are 100 randomly generated points with a uniform distribution, and on the right are points with a standard normal distribution. The black points on the x-axis represent X_i, and the blue lines connect $(X_i, 0)$ to $(X_i, g(X_i))$

$$F = \int_{z_0}^{z_1} \int_{y_0}^{y_1} \int_{x_0}^{x_1} f(x, y, z)\, dx\, dy\, dz$$

Similar to the 1D integral, the Monte Carlo estimate of F is given by

$$\langle F^N \rangle = \frac{(x_1 - x_0)(y_1 - y_0)(z_1 - z_0)}{N} \sum_{i=1}^{n} f(\mathbf{X}_i)$$

Let's try to compute the following integral using the Monte Carlo method: approximating the volume under the plane $z = 8x + 6y$ over the rectangular region $R = [0, 1] \times [0, 2]$. The result is 20. The program is presented in Listing 10.3. The result with $N = 10^3$ is 19.88. The advantage of this method is that even for a 1000-dimensional integral, we only need approximately this number of N. This is in stark contrast to numerical methods like Gauss or Simpson's rule, where the required N would increase significantly for higher dimensional integrals.

Listing 10.3 Monte Carlo method for double integrals.

```
using Distributions
function MC_2D_integral(f,a,b,c,d,N)
  uniform_dist1 = Uniform(a, b)
  uniform_dist2 = Uniform(c, d)
  X             = rand(uniform_dist1, N)
  Y             = rand(uniform_dist2, N)
  area          = (b-a)*(d-c)
  return area * mean(f.(X, Y))
end
MC_2D_integral((x,y)->8x+6y,0,1,0,2,10^3)
```

10.5 Solving Nonlinear Equations

Solving nonlinear equations of the form $f(x) = 0$ is a common problem in applied mathematics. Sometimes it is referred to as root finding. Of course we're interested in systematic methods that can solve any equation using some algorithm. Herein I discuss just some of them so that we know the general principles of these methods.

In Sect. 2.11, we have discussed the fixed point iteration method to solve nonlinear equations. The algorithm is simple, but we do not know why it works. Recall that to solve the equation $f(x) = 0$, we convert it to a fixed point problem, i.e., that of the form $x = g(x)$.[15] Starting with the initial x_0, the fixed point iterations are then given by

[15] One way is to cast $f(x) = 0$ as $x = x + f(x)$. But note that this form might not work. Herein I skip the discussion on how to choose $g(x)$ that guarantee the success of the method.

$$x_{n+1} = g(x_n), \quad n = 0, 1, 2, \ldots$$

And we continue until we arrive at the solution α (i.e., $\alpha = g(\alpha)$). Practically it is when the so-called stopping condition $|x_n - x_{n-1}| < \epsilon$ is met, where ϵ is a small positive number controlling the tolerance. Then the iterations stop.

To motivate the discussion, let's use the fixed point method to solve the following equations:

$$\begin{aligned} E1 &: x = 1 + 0.5 \sin x \\ E2 &: x = 3 + 2.0 \sin x \end{aligned} \tag{10.5.1}$$

A simple plotting shows that both equations have one solution. But if we use the fixed point iterations we see that, while it works for the first equation (the iterates x_n converge to the correct solution), it does not work for the second (Fig. 10.12). The question is why?

Now consider this equation $x = g(x)$ with one solution denoted by α. We consider the interval $[a, b]$ that contains α. Now, as always, to analyze the method, we study the error $e_{n+1} = \alpha - x_{n+1}$. The error is certainly a sequel, simply a list of many numbers, and if this sequence converges to zero when n is getting larger, then the method works. We compute the error now:

$$\alpha - x_{n+1} = g(\alpha) - f(x_n)$$

The mean value theorem (Sect. 4.11.2) gives us a way to understand $f(\alpha) - f(x_n)$:

$$g(\alpha) - g(x_n) = g'(\xi_n)(\alpha - x_n), \quad \xi_n \in [\alpha, x_n]$$

Let's denote by λ the maximum of $|f'(x)|$ for $x \in [a, b]$. Thus,

$$\alpha - x_{n+1} \le \lambda(\alpha - x_n) \iff \boxed{|\alpha - x_{n+1}| \le \lambda|(\alpha - x_n)|}$$

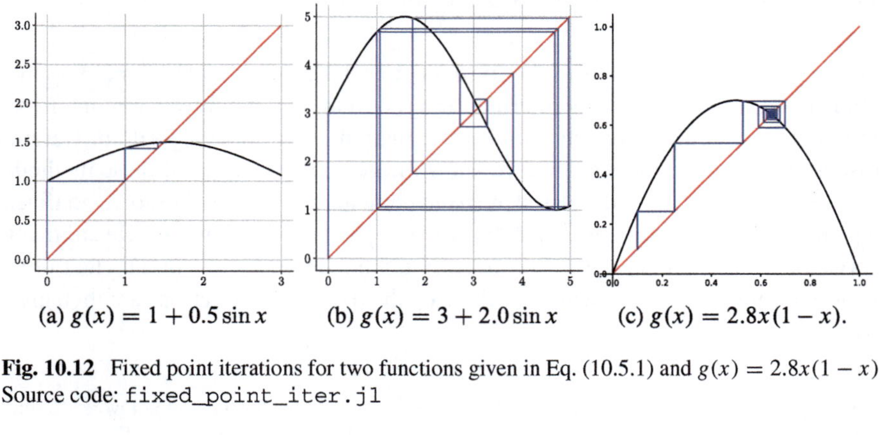

(a) $g(x) = 1 + 0.5 \sin x$ (b) $g(x) = 3 + 2.0 \sin x$ (c) $g(x) = 2.8x(1 - x)$.

Fig. 10.12 Fixed point iterations for two functions given in Eq. (10.5.1) and $g(x) = 2.8x(1 - x)$. Source code: `fixed_point_iter.jl`

Table 10.8 Fixed point iterations for $g(x) = 1 + 0.5 \sin x$ with $x_0 = 0$ and $\epsilon = 10^{-6}$

n	x_n	$\lvert\alpha - x_n\rvert$	$\lvert\alpha - x_{n+1}\rvert / \lvert\alpha - x_n\rvert$	$x_n - x_{n-1} / x_{n-1} - x_{n-2}$
0	0.0000000000000	1.4987011335180	–	–
1	1.0000000000000	0.4987011335180	0.3327555590402	–
2	1.4207354924039	0.0779656411141	0.1563374050587	0.4207354924039
3	1.4943809925643	0.0043201409537	0.0554108308730	0.1750399039063
6	1.4987009254070	0.0000002081110	0.0360178432695	0.0360577760154
7	1.4987011260224	0.0000000074956	0.0360171325034	0.0360178698262
8	1.4987011332479	0.0000000002701	0.0360359025969	0.0360164311995

The boxed equation tells us: the distance to the fixed point α is shrinking every iteration. And that's exactly what we see in Fig. 10.12a or Table 10.8. The boxed equation can be used to show that

$$|\alpha - x_n| \leq \lambda^n |\alpha - x_0|$$

If $\lambda \leq 1$, then the red term, e.g., λ^n will vanish, thus the error $|\alpha - x_n|$ is vanishing too, and voilà, the iterates x_n converge to the solution.

Is it possible to find λ? From Table 10.8, we can see that $\lambda \approx 0.0361$. It turns out that $\lambda = g'(\alpha) = g'(1.4987011332479)$. The reason is as follows. The MVT gives us

$$\alpha - x_{n+1} = g(\alpha) - g(x_n) = g'(\xi_n)(\alpha - x_n), \quad \xi_n \in [\alpha, x_n]$$

Thus,

$$\frac{|\alpha - x_{n+1}|}{|\alpha - x_n|} = |g'(\xi_n)| \Longrightarrow \lim_{n\to\infty} \frac{|\alpha - x_{n+1}|}{|\alpha - x_n|} = \lim_{n\to\infty} |g'(\xi_n)| \tag{10.5.2}$$

As $n \to \infty$, $x_n \to \alpha$, hence $\xi_n \to \alpha$, and as $g'(x)$ is a continuous function, $\lim_{n\to\infty} |g'(\xi_n)| = |g'(\alpha)|$. With $g(x) = 1 + 0.5 \sin x$, $g'(\alpha) = 0.5 \cos(1.4987011332479) = 0.0360$. The analysis confirms the numerical experiment shown in Table 10.8.

If we look at Fig. 10.12a and c we see that, although the iterates converge to the solution, the way they approach the solution is different. In Fig. 10.12c, the iterates approach the solution from both sides, thus making a cobweb. On the other hand, in Fig. 10.12a, the iterates approach the solution from only one side, making a staircase picture. Can we explain this? Of course yes. Implied by Eq. (10.5.2) is, close to α we have

$$\alpha - x_{n+1} \approx g'(\alpha)(\alpha - x_n), \quad \text{or} \quad \frac{\alpha - x_{n+1}}{\alpha - x_n} \approx g'(\alpha)$$

This equation says that the ratio $\alpha - x_{n+1} / \alpha - x_n$ has the same sign with $g'(\alpha)$. When $g'(\alpha)$ is negative, $\alpha - x_{n+1}$ and $\alpha - x_n$ are of different sign; the iterates oscillate between α. Indeed, for $g(x) = 2.8x(1 - x)$, $g'(x) = 2.8 - 5.6x < 0$ at the solution.

Now, we can also explain why the method did not work for $g(x) = 3 + 2\sin x$. This is because $|g'(\alpha)| > 1$. When this is the case, from $|\alpha - x_{n+1}| \approx |g'(\alpha)||(\alpha - x_n)|$, we see that the errors will increase as we approach the root rather than decrease in size. This is what we see in Fig. 10.12b. Precisely $g'(\alpha) = -1.99$, negative: that's why we're seeing a cobweb.

10.6 Numerical Solution of Ordinary Differential Equations

This section presents some common numerical methods (such as Euler's method) to solve either a single ordinary differential equation or a system of ordinary differential equations.[16] It is with the power of these methods that Katherine Johnson[17] helped to put men on the moon.

We begin with the simplest method—the Euler method (Sect. 10.6.1)—for first-order ODEs. Next, we discuss this method for second-order ODEs (e.g., equations of motions of harmonic oscillators and of planets orbiting the Sun) in Sect. 10.6.2. Albeit simple, the Euler method does not conserve energies, it therefore is bad for modeling the long-term behavior of oscillatory systems. Thus, we need a better method and one of them is the Euler-Aspel-Cromer method presented in Sect. 10.6.3. Having a good numerical method, we then apply it to the Kepler problem, i.e., we solve the Sun-Earth problem (Sect. 10.6.4). For what? To rediscover for ourselves that planets do indeed go around the Sun in elliptical orbits. And high school students can achieve that because the maths behind all of this is simple. In a logical development, we study three-body and N-body problems in Sect. 10.6.5. And in Sect. 10.6.6 we discuss again the simple pendulum but with large amplitude; we shall observe chaotic motion in this simple mechanical system.

Although Euler's method and related variants are simple and good, they are only first-order methods (i.e., the accuracy is low), I present a second-order method in Sect. 10.6.7. That is the Verlet method—a very popular method used to solve Newton's equations of motions, i.e., $F = ma$. Section 10.6.8 presents an analysis of the Euler method to answer questions such as what is the accuracy of the method.

---

[16] Refer to Chap. 7 for an introduction to these ODEs.

[17] Katherine Johnson (26 August 1918–24 February 2020) was an American mathematician whose calculations of orbital mechanics as a NASA employee were critical to the success of the first and subsequent U.S. crewed spaceflights. During her 33-year career at NASA and its predecessor, she earned a reputation for mastering complex manual calculations and helped pioneer the use of computers to perform the tasks. The space agency noted her "historical role as one of the first African-American women to work as a NASA scientist".

10.6.1 Euler's Method: First-Order ODE

To introduce Euler' method, let's consider the following first-order ODE

$$\dot{x} = f(x, t), \quad x(0) = x_0 \tag{10.6.1}$$

We can think of the above equation as the velocity of an object. Suppose that at a given time t the object has a certain position $x(t)$. What is the position at a slightly later time $t + \epsilon$? (ϵ is referred to as the time step). If we can answer this question we have solved Eq. (10.6.1), then we can start with the initial position x_0 and compute how it changes for the first instant ϵ, the next instant 2ϵ, and so on.

Now, if ϵ is small, we can compute the velocity as the averaged velocity[18]

$$\dot{x} = \frac{x(t+\epsilon) - x(t)}{\epsilon} \tag{10.6.2}$$

With that $\dot{x}$ being substituted into Eq. (10.6.1), we can get $x(t+\epsilon)$:

$$\frac{x(t+\epsilon) - x(t)}{\epsilon} = f(x, t) \Longrightarrow \boxed{x(t+\epsilon) = x(t) + \epsilon f(x, t)} \tag{10.6.3}$$

The boxed equation, which is the Euler method, enables the solution $x(t)$ to advance or March in time starting from $x(0)$. If you use Euler's method with small ϵ you will find that it works nicely (just try it with some first-order ODE). We rush now to second-order ODEs which are more fun.

But how small is small for ϵ? Does the numerical solution converge to the exact solution when ϵ goes to zero? What is the accuracy of the method? Those are questions that mathematicians seek answer for. For now, let's have fun first and in Sect. 10.6.8 we shall try to answer those questions. That's how scientists and engineers approach a problem.

10.6.2 Euler's Method: Second-Order ODE

As a typical second-order ODE, let's consider the simple harmonic oscillator with mass m, spring k, and damping b (Sect. 7.8):

$$m\ddot{x} + b\dot{x} + kx = 0, \quad x(0) = x_0, \quad \dot{x}(0) = v_0 \tag{10.6.4}$$

Now, introducing the velocity $v = \dot{x}$, the above equation is rewritten as

---

[18] Or if you like you can say that we are using the forward difference formula for the first derivative of $x(t)$. They are equivalent.

$$\dot{v} = -\frac{b}{m}v - \frac{k}{m}x := F(v, x) \tag{10.6.5}$$

Using the Euler method, that is, the boxed equation in Eq. (10.6.3), for the position equation $\dot{x} = v$ and the velocity equation $\dot{v} = F$, we obtain

$$\begin{aligned} x(t+\epsilon) &= x(t) + \epsilon v(t) \\ v(t+\epsilon) &= v(t) + \epsilon F(v(t), x(t)) \end{aligned} \tag{10.6.6}$$

The Euler method is easy to program. Usually it works nicely but for some problems it performs badly, and simple harmonic oscillation is one of them (Fig. 10.13). Input data used: $k = m = 1$, $x_0 = 1$, $v_0 = 0$, and $b = 0$ (i.e., no damping), the total time is three periods and time step $\epsilon = 0.01$. The plot of $x(t)$ shows that the amplitude of the oscillation keeps increasing (Fig. 10.13a). This means that energies also increase, and thus energy conservation is violated. Thus, the phase portrait is no longer a nice circle[19] (Fig. 10.13b). The orange is the exact phase portrait.

To understand what went wrong, we need a better notation. Instead of writing $x(n\epsilon)$, we write x_n. Thus the subscript n is used to indicate the time when a certain term is evaluated; the discrete time events are $t_n = n\epsilon$ for $n = 0, 1, 2, \ldots$ With the new notation, Eq. (10.6.6) becomes

$$\begin{aligned} x_{n+1} &= x_n + \epsilon v_n \\ v_{n+1} &= v_n + \epsilon F_n \end{aligned} \tag{10.6.7}$$

As the total energy is wrong, we analyze it. At two iterations or time steps t_n and t_{n+1}, the total energies are (without loss of generality I used $m = k = 1$)

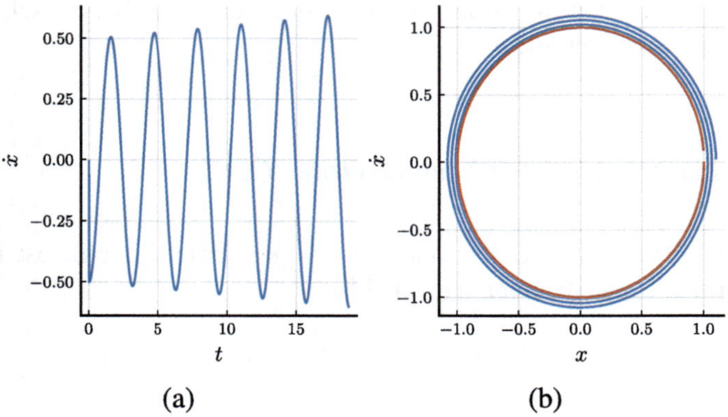

Fig. 10.13 Euler's method does not conserve: simple harmonic oscillation problem

[19] Refer to Fig. 7.8 and the related discussion if phase portrait is not clear.

$$\begin{aligned} E_n &= \frac{1}{2}v_n^2 + \frac{1}{2}x_n^2 \\ E_{n+1} &= \frac{1}{2}v_{n+1}^2 + \frac{1}{2}x_{n+1}^2 \end{aligned} \tag{10.6.8}$$

Now using Eq. (10.6.7), we compute E_{n+1}:

$$E_{n+1} = \frac{1}{2}(v_n + \epsilon F_n)^2 + \frac{1}{2}(x_n + \epsilon v_n)^2 = E_n + \epsilon F_n v_n + \frac{\epsilon^2}{2}F_n^2 + \epsilon x_n v_n + \frac{\epsilon^2}{2}v_n^2 \tag{10.6.9}$$

Noting that $F_n = -x_n$, thus the change in total energy is

$$\Delta E_n := E_{n+1} - E_n = \epsilon^2\left(\frac{1}{2}x_n^2 + \frac{1}{2}v_n^2\right) > 0 \tag{10.6.10}$$

And that's why the numerical total energy is increasing and finally it will blow up the computations.

10.6.3 Euler-Aspel-Cromer Method: Better Energy Conservation

The method that we now refer to as the Euler-Cromer method was discovered quite by accident. Around 1980, Abby Aspel (who at the time was a high school student) correctly coded up the Euler method for the Kepler problem. Thinking the resulting inaccurate model was caused by an error in her code, *she interchanged two lines in her program*, and the model seemed to work. Abby accidentally stumbled upon the method, given for our problem of SHO as follows:

$$\begin{aligned} v_{n+1} &= v_n + \epsilon F_n \\ x_{n+1} &= x_n + \epsilon v_{n+1} \end{aligned} \tag{10.6.11}$$

The only change is in the red term, instead of using v_n, now v_{n+1} is used. If you modify the code (very slightly) and rerun the SHO problem, you will see that the results are very good. Cromer in his paper entitled *Stable solutions using the Euler approximation* (so Cromer did not call his method Cromer's method and he gave credit to Aspel even though in a footnote) presented a mathematical analysis of why the method works.

The change in total energy is now given by

$$\Delta E_n = \epsilon^2\left(\frac{1}{2}v_n^2 - \frac{1}{2}x_n^2\right) - \epsilon^3 v_n x_n \tag{10.6.12}$$

10.6.4 Solving Kepler's Problem Numerically

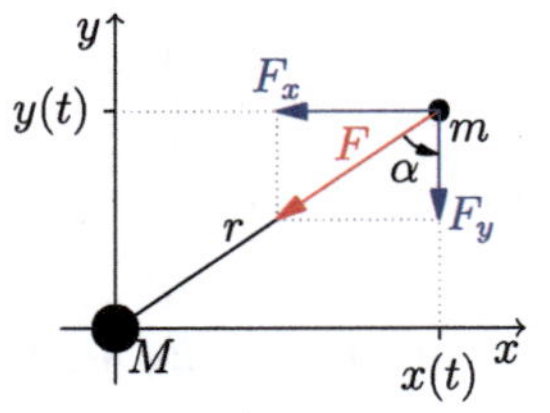

Now we have a powerful tool—the Euler-Aspel-Cromer method—to solve ordinary differential equations. Let's use it to analyze the motion of a planet around the sun. We hope that we would discover the elliptical orbit that Newton did. But we use a computer to do that. This is known as the 2-body problem. With that name we certainly have 3-body problems and generally N-body problems. Note that even though there exists closed-form solutions to the 2-body problem, there is no such formula for N-body problems when $N > 2$. Numerical methods such as Euler's method can solve N-body problems.

Let's use a Cartesian coordinate system with x being the horizontal and y the vertical axis. The sun is at the origin of this system. The position of a planet is $(x(t), y(t))$ at time t. Of course, we are going to use Newton's 2nd law and his theory of gravitation. When the planet is at a distance $r = \sqrt{x^2 + y^2}$ from the sun, it is pulled by the sun with a gravitational force of magnitude

$$F = \frac{GMm}{r^2}$$

We decompose this force into two components F_x and F_y ($F_x = -F \sin\alpha$, $\sin\alpha = {}^x/_r$ see above figure):

$$F_x = -F\frac{x}{r} = -\frac{GMm}{r^3}x, \quad F_y = -F\frac{y}{r} = -\frac{GMm}{r^3}y$$

Newton's 2nd law now gives us

$$m\frac{dv_x}{dt} = -\frac{GMm}{r^3}x, \quad m\frac{dv_y}{dt} = -\frac{GMm}{r^3}y \tag{10.6.13}$$

We have two ODEs, not one. But that's no problem. Don't forget that $r = \sqrt{x^2 + y^2}$. Using the Euler-Aspel-Cromer method, we have (as the mass of the Sun is too big, it is assumed that the Sun is stationary)

$$
\begin{aligned}
r_n &= \sqrt{x_n^2 + y_n^2} \\
v_{x,n+1} &= v_{x,n} + \epsilon \left(-\frac{GM}{r_n^3} x_n \right) \\
v_{y,n+1} &= v_{y,n} + \epsilon \left(-\frac{GM}{r_n^3} y_n \right) \\
x_{n+1} &= x_n + \epsilon v_{x,n+1} \\
y_{n+1} &= y_n + \epsilon v_{y,n+1}
\end{aligned}
\tag{10.6.14}
$$

with the initial conditions (x_0, y_0) and (v_{x0}, y_{y0}), to be discussed shortly. Remark: The notation got a bit ugly now: $v_{x,n+1}$ means the $x-$component of the velocity at time step $n + 1$.

Before we can run the code, there is the matter of choice of units. As the radius of Earth's orbit around the sun is about 1.5×10^{11}m, a graph showing this orbit would have labels of 1×10^{11}m, 2×10^{11}m, etc.., which is awkward. It is much more convenient to use astronomical units, AU, which are defined as follows. One astronomical unit of length (i.e., 1 AU) is the average distance between the Sun and the Earth, which is about 1.5×10^{11}m. For time, it is convenient to measure it in years. What is then the unit of mass?

Recall that the Earth's orbit is, to a very good approximation, circular. Thus, there must be a force equal to $M_E v^2/r$ $(r = 1$ AU), where v is the Earth's speed which is equal to $2\pi r/(1 \text{ yr}) = 2\pi$ AU/yr. Thus, we have

$$
\frac{M_E v^2}{r} = \frac{GMM_E}{r^2} \Longrightarrow GM = v^2 r = 4\pi^2 \text{ AU}^3/\text{yr}^2
$$

Now, we discuss the initial positions and velocities for Mercury (as we want to see an ellipse). Using astronomical data we know that the eccentricity of the elliptical orbit for Mercury is $e = 0.206$, and the radius (or semi-major axis) $a = 0.39$ AU. For the simulation, we assume that the initial position of Mercury is at the aphelion $(x_0, y_0) = (r_1, 0)$ with $r_1 = a(1 + e)$ (check Sect. 4.13.2 if something not clear). The initial velocity is $(0, v_1)$. How to compute this v_1? We need two equations: angular momentum conservation and energy conservation evaluated at two points; these two equations involve two unknown velocities v_1 and v_2. The angular momentum is $r_x p_y - r_y p_x$, evaluated at two points $(r_1, 0)$ and $(0, r_2)$:

$$
v_1 r_1 = v_2 b \Longrightarrow v_2 = \frac{v_1 r_1}{b}, \quad b = a\sqrt{1 - e^2}
$$

With m being the mass of Mercury and M the mass of the Sun, conservation of total energy provides us the second equation:

$$
-\frac{GMm}{r_1} + \frac{1}{2} m v_1^2 = -\frac{GMm}{r_2} + \frac{1}{2} m v_2^2
$$

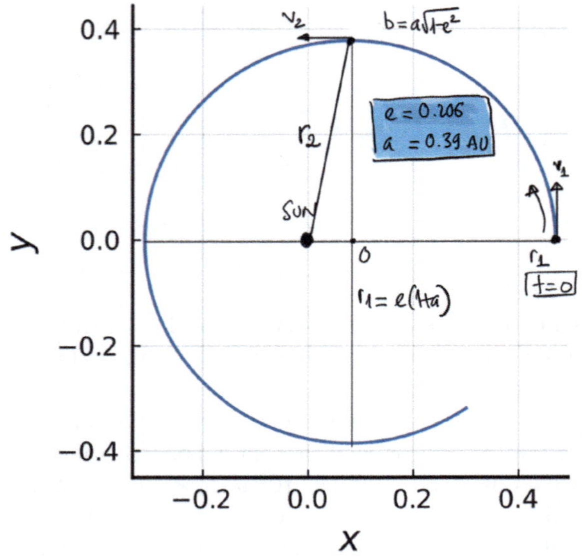

Fig. 10.14 Mercury elliptical orbit

Solving these two equations for v_1, noting that $r_2 = a$, we get

$$v_1 = \sqrt{\frac{GM}{a}\frac{1-e}{1+e}}$$

Now, we can really let the Mercury go! And with the Euler-Aspel-Cromer method and Newton's laws, we are able to get the elliptical orbit of planetary motion (Fig. 10.14). We can determine the period T (how?), etc. Applying the same method for other planets we can also discover Kepler's third law: for each planet just compute $T/a^{3/2}$ and you will see that this quantity is approximately one (recall that Kepler told us that this constant should be $k = {}^{2\pi}/\sqrt{GM} = 1$). We can also discover the 2nd Kepler's law.

10.6.5 Three-Body Problems and N-Body Problems

It is logical that after solving the two-body problem, we solve the three-body problem. Newton did that. In Proposition 66 of Book 1 of the Principia, and its 22 Corollaries, Newton took the first steps in the definition and study of the problem of the movements of three massive bodies subject to their mutually perturbing gravitational attractions. In Propositions 25–35 of Book 3, Newton also took the first steps in applying his results of Proposition 66 to the lunar theory, the motion of the Moon under the gravitational influence of Earth and the Sun. Newton did not succeed just because, as Poincaré pointed out many years later, there is no closed-form solution. If those solutions existed Newton would be able to discover them.

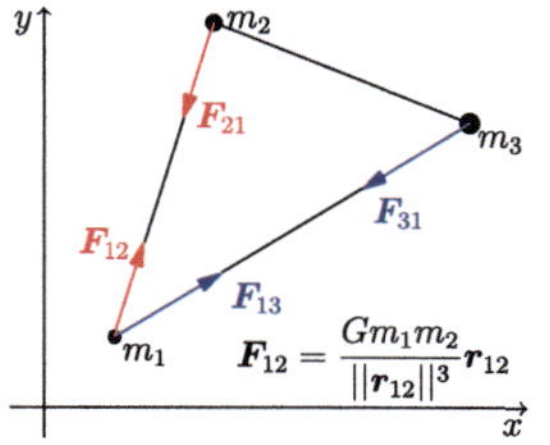

Herein, we solve the $N-$body problem using numerical method. We use the three-body problem as an example to set up the equations, but the code is written for N bodies. It is more convenient to use vectors now, because one vectorial equation replaces two normal equations: we save time. Consider now three bodies of masses m_1, m_2, m_3, their positions are $\boldsymbol{r}_i(t)$ and their velocities are $\boldsymbol{v}_i(t)$, $i = 1, 2, 3$. Now focusing on mass m_1, the forces acting on it are ($||\boldsymbol{r}||$ means the Euclidean length of $\boldsymbol{r}$)

$$\boldsymbol{F}_1 = \boldsymbol{F}_{12} + \boldsymbol{F}_{13}, \quad \boldsymbol{F}_{1j} = \frac{Gm_1m_j}{||\boldsymbol{r}_{1j}||^3}\boldsymbol{r}_{1j}, \quad \boldsymbol{r}_{1j} = \boldsymbol{r}_j - \boldsymbol{r}_1, \quad (||\boldsymbol{x}|| = \sqrt{x_1^2 + x_2^2})$$

Now the dynamical equations for m_1 are

$$m_1\frac{d\boldsymbol{v}_1}{dt} = \boldsymbol{F}_1, \quad \frac{d\boldsymbol{r}_1}{dt} = \boldsymbol{v}_1$$

Using the Euler-Aspel-Cromer method, we update the velocity and position for mass m_1:

$$\boldsymbol{v}_{1,n+1} = \boldsymbol{v}_{1,n} + \epsilon\boldsymbol{F}_{1,n}, \quad \boldsymbol{r}_{1,n+1} = \boldsymbol{r}_{1,n} + \epsilon\boldsymbol{v}_{1,n+1}$$

Then we do the same thing for the other two masses. That's it. It's time to generalize to N bodies. For body i, we do:

$$\begin{aligned}
\boldsymbol{F}_{i,n} &= \sum_{j=1, j\neq i}^{N}\frac{Gm_j}{||\boldsymbol{r}_{ij}||^3}\boldsymbol{r}_{ij}, \quad \boldsymbol{r}_{ij} = \boldsymbol{r}_{j,n} - \boldsymbol{r}_{i,n} \\
\boldsymbol{v}_{i,n+1} &= \boldsymbol{v}_{i,n} + \epsilon\boldsymbol{F}_{i,n} \\
\boldsymbol{r}_{i,n+1} &= \boldsymbol{r}_{i,n} + \epsilon\boldsymbol{v}_{i,n+1}
\end{aligned} \tag{10.6.15}$$

Let's have fun with this. From https://en.wikipedia.org/wiki/Three-body_problem#HistoryWikipedia page on three-body problems, I obtained the following initial conditions:

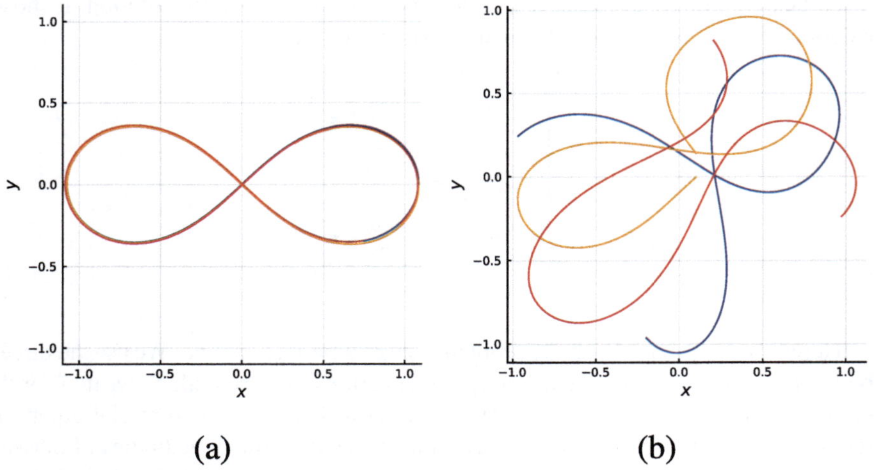

Fig. 10.15 Three-body problems solved with Euler-Aspel-Cromer method

$$\boldsymbol{r}_1(0) = -\boldsymbol{r}_3(0) = (-0.97000436, 0.24308753); \quad \boldsymbol{r}_2(0) = (0, 0)$$
$$\boldsymbol{v}_1(0) = \boldsymbol{v}_3(0) = (0.4662036850, 0.4323657300); \boldsymbol{v}_2(0) = (-0.93240737, -0.86473146)$$

And with that[20] we get the beautiful figure-eight in Fig. 10.15a with equal masses (I used $m_1 = m_2 = m_3 = 1$ and $G = 1$). You can go to the mentioned Wikipedia page to see the animation. Now with mass m_2 slightly changed to $\boldsymbol{r}_2(0) = (0.1, 0)$ instead, we get Fig. 10.15b. How about solution time? With a time step $\epsilon = 0.01$ and a total time of about 6 (whatever unit it is), that is, 600 iterations or steps, the code runtime is about 42 s including generation of animations on a 16 GB RAM Mac mini with Apple M1 chip.

10.6.6 Simple Pendulum: Large Amplitude and Chaos

Having a powerful numerical method to solve ODEs, we wow we will *have fun* by allowing large amplitude, air resistance $q\dot{\theta}$, and a forcing term $F_d \sin \omega_d t$ in a simple pendulum. For simplicity, let's consider frictional force proportional to velocity, denoted as $-q\dot{\theta}$ (negative because it acts as a resisting force to motion). And, let's choose a driving force with a fixed amplitude A and periodicity with angular frequency ω. Thus, the oscillation equation is written as

$$\boxed{ml\ddot{\theta} = -mg \sin \theta - q\dot{\theta} + A \cos \omega t} \tag{10.6.16}$$

---

[20] The program is presented in Sect. A.7.

This is a second-order nonlinear differential equation with the variable being the position of the mass blob m—that is, $\theta(t)$; second-order because it involves $\ddot{\theta}$, and nonlinear due to the presence of $\sin\theta$. In most cases, nobody can solve such equations exactly. Euler had to exclaim that *analysis has abandoned us!*

This led to the emergence of the famous approximation $\sin\theta \approx \theta$. However, in doing so, we unintentionally lost the opportunity to appreciate the beauty and richness that nonlinear equations bring. Enrico Fermi had to exclaim, "The Bible doesn't say that all the laws of nature must be expressed linearly!"

Equation 10.6.16 requires six parameters: m, l, g, q, A, and ω. When we change the unit system, these six parameters change values. Physicists have a clever way to avoid this nuisance. They normalize the equation! First, we normalize the time variable t by replacing it with $\hat{t}$ such that $\hat{t} = {}^{t}/_{t_c}$. Clearly, from the definition of $\hat{t}$, it is a dimensionless quantity (i.e., without units). Next, we need to replace $\dot{\theta}$ with ${}^{d\theta}/_{d\hat{t}}$ since we are no longer dealing with t:

$$\frac{d\theta}{dt} = \frac{d\theta}{d\hat{t}}\frac{d\hat{t}}{dt} = \frac{1}{t_c}\frac{d\theta}{d\hat{t}}$$

Similarly, the second derivative of θ with respect to t becomes

$$\frac{d^2\theta}{dt^2} = \frac{d}{dt}\left(\frac{d\theta}{dt}\right) = \frac{d}{dt}\left(\frac{1}{t_c}\frac{d\theta}{d\hat{t}}\right) = \frac{1}{t_c^2}\frac{d^2\theta}{d\hat{t}^2}$$

Substituting this into Eq. (10.6.16), we obtain

$$\frac{ml}{t_c^2}\frac{d^2\theta}{d\hat{t}^2} = -mg\sin\theta - \frac{q}{t_c}\frac{d\theta}{d\hat{t}} + A\cos(\omega t_c \hat{t}) \tag{10.6.17}$$

All terms in the equation have dimensions of force. Therefore, we divide the equation by mg—a force:

$$\frac{l}{gt_c^2}\frac{d^2\theta}{d\hat{t}^2} = -\sin\theta - \frac{q}{mgt_c}\frac{d\theta}{d\hat{t}} + \frac{A}{mg}\cos(\omega t_c \hat{t})$$

Now we have to choose t_c; note that it must have dimensionless time. We choose $t_c = 1/\omega_0$. Why? Because then $\omega t_c \hat{t}$ will be dimensionless. Therefore, $t_c^2 = l/g$. The equation becomes

$$\frac{d^2\theta}{d\hat{t}^2} = -\sin\theta - \frac{q\omega_0}{mg}\frac{d\theta}{d\hat{t}} + \frac{A}{mg}\cos\left(\frac{\omega}{\omega_0}\hat{t}\right)$$

We introduce the following dimensionless quantities:

$$Q = \frac{mg}{q\omega_0}, \quad \hat{A} = \frac{A}{mg}, \quad \hat{\omega} = \frac{\omega}{\omega_0} \tag{10.6.18}$$

And the final oscillation equation takes the form:

$$\boxed{\frac{d^2\theta}{d\hat{t}^2} = -\sin\theta - \frac{1}{Q}\frac{d\theta}{d\hat{t}} + \hat{A}\cos(\hat{\omega}\hat{t})} \tag{10.6.19}$$

Now we only need three parameters for the problem: Q, $\hat{A}$, and $\hat{\omega}$.

However, no one can solve, accurately, Eq. 10.6.19, a nonlinear differential equation. No one! Just like no one can accurately solve equations of degree 5 or higher. Nevertheless, we can easily solve it using numerical methods. First, we transform a second-order differential equation into two first-order differential equations, introducing the angular velocity $\omega = \dot{\theta}$:

$$\begin{aligned} \dot{\omega} &= -\sin\theta - \frac{1}{Q}\dot{\theta} + \hat{A}\cos(\hat{\omega}t) \\ \dot{\theta} &= \omega \end{aligned} \tag{10.6.20}$$

Then, we use the Euler-Cromer method, as presented in Sect. 10.6.3, to solve this system of equations. The algorithm is simple, outlined by the following equations:

$$\begin{aligned} F_n &= -\sin\theta_n - \frac{1}{Q}\omega_n + \hat{A}\cos(\hat{\omega}t) \\ \omega_{n+1} &= \omega_n + \epsilon F_n \\ \theta_{n+1} &= \theta_n + \epsilon\omega_{n+1} \end{aligned} \tag{10.6.21}$$

The programming part is similar to how we solve the Kepler problem, as shown in Sect. 10.6.4, so I won't introduce the code here. In Fig. 10.16a, the results of the program run with inputs $A = 0$, $Q = 2000$, $\hat{\omega} = 1$ (no driving force, and negligible damping force treated as 0) and initial conditions $\theta_0 = 0.1, \omega_0 = 0$ are depicted. The simulation runs for $T = 2$ time units, with a time step of $\epsilon = T/1000$. We observe harmonic oscillations with a circular trajectory. In Fig. 10.16a, we have the case of gradually damping oscillations with $Q = 20$ and $T = 10$ (10 oscillation cycles). The trajectory forms a spiral, converging to the point with coordinates (0, 0), indicating a stationary position of the pendulum.

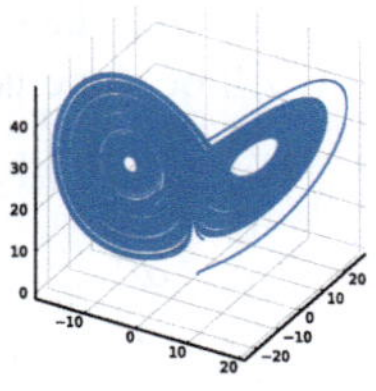

The results presented in Fig. 10.16 are something that all high school students know and anyone who has seen a swing (not being pushed) knows. The surprising part is when we consider the impact of an external force. For instance, with $A = 1.15$, $Q = 2$, $\hat{\omega} = 2/3$, the oscillation of the pendulum will be as shown in Fig. 10.17.

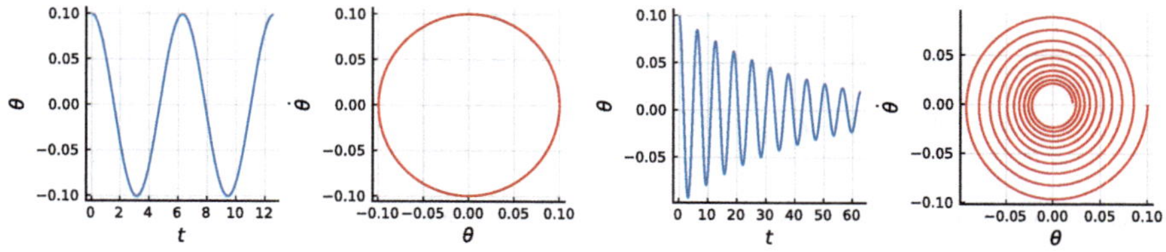

(a) $A = 0, Q = 2000, \hat{\omega} = 1$ (b) $A = 0, Q = 20, \hat{\omega} = 1$

Fig. 10.16 Single pendulum without external force: **a** perpetual harmonic oscillation and **b** gradually damping oscillation. In each figure, the left side shows the $\theta - t$ graph, and the right side shows the $\theta - \dot{\theta}$ graph, i.e., the phase diagram

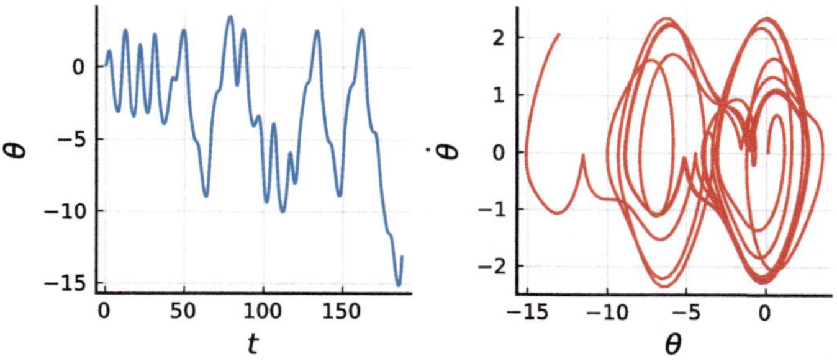

Fig. 10.17 Single pendulum with $A = 1.15$, $Q = 2$, $\hat{\omega} = 2/3$

This is called a chaotic motion, a phenomenon discovered by Edward Norton Lorenz (1917–2008). Lorenz was an American mathematician and meteorologist, described as "a genius with the soul of an artist" by his close friend and collaborator, Jule Charney. In 1961, while working at MIT, Lorenz used a simple computer, a Royal McBee LGP-30, to simulate weather models (for weather prediction). Everything was going smoothly until one fine day, he observed something similar to what we are seeing with a simple pendulum (see figure, which probably gave us the word "butterfly effect"). The story of chaos was told by James Gleick in his bestselling book *Chaos: Making a New Science*.

10.6.7 Verlet's Method

As Euler's method is only a first-order method, its accuracy is low (see Sect. 10.6.8 to understand what it means). Now I present a second-order method, the Verlet method.[21] The Verlet method is a popular method to integrate Newton's equations of motion $\ddot{x} = f(x, t)$. We begin with Taylor expansions:

[21] It is named after Loup Verlet (1931–2019), a French physicist who pioneered the computer simulation of molecular dynamics models. In a famous 1967 paper, he used what is now known as Verlet integration (a method for the numerical integration of equations of motion) and the Verlet

$$\begin{aligned} x(t+\epsilon) &= x(t) + \dot{x}(t)\epsilon + \frac{\ddot{x}(t)}{2}\epsilon^2 + \frac{\dddot{x}(t)}{3!}\epsilon^3 \\ x(t-\epsilon) &= x(t) - \dot{x}(t)\epsilon + \frac{\ddot{x}(t)}{2}\epsilon^2 - \frac{\dddot{x}(t)}{3!}\epsilon^3 \end{aligned} \tag{10.6.22}$$

Adding and subtracting these two equations, we obtain

$$\begin{aligned} x(t+\epsilon) + x(t-\epsilon) &= 2x(t) + \ddot{x}(t)\epsilon^2 \\ x(t+\epsilon) - x(t-\epsilon) &= 2\dot{x}(t)\epsilon \end{aligned} \tag{10.6.23}$$

And from that, we obtain the Verlet method[22]

$$\begin{aligned} x(t+\epsilon) &= 2x(t) - x(t-\epsilon) + \ddot{x}(t)\epsilon^2 \\ \dot{x}(t) &= \frac{x(t+\epsilon) - x(t-\epsilon)}{2\epsilon} \end{aligned} \tag{10.6.24}$$

We can see that the position update requires positions at previous two time steps (i.e., $x(t-\epsilon)$ and $x(t)$). Thus the Verlet method is a *two-step method* and furthermore it is not self-starting. At $t = 0$, we need $x(-\epsilon)$. The velocities are not required in the position update, but often they are necessary for the calculation of certain physical quantities like the kinetic energy. That is where the second equation comes in. Due to the blue term in the second equation we will have problem with round of errors. What is more, we have to store the position at three steps $x(t-\epsilon)$, $x(t)$, and $x(t+\epsilon)$.

A mathematically equivalent algorithm known as Velocity Verlet was developed to solve these issues. The Velocity Verlet method is[23]

$$\begin{aligned} x(t+\epsilon) &= x(t) + \dot{x}(t)\epsilon + \frac{1}{2}\ddot{x}(t)\epsilon^2 \\ \dot{x}(t+\epsilon) &= \dot{x}(t) + \left(\frac{\ddot{x}(t) + \ddot{x}(t+\epsilon)}{2}\right)\epsilon \end{aligned} \tag{10.6.25}$$

The first equation is obtained by eliminating $x(t-\epsilon)$ in Eq. (10.6.24): substituting that term obtained from the second into the first. The derivation of the velocity update is as follows:

list (a data structure that keeps track of each molecule's immediate neighbors in order to speed computer calculations of molecule-to-molecule interactions).

22 The algorithm was first used in 1791 by Delambre and has been rediscovered many times since then. It was also used by Cowell and Crommelin in 1909 to compute the orbit of Halley's Comet, and by Carl Størmer in 1907 to study the trajectories of electrical particles in a magnetic field (hence it is also called Störmer's method).

23 Note that as the velocity update requires the acceleration at $t + \epsilon$, the Verlet method cannot be used for problems in which the force depends on the velocity. For example, it cannot be used to solve damped harmonic oscillation problems.

$$\dot{x}(t+\epsilon) = \frac{x(t+2\epsilon) - x(t)}{2\epsilon}$$
$$x(t+2\epsilon) = x(t+\epsilon) + \dot{x}(t+\epsilon)\epsilon + \frac{1}{2}\ddot{x}(t+\epsilon)\epsilon^2$$
$$x(t+\epsilon) = x(t) + \dot{x}(t)\epsilon + \frac{1}{2}\ddot{x}(t)\epsilon^2$$

10.6.8 Analysis of Euler's Method

Let's forget time and motion, all first-order ODEs are of the following general form:

$$y'(x) = f(x, y), \quad a \le x \le b, \quad y(a) = y_0 \tag{10.6.26}$$

where the independent variable x lies within a and b. And our goal is to study the accuracy of Euler's method. Let's start with one example that we know the exact solution (from that we can calculate the error in Euler's method):

$$y' = xy, \quad y(0) = 0.1 \Longrightarrow y(x) = 0.1e^{\frac{x^2}{2}} \tag{10.6.27}$$

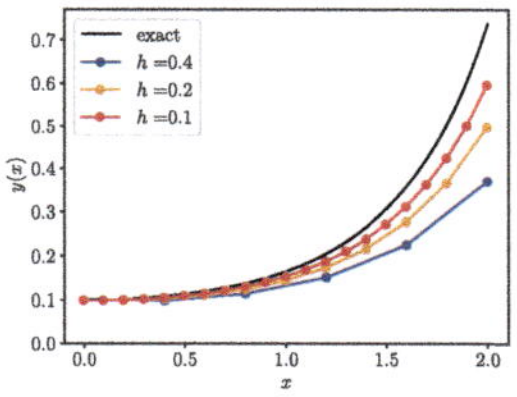

We solve this problem using Euler's method with different step sizes $h = 0.4, 0.2, 0.1$. And we plot these numerical solutions with the exact solution in one plot. In this way, we can understand the behavior of the method. From the results shown in the figure, we observe that the numerical solutions get closer to the exact one when the step size h is getting smaller. The second observation is that the error is getting bigger as x increases. Our problem is now to quantify the error and show that it is getting smaller and smaller when we reduce h.

Now, using Taylor's theorem we can write $y(x+h)$ as (noting that $y' = f$)

$$y(x+h) = y(x) + y'(x)h + \frac{y''(\xi)}{2}h^2 = y(x) + f(x, y)h + \frac{y''(\xi)}{2}h^2, \ \xi \in [x, x+h] \tag{10.6.28}$$

Up to now, we have been working with the exact solution $y(x)$. Now comes Euler, with the approximate solution. To differentiate the exact and approximate solution, the latter is denoted by $\tilde{y}(x)$. At $x + h$, Euler's approximate solution is

$$\tilde{y}(x+h) = \tilde{y}(x) + f(x, y)h \tag{10.6.29}$$

Putting the exact solutions and Euler's solution together, we get

$$\begin{aligned} y_{n+1} &= y_n + f(x_n, y_n)h + \frac{y''(\xi)}{2}h^2 \\ \tilde{y}_{n+1} &= \tilde{y}_n + f(x_n, \tilde{y}_n)h \end{aligned} \tag{10.6.30}$$

With that we can calculate the error, which is the difference between the exact solution and the numerical solution, that is, $E_{n+1} := y_{n+1} - \tilde{y}_{n+1}$. Subtracting the first from the second in Eq. (10.6.30), we get E_{n+1} as

$$E_{n+1} = E_n + \left[f(x_n, y_n) - f(x_n, \tilde{y}_n)\right]h + \frac{1}{2}y''(\xi)h^2 \tag{10.6.31}$$

The error consists of two parts (assume that rounding error is zero): the first part is the local truncation error—occurs when we neglected the red term—this error is $\mathcal{O}(h^2)$—and the second part is related to the blue term.

Now that we have an expression for the error, we need to find an upper bound for it, i.e., $|E_n| \le \square$. Note that for the error we're interested in its magnitude only, thus we need $|E_{n+1}|$. And the triangle inequality Eq. (2.20.12) enables us to write

$$|E_{n+1}| \le |E_n| + |f(x_n, y_n) - f(x_n, \tilde{y}_n)|h + \frac{1}{2}|y''(\xi)|h^2 \tag{10.6.32}$$

To proceed, we need to introduce some assumptions. The first one is

$$|f(x, y_1) - f(x, y_2)| \le L|y_1 - y_2| \tag{10.6.33}$$

$$\beta = \max \frac{1}{2}y''(x) \quad \text{for } x \in [a, b] \tag{10.6.34}$$

With these conditions, Eq. (10.6.32) is simplified to

$$|E_{n+1}| \le \alpha|E_n| + \beta h^2, \quad \alpha := 1 + hL \tag{10.6.35}$$

What we do with this equation? Start with E_0, which is assumed to be zero, we compute E_1, then E_2 and so on:

$$\begin{aligned} n = 0 &: |E_1| \le \alpha|E_0| + \beta h^2 = \beta h^2, \quad (|E_0| = 0) \\ n = 1 &: |E_2| \le \alpha|E_1| + \beta h^2 \le (1 + \alpha)\beta h^2 \\ n = 2 &: |E_3| \le \alpha|E_2| + \beta h^2 \le (1 + \alpha + \alpha^2)\beta h^2 \end{aligned}$$

We can see a pattern here, and that pattern gives us $|E_{n+1}|$ (recall that $\alpha = 1 + hL$):

$$|E_{n+1}| \le \frac{\alpha^n - 1}{\alpha - 1}\beta h^2 = \frac{(1 + hL)^n - 1}{L}\beta h \tag{10.6.36}$$

This equation gives a bound for $|E_n|$ in terms of h, L, β, and n. Note that for a fixed h, this error bound increases with increasing n. This is in agreement with the example of $y' = xy$ that we considered at the beginning of the section.

With this inequality $(1 + hL)^n \leq e^{nhL}$ and $nh \leq b - a$, we then have

$$|E_{n+1}| \leq \frac{e^{nhL} - 1}{L}\beta h \leq \frac{e^{(b-a)L} - 1}{L}\beta h := Kh \tag{10.6.37}$$

We have just showed that the error at time step n is proportional to h with the proportionality constant K depending on L, β and the time interval $b - a$. With this result, we're now able to talk about *the error of Euler's method*: it is defined as the maximum of $|E_n|$ over all the time steps:

$$E := \max_n |E_n| \leq Kh \Longrightarrow E = \mathcal{O}(h) \Longrightarrow \lim_{h \to 0} E = 0 \tag{10.6.38}$$

10.7 Numerical Solution of Partial Differential Equations

As discussed in Chap. 7, engineers and scientists and mathematicians resort to partial differential equations when they need to describe a complex phenomenon. The problem is that partial differential equations—as essential and ubiquitous as they are in science and engineering—are notoriously difficult to solve, if they can be solved at all.

Numerical method for partial differential equations is the branch of numerical analysis that studies the numerical solution of partial differential equations (PDEs). Common methods are finite difference method (FDM), finite volume method (FVM), finite element method (FEM), spectral methods, meshfree methods, etc. The field is simply huge and I do not have time to learn all of them. The finite difference method is often regarded as the simplest method to learn and use. This section is a brief introduction to the FDM.

10.7.1 Finite Difference for the 1D Heat Equation: Explicit Schemes

We now solve the 1D heat equation $\theta_t = \kappa^2\theta_{xx}$ for $\theta(x, t)$, with $0 \leq x \leq L$ and $0 \leq t \leq T$, using finite differences. The idea is simple as we simply follow Euler in approximating the derivatives by some finite difference formula. Thus, we construct a grid (or lattice or mesh or whatever you want to call it) of points in the 2D xt plane. A point on this grid is labeled by (i, n), which means that the spatial coordinate is $i\Delta x$ and the temporal coordinate is $n\Delta t$, where Δx is the grid spatial size and Δt is the time step. Such a grid is given in Fig. 10.18 (left). Note that this is a *uniform grid* in

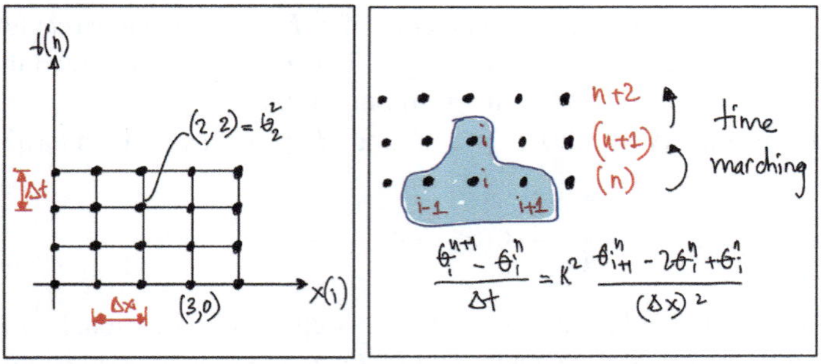

Fig. 10.18 A 2D (uniform) finite difference grid: the space $[0, L]$ is discretized by N points

which Δx and Δt are constant. (But nothing can prevent us from using non-uniform grids.) With such a grid, the temperature is only available at the points; for example, at point (i, n), the temperature is θ_i^n: the subscript is for space and the superscript is for time.

If, by any way, we can transform the PDE $\theta_t = \kappa^2 \theta_{xx}$ into a system of algebraic equations containing all θ_i^{n+1} given that the temperature at all grid point is known at the previous time step n, then we're done. This is so because we can start with $\theta_i^0, i = 0, 1, 2, \ldots$, compute θ_i^1, then θ_i^2, marching in time just as we do with Euler's method to solve ODEs.

To start simple, we use the forward difference for the time partial derivative θ_t evaluated at grid point (i, n):

$$\left(\frac{\partial \theta}{\partial t}\right)_i^n = \frac{\theta_i^{n+1} - \theta_i^n}{\Delta t} + \mathcal{O}(\Delta t) \tag{10.7.1}$$

and a central difference for the spatial second-order derivative θ_{xx} evaluated at grid point (i, n):

$$\left(\frac{\partial^2 \theta}{\partial x^2}\right)_i^n = \frac{\theta_{i+1}^n - 2\theta_i^n + \theta_{i-1}^n}{(\Delta x)^2} + \mathcal{O}((\Delta x)^2) \tag{10.7.2}$$

Substituting Eqs. (10.7.1) and (10.7.2) into the heat equation (after removing the high-order terms of course), we get the following equation:

$$\frac{\theta_i^{n+1} - \theta_i^n}{\Delta t} = \kappa^2 \frac{\theta_{i+1}^n - 2\theta_i^n + \theta_{i-1}^n}{(\Delta x)^2}, \quad i = 1, \ldots, N-2 \tag{10.7.3}$$

This is called a *finite difference equation* for the heat equation. Note that, there are $N-2$ such equations for N unknowns θ_i^{n+1} for $i = 0, 1, \ldots, N-1$ as the temperature is known at time n. But do not worry we have two equations coming from the boundary conditions. Another note, Eq. (10.7.3) is just one specific type of finite difference equation for the heat equation. We can develop other FD equations, e.g., if we use the backward difference for θ_t instead of the forward difference in Eq.

(10.7.1). But, there is something nice about Eq. (10.7.3). In that equation, for any i, there is only one unknown θ_i^{n+1}. So, we can solve for it easily:

$$\theta_i^{n+1} = \theta_i^n + \kappa^2 \frac{\Delta t}{(\Delta x)^2} \left[\theta_{i+1}^n - 2\theta_i^n + \theta_{i-1}^n\right], \quad i = 1, \ldots, N-2 \qquad (10.7.4)$$

This equation is called a *computational molecule or stencil* and plotted in Fig. 10.18 (right). And this finite difference method is known as the Forward Time Centered Space or FTCS method. What is more, it is an *explicit* method. It is so called because to determine θ_i^{n+1}, we do not have to solve any system of equations. Equation (10.7.4) provides an explicit formula to quickly compute θ_i^{n+1}. There are explicit methods, just because there are implicit ones. And the next section presents one implicit method.

10.7.2 Finite Difference for the 1D Heat Equation: Implicit Schemes

This section presents an implicit FDM using the backward Euler for θ_t. Let's use a simple ODE to demonstrate the difference between explicit and implicit methods: solving $\dot{x} = \sin x$ with $x(0) = x_0$. Using the backward difference formula (Sect. 10.2.1), we can write

$$\dot{x} = \frac{x_n - x_{n-1}}{\Delta t} \implies \boxed{\frac{x_n - x_{n-1}}{\Delta t} = \sin x_n}$$

Obviously to solve for x_n with x_{n-1} known we have to solve the boxed equation, which is a nonlinear equation. This is an implicit method which involves the solution of a nonlinear equation. On the contrary, an explicit method does not need to solve any equation; see Eq. (10.7.4) for example. So, you might be thinking we should not then use implicit methods. But that's not the whole story, otherwise the backward Euler's method would not have been developed.

Getting back to the heat equation, now we write θ_t as

$$\left(\frac{\partial \theta}{\partial t}\right)_i^n = \frac{\theta_i^n - \theta_i^{n-1}}{\Delta t} + \mathcal{O}(\Delta t) \qquad (10.7.5)$$

Substituting Eq. (10.7.5) and (10.7.2) into the heat equation, we get the following equation:

$$\frac{\theta_i^n - \theta_i^{n-1}}{\Delta t} = \kappa^2 \frac{\theta_{i+1}^n - 2\theta_i^n + \theta_{i-1}^n}{(\Delta x)^2}, \quad i = 1, \ldots, N-2 \qquad (10.7.6)$$

And we have obtained the Backward Time Centered Space (BTCS) difference method for the heat equation. In the above equation only the red term is known, and thus we cannot solve it equation by equation. Instead we have to assemble all the equations into $\mathbf{A}\boldsymbol{x} = \boldsymbol{b}$ and solve this system of linear equations once for all θ_i^n. To get the matrix $\mathbf{A}$, we just need to rewrite Eq. (10.7.6) in which we separate the knowns (in the RHS of the equation) and the unknowns[24]:

$$-s\theta_{i-1}^n + (1+2s)\theta_i^n - s\theta_{i+1}^n = \theta_i^{n-1}, \quad i = 1, \ldots, N-2, \quad s := \frac{\kappa^2 \Delta t}{(\Delta x)^2} \tag{10.7.7}$$

Noting that each equation involves only three unknowns at points $i-1, i$, and $i+1$, thus, when we assemble all the equations from all the nodes, we get a tridiagonal matrix $\mathbf{A}$. For example, if we have six points (i.e., $N = 6$), we will have (the first and last row come from the boundary conditions $\theta_{0/5}^n = \theta_{0/5}^*$):

$$\begin{bmatrix} 1 & 0 & 0 & 0 & 0 & 0 \\ -s & 1+2s & -s & 0 & 0 & 0 \\ 0 & -s & 1+2s & -s & 0 & 0 \\ 0 & 0 & -s & 1+2s & -s & 0 \\ 0 & 0 & 0 & -s & 1+2s & -s \\ 0 & 0 & 0 & 0 & 0 & 1 \end{bmatrix} \begin{bmatrix} \theta_0^n \\ \theta_1^n \\ \theta_2^n \\ \theta_3^n \\ \theta_4^n \\ \theta_5^n \end{bmatrix} = \begin{bmatrix} \theta_0^* \\ \theta_1^{n-1} \\ \theta_2^{n-1} \\ \theta_3^{n-1} \\ \theta_4^{n-1} \\ \theta_5^* \end{bmatrix} \tag{10.7.8}$$

To see more clearly the pattern of the matrix, we need to have a bigger matrix. For example, with 100 points, we have the matrix shown in Fig. 10.19; the one on the left shows the entire matrix and the right figure shows only the first ten rows/cols.

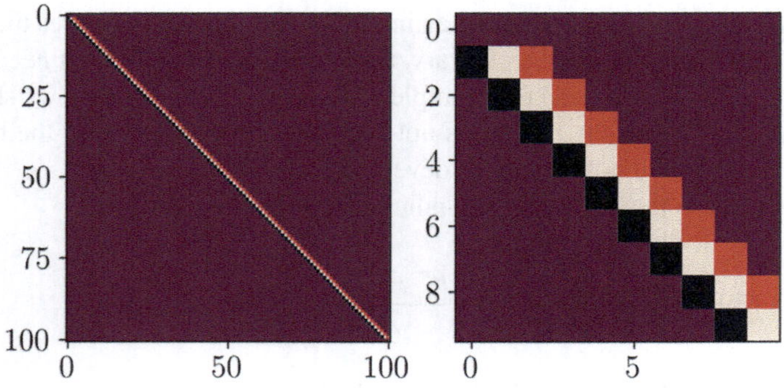

Fig. 10.19 A tridiagonal matrix resulting from the FDM for the heat equation: obtained using the function `imshow` in `matplotlib`. A tridiagonal matrix is a band matrix that has non-zero elements only on the main diagonal, the subdiagonal/lower diagonal (the first diagonal below this), and the supdiagonal/upper diagonal (the first diagonal above the main diagonal). Check the source file `heat_btcs.jl` for detail

[24] This finite difference equation appeared for the first time in 1924 in a paper of Erhard Schmidt.

Equation (10.7.8) is obviously of the form $\mathbf{A}\boldsymbol{x} = \boldsymbol{b}$ and without knowing it beforehand we are back to linear algebra! We need techniques from that field to have a fast method to solve this system. But we do not delve into that topic here. We just use a linear algebra library to do that so that we can focus on the PDE (and the physics we're interested in).

It is obvious that the BTCS finite difference method is an *implicit method* as we have to solve a system of (linear) equations to determine the temperature at all the nodes at a given time. What are then the pros/cons of implicit methods compared with explicit methods? The next section gives an answer to that question.

10.7.3 Implicit Versus Explicit Methods: Stability Analysis

Briefly, explicit methods are easy to code but they are unstable, if not used properly. On the other hand, implicit methods are harder to code but are stable. To see what stability means, let's consider a simple ODE:

$$\dot{x} = -kx, \quad x(0) = x_0, \quad k > 0 \Longrightarrow x(t) = x_0 \exp(-kt)$$

The exact solution is a decaying exponential function: as $t \to \infty$ the solution approaches x_0. Now, we use the forward Euler method to have

$$\frac{x_{n+1} - x_n}{\Delta t} = -kx_n \Longrightarrow x_{n+1} = (1 - k\Delta t)x_n \Longrightarrow \boxed{x_n = (1 - k\Delta t)^n x_0}$$

which is an explicit method, nothing can be simpler. However, it is easy to see that the solution depends heavily on the value of Δt. In Fig. 10.20, the exact solution is plotted with two numerical solutions—one with Δt such that $|1 - k\Delta t| \le 1$ (blue curve) and one with $|1 - k\Delta t| > 1$ (red curve). The red curve is absolutely wrong; that solution grows to infinity and blows up our computer! That is *numerical instability*. As the method gives stable solution only for $\Delta t \le 2/k$, the method is said to be *conditionally stable*.

von Neumann stability analysis is a procedure used to check the stability of finite difference schemes as applied to linear partial differential equations. The analysis is based on the Fourier decomposition of numerical error and was developed at Los Alamos National Laboratory after having been briefly described in a 1947 article by British researchers Crank and Nicolson. Later, the method was given a more rigorous treatment in an article by John von Neumann.

Let's denote by A the exact solution to the heat equation (i.e., $\theta_t = \alpha\theta_{xx}$), by D the exact solution to the finite difference equation corresponding to the heat equation. For example, if we consider the FTBS method, then D is the exact solution to the following equation:

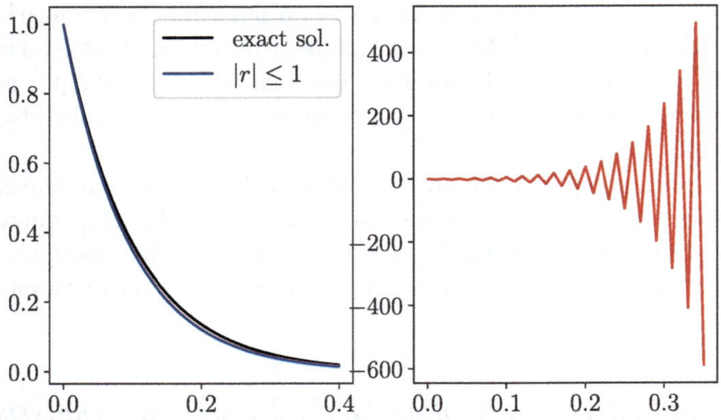

Fig. 10.20 Demonstration of numerical stability in solving ODEs using finite difference methods

$$\frac{\theta_i^{n+1} - \theta_i^n}{\Delta t} = \alpha \frac{\theta_{i+1}^n - 2\theta_i^n + \theta_{i-1}^n}{(\Delta x)^2}, \quad i = 1, \ldots, N-2 \tag{10.7.9}$$

This exact solution was obtained if our computer has no round-off errors (which is not reality). That's why we have another solution, N, which is the actual solution to Eq. (10.7.9) that we obtain from our computer. Now, we can define some errors:

$$\begin{aligned} \text{discretization error} &= A - D \\ \text{round-off error } \epsilon &= N - D \Longrightarrow N = \epsilon + D \end{aligned} \tag{10.7.10}$$

The stability of numerical schemes is closely associated with numerical error. A finite difference scheme is stable if the errors made at one time step of the calculation do not cause the errors to be magnified as the computations are continued. Thus the plan is now to study how ϵ behaves. We are going to show that the error is also a solution of Eq. (10.7.9). The proof is simply algebraic. Indeed, as N is the solution to Eq. (10.7.9), we have

$$\frac{\epsilon_i^{n+1} - \epsilon_i^n}{\Delta t} + \frac{D_i^{n+1} - D_i^n}{\Delta t} = \alpha \frac{\epsilon_{i+1}^n - 2\epsilon_i^n + \epsilon_{i-1}^n}{(\Delta x)^2} + \alpha \frac{D_{i+1}^n - 2D_i^n + D_{i-1}^n}{(\Delta x)^2}$$

where the red terms cancel each other leading to

$$\frac{\epsilon_i^{n+1} - \epsilon_i^n}{\Delta t} = \alpha \frac{\epsilon_{i+1}^n - 2\epsilon_i^n + \epsilon_{i-1}^n}{(\Delta x)^2} \tag{10.7.11}$$

Now comes the surprise: the error $\epsilon(x, t)$ is decomposed into a Fourier series in a complex form[25]:

[25] Check Sect. 4.19.3 if this is not clear.

$$\epsilon(x,t) = \sum_{n=-\infty}^{\infty} c_n e^{in2\pi x/L} = \sum_{n=-\infty}^{\infty} c_n e^{ik_n x}, \quad c_n = e^{at} \tag{10.7.12}$$

Instead of considering the whole series, we focus on just one term. That is $\epsilon(x,t) = e^{at}e^{ik_n x}$. With that and Eq. (10.7.11), we can obtain the following:

$$\frac{e^{a\Delta t} - 1}{\alpha \Delta t} = \frac{e^{ik_n \Delta x} - 2 + e^{-ik_n \Delta x}}{(\Delta x)^2} \tag{10.7.13}$$

and this allows us to determine the ratio of the error at two consecutive time steps $\epsilon_i^{n+1}/\epsilon_i^n$:

$$\begin{aligned}
\frac{\epsilon_i^{n+1}}{\epsilon_i^n} = e^{a\Delta t} &= 1 + \frac{\alpha \Delta t}{(\Delta x)^2}(e^{ik_n \Delta x} - 2 + e^{-ik_n \Delta x}) \text{ (10.7.13)} \\
&= 1 + \frac{\alpha \Delta t}{(\Delta x)^2}(2\cos k_n \Delta x - 2) \\
&= 1 - \frac{4\alpha \Delta t}{(\Delta x)^2}\sin^2 \frac{k_n \Delta x}{2}
\end{aligned} \tag{10.7.14}$$

The last two steps are purely algebraic. It is interesting that trigonometry identities play a role in the context of numerical solutions of the heat equation, isn't it?

We do not want the error to grow, so we're interested in when the following inequality holds $\left|\epsilon_i^{n+1}/\epsilon_i^n\right| \leq 1$. With Eq. (10.7.14), this condition becomes

$$\left|1 - \frac{4\alpha \Delta t}{(\Delta x)^2}\sin^2 \frac{k_n \Delta x}{2}\right| \leq 1 \Longrightarrow \frac{2\alpha \Delta t}{(\Delta x)^2}\sin^2 \frac{k_n \Delta x}{2} \leq 1 \Longrightarrow \boxed{\frac{\alpha \Delta t}{(\Delta x)^2} \leq \frac{1}{2}} \tag{10.7.15}$$

The boxed equation gives the stability requirement for the FTCS scheme as applied to one-dimensional heat equation. It says that for a given Δx, the allowed value of Δt must be small enough to satisfy the boxed equation.[26]

10.7.4 Analytical Solutions Versus Numerical Solutions

To test the implementation of various FD schemes for the heat equation and also to demonstrate the differences between analytical solutions and numerical solutions, let's solve a specific heat conduction problem:

$$\begin{aligned}
\frac{\partial \theta}{\partial t} = 0.1^2 \frac{\partial^2 \theta}{\partial x^2} \quad & 0 < x < 1, \ t > 0 \\
\theta(x, 0) = 1 \quad & 0 \leq x \leq 1 \\
\theta(0, t) = 0, \ \theta(1, t) = 0 \quad & t > 0
\end{aligned}$$

[26] One example to see how small the time step must be: $\alpha = 1$, $\Delta x = 0.1$, then $\Delta t \leq 0.05$.

Table 10.9 Numerical solutions are given in a tabular format: each row corresponds with a time step

	0.0	0.1	0.2	0.3	0.4	0.5	0.6	0.7	0.8	0.9	1.0
0.0	1.0	1.0	1.0	1.0	1.0	1.0	1.0	1.0	1.0	1.0	1.0
4.0	0.0	0.816	0.973	0.996	0.999	0.999	0.999	0.996	0.973	0.816	0.0
7.5	0.0	0.188	0.358	0.492	0.578	0.607	0.578	0.492	0.358	0.188	0.0

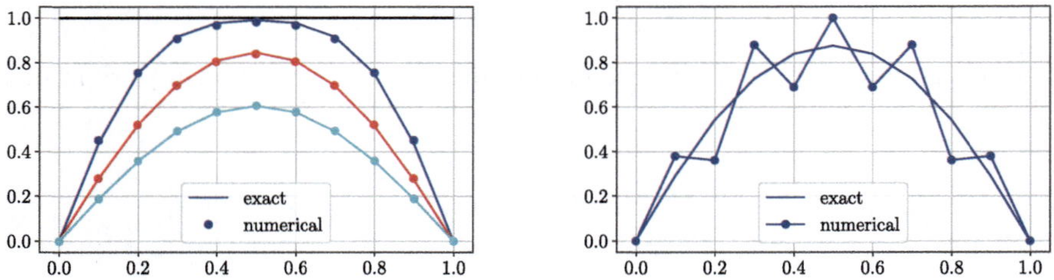

(a) BTCS (b) FTCS

Fig. 10.21 Analytical versus numerical solution of the heat equation. Ten terms are used in Eq. (10.7.16). For the FTCS scheme, a time step slightly larger than the upper limit in Eq. (10.7.15) was used. Thus, the solution shows instability. For later time steps, the numerical solution blew up

The exact solution in Eq. (7.9.14) for this specific problem is

$$\theta(x,t) = \sum_{n=1,3,\ldots}^{\infty} B_n e^{-(n\pi\kappa)^2 t} \sin(n\pi x), \quad B_n = \frac{4\pi}{n} \tag{10.7.16}$$

whereas a (part of) numerical solution is shown in Table 10.9. An analytical solution allows us to compute the solution at any point (in the domain). On the other hand, we only have the numerical solutions at some points (at the nodes). The analytical solution can tell us how the parameters (e.g., κ here) affect the solution. The numerical solutions are obtained only for a specific value of the parameters.

Now is the time for code verification. The results in Fig. 10.21 indicate that the implementation is correct and it also confirms the von Neumann stability analysis.

10.7.5 Finite Difference for the 1D Wave Equation

We now move on to solving the 1D wave equation $u_{tt} = c^2 u_{xx}$ for $u(x,t)$ also using finite differences. One simple explicit method is to adopt a central finite difference for both u_{tt} and u_{xx}. Now using Eq. (10.2.5), we can compute u_{tt} and u_{xx} as

$$\left(\frac{\partial^2 u}{\partial t^2}\right)_i^n = \frac{u_i^{n+1} - 2u_i^n + u_i^{n-1}}{(\Delta t)^2} + \mathcal{O}((\Delta x)^2)$$
$$\left(\frac{\partial^2 u}{\partial x^2}\right)_i^n = \frac{u_{i+1}^n - 2u_i^n + u_{i-1}^n}{(\Delta x)^2} + \mathcal{O}((\Delta x)^2) \tag{10.7.17}$$

Substituting these into the wave equation we obtain with $r := \frac{\Delta t}{\Delta x}c$

$$u_i^{n+1} = 2(1-r^2)u_i^n - u_i^{n-1} + r^2\left[u_{i+1}^n + u_{i-1}^n\right] \tag{10.7.18}$$

which allows us to determine u_i^{n+1} explicitly, i.e., without solving any system of equations. This is referred to as the Centered in Time and Space (CTCS) FD scheme. Now that we know of the von Neumann stability analysis, we can carry out such an analysis to see the condition on the time step Δt:

$$\frac{\epsilon_i^{n+1}}{\epsilon_i^n} = g, \quad \boxed{g^2 - 2\beta g + 1 = 0}, \quad \beta = 1 - 2r^2 \sin^2\frac{k_n \Delta x}{2}$$

Solving the boxed equation we obtain g as

$$g_{1,2} = \beta \pm \sqrt{\beta^2 - 1}$$

Note that g can be a complex number and we need $|g| \le 1$ so that our method is stable. And this requires that $|\beta| \le 1$. In this case, we can write g as

$$g_{1,2} = \beta \pm i\sqrt{1-\beta^2} \Longrightarrow |g| = 1$$

So, we conclude that the method is stable as long as $|\beta| \le 1$, or (Fig. 10.22)

$$\left|1 - 2r^2\sin^2\frac{k_n\Delta x}{2}\right| \le 1 \Longrightarrow r := \boxed{\frac{\Delta t}{\Delta x}c \le 1}. \tag{10.7.19}$$

10.7.6 Finite Difference for the 2D Heat Equation: Explicit Method

Here, we will use the FDM to solve the 2D heat equation given by

$$\frac{\partial \theta}{\partial t} = \kappa^2\left(\frac{\partial^2\theta}{\partial x^2} + \frac{\partial^2\theta}{\partial y^2}\right), \quad \theta = \theta(x, y, t) \tag{10.7.20}$$

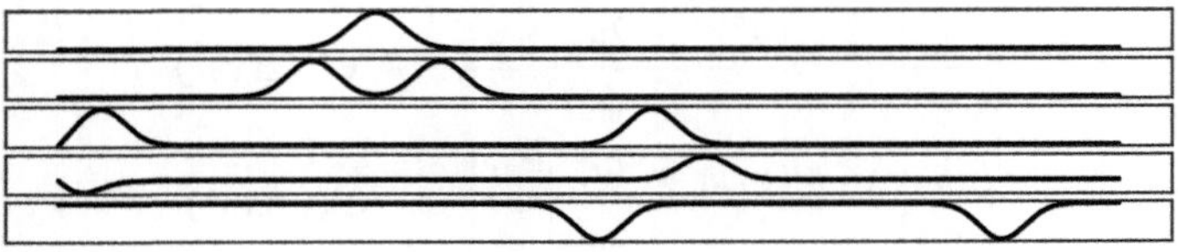

Fig. 10.22 Waves propagating on a string with fixed ends. The data are: $c = 300\,\text{m/s}$, $L = 1\,\text{m}$, $\Delta x = 0.01\,\text{m}$, $\Delta t = \Delta x/c$. The initial string shape is given at the top, which is a Gaussian pluck $u(x, 0) = \exp(-k(x - x_0)^2)$ with $x_0 = 0.3\,\text{m}$ and $k = 1000\,1/\text{m}^2$. The wave is split into two wavepackets (pulses) which travel in opposite directions (second and third figures). This is consistent with d' Alembert solution in Eq. (7.10.6). The left pulse reaches the left end and reflected, this reflection inverts the pulse so that its displacement is now negative (fourth figure). Meanwhile, the right pulse keeps going to the right, reaches the fixed end, reflected, and inverted

defined over the domain $D = [0, L_x] \times [0, L_y]$ (i.e., a rectangular region of size $L_x \times L_y$).

We discretize space and time as usual. The domain (space) of the problem—the rectangle D—will be replaced by a 2D grid similar to the left side of Fig. 10.18 (instead of the t-axis, now it's the y-axis); the grid has sizes Δx and Δy. Time is also discretized by Δt; imagine stacking the spatial grids on top of each other, separated by Δt. Thus, a point in D and at a certain time t will have coordinates (i, j, n), meaning $(i\Delta x, j\Delta y, n\Delta t)$ with $i, j, n = 0, 1, 2, \ldots$

We still use forward difference for the time derivative (θ_t) evaluated at the grid point (i, j, n):

$$\left(\frac{\partial \theta}{\partial t}\right)_{i,j}^n = \frac{\theta_{i,j}^{n+1} - \theta_{i,j}^n}{\Delta t} \tag{10.7.21}$$

and central difference for the second spatial derivative evaluated at the grid point (i, j, n):

$$\begin{aligned} \left(\frac{\partial^2 \theta}{\partial x^2}\right)_{i,j}^n &= \frac{\theta_{i+1,j}^n - 2\theta_{i,j}^n + \theta_{i-1,j}^n}{(\Delta x)^2} \\ \left(\frac{\partial^2 \theta}{\partial y^2}\right)_{i,j}^n &= \frac{\theta_{i,j+1}^n - 2\theta_{i,j}^n + \theta_{i,j-1}^n}{(\Delta y)^2} \end{aligned} \tag{10.7.22}$$

Substituting Eqs. (10.7.21) and (10.7.22) into the heat equation Eq. (10.7.20), we have the following equation (for simplicity, let's assume $\Delta x = \Delta y$)

$$\begin{aligned} \theta_{i,j}^{n+1} &= \theta_{i,j}^n + s\left(\theta_{i+1,j}^n + \theta_{i-1,j}^n + \theta_{i,j+1}^n + \theta_{i,j-1}^n - 4\theta_{i,j}^n\right), \quad (i, j = 1, \ldots, N-2), \\ s &:= \frac{\kappa^2 \Delta t}{(\Delta x)^2} \end{aligned}$$

Looking at this formula, we can see three loops: the outermost loop for time ($n = 0, 1, \ldots$), and the next two loops for grid points (i, j). The right-hand side of the formula is the known value, and we update the quantity on the left.

10.7.7 Turing Model

Sometimes it is the people no one can imagine anything of who do the things no one can imagine. *Alan Turing*

On 25 May 2011, President Barack Obama, while addressing the UK Parliament, mentioned only three names when referring to contributors to British science. Just three. They were Isaac Newton, Charles Darwin, and Alan Turing (1912–1954). Why did Obama choose Turing? First, Turing is considered the father of computer science. Second, Turing played a crucial role in deciphering German messages during World War II, helping to shorten this brutal war by about 2 years and saving around 14 million lives.

Yet, I only learned about Turing in 2014 when the movie *The Imitation Game* was released.[27] Knowing and forgetting, as it was the time for survival. Now, with more leisure, I delve into Turing, and that's the reason for this section. This section presents a mathematical model developed by Turing. It is related to mathematics, chemistry, biology, and programming; hence, it is an intriguing topic.

The Turing model is a concept introduced by Turing in his last paper in 1952, just 2 years before his sudden death. It is titled *The Chemical Basis of Morphogenesis*. The paper describes how patterns in nature, such as stripes and spots (Fig. 10.23), can naturally and independently emerge from a homogeneous, uniform state. Let's

Fig. 10.23 .

[27] This is a historical drama biopic based on the 1983 biography *Alan Turing: The Enigma* by Andrew Hodges, a British mathematician, and writer. Incidentally, here are some good movies about mathematics and science: *Oppenheimer*, *The Theory of Everything*, *Good Will Hunting*.

explore how the world's most famous codebreaker decoded the secrets of nature's beauty.

The final equation of the Turing model is quite simple, and, in fact, you have encountered a part of it. Where? In the heat equation! The heat equation, $\frac{\partial \theta}{\partial t} = \kappa^2 \frac{\partial^2 \theta}{\partial x^2}$, developed by Fourier, has a more accurate name: the diffusion equation. In reality, the equation we are considering does not have a general form; the general form of the heat/diffusion equation is given by

$$\frac{\partial \theta}{\partial t} = \kappa^2 \frac{\partial^2 \theta}{\partial x^2} + f(\theta) \tag{10.7.23}$$

where the term $f(\theta)$ is a function of θ describing a heat source (if it is a heat problem). Turing's equation is exactly like this, with the twist that now in the game, there are at least two characters. Limiting to the two-dimensional problem, these two characters are $u(x, y, t)$ and $v(x, y, t)$. The Turing equation is called the reaction-diffusion equation and has the following form:

$$\begin{aligned} \frac{\partial u}{\partial t} &= D_u \left(\frac{\partial^2 u}{\partial x^2} + \frac{\partial^2 u}{\partial y^2} \right) + R_u(u, v) = D_u \Delta u + R_u(u, v) \\ \frac{\partial v}{\partial t} &= D_v \left(\frac{\partial^2 v}{\partial x^2} + \frac{\partial^2 v}{\partial y^2} \right) + R_v(u, v) = D_v \Delta v + R_v(u, v) \end{aligned} \tag{10.7.24}$$

The variables $u(x, y, t)$ represent the concentration of a substance favoring skin color formation, while $v(x, y, t)$ represents another substance reacting with the first and hindering skin color formation. The diffusion coefficients of u and v are D_u and D_v, respectively. The functions R_u and R_v determine how the concentration changes due to chemical reactions between substances.

Many models originate from Eq. (10.7.24); each model differs only in the form of R_u and R_v. Here are two popular equations:

$$\begin{aligned} &\text{Gray-Scott:} && \frac{\partial u}{\partial t} = D_u \Delta u - uv^2 + f(1-u), ;; \frac{\partial v}{\partial t} = D_v \Delta v + uv^2 + (f+k)v \\ &\text{Fitzhugh-Nagumo:} && \frac{\partial u}{\partial t} = D_u \Delta u + u - u^3 - v + \alpha, ;; \frac{\partial v}{\partial t} = D_v \Delta v + \beta(u - v) \end{aligned} \tag{10.7.25}$$

In the Gray-Scott equation, f is the supply rate of u, and k is the removal/kill rate of v. The two substances interact with each other through uv^2 (specifically from a chemical reaction).

We will use the finite difference method explained in Sect. 10.7.6 to solve the system of equations given in Eq. (10.7.25) on the domain $[0, L] \times [0, L]$. This domain will be discretized into an $N \times N$ grid, where each grid point will have values u and v. Thus, we have a three-dimensional array u with sizes $(N, N, 2)$ where $u[:, :, 1]$ corresponds to $u(x, y)$, and $u[:, :, 2]$ corresponds to $v(x, y, t)$. We apply periodic boundary conditions for u and v:

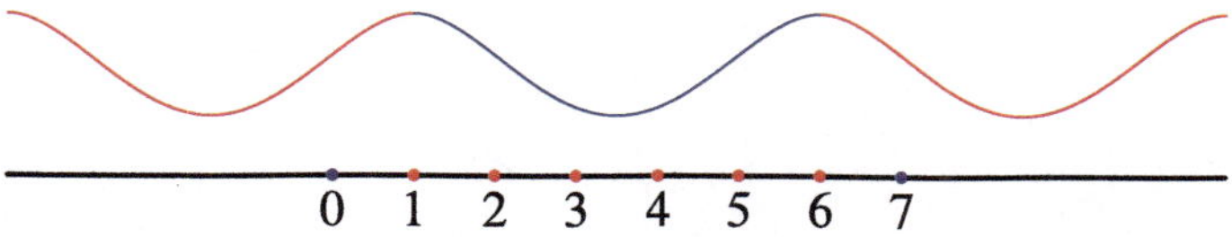

Fig. 10.24 Periodic boundary conditions: illustrated for a one-dimensional problem. This corresponds to row 5 in Listing 10.4

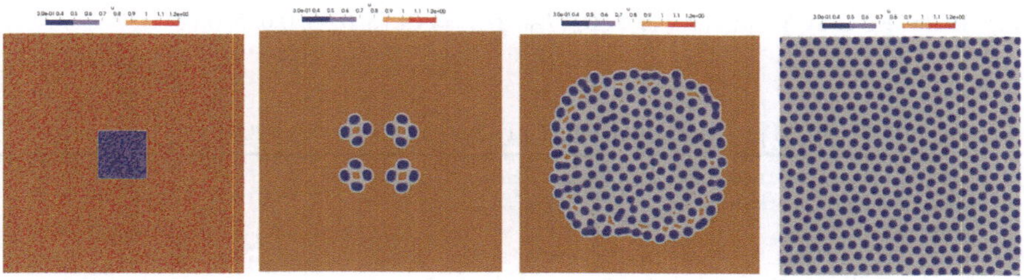

Fig. 10.25 PT Gray-Scott: $u(x, y, t)$: $(D_u, D_v, f, k) = (0.16, 0.08, 0.035, 0.065)$. The left most figure is for $t = 0$ and the right most figure is for $t = 20000$

$$u(x + L, y, t) = u(x, y, t) \; u(x, y + L, t) \quad = u(x, y, t) \tag{10.7.26}$$

(only written for u). Periodic boundary conditions are a way to reduce computational costs: instead of solving the problem over a large domain, we consider this domain as a floor tiled with identical tiles. Now we only need to consider one tile, with periodic boundary conditions. When calculating Δu for $i = 1$, we will encounter $u(0, j)$, and Fig. 10.24 shows the value for this: $u(5, j)$.

Although we can improve the 2D heat problem-solving program, and young readers are encouraged to do so, here I will code in a more professional way. I will use the `DifferentialEquations.jl` library because it provides almost all the solvers for differential equations (so far we only know about the Euler method, which is not the best method). We will use the finite difference method to discretize the Laplace operator, and then we will have a system of differential equations for $u[i, j, 1]$ and $u[i, j, 2]$. Using any solver provided by `DifferentialEquations.jl`, we will solve this system. The program is presented in Listing 10.4. In line 26, instead of using the `Euler()` solver, we can use any solver provided by `DifferentialEquations.jl`.

Here are some results from the program shown in Figs. 10.25 and 10.26. It's strange; we see spots and stripes from the initially random conditions. Who could have guessed that we would have such beautiful patterns? If we pay attention to the parameters D_u, D_v, k, f, we notice that even a slight change can give us completely different results!

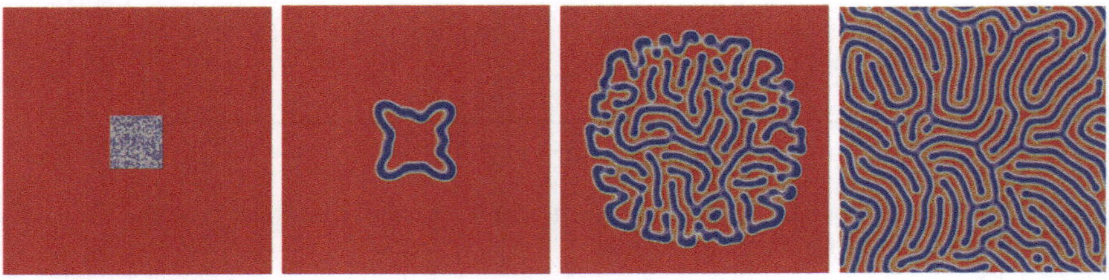

Fig. 10.26 PT Gray-Scott: $u(x, y, t)$: $(D_u, D_v, f, k) = (0.16, 0.08, 0.035, 0.06)$

Listing 10.4 The FDM program solving the Gray-Scott equation. $L = 200$.

```
using DifferentialEquations, LinearAlgebra, SparseArrays
using Random, Distributions # for initial conditions
const N   = 200
const xyd = range(0, stop = 200, length = N) # [0,L] into N-1 intervals
limit(a, N) = a == N + 1 ? 2 : a == 0 ? N - 1 : a  # periodic condition
function reaction_diffusion_Gray_Scott(du, u, p, t)
    # du = 3D array (N,N,2) => du(:,:,1) = du/dt, du(:,:,2) = dv/dt
    Da, Db, f, k, dx = p  # get the parameters stored in 'p'
    Da = Da / dx^2
    Db = Db / dx^2
    @inbounds for I in CartesianIndices((N, N))
        i, j = Tuple(I) # if needed: x, y = xyd[I[1]], xyd[I[2]]
        ip1,im1,jp1,jm1 = limit(i+1,N), limit(i-1,N), limit(j+1,N), limit(j-1,N)
        du[i,j,1] = Da * (u[im1,j,1] + u[ip1,j,1] + u[i,jp1,1]
        + u[i,jm1,1] - 4u[i,j,1]) - u[i,j,1] * u[i,j,2]^2 + f*(1 - u[i,j,1] )

        du[i,j,2] = Db * (u[im1,j,2] + u[ip1,j,2] + u[i,jp1,2]
        + u[i,jm1,2] - 4u[i, j,2]) + u[i,j,1] * u[i,j,2]^2 - (f+k) * u[i,j,2]
    end
end
p        = (0.16, 0.08, 0.035, 0.065, step(xyd))
tspan    = (0.0, 20000.0)   # time domain
u0       = init_gray_scott(xyd)  # build initial values for u,v
prob_ode = ODEProblem(reaction_diffusion_Gray_Scott, u0, tspan, p)
sol      = solve(prob_ode, Euler(), dt=.1,saveat = 100.);
```

10.8 Numerical Optimization

"Optimization" comes from the same root as "optimal", which means best. When you optimize something, you are "making it best". Of course when we write optimization we mean mathematical optimization. And we only discuss continuous optimization in the sense that we can use calculus. And numerical optimization refers to algorithms that are used to solve continuous optimization problems numerically (approximately). Those algorithms are developed to solve large-scale industry optimization problems.

Basically we have an objective function $f(\boldsymbol{x})$ of multiple variables $\boldsymbol{x} = (x_1, \ldots, x_n)$ where n is really big (e.g., millions). These variables are the inputs—things that you can control. Usually the inputs subject to some constraints, which are equations that place limits on how big or small some variables can get. There are two types of constraints: equality and inequality constraints. Equality constraints are usually noted $h_n(\boldsymbol{x})$ and inequality constraints are noted $g_n(\boldsymbol{x})$.

When there are constraints we are dealing with a constrained optimization problem. Otherwise, we have an unconstrained optimization problem.

Optimization is now a big branch of applied mathematics with a wide range of applications. This section is just a very brief introduction to the gradient descent method, which is commonly used to solve optimization problems.

Consider the problem of finding the minimum of the function $f(x_1, \ldots, x_n)$. The gradient descent method is an iterative method to solve this optimization problem. The idea is simple: starting at an initial point $\boldsymbol{x}_0$ we find the next point $\boldsymbol{x}_1$ that decreases the function as much as possible. Knowing that the gradient of f (∇f) is the direction of steepest ascent,[28] the direction of steepest descent is simply $-\nabla f$. But that only the direction, we need to know the step size γ which tells us how much to go in that $-\nabla f$ direction. Thus, the new solution is $\boldsymbol{x}_1 = \boldsymbol{x}_0 - \gamma \nabla f$, and from that we continue with $\boldsymbol{x}_2, \boldsymbol{x}_3, \ldots$. The algorithm is then (very simple, easy to code)

$$\boldsymbol{x}_{k+1} = \boldsymbol{x}_k - \gamma_k \nabla f(\boldsymbol{x}_k), \quad k = 0, 1, 2, \ldots$$

Let's consider one example to see how the value of γ affects the performance of the method.

Example 10.1 We're going to minimize the following quadratic function:

$$f(x_1, x_2) = \left(\frac{3}{4}x_1 - \frac{3}{2}\right)^2 + (x_2 - 2)^2 + \frac{x_1 x_2}{4}, \quad \nabla f = \left(\frac{9}{8}x_1 - \frac{9}{4} + \frac{1}{4}x_2, 2x_2 - 4 + \frac{1}{4}x_1\right)$$

The exact solution is $(1.6, 1.8)$. The source code is in `gradient_descent_example.jl`. The initial x_0 is $(5, 4)$ and various γ are used.

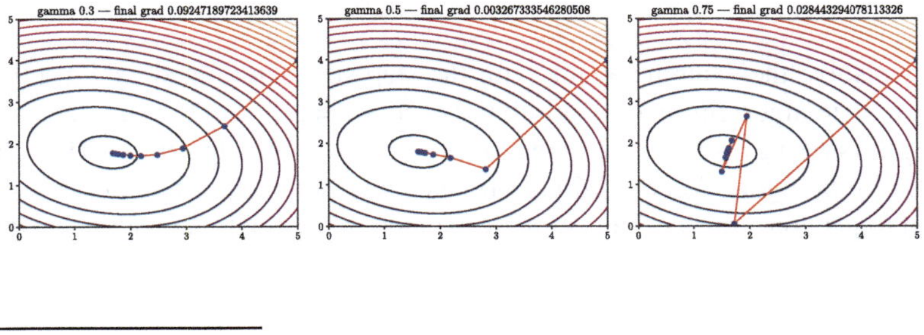

[28] Check Sect. 6.5 if this is not clear.

Two observations can be made: (1) each step indeed takes us towards the solution (i.e., decreasing the function f) and (2) we need to find a good value for γ to have a fast method.

The specific function considered in the above example belongs to a general quadratic function of the following form:

$$f(\boldsymbol{x}) = \frac{1}{2}\boldsymbol{x}^\top \mathbf{A}\boldsymbol{x} - \boldsymbol{b}^\top \boldsymbol{x} + c$$

where $\mathbf{A}$ is symmetric and positive definite.[29] In what follows we consider $c = 0$ for simplicity as it does not affect the solution but only the minimum value of f. Note that due to the positive definiteness of $\mathbf{A}$, the shape of f is like a bowl (think of the simple function $0.5ax^2 - bx$ with $a > 0$ or Fig. 6.9). Thus, there is only one minimum. The gradient of f is $\nabla f = \mathbf{A}\boldsymbol{x} - \boldsymbol{b}$.[30] For this case, we can find γ_k exactly. The idea is: choose γ_k such that $f(\boldsymbol{x}_{k+1})$ is minimized. This is simply an one-dimensional optimization problem. Let's consider the following function:

$$g(\gamma_k) = f(\boldsymbol{x}_{k+1}) = \frac{1}{2}\left(\boldsymbol{x}_k - \gamma_k \nabla f(\boldsymbol{x}_k)\right)^\top \mathbf{A}\left(\boldsymbol{x}_k - \gamma_k \nabla f(\boldsymbol{x}_k)\right) - \boldsymbol{b}^\top\left(\boldsymbol{x}_k - \gamma_k \nabla f(\boldsymbol{x}_k)\right)$$

which is nothing but a quadratic function $g(\gamma_k) = a\gamma_k^2 + d\gamma_k + e$ with

$$a = \frac{1}{2}\nabla f(\boldsymbol{x}_k)^\top \mathbf{A}\nabla f(\boldsymbol{x}_k), \; d = (\boldsymbol{b}^\top - \boldsymbol{x}_k^\top \mathbf{A})\nabla f(\boldsymbol{x}_k) = -\nabla f(\boldsymbol{x}_k)^\top \nabla f(\boldsymbol{x}_k)$$

Thus, γ_k is given by

$$\gamma_k = -\frac{d}{2a} = \frac{\nabla f(\boldsymbol{x}_k)^\top \nabla f(\boldsymbol{x}_k)}{\nabla f(\boldsymbol{x}_k)^\top \mathbf{A}\nabla f(\boldsymbol{x}_k)} \tag{10.8.1}$$

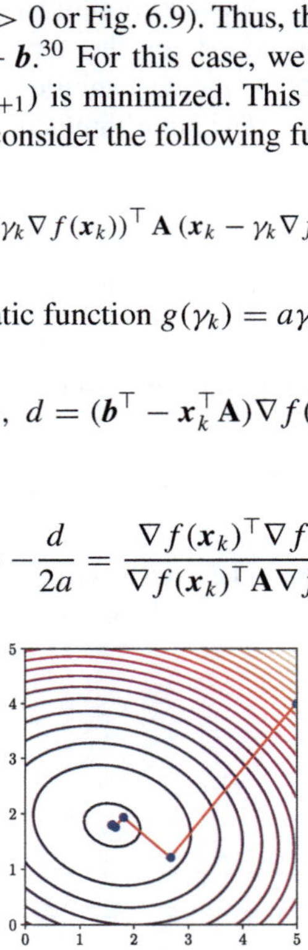

Now, we solve the example again, with the step size determined by Eq. (10.8.1). The complete algorithm is in Algorithm 2 together with a `Julia` implementation (noting the striking resemblance of the `Julia` code with the algorithm). The solution is given in the figure. Very good performance was obtained: a few iterations were needed to get convergence. But we notice something special: the path generated by

[29] See Sect. 9.10.6.

[30] See Sect. 10.9.2 for a proof of this.

$\{\boldsymbol{x}_k\}$ is zig-zag. We have the so-called zig-zag theorem. It goes like this: Let $\{\boldsymbol{x}_k\}$ be the sequence generated by the steepest descent algorithm. Then, for all k, $\boldsymbol{x}_{k+1} - \boldsymbol{x}_k$ is orthogonal to $\boldsymbol{x}_{k+2} - \boldsymbol{x}_{k+1}$.

Proof Of course to prove the orthogonality of two vectors, we need to show its dot product is zero. We have

$$\begin{cases} \boldsymbol{x}_{k+1} = \boldsymbol{x}_k - \gamma_k \nabla f(\boldsymbol{x}_k) \\ \boldsymbol{x}_{k+2} = \boldsymbol{x}_{k+1} - \gamma_{k+1} \nabla f(\boldsymbol{x}_{k+1}) \end{cases} \Longrightarrow (\boldsymbol{x}_{k+1} - \boldsymbol{x}_k, \boldsymbol{x}_{k+2} - \boldsymbol{x}_{k+1}) = \gamma_k \gamma_{k+1} (\nabla f(\boldsymbol{x}_k), \nabla f(\boldsymbol{x}_{k+1}))$$

As γ_k is the minimizer of $f(\boldsymbol{x}_k - \gamma_k \nabla f(\boldsymbol{x}_k))$, we have ${}^{df}/{}_{d\gamma} = 0$. Using the chain rule, this derivative is computed as

$$\frac{df}{d\gamma} = \nabla f(\boldsymbol{x}_k - \gamma_k \nabla f(\boldsymbol{x}_k)) \cdot \nabla f(\boldsymbol{x}_k) = (\nabla f(\boldsymbol{x}_{k+1}), -\nabla f(\boldsymbol{x}_k))$$

This derivative is zero leads to the dot product $(\nabla f(\boldsymbol{x}_{k+1}), -\nabla f(\boldsymbol{x}_k))$ being zero which results in $(\boldsymbol{x}_{k+1} - \boldsymbol{x}_k, \boldsymbol{x}_{k+2} - \boldsymbol{x}_{k+1}) = 0$. ■

Obviously a zig-zag path is not the shortest path, so the gradient descent is not a very fast method. This will be proved later using a convergence analysis and if we zoom in to look more closely at the path we see that we follow some direction that was taken earlier. In other words, there exist $\nabla f(\boldsymbol{x}_i)$ and $\nabla f(\boldsymbol{x}_j)$ which are parallel. This observation will lead to a better method: the conjugate gradient method, to be presented in Sect. 10.9.2.

Algorithm 2 Gradient descent algorithm (exact line search).

Inputs: $\mathbf{A}$, $\boldsymbol{b}$, $\boldsymbol{x}_0$, the tolerance ϵ
Outputs: the solution $\boldsymbol{x}$
$\boldsymbol{x} = \boldsymbol{x}_0$
$\nabla f = \mathbf{A}\boldsymbol{x} - \boldsymbol{b}$ ▷ gradient of f
while $\|\nabla f\| > \epsilon$ **do**
  $\gamma = \frac{\nabla f^\top \nabla f}{\nabla f^\top \mathbf{A} \nabla f}$ ▷ step size
  $\boldsymbol{x} = \boldsymbol{x} - \gamma \nabla f$ ▷ update $\boldsymbol{x}$
  $\nabla f = \mathbf{A}\boldsymbol{x} - \boldsymbol{b}$
end while

```
function gradient_descent_exact_convergence(A,b,x0,tol)
    x          = x0
    ∇f         = A*x-b
    # actual gradient descent algorithm
    while (  norm( ∇f ) > tol )
        γ          = ∇f'*∇f/(∇f'*A*∇f)
        x          = x - γ * ∇f
        ∇f         = A*x-b
    end
    return x
end

A=[9/8 1/4;1/4 2]
b=[9/4,4]
x0=[5,4]
tol=1e-6
x=gradient_descent_exact_convergence(A,b,x0,tol)
```

Convergence analysis. The gradient descent method generates a sequence $\{\boldsymbol{x}_k\}$ that converges towards $\boldsymbol{x}$—the solution. We have seen one numerical evidence of that. And we need a proof. Then, what is the convergence rate (of the method) that tells us how fast we go from $\boldsymbol{x}_0$ to $\boldsymbol{x}$. Certainly, this rate of convergence is evaluated using error function $E(\boldsymbol{x})$:

$$E : \mathbb{R}^n \to \mathbb{R} \text{ such that } E(\boldsymbol{x}) \geq 0 \ \forall \boldsymbol{x} \in \mathbb{R}^n$$

One choice is to define the error $\boldsymbol{e}_k = \boldsymbol{x}_k - \boldsymbol{x}$, then define E as the following energy norm:

$$E(\boldsymbol{x}_k) = \left(\boldsymbol{e}_k^\top \mathbf{A} \boldsymbol{e}_k\right)^{1/2}$$

We need the formula for updating $\boldsymbol{e}_k$, it satisfies the same equation as for $\boldsymbol{x}_k$:

$$\boldsymbol{x}_{k+1} - \boldsymbol{x} = (\boldsymbol{x}_k - \boldsymbol{x}) - \gamma_k \nabla f(\boldsymbol{x}_k) \Longleftrightarrow \boldsymbol{e}_{k+1} = \boldsymbol{e}_k - \gamma_k \nabla f(\boldsymbol{x}_k)$$

Now, we can compute $E(\boldsymbol{x}_{k+1})$ by considering its square, and relating it to $E(\boldsymbol{x}_k)$:

$$\begin{aligned}
[E(\boldsymbol{x}_{k+1})]^2 &= \boldsymbol{e}_{k+1}^\top \mathbf{A} \boldsymbol{e}_{k+1} \\
&= (\boldsymbol{e}_k^\top - \gamma_k \nabla f(\boldsymbol{x}_k)^\top)\mathbf{A}(\boldsymbol{e}_k - \gamma_k \nabla f(\boldsymbol{x}_k)) \\
&= \boldsymbol{e}_k^\top \mathbf{A} \boldsymbol{e}_k - 2\gamma_k \nabla f(\boldsymbol{x}_k)^\top \mathbf{A} \boldsymbol{e}_k + \gamma_k^2 \nabla f(\boldsymbol{x}_k)^\top)\mathbf{A} \nabla f(\boldsymbol{x}_k) \\
&= [E(\boldsymbol{x}_k)]^2 - \frac{[\nabla f(\boldsymbol{x}_k)^\top \nabla f(\boldsymbol{x}_k)]^2}{\nabla f(\boldsymbol{x}_k)^\top \mathbf{A} \nabla f(\boldsymbol{x}_k)} \\
&= [E(\boldsymbol{x}_k)]^2 \left[1 - \frac{[\nabla f(\boldsymbol{x}_k)^\top \nabla f(\boldsymbol{x}_k)]^2}{(\nabla f(\boldsymbol{x}_k)^\top \mathbf{A} \nabla f(\boldsymbol{x}_k))(\boldsymbol{e}_k^\top \mathbf{A} \boldsymbol{e}_k)}\right]
\end{aligned} \tag{10.8.2}$$

Now comes the magic of eigenvectors and eigenvalues. As $\mathbf{A}$ is a real symmetric matrix, it has n independent orthonormal eigenvectors $\boldsymbol{v}_i$ and n positive real eigenvalues λ_i, and we use them as a basis of $\mathbb{R}^n$ to express the error—which is a vector in $\mathbb{R}^n$–$\boldsymbol{e}_k$ as

$$\boldsymbol{e}_k = \sum_{i=1}^{n} \xi_i \boldsymbol{v}_i \tag{10.8.3}$$

With that it is possible to compute different terms in the last expression of Eq. (10.8.2). We start with

$$\nabla f(\boldsymbol{x}_k) = \mathbf{A}\boldsymbol{x}_k - \boldsymbol{b} = \mathbf{A}\boldsymbol{e}_k = \mathbf{A}\sum_{i=1}^{n} \xi_i \boldsymbol{v}_i = \sum_{i=1}^{n} \xi_i \lambda_i \boldsymbol{v}_i \tag{10.8.4}$$

Thus,

$$\begin{aligned}
\nabla f(\boldsymbol{x}_k)^\top \nabla f(\boldsymbol{x}_k) &= \left(\sum_{i=1}^{n} \xi_i \lambda_i \boldsymbol{v}_i\right) \cdot \left(\sum_{j=1}^{n} \xi_j \lambda_j \boldsymbol{v}_j\right) = \sum_{i=1}^{n} \xi_i^2 \lambda_i^2 \\
\nabla f(\boldsymbol{x}_k)^\top \mathbf{A} \nabla f(\boldsymbol{x}_k) &= \left(\sum_{i=1}^{n} \xi_i \lambda_i \boldsymbol{v}_i\right) \cdot \left(\sum_{j=1}^{n} \xi_j \lambda_j^2 \boldsymbol{v}_j\right) = \sum_{i=1}^{n} \xi_i^2 \lambda_i^3
\end{aligned} \tag{10.8.5}$$

And in the same manner, we can compute $\boldsymbol{e}_k^\top \mathbf{A}\boldsymbol{e}_k$ as

$$\boldsymbol{e}_k^\top \mathbf{A}\boldsymbol{e}_k = \left(\sum_{i=1}^{n} \xi_i \boldsymbol{v}_i\right) \cdot \left(\sum_{j=1}^{n} \xi_j \lambda_j \boldsymbol{v}_j\right) = \sum_{i=1}^{n} \xi_i^2 \lambda_i \tag{10.8.6}$$

With all the intermediate results, from Eq. (10.8.2), we can finally get

$$[E(\boldsymbol{x}_{k+1})]^2 = [E(\boldsymbol{x}_k)]^2 \left[1 - \frac{[\xi_i^2 \lambda_i^2]^2}{(\xi_i^2 \lambda_i^3)(\xi_i^2 \lambda_i)}\right] \tag{10.8.7}$$

We have finished the error analysis: we have now a formula relating the error in step $k+1$ to step k. The convergence depends on the number in the brackets. For details, please consult the very readable paper *An Introduction to the Conjugate Gradient Method Without the Agonizing Pain* by Jonathan Richard Shewchuk.

10.9 Numerical Linear Algebra

This section is a brief introduction to the field of numerical linear algebra, in which approximate and efficient methods to solve large systems of linear equations are the focus.

10.9.1 Iterative Methods to Solve a System of Linear Equations

In Sect. 3.7, we have met the Persian astronomer al-Kashi, who computed $x = \sin 1°$ *iteratively* from a trigonometry identity:

$$\sin 3° = 3 \sin 1° - 4 \sin^3 1° \Longrightarrow x = \frac{\sin 3° + 4x^3}{3}$$

We are going to do the same thing but for $\mathbf{A}\boldsymbol{x} = \boldsymbol{b}$: we split the matrix into two matrices $\mathbf{A} = \mathbf{S} - \mathbf{T}$, then the system becomes $(\mathbf{S} - \mathbf{T})\boldsymbol{x} = \boldsymbol{b}$ or $\mathbf{S}\boldsymbol{x} = \mathbf{T}\boldsymbol{x} + \boldsymbol{b}$. Then, following al-Kashi, we solve this system iteratively, starting from $\boldsymbol{x}_0$ we get $\boldsymbol{x}_1$, and from $\boldsymbol{x}_1$ we obtain $\boldsymbol{x}_2$ and so on:

$$\mathbf{S}\boldsymbol{x}_{k+1} = \mathbf{T}\boldsymbol{x}_k + \boldsymbol{b}, \quad k = 0, 1, 2, \dots \tag{10.9.1}$$

Thus, instead of solving $\mathbf{A}\boldsymbol{x} = \boldsymbol{b}$ directly using Gaussian elimination method, we're adopting an iterative method.

It is obvious that we need to select $\mathbf{S}$ in a way that

(a) Equation (10.9.1) is solved easily (or fast).
(b) The difference (or error) $\boldsymbol{x} - \boldsymbol{x}_k$ should go quickly to zero. To get an expression for this difference, subtracting Eq. (10.9.1) from $\mathbf{S}\boldsymbol{x} = \mathbf{T}\boldsymbol{x} + \boldsymbol{b}$:

$$\mathbf{S}\boldsymbol{e}_{k+1} = \mathbf{T}\boldsymbol{e}_k \Longrightarrow \boxed{\boldsymbol{e}_{k+1} = \mathbf{S}^{-1}\mathbf{T}\boldsymbol{e}_k}$$

The matrix $\mathbf{B} = \mathbf{S}^{-1}\mathbf{T}$ controls the convergence rate of the method.

To demonstrate iterative methods for $\mathbf{A}\boldsymbol{x} = \boldsymbol{b}$, we first consider the following methods:

Jacobi method: **S** is the diagonal part of **A**
Gauss-Seidel method: **S** is the lower triangular part of **A**(diagonal included)

Example 10.2 Consider the following system, with solution:

$$\begin{bmatrix} +2 & -1 \\ -1 & +2 \end{bmatrix} \begin{bmatrix} x \\ y \end{bmatrix} = \begin{bmatrix} +4 \\ -2 \end{bmatrix} \quad \text{has the solution} \begin{bmatrix} x \\ y \end{bmatrix} = \begin{bmatrix} 2 \\ 0 \end{bmatrix}$$

The Jacobi iterations are

$$\begin{bmatrix} 2 & 0 \\ 0 & 2 \end{bmatrix} \boldsymbol{x}_{k+1} = \begin{bmatrix} 0 & 1 \\ 1 & 0 \end{bmatrix} \boldsymbol{x}_k + \begin{bmatrix} +4 \\ -2 \end{bmatrix}, \quad \begin{cases} 2x_{k+1} = y_k + 4 \\ 2y_{k+1} = x_k - 2 \end{cases}$$

And the Gauss-Seidel iterations are

$$\begin{bmatrix} +2 & 0 \\ -1 & 2 \end{bmatrix} \boldsymbol{x}_{k+1} = \begin{bmatrix} 0 & 1 \\ 0 & 0 \end{bmatrix} \boldsymbol{x}_k + \begin{bmatrix} +4 \\ -2 \end{bmatrix}, \quad \begin{cases} 2x_{k+1} = y_k + 4 \\ 2y_{k+1} = x_{k+1} - 2 \end{cases}$$

Jacobi method		**Gauss-Seidel method**	
x_k	y_k	x_k	y_k
0.000	0.000000	0.000000000	−1.000000000
2.000	−1.000000	1.500000000	−0.250000000
1.500	0.000000	1.875000000	−0.062500000
2.000	−0.250000	1.968750000	−0.015625000

10.9.2 Solving Ax = b as a Minimization Problem

It is easy to see that the minimum point of the quadratic function $f(x) = 1/2ax^2 - bx + c$ (with $a > 0$) is x such that $ax = b$. By analogy, the minimum point of the following quadratic function where $\boldsymbol{x}, \boldsymbol{b} \in \mathbb{R}^n$

$$f(\boldsymbol{x}) = \frac{1}{2}\boldsymbol{x}^\top \mathbf{A}\boldsymbol{x} - \boldsymbol{b}^\top \boldsymbol{x} + c$$

is the solution to the linear system $\mathbf{A}\boldsymbol{x} = \boldsymbol{b}$ when $\mathbf{A}$ is symmetric and positive definite (similar to the fact that $a > 0$).

Proof We need to prove this $d/dx(\boldsymbol{x}^\top \mathbf{A}\boldsymbol{x}) = (\mathbf{A} + \mathbf{A}^\top)\boldsymbol{x}$. Indeed, $g(\boldsymbol{x}) = \boldsymbol{x}^\top \mathbf{A}\boldsymbol{x} = A_{ij}x_i x_j$, now take the derivative of $g(\boldsymbol{x})$ with respect to x_k,

$$\begin{aligned}
\frac{\partial x_k}{\partial g(\boldsymbol{x})} &= A_{ij}\frac{\partial x_i}{\partial x_k}x_j + A_{ij}\frac{\partial x_j}{\partial x_k}x_i = A_{ij}\frac{\partial x_i}{\partial x_k}x_j + A_{ij}\frac{\partial x_j}{\partial x_k}x_i \\
&= A_{ij}\delta_{ik}x_j + A_{ij}\delta_{jk}x_i = A_{kj}x_j + A_{ik}x_i
\end{aligned}$$

which is the kth component of $(\mathbf{A} + \mathbf{A}^\top)\boldsymbol{x}$. Doing something similar, we get the derivative of $\boldsymbol{b}^\top \boldsymbol{x}$ is $\boldsymbol{b}$.[31] Hence, the derivative of $f(\boldsymbol{x})$ is $\mathbf{A}\boldsymbol{x} - \boldsymbol{b}$. Setting this derivative to zero, we get the linear system $\mathbf{A}\boldsymbol{x} = \boldsymbol{b}$. ■

So, facing a large sparse linear system $\mathbf{A}\boldsymbol{x} = \boldsymbol{b}$, we do not solve it directly, but we find the minimum of the function $f(\boldsymbol{x})$. Why we do that? Intuitively finding a minimum of a nice function such as $f(x)$, which geometrically is a bowl, seems easy. We just need to start somewhere on the bowl and moving downhill, more often we will hit home: the bottom of the bowl. We have actually seen such a method: the gradient descent method. There exits many other methods, such as the conjugate gradient method.

[31] Noting that $\mathbf{A} = \mathbf{A}^\top$ as $\mathbf{A}$ is symmetric.

Appendix A
Codes

Leslie Lamport Coding is to programming as typing is to writing.

To encourage young students to learn coding and also to demonstrate the important role of coding in mathematics, engineering, and sciences, in this book, I have used many small programs to do some tedious (or boring) calculations. In this appendix, I provide some snippets of these programs so that young people can learn programming while learning maths/physics.

There are so many programming languages and I have selected `Julia` for two main reasons. First, it is open source (so we can *use it for free* and we can see its source code if we find that needed). Second, it is *easy to use*. For young students, the fact that a programming language is free is obviously important. The second reason—being easy to use—is more important as we use a programming language just as a tool; our main purpose is doing mathematics (or physics). Of course you can use *Python*; it is also free and easy to use and popular. The reason I have opted for *Julia* was to force me to learn this new language; I forced myself to go outside of my comfort zone, only then I could find something unexpected. There is actually another reason, although irrelevant here, is that `Julia` codes run faster than `Python` ones. Moreover, it is possible to use `Python` and `R`[1] in `Julia`.

It is worthy noting that our aim is to learn coding to use it to solve mathematical problems. We do not want to learn coding to write software for general use; that is a completely story. And that is why I do not spend time (for time is limited) learning how to make graphical user interfaces (GUI), and do not learn coding with languages such as *Visual Basic, Delphi*, and so on.

In the text, if there is certain amount of boring calculations (e.g., a table of partial sums of an infinite series), certainly I have used a small `Julia` program to do that job. And I have provided links to the code given in this appendix. Now, in the code snippets, I provide the link back to the associated text in the book.

[1] R is a free software environment for statistical computing and graphics. It compiles and runs on a wide variety of UNIX platforms, Windows, and MacOS.

V. P. Nguyen, *Mathematics for Engineers and Scientists*,
https://doi.org/10.1007/978-981-97-4631-6

Fig. A.1 Baby programming

To reduce the thickness of the book, all other codes, which are not given in the text, are put in GitHub[2] at this address.

As this appendix is not a tutorial to `Julia`, I just provide the codes and provide little explanations. I would like to emphasize that programming or coding is not hard. This is true, as kids can do it (Fig. A.1). A program is simply a set of instructions written by you to demand a computer to perform a certain task. These instructions are written in a programming language with vocabularies such as *for, while, if, and function*. All we need to do are: (1) write down what we need to achieve in clear steps (in your own mother language) and (2) translate these steps to the programming language we've selected. That's it!

And by the way, we do not have to memorize anything (such as should I write *for* $i=1$ to 10 or *for* $i=$ 1:10?). Google is our best friend. If you're not sure about anything, just Google. And with time all these *for/while/if*... will reside in your brain without you knowing it! Albert Einstein once said "Never memorize something that you can look up".

Having said that, programming becomes, however, harder when we have to write optimized codes; codes that are very fast to execute and codes that can be easy to maintain (i.e., easy to modify to add new codes). But that is beyond the scope of this book.

When you have known a language and enjoy the process, learn a new one or even more. And it is beneficial to read lots of code written by others, it is one of

[2] GitHub is a website and cloud-based service that helps developers store and manage their code, as well as track and control changes to their code.

the best ways to become a better programmer. This is similar to writers who read Shakespeare, Mark Twain, and Hemingway.

A.1 Algebra and Calculus

In this section, I present some codes used in Chap. 2. Not all programs are presented as a few representative ones are sufficient to demonstrate the basics of programming.

Listing A.1 is a code to compute the square root of any positive number S using the formula $x_{n+1} = 0.5(x_n + S/x_n)$ starting with an initial value x_0, see Sect. 2.9.3. This code is representative for any iteration-based formula. Note that x is replaced by the new value, so intermediate x's are not recorded.

Listing A.1 Computing the square root of a positive real number S. `Julia` built in functions are in **blue heavy bold font**

```
function square_root(S,x0,epsilon)
  x  = x0                # x is for x_{n+1} in our formula
  while (true)           # do the iterations, a loop without knowing the # of iterations
    x = 0.5 * ( x + S/x )
    if (abs(x*x-S) < epsilon) break end # if x is accurate enough, stop
  end
  return x
end
```

Listing A.2 is the code to compute the partial sums of a geometric series $\sum_{i=1}^{n} 1/2^i$. The code is typical for calculating a sum of n terms. We initialize the sum to zero, and using a for loop to add one term to the sum each time. Listing A.3 is a similar code, but for the Taylor series of the sine function $\sin x = \sum_{i=1}^{\infty}(-1)^{i-1} 1/(2i-1)! x^{2i-1}$, see Sect. 4.15.6. The code introduces the use of the `factorial(n)` function to compute $n!$ Note that we have to use big numbers as $n!$ is very large for large n.

Listing A.4 is the program to check whether a natural number is a factorion. Having such a function, we just need to sweep over, let say the first 100 000 numbers and check every number if it is a factorion. We provide two solutions: one using the built-in `Julia`'s function `digits` to get the digits of an integer. This solution is a lazy one. The second solution does not use that function. Only then, we're forced to work out how to get the digits of a number. Let's say the number is 3 258, we can get the digits starting from the first one (and get 3, 2, 5, 8) or we can start from the last digit (to get 8, 5, 2, 3). The second option is easier because $8 = 3258\%10$ (the last digit is the remainder of the division of the given number with 10). Once we have already got the last digit, we do not need it, so we just need to remove it; $325 = div(3258, 10)$; that is, 325 is the result of the integer division of 3258 with 10.

Listing A.2 Partial sum of geometric series $\sum_{i=1}^{n} 1/2^i$. Also produces directly Table 2.12

```
using PrettyTables                # you have to install this package first
function geometric_series(n)      # make a function named 'geometric_series' with 1 input
  S = 0.
  for k=1:n                       # using 'for' for loops with known number of iterations
    S += 1/2^k                    # S += ... is short for S = S + ...
  end
  return S
end
data = zeros(20,2)                # this is an array of 20 rows and 2 columns
for i=1:20                        # produce 20 rows in Table 2.10
  S         = geometric_series(i)
  data[i,1] = i                   # row 'i', first col is 'i'
  data[i,2] = S                   #          second col is S
end
pretty_table(data, ["n", "S"]) # print the table to terminal
```

Listing A.3 Calculating $\sin x$ using the sine series $\sin x = \sum_{i=1}^{\infty} (-1)^{i-1} 1/(2i-1)! x^{2i-1}$

```
using PrettyTables  # you have to install this package first
function sinx_series(x,n)
  A       = 0.
  for i=1:n
    A  += (-1)^(i-1) * x^(2*i-1) / factorial(big(2*i-1))
  end
  return A
end
# compute sin series with n=1,2,...,10 terms
data = zeros(10,2)
for i=1:size(data,1)
  S1        = sinx_series(pi/4,i)
  data[i,1] = i
  data[i,2] = S1
end
pretty_table(data, ["n", "S1"], backend = :latex,formatters = ft_printf("
```

Listing A.5 is the code for the calculation of $s_n = \prod_{k=0}^{n} \binom{n}{k}$ that is the product of all the binomial coefficients. The idea is the same as the calculation of a sum but we need to initialize the result to 1 (instead of 0). We use `Julia` built-in function `binomial` to compute $\binom{n}{k}$.

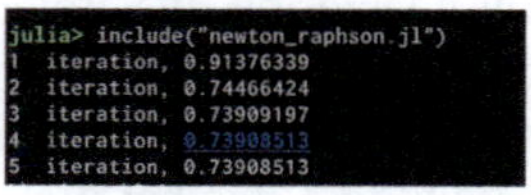

In Listing A.6, I present a code that implements the Newton-Raphson method for solving $f(x) = 0$. (See Sect. 4.5.4.) As it uses iterations, the code is similar to Listing A.1. Instead of calculating the first derivative of $f(x)$ directly, I used a central difference for this. I also introduced an increment variable i to count the number of iterations required to get the solution. The function was then applied to solving the equation $\cos x - x = 0$.

Listing A.4 Checking if an integer is a factorion (Sect. 2.26.2)

```
function is_factorion_version1(n)
  s = 0
  digits_n = digits(n)
  for i=1:length(digits_n)
    s += factorial(big(digits_n[i]))
  end
  return ( s == n )
end
function is_factorion_version2(n)
  s  = 0
  n0 = n                  # kepp the original number as we modify n
  while ( n > 0 )         # the loop stops when n = 0
    x  = n
    s += factorial(big(x))
    n  = div(n,10)        # remove the last digit
  end
  return ( s == n0 )
end
```

Listing A.5 $s_n = \prod_{k=0}^{n} \binom{n}{k} = \prod_{k=0}^{n} n!/(n-k)!k!$. See Pascal triangle and number e, Sect. 2.28.3

```
function sn(n)
 product=1.0
 for k=0:n
  product *= binomial(big(n),k)
 end
 return product
end
```

Listing A.6 Newton-Raphson method to solve $f(x) = 0$ using central difference for derivative

```
function newton_raphson(f,x0,epsilon)
  x  = x0
  i  = 0
  while ( true )
     i   += 1
     derx = (f(x0+1e-5)-f(x0-1e-5)) / (2e-5)
     x    = x0 - f(x0)/derx
     @printf "
     if ( abs(x-x0) < epsilon ) break end
     x0   = x
  end
end
f(x)   = cos(x) - x      # short functions
newton_raphson(f,0.1,1e-6)
```

Listing A.7 implements three functions used to generate Newton fractals shown in Fig. 1.3. The first function is the standard Newton-Raphson method, but the input is a function of a single complex variable. The second function `get_root_index` is to return the position of a root r in the list of all roots of the equation $f(z) = 0$. This function uses the built-in function `isapprox` to check the equality of two numbers.[3] The final function `plot_newton_fractal` loops over a grid of $n \times n$ points within the domain $[x_{\min}, x_{\max}]^2$, for each point (x, y), a complex variable $z_0 = x + iy$ is made and inserted to the function `newton` to find a root r. Then, it finds the position of r in the list `roots`. And finally it updates the matrix `m` accordingly. We used the code with the function $f(z) = z^4 - 1$, but you're encouraged to play with $f(z) = z^{12} - 1$.

Listing A.7 Newton fractals

```
function newton(z0,f,fprime;max_iter=1000)
  z = z0
  for i=1:max_iter
    dz = f(z)/fprime(z)
    if abs(dz) < TOL return z end
    z -= dz
  end
  return 0
end
function get_root_index(roots,r)
    i = 0
    for i=1:length(roots)
      if isapprox(roots[i],r)  # if r equals roots[i]
         return i
      end
    end
    if i == 0                  # if root r is not yet found, add to roots
      append!(roots,r)         # equivalent to: roots = [roots;r]
      return length(roots)
    end
end
function plot_newton_fractal(f,fprime;n=200,domain=(-1,1,-1,1))
  roots = []  # initialize roots to be empty
  m     = zeros(n,n)
  xmin,xmax,ymin,ymax = domain
  xarrays = range(xmin,xmax,length=n)  # range(0,1,5) => {0,0.25,0.5,0.75,1}
  yarrays = range(ymin,ymax,length=n)
  for (ix,x) in enumerate(xarrays)     # ix=1,2,... and x = xarrays[1], xarrays[2]...
    for (iy,y) in enumerate(yarrays)
       z0 = x + y * im        # 'im' is Julia for i
       r  = newton(z0,f,fprime)
       if r  != 0
           ir = get_root_index(roots,r) # roots be updated
           m[iy,ix] = ir
       end
    end
  end
  return m
end
# a concrete function f(z)
f(z)       = z^4 - 1; fprime(z) = 4*z^3
domain     = (0.414,0.445,0.414,0.445)
m          = plot_newton_fractal(f,fprime,n=500,domain=domain)
myplot     = spy(m,Scale.ContinuousColorScale(p -> get(ColorSchemes.rainbow, p)))
```

[3] We should never check the equality of real/complex numbers by checking $a == b$; instead we should check $|a - b| < \epsilon$, where ϵ is a small positive number. In other words, $0.99998 = 1.00001 = 1$ according to a computer. The built-in function is an optimal implementation of this check.

A.2 Recursion

In Sect. 2.10, we have met the Fibonacci numbers:

$$F_n = F_{n-1} + F_{n-2}, \quad n \geq 2, \quad F_0 = F_1 = 1 \tag{A.1}$$

To compute $F(n)$, we need to use the recursive relation in Eq. (A.1). Listing A.8 is the `Julia` implementation of Eq. (A.1). What is special about this "fibonacci" function? Inside the definition of that function we call it (with smaller values of n). The process in which a function calls itself directly or indirectly is called recursion and the corresponding function is called a recursive function.

The case $n = 0$ or $n = 1$ is called *the base case* of a recursive function. This is the case that we know the answer to, thus it can be solved without any more recursive calls. The base case is what stops the recursion from continuing on forever (i.e., infinite loop). *Every recursive function must have at least one base case* (many functions have more than one).

Listing A.8 Fibonacci numbers implemented as a recursive function

```
function fibonacci(n)
  if ( n==0  n==1 )
     return 1
  else
    return fibonacci(n-2) + fibonacci(n-1)
  end
end
```

Sometimes the problem does not appear to be recursive. Thus, to master recursion we must first find out how to think recursively. For example, consider the problem of computing the sum of the first n integers. Using recursion, we do this:

$$S(n) = 1 + 2 + \cdots + n = \underbrace{1 + 2 + \cdots + (n-1)}_{S(n-1)} + n$$

We also need the base case, which is obviously $S(1) = 1$. Now we can implement this in `Julia` as in Listing A.9.

A.3 Numerical Integration

Listing A.10 presents an implementation of the Simpson quadrature rule for $\int_a^b f(x)dx$. The integration interval can be divided into n equal parts. Using this as a template you can program other quadrature rules such as trapezoidal or Gauss rule.

Listing A.9 Sum of the first n integers implemented as a recursive function

```
function sum_first_integers(n)
  if ( n==1 )
     return 1
  else
    return sum_first_integers(n-1) + n
  end
end
```

Listing A.10 Simpson's quadrature for $\int_a^b f(x)dx$

```
using PrettyTables
function simpson_quad(f,a,b,n)
  A        = 0.
  deltan   = (b-a)/n
  deltax6  = deltan/6
  for i=1:n
    fa  = f(a+(i-1)*deltan)
    fb  = f(a+i*deltan)
    fm  = f(a+i*deltan-deltan/2)
    A   += fa + 4*fm + fb
  end
  return A*deltax6
end
fx4(x)  = x^4
I = simpson_quad(fx4,0,1,10)
```

A.4 Harmonic Oscillations

Listing A.11 is used to generate Fig. 7.9 about a weakly damped harmonic simple oscillator. This code is presented to demonstrate how to generate graphs from formula. Assuming that the formula is $x(t) = f(t)$, and $0 \leq t \leq 50$, then we generate a large number of points in this interval, and for each t_i, we compute $x(t_i)$. Then we plot the points $(t_i, x(t_i))$, these points are joined by a line and thus we have a smooth curve. This is achieved using the `Plots` package.

A.5 Polynomial Interpolation

Listing A.12 implements the Lagrange interpolation (Sect. 10.3.1). First a function is coded to compute the Lagrange interpolation function $l_i(x)$

$$l_i(x) = \prod_{\substack{j=0 \\ j \neq i}}^{n} \frac{x - x_j}{x_i - x_j}$$

Listing A.11 Weakly damped oscillation, Fig. 7.9

```
using Plots
using LaTeXStrings

omega0, beta, x0, v0 = 1.0, 0.05, 1.0, 3.0

omega0d = sqrt(omega0^2-beta^2)
theta   = atan(-(v0+beta*x0)/(omega0d*x0))
C       = x0 / cos(theta)

ta      = 0:0.1:50             # time domain divided in many points: 0,0.1,0.2,...
xt      = zeros(length(ta))    # length(ta) returns the size of the vector 'ta'
b1      = zeros(length(ta))    # zeros(10) => a vector of 10 elems, all are 0.
b2      = zeros(length(ta))

for i=1:length(ta)            # loop over t_i, compute x(t_i), ...
  t     = ta[i]
  xt[i] =  C*exp(-beta*t)*cos(omega0d*t+theta)
  b1[i] =  C*exp(-beta*t)
  b2[i] = -C*exp(-beta*t)
end
# generating the graphs, plot!() is to add another plot on the existing plot
pyplot()
p=plot(ta,xt,legend=false,size=(250,250))
plot!(ta,b1,legend=false,color="red", linestyle = :dash)
plot!(ta,b2,legend=false,color="red", linestyle = :dash)
xlabel!(L"t"), ylabel!(L"x(t)")
```

Then the Lagrange interpolation formula is programmed:

$$y(x) = \sum_{i=0}^{n} l_i(x) y_i$$

To generate Fig. 10.2, many points in [0, 6] are generated, and for each point x_i compute $y(x_i)$, then we plot the points $(x_i, y(x_i))$.

A.6 Probability

Monte Carlo for pi. I show in Listing A.13 the code that implements the Monte Carlo method for calculating π. This is the code used to generate Table 5.3 and Fig. 5.2 (this part of the code is not shown for brevity). It also presents how to work with arrays of unknown size (line 4 for the array `points2` as we do not in advance how many points will be inside the circle). In line 13, we add one row to this array. Final note, this function returns multiple values put in a tuple (line 16).

Listing A.12 Lagrange interpolation

```
# ith Lagrange basis, i =1,2,3,...
function lagrange_basis_i(i,data_x,x) # l_i(x)
  li = 1.0
  xi = data_x[i]
  for j=1:length(data_x)
    if j != i
      xj = data_x[j]
      li *= (x-xj)/(xi-xj)
    end
  end
  return li
end
function lagrange_interpolation(data_x,data_y,x)
  fx = 0.
  for i=1:length(data_x)
    li  = lagrange_basis_i(i,data_x,x)
    fx += li * data_y[i]
  end
  return fx
end
data_x=[0 1 2 3 4 5 6] # data points
data_y=[0 0.8415 0.9093 0.1411 -0.7568 -0.9589 -0.2794]
ta    = 0:0.1:6         # points where Lagrange interpolating function is drawn
func  = zeros(length(ta))
for i=1:length(ta)
  func[i] =    lagrange_interpolation(data_x,data_y,ta[i])
end
```

Listing A.13 Monte-Carlo method for calculating π

```
function monte_carlo_pi(n)
  inside  = 0                             # counter for points inside the circle
  points1 = zeros(n,2)                    # all generated points inside/outside
  points2 = Array{Float64}(undef, 0, 2) # points inside the circle. points1/2 for plot
                                          # 0 row, 2 cols, initial value: anything
  for i=1:n
    x             = rand()                # random number in [0,1]
    y             = rand()
    points1[i,:] = [x,y]                  # points1[i,1] = x, points1[i,2] = y
    if ( x^2 + y^2 <= 1. )
      inside    += 1
      points2     = [points2;[x y]]       # add [x y] to points2, one row a time
    end
  end
  return (4*(inside/n),points1,points2) # return a tuple (pi,points1,points2)
end
```

In Listing A.14, I present another implementation, which is much shorter using *list comprehension*.[4] In one line (line 3) all n points in $[0, 1]^2$ are generated. In line 4, we

[4] A list comprehension is a syntactic construct for creating a list based on existing lists. It follows the form of the mathematical set-builder notation (set comprehension). For example, $S = \{2 \cdot x : x \in \mathbb{N},\ x^2 > 3\}$.

get all the points inside the unit circle using the `filter` function[7] and an anonymous predicate (x− > norm(x) <= 1). The `norm` function, from the `LinearAlgebra` package, is for $\sqrt{x^2 + y^2}$.

Listing A.14 Monte-Carlo method for calculating π: shorter version

```
using LinearAlgebra
function monte_carlo_pi(n)
  points1  = [(rand(),rand()) for _ in 1:n]
  points2  = filter(x -> norm(x) <= 1., points1) # filter out elements : x^2+y^2 <=1
  return (4*(length(points2)/n),points1,points2)
end
# note that points1/2 are not matrices of nx2, but vectors of tuples
# to get the x-coords, we have to do: first.(points1)
# for example, to plot the points, do:
plot(first.(points1),last.(points1),"ro")
```

Computer experiment of tossing a coin. When we toss a coin we either get a head or a tail. In our virtual coin tossing experiment, we generate a random integer number within [1, 2], and we assign one to head and two to tail. We repeat this for n times and count the number of heads and tails. Listing A.15 is the resulting code. The code introduces the `rand` function to generate random numbers.

Using list comprehension we can have a shorter implementation shown in Listing A.16.

Listing A.15 Virtual experiment of tossing a coin in `Julia`

```
function tossing_a_coin(n)
  head_count = 0
  tail_count = 0
  for i=1:n
    result = rand(1:2) # 1: Head and 2: Tail
    if ( result == 1 ) head_count += 1 end
    if ( result == 2 ) tail_count += 1 end
  end
  return (head_count, tail_count)
end
data = zeros(5,3)
data[:,1] = [10,100,1000,2000,10000]
for i=1:size(data,1)
   h,t        = tossing_a_coin(data[i,1])
   data[i,2] = h/data[i,1]
   data[i,3] = t/data[i,1]
end
```

[7] Filter is a higher order function that processes a data structure (usually a list) in some order to produce a new data structure containing exactly those elements of the original data structure for which a given predicate returns the Boolean value true.

Listing A.16 Virtual experiment of tossing a coin in `Julia`: list comprehension based implementation

```
function tossing_a_coin(n)
  coin=[ rand(1:2)  for _ in 1:n]
  return (sum(coin .== 1), sum(coin .== 2))
end
```

Birthday problem. Now we present an implementation of the birthday problem. The procedure is: we repeat the following steps N times where N is a large counting number:

- Collect birthdays of n persons; this can be done with [rand(1:365) for _ in 1:n].
- Count the number of occurrences of the above birthdays array; for example, with three persons, we can have $\{1, 2, 2\}$, and after the counting we get $\{1, 2\}$ (there is shared birthday), or $\{4, 5, 6\}$ with no duplicated elements, we get $\{1, 1, 1\}$ thus no shared birthday.
- If there is shared birthday we return a true.

Refer to Listing A.17 for the code.

Listing A.17 List comprehension based implementations

```
# roll 2 six-sided dice N times, compute P(even number)
faces = 1:6
dice  = sum ([ iseven(rand(faces) + rand(faces)) for _ in 1:N ]) / N
# birthday problem
using StatsBase # for counts function
function birthday_event(n)
  birthday_a_year       = 1:365
  birthdays_n_pers      = [rand(birthday_a_year) for _ in 1:n]
  birthdays_occurences = counts(birthdays_n_pers)
  return maximum(birthdays_occurences) > 1
end
# counts(x): Count the number of times each value in x occurs
N = 10^5

function birthday_experiment(n)
    return sum([birthday_event(n) for _ in 1:N])/N
end

println("Probability ofr 23: $(birthday_experiment(23))")
```

Distributions.jl is a Julia package for probability distributions and associated functions. Listing A.18 presents a brief summary of some common functions.

The code in Listing A.19 is used to illustrate graphically the central limit theorem. The code generates n uniformly distributed variables (i.e., $X_1, X_2, \ldots, X_n$). Then it computes the mean of X_is, that is, $Y = (X_1 + \cdots + X_n)/n$. And this is done for a large number of times ($N = 2 \times 10^4$ for example). Then, a histogram of the vector of these N means is plotted (lines 7–8). What we get is Fig. 5.17a.

Listing A.18 Illustration of the `Distributions` package

```
using Distributions
xGrid          = -5:.01:5                         # where the PDF evaluated
normal_dist    = Normal(0,1)                      # make a normal RV X: mu=0,sigma=1
normalPDF(z)   = pdf(normal_dist,z)               # return the PDF of X at z
plot(xGrid,normalPDF.(xGrid),color="red")         # plot the PDF of X
xbar           = mean(normal_dist)                # get the mean of X
sig            = std(normal_dist)                 # get the SD of X
P              = cdf(normal_dist,0.6)             # get the CDF of X at 0.6: 0.7257
```

Listing A.19 Illustration of the central limit theorem (Fig. 5.17)

```
using Distributions, Random
n, N = 5, 20000
dist = Uniform(1.,2.)                          # X: uniform RV a=1,b=2
data = [mean(rand(dist,n)) for _ in 1:N]       # means of X_i, i=1:n
lb   = minimum(data), ub   = maximum(data)
nb   = Int( floor( (ub-lb)/width ) )           # width=bin width
fig , ax = plt.subplots(1, 1, figsize=set_size())
ax.hist(data, bins=nb,align="left", rwidth=0.9,density=1)
```

A.7 N-Body Problems

This section presents the program to solve the three-body problems discussed in Sect. 10.6.5. The program works for an arbitrary number of bodies. It consists of three parts: the input part is given in Listing A.20, the solution part in Listing A.21, and post-processing in Listing A.22. The code for the solution phase is pretty identical to the maths (Eq. (10.6.15)). Note that I do not care for efficiency: for example, the force between body i and body j is computed twice.

Listing A.20 $N-$body problem solved with Euler-Cromer's method: part I

```
using LinearAlgebra, Printf, Plots
time       = 6.                              # total simulation time
dt         = 0.01                            # time step
stepCount  = Int32(floor(time/dt))           # number of time steps
N          = 3                               # number of bodies
mass       = zeros(N)                        # mass vector
pos        = zeros(2,N,stepCount)            # position matrix (all bodies, all steps)
vel        = zeros(2,N,stepCount)            # velocity matrix
mass[1]   = 1.; mass[2]   = 1.; mass[3]   = 1.;
# initial conditions
pos[:,1,1] = ... pos[:,3,1] = ... pos[:,2,1] = ...
vel[:,1,1] = ... vel[:,3,1] = ... vel[:,2,1] = ...
function force(ri,rj,mj)
  rij = rj - ri
  d   = norm(rij)
  return (G*mj/d^3)* rij
end
```

Listing A.21 $N-$body problem solved with Euler-Cromer's method: part II

```
for n=1:stepCount-1
    for i = 1: N                              # loop over the bodies
      ri = pos[:,i,n]                         # position vector of body 'i' at time n
      fi = zeros(2)                           # compute force acting on 'i'
      for j = 1:N
        if ( j != i )
          rj  = pos[:,j,n]                    # position vector of body 'j' at time n
          mj  = mass[j]                       # mass of body 'j'
          fij = force(ri,rj,mj)               # call the force function
          fi += fij                           # add force of 'j' on 'i'
        end
      end
      vel[:,i,n+1] = vel[:,i,n]+dt*fi         # update velocity of body 'i'
      pos[:,i,n+1] = pos[:,i,n]+dt*vel[:,i,n+1] # update position of body 'i'
    end
end
```

Listing A.22 $N-$body problem solved with Euler-Cromer's method: part III

```
    colors      = [:blue,:orange,:red,:yellow]
    anim = @animate for n in 1:stepCount
    plot(;size=(400,400), axisratio=:equal, legend=false)
    xlims!(-1.1,1.1)
    ylims!(-1.1,1.1)
    scatter!(pos[1,:,n],pos[2,:,n],axisratio=:equal)  # plot three masses
    # plot the trajectory of three masses upto time n
    plot!(pos[1,1,1:n],pos[2,1,1:n],axisratio=:equal,color=colors[1])
    plot!(pos[1,2,1:n],pos[2,2,1:n],axisratio=:equal,color=colors[2])
    plot!(pos[1,3,1:n],pos[2,3,1:n],axisratio=:equal,color=colors[3])
    end
    gif(anim, "three-body.gif", fps=30) # fps = frames per second
```

A.8 Working with Images

`Images.jl` is a package that you need to manipulate images programmatically. When testing ideas or just following along with the documentation of `Images.jl`, it can be useful to have some images to work with. The `TestImages.jl` package bundles several standard images for you.

Listing A.23 is the code used to do an SVD image compression. The result of the code was given in Fig. 9.29. In the code I used the `map` function. In many programming languages, map is the name of a higher order function that applies a given function to each element of a collection, e.g., a list or set, returning the results in a collection of the same type. Listing A.24 demonstrates the use of map.

Listing A.23 Image compression using SVD in `Julia`

```
using Images, TestImages, LinearAlgebra
img0 = float.(testimage("mandrill")) # load the image (512,512) matrix
img  = Gray.(img0)                   # convert to grayscale
function rank_approx(F::SVD, k)
    U, S, V = F
    M       = U[:, 1:k] * Diagonal(S[1:k]) * V[:, 1:k]'
    clamp01!(M)
end
# svd(img) returns SVD{Float32, Float32, Matrix{Float32}}
imgs = [rank_approx(svd(img), k)  for k in (10,50,100)]
imgs = mosaicview(img, imgs...; nrow=1, npad=10)
save("compression-svd.png",imgs)
```

Listing A.24 Use of the `Julia` map function

```
map((x) -> x * 2, [1, 2, 3])      # => [2,4,6]
map(+, [1, 2, 3], [10, 20, 30])   # => [11,22,33]
imgs = map((10, 50, 100)) do k
    rank_approx(svd(img), k)
end
```

A.9 Reinventing the Wheel

Reinventing the wheel means "to work on an something that already exists". Thus, I have reinvented the wheel as I have reimplemented Simpson's quadrature, the Newton-Raphson method, Lagrange interpolation, and so on. There are excellent implementations of them, which are much better than my implementation, provided as "packages" or "libraries". When we learn something we should reinvent the wheel as it is usually the best way to understand something. But for real work, use libraries. Go to https://julialang.org for a list of packages available in `Julia`.

A.10 Computer Algebra System

Herein a summary of the library `SymPy` is given in Listing A.25 for reference. For a more complete documentation, I refer to the website.

A.11 Computer Graphics with `Processing`

We have seen some stunning fractals in Figs. 1.2, 1.5, and 1.6. Herein I present the code used to make them. I use `processing` for this task. We will create Fig. 1.5.

Listing A.25 `SymPy`

```
using SymPy
@vars x
# derivatives and plot
f   = 1 / ( 1 + 25 * x * x)
f2  = diff(f,x,2)                    # 2nd derivative of f
f6  = diff(f,x,6)                    # 6th derivative of f
xgrid=-1:0.01:1.0
yh6=[f6.subs(x,xi) for xi in xgrid]     # evaluate f6 at xgrid
plt.plot(xgrid,yh6,color="black","-",linewidth=1.,label="6th derivative")
# integrals
J   = integrate(f, (x, a, b))        # integral of f, from a to b
# limits
limit(sin(x)/x, x, 0)
limit( (x+1/x)^x, x,oo )             # oo for infinity
# series expansion
f.series(x, x0, n)                   # Taylor series around x0 n terms
# partial fraction decomposition
apart(f)
```

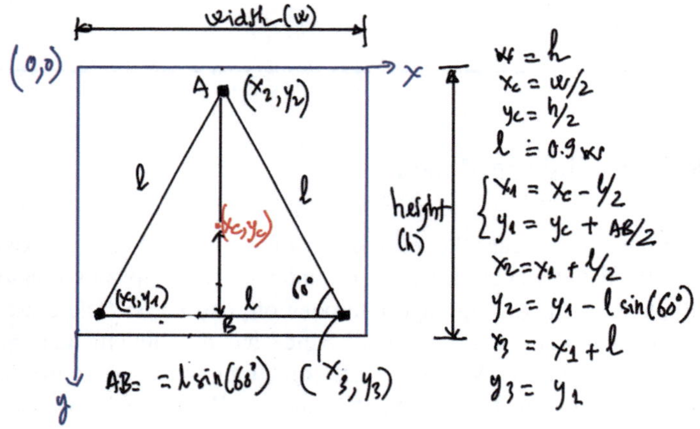

Fig. A.2 Canvas and its coordinate system in `processing`

The first thing is to define the canvas (think of it as the paper on which you draw, just that this paper is part of your computer screen). This can be done with the `setup` function shown in Listing A.26. On this canvas, a Cartesian coordinate system is setup with the origin at the top left corner, x-axis goes to the right, and y-axis goes down (Fig. A.2) (a bit weird). Then, we select the size of our equilateral triangle (the biggest triangle) l (l a bit smaller than the width of the canvas is fine). The center of this triangle is (x_c, y_c). Next, we determine (x_1, y_1)—the coords of the lower right vertex of the triangle. We need a function to draw a triangle given its lower left corner and its length, thus we wrote the function "tri" in Listing A.27.

Listing A.26 `setup` in `processing`

```
void setup() {
  size(900, 900);          // size of the canvas
  noStroke();              // no stroke (outline) around triangles
  fill(50);                // fill color for the triangles, black
}
```

Listing A.27 Draw a triangle with the lower left corner and side

```
void tri(float x, float y, float l) {
    triangle(x, y, x + l/2, y - sin(PI/3) * l, x + l, y);
}
```

Now, we study the problem carefully. The process is: start with an equilateral triangle. Subdivide it into four smaller congruent equilateral triangles and remove the central triangle. Repeat step 2 with each of the remaining smaller triangles infinitely. Of course we do not divide the triangles infinitely, but for a finite number of times denoted by n. Note also that subdividing the biggest triangle by four smaller triangles and removing the central one is equivalent to drawing three smaller triangles.

Now, if $n = 1$ we just draw the biggest triangle, which is straightforward. For $n = 2$ we need to draw three triangles. This is illustrated in Fig. A.3. We're now ready to write the main function called "divide", the code is in Listing A.28. The base case is $n = 1$ and if $n = 2$ we call this function again with l replaced by $l/2$ (smaller triangles) and n replaced by $n - 1$, which is one, and thus three $l/2$ sub-triangles are created. Finally, put the divide function inside the `processing` built-in function `draw` as shown in Listing A.29.

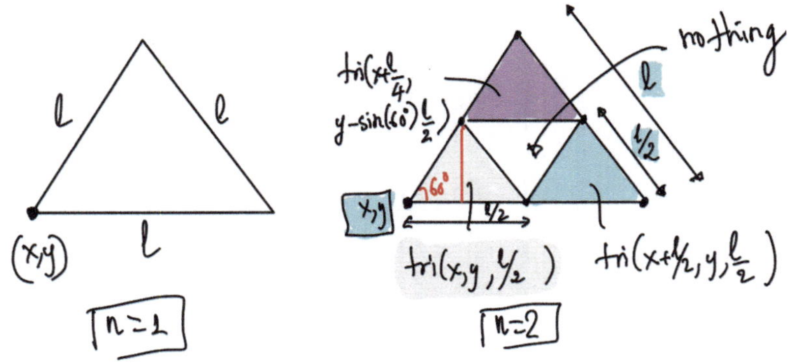

Fig. A.3 .

Listing A.28 The function "divide" as a recursive function

```
void divide(float x, float y, float l, int n) {
    if(n == 1) {
        tri(x, y, l);
    } else {
        divide(x,       y,                      l/2, n-1);
        divide(x + l/2, y,                      l/2, n-1);
        divide(x + l/4, y - sin(PI/3) * l/2, l/2, n-1);
    }
}
```

Listing A.29 Put the drawing functions inside the `draw` function

```
void draw() {
  background(255);  // background color
  divide(x1, y1, l, 3);
}
```

For more on `processing`, you can check out this YouTube channel.

Appendix B
Data Science with Julia

B.1 Introduction to DataFrames.jl

Data comes mostly in a tabular format. By tabular, we mean that the data consists of a table containing rows and columns. The rows denote observations while columns denote variables.

Comma-separated value (CSV) files are very effective way to store tables. A CSV file does exactly what the name indicates it does, namely, storing values by separating them using commas.

We are going to use two packages, namely, `CSV.jl` and `DataFrames.jl`. The former to read CSV files (that contain the data we want to analyze) and the latter is to store this data as a table format. See line 10 of Listing B.1.

Listing B.1 Reading a CSV file and creating a DataFrame

```
1  using DataFrames
2  using CSV
3  using PyCall
4  
5  plt = pyimport("matplotlib.pyplot")
6  sns = pyimport("seaborn")
7  
8  sns.set_style("ticks")
9  
10 train =  DataFrame(CSV.File("Pearson.csv"))
11 size(train)
12 names(train)
13 first(train,5)
14 train[!,:Father]
15 col = train[,:Father]
16 train[train.Father .> 70,:]
17 
18 fig , ax = plt.subplots(1, 1, figsize=(5,5))
19 ax.hist(train[!,:Father],bins=18,density=true)
20 plt.xlabel("Height")
21 plt.ylabel("Proportion of observations per unit bin")
```


V. P. Nguyen, *Mathematics for Engineers and Scientists*,
https://doi.org/10.1007/978-981-97-4631-6

References

1. D. Acheson, *The Wonder Book of Geometry: A Mathematical Story* (OUP Oxford, 2020). ISBN 9780192585387, 9780198846383
2. L. Alcock, *Mathematics Rebooted: A Fresh Approach to Understanding*, 1st edn. (Oxford University Press, 2017). ISBN 0198803796, 9780198803799
3. G. Wanner (auth.), A. Ostermann, in *Geometry By Its History*. Undergraduate Texts in Mathematics, 1st edn. (Springer, Berlin, 2012). ISBN 3642291627, 9783642291623
4. J. Anderson, *Computational Fluid Dynamics: The Basic and Applications*, 1st edn. (McGraw-Hill Science/Engineering/Math, 1995). ISBN 9780070016859, 0-07-001685-2, 0-07-001685-2
5. H.H. Goldstine, in *A History of the Calculus of Variations from the 17th through the 19th Century*. Studies in the History of Mathematics and Physical Sciences, vol. 5, 1st edn. (Springer, New York, 1980). ISBN 9781461381082; 1461381088; 9781461381068; 1461381061
6. W.W. Rouse Ball, *A Short Account of the History of Mathematics*. Michigan Historical Reprint (Scholarly Publishing Office, University of Michigan Library, 2005). ISBN 1418185272, 9781418185275
7. E.T. Bell, *Men of Mathematics: The Lives and Achievements of the Great Mathematicians from Zeno to Poincaré* (Touchstone, 1986). ISBN 0-671-62818-6, 978-1-4767-8425-0
8. A. Bellos, *Alex's Adventure in Numberland* (Bloomsbury Publishing PLC)
9. J. Borwein, D. Bailey, *Mathematics by Experiment: Plausible Reasoning in the 21st Century*, 2nd edn. (A K Peters/CRC Press, 2008). ISBN 1568814429, 9781568814421
10. U.C. Merzbach, C.B. Boyer, *A History of Mathematics* (Wiley, New York, 2011). ISBN 0470525487, 9780470525487
11. G. Van Brummelen, *Heavenly Mathematics: The Forgotten Art of Spherical Trigonometry* (Princeton, 2012). ISBN 0691148929 978-0691148922
12. D.N. Burghes, M.S. Borrie, *Modelling with Differential Equations* (Publisher, Mathematics and its Applications. Ellis Horwood Ltd., 1981). ISBN 0853122865; 9780853122869
13. F. Cajori, *A History of Mathematical Notations* (Dover Publications, 1993). ISBN 9780486677668, 0486677664
14. J. Coopersmith, *The Lazy Universe. An Introduction to the Principle of Least Action*, 1st edn. (Oxford University Press, 2017). ISBN 978-0-19-874304-0, 0198743041
15. R. Courant, H. Robbins, I. Stewart, *What is Mathematics?: An Elementary Approach to Ideas and Methods*, 2nd edn. (Oxford University Press, 1996). ISBN 0195105192, 9780195105193
16. J. Cummings, *Proofs: A Long-Form Mathematics Textbook (The Long-Form Math Textbook Series)* (Independently Published, 2021). ISBN 9798595265973

V. P. Nguyen, *Mathematics for Engineers and Scientists*,
https://doi.org/10.1007/978-981-97-4631-6

17. K. Devlin, *The Unfinished Game: Pascal, Fermat and the Letters*, 1st edn. (Basic Books, 2008). ISBN 0465009107, 9780465009107, 9780786726325
18. W. Dunham, *Journey through Genius: The Great Theorems of Mathematics* (1991). ISBN 9780140147391, 9780471500308, 014014739X, 0471500305
19. W. Dunham, *The Mathematical Universe: An Alphabetical Journey Through the Great Proofs, Problems, and Persons* (Wiley, New York, 1994). ISBN 9780471536567, 0471536563
20. W. Dunham, *Euler: The Master of Us All*, vol. 22 (American Mathematical Society, 2022)
21. C.H. Edwards, *The Historical Development of the Calculus* (Springer, Berlin, 1979). ISBN 3540904360, 9783540904366
22. J.S. Farlow, *Partial Differential Equations for Scientists and Engineers* (Courier Dover Publications, 1993). ISBN 048667620X, 9780486676203
23. R.P. Feynman, *The Feynman Lectures on Physics 3 Volume Set) Set v*, vol. 1–3 (Addison Wesley Longman, 1970). ISBN 0201021153, 9780201021158
24. G. Strang, *Introduction to Linear Algebra*, 5th edn. (Wellesley-Cambridge Press, U.S, 2021)
25. M.J. Gander, G. Wanner, From euler, ritz, and galerkin to modern computing. Siam Rev. **54**(4), 627–666 (2012) <error l="308" c="Invalid command: paragraph not started." />
26. C. Geuzaine, J.-F. Remacle, Gmsh: A 3-d finite element mesh generator with built-in pre-and post-processing facilities. Int. J. Numer. Methods Eng. **79**(11), 1309–1331 (2009)
27. N.J. Giordano, H. Nakanishi, *Computational Physics*, 2nd edn. (Addison-Wesley, 2005). ISBN 0131469908; 9780131469907
28. A. Hald, *A History of Probability and Statistics and Their Applications before 1750 (Wiley Series in Probability and Statistics)*, 1st edn. (Wiley-Interscience, 2003). ISBN 0471471291, 9780471471295
29. R. Hamming, *Numerical Methods for Scientists and Engineers*, 2nd edn. (Dover, 1987). ISBN 9780486652412, 0486652416
30. J. Havil, *Gamma: Exploring Euler's Constant*, Illustrated edition (Princeton Science Librar, Princeton University Press, 2009). ISBN 9780691141336, 0691141339, 0691099839, 9780691099835
31. B. Hopkins, R.J. Wilson, The truth about königsberg. College Math. J. **35**(3), 198–207 (2004)
32. E. Isaacson, H.B. Keller, *Analysis of Numerical Methods* (Dover Publications, 1994). ISBN 9780486680293, 0486680290
33. V.J. Katz, *A History of Mathematics*, 3rd edn. (Pearson, 2008). ISBN 0321387007, 9780321387004
34. D. Kleppner, R. Kolenkow, *An Introduction To Mechanics*, 1st edn. (McGraw-Hill, 1973). ISBN 0070350485, 9780070350489
35. M Kline, *Mathematical Thought From Ancient to Modern Times I* (Oxford University Press, 1972)
36. M. Kline, *Calculus: An Intuitive and Physical Approach* (Wiley, New York, 1967). ISBN 9780471023968, 0471023965
37. M. Kline, *Mathematics for the Nonmathematician (Dover books explaining science)*. Dover books explaining science, Illustrated edition (Dover Publications, 1985). ISBN 0486248232, 9780486248233, 048646329X, 9780486463292
38. C. Lanczos, *The Variational Principles of Mechanics* (1957)
39. S. Lang, *Math: Encounters with High School Students* (Springer, Berlin, 1985). ISBN 9780387961293, 0387961291
40. H.P. Langtangen, S. Linge, in *Finite Difference Computing with PDEs: A Modern Software Approach*, Texts in Computational Science and Engineering, vol. 16, 1st edn. (Springer International Publishing, 2017). ISBN 978-3-319-55455-6, 978-3-319-55456-3
41. P. Lockhart, *A Mathematician's Lament: How School Cheats Us Out of Our Most Fascinating and Imaginative Art Form* (Bellevue Literary Press, 2009). ISBN 1934137170, 9781934137178
42. P. Lockhart, *Measurement* (Belknap Press, 2012). ISBN 0674057554, 9780674057555
43. E. Maor, *To Infinity and Beyond: A Cultural History of the Infinite*, Illustrated edition (Princeton University Press, 1991). ISBN 9780691025117, 0691025118

44. E. Maor, *Trigonometric Delights* (Princeton University Press, 1998). ISBN 9780691057545, 9780691095417, 0691057540, 0691095418
45. J.E. Marsden, A. Tromba, *Vector Calculus*, 5th edn. (W.H. Freeman, 2003). ISBN 9780716749929; 0716749920
46. P.J. Nahin, *An Imaginary Tale: The Story of Square Root of–1*, pup edition (Princeton University Press, 1998). ISBN 0691027951, 9780691027951, 9780691127989, 0691127980
47. P.J. Nahin, *Dr. Euler's Fabulous Formula: Cures Many Mathematical Ills* (Princeton University Press, 2006). ISBN 0691118221, 9780691118222
48. P.J. Nahin, *When Least Is Best: How Mathematicians Discovered Many Clever Ways to Make Things as Small (or as Large) as Possible* (Princeton University Press, 2007). ISBN 0691130523, 9780691130521
49. P.J. Nahin, in *Inside Interesting, Integrals: A Collection of Sneaky Tricks, Sly Substitutions, and Numerous Other Stupendously Clever, Awesomely Wicked*, Undergraduate Lecture Notes in Physics, 2015th edn. (Springer, Berlin, 2014). ISBN 1493912763, 9781493912766
50. P.J. Nahin, *Hot Molecules, Cold Electrons: From the Mathematics of Heat to the Development of the Trans-Atlantic* (Princeton University Press, 2020). ISBN 9780691191720; 0691191727
51. R.B. Nelsen, *Proofs without Words: Exercises in Visual Thinking*, vol. 1 (MAA, 1993)
52. I.M. Niven, in *Numbers: Rational and Irrational*. New Mathematical Library, Random house edition (Mathematical Assn of America, 1961). ISBN 9780883856017, 0883856018
53. G. Polya, *How to Solve It; A New Aspect of Mathematical Method*. Princeton Paperbacks, vol. 246, 2nd edn. (Princeton University Press, 1971). ISBN 9780691023564, 9780691080970, 0691023565, 0691080976
54. D. Poole, *Linear Algebra* (A Modern Introduction, Brooks Cole, 2005). 0534998453, 9780534998455
55. D.S. Richeson, *Euler's Gem: The Polyhedron Formula and The Birth of Topology*, 1st edn. (Princeton University Press, 2008). ISBN 0691126771, 9780691126777
56. D.S. Richeson, *Tales of Impossibility: The 2000-Year Quest to Solve the Mathematical Problems of Antiquity* (Princeton University Press, 2019). ISBN 9780691192963
57. S.M. Ross, *A First Course in Probability*, 5th edn. (Prentice Hall, 1998). ISBN 0137463146, 9780137463145
58. H.M. Schey, *Div, Grad, Curl, and All That: An Informal Text on Vector Calculus*, 4th edn. (W. W. Norton & Company, 2005). ISBN 0393925161, 9780393925166
59. I. Stewart, in *Letters to a Young Mathematician*. Art of Mentoring (Basic Books, a member of the Perseus Books Group, 2006). ISBN 0465082319, 9780465082315, 0465082327, 9780465082322
60. I. Stewart, *Why Beauty Is Truth: The History of Symmetry* (Basic Books, 2007a). ISBN 046508236X, 9780465082360
61. J. Stewart, in *Calculus: Early Transcendentals*. Stewart's Calculus Series, 6th edn. (Brooks Cole, 2007b). ISBN 0495011665, 9780495011668
62. S.M. Stigler, *The History of Statistics: The Measurement of Uncertainty Before 1900*, Illustrated edition (Belknap Press, 1986). ISBN 0674403401, 9780674403406
63. J. Stillwell, in *The Four Pillars of Geometry*. Undergraduate texts in mathematics, 1st edn. (Springer, 2005). ISBN 9780387255309, 0387255303
64. J. Stillwell, *Mathematics and Its History*. Undergraduate Texts in Mathematics, 3rd edn. (Springer, New York, 2010). ISBN 144196052X, 9781441960528
65. G. Strang, *Calculus*, 2nd edn. (Wellesley College, 1991). ISBN 0961408820, 9780961408824
66. G. Strang, *Linear Algebra and Learning from Data*, vol. 1, 1st edn. (Wesley-Cambridge Press, 2019). ISBN 0692196382, 9780692196380
67. S. Strogatz, *Infinite Powers: How Calculus Reveals the Secrets of the Universe* (Houghton Mifflin Harcourt, 2019)
68. R.J. Taylor, *Classical Mechanics* (University Science Books, 2005). ISBN 189138922X, 9781891389221
69. L.N. Trefethen, *Approximation Theory and Approximation Practice* (2013)
70. P. Zeitz, *The Art and Craft of Problem Solving*, 2nd edn. (Wiley, New York, 2007). ISBN 9780471789017, 0471789011

Index

V. P. Nguyen, *Mathematics for Engineers and Scientists*,
https://doi.org/10.1007/978-981-97-4631-6

GPSR Compliance

The European Union's (EU) General Product Safety Regulation (GPSR) is a set of rules that requires consumer products to be safe and our obligations to ensure this.

If you have any concerns about our products, you can contact us on ProductSafety@springernature.com

In case Publisher is established outside the EU, the EU authorized representative is:

Springer Nature Customer Service Center GmbH
Europaplatz 3
69115 Heidelberg, Germany

Batch number: 10725823

Printed by Printforce, the Netherlands